SIGNAL ANALYSIS

IEEE Press
445 Hoes Lane
Piscataway, NJ 08854

SIGNAL ANALYSIS

TIME, FREQUENCY, SCALE, AND STRUCTURE

Ronald L. Allen
Duncan W. Mills

IEEE PRESS

A John Wiley & Sons, Inc., Publication

Published simultaneously in Canada.

Library of Congress Cataloging-in-Publication Data is available.

ISBN: 0-471-23441-9

Printed in the United States of America

10 9 8 7 6 5 4 3 2

To Beverley and to the memory of my parents, Mary and R.L. (Kelley).

R.L.A.

To those yet born, who will in some manner—large or small—benefit from the technology and principles described here. To the reader, who will contribute to making this happen.

D.W.M.

CONTENT

PREFACE

This text provides a complete introduction to signal analysis. Inclusion of fundamental ideas—analog and discrete signals, linear systems, Fourier transforms, and sampling theory—makes it suitable for introductory courses, self-study, and refreshers in the discipline. But along with these basics, *Signal Analysis: Time, Frequency, Scale, and Structure* gives a running tutorial on functional analysis—the mathematical concepts that generalize linear algebra and underlie signal theory. While the advanced mathematics can be skimmed, readers who absorb the material will be prepared for latter chapters that explain modern mixed-domain signal analysis: Short-time Fourier (Gabor) and wavelet transforms.

Quite early in the presentation, *Signal Analysis* surveys methods for edge detection, segmentation, texture identification, template matching, and pattern recognition. Typically, these are only covered in image processing or computer vision books. Indeed, the fourth chapter might seem like a detour to some readers. But the techniques are essential to one-dimensional signal analysis as well. Soon after learning the rudiments of systems and convolutions, students are invited to apply the ideas to make a computer understand a signal. Does it contain anything significant, expected, or unanticipated? Where are the significant parts of the signal? What are its local features, where are their boundaries, and what is their structure? The difficulties inherent in understanding a signal become apparent, as does the need for a comprehensive approach to signal frequency. This leads to the chapters on the frequency domain. Various continous and discrete Fourier transforms make their appearance. Their application, in turn, proves to be problematic for signals with transients, localized frequency components, and features of varying scale. The text delves into the new analytical tools—some discovered only in the last 20 years—for such signals. Time-frequency and time-scale transforms, their underlying mathematical theory, their limitations, how they differently reveal signal structure, and their promising applications complete the book. So the highlights of this book are:

- The signal analysis perspective;
- The tutorial material on advanced mathematics—in particular function spaces, cast in signal processing terms;
- The coverage of the latest mixed domain analysis methods.

We thought that there is a clear need for a text that begins at a basic level while taking a *signal analysis* as opposed to *signal processing* perspective on applications.

The goal of signal analysis is to arrive at a structural description of a signal so that later high-level algorithms can interpret its content. This differs from signal processing *per se*, which only seeks to modify the input signal, without changing its fundamental nature as a one-dimensional sequence of numerical values. From this viewpoint, signal analysis stands within the scope of artificial intelligence. Many modern technologies demand its skills. Human–computer interaction, voice recognition, industrial process control, seismology, bioinformatics, and medicine are examples.

Signal Analysis provides the abstract mathematics and functional analysis which is missing from the backgrounds of many readers, especially undergraduate science and engineering students and professional engineers. The reader can begin comfortably with the basic ideas. The book gradually dispenses the mathematics of Hilbert spaces, complex analysis, disributions, modern integration theory, random signals, and analog Fourier transforms; the less mathematically adept reader is not overwhelmed with hard analysis. There has been no easy route from standard signal processing texts to the latest treatises on wavelets, Gabor transforms, and the like. The gap must be spanned with knowledge of advanced mathematics. And this has been a problem for too many engineering students, classically-educated applied researchers, and practising engineers. We hope that *Signal Analysis* removes the obstacles. It has the signal processing fundamentals, the signal analysis perspective, the mathematics, and the bridge from all of these to crucial developments that began in the mid-1980s.

The last three chapters of this book cover the latest mixed-domain transform methods: Gabor transforms, wavelets, multiresolution analysis, frames, and their applications. Researchers who need to keep abreast of the advances that are revolutionizing their discipline will find a complete introductory treatment of time-frequency and time-scale transforms in the book. We prove the Balian-Low theorem, which pinpoints a limitation on short-time Fourier representations. We had envisioned a much wider scope for mixed-domain applications. Ultimately, the publication schedule and the explosive growth of the field prevented us from achieving a thorough coverage of all principal algorithms and applications—what might have been a fourth highlight of the book. The last chapter explains briefly how to use the new methods in applications, contrasts them with time domain tactics, and contains further references to the research literature.

Enough material exists for a year-long university course in signal processing and analysis. Instructors who have students captive for two semesters may cover the chapters in order. When a single semester must suffice, Chapters 1–3, 5, 7, 8, and 9 comprise the core ideas. We recommend at least the sections on segmentation and thresholding in Chapter 4. After some programming experiments, the students will see how hard it is to make computers do what we humans take for granted. The instructor should adjust the pace according to the students' preparation. For instance, if a system theory course is prerequisite—as is typical in the undergraduate engineering curriculum—then the theoretical treatments of signal spaces, the Dirac delta, and the Fourier transforms are appropriate. An advanced course can pick up the mathematical theory, the pattern recognition material in

Chapter 4, the generalized Fourier transform in Chapter 6, and the analog filter designs in Chapter 9. But the second semester work should move quickly to and concentrate upon Chapters 10–12. This equips the students for reading the research literature.

RONALD L. ALLEN
San José, California

DUNCAN W. MILLS
Mountain View, California

ACKNOWLEDGMENTS

We would like to thank the editorial and production staffs on John Wiley and Sons and IEEE Press for their efficiency, courtesy, patience, and professionalism while we wrote this book. We are especially grateful to Marilyn G. Catis and Anthony VenGraitis of IEEE Press for handling incremental submissions, managing reviews, and providing general support over the years. We are grateful to Beverley Andalora for help with the figures, to William Parker of Philips Speech Recognition Systems for providing digital speech samples, and to KLA-Tencor Corporation for reflectometry and scanning electron microscopy data samples.

RONALD L. ALLEN

DUNCAN W. MILLS

CHAPTER 1

Signals: Analog, Discrete, and Digital

Analog, discrete, and digital signals are the raw material of signal processing and analysis. Natural processes, whether dependent upon or independent of human control, generate analog signals; they occur in a continuous fashion over an interval of time or space. The mathematical model of an analog signal is a function defined over a part of the real number line. Analog signal conditioning uses conventional electronic circuitry to acquire, amplify, filter, and transmit these signals. At some point, digital processing may take place; today, this is almost always necessary. Perhaps the application requires superior noise immunity. Intricate processing steps are also easier to implement on digital computers. Furthermore, it is easier to improve and correct computerized algorithms than systems comprised of hard-wired analog components. Whatever the rationale for digital processing, the analog signal is captured, stored momentarily, and then converted to digital form. In contrast to an analog signal, a discrete signal has values only at isolated points. Its mathematical representation is a function on the integers; this is a fundamental difference. When the signal values are of finite precision, so that they can be stored in the registers of a computer, then the discrete signal is more precisely known as a digital signal. Digital signals thus come from sampling an analog signal, and—although there is such a thing as an analog computer—nowadays digital machines perform almost all analytical computations on discrete signal data.

This has not, of course, always been the case; only recently have discrete techniques come to dominate signal processing. The reasons for this are both theoretical and practical.

On the practical side, nineteenth century inventions for transmitting words, the telegraph and the telephone—written and spoken language, respectively—mark the beginnings of engineered signal generation and interpretation technologies. Mathematics that supports signal processing began long ago, of course. But only in the nineteenth century did signal theory begin to distinguish itself as a technical, engineering, and scientific pursuit separate from pure mathematics. Until then, scientists did not see mathematical entities—polynomials, sinusoids, and exponential functions, for example—as sequences of symbols or carriers of information. They were envisioned instead as ideal shapes, motions, patterns, or models of natural processes.

Signal Analysis: Time, Frequency, Scale, and Structure, by Ronald L. Allen and Duncan W. Mills
ISBN: 0-471-23441-9

The development of electromagnetic theory and the growth of electrical and electronic communications technologies began to divide these sciences. The functions of mathematics came to be studied as bearing information, requiring modification to be useful, suitable for interpretation, and having a meaning. The life story of this new discipline—signal processing, communications, signal analysis, and information theory—would follow a curious and ironic path. Electromagnetic waves consist of coupled electric and magnetic fields that oscillate in a sinusoidal pattern and are perpendicular to one another and to their direction of propagation. Fourier discovered that very general classes of functions, even those containing discontinuities, could be represented by sums of sinusoidal functions, now called a Fourier series [1]. This surprising insight, together with the great advances in analog communication methods at the beginning of the twentieth century, captured the most attention from scientists and engineers.

Research efforts into discrete techniques were producing important results, even as the analog age of signal processing and communication technology charged ahead. Discrete Fourier series calculations were widely understood, but seldom carried out; they demanded quite a bit of labor with pencil and paper. The first theoretical links between analog and discrete signals were found in the 1920s by Nyquist,[1] in the course of research on optimal telegraphic transmission mechanisms [2]. Shannon[2] built upon Nyquist's discovery with his famous sampling theorem [3]. He also proved something to be feasible that no one else even thought possible: error-free digital communication over noisy channels. Soon thereafter, in the late 1940s, digital computers began to appear. These early monsters were capable of performing signal processing operations, but their speed remained too slow for some of the most important computations in signal processing—the discrete versions of the Fourier series. All this changed two decades later when Cooley and Tukey disclosed their fast Fourier transform (FFT) algorithm to an eager computing public [4–6]. Digital computations of Fourier's series were now practical on real-time signal data, and in the following years digital methods would proliferate. At the present time, digital systems have supplanted much analog circuitry, and they are the core of almost all signal processing and analysis systems. Analog techniques handle only the early signal input, output, and conditioning chores.

There are a variety of texts available covering signal processing. Modern introductory systems and signal processing texts cover both analog and discrete theory [7–11]. Many reflect the shift to discrete methods that began with the discovery of the FFT and was fueled by the ever-increasing power of computing machines. These often concentrate on discrete techniques and presuppose a background in analog

[1]As a teenager, Harry Nyquist (1887–1976) emigrated from Sweden to the United States. Among his many contributions to signal and communication theory, he studied the relationship between analog signals and discrete signals extracted from them. The term *Nyquist rate* refers to the sampling frequency necessary for reconstructing an analog signal from its discrete samples.

[2]Claude E. Shannon (1916–2001) founded the modern discipline of information theory. He detailed the affinity between Boolean logic and electrical circuits in his 1937 Masters thesis at the Massachusetts Institute of Technology. Later, at Bell Laboratories, he developed the theory of reliable communication, of which the sampling theorem remains a cornerstone.

signal processing [12–15]. Again, there is a distinction between discrete and digital signals. Discrete signals are theoretical entities, derived by taking instantaneous—and therefore exact—samples from analog signals. They might assume irrational values at some time instants, and the range of their values might be infinite. Hence, a digital computer, whose memory elements only hold limited precision values, can only process those discrete signals whose values are finite in number and finite in their precision—digital signals. Early texts on discrete signal processing sometimes blurred the distinction between the two types of signals, though some further editions have adopted the more precise terminology. Noteworthy, however, are the burgeoning applications of digital signal processing integrated circuits: digital telephony, modems, mobile radio, digital control systems, and digital video to name a few. The first high-definition television (HDTV) systems were analog; but later, superior HDTV technologies have relied upon digital techniques. This technology has created a true digital signal processing literature, comprised of the technical manuals for various DSP chips, their application notes, and general treatments on fast algorithms for real-time signal processing and analysis applications on digital signal processors [16–21]. Some of our later examples and applications offer some observations on architectures appropriate for signal processing, special instruction sets, and fast algorithms suitable for DSP implementation.

This chapter introduces signals and the mathematical tools needed to work with them. Everyone should review this chapter's first six sections. This first chapter combines discussions of analog signals, discrete signals, digital signals, and the methods to transition from one of these realms to another. All that it requires of the reader is a familiarity with calculus. There are a wide variety of examples. They illustrate basic signal concepts, filtering methods, and some easily understood, albeit limited, techniques for signal interpretation. The first section introduces the terminology of signal processing, the conventional architecture of signal processing systems, and the notions of analog, discrete, and digital signals. It describes signals in terms of mathematical models—functions of a single real or integral variable. A specification of a sequence of numerical values ordered by time or some other spatial dimension is a time domain description of a signal. There are other approaches to signal description: the frequency and scale domains, as well as some—relatively recent—methods for combining them with the time domain description. Sections 1.2 and 1.3 cover the two basic signal families: analog and discrete, respectively. Many of the signals used as examples come from conventional algebra and analysis.

The discussion gets progressively more formal. Section 1.4 covers sampling and interpolation. Sampling picks a discrete signal from an analog source, and interpolation works the other way, restoring the gaps between discrete samples to fashion an analog signal from a discrete signal. By way of these operations, signals pass from the analog world into the discrete world and vice versa. Section 1.5 covers periodicity, and foremost among these signals is the class of sinusoids. These signals are the fundamental tools for constructing a frequency domain description of a signal. There are many special classes of signals that we need to consider, and Section 1.6 quickly collects them and discusses their properties. We will of course expand upon and deepen our understanding of these special types of signals throughout the book. Readers with signal processing backgrounds may quickly scan this material; however, those with little prior work in this area might well linger over these parts.

The last two sections cover some of the mathematics that arises in the detailed study of signals. The complex number system is essential for characterizing the timing relationships in signals and their frequency content. Section 1.7 explains why complex numbers are useful for signal processing and exposes some of their unique properties. Random signals are described in Section 1.8. Their application is to model the unpredictability in natural signals, both analog and discrete. Readers with a strong mathematics background may wish to skim the chapter for the special signal processing terminology and skip Sections 1.7 and 1.8. These sections can also be omitted from a first reading of the text.

A summary, a list of references, and a problem set complete the chapter. The summary provides supplemental historical notes. It also identifies some software resources and publicly available data sets. The references point out other introductory texts, reviews, and surveys from periodicals, as well as some of the recent research.

1.1 INTRODUCTION TO SIGNALS

There are several standpoints from which to study signal analysis problems: empirical, technical, and theoretical. This chapter uses all of them. We present lots of examples, and we will return to them often as we continue to develop methods for their processing and interpretation. After practical applications of signal processing and analysis, we introduce some basic terminology, goals, and strategies.

Our early methods will be largely experimental. It will be often be difficult to decide upon the best approach in an application; this is the limitation of an intuitive approach. But there will also be opportunities for making technical observations about the right mathematical tool or technique when engaged in a practical signal analysis problem. Mathematical tools for describing signals and their characteristics will continue to illuminate this technical side to our work. Finally, some abstract considerations will arise at the end of the chapter when we consider complex numbers and random signal theory. Right now, however, we seek only to spotlight some practical and technical issues related to signal processing and analysis applications. This will provide the motivation for building a significant theoretical apparatus in the sequel.

1.1.1 Basic Concepts

Signals are symbols or values that appear in some order, and they are familiar entities from science, technology, society, and life. Examples fit easily into these categories: radio-frequency emissions from a distant quasar; telegraph, telephone, and television transmissions; people speaking to one another, using hand gestures; raising a sequence of flags upon a ship's mast; the echolocation chirp of animals such as bats and dolphins; nerve impulses to muscles; and the sensation of light patterns striking the eye. Some of these signal values are quantifiable; the phenomenon is a measurable quantity, and its evolution is ordered by time or distance. Thus, a residential telephone signal's value is known by measuring the voltage across the pair of wires that comprise the circuit. Sound waves are longitudinal and produce minute, but measurable, pressure variations on a listener's eardrum. On the other hand, some signals appear to have a representation that is at root not quantifiable, but rather symbolic. Thus, most people would grant that sign language gestures, maritime signal flags, and even ASCII text could be considered signals, albeit of a symbolic nature.

Let us for the moment concentrate on signals with quantifiable values. These are the traditional mathematical signal models, and a rich mathematical theory is available for studying them. We will consider signals that assume symbolic values, too, but, unlike signals with quantifiable values, these entities are better described by relational mathematical structures, such as graphs.

Now, if the signal is a continuously occurring phenomenon, then we can represent it as a function of a time variable t; thus, $x(t)$ is the value of signal x at time t. We understand the units of measurement of $x(t)$ implicitly. The signal might vary with some other spatial dimension other than time, but in any case, we can suppose that its domain is a subset of the real numbers. We then say that $x(t)$ is an *analog signal*. Analog signal values are read from conventional indicating devices or scientific instruments, such as oscilloscopes, dial gauges, strip charts, and so forth.

An example of an analog signal is the seismogram, which records the shaking motion of the ground during an earthquake. A precision instrument, called a *seismograph*, measures ground displacements on the order of a micron (10^{-6} m) and produces the seismogram on a paper strip chart attached to a rotating drum. Figure 1.1 shows the record of the Loma Prieta earthquake, centered in the Santa Cruz mountains of northern California, which struck the San Francisco Bay area on 18 October 1989.

Seismologists analyze such a signal in several ways. The total deflection of the pen across the chart is useful in determining the temblor's magnitude. Seismograms register three important types of waves: the *primary*, or *P waves*; the *secondary*, or *S waves*; and the *surface waves*. P waves arrive first, and they are compressive, so their direction of motion aligns with the wave front propagation [22]. The transverse S waves follow. They oscillate perpendicular to the direction of propagation. Finally, the large, sweeping surface waves appear on the trace.

This simple example illustrates processing and analysis concepts. Processing the seismogram signal is useful to remove noise. Noise can be minute ground motions from human activity (construction activity, heavy machinery, vehicles, and the like), or it may arise from natural processes, such as waves hitting the beach. Whatever the source, an important signal processing operation is to smooth out these minute ripples in the seismogram trace so as to better detect the occurrence of the initial indications of a seismic event, the P waves. They typically manifest themselves as seismometer needle motions above some threshold value. Then the analysis problem of finding when the S waves begin is posed. Figure 1.1 shows the result of a signal analysis; it slices the Loma Prieta seismogram into its three constituent wave

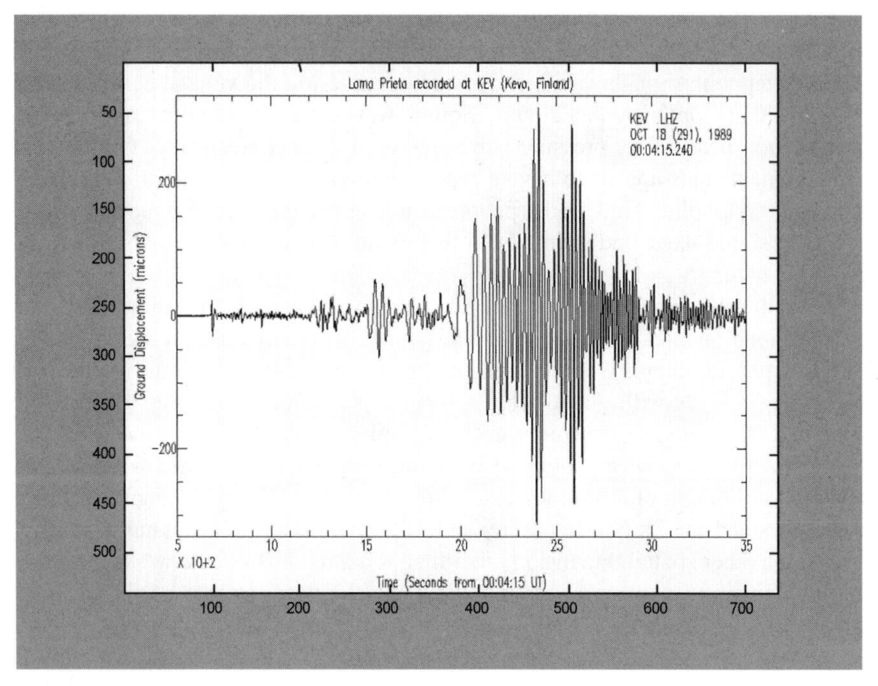

Fig. 1.1. Seismogram of the magnitude 7.1 Loma Prieta earthquake, recorded by a seismometer at Kevo, Finland. The first wiggle—some eight minutes after the actual event—marks the beginning of the low-magnitude P waves. The S waves arrive at approximately t = 1200 s, and the large sweeping surface waves begin near t = 2000 s.

trains. This type of signal analysis can be performed by inspection on analog seismograms.

Now, the time interval between the arrival of the P and S waves is critical. These undulations are simultaneously created at the earthquake's epicenter; however, they travel at different, but known, average speeds through the earth. Thus, if an analysis of the seismogram can reveal the time that these distinct wave trains arrive, then the time difference can be used to measure the distance from the instrument to the earthquake's epicenter. Reports from three separate seismological stations are sufficient to locate the epicenter. Analyzing smaller earthquakes is also important. Their location and the frequency of their occurrence may foretell a larger temblor [23]. Further, soundings in the earth are indicative of the underlying geological strata; seismologists use such methods to locate oil deposits, for example [24]. Other similar applications include the detection of nuclear arms detonations and avalanches. For all of these reasons—scientific, economic, and public safety—seismic signal intepretation is one of the most important areas in signal analysis and one of the areas in which new methods of signal analysis have been pioneered. These further signal interpretation tasks are more troublesome for human interpreters. The signal behavior that distinguishes a small earthquake from a distant nuclear detonation is not apparent. This demands thorough computerized analysis.

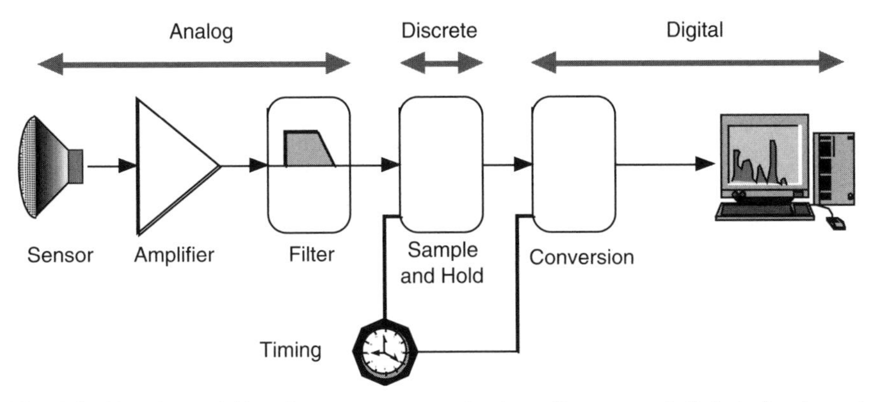

Fig. 1.2. Signal acquisition into a computer. Analog, discrete, and digital signals each occur—at least in principle—within such a system.

Suppose, therefore, that the signal is a discrete phenomenon, so that it occurs only at separate time instants or distance intervals and not continuously. Then we represent it as a function on a subset of the integers $x(n)$ and we identify $x(n)$ as a *discrete signal*. Furthermore, some discrete signals may have only a limited range of values. Their measurable values can be stored in the memory cells of a digital computer. The discrete signals that satisfy this further constraint are called *digital signals*.

Each of these three types of signals occurs at some stage in a conventional computerized signal acquisition system (Figure 1.2). Analog signals arise from some quantifiable, real-world process. The signal arrives at an interface card attached to the computer's input–output bus.

There are generally some signal amplification and conditioning components, all analog, at the system's front end. At the sample and hold circuit, a momentary storage component—a capacitor, for example—holds the signal value for a small time interval. The sampling occurs at regular intervals, which are set by a timer. Thus, the sequence of quantities appearing in the sample and hold device represents the discrete form of the signal. While the measurable quantity remains in the sample and hold unit, a digitization device composes its binary representation. The extracted value is moved into a digital acquisition register of finite length, thereby completing the analog-to-digital conversion process. The computer's signal processing software or its input–output driver reads the digital signal value out of the acquisition register, across the input–output bus, and into main memory. The computer itself may be a conventional general-purpose machine, such as a personal computer, an engineering workstation, or a mainframe computer. Or the processor may be one of the many special purpose *digital signal processors* (DSPs) now available. These are now a popular design choice in signal processing and analysis systems, especially those with strict execution time constraints.

Some natural processes generate more than one measurable quantity as a function of time. Each such quantity can be regarded as a separate signal, in which case

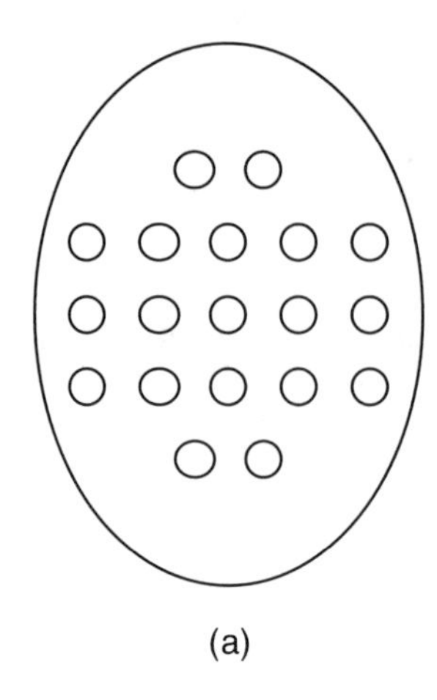

(a)

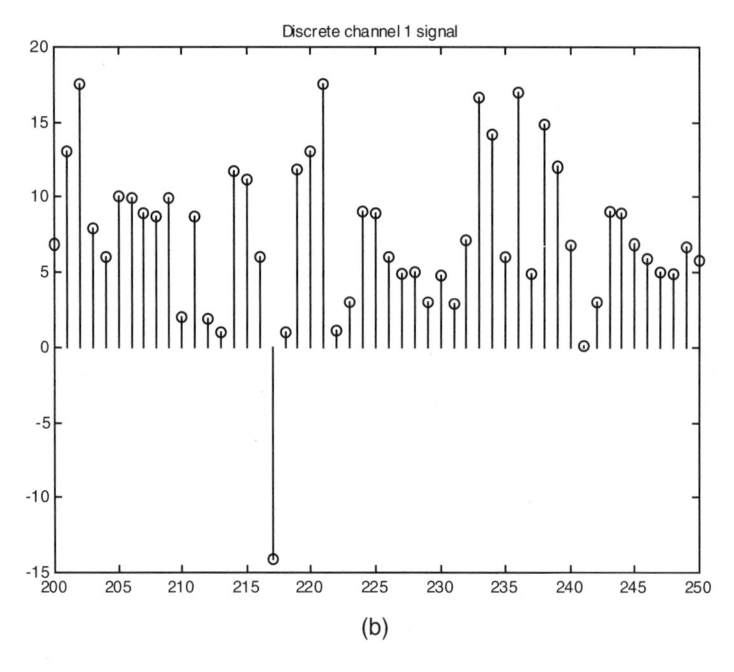

(b)

Fig. 1.3. A multichannel signal: The electroencephalogram (EEG) taken from a healthy young person, with eyes open. The standard EEG sensor arrangement consists of 19 electrodes (a). Discrete data points of channel one (b). Panels (c) and (d) show the complete traces for the first two channels, $x_1(n)$ and $x_2(n)$. These traces span an eight second time interval: 1024 samples. Note the jaggedness superimposed on gentler wavy patterns. The EEG varies according to whether the patient's eyes are open and according to the health of the individual; markedly different EEG traces typify, for example, Alzheimer's disease.

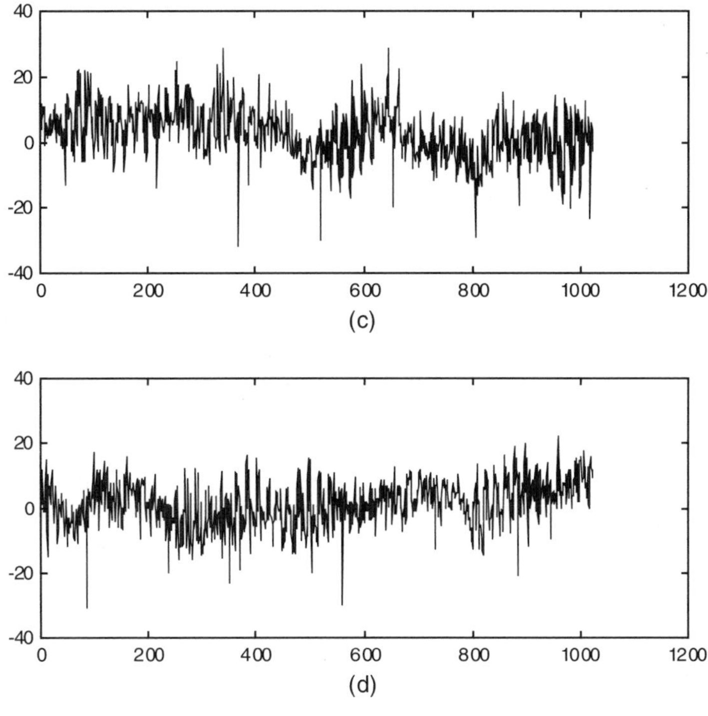

Fig. 1.3 (*Continued*)

they are all functions of the same independent variable with the same domain. Alternatively, it may be technically useful to maintain the multiple quantities together as a vector. This is called a *multichannel* signal. We use boldface letters to denote multichannel signals. Thus, if $\boldsymbol{x}$ is analog and has N channels, then $\boldsymbol{x}(t) = (x_1(t), x_2(t), \ldots, x_N(t))$, where the analog $x_i(t)$ are called the *component* or *channel* signals. Similarly, if $\boldsymbol{x}$ is discrete and has N channels, then $\boldsymbol{x}(n) = (x_1(n), x_2(n), \ldots, x_N(n))$.

One biomedical signal that is useful in diagnosing brain injuries, mental illness, and conditions such as Alzheimer's disease is the *electroencephalogram* (EEG) [25], a multichannel signal. It records electrical potential differences, or voltages, that arise from the interactions of massive numbers of neurons in different parts of the brain. For an EEG, 19 electrodes are attached from the front to the back of the scalp, in a two–five–five–five–two arrangement (Figure 1.3).

The EEG traces in Figure 1.3 are in fact digital signals, acquired one sample every 7.8 ms, or at a sampling frequency of 128 Hz. The signal appears to be continuous in nature, but this is due to the close spacing of the samples and linear interpolation by the plotting package.

Another variation on the nature of signals is that they may be functions of more than one independent variable. For example, we might measure air temperature as a function of height: $T(h)$ is an analog signal. But if we consider that the variation may occur along a north-to-south line as well, then the temperature depends upon a distance measure x as well: $T(x, h)$. Finally, over an area with location coordinates (x, y), the air temperature is a continuous function of three variables $T(x, y, h)$. When a signal has more than one independent variable, then it is a *multidimensional* signal. We usually think of an "image" as recording light intesity measurements of a scene, but multidimensional signals—especially those with two or three independent variables—are usually called *images*. Images may be discrete too. Temperature readings taken at kilometer intervals on the ground and in the air produce a discrete signal $T(m, n, k)$. A discrete signal is a sequence of numerical values, whereas an image is an array of numerical values. Two-dimensional image elements, especially those that represent light intensity values, are called *pixels*, an acronym for *picture elements*. Occasionally, one encounters the term *voxel*, which is a three-dimensional signal value, or a *volume element*.

An area of multidimensional signal processing and analysis of considerable importance is the intepretation of images of landscapes acquired by satellites and high altitude aircraft. Figure 1.4. shows some examples. Typical tasks are to automatically distinguish land from sea; determine the amount and extent of sea ice; distinguish agricultural land, urban areas, and forests; and, within the agricultural regions, recognize various crop types. These are remote sensing applications.

Processing two-dimensional signals is more commonly called picture or image processing, and the task of interpreting an image is called image analysis or computer vision. Many researchers are involved in robotics, where their efforts couple computer vision ideas with manipulation of the environment by a vision-based machine. Consequently, there is a vast, overlapping literature on image processing [26–28], computer vision [29–31], and robotics [32].

Our subject, signal analysis, concentrates on the mathematical foundations, processing, and especially the intepretation of one-dimensional, single-valued signals. Generally, we may select a single channel of a multichannel signal for consideration; but we do not tackle problems specific to multichannel signal interpretation. Likewise, we do not delve deeply into image processing and analysis. Certain images do arise, so it turns out, in several important techniques for analyzing signals. Sometimes a daunting one-dimensional problem can be turned into a tractable two-dimensional task. Thus, we prefer to pursue the one-dimensional problem into the multidimensional realm only to the point of acknowledging that a straightforward image analysis will produce the intepretation we seek.

So far we have introduced the basic concepts of signal theory, and we have considered some examples: analog, discrete, multichannel, and multidimensional signals. In each case we describe the signals as sequences of numerical values, or as a function of an independent time or other spatial dimension variable. This constitutes a time-domain description of a signal. From this perspective, we can display a signal, process it to produce another signal, and describe its significant features.

(a) Agricultural area; (b) Forested region;

(c) Ice at sea; (d) Urban area.

Fig. 1.4. Aerial scenes. Distinguishing terrain types is a typical problem of image analysis, the interpretation of two-dimensional signals. Some problems, however, admit a one-dimensional solution. A sample line through an image is in fact a signal, and it is therefore suitable for one-dimensional techniques. (a) Agricultural area. (b) Forested region. (c) Ice at sea. (d) Urban area.

1.1.2 Time-Domain Description of Signals

Since time flows continuously and irreversibly, it is natural to describe sequential signal values as given by a time ordering. This is often, but not always, the case; many signals depend upon a distance measure. It is also possible, and sometimes a very important analytical step, to consider signals as given by order of a salient event. Conceiving the signal this way makes the dependent variable—the signal value—a function of time, distance, or some other quantity indicated between successive events. Whether the independent variable is time, some other spatial dimension, or a counting of events, when we represent and discuss a signal in terms of its ordered values, we call this the *time-domain* description of a signal.

Note that a precise time-domain description may elude us, and it may not even be possible to specify a signal's values. A fundamentally unknowable or random process is the source of such signals. It is important to develop methods for handling the randomness inherent in signals. Techniques that presuppose a theory of signal randomness are the topic of the final section of the chapter.

Next we look further into two application areas we have already touched upon: biophysical and geophysical signals. Signals from representative applications in these two areas readily illustrate the time-domain description of signals.

1.1.2.1 Electrocardiogram Interpretation. Electrocardiology is one of the earliest techniques in biomedicine. It also remains one of the most important. The excitation and recovery of the heart muscles cause small electrical potentials, or voltages, on the order of a millivolt, within the body and measurable on the skin. Cardiologists observe the regularity and shape of this voltage signal to diagnose heart conditions resulting from disease, abnormality, or injury. Examples include cardiac dysrhythmia and fibrillation, narrowing of the coronary arteries, and enlargement of the heart [33]. Automatic interpretation of ECGs is useful for many aspects of clinical and emergency medicine: remote monitoring, as a diagnostic aid when skilled cardiac care personnel are unavailable, and as a surgical decision support tool.

A modern electrocardiogram (ECG or EKG) contains traces of the voltages from 12 leads, which in biomedical parlance refers to a configuration of electrodes attached to the body [34]. Refer to Figure 1.5. The voltage between the arms is Lead I, Lead II is the potential between the right arm and left leg, and Lead III reads between the left arm and leg. The WCT is a common point that is formed by connecting the three limb electrodes through weighting resistors. Lead aVL measures potential difference between the left arm and the WCT. Similarly, lead aVR is the voltage between the right arm and the WCT. Lead aVF is between the left leg and the WCT. Finally, six more electrodes are fixed upon the chest, around the heart. Leads V1 through V6 measure the voltages between these sensors and the WCT. This circuit

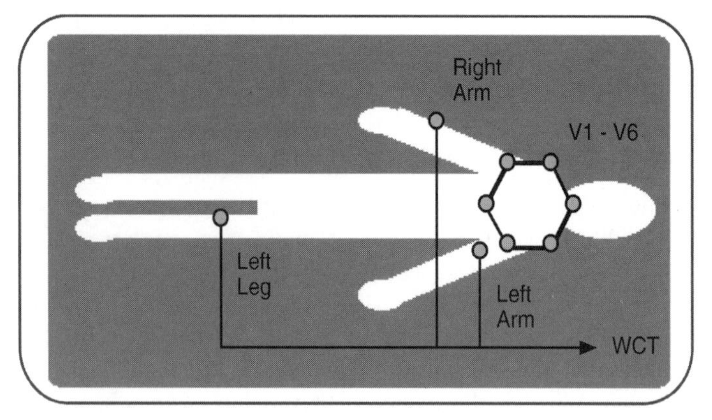

Fig. 1.5. The standard ECG configuration produces 12 signals from various electrodes attached to the subject's chest, arms, and leg.

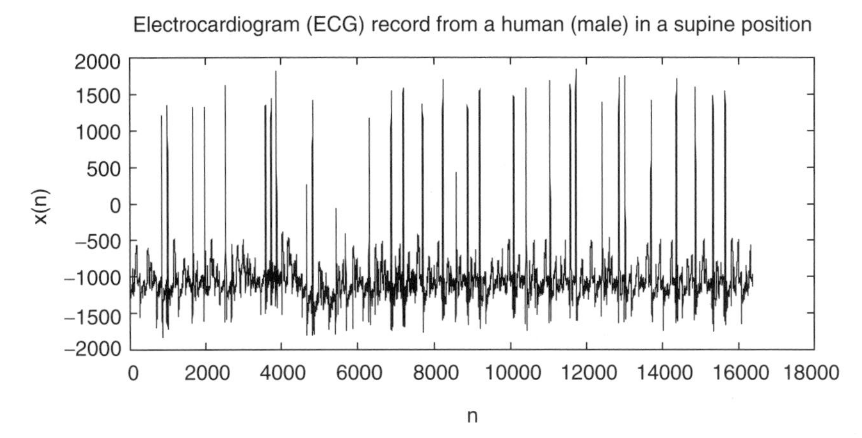

Fig. 1.6. One lead of an ECG: A human male in supine position. The sampling rate is 1 kHz, and the samples are digitized at 12 bits per sample. The irregularity of the heartbeat is evident.

arrangement is complicated; in fact, it is redundant. Redundancy provides for situations where a lead produces a poor signal and allows some cross-checking of the readings. Interpretation of 12-lead ECGs requires considerable training, experience, and expert judgment.

What does an ECG trace look like? Figure 1.6 shows an ECG trace from a single lead. Generally, an ECG has three discernible pulses: the P wave, the QRS complex, and the T wave. The P wave occurs upon excitation of the auricles of the heart, when they draw in blood from the body and lungs. The large-magnitude QRS complex occurs during the contraction of the vertricles as they contract to pump blood out of the heart. The Q and S waves are negative pulses, and the R wave is a positive pulse. The T wave arises during repolarization of the ventricles. The ECG signal is originally analog in nature; it is the continuous record of voltages produce across the various leads supported by the instrument. We could attach a millivoltmeter across an electrode pair and watch the needle jerk back and forth. Visualizing the signal's shape is easier with an oscilloscope, of course, because the instrument records the trace on its cathode ray tube. Both of these instruments display analog waveforms. If we could read the oscilloscope's output at regular time instants with perfect precision, then we would have—in principle, at least—a discrete representation of the ECG. But for computer display and automatic interpretation, the analog signal must be converted to digital form. In fact, Figure 1.6 is the result of such a digitization. The signal $v(n)$ appears continuous due to the large number of samples and the interpolating lines drawn by the graphics package that produced the illustration.

Interpreting ECGs is often difficult, especially in abnormal traces. A wide literature describing the 12-lead ECG exists. There are many guides to help technicians, nurses, and physicians use it to diagnose heart conditions. Signal processing and analysis of ECGs is a very active research area. Reports on new techniques, algorithms, and comparison studies continue to appear in the biomedical engineering and signal analysis literature [35].

One technical problem in ECG interpretation is to assess the regularity of the heart beat. As a time-domain signal description problem, this involves finding the separation between peaks of the QRS complex (Figure 1.6). Large time variations between peaks indicates dysrhythmia. If the time difference between two peaks, $v(n_1)$ and $v(n_0)$, is $\Delta_T = n_1 - n_0$, then the instantaneous heart rate becomes $60(\Delta_T)^{-1}$ beats/m. For the sample in Figure 1.6, this crude computation will, however, produce a wildly varying value of doubtful diagnostic use. The application calls for some kind of averaging and summary statistics, such as a report of the standard deviation of the running heart rate, to monitor the dysrhythmia.

There remains the technical problem of how to find the time location of QRS peaks. For an ideal QRS pulse, this is not too difficult, but the signal analysis algorithms must handle noise in the ECG trace. Now, because of the noise in the ECG signal, there are many local extrema. Evidently, the QRS complexes represent signal features that have inordinately high magnitudes; they are mountains above the forest of small-scale artifacts. So, to locate the peak of a QRS pulse, we might select a threshold M that is bigger than the small artifacts and smaller than the QRS peaks. We then deem any maximal, contiguous set of values $S = \{(n, v(n)): v(n) > M\}$ to be a QRS complex. Such regions will be disjoint. After finding the maximal value inside each such QRS complex, we can calculate Δ_T between each pair of maxima and give a running heart rate estimate. The task of dividing the signal up into disjoint regions, such as for the QRS pulses, is called signal *segmentation*. Chapter 4 explores this time domain procedure more thoroughly.

When there is poor heart rhythm, the QRS pulses may be jagged, misshapen, truncated, or irregulary spaced. A close inspection of the trace in Figure 1.7 seems to reveal this very phenomenon. In fact, one type of ventricular disorder that is

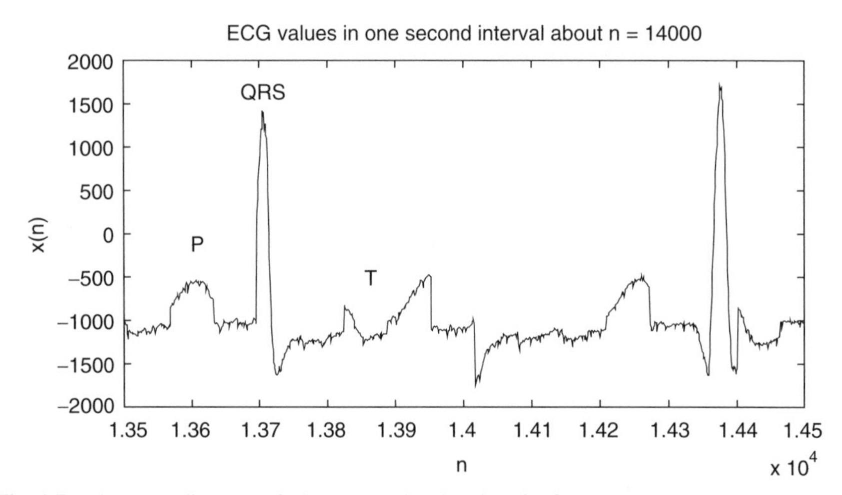

Fig. 1.7. Electrocardiogram of a human male, showing the fundametal waves. The 1-*s* time span around sample n = 14,000 is shown for the ECG of Figure 1.6. Note the locations of the P wave, the QRS complex, and—possibly—the T wave. Is there a broken P wave and a missing QRS pulse near the central time instant?

detectable in the ECG, provided that it employs a sufficiently high sampling rate, is *splintering* of the QRS complex. In this abnormal condition, the QRS consists of many closely spaced positive and negative transitions rather than a single, strong pulse. Note that in any ECG, there is a significant amount of signal noise. This too is clearly visible in the present example. Good peak detection and pulse location, especially for the smaller P and T waves, often require some data smoothing method. Averaging the signal values produces a smoother signal $w(n)$:

$$w(n) = \frac{1}{3}[v(n-1) + v(n) + v(n+1)]. \tag{1.1}$$

The particular formula (1.1) for processing the raw ECG signal to produce a less noisy $w(n)$ is called *moving average smoothing* or *moving average filtering*. This is a typical, almost ubiquitous signal processing operation. Equation (1.1) performs averaging within a symmetric window of width three about $v(n)$. Wider windows are possible and often useful. A window that is too wide can destroy signal features that bear on interpretation. Making a robust application requires judgment and experimentation.

Real-time smoothing operations require asymmetric windows. The underlying reason is that a symmetric smoothing window supposes knowledge of future signal values, such as $v(n + 1)$. To wit, as the computer monitoring system acquires each new ECG value $v(n)$, it can calculate the average of the last three values:

$$w(n) = \frac{1}{3}[v(n-2) + v(n-1) + v(n)]; \tag{1.2}$$

but at time instant n, it cannot possibly know the value of $v(n+1)$, which is necessary for calculating (1.1). If the smoothing operation occurs offline, after the entire set of signal values of interest has already been acquired and stored, then the whole range of signal values is accessible by the computer, and calculation (1.1) is, of course, feasible. When smoothing operations must procede in lockstep with acquisition operations, however, smoothing windows that look backward in time (1.2) must be applied.

Yet another method from removing noise from signals is to produce a signal whose values are the median of a window of raw input values. Thus, we might assign

$$w(n) = \text{Median}\{v(n-2), v(n-1), v(n), v(n+1), v(n+2)\} \tag{1.3}$$

so that $w(n)$ is the input value that lies closest to the middle of the range of five values around $v(n)$. A median filter tends to be superior to a moving average filter when the task is to remove isolated, large-magnitude spikes from a source signal. There are many variants. In general, smoothing is a common early processing step in signal analysis systems. In the present application, smoothing reduces the jagged noise in the ECG trace and improves the estimate of the QRS peak's location.

Contemplating the above algorithms for finding QRS peaks, smoothing the raw data, and estimating the instantaneous heart rate, we can note a variety of design choices. For example, how many values should we average to smooth the data? A span too small will fail to blur the jagged, noisy regions of the signal. A span too large may erode some of the QRS peaks. How should the threshold for segmenting QRS pulses be chosen? Again, an algorithm using values too small will falsely identify noisy bumps as QRS pulses. On the other hand, if the threshold values chosen are too large, then valid QRS complexes will be missed. Either circumstance will cause the application to fail. Can the thresholds be chosen automatically? The chemistry of the subject's skin could change while the leads are attached. This can cause the signal as a whole to trend up or down over time, with the result that the original threshold no longer works. Is there a way to adapt the threshold as the signal average changes so that QRS pulses remain detectable? These are but a few of the problems and tradeoffs involved in time domain signal processing and analysis.

Now we have illustrated some of the fundamental concepts of signal theory and, through the present example, have clarified the distinction between signal processing and analysis. Filtering for noise removal is a processing task. Signal averaging may serve our purposes, but it tends to smear isolated transients into what may be a quite different overall signal trend. Evidently, one aberrent upward spike can, after smoothing, assume the shape of a QRS pulse. An alternative that addresses this concern is median filtering. In either case—moving average or median filtering—the algorithm designer must still decide how wide to make the filters and discover the proper numerical values for thresholding the smoothed signal. Despite the analytical obstacles posed by signal noise and jagged shape, because of its prominence, the QRS complex is easier to characterize than the P and T waves.

There are alternative signal features that can serve as indicators of QRS complex location. We can locate the positive or negative transitions of QRS pulses, for example. Then the midpoint between the edges marks the center of each pulse, and the distance between these centers determines the instantaneous heart rate. This changes the technical problem from one of finding a local signal maximum to one of finding the positive- or negative-transition edges that bound the QRS complexes. Signal analysis, in fact, often revolves around edge detection. A useful indicator of edge presence is the discrete derivative, and a simple threshold operation identifies the significant changes.

1.1.2.2 Geothermal Measurements. Let us investigate an edge detection problem from geophysics. Ground temperature generally increases with depth. This variation is not as pronounced as the air temperature fluctuations or biophysical signals, to be sure, but local differences emerge due to the geological and volcanic history of the spot, thermal conductivity of the underlying rock strata, and even the amount of radioactivity. Mapping changes in ground termperature are important in the search for geothermal energy resources and are a supplementary indication of the underlying geological structures. If we plot temperature versus depth, we have a

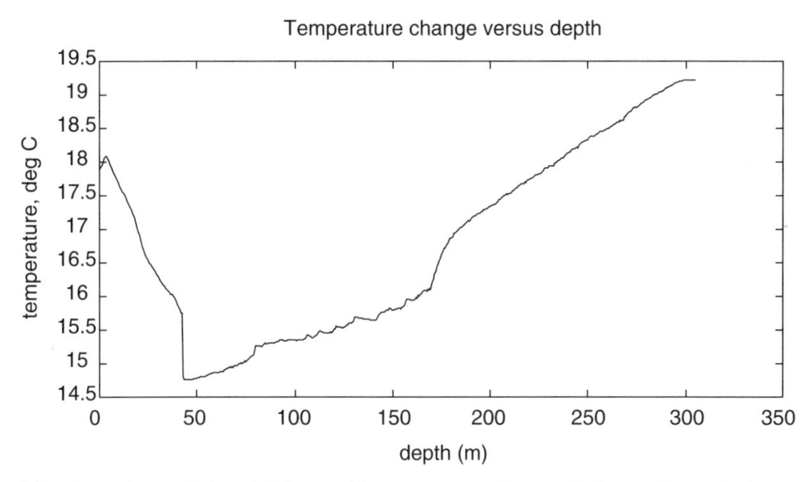

Fig. 1.8. A geothermal signal. The earth's temperature is sampled at various depths to produce a discrete signal with a spatially independent variable.

signal—the *geothermal gradient*—that is a function of distance, not time. It ramps up about 10°C per kilometer of depth and is a primary indicator for geothermal prospecting. In general, the geothermal gradient is higher for oceanic than for continental crust. Some 5% of the area of the United States has a gradient in the neighborhood of 40°C per kilometer of depth and has potential for use in geothermal power generation.

Mathematically, the geothermal gradient is the derivative of the signal with respect to its independent variable, which in this case measures depth into the earth. A very steep overall gradient may promise a geothermal energy source. A localized large magnitude gradient, or edge, in the temperature profile marks a geological artifact, such as a fracture zone. An example of the variation in ground temperature as one digs into the earth is shown in Figure 1.8.

The above data come from the second of four wells drilled on the Georgia–South Carolina border, in the eastern United States, in 1985 [36]. The temperature first declines with depth, which is typical, and then warmth from the earth's interior appears. Notice the large-magnitude positive gradients at approximately 80 and 175 m; these correspond to fracture zones. Large magnitude deviations often represent physically significant phenomena, and therein lies the importance of reliable methods for detecting, locating, and interpreting signal edges. Finding such large deviations in signal values is once again a time-domain signal analysis problem.

Suppose the analog ground temperature signal is $g(s)$, where s is depth into the earth. We seek large values of the derivative $g'(s) = dg/ds$. Approximating the derivative is possible once the data are digitized. We select a sampling interval $D > 0$ and set $x(n) = g(nD)$; then $x'(n) = x(n+1) - x(n-1)$ approximates the geothermal gradient at depth nD meters. It is further necessary to identify a threshold M for what constitutes a significant geothermal gradient. Threshold selection may rely upon expert scientific knowledge. A geophysicist might suggest significant gradients for the region. If we collect some statistics on temperature gradients, then the outlying values may be candidates for threshold selection. Again, there are local variations in the temperature profile, and noise does intrude into the signal acquisition apparatus. Hence, preliminary signal smoothing may once again be useful. Toward this end, we may also employ discrete derivative formulas that use more signal values:

$$x'(n) = \frac{1}{12}[x(n-2) - 8x(n-1) + 8x(n+1) - x(n+2)]. \tag{1.4}$$

Standard numerical analysis texts provide many alternatives [37]. Among the problems at the chapter's end are several edge detection applications. They weigh some of the alternatives for filtering, threshold selection, and finding extrema.

For now, let us remark that the edges in the ECG signal (Figure 1.6) are far steeper than the edges in the geothermal trace (Figure 1.8). The upshot is that the signal analyst must tailor the discrete derivative methods to the data at hand. Developing methods for edge detection that are robust with respect to sharp local variation of the signal features proves to be a formidable task. Time-domain methods, such as we consider here, are usually appropriate for edge detection problems. There comes a point, nonetheless, when the variety of edge shapes, the background noise in the source signals, and the diverse gradients cause problems for simple time domain techniques. In recent years, researchers have turned to edge detection algorithms that incorporate a notion of the size or scale of the signal features. Chapter 4 has more to say about time domain signal analysis and edge detection, in particular. The later chapters round out the story.

1.1.3 Analysis in the Time-Frequency Plane

What about signals whose values are symbolic rather than numeric? In ordinary usage, we consider sequences of signs to be signals. Thus, we deem the display of flags on a ship's mast, a series of hand gestures between baseball players, DNA codes, and, in general, any sequence of codes to all be "signals." We have already taken note of such usages. And this is an important idea, but we shall not call such a symbolic sequence a signal, reserving for that term a narrow scientific definition as an ordered set of numbers. Instead, we shall define a sequence of abstract symbols to be a structural interpretation of a signal.

It is in fact the conversion of an ordered set of numerical values into a sequence of symbols that constitutes a signal interpretation or analysis. Thus, a microphone receives a logitudinal compressive sound wave and converts it into electrical impulses, thereby creating an analog signal. If the analog speech signal is digitized, processed, and analyzed by a speech recognition engine, then the output in the form of ASCII text characters is a symbolic sequence that interprets, analyzes, or assigns meaning to the signal. The final result may be just the words that were uttered. But, more likely, the speech interpretation algorithms will generate a variety of intermediate representations of the signal's structure. It is common to build a large hierarchy of interpretations: isolated utterances; candidate individual word sounds within the utterances; possible word recognition results; refinements from grammatical rules and application context; and, finally, a structural result.

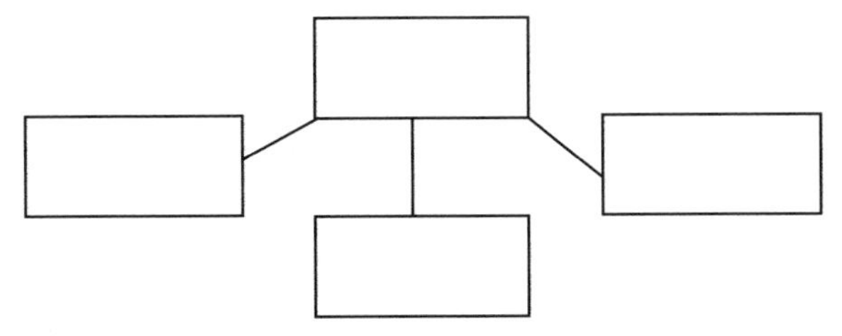

Fig. 1.9. Elementary graph structure for seismograms. One key analytical parameter is the time interval between the P waves and the S waves.

This framework applies to the applications covered in this section. A simple sequence of symbols representing the seismometer background, P waves, S waves, and surface waves may be the outcome of a structural analysis of a seismic signal (Figure 1.9).

The nodes of such a structure may have further information attached to them. For instance, the time-domain extent of the region, a confidence measure, or other analytical signal features can be inserted into the node data structure. Finding signal edges is often the prelude to a structural description of a signal. Figure 1.10

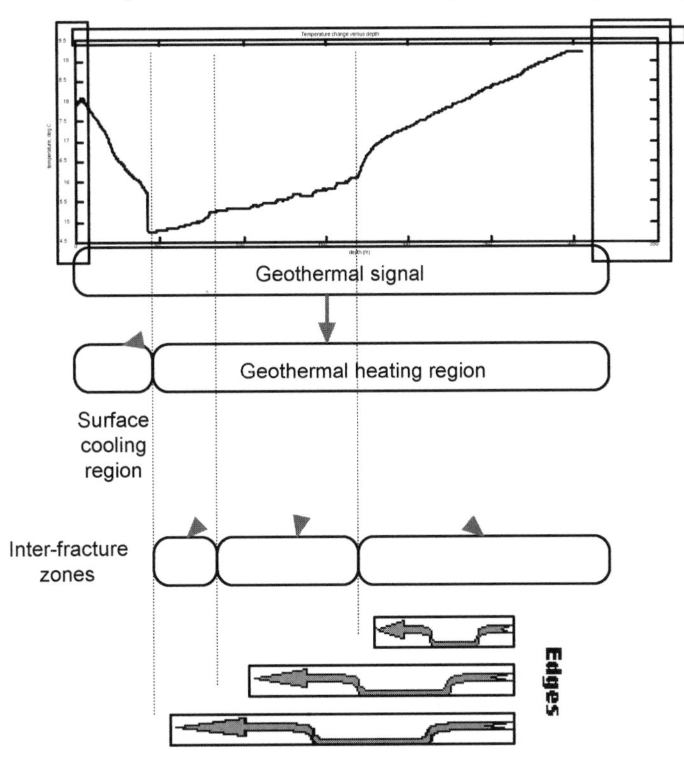

Fig. 1.10. Hypothetical geothermal signal structure. The root note of the interpretive structure represents the entire time-domain signal. Surface strata exhibit a cooling trend. Thereafter, geothermal heating effects are evident. Edges within the geothermal heating region indicate narrow fracture zones.

illustrates the decomposition of the geothermal profile from Figure 1.8 into a relational structure.

For many signal analysis problems, more or less flat relational structures that divide the signal domain into distinct regions are sufficient. Applications such as natural language understanding require more complicated, often hierarchical graph structures. Root nodes describe the coarse features and general subregions of the signal. Applying specialized algorithms to these distinct regions decomposes them further. Some regions may be deleted, further subdivided, or merged with their neighbors. Finally, the resulting graph structure can be compared with existing structural models or passed on to higher-level artificial intelligence applications.

1.1.4 Other Domains: Frequency and Scale

While we can achieve some success in processing and analyzing signals with elementary time-domain techniques, applied scientists regularly encounter applications demanding more sophisticated treatment. Thinking for a moment about the seismogram examples, we considered one aspect of their interpretation: finding the time difference between the arrival of the P and S waves. But how can one distinguish between the two wave sets? The distinction between them, which analysis algorithms must find, is in their oscillatory behavior and the magnitude of the oscillations. There is no monotone edge, such as characterized the geothermal signal. Rather, there is a change in the repetitiveness and the sweep of the seismograph needle's wiggling. When the oscillatory nature of a signal concerns us, then we are interested in its periodicity—or in other words, the reciprocal of period, the *frequency*.

Frequency-domain signal descriptions decompose the source signals into sinusoidal components. This strategy does improve upon pure time domain methods, given the appropriate application. A frequency-domain description uses some set of sinusoidal signals as a basis for describing a signal. The frequency of the sinusoid that most closely matches the signal is the principal frequency component of the signal. We can delete this principal frequency component from the source signal to get a difference signal. Then, we iterate. The first difference signal is further frequency analyzed to get a secondary periodic component and, of course, a second difference signal. The sinusoidal component identification and extraction continue until the difference signal consists of nothing but small magnitude, patternless, random perturbations—noise. This is a familiar procedure. It is just like the elementary linear algebra problem of finding the expansion coefficients of a given vector in terms of a basis set.

Thus, a frequency-domain approach is suitable for distinguishing the P waves from the S waves in seismogram interpretation. But, there is a caveat. We cannot apply the sinusoidal signal extraction to the whole signal, but rather only to small pieces of the signal. When the frequency components change radically on the separate, incoming small signal pieces, then the onset of the S waves must be at hand. The subtlety is to decide how to size the small signal pieces that will be subject to frequency analysis. If the seismographic station is far away, then the time interval between the initial P waves and the later S waves is large, and fairly large subintervals should suffice. If the seismographic station is close to the earthquake epicenter, on the other hand, then the algorithm must use very small pieces, or it will miss the short P wave region of the motion entirely. But if the pieces are made too small, then they may contain too few discrete samples for us to perform a frequency analysis. There is no way to know whether a temblor that has not happened yet will be close or far away. And the dilemma is how to size the signal subintervals in order to analyze all earthquakes, near and far, and all possible frequency ranges for the S and P waves.

It turns out that although such a frequency-domain approach as we describe is adequate for seismic signals, the strategy has proven to be problematic for the interpretation of electrocardiograms. The waves in abnormal ECGs are sometimes too variable for successful frequency-domain description and analysis.

Enter the notion of a *scale-domain* signal description. A scale-domain description of a signal breaks it into similarly shaped signal fragments of varying sizes. Problems that involve the time-domain size of signal features tend to favor this type of representation. For example, a scale-based analysis can offer improvements in electrocardiogram analysis; in this field it is a popular redoubt for researchers that have experimented with time domain methods, then frequency-domain methods, and still find only partial success in interpreting ECGs.

We shall also illustrate the ideas of frequency- and scale-domain descriptions in this first chapter. A complete understanding of the methods of frequency- and scale-domain descriptions requires a considerable mathematical expertise. The next two sections provide some formal definitions and a variety of mathematical examples of signals. The kinds of functions that one normally studies in algebra, calculus, and mathematical analysis are quite different from the ones at the center of signal theory. Functions representing signals are often discontinuous; they tend to be irregularly shaped, blocky, spiky, and altogether more ragged than the smooth and elegant entities of pure mathematics.

1.2 ANALOG SIGNALS

At the scale of objects immediately present to human consciousness and at the macroscopic scale of conventional science and technology, measurable phenomena tend to be continuous in nature. Hence, the raw signals that issue from nature—temperatures, pressures, voltages, flows, velocities, and so on—are commonly measured through analog instruments. In order to study such real-world signals, engineers and scientists model them with mathematical functions of a real variable. This strategy brings the power and precision of mathematical analysis to bear on engineering questions and problems that concern the acquisition, transmission, interpretation, and utilization of natural streams of numbers (i.e., signals).

Now, at a very small scale, in contrast to our perceived macroscopic world, natural processes are more discrete and quantized. The energy of electromagnetic radiation exists in the form of individual quanta with energy $E = h/\lambda$, where h is

Planck's constant,[3] and λ is the wavelength of the radiation. Phenomena that we normally conceive of possessing wave properties exhibit certain particle-like behaviors. On the other hand, elementary bits of matter, electrons for instance, may also reveal certain wave-like aspects. The quantization of nature at the subatomic and atomic levels leads to discrete interactions at the molecular level. Lumping ever greater numbers of discretized interactions together, overall statistics take priority over particular interactions, and the continuous nature of the laws of nature at a large scale then become apparent.[4] Though nature is indeed discrete at the microlevel, the historical beginnings of common sense, engineering, and scientific endeavor involve reasoning with continuously measurable phenomena. Only recently, within the last century have the quantized nature of the interactions of matter and energy become known. And only quite recently, within our own lifetimes, have machines become available to us—digital computers—that require for their application the discretization of their continuous input data.

1.2.1 Definitions and Notation

Analog signal theory proceeds directly from the analysis of functions of a real variable. This material is familiar from introductory calculus courses. Historically, it also precedes the development of discrete signal theory. And this is a curious circumstance, because the formal development of analog signal theory is far more subtle—some would no doubt insist the right term is perilous—than discrete time signal processing and analysis.

Definition (Analog Signals). An *analog signal* is a function $x: \mathbb{R} \to \mathbb{R}$, where $\mathbb{R}$ is the set of real numbers, and $x(t)$ is the signal value at time t. A *complex-valued* analog signal is a function $x: \mathbb{R} \to \mathbb{C}$. Thus, $x(t) = x_r(t) + jx_i(t)$, where $x_r(t)$ is the real part of $x(t)$; $x_i(t)$ is the imaginary part of $x(t)$; both of these are real-valued signals; and $j^2 = -1$.

Thus, we simply identify analog signals with functions of a real variable. Ordinarily, analog signals, such the temperature of an oven varying over time, take on real values. In other cases, where signal timing relationships come into question, or the frequency content of signals is an issue, complex-valued signals are often used. We will work with both real- and complex-valued signals in this section. Section 1.7 considers the complex number system, complex-valued signals, and the mathematics of complex numbers in more detail. Complex-valued signals arise primarily in the study of signal frequency.

[3]To account for the observation that the maximum velocity of electrons dislodged from materials depended on the frequency of incident light, Max Planck (1858–1947) conjectured that radiant energy consists of discrete packets, called *photons* or *quanta*, thus discovering the quantum theory.

[4]This process draws the attention of philosophers (N. Hartmann, *New Ways of Ontology*, translator R. C. Kuhn, Chicago: Henry Regnery, 1953) and scientists alike (W. Zurek, "Decoherence and the transition from quantum to classical," *Physics Today*, vol. 44, no. 10, pp. 36–44, October 1991).

Of course, the independent variable of an analog signal does not have to be a time variable. The pneumatic valve of a bicycle tire follows a sinusoidal course in height above ground as the rider moves down the street. In this case the analog signal is a function of distance ridden rather than time passed. And the geothermal gradient noted in the previous section is an example of a signal that is a function of depth in the earth's crust.

It is possible to generalize the above definition to include multichannel signals that take values in $\mathbb{R}^n$, $n \geq 2$. This is a straightforward generalization for all of the theory that we develop. Another way to generalize to higher dimensionality is to consider signals with domains contained in $\mathbb{R}^n$, $n \geq 2$. This is the discipline of image processing, at least for $n = 2$, 3, and 4. As a generalization of signal processing, it is not so straightforward as multichannel theory; the extra dimension in the independent signal variable leads to complications in signal interpretation and imposes severe memory and execution time burdens for computer-based applications.

We should like to point out that modeling natural signals with mathematical functions is an inherently flawed step; many functions do not correspond to any real-world signal. Mathematical functions can have nonzero values for arbitrarily large values of their independent variable, whereas in reality, such signals are impossible; every signal must have a finite past and eventually decay to nothing. To suppose otherwise would imply that the natural phenomenon giving rise to the signal could supply energy indefinitely. We can further imagine that some natural signals containing random noise cannot be exactly characterized by a mathematical rule associating one independent variable with another dependent variable.

But, is it acceptable to model real-world signals with mathematical models that eventually diminish to zero? This seems unsatisfactory. A real-world signal may decay at such a slow rate that in choosing a function for its mathematical model we are not sure where to say the function's values are all zero. Thus, we should prefer a theory of signals that allows signals to continue forever, perhaps diminishing at an allowable rate. If our signal theory accomodates such models, then we have every assurance that it can account for the wildest natural signal that the real world can offer. We will indeed pursue this goal, beginning in this first chapter. With persistence, we shall see that natural signals do have mathematical models that reflect the essential nature of the real-world phenomenon and yet are not limited to be zero within finite intervals. We shall find as well that the notion of randomness within a real-world signal can be accommodated within a mathematical framework.

1.2.2 Examples

The basic functions of mathematical analysis, known from algebra and calculus, furnish many elementary signal models. Because of this, it is common to mix the terms "signal" and "function." We may specify an analog signal from a formula that relates independent variable values with dependent variable values. Sometimes the formula can be given in closed form as a single equation defining the signal values. We may also specify other signals by defining them piecewise on their domain. Some functions may best be described by a geometric definition. Still other functions representing analog signals may be more convenient to sketch rather than specify mathematically.

1.2.2.1 *Polynomial, Rational, and Algebraic Signals.* Consider, for example, the *polynomial* signal,

$$x(t) = \sum_{k=0}^{N} a_k t^k . \tag{1.5}$$

$x(t)$ has derivatives of all orders and is continuous, along with all of its derivatives. It is quite unlike any of nature's signals, since its magnitude, $|x(t)|$, will approach infinity as $|t|$ becomes large. These signals are familiar from elementary algebra, where students find their roots and plot their graphs in the Cartesian plane. The domain of a polynomial $p(t)$ can be divided into disjoint regions of concavity: concave upward, where the second derivative is positive; concave downward, where the second derivative is negative; and regions of no concavity, where the second derivative is zero, and $p(t)$ is therefore a line. If the domain of a polynomial $p(t)$ containsan interval $a < t < b$ where $\frac{d^2}{dt^2}p(t) = 0$ for all $t \in (a, b)$, then $p(t)$ is a line. However familiar and natural the polynomials may be, they are not the signal family with which we are most intimately concerned in signal processing. Their behavior for large $|t|$ is the problem. We prefer mathematical functions that more closely resemble the kind of signals that occur in nature: Signals $x(t)$ which, as $|t|$ gets large, the signal either approaches a constant, oscillates, or decays to zero. Indeed, we expend quite an effort in Chapter 2 to discover signal families—called *function* or *signal spaces*—which are faithful models of natural signals.

The concavity of a signal is a very important concept in certain signal analysis applications. Years ago, the psychologist F. Attneave [38] noted that a scattering of simple curves suffices to convey the idea of a complex shape—for instance, a cat. Later, computer vision researchers developed the idea of assemblages of simple, oriented edges into complete theories of low-level image understanding [39–41]. Perhaps the most influential among them was David Marr, who conjectured that understanding a scene depends upon the extraction of edge information [39] over a range of visual resolutions from coarse to fine multiple scales. Marr challenged computer vision researchers to find processing and analysis paradigms within biological vision and apply them to machine vision. Researchers investigated the applications of concavity and convexity information at many different scales. Thus, an intricate shape might resolve into an intricate pattern at a fine scale, but at a coarser scale might appear to be just a tree. How this can be done, and how signals can be smoothed into larger regions of convexity and concavity without increasing the number of differently curved regions, is the topic of scale-space analysis [42,43]. We have already touched upon some of these ideas in our discussion of edges of the QRS complex of an electrocardiogram trace and in our discussion of the geothermal gradient. There the scale of an edge corresponded to the number of points incorporated in the discrete derivative computation. This is precisely the notion we are trying to illustrate, since the scale of an edge is a measure of its time-domain extent. Describing signal features by their scale is most satisfactorily accomplished using special classes of signals (Section 1.6). At the root of all of this deep theory, however, are the basic calculus notion of the sign of the second derivative and the intuitive and simple polynomial examples.

Besides motivating the notions of convexity and concavity as component building blocks for more complicated shapes, polynomials are also useful in signal theory as interpolating functions. The theory of splines generalizes linear interpolation. It is one approach to the modern theory of wavelet transforms. Interpolating the values of a discrete signal with continuous polynomial sections—connecting the dots, so to speak—is the opposite process to sampling a continuous-domain signal.

If $p(t)$ and $q(t)$ are polynomials, then $x(t) = p(t)/q(t)$ is a *rational* function. Signals modeled by rational functions need to have provisions made in their definitions for the times t_0 when $q(t_0) = 0$. If, when this is the case, $p(t_0) = 0$ also, then it is possible that the limit,

$$\lim_{t \to t_0} \frac{p(t)}{q(t)} = r_0 = x(t_0), \tag{1.6}$$

exists and can be taken to be $x(t_0)$. This limit does exist when the order of the zero of $p(t)$ at $t = t_0$ is at least the order of the zero of $q(t)$ at $t = t_0$.

Signals that involve a rational exponent of the time variable, such as $x(t) = t^{1/2}$, are called *algebraic* signals. There are often problems with the domains of such signals; to the point, $t^{1/2}$ does not take values on the negative real numbers. Consequently, we must usually partition the domain of such signals and define the signal piecewise. One tool for this is the upcoming unit step signal $u(t)$.

1.2.2.2 Sinusoids. A more real-to-life example is a *sinusoidal* signal, such as $\sin(t)$ or $\cos(t)$. Of course, the mathematician's sinusoidal signals are synthetic, ideal creations. They undulate forever, whereas natural periodic motion eventually deteriorates. Both $\sin(t)$ and $\cos(t)$ are differentiable: $\frac{d}{dt}\sin(t) = \cos(t)$ and $\frac{d}{dt}\cos(t) = -\sin(t)$. From this it follows that both have derivatives of all orders and have Taylor[5] series expansions about the origin:

$$\sin(t) = t - \frac{t^3}{3!} + \frac{t^5}{5!} - \frac{t^7}{7!} + \cdots \tag{1.7a}$$

$$\cos(t) = 1 - \frac{t^2}{2!} + \frac{t^4}{4!} - \frac{t^6}{6!} + \cdots . \tag{1.7b}$$

[5]The idea is due to Brook Taylor (1685–1731), an English mathematician, who—together with many others of his day—sought to provide rigorous underpinnings for Newton's calculus.

So, while $\sin(t)$ and $\cos(t)$ are most intuitively described by the coordinates of a point on the unit circle, there are also formulas (1.7a)–(1.7b) that define them. In fact, the Taylor series, where it is valid for a function $x(t)$ on some interval $a < t < b$ of the real line, shows that a function is the limit of a sequence of polynomials: $x(a)$, $x(a) + x^{(1)}(a)(t-a)$, $x(a) + x^{(1)}(t-a) + x^{(2)}(t-a)^2/2!, \ldots$, where we denote the nth-order derivative of $x(t)$ by $x^{(n)}(t)$.

The observation that $\sin(t)$ and $\cos(t)$ have a Taylor series representation (1.7a)–(1.7b) inspires what will become one of our driving principles. The polynomial signals may not be very lifelike, when we consider that naturally occurring signals will tend to wiggle and then diminish. But sequences of polynomials, taken to a limit, converge to the sinusoidal signals. The nature of the elements is completely changed by the limiting process. This underscores the importance of convergent sequences of signals, and throughout our exposition we will always be alert to examine the possibility of taking signal limits. Limit processes constitute a very powerful means for definining fundamentally new types of signals.

From their geometric definition on the unit circle, the sine and cosine signals are periodic; $\sin(t + 2\pi) = \sin(t)$ and $\cos(t + 2\pi) = \cos(t)$ for all $t \in \mathbb{R}$. We can use the trigonometric formulas for $\sin(s + t)$ and $\cos(s + t)$, the limit $\frac{\sin(t)}{t} \to 1$ as $t \to 0$, and the limit $\frac{\cos(t) - 1}{t} \to 0$ as $t \to 0$ to discover the derivatives and hence the Taylor series. Alternatively, we can define $\sin(t)$ and $\cos(t)$ by (1.7a)–(1.7b), whence we derive the addition formulas; define π as the unique point $1 < \pi/2 < 2$, where $\cos(t) = 0$; and, finally, show the periodicity of sine and cosine [44].

1.2.2.3 Exponentials.

Exponential signals are of the form

$$x(t) = Ce^{at}, \tag{1.8}$$

where C and a are constants, and e is the real number b for which the exponential $x(t) = b^t$ has derivative $\frac{d}{dt}x(t) = 1$ for $t = 0$. For $C = a = 1$, we often write $x(t) =$ $\exp(t)$. The derivative of $\exp(t)$ is itself. This leads to the Taylor series expansion about $t = 0$:

$$e = \exp(t) = 1 + t + \frac{t^2}{2!} + \frac{t^3}{3!} + \cdots = \sum_{k=0}^{\infty} \frac{t^k}{k!}. \tag{1.9}$$

Notice once more that a polynomial limit process creates a signal of a completely different genus. Instead of a periodic signal, the limit in (1.9) grows rapidly as $t \to \infty$ and decays rapidly as $t \to -\infty$.

If $C > 0$ and $a > 0$ in (1.9), then the graph of the exponential signal is an ever-increasing curve for $t > 0$ and an ever-decaying curve for $t < 0$. Since it has non-zero derivatives of arbitrarily high orders, such an exponential grows faster than any polynomial for positive time values. For $a < 0$, the graph of the exponential reflects across the y-axis (Figure 1.11).

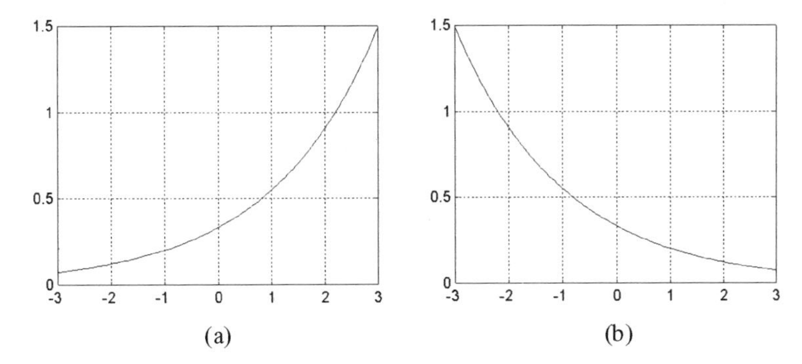

Fig. 1.11. Analog exponential signals. Panel (a) shows the exponential exp(t/2)/3, and (b) is its reflection across the y-axis, exp($-t$/2)3.

A particularly valuable relation for signal theory is the *Laplace*[6] *identity*; we take the exponent in (1.9) to be purely imaginary:

$$e^{js} = \exp(js) = \cos(s) + j\sin(s), \tag{1.10}$$

where s is real. Why this is true can be seen from the unit circle in the complex plane (Figure 1.12) and by examining the expansion (1.10) of e^{js} in the series (1.9). First, substitute js for t in the expansion (1.9). Next, group the real and

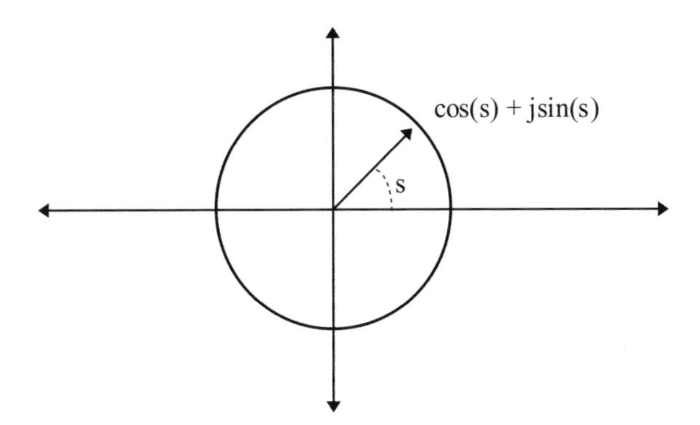

Fig. 1.12. Laplace's relation on the unit circle of the complex plane. By comparing Taylor series expansions, we find $e^{js} = \cos(s) + j\sin(s)$, and this corresponds to a point at arc distance s, counterclockwise on the unit circle from the positive x-axis.

[6]Pierre Simon Laplace (1749–1827), a French mathematician, physicist, and astronomer, theorized (along with German philosopher Immanuel Kant) that the solar system coalesced from a rotating gas cloud. The Laplace transform (Chapter 9) is named for him.

imaginary terms together. Observe that the sine and cosine Taylor series are in fact intermixed into the expansion of e^{js}, just as (1.10) expresses. The Laplace identity generalizes to any complex exponent $x + jy$: $e^{x+jy} = e^{x}[\cos(y) + j\sin(y)]$, where x and y are real. This is the most important formula in basic algebra.

The exponential signal is important in solving differential equations, such as arise from the study of heat transport and electromagnetism. For instance, the *heat diffusion equation* describes the propagation of heat $T(t, s)$ along a straight wire at time t and distance s from the end of the wire:

$$\frac{\partial T}{\partial t} = D\frac{\partial^2 T}{\partial s^2}, \tag{1.11}$$

where D is the *diffusion constant*. Solutions to (1.11) are $T(t, s) = e^{-\lambda t}e^{-jks}$, where λ and k are such that $D = \lambda/k^2$. The diffusion equation will make an unexpected appearance in Chapter 4 when we consider how to smooth a signal so that new regions of concavity do not appear as the smoothing progresses. Now, in electromagnetism, the electric and magnetic fields are vectors, $\boldsymbol{E}$ and $\boldsymbol{H}$, respectively, that depend upon one another. Maxwell's equations[7] for a vacuum describe this interaction in terms of space and time derivatives of the field vectors as follows:

$$\nabla \times \boldsymbol{E} = -\mu_0 \frac{\partial \boldsymbol{H}}{\partial t}, \tag{1.12a}$$

$$\nabla \times \boldsymbol{H} = \varepsilon_0 \frac{\partial \boldsymbol{E}}{\partial t}, \tag{1.12b}$$

$$\nabla \bullet \boldsymbol{E} = \nabla \bullet \boldsymbol{H} = 0. \tag{1.12c}$$

Equations (1.12a)–(1.12b) tell us that the curl of each field is proportional to the time derivative of the other field. The zero divergences in (1.12c) hold when there is no charge present. Constants μ_0 and ε_0 are the *magnetic permeability* and *electric permittivity* of space, respectively. By taking a second curl in (1.12a) and a second time derivative in (1.12b), separate equations in $\boldsymbol{E}$ and $\boldsymbol{H}$ result; for example, the electric field must satisfy

$$\nabla^2 \boldsymbol{E} = \mu_0 \varepsilon_0 \frac{\partial^2 \boldsymbol{E}}{\partial t^2}. \tag{1.13a}$$

For one spatial dimension, this becomes

$$\frac{\partial^2 \boldsymbol{E}}{\partial s^2} = \mu_0 \varepsilon_0 \frac{\partial^2 \boldsymbol{E}}{\partial t^2}. \tag{1.13b}$$

[7]Scottish physicist James Clerk Maxwell (1831–1879) is known best for the electromagnetic theory, but he also had a significant hand in the mechanical theory of heat.

Solutions to (1.13b) are sinusoids of the form $E(t, s) = A\cos(bs - \omega t)$, where $(b/\omega)^2 = \mu_0\varepsilon_0$, and b, ω, and A are constants.

Another signal of great importance in mathematics, statistics, engineering, and science is the *Gaussian.*[8]

Definition (Analog Gaussian). The *analog Gaussian* signal of mean μ and standard deviation σ is

$$g_{\mu,\sigma}(t) = \frac{1}{\sigma\sqrt{2\pi}} e^{-\frac{(t-\mu)^2}{2\sigma^2}}. \tag{1.14}$$

These terms are from statistics (Section 1.7). For now, however, let us note that the Gaussian $g_{\mu,\sigma}(t)$ can be integrated over the entire real line. Indeed, since (1.14) is always symmetric about the line $t = \mu$, we may take $\mu = 0$. The trick is to work out the square of the integral, relying on Fubini's theorem to turn the consequent iterated integral into a double integral:

$$\left(\int_{-\infty}^{\infty} g_{0,\sigma}(t)\,dt\right)^2 = \left(\frac{1}{\sqrt{2\pi}\sigma}\int_{-\infty}^{\infty} e^{-\frac{t^2}{2\sigma^2}}\,dt\right)\left(\frac{1}{\sqrt{2\pi}\sigma}\int_{-\infty}^{\infty} e^{-\frac{s^2}{2\sigma^2}}\,ds\right)$$
$$= \frac{1}{2\pi\sigma^2}\int_{-\infty}^{\infty}\int_{-\infty}^{\infty} e^{-\frac{s^2+t^2}{2\sigma^2}}\,dt\,ds \tag{1.15}$$

Changing to polar coordinates cracks the hard integral on the right-hand side of (1.15): $r^2 = t^2 + s^2$ and $dtds = rdrd\theta$. Hence,

$$\left(\int_{-\infty}^{\infty} g_{0,\sigma}(t)\,dt\right)^2 = \frac{1}{2\pi\sigma^2}\int_0^{2\pi}\int_0^{\infty} e^{-\frac{r^2}{2\sigma^2}}\,r\,dr\,d\theta = \frac{1}{2\pi}\int_0^{2\pi}\left(-e^{-\frac{r^2}{2\sigma^2}}\Bigg|_0^{\infty}\right)d\theta \tag{1.16}$$

and we have

$$\left(\int_{-\infty}^{\infty} g_{0,\sigma}(t)\,dt\right)^2 = \frac{1}{2\pi}\int_0^{2\pi}(0-(-1))\,d\theta = \frac{1}{2\pi}(\theta)\Big|_0^{2\pi} = 1. \tag{1.17}$$

[8]Karl Friedrich Gauss (1777–1855) is a renowned German mathematician, physicist, and astronomer.

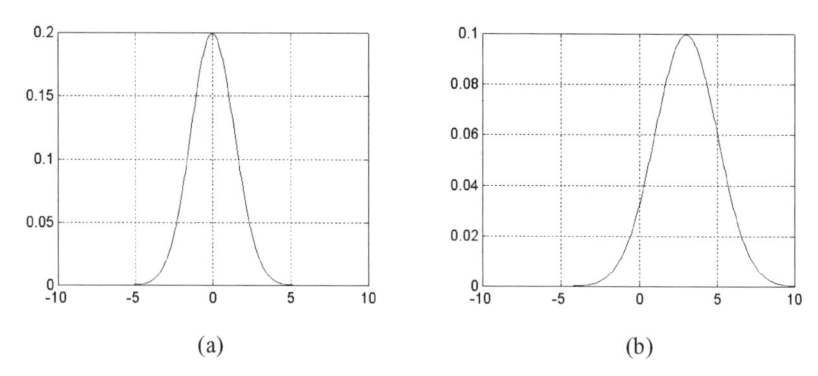

Fig. 1.13. Gaussian signals. The pulse on the left is $g_{0,2}(t)$, the Gaussian with mean μ equals 0, and the standard deviation σ equals 2. The pulse $g_{3,4}(t)$ is on the right. It has a wider spread than $g_{0,2}(t)$, and takes its smaller maximum value at $t = 3$.

Thus, there is unit area under the Gaussian curve. The Gaussian $g_{\mu,\sigma}(t)$ (1.14) is a bell-shaped curve (Figure 1.13), peaking at $t = \mu$, and symmetric about this line. The Gaussian decays forever as $|t| \to \infty$, but $g_{\mu,\sigma}(t) > 0$ for any $t \in \mathbb{R}$.

We define

$$g(t) = g_{0,1}(t) = \frac{1}{\sqrt{2\pi}} e^{-\frac{t^2}{2}}, \tag{1.18}$$

so that any Gaussian (1.14) is a scaled, shifted, and dilated version of $g(t)$:

$$g_{\mu,\sigma}(t) = \frac{1}{\sigma\sqrt{2\pi}} e^{-\frac{(t-\mu)^2}{2\sigma^2}} = \frac{1}{\sigma} g\left(\frac{t-\mu}{\sigma}\right). \tag{1.19}$$

The multiplying factor $(1/\sigma)$ governs the *scaling*, which may increase or decrease the height of the Gaussian. The same factor inside $g((t-\mu)/\sigma)$ *dilates* the Gaussian; it adjusts the spread of the bell curve according to the scale factor so as to preserve the unit area property. The peak of the bell shifts by the mean μ.

If we multiply a complex exponential $\exp(-j\omega t)$ by a Gaussian function, we get what is known as a *Gabor*[9] *elementary* function or signal [45].

[9]Dennis Gabor (1900–1979) analyzed these pulse-like signals in his 1946 study of optimal time and frequency signal representations. He is more famous outside the signal analysis discipline for having won the Nobel prize by inventing holography.

Definition (Gabor Elementary Functions). The *Gabor elementary function* $G_{\mu,\sigma,\omega}(t)$ is

$$G_{\mu,\,\sigma,\,\omega}(t) = g_{\mu,\,\sigma}(t)e^{j\omega t}. \tag{1.20}$$

Note that the real part of the Gabor elementary function $G_{\mu,\sigma,\omega}(t)$ in (1.20) is a cosine-like undulation in a Gaussian envelope. The imaginary part is a sine-like curve in a Gaussian envelope of the same shape (Figure 1.14). The time-frequency Gabor transform (Chapter 10) is based on Gabor elementary functions.

Interest in these signals surged in the mid-1980s when psychophysicists noticed that the modeled some aspects of the brain's visual processing. In particular, the receptive fields of adjacent neurons in the visual cortex seem to have profiles that resemble the real and imaginary parts of the Gabor elementary function. A controversy ensued, and researchers—electrical engineers, computer scientists, physiologists, and psychologists—armed with the techniques of mixed-domain signal decomposition continue to investigate and debate the mechanisms of animal visual perception [46, 47].

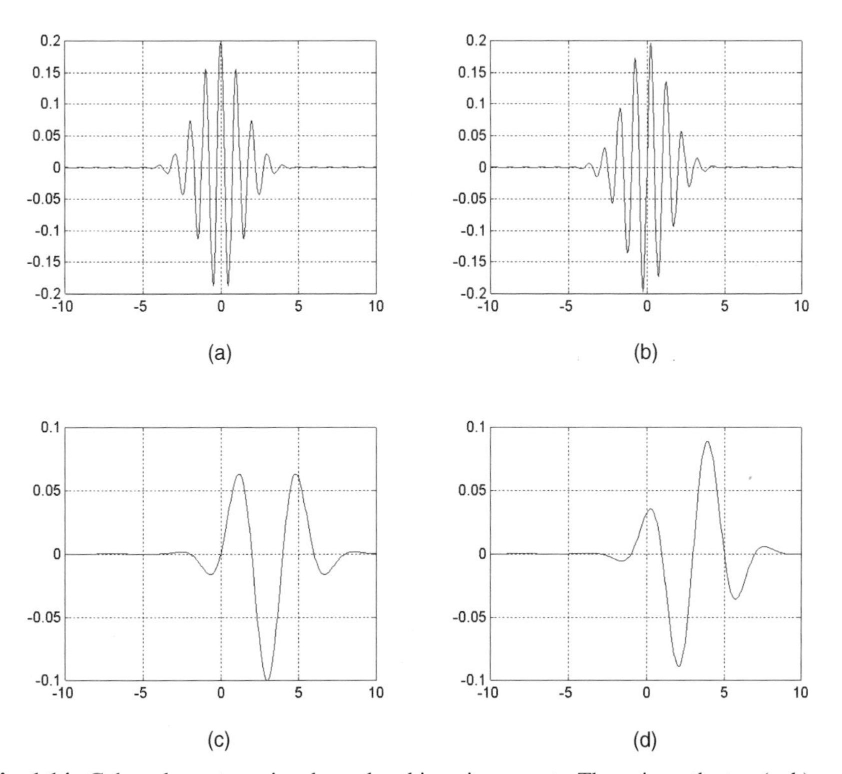

Fig. 1.14. Gabor elementary signals, real and imaginary parts. The pair on the top (a, b) are the real and imaginary parts of $g_{0,2}(t)\exp(j2\pi t)$. Below (c, d) is the Gabor pulse $G_{3,4,.5\pi}$. Note that if two Gabor elementary signals have the same sinusoidal frequency, but occupy Gaussian envelopes of different variances, then they have fundamentally different shapes.

1.2.3 Special Analog Signals

Several of the analog signal examples above are familiar from elementary algebra and calculus. Others, perhaps the Gabor elementary functions, are probably unfamiliar until one begins the formal study of signal processing and analysis. Some very simple analog signals play pivotal roles in the theoretical development.

1.2.3.1 Unit Step. We introduce the unit step and some closely related signals. The unit step signal (Figure 1.15) finds use in chopping up analog signals. It is also a building block for signals that consist of rectangular shapes and square pulses.

Definition (Unit Step). The *unit step* signal $u(t)$ is defined:

$$u(t) = \begin{cases} 1 & \text{if } t \geq 0, \\ 0 & \text{if } t < 0. \end{cases} \tag{1.21}$$

To chop up a signal using $u(t)$, we take the product $y(t) = x(t)u(t - c)$ for some $c \in \mathbb{R}$. The nonzero portion of $y(t)$ has some desired characteristic. Typically, this is how we zero-out the nonintegrable parts of signals such as $x(t) = t^{-2}$.

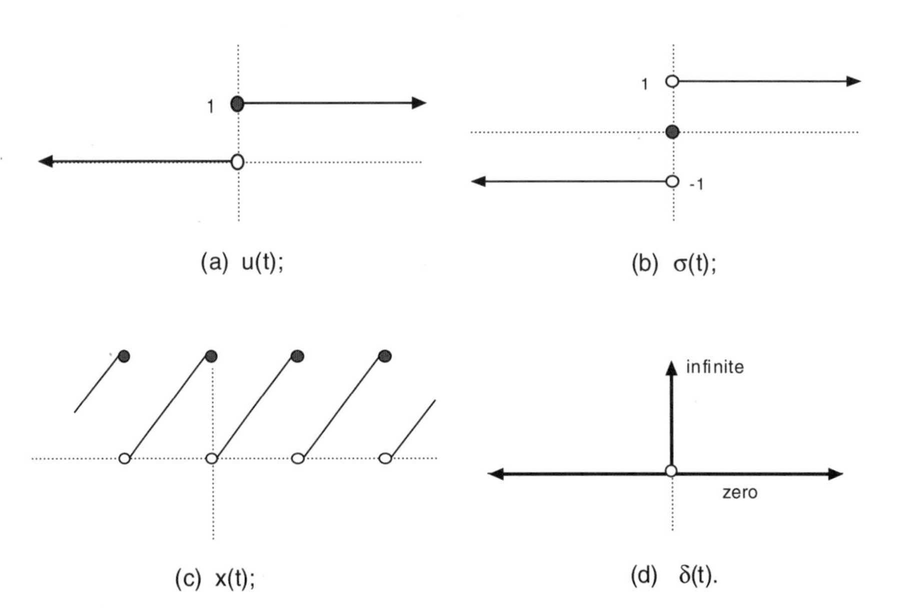

Fig. 1.15. Special Utility signals. (a) $u(t)$. (b) $\sigma(t)$. (c) $x(t)$. (d) $\delta(t)$. The unit step (a), signum (b), and sawtooth (c) are useful for constructing other signals and modeling their discontinuities. The Dirac delta (d) is "infinitely high" at $t = 0$ and zero otherwise; thus, it is not a bona fide analog signal. Chapters 3 and 5 provide the mathematical underpinnings of a valid, formal treatment of $\delta(t)$.

Definition (Signum). The *signum* signal $\sigma(t)$ is a cousin to the unit step:

$$\sigma(t) = \begin{cases} 1 & \text{if } t > 0, \\ 0 & \text{if } t = 0, \\ -1 & \text{if } t < 0. \end{cases} \tag{1.22}$$

Definition (Sawtooth). A sawtooth signal is a piecewise linear signal (Figure 1.15). For example, the infinite sawtooth $x(t)$ is

$$x(t) = \begin{cases} t & \text{if } t \geq 0, \\ 0 & \text{if } t < 0. \end{cases} \tag{1.23}$$

1.2.3.2 Dirac Delta. The *Dirac*[10] *delta* is really more of a fiction than a function. Nonetheless, it is a useful fiction. It can be made mathematically precise without losing its utility, and its informal development is familiar to many scientists and engineers.

For $n > 0$ let us define a sequence of analog signals $\delta_n(t)$:

$$\delta_n(t) = \begin{cases} \frac{n}{2} & \text{if } t \in \left[-\frac{1}{n}, \frac{1}{n}\right], \\ 0 & \text{if otherwise}. \end{cases} \tag{1.24}$$

The signals (1.24) are increasingly tall square spikes centered around the origin. Consider a general analog signal $x(t)$ and the integral over $\mathbb{R}$ of $x(t)\delta_n(t)$:

$$\int_{-\infty}^{\infty} x(t)\delta_n(t)\,dt = \int_{-1/n}^{1/n} x(t)\frac{n}{2}\,dt = \frac{1}{1/n-(-1/n)}\int_{-1/n}^{1/n} x(t)\,dt \tag{1.25}$$

The last term in (1.25) is the average value of $x(t)$ over $[-1/n, 1/n]$. As $n \rightarrow \infty$,

$$\lim_{n\rightarrow\infty}\int_{-\infty}^{\infty} x(t)\delta_n(t)\,dt = x(0). \tag{1.26}$$

The casual thought is to let $\delta(t)$ be the limit of the sequence $\{\delta_n(t): n > 0\}$ and conclude that the limit operation (1.26) can be moved inside the integral (Figure 1.16):

$$\lim_{n\rightarrow\infty}\int_{-\infty}^{\infty} x(t)\delta_n(t)\,dt = \int_{-\infty}^{\infty} x(t)\lim_{n\rightarrow\infty}\delta_n(t)\,dt = \int_{-\infty}^{\infty} x(t)\delta(t)\,dt = x(0). \tag{1.27}$$

[10]British physicist Paul Adrian Maurice Dirac (1902–1984) developed the theory of quantum electrodynamics. He received the Nobel prize in 1933.

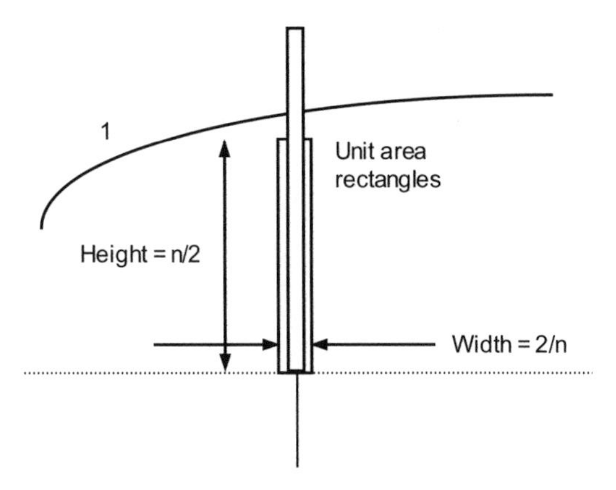

Fig. 1.16. Informal conception of the Dirac delta function. It is useful to think of $\delta(t)$ as the limit of a sequence of rectangles growing higher and narrower.

This idea is fundamentally mistaken, however. There is no pointwise limit of the sequence $\delta_n(t)$ at $t = 0$, and the limit of this signal sequence does not exist. The interchange of limit operations attempted in (1.27) is invalid. It is perhaps best to think of the final integral in (1.27) as an abbreviation for the valid limit operation in (1.26).

The Dirac delta can be shifted to any point t_0 in the domain of signal $x(t)$ and a similar argument applied. This gives the informal *sifting property* of the Dirac delta:

$$\int_{-\infty}^{\infty} x(t)\delta(t-t_0)\,dt = x(t_0). \tag{1.28}$$

Again, mathematical prudence suggests that we think of the sifting property as special way of writing a limit of integrals. We can add another story to this mythology: The Dirac delta is the derivative of the unit step $u(t)$. Let $n > 0$ and consider the following sequence of continuous signals $u_n(t)$ approximating the unit step.

$$u_n(t) = \begin{cases} 1 & \text{if } \frac{1}{n} < t, \\ \frac{(nt+1)}{2} & \text{if } -\frac{1}{n} \le t \le \frac{1}{n}, \\ 0 & \text{if } t < -\frac{1}{n}. \end{cases} \tag{1.29}$$

Note that as $n \to \infty$, $u_n(t) \to u(t)$ for all $t \neq 0$. Also, for all $t \notin [-1/n, 1/n]$,

$$\frac{d}{dt}u_n(t) = \delta_n(t). \tag{1.30}$$

We set aside our mathematical qualms and take limits as $n \to \infty$ of both sides of (1.30). The *derivative property* of the unit step results:

$$\lim_{n \to \infty} \frac{d}{dt} u_n(t) = \frac{d}{dt} \lim_{n \to \infty} u_n(t) = \frac{d}{dt} u(t) = \lim_{n \to \infty} \delta_n(t) = \delta(t). \qquad (1.31)$$

The convergence of a sequence of functions must be *uniform* in order that interchange of limit operations, such as (1.27) and (1.30), be valid. Advanced calculus texts cover this theory [44]. The mathematical theory of *distributions* [48, 49] provides a rigorous foundation for the idea of a Dirac delta, as well as the sifting and derivative properties.

1.3 DISCRETE SIGNALS

Now that we have looked at some functions that serve as models for real-world analog signals, let us assume that we have a method for acquiring samples. Depending upon the nature of the analog signal, this may be easy or difficult. To get a discrete signal that represents the hourly air temperature, noting the reading on a thermometer is sufficient. Air temperature varies so slowly that hand recording of values works just fine. Rapidly changing analog signals, in contrast, require faster sampling methods. To acquire digital samples over one million times per second is not at all easy and demands sophisticated electronic design.

In signal processing, both analog and digital signals play critical roles. The signal acquisition process takes place at the system's front end. These are electronic components connected to some sort of transducer: a microphone, for instance. An analog value is stored momentarily while the digitization takes place. This sample-and-hold operation represents a discrete signal. An analog-to-digital converter turns the stored sample into a digital format for computer manipulation. We will, however, not ordinarily deal with digital signals, because the limitation on numerical precision that digital form implies makes the theoretical development too awkward. Thus, the discrete signal—actually an abstraction of the key properties of digital signals that are necessary for mathematical simplicity and flexibility—turns out to be the most convenient theoretical model for real-life digital signals.

1.3.1 Definitions and Notation

Unlike analog signals, which have a continuous domain, the set of real numbers $\mathbb{R}$, discrete signals take values on the set of integers $\mathbb{Z}$. Each integer n in the domain of x represents a time instant at which the signal has a value $x(n)$. Expressions such as $x(2/3)$ make no sense for discrete signals; the function is not even defined there.

Definition (Discrete and Digital Signals). A *discrete-time* (or simply *discrete*) signal is a real-valued function $x: \mathbb{Z} \to \mathbb{R}$. $x(n)$ is the signal value at time instant n. A *digital* signal is an integer-valued function $x: \mathbb{Z} \to [-N, N]$, with domain $\mathbb{Z}$, $N \in \mathbb{Z}$, and $N > 0$. A complex-valued discrete-time signal is a function $x: \mathbb{Z} \to \mathbb{C}$, with domain $\mathbb{Z}$ and range included in the complex numbers $\mathbb{C}$.

Digital signals constitute a special class within the discrete signals. Because they can take on only a finite number of output values in the dependent variable, digital signals are rarely at the center of signal theory analyses. It is awkward to limit signal values to a finite set of integers, especially when arithmetic operations are performed on the signal values. Amplification is an example. What happens when the amplified value exceeds the maximum digital value? This is saturation, a very real problem for discrete signal processing systems. Some approach for avoiding saturation and some policy for handling it when it does occur must enter into the design considerations for engineered systems. To understand the theory of signals, however, it is far simpler to work with real-valued signals that may become arbitrarily small, arbitrarily large negative, and arbitrarily large positive. It is simply assumed that a real machine implementing signal operations would have a sufficiently high dynamic range within its arithmetic registers.

Notation. We use variable names such as "n", "m", and "k" for the independent variables of discrete signals. We prefer that analog signal independent variables have names such as "t" and "s". This is a tradition many readers will be comfortable with from Fortran computer programming. On those occasions when the discussion involves a sampling operation, and we want to use like names for the analog source and discrete result, we will subscript the continuous-domain signal: $x_a(t)$ is the analog source, and $x(n) = x_a(nT)$ is the discrete signal obtained from $x_a(t)$ by taking values every T time units.

With discrete-time signals we can tabulate or list signal values—for example, $x(n) = [3, 2, 1, \underline{5}, 1, 2, 3]$ (Figure 1.17). The square brackets signify that this is a discrete signal definition, rather than a set of integers. We must specify where the independent variable's zero time instant falls in the list. In this case, the value at $n = 0$ is

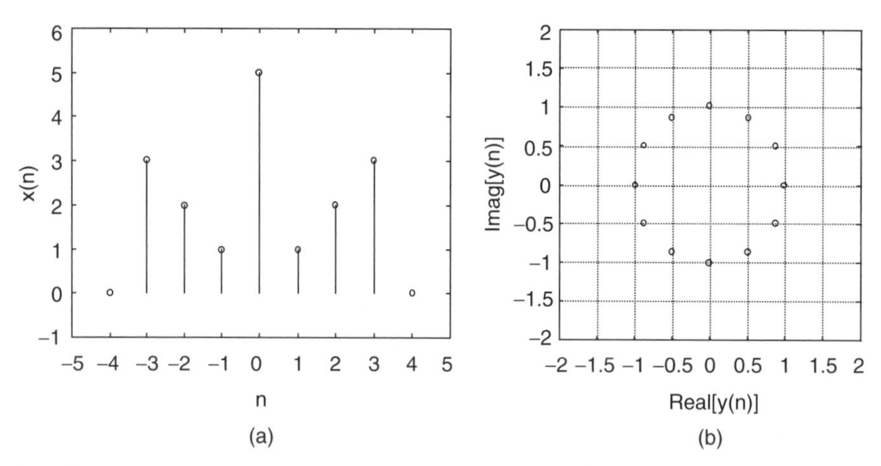

Fig. 1.17. Discrete signals. Panel (a) shows the signal $x(n) = [3, 2, 1, \underline{5}, 1, 2, 3]$. Signals may also be complex-valued, in which case their graphs are plotted in the complex plane. In (b), points of the signal $y(n) = \cos(n\pi/6) + j\sin(n\pi/6)$ are shown as pairs, (Real[$y(n)$], Imag[$y(n)$]).

underlined, and $x(0) = 5$, $x(-1) = 1$, $x(1) = 1$, and so on. For time instants not shown, signal values are zero. Thus, for discrete signals with an infinite number of nonzero values, we must provide a formula or rule that relates time instances to signal values, just as with analog signals.

1.3.2 Examples

We can derive straightforward discrete equivalents from the examples of analog signals above. A few curious properties relating to periodicity and derivatives arise.

1.3.2.1 Polynomials and Kindred Signals. There are discrete polynomial, rational, and algebraic signals. Discrete polynomials have the form

$$x(n) = \sum_{k=0}^{N} a_k n^k. \tag{1.32}$$

We cannot form the instantaneous derivative of $x(n)$ in (1.32) as with analog signals; instead discrete approximations must suffice. A variety of sometimes useful, but often problematic, notions of discrete derivatives do exist. For example, the left-hand discrete derivative of (1.32) is defined by $x_{\text{left}}(n) = x(n) - x(n-1)$. And a right-hand derivative exists too: $x_{\text{right}}(n) = x(n+1) - x(n)$. We can continue taking discrete derivatives of discrete derivatives. Note that there is no worry over the existence of a limit, since we are dividing the difference of successive signal values by the distance between them, and that can be no smaller than unity. Thus, discrete signals have (discrete) derivatives of all orders.

The domain of a polynomial $p(n)$ can be divided into disjoint regions of concavity: concave upward, where the second discrete derivative is positive; concave downward, where the second discrete derivative is negative; and regions of no concavity, where the second discrete derivative is zero, and $p(n)$ is therefore a set of dots on a line. Here is a first example, by the way, of how different analog signals can be from their discretely sampled versions. In the case of nonlinear analog polynomials, inflection points are always isolated. For discrete polynomials, though, there can be whole multiple point segments where the second derivative is zero.

If $p(n)$ and $q(n)$ are polynomials, then $x(n) = p(n)/q(n)$ is a *discrete rational* function. Signals modeled by discrete rational functions need to have provisions made in their definitions for the times n_0 when $q(n_0) = 0$. If, when this is the case, $p(n_0) = 0$ also, then it is necessary to separately specify the value of $x(n_0)$. There is no possibility of resorting to a limit procedure on $x(n)$ for a signal value, as with analog signals. Of course, if both $p(n)$ and $q(n)$ derive via sampling from analog ancestors, then such a limit, if it exists, could serve as the missing datum for $x(n_0)$.

Signals that involve a rational exponent of the time variable, such as $x(n) = n^{1/2}$, are called *discrete algebraic* signals. Again, there are problems with the domains of such signals; $n^{1/2}$ does not take values on the negative real numbers, for example. Consequently, we must usually partition the domain of such signals and define the signal piecewise.

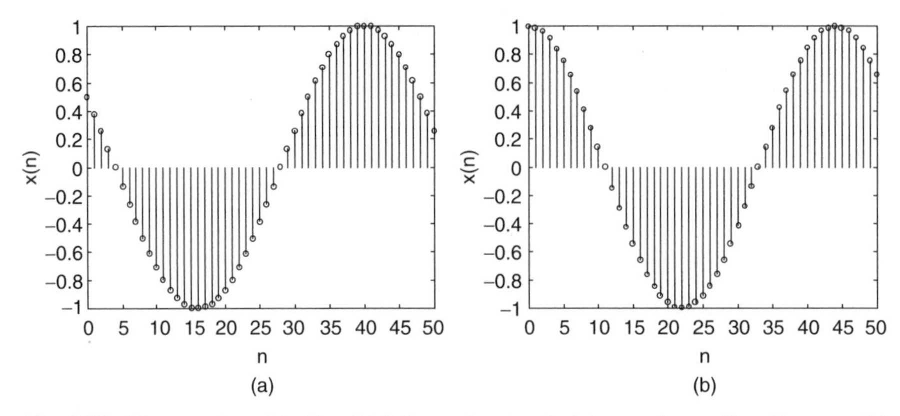

Fig. 1.18. Discrete sinusoids. Panel (a) shows the signal $x(n) = \cos(\omega n + \phi)$, with $\omega = \pi/24$ and $\phi = \pi/3$. Signal $x(n)$ has period $T = 48$. Panel (b,) on the other hand, shows the signal $y(n) = \cos(\omega n + \phi)$, with $\omega = 1/7$. It is not periodic.

1.3.2.2 *Sinusoids.* Discrete sinusoidal signals, such as $\sin(\omega n)$ or $\cos(\omega n)$, arise from sampling analog sinusoids (Figure 1.18). The function $\sin(\omega n + \phi)$ is the *discrete sine* function of *radial frequency* ω and phase ϕ. We will often work with $\cos(\omega n + \phi)$—and call it a sinusoid also—instead of the sine function. Note that—somewhat counter to intuition—discrete sinusoids may not be periodic! The periodicity depends upon the value of ω in $\cos(\omega n + \phi)$. We will study this nuance later, in Section 1.5.

1.3.2.3 *Exponentials.* Discrete exponential functions take the form

$$x(n) = Ce^{an} = C\exp(an), \tag{1.33}$$

where C and a are constants. Discrete exponentials (Figure 1.19) are used in frequency domain signal analysis (Chapters 7–9).

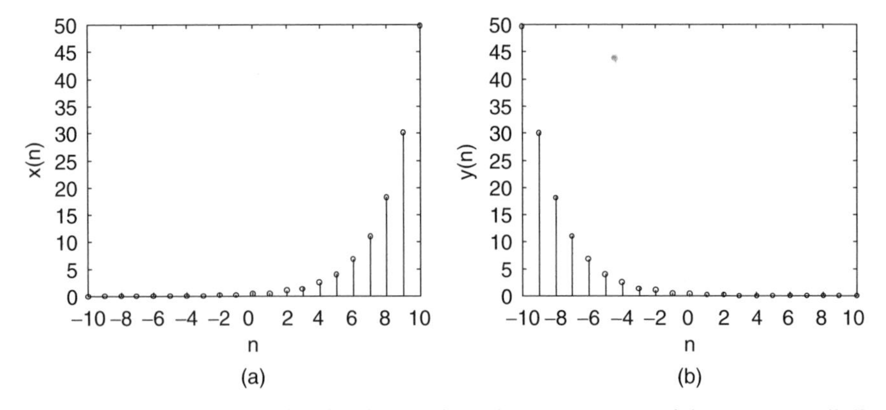

Fig. 1.19. Discrete exponential signals. Panel (a) shows the exponential $x(n) = \exp(n/2)/3$, and (b) is its reflection across the y-axis, $y(n) = \exp(-n/2)/3$.

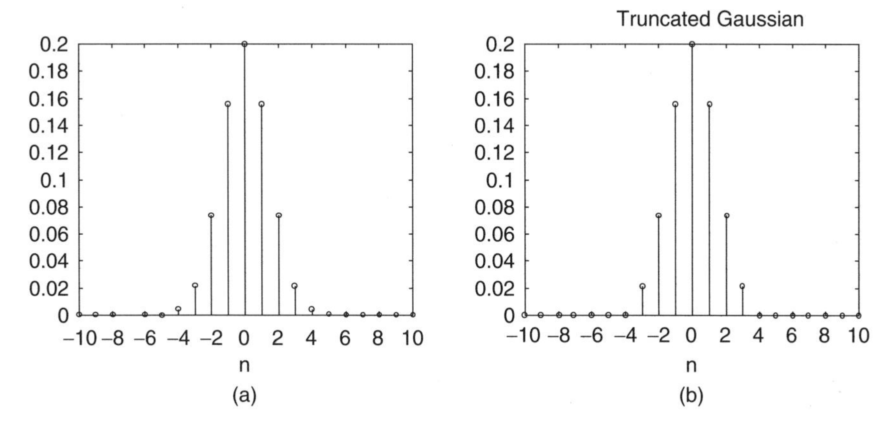

Fig. 1.20. Discrete Gaussian signals. The discrete pulse in (a) is $g_{0,2}(n)$, the Gaussian with mean μ equals 0 and standard deviation σ equals 2. Panel (b) illustrates a typical truncated Gaussian pulse.

The *discrete Gaussian* signal is

$$g_{\mu,\sigma}(n) = \frac{1}{\sigma\sqrt{2\pi}} e^{-\frac{(n-\mu)^2}{2\sigma^2}}. \tag{1.34}$$

Truncating the discrete Gaussian so that it is zero outside of some interval $[-N, N]$ is a discrete signal processing commonplace (Figure 1.20). This assumes that $g_{\mu,\sigma}(n)$ is small for $|n| > N$. As a signal with a finite number of nonzero values, the truncated discrete Gaussian serves as a noise removal filter. We can use it instead of the moving average filter (Section 1.1), giving preference to local signal values, for example. Also, for noise removal it makes sense to normalize the nonzero values so that their sum is unity. This preserves the average value of the raw signal.

There are also discrete versions of the Gabor elementary functions:

$$G_{\mu,\sigma,\omega}(n) = g_{\mu,\sigma}(n) e^{j\omega n}. \tag{1.35}$$

1.3.3 Special Discrete Signals

Discrete delta and unit step present no theoretical difficulties.

Definition (Discrete Delta). The *discrete delta* or *impulse* signal $\delta(n)$ is

$$\delta(n) = \begin{cases} 1 & \text{if } n = 0, \\ 0 & \text{if } n \neq 0. \end{cases} \tag{1.36}$$

There is a *sifting property* for the discrete impulse:

$$\sum_{n=-\infty}^{\infty} x(n)\delta(n-k) = x(k). \tag{1.37}$$

Discrete summation replaces the analog integral; this will become familiar.

Definition (Discrete Unit Step). The *unit step* signal $u(n)$ is

$$u(n) = \begin{cases} 1 & \text{if } n \geq 0, \\ 0 & \text{if } n < 0. \end{cases} \tag{1.38}$$

Note that if $m > k$, then $b(n) = u(n-k) - u(n-m)$ is a square pulse of unit height on $[k, m-1]$. Products $s(n)b(n)$ extract a chunk of the original discrete signal $s(n)$. And translated copies of $u(n)$ are handy for creating new signals on the positive or negative side of a signal: $y(n) = x(n)u(n-k)$.

1.4 SAMPLING AND INTERPOLATION

Sampling and interpolation take us back and forth between the analog and digital worlds. Sampling converts an analog signal into a digital signal. The procedure is straightforward: Take the values of the analog source at regular intervals. Interpolation converts a discrete signal into an analog signal. Its procedure is almost as easy: make some assumption about the signal between known values—linearity for instance—and fill them in accordingly. In the sampling process, much of the analog signal's information appears to be lost forever, because an infinite number of signal values are thrown away between successive sampling instants. On the other hand, interpolation appears to make some assumptions about what the discrete signal ought to look like between samples, when, in fact, the discrete signal says nothing about signal behavior between samples. Both operations would appear to be fundamentally flawed.

Nevertheless, we shall eventually find conditions upon analog signals that allow us to reconstruct them exactly from their samples. This was the discovery of Nyquist [2] and Shannon [3] (Chapter 7).

1.4.1 Introduction

This section explains the basic ideas of signal sampling: Sampling interval, sampling frequency, and quantization. The *sampling interval* is the time (or other spatial dimension measure) between samples. For a time signal, the *sampling frequency* is measured in hertz (Hz); it is the reciprocal of the sampling interval, measured in seconds (s). If the signal is a distance signal, on the other hand, with the sampling interval given in meters, then the sampling frequency is in units of $(\text{meters})^{-1}$

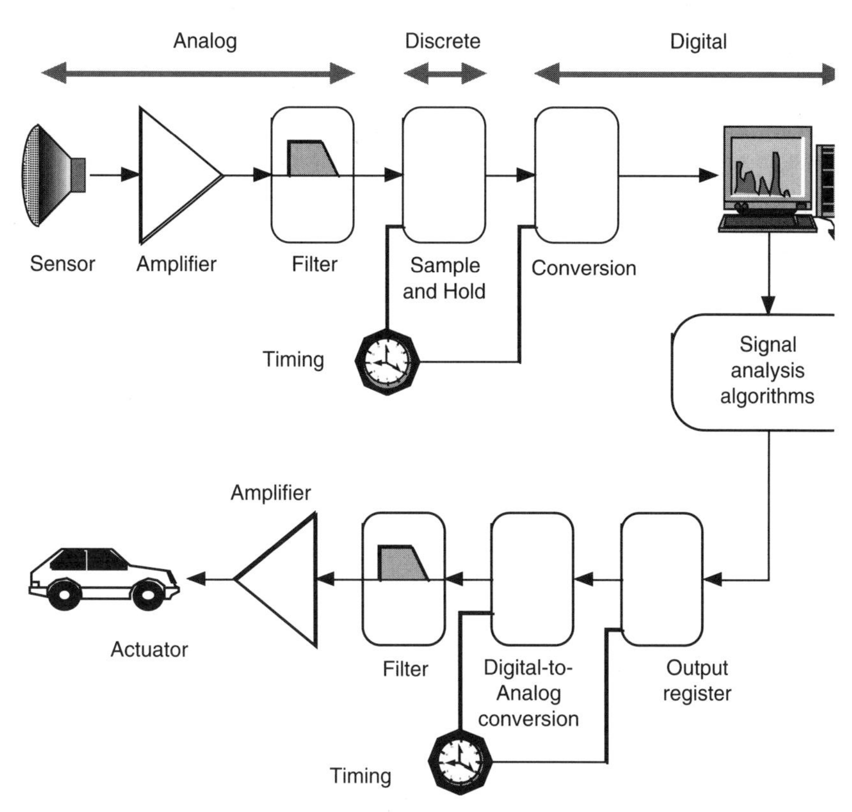

Fig. 1.21. Analog-to-digital conversion. Conversion of analog signals to digital signals requires several steps. Once digitized, algorithms running on the computer can analyze the signal. There may exist a closed loop with the analog world. For example, a digital output signal is converted back into analog form to control an actuator, such as anti-skid brakes on an automobile.

Discrete signals are more convenient for theoretical work, but for computer processing only a finite number of bits can represent the value in binary form. The signal must be digitized, or, in other words, the signal values must be *quantized.* By squeezing the signal value into an N-bit register, some fraction of the true signal value is lost, resulting in a *quantization error.* The number of possible digital signal values is called the *dynamic range* of the conversion.

If $x_a(t)$ is an analog signal, then $x(n) = x_a(n)$ defines a discrete signal. The time interval between $x(n)$ values is unity. We can also take more widely or narrowly spaced samples from $x_a(t)$: $x(n) = x_a(nT)$, where $T > 0$. In an actual system, electronic clock circuits set the sampling rate (Figure 1.21).

An N-bit register can hold non-negative digital values from 0 to $2^N - 1$. The smallest value is present when all bits are clear, and the largest value is when all bits are set. The two's complement representation of a digital value most common for storing signed digital signal values. Suppose there are N bits available in the input register, and the quantized signal's bit values are b_{N-1}, b_{N-2}, ..., b_1, b_0. Then, the digital value is

$$D = -b_{N-1}2^{N-1} + b_{N-2}2^{N-2} + \dots + b_2 2^2 + b_1 2^1 + b_0 2^0. \tag{1.39}$$

In this form, a register full of zeros represents a digital zero value; a single bit in the low-order position, $b_0 = 1$, represents unity; and a register having all bits set contains -1. The dynamic range of an N-bit register is 2^N.

There are several popular analog-to-digital converter (ADC) designs: Successive approximation, flash, dual-slope integration, and sigma–delta. The popular successive approximation converter operates like a balance beam scale. Starting at half of the digital maximum, it sets a bit, converts the tentative digital value to analog, and compares the analog equivalent to the analog input value. If the analog value is less than the converted digital guess, then the bit remains set; otherwise, the bit is cleared. The process continues with the next highest bit position in succession until all bits are tested against the input value. Thus, it adds and removes half-gram weights, quarter-gram weights, and so forth, until it balances the two pans on the beam. Successive approximation converters are accurate, slow, and common. A flash converter implements a whole bank of analog comparators. These devices are fast, nowadays operating at sampling rates of over 250 MHz. However, they have a restricted dynamic range. Dual-slope integration devices are slower, but offer better noise rejection. The sigma–delta converters represent a good design compromise. These units can digitize to over 20 bits and push sampling rates to almost 100 MHz.

1.4.2 Sampling Sinusoidal Signals

Let us consider sampling a sinusoid, $x_a(t) = \cos(\omega t)$, as in Figure 1.22. We sample it at a variety of rates T: $x(n) = x_a(nT)$. For high sampling rates, the discrete result resembles the analog original. But as the sampling interval widens, the resemblance fades. Eventually, we cannot know whether the original analog signal, or, possibly, one of much lower frequency was the analog source for $x(n)$.

To answer the simple question—what conditions can we impose on an analog signal $x_a(t)$ in order to recover it from discrete samples $x(n)$?—requires that we develop both the analog and discrete Fourier transform theory (Chapters 5–7).

1.4.3 Interpolation

Why reconstruct an analog signal from discrete samples? Perhaps the discrete signal is the original form in which a measurement comes to us. This is the case with the geothermal signals we considered in Section 1.1. There, we were given temperature values taken at regular intervals of depth into the earth. It may be of interest—especially when the intervals between samples are very wide or irregular—to provide estimates of the missing, intermediate values. Also, some engineered systems use digital-to-analog converters to take a discrete signal back out into the analog world again. Digital communication and entertainment devices come to mind. So, there is an impetus to better understand and improve upon the analog conversion process.

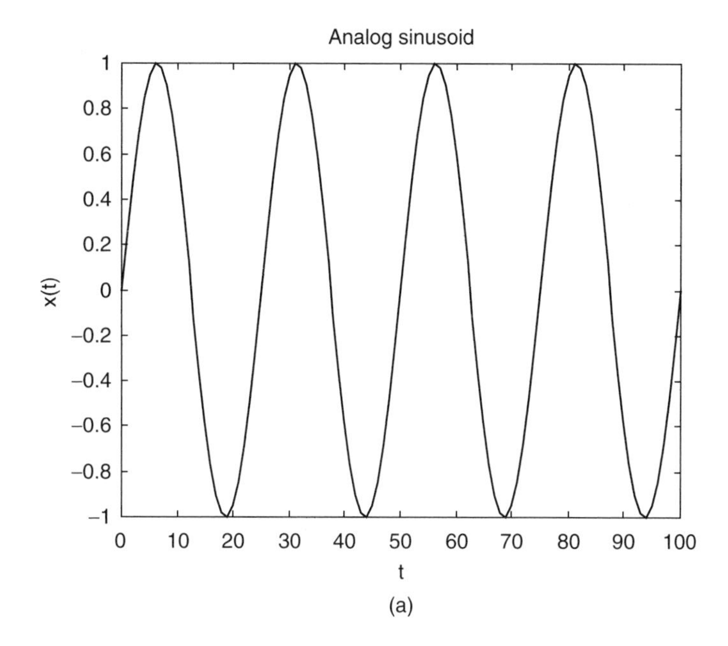

(a)

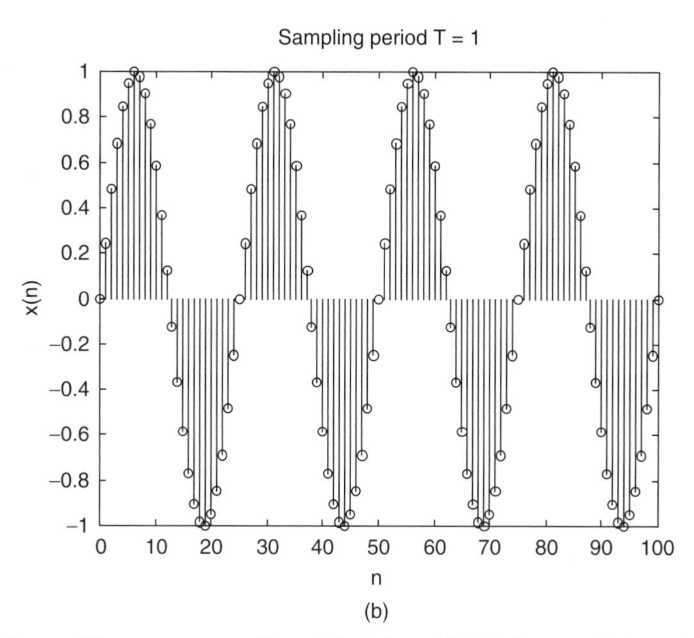

(b)

Fig. 1.22. Impossible to reconstruct. The original sinusoid (a) is first sampled at unit intervals (b). Sampling at a slower rate (c) suggests the same original $x(t)$ But when the rate falls, lower-frequency analog sinusoids could be the original signal (d).

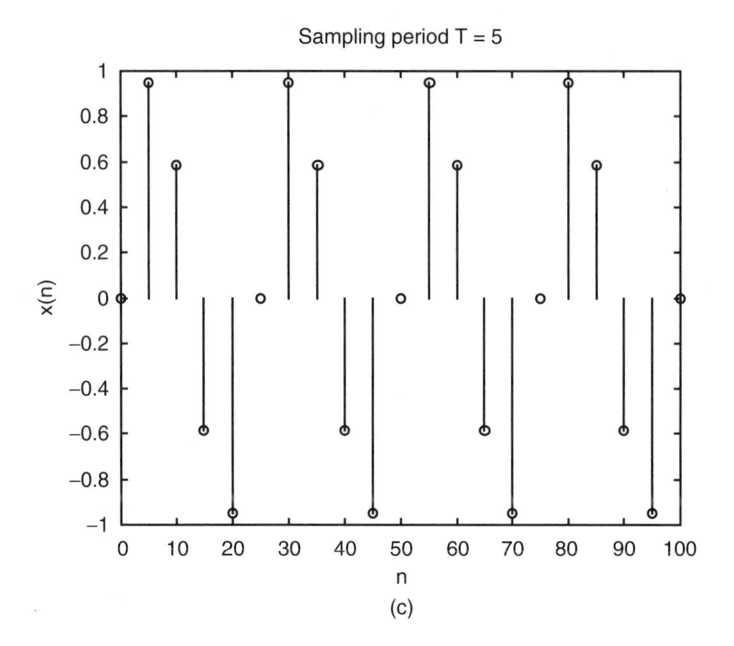

(c)

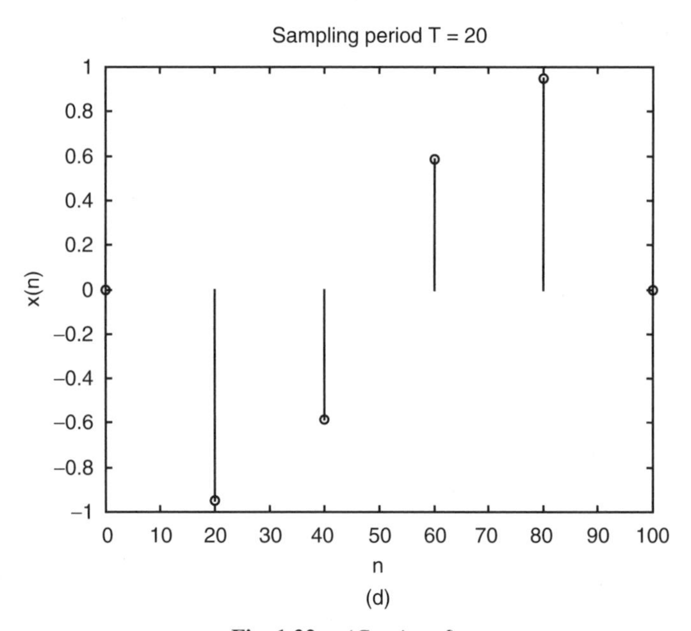

(d)

Fig. 1.22 *(Continued)*

1.4.3.1 Linear. Perhaps the simplest method of constructing an analog signal from discrete values is to use *linear interpolation* (Figure 1.23). Let $y_n = x(n)$, and define

$$x(t) = y_n + (y_{n+1} - y_n)(t - n) \tag{1.40}$$

for $t \in (n, n+1)$. The given samples are called *knots*, as if we were tying short sections of rope together. In this case, the analog signal passes through the knots. Observe that this scheme leaves corners—discontinuities in the first derivative—at the knots. The analog signal constructed from discrete samples via linear interpolation may therefore be unrealistic; nature's signals are usually smooth.

1.4.3.2 Polynomial. Smooth interpolations are possible with quadratic and higher-order polynomial interpolation.

Theorem (Lagrange[11] Interpolation). There is a unique polynomial $p(t)$ of degree $N > 0$ whose graph $(t, p(t))$ contains the distinct points $P_k = (n_k, x(n_k))$ for

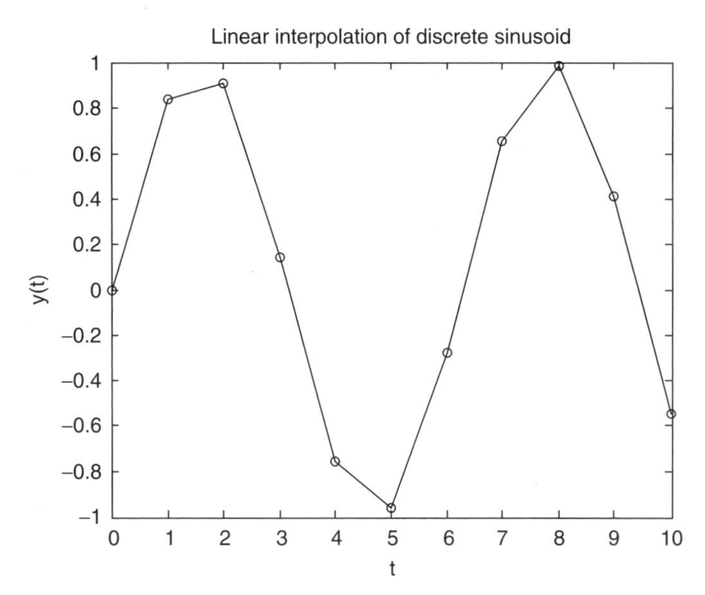

Fig. 1.23. Linear interpolation. Consider the signal $x(n) = \sin(n)$. Linear interpolation of this discrete sinusoid produces a jagged analog result.

[11]Joseph Louis Lagrange (1736–1813)—professor at Turin, Berlin, and Paris—was, with Euler, one of the great number theorists of the eighteenth century.

$0 \le k \le N$:

$$p(t) = \sum_{k=0}^{N} x(n_k) \prod_{\substack{m=0, \\ m \ne k}}^{N} \frac{(t-n_m)}{(n_k-n_m)} \tag{1.41}$$

Proof: Clearly $p(t)$ is of degree N and its graph passes through the points P_k. If $q(t)$ is another polynomial like this, then $d(t) = p(t) - q(t)$ is of degree N and has zeros at the $N + 1$ places: n_k, $0 \le k \le N$. Since a nonzero polynomial of degree N has at most N roots, the difference $d(t)$ must be identically zero. ■

Lagrange interpolation is not completely satisfactory. If N is large, then the method depends on a quotient of products of many terms; it is thereby subject to numerical round-off errors in digital computation. Furthermore, it is only valid for a restricted interval of points in a discrete signal. We may delete some points, since the interval between successive knots does not need to be unity in (1.41). However, we are still left with a polynomial approximation to a small part of the signal. If the signal is zero outside an interval, then however well the interpolant matches the nonzero signal region, $p(t)$ still grows large in magnitude with $|t|$.

An alternative is to compute quadratic polynomials on sequential triples of points. This provides some smoothness, but at every other knot, there is the possibility of a discontinuity in the derivative of the interpolants. Another problem is that the interpolation results vary depending on the point at which one starts choosing triples. To the point, how should we smooth the unit step signal $u(n)$ with interpolating quadratics? The interpolants are lines, except near the origin. If we begin with the triple $(-2, -1, 0)$ and fit it with a quadratic polynomial, then our first interpolant will be concave up. If, on the other hand, we begin interpolating with the triad $(-1, 0, 1)$, then our first quadratic approximation will be concave down. The exercises explore this and other questions entailed by quadratic interpolation. There is, however, a better way.

1.4.4 Cubic Splines

Perhaps the best method to make an analog signal out of discrete samples is to interpolate with spline[12] functions. The idea is to use a cubic polynomial between each successive pair of knots, $(n_k, x(n_k))$ and $(n_{k+1}, x(n_{k+1}))$ and yet match the first derivatives of cubics on either side of a knot.

To understand how this might work, let us first allow that the distance between known points need not be unity. Perhaps the system that collects the discrete samples cannot guarantee a regular sampling interval. The irregular sampling

[12]Architects and engineers once used a jointed, flexible ruler—known as a *spline*—to draw curves. These tools were made from sections of wood or plastic, attached together by brass rivets; modern splines, however, are almost always made of software.

notwithstanding, it is desirable to compose an analog signal that models the discrete data as a continuous, naturally occurring phenomenon. Later, it may even be useful to sample the analog model at regular intervals. So we assume that there are $N+1$ data points $(n_k, x(n_k))$, $0 \le k \le N$. Next, we set $\Delta_k = n_{k+1} - n_k$, $y_k = x(n_k)$, and we consider a polynomial $p_k(t)$ between the knots (n_k, y_k) and (n_{k+1}, y_{k+1}), for $0 \le k < N$. If the polynomial is quadratic or of higher degree and it contains the knots, then we need additional conditions to specify it. We prefer no sharp corners at the knots; thus, let us also stipulate that the derivatives of successive polyomials, $p_k'(t)$ and $p_{k+1}'(t)$, agree on their common knots: $p_k'(n_{k+1}) = p_{k+1}'(n_{k+1})$. Now there are four conditions on any interpolant: It must pass through two given knots and agree with its neighbors on endpoint derivatives. This suggests a cubic, since there are four unknowns. Readers might suspect that something more is necessary, because a condition on a polynomial's derivative is much less restrictive than requiring it to contain a given point.

Indeed, we need two further conditions on the second derivative in order to uniquely determine the interpolating polynomial. This reduces the search for a set of interpolating cubics to a set of linear equations. The two supplementary conditions are that we must have continuity of the second derivatives at knots, and we must specify second derivative values at the endpoints, (n_0, y_0) and (n_N, y_N). Then the equations are solvable [37, 50].

We write the interpolant on the interval $[n_k, n_{k+1}]$ in the form

$$p_k(t) = a_k(t-n_k)^3 + b_k(t-n_k)^2 + c_k(t-n_k) + y_k. \tag{1.42}$$

Then the derivative is

$$p_k'(t) = 3a_k(t-n_k)^2 + 2b_k(t-n_k) + c_k, \tag{1.43}$$

and the second derivative is

$$p_k''(t) = 6a_k(t-n_k) + 2b_k. \tag{1.44}$$

We define $D_k = p_k''(n_k)$ and $E_k = p_k''(n_{k+1})$. From (1.44),

$$D_k = 2b_k \tag{1.45}$$

and

$$E_k = 6a_k(n_{k+1} - n_k) + 2b_k = 6a_k\Delta_k + 2b_k, \tag{1.46}$$

with $\Delta_k = n_{k+1} - n_k$. Thus, we can express b_k and a_k in terms of Δ_k, which is known, and D_k and E_k, which are as yet unknown. Using (1.42), we can write c_k as follows:

$$c_k = \frac{y_{k+1} - y_k}{\Delta_k} - a_k\Delta_k^2 - b_k\Delta_k. \tag{1.47}$$

For $0 \le k < N$, (1.45)–(1.47) imply

$$c_k = \frac{y_{k+1} - y_k}{\Delta_k} - \frac{E_k - D_k}{6}\Delta_k - \frac{D_k}{2}\Delta_k = \frac{y_{k+1} - y_k}{\Delta_k} - \frac{\Delta_k}{6}(E_k + 2D_k). \tag{1.48}$$

To make a system of linear equations, we need to express this in terms of the second derivatives, D_k and E_k. The derivatives, $p_k'(t)$ and $p_{k+1}'(t)$, are equal for $t = n_{k+1}$; this is a required property of the interpolating cubics. Hence,

$$p_k'(n_{k+1}) = 3a_k(\Delta_k)^2 + 2b_k(\Delta_k) + c_k = p_{k+1}'(n_{k+1}) = c_{k+1}. \tag{1.49}$$

Inserting the expressions for c_{k+1}, c_k, b_k, and a_k in terms of second derivatives gives

$$6\frac{y_{k+2} - y_{k+1}}{\Delta_{k+1}} - 6\frac{y_{k+1} - y_k}{\Delta_k} = \Delta_k(2E_k + D_k) + \Delta_{k+1}(2D_{k+1} + E_{k+1}). \tag{1.50}$$

Now, $E_k = p_k''(n_{k+1}) = D_{k+1} = p_{k+1}''(n_{k+1})$, by invoking the continuity assumption on second derivatives. We also set $E_{N-1} = D_N$, producing a linear equation in D_k, D_{k+1}, and D_{k+2}:

$$6\frac{y_{k+2} - y_{k+1}}{\Delta_{k+1}} - 6\frac{y_{k+1} - y_k}{\Delta_k} = \Delta_k(2D_{k+1} + D_k) + \Delta_{k+1}(2D_{k+1} + D_{k+2}). \tag{1.51}$$

This system of equations has $N + 1$ variables, $D_0, D_1, \ldots, D_N$. Unfortunately, (1.51) has only $N - 1$ equations, for $k = 0, 1, \ldots, N - 2$. Let us lay them out as follows:

$$\begin{bmatrix} \Delta_0 & 2(\Delta_0+\Delta_1) & \Delta_1 & 0 & 0 & \cdots & 0 \\ 0 & \Delta_1 & 2(\Delta_1+\Delta_2) & \Delta_2 & 0 & \cdots & 0 \\ 0 & 0 & \Delta_2 & \cdots & \cdots & \cdots & 0 \\ 0 & 0 & 0 & \cdots & \cdots & \cdots & \cdots \\ \cdots & \cdots & \cdots & \cdots & \cdots & \Delta_{N-2} & 0 \\ 0 & 0 & 0 & 0 & \Delta_{N-2} & 2(\Delta_{N-2}+\Delta_{N-1}) & \Delta_{N-1} \end{bmatrix} \begin{bmatrix} D_0 \\ D_1 \\ D_2 \\ D_4 \\ \cdots \\ D_N \end{bmatrix}$$

$$= 6\begin{bmatrix} \frac{y_2 - y_1}{\Delta_1} - \frac{y_1 - y_0}{\Delta_0} \\ \frac{y_3 - y_2}{\Delta_2} - \frac{y_2 - y_1}{\Delta_1} \\ \cdots \\ \cdots \\ \cdots \\ \frac{y_N - y_{N-1}}{\Delta_{N-1}} - \frac{y_{N-1} - y_{N-2}}{\Delta_{N-2}} \end{bmatrix}. \tag{1.52}$$

From linear algebra, the system (1.52) may have no solution or multiple solutions [51, 52]. It has a unique solution only if the number of variables equals the number of equations. Then there is a solution if and only if the rows of the coefficient matrix—and consequently its columns—are linearly independent. Thus, we must reduce the number of variables by a pair, and this is where the final condition on second derivatives applies.

We specify values for D_0 and D_N. The most common choice is to set $D_0 = D_N = 0$; this gives the so-called *natural spline* along the knots (n_0, y_0), (n_1, y_1), ..., (n_N, y_N). The coefficient matrix of the linear system (1.52) loses its first and last columns, simplifying to the symmetric system (1.53). Other choices for D_0 and D_N exist and are often recommended [37, 50]. It remains to show that (1.53) always has a solution.

$$\begin{bmatrix} 2(\Delta_0+\Delta_1) & \Delta_1 & 0 & 0 & 0 & \dots & 0 \\ \Delta_1 & 2(\Delta_1+\Delta_2) & \Delta_2 & 0 & 0 & \dots & 0 \\ 0 & \Delta_2 & 2(\Delta_2+\Delta_3) & \Delta_3 & \dots & \dots & 0 \\ 0 & 0 & \Delta_3 & \dots & \dots & \dots & \dots \\ \dots & \dots & \dots & \dots & \dots & \dots & \Delta_{N-2} \\ 0 & 0 & \dots & \dots & 0 & \Delta_{N-2} & 2(\Delta_{N-2}+\Delta_{N-1}) \end{bmatrix} \begin{bmatrix} D_1 \\ D_2 \\ D_3 \\ D_4 \\ \dots \\ D_{N-1} \end{bmatrix}$$

$$= 6\begin{bmatrix} \frac{y_2-y_1}{\Delta_1} - \frac{y_1-y_0}{\Delta_0} \\ \frac{y_3-y_2}{\Delta_2} - \frac{y_2-y_1}{\Delta_1} \\ \dots \\ \dots \\ \dots \\ \frac{y_N-y_{N-1}}{\Delta_{N-1}} - \frac{y_{N-1}-y_{N-2}}{\Delta_{N-2}} \end{bmatrix}. \tag{1.53}$$

Theorem (Existence of Natural Splines). Suppose the points (n_0, y_0), (n_1, y_1), ..., (n_N, y_N) are given and $n_0 < n_1 < \dots < n_N$. Let $\Delta_k = n_{k+1} - n_k$. Then the system $A\boldsymbol{v} = \boldsymbol{y}$ in (1.53) has a solution $\boldsymbol{v} = [D_1, D_2, \dots, D_{N-1}]^T$.

Proof: Gaussian elimination solves the system, using row operations to convert $\boldsymbol{A} = [A_{r,c}]$ into an upper-triangular matrix. The elements on the diagonal of the coefficient matrix are called the *pivots*. The first pivot is $P_1 = 2\Delta_0 + 2\Delta_1$, which is *positive*. We multiply the first row of $\boldsymbol{A}$ by the factor $f_1 = \frac{-\Delta_1}{2\Delta_0+2\Delta_1}$ and add it to the second row, thereby annihilating $A_{2,1}$. A second pivot P_2 appears in place of $A_{2,2}$:

$$\begin{aligned} P_2 &= 2\Delta_1 + 2\Delta_2 + \Delta_1 f_1 = \Delta_1 \frac{-\Delta_1}{(2\Delta_0 + 2\Delta_1)} + 2\Delta_1 + 2\Delta_2 \\ &= \frac{3\Delta_1^2 + 4\Delta_0\Delta_1}{(2\Delta_0 + 2\Delta_1)} + 2\Delta_2. \end{aligned} \tag{1.54}$$

We update the vector $\boldsymbol{y}$ according to the row operation as well. Notice that $P_2 > 2\Delta_2 > 0$. The process produces another positive pivot. Indeed, the algorithm continues to produce positive pivots P_r. These are more than double the coefficient $A_{r+1,r}$, which the next row operation will annihilate. Thus, this process will eventually produce an upper-triangular matrix. We can find the solution to (1.53) by back substitution, beginning with D_{N-1} on the upper-triangular result. ■

Figure 1.24 shows how nicely cubic spline interpolation works on a discrete sinusoid. Besides their value for reconstructing analog signals from discrete samples, splines are important for building multiresolution signal decompositions that support modern wavelet theory [53] (Chapters 11 and 12).

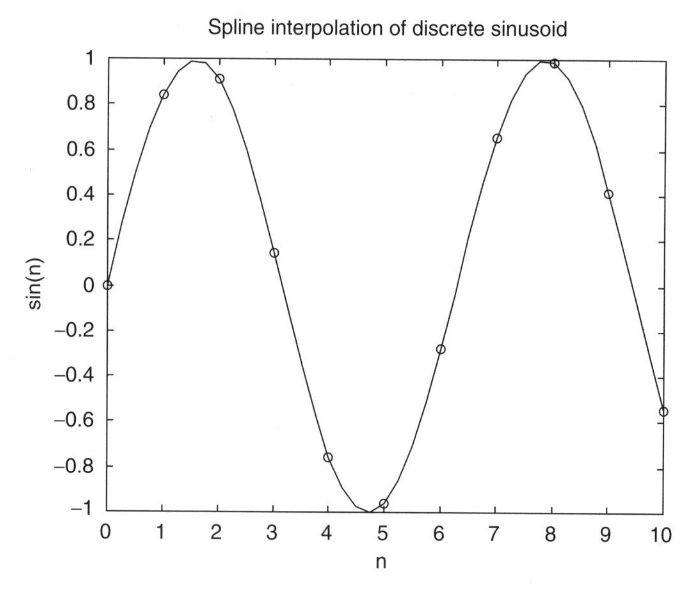

Fig. 1.24. Cubic spline interpolation. Again, discrete samples of the signal $x(n) = \sin(n)$ are used for the knots. Cubic spline interpolation offers a smooth model of the undulations and clearly captures the sinusoidal behavior of the original analog signal.

1.5 PERIODIC SIGNALS

Periodic signals, whether analog or discrete, repeat their values over intervals. The most familar ones are sinusoids. These signals arise from the mechanics of circular motion, in electric and magnetic interactions, and they are found in many natural phenomena. For instance, we considered the solution of Maxwell's equations, which describe the relation between the electric and magnetic fields. There we showed how to derive from the field equations a set of differential equations whose solution involves sinsusoidal functions. Radio waves propagate through empty space as electric and magnetic sinusoids at right angles to one another.

1.5.1 Fundamental Period and Frequency

The interval over which a signal repeats itself is its period, and the reciprocal of its period is its frequency. Of course, if a signal repeats itself over an interval, then it also repeats itself over any positive integral multiple of that interval; we must characterize a periodic signal by the smallest such interval of repetition.

Definition (Periodicity). An analog signal $x(t)$ is *periodic* if there is a $T > 0$ with $x(t + T) = x(t)$ for all t. A discrete signal $x(n)$ is periodic if there is an integer $N > 0$ with $x(n) = x(n + N)$ for all n. The smallest value for which a signal is periodic is called the *fundamental period.*

Definition (Analog Sinusoid). The signal

$$x(t) = A\cos(\Omega t + \phi) \tag{1.55}$$

is an *analog sinusoid.* A is the *amplitude* of $x(t)$, which gives its maximum value; Ω is its *frequency* in radians per second; and ϕ is its *phase* in radians. $\Omega = 2\pi F$, where F is the frequency in hertz.

Example. If $x(t) = A\cos(\Omega t + \phi)$, then $x(t) = x(t + \frac{2\pi}{\Omega})$, from the 2π-periodicity of the cosine function. So $T = \frac{2\pi}{\Omega} = \frac{1}{F}$ is the fundamental period of $x(t)$.

The sinusodal signals in nature are, to be sure, never the perfect sinusoid that our mathematical models suggest. Electromagnetic propagation through space comes close to the ideal, but always present are traces of matter, interference from other radiation sources, and the minute effects of gravity. Noise corrupts many of the phenomena that fall under our analytical eye, and often the phenomena are only vaguely sinusoidal. An example is the periodic trends of solar activity—in particular, the 11-year sunspot cycle, which we consider in more detail in the next section. But signal noise is only one aspect of nature's refusal to strictly obey our mathematical forumulas.

Natural sinusoidal signals decay. Thus, for it to be a faithful mathematical model of a naturally occurring signal, the amplitude of the sinusoid (1.55) should decrease. Its fidelity to nature's processes improves with a time-varying amplitude:

$$x(t) = A(t)\cos(\Omega t + \phi). \tag{1.56}$$

The earliest radio telephony technique, *amplitude modulation* (AM), makes use of this idea. The AM radio wave has a constant *carrier frequency*, $F = \frac{\Omega}{2\pi}$ Hz, but its amplitude $A(t)$ is made to vary with the transmitted signal. Electronic circuits on the receiving end tune to the carrier frequency. The amplitude cannot jump up and down so quickly that it alters the carrier frequency, so AM is feasible only if F greatly exceeds the frequency of the superimposed amplitude modulation. This works for common AM content—voice and music—since their highest useful frequencies are about 8 and 25 kHz, respectively. In fact, limiting voice frequencies to only 4 kHz produces a very lifelike voice audio, suitable for telephony. Accordingly, the AM radio band, 550 kHz to 1600 kHz, is set well above these values. The signal looks like a sine wave whose envelope (the curve that follows local signal maxima) matches the transmitted speech or music.

Natural and engineered systems also vary the frequency value in (1.55). The basic *frequency-modulated* (FM) signal is the chirp, wherein the frequency increases linearly. Animals—birds, dolphins, and whales, for example—use frequency varying signals for communication. Other animals, such as bats, use chirps for echolocation. Some natural languages, such as Chinese, use changing tones as a critical indication of word meaning. In other languages, such as English and Russian, it plays only an ancillary role, helping to indicate whether a sentence is a question or a statement. Thus, we consider signals of the form

$$x(t) = A\cos(2\pi F(t) + \phi), \tag{1.57}$$

where $F(t)$ need not be linear. An FM signal (1.57) is not a true sinusoid, but it provides the analyst with a different kind of signal model, suitable for situations where the frequency is not constant over the time region of interest. Applications that rely on FM signals include such systems as radars, sonars, seismic prospecting systems, and, of course, communication systems.

A *phase-modulated* signal is of the form

$$x(t) = A\cos(2\pi F + \phi(t)). \tag{1.58}$$

There is a close relation between phase and frequency modulation, namely, that the derivative of the phase function $\phi(t)$ in (1.58) is the *instantaneous frequency* of the signal $x(t)$ [54, 55]. The idea of instantaeous frequency is that there is a sinusoid that best resembles $x(t)$ at time t. It arose as recently as the late 1930s in the context of FM communication systems design, and its physical meaning has been the subject of some controversy [56]. If we fix F in (1.58) and allow $\phi(t)$ to vary, then the

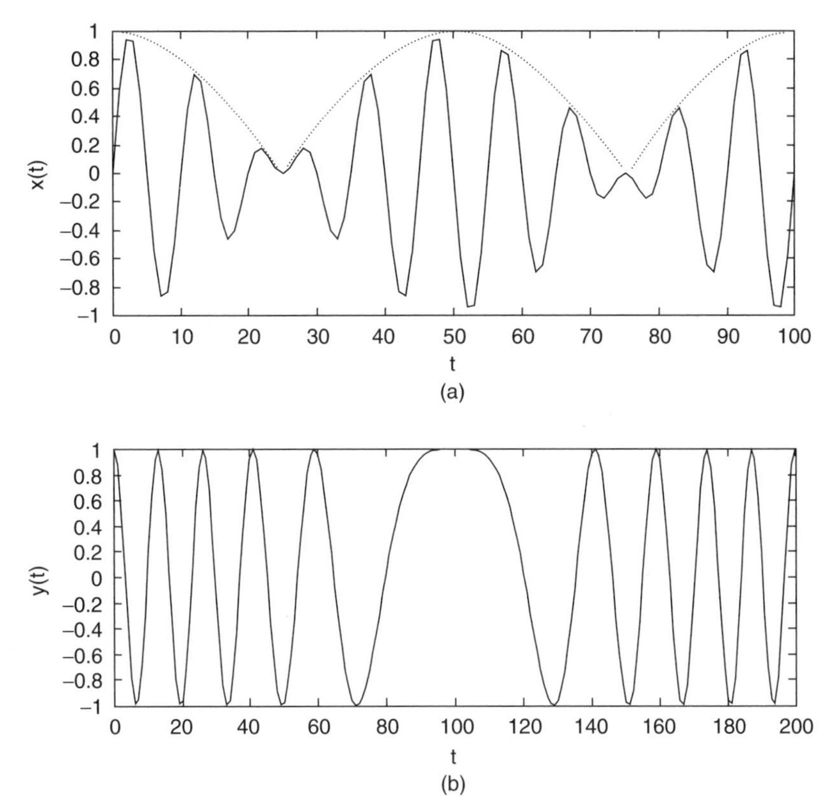

Fig. 1.25. AM and FM signals. Panel (a) shows a sinusoidal carrier modulated by a sinusoidal signal. The information-bearing part of the signal is given by the envelope of the signal, shown by dotted lines. In (b), a simple FM signal with the frequency varying sinusoidally is shown. Note that the oscillations bunch up and spread out as time passes, indicating rising and falling signal frequency, respectively.

frequency of the cosine wave changes. Over Δt seconds, the radial frequency of $x(t)$ changes by amount $[\Omega + \phi(t + \Delta t)] - [\Omega + \phi(t)] = \phi(t + \Delta t) - \phi(t)$, where $\Omega = 2\pi f$. The average change in Hertz frequency over this time interval is $\frac{\phi(t + \Delta t) - \phi(t)}{2\pi\Delta t}$. As $\Delta t \to 0$, this value becomes the derivative $\frac{d}{dt}\phi(t)$, the instantaneous frequency of the phase modulated signal $x(t)$.

If it seems odd that the derivative of phase is the signal frequency, then perhaps thinking about the familiar Doppler[13] effect can help reveal the connection. Suppose a train whistle makes a pure sinusoidal tone. If the train is standing still, then someone within earshot hears a sound of pure tone that varies neither in amplitude nor pitch. If the train moves while the whistle continues to blow, however, then the

[13]Austrian physicist Christian Doppler (1803–1853) discovered and described this phenomenon, first experimenting with trumpeters on a freight train.

tone changes. Coming toward us, the train whistle mechanically reproduces the same blast of air through an orifice, but the signal that we hear is different. The pitch increases as the train comes toward us. That means the signal frequency is increasing, but all that it takes to accomplish that is to move the train. In other words, a change in the phase of the whistle signal results in a different frequency in sound produced. A similar effect occurs in astronomical signals with the red shift of optical spectral lines from distant galaxies. In fact, the further they are away from us, the more their frequency is shifted. This means that the further they are from earth, the more rapidly is the phase changing. Objects further away move away faster. This led Hubble[14] to conclude that the universe is expanding, and the galaxies are spreading apart as do inked dots on an inflating balloon.

Some signals, natural and synthetic, are superpositions of sinusoids. In speech analysis, for example, it is often possible to model vowels as the sum of two sinusoidal components, called *formants*:

$$x(t) = x_1(t) + x_2(t) = A_1\cos(\Omega_1 t + \phi_1) + A_2\cos(\Omega_2 t + \phi_2). \tag{1.59}$$

Generally, $x(t)$ in (1.59) is not sinusoidal, unless $\Omega_1 = \Omega_2 = \Omega$. A geometric argument demonstrates this. If the radial frequencies of the sinusoidal components are equal, then the vectors $\boldsymbol{v}_1 = (A_1\cos(\Omega t + \phi_1), A_1\sin(\Omega t + \phi_1))$ and $\boldsymbol{v}_2 = (A_2\cos(\Omega t + \phi_2), A_2\sin(\Omega t + \phi_2))$ rotate around the origin at equal speeds. This forms a parallelogram structure, rotating about the origin at the same speed as $\boldsymbol{v}_1$ and $\boldsymbol{v}_2$ (Figure 1.26), namely Ω radians per unit time. The x-coordinates of $\boldsymbol{v}_1$ and

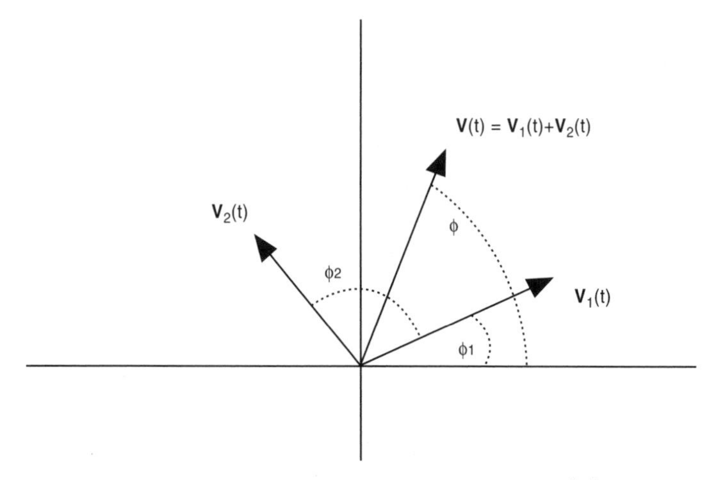

Fig. 1.26. Sinusoidal summation. Vector $\boldsymbol{v} = \boldsymbol{v}_1 + \boldsymbol{v}_2$ has length $\|\boldsymbol{v}\| = A$. Its x-coordinate, as a function of t, is a sinusoid of radial frequency Ω and phase $\phi = (\phi_2 - \phi_1)/2$.

[14]Working at the Mount Wilson Observatory near Los Angeles, Edwin Powell Hubble (1889–1953) discovered that galaxies are islands of stars in the vast sea of space, the red shift relates velocity to distance, and the universe itself expands in all directions.

v_2 are the values of $x_1(t)$ and $x_2(t)$, respectively. The sum $x(t)$ is the x-coordinate of $v_1 + v_2$. Now, $\|v_1 + v_2\| = \|v\| = A$, where

$$\begin{aligned} A^2 &= A_1^2 + A_2^2 + 2A_1A_2\cos(\Omega t + \phi_1)\cos(\Omega t + \phi_2) \\ &\quad + 2A_1A_2\sin(\Omega t + \phi_1)\cos(\Omega t + \phi_2) \\ &= A_1^2 + A_2^2 + 2A_1A_2\cos(\phi_1 - \phi_2). \end{aligned} \tag{1.60}$$

We also see from Figure 1.26 that the sum lies half way between v_1 and v_2. Thus, the phase of **v** is $\phi = \frac{\phi_2 - \phi_1}{2}$, and we have $x(t) = A\cos(\Omega t + \phi)$.

1.5.2 Discrete Signal Frequency

Due to the gap between successive signal values, discrete periodic signals have several properties that distinguish them from analog periodic waveforms:

(i) Discrete periodic signals have lower limits on their period; it makes no sense to have a discrete signal with period less than unity, because the discrete world does not even define signals at intervals smaller than unity.

(ii) A discrete signal with unit period is constant.

(iii) For sinusoids, the restriction to unit periods or more means that they have a maximum frequency: $|\Omega| = \pi$.

(iv) Not all sinusoids are periodic; periodicity only obtains when the frequency of the sampled signal is matched to the sampling interval.

This section covers these idiosyncrasies.

Proposition (Discrete Period). The smallest period for a discrete signal is $T = 1$. The largest frequency for a discrete sinusoid is $|\Omega| = \pi$, or equivalently, $|F| = 1$, where $\Omega = 2\pi F$ is the frequency in radians per sample.

Proof: Exercise. ■

Proposition (Periodicity of Discrete Sinusoids). Discrete sinusoid $x(n) = A\cos(\Omega n + \phi)$, $A \neq 0$, is periodic if and only if $\Omega = 2\pi p$, where $p \in \mathbb{Q}$, the rational numbers.

Proof: First, suppose that $\Omega = 2\pi p$, where $p \in \mathbb{Q}$. Let $p = m/k$ where $m, k \in \mathbb{N}$. If $m = 0$, then $x(n)$ is periodic; in fact, it is constant. Therefore, suppose $m \neq 0$ and choose $N = |k/m|$. Then,

$$x(n+N) = A\cos\left(\Omega n + \Omega\left|\frac{k}{m}\right| + \phi\right) = A\cos\left(\Omega n + 2\pi\frac{m}{k}\left|\frac{k}{m}\right| + \phi\right)$$

$$= A\cos(\Omega n + 2\pi(\pm 1) + \phi) = A\cos(\Omega n + \phi) = x(n)$$

by the 2π-periodicity of the cosine function. Thus, $x(n)$ is periodic with period N. Conversely, suppose that for some $N > 0$, $x(n + N) = x(n)$ for all n. Then, $A\cos(\Omega n + \phi) = A\cos(\Omega n + \Omega N + \phi)$. Since $A \neq 0$, we must have

$\cos(\Omega n + \phi) = \cos((\Omega n + \phi) + \Omega N)$. And, since cosine can only assume the same values on intervals that are integral multiples of π, we must have $\Omega N = m\pi$ for some $m \in \mathbb{N}$. Then, $\Omega = m\pi/N$, so that Ω is a rational multiple of π. ■

Let us reinforce this idea. Suppose that $x(n) = x_a(Tn)$, where $x_a(t) = A\cos(\Omega t + \phi)$, with $A \neq 0$. Then $x_a(t)$ is an analog periodic signal. But $x(n)$ is not necessarily periodic. Indeed, $x(n) = A\cos(\Omega nT + \phi)$, so by the proposition, $x(n)$ is periodic only if ΩT is a rational multiple of π. Also, the discrete sinusoid $x(n) = \cos(2\pi fn)$ is periodic if and only if the frequency $f \in \mathbb{Q}$. The analog signal $s(t) = \cos(2\pi ft)$, $f > 0$, is always periodic with period $1/f$. But if $f = m/k$, with $m, k \in \mathbb{N}$, and m and k are relatively prime, then $x(n)$ has period k, not $1/f$. It takes time to get used to the odd habits of discrete sinusoids.

1.5.3 Frequency Domain

Having introduced analog and discrete sinusoids, fundamental period, and sinusoidal frequency, let us explain what it means to give a frequency-domain description of a signal. We already know that signals from nature and technology are not always pure sinusoids. Sometimes a process involves superpositions of sinusoids. The signal amplitude may vary too, and this behavior may be critical to system understanding. A variant of the pure sinusoid, the amplitude-modulated sine wave, models this situation. Another possibility is a variable frequency characteristic in the signal, and the frequency-modulated sine wave model accounts for it. There is also phase modulation, such as produced by a moving signal source. Finally, we must always be cognizant of, prepare for, and accommodate noise within the signal. How can we apply ordinary sinusoids to the study of these diverse signal processing and analysis applications?

1.5.3.1 Signal Decomposition. Many natural and synthetic signals contain regular oscillatory components. The purpose of a frequency-domain description of a signal is to identify these components. The most familar tool for aiding in identifying periodicities in signals are the sinusoidal signals, sin(t) and cos(t). Thus, a frequency-domain description presupposes that the signal to be analyzed consists of a sum of a few sinusoidal components. Perhaps the sum of sinusoids does not exactly capture the signal values, but what is left over may be deemed noise or background. We consider two examples: sunspot counts and speech. If we can identify some simple sinusoidal components, then a frequency-domain description offers a much simpler signal description. For instead of needing a great number of time domain values to define the signal, we need only a few triplets of real numbers—the amplitude, frequency, and phase of each substantive component—in order to capture the essence of the signal.

Consider the signal $x(n)$ of Figure 1.27, which consists of a series of irregular pulses. There appears to be no rule or regularity of the values that would allow us to describe it by more that a listing of its time-domain values. We note that certain regions of the signal appear to be purely sinusoidal, but the juxtaposition of unfamiliar, oddly shaped pulses upsets this local pattern.

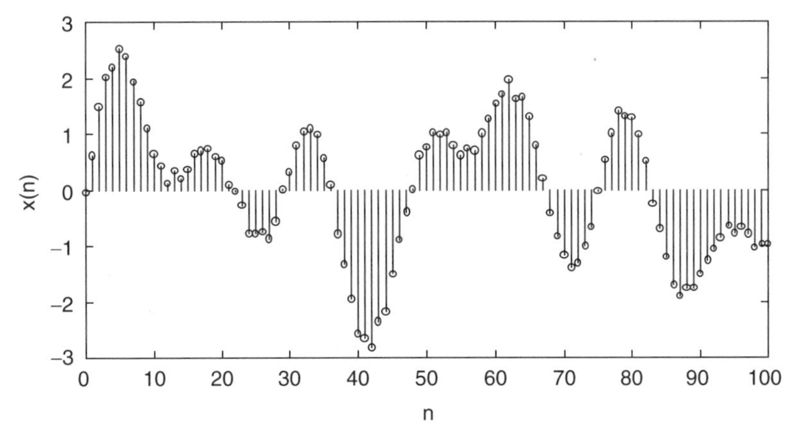

Fig. 1.27. Efficiency of freqency-domain signal description. An irregular signal, appearing to lack any regular description beyond a listing of its time-domain values, turns out to be the sum of three sinusoids and a small background noise signal.

The signal does, however, have a concise description as a sum of three sinusoids and a small noise component:

$$x(n) = \sin\left(\frac{2\pi n}{15}\right) + \sin\left(\frac{2\pi n}{25}\right) + \sin\left(\frac{2\pi n}{50}\right) + N(n)\,. \tag{1.61}$$

Thus, signal description, and hence the further analysis of signals, is often made more powerful by representing signals in terms of an established set of prototype signals. For a frequency domain signal description, sinusoidal models are appropriate. Some background noise may elude an immediate description in terms of sinusoidal components. Depending on the application, of course, this residual signal may be negligible because it has a much lower magnitude than the source signal of interest. If the sinusoidal trends are localized within a signal, the arithmetic of superposition may allow us to describe them even with sinusoids of longer duration, as Figure 1.27 illustrates.

1.5.3.2 Sunspots. One periodic phenomenon with which we are familiar is the sunspot cycle, an example of a natural periodic trend. Sunspots appear regularly on the sun's surface as clusters of dark spots and, at their highest intensities, disrupt high-frequency radio communication. Scientists have scrutinized sunspot activity, because, at its height during the 11-year cycle, it occasionally hampers very high frequency radio communication. Until recently, the mechanism responsible for this phenomenon was not known. With temperature of 3800 °K, they are considerably cooler than the rest of the sun's surface, which has an average temperature of some 5400 °K.

Ancient people sometimes observed dark spots on the solar disk when it was obscured by fog, mist, or smoke. Now we check for them with a simple telescope

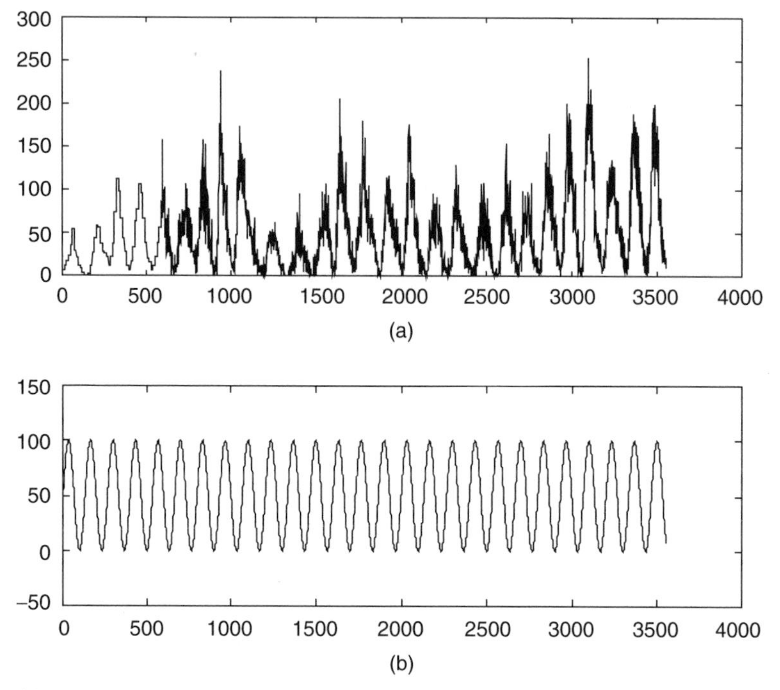

Fig. 1.28. Wolf sunspot numbers. Panel (a) plots the time-domain values of $w(n) = 10G(n) + S(n)$ for each month from 1700 through 1995. We compare the oscillation with sinusoids, for example, when period $T = 11.1$ years as in (b).

that projects the sun's image onto a white plate. Galileo[15] was the first to do so. His observations upset the prevailing dogma of seventeenth century Europe insisting that the sun was a perfect disk. Standardized sunspot reports began in the mid-1700s, and the earlier values given in our data plots (Figure 1.28) are assumptions based on informal observations.

Sunspot activity can be formulated as a discrete signal by counting the number of groups of sunspots. In 1848, the Swiss astronomer, Johann Rudolph Wolf, introduced a daily measurement of sunspot number. His method, which is still used today, counts the total number of spots visible on the face of the sun and the number of groups into which they cluster, because neither quantity alone satisfactorily measures sunspot activity. The *Wolf*[16] *sunspot number* is $w(n) = 10G(n) + S(n)$, where $G(n)$ is the average number of sunspot groups and $S(n)$ is the average number of spots. Individual observational results do vary greatly, however, since the measurement strongly depends on interpretation, experience, and the stability of the earth's atmosphere

[15]In addition to finding solar blemishes in 1610, Galileo Galilei (1564–1642) used his telescope to resolve the Milky Way into faint stars and, with his discovery of the phases of Venus, confirmed Copernicus's heliocentric theory.

[16]After Swiss astronomer Johann Rudolph Wolf (1816–1893).

above the observing site. The use of the earth as a platform from which to record these numbers contributes to their variability, too, because the sun rotates and the evolving spot groups are distributed unevenly across solar longitudes. To compensate for these limitations, each daily international number is computed as a weighted average of measurements made from a network of cooperating observatories.

One way to elucidate a frequency-domain description of the sunspot signal $w(n)$ is to compare it with a sinusoidal signal. For example, we can align sinusoids of varying frequency with $w(n)$, as shown in Figure 1.28b. Thus, the sinusoids are models of an ideal sunspot cycle. This ideal does not match reality perfectly, of course, but by pursuing the mathematical comparison between the trignonometric model and the raw data, we can get a primitive frequency-domain description of the sunspot cycle. What we want to derive in a frequency-domain description of a signal is some kind of quantification of how much a signal resembles model sinusoids of various frequencies. In other words, we seek the relative weighting of supposed frequency components within the signal. Thinking of the signal values as a very long vector, we can compute the inner product of $w(n)$ with the unit vectors whose values are given by sinusoids

$$s(n) = 50 + 50\sin\left(\frac{2\pi n}{12T}\right), \tag{1.62}$$

where T varies from 0.1 year to 16 years. Then we compute the difference $e_T(n) = w(n) - s(n)$, an error term which varies with periodicity of the sinusoid (1.62) determined by T. Now, we take evalutate the norm of the vector $e_T(n)$, which has length $12 \times 296 = 3552 : \|e_T\|$. If we plot the norm of the error vectors with respect to the supposed period T of the sinusoidal model, we see that there is a pronounced minimum near $T = 11$ (Figure 1.29).

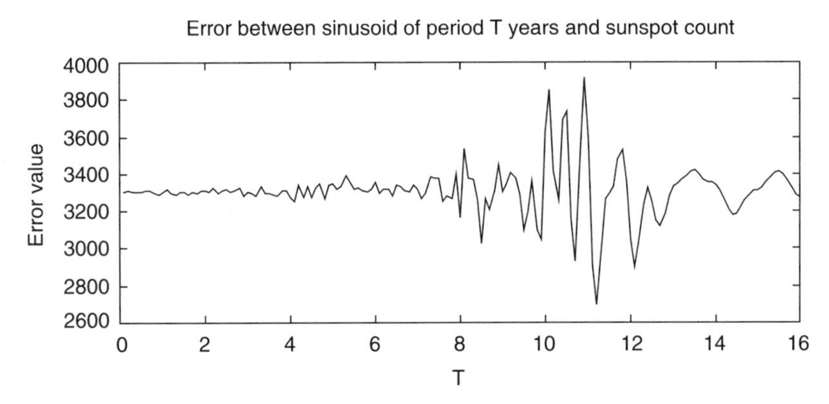

Fig. 1.29. Comparing sinusoids with sunspot numbers. The minimal error between such sinusoids and the sunspot oscillations occurs just above $T = 11$. Note that there are other local minima both below and above $T = 11$. In point of fact, sometimes the sunspot cycle peaks a few months early and sometimes a few months later.

Note that we have exploited a few specific facts about the sunspot numbers in this analysis. In particular, we have not worried about the relative position of the sinusoidal model. That is, we have not taken the relative phases of the sinusoid $s(n)$ and the discrete sunspot signal $w(n)$ into account. Could a slightly shifted sinusoid result in a smaller error term? This is indeed quite a likely possibility, and we avoid it for two reasons. First, it turns the minimization of the error term into a two-dimensional problem. This is a practical, application-oriented discussion. Although we want to explain the technical issues that arise in signal analysis problems, we do not want to stray into two-dimensional problems. Our focus is one-dimensional—signal analysis, not image analysis—problems. In fact, with more powerful tools, such as the discrete Fourier transform (Chapter 7), we we can achieve a frequency domain analysis in one dimension that handles the relative phase problem.

1.5.3.3 Speech. Scientists and engineers have long attempted to build commercial speech recognition products. Such products now exist, but their applicability remains limited. Computers are so fast, mathematics so rich, and the investigations so deep: How can there be a failure to achieve? The answers seem to lie in the fundamental differences between how signal analyzing computers and human minds—or animal minds in general, for that matter—process the acoustical signals they acquire. The biological systems process data in larger chunks with a greater application of top-down, goal-directed information than is presently possible with present signal analysis and artificial intelligence techniques.

An interesting contrast to speech recognition is speech generation. Speech synthesis is in some sense the opposite of recognition, since it begins with a structural description of a signal—an ASCII text string, for instance—and generates speech sounds therefrom. Speech synthesis technology has come very far, and now at the turn of the century it is found in all kinds of commercial systems: telephones, home appliances, toys, personal computer interfaces, and automobiles. This illustrates the fundamental asymmetry between signal synthesis and analysis. Speech recognition systems have become increasingly sophisticated, some capable of handling large vocabularies [57–59]. In recent years, some of the recognition systems have begun to rely on artificial neural networks, which mimic the processing capabilities of biological signal and image understanding systems [60].

To begin to understand this fundamental difference and some of the daunting problems faced by speech recognition researchers, let us consider an example of digitized voice (Figure 1.30). Linguists typically classify speech events according to whether the vocal cords vibrate during the pronunciation of a speech sound, called a *phone* [61]. Phones are speech fragments. They are realizations of the basic, abstract components of a natural language, called *phonemes*. Not all natural languages have the same phonemes. Speakers of one language sometimes have extreme difficulties hearing and saying phonemes of a foreign tongue. And within a given language, some phonemes are more prevalent than others. The most common strategy for speech recognition technology is to break apart a digital speech sample into separate phones and then identify phonemes among

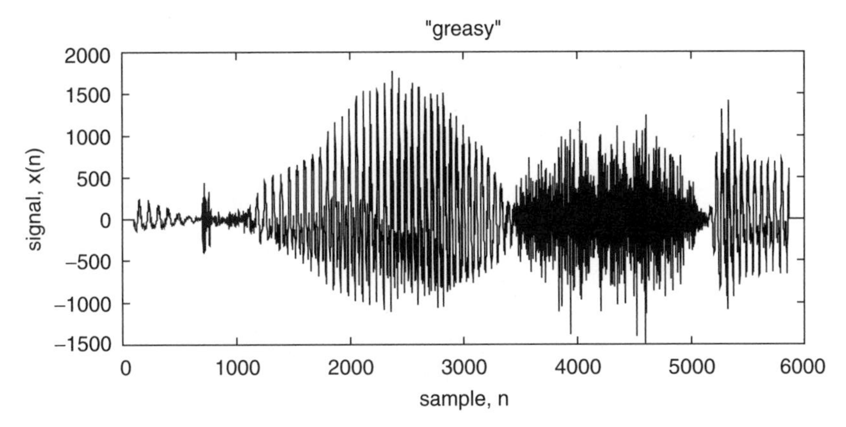

Fig. 1.30. A digitized voice signal, the word "greasy". The sampling rate is 16 kHz. It is difficult, even to the trained eye, to recognize the spoken content of a signal from its time-domain signal values.

them [59, 62]. The speech fragment of Figure 1.30 contains five phonemes. Sometimes the preliminary sequence of phonemes makes no sense; a sophisticated algorithm may merge or further segment some of the phones for a better result. A higher-level process uses the phoneme stream to extract whole words. Perhaps some posited word does not fit the application's context. Still higher-level algorithms—and at this point the application has removed itself from signal processing and analysis proper to the realm of artificial intelligence—may substitute one phoneme for another to improve the interpretation. This is called *contextual analysis*. Although we will consider speech analysis in more detail in later chapters, topics such as contextual analysis blend into artificial intelligence, and are outside the scope of our presentation. It is nevertheless interesting to note that computers are generally better than humans at recognizing individual phonemes, while humans are far superior when recognizing complete words [63], a hint of the power of contextual methods.

Linguists separate phonemes into two categories: *voiced* and *unvoiced*, according to whether the vocal cords vibrate or not, respectively. Vowels are voiced, and it turns out that a frequency-domain description helps to detect the presence of a vowel sound. Vowel sounds typically contain two sinusoidal components, and one important early step in speech processing is to determine the frequency components of the signal. We can see this in a digital speech sample of a vowel (Figure 1.31). There are clearly two trends of oscillatory behavior in the signal.

Thus, depending upon the relative strength of the two components and upon their actual frequency, a frequency domain description can identify this phone as the /u/ phoneme. There are complicating factors of course. The frequency components change with the gender of the speaker. And noise may corrupt the analysis. Nevertheless, a frequency-domain description is an important beginning point in phoneme recognition.

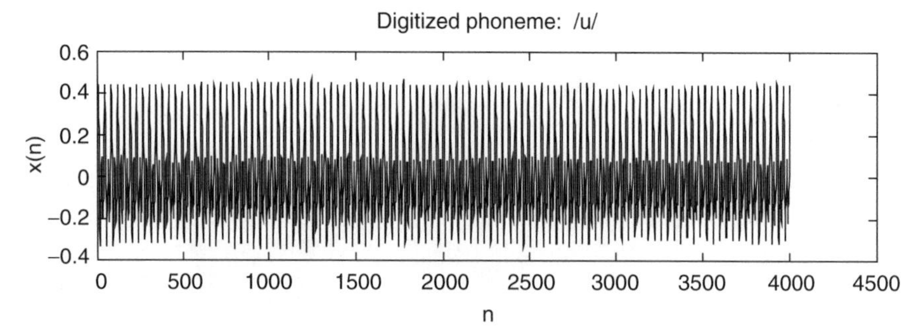

Fig. 1.31. A vowel phoneme, the sound "*u*", sampled at 8 kHz. Note the two sinusoids in the time-domain trace. The lower frequency component has a higher amplitude. Principal frequency components are at approximately 424 Hz and at 212 Hz.

1.5.4 Time and Frequency Combined

Until we develop the formal theory for the frequency analysis of signals—the Fourier transform, in particular—we continue to illustrate signals in terms of their time-domain values. We can nonetheless identify the oscillatory components by inspection. Parts of a given signal resemble a sinusoid of a given frequency. Sometimes the oscillations continue throughout the time span of the signal. For such signals, the Fourier transform (Chapters 5 and 6, and introduced from a practical standpoint in Chapter 4) is the appropriate tool.

Sometimes, however, the oscillatory components die out. This makes the game interesting, because our ordinary sinusoidal models continue oscillating forever. We can arbitrarily limit the time domain extent for our sinusoidal models, and this happens to be effective for applications such as speech analysis. Special mathematical tools exist that can decompose a signal into a form that exposes its frequency components within distinct time intervals. Among these tools is the Gabor transform, one of several time-frequency transforms that we explore in Chapter 10. Until the theoretical groundwork is laid for understanding these transforms, however, we must content ourselves with intuitive methods for describing the frequency content of signals.

Let us also remark that frequency-domain methods have been extensively explored in electrocardiography and electroencephalography—applications we considered at the beginning of the chapter. This seems a natural approach for the ECG, since the heartbeat is a regular pulse. Perhaps surprisingly, for heartbeat irregularities, frequency-domain techniques have been found to be problematic. In EEG work, the detection of certain transients and their development into regular waves is important for diagnosing epilepsy. And here again frequency-domain tools—even those that employ an explicit mixed time-frequency analysis strategy—do not address all of the difficulties [35, 64–66]. Like problems arise in seismic signal interpretation [67, 68]. In fact, problems in working with time-frequency analysis methods for seismic soundings analysis in petroleum prospecting led to the discovery of the scale-based wavelet transform in the mid-1980s [69].

Researchers have thus begun to investigate methods that employ signal shape and scale as a tool, rather than frequency. Unlike the sinusoidal components that are the basis for a frequency-domain signal description, the components for a scale-based description are limited in their time-domain extent. The next section considers several special signal classes, among them several types which are time-limited. Of these, the finite-energy signals are particularly attractive, as later chapters demonstrate, for signal descriptions based on scale.

1.6 SPECIAL SIGNAL CLASSES

This section covers some special signal classes: finitely supported signals, even and odd signals, absolutely summable signals, finite energy signals, and finite average power signals. The finite-energy signals are by far the most important. This signal class has a particularly elegant structure. The finite energy signals are usually at the center of theoretical signal processing and analysis discussions.

It is from such signal families that the notion of a scale-domain description of a signal arises. A scale-domain description decomposes a signal into parts based on shape over a given length of time. All of the parts contain the same shape, even though the time-domain extent of the shape element varies. In order for a shape element to be so localized, the component signal must eventually die out; it becomes zero, or at least effectively zero. Thus, signals that oscillate forever, as do the sinusoids, do not directly serve a scale-domain analysis. Signals that diminish near infinity, such as Gaussians, Gabor elementary functions, and the like, are used for scale-domain description.

1.6.1 Basic Classes

Useful distinguishing properties of signals are their symmetry and their behavior near infinity.

1.6.1.1 Even and Odd Signals. One of the important characteristics of a signal is its symmetry. Symmetries allow us to simplify the description of a signal; we only need to know about the shape of the signal over some restricted domain. Uncovering symmetries can also be a first step to decomposing a signal into constituent parts. For brevity, this section primarily discusses discrete signals, but for analog signals, similar definitions and properties follow easily.

Definition (Symmetry, Even and Odd Signals). A discrete signal $x(n)$ is *symmetric* about the time instant $n = p$ if $x(p + n) = x(p - n)$ for all $n \in \mathbb{Z}$. And $x(n)$ is *antisymmetric* about the time instant p if $x(p + n) = -x(p - n)$ for all nonzero $n \in \mathbb{Z}$. A discrete signal $x(n)$ is *even* if it is symmetric about $n = 0$. Similarly, if $x(n)$ is antisymmetric about $n = 0$, then x is *odd*.

Corresponding definitions exist for symmetries of analog signals $x(t)$.

Definition (Even and Odd Part of Signals). Let $x(n)$ be a discrete signal. Then the *even part* of $x(n)$ is

$$x_e(n) = \frac{x(n) + x(-n)}{2}. \tag{1.63a}$$

The *odd part* of $x(n)$ is

$$x_o(n) = \frac{x(n) - x(-n)}{2}. \tag{1.63b}$$

There are corresponding definitions for the even and odd parts of analog signals as well.

Proposition (Even/Odd Decomposition). If $x(n)$ is a discrete signal, then

(i) $x_e(n)$ is even;
(ii) $x_o(n)$ is odd;
(iii) $x(n) = x_e(n) + x_o(n)$.

Proof: Exercise. ■

Examples. $\sin(t)$ is odd; $\cos(t)$ is even; and the Gaussian, $g_{\mu,\sigma}(t)$ of mean μ and standard deviation σ (1.14), is symmetric about μ.

Of course, some signals are neither even nor odd. For complex-valued signals, we often look at the real and imaginary components for even and odd symmetries.

1.6.1.2 Finitely Supported Signals. The set of time values over which a signal x is nonzero is called the *support* of x. *Finitely supported* signals are zero outside some finite interval. For analog signals, a related concept is also useful—compact support.

Definition (Finite Support). A discrete signal $x(n)$ is *finitely supported* if there are integers $M < N$ such that $x(n) = 0$ for $n < M$ and $n > N$.

If $x(n)$ is finitely supported, then it can be specified via square brackets notation: $x = [k_M, \ldots, \underline{kd_0}, \ldots, k_N]$, where $x(n) = k_n$ and $M \leq 0 \leq N$.

For analog signals, we define the concept of finite support as we do with discrete signals; that is, $x(t)$ is of *finite support* if it is zero outside some interval $[a, b]$ on the real line. It turns out that our analog theory will need more specialized concepts from the topology of the real number line [44, 70].

Definition (Open and Closed Sets, Open Covering, Compactness). A set $S \subseteq \mathbb{R}$ is *open* if for every $s \in S$, there is an open interval *(a, b)* such that $s \in (a, b) \subseteq S$. A set is *closed* if its complement is open. An *open covering* of a $S \subseteq \mathbb{R}$ is a family of open sets $\{O_n \mid n \in \mathbb{N}\}$ such that $\bigcup_{n=0}^{\infty} O_n \supseteq S$. Finally, a set$S \subseteq \mathbb{R}$ is *compact* if for every open covering of *S*, $\{O_n \mid n \in \mathbb{N}\}$, there is a finite subset that also contains S:

$$\bigcup_{n=0}^{N} O_n \supseteq S, \tag{1.64}$$

for some $\{n \in \mathbb{N}\}$.

Definition (Compact Support). An analog signal *x*(*t*) has *compact support* if $\{t \in \mathbb{R} \mid x(t) \neq 0\}$ is compact.

It is easy to show that a (finite) sum of finitely supported discrete signals is still of finite support; that is, the class of finitely supported signals is closed under addition. We will explore this and other operations on signals, as well as the associated closure properties in Chapters 2 and 3. The following theorem connects the idea of compact support for analog signals to the analogous concept of finite support for discrete signals [44, 70].

Theorem (Heine–Borel). $S \subseteq \mathbb{R}$ is compact if and only if it is closed and contained within some finite interval [*a*, *b*] (that is, it is *bounded*).

Proof: The exercises outline the proof. ■

1.6.2 Summable and Integrable Signals

Compact support is a very strong constraint on a signal. This section introduces the classes of absolutely summable (discrete) and absolutely integrable (analog) signals. Their decay is sufficiently fast so that they are often neglible for large time values. Their interesting values are concentrated near the origin, and we can consider them as having localized shape.

Definition (Absolutely Summable Signals). A discrete signal *x*(*n*) is *absolutely summable* (or simply *summable*) if the sum of its absolute values is finite:

$$\sum_{n=-\infty}^{\infty} |x(n)| < \infty . \tag{1.65}$$

Another notation for this family of discrete signals is l^1. Finite support implies absolutely summability.

Definition (Absolutely Integrable Signals). A signal $x(t)$ is *absolutely integrable* (or simply *integrable*) if the integral of its absolute value over $\mathbb{R}$ is finite:

$$\int_{-\infty}^{\infty} |x(t)|\, dt < \infty . \quad (1.66)$$

Other notations for this analog signal family are L^1 or $L^1[\mathbb{R}]$. Signals that are integrable of an interval $[a, b]$ are in $L^1[a, b]$. They satisfy

$$\int_{a}^{b} |x(t)|\, dt < \infty . \quad (1.67)$$

1.6.3 Finite-Energy Signals

The most important signal classes are the discrete and analog finite energy signals.

Definition (Finite–Energy Discrete Signals). A discrete signal $x(n)$ has *finite energy* or is *square-summable* if

$$\sum_{n=-\infty}^{\infty} |x(n)|^2 < \infty . \quad (1.68)$$

Another notation for this family of discrete signals is l^2. Note that a discrete signal that is absolutely summable must also be finite energy. We require the square of the absolute value $|x(n)|^2$ in (1.68) to accomodate complex-valued signals.

Definition (Finite-Energy Analog Signals). An analog signal $x(t)$ is *finite-energy* (or *square-integrable*) if

$$\int_{-\infty}^{\infty} |x(t)|^2\, dt < \infty. \quad (1.69)$$

Alternative names for this family are L^2 or $L^2[\mathbb{R}]$. $L^1[a, b]$ signals satisfy

$$\int_{a}^{b} |x(t)|^2\, dt < \infty. \quad (1.70)$$

The term "finite-energy" has a physical meaning. The amount of energy required to generate a real-world signal is proportional to the total squares of its values. In classical electromagnetic theory, for example, a radio wave carries energy that is proportional to the sum of the squares of its electric and magnetic fields integrated over the empty space through which the fields propagate.

Discrete and analog finite-energy signals are central to the later theoretical development. The next two chapters generalize the concept of a vector space to infinite dimensions. A discrete signal is like a vector that is infinitely long in both positive and negative directions. We need to justify mathematical operations on signals so that we can study the processes that operate upon them in either nature or in engineered systems. The goal is to find classes of signals that allow infinite support, yet possess all of the handy operations that vector space theory gives us: signal sums, scalar multiplication, inner (dot) product, norms, and so forth.

1.6.4 Scale Description

Only recently have we come to understand the advantages of analyzing signals by the size of their time-domain features. Before the mid-1980s, signal descriptions using frequency content predominated. Sometimes the frequency description was localized, but sometimes these methods break down. Other applications naturally invite an analysis in terms of the feature scales. At a coarse scale, only large features of the signal are evident. In a speech recognition application, for example, one does not perform a frequency decomposition or further try to identify phonemes if a coarse-scale inspection of the signal reveals only the presence of low-level background noise. At a finer scale, algorithms separate words. And at even higher resolution, the words may be segmented into phonemes that are finally subjected to recognition efforts. Although a frequency-domain analysis is necessary to identify phonemes, therefore, some kind of scale-domain analysis may be appropriate for the initial decomposition of signal.

Figure 1.32 shows an example from image analysis. One-dimensional analysis is possible by extracting lines from the image. In fact, many image analysis applications approach the early segmentation steps by using one-dimensional methods at a series of coarse scales. The time-consuming, two-dimensional analysis is thus postponed as long as possible.

1.6.5 Scale and Structure

Signal description at many scales is one of the most powerful methods for exposing a signal's structure. Of course, a simple parsing of a signal into time-domain subsets that do and do not contain useful signal represents a structural decomposition. However, when this type of signal breakdown is presented a different scales, then an artificial intelligence algorithm can home in on areas of interest, perform some goal-directed interpretation, and proceed—based upon the coarse scale results—to focus on minute details that were ignored previously. Thus, the structural description of a signal resembles a tree, and this tree, properly constructed, becomes a guide for the interpretation of the signal by high-level algorithms.

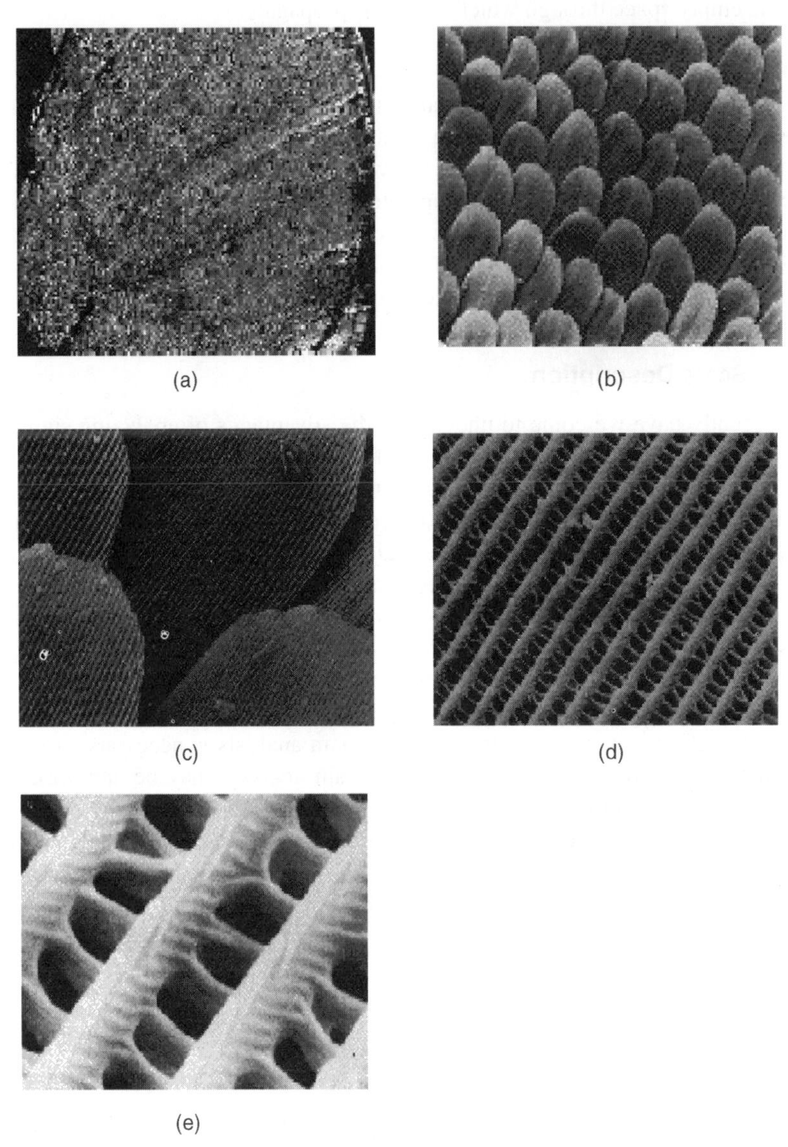

Fig. 1.32. A butterfly and a scanning electron microscope (SEM) teach us the concept of scale. The first image (a), which looks like a butterfly wing is taken at a magnification of 9×. (b) The SEM, at a power of 330×, reveals a scaly pattern. (c) At 1700× the scales appear to possess striations. (d) This confirms the existence of striations at 8500× and hints of small-scale integuments between the principal linear structures. (e) This exposes both the coarse-scale striations and the fine-scale integuments between them at 40,000× magnification.

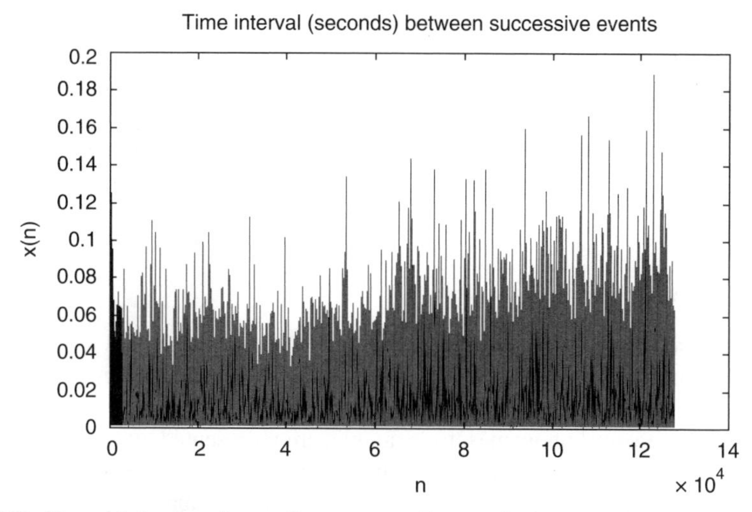

Fig. 1.33. Fractal behavior of an auditory neuron. From a single neuron in the hearing pathways of an anesthesized cat, researchers recorded the time intervals between successive neural discharge events in the presence of a tone stimulus. Note in particular that the independent variable represents not the flow of time or an ordering by distance but rather an enumeration of successive neural discharge events.

Two methods of structural description rely on signal shape models:

(i) Using self-similar shape models over a range of scales
(ii) Using a library of special shapes

Applications using self-similar shape models have begun to draw the attention of researchers in recent years. This approach makes possible a fractal analysis of a signal pattern. A fractal application attempts to understand to what degree the same shape occurs within a signal at different scales.

Sometimes, naturally occurring signals unexpectedly reveal fractal properties (Figure 1.33). This signal monitors neurons in the auditory pathway of an anesthetized cat. The signal's independent variable is not temporally dimensioned. Rather, each signal instant represents the next neural event, and the value of the signal is the time interval between such events. This twist in the conventional way of sketching a signal is key to arriving at the fractal behavior of the neuron.

Representing the signal in this special way provides insight into the nature of the neural process [71]. The discharge pattern reveals fractal properties; its behavior at large scales (in this case, over a span of many discharge events) resembles its behavior over small scales. Recently, signal analysts have brought their latest tools—among them the wavelet transform (Chapter 11), which breaks down a signal according to its frequency content within a telescoping family of scales—to bear on fractal analysis problems [53, 69].

1.7 SIGNALS AND COMPLEX NUMBERS

Complex numbers are useful for two important areas of signal theory:

(i) Computation of timing relationships (phase) between signals;
(ii) Studying the frequency content of signals.

1.7.1 Introduction

To understand why complex numbers are computationally useful, let us consider the superposition of two sinusoidal waves. Earlier, we considered this important case. It occurs in electrical circuits, where two voltages are summed together; in speech recognition, where vowel phonemes, for example, are represented by a sinusoidal sum; and in optics, where two optical wavefronts combine and interfere with one another to produce an interference pattern. Introducing complex numbers into the mathematical description of signal phenomena makes the analysis much more tractable [72].

Let $x(t) = x_1(t) + x_2(t) = A_1\cos(\Omega_1 t + \phi_1) + A_2\cos(\Omega_2 t + \phi_2)$. An awkward geometric argument showed earlier that if $\Omega_1 = \Omega_2 = \Omega$, then $x(t)$ remains sinusoidal. Why should a purely algebraic result demand a proof idea based on rotating parallelograms? Complex numbers make it easier. We let $x_1(t)$ and $x_2(t)$ be the real parts of the complex exponentials: $x_1(t) = \mathrm{Real}[z_1(t)]$ and $x_2(t) = \mathrm{Real}[z_2(t)]$, where

$$z_1(t) = A_1\exp(j[\Omega t + \phi_1]), \tag{1.71a}$$

$$z_2(t) = A_2\exp(j[\Omega t + \phi_2]). \tag{1.71b}$$

Then $x(t) = \mathrm{Real}[z_1(t) + z_2(t)] = \mathrm{Real}[z(t)]$. We calculate

$$\begin{aligned} z(t) &= A_1\exp(j[\Omega t + \phi_1]) + A_2\exp(j[\Omega t + \phi_2]) \\ &= \exp(j\Omega t)[A_1\exp(j\phi_1) + A_2\exp(j\phi_2)] \end{aligned} \tag{1.72}$$

Notice that the sum (1.72) has radial frequency Ω radians per second; the $\exp(j\Omega t)$ term is the only one with a time dependence. To calculate $\|z(t)\|$, note that $\|e^{j\Omega t}\| = 1$, and so $z(t) = \|A_1\exp(j\phi_1) + A_2\exp(j\phi_2)\|$. As before, we find

$$\|z(t)\|^2 = A_1^2 + A_2^2 + 2A_1A_2\cos(\phi_2 - \phi_1). \tag{1.73}$$

Thus, complex arithmetic takes care of the phase term for us, and this is one reason why complex arithmetic figures in signal theory. Of course, we understand that only the real part of the complex-valued signal model corresponds to any physical reality.

Now let us consider complex-valued functions, $f: \mathbb{C} \to \mathbb{C}$, and develop the ideas of calculus for them. Such functions may at first glance appear to bear a strong resemblance to signals defined on pairs of real numbers: $f: \mathbb{R} \times \mathbb{R} \to \mathbb{C}$, for example. To pursue the idea for a moment, we can let $z = x + jy$, where $j^2 = -1$. Then, $f(z) = f(x + jy)$, and f is a function of the real pair (x, y). The function f is complex-valued, of course, but we can take its real and imaginary parts: $f(z) = \text{Real}[f(x + jy)] + j\mathit{Imag}[f(x + jy)]$. Now $f(z)$ looks suspiciously like a sum of two multidimensional signals—a sum of two images—one of which is scaled by the imaginary square root of -1. If we were to define differentiation of $f(z)$ with respect to x and y, where $z = x + jy$, then this would indeed be the case; our theory of complex analysis would look a lot like ordinary real analysis. But when we define differentiation with respect to the complex variable z, what a huge difference it makes! A seemingly innocuous distinction about how to define differentiation makes the calculus of complex variables rich, novel, and powerful.

1.7.2 Analytic Functions

The existence of a derivative is a very special and far-reaching property for a complex function $f(z)$.

Definition (Differentiation, Derivative). Let $S \subseteq \mathbb{C}$ and $f: S \to \mathbb{C}$. Then f is *differentiable* at a point $z \in S$ if the limit,

$$f'(z) = \lim_{w \to z} \frac{f(w) - f(z)}{w - z} \tag{1.74}$$

exists. As in calculus, the limit $f'(z) = \frac{d}{dz}f(z)$ is called the *derivative* of f at z.

Definition (Analyticity). Let $w \in \mathbb{C}$. If there is an $R > 0$ such that $f(z)$ is differentiable for all at all z such that $|w - z| < R$, then f is *analytic* at w. If $f(z)$ is analytic at every $w \in S$, then $f(z)$ is analytic in S.

Proposition (Differentiation). Let f and g be differentiable at $z \in \mathbb{C}$. Then

(i) f is continuous at z.
(ii) If $c \in \mathbb{C}$, then cf is differentiable at z, and $(cf)'(z) = cf'(z)$.
(iii) $f+g$ is differentiable at z, and $(f+g)'(z) = f'(z) + g'(z)$.
(iv) fg is differentiable at z, and $(fg)'(z) = fg'(z) + f'g(z)$.
(v) If $g(z) \neq 0$, then f/g is differentiable at z, and

$$(f/g)'(z) = \frac{g(z)f'(z) - f(z)g'(z)}{g^2(z)}. \tag{1.75}$$

Proof: As in calculus of a real variable [44]; also see complex analysis texts [73–75]. ■

Proposition (Chain Rule). Let f be differentiable at $z \in \mathbb{C}$ and let g be differentiable at $f(z)$. Then the composition of the two functions, $(g \circ f)(z) = g(f(z))$, is also differentiable, and $(g \circ f)'(z) = g'(f(z))f'(z)$.

Proof: Refer to Refs. [44] and [73–75]. ■

Power series of a complex variable are useful for the study of discrete *systems*, signal processing entities that modify discrete signals. A system takes a signal as an input and produces another signal as an output. A special complex power series, called the *z-transform* of a signal, is studied in Chapters 8 and 9.

Definition (Power and Laurent Series). A *complex power series* is a sum of scaled powers of the complex variable z:

$$\sum_{n=0}^{\infty} a_n z^n , \tag{1.76a}$$

where the a_n are (possibly complex) coefficients. Sometimes we expand a complex power series about a point $w \in \mathbb{C}$:

$$\sum_{n=0}^{\infty} a_n (z-w)^n . \tag{1.76b}$$

A *Laurent series* is two-sided:

$$\sum_{n=-\infty}^{\infty} a_n z^n \tag{1.77a}$$

and can be expanded about $w \in \mathbb{C}$:

$$\sum_{n=-\infty}^{\infty} a_n (z-w)^n . \tag{1.77b}$$

[The z-transform of $x(n)$ is in fact a Laurent expansion on the values of the discrete signal: $x(n) = a_n$ in (1.77a).]

We define special complex functions in terms of power series:

$$\sin(z) = z - \frac{z^3}{3!} + \frac{z^5}{5!} - \frac{z^7}{7!} + \cdots ; \tag{1.78a}$$

$$\cos(z) = 1 - \frac{z^2}{2!} + \frac{z^4}{4!} - \frac{z^6}{6!} + \cdots ; \tag{1.78b}$$

and the most important function in mathematics,

$$\exp(z) = 1 + z + \frac{z^2}{2!} + \frac{z^3}{3!} + \frac{z^4}{4!} + \cdots = e^z. \tag{1.79}$$

Their convergence criteria are similar to those of real power series. The following theorem says that a convergent power series is differentiable and its derivative may be computed by termwise differentiation.

Proposition (Power Series Differentiation). Suppose that

$$p(z) = \sum_{n=0}^{\infty} a_n (z-w)^n \tag{1.80}$$

converges in $S = \{z : |z-w| < R\}$. Then $p(z)$ is analytic (differentiable at every point) inside S, and

$$p'(z) = \sum_{n=1}^{\infty} n a_n (z-w)^{n-1}. \tag{1.81}$$

Proof: References [44] and [73–75]. ■

The next theorem suggests that complex function calculus is very different from conventional real variable theory [73–75].

Theorem (Cauchy–Riemann Equations). Suppose that $f(z) = u(x, y) + jv(x, y)$, where u and v are the real and imaginary parts, respectively, of $f(z)$. If f is differentiable at $z = w$, then the partial derivative $\partial u/\partial x$, $\partial u/\partial y$, $\partial v/\partial x$, and $\partial v/\partial y$ all exist; furthermore,

$$\frac{\partial u}{\partial x}(w) = \frac{\partial v}{\partial y}(w), \tag{1.82a}$$

$$\frac{\partial u}{\partial y}(w) = \frac{\partial v}{\partial x}(w). \tag{1.82b}$$

Proof: We can compute the derivative $f'(w)$ in two different ways. We can approach w along the real axis or along the imaginary axis. Thus, we see that the following limit gives $f'(w)$ when we approach w from values $w + h$, where h is real:

$$\lim_{h\to 0} \frac{f(w+h)-f(w)}{h} = \lim_{h\to 0} \frac{u(x+h,y)-u(x,y)}{h} + j \lim_{h\to 0} \frac{v(x+h,y)-v(x,y)}{h}. \tag{1.83a}$$

But $f'(w)$ can also be computed by taking the limit along the imaginary axis; we now approach w from values $w+jk$, where k is real. Consequently,

$$\lim_{k \to 0} \frac{f(w+jk)-f(w)}{jk} = \lim_{k \to 0} \frac{u(x, y+k)-u(x, y)}{jk} + j \lim_{k \to 0} \frac{v(x, y+k)-v(x, y)}{jk}. \tag{1.83b}$$

The limit (1.83a) is

$$f'(w) = \frac{\partial u}{\partial x}(x, y) + j\left[\frac{\partial v}{\partial x}(x, y)\right], \tag{1.84a}$$

whereas (1.83b) is

$$f'(w) = -j\left[\frac{\partial u}{\partial y}(x, y)\right] + \frac{\partial v}{\partial y}(x, y). \tag{1.84b}$$

Since $f'(w)$ has to be equal to both limits, the only way to reconcile (1.84a) and (1.84b) is to equate their real and imaginary parts, which gives (1.82a)–(1.82b). ■

Remark. The Cauchy–Riemann equations imply that some surprisingly simple complex functions, $f(z) = f(x + jy) = x - jy$, for example, are not differentiable.

The converse to the theorem requires an additional criterion on the partial derivatives, namely that they be continuous.

Theorem (Cauchy–Riemann Converse). Let $f(z) = u(x, y) + jv(x, y)$, where u and v are the real and imaginary parts, respectively, of $f(z)$. Furthermore, let the partial derivatives $\partial u/\partial x$, $\partial u/\partial y$, $\partial v/\partial x$, and $\partial v/\partial y$ all exist and be continuous and satisfy the Cauchy–Riemann equations (1.82a)–(1.82b) at $z = w$. Then $f'(w)$ exists.

Proof: Not too difficult [73–75]. ■

Corollary. Let $f(z) = u(x, y) + jv(x, y)$, where u and v are the real and imaginary parts, respectively, of $f(z)$. If $f'(w)$ and $f''(w)$ exist, then the partial derivatives of u and v obey the Laplace equation:

$$\frac{\partial^2 \phi}{\partial x^2} + \frac{\partial^2 \phi}{\partial y^2} = 0. \tag{1.85}$$

Proof: By the Cauchy–Riemann theorem, u and v satisfy (1.82a)–(1.82b). Applying the theorem again to the deriviatives, and using the fact from calculus that mixed partial derivatives are equal where they are continuous, we find

$$\frac{\partial}{\partial x}\frac{\partial u}{\partial x}(w) = \frac{\partial}{\partial x}\frac{\partial v}{\partial y}(w) = \frac{\partial}{\partial y}\frac{\partial v}{\partial x}(w) = -\frac{\partial}{\partial y}\frac{\partial u}{\partial y}(w) \tag{1.86a}$$

Similarly,

$$\frac{\partial}{\partial y}\frac{\partial v}{\partial y}(w) = -\frac{\partial}{\partial x}\frac{\partial v}{\partial x}(w). \tag{1.86b}$$

The Laplace equations for both u and v follow. ■

This is an intriguing result. Complex differentiability leads to a second-order partial differential equation. That is, if a function $f(z)$ is twice differentiable, then it is harmonic in a set $S \subseteq \mathbb{C}$. Thus, complex differentiation is already seen to be a much more restricted condition on a function than real differentiation. Laplace's equation appears in many applications of physics and mechanics: heat conduction, gravitation, current flow, and fluid flow, to name a few. The import of the corollary is that complex functions are a key tool for understanding such physical systems. For applications to the theory of fluid flow, for example, see Ref. 75.

Even stronger results are provable. The next section outlines the development of complex integration theory. It seems quite backwards to prove theorems about differentiation by means of integration theory; but in the exotic realm of complex analysis, that is exactly the course we follow. Using contour integration in the complex plane, it is possible to prove that an analytic function (differentiable in a region) has continuous derivatives of all orders. That is, every analytic function expands in a Taylor series.

1.7.3 Complex Integration

This section continues our sweep through complex analysis, turning now to integration in the complex plane. Given the results of the previous section, one might imagine that complex integration should also have special properties unlike anything in real analysis. Such readers will not be disappointed; the theory of complex integration is even more amazing than differentiation.

Definition (Contour). A *curve* in the complex plane is a function $s : [a, b] \rightarrow \mathbb{C}$, where $[a, b] \subset \mathbb{R}$. We say that s *parameterizes* its range. If the real and imaginary parts of $s(t)$ are continuously differentiable, then s is called an *arc*. If $s(a) = s(b)$, then the curve s is *closed*. And if $s(t_1) = s(t_2)$ on (a, b) implies $t_1 = t_2$, then the curve s is *simple*. A sequence of arcs $\{s_n(t) : [a_n, b_n] \rightarrow C : 1 \le n \le N\}$ is a *contour* if $s_n(b_n) = s_{n+1}(a_{n+1})$, for $n = 1, 2, \ldots, N-1$.

Remarks. A curve is a complex-valued analog signal, defined on a closed interval of the real line. An arc is a continuously differentiable, complex-valued analog signal. A simple curve does not intersect itself, save at its endpoints. We often denote an arc in the complex plane by its range, $C = \{z : z = s(t), \text{ for some } a \le t \le b\}$, and the defining curve function is implicit. Our purpose is to define integration along a contour [76].

Definition (Contour Integral). If the complex function $f(z)$ is continuous in a region containing an arc C, then the contour integral of f over C is defined by

$$\oint_C f(z)\, dz = \int_a^b f[s(t)]s'(t)\, dt\,, \tag{1.87}$$

where $s(t)$ is the function that parameterizes C.

Since $f(z)$, $s(t)$, and $s'(t)$ are all continuous, the integrand in (1.87) is Riemann integrable. The function $f[s(t)]s'(t)$ is complex-valued; we therefore perform the real integration (that is, with respect to t) twice, once for the real part and once for the imaginary part of the integrand. Observe that the change of integration variable, $z = s(t)$ and $dz = s'(t)\, dt$, converts the integral's definition with respect to z in (1.87) to one with respect to t.

The main result of this section is Cauchy's integral theorem. There is an interpretation of contour integration that provides an intuitive link to the familiar theory of integration from calculus and an informal argument for the theorem [77]. Readers seeking rigor and details will find them in the standard texts [73–76]. We first consider the case where C is a circle around the origin, which is a simple, closed arc. Then we shall argue the extension to general arcs, by supposing the arcs to be the limit of a local tiling of the region by adjacent triangles. From this, the extension to contours, which are a sequence of arcs, follows directly.

Theorem (Cauchy Integral for a Circle). Suppose $f(z)$ is analytic in a region containing the closed circle C, with radius R and center $z = (0, 0)$. Then,

$$\frac{1}{2\pi j}\oint_C z^m dz = \begin{cases} 0 & \text{if } m \ne 1, \\ 1 & \text{if } m = -1. \end{cases} \tag{1.88}$$

Proof: In calculus courses [44], Riemann integrals are the limits of Riemann sums. For the case of a contour integral, such a sum is a limit:

$$\oint_C f(z)dz = \lim_{N \to \infty} \sum_{n=1}^{N} f(w_n)[z_{n+1} - z_n]\,. \tag{1.89}$$

where $s : [a, b] \to \mathbb{C}$ parameterizes the arc C; $a = t_1 < t_2 < \cdots < t_N < t_{N+1} = b$ partitions $[a, b]$; $z_n = s(t_n)$; and $w_n = s(t)$ for some $t \in [t_n, t_{n+1}]$. Suppose further that we select the t_n so that $|z_{n+1} - z_n| = \varepsilon_N = \text{Length}(C)/N = L_C/N$. Then we have

$$\oint_C f(z)dz = \lim_{N\to\infty} \sum_{n=1}^{N} f(w_n)\frac{[z_{n+1}-z_n]}{|z_{n+1}-z_n|}|z_{n+1}-z_n|$$
$$= \lim_{N\to\infty} \varepsilon_N \sum_{n=1}^{N} f(w_n)\frac{[z_{n+1}-z_n]}{|z_{n+1}-z_n|} = L_C \lim_{N\to\infty}\frac{1}{N}\sum_{n=1}^{N} f(w_n)\frac{[z_{n+1}-z_n]}{|z_{n+1}-z_n|}. \tag{1.90}$$

Note that as $N \to \infty$, we have $w_n \to z_n$, and $(z_{n+1}-z_n)/|z_{n+1}-z_n|$ approaches a complex value whose real and imaginary parts are the components of the unit tangent vector to C at z_n, $\boldsymbol{T}(z_n)$. Since C has radius R, $\boldsymbol{T}(z_n) = jz/R$ and $L_C = 2\pi R$. Therefore, the final sum in (1.90) approaches $L_C \times \{\text{average over } C \text{ of } f(z)\boldsymbol{T}(z)\}$. We conclude

$$\frac{1}{L_C}\oint_C f(z)\, dz = \mathrm{Avg}_{z\in C}[f(z)\boldsymbol{T}(z)]. \tag{1.91}$$

Now suppose $m = -1$, so that $f(z) = z^{-1}$. Then

$$\frac{1}{2\pi R}\oint_C f(z)\, dz = \mathrm{Avg}_{z\in C}\left[\frac{\boldsymbol{T}(z)}{z}\right] = \mathrm{Avg}_{z\in C}\left[\frac{jz}{Rz}\right] = \mathrm{Avg}_{z\in C}\left[\frac{j}{R}\right] = \frac{j}{R}. \tag{1.92}$$

To show the other possibility in (1.88), we let $m \neq -1$ and find

$$\frac{1}{2\pi R}\oint_C f(z)\, dz = \mathrm{Avg}_{z\in C}\left[\frac{jz^{m+1}}{R}\right] = \frac{j}{R}\mathrm{Avg}_{z\in C}[z^{m+1}]. \tag{1.93}$$

But, the average of all values z^{m+1} over the circle $|z| = R$ is zero, which demonstrates the second possibility of (1.88) and concludes the proof. ■

Note that the informal limit (1.89) is very like the standard calculus formulation of the Riemann integral. The definition of the contour integral is thus a plausible generalization to complex-valued functions.

We will apply this result in Chapter 8 to derive one form of the inverse z-transform. This produces discrete signal values $x(n)$ from the complex function $X(z)$ according to the rule:

$$x(n) = \frac{1}{2\pi j}\oint_C X(z)z^{n-1}dz. \tag{1.94}$$

The Cauchy residue theorem leads to the following concepts.

Definition (Poles and Zeros). A complex function $f(z)$ has a *pole* of order k at $z = p$ if there is a $g(z)$ such that $f(z) = g(z)/(z-p)^k$, $g(z)$ is analytic in an open set containing $z = p$, and $g(p) \neq 0$. We say that $f(z)$ has a *zero* of order k at $z = p$ if there is a $g(z)$ such that $f(z) = g(z)(z-p)^k$, $g(z)$ is analytic in a region about $z = p$, and $g(p) \neq 0$.

Definition (Residue). The *residue* of $f(z)$ at the pole $z = p$ is given by

$$\operatorname{Res}(f(z), p) = \begin{cases} \left[\dfrac{1}{(k-1)!} f^{(k-1)}(p)\right] & \text{if } p \in \text{Interior}(C), \\ 0 & \text{if otherwise,} \end{cases} \tag{1.95}$$

where k is the order of the pole.

Theorem (Cauchy Residue). Assume that $f(z)$ is a complex function, which is analytic on and within a curve C; $a \notin C$; and $f(z)$ is finite (has no pole) at $z = a$. Then

$$\frac{1}{2\pi j} \oint_C \frac{f(z)}{(z-a)^{m-1}} dz = \begin{cases} \left[\dfrac{1}{(m-1)!} f^{(m-1)}(a)\right] & \text{if } a \in \text{Interior}(C), \\ 0 & \text{if otherwise.} \end{cases} \tag{1.96}$$

More generally, we state the following theorem.

Theorem (Cauchy Residue, General Case). Assume that C is a simple, closed curve; $a_m \notin C$ for $1 \le m \le M$; and $f(z)$ is analytic on and within C, except for poles at each of the a_m. Then

$$\frac{1}{2\pi j} \oint_C f(z)\, dz = \sum_{m=1}^{M} \operatorname{Res}(f(z), a_m), \tag{1.97}$$

Proof: References 73–76. ■

1.8 RANDOM SIGNALS AND NOISE

Up until now, we have assumed a close link between mathematical formulas or explicit rules and our signal values. Naturally occurring signals are inevitably corrupted by some random noise, and we have yet to capture this aspect of signal processing in our mathematical models. To incorporate randomness and make the models more realistic, we need more theory.

We therefore distinguish between *random* signals and *deterministic* signals. Deterministic signals are those whose values are completely specified in terms of their independent variable; their exact time domain description is possible. The signal may be discrete or continuous in nature, but as long as there is a rule or formula that relates an independent variable value to a corresponding signal value, then the signal is deterministic. In contrast, a random signal is one whose values are not known in terms of the value of its independent variable. It is best to think of time-dependent signals to understand this. For a random signal, we cannot know the value of the signal in advance; however, once we measure the signal at a particular time instant, only then do we know its value. Deterministic signals are good for carrying information, because we can reliably insert and extract the information we need to move in a reliable fashion. Nature is not kind, however, to our designs. A random component—for example, a measurement error, digitization error, or thermal noise—corrupts the deterministic signal and makes recovery of the signal information more difficult.

This situation often confronts electrical communication engineers. There are many sources of noise on telephone circuits, for example. If the circuits are physically close, electromagnetic coupling between them occurs. Faint, but altogether annoying, voices will interfere with a conversation. One might argue that this is really a deterministic interference: someone else is deliberately talking, and indeed, if the coupling is strong enough, the other coversation can be understood. However, it is in general impossible to predict when this will occur, if at all, and telephony engineers allow for its possibility by considering models of random signal interference within their designs. Thermal noise from the random motion of electrons in conductors is truly random. It is generally negligible. But it becomes significant when the information-bearing signals are quite weak, such as at the receiving end of a long line or wireless link.

An important signal analysis problem arises in communication system design. In a conversation, a person speaks about 35% of the time. Even allowing that there are two persons talking and that both may speak at once, there is still time available on their communication channel when nobody speaks. If circuitry or algorithms can detect such episodes, the channel can be reused by quickly switching in another conversation. The key idea is to distinguish voice signals from the channel's background noise. There is one quirk: When the conversation is broken, the telephone line sounds dead; one listener or the other invariably asks, "Are you still there?" In order to not distress subscribers when the equipment seizes their channel in this manner, telephone companies actually synthesize noise for both ends of the conversation; it sounds like the connection still exists when, in fact, it has been momentarily broken for reuse by a third party. This is called *comfort noise* generation. A further problem in digital telephony is to estimate the background noise level on a voice circuit so that the equipment can synthesize equivalent noise at just the right time.

Now let us provide some foundation for using random signals in our development. Our treatment is quite compact; we assume the reader is at least partly familiar with the material. Readers can gain a deeper appreciation for discrete and continuous probability space theory from standard introductory texts [78–81]. Random signal theory is covered by general signal processing texts [13, 14] and by books that specialize in the treatment of random signals [82, 83].

1.8.1 Probability Theory

This section introduces the basic principles and underlying definitions of probability theory, material that should already be familiar to most readers.

Consider the noise in the 12-lead electrocardiogram signal. Close inspection of its trace shows small magnitude jaggedness, roughness of texture, and spiky artifacts. Variations in the patient's physical condition and skin chemistry, imperfections in the sensors, and flaws in the electronic signal conditioning equipment impose an element of randomness and unknowability on the ECG's value at any time. We cannot know the exact voltage across one of the ECG leads in advance of the measurement. Hence, at any time t, the voltage across a chosen ECG lead $v(t)$ is a *random variable*. All of the possible activities of ECG signal acquisition constitute the *sample space*. An *event* is a subset of the the sample space. For instance, recording the ECG signal at a moment in time is an event. We assign numerical likelihoods or probabilities to the ECG signal acquisition events.

1.8.1.1 Basic Concepts and Definitions. In order that probability and random signal theory work correctly, the events must obey certain rules for separating and combining them.

Definition (Algebra and σ-Algebra). An *algebra* over a set Ω is a collection of subsets of Ω, $\Sigma \subseteq \wp(\Sigma) = \{A : A \subseteq \Omega\}$, with the following properties:

(i) The empty set is in Σ: $\varnothing \in \Sigma$.
(ii) If $A \in \Sigma$, then the complement of A is in Σ: $A' \in \Sigma$.
(iii) If $A, B \in \Sigma$, then $A \cup B \in \Sigma$.

A *σ-algebra* over a set Ω is an algebra Σ with a further property:

(iv) If $A_n \in \Sigma$ for all $n \in \mathbb{N}$, then their union is in Σ:

$$\bigcup_{n=0}^{\infty} A_n \in \Sigma. \tag{1.98}$$

It is easy to verify that in an algebra Σ, $\Omega \in \Sigma$, the union of any finite set of its elements is still in Σ, and Σ is closed under finite intersections. A σ-algebra is also closed under the intersection of infinite families of elements as in (1.98).

The probability measure must have certain mathematical properties.

Definition (Probability Measure). A *probability measure* on a σ-algebra Σ over Ω is a function $P: \Sigma \to [0, 1]$ such that

(i) $P(\Omega) = 1$;
(ii) P sums on disjoints unions; that is, if $\{A_n : n \in I\} \subseteq \Sigma$, where $I \subseteq \mathbb{N}$, and $A_n \cap A_m = \varnothing$, when $n \neq m$, then

$$P\left(\bigcup_{n \in I} A_n\right) = \sum_{n \in I} P(A_n). \tag{1.99}$$

Definition (Probability Space). A *probability space* is an ordered triple (Ω, Σ, P), where Ω is a set of experimental outcomes, called the *sample space*; Σ is a σ-algebra over Ω, the elements of which are called *events*; and P is a probability measure on Σ. The event $\varnothing$ is called the *impossible* event, and the event Ω is called the *certain* event.

Alternative approaches to probability exist. The earliest theories are drawn from the experiments of early gambler-mathematicians, such as Cardano and Pascal.[17] Their dice and card games, run through many cycles—sometimes to the point of financial ruin of the investigator—inspired an alternative definition of probability. It is the value given by the limiting ratio of the number of times the event occurs divided by the number of times the experiment has been tried:

$$P(X) = \lim_{n \to \infty} \frac{O_{X,n}}{n}, \tag{1.100}$$

where $O_{X,n}$ is the number of observations through n trials where X occurred. This intuition serves as a foundation for probability. The exercises supply some flavor of the theoretical development. More widely accepted, however, is the axiomatic approach we follow here. Soviet mathematicians—notably Kolmogorov[18]—pioneered this approach in the 1930s. Through William Feller's classic treatise [79] the axiomatic development became popular outside the Soviet Union. Most readers are probably familiar with this material; those who require a complete treatment will find [78–81] helpful.

1.8.1.2 Conditional Probability.

Conditional probability describes experiments where the probability of one event is linked to the occurrence of another.

Definition (Conditional Probability, Independence). Suppose A and B are two events. The probability that A will occur, given that B has occurred, is defined as

$$P(A \mid B) = \frac{P(A \cap B)}{P(B)}. \tag{1.101}$$

The quotient $P(A|B)$ is called the *conditional probability* of event A given B.

[17]Girolamo Cardano (1501–1576) led a scandalous life as a gambler, but learned enough to found the theory of probability decades before Fermat and Pascal. Blaise Pascal (1623–1662) was a French mathematician and philosopher. See O. Ore, *Cardano, The Gambling Scholar*, New York: Dover, 1953; also, O. Ore, Pascal and the invention of probability theory, *American Mathematical Monthly*, vol. 67, pp. 409–419, 1960.

[18]Andrei Nikolaevich Kolmogorov (1903–1987) became professor of mathematics at Moscow University in 1931. His foundational treatise on probability theory appeared in 1933.

B must occur with nonzero probability for the conditional probability $P(A|B)$ to be defined.

Definition (Independent Events). Suppose A and B are two events. If $P(A|B) = P(A)P(B)$, then A and B are said to be *independent* events.

Proposition. If A and B are independent, then

(i) A and $\sim B$ are independent;
(ii) $\sim A$ and $\sim B$ are independent.

Proof: Exercise. ■

Proposition (Total Probability). Suppose $\{B_n: 1 \le n \le N\}$ is a partition of Ω and $P(B_n) > 0$ for all n. Then for any A,

$$P(A) = \sum_{n=1}^{N} P(B_n)P(A \mid B_n). \tag{1.102}$$

Proof: $A = (A \cap B_1) \cup (A \cap B_2) \cup \ldots \cup (A \cap B_N)$, which is a disjoint union. The definition of conditional probability entails

$$P(A) = \sum_{n=1}^{N} P(A \cap B_n) = \sum_{n=1}^{N} P(B_n)P(A \mid B_n). \tag{1.103}$$

■

1.8.1.3 Bayes's Theorem. An important consequence of the total probability property, known as Bayes's[19] theorem, is central to a popular pattern classification scheme (Chapter 4).

Theorem (Bayes's). Suppose $\{C_n: 1 \le n \le N\}$ is a partition of Ω and $P(C_n) > 0$ for all n. If $P(A) > 0$ and $1 \le k \le N$

$$P(C_k \mid \mathrm{A}) = \frac{P(C_k)P(A \mid C_k)}{\sum_{n=1}^{N} P(C_n)P(A \mid C_n)}. \tag{1.104}$$

[19]A friend of Thomas Bayes (1703–1761) published the Nonconformist minister's theorem in a 1764 paper before the Royal Society of London.

Proof: The definition of conditional probability implies

$$P(C_k \mid A) = \frac{P(A \cap C_k)}{P(A)} = \frac{P(C_k)P(A \mid C_k)}{P(A)} = \frac{P(C_k)P(A \mid C_k)}{\sum_{n=1}^{N} P(C_n)P(A \mid C_n)}. \quad (1.105)$$

■

Example (Phoneme Classification). Consider the application of Bayes's theorem to a phoneme classification system. Phonemes fall into a fixed number of classes, $C_1, C_2, \ldots, C_N$, given by the application domain. There are also a set of signal features that the application computes for each candidate phoneme. Let us suppose that there are M features, $A_1, A_2, \ldots, A_M$, and the application design is so well done that, for any phoneme-bearing signal, it is possible to both reliably distinguish the phonemes from one another and to assign one of the classes A_m as the principal feature of the signal. A typical feature might be a set of sinusoidal frequencies (formants) that dominate the energy contained in the signal. In any case, we are interested in the phoneme class C_n to which a given input signal belongs. Suppose that the dominant feature is $A = A_m$. We calculate each of the probabilities: $P(C_1|A)$, $P(C_2|A), \ldots, P(C_N|A)$. The highest of these probabilities is the answer—the *Bayes classification*.

How can we calculate these N probabilities? Evidently, we must know $P(C_n)$ for each n. But any of the features might be the dominant one within a signal. Therefore, we must know $P(A_m|C_n)$ for each m and n. And, finally, we must know $P(A_m)$ for each m. A working Bayes classifier requires many probabilities to be known in advance. It is possible to develop these statistics, however, a step called the classifier *training phase*. We gather a large, representative body of speech for the application. If we classify the phonemes manually, in an offline effort, then the relative frequencies of each phoneme can be used in the real-time application. This gives us $P(C_n)$, $1 \le n \le N$. Once we identify a phoneme's class, then we find its predominant feature. For each phoneme C_n, we calculate the number of times that feature A_m turns out to be its predominant feature, which approximates $P(A_m|C_n)$. Lastly, we compute the number of times that each feature is dominant and thus estimate $P(A_m)$. Now all of the numbers are available from the training phase to support the execution of the phoneme classifier on actual data. The more sample phonemes we process and the more genuinely the training data reflects the actual application sources, the better should be our probability estimates.

It is unfortunately often the case that one cannot discover any predominant feature from a set of signal data. What we usually encounter is a feature vector $\boldsymbol{a} = (a_1, a_2, \ldots, a_M)$, where the a_m represent numerical values or scores indicating the presence of each feature A_m. We can compute the probability of a vector of features, but that can only be done after a little more development.

1.8.2 Random Variables

A *random variable* is a function that maps events to numerical values.

Definition (Random Variable). Suppose that (Ω, Σ, P) is a probability space. A *random variable* x on Ω is a function $x : \Omega \to \mathbb{R}$, such that for all $r \in \mathbb{R}$, $\{\omega \in \mathbb{R}: x(\omega) \le r\} \in \Sigma$.

Notation. $x \le r$ or $\{x \le r\}$ is standard for the event $\{\omega \in \mathbb{R}: x(\omega) \le r\} \in \Sigma$. Similarly, we write $x > r$, $x = r$, $r < x \le s$, and so on. Using the properties of a σ-algebra, we can show these too are events in S. It is also possible to consider complex-valued random variables, $z : \Omega \to \mathbb{C}$.

Definition (Distribution Function). Suppose that (Ω, Σ, P) is a probability space and x is a random variable $x : \Omega \to \mathbb{R}$. Then the *probability distribution function*, or simply the *distribution function*, for **x** is defined by $F_x(r) = P(x \le r)$.

Since there is no ordering relation on the complex numbers, there is no distribution function for a complex-valued random variable. However, we can consider distribution functions of the real and imaginary part combined; this topic is explored later via the concept of multivariate distributions.

Proposition (Distribution Function Properties). Let $x : \Omega \to \mathbb{R}$ be a random variable in the probability space (Ω, Σ, P), and let $F_x(r)$ be its distribution function. Then the following properties hold:

(i) If $r < s$, then $F_x(r) \le F_x(s)$.
(ii) $\lim\limits_{r \to \infty} F_x(r) = 1$ and $\lim\limits_{x \to \infty} F_x(r) = 0$.
(iii) $P(x > r) = 1 - F_x(r)$.
(iv) $P(r < x \le s) = F_x(s) - F_x(r)$.
(v) $P(x = r) = F_x(r) - \lim\limits_{s > 0,\, s \to 0} F_x(r - s)$.
(vi) $P(r \le x \le s) = F_x(s) - \lim\limits_{t > 0,\, t \to 0} F_x(r - t)$.
(vii) If $F_x(r)$ is a continuous function of r, then $P(x = r) = 0$ for all r.

Proof: Exercise [81]. ■

The proposition's first statement (i) is a *monotonicity* property.

The distribution function of a random variable may be computed by experiment or may be assumed to obey a given mathematical rule. Special mathematical properties are often assumed for the distribution function; this facilitates mathematical investigations into the behavior of the random variable. One common assumption is that the distribution function is differentiable. This motivates the next definition.

Definition (Density Function). Suppose that (Ω, Σ, P) is a probability space and $\boldsymbol{x}$ is a random variable on Ω. If $F_x(r)$ is differentiable, then the derivative with respect to r of $F_x(r)$, denoted with a lowercase letter f,

$$f_{\boldsymbol{x}}(r) = \frac{d}{dr}F_{\boldsymbol{x}}(r), \tag{1.106}$$

is called the *probability density* function or simply the *density* function of x.

Only functions with specific properties can be density functions. The exercises explore some specific cases.

Proposition (Density Function Properties). Let $\boldsymbol{x} : \Omega \to \mathbb{R}$ be a random variable in the probability space (Ω, Σ, P) with distribution function $F_{\boldsymbol{x}}(r)$. Then

(i) $0 \leq f_{\boldsymbol{x}}(r)$ for all $r \in \mathbb{R}$.

(ii)

$$\int_{-\infty}^{\infty} f_x(t)\, dt = 1. \tag{1.107}$$

(iii)

$$F_{\boldsymbol{x}}(r) = \int_{-\infty}^{r} f_x(t)\, dt. \tag{1.108}$$

(iv)

$$P(r < x \leq s) = F_x(s) - F_x(r) = \int_{s}^{r} f_x(t)\, dt. \tag{1.109}$$

Proof: Property (i) follows from the monotonicity property of the distribution function. Property (iv) follows from the fundamental theorem of calculus [44], where we let the lower limit of the integral pass to infinity in the limit. Properties (ii) and (ii) derive from (iv) via the distribution function limit properties. ■

In the proposition, (i) and (ii) are the conditions that a general function $f: \mathbb{R} \to \mathbb{R}$ must satisfy in order to be a density function. One may also prove an existence theorem that constructs a random variable from such a density function [81]. Random variables divide into two classes: discrete and continuous, based on the continuity of the distribution function. (There is also a *mixed distribution* that has aspects of both, but it is outside our scope.)

1.8.2.1 Discrete Random Variables. Discrete random variables prevail within discrete signal theory.

Definition (Discrete Random Variable). The random variable $\boldsymbol{x}$ is *discrete* if its distribution function is a step function.

In this case, there is a set $M = \{r_n: n \in \mathbb{Z}\}$, such that $m < n$ implies $r_m < r_n$, the set of half-open intervals $[r_m, r_n)$ partition $\mathbb{R}$, and $F_{\boldsymbol{x}}(r)$ is constant on each $[r_m, r_n)$.

Proposition (Discrete Random Variable Characterization). Let $\boldsymbol{x}$ be a random variable in the probability space (Ω, Σ, P) with distribution function $F_{\boldsymbol{x}}(r)$. Set $M = \{r \in \mathbb{R}: \mathrm{P(x = r)} > 0\}$. Then, $\boldsymbol{x}$ is discrete if and only if

$$\sum_{r \in M} P_x(\mathrm{x = r}) = 1. \tag{1.110}$$

Proof: By the definition, we see that $P(\boldsymbol{x} \le r_n) = P(\boldsymbol{x} < r_{n+1})$. This occurs if and only if $P(r_n \le \boldsymbol{x} \le r_{n+1}) = P(\boldsymbol{x} = r_n)$. Therefore the sum (1.110) is

$$\sum_{r \in M} P_x(\mathrm{x = r}) = \lim_{r \to \infty} F_{\boldsymbol{x}}(r) + \lim_{r \to -\infty} F_{\boldsymbol{x}}(r) = 1 \tag{1.111}$$

by the distribution function properties. ■

If the random variable $\boldsymbol{x}$ is discrete, then the $F_{\boldsymbol{x}}(r)$ step heights approach zero as $r \to -\infty$ and approach unity as $r \to \infty$. Because a step function is not differentiable, we cannot define a density function for a discrete random variable as in the previous section. However, we can separately define the density function for a discrete random variable as discrete impulses corresponding to the transition points between steps.

Definition (Discrete Density Function). Suppose the random variable $\boldsymbol{x}$ is discrete, and its distribution function $F_x(r)$ is constant on half-open intervals $[r_n, r_m)$ that partition R. Its density function $f_x(r)$ is defined:

$$f_x(r) = \begin{cases} F_x(r_{n+1}) - F_x(r_n) & \text{if } r = r_n, \\ 0 & \text{if otherwise.} \end{cases} \tag{1.112}$$

Example (Dice). Consider an experiment where two fair dice are thrown, such as at a Las Vegas craps table. Each die shows one to six dots. The probability of any roll on one die is, given honest dice, 1/6. The throw's total is the sum, a random variable $\boldsymbol{x}$. The values of $\boldsymbol{x}$ can be 2, 12, or any natural number in between. There are 36 possible rolls, and the probability of the event that either 2 or 12 is rolled is 1/36. Lucky seven is the most common event—with probability 6/36—as it occurs through the following tosses: (1, 6), (2, 5), (3, 4), (4, 3), (5, 2), or (6, 1). Figure 1.34 shows the distribution function and the density functions for the dice toss.

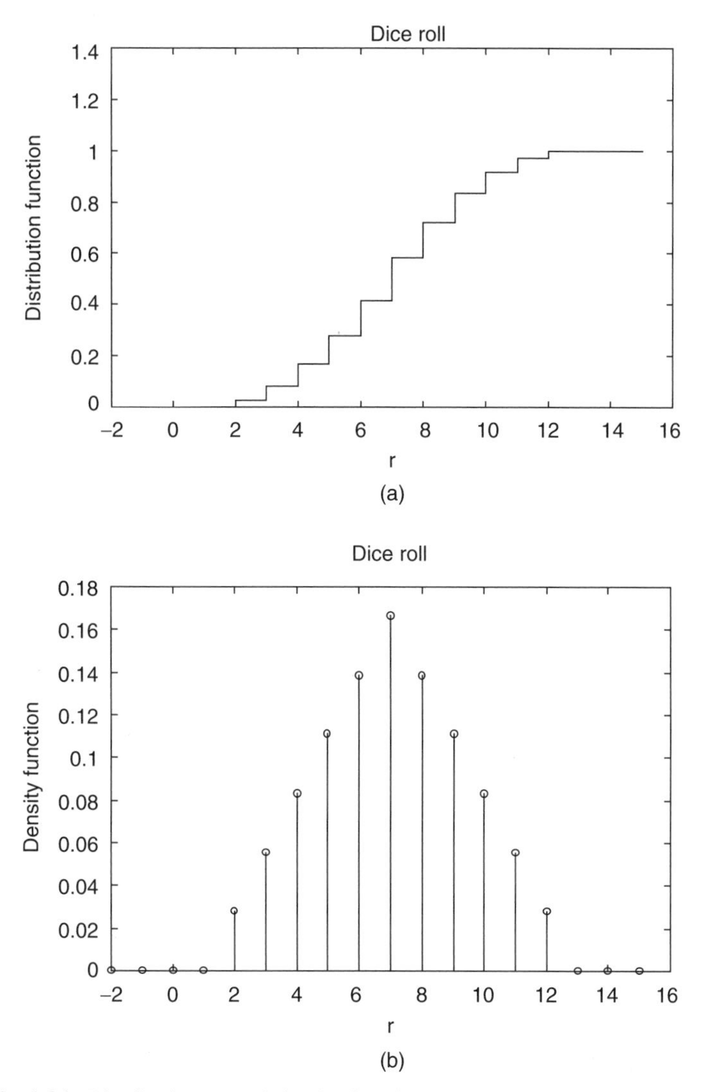

Fig. 1.34. Distribution (a) and density functions (b) for tossing a pair of dice.

Discrete signal theory commonly assumes a density or distribution function for a random variable. If the density or distribution is unknown, it can be measured, of course, but that is sometimes impractical. Instead, one typically approximates it by a distribution that has tractable mathematical properties.

Definition (Binomial Distribution). Suppose that the discrete random variable x has $\mathrm{Range}(\boldsymbol{x}) \subseteq \{0, 1, 2, \ldots, n\}$. Then $\boldsymbol{x}$ has a *binomial distribution* of order n if there are non-negative values p and q, such that $p + q = 1$, and

$$P(\boldsymbol{x} = k) = \binom{n}{k} p^k q^{n-k}. \tag{1.113}$$

Definition (Poisson[20] Distribution). Suppose the discrete random variable $\boldsymbol{x}$ has $\mathrm{Range}(\boldsymbol{x}) \subseteq \{0, 1, 2, \ldots, n\}$ and $a > 0$. Then $\boldsymbol{x}$ has a *Poisson distribution* with parameter a if

$$P(\boldsymbol{x} = k) = \frac{a^k}{k!} e^{-a}. \tag{1.114}$$

We cannot know the value that a random variable will assume on an event before the event occurs. However, we may know enough about the trend of the random variable to be able to specify its average value over time and how well grouped about its average the random values tend to be. There are a variety of parameters associated with a random variable; these we calculate from its distribution or density functions. The most important of these are the mean and standard deviation.

Definition (Discrete Mean). If the random variable $\boldsymbol{x}$ is discrete and $M = \{r \in \mathbb{R}: P(\boldsymbol{x} = r) > 0\}$, then the *mean* or *expectation* of $\boldsymbol{x}$, written $E[x]$, is

$$E(x) = \sum_{r \in M} rP(\boldsymbol{x} = r). \tag{1.115}$$

Definition (Discrete Variance, Standard Deviation). Let the random variable $\boldsymbol{x}$ be discrete, $M = \{r \in \mathbb{R}: P(\boldsymbol{x} = r) > 0\}$, and $\mu = E[\boldsymbol{x}]$. Then the *variance* of $\boldsymbol{x}$, σ_x^2, is

$$\sigma_x^2 = \sum_{r \in M} (r - \mu)^2 P(\boldsymbol{x} = r). \tag{1.116}$$

The *standard deviation* of $\boldsymbol{x}$ is the square root of the variance: σ_x.

1.8.2.2 Continuous Random Variables. The distribution function may have no steps.

Definition (Continuous Random Variable). The random variable $\boldsymbol{x}$ is *continuous* if its distribution function $F_{\boldsymbol{x}}(r)$ is continuous.

Proposition (Continuous Random Variable Characterization). Let x be a continuous random variable in the probability space (Ω, Σ, P) with distribution function $F_{\boldsymbol{x}}(r)$. Then, $P(\boldsymbol{x} = r) = 0$ for all $r \in \mathbb{R}$.

[20]This distribution was first described in 1837 by French mathematician Siméon-Denis Poisson (1781–1840).

Proof: By continuity $F_x(r) = \lim_{s>0,\, s\to 0} F_x(r-s)$. But the distribution function properties entail that $P(\boldsymbol{x} = r) = F_x(r) - \lim_{s>0,\, s\to 0} F_x(r-s)$. So, $P(\boldsymbol{x} = r) = 0$.

■

Assuming a particular form of the density function and then integrating it to get a distribution is common in analytical work. The only restrictions are that the density function must be non-negative and that its integral over the entire real line be unity. This implies that density functions for continuous random variables are in fact absolutely integrable. There are many distribution functions useful for analog signal theory, but the normal or Gaussian distribution is of paramount importance.

Definition (Normal Distribution). The random variable x is *normally* or *Gaussian* distributed if its probability density function is of the form

$$f_x(r) = \frac{1}{\sigma\sqrt{2\pi}}\exp\left(-\frac{(r-\mu)^2}{2\sigma^2}\right). \tag{1.117}$$

where μ and σ are the mean and standard deviation of the Gaussian (1.117), respectively.

Definition (Exponential Distribution). The random variable $\boldsymbol{x}$ has an *exponential distribution* with parameter $a > 0$ if its density function is of the form

$$f_x(r) = \begin{cases} a\exp(-ar) & \text{if } r > 0, \\ 0 & \text{if } r \le 0. \end{cases} \tag{1.118}$$

Definition (Gamma Distribution). The random variable x has a *gamma distribution* with *scale* parameter $a > 0$ and *shape* parameter $b > 0$ if its density function is of the form

$$f_x(r) = \begin{cases} \dfrac{a^b r^{b-1}\exp(-ar)}{\Gamma(b)} & \text{if } r > 0, \\ 0 & \text{if } r \le 0, \end{cases} \tag{1.119}$$

where $\Gamma(t)$ is the gamma function [80]:

$$\Gamma(t) = \int_0^\infty s^{t-1}\exp(-s)\, ds, \tag{1.120}$$

defined for $t > 0$.

Definition (Continuous Mean). If the random variable $\boldsymbol{x}$ is continuous and has density function $f_x(r)$ and if $xf_x(r)$ is in $L^1(\mathbb{R})$, then the *mean* or *expectation* of $\boldsymbol{x}$, written $E[\boldsymbol{x}]$, is

$$E(x) = \int_{-\infty}^{\infty} r f_x(r)\, dr. \tag{1.121}$$

Definition (Continuous Variance, Standard Deviation). Suppose that the random variable $\boldsymbol{x}$ is continuous, $x f_x(r)$ is in $L^1(\mathbb{R})$, $\mu = E[\boldsymbol{x}]$, and $x^2 f_x(r)$ is in $L^1(\mathbb{R})$. Then the *variance* of $\boldsymbol{x}$, σ_x^2, is

$$\sigma_x^2 = \int_{-\infty}^{\infty} (r-\mu)^2 f_x(r)\, dr. \tag{1.122}$$

1.8.2.3 Multivariate Distributions. This section considers the description of random vectors, entities that consist of two or more random components. Much of the development follows from a direct, albeit somewhat messy, extension of the ideas from the single random variables.

Definition (Multivariate Distributions). Let $\boldsymbol{x}$ and $\boldsymbol{y}$ be random variables in the probability space (Ω, Σ, P). Their joint distribution function is defined by $F_{\boldsymbol{x},\boldsymbol{y}}(r, s) = P[(\boldsymbol{x} \le r) \cap (\boldsymbol{y} \le s)]$. This generalizes to an arbitrary finite number of random variables, $\boldsymbol{r} = (\boldsymbol{r}_1, \boldsymbol{r}_2, \ldots, \boldsymbol{r}_M)$. For continuous random variables, the *joint density* of $\boldsymbol{x}$ and $\boldsymbol{y}$ is

$$f_{x,y}(r, s) = \frac{\partial^2}{\partial x \partial y} F_{x,y}(r, s). \tag{1.123}$$

We can define joint probability density functions for families of random variables too. This requires vector and matrix formulations in order to preserve the properties of density and distribution functions. For example, for the multivariate normal density, we have the following definition.

Definition (Joint Normal Density). Suppose that $X = (\boldsymbol{x}_1, \boldsymbol{x}_2, \ldots, \boldsymbol{x}_M)$ is a vector of M random variables on the probability space (Ω, Θ, P). We define the *joint normal density* function $f_X(r)$ by

$$f(r) = \frac{\exp\left[-\frac{1}{2}(\boldsymbol{r}-\mu)^T \Sigma^{-1} (\boldsymbol{r}-\mu)\right]}{\sqrt{\det(\Sigma)(2\pi)^M}}, \tag{1.124}$$

where $\boldsymbol{r} = (\boldsymbol{r}_1, \boldsymbol{r}_2, \ldots, \boldsymbol{r}_M)$ is a vector of length M; $\mu = (E[\boldsymbol{x}_1], E[\boldsymbol{x}_2], \ldots, E[\boldsymbol{x}_M])$ is the vector of means; $(\boldsymbol{r} - \mu)^T$ is the transpose of $\boldsymbol{r} - \mu$; Σ is the $M \times M$ covariance matrix for X, $\Sigma = [\sigma_{m,n}] = E[(x_m - \mu_m)(x_n - \mu_n)]$; $\det(\Sigma)$ is its determinant; and its inverse is Σ^{-1}.

Earlier we considered how to apply Bayes's theorem to the problem of signal classification. However, we noted that it is not easy to distinguish signals by one feature alone, and our one-dimensional statistical classification breaks down. Now let's consider how to use statistical information about feature vectors and classes of signals to develop statistical discriminant functions. Suppose that we know the *a priori* probability of occurrence of each of the classes C_k, $P(C_k)$. Suppose further that for each class C_k we know the probability density function for the feature vector $\boldsymbol{v}$, $p(\boldsymbol{v}|C_k)$. The conditional probability $P(\boldsymbol{v}|C_k)$ provides the likelihood that class k is present, given that the input signal has feature vector $\boldsymbol{v}$. If we could compute $P(C_k|\boldsymbol{v})$ for each C_k and $\boldsymbol{v}$, then this would constitute a statistical basis for selecting one class over another for categorizing the input signal f.

We can restate Bayes's theorem for the multivariate case as follows.

Theorem (Multivariate Bayes). Suppose that for K signal classes C_k we know the *a priori* probability of occurrence of each of the classes $P(C_k)$ and the probability density function for the feature vector $\boldsymbol{v}$, $P(\boldsymbol{v}|C_k)$. Then

$$P(C_k \mid \boldsymbol{v}) = \frac{p(\boldsymbol{v} \mid C_k)P(C_k)}{p(\boldsymbol{v})} = \frac{p(\boldsymbol{v} \mid C_k)P(C_k)}{\sum_{i=1}^{K} p(\boldsymbol{v} \mid C_i)P(C_i)}, \tag{1.125}$$

where $p(\boldsymbol{v})$ is the probability density function for feature vector $\boldsymbol{v}$.

1.8.3 Random Signals

The ideas of random variables, their distribution and density functions, and the principal parameters that describe them are the basis for a definition of a random signal.

When we say that a signal is random, that is not to say that we know nothing of its values; in fact, we might know that the value of the signal at a time instant is almost certain to be in a given range. We might know that the signal remains, at other times, in some other range. It should be possible to provide a table that specifies the possible ranges of the signal and furnishes rough measures for how likely the signal value is to fall within that range. Every time the signal is measured or evaluated at a time, the signal is different, but we have an approximate idea of how these values behave. We can find one set of signal values, one instance of the random signal, but the next instance will differ. Thus, our concept of a random signal is embodied by a family of signals, and each member of the family represents a possible measurement of the signal over its domain. In probabilty theory, this is known as a *random* or *stochastic process.*

Definition (Random Signal). Suppose that (Ω, Σ, P) is a probability space. Let $X = \{x(r): t \in T\}$ be a family of random variables on (Ω, Σ, P) indexed by the set T. Then X is a *stochastic process* or *random signal*. If the index set T is the integers, then we say that X is a *discrete* random signal. If T is the real numbers, then we call X an *analog* random signal.

1.9 SUMMARY

There are two distinct signal categories: those with a continuous independent variable and those with a discrete independent variable. Natural signals are generally analog, and they become discrete—or more precisely, digital—by means of a conversion apparatus or by the way in which they are collected. The electrocardiogram source is analog, and it could be displayed in such a mode on an oscilloscope, for example. But nowadays it is often digitized for computer analysis. Such digital signals have a finite range. They are the true objects of digital signal processing on computers, but they are awkward for theoretical development. The temperature of the earth is a continuous function of depth, and it could be continuously recorded on a strip chart. But since the changes in the geothermal signal are so slow, it is more practical to collect isolated measurements. It is therefore discrete from the beginning. We rely on mathematical model for signal theory: continuous time functions, defined on the real numbers, for analog signals, and discrete time functions, defined on the integers, for discrete signals.

There is also a notion of the units of the interval between numerical signal values; This is called the independent variable. It is often a time variable, measured in seconds, minutes, hours, and so on, and this is natural, because time of occurrence provides a strict, irreversible ordering of events. So often are signals based on time that we get imprecise and routinely speak in temporal terms of the independent variable. On occasion, the independent variable that defines earlier and later signal values is a distance measure. The geothermal signal has an independent variable typically measured in meters, or even kilometers, of depth into the earth. Despite the fact that the independent variable is a distance measure, we often casually refer to the list of the signal's values as its "time-domain" specification.

The dependent variable of the signal is generally a real value for analog and discrete signals, and it is an integral value for digital signals. These are the signal values, and we stipulate that they assume numerical values, so that we can apply mathematical methods to study them. So the terminology here follows that from the mathematical notions of the independent and dependent variable for a mathematical function. We reserve the idea of a sequence of symbols, which is sometimes called a signal in ordinary language, for our concept of a signal interpretation.

We are concerned mainly with signals that have a one-dimensional independent and dependent variables. It is possible for a signal's dependent measurement to depend on multiple independent measurements. Image processing performs conditioning operations on two-dimensional signals. Computer vision analyzes multidimensional signals and produces a structural description or interpretation of them. We distinguish between single channel and multiple channel signals. If a signal produces a one-dimensional value, then it is single channel. An example is the temperature versus depth measurement. Signals that generate multidimensional values are called multichannel signals. An example is the 12-lead ECG. Multichannel signals have multidimensional range values. They arise in many applications, but we confine our discussions primarily to single-channel signals and refer interested readers to the more specialized literature on sensor fusion.

1.9.1 Historical Notes

A popular introduction to signal theory with historical background is Ref. 84. One of the latest discoveries in signal analysis is wavelet theory—a relatively recent and exciting approach for time-scale analysis [85]. An overview of wavelets and related time-frequency transforms is Ref. 69.

Several practical inventions spurred the early development of signal processing and analysis. The telegraph, invented by the American portrait painter Samuel F. B. Morse (1791–1872) in 1835, transmitted messages comprised of a sequence of isolated pulse patterns, or symbols, standing for letters of the English alphabet. The symbols themselves were (and still are, for the Morse code has been amended slightly and standardized internationally) a finite sequence of short and long electrical pulses, called dots and dashes, respectively. Shorter symbols represent the more prevalent English letters. For example, single dot and single dash represent the most common English letters, E and T, respectively. Morse's signaling scheme is an essentially discrete coding, since there is no continuous transition between either the full symbols or the component dots and dashes. Moreover, as a means of communication it could be considered to be digital, since the code elements are finite in number. But it would eventually be supplanted by analog communication technologies—the telephone and voice radio—which relied on a continuously varying representation of natural language.

Alexander Graham Bell (1847–1922), the famed U.S. inventor and educator of the deaf, discovered the telephone in the course of his experiments, undertaken in the mid-1870s, to improve the telegraph. The telegraph carried a single signal on a single pair of conductors. Bell sought to multiplex several noninterfering telegraphic messages onto a single circuit. The economic advantages of Bell's Harmonic Telegraph would have been tremendous, but the results were modest. Instead, Bell happened upon a technique for continuously converting human voices into electrical current variations and accurately reproducing the voice sounds at a remote location. Bell patented the telephone less than a year later, in March 1876; verified the concept six months later in sustained conversations between Boston and Cambridgeport, Massachusetts; and built the first commerical telephone exhange, at New Haven, Connecticut in January 1878. Bell's patent application points to the analog nature of telephony as clearly distinguishing it from discrete telegraphy.

Wireless telegraphy—and eventually wireless telephony—were the fruit of persistent efforts by yet another scientific layperson, Guglielmo Marconi (1874–1937). The young Italian inventor was aware of both J. C. Maxwell's theory of electromagnetic waves[21]and H. R. Hertz's demonstration[22] of precisely this radiation with a spark coil transmitter and wire loop receiver. But Hertz's apparatus was too weak for practical use. Marconi's improvements—a telegraph key to control the firing of the spark gap, a long wire antenna and earth ground for greater signal strength, and

[21]Scottish physicist James Clerk Maxwell (1831–1879) announced his electromagnetic field theory to a skeptical scientific community in 1864.

[22]With a small spark coil, German physicist Heinrich Rudolf Hertz (1857–1894) generated the first electromagnetic waves at the University of Karlsruhe and verified Maxwell's theory.

the crystal detector for improved reception—enabled him to demonstrate radio telegraphy in 1895. Unheralded in his native Italy, Marconi took his technology to England, received a patent in 1896, formed his own company a year later, and received the Nobel prize in 1909. These techniques could only serve discrete modes of telecommunication, however. Analog communication awaited further improvements in radio communication, in particular the radio-frequency alternator.

These practical advances in analog technologies were complemented by the discovery of the richness within Jean-Baptiste Fourier's discovery, long past, that even signals containing discontinuities could be represented by sums of smoothly undulating sinusoids. Fourier developed his theory for the purpose of studying heat propagation. In particular, it remains a principal tool for solving the differential equations governing such phenomena as heat conduction, Fourier's original problem [1]. Thus, at the turn of the last century, the most important signal technologies and the most important signal theories revolved around analog methods.

Theory would not link the analog and discrete realms of signal processing until the early twentieth century, when Nyquist [2], Shannon [3], and Vladimir Kotelnikov[23] developed the sampling theory. Nyquist's original research article focused on telegraphy, and it established a first theoretical link between discrete and analog communication methods. In particular, he showed that a continuous domain signal containing but a limited variety of frequencies could be captured and regenerated with a discrete signal. But analog practice and analog theory ruled supreme, and Nyquist's contribution was largely overlooked. Only when Shannon proved that error-free digital communication—even in the presence of noise—was possible did the attention of scientists and engineers turn once more to discrete modes of communication. The contributions of Nyquist and Shannon did firmly establish signal theory as a distinct scientific and engineering discipline. Both analog and discrete signal theory were soundly fixed upon mathematical foundations and shared a link through the Shannon–Nyquist results.

One seemingly insurmountable problem remained. The frequency analysis of analog signals was possible using conventional analog instruments such as a frequency analyzer. But discrete signal frequencies could not be calculated fast enough to keep pace with the arrival of discrete values to a processing apparatus. Therefore, although mathematicians developed a considerable complement of tools for understanding discrete signals, engineers remained preoccupied with analog tools which could handle their signals in real time.

The discovery of the fast Fourier transform (FFT) by J. W. Cooley and J. W. Tukey in 1965 shattered the analog tradition in signal processing. By eliminating duplicate computations in the DFT, it became possible to produce the frequency spectrum of a signal with N data points in $N\log_2 N$ operations; real-time digital signal spectrum analysis became feasible [4–6].

[23]Vladimir A. Kotelnikov (1908–), a Russian communications engineer, independently discovered the sampling theorem in 1933. His work was largely unknown outside the Soviet Union.

1.9.2 Resources

A vast array of resources—commercial products, inexpensive shareware applications, and public-domain software—are available nowadays for studying signal theory. Researchers, university laboratories, private firms, and interested individuals have also made available signal and image processing data sets. Some of these have become standards for experimentation, algorithm development, and performance comparisons.

1.9.2.1 Signal Processing Tools. Commercial packages used for this chapter's examples include:

- *Matlab*, available from The MathWorks, Inc., 24 Prime Park Way, Natick, MA, 01760, USA [86].
- *Mathematica*, available from Wolfram Research, Inc., 100 Trade Center Drive, Champaign, IL, 61820, USA [87].

Public-domain packages include the following:

- *Wavelab*, which uses Matlab and includes several popular research data sets, available free of charge from Stanford University: `http://playfair.stanford.edu/~wavelab`.
- *Khoros*, available free of charge via anonymous ftp from the University of New Mexico: `ftp.eece.unm.edu` [88].

1.9.2.2 Data. There are many data sets available over the internet, including several smaller data archives, maintained by individual researchers, tapped for examples in this chapter.

Among the larger repositories are the following:

- Rice University, Houston, TX, in conjunction with the Institute of Electrical and Electronic Engineers (IEEE), supports the Signal Processing Information Base (SPIB): `http://spib.rice.edu/spib.html`. SPIB contains a variety of signal and image data sets, several of which found their way into the examples of this text.
- The University of California at Irvine, Irvine, CA supports a machine intelligence database.

Every effort has been made to use example data sets that are available to the reader. Readers should be able to find this chapter's signal data examples within the public domain. Figure 1.1 is on the web site of the Princeton Earth Physics Project (`http://www.gns.cri.nz/quaketrackers/curr/seismic_waves.htm`). The EEG signals of Figure 1.3 are from Krishna Nayak's Florida State University web site, `http://www.scri.fsu.edu`. The aerial scenes of Figure 1.4 are from the Danish Center for Remote Sensing, `http://www.dcrs.dk`. The ECG signal of Figure 1.6 is from SPIB. The geothermal data of Figure 1.8 comes from the Appalachian Deep Core Hole project and is available at the ADCOH web site [36]. The auditory neuron pulse train data are from SPIB.

1.9.3 Looking Forward

Now that we have introduced the basic raw material, signals, we proceed in Chapters 2 and 3 to introduce the machinery, systems. The term "system" is a very broad term, but in signal theory it is used in a quite specific sense. A *system* is the mathematical entity that accomplishes signal processing; it takes a signal as input and produces a signal as output. A system is a function that operates on signals.

An understanding of signals requires ideas from basic mathematics, algebra, calculus, a dose of complex analysis, and some random variable theory. In contrast, a firm understanding of the ideas of systems—the mechanisms that convert one signal into another, signal processing in other words—depends upon ideas from advanced mathematical analysis. In particular, we must draw upon the concepts of functional analysis—especially Hilbert space theory—topics normally taught at the university graduate mathematics level. For practical-minded scientists and engineers, this seems ominous. But the good news is that this development is straightforward for discrete signals. Thus, in Chapter 2 we concentrate exclusively on discrete signal spaces, of which discrete Hilbert spaces are a special case.

To most of us, the mastery of analog signal processing theory comes less readily than a thorough understanding of discrete theory. Readers need to understand both developments, even though the analog theory is more mathematically involved. However, scientists, applied mathematicians, and engineers who are looking further toward modern mixed-domain signal processing methods need a good foundation in signal spaces and an advanced presentation of analog signal analysis. Chapter 3 presents the prerequisite background in continuous-domain signal spaces.

1.9.4 Guide to Problems

All of the chapters provide problems. They range in difficulty from simple exercises that recall basic ideas from the text to more complicated problems that extend and develop the chapter's material. Some of them are outlines of research projects that may involve several weeks of work. The student may need to make simplifying assumptions, discover constraints, and—quite likely—will not arrive at a once-and-for-all answer to the problems posed.

REFERENCES

1. J.-B. J. Fourier, *The Analytical Theory of Heat*, translated by A. Freeman, New York: Dover, 1955.
2. H. Nyquist, Certain topics in telegraph transmission theory, *Transactions of the AIEE*, vol. 47, pp. 617–644, 1928.
3. C. E. Shannon, A mathematical theory of communication, *Bell Systems Technical Journal*, vol. 27, pp. 379–423 and pp. 623–656, 1948.
4. J. W. Cooley and J. W. Tukey, An algorithm for the machine calculation of complex Fourier series," *Mathematics of Computation*, vol. 19, pp. 297–301, April 1965.
5. J. W. Cooley, P. A. Lewis, and P. D. Welch, Historical notes on the fast Fourier transform, *IEEE Transactions on Audio and Electroacoustics*, vol AU-15, pp. 76–79, June 1967.
6. J. W. Cooley, How the FFT gained acceptance, *IEEE SP Magazine*, pp. 10–13, January 1992.
7. H. Baher, *Analog and Digital Signal Processing*, New York: Wiley, 1990.
8. J. A. Cadzow and H. F. van Landingham, *Signals, Systems, and Transforms*, Englewood Cliffs, N J: Prentice-Hall, 1989.
9. L. B. Jackson, *Signals, Systems, and Transforms*, Reading, MA: Addison-Wesley, 1991.
10. A. V. Oppenheim, A. S. Willsky, and S. H. Nawab, *Signals and Systems*, 2nd ed., Englewood Cliffs, NJ: Prentice-Hall, 1989.
11. R. E. Ziemer, W. H. Tranter, and D. R. Fannin, *Signals and Systems*: *Continuous and Discrete*, New York: Macmillan, 1989.
12. L. B. Jackson, *Digital Filters and Signal Processing*, Boston: Kluwer Academic Publishers, 1989.
13. A. V. Oppenheim and R. W. Shafer, *Discrete-Time Signal Processing*, Englewood Cliffs, NJ: Prentice-Hall, 1989.
14. J. G. Proakis and D. G. Manolakis, *Digital Signal Processing: Principles, Algorithms, and Applications*, 2nd ed., New York: Macmillan, 1992.
15. L. R. Rabiner and B. Gold, *Theory and Application of Digital Signal Processing*, Englewood Cliffs, NJ: Prentice-Hall, 1975.
16. K.-S. Lin, ed., *Digital Signal Processing Applications with the TMS320 Family*, vol. 1, Dallas, TX: Texas Instruments, 1989.
17. R. J. Simpson, *Digital Signal Processing Using the Motorola DSP Family*, Englewood Cliffs, NJ: Prentice-Hall, 1994.
18. Motorola, Inc., *DSP56000/DSP56001 Digital Signal Processor User's Manual*, Phoenix, AZ: Motorola Literature Distribution, 1990.
19. R. J. Higgins, *Digital Signal Processing in VLSI*, Englewood Cliffs, NJ: Prentice-Hall, 1990.
20. A. Mar, ed., *Digital Signal Processing Applications Using the ADSP-2100 Family*, Englewood Cliffs, NJ: Prentice-Hall, 1990.
21. V. K. Madisetti, *VLSI Digital Signal Processors: An Introduction to Rapid Prototyping and Design Synthesis*, Piscataway, NJ: IEEE Press, 1995.
22. B. A. Bolt, *Earthquakes and Geologic Discovery*, New York: Scientific American Library, 1990.

23. O. Kulhanek, *Anatomy of Seismograms*, Amsterdam: Elsevier, 1990.
24. J. F. Claerbout, *Fundamentals of Geophysical Data Processing: With Applications to Petroleum Prospecting*, Boston: Blackwell Scientific, 1985.
25. M. Akay, *Biomedical Signal Processing*, San Diego, CA: Academic Press, 1994.
26. J. C. Russ, *The Image Processing Handbook*, Boca Raton, FL: CRC Press, 1995.
27. A. Rosenfeld and A. C. Kak, *Digital Picture Processing*, vols. 1 and 2, Orlando, FL: Academic Press, 1982.
28. A. K. Jain, *Fundamentals of Digital Image Processing*, Englewood Cliffs, NJ: Prentice-Hall, 1989.
29. R. J. Schalkoff, *Digital Image Processing and Computer Vision*, New York: Wiley, 1989.
30. D. H. Ballard and C. M. Brown, *Computer Vision*, Englewood Cliffs, NJ: Prentice-Hall, 1982.
31. R. M. Haralick and L. G. Shapiro, *Computer and Robot Vision*, vols. 1 and 2, New York: Addison-Wesley, 1992.
32. B. K. P. Horn, *Robot Vision*, Cambridge, MA: MIT Press, 1986.
33. M. J. Goldman, *Principles of Clinical Electrocardiography*, Los Altos, CA: Lange Medical Publications, 1986.
34. D. B. Geselowitz, On the theory of the electrocardiogram, *Proceedings of the IEEE*, vol. 77, no. 6, pp. 857–872, June 1989.
35. M. Unser and A. Aldroubi, A review of wavelets in biomedical applications, *Proceedings of the IEEE*, vol. 84, no. 4, pp. 626–638, April 1996.
36. J. K. Costain and E. R. Decker, Heat flow at the proposed Appalachian Ultradeep Core Hole (ADCOH) site: Tectonic implications, *Geophysical Research Letters*, vol. 14, no. 3, pp. 252–255, 1987.
37. C .F. Gerald and P. O. Wheatley, *Applied Numerical Analysis*, Reading, MA: Addison-Wesley, 1990.
38. F. Attneave, Some informational aspects of visual perception, *Psychological Review*, vol. 61, pp. 183–193, 1954.
39. D. Marr, *Vision*, New York: W. H. Freeman and Company, 1982.
40. H. Asada and M. Brady, The curvature primal sketch, *IEEE Transactions on Pattern Analysis and Machine Intelligence*, vol. PAMI-8, no. 1, pp. 2–14, January 1986.
41. I. Biedermann, Human image understanding: Recent research and a theory," *Computer Vision, Graphics, and Image Processing*, vol. 32, pp. 29–73, 1985.
42. A. P. Witkin, Scale-space filtering, *Proceedings of the 8th International Joint Conference on Artificial Intelligence*, Karlsruhe, W. Germany, 1983. See also A. P. Witkin, Scale-space filtering, in *From Pixels to Predicates*, A. P. Pentland, ed., Norwood, NJ: Ablex, 1986.
43. T. Lindeberg, Scale space for discrete signals, *IEEE Transactions on Pattern Analysis and Machine Intelligence*, vol. 12, no. 3, pp. 234–254, March 1990.
44. M. Rosenlicht, *Introduction to Analysis*, New York: Dover, 1978.
45. D. Gabor, Theory of communication, *Journal of the Institute of Electrical Engineers*, vol. 93, pp. 429–457, 1946.
46. D. A. Pollen and S. F. Ronner, Visual cortical neurons as localized spatial frequency filters, *IEEE Transactions on Systems, Man, and Cybernetics*, vol. SMC-13, no. 5, pp. 907–916, September–October 1983.

47. J. J. Kulikowski, S. Marcelja, and P. O. Bishop, Theory of spatial position and spatial frequency relations in the receptive fields of simple cells in the visual cortex, *Biological Cybernetics*, vol. 43, pp. 187–198, 1982.
48. A. H. Zemanian, *Distribution Theory and Transform Analysis*, New York: Dover, 1965.
49. M. J. Lighthill, *Introduction to Fourier Analysis and Generalized Functions*, New York: Cambridge University Press, 1958.
50. J. Stoer and R. Bulirsch, *Introduction to Numerical Analysis*, 2nd ed., New York: Springer-Verlag, 1993.
51. S. Lang, *Linear Algebra*, Reading, MA: Addison-Wesley, 1968.
52. G. Strang, *Linear Algebra and Its Applications*, 2nd ed., New York: Academic Press, 1980.
53. S. G. Mallat, A theory for multiresolution signal decomposition: The wavelet representation, *IEEE Transactions on Pattern Analysis and Machine Intelligence*, vol. 11, no. 7, pp. 674–693, July 1989.
54. J. Carson and T. Fry, Variable frequency electric circuit theory with applications to the theory of frequency modulation, *Bell Systems Technical Journal*, vol. 16, pp. 513–540, 1937.
55. B. Van der Pol, The fundamental principles of frequency modulation, *Proceedings of the IEE*, vol. 93, pp. 153–158, 1946.
56. J. Shekel, Instantaneous frequency, *Proceedings of the IRE*, vol. 41, p. 548, 1953.
57. F. Jelinek, Continuous speech recognition by statistical methods, *Proceedings of the IEEE*, vol. 64, no. 4, pp. 532–556, April 1976.
58. S. Young, A review of large-vocabulary continuous-speech recognition, *IEEE Signal Processing Magazine*, vol. 13, no. 5, pp. 45–57, September 1996.
59. L. Rabiner and B.-H. Juang, *Fundamentals of Speech Recognition*, Englewood Cliffs, NJ: Prentice-Hall, 1993.
60. N. Morgan and H. Bourlard, Continuous speech recognition, *IEEE Signal Processing Magazine*, vol. 12, no. 3, pp. 25–42, May 1995.
61. B. Malmberg, *Phonetics*, New York: Dover, 1963.
62. J. L. Flanagan, *Speech Analysis, Synthesis, and Perception*, 2nd ed., New York: Springer-Verlag, 1972.
63. N. Deshmukh, R. J. Duncan, A. Ganapathiraju, and J. Picone, Benchmarking human performance for continuous speech recognition, *Proceedings of the Fourth International Conference on Spoken Language Processing*, Philadelphia, pp. SUP1-SUP10, October 1996.
64. J. R. Cox, Jr., F. M. Nolle, and R. M. Arthur, Digital analysis of electroencephalogram, the blood pressure wave, and the electrocardiogram, *Proceedings of the IEEE*, vol. 60, pp. 1137–1164, 1972.
65. L. Khadra, M. Matalgah, B. El-Asir, and S. Mawagdeh, Representation of ECG-late potentials in the time frequency plane, *Journal of Medical Engineering and Technology*, vol. 17, no. 6, pp. 228–231, 1993.
66. F. B. Tuteur, Wavelet transformations in signal detection, in *Wavelets: Time-Frequency Methods and Phase Space*, J. M. Combes, A. Grossmann, and P. Tchamitchian, eds., 2nd ed., Berlin: Springer-Verlag, pp. 132–138, 1990.

67. G. Olmo and L. Lo Presti, Applications of the wavelet transform for seismic activity monitoring, in *Wavelets: Theory, Applications, and Applications*, C. K. Chui, L. Montefusco, and L. Puccio, eds., San Diego, CA: Academic Press, pp. 561–572, 1994.
68. J. L. Larsonneur and J. Morlet, Wavelets and seismic interpretation, in *Wavelets: Time-Frequency Methods and Phase Space*, J. M. Combes, A. Grossmann, and P. Tchamitchian, eds., 2nd ed., Berlin: Springer-Verlag, pp. 126–131, 1990.
69. Y. Meyer, *Wavelets: Algorithms and Applications*, Philadelphia: SIAM, 1993.
70. A. N. Kolmogorov and S. V. Fomin, *Introductory Real Analysis*, New York: Dover, 1975.
71. M. C. Teich, D. H. Johnson, A. R. Kumar, and R. Turcott, Fractional power law behavior of single units in the lower auditory system, *Hearing Research*, vol. 46, pp. 41–52, May 1990.
72. R. P. Feynman, R. B. Leighton, and M. Sands, *The Feynman Lectures on Physics*, vol. 1, Reading, MA: Addison-Wesley, 1977.
73. L. V. Ahlfors, *Complex Analysis*, 2nd ed., New York: McGraw-Hill, 1966.
74. E. Hille, *Analytic Function Theory*, vol. 1, Waltam, MA: Blaisdell, 1959.
75. N. Levinson and R. M. Redheffer, *Complex Variables*, San Francisco: Holden-Day, 1970.
76. R. Beals, *Advanced Mathematical Analysis*, New York: Springer-Verlag, 1987.
77. A. Gluchoff, A simple interpretation of the complex contour integral, *Teaching of Mathematics*, pp. 641–644, August–September 1991.
78. A. O. Allen, *Probability, Statistics, and Queueing Theory with Computer Scinece Applications*, Boston: Academic, 1990.
79. W. Feller, *An Introduction to Probability Theory and Its Applications*, New York: Wiley, 1968.
80. E. Parzen, *Modern Probability Theory and Its Applications*, New York: Wiley, 1960.
81. A. Papoulis, *Probability, Random Variables, and Stochastic Processes*, New York: McGraw-Hill, 1984.
82. R. E. Mortensen, *Random Signals and Systems*, New York: Wiley, 1984.
83. W. B. Davenport and W. L. Root, *An Introduction to the Theory of Random Signals and Noise*, New York: McGraw-Hill, 1958.
84. J. R. Pierce and A. M. Noll, *Signals: The Science of Telecommunications*, New York: Scientific American Library, 1990.
85. B. B. Hubbard, *The World According to Wavelets*, Wellesley, MA: A. K. Peters, 1996.
86. A. Biran and M. Breiner, *Matlab for Engineers*, Harlow, England: Addison-Wesley, 1995.
87. S. Wolfram, *The Mathematica Book*, 3rd ed., Cambridge, UK: Cambridge University Press, 1996.
88. K. Konstantinides and J. R. Rasure, The Khoros software development environment for image and signal processing, *IEEE Transactions on Image Processing*, vol. 3, no. 3, pp. 243–252, May 1994.

PROBLEMS

1. Which of the following signals are analog, discrete, or digital? Explain.

(a) The temperature reading on a mercury thermometer, as a function of height, attached to a rising weather balloon.

(b) The time interval, given by a mechanical clock, between arriving customers at a bank teller's window.

(c) The number of customers that have been serviced at a bank teller's window, as recorded at fifteen minutes intervals throughout the workday.

2. Which of the following constitute time domain, frequency domain, or scale domain descriptions of a signal? Explain.

(a) A listing of the percentages of 2-kHz, 4-kHz, 8-kHz, and 16-kHz tones in a ten second long tape recording of music.

(b) The atmospheric pressure readings reported from a weather balloon, as it rises above the earth.

(c) From a digital electrocardiogram, the number of QRS pulses that extend for 5, 10, 15, 20, 25, and 30 ms.

3. Sketch the following signals derived from the unit step $u(t)$:

(a) $u(t-1)$

(b) $u(t+2)$

(c) $u(-t)$

(d) $u(-t-1)$

(e) $u(-t+2)$

(f) $u(t-2) - u(t-8)$

4. Sketch the following signals derived from the discrete unit step $u(n)$:

(a) $u(n-4)$

(b) $u(n+3)$

(c) $u(-n)$

(d) $u(-n-3)$

(e) $u(-n+3)$

(f) $u(n-2) - u(n-8)$

(g) $u(n+6) - u(n-3)$

5. Describe the difference between the graphs of a signal $x(t)$; the shifted version of $x(t)$, $y(t) = x(t-c)$; and the reflected and shifted version, $z(t) = x(-t-c)$. Consider all cases for $c > 0$, $c < 0$, and $c = 0$.

6. Suppose that an N-bit register stores non-negative digital values ranging from 0 (all bits clear) to all bits set. The value of bit b_n is 2^n, $n = 0, 1, ..., N-1$. Show that the largest possible value is $2^N - 1$.

7. Consider the two's complement representation of a digital value in an N-bit register. If the bits are $b_{N-1}, b_{N-2}, ..., b_1, b_0$, then the digital value is $-b_{N-1}2^{N-1} + b_{N-2}2^{N-2} + \cdots + b_2 2^2 + b_1 2^1 + b_0 2^0$.

(a) Find the largest positive value and give its bit values.

(b) Find the most negative value and give its bit values.

(c) Show that the dynamic range is 2^N.

8. Suppose that an N-bit register uses the most significant bit b_{N-1} as a sign bit: If $b_{N-1} = 1$, then the value is -1 times the value in first $N-1$ bits; otherwise the value is positive, 1 times the value in first $N-1$ bits. The remaining $N-1$ bits store a value as in Problem 6.
   (a) Again, find the largest possible positive value and the most negative value.
   (b) What is the dynamic range for this type of digital storage register? Explain the result.

9. Suppose discrete signal $x(n)$ is known at distinct points $(n_k, x(n_k)) = (n_k, y_k)$, where $0 \le k \le N$. Suppose too that there are interpolating cubic polynomials over the $[n_k, n_{k+1}]$:

$$p(t) = a_k(t-n_k)^3 + b_k(t-n_k)^2 + c_k(t-n_k) + d_k. \tag{1.126}$$

   (a) If the interpolants passes through the knots (n_k, y_k), then show $y_k = d_k$.
   (b) Compute the derivatives, $p_k'(t)$ and $p_k''(t)$ for each k, and show that if $D_k = p_k''(n_k)$ and $E_k = p_k''(n_{k+1})$, then a_k and b_k can be written in terms of D_k and E_k.
   (c) Suppose that for some k, we know both D_k and E_k. Show that we can then give the coefficients of the interpolating cubic, $p_k(t)$ on $[n_k, n_{k+1}]$.

10. Let $x(t) = 5\sin(2400t + 400)$, where t is a (real) time value in seconds. Give:
   (a) The amplitude of x
   (b) The phase of x
   (c) The frequency of x in Hz (cycles/second)
   (d) The frequency of x in radians/second
   (e) The period of x

11. Consider a discrete signal $x(n) = A\cos(\Omega n + \phi)$ for which there is an $N>0$ with $x(n) = x(n+N)$ for all n.
   (a) Explain why the smallest period for all discrete signals is $N=1$, but there is no such lowest possible period for the class of analog signals.
   (b) Show that if $x(n)$ is a sinusoid, then the largest frequency it can have is $|\Omega| = \pi$ or, equivalently, $|F| = 1$, where $\Omega = 2\pi F$.

12. Let $s(n) = -8\cos\left(\frac{21}{2}n + 3\right)$ be a discrete signal. Find the following:
   (a) The amplitude of s
   (b) The phase of s
   (c) The frequency of s in radians/sample
   (d) The frequency of s in Hz (cycles/sample)
   (e) Does s have a period? Why?

13. Find the frequency of the following discrete signals. Which ones are even, odd, finitely supported? Which ones are equal?

(a) $a(n) = 5\cos\left(n\frac{\pi}{4}\right)$.

(b) $b(n) = 5\cos\left(-n\frac{\pi}{4}\right)$.

(c) $c(n) = 5\sin\left(n\frac{\pi}{4}\right)$.

(d) $d(n) = 5\sin\left(-n\frac{\pi}{4}\right)$.

14. Prove that a signal decomposes into its even and odd parts. If $x(n)$ is a discrete signal, then show that:

(a) $x_e(n)$ is even.

(b) $x_o(n)$ is odd.

(c) $x(n) = x_e(n) + x_o(n)$.

15. Consider the signal $x(n) = [3, 2, 1, -1, -1, -\underline{1}, 0, 1, 2]$. Write $x(n)$ as a sum of even and odd discrete functions.

16. Show the following:

(a) $\sin(t)$ is odd.

(b) $\cos(t)$ is even.

(c) $g_{\mu,\sigma}(t)$ of mean μ and standard deviation σ (1.14) is symmetric about μ.

(d) Suppose a polynomial $x(t)$ is even; what can you say about $x(t)$? Explain.

(e) Suppose a polynomial $x(t)$ is odd; what can you say about $x(t)$? Explain.

(f) Show that the norm of the Gabor elementary function $\|G_{\mu,\sigma,\omega}(t)\|$ (1.20) is even.

(g) Characterize the real and imaginary parts of $G_{\mu,\sigma,\omega}(t)$ as even or odd.

17. Show that rational signal $x(t) = 1/t$ is neither integrable nor square-integrable in the positive real half-line $\{t : t > 0\}$. Show that $s(t) = t^{-2}$, however, is integrable for $\{t : t > 1\}$.

18. Show that $f(z) = f(x + jy) = x - jy$, the complex conjugate function, is not differentiable at a general point $z \in \mathbb{C}$.

19. Suppose that $f(z) = z$ and C is the straight line arc from a point u to point v in the complex plane.

(a) Find the contour integral

$$\oint_C f(z)\,dz. \tag{1.127}$$

(b) Suppose that $f(z) = z^{-1}$; again evaluate the contour integral in part (a); what assumptions must be made? Explain.

20. Suppose Σ is an algebra over a set Ω.

(a) Show that $\Omega \in \Sigma$.

(b) Show that Σ is closed under finite unions; that is, show that a finite union of elements of Σ is still in Σ.

(c) Show that Σ is closed under finite intersections.

(d) Supposing that Σ is a σ-algebra as well and that $S_n \in \Sigma$ for all natural numbers $n \in \mathbb{N}$, show that

$$\bigcap_{n \in \mathbb{N}} S_n \in \Sigma. \tag{1.128}$$

21. Suppose that (Ω, Σ, P) is a probability space. Let S and T be events in Σ. Show the following:

(a) $P(\varnothing) = 0$.

(b) $P(S) = 1 - P(S')$, where S' is the complement of S inside Ω.

(c) If $S \subseteq T$, then $P(S) \leq P(T)$.

(d) $P(S \cup T) = P(S) + P(T) - P(S \cap T)$.

22. Suppose that Ω is a set.

(a) What is the smallest algebra over Ω?

(b) What is the largest algebra over Ω?

(c) Find an example set Ω and an algebra Σ over Ω that is not a σ-algebra;

(d) Suppose that every algebra over Ω is also a σ-algebra. What can you say about Ω? Explain.

23. If A and B are independent, show that

(a) A and $\sim B$ are independent.

(b) $\sim A$ and $\sim B$ are independent.

24. Let $\boldsymbol{x}$ be a random variable and let r and s be real numbers. Then, by the definition of a random variable, the set $\{\omega \in \Omega : \boldsymbol{x}(\omega) \leq r\}$ is an event. Provide definitions for the following and show that they must be events:

(a) $\boldsymbol{x} > r$.

(b) $r < \boldsymbol{x} \leq s$.

(c) $\boldsymbol{x} = r$.

25. Find constants A, B, C, and D so that the following are probability density functions:

(a) $x(n) = A \times [\underline{4}, 3, 2, 1]$.

(b) $f(r) = B[u(r) - u(r-2)]$, where $u(t)$ is the analog unit step signal.

(c) The Rayleigh density function is

$$f(r) = \begin{cases} Cr\exp\left(-\frac{r^2}{2}\right) & \text{if } r \geq 0, \\ 0 & \text{if } r < 0. \end{cases} \tag{1.129}$$

(d) $f(r)$ is defined as follows:

$$f(r) = \begin{cases} \dfrac{D}{\sqrt{1-r^2}} & \text{if } |r| < 1, \\ 0 & \text{if } r \geq 1. \end{cases} \tag{1.130}$$

The following problems are more involved and, in some cases, expand upon ideas in the text.

26. Let $x(t) = A\cos(\Omega t)$ and $y(t) = A\cos(\Phi t)$ be continuous-domain (analog) signals. Find conditions for A, B, Ω, and Φ so that the following statement is true, and then prove it: If $x(t) = y(t)$ for all t, then $A = B$ and $\Omega = \Phi$.

27. Explain the following statement: There is a unique discrete sinusoid $x(n)$ with radial frequency $|\omega| \leq \pi$.

28. The following steps show that the support of a signal $x(t)$ is compact if and only if its support is both closed and bounded [44, 70].

(a) Show that a convergent set of points in a closed set S converges to a point in S.

(b) Prove that a compact $S \subset \mathbb{R}$ is bounded.

(c) Show that a compact $S \subset \mathbb{R}$ has at least one cluster point; that is, there is a t in S such that any open interval (a, b) containing t contains infinitely many points of S.

(d) Using (a) and (b), show that a compact set is closed.

(e) If $r > 0$ and $S \subset \mathbb{R}$ is bounded, show S is contained in the union of a finite number of closed intervals of length r.

(f) Show that if $S \subset \mathbb{R}$ is closed and bounded, then S is compact.

29. The *average power* of the discrete signal $x(n)$ is defined by

$$P_x = \lim_{N \to \infty} \left[\frac{1}{2N+1} \sum_{n=-n}^{n} |x(n)|^2 \right]. \tag{1.131}$$

If the limit defining P_x exists, then we say that $x(n)$ has *finite average power.* Show the following.

(a) An exponential signal $x(n) = Ae^{j\omega n}$, where A is real and nonzero, has finite average power, but not finite energy.

(b) If $x(n)$ is periodic and $x(n)$ is non-zero, then $x(n)$ is neither absolutely summable nor square summable.

(c) If $x(n)$ is periodic, then $x(n)$ has finite average power.

30. Show that under any of the following conditions, the differentiable function $f(z)$ must be constant on $\mathbb{C}$.

(a) Real$[f(z)]$ is constant.

(b) Imag$[f(z)]$ is constant.

(c) $|f(z)|$ is constant.

31. Show that a discrete polynomial $p(k)$ may have consecutive points, k_0, k_1, ..., and so on, where the discrete second derivative is zero.

(a) For a given degree n, what is the limit, if any, on the number of consecutive points where the discrete second derivative is zero? Explain.

(b) For a discrete polynomial $p(k)$ of degree n, find formulas for the first, second, and third derivatives of $p(k)$.

(c) Show that a polynomial $p(t)$ of degree $n > 1$ has only isolated points, t_0, t_1, ..., t_N, where the second derivative is zero. What is N?

32. Prove the proposition on distribution function properties of Section 1.8.2 [81].

33. Suppose the discrete random variable $\boldsymbol{x}$ has a binomial distribution (1.113).

(a) Find the density function $f_{\boldsymbol{x}}(r)$.

(b) Find the distribution function $F_{\boldsymbol{x}}(r)$.

(c) Find the mean $E[\boldsymbol{x}]$.

(d) Find the variance $(\sigma_{\boldsymbol{x}})^2$.

(e) Discuss the case where p or q is zero in (1.113).

34. Suppose the discrete random variable $\mathbf{x}$ has a Poisson distribution with parameter $a > 0$ (1.114).

(a) Find the density function $f_{\boldsymbol{x}}(r)$.

(b) Find the distribution function $F_{\boldsymbol{x}}(r)$.

(c) Find the mean $E[\boldsymbol{x}]$.

(d) Find the variance $(\sigma_{\boldsymbol{x}})^2$.

The next several problems consider electrocardiogram processing and analysis.

35. Develop algorithms for calculating the running heart rate from a single ECG lead.

(a) Obtain the ECG trace of Figure 1.6 from the Signal Processing Information Base (see Section 1.9.2.2). Plot the data set using a standard graphing package or spreadsheet application. For example, in Matlab, execute the command lines: `load ecg.txt; plot (ecg)`. As an alternative, develop C or C++ code to load the file, plot the signal, and print out the time-domain values. Identify the QRS complexes and give a threshold value M which allows you to separate QRS pulses from noise and other cardiac events.

(b) Give an algorithm that reads the data sequentially; identifies the beginning of a QRS complex using the threshold M from (a); identifies the end of the QRS pulse; and finds the maximum value over the QRS event just determined.

(c) Suppose two successive QRS pulse maxima are located at n_0 and n_1, where $n_1 > n_0$. Let the sampling interval be T seconds. Find the elapsed time (seconds) between the two maxima, $v(n_1)$ and $v(n_0)$. Give a formula for the heart rate from this single interval; let us call this value $H(n_1)$.

(d) Critique the algorithm for instantaneous heart rate above. Explain any assumptions you have made in the algorithm. Calculate the instantaneous heart rate $H(n_i)$, for all successive pairs of QRS pulses beginning at n_{i-1}. Plot this $H(n_i)$ value over the entire span of the signal. What do you observe? What if the threshold you choose in (a) is different? How does this affect your running heart rate value?

(e) Suppose that the running heart rate is computed as the average of the last several $H(n_i)$ values—for example, $H_3(n_i) = [H(n_i) + H(n_{i-1}) + H(n_{i-2})]/3$. Is the instantaneous heart rate readout better? Is there a practical limit to how many past values you should average?

36. Explore the usefulness of signal averaging when computing the instantaneous heart rate.

(a) Use a symmetric moving average filter on the raw ECG trace:

$$w(n) = \frac{1}{3}[v(n-1) + v(n) + v(n+1)]. \tag{1.132}$$

Calculate the running heart rates as in the previous problem using $w(n)$ instead of $v(n)$.

(b) How does an asymmetric smoothing filter,

$$w(n) = \frac{1}{3}[v(n) + v(n-1) + v(n-2)], \tag{1.133}$$

affect the results? Explain.

(c) Sketch an application scenario which might require an asymmetric filter.

(d) Try symmetric moving average filters of widths five and seven for the task of part (a). Graph the resulting ECG traces. Are the results improved? Is the appearance of the signal markedly different?

(e) Why do signal analysts use symmetric filters with an odd number of terms?

(f) When smoothing a signal, such as the ECG trace $v(n)$, would it be useful to weight the signal values according to how close they are to the most recently acquired datum? Contemplate filters of the form

$$w(n) = \frac{1}{4}v(n-1) + \frac{1}{2}v(n) + \frac{1}{4}v(n+1), \tag{1.134}$$

and discuss their practicality.

(g) Why do we choose the weighting coefficients in the equation of part (f) to have unit sum? Explain.

(h) Finally, consider weighted filter coefficients for asymmetric filters. How might these be chosen, and what is the motivation for so doing? Provide examples and explain them.

37. Develop algorithms for alerting medical personnel to the presence of cardiac dysrhythmia. A statistical measure of the variability of numerical data is the standard deviation,

(a) What is the average heart rate over the entire span of the ECG trace, once again using the distance between QRS pulse peaks as the basis for computing the instantaneous heart rate.

(b) Calculate the standard deviation of time intervals between QRS pulses. How many pulses are necessary for meaningful dysrhythmia computations?

38. Find algorithms for detecting the presence of the P wave and T wave in an ECG. One approach is to again identify QRS pulses and then locate the P wave pulse prior to the detected QRS complex and the T wave pulse subsequent to the detected QRS complex.

(a) Find the presence of a QRS pulse using a threshold method as before. That is, a QRS pulse is indicated by signal values above a threshold T_q.

(b) However, to locate P and T waves adjacent to the QRS complex, we must develop an algorithm for finding the time domain extent, that is the *scale*, of QRS pulses in the ECG trace. Develop an algorithm that segments the signal into the QRS pulse regions and non-QRS pulse regions. How do you handle the problem of noise that might split a QRS region? Is the method robust to extremely jagged QRS pulses—that is, *splintering* of the QRS complex?

(c) Show how a second, smaller threshold T_p can be used to find the P wave prior to the QRS complex. Similarly, a third threshold T_t can be used to find the T wave after the falling edge of the QRS complex.

(d) Should the thresholds T_p and T_t be global constants, or should they be chosen according to the signal levels of the analog and discrete signal acquisition procedures? Explain.

CHAPTER 2

Discrete Systems and Signal Spaces

The first chapter introduced many different sorts of signals—simple, complicated, interesting, boring, synthetic, natural, clean, noisy, analog, discrete—the initial mixed stock of fuel for the signal analysis machinery that we are going to build. But few signals are ends in themselves. In an audio system, for example, the end result is music to our ears. But that longitudinal, compressive signal is only the last product of many transformations of many representations of the sound on its path from compact disc to cochlea. It begins as a stretch of some six billion tiny pits on a compact disc. A laser light strobes the disc, with its reflection forming a pair of 16-bit sound magnitude values 44,000 times each second. This digital technique, known as *Pulse Code Modulation* (PCM), provides an extremely accurate musical tone rendition. Filtering circuits remove undesirable artifacts from the digital signal. The digital signal is converted into analog form, filtered again, amplified, bandpass filtered through the graphic equalizer, amplified again, and finally delivered strong and clean to the speakers. The superior sound quality of digital technology has the drawback of introducing some distortion at higher frequencies. Filtering circuits get rid of some distortion. Since we are human, though, we cannot hear the false notes; such interference occurs at frequencies above 22 kHz, essentially outside our audio range. Of the many forms the music signal takes through the stereo equipment, all but the last are transitional, intended for further conversion, correction, and enhancement. Indeed, the final output is pleasant only because the design of the system incorporates many special intermediate processing steps. The abstract notion of taking an input signal, performing an operation on it, and obtaining an output is called a *system*. Chapter 2 covers systems for discrete signals.

As a mathematical entity, a system is analogous to a vector-valued function on vectors, except that, of course, the "vectors" have an infinite extent. Signal processing systems may require a single signal, a pair of signals, or more for their inputs. We shall develop theory primarily for systems that input and output a single signal. Later (Chapter 4) we consider operations that accept a signal as an input but fundamentally change it or break it down somehow to produce an output that is not a signal. For example, the output could be a structural description, an interpretation, or

Signal Analysis: Time, Frequency, Scale, and Structure, by Ronald L. Allen and Duncan W. Mills
ISBN: 0-471-23441-9

just a number that indicates the type of the signal. We will call this a *signal analysis system* in order to distinguish it from the present case of *signal processing systems*.

Examples of systems abound in electronic communication technology: amplifiers, attenuators, modulators, demodulators, coders, decoders, and so on. A radio receiver is a system. It consists of a sequence of systems, each accepting an input from an earlier system, performing a particular operation on the signal, and passing its output on to a subsequent system. The entire cascade of processing steps converts the minute voltage induced at the antenna into sound waves at the loudspeaker. In modern audio technology, more and more of the processing stages operate on digital signal information. The compact disc player is a wonderful example, embodying many of the systems that we cover in this chapter. Its processed signals take many forms: discrete engravings on the disc, light pulses, digital encodings, analog voltages, vibration of the loudspeaker membrane, and finally sound waves.

Although systems that process on digital signals may not be the first to come to mind—and they are certainly not the first to have been developed in electronic signal conditioning applications—it turns out that their mathematical description is much simpler. We shall cover the two subjects separately, beginning here with the realm that is easier realm to conquer: discrete signal spaces and systems. Many signal processing treatments combine the introduction of discrete and continuous time systems [1–4]. Chapter 3 covers the subtler theory of analog systems.

2.1 OPERATIONS ON SIGNALS

This section explores the idea of a discrete system, which performs an operation on signals. To help classify systems, we define special properties of systems, provide examples, and prove some basic theorems about them. The proofs at this level are straightforward.

A discrete system is a function that maps discrete signals to discrete signals. Systems may be defined by rules relating input signals to output signals. For example, the rule $y(n) = 2x(n)$ governs an *amplifier* system. This system multiplies each input value $x(n)$ by a constant $A \geq 1$. If H is a system and $x(n)$ is a signal, then $y = H(x)$ is the output of the system H, given the input signal x. More compact and common, when clarity permits, is $y = Hx$. To highlight the independent variable of the signals we can also say $y(n) = H(x(n))$. But there should be no misunderstanding: The system H operates on the whole signal x not its individual values $x(n)$, found at time instants n. Signal flow diagrams, with arrows and boxes, are good for visualizing signal processing operations (Figure 2.1).

Not every input signal to a given system produces a valid output signal. Recall that a function on the real numbers might not have all real numbers in its domain. An example is $f(t) = \sqrt{t}$ with $\mathrm{Domain}(f) = \{t \in \mathbb{R}: t \geq 0\}$. Consider now the *accumulator* system $y = Hx$ defined by the rule

$$y(n) = \sum_{k=-\infty}^{n} x(k). \tag{2.1}$$

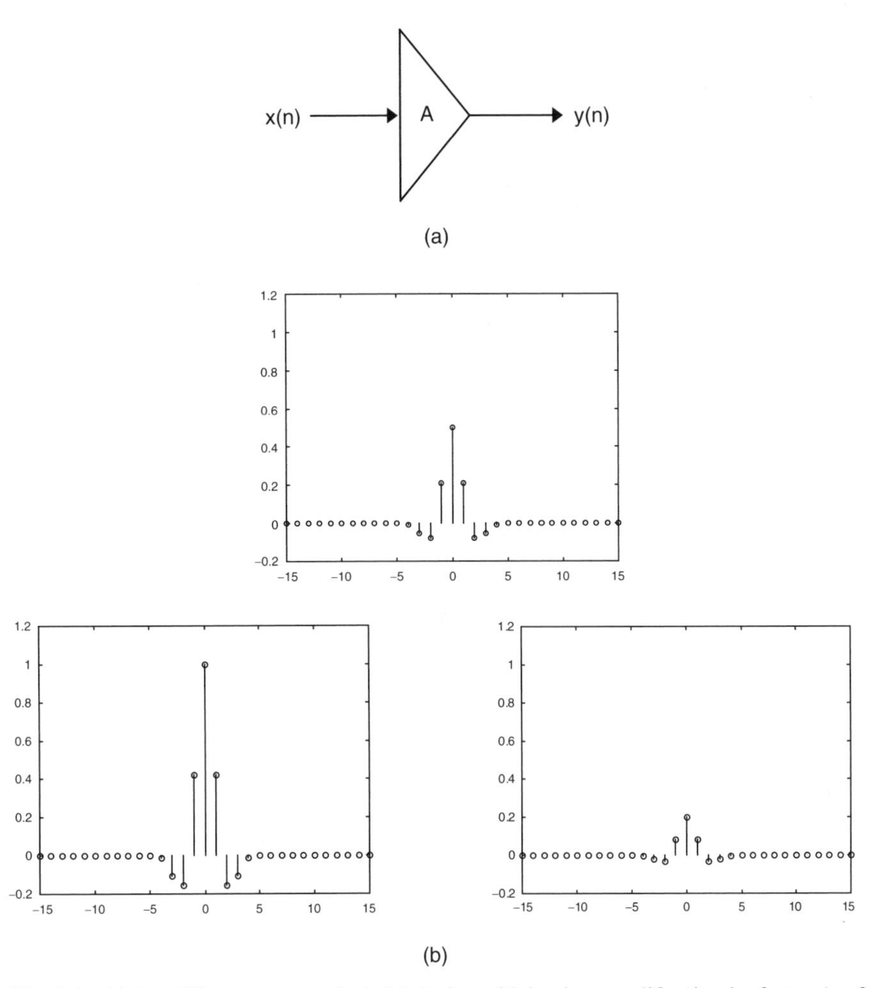

Fig. 2.1. (a) Amplifier system symbol. (b) A sinusoidal pulse, amplification by factor $A = 2$, and attenuation by factor $A = 0.4$.

With $x(n) = 1$ for all n, H has no output. So, as with functions on the real numbers, it is best to think of a system as a partial function on discrete signals.

2.1.1 Operations on Signals and Discrete Systems

There are many types of operations on signals and the particular cases that happen to be discrete systems. We list a variety of cases, some quite simple. But it will turn out that many types of discrete systems decompose into such simple system components, just as individual signals break down into sums of shifted impulses.

Definition (Discrete System). A *discrete system* H is a partial function from the set of all discrete signals to itself. If $y(n)$ is the signal output by H from the input $x(n)$, then $y = Hx$ or $y = H(x)$. It is common to call y the *response* of the system H to input

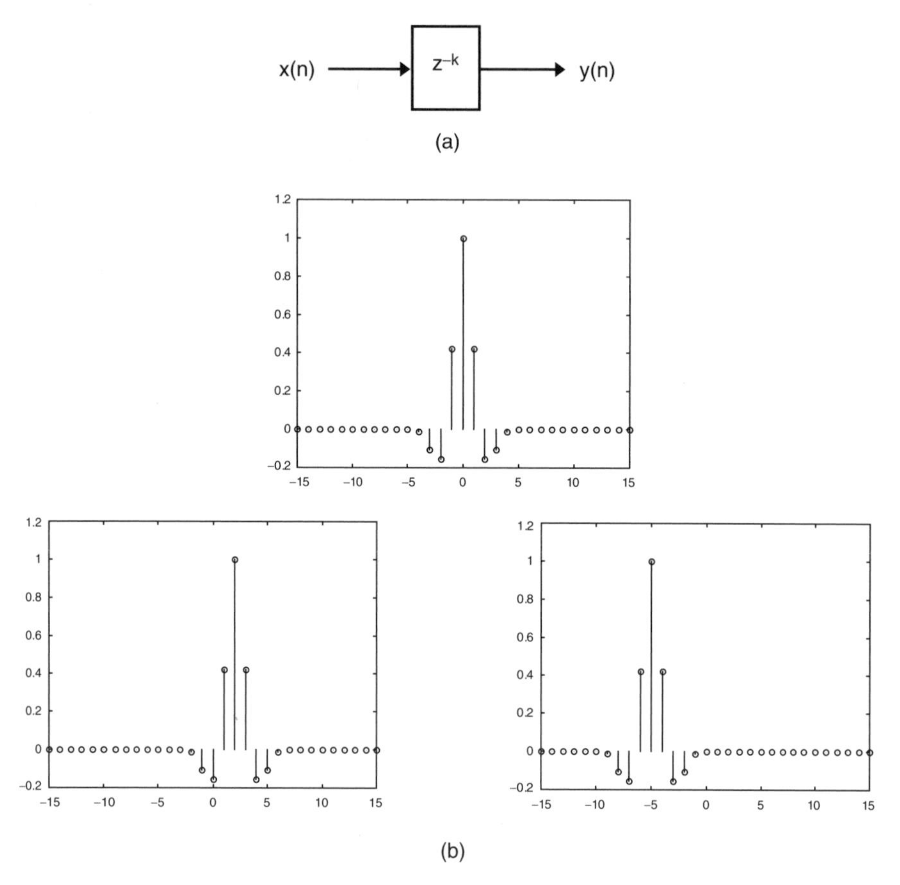

Fig. 2.2. Translation systems. (a) Diagram for a translation system, $y(n) = x(n-k)$. (b) discrete sinusoidal pulse within a Gaussian envelope, $x(n)$; $x(n-k)$, with $k = 2$; and $x(n-k)$ with $k = -5$.

x. The set of signals x for which some $y = Hx$ is the *domain* of the system H. The set of signals y for which $y = Hx$ for some signal x is the *range* of H.

One simple signal operation is to multiply each signal value by a constant (Figure 2.1). If H is the system and $y = Hx$, then the output values $y(n)$ are related to input values by $y(n) = Ax(n)$. This operation *inverts* the input signal when $A < 0$. When $|A| > 1$, the system *amplifies* the input. When $|A| < 1$, the system *attenuates* the input signal. This system is also referred to as a *scaling* system. (Unfortunately, another type of system, one that dilates a signal by distorting its independent variable, is also called a scaling system. Both appellations are widespread, but the two notions are so different that the context is usually enough to avoid confusion.)

The domain of an amplification system is all discrete signals. Except for the case $A = 0$, the range of all amplification systems is all discrete signals.

Another basic signal operation is to delay or advance its values (Figure 2.2). Thus, if $x(n)$ is an input signal, this system produces an output signal $y(n) = x(n - k)$

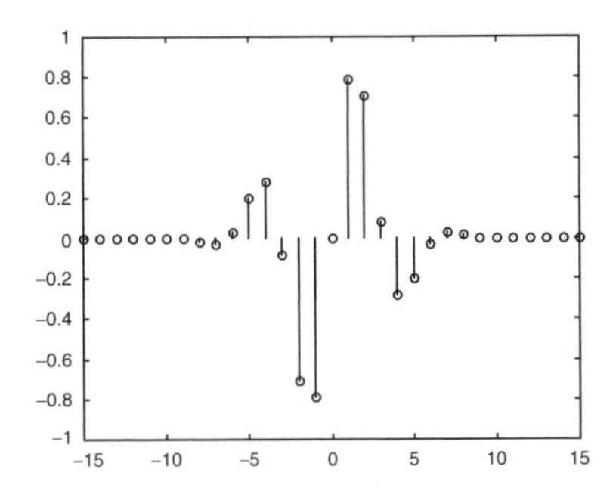
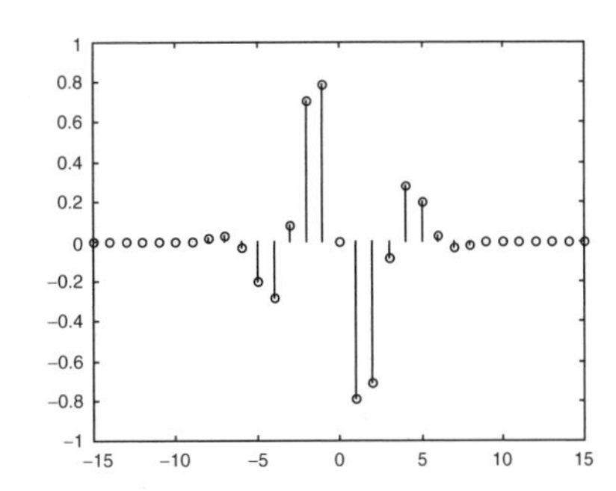

Fig. 2.3. Time reversal $y(n) = x(-n)$. There does not seem to be a conventional block diagram symbol for time reversal, perhaps owing to its physical impossibility for time-based signals.

for some integer k. This is called a *time shift* when the signal's independent variable indicates a time instant or a translation when the independent variable stands for a non-time quantity, such as distance. If $k > 0$, then the shift operation is a *delay.* If $k < 0$, then this system *advances* the signal. The diagram notation z^{-k} is inspired by the notions of the impulse response of the translation system, which we shall discover later in this chapter, and the z-transform (Chapter 8).

The set of all translates of a signal is closed under the translation operation. This system is also commutative; the order of two successive translations of a signal does not matter. Translations cause no domain and range problems. If T is a translation system, then $Domain(T)$ = Range(T) = $\{s(n)$: *s is a discrete signal*$\}$.

Signal *reflection* reverses the order of signal values: $y(n) = x(-n)$. For time signals, we will call this a *time reversal* system (Figure 2.3). It flips the signal values $x(n)$ around the instant $n = 0$. Note that the reflection and translation operations do not commute with one another. If H is a reflection system and G is a translation $y(n) = x(n-k)$, then $H(Gx) \neq G(Hx)$ for all x unless $k = 0$. Notice also that we are careful to say "for all x" in this property. In order for two systems to be identical, it is necessary that their outputs are identical when their inputs are identical. It is not enough that the system outputs coincide for a particular input signal.

Signal *addition* or *summation* adds a given signal to the input, $y(n) = x(n) + x_0(n)$, where $x_0(n)$ is a fixed signal associated with the system H (Figure 2.4). If we allow systems with two inputs, then we can define $y = H(v,w) = v + w$.

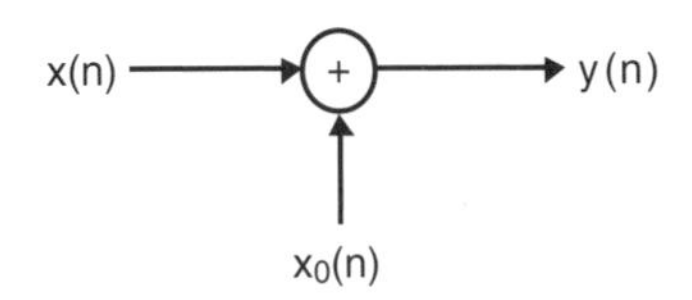

Fig. 2.4. System $y = Hx$ adds another (fixed) signal to the input.

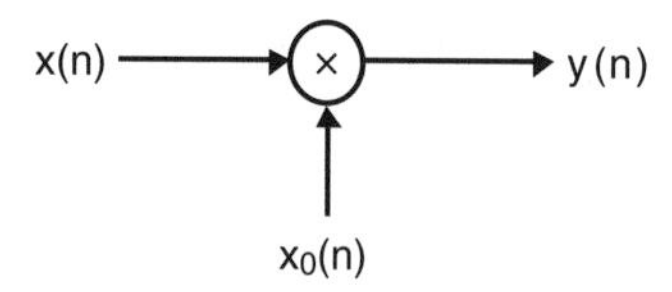

Fig. 2.5. System $y = Hx$ multiplies the input term-by-term with another (fixed) signal. This also called *modulation*, especially in communication theory.

Another type of system, called a *multiplier* or *modulator*, forms the termwise product of the input, $y(n) = x(n)x_0(n)$, where $x_0(n)$ is a fixed signal associated with the system H (Figure 2.5). Note carefully that the product system is *not* written with an asterisk operator $y = x{*}h$. This is the notation for convolution, a more important signal operation, which we will cover below.

The *correlation* of two signals is the sum of their termwise products:

$$C = \sum_{n=-\infty}^{\infty} x(n)\overline{y(n)}. \tag{2.2}$$

Signals $x(n)$ and $y(n)$ may be complex-valued; in this case, we take the complex conjugate of the second operand. The correlation of a signal with itself, the *autocorrelation*, will then always be a non-negative real number. In (2.2) the sum is infinite, so the limit may not exist. Also note that (2.2) does not define a system, because the output is a number, not a signal. When we study abstract signal spaces later, we will call correlation the *inner product* of the two signals. It is a challenge to find classes of signals, not necessarily having finite support, for which the inner product always exists.

The *cross-correlation system* is defined by the input–output relation

$$y(n) = (x \circ h)(n) = \sum_{k=-\infty}^{\infty} x(k)h(n+k). \tag{2.3}$$

In (2.3) the signal $h(n)$ is translated before the sum of products correlation is computed for each $y(n)$. If the signals are complex-valued, then we use the complex conjugate of $h(n)$:

$$y(n) = (x \circ h)(n) = \sum_{k=-\infty}^{\infty} x(k)\overline{h(n+k)}. \tag{2.4}$$

This makes the autocorrelation have a non-negative real value for $n = 0$; if $x = h$,

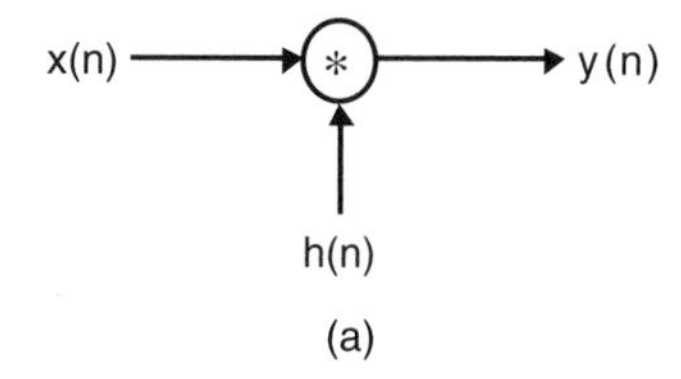

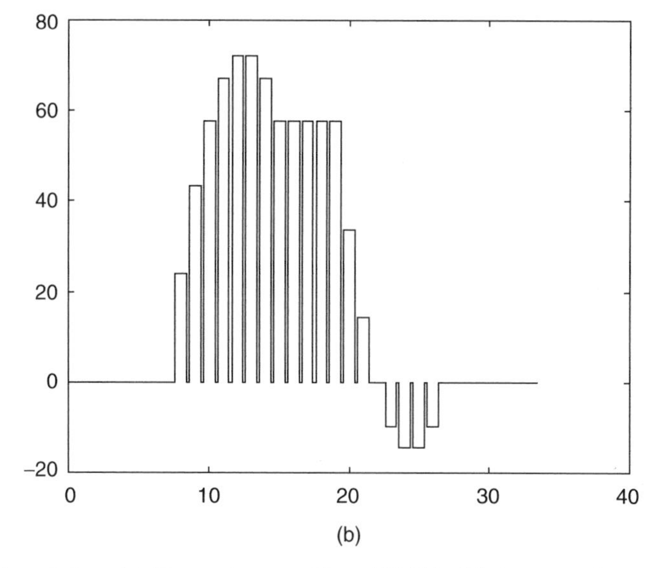

Fig. 2.6. Convolution. (a) The system $y = Hx = (h*x)(n)$. (b) An example convolving two finitely supported signals a square pulse $h(n) = 4.8[u(n-8) - u(n-20)]$ and a triangular pulse $x(n) = (6-n)[u(n-1) - u(n-9)]$.

$$y(0) = (x \circ h)(0) = \sum_{k=-\infty}^{\infty} x(k)\overline{x(nk)} = \|x\|_2^2, \tag{2.5}$$

the square of the l^2 norm of x, which will become quite important later.

Convolution seems strange at first glance—a combination of reflection, translation, and correlation:

$$y(n) = (x * h)(n) = \sum_{k=-\infty}^{\infty} x(k)h(n-k). \tag{2.6}$$

But this operation lies at the heart of linear translation invariant system theory, transforms, and filtering. As Figure 2.6 shows, in convolution one signal is flipped and then shifted relative to the other. At each shift position, a new $y(n)$ is calculated as the sum of products.

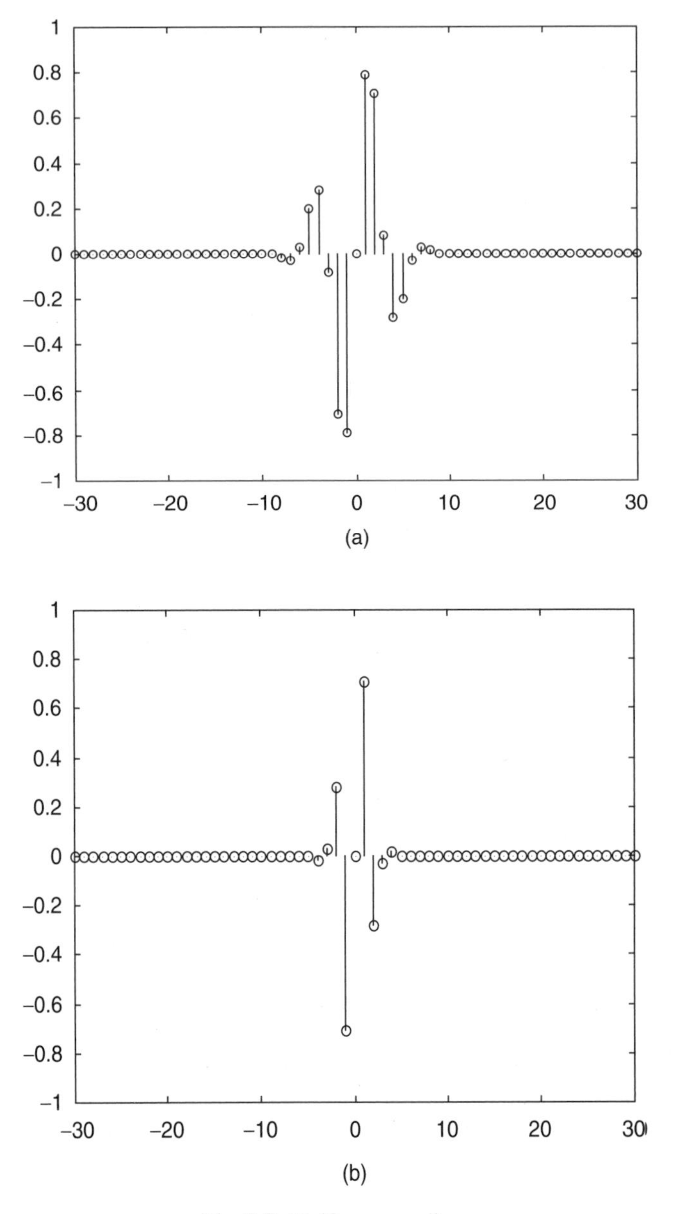

Fig. 2.7. Uniform sampling.

A *subsampling* or *downsampling* system skips over samples of the input signal: $y(n) = x(kn)$, where $k > 0$ is the *sampling interval*. This is *uniform* sampling (Figure 2.7). In nonuniform sampling, the intervals between samples vary. Both forms of signal sampling have proven handy in applications.

Sampling is useful when the content of a signal contains less information than the density of the samples warrants. In uniform sampling, the relevant signal information is adequately conveyed by every kth sample of the input. Thus, subsampling is a preliminary signal compression step. Scientists and engineers, attempting to squeeze every bit of digital information through narrow bandwidth communication channels, have been seeking better ways to compress signals and images in recent years [5]. Technologies such as network video and cellular radio communications hinge on the efficiency and integrity of the compression operations. Also, in signal and image analysis applications, we can filter signals and subsample them at multiple rates for coarse-to-fine recognition. One outcome of all the research is that if the filtering and subsampling operations are judiciously chosen, then the sampled signals are adequate for exact reconstruction of the original signal. Progressive transmission is therefore feasible. Moreover, there is an unexpected, intimate connection between multirate signal sampling and the modern theory of time-scale transforms, or wavelet analysis [6]. Later chapters detail these aspects of signal subsampling operations. Nonuniform sampling is useful when some local regions of a signal must be more carefully preserved in the sampled output. Such systems have also become the focus of modern research efforts [7].

An *upsampling* operation (Figure 2.8) inserts extra values between input samples to produce an output signal. Let $k > 0$ be an integer. Then we form the upsampled output signal $y(n)$ from input signal $x(n)$ by

$$y(n) = \begin{cases} x\left(\frac{n}{k}\right) & \text{if } \frac{n}{k} \in \mathbb{Z}, \\ 0 & \text{if otherwise.} \end{cases} \tag{2.7}$$

A *multiplexer* merges two signals together:

$$y(n) = \begin{cases} x_0(n) & \text{if } n \text{ is even,} \\ x_1(n) & \text{if } n \text{ is odd.} \end{cases} \tag{2.8}$$

A related system, the *demultiplexer*, accepts a single signal $x(n)$ as an input and produces two signals on output, $y_0(n)$ and $y_1(n)$. It also is possible to multiplex and demultiplex more than two signals at a time. These are important systems for communications engineering.

Thresholding is an utterly simple operation, ubiquitous as well as notorious in the signal analysis literature. Given a threshold value T, this system segments a signal:

$$y(n) = \begin{cases} 1 & \text{if } x(n) \geq T, \\ 0 & \text{if } x(n) < T. \end{cases} \tag{2.9}$$

The threshold value T can be any real number; however, it is usually positive and a thresholding system usually takes non-negative real-valued signals as inputs. If the input signal takes on negative or complex values, then it may make sense to first produce its magnitude $y(n) = |x(n)|$ before thresholding.

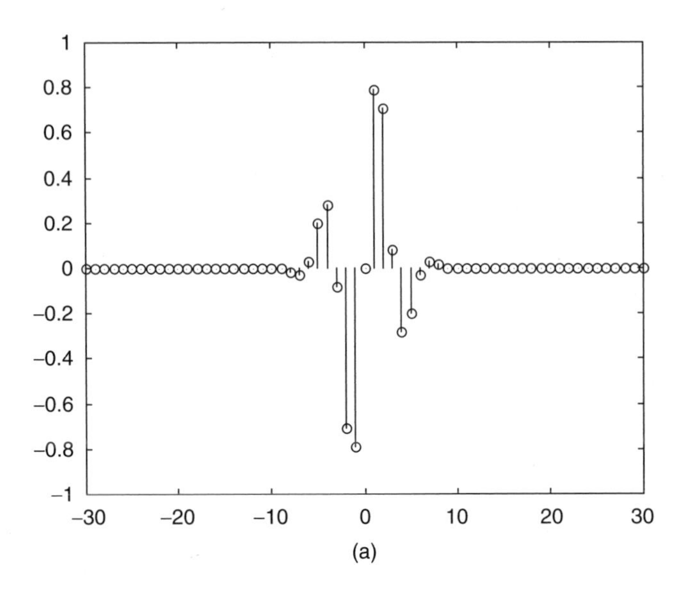

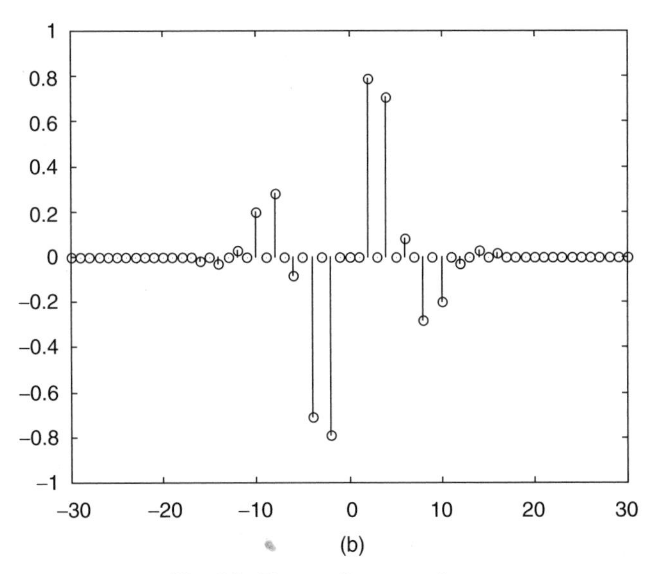

Fig. 2.8. Upsampling operation.

The thresholding operation of (2.9) is an elementary signal analysis system. Input signals $x(n)$ are labeled by the output signal $y(n)$ which takes on two values, 0 or 1. Thus, 1 indicates the presence of meaningful information, and 0 indicates the presence of nonmeaningful information in $x(n)$. When we say that the thresholding operation *segments* the signal into useful and nonuseful portions, we mean that the system partitions the input signal's domain. Typically, we use thresholding as a crude method for removing background noise from signals. A problem with brute-force thresholding is that signals may contain noise impulses that are as large in magnitude as some meaningful signal values. Furthermore, the noise magnitude may be as large as the interesting signal values and thresholding therefore fails completely. Chapter 4 elaborates on thresholding subtleties.

Threshold selection dramatically affects the segmentation. Thresholding usually follows the filtering or transforming of an input signal. The trick is choosing the proper threshold so that the output binary image correctly marks signal regions. To gain some appreciation of this, let's consider a thresholding problem on a two-dimensional discrete signal—that is an 8-bit, 256 gray scales image. Figure 2.9 shows the original image, a parcel delivery service's shipping label. A very thoroughly studied image analysis problem is to interpret the text on the label so that automated systems can handle and route the parcel to its destination. Optical character recognition (OCR) systems read the individual characters. Since almost all current OCR engines accept only binary image data, an essential first step is to convert this gray scale image to binary form [8, 9]. Although the image itself is fairly legible, only the most careful selection of a single threshold for the picture suffices to correctly binarize it.

The *accumulator* system, $y = Hx$, is given by

$$y(n) = \sum_{k=-\infty}^{n} x(k). \tag{2.10}$$

The accumulator outputs a value that is the sum of all input values to the present signal instant. Any signal with finite support is in the domain of the accumulator system. But, as already noted, not all signals are in the accumulator's domain. If a signal is absolutely summable (Section 1.6.2), then it is in the domain of the accumulator. Some finite-energy signals are not in the domain of the accumulator. An example that cannot be fed into an accumulator is the finite-energy signal

$$x(n) = \begin{cases} 0 & \text{if } n \geq 0, \\ -\dfrac{1}{n} & \text{if } n < 0. \end{cases} \tag{2.11}$$

A system may extract the nearest integer to a signal value. This called a *digitizer* because in principle a finite-length storage register can hold the integral values produced from a bounded input signal. There are two variants. One kind of digitizer produces a signal that contains the integral *ceiling* of the input,

$$y(n) = \lceil x(n) \rceil = \text{least integer} \geq x(n). \tag{2.12}$$

The other variant produces the integral *floor* of the input,

$$y(n) = \lfloor x(n) \rfloor = \text{greatest integer} \le x(n). \tag{2.13}$$

The moving average system, $y = Hx$, is given by

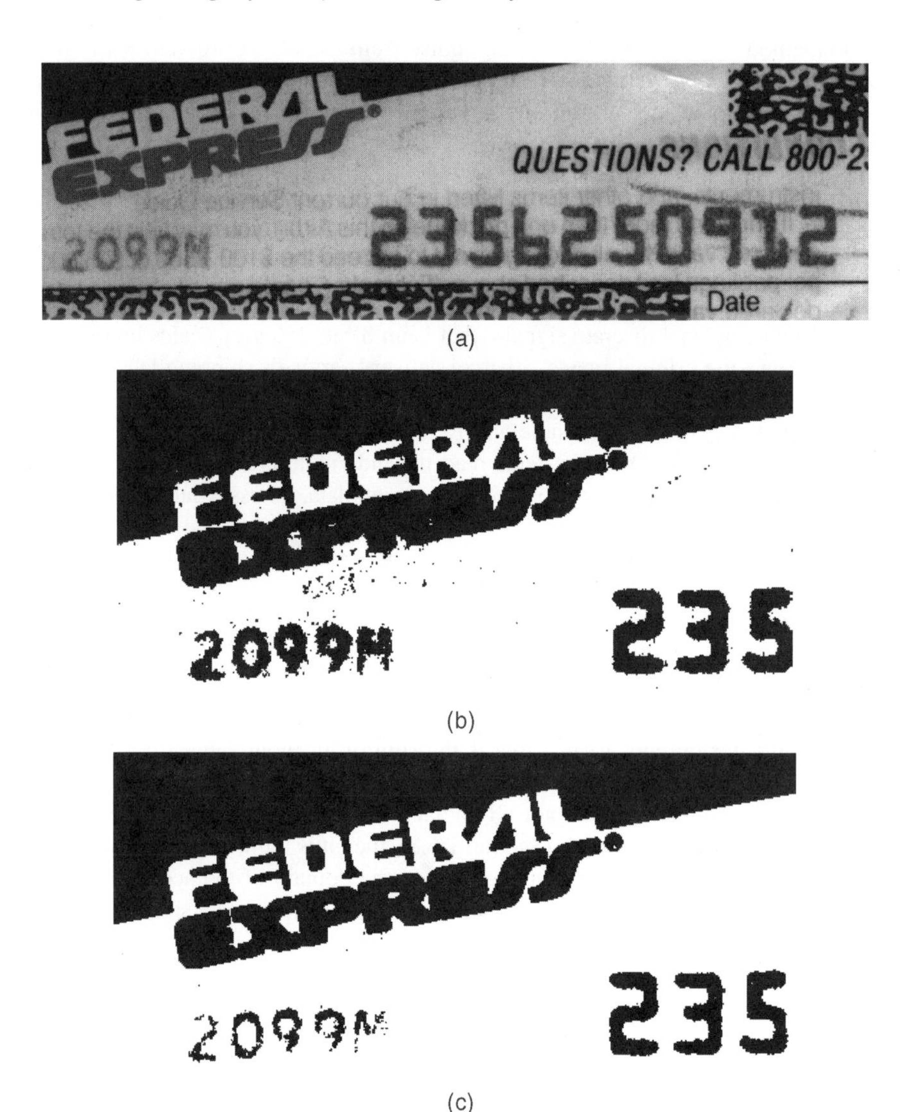

Fig. 2.9. Thresholding. An 8-bit, 256 gray scales image of a shipping label (a). First threshold applied (b). Here the threshold results in an image that is too dark. The company logo is only marginally readable. With a different threshold (c), the logo improves, but the shipping code beneath begins to erode. It is difficult to pick an appropriate global threshold for the image.

$$y(n) = \frac{1}{2N+1}\sum_{k=-N}^{N} x(k). \tag{2.14}$$

This system outputs a signal that averages the $2N + 1$ values around each $x(n)$. This smoothes the input signal, and it is commonly used to improve the signal-to-noise ratio of raw signals in data acquisition systems.

2.1.2 Operations on Systems

We can build more complicated systems by combining the systems, taking the output of one system and using it as an input to another.

Definition (System Composition). Let H and G be systems. Then the *composite system* GH is the system defined by $y(n) = G(H(x(n)))$. GH is called the *composition* or the *cascade* of G with H.

Remark. In general, the order of system composition is important. Many signal operations—however intuitive and simple they may seem—may not commute with one another. In particular, note again that the shift operation does not commute with reflection. If G is a shift operation, $G(x(n)) = x(n - k)$, k is nonzero, and H is a reflection, then $GH \neq HG$.

2.1.3 Types of Systems

Now let us categorize the various systems we have considered. These characterizations will prove useful for understanding the behavior of a system in an application. Some systems, because of the form of their definition, may seem impossible to implement on a computer. For example, the accumulator (2.9) is defined with an infinite sum that uses every earlier input signal value. We have observed already that it is an example of a system in which not every input produces a meaningful output. Do we need an infinite set of memory locations to store input signal values? No, we can still implement an accumulator on a computer by exploiting the recurrence relation $y(n) = y(n - 1) + x(n)$. It is only necessary to know an initial value of $y(n)$ at some past time instant $n = n_0$. The accumulator belongs to a class of systems—called *recursive* systems—that can be implemented by reusing output values from previous time instants.

Now we develop some basic categories for systems: static and dynamic systems, recursive systems, and causal systems.

We can start by distinguishing systems according to whether they require computer memory for signal values at other than the current instant for their implementation. Let H be a system $y = Hx$. H is *static* if $y(n)$ can always be found from the current value of input signal $x(n)$ and n itself. That is, H is *static* when $y(n) = F(x(n))$ for some defining function or rule F for H. H is *dynamic* when it must use values $x(m)$ or $y(m)$ for $m \neq n$ in order to specify $y(n)$.

A system that depends on future values of the input signal $x(n)$ is dynamic too. This seems impossible, if one thinks only of signals that have a time-related independent variable. But signals can independent variables of distance, for example. So the values of $x(n + k)$, $k > 0$, that need to be known to compute $y(n)$ are just values of the input signal in a different direction. Systems for two-dimensional signals (i.e., images) are in fact a very widely studied case of signals whose values may depend on "future" values of the input.

Example. The accumulator is a dynamic system because, for a general input $x(n)$, it cannot be implemented without knowing either

(i) $x(n)$ and all previous values of the input signal; or
(ii) for some $k > 1$, $y(n - k)$ and all $x(n - k + p)$ for $p = 0, ..., k$.

Dynamic systems require memory units to store previous values of the input signal. So static systems are also commonly called *memoryless* systems.

A concept related to the static versus dynamic distinction is that of a *recursive* system. Pondering the accumulator system once more, we note that this dynamic system cannot be implemented with a finite set of memory elements that only contain previous values of $x(n)$. Let H be a system, $y = Hx$. H is *nonrecursive* if there is an $M > 0$, such that $y(n)$ can always be found as a function of $x(n), x(n - 1), ..., x(n - M)$. If $y(n)$ depends on $y(n - 1), y(n - 2), ..., y(n - N)$, for some $N > 0$, and perhaps upon $x(n - 1), ..., x(n - M)$, for some M, then H is *recursive*.

A system $y = Hx$ is *causal* if $y(n)$ can always be computed from present and past inputs $x(n), x(n - 1), x(n - 2), ...$. Real, engineered systems for time-based signals must always be causal, and, if for no other reason, causality is important. Nevertheless, where the signals are not functions of discrete time variables, noncausal signals are acceptable. A nonrecursive system is causal, but the converse is not true.

Examples (System Causality)

(i) $y(n) = x(n) + x(n - 1) + x(n - 2)$ is causal and nonrecursive.
(ii) $y(n) = x(n) + x(n + 1)$ is not causal.
(iii) $y(n) = x(2 - n)$ is noncausal.
(iv) $y(n) = x(|n|)$ is noncausal.
(v) The accumulator (2.10) is causal.
(vi) The moving average system (2.14) is not causal.

2.2 LINEAR SYSTEMS

Linearity prevails in many signal processing systems. It is desirable in entertainment audio systems, for example. Underlying the concept of linearity are two ideas, and they are both straightforward:

(i) When the input signal to the system is larger (or smaller) in amplitude, then the output signal from the system produces is proportionally larger (or smaller). This is the *scaling* property of linearity. In other words, if a signal is amplified or attenuated and then input to a linear system, then the output is a signal that is amplified or attenuated by the same amount.

(ii) Furthermore, if two signals are added together before input, then the result is just the sum of the outputs that would occur if each input component were passed through the system independently. This is the *superposition* property.

Obviously, technology cannot produce a truly linear system; there is a range within which the linearity of a system can be assumed. Real systems add noise to any signal. When the input signal becomes too small, the output may disappear into the noise. The input could become so large intensity that the output is distorted. Worse, the system may fail if subjected to huge input signals. Practical, nearly linear systems are possible, however, and engineers have discovered some clever techniques to make signal amplification as linear as possible.

When amplification factors must be large, the nonlinearity of the circuit components—vacuum tubes, transistors, resistors, capacitors, inductors, and so on—becomes more and more of a factor affecting the output signal. The discrete components lose their linearity at higher power levels. A change in the output proportional to the change in the input becomes very difficult to maintain for large amplification ratios. Strong amplification is essential, however, if signals must travel long distances from transmitter to repeater to receiver. One way to lessen the distortion by amplification components is to feed back a tiny fraction of the output signal, invert it, and add it to the input signal. Subject to a sharp attenuation, the feedback signal remains much closer to true linearity. When the output varies from the amplification setting, the input biases in the opposite direction. This restores the amplified signal to the required magnitude. An illustration helps to clarify the concept (Figure 2.10).

From Figure 2.10 we can see that the output signal is $y(n) = A(x(n) - By(n-1))$. Assuming that the output is relatively stable, so that $y(n) \approx y(n-1)$, we can express

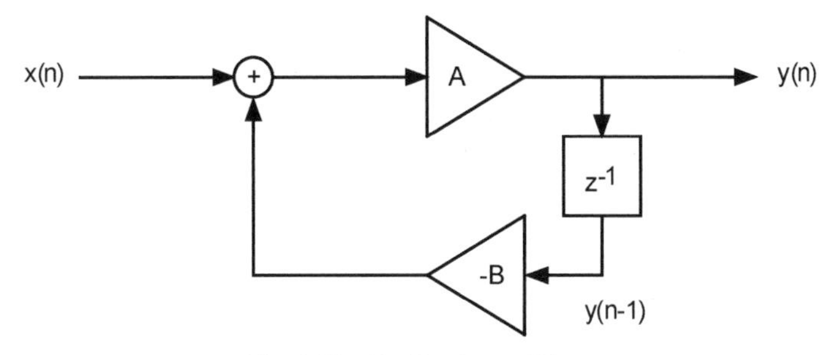

Fig. 2.10. Feedback amplifier.

the system gain as follows:

$$y(n) = \frac{A}{AB+1}x(n) = \frac{1}{B+A^{-1}}x(n). \tag{2.15}$$

Evidently, if the amplification factor A is large, then A^{-1} is small, and the system gain depends mostly on the inverse of the attenuation factor in the feedback. It is approximately B^{-1}. Due to its low power levels, the feedback circuit is inherently more linear, and it improves the overall linearity of the system immensely. Achieving good linearity was an early stumbling block in the development of the telephone system.[1]

Since many important results and characterizations about systems follow from the assumption of linearity, it is a central object of our study in signal processing and analysis.

2.2.1 Properties

Let us formalize these ideas and explore the properties of linear systems.

Definition (Linear System). Let $y = Hx$, and let A be a scalar (real or complex number). The system H is *linear* if it obeys the scaling and superposition properties:

(i) *Scaling*:

$$H(Ax) = AH(x). \tag{2.16}$$

(ii) *Superposition*:

$$H(x+y) = Hx + Hy. \tag{2.17}$$

There is a useful criterion for system linearity.

Proposition (Linearity Criterion). If system H is linear, $x(n) = 0$ for all n, and $y = Hx$, then $y(n) = 0$ for all n also.

Proof: Although it seems tricky to many who are not familiar with this type of argument, the proof is quite simple. If $x(n) = 0$ for all n, then $x(n) = x(n) + x(n)$. Hence, $y(n) = H(x(n) + x(n)) = H(x(n))$. But $H(x + x) = Hx + Hx$ by superposition. So $Hx + Hx = Hx$; the consequence, subtracting Hx from both sides of this equation, is that $0 = Hx = y$. That is, $y(n)$ is the zero signal, and the criterion is proven. Note,

[1]The big technical breakthrough came when a Bell Laboratories engineer, Harold Black, had a flash of insight on the ferry from Hoboken, New Jersey, to work in New York City on a summer morning in 1927 [J. R. Pierce and A. M. Noll, *Signals: The Science of Telecommunications*, New York: Scientific American Library, 1990].

by the way, that this last equality is an equality of signals and that the "0" in the equation is really the signal $y(n)$ that is zero for all time instants n. ■

Examples. To show a system is linear, we must check that it satisfies both the scaling and superposition properties. The linearity criterion is helpful for exposing nonlinear systems.

(i) The system $y(n) = 5 + x(n)$ is nonlinear, an easy application of the criterion.

(ii) The system $y(n) = x(n)x(n)$ is nonlinear. The criterion does not help with this example. However, the system violates both the scaling and superposition properties.

(iii) The reflection system $y(n) = x(-n)$ is linear.

(iv) The system $y(n) = \cos(n)x(n)$ is linear. Note this example. The idea of linearity is that systems are linear in their input signals. There may be nonlinear functions, such as $\cos(t)$, involved in the definition of the system relation. Nevertheless, as long as the overall relation between input and output signals obeys both scaling and superposition (and this one does indeed), then the system is linear.

2.2.2 Decomposition

A very useful mathematical technique is to resolve complicated entities into simpler components. For example, the Taylor series from calculus resolves differentiable functions into sums of polynomials. Linear systems can defy initial analysis because of their apparent complexity, and our goal is to break them down into elementary systems more amenable to study. There are two steps:

(i) We first break down signals into sums of scaled, shifted unit impulse signals.

(ii) This decomposition can it turn be used to resolve the outputs of linear systems into sums of simple component outputs.

Proposition (Decomposition). Let $x(n)$ be a discrete signal and define the constants $c_k = x(k)$. Then $x(n)$ can be decomposed into a sum of scaled, shifted impulse signals as follows:

$$x(n) = \sum_{k=-\infty}^{\infty} c_k \delta(n-k). \tag{2.18}$$

Proof: Let $w(n)$ be the right-hand side of (2.18). Then $w(k) = c_k$, since all of the terms $\delta(n-k)$ are zero, unless $n = k$. But this is just $x(k)$ by the definition of the constants c_k. So $w(k) = x(k)$ for all k, completing the proof. ■

The next definition prepares us for a theorem that characterizes discrete linear systems. From the examples above, linear systems can come in quite a few varieties. Sometimes the system's linear nature is nonintuitive. However, any general linear system is completely known by its behavior on one type of input signal: a shifted impulse. (Our proof is informal, in that we assume that scaling and superposition apply to infinite sums and that a convergent series can be rearranged without affecting its limit.)

Definition (Output of Shifted Input). Let H be a linear system and $y = Hx$. Define $y(n, k) = H(x(n-k))$ and $h(n, k) = H(\delta(n-k))$, where δ is the discrete unit impulse signal.

Theorem (Linearity Characterization). Let H be a linear system, $x(n)$ a signal, $c_k = x(k)$, and $y = Hx$. Then

$$y(n) = \sum_{k=-\infty}^{\infty} c_k h(n, k). \tag{2.19}$$

Proof: By the decomposition of discrete signals into impulses, we know that $x(n) = \sum_{k=-\infty}^{\infty} c_k \delta(n-k)$. So with $y(n) = H(x(n))$, it follows from superposition that

$$\begin{aligned} y(n) &= H\left(\sum_{k=-\infty}^{\infty} c_k \delta(n-k)\right) \\ &= H\left(\sum_{k=1}^{\infty} c_k \delta(n-k)\right) + H(c_0 \delta(n)) + H\left(\sum_{k=-\infty}^{-1} c_k \delta(n-k)\right). \end{aligned} \tag{2.20}$$

And then by the scaling property applied to the middle term of (2.20),

$$\begin{aligned} y(n) &= H\left(\sum_{k=-\infty}^{\infty} c_k \delta(n-k)\right) \\ &= H\left(\sum_{k=1}^{\infty} c_k \delta(n-k)\right) + c_0 h(n, 0) + H\left(\sum_{k=-\infty}^{-1} c_k \delta(n-k)\right). \end{aligned} \tag{2.21}$$

Repeatedly using the linearity properties to break out middle terms in (2.21) gives the desired result. ■

2.3 TRANSLATION INVARIANT SYSTEMS

In many real-life systems, the outputs do not depend on the absolute time at which the inputs were done. For example, a compact disc player is a reasonably good time-invariant system. The music played from a compact disc on Monday will not differ substantially when the same disc played on Tuesday. Loosely put, time-invariant systems produce identically shifted outputs from shifted inputs. Since not all systems, of course, are time-based, we prefer to call such systems *translation-invariant.*

To formalize this idea, we need to carefully distinguish between the following:

- The signal $y(n, k)$ is the output of a system, given a delay (or advance, depending on the sign of k) of the input signal $x(n)$ by k time units.
- On the other hand, $y(n - k)$ is the signal obtained by finding $y = Hx$ and then shifting the resulting output signal, $y(n)$, by k.

These two results may not be the same. An example is the system $y(n) = x(n) + n$. It is easy to find input signals $x(n)$ which for this system give $y(n-k) \neq y(n, k)$. Hence, the important definition:

Definition (Translation-Invariant Systems). Let H be the system $y = Hx$. Then, if for all signals $x(n)$ in $Domain(H)$, if $y(n, k) = H(x(n - k)) = y(n - k)$, then H is *translation-invariant.* Another term is *shift-invariant.* For time-based signals, it is common to say *time-invariant.*

To show a system is translation-invariant, we must compare the shifted output, $y(n - k)$, with the output from the shifted input, $H(x(n - k))$. If these are equal for all signals in the system's domain, then H is indeed translation-invariant. It is very easy to confuse these two situations, especially for readers new to signal processing. But there is a simple, effective technique for showing translation-invariance: we rename the input signal after it is shifted, hiding the shift amount.

Examples (Translation-Invariance). Let us try this technique on the systems we checked for linearity earlier.

(i) The system $y(n) = 5 + x(n)$ is translation-invariant. If $y = Hx$, then the output shifted by k is $y(n - k) = 5 + x(n - k)$. We just substitute "$n - k$" for "n" at each instance in the defining rule for the system's input–output relation. What if we translate the input by k units also? This produces $x(n - k)$. We rename it $w(n) = x(n - k)$. This hides the shift amount within the new name for the input signal. It is easy to see what H does to any signal, be it named x, w, u, v, or whatever. H adds the constant 5 to each signal value. Hence we see $y(n, k) = H(w(n)) = 5 + w(n)$. Now we put the expression for x in terms of w back into the expression: $y(n, k) = 5 + x(n - k)$. So $y(n - k) = y(n, k)$, and the system is translation-invariant.

(ii) The system $y(n) = x(n)x(n)$ is translation-invariant. The output shifted by k is $y(n - k) = x(n - k)x(n - k)$. We rename once more the shifted input $w(n) = x(n - k)$. Then $y(n, k) = H(w(n)) = w(n)w(n) = x(n - k)x(n - k)$. Again, the system is translation-invariant.

(iii) The reflection system $y(n) = x(-n)$ is not translation-invariant. The shifted output in this case is $y(n - k) = x(-(n - k)) = x(k - n)$. We make a new name for the shifted input, $w(n) = x(n - k)$. Then $y(n, k) = H(w(n)) = w(-n)$. But $w(-n) = x(-n - k)$. Thus, $y(n - k) \neq y(n, k)$ in general. In particular, we can take $x(n) = \delta(n)$, the unit impulse signal, and $k = 1$. Then $y(n - k) = \delta(1 - n)$, although $y(n, k) = \delta(-1 - n)$.

(iv) The system $y(n) = \cos(n)x(n)$ is not translation-invariant. The shifted output in this case is $y(n - k) = \cos(n - k)x(n - k)$. Once more, we rename the shifted input $w(n) = x(n - k)$. Then $y(n, k) = H(w(n)) = \cos(n)w(n) = \cos(n)x(n - k)$. Again, $y(n - k) \neq y(n, k)$ in general.

2.4 CONVOLUTIONAL SYSTEMS

The most important systems obey both the linearity and translation-invariance properties. Many physical systems are practically linear and practically translation-invariant. If these special properties can be assumed for a physical system, then the analysis of that system simplifies tremendously. The key relationship, it turns out, is that linear translation-invariant systems are fully characterized by one signal associated with the system, the impulse response. By way of contrast, think again of the characterization of linear systems that was given in Section 2.2. For a linear system, there is a characterization of the system's outputs as sums of scaled signals $h(n, k)$. However, there are in general an infinite number of $h(n, k)$ components. This infinite set reduces to just one signal if the system is translation-invariant too.

2.4.1 Linear, Translation-Invariant Systems

This is the most important type of system in basic signal processing theory.

Definition (LTI System). A system H that is both linear and translation-invariant is called an *LTI system*. When signals are functions of an independent time variable, we may say *linear, time-invariant*, but the abbreviation is the same. Some authors use the term *shift-invariant* and refer to *LSI* systems.

Definition (Impulse Response). Let H be an LTI system and $y = Hx$. Then we define the impulse response of the system as $h(n) = H(\delta(n))$, where δ is the discrete unit impulse signal.

Theorem (Convolution). Let H be an LTI system, $y = Hx$, and $h = H\delta$. Then

$$y(n) = \sum_{k=-\infty}^{\infty} x(k)h(n-k). \tag{2.22}$$

Proof: With H linear, there follows the decomposition

$$y(n) = \sum_{k=-\infty}^{\infty} x(k)h(n,k) = \sum_{k=-\infty}^{\infty} x(k)H(\delta(n-k)). \tag{2.23}$$

But since H is translation-invariant, $h(n-k) = H(\delta(n-k))$. Inserting this into (2.23) proves the theorem. ■

Remarks. Note that (2.22) is the convolution of the input signal $x(n)$ and the impulse response $\delta(n)$ of the LTI system H: $y(n) = (x*h)(n)$. For this reason, we call LTI systems *convolutional.* Although it is a simple result, the importance of the theorem cannot be overemphasized. In order to understand a system, we must know how it operates on its input signals. This could be extremely complicated. But for LTI systems, all we need to do is find one signal—the system's impulse response. Then, for any input, the output can be computed by convolving the input signal and the impulse response. There is some more theory to cover, but let us wait and give an application of the theorem.

Example (System Determination). Consider the LTI system $y = Hx$ for which only two test cases of input–output signal pairs are known: $y_1 = Hx_1$ and $y_2 = Hx_2$, where $y_1 = [-2, 2, \underline{0}, 2, -2]$, $x_1 = [2, \underline{2}, 2]$, $y_2 = [1, -2, \underline{2}, -2, 1]$, $x_2 = [-1, \underline{0}, -1]$. (Recall from Chapter 1 the square brackets notation: $x = [n_{-M}, ..., n_{-1}, \underline{n_0}, n_1, ..., n_N]$ is the finitely supported signal on the interval $[M, N]$ with $x(0) = n_0$, $x(-1) = n_{-1}$, $x(1) = n_1$, and so on. The underscored value indicates the zero time instant value.) The problem is to find the output $y = Hx$ when x is the ramp signal $x = [1, \underline{2}, 3]$. An inspection of the signal pairs reveals that $2\delta(n) = x_1 + 2x_2$. Thus, $2h = Hx_1 + 2Hx_2 = y_1 + 2y_2$ by the linearity of H. The impulse response of H must be the signal $(y_1 + 2y_2)/2 = h = [-1, \underline{2}, -1]$. Finding the impulse response is the key. Now the convolution theorem implies that the response of the system to $x(n)$ is the convolution $x*h$. So $y = [-1, 0, \underline{0}, 4, -3]$.

Incidentally, this example previews some of the methods we develop extensively in Chapter 4. Note that the output of the system takes zero values in the middle of the ramp input and large values at or near the edges of the ramp. In fact, the impulse response is known as a *discrete Laplacian operator.* It is an elementary example of a signal *edge detector*, a type of signal analysis system. They produce low magnitude outputs where the input signal is constant or uniformly changing, and they produce large magnitude outputs where the signal changes abruptly. We could proceed further to find a threshold value T for a signal analysis system that would mark the signal edges with nonzero values. In fact, even this easy example gives us a taste of the problems with picking threshold values. If we make T large, we will only detect the large step edge of the ramp input signal above. If we make T small enough to detect the beginning edge of the ramp, then there are two instants at which the final edge is detected.

2.4.2 Systems Defined by Difference Equations

A discrete difference equation can specify a system. Difference equations are the discrete equivalent of differential equations in continuous-time mathematical analysis, science, and engineering. Consider the LTI system $y = Hx$ where the input and output signals always satisfy a linear, constant-coefficient difference equation:

$$y(n) = \sum_{k=1}^{K} a_k y(n-k) + \sum_{m=0}^{M} b_m x(n-m). \tag{2.24}$$

The output $y(n)$ can be determined from the current input value, recent input values, and recent output values.

Example. Suppose the inputs and outputs of the LTI system H are related by

$$y(n) = ay(n-1) + x(n), \tag{2.25}$$

where $a \neq 0$. In order to characterize inputs and outputs of this system, it suffices to find the impulse response. There are, however, many signals that may be the impulse response of a system satisfying (2.25). If $h = H\delta$, is an input–output pair that satisfies (2.25), with $x = \delta$ and $y = h$, then a single known value of $h(n_0)$ determines all of $h(n)$. Let us say that $h(0) = c$. Then repeatedly applying (2.25) we can find $h(1) = ac + \delta(1) = ac$; $h(2) = a^2c$; and, for non-negative k, $h(k) = a^k c$. Furthermore, by writing the equation for $y(n-1)$ in terms of $y(n)$ and $x(n)$, we have

$$y(n-1) = \frac{y(n) - x(n)}{a}. \tag{2.26}$$

Working with (2.26) from the initial known value $h(0) = c$ gives $h(k) = a^k(c-1)$, for $k < 0$. So

$$h(n) = \begin{cases} a^n c & \text{if } n \geq 0, \\ a^n(c-1) & \text{if } n < 0. \end{cases} \tag{2.27}$$

From the convolution relation, we now know exactly what LTI systems satisfy the difference equation (2.25). It can be shown that if the signal pair (δ, h) satisfies the difference equation (2.24), then the pair $(x, h*x)$ also satisfies (2.24).

Example. Not all solutions (x, y) of (2.24) are an input–output pair for an LTI system. To see this, let $a \neq 0$ and consider the *homogeneous* linear difference equation

$$y(n) - ay(n-1) = 0. \tag{2.28}$$

Clearly, any signal of the form $y(n) = da^n$, where d is a constant, satisfies the homogeneous equation. If (x, y) is a solution pair for (2.25), and y_h is a solution of the homogeneous equation, then $(x, y + y_h)$ is yet another solution of (2.25). We know by the linearity criterion (Section 2.2) that for an LTI system, the only possible input–output pair (x, y) when $x(n) = 0$ for all n is $y(n) = 0$ for all n. In particular, $(0, da^n)$ is a solution of (2.25), but not an input–output pair on any LTI system.

2.4.3 Convolution Properties

Although it seems at first glance to be a very odd operation on signals, convolution is closely related to two quite natural conditions on systems: linearity and translation-invariance. Convolution in fact enjoys a number of algebraic properties: associativity, commutativity, and distributivity.

Proposition (Convolution Properties). Let x, y, and z be discrete signals. Then

(i) (*Associativity*) $x*(y*z) = x*(y*z)$.
(ii) (*Commutativity*) $x*y = y*x$.
(iii) (*Distributivity*) $x*(y+z) = x*y + x*z$.

Proof: We begin with associativity. Let $w = y*z$. Then, by the definition of convolution, we have $[x*(y*z)](n) = (x*w)(n)$. But, x convolved with w is

$$\sum_{k=-\infty}^{\infty} x(k)w(n-k) = \sum_{k=-\infty}^{\infty} x(k) \sum_{l=-\infty}^{\infty} y(l)z((n-k)-l)$$
$$= \sum_{l=-\infty}^{\infty} \sum_{k=-\infty}^{\infty} x(k)y(l)z((n-k)-l). \tag{2.29}$$

Let $p = k + l$ so that $l = p - k$, and note that $p \to \pm\infty$ as $l \to \pm\infty$. Then make the change of summation in (2.29):

$$(x*w)(n) = (x*(y*z))(n) = \sum_{p=-\infty}^{\infty} \sum_{k=-\infty}^{\infty} x(k)y(p-k)z(n-p)$$
$$= \sum_{p=-\infty}^{\infty} \left(\sum_{k=-\infty}^{\infty} x(k)y(p-k) \right) z(n-p)$$
$$= \sum_{p=-\infty}^{\infty} (x*y)(p)z(n-p) = ((x*y)*z)(n). \tag{2.30}$$

Commutativity is likewise a matter of juggling summations:

$$(x * y)(n) = \sum_{k=-\infty}^{\infty} x(k)y(n-k) = \sum_{k=-\infty}^{\infty} y(n-k)x(k)$$
$$= \sum_{k=\infty}^{-\infty} y(n-k)x(k) = \sum_{p=-\infty}^{\infty} y(p)x(n-p) = (y * x)(n). \tag{2.31}$$

Finally, distributivity is the easiest property, since

$$(x * (y+z))(n) = \sum_{k=-\infty}^{\infty} x(k)(y+z)(n-k) = \sum_{k=-\infty}^{\infty} [x(k)y(n-k) + x(k)z(n-k)]$$
$$= \sum_{k=-\infty}^{\infty} x(k)y(n-k) + \sum_{k=-\infty}^{\infty} x(k)z(n-k)$$
$$= (x * y)(n) + (x * z)(n). \tag{2.32}$$

This completes the theorem. ■

The convolution theorem has a converse, which means that convolution with the impulse response characterizes LTI systems.

Theorem (Convolution Converse). Let $h(n)$ be a discrete signal and H be the system defined by $y = Hx = x*h$. Then H is LTI and $h = H\delta$.

Proof: Superposition follows from the distributive property. The scaling property of linearity is straightforward (exercise). To see that H is translation-invariant, note that the shifted output $y(n-l)$ is given by

$$y(n-l) = \sum_{k=-\infty}^{\infty} x(k)h((n-l)-k). \tag{2.33}$$

We compare this to the response of the system to the shifted input, $w(n) = x(n-l)$:

$$H(x(n-l)) = (Hw)(n) = \sum_{k=-\infty}^{\infty} w(k)h(n-k) = \sum_{k=-\infty}^{\infty} x(k-l)h(n-k)$$
$$= \sum_{p=-\infty}^{\infty} x(p)h(n-(p+l)) = \sum_{p=-\infty}^{\infty} x(p)h((n-l)-p) = y(n-l). \tag{2.34}$$

So H is indeed translation-invariant. It is left to the reader to verify that $h = H\delta$. ■

Definition (FIR and IIR Systems). An LTI system H having an impulse response $h = H\delta$ with finite support is called a *finite impulse response* (*FIR*) system. If the LTI system H is not FIR, then it is called *infinite impulse response* (*IIR*).

Remarks. FIR systems are defined by a convolution sum that may be computed for any input signal. IIR systems will have some signals that are not in their domain. IIR systems may nevertheless have particularly simple implementations. This is the case when the system can be implemented via a difference equation, where previous known output values are stored in memory for the computation of current response values. Indeed, an entire theory of signal processing with IIR systems and their compact implementation on digital computers has been developed and is covered in signal processing texts [10].

2.4.4 Application: Echo Cancellation in Digital Telephony

In telephone systems, especially those that include digital links (almost all intermediate- and long-distance circuits in a modern system), *echo* is a persistent problem. Without some special equipment—either echo suppressors or echo cancellers—a speaker can hear a replica of his or her own voice. This section looks at an important application of the theory of convolution and LTI systems for constructing effective echo cancellers on digital telephone circuits.

Echo arises at the connection between two-wire telephone circuits, such as found in a typical residential system, and four-wire circuits, that are used in long-haul circuits. The telephone circuits at the subscriber's site rely on two wires to carry the near-end and far-end speakers' voices and an earth ground as the common conductor between the two circuit paths. The earth ground is noisy, however, and for long-distance circuits, quite unacceptable. Good noise immunity requires a four-wire circuit. It contains separate two-wire paths for the far-end and near-end voices. A device called a *hybrid transformer*, or simply a *hybrid*, effects the transition between the two systems [11]. Were it an ideal device, the hybrid would convert all of the energy in a signal from the far-end speaker into energy on the near-end two-wire circuit. Instead, some of the energy leaks through the hybrid (Figure 2.11) into the circuit that carries the near-end voice outbound to the far-end speaker. The result

Fig. 2.11. Impedance mismatches in the four-wire to two-wire hybrid transformer allow an echo signal to pass into the speech signal from the near-end speaker. The result is that the far-end speaker hears an echo.

is that far-end speakers hear echoes of their own voices. Since the system design is often symmetrical, the echo problem is symmetrical too, and near-end speakers also suffer annoying echoes. The solution is to employ an *echo suppression* or *echo cancellation* device.

Echo suppression is the older of the two approaches. Long-haul telephone circuits are typically digital and, in the case of the common North American T1 standard, multiplex upwards of 24 digital signals on a single circuit. Since the echo removal is most economically viable at telephone service provider central offices, the echo removal equipment must also be digital in nature. A digital echo suppressor extracts samples from the far-end and near-end digital voice circuits. It compares the magnitudes of the two signals, generally using a threshold on the difference in signal amplitudes. It opens the near-end to far-end voice path when there is sufficient far-end speech detected to cause an echo through the hybrid, but insufficient near-end speech to warrant maintaining the circuit so that the two speakers talk at once. (This situation, called *double-talk* in telephone engineering parlance, occurs some 30% of the time during a typical conversation.) Thus, let $T > 0$ be a threshold parameter, and suppose that M far-end samples and N near-end samples are compared to decide a suppression action. Let $x(n)$ and $s(n)$ be the far- and near-end digital signals, respectively. If $|x(n)| + |x(n-1)| + \cdots + |x(n-M+1)| > T(|s(n)| + |s(n-1)| + \cdots + |s(n-N+1)|)$ at time instant n, then the suppressor mutes the near-end speaker's voice. Now, this may seem crude.

And echo suppression truly is crude. When both people speak, there is echo, but it is less noticeable. The suppressor activates only when the near-end speaker stops talking. Hence, unless the threshold T, the far-end window size M, and the near-end window size N are carefully chosen, the suppressor haphazardly interrupts the near-end signal and makes the resultant voice at the far-end sound choppy. Even granting that these parameters are correct for given circuit conditions, there is no guarantee that system component performances will not drift, or that one speaker will be unusually loud, or that the other will be unusually soft-voiced. Moreover, the suppressor design ought to provide a noise matching capability, whereby it substitutes *comfort noise* during periods of voice interruption; otherwise, from the utter silence, the far-end listener also gets the disturbing impression that the circuit is being repeatedly broken and reestablished. Chapter 4 will study some methods for updating such parameters to maintain good echo suppressor performance. For now, however, we wish to turn to the echo canceller, a more reliable and also more modern alternative.

An echo canceller builds a signal model of the echo which slips through the hybrid. It then subtracts the model signal from the near-end speaker's outbound voice signal (Figure 2.12). How can this work? Note that the hybrid is an approximately linear device. If the far-end speaker's signal $x(n)$ is louder, then the echo gets proportionally louder. Also, since telephone circuit transmission characteristics do not change much over the time of a telephone call, the echo that the far-end signal produces at one moment is approximately the same as it produces at another moment. That is, the hybrid is very nearly translation-invariant. This provides an opportunity to invoke the convolution theorem for LTI systems.

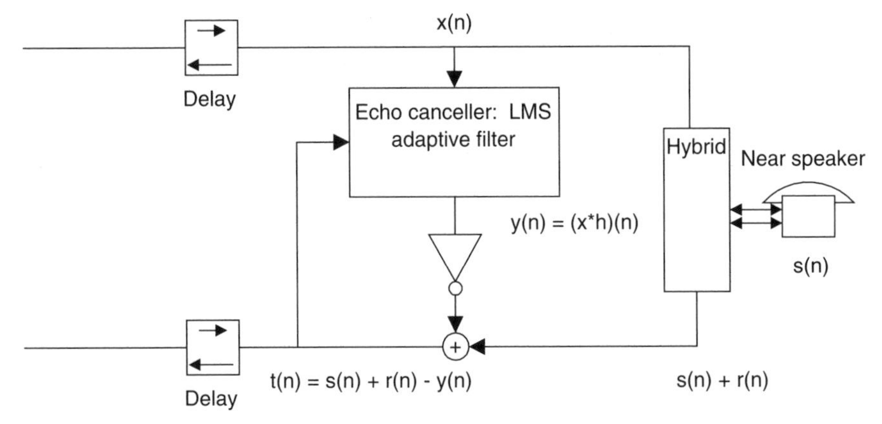

Fig. 2.12. An echo canceller holds a set of coefficients from which it can approximately model the echo signal $r(n)$. Assuming that the hybrid leakage is both linear and translation-invariant, we can apply the convolution theorem for LTI systems. The canceller stores the impulse response of the inbound far-end voice signal $h(n)$ and computes the echo as $y(n) = (x*h)(n)$. Subtracting $y(n)$ from the near-end speech and echo signal, $s(n) + r(n)$, gives the transmitted digital voice signal $t(n) = s(n) + r(n) - y(n)$.

Let H be the hybrid system, so that $r = Hx$ is LTI, where $x(n)$ is the inbound signal from the far-end speaker, and $r(n)$ is the echo through the hybrid. Suppose that the canceller stores the impulse response of the inbound far-end voice signal $h(n)$. Then it can approximate the echo as $y(n) = (x*h)(n)$, and this is the crux of the design. Subtracting $y(n)$ from the near-end speech and echo signal, $s(n) + r(n)$, gives the transmitted digital voice signal $t(n) = s(n) + r(n) - y(n)$ with echo largely removed. Digital signal processors can perform these convolution and subtraction steps in real time on digital telephony circuits [12]. Alternatively, echo cancellation can be implemented in application-specific integrated circuits [13].

An intriguing problem is how to establish the echo impulse response coefficients $h(k)$. The echo may change based on the connection and disconnection of equipment on the near-end circuit, including the two-wire drop to the subscriber. Thus, it is useful to allow $h(k)$ to change slowly over time. We can allow $h(k)$ to adapt so as to minimize the residual error signal, $e(n) = r(n) - y(n)$, that occurs when the near-end speaker is silent, $s(n) = 0$. Suppose that discrete values up to time instant n have been received and the coefficients must be adjusted so that the energy of the error $e(n)$ is a minimum. The energy of the error signal's last sample is $e^2(n)$. Viewing $e^2(n)$ as a function of the coefficients $h(k)$, the maximum decrease in the error is in the direction of the negative gradient, given by the vector with components

$$-\nabla e^2(n)_k = -\frac{\partial}{\partial h(k)}(e^2(n)) = 2r(n)x(n-k). \tag{2.35}$$

To smoothly converge on a good set of $h(k)$ values, it is best to use some proportion parameter η; thus, we adjust $h(k)$ by adding $2\eta r(n)x(n-k)$. This is an adaptive

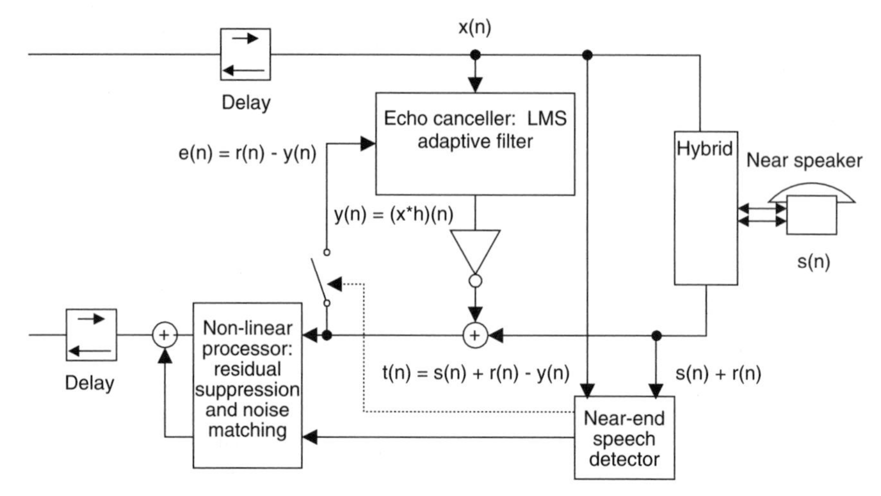

Fig. 2.13. An advanced echo canceller implements some of the design ideas from echo suppressors. The nonlinear processor mutes residual echo (the signal $t(n)$ after cancellation) when it exceeds a threshold. Noise matching supplies comfort noise to the outbound circuit so that the far-end listener does not suspect a dead line.

filtering algorithm known in the research literature as the *least mean squares* (*LMS*) *algorithm* [14].

An improvement in echo cancellation results from implementing some of the ideas from echo suppressor designs. A component commonly called the nonlinear processor (NLP) performs muting and noise insertion in the advanced canceller. Referring to Figure 2.13, note that the NLP mutes residual echo $t(n)$ when it falls below a small threshold and there is no near-end speech. NLP noise matching supplies comfort noise to the outbound circuit so the far-end listener does not suspect a dead line. Finally, the adaptation of the coefficients $h(n)$ should cease during periods of double-talk; otherwise, just as does genuine echo, the large magnitude near-end speech will be seen as echo by the canceller. The upshot is that the $h(n)$ coefficients will diverge from their proper settings, and a burst of echo will occur when the near speaker stops talking.

2.5 THE l^p SIGNAL SPACES

The remainder of the chapter develops the mathematical foundation for discrete signal processing theory. Vector calculus relies on such a foundation. But it is so intuitive, however, that many students do not even realize that the finite-dimensional vector spaces, $\mathbb{R}^n$ and $\mathbb{C}^n$, underlie multivariable calculus. The theories of differentiation, integration, limits, and series easily generalize to work with vectors instead of just scalars. Our problems are much more complicated now, since the objects of our theory—signals—are like vectors that are infinitely long and extend in two directions. We cannot, for example, carelessly generalize the dot product from finite dimensional vector spaces to signals; the resulting sum may not converge.

2.5.1 l^p Signals

What signal operations must our formal framework support? We have discovered that convolution operations play a pivotal role in the study of LTI systems. Since the summation in this important operation is infinite, we may ask under what circumstances one can compute the limit which it represents. It is clear that if the system is FIR, then the convolution sum is computable for any input signal. From a practical perspective, every signal encountered in engineering is finitely supported and cannot continue with nonzero values forever. However, it is reassuring to know that this is not a fundamental limitation of the theory we develop for signals. By looking a little deeper into the mathematics of signals and operations on them, we can in fact find classes of signals, not finitely supported, that allow us to compute convolutions for IIR systems.

Definition (l^p Signal Spaces). Let $p \geq 1$ be a real number. Then l^p is the set of all real-valued (or complex-valued) signals $x(n)$ such that

$$\sum_{n=-\infty}^{\infty} |x(n)|^p < \infty. \tag{2.36}$$

We sometimes call the l^p signals *p-summable*. If $x(n)$ is in l^p, then its *l^p-norm* is

$$\|x(n)\|_p = \left(\sum_{n=-\infty}^{\infty} |x(n)|^p \right)^{\frac{1}{p}}. \tag{2.37}$$

There is an l^p *distance measure* as well: $d_p(x, y) = \|x - y\|_p$. There is a special case of l^p signals, the bounded signals. A discrete signal $x(n)$ is *bounded* if there is a positive real number M_x such that for all n,

$$|x(n)| \leq M_x. \tag{2.38}$$

It is customary to denote the class of bounded signals as l^∞. (This notation allows an elegant formulation of an upcoming theorem.) Finally, we define the l^∞ *norm* of a signal to be its least upper bound:

$$\|x(n)\|_\infty = \min \{M: x(n) \leq M \text{ for all } n\}. \tag{2.39}$$

The l^p signal classes share a number of important properties that we need in order to do signal processing within them. This and the next section comprise a tutorial on the mathematical discipline known as *functional analysis* [15], which studies the properties of mathematical functions from a geometric standpoint. It is a generalization of finite-dimensional real and complex vector space theory. In fact, we already know some examples of l^p spaces.

Example (Finite-Energy Signals). The l^2 signals are precisely the finite energy, or square-summable discrete signals. The l^2 norm is also the square root of the energy of the signal. We will find that l^2 is a standout l^p space, since it supports an inner product relation that generalizes the dot product on signals. There are signal spaces for finite energy analog signals too (Chapter 3). Furthermore, we will find in Chapter 7 that it is possible to build a complete theory for the frequency analysis of l^2 signals [16]. Chapter 11 explains the recent theory of multiresolution analysis for finite energy signals—a recent advance in signal analysis and one kind of signal decomposition based on wavelets [17].

Example (Absolutely Summable Signals). The l^1 signal space is the set of absolutely summable signals. It too enjoys a complete signal frequency theory. Interestingly, for any p, $1 \le p \le 2$, there is an associated frequency analysis theory for the space of l^p signals [16]. Since signal processing and analysis uses mainly l^1 and l^2, however, we will not elaborate the Fourier transform theory for l^p, $1 < p < 2$.

2.5.2 Stable Systems

The notion of the l^p signal spaces applies readily to study of stable systems.

Definition (Stability). The system H is *stable* if $y = Hx$ is bounded whenever x is bounded. Another term for stable is *bounded input–bounded output* (*BIBO*).

Thus, the response of a stable system to an l^∞ input signal is still an l^∞ signal. The next theorem is useful for discovering whether or not an LTI system is stable. This result depends on an important property of the real number system: Every sequence that has a upper bound has a least upper bound [18].

Theorem (Stability Characterization). An LTI system H is stable if and only if its impulse response, $h = H\delta$, is in l^1 (absolutely summable).

Proof: First, suppose $h = H\delta$ is in l^1, and $x(n)$ is an l^∞ input signal with $|x(n)| \le M$. Let $y = Hx$. Then,

$$|y(n)| = \left|\sum_{k=-\infty}^{\infty} h(k)x(n-k)\right| \le \sum_{k=-\infty}^{\infty} |h(k)||x(n-k)| \le \sum_{k=-\infty}^{\infty} |h(k)|M$$
$$= M\sum_{k=-\infty}^{\infty} |h(k)| = M\|h\|_1 < \infty. \tag{2.40}$$

So $y(n)$ is bounded, proving that H is stable. Conversely, suppose the system H is stable, but the impulse response $h(n)$ is not in l^1. Now, the expression (2.37) for the l^1-norm of h is in fact a limit operation on a monotonically increasing sequence of sums. This sequence is either bounded or it is not. If it is bounded, then it would have a least upper bound, which must be the limit of the infinite sum (2.37). But we are assuming this limit does not exist. Thus, the sum must in fact be unbounded. And the only way that the limit cannot exist is if it diverges to infinity. That is,

$$\|h\|_1 = \sum_{k=-\infty}^{\infty} |h(k)| = \infty. \tag{2.41}$$

Let us consider the bounded input signal $x(n)$ defined

$$x(n) = \begin{cases} \dfrac{h(-n)}{|h(-n)|} & \text{if } h(-n) \neq 0, \\ 0 & \text{if } h(-n) = 0. \end{cases} \tag{2.42}$$

What is the response of our supposedly stable system to the signal $x(n)$? The convolution theorem for LTI systems tells us that we can find, for example, $y(0)$ to be

$$y(0) = \sum_{k=-\infty}^{\infty} h(k)x(0-k) = \sum_{k=-\infty}^{\infty} h(k)\frac{h(k)}{|h(k)|} = \sum_{k=-\infty}^{\infty} |h(k)| = \|h\|_1 = \infty. \tag{2.43}$$

This shows that y is unbounded, so that H is not stable, contradicting the assumption. Consequently, it must be the case that h is in l^1. ■

2.5.3 Toward Abstract Signal Spaces

One of the first things to verify is the closure of the l^p spaces under certain arithmetic or algebraic signal operations. This involves two steps:

(i) Verifying that the result of the operation is still a signal; that is, we can compute the value of the result at any time instant (if one of the signal values becomes infinite, then it is not a signal).

(ii) Verifying that the resulting signal is in the signal space of the operands.

2.5.3.1 Closure Properties.

The closure property for a signal operation shows that we can process l^p space signals through systems that are defined by the given operation. For example, the proof of the following closure proposition is easy and left as an exercise. What it shows is that an l^p signal can be fed into an amplifier system and yet it remains an l^p signal. Similarly, a delay or advance system preserves the l^p nature of its input signals.

Proposition (Closure of l^p Spaces). Let $x(n)$ be a signal in l^p, $1 \le p \le \infty$, let c be a real (or complex) number, and let k be an integer. Then:

(i) $cx(n)$ is in l^p and $\|cx\|_p = |c|\,\|x\|_p$.
(ii) $x(k-n)$ is in l^p and $\|x(k-n)\|_p = \|x(n)\|_p$.

Proof: Exercise. ■

Proposition (Closure of l^1 Spaces). Let $x(n)$ and $y(n)$ be l^1 signals. Then $w(n) = x(n) + y(n)$ is in l^1 also.

Proof: To show that $\|w(n)\|_1$ is finite, we only need to generalize the *triangle inequality* from arithmetic, $|x+y| \le |x| + |y|$, to infinite sums, and this is straightforward. ■

Proposition (Closure of l^∞ Spaces). Let $x(n)$ and $y(n)$ be l^∞ signals. Then $w(n) = x(n) + y(n)$ is in l^∞ also.

Proof: $\|w(n)\|_\infty$ is the least upper bound of $\{|w(n)|: n \text{ an integer}\}$. The arithmetic triangle inequality extends to infinite sets for upper bounds as well. ■

2.5.3.2 Vector Spaces. Before getting into the thick of abstract signal spaces, let us review the properties of a vector V space over the real numbers $\mathbb{R}$. There is a zero vector. Vectors may be added, and this addition is commutative. Vectors have additive inverses. Vectors can be multiplied by scalars, that is, elements of $\mathbb{R}$. Also, there are distributive and associative rules for scalar multiplication of vectors. One vector space is the real numbers over the real numbers: not very provocative. More interesting is the space with vectors taken from $\mathbb{R} \times \mathbb{R} = \{(a, b): a, b \in \mathbb{R}\}$. This set of real ordered pairs is called the *Cartesian product*, after Descartes,[2] but the concept was also pioneered by Fermat.[3] In general, we can take the Cartesian

[2]Forming a set of ordered pairs into a structure that combines algebraic and geometric concepts originates with Rene Descartes (1596–1650) and Pierre Fermat (1601–1665) [D. Struik, ed., *A Source Book in Mathematics,* 1200–1800, Princeton, NJ: Princeton University Press, 1986]. Descartes, among other things, invented the current notation for algebraic equations; developed coordinate geometry; inaugurated, in his *Meditations*, the epoch of modern philosophy; and was subject to a pointed critique by his pupil, Queen Christina of Sweden.

[3]Fermat is famous for notes he jotted down while perusing Diophantus's *Arithmetic*. One was a claim to the discovery of a proof for his Last Theorem: There are no nonzero whole number solutions of $x^n + y^n = z^n$ for $n > 1$. Historians of science, mathematicians, and the lay public for that matter have come to doubt that the Toulouse lawyer had a "marvelous demonstration," easily inserted but for the tight margins left by the printer. Three centuries after Fermat's teasing marginalia, A. Wiles of Princeton University followed an unexpected series of deep results from mathematicians around the world with his own six-year assault and produced a convincing proof of Fermat's Last Theorem [K. Devlin, *Mathematics: The Science of Patterns*, New York: Scientific American Library, 1994.]

product any positive number of times. The vector space exemplar is in fact the set of ordered n-tuples of real numbers $(a_1, a_2, \ldots, a_n) \in \mathbb{R}^n$. This space is called *Euclidean n-space.*[4]

Ideas of linear combination, span, linear independence, basis, and dimension are important; we will generalize these notions as we continue to lay our foundation for signal analysis. When a set of vectors $S = \{u_i: i \in I\}$ *spans* V, then each vector $v \in V$ is a *linear combination* of some of the u_i: there are real numbers a_i such that $v = a_1 u_1 + \cdots + a_N u_N$. If S is finite and spans V, then there is a *linearly independent* subset of S that spans V. In other words, there is some $B \subseteq S$, $B = \{b_i: 1 \le i \le N\}$, B spans V, and no nontrivial linear combination of the $\mathbf{b_i}$ is the zero vector: $\mathbf{0} = a_1\mathbf{u}_1 + \cdots + a_N\mathbf{u}_N$ implies $a_i = 0$. A spanning, linearly independent set is called a *basis* for V. If V has a finite basis, then every basis for V contains the same number of vectors—the *dimension* of V. A vector in Euclidean n-space has a *norm*, or *length*, which is the square root of the sum of the squares of its elements.

There is also an *inner product*, or *dot product*, for $\mathbb{R}^n$: $\langle a, b\rangle = \sum_{i=1}^{n} a_i b_i = a \cdot b$.

The dot product is a means of comparing two vectors via the relation $\langle a, b\rangle = \|a\|\,\|b\|\cos\theta$, where θ is the angle between the two vectors.

The vector space may also be defined over the complex numbers $\mathbb{C}$ in which casewe call it a *complex vector space*. The set of n-tuples of complex numbers $\mathbb{C}^n$ is called *unitary n-space*, an n-dimensional complex vector space. All of the vector space definitions and properties carry directly over from real to complex vector spaces, except for those associated with the inner product. We have to define $\langle a, b\rangle = \sum_{i=1}^{n} a_i \overline{b_i} = a \cdot b$, where $\overline{c}$ is the complex conjugate. A classic reference on vector spaces (and modern algebra all the way up to the unsolvability of quintic polynomials by radicals $\sqrt[n]{r}$, roots of order n) is Ref. 19.

2.5.3.3 Metric Spaces. A *metric space* is an abstract space that incorporates into its definition only the notion of a distance measure between elements.

Definition (Metric Space). Suppose that M is a set and d maps pairs of elements of M into the real numbers, $d: M \times M \to \mathbb{R}$. Then M is a *metric space* with *metric*, or *distance measure d*, if:

(i) $d(u, v) \ge 0$ for all u, v in M.

(ii) $d(u, v) = 0$ if and only if $u = v$.

(iii) $d(u, v) = d(v, u)$ for all u, v in M.

(iv) For any u, v, and w in M, $d(u, v) \le d(u, w) + d(w, v)$.

[4]The ancient Greek mathematician Euclid (ca. 300 B.C.) was (probably) educated at Plato's Academy in Athens, compiled the *Elements*, and founded a school at Alexandria, Egypt.

Items (i) and (ii) are known as the *positive-definiteness* conditions; (iii) is a *symmetry* condition; and (iv) is the *triangle inequality*, analogous to the distances along sides of a triangle.

Example (Euclidean, Unitary Metric Spaces). Clearly the Euclidean and unitary spaces are metric spaces. We can simply take the metric to be the Euclidean norm: $d(\mathbf{u}, \mathbf{v}) = \|\mathbf{u} - \mathbf{v}\|$.

Example (City Block Metric). But other distance measures are possible, too. For example, if we set $d((u_1, v_1), (u_2, v_2)) = |u_1 - u_2| + |v_1 - v_2|$, then this is a metric on $\mathbb{R}^2$, called the *city block distance*. This is easy to check and left as an exercise. Note that the same set of elements can underlie a different metric space, depending upon the particular distance measure chosen, as the city block distance shows. Thus, it is common to write a metric space as an ordered pair (M, d), where M is the set of elements of the space, and d is the distance measure.

The triangle inequality is a crucial property. It allows us to form groups of metric space elements, all centered around a single element. Thus, it is the mathematical foundation for the notion of the proximity of one element to another. It makes the notion of distance make sense: You can jump from u to v directly, or you can jump twice, once to w and thence to v. But since you end up in the same place, namely at element v, a double jump should not be a shorter overall trip.

We would like to compare two signals for similarity—for instance, to match one signal against another (Chapter 4). One way to do this is to subtract the signal values from each other and calculate the size or magnitude of their difference. We can't easily adapt our inner product and norm definitions from Euclidean and unitary spaces to signals because signals contain an infinite number of components. Intuitively, a finitely supported signal could have some norm like a finite-dimensional vector. But what about other signals? The inner product sum we are tempted to write for a discrete signal becomes an infinite sum if the signal is not finitely supported. When does this sum converge? What about bases and dimension? Can the span, basis, and dimension notions extend to encompass discrete signals too?

2.5.4 Normed Spaces

This section introduces the *normed space*, which combines the ideas of vector spaces and metric spaces.

Definition (Normed Space). A *normed space*, or *normed linear space*, is a vector space V with a norm $\|v\|$ such that for any u and v in V,

(i) $\|v\|$ is real and $\|v\| \geq 0$.

(ii) $\|v\| = 0$ if and only if $v = \mathbf{0}$, the zero vector in V.

(iii) $\|av\| = |a|\ \|v\|$, for all real numbers a.
(iv) $\|u + v\| \le \|u\| + \|v\|$.

If $S \subseteq V$ is a vector subspace of V, then we may define a norm on S by just taking the norm of V restricted to S. Then S becomes a normed space too. S is called a *normed subspace* of V.

We adopt Euclidean and unitary vector spaces as our inspiration. The goal is to take the abstract properties we need for signals from them and define a special type of vector space that has at least a norm. Now, we may not be able to make the class of all signals into a normed space; there does not appear to be any sensible way to define a norm for a general signal with infinite support. One might try to define a norm on a restricted set of signals in the same way as we define the norm of a vector in $\mathbb{R}^3$. The l^2 signals have such a norm defined for them. The problem is that we do not yet know whether the l^2 signals form a vector space. In particular, we must show that a sum of l^2 signals is still an l^2 signal (additive closure). Now, we have already shown that the l^1 signals with the norm $\|x\|_1$ do form a normed space. Thus, our strategy is to work out the specific properties we need for signal theory, specify an abstract space with these traits, and then discover those concrete classes of signals that fulfill our axiom system's requirements. This strategy has proven quite successful in applied mathematics.[5] The discipline of *functional analysis* provides the tools we need [15, 20–23]. There is a complete history as well [24]. We start with a lemma about conjugate exponents [15].

Definition (Conjugate Exponents). Let $p > 1$. If $p^{-1} + q^{-1} = 1$, then q is a *conjugate exponent* of p. For $p = 1$, the conjugate exponent of p is $q = \infty$.

Let us collect a few simple facts about conjugate exponents.

Proposition (Conjugate Exponent Properties). Let p and q be conjugate exponents. Then

(i) $(p + q)/pq = 1$.
(ii) $pq = p + q$.
(iii) $(p - 1)(q - 1) = 1$.
(iv) $(p - 1)^{-1} = q - 1$.
(v) If $u = tp^{-1}$, then $t = uq^{-1}$.

[5]Several mathematicians—among them Wiener, Hahn, and Banach—simultaneously and independently worked out the concept and properties for a normed space in the 1920s. The discipline of functional analysis grew quickly. It incorporated the previous results of Hölder and Minkowski, found applications in quantum mechanics, helped unify the study of differential equations, and, within a decade, was the subject of general treatises and reviews.

Lemma (Conjugate Exponents). Let $a > 0$ and $b > 0$ be real numbers. Let p and q be conjugate exponents. Then $ab \le p^{-1}a^p + q^{-1}b^q$.

Proof: The trick is to see that the inequality statement of the lemma reduces to a geometric argument about the areas of regions bounded by the curve $u = tp^{-1}$. Note that

$$\frac{a^p}{p} = \int_0^a t^{p-1}\,dt, \tag{2.44}$$

$$\frac{b^q}{q} = \int_0^b u^{q-1}\,du. \tag{2.45}$$

Definite integrals (2.44) and (2.45) are areas bounded by the curve and the t- and u-axis, respectively (Figure 2.14). The sum of these areas is not smaller than the area of the rectangle defined by (0,0) and (a,b) in any case. ■

Our first nontrivial closure result on l^p spaces, Hölder's[6] inequality [15], shows that it is possible to form the product of signals from conjugate l^p spaces.

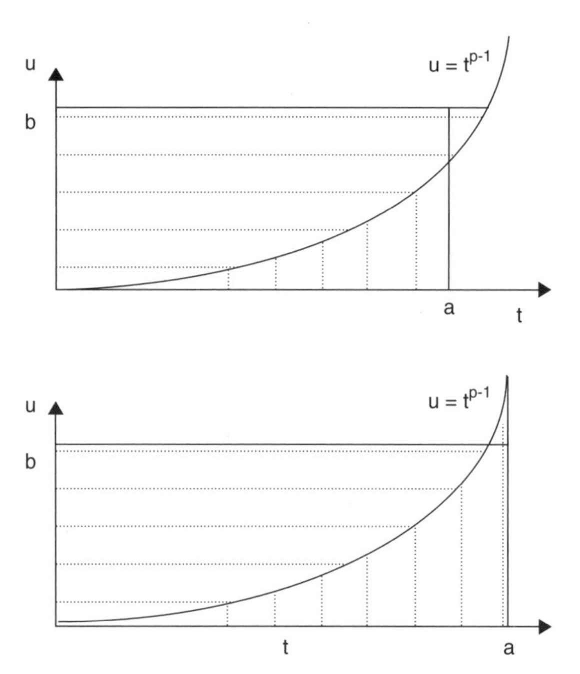

Fig. 2.14. There are two cases: Either point (a, b) is below the curve (bottom) or above the curve (top). In either case the area of the rectangle determined by the origin and (a, b) has a smaller area than the total area of regions bounded by the curve and the t- and u-axes.

[6]German mathematician Otto Ludwig Hölder (1859–1937) discovered the relation in 1884.

Theorem (Hölder's Inequality). Let $x(n)$ be in l^p and let $h(n)$ in l^q, where p and q are conjugate exponents. Then $y(n) = x(n)h(n)$ is in l^1 and $\|y\|_1 \le \|x\|_p \|h\|_q$.

Proof: Since the inequality clearly holds if $\|x\|_p = 0$ or $\|h\|_q = 0$, we assume that these signals are not identically 0. Next, let $x'(n) = \frac{x(n)}{\|x\|_p}$ and $h'(n) = \frac{h(n)}{\|h\|_q}$, and set $y'(n) = x'(n)h'(n)$. By the lemma,

$$|x'(n)||h'(n)| \le \frac{|x'(n)|^p}{p} + \frac{|h(n)|^q}{q}. \tag{2.46}$$

Putting (2.46) into the expression for $\|y'\|_1$, we find that

$$\begin{aligned} \|y'\|_1 &= \sum_{n=-\infty}^{\infty} |x'(n)||h'(n)| \le \sum_{n=-\infty}^{\infty} \frac{|x'(n)|^p}{p} + \sum_{n=-\infty}^{\infty} \frac{|h'(n)|^q}{q} \\ &= \frac{1}{p} \sum_{n=-\infty}^{\infty} \frac{|x(n)|^p}{\|x\|_p} + \frac{1}{q} \sum_{n=-\infty}^{\infty} \frac{|h(n)|^q}{\|h\|_q} = \frac{1}{p} + \frac{1}{q} = 1. \end{aligned} \tag{2.47}$$

Hence,

$$\|y'\|_1 = \frac{\|xh\|_1}{\|x\|_p \|h\|_q} = \frac{\|y\|_1}{\|x\|_p \|h\|_q} \le 1 \tag{2.48}$$

and the Hölder inequality follows. ■

Remarks. From the Hölder inequality, it is easy to show that there are IIR systems that have large classes of infinitely supported signals in their domains. This shows that our theory of signals can cope with more than FIR systems. In particular, the space of finite-energy signals is contained within the domain of any LTI system with a finite-energy impulse response.

Theorem (Domain of l^q Impulse Response Systems). Suppose H is an LTI system and $h = H\delta$ is in l^q. Then any $x(n)$ in l^p, where p and q are conjugate exponents, is in *Domain*(H).

Proof: By the convolution theorem for LTI systems, the response of H to input x is $y = x * h$. Thus,

$$|y(n)| = \left| \sum_{k=-\infty}^{\infty} x(k)h(n-k) \right| \le \sum_{k=-\infty}^{\infty} |x(k)||h(n-k)|. \tag{2.49}$$

Since the q-summable spaces are closed under translation and reflection, $h(n-k)$ is in l^q for all n. So the Hölder inequality implies that the product signal $w_n(k) = x(k)h(n-k)$ is in l^1. Computing $\|w_n\|_1$, we see that it is just the right-hand side of (2.49). Therefore, $|y(n)| \le \|w_n\|_1$ for all n. The convolution sum (2.49) converges. ■

Corollary (Cauchy–Schwarz[7] Inequality). Suppose $x(n)$ and $h(n)$ are in l^2. Then the product $y(n) = x(n)h(n)$ is in l^1 and

$$\|y\|_1 = \sum_{n=-\infty}^{\infty} |x(n)||h(n)| \le \sqrt{\sum_{n=-\infty}^{\infty} |x(n)|^2}\sqrt{\sum_{n=-\infty}^{\infty} |h(n)|^2} = \|x\|_2\|h\|_2. \quad (2.50)$$

Proof: Take $p = q = 2$ in the Hölder inequality. ∎

Theorem (Minkowski[8] Inequality). If x and y are in l^p, then $w = x + y$ is in l^p, and $\|w\|_p \le \|x\|_p + \|y\|_p$.

Proof: Assume $p > 1$. Then,

$$|w(n)|^p = |w(n)|^{p-1}|x(n) + y(n)| \le |w(n)|^{p-1}(|x(n)| + |y(n)|). \quad (2.51)$$

Summing over all n gives

$$\sum_{n=-\infty}^{\infty} |w(n)|^p \le \sum_{n=-\infty}^{\infty} |x(n)||w(n)|^{p-1} + \sum_{n=-\infty}^{\infty} |y(n)||w(n)|^{p-1}. \quad (2.52)$$

The Hölder inequality applies to the first sum on the right-hand side of (2.52) with $q = p/(p-1)$ as follows:

$$\sum_{n=-\infty}^{\infty} |x(n)||w(n)|^{p-1} \le \left(\sum_{n=-\infty}^{\infty} |x(n)|^p\right)^{\frac{1}{p}}\left(\sum_{n=-\infty}^{\infty} [|w(n)|^{p-1}]^q\right)^{\frac{1}{q}}$$
$$= \left(\sum_{n=-\infty}^{\infty} |x(n)|^p\right)^{\frac{1}{p}}\left(\sum_{n=-\infty}^{\infty} |w(n)|^p\right)^{\frac{1}{q}}. \quad (2.53)$$

Similarly, for the second sum of the right-hand side of (2.52), we have

$$\sum_{n=-\infty}^{\infty} |y(n)||w(n)|^{p-1} \le \left(\sum_{n=-\infty}^{\infty} |y(n)|^p\right)^{\frac{1}{p}}\left(\sum_{n=-\infty}^{\infty} |w(n)|^p\right)^{\frac{1}{q}}. \quad (2.54)$$

Putting (2.52)–(2.54) together gives

$$\sum_{n=-\infty}^{\infty} |w(n)|^p \le \left[\left(\sum_{n=-\infty}^{\infty} |x(n)|^p\right)^{\frac{1}{p}} + \left(\sum_{n=-\infty}^{\infty} |y(n)|^p\right)^{\frac{1}{p}}\right]\left(\sum_{n=-\infty}^{\infty} |w(n)|^p\right)^{\frac{1}{q}}. \quad (2.55)$$

[7]After French mathematician Augustin-Louis Cauchy (1789–1857) and German mathematician Hermann Amandus Schwarz (1843–1921).

[8]Although he was born in what is now Lithuania, Hermann Minkowski (1864–1909) spent his academic career at German universities. He studied physics as well as pure mathematics. He was one of the first to propose a space-time continuum for relativity theory.

Finally, (2.55) entails

$$\frac{\sum\limits_{n=-\infty}^{\infty}|w(n)|^p}{\left(\sum\limits_{n=-\infty}^{\infty}|w(n)|^p\right)^{\frac{1}{q}}} = \left(\sum_{n=-\infty}^{\infty}|w(n)|^p\right)^{1-\frac{1}{q}} = \left(\sum_{n=-\infty}^{\infty}|w(n)|^p\right)^{\frac{1}{p}}$$

$$= \|w\|_p \leq \left(\sum_{n=-\infty}^{\infty}|x(n)|^p\right)^{\frac{1}{p}} + \left(\sum_{n=-\infty}^{\infty}|y(n)|^p\right)^{\frac{1}{p}} = \|x\|_p + \|y\|_p, \tag{2.56}$$

completing the proof. ■

Now, it's easy to check that the norms associated with the l^p signal spaces, $\|x\|_p$, are in fact norms under the above abstract definition of a normed linear space:

Theorem (l^p Spaces Characterization). The l^p spaces are normed spaces for $1 \leq p \leq \infty$. ■

So far we find that the l^p spaces support several needed signal operations: addition, scalar multiplication, and convolution. Sometimes the result is not in the same class as the operands, but it is still in another related class; this is not ideal, but at least we can work with the result. Now let us try to incorporate the idea of signal convergence—limits of sequences of signals—into our formal signal theory.

2.5.5 Banach Spaces

In signal processing and analysis, we often consider sequences of signals $\{x_k(n)\}$. For example, the sequence could be a series of transformations of a source signal. It is of interest to know whether the sequence of signals converges to another limiting signal. In particular, we are concerned with the convergence of sequences of signals in l^p spaces, since we have already shown them to obey special closure properties, and they have a close connection with such signal processing ideas as stability. We have also shown that the l^p signal spaces are normed spaces. We cannot expect every sequence of signals to converge; after all, not every sequence of real numbers converges. However, we recall from calculus the Cauchy condition for convergence: A sequence of real (or complex) numbers $\{a_k\}$ is a *Cauchy sequence* if for every $\varepsilon > 0$, there is an $N > 0$ such that for k, $l > N$, $|a_k - a_l| < \varepsilon$. Informally, this means that if we wait long enough, the numbers in the sequence will remain arbitrarily close together. An essential property of the real line is that every Cauchy sequence of converges to a limit [18]. If we have an analogous property for signals in a normed space, then we call the signal space a Banach[9] space [15].

Definition (Banach Space). A Banach space B is a normed space that is *complete*. That is, any sequence $\{x_k(n)\}$ of signals in B that is a Cauchy sequence converges in B to a signal $x(n)$ also in B. Note that $\{x_k(n)\}$ is a *Cauchy sequence* if for every $\varepsilon > 0$, there is an $N > 0$ so that whenever $k, l > N$, we have $\|x_k - x_l\| < \varepsilon$. If $S \subseteq B$ is a complete normed subspace of B, then we call S a *Banach subspace* of B.

Theorem (Completeness of l^p Spaces). The l^p spaces are complete, $1 \le p \le \infty$.

Proof: The exercises sketch the proofs. ■

Banach spaces have historically proven difficult to analyze, evidently due to the lack of an inner product relation. Breakthrough research has of late cleared up some of the mysteries of these abstract spaces and revealed surprising structure. The area remains one of intense mathematical research activity. For signal theory, we need more analytical power than what Banach spaces furnish. In particular, we need some theoretical framework for establishing the similarity or dissimilarity of two signals—we need to augment our abstract signal theory space with an inner product relation.

Examples

(i) An example of a Banach subspace is l^1, which is a subspace of all l^p $1 \le p \le \infty$.

(ii) The set of signals in l^p that are zero on a nonempty subset $Y \subseteq \mathbb{Z}$ is easily shown to be a Banach space. This is a proper subspace of l^p for all p.

(iii) The normed subspace of l^p that consists of all finitely supported p-summable signals is not a Banach subspace. There is a sequence of finitely supported signals that is a Cauchy sequence (and therefore converges inside l^p) but does not converge to a finitely supported signal.

Recall from calculus the ideas of open and closed subsets of the real line. A set $S \subseteq \mathbb{R}$ is *open* if for every point p in S, there is $\varepsilon > 0$ such that $Ball(p, \varepsilon) = \{x \in \mathbb{R}: |x - p| < \varepsilon\} \subseteq S$. That is, every point p of S is contained in an open ball that is contained in S. A set $S \subseteq \mathbb{R}$ is *closed* if its complement is open. Let V be a normed space. Then a set $S \subseteq V$ is *open* if for every point p in S, there is $\varepsilon > 0$ such that $Ball(p, \varepsilon) = \{x \in V: \|x - p\| < \varepsilon\} \subseteq S$. That is, every point p of S is contained in an open ball that is contained in S.

Theorem (Banach Subspace Characterization). Let B be a Banach space and S a normed subspace of B. Then S is a Banach subspace if and only if S is closed in B.

[9]The Polish mathematician S. Banach (1892–1945) developed so much of the initial theory of complete normed spaces that the structure is named after him. Banach published one of the first texts on functional analysis in the early 1930s.

Proof: First suppose that S is a Banach subspace. We need to show that S is closed in B. Let $p \in B$, and $p \notin S$. We claim that there is an $\varepsilon > 0$ such that the open ball $Ball(p, \varepsilon) \cap S = \varnothing$. If not, then for any integer $n > 0$, there is a point $s_n \in S$ that is within the ball $Ball(p, 1/n)$. The sequence $\{s_n: n > 0\}$ is a Cauchy sequence in S. Since S is Banach, this sequence converges to $s \in S$. However, this means we must have $s = p$, showing that $p \in S$, a contradiction.

Conversely, suppose that S is closed and $\{s_n: n > 0\}$ is a Cauchy sequence in S. We need to show that $\{s_n\}$ converges to an element in S. The sequence is still a Cauchy sequence in all of B; the sequence converges to $p \in B$. We claim $p \in S$. If not, then since p is in the complement of S and S is closed, there must be an $\varepsilon > 0$ and an open ball $Ball(p, \varepsilon) \subseteq S'$. This contradicts the fact that $d(s_n, s) \to 0$, proving the claim and the theorem. ■

2.6 INNER PRODUCT SPACES

Inner product spaces have a binary operator for measuring the similarity of two elements. Remember that in linear algebra and vector calculus over the normed spaces $\mathbb{R}^n$, the inner (or dot) product operation has a distinct geometric interpretation. We use it to find the angle between two vectors in $\mathbb{R}^n$: $\langle u, v\rangle = \|u\|\|v\|\cos(\theta)$, where θ is the angle between the vectors. From the inner product, we define the notion of orthogonality and of an orthogonal set of basis elements. Orthogonal bases are important because they furnish a very easy set of computations for decomposing general elements of the space.

2.6.1 Definitions and Examples

Once again, we abstract the desired properties from the Euclidean and unitary spaces.

Definition (Inner Product Space). An *inner product space I* is a vector space with an inner product defined on it. The inner product, written with brackets notation $\langle x, y\rangle$, can be real- or complex-valued, according to whether I is a real or complex vector space, respectively. The inner product satisfies these five rules:

(i) $0 \le \langle x, x\rangle$ for all $x \in I$.

(ii) For all $x \in I, 0 = \langle x, x\rangle$ if and only if $x = 0$ (that is, x is the zero vector).

(iii) For all $x, y \in I$, $\langle x, y\rangle = \overline{\langle y, x\rangle}$, where $\bar{c}$ is the complex conjugate of c.

(iv) For all $c \in \mathbb{C}$ (or just $\mathbb{R}$, if I is a real inner product space) and all $x, y \in I$, $\langle cx, y\rangle = c\langle x, y\rangle$.

(v) For all $w, x, y \in I$, $\langle w + x, y\rangle = \langle w, y\rangle + \langle x, y\rangle$.

If $S \subseteq I$, then S becomes an inner product space by taking its inner product to be the inner product of I restricted to S. We call S an *inner product subspace* of I.

Remarks. Note that the inner product is linear in the first component, but, when the inner product spaces are complex, it is *conjugate linear* in the second component. When the definition speaks of "vectors," these are understood to be abstract elements; they could, for example, be infinitely long or functions from the integers to the real numbers (discrete signals).

Examples (Inner Product Spaces)

(i) The normed space $\mathbb{R}^n$, Euclidean n-space, with inner product defined $\langle x, y\rangle = \sum_{k=1}^{n} x_k y_k$ is a real inner product space. This space is familiar from linear algebra and vector calculus.

(ii) The normed space $\mathbb{C}^n$ with inner product defined $\langle x, y\rangle = \sum_{k=1}^{n} x_k \overline{y_k}$ is a complex inner product space. (We take complex conjugates in the definition so that we can define $\|x\|$ on $\mathbb{C}^n$ from the inner product.)

(iii) The signal space l^2 is an inner product space when we define its inner product

$$\langle x, y\rangle = \sum_{n=-\infty}^{\infty} x(n)\overline{y(n)}. \tag{2.57}$$

Remarks. Notice that the Cauchy–Schwarz result (2.50) implies convergence of (2.57). Furthermore, since we know that the l^p spaces are Banach spaces and therefore complete, l^2 is our first example of a *Hilbert space*[10] (Figure 2.15).

The ideas underlying Banach and Hilbert spaces are central to understanding the latest developments in signal analysis: time-frequency transforms, time-scale transforms, and frames (Chapters 10–12).

[10]D. Hilbert studied the special case of l^2 around 1900. Later, in a landmark 1910 paper, F. Riesz generalized the concept and defined the l^p spaces we know today. The Cauchy–Schwarz inequality was known for discrete finite sums (i.e., discrete signals with finite support) by A. Cauchy in the early nineteenth century. H. Schwarz proved the analagous result for continuous signals (we see this in the next chapter, when the summations become integrals) and used it well in a prominent 1885 paper on minimal surfaces. V. Buniakowski had in fact already discovered Schwarz's integral form of the inequality in 1859, but his result drew little attention. O. Hölder published a paper containing his inequality in 1889. H. Minkowski disclosed the inequality that now bears his name as late as 1896; it was, however, restricted to finite sums. Riesz's 1910 paper would extend both the Hölder and Minkowski results to analog signals, for which integrals replace the discrete sums.

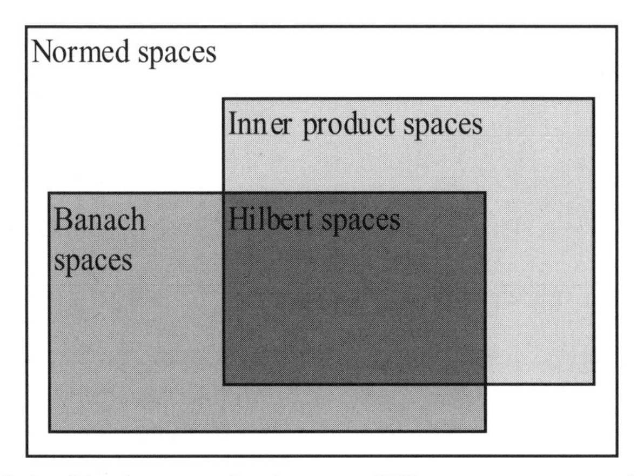

Fig. 2.15. Relationships between signal spaces. Hilbert spaces are precisely the inner prodct spaces that are also Banach spaces.

2.6.2 Norm and Metric

An inner product space I has a natural *norm* associated with its inner product: $\|x\| = \sqrt{\langle x, x\rangle}$. With this definition, the triangle inequality follows from the Cauchy–Schwarz relation, and the other normed space properties follow easily. There is also a natural distance measure: $d(x,y) = \|x - y\|$.

Theorem (Cauchy–Schwarz Inequality for Inner Product Spaces). Let I be an inner product space and $u,v \in I$. Then

$$|\langle u, v\rangle| \leq \|u\|\,\|v\| \tag{2.58}$$

Furthermore, $|\langle u, v\rangle| = \|u\|\,\|v\|$ if and only if u and v are linearly dependent.

Proof: First, suppose that u and v are linearly dependent. If, say, $v = 0$, then both sides of (2.58) are zero. Also, if $u = cv$ for some scalar (real or complex number) c, then $|\langle u, v\rangle| = c\|v\|^2 = \|u\|\,\|v\|$, proving (2.58) with equality. Next, let us show that there is strict inequality in (2.58) if u and v are linearly independent. We resort to a standard trick. By linear independence, $u + cv \neq 0$ for any scalar c. Hence,

$$0 < \|u - cv\|^2 = \langle u, u\rangle - \bar{c}\langle u, v\rangle - c\langle v, u\rangle - |c|^2\langle v, v\rangle \tag{2.59}$$

for any c. In particular, by taking $c = \frac{\overline{\langle v, u\rangle}}{\langle v, v\rangle}$ in (2.59),

$$0 < \|u\|^2 - \frac{|\langle u, v\rangle|^2}{\|v\|^2}. \tag{2.60}$$

The Cauchy–Schwarz inequality follows. ■

We can now show that with $\|x\| = \sqrt{\langle x, x\rangle}$, I is a normed space. All of the properties of a normed space are simple, except for the triangle inequality. By expanding the inner products in $\|u + v\|^2$, it turns out that

$$\|u+v\|^2 = \|u\|^2 + \|v\|^2 + 2Real(\langle u, v\rangle) \le \|u\|^2 + \|v\|^2 + 2|\langle u, v\rangle|, \tag{2.61}$$

where $Real(c)$ is the real part of $c \in \mathbb{C}$. Applying Cauchy–Schwarz to (2.61),

$$\|u+v\|^2 \le \|u\|^2 + \|v\|^2 + 2\|u\|\|v\| = (\|u\| + \|v\|)^2, \tag{2.62}$$

and we have shown the triangle inequality for the norm.

The natural distance measure $d(x,y) = \|x - y\|$ is indeed a metric. That is, for all $x, y \in I$ we have $d(x, y) \ge 0$. Also, $d(x, y) = 0$ if and only if $x = y$. Symmetry exists: $d(x, y) = d(y, x)$. And the triangle inequality holds: Given $z \in I$, $d(x, y) \le d(x, z) + d(z, y)$. We use the distance metric to define convergent sequences in I: If $\{x_n\}$ is a sequence in I, and $d(x_n, x) \to 0$ as $n \to \infty$, then we say $x_n \to x$. The inner product in I is continuous (exercise), another corollary of the Cauchy–Schwarz inequality.

Proposition (Parallelogram Rule). Let I be an inner product space, let x and y be elements of I, and let $\|x\| = \sqrt{\langle x, x\rangle}$ be the the inner product space norm. Then $\|x + y\|^2 + \|x - y\|^2 = 2\|x\|^2 + 2\|y\|^2$.

Proof: Expanding the norms in terms of their definition by the inner product gives

$$\|x+y\|^2 = \|x\|^2 + \langle x, y\rangle + \langle y, x\rangle + \|y\|^2, \tag{2.63}$$

$$\|x-y\|^2 = \|x\|^2 - \langle x, y\rangle - \langle y, x\rangle + \|y\|^2. \tag{2.64}$$

Adding (2.63) and (2.64) together gives the rule. ∎

The reason for the rule's name lies in a nice geometric interpretation (Figure 2.16). The parallelogram rule is an abstract signal space equivalent of an elementary

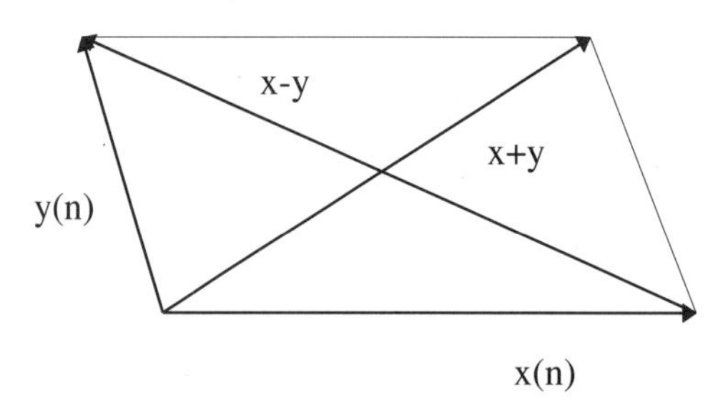

Fig. 2.16. Parallelogram rule. This simple geometric relationship between signals in an inner product space imposes a severe constraint on the l^p spaces. Only subspaces of l^2 support an inner product relation for our signal theory.

property from plane geometry. This shows the formal and conceptual power of the function space approach to signal theory. We can derive the algebraic rule from blind manipulations, and we can also resort to geometric constructs for insights into the relationships between the abstractions.

There is a negative consequence of the parallelogram rule: Except for l^2, none of the l^p spaces support an inner product definition that is related to the norm, $\| \|_p$. To verify this, consider the signals $x = [\underline{0}, 1]$, $y = [\underline{1}, 0]$, and compute $\|x\|_p^2 = \|y\|_p^2 = 1$ and $\|x+y\|_p^2 = \|x-y\|_p^2 = 2^{2/p}$. By the rule, we must have $2 = 2^{2/p}$, so that $p = 2$. The consequence is that although we have developed a certain body of theory for p-summable signals, found closure rules for basic operations like signal summation, amplification, and convolution, and shown that Cauchy sequences of p-summable signals converge, we cannot have an inner product relation for comparing signals unless $p = 2$. One might worry that l^2 is a signal analysis monoculture—it lacks the diversity of examples, counterexamples, classes, and conditions that we need to run and maintain the signal analysis economy. The good news is that the square-summable signal space is quite rich and that we can find in its recesses the exceptional signals we happen to need.

2.6.3 Orthogonality

The inner product relation is the key to decomposing a signal into a set of simpler components and for characterizing subspaces of signals. The pivotal concept for inner product signal spaces, as with simple Euclidean and unitary vector spaces, is orthogonality.

2.6.3.1 Definition and Examples.

If we consider inner product spaces of discrete signals, then the inner product is a measure of the similarity of two signals. Signals are similar to themselves, and so nonzero signals have a positive inner product with themselves: $\langle u, u \rangle > 0$. Two signals that are not at all alike, following this intuition, have zero inner product with each other: $\langle u, u \rangle = 0$.

Definition (Orthogonality). In an inner product space I, two elements—u and v— are *orthogonal* if $\langle u, v \rangle = 0$; in this case, we write $u \perp v$. If u is orthogonal to every element of a set S, we write $u \perp S$. The set of all u in I such that for all s in S, we have $u \perp s$ is called the *orthogonal complement* of S; it is written $S^{\perp}$. A set S of nonzero elements of I is an *orthogonal* set if $u \perp v$ for u, v in S, $u \neq v$. If S is an orthogonal set such that $\|u\| = 1$ for all u in S, then S is an *orthonormal* set.

Example (Unit Impulses). The shifted unit impulse signals $S = \{\delta(n-k)\}$ form an orthonormal set in the inner product space l^2.

Notice that we do not say that the unit impulse signals are a *basis* for the square-summable signals. In finite-dimensional vector spaces, bases span the entire space; thus, every vector is a linear combination of a finite number of elements from the basis set. This is not true for general l^2 signals and the shifted unit impulses. No finite set of shifted unit impulses can span the whole Hilbert space. Moreover, some signals—those without finite support—cannot be a linear combination of a finite number of shifted unit impulses. While we were able to smoothly migrate most of the ideas from finite-dimensional vector spaces to inner product spaces, we now find that the concept of a basis did not fare so well. The problem is that general inner product spaces (among them our remaining l^p space, l^2) may have an "infinite" dimension.

We must loosen up the old idea of basis considerably. Let us reflect for a moment on the shifted unit impulses and a signal $x \in l^2$. Since $\|x\|_2 < \infty$, given $\varepsilon > 0$, we can find $N > 0$, so that the energy of the outer fringes of the signal—$x(n)$ values for $n > N$—have energy less than ε. This means that some linear combination of shifted impulses comes arbitrarily close to x in l^2 norm. Here then is where we loosen the basis concept so that it works for inner product spaces. We allow that a linear combination come arbitrarily close to signal x. This is the key idea of *completeness*: A set of elements S is *complete* or *total* if every element of the space I is the limit of a sequence of elements from the linear span of S. We also say that S is *dense* in I. The term "complete" for this idea is terrible, but standard. In a space with a distance measure, "complete" means that every Cauchy sequence converges to an element in the space. For example, the real number line is complete this sense. Now we use the same term for something different, namely the existence of a Cauchy sequence of linear combinations. The better term is "total" [15]. The only good advice is to pay close attention to the context; if the discussion is about bases, then "complete" probably signifies this new sense. So we could stipulate that a *basis* for I is a linearly independent set of signals whose linear span is dense in I.

Should we also assert that the shifted unit samples are a basis for the square-integrable signals? Notice that we are talking here about convergence in a general inner product space. Some inner product spaces are not complete. So, we will postpone the proper definition of completeness and the generalization of the basis concept that we need until we develop more inner product space theory.

Example (Rademacher[11] Signals). Consider the signals $e_0 = [\underline{1}, -1]$, $e_1 = [\underline{1}, 1, -1, -1]$, $e_2 = [\underline{1}, 1, 1, 1, -1, -1\ -1, -1]$, and so on. These signals are orthogonal. Notice that we can shift e_0 by multiples of 2, and the result is still orthogonal to all of the e_i. And we can delay e_1 by $4k$ for some integer k, and it is still orthogonal to all of the e_i and all of the shifted versions of e_0. Let us continue to use $e_{i,k}$ to denote the signal e_i delayed by amount k. Then the set $\{e_{i,k}$: *i is a natural number and* $k = m^{2i+1}$, *for some m*$\}$ is orthogonal. Notice that the sum of signal values of a linear combination

[11]Although not Jewish, but rather a pacifist, German number theorist and analyst Hans Rademacher (1892–1969) was forced from his professorship at Breslau in 1934 by the Nazi regime and took refuge in the United States.

of Rademacher signals is always zero. Hence, the Rademacher signals are not complete. The signal $\delta(n)$, for instance, is not in the closure of the linear span of the Rademacher signals. Note too that the Rademacher signals are not an orthonormal set. But we can orthonormalize them by dividing each $e_{i,k}$ by its norm, 2^{i+1}.

We will use the notion of orthogonality extensively. Note first of all that this formal definition of orthogonality conforms to our geometric intuition. The Pythagoras[12] relation for a right triangle states that the square of the length of the hypoteneuse is equal to the sum of the squares of the other two sides. It is easy to show that $u \perp v$ in an inner product space entails $\|u\|^2 + \|v\|^2 = \|u+v\|^2$. This also generalizes to an arbitrary finite orthogonal set. Other familiar inner product properties from the realm of vector spaces reappear as well:

(i) If $x \perp u_i$ for $i = 1, ..., n$, then x is orthogonal to any linear combination of the u_i.
(ii) If $x \perp u_i$ for i in the natural numbers, and $u_i \to u$, then $x \perp u$.
(iii) If $x \perp S$ (where S is an orthogonal set), u_i is in S, and $u_i \to u$, then $x \perp u$.

Orthonormal sets can be found that span the inner product signal spaces I that we commonly use to model our discrete (and later, analog) signals. The idea of spanning an inner product space generalizes the same notion for finite-dimensional vector spaces. We are often interested in decomposing a signal $x(n)$ into linear combination of simpler signals $\{u_0, u_1, ...\}$. That is, we seek scalars c_k such that $x(n) = \sum c_k u_k(n)$. If the family $\{c_k\}$ is finite, we say that $x(n)$ is in the *linear span* of $\{u_0, u_1, ...\}$. Orthonormal sets are handy for the decomposition because the scalars c_k are particularly easy to find. If $x(n) = \sum c_k u_k(n)$ and the u_k are orthonormal, then $c_k = \langle x, u_k \rangle$. For example, the shifted unit impulse signals $S = \{\delta(n - k)\}$ form an orthonormal set in the inner product space l^2. Decomposing $x(n)$ on the orthonormal shifted unit impulses is trivial: $c_k = x(k)$. The unit impulses are not a very informative decomposition of a discrete signal, however, because they do not provide any more information about the signal than its values contain at time instants. The problem of signal decomposition becomes much more interesting and useful when the composition elements become complicated. Each u_k then encapsulates more elaborate information about $x(n)$ within the decomposition. We may also interpret $|c_k|$ as a measure of how much alike are $x(n)$ and $u_k(n)$.

Example (Discrete Fourier Transform). To illustrate a nontrivial orthonormal decomposition, let $N > 0$ and consider the windowed exponential signals

$$u(n) = \frac{\exp\left(2\pi jk\frac{n}{N}\right)}{\sqrt{N}}[u(n) - u(n-N)], \tag{2.65}$$

[12]This property of right triangles was known to the Babylonians long before the mystic Greek number theorist Pythagoras (fl. ca. 510 B.C.) [B. L. van der Waerden, *Science Awakening*, translator. A. Dresden, Groningen, Holland: Nordhoff, 1954].

where $u(n)$ is the unit step signal. They form an orthonormal set on $[0, N-1]$. Suppose $x(n)$ also has support in $[0, N-1]$. We will show later that $x(n)$ is in the linear span of $\{u_k(n): 0 \le k \le N-1\}$ (Chapter 7). Since $x, u_1, \ldots, u_{N-1}$ are in l^2, we have

$$x(n) = \sum_{k=0}^{N-1} c_k u_k = \sum_{k=0}^{N-1} \langle x, u_k \rangle u_k . \tag{2.66}$$

This decomposes $x(n)$ into a sum of scaled frequency components; we have, in fact, quite easily discovered the *discrete Fourier transform* (DFT) using a bit of inner product space theory. Fourier transforms in signal processing are a class of signal operations that resolve an analog or discrete signal into its frequency components—sinusoids or exponentials. The components may be called "Fourier components" when the underlying orthonormal set is not made up of sinusoidal or exponential components. Thus we have the definition: If $\{u_k(n): 0 \le k \le N-1\}$ is an orthonormal set and x is a signal in an inner product space, then $c_k = \langle x, u_k \rangle$ is the kth Fourier coefficient of $x(n)$ with respect to the $\{u_k\}$. If $x(n) = \sum c_k u_k$, then we say that x is represented by a *Fourier series* in the $\{u_k\}$.

Note that the DFT system is linear, but, owing to the fixed decomposition window, not translation-invariant. There are a great many other properties and applications of the DFT (Chapter 7).

2.6.3.2 *Bessel's Inequality.* Geometric intuition about inner product spaces can tell us how we might use Fourier coefficients to characterize a signal. From (2.66) we can see that each Fourier coefficient indicates how much of each $u_k(n)$ there is in the signal $x(n)$. If $0 = \langle x, u_k \rangle$, then there is nothing like $u_k(n)$ in $x(n)$; if $|\langle x, u_k \rangle|$ is large, it means that there is an important u_k-like element in $x(n)$; and when $|\langle x, u_k \rangle|$ is large and the rest of the Fourier coefficients are small, it means that as a signal $x(n)$ has a significant similarity to $u_k(n)$. Given an orthonormal set $S = \{u_k\}$ and a signal $x(n)$, what do the Fourier coefficients of $x(n)$ with respect to S look like? It is clear from (2.66) that when $\|x(n)\|$ is large, then the norms of the Fourier coefficients also become large. How large can the Fourier coefficients be with respect to $\|x(n)\|$? If the set S is infinite, are most of the Fourier coefficients zero? Is it possible for the Fourier coefficients of $x(n)$ to be arbitrarily large? Perhaps the $\langle x, u_k \rangle \to 0$ as $k \to \infty$? If not, then there is an $\varepsilon > 0$, such that for any $N > 0$, there is a $k > N$ with $|\langle x, u_k \rangle| > \varepsilon$. In other words, can the significant Fourier components in $x(n)$ with respect to S go on forever? The next set of results, leading to Bessel's theorem for inner product spaces, helps to answer these questions.

Consider a signal $x(n)$ and an orthonormal set $S = \{u_k(n): 0 \le k < N\}$ in an inner product space I. $x(n)$ may be in the linear span of S, in which case the Fourier coefficients tell us the degree of similarity of $x(n)$ to each of the elements u_k. But if $x(n)$ is not in the linear span of S, then we might try to find the $y(n)$ in the linear span of S that is the closest signal to $x(n)$. In other words, if S cannot give us an exact breakdown of $x(n)$, what is the best model of $x(n)$ that S can provide? Let c_k be complex numbers, and let $y = \sum c_k u_k$. $y(n)$ is a linear combination of elements of S which we want to be close to $x(n)$. Then, after some inner product algebra and help from Pythagoras,

$$\|x-y\|^2 = \left\langle x - \sum c_k u_k, x - \sum c_k u_k \right\rangle = \|x\|^2 - \sum |c_k - \langle u_k, x\rangle|^2 - \sum |\langle u_k, x\rangle|^2. \tag{2.67}$$

By varying the c_k, $y(n)$ becomes any general signal in the linear span of S. The minimum distance between $x(n)$ and $y(n)$ occurs when the middle term of (2.67) is zero: $c_k = \langle u_k, x\rangle$. With this choice of $\{c_k\}$, $y(n)$ is the best model S can provide for $x(n)$.

We can apply this last result to answer the question about the magnitude of Fourier coefficients for a signal $x(n)$ in an inner product space. We would like to find an orthogonal set S so that its linear span contains every element of the inner product space. Then, we might characterize a signal $x(n)$ by its Fourier coefficients with respect to S. Unfortunately, as with the Rademacher signals, the linear span of an orthonormal set may not include the entire inner product space. Nevertheless, it is still possible to derive Bessel's inequality for inner product spaces.

Theorem (Inner Product Space Bessel[13] Inequality). Let I be an inner product space, let $S = \{u_k\}$ an orthonormal family of signals in I, and let $x \in I$. Then

$$\|x\|^2 \geq \sum |\langle u_k, x\rangle|^2. \tag{2.68}$$

Proof: Proceeding from (2.67), we set $c_k = \langle u_k, x\rangle$. Then, $\sum |c_k - \langle u_k, x\rangle|^2 = 0$, and $\|x-y\|^2 = \|x\|^2 - \sum |\langle u_k, x\rangle|^2$. But $\|x-y\|^2 \geq 0$. Thus, $\|x\|^2 - \sum |\langle u_k, x\rangle|^2 \geq 0$ and (2.68) follows. ■

2.6.3.3 Summary. We use the inner product relation as a measure of the similarity of signals, just as we do with finite-dimensional vectors. Orthonormal families of signals $S = \{u^k\}$ are especially convenient for decomposing signals $x(n)$, since the coefficients of the decomposition sum are readily computed as the inner product $c_k = \langle u_k, x\rangle$. It may well be that $x(n)$ cannot be expressed as a sum (possibly infinite) of elements of S; nevertheless, we can find the coefficients c_k that give us the closest signal to $x(n)$. From Bessel's relation, we see that the Fourier coefficients for a general signal $x(n)$ with respect to an orthonormal family S are bounded by $\|x\|$. One does not find a monstrous Fourier coefficient unless the original signal itself is monstrous. Moreover, the sum in (2.68) could involve an infinite number of nonzero terms. Then the fact that the sum converges indicates that the Fourier coefficients must eventually become arbitrarily small. No signal has Fourier coefficients with respect to S that get arbitrarily large. When S is infinite, every signal has Fourier coefficients $c_k = \langle u_k, x\rangle$ such that $c_k \to 0$ as $k \to \infty$. Bessel's inequality guarantees that the Fourier coefficients for $x(n)$ are well-behaved.

[13]After German mathematician and astronomer Friedrich Wilhelm Bessel (1794–1846).

Under what conditions does $x = \sum c_k u_k$, with $c_k = \langle u_k, x\rangle$? That is, we are interested in whether $x(n)$ has a Fourier series representation with respect to the $\{u_k\}$. If the orthonormal set S is finite and x is in the linear span of S, then this is true. If, on the other hand, S is infinite, then the sum becomes (possibly) infinite and the problem becomes whether the limit that formally defines this summation exists. Recall that calculus explains convergence of infinite sums in terms of Cauchy sequences of partial series sums. Thus, $\sum a_k = a$, if $\lim A_n = a$, where $A_n = \sum_{k=1}^{n} a_k$. If every Cauchy sequence has a limit, then the abstract space is called *complete*. Banach spaces are normed spaces that are complete. If we add completeness to the required properties of an inner product space, then what we get is the abstract structure known as a *Hilbert space*—one of the most important tools in applied mathematics, signal analysis, and physics.

2.7 HILBERT SPACES

In addition to the many excellent treatments of inner product and Hilbert spaces in functional analysis treatises, Hilbert space theory is found in specialized, introductory texts [25, 26].

2.7.1 Definitions and Examples

Definition (Hilbert Space). A *Hilbert space* is a complete inner product space. If $S \subseteq H$ is an inner product subspace of H and every Cauchy sequence of elements of S converges to an element of S, then S is a *Hilbert subspace* of H.

Recall that $\{x_k(n)\}$ is a Cauchy sequence if for every $\varepsilon > 0$, there is an $N > 0$ so that whenever $k, l > N$, we have $d(x_k, x_l) = \|x_k - x_l\| < \varepsilon$. Completeness means that every Cauchy sequence of signals in the space converges to a signal also in that space. Since we are working with inner product spaces, this norm must be interpreted as the inner product space norm. The least special of all the spaces is the normed space. Within its class, and distinct from one another, are the inner product and Banach spaces. The Banach spaces that are blessed with an inner product are the Hilbert spaces (Figure 2.15).

Examples (Hilbert Spaces). The following are Hilbert spaces:

(i) l^2 with the inner product defined

$$\langle x, y\rangle = \sum_{n=-\infty}^{\infty} x(n)\overline{y(n)}. \tag{2.69}$$

(ii) The set of signals in l^2 that are zero on a nonempty subset $Y \subseteq \mathbb{Z}$.

(iii) The inner product space $\mathbb{R}^n$, Euclidean n-space, with the standard dot product.

(iv) Similarly, the unitary space $\mathbb{R}^n$ with the standard inner product is a complex Hilbert space.

(v) Consider some subset of the shifted unit impulse signals $S = \{\delta(n-k)\}$. The linear span of S is an inner product subspace of l^2. If we take the set of limit points of Cauchy sequences of the linear span of S, then we get a Hilbert subspace of l^2. These subspaces are identical to those of (ii).

2.7.2 Decomposition and Direct Sums

The notions of orthogonality, basis, and subspace are interlinked within Hilbert space theory. The results in this section will show that l^2 Hilbert space looks very much like an "infinite-dimensional" extension of our finite-dimensional Euclidean and unitary n-spaces.

2.7.2.1 Subspace Decomposition. The following theorem is basic.

Theorem (Hilbert Space Decomposition). Let H be a Hilbert space, let X be a Hilbert subspace of H, and let $Y = X^{\perp}$ be the orthogonal complement of X in H. Then for any h in H, $h = x + y$, where $x \in X$ and $y \in Y$.

Proof: Let $h \in H$ and consider the distance from the subspace X to h. This number, call it $\delta = d(h, X)$, is the greatest lower bound of $\{\|x-h\|: x \text{ in } X\}$. Since we can find elements $x \in X$ whose distance to h differs by an arbitrarily small value from δ, there must be a sequence $\{x_n: x_n \text{ in } X, n > 0\}$ with $\|x_n - h\| < 1/n + \delta$.

We claim that $\{x_n\}$ is a Cauchy sequence in X. By applying the parallelogram rule to $x_n - h$ and $x_m - h$, we have

$$\|(x_n-h)+(x_m-h)\|^2 + \|x_n - x_m\|^2 = 2\|(x_n-h)\|^2 + 2\|(x_m-h)\|^2. \tag{2.70}$$

Since X is closed under addition and scalar multiplication, we have $\frac{x_n+x_m}{2} \in X$, and therefore

$$\|(x_n-h)+(x_m-h)\| = \|x_n + x_m - 2h\| = 2\left\|\frac{x_n+x_m}{2} - 2\right\| \geq 2\delta. \tag{2.71}$$

Putting the inequality (2.71) into (2.70) and rearranging gives

$$\|x_n - x_m\|^2 \leq 2\left(\delta + \frac{1}{n}\right) + 2\left(\delta + \frac{1}{m}\right) - \|x_n - h + x_m - h\|^2. \tag{2.72}$$

Consequently,

$$\|x_n - x_m\|^2 \le 4\left(\frac{\delta}{n}\right) + \frac{2}{n^2} + 4\left(\frac{\delta}{m}\right) + \frac{2}{m^2}, \tag{2.73}$$

which shows $\{x_n\}$ is Cauchy, as claimed. Since X is complete and contains this sequence, there must be a limit point: $x_n \to x \in X$. Let $y = h - x$. We claim that $y \in X^{\perp} = Y$. To prove this claim, let $0 \ne w \in X$. We must show that $\langle y, w\rangle = 0$. First note that since $\delta = d(h, X)$ and $x \in X$, we have

$$\delta \le \|y\| = \|h - x\| = \left\|h - \lim_{k \to \infty} x_k\right\| \le \frac{1}{n} + \delta \tag{2.74}$$

for all n. Thus, $\delta = \|y\| = \langle y, y\rangle^{1/2}$. Next, let a be a scalar. Closure properties of X imply that $x + aw \in X$ so that $\|y - aw\| = \|h - (x + aw)\| \ge \delta$. Expanding this last inequality in terms of the inner product on H gives

$$\langle y, y\rangle - \bar{a}\langle y, w\rangle - a\langle w, y\rangle + |a|^2\langle w, w\rangle \ge \delta^2. \tag{2.75}$$

Because $\langle y, y\rangle = \delta^2$, we can simplify (2.75) and then take $a = \frac{\langle y, w\rangle}{\langle w, w\rangle}$, which greatly simplifies to produce $|\langle y, w\rangle|^2 \le 0$. This must mean that $0 = \langle y, w\rangle$, $y \perp w$, and $y \in X^{\perp}$. This proves the last claim and completes the proof. ■

Let us list a few facts that follow from the theorem:

(i) The norm of element y is precisely the distance from h to the subspace X, $\|y\| = \delta = d(h, X)$. This was shown in the course of the proof.
(ii) Also, the decomposition of $h = x + y$ is unique (exercise).
(iii) One last corollary is that a set S of elements in H is complete (that is, the closure of its linear span is all of H) if and only if the only element that is orthogonal to all elements of S is the zero element.

Definition (Direct Sum, Projection). Suppose that H is a Hilbert space with $X, Y \subseteq H$. Then H is the *direct sum* of X and Y if every $h \in H$ is a unique sum of a signal in X and a signal in Y: $h = x + y$. We write $H = X \oplus Y$ and, in this case, if $h = x + y$ with $x \in X$ and $y \in Y$, we say that x is the *projection* of h onto X and y is the projection of h onto Y.

The decomposition theorem tells us that a Hilbert space is the direct sum of any Hilbert subspace and its orthogonal complement. The direct sum decomposition of a Hilbert space leads naturally to a linear system.

Definition (Projection System). Let H be a Hilbert space and X a Hilbert subspace of H. The *projection from* H *to* X is the mapping $T: H \to X$ defined by $T(h) = x$, where $h = x + y$, with $y \in X^{\perp}$.

Remark. This definition makes sense (in other words, the mapping is *well-defined*) because there is a unique $x \in X$ that can be associated with any $h \in H$.

2.7.2.2 Convergence Criterion. Combining the decomposition system with orthogonality gives Hilbert space theory much of the power it has for application in signal analysis. Consider an orthonormal set of signals $\{u_k(n)\}$ in a Hilbert space H, a set of scalars $\{a_k\}$, and the sum

$$x(n) = \sum_{k=-\infty}^{\infty} a_k u_k(n). \tag{2.76}$$

This sum may or may not converge in H. If the sequence of partial sums $\{s_N(n)\}$

$$s_N(n) = \sum_{k=-N}^{N} a_k u_k(n). \tag{2.77}$$

is a Cauchy sequence, then (2.76) has a limit.

Theorem (Series Convergence Criterion). The sum (2.76) converges in Hilbert space H if and only if the signal $a(n) = a_n$ is in l^2.

Proof: Let $N > M$ and take the difference $s_N - s_M$ in (2.77). Then,

$$\|s_N - s_M\|^2 = |a_{-N}|^2 + |a_{-N+1}|^2 + \dots + |a_{-M-1}|^2 + |a_{M+1}|^2 + \dots + |a_N|^2, \tag{2.78}$$

because of the orthonormality of the $\{u_k(n)\}$ signal family. Thus, $d(s_N, s_M)$ tends to zero if and only if the sums of squares of the $|a_n|$ tend to zero. ∎

Note too that if (2.76) converges, then the a_n are the Fourier coefficients of x with respect to the orthonormal family $\{u_k(n)\}$. This follows from taking the inner product of x with a typical u_k:

$$\langle x, u_k \rangle = \left\langle \sum_{i=-\infty}^{\infty} a_i u_i(n), u_k \right\rangle = \left\langle \lim_{N \to \infty} \sum_{i=-N}^{N} a_i u_i(n), u_k \right\rangle = a_k \langle u_k, u_k \rangle = a_k. \tag{2.79}$$

Therefore,

$$x(n) = \sum_{k=-\infty}^{\infty} \langle x, u_k \rangle u_k(n). \tag{2.80}$$

Thus, if the orthonormal family $\{u_k(n)\}$ is complete, then any $x(n)$ in H can be written as a limit of partial sums, and the representation (2.80) holds.

The theorem shows that there is a surprisingly close relationship between a general Hilbert space and the square-summable sequences l^2.

Orthonormal families and inner products are powerful tools for finding the significant components within signals. When does a Hilbert space have a complete orthonormal family? It turns out that every Hilbert space has a complete orthonormal family, a result that we will explain in a moment. There is also a method whereby any linearly independent set of signals in an inner product space can be converted into an orthonormal family.

2.7.2.3 Orthogonalization.

Let us begin by showing that there is an algorithm, called *Gram–Schmidt*[14] *orthogonalization*, for converting a linearly independent set of signals into an orthormal family. Many readers will recognize the procedure from linear algebra.

Theorem (Gram–Schmidt Orthogonalization). Let H be a Hilbert space containing a linearly independent family $\{u_n\}$. Then there is an orthonormal family $\{v_n\}$ with each v_n in the linear span of $\{u_k: 0 \le k \le n\}$.

Proof: The proof is by induction on n. For $n = 0$, we can take $v_0 = \frac{u_0}{\|u_0\|}$. Now suppose that the algorithm works for $n = 0, 1, ..., k$. We want to show that the orthonormal elements can be expanded one more time, for $n = k + 1$. Let U be the subspace of H that consists of the linear span of $\{u_0, u_1, ..., u_k\}$. This is a Hilbert subspace; for instance, it is closed and therefore complete. Let $V = U^{\perp}$. By linear independence, u_{k+1} is not in U. This means that in the unique decomposition $u_{k+1} = u + v$, with u in U and v in V, we must have $v \neq 0$, the zero signal. If we set $v_{k+1} = \frac{v}{\|v\|}$, then $\|v_{k+1}\| = 1$; $v_{k+1} \in U^{\perp}$; and, because $v = u_{k+1} - u$, v_{k+1} is in the linear span of $\{u_i: 0 \le i \le k+1\}$. ■

It is easier to find linearly independent than fully orthogonal signal families. So the Gram–Schmidt method is useful. The Gram–Schmidt procedure shows that if the linearly independent family is complete, then the algorithm converts it into a complete, orthonomal family.

[14]Erhard Schmidt (1876–1959), to whom the algorithm had been attributed, was Hilbert's student. Schmidt specified the algorithm in 1907. But it was discovered later that Jorgen Pedersen Gram (1850–1916) of Denmark had resorted to the same technique during his groundbreaking 1883 study on least squares approximation problems.

2.7.3 Orthonormal Bases

We now show how to build complete orthonormal families of signals in Hilbert space. That is, we want every element in the space to be approximated arbitrarily well by some linear combination of signals from the orthonormal family. Euclidean and unitary n-dimensional vector spaces all have orthonormal bases. This is a central idea in linear algebra. We are close to having shown the existence of orthonormal bases for general Hilbert spaces, too. But to get there with the Gram–Schmidt algorithm, we need to start with a complete (total) linearly independent family of signals. At this point, it is not clear that a general Hilbert space should even have a total linearly independent set.

Definition (Orthonormal Basis). In a Hilbert space, a complete orthonormal set is called an *orthonormal basis*.

We have already observed that the shifted unit sample signals are an orthonormal basis for the Hilbert space l^2. Remember the important distinction between this looser concept of basis and that for the finite-dimensional Euclidean and unitary spaces. In the cases of $\mathbb{R}^n$ and $\mathbb{C}^n$, the bases span the entire space. For some Hilbert spaces, however—and l^2 is a fine example—the linear combinations of the orthonormal basis signals only come arbitrarily close in norm to some signals.

2.7.3.1 Set Theoretic Preliminaries. There are some mathematical subtleties involved in showing that every Hilbert space has an orthonormal basis. The notions we need hinge on some fundamental results from mathematical set theory. A very readable introduction to these ideas is [27]. Most readers are probably aware that there are different orders of infinity in mathematics. (Those that are not may be in for a shock.) The number of points on a line (i.e., the set of real numbers) is a larger infinity than the natural numbers, because $\mathbb{R}$ cannot be placed in a one-to-one correspondence with $\mathbb{N}$. We say that two sets between which a one-to-one map exists have the same *cardinality*. The notation for the cardinality of a set X is $|X|$. In fact, the natural numbers, the integers, the rational numbers, and even all the real numbers which are roots of rational polynomials have the same cardinality, $|\mathbb{N}|$. They are called *countable* sets, because there is a one-to-one and onto map from $\mathbb{N}$, the counting set, to each of them. The real numbers are an *uncountable* set. Also uncountable is the set of subsets of the natural numbers, called the *power set* of $\mathbb{N}$, written $\mathcal{P}(\mathbb{N})$. It turns out that $|\mathcal{P}(\mathbb{N})| = |\mathbb{R}|$. The discovery of different orders of infinity—different cardinalities—is due to Cantor.[15]

[15]Georg Cantor (1845–1918) worked himself to the point of physical, emotional, and mental exhaustion trying to demonstrate the *continuum hypothesis*: there is no cardinality of sets in between $|\mathcal{N}|$ and $|\mathcal{R}|$. He retreated from set theory to an asylum, but never proved or disproved the continuum hypothesis. It is a good thing, too. In 1963, Paul Cohen proved that the continuum hypothesis is independent of the usual axioms of set theory; it can be neither proved nor disproved! [K. Devlin, *Mathematics: The Science of Patterns*, New York: Scientific American Library, 1994.]

Some basic facts about countable sets are as follows (exercises):

(i) The Cartesian product $X \times Y$ of two countable sets is countable.

(ii) The Cartesian product $X_1 \times X_2 \times \cdots \times X_n$ of a finite number of countable sets is countable.

(iii) A countable union of countable sets is countable.

(iv) The set that consists of all finite subsets of a countable set is countable.

(v) The set of all subsets of a set X always has a larger cardinality than X; in other words, $|X| < |\mathcal{P}(X)|$.

Observe carefully that indexing notation presupposes a one-to-one, onto map from the indexing set to the indexed set. Suppose X is a countable set—for example, the set of shifted impulses, $X = \{\delta(n-k)$: k an integer$\}$. We can index X by $\mathbb{N}$ with the map $f(k) = \delta(n-k)$. Trivially, f is a one-to-one and onto map of $\mathbb{N}$ to X. Now let $Y = \{a\delta(n)$: a is a real number$\}$ be the set of amplified unit impulse signals. It is impossible to index Y with the natural numbers, because Y has the same cardinality as the real line. Instead, if it is necessary to index such a collection, we must pick an indexing set that has the same cardinality as Y.

2.7.3.2 Separability. We draw upon these set theoretic ideas in order to show that every Hilbert space has an orthonormal basis. In particular, we need to bring the notion of cardinality into the discussion of Hilbert space and to invoke another concept from set theory—the Axiom of Choice.

Definition (Separable Hilbert Space). A Hilbert space is *separable* if it contains a countable dense set.

Notice that l^2 is a separable Hilbert space. The set of shifted impulse signals is an orthonormal basis for l^2. Now the set of all scalar multiples of linear combinations of the shifted impulses is not countable, because there are an uncountable number of magnitude values possible. However, we can get arbitrarily close to a linear combination of shifted impulses with a linear combination that has rational coefficients. There are a countable number of rationals. The set of finite sequences of rationals is therefore countable. Thus, the set of linear combinations of shifted impulses with rational coefficients is a countable dense subset of l^2.

Let's continue this line of reasoning and assume that we have a countable dense subset S of a Hilbert space H. We wish to fashion S into an orthonormal basis. We may write the dense family using the natural numbers as an indexing set: $S = \{s_n$: n is in $\mathbb{N}\}$. If S is a linearly independent family, then the Gram–Schmidt procedure applies, and we can construct from S an orthonormal family that is still dense in H. Thus, in this case, H has a countable basis. If S is has some linear dependency, we can pick the first element of S, call it d_0, that is a linear combination of the previous ones. We delete d_0 from S to form S_0, which still has the same linear span, and hence is just as dense as S in H. Continue this process for all natural numbers, finding d_{n+1} and cutting it from S_n to produce S_{n+1}. The result is a linearly independent set, S_ω. If S_ω is not linearly independent, then there is an element that is a linear combination of the others; call it t. We see immediately a contradiction, because t had to be chosen from the orthogonal complement of the elements that we chose before it and because the elements that were chosen later had to be orthogonal—and therefore linearly independent—to t. We note as well that S_ω has a linear span which is dense in H; and, using the Gram–Schmidt algorithm, it can be sculpted into an orthonormal basis for H.

Without separability of the Hilbert space H, the above argument breaks down. We could begin an attack on the problem by assuming a dense subset $S \subseteq H$. But what subsets, other than H itself, can we assume for a general, abstract Hilbert space? Examining the separable case's argument more closely, we see that we really built up a linearly independent basis incrementally, beginning from the bottom $s_0 \in S$. Here we can begin with some nonzero element of H, call it s_a, where we index by some other set A that has sufficient cardinality to completely index the orthonormal set we construct. If the linear span of $\{s_a\}$ includes all of H, then we are done; otherwise, there is an element in the orthogonal complement of the Hilbert subspace spanned by $\{s_a\}$. Call this element s_b. Then $\{s_a, s_b\}$ is a linearly independent set in H. Continue the process: Check whether the current set of linearly independent elements has a dense linear span; if not, select a vector from the orthogonal complement, and add this vector to the linearly independent family. In the induction procedure for the case of a separable H, the ultimate completion of the construction was evident. Without completion a contradiction arises. For if our "continuation" on the natural numbers does not work, can we find a least element that is a linear combination of the others, leading to a contradiction. But how can we find a "least" element of the index set A in the nonseparable case? We do not even know of an ordering for A. Thus there is a stumbling block in showing the existence of an orthonormal basis for a nonseparable Hilbert space.

2.7.3.3 Existence. The key is an axiom from set theory, called the *Axiom of Choice*, and one of its related formulations, called *Zorn's lemma.*[16] The Axiom of Choice states that the Cartesian product of a family of sets $\{S_a : a \in A\}$ is not empty. That is, $P = \{(s_a, s_b, s_c, \ldots) : s_a \in S_a, s_b \in S_b, \ldots\}$ has at least one element. The existence of an element in P means that there is a way to simultaneously choose one element from each of the sets S_a of the collection S. Zorn's lemma seems to say nothing like this. The lemma states that if a family of sets $S = \{S_a : a \in A\}$ has the property that for every chain $S_a \subseteq S_b \subseteq \ldots$ of sets in S, there is a T in S that is a superset of each of the chain elements, then S itself has an element that is contained properly in no other element of S; that is, S has a *maximal* set. Most people are inclined to think that the Axiom of Choice is obviously true and that Zorn's lemma is very suspicious, if not an outright fiction. On the contrary: Zorn's lemma is true if and only if the Axiom of Choice is true [27].

[16]Algebraist Max Zorn (1906–1993) used his *maximal set principle* in a 1935 paper.

Let us return now to our problem of constructing a dense linearly independent set in a Hilbert space H and apply the Zorn's lemma formulation of the Axiom of Choice. In a Hilbert space, the union of any chain of linearly independent subsets is also linearly independent. Thus, H must have, by Zorn, a maximal linearly independent set S. We claim that K, the linear span of S, is dense. Suppose not. Now K is a Hilbert subspace. So there is a vector v in the orthogonal complement to K. Contradiction is imminent. The set $S \cup \{v\}$ is linearly independent and properly includes S; this is impossible since S was selected to be maximal. So S must be complete (total). Its linear span is dense in H. Now we apply the Gram–Schmidt procedure to S. One final obstacle remains. We showed the Gram–Schmidt algorithm while using the natural numbers as an index set, and thus implicitly assumed a countable collection! We must not assume this now. Instead we apply Zorn's lemma to the Gram–Schmidt procedure, finding a maximal orthonormal set with same span as S. We have, with the aid of some set theory, finally shown the following.

Theorem (Existence of Orthonormal Bases). Every Hilbert space contains an orthonormal basis. ■

If the Hilbert space is spanned by a finite set of signals, then the orthonormal basis has a finite number of elements. Examples of finite-dimensional Hilbert spaces are the familiar Euclidean and unitary spaces. If the Hilbert space is separable, but is not spanned by a finite set, then it has a countably infinite orthonormal basis. Lastly, there are cases of Hilbert spaces which are not separable.

2.7.3.4 Fourier Series. Let us complete this chapter with a theorem that wraps up many of the ideas of discrete signal spaces: orthonormal bases, Fourier coefficients, and completeness.

Theorem (Fourier Series Representation). Let H be a Hilbert space and let $S = \{u_a : a \in A\}$ be an orthonormal family in H. Then,

(i) Any $x \in H$ has at most countably many nonzero Fourier coefficients with respect to the u_a.

(ii) S is complete (its linear span is dense in H) if and only if for all signals $x \in H$ we have

$$\|x\|^2 = \sum_{a \in A} |\langle x, u_a \rangle|^2, \tag{2.81}$$

where the sum is taken over all a, such that the Fourier coefficient of x with respect to u_a is not zero.

(iii) (Riesz–Fischer Theorem[17]) If $\{c_a : a \in A\}$ is a set of scalars such that

[17]Hungarian Frigyes Riesz (1880–1956) and Austrian Ernst Sigismund Fischer (1875–1954) arrived at this result independently in 1907 [22].

$$\sum_{a \in A} |c_a|^2 < \infty, \tag{2.82}$$

then there is a unique x in H such that $\langle x, u_a \rangle = c_a$, and

$$x = \sum_{a \in A} c_a u_a. \tag{2.83}$$

Proof: We have already used most of the proof ideas in previous results.

(i) The set of nonzero Fourier coefficients of x with respect to the u_a is the same as the set of Fourier coefficients that are greater than $1/n$ for some integer n. Since there can only be finitely many Fourier coefficients that are greater than $1/n$, we must have a countable union of finite sets, which is still countable. Therefore, there may only be a countable number of $\langle x, u_a \rangle \neq 0$.

(ii) Suppose first that S is complete and $x \in H$. Since there can be at most a countably infinite number of nonzero Fourier coefficients, it is possible to form the series sum,

$$s = \sum_{a \in A} \langle x, u_a \rangle u_a. \tag{2.84}$$

This sum converges by the Bessel inequality for inner product spaces. Consider $t = s - x$. It is easy to see that $t \in S^{\perp}$ by taking the inner product of t with each $u_a \in S$. But since S is complete, this means that there can be no nonzero element in its orthogonal complement; in other words, $t = 0$ and $s = x$. Now, since $\langle u_a, u_b \rangle \neq 0$ when $a \neq b$, we see that

$$\|x\|^2 = \langle x, x \rangle = \left\langle \sum_{a \in A} \langle x, u_a \rangle u_a, \sum_{a \in A} \langle x, u_a \rangle u_a \right\rangle = \sum_{a \in A} \langle x, u_a \rangle \overline{\langle x, u_a \rangle}. \tag{2.85}$$

Next, suppose that the relation (2.81) holds for all x. Assume for the sake of contradiction that S is not complete. Then by the Hilbert space decomposition theorem, we know that there is some nonzero $x \in S^{\perp}$. This means that $\langle x, u_a \rangle = 0$ for all u_a and that the sum (2.81) is zero. The contradiction is that now we must have $x = 0$, the zero signal.

(iii) If $\{c_a: a \in A\}$ is a set of scalars such that (2.82) holds, then at most a countable number of them can be nonzero. This follows from an argument similar to the proof of (i). Since we have a countable collection in (2.82), we may use the Hilbert space series convergence criterion, which was stated (implicitly at that point in the text) for a countable collection. ■

An extremely powerful technique for specifying discrete systems follows from these results. Given a Hilbert space, we can find an orthonormal basis for it. In the case of a separable Hilbert space, there is an iterative procedure to find a linearly independent family and orthogonalize it using the Gram–Schmidt algorithm. If the

Hilbert space is not separable, then we do not have such a construction. But the existence of the orthonormal basis $U = \{u_a : a \in A\}$ is still guaranteed by Zorn's lemma. Now suppose we use the orthonormal basis to analyze a signal. Certain of the basis elements, $V = \{v_b : \text{b} \in B \subseteq A\}$, have features we seek in general signals, x. We form the linear system $T(x) = y$, defined by

$$Tx = y = \sum_{b \in B} \langle x, v_b \rangle v_b. \tag{2.86}$$

Now the signal y is that part of x that resembles the critical basis elements. Since the theorem guarantees that we can expand any general element in terms of the orthonormal basis U, we know that the sum (2.86) converges. We can tune our linear system to provide precisely the characteristics we wish to preserve in or remove from signal x by selecting the appropriate orthonormal basis elements. Once the output $y = Tx$ is found, we can find the features we desire in x more easily in y. Also, y may prove that x is desirable in some way because it has a large norm; that is $\|x\| \approx \|y\|$. And, continuing this reasoning, y may prove that x is quite undesirable because $\|y\|$ is small.

In its many guises, we will be pursuing this idea for the remainder of the book.

2.8 SUMMARY

This chapter began with a practical—perhaps even naïve—exploration of the types of operations that one can perform on signals. Many of these simple systems will arise again and again as we develop methods for processing and interpreting discrete and continuous signals. The later chapters will demonstrate that the most important type of system we have identified so far is the linear, time-invariant system. In fact, the importance of the characterization result, the convolution theorem for LTI systems, cannot be overemphasized. This simple result underlies almost all of our subsequent work. Some of the most important concepts in signal filtering and frequency analysis depend directly on this result.

Our explorations acquire quite a bit of mathematical sophistication, however, when we investigate the closure properties of our naively formulated signal processing systems. We needed some good answers for what types of signals can be used with certain operations. It seems obvious enough that we would like to be able to sum any two signals that we consider, and this is clearly feasible for finitely supported signals. For other signals, however, this simple summing problem is not so swiftly answered. We need a formal mathematical framework for signal processing and analysis. Inspired by basic vector space properties, we began a search for the mathematical underpinnings of signal theory with the idea of a normed space. The l^p Banach spaces conveniently generalize some natural signal families that we first encountered in Chapter 1. Moreover, these spaces are an adequate realm for developing the theory of signals, stable systems, closure, convolution, and convergence of signals.

Unfortunately, except for l^2, none of the l^p spaces support an inner product definition that is related to the norm, $\| \|_p$. This is a profoundly negative result. But it once again shows the unique nature of the l^2 space. Of the l^p Banach spaces, only l^2 can be equipped with an inner product that makes it into a Hilbert space. This explains why finite-energy signals are so often the focus of signal theory. Only the l^2 Hilbert space, or one of its closed subspaces, has all of the features from Euclidean vector spaces that we find so essential for studying signals and systems.

We see that all Hilbert spaces have orthonormal bases, whether they are finite, countable, or uncountable. Furthermore, a close link exists between orthonormal bases for Hilbert spaces and linear systems that map one signal to another yet retain only desirable properties of the input. We will see in the sequel that it is possible to find special orthonormal bases that provide for the efficient extraction of special characteristics of signals, help us to find certain frequency and scale components of a signal, and, finally, allow us to discover the structure and analyze a signal.

REFERENCES

1. A. V. Oppenheim, A. S. Willsky, and S. H. Nawab, *Signals and Systems*, 2nd ed., Upper Saddle River, NJ: Prentice-Hall, 1997.
2. J. A. Cadzow and H. F. Van Landingham, *Signals, Systems, and Transforms*, Englewood Cliffs, NJ: Prentice-Hall, 1983.
3. H. Baher, *Analog and Dignal Signal Processing*, New York: Wiley, 1990.
4. R. E. Ziemer, W. H. Tranter, and D. R. Fannin, *Signals and Systems: Continouous and Discrete*, 3rd ed., New York: Macmillan, 1993.
5. P. P. Vaidyanathan, Multirate digital filters, filter banks, polyphase networks, and applications: A tutorial, *Proceedings of the IEEE*, vol. 78, pp. 56–93, January 1990.
6. O. Rioul and M. Vetterli, Wavelets and signal processing, *IEEE SP Magazine*, pp. 14–38, October 1991.
7. M. R. K. Khansari and A. Leon-Garcia, Subband decomposition of signals with generalized sampling, *IEEE Transactions on Signal Processing*, vol. 41, pp. 3365–3376, December 1993.
8. M. Kamel and A. Zhao, Extraction of binary character/graphics images from grayscale document images, *CVGIP: Graphical Models and Image Processing*, vol. 55, no. 3, pp. 203–217, 1993.
9. P. K. Sahoo, S. Soltani, and A. K. C. Wong, A survey of thresholding techniques, *Computer Vision, Graphics, and Image Processing*, vol. 41, pp. 233–260, 1988.
10. J. G. Proakis and D. G. Manolakis, *Digital Signal Processing*, 2nd ed., New York: Macmillan, 1992.
11. L. W. Couch, *Digital and Analog Communication Systems*, 4th ed., Upper Saddle River, NJ: Prentice-Hall, 1993.
12. D. Messerschmitt, D. Hedberg, C. Cole, A. Haoui, and P. Winsip, Digital voice echo canceller with a TMS32020, *Digital Signal Processing Applications with the TMS320 Family: Theory, Applications, and Implementations*, Dallas: Texas Instruments, Inc., pp. 415–454, 1989.

13. D. L. Duttweiler and Y. S. Chen, A single-chip VLSI echo canceller, *Bell System Technical Journal*, vol. 59, no. 2, pp. 149–160, February 1980.
14. B. Widrow, J. Kaunitz, J. Glover, and C. Williams, Adaptive noise cancelling: Principles and applications, *Proceedings of the IEEE*, vol. 63, pp. 1692–1716, December 1975.
15. E. Kreysig, *Introductory Functional Analysis with Applications*, New York: Wiley, 1989.
16. D. C. Champeney, *A Handbook of Fourier Theorems*, Cambridge: Cambridge University Press, 1987.
17. S. Mallat, A theory for multiresolution signal decomposition: The wavelet decomposition, *IEEE Transactions on Pattern Analysis and Machine Intelligence*, vol. 11, pp. 674–693, July 1989.
18. M. Rosenlicht, *An Introduction to Analysis*, New York: Dover, 1978.
19. G. Birkhoff and S. MacLane, *A Survey of Modern Algebra*, New York: Macmillan, 1965.
20. K. Yosida, *Functional Analysis*, 6th ed., Berlin: Springer-Verlag, 1980.
21. W. F. Rudin, *Functional Analysis*, 2nd ed., New York: McGraw-Hill, 1991.
22. F. Riesz and B. Sz.-Nagy, *Functional Analysis*, New York: Dover, 1990.
23. L. Kantorovich and G. Akilov, *Functional Analysis*, Oxford: Pergamon, 1982.
24. J. Dieudonne, *History of Functional Analysis*, Amsterdam: North-Holland, 1981.
25. N. Young, *An Introduction to Hilbert Space*, Cambridge: Cambridge University Press, 1968.
26. L. Debnath and P. Mikusinski, *Introduction to Hilbert Spaces with Applications*, 2nd ed., San Diego, CA: Academic Press, 1999.
27. P. Halmos, *Naive Set Theory*, New York: Van Nostrand, 1960.

PROBLEMS

1. Find the domain and range of the following systems:

(a) The amplifier system: $y(n) = Ax(n)$.

(b) A translation system: $y(n) = x(n - k)$.

(c) The discrete system on real-valued signals, $y(n) = x(n)^{1/2}$.

(d) The discrete system on complex-valued signals, $y(n) = x(n)^{1/2}$.

(e) An adder: $y(n) = x(n) + x_0(n)$.

(f) Termwise multiplication (modulation): $y(n) = x(n)x_0(n)$.

(g) Convolution: $y(n) = x(n)*h(n)$.

(h) Accumulator: $y(n) = x(n) + y(n-1)$.

2. Consider the LTI system $y = Hx$ that satisfies a linear, constant-coefficient difference equation

$$y(n) = \sum_{k=1}^{K} a_k y(n-k) + \sum_{m=0}^{M} b_m x(n-m). \tag{2.87}$$

Show that any K successive values of the output $h = H\delta$ are sufficient to characterize the system.

3. Consider an LTI system $y = Hx$ that satisfies the difference equation (2.87).

(a) Give the homogeneous equation corresponding to (2.87).

(b) Show that if (x, y) is a solution pair for (2.87) and y_h is a solution of its homogeneous equation, then $(x, y + y_h)$ is a solution of the difference equation.

(c) Show that if (x, y_1) and (x, y_2) are solution pairs for (2.87), then $y_1 - y_2$ is a solution to the homogeneous equation in (a).

4. Consider the LTI system $y = Hx$ that satisfies a linear, constant-coefficient difference equation (2.87). Prove that if the signal pair (δ, h) satisfies the difference equation and $y = x*h$, then the pair (x, y) also satisfies the difference equation.

5. Prove the converse of the convolution theorem for LTI Systems: Let $h(n)$ be a discrete signal and H be the system defined by $y = Hx = x*h$. Then H is LTI and $h = H\delta$.

6. Suppose $x(n)$ is in l^p, $1 \le p \le \infty$. Let c be a scalar (real or complex number) and let k be an integer. Show the following closure rules:

(a) $cx(n)$ is in l^p and $\|cx\|_p = |c|\ \|x\|_p$.

(b) $x(k - n)$ is in l^p and $\|x(k-n)\|_p = \|x(n)\|_p$.

7. Show that the signal space l^p is a normed space. The triangle inequality of the norm is proven by Minkowski's inequality. It remains to show the following:

(a) $\|x\|_p \ge 0$ for all x.

(b) $\|x\|p = 0$ if and only if $x(n) = 0$ for all n.

(c) $\|ax\|_p = |a|\ \|x\|_p$ for all scalars a and all signals $x(n)$.

8. Let p and q be conjugate exponents. Show the following:

(a) $(p + q)/pq = 1$.

(b) $pq = p + q$.

(c) $(p - 1)(q - 1) = 1$.

(d) $(p - 1)^{-1} = q - 1$.

(e) If $u = t^{p-1}$, then $t = uq^{-1}$.

9. Show that the l^p spaces are complete, $1 \le p < \infty$. Let $\{x_k(n)\}$ be a Cauchy sequence of signals in l^p.

(a) Show that for any integer n, the values of the signals in the sequence at time instant n are a Cauchy sequence. That is, with n fixed, the sequence of scalars $\{x_k(n)$: k an integer$\}$ is a Cauchy sequence.

(b) Since the real (complex) numbers are complete, we can fix n, and take the limit

$$c_n = \lim_{k \to \infty} x_k(n). \tag{2.88}$$

Show that the signal defined by $x(n) = c_n$ is in l^p.

(c) Show that

$$\lim_{k \to \infty} \|x_k - x\|_p = 0, \tag{2.89}$$

so that the signals x_k converge to x in the l^p distance measure d_p.

10. Show that the l^∞ signal space is complete.

11. Let I be an inner product space. Show that the inner product is continuous in I; that is if $x_n \to x$ and $y_n \to y$, then $\langle x_n, y_n \rangle \to \langle x, y \rangle$.

12. Show that orthogonal signals in an inner product space are linearly independent.

13. Let I be an inner product space and $d(u, v) = \|u - v\|$ be its distance measure. Show that with the distance measure $d(u, v)$, I is a metric space:

(a) $d(u, v) \geq 0$ for all u, v.

(b) $d(u, v) = 0$ if and only if $u = v$.

(c) $d(u, v) = d(v, u)$ for all u, v.

(d) For any w, $d(u, v) \leq d(u, w) + d(w, v)$.

14. Show that the *discrete Euclidean* space $\mathbb{Z}^n = \{(k_1, k_2, \cdots, k_n) \mid k_i \textit{ is in } \mathbb{Z}\}$ is a metric space. Is it a normed linear space? Explain.

15. Show that if for the Euclidean space $\mathbb{R}^n$, we define the metric $d((u_1, u_2, ..., u_n), (v_1, v_2, \cdots, v_n)) = |u_1 - v_1| + |u_2 - v_2| + \cdots + |u_n - v_n|$, then $(\mathbb{R}^n, d)$ is a metric space.

16. Show that the following sets are countable.

(a) The integers $\mathbb{Z}$ by arranging them in two rows:

0, 2, 4, 6, ...

1, 3, 5, 7, ...

and enumerating them with a zigzag traversal.

(b) All ordered pairs in the Cartesian product $\mathbb{N} \times \mathbb{N}$.

(c) The rational numbers $\mathbb{Q}$.

(d) All ordered k-tuples of a countable set X.

(e) Any countable union of countable sets.

CHAPTER 3

Analog Systems and Signal Spaces

This chapter extends linear systems and Hilbert space ideas to continuous domain signals, filling the gap Chapter 2 left conspicuous. Indeed, noting that an integral over the real line displaces an integral summation, the definitions, theorems, and examples are quite similar in form to their discrete-world cousins. We mainly verify that after replacing summations with integrations we can still construct analog signal spaces that support signal theory.

The initial presentation is informal, not rigorous, and takes a quicker pace. We require normed vector space operations for analog signals: the capability to add, scalar multiply, and measure the size of a signal. The signal spaces should also support limit operations, which imply that arbitrarily precise signal approximations are possible; we find that we can construct analog Banach spaces, too. More important is the measure of similarity between two analog signals—the inner product relation—and we take some care in showing that our abstract structures survive the transition to the analog world. There is an analog Hilbert space theory, for which many of the purely algebraic results of Chapter 2 remain valid. This is convenient, because we can simply quote the same results for analog signals. Hilbert spaces, principally represented by the discrete and analog finite energy signals, will prove to be the most important abstract structure in the sequel. Introductory texts that cover analog signal theory include Refs. 1–5.

The last two sections are optional reading; they supply rigor to the analog theory. One might hope that it is only necessary to replace the infinite summations with infinite integrations, but subtle problems thwart this optimistic scheme. The Riemann integral, familiar from college calculus, cannot handle signals with an infinite number of discontinuities, for example. Its behavior under limit operations is also problematic. Mathematicians faced this same problem at the end of the nineteenth century when they originally developed function space theories. The solution they found—the Lebesgue integral—works on exotic signals, has good limit operation properties, and is identical to the Riemann integral on piecewise continuous functions. Lebesgue measure and integration theory is covered in Section 3.4. In another area, the discrete-world results do not straightforwardly generalize to the analog world: There is no bona fide continuous-time delta function. The intuitive

Signal Analysis: Time, Frequency, Scale, and Structure, by Ronald L. Allen and Duncan W. Mills
ISBN: 0-471-23441-9

treatment, which we offer to begin with, does leave the theory of linear, translation-invariant systems with a glaring hole. It took mathematicians some time to put forward a theoretically sound alternative to the informal delta function concept as well. The concept of a distribution was worked out in the 1930s, and we introduce the theory in Section 3.5.

The chapter contains two important applications. The first, called the matched filter, uses the ideas of inner product and orthogonalization to construct an optimal detector for an analog waveform. The second application introduces the idea of a frame. Frames generalize the concept of a basis, and we show that they are the basic tool for numerically stable pattern detection using a family of signal models. The next chapter applies matched filters to the problem of recognizing signal shapes. Chapter 10 develops frame theory further in the context of time-frequency signal transforms.

3.1 ANALOG SYSTEMS

This section introduces operations on analog signals. Analog signals have a continuous rather than discrete independent time-domain variable. Analog systems operate on analog systems, and among them we find the familiar amplifiers, attenuators, summers, and so on. This section is a quick read for readers who are already familiar with analog systems.

3.1.1 Operations on Analog Signals

For almost every discrete system there corresponds an analog system. We can skim quickly over these ideas, so similar they are to the discrete-world development in the previous chapter.

Definition (Analog System). An analog system H is a partial function from the set of all analog signals to itself. If $x(t)$ is a signal and $y(t)$ is the signal output by H from the input $x(t)$, then $y = Hx$, $y(t) = (Hx)(t)$, or $y(t) = H(x(t))$. As with discrete signals, we call $y(t)$ the response of the system H to input $x(t)$. The set of signals $x(t)$ for which some $y = Hx$ is determined is the domain of the system H. The set of signals y for which $y = Hx$ for some signal x is the range of H.

3.1.2 Extensions to the Analog World

Let us begin by enumerating some common, but not unimportant, analog operations that pose no theoretical problems. As with discrete signals, the operations of scaling (in the sense of amplification and attenuation) and translation (or shifting) an analog signal are central.

We may amplify or attenuate an analog signal by multiplying its values by a constant:

$$y = Ax(t). \tag{3.1}$$

This is sometimes called a scaling operation, which introduces a possible confusion with the notion of dilation $y(t) = x(At)$, which is also called "scaling." The scaling operation inverts the input signal when $A < 0$, amplifies the signal when $|A| > 1$, and attenuates the signal when $|A| < 1$. The domain of a scaling system is all analog signals, as is the range, as long as $A = 0$.

An analog signal may be translated or shifted by any real time value:

$$y = x(t - t_0). \tag{3.2}$$

When $t_0 > 0$ the translation is a delay, and when $t_0 < 0$ the system can more precisely be called an advance. As in the discrete world, translations cause no domain and range problems. If T is an analog time shift, then $\mathrm{Dom}(T) = \mathrm{Ran}(T) = \{s(t)$: s is an analog signal$\}$.

Analog signal reflection reverses the order of signal values: $y(t) = x(-\mathrm{t})$. For analog time signals, this time reversal system reflects the signal values $x(t)$ around the time $t = 0$. As with discrete signals, the reflection and translation operations do not commute.

The basic arithmetic operations on signals exist for the analog world as well. Signal addition or summation adds a given signal to the input, $y(t) = x(t) + x_0(t)$, where $x_0(t)$ is a fixed signal associated with the system H. We can also consider the system that takes the termwise product of a given signal with the input, $y(t) = x(t)x_0(t)$.

One benefit of a continuous-domain variable is that analog signals allow some operations that were impossible or at least problematic in the discrete world.

Dilation always works in the analog world. We can form $y(t) = x(at)$ whatever the value of $a \in \mathbb{R}$. The corresponding discrete operation, $y(n) = x(bn)$, works nicely only if $b \in \mathbb{Z}$ and $|b| \geq 1$; when $0 < |b| < 1$ and $b \in \mathbb{Q}$ we have to create special values (typically zero) for those $y(n)$ for which $b \in \mathbb{Z}$. As noted earlier, dilation is often called scaling, because it changes the scale of a signal. Dilation enlarges or shrinks signal features according to whether $|a| < 1$ or $|a| > 1$, respectively.

Another analog operation is differentiation. If it is smooth enough, we can take the derivative of an analog signal:

$$y(t) = \frac{dx}{dt} = x'(t). \tag{3.3}$$

If the signal $x(t)$ is only piecewise differentiable in (3.3), then we can assign some other value to $y(t)$ at the points of nondifferentiability.

3.1.3 Cross-Correlation, Autocorrelation, and Convolution

The correlation and convolution operations depend on signal integrals. In the discrete world, systems that implement these operations have input–output relations that involve infinite summations over the integers. In continuous-domain signal processing, the corresponding operations rely on integrations over the entire real line.

These do pose some theoretical problems—just as did the infinite summations reminiscent of Chapter 2; we shall address them later when we consider signal spaces of analog signals.

The analog *convolution* operation is once again denoted by the * operator: $y = x*h$. We define:

$$y(t) = (x*h)(t) = \int_{-\infty}^{\infty} x(s)h(t-s)\, ds. \tag{3.4}$$

The *cross-correlation* system is defined by the rule $y = x^{\circ}h$, where

$$y(t) = (x^{\circ}h)(t) = \int_{-\infty}^{\infty} x(s)h(t+s)\, ds. \tag{3.5}$$

The analog *autocorrelation* operation on a signal is $y = x^{\circ}x$, and when the signals are complex-valued, we use the complex conjugate of the kernel function $h(t)$:

$$y(t) = (x^{\circ}h)(t) = \int_{-\infty}^{\infty} x(s)\overline{h(t+s)}\, ds. \tag{3.6}$$

The autocorrelation is defined by $y(t) = (x^{\circ}x)(t)$. One of the applications of functional analysis ideas to signal processing, which we shall provide below, is to show the existence of the correlation and autocorrelation functions for square-integrable signals $x(t)$ and $h(t)$.

We can show that linear translation invariant analog systems are again characterized by the convolution operation. This is not as easy as it was back in the discrete realm. We have no analog signal that corresponds to the discrete impulse, and discovering the right generalization demands that we invent an entirely new theory: distributions.

3.1.4 Miscellaneous Operations

Let us briefly survey other useful analog operations.

A subsampling or downsampling system continuously expands or contracts an analog signal: $y(t) = x(at)$, where $a > 0$ is the scale or dilation factor. Tedious as it is to say, we once more have a terminology conflict; the term "scale" also commonly refers in the signal theory literature to the operation of amplifying or attenuating a signal: $y(t) = ax(t)$.

Analog thresholding is a just as simple as in the discrete world:

$$y(t) = \begin{cases} 1 & \text{if } x(t) \geq T, \\ 0 & \text{if } x(t) < T. \end{cases} \tag{3.7}$$

The accumulator system, $y = Hx$, is given by

$$y(t) = \int_{-\infty}^{t} x(s)\, ds. \tag{3.8}$$

The accumulator outputs a value that is the sum of all input values to the present signal instant. As already noted, not all signals are in the domain of an accumulator system. The exercises explore some of these ideas further.

The moving average system is given by

$$y(t) = \frac{1}{2a} \int_{t-a}^{t+a} x(s)\, ds, \tag{3.9}$$

where $a > 0$. This system averages $x(s)$ around in an inteval of width $2a$ to output $y(t)$.

3.2 CONVOLUTION AND ANALOG LTI SYSTEMS

The characterization of a linear, translation-invariant analog system as one given by the convolution operation holds for the case of continuous domain signals too. We take aim at this idea right away. But, the notion of an impulse response—so elementary it is an embarassment within discrete system theory—does not come to us so readily in the analog world. It is not an understatement to say that the proper explication of the analog delta requires supplementary theory; what it demands is a complete alternative conceptualization of the mathematical representation of analog signals. We will offer an informal definition for the moment, and this might be the prudent stopping point for first-time readers. We shall postpone the more abstruse development, known as distribution theory, until Section 3.5.

3.2.1 Linearity and Translation-Invariance

Analog systems can be classified much like discrete systems. The important discrete signal definitions of linearity and translation- (or shift- or time-) invariance extend readily to the analog world.

Definition (Linear System). An analog system H is linear if $H(ax) = aH(x)$ and $H(x + y) = H(x) + H(y)$. Often the system function notation drops the parentheses; thus, we write Hx instead of $H(x)$. $H(x)$ is a signal, a function of a time variable, and so we use the notation $y(t) = (Hx)(t)$ to include the independent variable of the output signal.

Definition (Translation-Invariant). An analog system H is translation-invariant if whenever $y = Hx$ and $s(t) = x(t - a)$, then $H(s) = y(t - a)$.

A linear system obeys the principles of scaling and superposition. When a system is translation-invariant, then the output of the shifted input is precisely the shifted output.

Definition (LTI System). An LTI system is both linear and translation-invariant.

Let us consider some examples.

Example. Let the system $y = Hx$ be given by $y(t) = x(t)\cos(t)$. The cosine term is a nonlinear distraction, but this system is linear. Indeed, $H(x_1 + x_2)(t) = [x_1(t) + x_2(t)]\cos(t) = x_1(t)\cos(t) + x_2(t)\cos(t) = H(x_1)(t) + H(x_2)(t)$. Also, $H(ax)(t) = [ax(t)]\cos(t) = a[x(t)\cos(t)] = a(Hx)(t)$.

Example. Let $y = Hx$ by given by $y(t) = tx(t)$. Then H is not translation-invariant. The decision about whether a system is or is not translation-invariant can sometimes bedevil signal processing students. The key idea is to hide the shift amount inside a new signal's definition: Let $w(t) = x(t - a)$. Then $w(t)$ is the shifted input signal. $(Hw)(t) = tw(t)$ by definition of the system H. But $tw(t) = tx(t - a)$. Is this the shifted output? Well, the shifted output is $y(t - a) = (t - a)x(t - a)$. In general, this will not equal $tx(t - a)$, so the system H is not shift-invariant.

Example. Let $y = Hx$, where $y(t) = x^2(t) + 8$. This system is translation-invariant. Again, let $w(t) = x(t - a)$, so that $(Hw)(t) = w^2(t) + 8 = x(t - a)x(t - a) + 8$ is the output of the translated input signal. Is this the translated output signal? Yes, because $y(t - a) = x^2(t - a) + 8 = (Hw)(t)$.

Example (Moving Average). Let $T > 0$ and consider the system $y = Hx$:

$$y(t) = \int_{t-T}^{t+T} x(s)\, ds. \tag{3.10}$$

Then H is LTI. The integration is a linear operation, which is easy to show. So let us consider a translated input signal $w(t) = x(t - a)$. Then $(Hw)(t)$ is

$$\int_{t-T}^{t+T} w(s)\, ds = \int_{t-T}^{t+T} x(s-a)\, ds = \int_{t-T-a}^{t+T-a} x(u)\, du, \tag{3.11}$$

where we have changed the integration variable with $u = s - a$. But note that the shifted output is

$$y(t-a) = \int_{t-a-T}^{t-a+T} x(s)\, ds, \tag{3.12}$$

and (3.12) is identical to (3.11).

3.2.2 LTI Systems, Impulse Response, and Convolution

Putting aside mathematical formalities, it is possible to characterize analog linear, translation-invariant systems by convolution of the input signal with the system impulse response.

3.2.2.1 Analog Delta and Impulse Response. Let us begin by developing the idea of the analog delta function, or Dirac[1] delta, $\delta(t)$. This signal should—like the discrete delta, $\delta(n)$—be zero everywhere except at time $t = 0$. Discrete convolution is a discrete sum, so $\delta(0) = 1$ suffices for sifting out values of discrete siganls $x(n)$. Analog convolution is an integral, and if $\delta(t)$ is a signal which is nonzero only at $t = 0$, then the integral of any integrand $x(s)\delta(t - s)$ in the convolution integral (3.4) is zero. Consequently, it is conventional to imagine the analog impulse signal as being infinite at $t = 0$ and zero otherwise; informally, then,

$$\delta(t) = \begin{cases} \infty & \text{if } t \neq 0, \\ 0 & \text{if otherwise.} \end{cases} \tag{3.13}$$

Another way to define the analog delta function is through the following convolutional identity:

Sifting Property. The analog impulse is the signal for which, given analog signal $x(t)$,

$$x(t) = (x*\delta)(t) = \int_{-\infty}^{\infty} x(s)\delta(t-s)\,ds. \tag{3.14}$$

No signal satisfying (3.13) or having the property (3.14) exists, however. The choice seems to be between the Scylla of an impossible function or the Charybdis of an incorrect integration.

To escape this quandary, let us try to approximate the ideal, unattainable analog sifting property by a local average. Let $\delta_n(t)$ be defined for $n > 0$ by

$$\delta_n(t) = \begin{cases} n & \text{if } t \in \left[-\frac{1}{2n}, \frac{1}{2n}\right], \\ 0 & \text{if otherwise.} \end{cases} \tag{3.15}$$

[1]P. A. M. Dirac (1902–1984) applied the delta function to the discontinuous energy states found in quantum mechanics (*The Principles of Quantum Mechanics*, Oxford: Clarendon, 1930). Born in England, Dirac studied electrical engineering and mathematics at Bristol and then Cambridge, respectively. He developed the relativistic theory of the electron and predicted the existence of the positron.

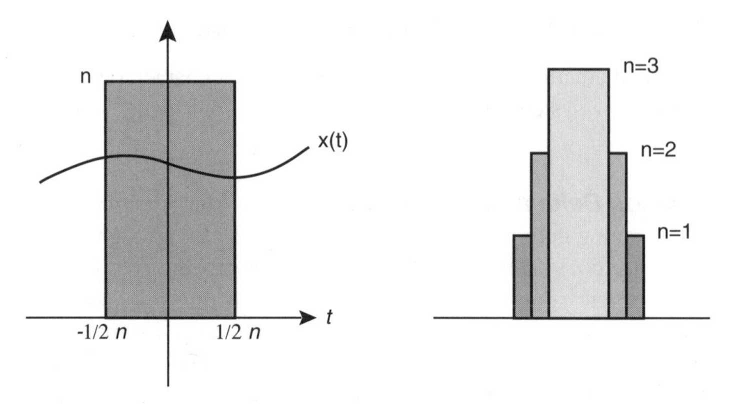

Fig. 3.1. Approximating the analog impulse, $\delta_n(t) = n$ on the interval of width $1/n$ around the origin.

Then, referring to Figure 3.1, $\delta_n(t)$ integrates to unity on the real line,

$$\int_{-\infty}^{\infty} \delta_n(t)\, dt = 1. \tag{3.16}$$

Furthermore, $\delta_n(t)$ provides a rudimentary sifting relationship,

$$\int_{-\infty}^{\infty} x(t)\delta_n(t)\, dt = \text{Average value of } x(t) \text{ on}\left[-\frac{1}{2n}, \frac{1}{2n}\right]. \tag{3.17}$$

To verify this, note that

$$\int_{-\infty}^{\infty} x(t)\delta_n(t)\, dt = n \int_{-1/2n}^{1/2n} x(t)\, dt, \tag{3.18}$$

while

$$\int_{a}^{b} x(t)\, dt = (b-a) \times \text{Average value of } x(t) \text{ on } \left[-\frac{1}{2n}, \frac{1}{2n}\right]. \tag{3.19}$$

Combining (3.18) and (3.19) proves that $\delta_n(t)$ has a sifting-like property. Note that as $n \to \infty$ the square impluse in Figure 3.1 grows higher and narrower, approximating an infinitely high spike. Under this limit the integral (3.17) is the average of $x(t)$ within an increasingly minute window around $t = 0$:

$$\lim_{n \to \infty} \int_{-\infty}^{\infty} x(t)\delta_n(t)\, dt = x(0). \tag{3.20}$$

This argument works whatever the location of the tall averaging rectangles defined by $\delta_n(t)$. We can conclude that

$$\lim_{n \to \infty} \int_{-\infty}^{\infty} x(s)\delta_n(t-s)\, ds = x(t). \tag{3.21}$$

An interchange of limit and integration in (3.21) gives

$$\lim_{n \to \infty} \int_{-\infty}^{\infty} x(s)\delta_n(t-s)\, ds = \int_{-\infty}^{\infty} x(s) \lim_{n \to \infty} \delta_n(t-s)\, ds. \tag{3.22}$$

Finally, and if we assume $\delta_n(t) \to \delta(t)$, that the square spikes converge to the Dirac delta, then the sifting property follows from (3.22).

Rigorously speaking, of course, there is no signal $\delta(t)$ to which the $\delta_n(t)$ converge, no such limit exists, and our interchange of limits is quite invalid [6]. It is possible to formally substantiate delta function theory with the theory of generalized functions and distributions [7–10]. Some applied mathematics texts take time to validate their use of Diracs [11–13]. In fact, the amendments follow fairly closely the informal motivation that we have provided above. Despite our momentary neglect of mathematical justification, at this point these ideas turn out to be very useful in analog signal theory; we shamelessly proceed to feed delta "functions" into linear, translation-invariant systems.

Definition (Impulse Response). Let H be an analog LTI system and $\delta(t)$ be the Dirac delta function. Then the impulse response of H is $h(t) = (H\delta)(t)$. Standard notation uses a lowercase "h" for the impulse response of the LTI system "H."

The next section applies these ideas toward a characterization of analog LTI systems.

3.2.2.2 LTI System Characterization. The most important idea in analog system theory is that a convolution operation—the input signal with the system's impulse response—characterizes LTI systems. We first rewrite an analog signal $x(t)$ as a scaled sum of shifted impulses. Why take a perfectly good—perhaps infinitely differentiable—signal and write it as a linear combination of these spikey, problematic components? This is how, in Chapter 2, we saw that convolution governs the input–output relation of an LTI discrete system.

To decompose $x(t)$ into Diracs, note that any integral is a limit of Riemann sums of decreasing width:

$$\int_{-\infty}^{\infty} f(s)ds = \lim_{Len(I) \to 0} \sum_{n=-\infty}^{\infty} f(a_n)Len(I_n), \tag{3.23}$$

where $I = \{I_n: n \in \mathbb{Z}\}$ is a partition of the real line, $I_n = [a_n, b_n]$, the length of interval I_n is $\mathrm{Len}(I_n) = b_n - a_n$, and $\mathrm{Len}(I) = \max\{\mathrm{Len}(I_n): n \in \mathbb{Z}\}$. If $f(s) = x(s)\delta(t - s)$, then

$$x(t) = \int_{-\infty}^{\infty} f(s)\, ds = \lim_{Len(I) \to 0} \sum_{n=-\infty}^{\infty} x(a_n)\delta(t - a_n)\, Len(I_n), \tag{3.24}$$

so that

$$\begin{aligned} y(t) = H(x)(t) &= H\left[\lim_{Len(I) \to 0} \sum_{n=-\infty}^{\infty} x(a_n)\delta(t - a_n) Len(I_n)\right] \\ &= \lim_{Len(I) \to 0} \sum_{n=-\infty}^{\infty} x(a_n) H[\delta(t - a_n)] Len(I_n) \end{aligned} \tag{3.25}$$

Besides blithely swapping the limit and system operators, (3.25) applies linearity twice: scaling with the factor $x(a_n)$ and superposition with the summation. By translation-invariance, $H[\delta(t - a_n)] = (H\delta)(t - a_n)$. The final summation above is an integral itself, precisely the LTI system characterization we have sought:

Theorem (Convolution for LTI Systems). Let H be an LTI system, $y = Hx$, and $h = Hd$. Then

$$y(t) = (x * h)(t) = \int_{-\infty}^{\infty} x(s) h(t - s)\, ds. \tag{3.26}$$

Now let us consider a basic application of the theorem to the study of stable and causal systems.

3.2.2.3 *Stable and Causal Systems.* The same ideas of stability and causality, known from discrete system theory, apply to analog systems.

Definition (Bounded). A signal $x(t)$ is *bounded* if there is a constant M such that $|x(t)| < M$ for all $t \in \mathbb{R}$.

Definition (Absolutely Integrable). A signal $x(t)$ is *absolutely integrable* if

$$\int_{-\infty}^{\infty} |x(t)|\, dt < \infty. \tag{3.27}$$

The equivalent concept in Chapter 2 is *absolutely summable*.

Definition (Stable System). If $y = Hx$ is a bounded output signal whenever $x(t)$ is a bounded input signal, then the system H is *stable*.

Proposition (Stability Characterization). The LTI system $y = Hx$ is stable if and only if its impulse response $h = H\delta$ is absolutely integrable.

Proof: Suppose h is absolutely integrable and $|x(t)| < M$ for all $t \in \mathbb{R}$. Then

$$|y(t)| = \left|\int_{-\infty}^{\infty} x(s)h(t-s)\,ds\right| \le \int_{-\infty}^{\infty} |x(s)h(t-s)|\,ds \le M\int_{-\infty}^{\infty} |h(t-s)|\,ds. \tag{3.28}$$

Since the final integral in (3.28) is finite, $y(t)$ is bounded and H must be stable. Conversely, suppose that H is stable but $h = H\delta$ is not absolutely integrable:

$$\int_{-\infty}^{\infty} |h(t)|\,dt = \infty; \tag{3.29}$$

we seek a contradiction. If

$$x(t) = \begin{cases} \dfrac{\overline{h(-t)}}{|h(-t)|} & \text{if } h(-t) \ne 0, \\ 0 & \text{if otherwise,} \end{cases} \tag{3.30}$$

where we are allowing for the situation that $h(t)$ is complex-valued, then $x(t)$ is clearly bounded. But convolution with $h(t)$ governs the input-output relation of H, and that is the key. We can write $y = Hx$ as

$$y(t) = \int_{-\infty}^{\infty} x(s)h(t-s)\,ds = \int_{-\infty}^{\infty} \frac{\overline{h(-s)}}{|h(-s)|}h(t-s)\,ds. \tag{3.31}$$

So $y(t)$ is the output of stable system H given bounded input $x(t)$. $y(t)$ should be bounded. What is $y(0)$? Well,

$$y(0) = \int_{-\infty}^{\infty} x(s)h(0-s)\,ds = \int_{-\infty}^{\infty} \frac{\overline{h(-s)}}{|h(-s)|}h(0-s)\,ds = \int_{-\infty}^{\infty} \frac{|h(-s)|^2}{|h(-s)|}\,ds$$
$$= \int_{-\infty}^{\infty} \frac{|h(-s)|^2}{|h(-s)|}\,ds = \int_{-\infty}^{\infty} |h(-s)|\,ds. \tag{3.32}$$

However, the final integral in (3.32) is infinite by (3.29). So $y(t)$ cannot be bounded, and this contradicts the converse's assumption. ■

To the discrete theory, there also corresponds an analog notion of causality. A system is causal if the future of the input signals have no bearing on their response.

Definition (Causal System). The system H is *causal* if $y = Hx$ can be found using only present and past values of $x(t)$.

Proposition (Causality Criterion). The LTI system $y = Hx$ is causal if its impulse response $h = H\delta$ satisfies $h(r) = 0$ for $r < 0$.

Proof: If $h(t - s) = 0$ for $s > t$, then jotting down the convolution integral shows

$$y(t) = \int_{-\infty}^{\infty} x(s)h(t-s)\, ds = \int_{-\infty}^{t} x(s)h(t-s)\, ds. \tag{3.33}$$

We have written $y(t)$ knowing only earlier $x(s)$ values, and so H must be causal. ■

Remark. For analog signals, the converse of the causality criterion is not true. The impulse response $h(t)$ could be nonzero at an isolated point, say $t = -2$, and so the convolution integral (3.33) does not depend on negative times.

Example. Let H be the system with impulse response

$$h(t) = \begin{cases} e^{-t} & \text{if } t \geq 0, \\ 1 & \text{if } t = -1, \\ 0 & \text{if otherwise.} \end{cases} \tag{3.34}$$

The system $y(t) = (x^*h)(t)$ is LTI, and H is causal, but it does not obey the converse of the causality criterion.

The causality criterion's lack of a converse distinguishes analog from discrete theory. In discrete system theory, an LTI system H is causal if and only if its impulse response $h(n)$ is zero for $n < 0$. We can improve on the proposition, by developing a stronger integral. Indeed, the fact that we might allow signals that have isolated finite impulses, or even a countably infinite number of finite impulse discontinuities points to the Riemann integral's inadequacy. This same defect troubled mathematicians in the early 1900s, when they first drew set theoretic concepts into analysis and began to demand limit operations from the integral. Modern measure and integration theory was the outcome, and we cover it in Section 3.4.

3.2.3 Convolution Properties

Convolution is the most important signal operation, since $y = h^*x$ gives the input–output relation for the LTI system H, where $h = H\delta$. This section develops basic theorems, these coming straight from the properties of the convolution integral.

Proposition (Linearity). The convolution operation is linear: $h^*(ax) = ah^*x$, and $h^*(x + y) = h^*x + h^*y$.

Proof: Easy by the linearity of integration (exercise). ■

Proposition (Translation-Invariance). The convolution operation is translation invariant: $h^*[x(t-a)] = (h^*x)(t-a)$.

Proof: It is helpful to hide the time shift in a signal with a new name. Let $w(t) = x(t-a)$, so that the translated input is $w(t)$. We are asking, what is the convolution of the shifted input? It is $(h^*w)(t)$. Well,

$$(h^*w)(t) = \int_{-\infty}^{\infty} h(s)w(t-s)\,ds = \int_{-\infty}^{\infty} h(s)x((t-a)-s)\,ds = (h^*x)(t-a), \tag{3.35}$$

which is the translation of the output by the same amount. ■

These last two propositions comprise a converse to the convolution theorem (3.26) for LTI systems: a system $y = h^*x$ is an LTI system. The next property shows that the order of LTI processing steps does not matter.

Proposition (Commutativity). The convolution operation is commutative: $x^*y = y^*x$.

Proof: Let $u = t - s$ for a change of integration variable. Then $du = -ds$, and

$$(x^*y)(t) = \int_{-\infty}^{\infty} x(s)y(t-s)\,ds = -\int_{\infty}^{-\infty} x(t-u)y(u)\,du = \int_{-\infty}^{\infty} x(t-u)y(u)\,du. \tag{3.36}$$

The last integral in (3.36) we recognize to be the convolution y^*x. ■

Proposition (Associativity). The convolution operation is associative: $h^*(x^*y) = h^*(x^*y)$.

Proof: Exercise. ■

Table 3.1 summarizes the above results.

TABLE 3.1. Convolution Properties

Signal Expression	Property Name
$(x^*y)(t) = \int_{-\infty}^{\infty} x(s)y(t-s)\,ds$	Definition
$h^*[ax+y] = ah^*x + h^*y$	Linearity
$h^*[x(t-a)] = (h^*x)(t-a)$	Translation- or shift-invariance
$x^*y = y^*x$	Commutativity
$h^*(x^*y) = (h^*x)^*y$	Associativity
$h^*(\delta(t-a)) = h(a)$	Sifting

3.2.4 Dirac Delta Properties

The Dirac delta function has some unusual properties. Although they can be rigorously formulated, they also follow from its informal description as a limit of ever higher and narrower rectangles. We maintain a heuristic approach.

Proposition. Let $u(t)$ be the unit step signal. Then $\delta(t) = \frac{d}{dt}[u(t)]$.

Proof: Consider the functions $u_n(t)$:

$$u_n(t) = \begin{cases} 1 & \text{if } t \geq \frac{1}{2n}, \\ 0 & \text{if } t \leq -\frac{1}{2n}, \\ nt + \frac{1}{2} & \text{if otherwise.} \end{cases} \tag{3.37}$$

Notice that the derivatives of $u_n(t)$ are precisely the $\delta_n(t)$ of (3.15) and that as $n \to \infty$ we have $u_n(t) \to u(t)$. Taking the liberty of assuming that in this case differentiation and the limit operations are interchangeable, we have

$$\delta(t) = \lim_{n \to \infty} \delta_n(t) = \lim_{n \to \infty} \frac{d}{dt} u_n(t) = \frac{d}{dt} \lim_{n \to \infty} u_n(t) = \frac{d}{dt} u(t). \tag{3.38}$$

■

Remarks. The above argument does stand mathematical rigor on its head. In differential calculus [6] we need to verify several conditions on a sequence of signals and their derivatives in order to conclude that the limit of the derivatives is the derivative of their limit. In particular, one must verify the following in some interval I around the origin:

(i) Each function in $\{x_n(t) \mid n \in \mathbb{N}\}$ is continuously differentiable on I.
(ii) There is an $a \in I$ such that $\{x_n(a)\}$ converges.
(iii) The sequence $\{x_n'(t)\}$ converges uniformly on I.

Uniform convergence of $\{y_n(t) \mid n \in \mathbb{N}\}$ means that for every $\varepsilon > 0$ there is an $N_\varepsilon > 0$ such that $m, n > N_\varepsilon$ implies $|y_n(t) - y_m(t)| < \varepsilon$ for all $t \in I$. The key distinction between uniform convergence and ordinary, or *pointwise*, convergence is that uniform convergence requires that an N_ε be found that pinches $y_n(t)$ and $y_m(t)$ together throughout the interval. If the N_ε has to depend on $t \in I$ (such as could happen if the derivatives of the $y_n(t)$ are not bounded on I), then we might find pointwise convergence exists, but that uniform convergence is lost. The exercises delve further into these ideas. Note also that a very important instance of this distinction occurs in Fourier series convergence, which we detail in Chapters 5 and 7. There we show that around a discontinuity in a signal $x(t)$ the Fourier series is pointwise but not uniformly convergent. Connected with this idea is the famous ringing artifact of Fourier series approximations, known as the *Gibbs phenomenon*.

The next property describes the scaling or time dilation behavior of $\delta(t)$. Needless to say, it looks quite weird at first glance.

Proposition (Scaling Property). Let $u(t)$ be the unit step signal and $a \in \mathbb{R}$, $a \neq 0$. Then

$$\delta(at) = \frac{1}{|a|}\delta(t). \tag{3.39}$$

Proof: Consider the rectangles approaching $\delta(t)$ as in Figure 3.1. If the $\delta_n(t)$ are dilated to form $\delta_n(at)$, then in order to maintain unit area, we must alter their height; this makes the integrals an average of $x(t)$. This intuition explains away the property's initial oddness. However, we can also argue for the proposition by changing integration variables: $s = at$. Assume first that $a > 0$; thus,

$$\int_{-\infty}^{\infty} x(t)\delta(at)\,dt = \int_{-\infty}^{\infty} x\left(\frac{s}{a}\right)\delta(s)\frac{1}{a}\,ds = \frac{1}{a}\int_{-\infty}^{\infty} x\left(\frac{s}{a}\right)\delta(s)\,ds = \frac{1}{a}x(0), \tag{3.40}$$

and $\delta(at)$ behaves just like $\frac{1}{|a|}\delta(t)$. If $a < 0$, then $s = at = -|a|t$, and

$$\int_{-\infty}^{\infty} x(t)\delta(at)\,dt = \int_{\infty}^{-\infty} x\left(\frac{s}{a}\right)\delta(s)\frac{-1}{|a|}\,ds = \frac{1}{|a|}\int_{-\infty}^{\infty} x\left(\frac{s}{a}\right)\delta(s)\,ds = \frac{1}{|a|}x(0). \tag{3.41}$$

Whatever the sign of the scaling parameter, (3.39) follows. ∎

The Dirac is even: $\delta(-t) = \delta(t)$, as follows from the Dirac scaling proposition. The next property uses this corollary to show how to sift out the signal derivative (Table 3.2).

TABLE 3.2. Dirac Delta Properties

Signal Expression	Property Name
$x(t) = \int_{-\infty}^{\infty} x(s)\delta(t-s)\,ds$	Sifting
$\delta(t) = \frac{d}{dt}[u(t)]$	Derivative of unit step
$\delta(at) = \frac{1}{\lvert a\rvert}\delta(t)$	Scaling
$\delta(t) = \delta(-t)$	Even
$\int_{-\infty}^{\infty} x(t)\frac{d}{dt}\delta(t)\,dt = -\frac{d}{dt}x(t)\Big\vert_{t=0}$	Derivative of Dirac

Proposition. Let $\delta(t)$ be the Dirac delta and $x(t)$ a signal. Then,

$$\int_{-\infty}^{\infty} x(t)\frac{d}{dt}\delta(t)\,dt = -\frac{d}{dt}(x(t))\bigg|_{t=0}. \tag{3.42}$$

Proof: We differentiate both sides of the sifting property equality to obtain

$$\frac{d}{dt}x(t) = \frac{d}{dt}\int_{-\infty}^{\infty} x(s)\delta(t-s)\,ds = \int_{-\infty}^{\infty} x(s)\frac{d}{dt}\delta(t-s)\,ds = \int_{-\infty}^{\infty} x(s)\frac{d}{dt}\delta(s-t)\,ds, \tag{3.43}$$

and the proposition follows by taking $t = 0$ in (3.43). ■

3.2.5 Splines

Splines are signals formed by interpolating discrete points with polynomials, and they are particularly useful for signal modeling and analysis. Sampling converts an analog signal into a discrete signal, and we can effect the reverse by stretching polynomials between isolated time instants. The term "spline" comes from a jointed wood or plastic tool of the same name used by architects to hand draw elongated, smooth curves for the design of a ship or building. Here we provide a brief overview of the basic splines, or *B-splines*. There are many other types of splines, but what now interests us in B-splines is their definition by repetitive convolution.

Definition (Spline). The analog signal $s(t)$ is a *spline* if there is a set of points $K = \{k_m: m \in \mathbb{Z}\}$ such that $s(t)$ is continuous and equal to a polynomial on each interval $[k_m, k_{m+1}]$. Elements of K are called *knots*. If $s(t)$ is a polynomial of degree n on each $[k_m, k_{m+1}]$, then $s(t)$ has degree n. A spline $s(t)$ of degree n is called *smooth* if it is $n - 1$ times continuously differentiable.

Splines are interesting examples of analog signals, because they are based on discrete samples—their set of knots—and yet they are very well-behaved between the knots. Linear interpolation is probably everyone's first thought to get from a discrete to an analog signal representation. But the sharp corners are problematic. Splines of small order, say $n = 2$ or $n = 3$, are a useful alternative; their smoothness is often precisely what we want to model a natural, continuously occurring process.

Splines accomplish function approximation in applied mathematics [14, 15] and shape generation in computer graphics [16, 17]. They have become very popular signal modeling tools in signal processing and analysis [18, 19] especially in connection with recent developments in wavelet theory [20]. A recent tutorial is Ref. 21.

For signal processing, splines with uniformly spaced knots $K = \{k_m: m \in \mathbb{Z}\}$ are most useful. The knots then represent samples of the spline $s(t)$ on intervals of length $T = k_1 - k_0$. We confine the discussion to smooth splines.

Definition (B-Spline). The zero-order B-spline $\beta_0(t)$ is given by

$$\beta_0(t) = \begin{cases} 1 & \text{if } -\frac{1}{2} < t < \frac{1}{2}, \\ \frac{1}{2} & \text{if } |t| = \frac{1}{2}, \\ 0 & \text{if otherwise.} \end{cases} \tag{3.44}$$

The *B-spline of order n*, $\beta_n(t)$, is

$$\beta_n(t) = \underbrace{\beta_0(t) * \beta_0(t) * \ldots \beta_0(t)}_{n\,+\,1 \text{ times}}. \tag{3.45}$$

The B-splines are isolated pulses. The zeroth-order B-spline is a square, $\beta_1(t)$ is a triangle, and higher-order functions are Gaussian-like creatures (Figure 3.2).

Before proving some theorems about splines, let us recall a few ideas from the topology of the real number system $\mathbb{R}$ [6]. Open sets on the real line have soft edges; $S \subset \mathbb{R}$ is *open* if every $s \in S$ is contained in an open interval $I = (a, b)$ that is completely within S: $s \in I \subset S$. Closed sets have hard edges: S is *closed* if its complement is open. Unions of open sets are open, and intersections of closed sets are closed. S is *bounded* if it is contained in some finite interval (a, b). A set of open sets $O = \{O_n\}$ is an *open covering* of S if $\bigcup_n O_n \supset S$. Set S is *compact* if for every open

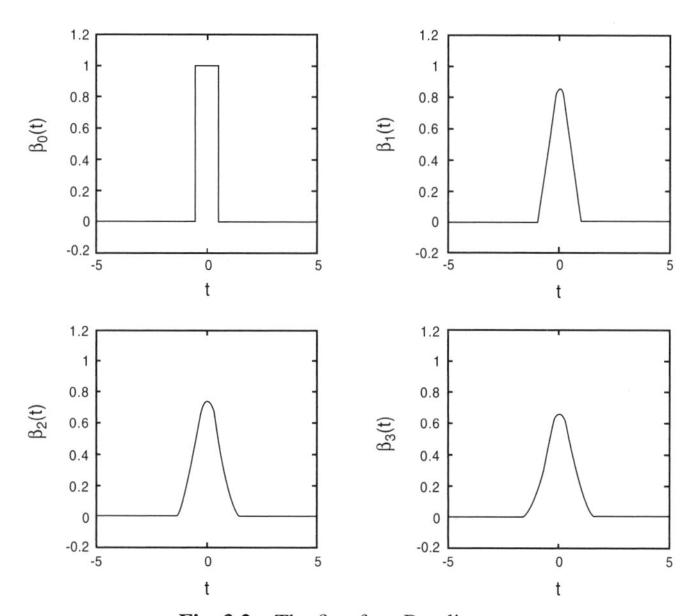

Fig. 3.2. The first four B-splines.

covering O of S there is a subset $P \subset O$ such that P is finite and P covers S. The famous Heine–Borel theorem[2] states that a set is compact if and only if it is closed and bounded [22, 23].

Definition (Support). Let $x(t)$ be an analog signal. Then $x(t)$ is *compactly supported* if $x(t) = 0$ outside some interval $[a, b]$. The *support* of $x(t)$, written Support(x), is the smallest closed set S such that $x(t) = 0$ if $t \notin S$.

Proposition. The n th-order B-spline $\beta_n(t)$ has compact support; that is, the smallest closed set which contains the domain over which set $\beta_n(t)$ is nonzero is closed and bounded.

Proof: By induction on the order (exercise). ■

The next result shows that general splines are linear combinations of shifted B-splines [14].

Theorem (Schoenberg). Let $x(t)$ be a spline having degree n and integral knots $K = \{m = k_m : m \in \mathbb{Z}\}$. Then there are constants c_m such that

$$s(t) = \sum_{m=-\infty}^{\infty} c_m \beta_n(t-m). \tag{3.46}$$

Proof: The proof is outside our present scope; we need Fourier transform tools (Chapters 5 and 6) for the arguments; interested readers will find some of the steps sketched in the latter chapter's problems. ■

The Schoenberg theorem shows that sums of simple shifts and scalings (amplification or attentuation) of the B-spline $\beta_n(t)$ will generate any spline of order n. Thus, the B-splines constitute a kind of signal atom. If signal $x(t)$ admits a spline model, then we can decompose the model into a sum of atoms. The relative abundance of each atom in the original signal at a time value $t = m$ is given by a simple coefficient c_m. This data reduction can be a great help to signal processing and analysis applications, such as filtering and pattern recognition. Notice too that as models of naturally occurring signals, B-splines have the realistic property that they decay to zero as time increases toward $\pm\infty$.

Once we develop more signal theory we shall explore B-spline ideas further. We require a more powerful tool: the full theory of signal frequency, which the Fourier transform provides. Looking forward, we will apply B-splines to construct windowed Fourier transforms (Chapter 10) and scaling functions for wavelet multiresolution analysis structures (Chapter 11).

[2]German analyst Heinrich Eduard Heine (1821–1881) first defined uniform continuity. Emile Borel (1871–1956) is mostly known for his contributions to topology and modern analysis. But he was also Minister of the French Navy (1925–1940) and received the Resistance Medal in 1945 for his efforts opposing the Vichy government during the war.

3.3 ANALOG SIGNAL SPACES

We must once again find sufficiently powerful mathematical structures to support the analog signal processing operations that we have been working with. For instance, convolution characterizes linear, translation-invariant analog systems. But how does one know that the convolution of two signals produces a genuine signal? Since convolution involves integration between infinite limits, there is ample room for doubt about the validity of the result. Are there classes of signals that are broad enough to capture the behavior of all the real-world signals we encounter and yet are closed under the convolution operation? We are thus once again confronted with the task of finding formal mathematical structures that support the operations we need to study systems that process analog signals.

3.3.1 L^p Spaces

Let us begin by defining the p-integrable signal spaces, the analog equivalents of the p-summable spaces studied in the previous chapter. Again, some aforementioned signal classes—such as the absolutely integrable, bounded, and finite-energy signals—fall into one of these families. The theory of signal frequency (Fourier analysis) in Chapters 5 and 7 uses these concepts extensively. These ideas also lie at the heart of recent developments in mixed-domain signal analysis: Gabor transforms, wavelets, and their applications, which comprise the final three chapters of the book.

3.3.1.1 Definition and Basic Properties. We begin with the definition of the p norm for analog signals. The norm is a measure of analog signal size. It looks a lot like the definition of the l^p norm for discrete signals, with an infinite integral replacing the infinite sum. Again, our signals may be real- or complex-valued.

Definition (L^p Norm or p Norm). If $x(t)$ is an analog signal and $p \geq 1$ is finite, then its L^p *norm* is

$$\|x\|_p = \left[\int_{-\infty}^{\infty} |x(t)|^p \, dt\right]^{\frac{1}{p}}, \tag{3.47}$$

if the Riemann integral exists. If $p = \infty$, then $\|x\|_\infty$ is the least upper bound of $\{|x(t)| \mid x \in \mathbb{R}\}$, if it exists. The L^p norm of $x(t)$ restricted to the interval $[a, b]$ is

$$\|x\|_{L^p[a,b]} = \left[\int_a^b |x(t)|^p \, dt\right]^{\frac{1}{p}}. \tag{3.48}$$

Other notations are $\|x\|_{p,\mathbb{R}}$ and $\|x\|_{p,[a,b]}$. The basic properties of the p-norm are the Hölder, Schwarz, and Minkowski inequalities; let us extend them to integrals.

Theorem (Hölder Inequality). Suppose $1 \leq p \leq \infty$ and p and q are conjugate exponents: $p^{-1} + q^{-1} = 1$. If $\|x\|_p < \infty$ and $\|y\|_q < \infty$, then $\|xy\|_1 \leq \|x\|_p \|y\|_q$.

Proof: Recall the technical lemma from Section 2.5.4: $ab \le p^{-1}a^p + q^{-1}b^q$, where p and q are conjugate exponents and $a, b > 0$. So by the same lemma,

$$\frac{|x(t)|}{\|x\|_p}\frac{|y(t)|}{\|y\|_q} \le \frac{|x(t)|^p}{p\|x\|_p^p} + \frac{|y(t)|^q}{q\|y\|_q^q}. \tag{3.49}$$

Integrating (3.49) on both sides of the inequality gives

$$\frac{1}{\|x\|_p\|y\|_q}\int_{-\infty}^{\infty}|x(t)||y(t)|\,dt \le \frac{1}{p\|x\|_p^p}\int_{-\infty}^{\infty}|x(t)|^p dt + \frac{1}{q\|y\|_q^q}\int_{-\infty}^{\infty}|y(t)|^q\,dt = \frac{1}{p}+\frac{1}{q} = 1. \tag{3.50}$$

The case $p = 1$ and $q = \infty$ is straightforward and left as an exercise. ■

Corollary (Schwarz Inequality). If $\|x\|_2 < \infty$ and $\|y\|_2 < \infty$, then $\|xy\|_1 \le \|x\|_2\|y\|_2$.

Proof: Because $p = q = 2$ are conjugate exponents. ■

Theorem (Minkowski Inequality). Let $1 \le p \le \infty$, $\|x\|_p < \infty$, and $\|y\|_p < \infty$. Then $\|x + y\|_p \le \|x\|_p + \|y\|_p$.

Proof: We prove the theorem for $1 < p < \infty$ and leave the remaining cases as exercises. Because $|x(t) + y(t)| \le |x(t)| + |y(t)|$ and $1 \le p$, we have $|x(t) + y(t)|^p \le (|x(t)| + |y(t)|)^p = (|x| + |y|)^{p-1}|x| + (|x| + |y|)^{p-1}|y|$. Integration gives

$$\int_{-\infty}^{\infty}|x(t)+y(t)|^p dt \le \int_{-\infty}^{\infty}(|x|+|y|)^p dt = \int_{-\infty}^{\infty}(|x|+|y|)^{p-1}|x|\,dt + \int_{-\infty}^{\infty}(|x|+|y|)^{p-1}|y|\,dt. \tag{3.51}$$

Let $q = \frac{p}{p-1}$ so that p and q are conjugate exponents. Hölder's inequality then applies to both integrals on the right-hand side of (3.51):

$$\int_{-\infty}^{\infty}|x|(|x|+|y|)^{p-1}dt = \int_{-\infty}^{\infty}|x|(|x|+|y|)^{\frac{p}{q}}dt \le \|x\|_p\left[\int_{-\infty}^{\infty}(|x|+|y|)^p dt\right]^{\frac{1}{q}} \tag{3.52a}$$

and

$$\int_{-\infty}^{\infty}|y|(|x|+|y|)^{p-1}dt = \int_{-\infty}^{\infty}|y|(|x|+|y|)^{\frac{p}{q}}dt \le \|y\|_p\left[\int_{-\infty}^{\infty}(|x|+|y|)^p dt\right]^{\frac{1}{q}}. \tag{3.52b}$$

If the term in square brackets is zero, the theorem is trivially true; we assme otherwise. Putting (3.52a) and (3.52b) together into (3.51) and dividing through by the square-bracketed term gives

$$\frac{\int_{-\infty}^{\infty} (|x| + |y|)^p dt}{\left[\int_{-\infty}^{\infty} (|x| + |y|)^p dt\right]^{\frac{1}{q}}} = \left[\int_{-\infty}^{\infty} (|x| + |y|)^p dt\right]^{p - \frac{1}{q}} \leq \|x\|_p + \|y\|_p . \tag{3.53}$$

Since $p - \frac{1}{q} = \frac{1}{p}$, the middle term in (3.53) is $\|\, |x| + |y| \,\|_p$. But $\|x + y\|_p \leq \|\, |x| + |y| \,\|_p$ and we are done. ■

Now we can define the principal abstract spaces for signal theory. The definition is provisional only, since it relies upon the Riemann integral for the idea that $\|x\|_p < \infty$. We offer two refinements in what follows.

Definition ($L^p(\mathbb{R})$, $L^p[a, b]$). Let $1 \leq p \leq \infty$. For $p < \infty$, the p-integrable space of analog signals or functions defined on the real numbers is $L^p(\mathbb{R}) = \{x(t) \mid x\colon \mathbb{R} \rightarrow \mathbb{K}$ and $\|x\|_p < \infty\}$, where $\|\cdot\|$ is the L^p norm and $\mathbb{K}$ is either the real numbers $\mathbb{R}$ or the complex numbers $\mathbb{C}$. Also if $a < b$, then $L^p[a, b] = \{x(t) \mid x\colon [a, b] \rightarrow \mathbb{K}$ and $\|x\|_{p,[a,b]} < \infty\}$. If $p = \infty$, then $L^\infty(\mathbb{R})$ and $L^\infty[a, b]$ are the bounded signals on $\mathbb{R}$ and $[a, b]$, respectively.

It is conventional to use uppercase letters for the analog p-integrable spaces and lowercase letters for the discrete p-summable signal spaces. It is also possible to consider half-infinite L^p spaces: $L^p(-\infty, a]$ and $L^p[a, +\infty)$.

These ideas have signal processing significance. The absolutely integrable signals can be used with other L^p signals under the convolution operation [12]. The following proposition tells us that as long as its impulse response $h = H\delta$ is absolutely integrable, an LTI system will produce an L^1 output from an L^1 input.

Proposition. If $x, h \in L^1(\mathbb{R})$, then $y = x * h \in L^1(\mathbb{R})$, and $\|y\|_1 \leq \|x\|_1 \|h\|_1$.

Proof

$$\int_{-\infty}^{\infty} |y(t)|\, dt = \int_{-\infty}^{\infty} \left| \int_{-\infty}^{\infty} x(s)h(t-s)\, ds \right| dt \leq \int_{-\infty}^{\infty} \int_{-\infty}^{\infty} |x(s)h(t-s)|\, ds dt. \tag{3.54}$$

From the Fubini–Tonelli theorem [24], if the two-dimensional integrand is either absolutely integrable or non-negative, then the double integral equals either iterated integral. Thus,

$$\|y\|_1 \le \int_{-\infty}^{\infty} |x(s)| \left[\int_{-\infty}^{\infty} |h(t-s)|\, dt \right] ds = \|x\|_1 \|h\|_1 . \tag{3.55}$$

■

The next proposition concerns the concept of uniform continuity, which readers may recall from calculus [6]. A function $y(t)$ is *uniformly continuous* means that for any $e > 0$ there is a $\delta > 0$ such that $|t - s| < \delta$ implies $|y(t) - y(s)| < \varepsilon$. The key idea is that for any $\varepsilon > 0$, it is possible to find a $\delta > 0$ that works for all time values. When the interval width δ must depend on $t \in \mathbb{R}$, then we may have ordinary continuity, but not necessarily uniform continuity. An example of a signal that is uniformly continuous on $\mathbb{R}$ is $\sin(t)$. A signal that is continuous, but not uniformly so, is t^2.

Proposition. If $x \in L^2(\mathbb{R})$ and $y = x \circ x$, then $|y(t)| \le \|x\|_2^2$. Furthermore, $y(t)$ is uniformly continuous.

Proof: We apply the Schwarz inequality, $\|fg\|_1 \le \|f\|_2 \|g\|_2$:

$$|y(t)| \le \int_{-\infty}^{\infty} |x(s)| \left|\overline{x(t+s)}\right| ds \le \|x(s)\|_2 \|x(t+s)\|_2 = \|x\|_2^2 . \tag{3.56}$$

To show uniform continuity, let us consider the magnitude $|y(t + \Delta t) - y(t)|$:

$$|y(t+\Delta t) - y(t)| \le \int_{-\infty}^{\infty} |x(s)| \left|\overline{x(t+\Delta t+s) - x(t+s)}\right| ds. \tag{3.57}$$

Invoking the Schwarz inequality on the right-hand side of (3.57) and changing the integration variable with $\tau = t + s$, we obtain

$$|y(t+\Delta t) - y(t)| \le \|x\|_2 \left[\int_{-\infty}^{\infty} |x(\tau+\Delta t) - x(\tau)|^2 d\tau \right]^{\frac{1}{2}} . \tag{3.58}$$

The limit, $\lim_{\Delta t \to 0} y(t + \Delta t) - y(t)$, concerns us. From integration theory—for instance, Ref. 24, p. 91—we know that

$$\lim_{\tau \to 0} \int_{-\infty}^{\infty} |x(\tau+\Delta t) - x(\tau)|\, d\tau = 0, \tag{3.59}$$

and since this limit does not depend upon t, $y(t)$ is uniform continuous. ■

Proposition. Let $1 \le p \le \infty$, $x \in L^p(\mathbb{R})$, and $h \in L^1(\mathbb{R})$. Then $y = x * h \in L^p(\mathbb{R})$, and $\|y\|_p \le \|x\|_p \|h\|_1$.

Proof: Exercise. ■

The criterion upon which the definition rides depends completely on whether $|x(t)|^p$ is integrable or not; that is, it depends how powerful an integral we can roust up. The Riemann integral, of college calculus renown, is good for functions that are piecewise continuous. Basic texts usually assume continuity of the integrand, but their theory generalizes easily to those functions having a finite number of discontinuities; it is only necessary to count the pieces and perform separate Riemann integrals on each segment. A refinement of the definition is possible, still based on Riemann integration. We make this refinement after we discover how to construct Banach spaces out of the L^p-normed linear spaces given by the first defintion of $\|x\|_p$. We issue the following warning: The Riemann integral will fail to serve our signal theoretic needs. We will see this as soon as we delve into the basic abstract signal structures: normed linear, Banach, inner product, and Hilbert spaces. The modern Lebesgue integral replaces it. Our final definition, which accords with modern practice, will provide the same finite integral criterion, but make use of the modern Lebesgue integral instead. This means that although the above definition will not change in form, when we interpret it in the light of Lebesgue's rather than Riemann's integral, the L^p spaces will admit a far wider class of signals.

3.3.1.2 Normed Linear Spaces.

A normed linear space allows basic signal operations such as summation and scalar multiplication (amplification or attenuation) and in addition provides a measure of signal size, the norm operator, written $\|\cdot\|$. Normed spaces can be made up of abstract elements, but generally we consider those that are sets of analog signals.

Definition (Norm, Normed Linear Space). Let X be a vector space of analog signals over $\mathbb{K}$ ($\mathbb{R}$ or $\mathbb{C}$). Then a *norm*, written $\|\cdot\|$, is a map $\|\cdot\|: X \to \mathbb{R}$ such that

(i) (Non-negative) $0 \le \|x\|$ for all $x \in X$.
(ii) (Zero) $\|x\| = 0$ if and only if $x(t) = 0$ for all t.
(iii) (Scalar multiplication) $\|ax\| = |c|\,\|x\|$ for every scalar $c \in \mathbb{K}$.
(iv) (Triangle inequality) $\|x + y\| \le \|x\| + \|y\|$.

If X is a vector space of analog signals and $\|\cdot\|$ is a norm on X, then $(X, \|\cdot\|)$ is a *normed linear space*. Other common terms are normed vector space or simply normed space.

One can show that the norm is a continuous map. That is, for any $x \in X$ and $\varepsilon > 0$ there is a $\delta > 0$ such that for all $y \in X$, $\|y - x\| < \delta$ implies $|\,\|y\| - \|x\|\,| < \varepsilon$. The algebraic operations, addition and scalar multiplication, are also continuous.

Example. Let $a < b$ and consider the set of continuous functions $x(t)$ on $[a, b]$. This space, denoted $C^0[a, b]$ is a normed linear space with the following norm: $\|x(t)\| = \sup\{|x(t)|: t \in [a, b]\}$. Since the closed interval $[a, b]$ is compact, it is closed and bounded (by the Heine–Borel theorem), and a continuous function therefore achieves a maximum [6]. Thus, the norm is well-defined.

Example. Let $C^0(\mathbb{R})$ the set of bounded continuous analog signals. This too is a normed linear space, given the supremum norm: $\|x(t)\| = \sup\{|x(t)|: t \in \mathbb{R}\}$.

Example. Different norms can be given for the same underlying set of signals, and this results in different normed vector spaces. For example, we can choose the energy of a continuous signal $x(t) \in C^0[a, b]$ as a norm:

$$\|x\| = E_x = \left[\int_a^b |x(t)|^2 dt\right]^{\frac{1}{2}}. \tag{3.60}$$

The next proposition ensures that the L^p-norm is indeed a norm for continuous signals.

Proposition. Let X be the set of continuous, p-integrable signals $x : \mathbb{R} \rightarrow \mathbb{K}$, where $\mathbb{K}$ is either $\mathbb{R}$ or $\mathbb{C}$. Then $\|x\|_p$ is a norm, and $(X, \|x\|_p)$ is a normed linear space.

Proof: The continuous signals are clearly an Abelian (commutative) group under addition, and obey the scalar multiplication rules for a vector space. Norm properties (i) and (iii) follow from basic integration theory. For (ii), note that if $x(t)$ is not identically zero, then there must be some t_0 such that $x(t_0) = \varepsilon \neq 0$. By continuity, in an interval $I = (a, b)$ about t_0, we must have $|x(t)| > \varepsilon/2$. But then the norm integral is at least $[\varepsilon(b - a)/2]^p > 0$. The last property follows from Minkowski's inequality for analog signals. ■

Proposition. Let X be the set of continuous, p-integrable signals x: $[a, b] \rightarrow \mathbb{K}$, where $a < b$. Then $(X, \|x\|_{p,[a, b]})$ is a normed linear space.

Proof: Exercise. ■

Must analog signal theory confine itself to continuous signals? Some important analog signals contain discontinuities, such as the unit step and square pulse signals, and we should have enough confidence in our theory to apply it to signals with an infinite number of discontinuities. Describing signal noise, for example, might demand just as much. The continuity assumption enforces the zero property of the norm, (ii) above; without presupposing continuity, signals that are zero except on a finite number of points, for example, violate (ii). The full spaces, $L^p(\mathbb{R})$ and $L^p[a,b]$, are not—from discussion so far—normed spaces.

A *metric space* derives naturally from a normed linear space. Recall from the Chapter 2 exercises that a *metric* $d(x, y)$ is a map from pairs of signals to real numbers. Its four properties are:

(i) $d(x, y) \geq 0$ for all x, y.
(ii) $d(x, y) = 0$ if and only if $x = y$.

(iii) $d(u, v) = d(v, u)$ for all u, v.
(iv) For any s, $d(x, y) \le d(x, s) + d(s, y)$.

Thus, the L^p-norm generates a metric. In signal analysis matching applications, a common application requirement is to develop a measure of match between candidate signals and prototype signals. The candidate signals are fed into the analysis system, and the prototypes are models or library elements which are expected among the inputs. The goal is to match candidates against prototypes, and it is typical to require that the match measure be a metric.

Mappings between normed spaces are also important in signal theory. Such maps abstractly model the idea of filtering a signal: signal-in and signal-out. When the normed spaces contain analog signals, or rather functions on the real number line, then such maps are precisely the analog systems covered earlier in the chapter. For applications we are often interested in linear maps.

Definition (Linear Operator). Let X and W be normed spaces over $\mathbb{K}$ ($\mathbb{R}$ or $\mathbb{C}$) and $T: X \to W$ such that for all $x, y \in X$ and any $a \in \mathbb{K}$ we have

(i) $T(x + y) = T(x) + T(y)$.
(ii) $T(ax) = aT(x)$.

Then T is called a *linear operator* or *linear map*. Dropping parentheses is widespread: $Tx \equiv T(x)$. If the operator's range is included in $\mathbb{R}$ or $\mathbb{C}$, then we more specifically call T a *linear functional*.

Proposition (Properties). Let X and W be normed spaces over $\mathbb{K}$ and let $T: X \to W$ be a linear operator. Then

(i) Range(T) is a normed linear subspace of W.
(ii) The null space of T, $\{x \in X \mid Tx = 0\}$ is a normed linear subspace of X.
(iii) The inverse map T^{-1}: Range(T) $\to X$ exists if and only if the null space of T is precisely $\{0\}$ in X.

Proof: Exercise. ■

Definition (Continuous Operator). Let X and W be normed spaces over $\mathbb{K}$ and $T: X \to W$. Then T is *continuous at x* if for any $\varepsilon > 0$ there is a $\delta > 0$ such that if $\|y - x\| < \delta$, then $\|Ty - Tx\| < \varepsilon$. T is *continuous* if it is continuous at every $x \in X$.

Definition (Norm, Bounded Linear Operator). Let X and W be normed spaces over $\mathbb{K}$ and $T: X \to W$ be a linear operator. Then we define the *norm* of T, written $\|T\|$, by $\|T\| = \sup\{\|Tx\|/\|x\|: x \in X, x \ne 0\}$. If $\|T\| < \infty$, then T is a *bounded linear operator*.

Theorem (Boundedness). Let X and W be normed spaces over $\mathbb{K}$ and $T: X \rightarrow W$ be a linear operator. Then the following are equivalent:

(i) T is bounded.
(ii) T is continuous.
(iii) T is continuous at $0 \in X$.

Proof: The easy part of this proof is not hard to spot: (ii) obviously implies (iii). Let us therefore assume continuity at zero (iii) and show that T is bounded. Let $\delta > 0$ such that $\|Tx - 0\| = \|Tx\| < 1$ when $\|x - 0\| = \|x\| < \delta$. Let $y \in X$ be nonzero. Then $\left\|\frac{\delta y}{2\|y\|}\right\| = \frac{\delta}{2} < \delta$, so that $\left\|\frac{T(\delta y)}{2\|y\|}\right\| < 1$ and $\|Ty\| < \frac{2\|y\|}{\delta}$; T is bounded. Now we assume T is bounded (i) and show continuity. Note that $\|Tx - Ty\| = \|T(x - y)\| \leq \|T\| \|x - y\|$, from which continuity follows. ■

The boundedness theorem seems strange at first glance. But what $\|T\| < \infty$ really says is that T amplifies signals by a limited amount. So there cannot be any sudden jumps in the range when there are only slight changes in the domain. Still, it might seem that a system could be continuous without being bounded, since it could allow no jumps but still amplify signals by arbitrarily large factors. The linearity assumption on T prevents this, however.

3.3.1.3 Banach Spaces. Analog Banach spaces are normed linear spaces for which every Cauchy sequence of signals converges to a limit signal also in the space. Again, after formal developments in the previous chapter, the main task here is to investigate how analog spaces using the L^p norm can be complete. Using familiar Riemann integration, we can solve this problem with an abstract mathematical technique: forming the completion of a given normed linear space. But this solution is unsatisfactory because it leaves us with an abstract Banach space whose elements are quite different in nature from the simple analog signals with which we began. Interested readers will find that the ultimate solution is to replace the Riemann with the modern Lebesgue integral.

Recall that a sequence of signals $\{x_n(t): n \in \mathbb{Z}\}$ is Cauchy when for all $\varepsilon > 0$ there is an N such that if $m, n > N$, then $\|x_m - x_n\| < \varepsilon$. Note that the definition depends on the choice of norm on the space X. That is, the signals get arbitrarily close to one another; as the sequence continues, signal perturbations become less and less significant—at least as far as the norm can measure. A signal $x(t)$ is the limit of a sequence $\{x_n(t)\}$ means that for any $\varepsilon > 0$ there is an $N > 0$ such that $n > N$ implies $\|x_n - x\| < \varepsilon$. A normed space X is complete if for any Cauchy sequence $\{x_n(t) : n \in \mathbb{N}\}$ there is an $x(t) \in X$ such that $x(t)$ is the limit of $\{x_n(t)\}$. A complete normed space is also called a Banach space.

In the previous section, we considered the continuous analog signals on the real line, or on an interval, and showed that with the L^p norm, they constituted normed linear spaces. Are they also Banach spaces? The answer is no, unfortunately;

Cauchy sequences of continuous signals may converge to signals that are not continuous, as the counterexample below illustrates.

Example. If $p < \infty$, then the signal space $(C^0[-1, 1], \|\cdot\|_{p,[-1,1]})$, consisting of all continuous signals on $[-1, 1]$ with the L^p norm, is not complete. The claim is secure if we can exhibit a Cauchy sequence of continuous functions that converges to a discontinuous function. A sequence that approximates the unit step on $[-1, 1]$ is

$$x_n(t) = \begin{cases} 0 & \text{if } t < -\frac{1}{n}, \\ \frac{tn}{2} + \frac{1}{2} & \text{if } -\frac{1}{n} \le t \le \frac{1}{n}, \\ 1 & \text{if } t > \frac{1}{n}. \end{cases} \tag{3.61}$$

The $\{x_n(t)\}$, shown in Figure 3.3, clearly converge pointwise. Indeed, for any $t_0 < 0$, $x_n(t_0) \to 0$; for any $t_0 > 0$, $x_n(t_0) \to 1$; and for $t_0 = 0$, $x_n(t_0) = 1/2$ for all n. Now, if we assume that $n < m$, then

$$\int_{-1}^{1} |x_m(t) - x_n(t)|^p dt = \int_{-1/n}^{1/n} |x_m(t) - x_n(t)|^p dt \le \frac{2}{n}\left(\frac{1}{2^p}\right). \tag{3.62}$$

The sequence is Cauchy, but converges to a discontinuous signal. The same reasoning applies to $L^p[a, b]$, where $a < b$, and to $L^p(\mathbb{R})$.

Example. Now consider $C^0[a, b]$ the set of bounded continuous analog signals on $[a, b]$ with the supremum or L^∞ norm: $\|x(t)\|_\infty = \sup\{|x(t)| : t \in [a, b]\}$. This space's norm avoids integration, so $(C^0[a, b], \|\cdot\|_\infty)$ earns Banach space status. To see this, note that if $\{x_n(t)\}$ is Cauchy and $\varepsilon > 0$, then there is an $N > 0$ such that for all $m, n > N$ we have $\|x_n - x_m\|_\infty < \varepsilon$. Fixing $t_0 \in [a, b]$, the sequence of real numbers $\{x_n(t_0)\}$ is Cauchy in $\mathbb{R}$. Calculus teaches that Cauchy sequences of real numbers converge to a limit in $\mathbb{R}$; for each $t \in [a, b]$, we may therefore set $x(t) = \lim_{n \to \infty} x_n(t)$. We claim $x(t)$ is continuous. Indeed, since the sequence $\{x_n(t)\}$ is Cauchy in the L^∞ norm, the sequence must converge not just pointwise, but uniformly to $x(t)$. That is, for any $\varepsilon > 0$, there is an $N > 0$ such that $m, n > N$ implies $|x_m(t) - x_n(t)| < \varepsilon$ for all $t \in [a, b]$. Uniformly convergent sequences of continuous functions converge to a continuous limit [6] $x(t)$ must therefore be continuous, and $C^0[a, b]$ is a Banach space.

Analog signal spaces seem to leave us in a quandary. We need continuity in order to achieve the basic properties of normed linear spaces, which provide a basic signal size function, namely the norm. Prodded by our intuition that worldly processes—at least at our own perceptual level of objects, forces between them, and their motions—are continuously defined, we might proceed to develop analog signal theory from

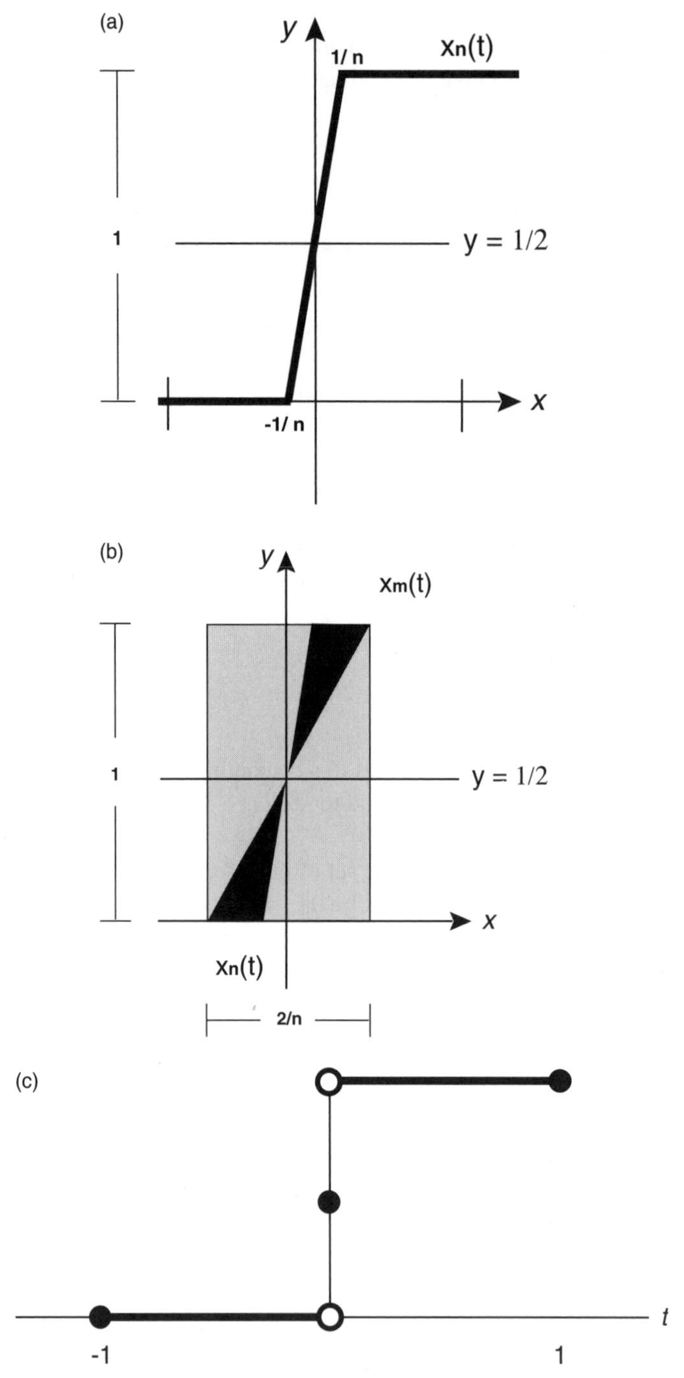

Fig. 3.3. (a) A Cauchy sequence of continuous signals in $L^p[-1, 1]$. (b) Detail of the difference between x_m and x_n. Assuming $n < m$, the signals differ by at most 1/2 within the rectangular region, which has width $2/n$. (c) Diagram showing discontinuous limit.

continuous signals alone. However, arbitrarily precise signal approximations depend on the mathematical theory of limits, which in turn begs the question of the completeness for our designated theoretical base, $(C^0[-1, 1], \|\cdot\|_p)$ for example. But the first above example shows that we cannot get completeness from families of continuous signals with the L^p-norm, where $p < \infty$.

What about working with the supremum norm, $\|\cdot\|_\infty$? This is a perfectly good norm for certain signal analysis applications. Some matching applications rely on it, for example. Nonetheless, the next section shows that L^∞ does not support an inner product. Inner products are crucial for much of our later development: Fourier analysis, windowed Fourier (Gabor) transforms, and wavelet transforms. Also many applications presuppose square-integrable physical quantities. It would appear that the supremum norm would confine signal theory to a narrow range of processing and analysis problems.

Let us persist: Can we only allow sequences that do converge? As the exercises explain, uniformly convergent sequences of continuous signals converge to a continuous limit. The problem is when the signal values, say $x_n(t_0)$, converges for every $t_0 \in [-1, 1]$, then we should expect that x_n itself converges to an allowable member of our signal space. Alternatively, since Cauchy sequences of continuous signals lead us to discontinuous entities, can we just incorporate into our foundational signal space the piecewise continuous signals that are p-integrable? We would allow signals with a finite number of discontinuities. The Riemann integral extends to them, and it serves the L^p-norm definition (3.47). We would have to give up one of the criteria for a normed linear space; signals would differ, but the norm of their difference would be zero. Frustratingly, this scheme fails, too, as the next example shows [25].

Example. Consider the signals $\{x_n(t)\}$ defined on $[-1, 1]$ defined by

$$x_n(t) = \lim_{m \to \infty} [\cos(n!\pi t)]^{2m} = \begin{cases} 1 & \text{if } n!t \in \mathbb{Z}, \\ 0 & \text{if otherwise.} \end{cases} \tag{3.63}$$

Note that $x_n(t)$ is zero at all but a finite number of points in $[-1, 1]$; $x_n(t)$ is Riemann-integrable, and its Riemann integral is zero on $[-1, 1]$. Also $\|x_n(t) - x_m(t)\|_{p,[-1,1]} = 0$, which means that the sequence $\{x_n(t)\}$ is Cauchy. It converges, however, to

$$x(t) = \begin{cases} 1 & \text{if } t \in \mathbb{Q}, \\ 0 & \text{if otherwise}. \end{cases} \tag{3.64}$$

which is not Riemann-integrable.

3.3.1.4 Constructing Banach Spaces. An abstract mathematical technique, called completing the normed space, eliminates most of the above aggravations. The completion of a normed linear space X, is a Banach space B having a subspace C isometric to X. Two normed spaces, M and N, are isometric means there is a one-to-one, onto map $f: M \to N$ such that f is a vector space isomorphism and for all $x \in M$, $\|x\|_M = \|f(x)\|_N$. Isometries preserve norms. The next theorem shows how to construct the completion of a general normed linear space.

Let us cover an algebraic idea which features in the completion result: equivalence relations [26].

Definition (Equivalence Relation). We say $a \sim b$ is an equivalence relation on a set S if it satisfies the following three properties:

(i) (Reflexive) For all $a \in S$, $a \sim a$.
(ii) (Symmetric) For all $a, b \in S$, if $a \sim b$, then $b \sim a$.
(iii) (Transitive) For all $a, b, c \in S$, if $a \sim b$ and $b \sim c$, then $a \sim c$.

Equivalence relations associate things in a collection that are similar in form.

Example. For example, we might consider ordered pairs of integers (m, n) with $n \neq 0$. If $p = (m, n)$ and $q = (i, k)$, then it is easy to check that the relation, $p \sim q$ if and only if $m \cdot k = n \cdot i$, is an equivalence relation. In fact, this is the equivalence relation-for different forms of the same rational number: $\frac{7}{8} = \frac{42}{48}$, for instance. Rational numbers themselves are not pairs of integers, but are represented by an infinite collection of such pairs. This is an important idea.

Definition (Equivalence Class). Let $a \sim b$ be an equivalence relation on a set S. The equivalence class of $a \in S$ is $[a] = \{b \in S \mid a \sim b\}$.

If $a \sim b$ is an equivalence relation on a set S, then the equivalence classes form a partition of S. Every $a \in S$ belongs to one and only one equivalence class.

Example (Rational Numbers). From the previous example we can let $S = \{(m, n) \mid m, n \in \mathbb{Z} \text{ and } n \neq 0\}$. Let $\mathbb{Q} = \{[q] \mid q \in S\}$. We can define addition and multiplication on elements of $\mathbb{Q}$. This is easy. If $p = (m, n)$ and $q = (i, k)$, then we define $[p] + [q] = [(M, N)]$, where $\frac{M}{N} = \frac{m}{n} + \frac{i}{k}$. Constructing a rational multiplication operator is simple, too. These steps construct the rational numbers from the integers [26].

But for signal theory, so what? Well, equivalence relations are precisely what we need to deal with the problem that arises when making a normed linear space based on the L^p norm. Just a moment ago, we noted that the zero property of the norm—namely, that $\|x\|_p = 0$ if and only if $x(t) = 0$ identically—compelled us to use continuous signals for normed linear spaces. But we use and need piecewise continuous entities, such as the unit step, in signal processing. Also, signals with point discontinuities are also useful for modeling noise spikes and the like. The strategy is to begin with piecewise discontinuous signals, but to assemble them into equivalence classes for making the normed linear space.

Example (Signal Equivalence Classes). Let $1 \le p < \infty$, and suppose that we have extended the Riemann integral to piecewise continuous analog signals (that is, having at most a finite number of discontinuities) on the interval $[a, b]$. Strictly speaking, this class of signals is not a normed linear space using $\|\cdot\|_p$; the zero property of the norm fails. However, we let $[x] = \{y(t)\colon \|y\|_p = \|x\|_p\}$ and define $\underline{L}^p[a, b] = \{[x] \mid x \in L^p[a, b]\}$. Evidently, we have identified all signals that differ from one another by a finite number of discontinuities as being essentially the same. That is what lumping them into equivalence classes accomplishes. Now we can define signal addition, scalar multiplication, and the norm on equivalence classes. This reinterpretation of $L^p[a, b]$ in terms of equivalence classes is often implicit in many signal processing treatments. The same idea applies to $L^p(\mathbb{R})$. The exercises explore these ideas in greater depth.

Now let us turn to the problem of constructing a Banach space from a given normed linear space [22, 27]. This leads to a refined definition for the L^p spaces.

Theorem (Completion of Normed Linear Space). Let X be a normed linear space. Then there is a Banach space B and an isometry $f\colon X \to B$ such that $C = \text{Range}(f)$ is dense in B.

Proof: Let $S = \{x_n\}$ and $T = \{y_n\}$ be Cauchy sequences in X. We define the relation $S \sim T$ if and only if $\lim_{n\to\infty}\|x_n - y_n\| = 0$. It is not hard to show this is an equivalence relation (exercise). Let $[S] = \{T\colon T \sim S\}$ and set $B = \{[S]\colon\ S = \{x_n\} \text{ is Cauchy in } X\}$. Let $x \in X$ and define $f(x) = [\{x_n\}]$, where $x_n = x$ for all n; the image $f(x)$ is a constant sequence. There are several things we must show.

With appropriate definitions of addition, scalar multiplication, and norm, we can make B into a normed linear space. If $S = \{x_n\}$, $T = \{y_n\} \in B$, then we define an additon operation on B by $[S] + [T] = [\{x_n + y_n\}]$. This works, but there is a slight technical problem. Many different Cauchy sequences $\{a_n\}$ can be the source for a single equivalence class, say $[S]$. We must show that the definition of additon does not depend on which sequences in the respective equivalence classes, S and T, are taken for defining the sums in $\{x_n + y_n\}$. So suppose $S = [\{a_n\}]$ and $T = [\{b_n\}]$ so that $\{a_n\} \sim \{x_n\}$ and $\{b_n\} \sim \{y_n\}$. We claim that $[\{x_n + y_n\}] = [\{a_n + b_n\}]$; that is, our addition operation is well-defined. Because $\|(x_n + y_n) - (a_n + b_n)\| = \|(x_n - a_n) + (y_n - b_n)\| \le \|x_n - a_n\| + \|y_n - b_n\|$, and both of these last terms approach zero, we must have $\{x_n + y_n\} \sim \{a_n + b_n\}$, proving the claim. We define scalar multiplication by $c[S] = [\{cx_n\}]$. It is straightforward to show that these definitions make B into a vector space. For the norm, we define $\|[S]\| = \lim_{n\to\infty}\|x_n\|$. Justifying the definition requires that the limit exists and that the definition is independent of the sequence chosen from the equivalence class $[S]$. We have

$$\big|\|x_n\| - \|x_m\|\big| \le \|x_n - x_m\| \tag{3.65}$$

by the triangle inequality. Since $\{x_n\}$ is Cauchy in X, so must $\{\|x_n\|\}$ be in $\mathbb{R}$. Next, suppose that some other sequence $\{a_n\}$ generates the same equivalence class: $[\{a_n\}] = [S]$. We need to show that $\lim_{n\to\infty}\|x_n\| = \lim_{n\to\infty}\|a_n\|$. In fact, we know that $[\{a_n\}] \sim [\{x_n\}]$, since they are in the same equivalence class. Thus, $\lim_{n\to\infty}\|x_n - a_n\| = 0$. Since $\|x_n - a_n\| \geq |\|x_n\| - \|a_a\||$, $\lim_{n\to\infty}[\|x_n\| - \|a_n\|] = 0$, and we have shown our second point necessary for a well-defined norm. Verifying the normed space properties remains, but it is straightforward and perhaps tedious.

Notice that the map f is a normed space isomorphism that preserves norms—an *isometry.* An *isomorphism* is one-to-one and onto, $f(x + y) = f(x) + f(y)$, and $f(cx) = cf(x)$ for scalars $c \in \mathbb{R}$ (or $\mathbb{C}$). In an isometry we also have $\|x\| = \|f(x)\|$.

Our major claim is that B is complete, but a convenient shortcut is to first show that Range(f) is dense in B. Given $[T] = [\{y_n\}] \in B$, we seek an $x \in X$ such that $f(x)$ is arbitrarily close to $[T]$. Since $\{y_n\}$ is Cauchy, for any $\varepsilon > 0$, there is an N_ε such that if $m, n > N_\varepsilon$, then $\|y_m - y_n\| < \varepsilon$. Let $k > N_\varepsilon$, and set $x = x_n = y_k$ for all $n \in \mathbb{N}$. Then $f(x) = [\{x_n\}]$, and $\|[\{y_n\}] - [\{x_n\}]\| = \lim_{n\to\infty}\|y_n - x_n\| = \lim_{n\to\infty}\|y_n - y_k\| \leq \varepsilon$. Since ε is arbitrary, $f(X)$ must be dense in B.

Finally, to show that B is complete, let $\{S_n\}$ be a Cauchy sequence in B; we have to find an $S \in B$ such that $\lim_{n\to\infty}S_n = S$. Since Range(f) is dense in B, there must exist $x_n \in X$ such that $\|f(x_n) - S_n\| < 1/(n + 1)$ for all $n \in \mathbb{N}$. We claim that $\{x_n\}$ is Cauchy in X, and if we set $S = [\{x_n\}]$, then $S = \lim_{n\to\infty}S_n$. Since f is an isometry,

$$\begin{aligned}\|x_n - x_m\| &= \|f(x_n) - f(x_m)\| = \|(f(x_n) - S_n) + (S_m - f(x_m)) + (S_n - S_m)\| \\ &\leq \|f(x_n) - S_n\| + \|S_m - f(x_m)\| + \|S_n - S_m\|.\end{aligned} \tag{3.66}$$

By the choice of $\{x_n\}$, the first two terms on the bottom of (3.66) are small for sufficiently large m, n. The final term in (3.66) is small too, since $\{S_n\}$ is Cauchy. Consequently, $\{x_n\}$ is Cauchy, and $S = [\{x_n\}]$ must be a bona fide element of B. Furthermore, note that $\|S_k - S\| \leq \|f(x_k) - S_k\| + \|f(x_k) - S\|$. Again, $\|f(x_k) - S_k\| < 1/(k + 1)$; we must attend to the second term. Let $y_n = x_k$ for all $n \in N$, so that $f(x_k) = [\{y_n\}]$. Then $\|f(x_k) - S\| = \|[\{y_n\}] - [\{x_n\}]\| = \lim_{n\to\infty}\|y_n - x_n\| = \lim_{n\to\infty}\|x_k - x_n\|$. But $\{x_n\}$ is Cauchy, so this last expression is also small for large n. ■

Corollary (Uniqueness). Let X be a normed linear space and suppose B is a Banach space with a dense subset C isometric to X. Then B is isometric to the completion of X.

Proof: Let f: $X \to C$ be the isometry and suppose $\overline{X}$ is the completion of X. Any element of B is a limit of a Cauchy sequence of elements in C: $b = \lim_{n\to\infty}c_n$. We can extend f to a map from $\overline{X}$ to B by $f(x) = b$, where $x = \lim_{n\to\infty}f^{-1}(c_n)$. We trust the further demonstration that this is an isometry to the reader (exercise). ■

The corollary justifies our referring to *the* completion of a normed linear space. Now we can refine the definition of the p-integrable signal spaces.

Definition ($L^p(\mathbb{R})$, $L^p[a, b]$). For $1 \le p < \infty$, $L^p(\mathbb{R})$ and $L^p[a, b]$ are the completions of the normed linear spaces consisting of the continuous, Riemann p-integrable analog signals on $\mathbb{R}$ and $[a, b]$, respectively.

So the completion theorem builds up Banach spaces from normed linear spaces having only a limited complement of signals—consisting, for instance, of just continuous signals. The problem with completion is that it provides no clear picture of what the elements in completed normed space look like. We do need to expand our realm of allowable signals because limit operations lead us beyond functions that are piecewise continuous. We also seek constructive and elemental descriptions of such functions and hope to avoid invoking abstract, indirect operations such as with the completion theorem. Can we accomplish so much and still preserve closure under limit operations?

The Lebesgue integral is the key concept. Modern integration theory removes almost every burden the Riemann integral imposes, but some readers may prefer to skip the purely mathematical development; the rest of the text is quite accessible without Lebesgue integration. So we postpone the formalities and turn instead to inner products, Hilbert spaces, and ideas on orthonormality and basis expansions that we need for our later analysis of signal frequency.

3.3.2 Inner Product and Hilbert Spaces

Inner product and Hilbert spaces provide many of the theoretical underpinnings for time domain signal pattern recognition applicatins and for the whole theory of signal frequency, or Fourier analysis.

3.3.2.1 Inner Product Spaces. An inner product space X is a vector space equipped with an inner product relation $\langle x, y\rangle$. The operation $\langle \cdot, \cdot\rangle$ takes pairs of elements in X and maps them to the real numbers or, more generally, the complex numbers. The algebraic content of Chapter 2's development is still valid; again, all we need to do is define the inner product for analog signals and verify that the properties of an abstract inner product space remain true.

Definition (Inner Product). Let $x(t)$ and $y(t)$ be real- or complex-valued analog signals. Then their inner product is $\langle x, y\rangle$:

$$\langle x, y\rangle = \int_{-\infty}^{\infty} x(t)\overline{y(t)}\,dt\,. \tag{3.67}$$

The inner product induces a norm, $\|x\| = (\langle x, x\rangle)^{1/2}$. So any inner product space thereby becomes a normed linear space. Readers mindful of Chapter 2's theorems will rightly suspect that the converse does not hold for analog signals. Recall that the inner product norm obeys the *parallelogram law*,

$$\|x+y\|^2 + \|x-y\|^2 = 2(\|x\|^2 + \|y\|^2)\,, \tag{3.68}$$

and the *polarization identity*,

$$4\langle x, y\rangle = \|x+y\|^2 - \|x-y\|^2 + j\|x+jy\|^2 - j\|x-jy\|^2. \tag{3.69}$$

It is not hard to show that the definition (3.67) satisfies the properties of an inner product. The main difficulty—as in the discrete world—is to find out for which abstract signal spaces the integral (3.67) actually exists. As with our earlier discrete results, we find that the space L^2 is special.

Example (Square-Integrable Signals). The spaces $L^2[a, b]$ and $L^2(\mathbb{R})$ are inner product spaces. Let $x(t)$ and $y(t)$ be real- or complex-valued analog signals. By the Schwarz inequality we know that if signals x and y are square-integrable, that is $\|x\|_2 < \infty$ and $\|y\|_2 < \infty$, then $\|xy\|_1 \le \|x\|_2\|y\|_2$. We must show that their inner product integral (3.67) exists. But,

$$|\langle x, y\rangle| = \left|\int_{-\infty}^{\infty} x(t)\overline{y(t)}\,dt\right| \le \int_{-\infty}^{\infty} \left|x(t)\overline{y(t)}\right|\,dt = \int_{-\infty}^{\infty} |x(t)||y(t)|\,dt = \|xy\|_1. \tag{3.70}$$

Schwarz's inequality shows the integration works. It states that if $\|x\|_2 < \infty$ and $\|y\|_2 < \infty$, then $\|xy\|_1 \le \|x\|_2\|y\|_2$. But (3.70) shows that $|\langle x, y\rangle| \le \|xy\|_1$. Requisite properties of an inner product are:

(i) $0 \le \langle x, x\rangle$ and $\langle x, x\rangle = 0$ if and only if $x(t) = 0$ for all t.
(ii) $\langle x + y, z\rangle = \langle x, z\rangle + \langle y, z\rangle$.
(iii) $\langle cx, y\rangle = c\langle x, y\rangle$, for any scalar c.
(iv) $\langle x, y\rangle = \overline{\langle y, x\rangle}$.

Their verification from (3.67) follows from the basic properties of Riemann integration (exercise).

Example (L^p spaces, $p \ne 2$). The spaces $L^p[a, b]$ and $L^p(\mathbb{R})$ are *not* inner product spaces with $\langle\cdot,\cdot\rangle$ defined in (3.67). Let $x(t)$ and $y(t)$ be the signals shown in Figure 3.4. Observe that $\|x\|_p = \|y\|_p = 2^{1/p}$, but $\|x + y\|_p = \|x - y\|_p = 2$. The parallelogram law holds in an inner product space, which for these signals implies $2 = 2^{2/p}$. This is only possible if $p = 2$.

3.3.2.2 Hilbert Spaces. An inner product space that is complete with respect to its induced norm is called a *Hilbert space*. All of the L^p signal spaces are Banach spaces, but only L^2 is an inner product space. So our generalization of linear algebra to encompass vectors that are infinitely long in both directions—that is, *signals*—succeeds but at the cost of eliminating all but an apparently narrow class of signals.

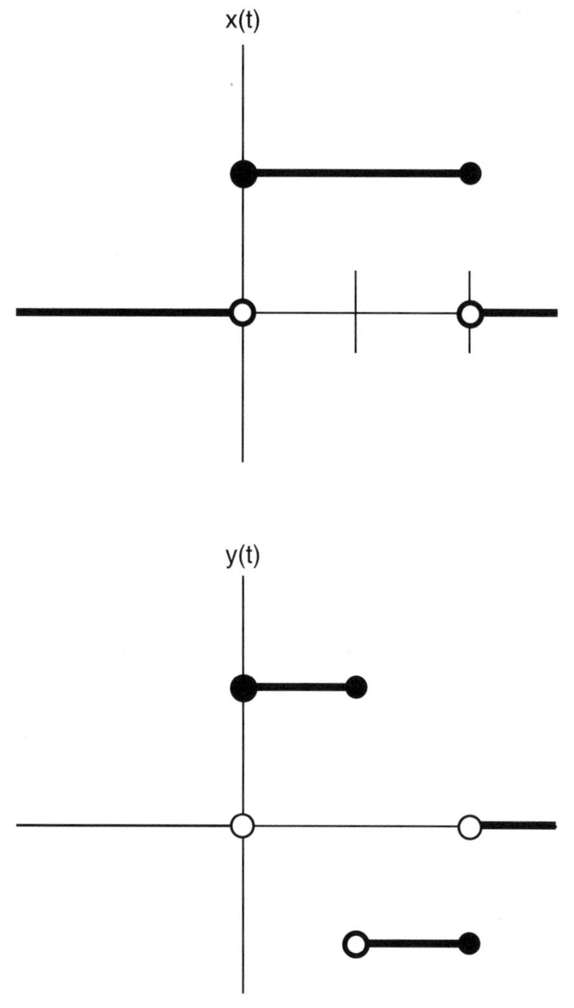

Fig. 3.4. Signals in the L^p Banach spaces: a square pulse $x(t)$ and a step signal $y(t)$.

Though it is true that square-integrable signals shall be our principal realm for signal theory, other classes do feature significantly in the sequel:

- Absolutely integrable signals, L^1;
- Bounded signals, L^∞;
- Certain subclasses of L^2, such as $L^1 \cap L^2$;
- Infinitely differentiable, rapidly decreasing signals.

It turns out that the L^1 signals constitute the best stepping-off point for constructing the Fourier transform (Chapter 5). This theory is the foundation of signal frequency analysis. With some extra effort, we can also handle L^2 signals, but we have to resort to limit operations to so extend the Fourier transform. Boundedness connects with absolute integrability in the theory of stable linear, translation-invariant systems: $y = Hx$ is stable when its impulse response $h = H\delta \in L^1(\mathbb{R})$. L^∞ is a basic, but useful, signal class. For example, if $f \in L^\infty$ and $g \in L^p$, then $fg \in L^p$.

Example (Square-Integrable Signals). The spaces $L^2[a, b]$ and $L^2(\mathbb{R})$ are Hilbert spaces. All of the L^p spaces are complete, and L^2 has an inner product that corresponds to its standard p-norm. These and their discrete cousin, $l^2(\mathbb{Z})$, the square-summable discrete signals, are the most important Hilbert spaces for signal theory.

Example (Absolutely and Square-Integrable Signals). The inner product on $L^2(\mathbb{R})$, restricted to those signals that are also absolutely integrable, furnishes a $\langle \cdot, \cdot \rangle$ operation for $L^1 \cap L^2$. We have to define $\|x\| = \langle x, x \rangle^{1/2}$ on this space. Note too that any Cauchy sequence of signals in $L^1 \cap L^2$ is still Cauchy in L^1. Thus, the sequence converges to an absolutely integrable limit. This sequence and its limit is also square-integrable, and so the limit is also in L^2. Thus, $L^1 \cap L^2$ is complete. It is easy to show that the restrictions of the signal addition and scalar multiplication operations to $L^1 \cap L^2$ are closed on that space. So $L^1 \cap L^2$ is a Hilbert space. We can say more: $L^1 \cap L^2$ is a *dense* subspace of L^2; for every square integrable signal $x(t)$ and any $\varepsilon > 0$, there is $y(t) \in L^1 \cap L^2$ such that $\|y - x\| < \varepsilon$. In fact, if $x(t) \in L^2$, then we may take

$$x_n(t) = \begin{cases} x(t) & \text{if } -n \le t \le n, \\ 0 & \text{if otherwise.} \end{cases} \tag{3.71}$$

Then $\lim_{n \to \infty} x_n = x$, and $x_n \in L^1(\mathbb{R})$ are absolutely integrable because they are compactly supported.

Example (Schwarz Space). The Schwarz space S is the class of infinitely differentiable, rapidly decreasing functions of a real variable [28]. *Infinitely differentiable* means that each $x(t) \in S$ has derivatives of all orders. Thus, $x(t)$ and its derivatives are all continuous. *Rapidly decreasing* means that $\lim_{t \to \infty} t^m x^{(n)}(t) = 0$ for all $m, n \in \mathbb{N}$, where $x^{(n)}(t) = d^n x / dt^n$. Examples of signals in S are the Gaussians of mean μ and standard deviation $\sigma > 0$:

$$g_{\mu, \sigma}(t) = \frac{1}{\sigma\sqrt{2\pi}} e^{-\frac{(t-\mu)^2}{2\sigma^2}}. \tag{3.72}$$

Examples of signals not in S include rational signals such as $(\sigma^2 + t^2)^{-1}$; these are not rapidly decreasing. The even decaying exponentials $\exp(-\sigma|t|)$ rapidly decrease, but fail differentiability, so they are not in the Schwarz class. The Schwarz space is a plausible candidate for the mathematical models of continuously defined naturally occurring signals: temperatures, voltages, pressures, elevations, and like quantitative phenomena. The Schwarz class strikes us as incredibly special and probably populated by very few signals. In fact, S is dense in both $L^1(\mathbb{R})$ and $L^2(\mathbb{R})$. To see this, it suffices to show that elements of S are arbitrarily close to square pulse functions. Since linear combinations of square pulses are precisely the step functions, and since step functions are dense in $L^1(\mathbb{R})$ and $L^2(\mathbb{R})$, this implies that S is dense as well. The trick is to blend the upper and lower ledge of a step together in an infinitely differentiable way [28]. The following function interpolates a unit step edge:

$$s(t) = \begin{cases} 0 & \text{if } t \le 0, \\ e^{-\frac{1}{t\exp\left[\frac{1}{t-1}\right]}} & \text{if } 0 < t < 1. \\ 1 & \text{if } 1 \le t. \end{cases} \tag{3.73}$$

Scalings and dilations of $s(t)$ interpolate a general step edge, and it is evident that arbitrary step functions are approximated to any precision by linear combinations of functions of the form (3.73).

There is a rich theory of linear operators and especially linear functionals on Hilbert spaces. Detailed explication of the ideas would take this presentation too far astray into abstract functional analysis; we refer the reader to the broad, excellent literature [13, 25, 26, 29] We shall be obliged to quote some of these results here, as elaborating some of our later signal analysis tools depends upon them.

Example (Inner Product). If we fix $h \in L^2(\mathbb{R})$, then the inner product $Tx = \langle x, h \rangle$ is a bounded linear functional with $\|T\| = \|h\|_2$. Indeed by Schwarz's inequality (3.70), $\|Tx\| = |\langle x, h \rangle| = \|xh\|_1 \le \|x\|_2 \|h\|_2$.

We have held up the inner product operation as the standard by which two signals may be compared. Is this right? It is conjugate linear and defined for square-integrable analog signals, which are desirable properties. We might well wonder whether any functional other than the inner product could better serve us as a tool for signal comparisons. The following theorem provides the answer.

Theorem (Riesz Representation). Let T be a bounded linear function on a Hilbert Space H. Then there is a unique $h \in H$ such that $Tx = \langle x, h \rangle$ for all $x \in H$. Futhermore, $\|T\| = \|h\|$.

Proof: See Ref. 29. ■

Inner products and bounded linear operators are very closely related. Using a generalization of the Riesz representation theorem, it is possible to show that every bounded linear Hilbert space operator $T : H \to K$ has a related map $S : K \to H$ which cross-couples the inner product.

Theorem. Let $T: H \to K$ be a bounded linear function on Hilbert space H and K. Then there is a bounded linear operator $S : K \to H$ such that:

(i) $\|T\| = \|S\|$.
(ii) For all $h \in H$ and $k \in K$, $\langle Th, k\rangle = \langle h, Sk\rangle$.

Proof: The idea is as follows. Let $k \in K$ and define the linear functional $L : H \to \mathbb{K}$ ($\mathbb{R}$ or $\mathbb{C}$) by $L(h) = \langle Th, k\rangle$. L is linear by the properties of the inner product. L is also bounded. The Riesz representation theorem applies, guaranteeing that there is a unique $g \in H$ such that $L(h) = \langle h, g\rangle$. Thus, we set $S(k) = g$. After verifying that S is linear and bounded, we see that it satisfies the two required properties [26]. ■

Definition (Adjoint). Let $T: H \to K$ be a bounded linear operator on Hilbert spaces H and K and S be the map identified by the previous theorem. Then S is called the *Hilbert adjoint operator* of T and is usually written $S = T^*$. If $H = K$ and $T^* = T$, then T is called *self-adjoint.*

Note that if T is self-adjoint, then $\langle Th, h\rangle = \langle h, Th\rangle$. So $\langle Th, h\rangle \in \mathbb{R}$ for all $h \in H$. This observation enables us to order self-adjoint operators.

Definition (Positive Operator, Ordering). A self-adjoint linear operator is *positive*, written $T \geq 0$, if for all $h \in H$, $0 \leq \langle Th, h\rangle$. If S and T are self-adjoint operators on a Hilbert space H with $T - S \geq 0$, then we say $S \leq T$.

We shall use ordered self-adjoint linear operators when we study frame theory in Section 3.3.4.

Finally, important special cases of Hilbert operators are those that are isometries.

Definition (Isometry). If $T: H \to K$ is linear operator on Hilbert spaces H and K and for all $g, h \in H$ we have $\langle Tg, Th\rangle = \langle g, h\rangle$, then T is an *isometry.*

3.3.2.3 Application: Constructing Optimal Detectors. As an application of functional analysis ideas to signal analysis, consider the problem of finding a known or prototype signal $p(t)$ within a given, candidate signal $x(t)$. The idea is to convolve a kernel $k(t)$ with the input: $y(t) = (x * k)(t)$. Where the response $y(t)$ has a maximum, then hopefully $x(t)$ closely resembles the model signal $p(t)$. How should we choose $k(t)$ to make this work?

A commonly used approach is to let $k(t) = p(-t)$, the prototype pattern's reflection [30, 31]. Then $y(t) = (x * k)(t) = (x \circ p)(t)$, the correlation of $x(t)$ and $p(t)$. Since

$$\int_{-\infty}^{\infty}\int_{-\infty}^{\infty} (x(t) - p(t))^2\, dt = \int_{-\infty}^{\infty}\int_{-\infty}^{\infty} x^2(t)dt + \int_{-\infty}^{\infty}\int_{-\infty}^{\infty} p^2(t)dt - 2\int_{-\infty}^{\infty}\int_{-\infty}^{\infty} x(t)p(t)\, dt, \quad (3.74)$$

we see that minimizing the energy of the difference of $x(t)$ and $p(t)$ is equivalent to maximizing their inner product as long as the two signals have constant 2-norms. It is easy to do this for the prototype; we use the normalized signal $\tilde{p}(t) = p(t)/\|p\|_2$ as the model signal, for example. This step is called normalization, and so the method is often called *normalized cross-correlation*. If the entire candidate $x(t)$ is available at the moment of comparison, such as for signals acquired offline or as functions of a nontemporal independent variable, then we can similarly normalize $x(t)$ and compare it to $\tilde{p}(t)$. If, on the other hand, $x(t)$ is acquired in real time, then the feasible analysis works on past fragments of $x(t)$.

The Schwarz inequality tells us that equality exists if and only if the candidate and prototype are constant multiples of one another. If we subtract the mean of each signal before computing the normalized cross-correlation, then the normalized cross-correlation has unit magnitude if and only if the signals are related by $x(t) = Ap(t) + B$, for some constants A, B. Since $y(t) = (x * k)(t) = (x \circ p)(t)$ attains a maximum response when the prototype $p(t)$ matches the candidate $x(t)$, this technique is also known as matched filtering. It can be shown that in the presence of a random additive white noise signal, the optimal detector for a known pattern is still given by the matched filter [32].

Many signal and image processing applications depend upon matched filtering. In speech processing, one problem is to minimize the effect of reverberation from the walls and furnishings within a room on the recorded sound. This is an echo cancellation problem where there may be multiple microphones. The idea is to filter each microphone's input by the reflected impulse response of the room system [33].

Normalized cross correlation can be computationally demanding. When it is applied to images, this is especially the case. Consequently, many applications in image-based pattern matching use coarse resolution matched filtering to develop a set of likely match locations of the template against the input image. Then, finer resolution versions of the template and original image are compared [34]. Correlation techniques are also one of the cornerstones of image motion analysis [35].

3.3.3 Orthonormal Bases

In Chapter 2 a close relationship was established between general Hilbert spaces and the space of square-summable discrete signals. Here we list discuss four different orthogonal basis sets:

- Exponential signals of the form $\exp(jnt)$, where $n \in \mathbb{N}$
- Closely related sinusoids, which are the real and imaginary parts of the exponentials
- Haar basis, which consists of translations and dilations of a single, simple step function
- Sinc functions of the form $s_n(t) = \frac{1}{\sqrt{\pi}}\frac{\sin(At - n\pi)}{At - n\pi}$, where $A > 0$ and $n \in \mathbb{N}$

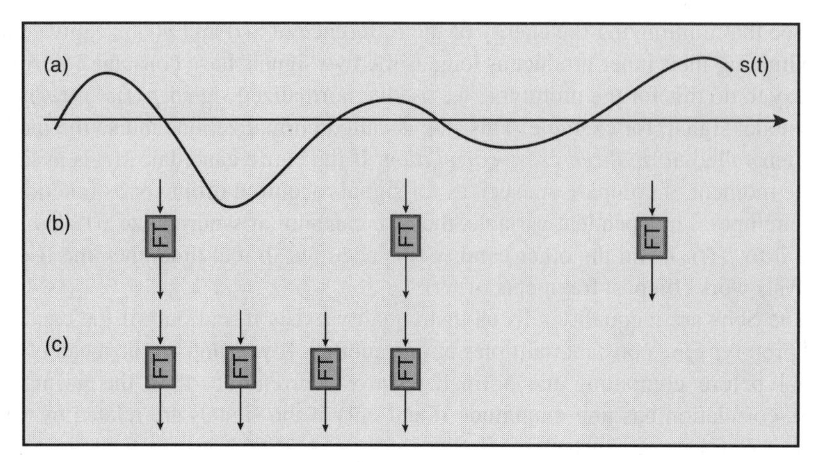

Fig. 3.5. Signal decomposition on an orthonormal basis (a). Sparse representations are better (b) than decompositions that spread signal energy across many different basis elements (c).

Orthonormal bases are fundamental signal identification tools. Given a signal $x(t)$, we project it onto the orthonormal basis set $\{e_n \mid n \in \mathbb{N}\}$ by taking the inner products $c_n = \langle x(t), e_n(t) \rangle$. Each c_n indicates the relative presence of the basis element e_n inside $x(t)$. The strategy is that—hopefully, at least—the set of coefficients $\{c_n \mid n \in \mathbb{N}\}$ is a simpler description of the original signal. If they are not, then we attempt to find an alternative basis set that better captures the character of anticipated input signals. We can say more, though. It works best for the signal recognition application if decomposition produces only a few significant c_n values for every typical $x(t)$ that we try to analyze. In other words, for the original signals $x(t)$ that we expect to feed into our analysis application, the energy of the decomposition coefficients is sparse and concentrated in a relative few values. On the other hand, dense decomposition coefficient sets make signal classification harder, because we cannot clearly distinguish which $e_n(t)$ factor most critically within $x(t)$. Figure 3.5 illustrates the idea. For signal identification, therefore, the upshot is that the statistics of the decomposition coefficients for typical system input signals are our guide for selecting an orthonormal basis set.

3.3.3.1 Exponentials. The most important basis for the L^2 Hilbert spaces is the exponential signals. We begin by considering the space $L^2[-\pi, \pi]$.

Let $\{e_n(t) \mid n \in \mathbb{N}\}$ be defined by $e_n(t) = (2\pi)^{-1/2}\exp(jnt)$. It can be easily shown that the $e_n(t)$ are indeed orthonormal (exercise). Similarly, if we set

$$e_n(t) = \frac{1}{\sqrt{b-a}} e^{2\pi jn\frac{t-a}{b-a}}, \tag{3.75}$$

then $B = \{e_n(t) \mid n \in \mathbb{N}\}$ is orthonormal in $L^2[a, b]$. We shall show in Chapter 5 that B is complete so that it is, in fact, a basis. Thus, for any square-integrable signal $x(t)$ on $[a, b]$ a linear combination of $\{e_n\}$ is arbitrarily close to $x(t)$; in other words,

$$x(t) = \sum_{n=-\infty}^{\infty} c_n e^{jnt} \tag{3.76}$$

for some constants $\{c_n \mid n \in \mathbb{N}\}$, called the Fourier series coefficients for $x(t)$. Note that c_n measures the similarity of $x(t)$ to the basis element $\exp(jnt)$: $\langle x(t), \exp(jnt)\rangle = c_n$ by orthonormality.

Now consider the case of $L^2(\mathbb{R})$. We can break up the real line into 2π-wide intervals $I_m = [(2m - 1)\pi, (2m + 1)\pi]$. Let X_m be the characteristic function on I_m, and set $e_{m,n}(t) = X_m(2\pi)^{-1/2}\exp(jnt)$. Then clearly $\{e_{m,n}(t) \mid m, n \in \mathbb{N}\}$ is an orthonormal basis for $L^2(\mathbb{R})$.

3.3.3.2 Sinusoids. There is an orthonormal basis for $L^2[-\pi, \pi]$ consisting entirely of sinusoids. Let us break up $e_n(t) = (2\pi)^{-1/2}\exp(jnt)$ into its real and imaginary parts using $\exp(jt) = \cos(t) = j\sin(t)$. We set $a_n = c_n + c_{-n}$ and $jb_n = c_{-n} - c_n$. Thus, (3.36) becomes

$$x(t) = \frac{a_0}{2} + \sum_{n=1}^{\infty} a_n \cos(nt) + \sum_{n=1} b_n \sin(nt), \tag{3.77}$$

and any $x(t) \in L^2[-\pi, \pi]$ can be expressed as a sum of sinusoids. Equation (3.77) shows that in addition to bona fide sinusoids on $[-\pi, \pi]$, we need one constant function to comprise a spanning set. That the sinusoids are also orthogonal follows from writing them in terms of exponentials:

$$\cos(t) = \frac{e^{jt} + e^{-jt}}{2} \tag{3.78a}$$

$$\sin(t) = \frac{e^{jt} - e^{-jt}}{2j} \tag{3.78b}$$

and using the orthonormality of the exponentials once more. As with exponentials, the sinusoids can be assembled interval-by-interval into an orthonormal basis for $L^2(\mathbb{R})$. The exercises further explore exponential and sinusoidal basis decomposition.

3.3.3.3 Haar Basis. The Haar[3] basis uses differences of shifted square pulses to form an orthonormal basis for square-integrable signals. It is a classic construction

[3]Hungarian mathematician Alfréd Haar (1885–1933) was Hilbert's graduate student at Göttingen. The results of his 1909 dissertation on orthogonal systems, including his famous basis set, were published a year later.

[36], dating from the early 1900s. It is also quite different in nature from the exponential and sinusoidal bases discussed above.

The sinusoidal basis elements consist of sinusoids whose frequencies are all integral multiples of one another—*harmonics*. As such, they all have different shapes. Thus, cos(t) follows one undulation on $[-\pi, \pi]$, and it looks like a shifted version of sin(t). Futhermore, $\cos(2t)$ and $\sin(2t)$ resemble one another as shapes, but they are certainly different from $\cos(t)$, $\sin(t)$, and any other basis elements of the form $\cos(nt)$ or $\sin(nt)$ where $n \neq 2$.

Haar's orthonormal family begins with a single step function, defined as follows:

$$h(t) = \begin{cases} 1 & \text{if } 0 \le t < \frac{1}{2}, \\ -1 & \text{if } \frac{1}{2} \le t \le 1, \\ 0 & \text{if otherwise.} \end{cases} \tag{3.79}$$

Nowadays $h(t)$ is called a *Haar wavelet*. Haar's basis contains dilations and translations of this single atomic step function (Figure 3.6). Indeed, if we set $H = \{h_{m,n}(t) = 2^{n/2}h(2^n t - m) \mid m, n \in \mathbb{N}\}$, then we claim that H is an orthonormal basis for $L^2(\mathbb{R})$.

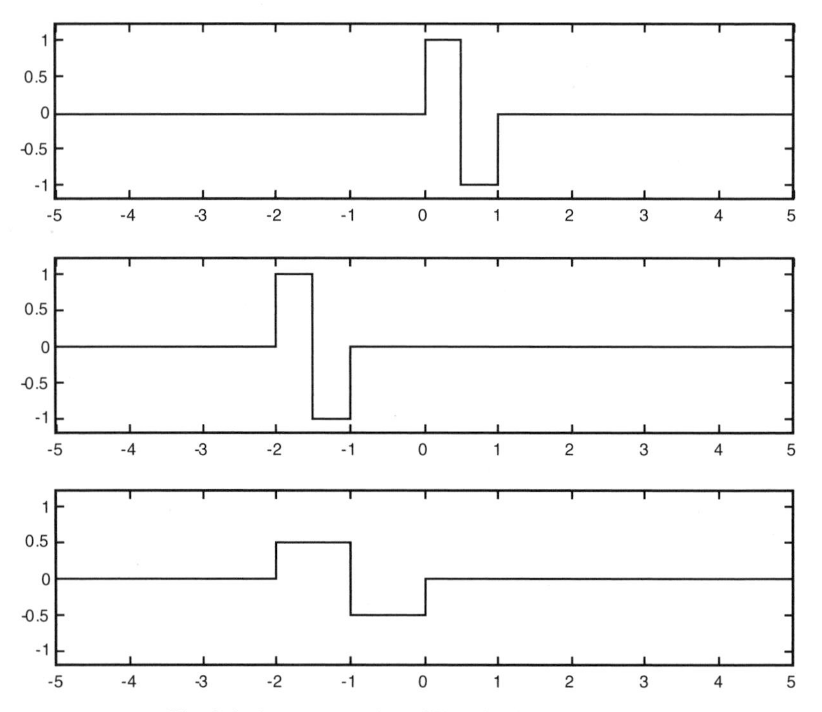

Fig. 3.6. Some examples of Haar basis elements.

Orthonormality is easy. If the supports of two unequal basis elements, $h_{m,n}(t)$ and $h_{p,q}(t)$, overlap, then we must have $n \neq q$; let us suppose $n > q$. Then the support of $h_{m,n}(t)$ will lie completely within one of the dyadic subintervals of Support $(h_{p,q}(t))$. The inner product must be zero: $\langle h_{m,n}(t), h_{p,q}(t)\rangle = 0$.

Completeness—that the closure of the span of $H = \{h_{m,n}(t) \mid m, n \in \mathbb{N}\}$ happens to be all of $L^2(\mathbb{R})$—is somewhat harder to show. Informally, any square-integrable signal can be approximated to arbitrary precision with a step function. Likewise, if we can approximate step functions with linear combinations of $\{h_{m,n}(t)\}$, then completeness follows. We first show that the unit square pulse, $s(t) = u(t-1) - u(t)$, is in Span(H). Indeed, one can check that for any $N \in \mathbb{N}$,

$$\sum_{n=-\infty}^{-1} 2^{\frac{n}{2}} h_{0,n}(t) = \begin{cases} 0 & \text{if } t \leq 0, \\ 1 & \text{if } 0 < t < 1, \\ 0 & \text{if } 2^N < t < 2^{N+1}. \end{cases} \tag{3.80}$$

So on all but a countably infinite set of points, the sum on the left-hand side of (3.80) is $s(t)$. When we study the Lebesgue integral, we shall see that these discontinuities do not affect the integral. Similarly, we can show that dyadic dilations and translations of $s(t)$ are in Span(H). Linear combinations of these dilations and translations can be arbitrarily close to a general step function, which justifies our completeness claim.

Practically, the importance of this is that it is good for finding one particular shape, which might manifest itself in different sizes or scales inside a signal. By contrast, the sinusoidal basis elements find various harmonics—distinct shapes—on the subintervals by which they divide the time domain. Of course, the elemental shape of the Haar basis is quite simple, but the basis elements are tuned to this shape at an infinite range of scales and an infinite set of positions. The two different types of basis in fact represent two different ways of analyzing signals. The sinusoidal and exponential bases find frequencies inside signals; they are *frequency-domain* tools for signal interpretation. The Haar basis finds a single atomic shape at different scales inside a signal, so it exemplifies *scale-domain* methods.

Recent books on wavelets [37, 38] cover Haar's basis. Until the mid-1980s, mathematicians believed that the only possible orthonormal bases for $L^2(\mathbb{R})$ which used dilations and translations of a single signal atom were those like Haar's construction where the atom has step discontinuities. The new wavelet theories, however, have completely toppled this long-standing conviction (Chapter 11).

3.3.3.4 *Sinc Functions.* Another important basis set consists of $\text{sinc}(t) = \sin(t)/t$ functions. In later chapters we shall have the tools in hand to show that the following family of functions is orthonormal: $\{s_n(t) \mid n \in \mathbb{N}\}$, where

$$s_n(t) = \frac{1}{\sqrt{\pi}} \frac{\sin(At - n\pi)}{At - n\pi} \tag{3.81}$$

and $A > 0$. We shall find that $\{s_n(t)\}$ defined in (3.81) spans an important Hilbert subspace of $L^2(\mathbb{R})$: a band-limited subspace that consists of signals whose spectral content lies entirely within a finite frequency range.

3.3.4 Frames

The concept of a frame generalizes the idea of an orthonormal basis in Hilbert space. Frame representations may be *overcomplete*. This means they may represent a signal in more than one way, and therefore they are not orthogonal. To understand why this might be useful and, indeed, why a signal analysis application based on frames can still be made workable, we have to reflect for a moment on the use of orthonormal bases for signal representation.

Many signal and image analysis applications need to recognize an unknown, candidate profile as an instance or combination of certain known prototypes. Probably the first approach that comes to mind is to decompose the candidate signal as a linear combination of the prototypical building blocks: $x(t) = \Sigma c_n e_n(t)$. The c_n represent the relative weight of each $e_n(t)$ in this linear signal combination. The application's desirable features are as follows:

(i) Incoming signals are uniquely represented by the coefficients $\langle x, e_n \rangle$; $x(t)$ does not have multiple identifying strings of weighting coefficients.

(ii) Two representations in terms of decomposition coefficients should permit a straightforward comparison of source signals for the differences between them.

(iii) Any original signal should be reconstructible from the stored coefficients; this is a completeness criterion.

(iv) Finally, the reconstruction of $x(t)$ from $\{c_n\}$ should be numerically stable; a small change in the coefficients results in a small change on the rebuilt signal.

These criteria suggest orthonormal bases. If we use an orthogonal set, such as the exponential functions $e_n(t) = e^{-jnt}$, then the four conditions hold in theory. The orthogonal character of the underlying special functions eases the computation of coefficients. However, a local change in the incoming signal leads to changes in the whole range of decomposition coefficients. Noise in signals, however sparse, drastically perturbs the stored digital form of the signal. This leads to practical difficulties in comparing two outwardly similar signals.

Sometimes, in an attempt to localize changes in signals to appropriate portions of the coefficient set, the decomposition functions are chosen to be windowed or "short-time" exponentials. These may take the form $e_{n,m}(t) = C_m e^{-jnt}$, where C_m is the characteristic function of the integral interval $[m, m + 1]$. The downfall of this tactic is that it adds high-frequency components to the decomposition. Relatively smooth signals decompose into sequences with unusually large coefficients for large values of n. This problem can be ameliorated by choosing a smoother window function—a Gaussian instead of a square window, for example. This becomes a representation in terms of the Gabor elementary functions, which first chapter introduced. But a deeper problem with windowed exponentials arises. It turns out that one cannot construct an orthonormal basis out of Gaussian-windowed exponentials. Chapter 10 will cover this famous result, known as the Balian–Low theorem. Nonetheless, it is possible to make *frames* out Gabor elementary functions, and this is a prime reason for the technique's recent popularity.

Earlier, however, we noted that the statistics of the decomposition coefficients are an important consideration. Sparse sets are better than dense sets. We form vectors out of the decomposition coefficient sets of system input signals. The vectors comprise a library of signal models. Sparse decomposition coefficients imply short vectors, and short vectors mean that the library more succinctly represents the essential aspects of our signal models.

Unfortunately, it quite often happens that orthonormal bases produce nonsparse coefficient sets on fairly simple signals. For example, a decaying pulse signal, $x(t) = \exp(-At^2)\cos(Bt)$, contains a local frequency component set by the width of the Gaussian envelope. Now, $x(t)$ can be represented by the exponential or sinusoidal basis sets, but far from the origin there will always be large weighting coefficients. These distant, high-frequency sinusoidal wiggles have to cancel one another in order to correctly represent the negligible amplitude of $x(t)$ in their neighborhood. Libraries based on such orthonormal bases can also be problematic; when the signal changes a little bit, many decomposition coefficients must change globally to effect just the right weighting to cancel and reinforce the right signal components.

One surprising result from contemporary signal analysis research is that overcomplete representations—in particular, the frame decomposition that we cover here—can help in constructing sparse signal representations. Special techniques, such as the matching pursuit algorithm have been devised to cope with the overcompleteness [39, 40]. Originating in the 1950s [41, 42], frames are now widely used in connection with the recent development of the theory of time-frequency and time-scale transforms [38, 43].

First-time readers may skip this section. Visitors to the latest research literature on time-frequency and time-scale transform methods will, however, commonly encounter frame theoretic discussions. We shall cover frames more thoroughly in Chapters 10–12.

3.3.4.1 Definition and Basic Properties.

This section defines the notion of a Hilbert space frame and provides some simple connections to the more specific and familiar concept of orthonormal basis. In what follows, we shall allow Hilbert spaces defined over the complex numbers.

Definition (Frame). Let $\{f_n: n \in \mathbb{Z}\}$ be signals in a Hilbert space H. If there are positive $A, B \in \mathbb{R}$ such that for all $x \in H$ we have

$$A\|x\|^2 \le \sum_{n=-\infty}^{\infty} |\langle x, f_n\rangle|^2 \le B\|x\|^2, \tag{3.82}$$

then the $\{f_n\}$ constitute a *frame* in H. The constants A and B are the *frame bounds*, *lower* and *upper*, respectively. The frame is *tight* if $A = B$. A frame is *exact* if it is no longer a frame following the deletion of a single element.

It is immediately clear that the frame condition (3.82) generalizes the idea of an orthonormal basis, makes the convergence unconditional, and ensures that a frame is a complete set (the closure of its linear span is all of H).

Proposition. Let $\{e_n: n \in \mathbb{Z}\}$ be an orthonormal basis for a Hilbert space H. Then $\{e_n\}$ is a tight exact frame having bounds $A = B = 1$.

Proof: Let $x \in H$. Parseval's relation for Hilbert spaces implies

$$\sum_{n=-\infty}^{\infty} |\langle x, e_n\rangle|^2 = \|x\|^2. \tag{3.83}$$

Therefore,

$$1 \cdot \|x\|^2 = \sum_{n=-\infty}^{\infty} |\langle x, e_n\rangle|^2 = 1 \cdot \|x\|^2, \tag{3.84}$$

which is precisely the frame condition (3.82), with $A = B = 1$. ■

Proposition (Unconditionality). Let $\{f_n : n \in \mathbb{Z}\}$ be a frame in a Hilbert space H. Then any rearrangement of the sequence $\{f_n\}$ is also a frame.

Proof: If $x \in H$, then $\{|\langle x, f_n\rangle|^2\}$ is a sequence of positive real numbers, and the convergence of the series

$$\sum_{n=-\infty}^{\infty} |\langle x, f_n\rangle|^2 \tag{3.85}$$

is absolute [6]. This means that the above sum converges to the same value under any rearrangement of the $\{f_n\}$. ■

Remark. Thus, we are free to renumber a frame with the natural numbers. Later, in Chapter 10, we shall find it convenient to use pairs of integers to index frame elements. This same idea is at work, as long as the index set has the same cardinality as the natural numbers.

Proposition (Completeness). Let $F = \{f_n : n \in \mathbb{Z}\}$ be a frame in a Hilbert space H. Then $\{f_n\}$ is a total set: $\overline{\text{Span}\,\{f_n\}} = H$.

Proof: Let $x \in H$, $\langle x, f_n \rangle = 0$ for all n, and $A > 0$ be the lower frame bound for F. By the definition of a frame,

$$0 \le A\|x\|^2 \le \sum_{n=-\infty}^{\infty} |\langle x, f_n \rangle|^2 = 0. \tag{3.86}$$

Equation (3.86) shows that $x = 0$. Since a subset X of H is complete if and only if no nonzero element is orthogonal to all elements of X, it must be the case that F is total in H. ∎

3.3.4.2 Examples. In the following examples, H is a Hilbert space, and $\{e_n: n \in \mathbb{N}\}$ is an orthonormal basis in H.

Example (Overcompleteness). Let $F = \{f_n\} = \{e_0, e_0, e_1, e_1, e_2, e_2, \ldots\}$. Then F is a tight frame with bounds $A = B = 2$. Of course, it is not an orthonormal basis, although the subsequence $\{f_{2n}: n \in \mathbb{N}\}$ is orthonormal. It is also not exact. Elements of H have multiple decompositions over the frame elements.

Example (Orthogonal, Yet Not a Frame). $F = \{f_n\} = \{e_0, e_1/2, e_2/3, \ldots\}$ is complete and orthogonal. However, it is not a frame, because it can have no positive lower frame bound. To see this, assume instead that F is a frame and let $A > 0$ be its lower bound. Let N be large enough that $N^{-2} < A$ and set $x = e_N$. Then, (3.82) gives

$$A = A\|x\|^2 \le \sum_n |\langle x, f_n \rangle|^2 = |\langle e_N, f_N \rangle|^2 = \frac{1}{N^2} < A, \tag{3.87}$$

a contradiction.

Example (Tight Frame). $F = \{f_n\} = \{e_0, 2^{-1/2}e_1, 2^{-1/2}e_1, 3^{-1/2}e_2, 3^{-1/2}e_2, 3^{-1/2}e_2, \ldots\}$ is a tight frame with bounds $A = B = 1$. F is not exact.

Example (Exact But Not Tight). $F = \{f_n\} = \{2e_0, e_1, e_2, \ldots\}$ is a frame, with lower bound $A = 1$ and upper bound $B = 2$. F is exact but not tight.

3.3.4.3 Frame Operator. There is a natural bounded linear operator associated with a frame. In fact, if the decomposition set is a frame, then the basic signal analysis system that associates signals in a Hilbert space with their decomposition coefficients is just such an operator. This section gives their definition and properties. While they are mathematically elegant and abstract, frame operators also factor critically in using overcomplete frames in signal representation.

Definition (Frame Operator). If $F = \{f_n(t): m, n \in \mathbb{Z}\}$ is a frame in a Hilbert space H, then the associated frame operator $T_F: H \to l^2(\mathbb{Z})$ is defined by

$$T_F(x)(n) = \langle x, f_n \rangle. \tag{3.88}$$

In other words, $y = T_F(x) = T_F x$ is the complex-valued function defined on the integers such that $y(n) = \langle x, f_n\rangle$. When the frame is clear by the context we may drop the subscript on the frame operator: $y = Tx$.

Proposition. If $F = \{f_n(t): n \in \mathbb{Z}\}$ is a frame in a Hilbert space H, then the associated frame operator T given by (3.88) is linear and bounded. Furthermore, if B is the upper frame bound, then $\|T\| \le B^{1/2}$.

Proof: Linearity is clear from inner product properties. If $x \in H$, then

$$\|Tx\| = \left(\sum_n \langle x, x_n\rangle \overline{\langle x, x_n\rangle}\right)^{\frac{1}{2}} = \left(\sum_n |\langle x, x_n\rangle|^2\right)^{\frac{1}{2}} \le (B\|x\|^2)^{\frac{1}{2}} = B^{\frac{1}{2}}\|x\| . \tag{3.89}$$

■

Definition (Adjoint Frame Operator). Let $F = \{f_n(t) : n \in \mathbb{Z}\}$ be a frame in a Hilbert space H and let T be its associated frame operator T (3.88). The *adjoint frame operator* $S : l^2(\mathbb{Z}) \to H$ is defined for $y(n) \in l^2(\mathbb{Z})$ by

$$S(y) = \sum_{n=-\infty}^{\infty} y(n) f_n. \tag{3.90}$$

Proposition. Let S be given by (3.90). Then S is the Hilbert adjoint operator with respect to frame operator T of F: $S = T^*$.

Proof: That S is linear is left as an exercise. Let $x \in H$, T the frame operator for $F = \{f_n(t) : n \in \mathbb{Z}\}$, and let $y = \{y_n : n \in \mathbb{Z}\}$ be some sequence in l^2. Then

$$\langle x, Sy\rangle = \left\langle x, \sum_n y_n x_n \right\rangle = \sum_n \overline{y_n} \langle x, x_n\rangle, \tag{3.91}$$

and also

$$\langle Tx, y\rangle = \sum_n \langle x, x_n\rangle \overline{y_n} = \sum_n \overline{y_n} \langle x, x_n\rangle . \tag{3.92}$$

Together, (3.91) and (3.92) show that T and S cross-couple the inner products of the two Hilbert spaces. Therefore, $S = T^*$. ■

The next theorem, one of the classic results on frames and frame operators, offers a characterization of frames [41]. It uses the idea of a positive operator. Recall that an operator T on a Hilbert space H is *positive*, written $T \ge 0$, when $\langle Tx, x\rangle \ge 0$ for all $x \in H$. Also, if S and T are operators on H, then $T \ge S$ means that $T - S \ge 0$. Positive operators are self-adjoint [26, 29].

Theorem (Frame Characterization). Let $F = \{f_n : n \in \mathbb{Z}\}$ be a sequence in a Hilbert space H; $A, B > 0$; and let I be the identity operator on H. Then F is a frame with lower and upper bounds A and B, respectively, if and only if the operator S defined by

$$Sx = \sum_n \langle x, f_n\rangle f_n \tag{3.93}$$

is a bounded linear operator with

$$AI \le S \le BI. \tag{3.94}$$

Proof: Assume that (3.93) defines S and (3.94) holds. By the definition of $\le$ for operators, for all $x \in H$ (3.94) implies

$$\langle AIx, x\rangle \le \langle Sx, x\rangle \le \langle BIx, x\rangle. \tag{3.95}$$

However,

$$\langle AIx, x\rangle = A\|x\|^2 \tag{3.96a}$$

and

$$\langle BIx, x\rangle = B\|x\|^2. \tag{3.96b}$$

The middle term of (3.95) is

$$\langle Sx, x\rangle = \left\langle \sum_n \langle x, f_n\rangle f_n, x\right\rangle = \sum_n \langle x, f_n\rangle \overline{\langle f_n, x\rangle} = \sum_n \left|\langle x, f_n\rangle\right|^2. \tag{3.97}$$

Together (3.96a), (3.96b), and (3.97) can be inserted into (3.95) to show that the frame condition is satisfied for F.

Conversely, suppose that F is a frame. We must first show that S is well-defined—that is, the series (3.93) converges. Now, the Schwarz inequality implies that the norm of any $z \in H$ is $\sup\{|\langle z, y\rangle| : y \in H \text{ and } \|y\| = 1\}$. Let s_N represent partial sums of the series (3.93):

$$s_N(x) = \sum_{n=-N}^{N} \langle x, f_n\rangle f_n. \tag{3.98}$$

When $M \le N$ the Schwarz inequality applies again:

$$\|s_N - s_M\|^2 = \sup_{\|y\| = 1}\left\{\left|\langle s_N - s_M, y\rangle\right|^2\right\} = \sup_{\|y\| = 1}\left\{\left|\sum_{n=M+1}^{N} \langle x, f_n\rangle \langle f_n, y\rangle\right|^2\right\}. \tag{3.99}$$

Algebra on the last term above gives

$$\|s_N - s_M\|^2 \le \sup_{\|y\| = 1}\left\{\sum_{n=M+1}^{N} \left|\langle x, f_n\rangle\right|^2 \sum_{n=M+1}^{N} \left|\langle f_n, y\rangle\right|^2\right\}. \tag{3.100}$$

By the frame condition,

$$\|s_N - s_M\|^2 \le \sup_{\|y\| = 1} \sum_{n=M+1}^{N} \left|\langle x, f_n\rangle\right|^2 B\|y\|^2 = B \sum_{n=M+1}^{N} \left|\langle x, f_n\rangle\right|^2, \tag{3.101}$$

and the final term in (3.101) must approach zero as $M, N \to \infty$. This shows that the sequence $\{s_N \mid N \in \mathbb{N}\}$ is Cauchy in the Hilbert space H. H is complete, so the $\{s_N\}$ converge, the series must converge, and the operator S is well-defined.

Similarly,

$$\|Sx\|^2 \leq \sup_{\|y\| = 1} |\langle Sx, y\rangle|^2, \tag{3.102}$$

which entails $\|S\| \leq B$. From the frame condition and (3.93) the operator ordering, $AI \leq S \leq BI$ follows immediately. ■

Remark. Notice that for a frame $F = \{f_n : n \in \mathbb{Z}\}$ the theorem's operator S is the composite T^*T, where T is the frame operator and T^* is its adjoint. The following corollaries provides further properties of T^*T.

Corollary. Let $F = \{f_n : n \in \mathbb{Z}\}$ be a frame in a Hilbert space H and let T be the frame operator. Then the map $S = T^*T : L^2(\mathbb{R}) \to L^2(\mathbb{R})$ given by

$$(T^*T)(x) = \sum_{n=-\infty}^{\infty} \langle x, f_n\rangle f_n. \tag{3.103}$$

is positive and invertible.

Proof: Let A be the lower frame bound for F. Since $AI \leq S$, $S - AI \geq 0$, by the definition of the operator ordering relation. Also, since $A > 0$, $S/A - I \geq 0$. A property of positive operators is that if an operator, say U, is positive, $U \geq 0$, then $U + I$ is invertible [27]. Therefore, $S/A - I + I = S/A$ is invertible. Clearly then, S is invertible. Moreover, adding a positive operator I to $S/A - I$ still gives a positive operator. This shows that S is indeed positive. ■

Corollary. Let $F = \{f_n : n \in \mathbb{Z}\}$ be a frame in a Hilbert space H; let A and B be the lower and upper frame bounds, respectively, for F; let T be the frame operator; and let $S = T^*T$ be given by the previous corollary. Then $I/B \leq S^{-1} \leq I/A$.

Proof: The previous corollary shows that S^{-1} exists. Since S^{-1} commutes with I and with S, and since $AI \leq S \leq BI$, it follows that $S^{-1}AI \leq S^{-1}S \leq S^{-1}BI$. Upon rearrangment, this yields $B^{-1}I \leq S^{-1} \leq A^{-1}I$. ■

These results allow us to define the concept of the dual frame. The dual frame is the key concept for applying frames in signal analysis applications.

Definition (Dual Frame). Let $F = \{f_n : n \in \mathbb{Z}\}$ be a frame in a Hilbert space H and T be the frame operator. We define the *dual frame* to F by applying the inverse of T^*T to frame elements:

$$\tilde{F} = \left\{(T^*T)^{-1}(f_n)\right\}_{n \in Z}. \tag{3.104}$$

Corollary. Let $F = \{f_n: n \in Z\}$ be a frame in a Hilbert space H; let A and B the lower and upper frame bounds, respectively, for F; let T be the frame operator; and let $S = T^*T$. Then the sequence $\{S^{-1}f_n \mid n \in \mathbb{Z}\}$ in H is a frame with lower bound B^{-1} and upper bound A^{-1}.

Proof: S^{-1} exists and is positive. Let $x \in H$ and note that

$$S^{-1}x = S^{-1}(SS^{-1}x) = S^{-1}\left(\sum_n \langle S^{-1}x, f_n\rangle f_n\right) = \sum_n \langle S^{-1}x, f_n\rangle S^{-1}f_n \qquad (3.105)$$

by the linearity and continuity of S^{-1}. Since every positive operator is self-adjoint (Hermitian) [29], S^{-1} is self-adjoint. Hence,

$$S^{-1}x = \sum_n \langle S^{-1}x, f_n\rangle S^{-1}f_n = \sum_n \langle x, S^{-1}f_n\rangle S^{-1}f_n . \qquad (3.106)$$

Notice that (3.106) is precisely the form that the operator S^{-1} takes in (3.93) of the frame characterization theorem. That $B^{-1}I \le S^{-1} \le A^{-1}I$ follows from an earlier corollary. Thus, the theorem's condition applies, and $\{S^{-1}f_n\}$ is a frame in H. ■

Corollary. Under the assumptions of the previous corollary, any $x \in H$ can be written

$$x = \sum_n \langle x, S^{-1}f_n\rangle f_n = \sum_n \langle x, f_n\rangle S^{-1}f_n. \qquad (3.107)$$

Proof: Using (3.105), (3.106), and $x = SS^{-1}x = S^{-1}Sx$, the result follows easily. ■

Corollary. Further assuming that the frame T is tight, we have $S = AI$, $S^{-1} = A^{-1}I$, and, if $x \in H$, then

$$x = A^{-1}\sum_n \langle x, f_n\rangle f_n. \qquad (3.108)$$

Proof: Clear from the definition of tightness and the preceding corollaries. ■

3.3.4.4 Application: Stable Modeling and Characterization.

These results shed light on our proposed requirements for a typical signal analysis system. Let us list what we know so far:

- The first requirement—that the representation be unique—was demonstrated not to hold for general frames by an easy counterexample.
- The second specification—that signal representations should permit a straightforward comparison of two incoming signals—is satisfied by the frame operator that maps signals to sequences of complex numbers allowing us to use the l^2 norm for comparing signals.

- The corollary (3.107) fulfills the requirement that the original signal should be reconstructible from the decomposition coefficients.
- The fourth requirement has been left rather vague: What does numerical instability mean?

We can understand numerical instability in terms of bounded operators. Let $F = \{f_n\}$ be a frame and $T = T_F$ its frame operator. If the inverse mapping T^{-1} is unbounded, then elements of l^2 of unit norm will be mapped back to elements of H having arbitrarily large norms. This is not at all desirable; signals of enormous power as well as signals of miniscule power will map to decomposition coefficients of small l^2-norm. *This is numerical instability.* The next result shows that frame-based signal decomposition realizes the fourth requirement of a signal analysis system.

Corollary. Let $F = \{f_n : n \in \mathbb{Z}\}$ be a frame in a Hilbert space H and let $T = T_F$ be the associated frame operator. Then the inverse T^{-1} exists and is bounded.

Proof: Let $S = T^*T$, as in the previous section. Then S^{-1} exists and $S^{-1}T^* = T^{-1}$ is the bounded inverse of T. Alternatively, (3.107) explicitly maps a square-summable sequence in $\mathbb{C}$ to H, and it inverts T. A straightforward calculation shows that the map is bounded with $\|T^{-1}\| \leq A^{-1/2}$. ■

Remarks. So the use of a frame decomposition for the signal analysis system allows a *stable reconstruction* of incoming signals from the coefficients obtained previously. In the case that the frame used in the signal processing system is tight, then the reconstruction is much simpler (3.108). We can reconstruct a signal from its decomposition coefficients using (3.108) alone; there is no need to invert $S = T^*T$ to get $S^{-1}f_n$ values.

We have substantiated all of the basic requirements of a signal analysis system, except for the first stipulation—that the coefficients of the decomposition be unique. The exercises elaborate some properties of exact frames that allow us to recover this uniqueness property. Briefly, if $F = \{f_n(t) : n \in \mathbb{Z}\}$ is a frame in a Hilbert space H, $T = T_F$ is its associated frame operator (3.88), and $S = T^*T$, then we know from the corollaries to the frame representation theorem that for any $x \in H$, if $a_n = \langle x, S^{-1}f_n \rangle$, then $x = \Sigma a_n f_n$. We can also show (exercises) that if there is some other representation of x, then it is no better than the the one we give in terms of the dual frame. That is, if there are $c_n \in C$ such that $x = \Sigma c_n f_n$, then

$$\sum_n |c_n|^2 = \sum_n |a_n|^2 + \sum_n |a_n - c_n|^2 ; \tag{3.109}$$

the representation by dual frame elements is the best in a least-squares sense. Later chapters (10–12) further cover frame representations and elaborate upon this idea.

To sum up. We began by listing the desired features of a signal analysis system. The notion of a frame can serve as a mathematical foundation for the decomposition, analysis, and reconstruction of signals. Orthonormal bases have very nice computational properties, but their application is often confounded by an undesirable practicality: The representations may well not be sparse. Frame theoretic approaches are a noteworthy alternative trend in recent signal analysis research, offering improved representation density over orthonormal bases [39, 40, 43].

3.4 MODERN INTEGRATION THEORY

The Lebesgue integral offers several theoretical advantages over the Riemann integral. The modern integral allows us to define Banach spaces directly, rather than in terms of the completion of a simpler space based on continuous analog signals. Also, the Lebesgue integral widens considerably the class of functions for which we can develop signal theory. It supports a powerful set of limit operations. Practically speaking, this means that we can develop powerful signal approximation techniques where signal and error magnitudes are based upon Lebesgue integrals (here we have in mind, of course, the L^p-norm).

This material has not been traditionally included in the university engineering and science curricula. Until recently, these disciplines could get along quite well without the mathematician's full toolkit. Mixed domain transform methods, Gabor and wavelet transforms in particular, have entered into widespread use in signal processing and analysis in the last several years. And carried with them has been an increased need for ideas from abstract functional analysis, Hilbert space techniques, and their mathematical underpinnings, among them the Lebesgue integral.

First-time readers and those who are content to build Banach spaces indirectly, by completing a given normed linear space, may elect to skip the material on the modern integral. Frankly, much of the sequel will still be quite understandable. Occasionally, we may worry that a signal is nonzero, but has Lebesgue integral zero; the mathematical term is that the signal is zero *almost everywhere* on an interval $[a, b]$. This is the standard, albeit homely, term for a function with so many zero points that its L^1-norm with respect to the Lebesgue integral is zero. We also say two signals are *equal almost everywhere* when their difference is zero almost everywhere.

Standard mathematical analysis texts cover measure and integration theory in far more detail and generality than we need here [24, 44, 45]. A tutorial is contained in Ref. 25. We cover the basic theoretical development only in these settings: measures on subsets of the real line and complex plane, real- and complex-valued functions, and their integrals. So restricted, this treatment follows the classic approach of Ref. 44.

Calculus defines the Riemann[4] integral as a limit of sums of areas of rectangles [6, 25]. Another approach uses trapezoids instead of rectangles; it offers somewhat

[4]Georg Friedrich Bernhard Riemann (1826–1866) studied under a number of great German mathematicians of the nineteenth century, including Gauss and Dirichlet. In 1857 he assumed a professorship at Göttingen. He contributed important results to complex variable theory, within which the Cauchy–Riemann equations are fundamental, and to non-Euclidean geometries, whose Riemannian manifolds Einstein much later appropriated for modern cosmology. Riemann is most widely known, ironically perhaps, for a relatively minor accomplishment—formalizing the definition of the conventional integral from calculus [R. Dedekind, Biography of Riemann, in H. Weber, ed., *Collected Works of Bernhard Riemann*, New York: Dover, 1953].

better numerical convergence. Either approach produces the same result. For rectangular sums, let us recall the definition.

Definition (Riemann Integral). Let $x(t)$ be continuous on the interval $I = [a, b]$; $a = t_0 < t_1 < \cdots < t_N = b$ partition I; and, for each subinterval, $I_k = [t_k, t_{k+1}]$, let r_k and s_k be the minimum and maximum values, respectively, of $x(t)$ on I_k. Then the lower and upper Riemann sums for $x(t)$ and I are

$$R_{x,I} = \sum_{k=1}^{N} r_k(t_k - t_{k-1}). \tag{3.110a}$$

$$S_{x,I} = \sum_{k=1}^{N} s_k(t_k - t_{k-1}). \tag{3.110b}$$

The Riemann integral is defined by

$$\int_a^b x(t)\,dt = \lim_{\Delta_I \to 0} R_{x,I}, \tag{3.110c}$$

where $\Delta_I = \max\{t_k - t_{k-1} \mid k = 1, 2, \cdots, N\}$.

Calculus proves that limit (3.110c) remains the same whether we use upper or lower Riemann sums. The height of the rectangle may indeed be the function value at any point within the domain interval. We have chosen to define the Riemann integral using the extreme cases, because, in fact, the modern Lebesgue integral uses sums from below similar to (3.110c).

The base of the rectangle or trapezoid is, of course, an interval. So the Riemann integral partitions the function *domain* and lets that partition determine the range values used for computing little areas. The insight of modern integration is this: Partition the *range*, not the domain, and then look at the sets in the function's domain that it maps to the range regions. There is a way to measure the area of these domain sets (next section), and then their area is weighted by the range values in much the same way as the Riemann integral. The difference seems simple. The difference seems inconsequential. But the implications are enormous.

3.4.1 Measure Theory

The standard approach to the Lebesgue integral is to develop a preliminary theory of the *measure* of a set. This generalizes the notion of simple interval length to a much wider class of sets. Although the Lebesgue integral can be defined without first building a foundation in measure theory (cf. Refs. 13 and 25), the idea of a measure is not difficult. Just as the Riemann integral is a limit of sums of areas, which are interval widths wide and function values high, the Lebesgue integral is a limit of sums of weighted measures—set measures scaled by function values.

Measurable sets, however, can be much more intricate than simple intervals. For example, the rational numbers $\mathbb{Q}$ is a measurable set, and its measure, or its *area* if you will, is zero. Furthermore, any countable set (i.e., a set that can be put into a one–one correspondence with the natural numbers $\mathbb{N}$) is measurable and has zero measure. The interval [0, 1] has unit measure, which is no doubt reassuring, and if we remove all the rational points from it, obtaining $[0, 1] \setminus \mathbb{Q}$, then the result is still measurable and still has unit measure. The rest of this section sketches the developments upon which these appealing ideas can be justified.

3.4.1.1 Rudiments of the Theory. Measure theory axiomatics are closely related to the ideas of a probability measure, which we covered in Chapter 1. We recall therefrom the concept of a σ-algebra Σ. These can be defined for abstract spaces, but we shall stick to sets of real or complex numbers, since these are the spaces for which we define analog signals. Let $\mathbb{K}$ be either $\mathbb{R}$ or $\mathbb{C}$. The four properties of Σ are:

(i) Elements of Σ are subsets of $\mathbb{R}$ (or $\mathbb{C}$): $\wp(\mathbb{K}) \supset \Sigma$.
(ii) The entire space is in Σ: $\mathbb{K} \in \Sigma$.
(iii) Closure under complement: If $A \in \Sigma$, then $\overline{A} = \{t \in \mathbb{K} \mid t \notin A\} \in \Sigma$.
(iv) Closure under countable union: If $\Sigma \supset \{A_n \mid n \in \mathbb{N}\}$, then $\bigcup_{n \in \mathrm{N}} A_n \in \Sigma$.

Examples. A couple of extreme examples of σ-algebras are as follows:

- $\Sigma_1 = \wp(\mathbb{R})$, the set of all subsets of $\mathbb{R}$.
- $\Sigma_0 = \{\varnothing, \mathbb{R}\}$.

There are equivalent examples for $\mathbb{C}$. We have a more interesting σ-algebra in mind, the Borel sets, and we will cover this class momentarily. Combining the closure under countable unions and complement rules, we can show (exercise) that σ-algebras are closed under countable intersections. Two basic concepts are that of a *measurable function*, which we interpret as a measurable signal, and of a *measure* itself, which is a map from a σ-algebra to the non-negative reals.

Definition (Measurable Function). A real- or complex-valued function $x(t)$ is measurable with respect to a σ-algebra Σ if $x^{-1}(A) = \{t \in \mathrm{R} \mid x(t) \in A\} \in \Sigma$ for all open sets A in $\mathbb{K}$.

Proposition. Let Σ be a σ-algebra and let $x : \mathbb{R} \to \mathbb{K}$ be a real- or complex-valued function. Let $\Theta = \{T \in \wp(\mathbb{K}) \mid x^{-1}(T) \in \Sigma\}$. Then Θ is a σ-algebra in $\mathbb{K}$.

Proof: Clearly, $\varnothing$ and $\mathbb{K} \in \Theta$. If $T \in \Theta$, then $S = x^{-1}(T) \in \Sigma$ and $\mathbb{R} \setminus S \in \Sigma$. But $\overline{T} = x(\mathbb{R} \setminus S)$, so $\mathbb{T} \in \Theta$. Finally, $x^{-1}(\cup_{n\in\mathbb{N}} T_n) = \cup_{n\in N} x^{-1}(T_n)$, so closure under countable unions holds as well. ■

The properties of a measure and limit theorems for the modern integral depend on an extension of the real numbers to include two infinite values: ∞ and $-\infty$. Formally, these are just symbols. Intuitively, however, ∞ is an abstract positive value that is larger than any than any real number and $-\infty$ is an abstract negative value that has larger magnitude than any real number. Let us first consider ∞. This value's arithmetic operations are limited. For example, $r + \infty = \infty$ for any $r \in$ R; $r \times \infty = \infty$ for any $r > 0$; if $r = 0$, then $r \times \infty = 0$; and addition and multiplication with ∞ are commutative. There is also a negative infinity element, $-\infty$, so that $r + (-\infty) = -\infty$ for any $r \in \mathbb{R}$. Furthermore, $r \times (-\infty) = -\infty$ for any $r > 0$; if $r = 0$, then $r \times (-\infty) = 0$; if $r < 0$, then $r \times \infty = -\infty$; and so on.

Note that subtraction, division, and cancellation operations only work with finite values. That is, $r - s$ is not defined if both r and s are infinite, and a similar restriction applies to r/s. If $rs = rt$, then we can conclude $s = t$ only if r is finite. A similar restriction applies to $r + s = r + t$.

We can also consider the extended real line, $\mathbb{R}^+ = \mathbb{R} \cup \{\infty\} \cup \{-\infty\}$. We consider $\mathbb{R}^+$ as having the additional open sets $(r, \infty]$ and $[-\infty, r)$ for any finite $r \in \mathbb{R}$. Of course, countable unions of open sets are open in the extended reals. Analog signals can be extended so that they take on infinite values at their singularities. Thus, a signal like $x(t) = t^{-2}$ is undefined at $t = 0$. In the extended reals, however, we may set $x(0) = \infty$. This makes the later limit theorems on modern integration (Section 3.4.3) into equalities. Note that we do not use the extended reals as the domain for analog signals. We only use the extension $\mathbb{R} \cup \{\infty\} \cup \{-\infty\}$ for defining the idea of a measure on σ-algebras and for extending the range of analog signals.

Definition (Measure). A *measure* on a σ-algebra Σ is a function $\mu : \Sigma \to [0, \infty]$ such that

$$\mu\left(\bigcup_{n=-\infty}^{\infty} A_n\right) = \sum_{n=-\infty}^{\infty} \mu(A_n) \tag{3.111}$$

whenever $\{A_n\}$ are pairwise disjoint.

Thus a measure is just like a probability measure, except that its values range in $[0, \infty]$ rather than in $[0, 1]$. A measure function gives a size value for a set. Thus, the measure might indicate the relative size of part of a signal's domain. We are also limiting the discussion to the real line, even though the ideas generalize to measures on σ-algebras in abstract metric spaces. Real analysis texts formalize these notions [24, 44, 45], but we prefer to limit the scope to just what we need for analog signal theory. Here are some easy examples.

Example (All or Nothing Measure). Let $\Sigma = \wp(\mathbb{R})$, and for $A \in \Sigma$ define $\mu(\varnothing) = 0$ and $\mu(A) = \infty$ if $A \neq \varnothing$. Then μ is a measure on Σ.

Example (Counting Measure). Again let $\Sigma = \wp(\mathbb{R})$, and for $A \in \Sigma$ define $\mu(A) = N$ if A contains exactly N elements and $\mu(A) = \infty$ otherwise. Then μ is a measure on Σ.

Proposition. Let μ be a measure on the σ-algebra Σ. Then

(i) (*Null set*) $\mu(\varnothing) = 0$.
(ii) (*Additivity*) If $A_p \cap A_q = \varnothing$ when $p \neq q$, then $\mu(A_1 \cup A_2 \cup \cdots \cup A_n) = \mu(A_1) + \mu(A_2) + \cdots + \mu(A_n)$.
(iii) (*Monotonicity*) If $B \supset A$, then $\mu(B) \geq \mu(A)$.

Proof: Similar to probability measure arguments. ■

3.4.1.2 Lebesgue Measurable Sets. There are lots of σ-algebras on the real line. Analog signal theory needs only the smallest σ-algebra that contains all the open sets in $\mathbb{R}$. We must show that such a smallest σ-algebra exists.

Theorem. There is a σ-algebra $\mathcal{B}$ on $\mathbb{R}$ such that:

(i) $\mathcal{B}$ contains all open subsets of $\mathbb{R}$.
(ii) If Σ is a σ-algebra containing all the open sets, then $\Sigma \supset \mathcal{B}$.

Proof: To begin with, the power set on the real line, $\wp(\mathbb{R})$, is itself is a σ-algebra and contains all open sets. Nonconstructively, we set $\mathcal{B}$ to be the intersection of all such σ-algebras. It is straightforward to show that this intersection is still a σ-algera. It is still a subset of $\wp(\mathbb{R})$. Since $\mathbb{R}$ must be in every σ-algebra, it is must be in the intersection of those containing the open sets. Closure under complement is also easy: if $A \in \Sigma$, where Σ is a any σ-algebra containing the open sets, then $\overline{A} \in \Sigma$; thus, $\overline{A}$ is in the intersection of all such σ-algebras. Finally, let $\mathcal{B} \supset \{A_n \mid n \in \mathbb{N}\}$. Then for all $n \in \mathbb{N}$, A_n is in every σ-algebra Σ that contains all the open sets in $\mathbb{R}$. Thus the countable family $\{A_n\}$ is a subset of every such σ-algebra. Hence $\bigcup_{n \in N} A_n$ is in each of these σ-algebras; and, consequently, $\bigcup_{n \in N} A_n$ is in the intersection $\mathcal{B}$. ■

Definition (Borel sets). The class of *Borel* or *Lebesgue measurable* sets is the smallest σ-algebra that contains every open set.

All the sets we normally use in signal theory are Borel sets. In fact, it takes a certain amount of craftiness to exhibit a set that is not Lebesgue measurable.

Example (Intervals). All of the open and closed sets, and the intervals (a, b) and $[a, b]$ in particular, are Lebesgue measurable. That closed sets are measurable follows from the σ-algebra's complement property. Also, since we take the half-infinite intervals $(a, \infty]$ and $[-\infty, a)$ to be open, these too are measurable. Finally, we can form countable unions involving these basic measurable sets. In the complex plane, open disks $\{z: |z| < r\}$, and closed disks $\{z: |z| \le r\}$ are measurable as are half-infinite sets $\{z: |z| > r\}$, and so on.

Example (Countable Sets). Any countable set is measurable. For example, real singletons $\{a\}$ are measurable because they are closed. So any countable union of singletons is measurable.

Proposition. If $x(t)$ is measurable and T is a Lebesgue measurable set in $\mathbb{K}$ ($\mathbb{R}$ or $\mathbb{C}$), then $x^{-1}(T) \in \mathcal{B}$; in other words, $x^{-1}(T)$ is Lebesgue measurable in $\mathbb{R}$.

Proof: Let $\Theta = \{T \in \wp(\mathbb{K}) \mid x^{-1}(T) \in \mathcal{B}\}$. Since $x(t)$ is measurable, Θ contains all the open sets in $\mathbb{K}$. In the previous section, we showed that Θ is a σ-algebra, so it must contain all the Lebesgue measurable sets. ∎

3.4.1.3 Lebesgue Measure. There are lots of possible measures on σ-algebras in $\mathbb{R}$ (or $\mathbb{C}$). Again, we need only one of them: the *Lebesgue measure*. It applies to the Lebesgue measurable, or Borel sets.

Definition (Open Covering). Let S be a set and $O = \{A_n \mid n \in \mathbb{N}\}$ be a family of open sets. If $S \subset \bigcup_{n \in N} A_n$, then O is an *open covering* for S.

Definition (Real Lebesgue Measure). Let $\mathcal{B}$ be the Lebesgue measurable sets on $\mathbb{R}$ and let the measure function μ be defined as follows:

(i) $\mu(a, b) = b - a$.

(ii) If S is an open set in R, then

$$\mu(S) = \inf_{S \subset \bigcup A_n} \left[\sum_{n=0}^{\infty} \mu(A_n) \right], \tag{3.112}$$

where the greatest lower bound is taken over all open coverings of S by intervals. The function μ is called the (real) *Lebesgue measure*.

Definition (Complex Lebesgue Measure). Let $\mathcal{B}$ be the Lebesgue measurable sets on $\mathbb{C}$. Let the measure function μ be defined as follows:

(i) If $B = \{z \in \mathbb{C} \mid |z - c| < r\}$, then $\mu(B) = \pi r^2$.

(ii) If S is an open set in $\mathbb{C}$, then

$$\mu(S) = \inf_{S \subset \bigcup B_n} \left[\sum_{n=0}^{\infty} \mu(B_n) \right], \tag{3.113}$$

where the greatest lower bound is taken over all open coverings of S by open balls. The function μ is called the (complex) *Lebesgue measure*.

Accepting that these definitions do produce functions that are indeed measures on the Borel sets, let us provide some examples of the measures of sets.

Example (Singletons). The measure of a singleton $\{a\}$ is zero. Let $\varepsilon > 0$. Then the single interval $I_\varepsilon = (a - \varepsilon, a + \varepsilon)$ covers $\{a\}$. The Lebesgue measure of I_ε is 2ε. Since ε was arbitrary and positive, the greatest lower bound of the lengths of all such intervals cannot be positive, so $\mu\{a\} = 0$.

Example (Intervals). The measure of a half-open interval $[a, b)$ is $b - a$. The measure of (a, b) is $b - a$, and the singleton $\{a\}$ has measure zero. Because (a, b) and $\{a\}$ are disjoint, $\mu(a, b) + \mu\{a\} = \mu[a, b) = b - a$. We also have $\mu(a, \infty] = \infty$; this is a consequence of the monotonicity property of a measure, since there are infinitely many disjoint intervals of unit length that are contained in a half-infinite interval.

Example (Countable Sets). Suppose A is a countable set $A = \{a_n \mid n \in \mathbb{N}\}$. Then for each $\varepsilon > 0$ we can find a set of intervals that covers A such that the sum of the lengths of the intervals is ε. For example, let $I_{n,\varepsilon} = (a_n - \varepsilon 2^{-n-2}, a_n + \varepsilon 2^{-n-2})$. Since ε is arbitrary, $\mu A = 0$.

Definition (Almost Everywhere). A property is said to hold *almost everywhere* on a measurable set A if the set of elements of A upon which is does not hold has Lebesgue measure zero.

Example (Nonmeasurable Set). To show that there are non-Lebesgue measurable sets, we recapitulate the example from Ref. 24. Consider the half-open unit interval $I = [0, 1)$, for which $\mu I = 1$. For $a, b \in I$, define $a \oplus b$ by

$$a \oplus b = \begin{cases} a + b & \text{if } a + b < 1, \\ a + b - 1 & \text{if } a + b \geq 1. \end{cases} \tag{3.114}$$

We can easily see that Lebesgue measure is translation invariant: If $\mu S = r$, then $\mu(S + a) = \mu\{s + a \mid s \in S\} = r$ for any $a \in \mathbb{R}$. Similarly, if we define $S \oplus a = \{s \oplus a \mid s \in S\}$, then $\mu(S \oplus a) = \mu(S)$. Now define an equivalence relation $a \sim b$ on I to mean $a - b \in \mathbb{Q}$, the rational numbers. Let $[a] = \{b \in I \mid a \sim b\}$ be the equivalence class of any $a \in I$ and $K = \{[a] \mid a \in I\}$. Define the set C to contain exactly one element from each equivalence class in K. Set theory's Axiom of Choice [46] ensures that set C exists. Since $\mathbb{Q}$ is countable, we can index $\mathbb{Q} \cap I$ by $\{q_n \mid n \in \mathbb{N}\}$, with $q_0 = 0$. We set $C_n = C \oplus q_n$. The properties of the C_n are as follows:

(i) $C_0 = C$.

(ii) The C_n are disjoint; for if $r \in C_m \cap C_n$, then $r = c + q_m = d + q_n$ for some c, $d \in C$; so $c \sim d$, and since C was a choice set containing exactly one element from disjoint equivalence classes in K, we must have $m = n$.

(iii) $I = \bigcup_{n \in \mathbb{N}} C_n$; if $r \in I$, then there is an $[a] \in K$ with $r \in [a]$ and $a \in C$; this implies $r \sim a$ or $r - a \in \mathbb{Q} \cap I$; but $\{q_n\}$ indexes such rational numbers, so $r - a = q_n$ for some $n \in \mathbb{N}$; thus, $r = a + q_n \in C_n$.

If C is Lebesgue measurable, then by the properties of the Lebesgue measure under translations, for all $n \in \mathbb{N}$, C_n is measurable and $\mu C = \mu C_n$.

Thus,

$$\mu(I) = \mu\Big(\bigcup_{n \in \mathbb{N}} C_n\Big) = \sum_{n=0}^{\infty} \mu(C_n) = \sum_{n=0}^{\infty} \mu C = \begin{cases} \infty & \text{if } \mu C > 0 \\ 0 & \text{if } \mu C = 0 \end{cases} \tag{3.115}$$

However, we know $\mu I = 1$, so that (3.115) is a contradiction; it must be the case that the choice set C is not Lebesgue measurable.

Our intuition might well suggest an easy generalization of the idea of an interval's length to a length measure for any subset of the real line. This last example has shown that the task demands some care. We cannot have the proposed properties of a measure and still be able to measure the size (length or area) of all sets. Some functions must be outside our theory of analog signals. The characteristic function for the choice set C in the above example is a case in point. Nevertheless, the class of measurable sets is quite large, and so too is the class of measurable functions. Let us turn to integration of measurable functions and apply the modern integral to signal theory. Here we shall see how these mathematical tools sharpen the definitions of the basic analog signal spaces. Moreover, these concepts will support our later development of signal approximation and transform techniques.

3.4.2 Lebesgue Integration

The key distinction between the Lebesgue and Riemann integrals is that the modern integral partitions the range of function values, whereas the classic integral partitions the domain of the function. A seemingly inconsequential difference at first glance, this insight is critical.

To illustrate the idea, consider the task of counting the supply of canned goods on a cabinet shelf. (Perhaps the reader lives in a seismically active region, such as this book's Californian authors, and is assessing household earthquake preparedness.) One way to do this is to iterate over the shelf, left to right, front to back, adding up the volume of each canned food item. No doubt, most people would count cans like this. And it is just how we Riemann integrate $x(t)$ by adding areas: for each interval $I = [a, b)$ in the domain partition, accumulate the area, say $x(a) \times (b - a)$; refine the partition; find a better Riemann sum; and continue to the limit. The other way to guage the stock of canned food is to first count all the cans of one size, say the small tins of tuna. Then proceed to the medium size soup cans. Next the large canned vegetables. Finally—and perhaps exhausting the reader's patience—we count the giant fruit juices. It might seem silly, but it works.

And this is also Lebesgue's insight for defining the integral. We slice up the range of $x(t)$, forming sums with the weighted measures $x(a) \times \mu[x^{-1}\{a\}]$, where $x^{-1}\{a\} = \{t \in \mathbb{R} \mid x(t) = a\}$ and μ is some measure of the size of $x^{-1}\{a\}$. Our remaining exposition omits many abstractions and details; mathematical treatises that excel in this area are readily available. Instead we seek a clear and simple statement of how the ideas on measure and measurable functions fit together to support the modern integral.

3.4.2.1 Simple Functions.

Like the Riemann integral, the Lebesgue integral is also a limit. It is the limit of integrals of so-called simple functions.

Definition (Simple Function). If $x : \mathbb{R} \to \mathbb{R}$ is Lebesgue measurable and Range(x) is finite, then $x(t)$ is called a simple function.

Every Lebesgue measurable function can be approximated from below by such simple functions [44]. This theorem's proof uses the stereotypical Lebesgue integration technique of partitioning the range of a measurable function.

Theorem (Approximation by Simple Functions). If $x(t)$ is measurable, then there is a sequence of simple functions $s_n(t)$ such that:

(i) If $n \le m$, then $|s_n(t)| \le |s_m(t)|$.
(ii) $|s_n(\mathrm{t})| \le |x(t)|$.
(iii) $\lim_{n\to\infty} s_n(t) = x(t)$.

Proof: Since we can split $x(t)$ into its negative and non-negative parts, and these are still measurable, we may assume that $x(t) \in [0, \infty]$ for all $t \in \mathbb{R}$. The idea is to break Range(x) into two parts: $S_n = [0, n)$ and $T_n = [n, \infty]$ for each $n > 0$. We further subdivide S_n into $n2^n$ subintervals of length 2^{-n}: $S_{m,n} = [m2^{-n}, (m + 1)2^{-n})$, for $0 \le m < n2^n$. Set $A_n = x^{-1}(S_n)$ and $B_n = x^{-1}(T_n)$. Define

$$s_n(t) = n\chi_{B_n}(t) + \sum_{m=0}^{n2^n - 1} \frac{m}{2^n}\chi_{B_n}(t), \tag{3.116}$$

where χ_S is the characteristic function on the set S. The simple functions (3.116) satisfy the conditions (i)–(iii). ■

3.4.2.2 Definition and Basic Properties. We define the modern integral in increments: first for non-negative functions, then for functions that go negative, and finally for complex-valued functions.

Definition (Lebesgue Integral, Non-negative Functions). Let μ the Lebesgue measure, let A be a Lebesgue measurable set in $\mathbb{R}$, and let $x : \mathbb{R} \to [0, \infty]$ be a measurable function. Then the Lebesgue integral with respect to μ of $x(t)$ over A is

$$\int_A x(t)\, d\mu = \sup_{0 \le s(t) \le x(t)} \left\{ \int_A s(t)\, d\mu \right\}, \tag{3.117}$$

where the functions $s(t)$ used to take the least upper bound in (3.117) are all simple.

Definition (Lebesgue Integral, Real-Valued Functions). Let μ the Lebesgue measure, let A be a Lebesgue measurable set in $\mathbb{R}$, and let $x : \mathbb{R} \to \mathbb{R}$ be a measurable function. Furthermore, let $x(t) = p(t) - n(t)$, where $p(t) > 0$ and $n(t) > 0$ for all $t \in \mathbb{R}$. Then the Lebesgue integral with respect to μ of $x(t)$ over A is

$$\int_A x(t)\, d\mu = \int_A p(t)\, d\mu - \int_A n(t)\, d\mu \tag{3.118}$$

as long as one of the integrals on the right-hand side of (3.118) is finite.

Remarks. By the elementary properties of Lebesgue measurable functions, the positive, negative, and zero parts of $x(t)$ are measurable functions. Note that the zero part of $x(t)$ does not contribute to the integral. In general, definitions and properties of the Lebesgue integral must assume that the subtractions of extended reals make sense, such as in (3.118). This assumption is implicit in what follows.

Definition (Lebesgue Integral, Complex-Valued Functions). Let μ the Lebesgue measure; let A be a Lebesgue measurable set in $\mathbb{R}$, let x: $\mathbb{R} \to \mathbb{C}$ be a measurable function, and let $x(t) = x_r(t) + jx_i(t)$, where $x_r(t)$ and $x_i(t)$ are real-valued for all $t \in \mathbb{R}$. Then the Lebesgue integral with respect to μ of $x(t)$ over A is

$$\int_A x(t)\, d\mu = \int_A x_r(t)\, d\mu + j\int_A x_i(t)\, d\mu . \tag{3.119}$$

The modern Lebesgue integral obeys all the rules one expects of an integral. It also agrees with the classic Riemann integral on piecewise continuous functions. Finally, it has superior limit operation properties.

Proposition (Linearity). Let μ the Lebesgue measure, let A be a Lebesgue measurable set in $\mathbb{R}$, and let $x, y : \mathbb{R} \to \mathbb{C}$ be measurable functions. Then,

(i) (Scaling) For any $c \in \mathbb{C}$, $\int_A cx(t)\, d\mu = c\int_A x(t)\, d\mu$.

(ii) (Superposition) $\int_A [x(t) + y(t)]\, d\mu = \int_A x(t)\, d\mu + \int_A y(t)\, d\mu$.

Proof: The integral of simple functions and the supremum are linear. ■

Remark. If $x(t) = 0$ for all $t \in \mathbb{R}$, then (even if $\mu A = \infty$) $\int_A x(t)\, d\mu = 0$ by (i).

Proposition. Let μ be the Lebesgue measure, let $[a, b]$ be an interval on $\mathbb{R}$, and let x: $\mathbb{R} \to \mathbb{C}$ be a piecewise continuous function. Then the Lebesgue and Riemann integrals of $x(t)$ are identical:

$$\int_{[a,b]} x(t)\, d\mu = \int_a^b x(t)\, dt. \tag{3.120}$$

Proof: $x(t)$ is both Riemann integrable and Lebesgue integrable. The Riemann integral, computed as a limit of lower rectangular Riemann sums (3.110a), is precisely a limit of simple function integrals. ■

Proposition (Domain Properties). Let μ be the Lebesgue measure; let A, B be Lebesgue measurable sets in $\mathbb{R}$; and let x: $\mathbb{R} \to \mathbb{C}$ be a measurable function. Then,

(i) (Subset) If $B \supset A$ and $x(t)$ is non-negative, then $\int_B x(t)\, d\mu \geq \int_A x(t)\, d\mu$.

(ii) (Union) $\int_{A \cup B} x(t)\, d\mu = \int_A x(t)\, d\mu + \int_B x(t)\, d\mu - \int_{A \cap B} x(t)\, d\mu$.

(iii) (Measure Zero Set) If $\mu(A) = 0$, then $\int_A x(t)\, d\mu = 0$.

Remark. Note that $\int_A x(t)\, d\mu = 0$ in (iii) even if $x(t) = \infty$ for all $t \in \mathbb{R}$.

Proposition (Integrand Properties). Let μ the Lebesgue measure; let A be Lebesgue measurable in $\mathbb{R}$; let $\chi_A(t)$ be the characteristic function on A; and let x, y: $\mathbb{R} \to \mathbb{C}$ be measurable functions. Then,

(i) (Monotonicity) If $y(t) \geq x(t) \geq 0$ for all $t \in \mathbb{R}$, then $\int_A y(t)\, d\mu \geq \int_A x(t)\, d\mu$.

(ii) (Characteristic Function) $\int_A x(t)\, d\mu = \int_{\mathbb{R}} x(t)\chi_A(t)\, d\mu$.

Proof: The proofs of these propositions follows from the definition of simple functions and integrals as limits thereof. ■

3.4.2.3 Limit Operations with Lebesgue's Integral.

The modern integral supports much more powerful limit operations than does the Riemann integral. We recall that sequence of functions can converge to a limit that is not Riemann integrable. In order to simplify the discussion, we offer the following theorems without detailed proofs; the interested reader can find them in treatises on modern analysis, [13, 24, 25, 44, 45].

Theorem (Monotone Convergence). Let μ be the Lebesgue measure, let A be a measurable set in $\mathbb{R}$, and x_n: $\mathbb{R} \to \mathbb{R}$ be Lebesgue measurable for $n \in \mathbb{N}$. If $0 \le x_n(t) \le x_m(t)$ for $n < m$ and $\lim_{n\to\infty} x_n(t) = x(t)$ for all $t \in A$, then $x(t)$ is measurable and

$$\lim_{n \to \infty} \int_A x_n(t)\, d\mu = \int_A x(t)\, d\mu . \tag{3.121}$$

Proof: $x(t)$ is measurable, because for any $r \in \mathbb{R}$,

$$x^{-1}(r, \infty] = \bigcup_{n=0}^{\infty} x_n^{-1}(r, \infty] . \tag{3.122}$$

Hence, $\lim_{n \to \infty} \int_A x_n(t)\, d\mu \le \int_A x(t)\, d\mu$. Arguing the inequality the other way [44] requires that we consider simple functions $s(t)$ such that $0 \le s(t) \le x(t)$ for all $t \in \mathbb{R}$. Let $0 < c < 1$ be constant and set $A_n = \{t \in \mathbb{R} | \, cs(t) \le x_n(t)\}$. Then, $A_{n+1} \supset A_n$ for all $n \in \mathbb{N}$, and $A = \bigcup_{n=0}^{\infty} A_N$. Thus,

$$c \int_{A_n} s(t)\, d\mu \le \int_A x_n(t)\, d\mu . \tag{3.123}$$

As $n \to \infty$ on the right-hand side of (3.123), we see

$$c \int_A s(t)\, d\mu \le \lim_{n \to \infty} \int_A x_n(t)\, d\mu , \tag{3.124a}$$

which is true for every $0 < c < 1$. Let $c \to 1$, so that

$$\int_A s(t)\, d\mu \le \lim_{n \to \infty} \int_A x_n(t)\, d\mu . \tag{3.124b}$$

But $s(t)$ can be any simple function bounding $x(t)$ below, so by the definition of Lebesgue integration we know $\lim_{n \to \infty} \int_A x_n(t)\, d\mu \ge \int_A x(t)\, d\mu$. ■

Corollary. Let μ the Lebesgue measure; let A be a measurable set in $\mathbb{R}$; let x_n: $\mathbb{R} \to \mathbb{R}$ be Lebesgue measurable for $n \in \mathbb{N}$; and, for all $t \in A$, suppose $\lim_{n\to\infty} x_n(t) = x(t)$. Then, $x(t)$ is measurable and

$$\lim_{n \to \infty} \int_A x_n(t)\, d\mu = \int_A x(t)\, d\mu . \tag{3.125}$$

Proof: Split $x(t)$ into negative and positive parts. ■

A similar result holds for complex-valued functions. The next corollary shows that we may interchange Lebesgue integration and series summation.

Corollary (Integral of Series). Let μ the Lebesgue measure; let A be a measurable set in $\mathbb{R}$; let x_n: $\mathbb{R} \to \mathbb{C}$ be Lebesgue measurable for $n \in \mathbb{N}$; and, for all $t \in A$, suppose $\Sigma_{n\to\infty} x_n(t) = x(t)$. Then,

$$\sum_{n=0}^{\infty} \int_A x_n(t)\, d\mu = \int_A x(t)\, d\mu. \tag{3.126}$$

Proof: Beginning with non-negative functions, apply the theorem to the partial sums in (3.126). Then extend the result to general real- and complex-valued functions. ■

The next theorem, Fatou's lemma,[5] relies on the idea of the lower limit of a sequence [23].

Definition (lim inf, lim sup). Let $A = \{a_n \mid n \in \mathbb{N}\}$ be a set of real numbers and let $A_N = \{a_n \mid n \ge \mathrm{N}\}$. Let $r_N = \inf A_N$ be the greatest lower bound of A_N in the extended real numbers $\mathbb{R} \cup \{\infty\} \cup \{-\infty\}$. Let $s_N = \sup A_N$ be the least upper bound of A_N in the extended real numbers. We define lim inf A and lim sup A by

$$\liminf \{a_n\} = \lim_{N\to\infty} r_N \tag{3.127a}$$

and

$$\limsup \{a_n\} = \lim_{N\to\infty} s_N. \tag{3.127b}$$

The main things anyone has to know are:

- $\liminf\{a_n\}$ is the smallest limit point in the sequence $\{a_n\}$;
- $\limsup\{a_n\}$ is the largest limit point in the sequence $\{a_n\}$;
- $\liminf\{a_n\} = \limsup\{a_n\}$ if and only if the sequence $\{a_n\}$ converges to some limit value $a = \lim_{n\to\infty}\{a_n\}$, in which case $a = \liminf\{a_n\} = \limsup\{a_n\}$;
- the upper and lower limits could be infinite, but they always exist;
- if $a_n \le b_n$ then $\liminf\{a_n\} \le \liminf\{b_n\}$;
- $\limsup\{-a_n\} = -\liminf\{a_n\}$.

Example. Consider the sequence $\{a_n\} = \{1^{-1}, 1, -1, 2^{-1}, 2, -2, 3^{-1}, 3, -3, \ldots\}$. This sequence has three limit points: $-\infty$, ∞, and 0. We have $\liminf\{a_n\} = -\infty$, $\limsup\{a_n\} = +\infty$, and $\lim_{n\to\infty}\{a_{3n}\} = 0$. Only this last sequence is a genuine Cauchy sequence, however.

Theorem (Fatou's Lemma). Let μ the Lebesgue measure; let A be a measurable set in $\mathbb{R}$; let x_n: $\mathbb{R} \to \mathbb{R}$ be Lebesgue measurable for $n \in \mathbb{N}$; and, for all $n \in \mathbb{N}$ and $t \in A$, $0 \le x_n(t) \le \infty$. Then,

$$\int_A \liminf[x_n(t)]\, d\mu \le \liminf \int_A x_n(t)\, d\mu. \tag{3.128}$$

[5]Pierre Fatou (1878–1929), mathematician and astronomer at the Paris Observatory.

Proof: Define

$$y_n(t) = \inf_{k \geq n} \{x_k(t)\}, \tag{3.129}$$

so that the $y_n(t)$ approach $\liminf x_n(t)$ as $n \to \infty$ and $y_n(t) \leq y_m(t)$ when $n < m$. Note that for all $n \in \mathbb{N}$, $y_n(t) \leq x_n(t)$. Consequently,

$$\int_A y_n(t)\, d\mu \leq \int_A x_n(t)\, d\mu, \tag{3.130}$$

and thus,

$$\liminf \int_A y_n(t)\, d\mu \leq \liminf \int_A x_n(t)\, d\mu. \tag{3.131}$$

The left-hand side of (3.131) draws our attention. Since $\lim\{y_n(t)\} = \liminf \{x_n(t)\}$ and $\{y_n(t)\}$ are monotone increasing, Lebesgue's monotone convergence theorem implies

$$\liminf \int_A y_n(t)\, d\mu = \lim \int_A y_n(t)\, d\mu = \int_A \lim[y_n(t)]\, d\mu = \int_A \liminf[x_n(t)]\, d\mu. \tag{3.132}$$

Combining (3.131) and (3.132) completes the proof. ∎

Theorem (Lebesgue's Dominated Convergence). Let μ be the Lebesgue measure, let A be a measurable set in $\mathbb{R}$, and let x_n: $\mathbb{R} \to \mathbb{C}$ be Lebesgue measurable for $n \in \mathbb{N}$, $\lim_{n\to\infty} x_n(t) = x(t)$ for all $t \in A$, and $|x_n(t)| \leq g(t) \in L^1(\mathbb{R})$. Then $x(t) \in L^1(\mathbb{R})$ and

$$\lim_{n \to \infty} \int_A x_n(t)\, d\mu = \int_A x(t)\, d\mu. \tag{3.133}$$

Proof: We borrow the proof from Ref. 44. Note first that limit $x(t)$ is a measurable function and it is dominated by $g(t)$, so $x(t) \in L^1(\mathbb{R})$. Next, we have $|x(t) - x_n(t)| \leq 2g(t)$ and we apply Fatou's lemma to the difference $2g(t) - |x_n(t) - x(t)|$:

$$2\int_A g(t) d\mu \leq \liminf_{n \to \infty} \int_A \{2g(t) - |x(t) - x_n(t)|\} d\mu. \tag{3.134a}$$

Manipulating the lower limit on the right-hand integral in (3.134a) gives

$$2\int_A g(t)\, d\mu \leq 2\int_A g(t)\, d\mu - \limsup_{n \to \infty} \int_A |x(t) - x_n(t)|\, d\mu. \tag{3.134b}$$

Subtracting $2\int_A g\, d\mu$ out of (3.134b), we have $\lim_{n\to\infty} \sup \int_A |x(t) - x_n(t)|\, d\mu \leq 0$ from which $\lim_{n\to\infty} \int_A |x(t) - x_n(t)|\, d\mu = 0$ and (3.133) follows. ∎

Corollary (Interchange of Limits). Let μ the Lebesgue measure, let A be a measurable set in $\mathbb{R}$, and let x_n: $\mathbb{R} \rightarrow \mathbb{C}$ be Lebesgue measurable for $n \in \mathbb{N}$. Suppose that

$$\sum_{n=-\infty}^{\infty} \int_A |x_n(t)| \, d\mu < \infty . \tag{3.135}$$

Then the series $\Sigma_n x_n(t) = x(t)$ for almost all $t \in A$, $x(t) \in L^1(\mathbb{R})$, and

$$\sum_{n=-\infty}^{\infty} \int_A x_n(t) \, d\mu = \int_A x(t) \, d\mu . \tag{3.136}$$

Proof: Apply the dominated convergence theorem to partial series sums. ■

3.4.2.4 Lebesgue Integrals in Signal Theory. To this point, our definitions of L^p signal spaces were defined abstractly as completions of more rudimentary spaces. The new integral helps avoid such stilted formulations. We define both the L^p norm and L^p signals spaces using the Lebesgue integral.

Definition (L^p, L^p norm). Let μ the Lebesgue measure and let A be a measurable set in $\mathbb{R}$. Then $L^p(A)$ is the set of all Lebesgue measurable signals $x(t)$ such that

$$\int_A |x(t)|^p d\mu < \infty . \tag{3.137}$$

We define $\|x\|_{p, A}$ to be

$$\|x\|_{p, A} = \left(\int_A |x(t)|^p d\mu \right)^{\frac{1}{p}} \tag{3.138}$$

when the integral (3.137) exists. In the case of $p = \infty$, we take $L^\infty(A)$ to be the set of all $x(t)$ for which there exists M_x with $|x(t)| < M_x$ almost everywhere on A.

Now we can recast the entire theory of L^p spaces using the modern integral. We must still identify L^p space elements with the equivalence class of all functions that differ only by a set of measure zero. The basic inequalities of Holder, Minkowski, and Schwarz still hold. For instance, Minkowski's inequality states that $\|x + y\|_p \leq \|x\|_p + \|y\|_p$, where the p-norm is defined by Lebesgue integral. The more powerful limit theorems of the modern integral, however, allow us to prove the following completeness result [25, 44, 45].

Theorem. For Lebesgue measurable A, the $L^p(A)$ spaces are complete, $1 \leq p \leq \infty$.

Proof: We leave the case $p = \infty$ as an exercise. Let $\{x_n(t)\}$ be Cauchy in $L^p(A)$ We can extract a subsequence $\{y_n(t)\}$ of $\{x_n(t)\}$ with $\|y_{n+1} - y_n\|_p < 2^{-n}$ for all n. We then define $f_n(t) = |y_1(t) - y_0(t)| + |y_2(t) - y_1(t)| + \cdots + |y_{n+1}(t) - y_n(t)|$ and $f(t) = \lim_{n\to\infty} f_n(t)$.

The choice of the subsequence and Minkowski's inequality together imply $\|f_n(t)\|_p < 1$. Invoking Fatou's lemma (3.128) on $\{[f_n(t)]^p\}$, we obtain

$$\int_A [f(t)]^p d\mu = \int_A \liminf [f_n(t)]^p d\mu \le \liminf \int_A [f_n(t)]^p d\mu \le \liminf \{1^p\} = 1. \tag{3.139}$$

This also shows $f(t) < \infty$ almost everywhere on A. If $x(t)$ is given by

$$x(t) = y_0(t) + \sum_{n=0}^{\infty} (y_{n+1}(t) - y_n(t)), \tag{3.140}$$

then the convergence of $f(t)$ guarantees that $x(t)$ converges absolutely almost everywhere on A. Thus, $x(t) = \lim_{n\to\infty} y_n(t)$, the Cauchy sequence $\{x_n(t)\}$ has a convergent subsequence, and so $\{x_n(t)\}$ must have the same limit. ■

The next result concerns two-dimensional integrals. These occur often, even in one-dimensional signal theory. The Fubini theorem[6] provides conditions under which iterated one-dimensional integrals—which arise when we apply successive integral operators—are equal to the associated two-dimensional integral [24]. Functions defined on $\mathbb{R} \times \mathbb{R}$ are really two-dimensional signals—*analog images*—and generally outside the scope of this book. However, in later chapters our signal transform operations will mutate a one-dimensional signal into a two-dimensional transform representation. Opportunties to apply the following result will abound.

Theorem (Fubini). Let $x(s, t)$ be a measurable function on a measurable subset $A \times B$ of the plane $\mathbb{R}^2$. If either of these conditions obtains,

(i) $x(s, t) \in L^1(A \times B)$,
(ii) $0 \le x(s, t)$ on $A \times B$,

then the order of integration may be interchanged:

$$\int_A \left[\int_B x(s, t)\, dt \right] ds = \int_B \left[\int_A x(s, t)\, ds \right] dt. \tag{3.141}$$

Proof: Refer to Ref. 24.

3.4.2.5 Differentiation. It remains to explain the concept of a derivative in the context of Lebesgue integration. Modern integration seems to do everything backwards. First, it defines the sums for integration in terms of range values rather than domain intervals. Then, unlike conventional calculus courses, it leaves out the intuitively easier differentiation theory until the end. Lastly, as we shall see below, it defines the derivative in terms of an integral.

[6]Guido Fubini (1879–1943) was a mathematics professor at Genoa and Turin until anti-Semitic decrees issued by Mussolini's fascist regime compelled him to retire. Fubini moved to the United States from Italy, taking a position at the Institute for Advanced Study in Princeton, New Jersey in 1939.

Definition (Derivative). Let $y(t)$ be a Lebesgue measurable function. Suppose $x(t)$ is a Lebesgue measurable function such that

$$y(t) = \int_a^t x \, d\mu , \tag{3.142}$$

almost everywhere on any interval $[a, b]$ that contains t. Then we say $x(t)$ is the *derivative* of $y(t)$. With the usual notations, we write $\frac{dy}{dt} = y'(t) = y^{(1)}(t) = x(t)$. Second and higher derivatives may be further defined, and the notations carry through as well.

All of the differentiation properties of conventional calculus check out under this definition.

3.5 DISTRIBUTIONS

Distributions extend the utility of Hilbert space to embrace certain useful quantities which are not classically defined in Riemannian calculus. The most celebrated example of a distribution is the *Dirac delta*, which played a seminal role in the development of quantum mechanics and has been extended to other areas of quantitative science [7–11, 47–49]. The Dirac delta is easy to apply but its development within the context of distributions is often obscured, particularly at the introductory level. This section develops the foundations of distribution theory with emphasis on the Dirac delta. The theory and definitions developed here are also the basis for the generalized Fourier transform of Chapter 6.

3.5.1 From Function to Functional

Quantitative science is concerned with generating numbers and the notion of a function as a mapping from the one set of complex numbers to another is well-established. The inner product in Hilbert space is another tool for generating physically relevant data, and this Hilbert space mapping is conveniently generalized by the concept of a *functional*.

Definition (Functional). Let $\phi(t)$ be a function belonging to the class of so-called *test functions* (to be defined shortly). If the inner product

$$\langle f(t), \phi(t) \rangle \equiv \int_{-\infty}^{\infty} f(t) \phi^*(t) \, dy \tag{3.143}$$

converges, the quantity $f(t)$ is a *functional* on the space of test functions.

The test functions are defined as follows.

Definition (Rapid Descent). A function is said to be *rapidly descending* if for each positive integer *N*, the product

$$t^N \cdot f(t) \tag{3.144}$$

remains bounded as $|t| \to \infty$. A test function $\phi(t)$ will be classified as *rapidly decreasing* if $\phi(t)$ and all its derivatives are rapidly decreasing.

The derivatives are included in this definition to ensure that the concept of rapid descent is closed under this operation. Exponentials such as e^{-t}, and Gaussians e^{-t^2} are rapidly decreasing. On the other hand, polynomials, the exponential e^t, and $\sin t$, $\cos t$ are not rapidly decreasing. (These will be categorized shortly.) Furthermore, rapidly decreasing functions are integrable.

The condition of rapid descent can be guaranteed for a large class of functions by forcing them vanish identically for all t outside some interval $[t_1, t_2]$, that is,

$$f(t) \to f(t) \cdot [u(t-t_1) - u(t-t_2)]. \tag{3.145}$$

These *test functions of compact support* are a subset of all test functions of rapid descent.

Remark. The test functions are our slaves; by stipulating that they decay sufficiently rapidly, we can ensure that (3.143) converges for a sufficiently broad class of functionals $f(t)$. In many discussions, particularly general theoretical treatment of functionals, the exact form of the test function is immaterial; it is merely a vehicle for ensuring that the inner products on the space of $\phi(t)$ converge. Often, all that is required is the knowledge that a test function behaves in a certain way under selected operations, such as differentiation, translation, scaling, and more advanced operations such as the Fourier transform (Chapters 5 and 6). Whenever possible, it is advantageous to work in the space of compact support test functions since that eliminates any questions as to whether its descent is sufficiently rapid; if $\phi(t)$ has compact support, the product $\phi(t)f(t)$ follows suit, admitting a large set of $f(t)$ for which (3.143) generates good data. However, in some advanced applications, such as the generalized Fourier transform (Chapter 6), we do not have the luxury of assuming that all test functions are compactly supported.

3.5.2 From Functional to Distribution

A distribution is a subset of the class of functionals with some additional (and physically reasonable) properties imposed.

3.5.2.1 Defintion and Classification

Definition (Distribution). Let $\phi(t)$ be a test function of rapid descent. A functional $f(t)$ is a distribution if it satisfies conditions of continuity and linearity:

(i) *Continuity.* Functional $f(t)$ is continuous if when the sequence $\phi_k(t)$ converges to zero in the space of test functions, then $\lim_{k \to \infty} \langle f(t), \phi_k(t) \rangle \to 0$.

(ii) *Linearity.* Functional $f(t)$ is linear if for all complex constants c_1 and c_2 and for $\psi(t)$ of rapid descent,

$$\langle f(t), c_1\phi(t) + c_2\psi(t) \rangle = c_1\langle f(t), \phi(t) \rangle + c_2\langle f(t), \psi(t) \rangle. \tag{3.146}$$

Linearity plays such a central role to signal analysis that any functional that does not belong to the class of distributions is not useful for our applications. Distributions of all types are sometimes referred to as *generalized functions.*

Definition (Equivalent Distributions). Two functionals $f(t)$ and $g(t)$ are equivalent if

$$\langle f(t), \phi(t) \rangle = \langle g(t), \phi(t) \rangle \tag{3.147}$$

for all test functions.

A distribution $f(t)$ that is equivalent to a classically defined function $f_0(t)$ so that

$$\langle f(t), \phi(t) \rangle = \langle f_0(t), \phi(t) \rangle \tag{3.148}$$

is termed a *regular distribution*. Distributions that have no direct expression as a standard function belong to the class of *singular distributions*. The most celebrated example of a singular distribution is the Dirac delta, which we will study in detail.

Distributions $f(t)$ defined on the space of the rapidly descending test functions are *tempered distributions* or *distributions of slow growth.* The concept of slow growth applies equally well to regular and singular distributions. For the former, it can be illustrated with familiar concepts:

Definition (Slow Increase). A function is said to be *slowly increasing*, *tempered*, or of *slow growth* if for some positive integer M, the product

$$t^{-M} \cdot f(t) \tag{3.149}$$

remains bounded as $|t| \to \infty$.

In essence, a slowly increasing function is one that can be tamed (tempered) by a sufficiently high power of t in (3.149). Examples include the polynomials, the sine and cosine, as well as $\sin t/t$.The exponential e^t grows too rapidly to be tempered.

Remarks. Functions of slow growth are not generally integrable, a fact that later hinders the description of their spectral content via the integral Fourier transform.

We will demonstrate this difficulty in Chapter 5 and remedy the situation by defining a generalized Fourier transform in Chapter 6; of utility will be many of the concepts developed here.

Note that the product of a function of slow growth $g(t)$ and a function of rapid descent $f(t)$ is a function of rapid descent. This is easily demonstrated. According to (3.144) and (3.149), there is some M for which

$$\upsilon \cdot g(t)f(t) \tag{3.150}$$

remains bounded as $t \to \infty$ for any positive integer $\upsilon \equiv N - M$. Therefore $h(t) \equiv f(t)g(t)$ decreases rapidly. Such products are therefore integrable. For signal analysis applications, distributions of slow growth and test functions of rapid descent are the most useful set of *dual spaces* in the distribution literature. For alternative spaces, see Ref. 10.

3.5.2.2 *Properties of Distributions.*

Many standard operations such as scaling, addition, and multiplication by constants have predictable effects on the inner product defining distributions. In selected cases these operations map distributions to distributions, giving us flexibility to add them, scale the independent variable, multiply by classically defined functions or constants, and take derivatives.

Proposition. Let $f(t)$ be a distribution. Then $\frac{df}{dt}$ is a functional which is continuous and linear.

Proof: By definition, a distribution $\frac{df}{dt}$ is a functional satisfying

$$\left\langle \frac{df}{dt}, \phi(t) \right\rangle = f(t)\phi(t)\Big|_{-\infty}^{\infty} - \left\langle f(t), \frac{d\phi}{dt} \right\rangle = -\left\langle f(t), \frac{d\phi}{dt} \right\rangle \tag{3.151}$$

Continuity is assured by noting

$$\lim_{k \to \infty} \left\langle \frac{df}{dt}, \phi_k(t) \right\rangle = -\lim_{k \to \infty} \left\langle f(t), \frac{d\phi_k}{dt} \right\rangle = 0, \tag{3.152}$$

which follows directly from the stipulations placed on the test function $\phi(t)$. Linearity is easy to establish and is left to the reader. ■

Derivatives of higher order follow in a similar manner. Consider the second derivative $\frac{d^2 f}{dt^2}$. Let $g(t) = f'(t)$, where the prime denotes differentiation. Then for $g'(t) \equiv f''(t)$ the derivative rule leads to

$$\int_{-\infty}^{\infty} g'(t)\phi(t)\,dt = -\int_{-\infty}^{\infty} g(t)\phi'(t)\,dt. \tag{3.153}$$

According to the conditions defining a test function, $\psi(t) \equiv \phi'(t)$ is also a bona fide test function, so we can reexpress the derivative rule for $g(t)$:

$$\int_{-\infty}^{\infty} g'(t)\phi(t)\,dt = -\int_{-\infty}^{\infty} g(t)\phi'(t)\,dt = -\int_{-\infty}^{\infty} g(t)\psi(t)\,dt. \tag{3.154}$$

But the last element in this equality can be further developed to

$$-\int_{-\infty}^{\infty} g(t)\psi(t)\,dt = -\int_{-\infty}^{\infty} f'(t)\psi(t)\,dt = -\left[\int_{-\infty}^{\infty} f'(t)\psi(t)\,dt\right] = \int_{-\infty}^{\infty} f(t)\psi'(t)\,dt \tag{3.155}$$

where we used the first derivative property for $f(t)$. Linking (3.154) and (3.155) leads to the desired result expressed in terms of the original test function $\phi(t)$:

$$\int_{-\infty}^{\infty} f''(t)\phi(t)\,dt = \int_{-\infty}^{\infty} f(t)\phi''(t)\,dt \tag{3.156}$$

This result generalizes to derivatives of all orders (exercise).

Proposition (Scaling). Let a be a constant. Then

$$\langle f(at),\phi(t)\rangle = \frac{1}{|a|}\left\langle f(t),\phi\left(\frac{t}{a}\right)\right\rangle \tag{3.157}$$

Proof: Left as an exercise. ■

Note that (3.157) does not *necessarily* imply that $f(at) = \frac{1}{|a|}f(t)$, although such equivalence may be obtained in special cases. If this is not clear, reconsider the definition of equivalence.

Proposition (Multiplication by Constant). If a is a constant, if follows that

$$\langle af(t),\phi(t)\rangle = a\langle f(t),\phi(t)\rangle. \tag{3.158}$$

Proof: Trivial. ■

In many signal analysis applications it is common to mix distributions and classically defined functions.

Proposition (Associativity). Let $f_0(t)$ be an infinitely differentiable regular distribution. Then

$$\langle f_0(t)f(t),\ \phi(t)\rangle = \langle f(t), f_0(t)\phi(t)\rangle. \tag{3.159}$$

Proof: Exercise. ■

Why must we stipulate that $f_0(t)$ be infinitely differentiable? In general, the product of two distributions is *not* defined unless at least one of them is an infinitely differentiable regular distribution. This associative law (3.159) could be established because $f_0(t)$ was a regular distribution and the product $f_0(t)\phi(t)$ has meaning as a test function.

Proposition (Derivative of Product). Let $f_0(t)$ be a regular distribution and let $f(t)$ be a distribution of arbitrary type. The derivative of their product is a distribution satisfying

$$\frac{d}{dt}[f_0(t)f(t)] = f(t)\frac{df_0}{dt} + f_0(t)\frac{df}{dt}. \tag{3.160}$$

Proof: Consider the distribution represented by the second term above, regrouping factors and then applying the derivative rule:

$$\int_{-\infty}^{\infty} f_0(t)\frac{df}{dt}\phi(t)\,dt = \int_{-\infty}^{\infty} \frac{df}{dt}\cdot f_0(t)\phi(t)\,dt = -\int_{-\infty}^{\infty} f(t)\cdot\frac{d}{dt}[f_0(t)\phi(t)]\,dt. \tag{3.161}$$

The derivative in the last equality can be unpacked using the classical product rule since both factors are regular functions,

$$-\int_{-\infty}^{\infty} f(t)\cdot\left[\phi(t)\frac{df_0}{dt} + f_0(t)\frac{d\phi}{dt}\right] dt = -\int_{-\infty}^{\infty} f(t)\cdot\phi(t)\frac{df_0}{dt}\,dt - \int_{-\infty}^{\infty} f(t)\cdot\frac{d\phi}{dt}f_0(t)\,dt. \tag{3.162}$$

The right-hand side can be rearranged by applying the derivative rule to the second term. This gives two terms with $\phi(t)$ acting as a test function:

$$-\int_{-\infty}^{\infty} f(t)\frac{df_0}{dt}\cdot\phi(t)\,dt + \int_{-\infty}^{\infty} \frac{d}{dt}[f(t)f_0(t)]\cdot\phi(t)\,dt. \tag{3.163}$$

Comparing (3.162) and (3.163) and applying the definition of equivalence gives

$$f_0(t)\frac{df}{dt} = -f(t)\frac{df_0}{dt} + \frac{d}{dt}[f(t)f_0(t)], \tag{3.164}$$

from which the desired result (3.161) follows. ■

In many applications it is convenient to scale the independent variable. Distributions admit a chain rule under differentiation.

Proposition (Chain Rule). If $f(t)$ is an arbitrary distribution, a is a constant, and $y \equiv at$, then

$$\frac{d}{dt}f(at) = a\frac{d}{dt}f(t). \tag{3.165}$$

Proof: Exercise. ■

3.5.3 The Dirac Delta

Certain signals $f(t)$ exhibit jump discontinuities. From a classical Riemannian perspective, the derivative df/dt is singular at the jump. The situation can be reassessed within the context of distributions. Consider the unit step $u(t)$. For a test function of rapid descent $\phi(t)$, integration by parts produces

$$\int_{-\infty}^{\infty} \frac{du}{dt}\phi(t)\, dt = u(t)\phi(t)\Big|_{-\infty}^{\infty} - \int_{-\infty}^{\infty} u(t)\frac{d\phi}{dt}\, dt = -\int_{-\infty}^{\infty} u(t)\frac{d\phi}{dt}\, dt. \tag{3.166}$$

This reduces to

$$-\int_{0}^{\infty} u(t)\frac{du}{dt}\, dt = -[\phi(\infty) - \phi(0)] = \phi(0), \tag{3.167}$$

so that

$$\int_{-\infty}^{\infty} \frac{du}{dt}\phi(t)\, dt = \phi(0). \tag{3.168}$$

The existence of $\frac{d\phi}{dt}$ is central to the preceding argument and is guaranteed by the definition of the test function. Consider the following definition.

Definition (Dirac Delta). The Dirac delta $\delta(t)$ is a functional that is equivalent to the derivative of the unit step,

$$\delta(t) \equiv \frac{du}{dt}. \tag{3.169}$$

From (3.168), we have

$$\langle \delta(t), \phi(t) \rangle = \left\langle \frac{du}{dt}, \phi(t) \right\rangle = \phi(0) \tag{3.170}$$

so the value returned by the distribution $\delta(t)$ is the value of the test function at the origin. It is straightforward to show that the Dirac delta satisfies the conditions of continuity and linearity; it therefore belongs to the class of functionals defined as distributions. Appearances are deceiving: despite its singular nature, the Dirac delta is a distribution of slow growth since it is defined on the (dual) space consisting of the test functions of rapid descent.

By simple substitution of variables we can generalize further:

$$\langle \delta(t-\tau), \phi(t) \rangle = \int_{-\infty}^{\infty} \frac{d}{dt}u(t-\tau)\phi(t)\, dt = \phi(\tau). \tag{3.171}$$

This establishes the Dirac delta as a *sifting* operator which returns the value of the test function at an arbitrary point $t = \tau$. In signal analysis, the process of sampling, whereby a the value of a function is determined and stored, is ideally represented by an inner product of the form (3.171).

Remark. Unfortunately, it has become common to refer to the Dirac delta as the "delta function." This typically requires apologies such as "the delta function is not a function", which can be confusing to the novice, and imply that the Dirac delta is the result of mathematical sleight of hand. The term delta distribution is more appropriate. The truth is simple: The delta function is not a *function*; it is a *functional* and if the foregoing discussion is understood thoroughly, the Dirac delta is (rightly) stripped of unnecessary mathematical mystique.

The sifting property described by (3.171) can be further refined through a test compact support on the interval $t \in [a, b]$. The relevant integration by parts,

$$\int_a^b \frac{du}{dt}\phi(t)\,du = u(t)\phi(t)\Big|_a^b - \int_a^b u(t)\frac{d\phi}{dt}\,dt, \tag{3.172}$$

takes specific values depending on the relative location of the discontinuity (in this case, located at $t = 0$ for convenience) and the interval on which the test function is supported. There are three cases:

(i) $a < 0 < b$:

$$\int_a^b \frac{du}{dt}\phi(t)\,du = u(t)\phi(t)\Big|_a^b - \int_0^b 1\cdot\frac{d\phi}{dt}\,dt = \phi(b) - [\phi(b) - \phi(0)] = \phi(0). \tag{3.173}$$

(ii) $a < b < 0$: Since the unit step is identically zero on this interval,

$$\int_a^b \frac{du}{dt}\phi(t)\,du = 0 - \int_a^b 0\cdot\frac{d\phi}{dt}\,dt = 0. \tag{3.174}$$

(iii) $b > a > 0$: on this interval, $u(t) = 1$, so that

$$\int_a^b \frac{du}{dt}\phi(t)\,du = \phi(t)\Big|_a^b - \int_a^b 1\cdot\frac{d\phi}{dt}\,dt = 0. \tag{3.175}$$

Remark. In general, it is not meaningful to assign a pointwise value to a distribution since by their nature they are defined by an integral over an interval specified by the test function. Cases (ii) and (iii) assert that $\delta(t)$ is identically zero on the interval $[a, b]$. The above arguments can be applied to the intervals $[-\infty, -\varepsilon]$ and $[\varepsilon, \infty]$ where ε is arbitrarily small, demonstrating that the Dirac delta is identically zero at all points along the real line except the origin, consistent with expected behavior of $\frac{du}{dt}$.

The general scaling law (3.157) relates two inner products. But when the distribution is a Dirac delta, there are further consequences.

Proposition (Scaling). If $a \neq 0$, then

$$\delta(at) = \frac{1}{|a|}\delta(t). \tag{3.176}$$

Proof: Note that

$$\int_{-\infty}^{\infty} \delta(at)\phi(t)\,dt = \frac{1}{|a|}\int_{-\infty}^{\infty} \delta(t)\phi\left(\frac{t}{a}\right)\,dt = \frac{1}{|a|}\phi(0). \tag{3.177}$$

The right-hand side can be expressed as

$$\frac{1}{|a|}\phi(0) = \frac{1}{|a|}\int_{-\infty}^{\infty} \delta(t)\phi(t)\,dt. \tag{3.178}$$

Applying the definition of equivalence leads to the scaling law (3.176). ■

The special case $a = -1$ leads to the relation $\delta(-t) = \delta(t)$, so the Dirac delta has even symmetry. Some useful relations follow from the application of (3.159) to the Dirac delta.

Proposition (Associative Property). We have

$$\delta(t)f_0(t) = \delta(t)f_0(0). \tag{3.179}$$

Proof: Since

$$\int_{-\infty}^{\infty} \delta(t)f_0(t)\phi(t)\,dt = f_0(0)\phi(0) \equiv \int_{-\infty}^{\infty} \delta(t)f_0(0)\phi(t)\,dt, \tag{3.180}$$

this establishes the equivalence (3.179). ■

Note this does *not* imply $f_0(t) = f_0(0)$, since division is not an operation that is naturally defined for arbitrary singular distributions such as $\delta(t)$. However, (3.179) leads to some interesting algebra, as shown in the following example.

Example. Suppose $f_0(t) = t$. According to (3.179),

$$t\delta(t) = 0 \tag{3.181}$$

for all t. This implies that if $f(t)$ and $g(t)$ are distributions, and $tg(t) = tf(t)$, then

$$g(t) = f(t) + a_0\delta(t), \tag{3.182}$$

where a_0 is constant. Note that (3.182) applies equally well to regular and singular distributions; it is central to establishing important relations involving the generalized Fourier transform in Chapter 6.

The inner product defining the derivative,

$$\int_{-\infty}^{\infty} \frac{df}{dt}\phi(t)\, dt = -\int_{-\infty}^{\infty} f(t)\frac{d\phi}{dt}\, dt, \tag{3.183}$$

also leads to a sifting property.

Proposition (Differentiation)

$$\int_{-\infty}^{\infty} \delta'(t-\tau)\phi(t)\, dt = -\phi'(\tau). \tag{3.184}$$

Proof: Using a prime to denote differentiation with respect to t, we obtain

$$\int_{-\infty}^{\infty} \delta'(t)\phi(t)\, dt = -\int_{-\infty}^{\infty} \delta(t)\phi'(t)\, dt = -\phi'(0). \tag{3.185}$$

For an arbitrary Dirac delta centered at $t = \tau$, this generalizes to (3.184). ■

3.5.4 Distributions and Convolution

The convolution operation is central to analyzing the output of linear systems. Since selected signals and system impulse responses may be expressed in terms of the Dirac delta, some of our applications may involve the convolution of two singular distributions, or singular and regular distributions. Given that the product of two singular distributions is not defined, it may come as an unexpected result to define a convolution operation. As before, we will base the development on an analogous result derived from Riemannian calculus.

First, consider the convolution of a distribution with a test function. This problem is straightforward. Let $f(t)$ be an arbitrary distribution. Then the convolution

$$(f*\phi)(u) = \int_{-\infty}^{\infty} f(t)\phi(u-t)\, dt = \int_{-\infty}^{\infty} f(u-t)\phi(t)\, dt \tag{3.186}$$

is a function in the variable u.

Next, consider the convolution of two test functions. If $\phi(t)$ and $\psi(t)$ are test functions, their convolution presents no difficulty:

$$(\phi*\psi)(u) \equiv \int_{-\infty}^{\infty} \phi(u-t)\psi(t)\, dt = \int_{-\infty}^{\infty} \phi(t)\psi(u-t)\, dt = (\psi*\phi)(u). \tag{3.187}$$

Remark. If $(\phi*\psi)(u)$ is to be a test function of compact support, both $\phi(t)$ and $\psi(t)$ must also be compactly supported. When this consideration impacts an important conclusion, it will be noted.

Now reconsider this convolution in terms of a Hilbert space inner product. If we define

$$\psi_{\text{ref}} \equiv \psi(-t), \tag{3.188}$$

then

$$(\phi*\psi_{\text{ref}})(u) = \int_{-\infty}^{\infty} \phi(t)\psi(u+t)\, dt. \tag{3.189}$$

This leads to a standard inner product, since

$$(\phi*\psi_{\text{ref}})(0) = \int_{-\infty}^{\infty} \phi(t)\psi(t)\, dt = \langle \phi(t), \psi(t) \rangle. \tag{3.190}$$

Because $(\phi*\psi)(t)$ is a test function, it follows that

$$\langle \phi*\psi, \eta \rangle = ((\phi*\psi)*\eta_{\text{ref}})(0) = (\phi*(\psi_{\text{ref}}*\eta)_{\text{ref}})(0). \tag{3.191}$$

However,

$$(\phi*(\psi_{\text{ref}}*\eta)_{\text{ref}})(0) = \langle \phi, \psi_{\text{ref}}*\eta \rangle. \tag{3.192}$$

Comparing the last two equations gives the desired result:

$$\langle \phi*\psi, \eta \rangle = \langle \phi, \psi_{\text{ref}}*\eta \rangle. \tag{3.193}$$

The purpose of this exercise was to allow the convolution to migrate to the right-hand side of the inner product. This leads naturally to a definition that embraces the convolution of two singular distributions.

Definition (Convolution of Distributions). Using (3.193) as a guide, if $f(t)$ and $g(t)$ are distributions of any type, including singular, their convolution is defined by

$$\langle f*g, \phi \rangle \equiv \langle f, g_{\text{ref}}*\phi \rangle. \tag{3.194}$$

Example (Dirac Delta). Given two delta distributions $f(t) = \delta(t-a)$ and $g(t) = \delta(t-b)$, we have

$$g_{\text{ref}} = \delta(-t-b) = \delta(-(t+b)) = \delta(t+b). \tag{3.195}$$

So

$$g_{\text{ref}}{}^{*}\phi = \int_{-\infty}^{\infty} \phi(u-t)\delta(t+b)\, dt = \phi(u+b), \tag{3.196}$$

and the relevant inner product (3.194) takes the form

$$\langle f^{*}g, \phi\rangle = \int_{-\infty}^{\infty} \delta(u-a)\phi(u+b)\, dt = \phi(a+b). \tag{3.197}$$

So the desired convolution is expressed,

$$f^{*}g = \delta(t-(a+b)). \tag{3.198}$$

Later developments will lead us derivatives of distributions. An importaint result involving convolution is the following. A similar result holds for standard analog signals (functions on the real line).

Proposition (Differentiation of Convolution). Let f and g be arbitrary distributions. Then,

$$\frac{d}{dt}(f^{*}g) = f^{*}\frac{dg}{dt} = \frac{df}{dt}{}^{*}g\,. \tag{3.199}$$

Proof: The proof of (3.199) requires the usual inner product setting applicable to distributions and is left as an exercise. (You may assume that convolution of distributions is commutative, which has been demonstrated for standard functions and is not difficult to prove in the present context.) ■

3.5.5 Distributions as a Limit of a Sequence

Another concept that carries over from function theory defines a functional as a limit of a sequence. For our purposes, this has two important consequences. First, limits of this type generate approximations to the Dirac delta, a convenient property that impacts the wavelet transform (Chapter 11). Second, such limits make predictions of the high-frequency behavior of pure oscillations (sinusoids) which are the foundations of Fourier analysis (Chapters 5 and 6). This section covers both topics.

Definition. Let $f(t)$ be a distribution of arbitrary type, and suppose $f_n(t)$ is a sequence in some parameter n. Then if

$$\lim_{n\to\infty} \langle f_n(t),\phi(t)\rangle \to \langle f(t),\phi(t)\rangle, \tag{3.200}$$

then the sequence $f_n(t)$ approaches $f(t)$ in this limit:

$$\lim_{n \to \infty} f_n(t) = f(t). \tag{3.201}$$

3.5.5.1 Approximate Identities for the Dirac Delta. The Dirac delta has been introduced as the derivative of the unit step, a functional which is identically zero along the entire real line except in an infinitesimal neighborhood of the origin. A surprising number of regular distributions—such as the square pulse, suitably scaled functions involving trigonometric functions, and another family generated by Gaussians—approach the Dirac delta when sufficiently scaled and squeezed in the limit in which the scaling factor becomes large. Intuitively, we need normalized, symmetric functions with maxima at the origin and that approach zero on other points along the real line as the appropriate limit (3.201) is taken. This leads to the following theorem.

Theorem (Approximating Identities). Let $f(t)$ be a regular distribution satisfying the criteria:

$$f(t) \in L^1(R), \tag{3.202}$$

$$\int_{-\infty}^{\infty} f(t)\, dt = 1. \tag{3.203}$$

Define

$$f_a(t) \equiv af(at) \tag{3.204}$$

and stipulate

$$\lim_{a \to \infty} \int_{\tilde{R}} |f_a(t)| = 0, \tag{3.205}$$

where $\tilde{R}$ denotes the real line minus a segment of radius ρ centered around the origin: $\tilde{R} \equiv R/[-\rho, \rho]$.

Then $\lim_{a \to \infty} f_a(t) = \delta(t)$, and $f_a(t)$ is said to be an *approximating identity.*

Proof: First,

$$|\langle f_a(t), \phi(t)\rangle - \langle \delta(t), \phi(t)\rangle| = \left| \int_{-\infty}^{\infty} af(at)\phi(t)\, dt - \phi(0) \right|. \tag{3.206}$$

From (3.203) a simple substitution of variables shows $\int_{-\infty}^{\infty} af(at)\, dt = 1$. We can thus recast the right-hand side of (3.206) into a more convenient form:

$$\left| \int_{-\infty}^{\infty} af(at)\phi(t)\, dt - \phi(0) \right| = \left| \int_{-\infty}^{\infty} af(at)\phi(t)\, dt - \int_{-\infty}^{\infty} af(at)\phi(0)\, dt \right|. \tag{3.207}$$

Dividing the real line into regions near and far from the origin, a change of variables $u \equiv at$ gives

$$\left| \int_{-\infty}^{\infty} f(u) \left[\phi\left(\frac{u}{a}\right) - \phi(0) \right] du \right| = I_1 + I_2, \tag{3.208}$$

where

$$I_1 = \left| \int_{-\rho/a}^{\rho/a} f(u) \left[\phi\left(\frac{u}{a}\right) - \phi(0) \right] du \right| \tag{3.209}$$

and

$$I_2 = \left| \int_{R/\left(\frac{-\rho}{a}, \frac{\rho}{a}\right)} f(u) \left[\phi\left(\frac{u}{a}\right) - \phi(0) \right] du \right|. \tag{3.210}$$

Since $\phi(t)$ is continuous, for some t suitably close to the origin, we have

$$\left| \phi\left(\frac{u}{a}\right) - \phi(0) \right| < \varepsilon. \tag{3.211}$$

Thus, I_1 is bounded above:

$$I_1 \le \varepsilon \int_{-\rho/a}^{\rho/a} |f(u)|\, du. \tag{3.212}$$

In the second integral, note that there exists some ρ such that

$$\int_{R/\left(\frac{-\rho}{a}, \frac{\rho}{a}\right)} |f(u)|\, du \le \varepsilon. \tag{3.213}$$

Equation (3.210) can be expressed as

$$I_2 \le \int_{R/\left(\frac{-\rho}{a}, \frac{\rho}{a}\right)} \phi\left(\frac{u}{a}\right) |f(u)|\, du - |\phi(0)| \int_{R/\left(\frac{-\rho}{a}, \frac{\rho}{a}\right)} |f(u)|\, du. \tag{3.214}$$

Furthermore, by definition, $\phi(t)$ is bounded above. Consequently,

$$I_2 \le \left| \phi\left(\frac{u}{a}\right) \right|_{\max} \int_{R/\left(\frac{-\rho}{a}, \frac{\rho}{a}\right)} |f(u)|\, du - |\phi(0)| \varepsilon, \tag{3.215}$$

which reduces to

$$I_2 \le \varepsilon \left[\left| \phi\left(\frac{u}{a}\right) \right|_{\max} - |\phi(0)| \right]. \tag{3.216}$$

Returning to (3.206), we have

$$\left|\langle f_a(t),\phi(t)\rangle - \langle\delta(t),\phi(t)\rangle\right| \le \varepsilon\left[\int\limits_{R/\left(\frac{-\rho}{a},\frac{\rho}{a}\right)} |f(u)|\, du + \left|\phi\left(\frac{u}{a}\right)\right|_{\max} - |\phi(0)|\right]. \quad (3.217)$$

In the limit of large a, we have

$$\lim_{a\to\infty} \varepsilon = 0, \quad (3.218)$$

so that

$$\lim_{a\to\infty}\left|\langle f_a(t),\phi(t)\rangle - \langle\delta(t),\phi(t)\rangle\right| = 0 \quad (3.219)$$

and we obtain the desired result. ■

The decay condition (3.205) is equivalent to

$$\lim_{t\to\infty} f(t) = 0, \quad (3.220)$$

which can be stated: for each $\lambda > 0$ there exists $T_\lambda > 0$ such that $|f(t)| < \lambda$ for all $|t| > T_\lambda$. This is equivalent to stating that if $\lambda/a > 0$ there exists $T_{\lambda/a} > 0$ such that $|f(at)| < \lambda/a$ for $|at| > T_{\lambda/a}$. The scaling required to convert $f(t)$ to an approximate identity implies that

$$|af(at)| = a|f(t)| < a\frac{\lambda}{a} < \lambda \quad (3.221)$$

so that

$$\lim_{|t|\to\infty} af(at) = 0. \quad (3.222)$$

Remark. Functions $f(t)$ that satisfy (3.202)–(3.205) are commonly called *weight functions*. They are of more than academic interest, since their localized atomistic character and ability to wrap themselves around a selected location make them useful for generating wavelets (Chapter 11). Such wavelets can zoom in on small-scale signal features. The class of weight functions is by no means small; it includes both uniformly and piecewise continuous functions.

There are several variations on these approximating identities. For example, (3.203) can be relaxed so that if the weight function has arbitrary finite area

$$\Gamma \equiv \int_{-\infty}^{\infty} f(t)\, dt, \quad (3.223)$$

then (3.125) reads

$$\lim_{a \to \infty} f_a(t) = \Gamma\delta(t). \tag{3.224}$$

The scaling of the amplitude can be modified so that

$$f_a(t) \equiv \sqrt{a} f(at) \tag{3.225}$$

is an approximating identity. The proofs are left as exercises. These approximating identities give the wavelet transform the ability to extract information about the local features in signals, as we will demonstrate in Chapter 11.

We consider several common weight functions below.

Example (Gaussian). Let

$$f(t) = A_0 e^{-t^2/2}, \tag{3.226}$$

where A_0 is a constant. The general Gaussian integral

$$\int_{-\infty}^{\infty} e^{-\alpha y^2} dy \equiv \sqrt{\frac{\pi}{\alpha}} \tag{3.227}$$

implies that

$$A_0 = \frac{1}{\sqrt{2\pi}} \tag{3.228}$$

if $f(t)$ is to have unit area. Note that the ability to normalize the area is proof that $f(t)$ is integrable, so two of the weight function criteria are satisfied. Figure 3.7 illustrates a Gaussian approximate identity

$$f_a(t) = \frac{a}{\sqrt{2\pi}} e^{-(at)^2/2} \tag{3.229}$$

for increasing values of a. The Gaussian is a powerful tool in the development of the continuous wavelet transform. We shall review its role as an approximating identity when we cover the small-scale resolution of the wavelet transform.

Example (Abel's Function). In this case the weight function permits an arbitrary positive-definite parameter β,

$$f(t) = A_0 \frac{\beta}{1 + \beta^2 t^2}. \tag{3.230}$$

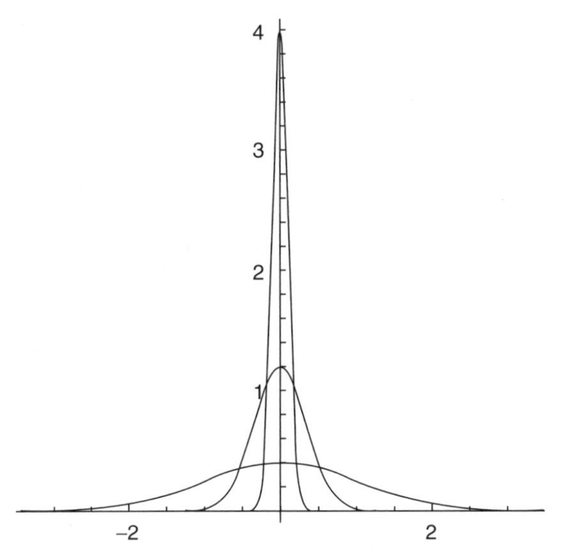

Fig. 3.7. The evolution of the scaled Gaussian (3.229) for $a = 1$ (widest), $a = 3$, and $a = 10$. In the limit $a \to \infty$, the scaled Gaussian approaches a Dirac delta.

It is left as an exercise to show that the condition of unit area requires

$$A_0 = \frac{1}{\pi}. \tag{3.231}$$

For this purpose it is convenient to use the tabulated integral,

$$\int_0^\infty \frac{x^{\mu-1}}{1+x^\nu}dx = \frac{\pi}{\nu}\csc\frac{\mu\pi}{\nu}, \tag{3.232}$$

where $Re\nu > Re\mu > 0$. The resulting approximate identity has the form

$$f_a(t) = \frac{a}{\pi}\frac{\beta}{1+\left(\frac{\beta}{a}\right)^2 t^2}. \tag{3.233}$$

Example (Decaying Exponential). This weight function is piecewise continuous, but fulfills the required criteria:

$$f(t) = A_0 e^{-|t|}. \tag{3.234}$$

It is left as an exercise to show that $A_0 = 1$, so that $f_a(t) = ae^{-|t/a|}$.

Example (Rectangular Pulse). The unit-area rectangular pulse described by

$$f(t) = \frac{1}{\sigma}[u(t) - u(t-\sigma)] \tag{3.235}$$

approximates a Dirac delta (exercise).

Remark. The weight functions described in this section embody properties that are sufficient to generate approximate identities. These conditions are by no means necessary. The family of sinc pulses

$$f(t) = \frac{1}{\pi}\frac{\sin t}{t} \tag{3.236}$$

does not qualify as a weight function (why not?), yet it can be shown that

$$\lim_{a \to \infty} \frac{1}{\pi}\frac{\sin at}{t} = \delta(t). \tag{3.237}$$

Note that the scale is applied only in the numerator. The properties of this approximate identity are explored in the exercises.

3.5.5.2 The Riemann–Lebesgue Lemma. Much of signal analysis involves the study of spectral content—the relative contributions from individual pure tones or oscillations at given frequency ω, represented by $\sin(\omega t)$ and $\cos(\omega t)$. The behavior of these oscillations as $\omega \to \infty$ is not predicted classically, but by treating the complex exponential $e^{j\omega t}$ as a regular distribution, the foregoing theory predicts that $\sin(\omega t)$ and $\cos(\omega t)$ vanish identically in the limit of high frequency. This result is useful in several subsequent developments.

Theorem (Riemann–Lebesgue Lemma). Let $f(t) \equiv e^{-j\omega t}$. If $\phi(t)$ is a test function of compact support on the interval $t \in [t_1, t_2]$, then

$$\lim_{\omega \to \infty} \int_{-\infty}^{\infty} f(t)\phi(t)\, dt = 0. \tag{3.238}$$

Proof: It is convenient to define a $g(t)$ such that

$$\frac{dg}{dt} \equiv f(t) = e^{-j\omega t} \tag{3.239}$$

and apply integration by parts to obtain

$$\int_{t_1}^{t_2} \frac{dg}{dt}\phi(t)\, dt = g(t)\phi(t)\Big|_{t_1}^{t_2} - \int_{t_1}^{t_2} g(t)\frac{d\phi}{dt}\, dt. \tag{3.240}$$

Then

$$\int_{t_1}^{t_2} e^{-j\omega t}\phi(t)\, dt = \frac{1}{j\omega}[\phi(t_2)e^{-j\omega t_2} - \phi(t_1)e^{-j\omega t_1}] + \frac{1}{j\omega}\int_{t_1}^{t_2} e^{-j\omega t}\frac{d\phi}{dt}\, dt. \tag{3.241}$$

By definition, the test function is bounded. Hence,

$$\left|\phi(t)e^{-j\omega t}\right| \le |\phi(t)|\left|e^{-j\omega t}\right| \le |\phi(t)|, \tag{3.242}$$

and the numerator of the first term in (3.241) is bounded. Consequently, in the limit $\omega \to 0$, this first term vanishes. The integral in the second term is also finite (it is the Fourier transform of $\frac{d\phi}{dt}$), and we defer proof of its boundedness until Chapter 5. Finally,

$$\lim_{\omega \to \infty} e^{-j\omega t} = 0. \tag{3.243}$$

By extension, the real and imaginary parts of this exponential ($\cos(\omega t)$ and $\sin(\omega t)$) also vanish identically in this limit. ∎

Remark. Test functions of compact support, as used herein, greatly simplify matters. When $(t_1, t_2) \to (-\infty, \infty)$, thus generalizing the result to all t, the same result obtains; we omit the complete proof, which is quite technical.

3.6 SUMMARY

Analog signal processing builds directly upon the foundation of continuous domain function theory afforded by calculus [6] and basic topology of sets of real numbers [22]. An analog system accepts an input signal and alters it to produce an output analog signal—a simple concept as long as the operations involve sums, scalar multiplications, translations, and so on. Linear, translation-invariant systems appear in nature and engineering. The notion is a mix of simple operations, and although there was a very straightforward theory of discrete LTI systems and discrete impulse response $\delta(n)$, replacing the n by a t stirs up serious complications.

Informal arguments deliver an analog equivalent of the convolution theorem, and they ought to be good enough justification for most readers. Others might worry about the analog impulse's scaling property and the supposition that an ordinary signal $x(t)$ decomposes into a linear combination of Diracs. But rigorous justification is possible by way of distribution theory [8–10].

Analog signals of certain classes, of which the L^p signals are especially convenient for signal processing, form elegant mathematical structures: normed spaces, Banach spaces, inner product, and Hilbert spaces. The inner product is a natural measure of similarity between signals. Completeness, or closure under limit operations, exists in Banach and Hilbert spaces and allows for incremental approximation of signals. Many of the results from discrete theory are purely algebraic in nature; we have been able to appropriate them once we show that analog signals—having rightly chosen a norm or inner product definition, of course—reside in one of the familiar abstract function spaces.

3.6.1 Historical Notes

Distribution theory dates from the mid-1930s with the work of S.L. Sobolev. Independent of the Soviet mathematician, L. Schwartz formalized the notion of a distribution and developed a rigorous delta function theory [7]. Schwartz's lectures are in fact the inspiration for Ref. 8.

Splines were first studied by Schoenberg in 1946 [14]. It is interesting to note that while his paper preceded Shannon's more famous results on sampling and signal reconstruction [50], the signal processing research community overlooked splines for many years.

Mathematicians developed frame theory in the early 1950s and used it to represent functions with Fourier series not involving the usual sinusoidal harmonics [41, 42]. Only some 30 years later did applied mathematicians, physicists, and engineers discover the applicability of the concept to signal analysis [43].

Details on the origins of signal theory, signal spaces, and Hilbert spaces can be found in mathematically oriented histories [51, 52]. Advanced treatments of functional analysis include Refs. 53 and 54.

Lebesgue's own presentation of his integral is given in Ref. 55. This book also contains a short biography of Lebesgue. Lebesgue's approach was the best of many competing approaches to replace the Riemannian integral [56, 57].

3.6.2 Looking Forward

The next chapter covers time-domain signal analysis. Taking representative signal interpretation problems, Chapter 4 attempts to solve them using the tools we have developed so far. The time-domain tools we can formulate, however, turn out to be deficient in analyzing signals that contain periodicities. This motivates the frequency-domain analysis of signals, beginning with the analog Fourier transforms in Chapter 5.

In Chapter 6, the formal development of the Dirac delta as a functional will help build a Fourier transform theory for periodic, constant, and otherwise untransformable analog signals.

3.6.3 Guide to Problems

This chapter includes many basic problems (1–46) that reinforce the ideas of the text. Some mathematical subtleties have been covered but casually in the main text, and the problems help the interested reader pursue and understand them. Of course, in working the problems, the student should feel free to consult the mathematical literature for help. The advanced problems (47–56) require broader, deeper investigation. For example, the last few problems advance the presentation on frames. Searching the literature should provide a number of approaches that have been considered for these problems. Indeed, investigators have not satisfactorily answered all of the questions posed.

REFERENCES

1. A. V. Oppenheim, A. S. Willsky, and S. H. Nawab, *Signals and Systems*, 2nd ed., Upper Saddle River, NJ: Prentice-Hall, 1997.
2. J. A. Cadzow and H. F. Van Landingham, *Signals, Systems, and Transforms*, Englewood Cliffs, New Jersey: Prentice-Hall, 1983.
3. H. Baher, *Analog and Dignal Signal Processing*, New York: Wiley, 1990.
4. R. E. Ziemer, W. H. Tranter, and D. R. Fannin, *Signals and Systems: Continouous and Discrete*, 3rd ed., New York: Macmillan, 1993.
5. L. B. Jackson, Signals, *Systems, and Transforms*, Reading, MA: Addison-Wesley, 1991.
6. M. Rosenlicht, *An Introduction to Analysis*, New York: Dover, 1978.
7. L. Schwartz, *Théorie des Distributions*, vols. I and II, Paris: Hermann, 1950–1951.
8. I. Halperin, *Introduction to the Theory of Distributions*, Toronto: University of Toronto Press, 1952.
9. M. J. Lighthill, *Introduction to Fourier Analysis and Generalized Functions*, New York: Cambridge University Press, 1958.
10. A. H. Zemanian, *Distribution Theory and Transform Analysis*, New York: Dover, 1965.
11. A. Friedman, *Generalized Functions and Partial Differential Equations*, Englewood Cliffs, NJ: Prentice-Hall, 1963.
12. E. M. Stein and G. Weiss, *Introduction to Fourier Analysis on Euclidean Spaces*, Princeton, NJ: Princeton University Press, 1971.
13. L. Debnath and P. Mikusinski, *Introduction to Hilbert Spaces with Applications*, 2nd ed., San Diego, CA: Academic Press, 1999.
14. I. J. Schoenberg, Contribution to the problem of approximation of equidistant data by analytic functions, *Quarterly of Applied Mathematics*, vol. 4, pp. 45–99, 112–141, 1946.
15. J. H. Ahlberg, E. N. Nilson, and J. L. Walsh, *The Theory of Splines and Their Applications*, New York: Academic Press, 1967.
16. D. Hearn and M. P. Baker, *Computer Graphics*, Englewood Cliffs, NJ: Prentice-Hall, 1986.
17. R. H. Bartels, J. C. Beatty, and B. A. Barsky, *Splines for Use in Computer Graphics*, Los Altos, CA: Morgan Kaufmann, 1987.
18. M. Unser, A. Aldroubi, and M. Eden, B-spline signal processing: Part I—theory, *IEEE Transactions on Signal Processing*, vol. 41, no. 2, pp. 821–833, February 1993.
19. M. Unser, A. Aldroubi, and M. Eden, B-spline signal processing: Part II—efficient design and applications, *IEEE Transactions on Signal Processing*, vol. 41, no. 2, pp. 834–848, February 1993.
20. S. Sakakibara, A practice of data smoothing by B-spline wavelets, in C. K. Chui, et al., eds., *Wavelets: Theory, Algorithms, and Applications*, San Diego, CA: Academic Press, pp. 179–196, 1994.
21. M. Unser, Splines: A perfect fit for signal and image processing, *IEEE Signal Processing Magazine*, vol. 16, no. 6, pp. 22–38, November 1999.
22. A. N. Kolmogorov and S. V. Fomin, *Introductory Real Analysis*, translator R. A. Silverman, New York: Dover, 1975.
23. R. Beals, *Advanced Mathematical Analysis*, New York: Springer-Verlag, 1987.

24. H. L. Royden, *Real Analysis*, 2nd ed., Toronto: Macmillan, 1968.
25. A. W. Naylor and G. R. Sell, *Linear Operator Theory in Engineering and Science*, New York: Springer-Verlag, 1982.
26. G. Birkhoff and S. MacLane, *A Survey of Modern Algebra*, New York: Macmillan, 1965.
27. E. Kreysig, *Introductory Functional Analysis with Applications*, New York: Wiley, 1989.
28. H. Dym and H. P. McKean, *Fourier Series and Integrals*, New York: Academic, 1972.
29. N. Young, *An Introduction to Hilbert Space*, Cambridge: Cambridge University Press, 1988.
30. A. Rosenfeld and A. C. Kak, *Digital Picture Processing*, 2nd ed., vol. II, Orlando, FL: Academic, 1982.
31. R. O. Duda and P. E. Hart, *Pattern Classification and Scene Analysis*, New York: Wiley, 1973.
32. S. Haykin, *Communication Systems*, 3rd ed., New York: Wiley, 1994.
33. J. L. Flanagan, A. C. Surendran, and E.-E. Jan, Spatially selective sound capture for speech and audio processing, *Speech Communication*, vol. 13, pp. 207–222, 1993.
34. F. Glazer, G. Reynolds, and P. Anandan, Scene matching by hierarchical correlation, *Proceedings of the IEEE Conference on Computer Vision and Pattern Recognition*, June 19–23, Washington, D.C., pp. 432–441, 1983.
35. P. J. Burt, C. Yen, and X. Xu, Local correlation measures for motion analysis: a comparative study, *Proceedings of the IEEE Conference on Pattern Recogniton and Image Processing*, pp. 269–274, 1982.
36. A. Haar, Zur theorie der orthogonalen Functionensysteme, *Mathematics Annalen*, vol. 69, pp. 331–371, 1910.
37. P. Wojtaszczyk, *A Mathematical Introduction to Wavelets*, Cambridge: Cambridge University Press, 1997.
38. I. Daubechies, *Ten Lectures on Wavelets*, Philadelphia: SIAM Press, 1992.
39. S. Mallat and Z. Zhang Matching pursuits with time-frequency dictionaries, *IEEE Transactions on Signal Processing*, vol. 41, no. 12, pp. 3397–3415, December 1993.
40. Goodwin and M. Vetterli, Matching pursuit and atomic signal models based on recursive filter banks, *IEEE Transactions on Signal Processing*, vol. 47, no. 7, pp. 1890–1902, July 1999.
41. R. J. Duffin and A. C. Schaeffler, A class of nonharmonic Fourier series, *Transactions of the American Mathematical Society*, vol. 72, pp. 341–366, 1952.
42. R. M. Young, *An Introduction to Nonharmonic Fourier Series*, New York: Academic Press, 1980.
43. I. Daubechies, A. Grossmann, and Y. Meyer, Painless nonorthogonal expansions, *Journal of Mathematical Physics*, vol. 27, pp. 1271–1283, 1986.
44. W. Rudin, *Real and Complex Analysis*, 2nd ed., New York: McGraw-Hill, 1974.
45. E. Hewitt and K. Stromberg, *Real and Abstract Analysis*, New York: Springer, 1991.
46. P. R. Halmos, *Naive Set Theory*, New York: Van Nostrand, 1960.
47. J. H. Richards and H. K. Youn, *Theory of Distributions: A Nontechnical Introduction*, New York: Cambridge University Press, 1990.
48. A. Papoulis, *The Fourier Integral and Its Application*, New York: McGraw-Hill, 1962.

49. N. Wiener, *The Fourier Integral and Certain of Its Applications*, New York: Dover Publications, 1958.

50. C. E. Shannon, Communication in the presence of noise, *Proceedings of the Institute of Radio Engineers*, vol. 37, pp. 10–21, 1949.

51. J. Dieudonne, *History of Functional Analysis*, Amsterdam: North-Holland, 1981.

52. J. Stillwell, *Mathematics and Its History*, New York: Springer-Verlag, 1989.

53. K. Yosida, *Functional Analysis*, 6th ed., Berlin: Springer-Verlag, 1980.

54. W. F. Rudin, *Functional Analysis*, 2nd ed., New York: McGraw-Hill, 1991.

55. H. L. Lebesgue, *Measure and the Integral*, edited, with a biography by K. O. May, San Francisco: Holden-Day, 1966.

56. J.-P. Pier, *Development of Mathematics: 1900–1950*, Basel: Birkhaeuser Verlag, 1994.

57. T. Hawkins, *Lebesgue's Theory of Integration: Its Origin and Development*, New York: Chelsea, 1970.

PROBLEMS

1. Find the domain and range of the following analog systems; if necessary, narrow the problem domain to a particular analog signal space.

(a) The amplifier (or attenuator, $|A| < 1$) system: $y(t) = Ax(t)$

(b) A translation system: $y(t) = x(t - a)$

(c) The system on real-valued signals, $y(t) = x(t)^{1/2}$

(d) The system on complex-valued signals, $y(t) = x(t)^{1/2}$

(e) The adder: $y(t) = x(t) + x_0(t)$

(f) Termwise multiplication: $y(t) = x(t) \times x_0(t)$

(g) Convolution: $y(t) = x(t)*h(t)$

(h) Accumulator:

$$y(t) = \int_{-\infty}^{t} x(s)\, ds. \tag{3.244}$$

2. Let $x(t)$ be an analog signal and let H be an analog accumulator system. Show that:

(a) If $x(t)$ has finite support, then $x(t)$ is in the domain of H.

(b) If $x(t)$ is absolutely integrable, then it is in the domain of H.

(c) There are finite energy signals are not in the domain of H. (*Hint*: Provide an example based on the signal $x(t) = t^{-1}$.)

3. Consider the systems in Problem 1.

(a) Which of them are linear?

(b) Which of them are translation-invariant?

(c) Which of them are stable?

4. Suppose that the analog systems H and G are LTI. Let $T = aH$ and $S = H + G$. Show that both T and S are LTI systems.

5. Which of the following systems are linear? translation-invariant? stable? causal?

(a) $y(t) = x(t/2)$

(b) $y(t) = x(2t)$

(c) $y(t) = x(0) + x(t)$

(d) $y(t) = x(-t)$

(e) The system given by

$$y(t) = \int_{-\infty}^{\infty} x(s)\cos(s)\,ds. \tag{3.245}$$

6. Consider the cross-correlation system $y = x^{\circ}h = Hx$, where

$$y(t) = (x^{\circ}h)(t) = \int_{-\infty}^{\infty} x(s)h(t+s)\,ds. \tag{3.246}$$

(a) Prove or disprove: H is linear.

(b) Prove or disprove: H is translation-invariant.

(c) Prove or disprove: H is stable.

(d) Is $h(t)$ the impulse response of H? Explain.

(e) Answer these same questions for the autocorrelation operation $y = x^{\circ}x$.

7. An analog LTI system H has impulse response $h(t) = 2e^{t}u(2-t)$, where $u(t)$ is the unit step signal. What is the response of H to $x(t) = u(t+3) - u(t-4)$?

8. Analog LTI system G has impulse response $g(t) = u(t+13) - u(t)$; $u(t)$ is the unit step. What is the response of G to $x(t) = u(t)e^{-t}$?

9. Analog LTI system K has impulse response $k(t) = \delta(t+1) + 2\delta(t-2)$, where $\delta(t)$ is the Dirac delta. What is K's response to $x(t) = u(t-1) - u(t)$?

10. Show that convolution is linear:

(a) $h*(ax) = ah*x$.

(b) $h*(x+y) = h*x + h*y$.

11. Show that if H and G are LTI systems, then $H(G(x)) = G(H(x))$.

12. Show that convolution is associative: $h*(x*y) = (h*x)*y$.

13. Give an alternative argument that $\delta(t) = u'(t)$, the derivative of the unit step [3].

(a) Let $x(t)$ be a signal; show that integration by parts implies

$$\int_{-\infty}^{\infty} x(t)\frac{d}{dt}u(t)\,dt = x(t)u(t)\Big|_{-\infty}^{\infty} - \int_{-\infty}^{\infty} x'(t)u(t)\,dt. \tag{3.247}$$

(b) Remove $u(t)$ from the final integrand in (3.247) and correct the limits of integration.

(c) Show that

$$\int_{-\infty}^{\infty} x(t)\frac{d}{dt}u(t)\,dt = x(0), \tag{3.248}$$

which is the sifting property once again; hence, $\delta(t) = \frac{d}{dt}u(t)$.

14. Suppose the function (an analog signal) $x(t)$ is uniformly continuous over some real interval $I = (a, b)$. This means that for every $\varepsilon > 0$, there is a $\Delta > 0$ such that if s, $t \in I$ with $|s - t| < \Delta$, then $|x(s) - x(t)| < \varepsilon$. We allow the cases that $a = -\infty$, $b = \infty$, or both.

(a) Show that the signal $\sin(t)$ is uniformly continuous on $\mathbb{R}$.

(b) Show that $\exp(t)$ is not uniformly continuous on $\mathbb{R}$.

(c) Show that $x(t) = t^2$ is uniformly continuous on any finite interval (a, b), but not on any unbounded interval ($a = -\infty$, $b = \infty$, or both).

(d) Show that $x(t) = \sqrt{|t|}$ is not uniformly continuous on any interval that includes the origin.

(e) Prove or disprove: If $x(t)$ is continuous and differentiable and has a bounded derivative in the interval (a, b), then $x(t)$ is uniformly continuous on (a, b).

15. Suppose the sequence $\{x_n(t) \in \mathbb{N}\}$ converges uniformly on some real time interval (a, b). That is, every $\varepsilon > 0$ there is an $N_\varepsilon > 0$ such that $m, n > N_\varepsilon$ implies that for all $t \in I$ we have $|y_n(t) - y_m(t)| < \varepsilon$.

(a) Show that if each $x_n(t)$ is continuous, then the limit$x(t) = \lim_{n \to \infty} x_n(t)$ is also continuous.

(b) Show that $x(t)$ may not be continuous if the convergence is not uniform.

(c) Prove or disprove: If each $x_n(t)$ is bounded, then $x(t)$ is also bounded.

16. This problem explores interchanging limit and integration operations [6]. Suppose the signal $x_n(t)$ is defined for $n > 0$ by

$$x_n(t) = \begin{cases} 4n^2 t & \text{if } 0 \le t \le \frac{1}{2n}, \\ 4n - 4n^2 t & \text{if } \frac{1}{2n} \le t \le \frac{1}{n}, \\ 0 & \text{if otherwise.} \end{cases} \tag{3.249}$$

(a) Show that $x_n(t)$ is continuous and integrates to unity on (0, 1).

$$\int_0^1 x_n(t)\,dt = 1 \tag{3.250}$$

(b) Let $x(t) = \lim_{n \to \infty} x_n(t)$. Show that $x(t) = 0$ for all $t \in \mathbb{R}$.

(c) Conclude that

$$\int_0^1 x(t)\, dt \neq \lim_{n \to \infty} \int_0^1 x_n(t)\, dt. \tag{3.251}$$

17. Let $\{x_n(t)\}$ be continuous and converge uniformly on the real interval $[a, b]$ to $x(t)$.

(a) Show that $x(t)$ is continuous and therefore Riemann integrable on $[a, b]$.

(b) Using the uniform continuity, find a bound for the integral of $x_n(t) - x(t)$.

(c) Finally, show that

$$\lim_{n \to \infty} \int_a^b x_n(t)\, dt = \int_a^b x(t)\, dt. \tag{3.252}$$

18. Prove (3.252) assuming that $\{x_n(t)\}$ converge uniformly and are Riemann-integrable (but not necessarily continuous) on $[a, b]$.

(a) Approximate $x_n(t)$ by step functions.

(b) Show that the limit $x(t)$ is Riemann-integrable on $[a, b]$.

(c) Conclude that the previous problem applies and (3.252) holds once again [6].

19. Suppose that the sequence $\{x_n(t) \in \mathbb{N}\}$ converges uniformly on some real interval $I = (a, b)$ to $x(t)$; that each $x_n(t)$ is continuously differentiable on I; that for some $c \in I$, $\{x_n(c)\}$ converges; and that the sequence $\{x_n'(t)\}$ converges uniformly on I.

(a) Show that for all $n \in \mathbb{N}$ and all $t \in I$,

$$x_n(t) - x_n(a) = \int_a^t x_n'(s)\, ds. \tag{3.253}$$

(b) By the previous problem, $\lim [x_n(t) - x_n(a)]$ exists and

$$\lim_{n \to \infty} [x_n(t) - x_n(a)] = \int_a^t \lim_{n \to \infty} x_n'(s)\, ds = x(t) - x(a). \tag{3.254}$$

(c) So by the fundamental theorem of calculus [6], we have

$$\lim_{n \to \infty} x_n'(t) = x(t). \tag{3.255}$$

20. Using informal arguments, show the following properties of the Dirac delta:

(a) $\delta(-t) = \delta(t)$.

(b)

$$\int_{-\infty}^{\infty} x(t)\frac{d}{dt}u(t)\,dt = x(0). \tag{3.256}$$

(c) Assuming the Dirac is differentiable and that interchange of differentiation and integration is permissable, show that

$$\int_{-\infty}^{\infty} x(t)\frac{d^n}{dt^n}\delta(t)\,dt = (-1)^n \frac{d^n}{dt^n}x(t)\Bigg|_{t=0}. \tag{3.257}$$

21. Show that the nth-order B–spline $\beta_n(t)$ has compact support.

22. Let $\beta_n(t)$ be the nth-order B-spline. Show the following [21]:

(a)

$$\frac{d}{dt}\beta_n(t) = \beta_{n-1}\left(t+\frac{1}{2}\right) - \beta_{n-1}\left(t-\frac{1}{2}\right). \tag{3.258}$$

(b)

$$\int_{-\infty}^{t} \beta_n(s)\,ds = \sum_{k=0}^{\infty} \beta_{n+1}\left(t-\frac{1}{2}-k\right). \tag{3.259}$$

23. Assuming that $\|x\|_1 < \infty$ and $\|y\|_\infty < \infty$, show that $\|xy\|_1 \le \|x\|_1\|y\|_\infty$.

24. If $\|x\|_\infty < \infty$ and $\|y\|_\infty < \infty$, show $\|x+y\|_\infty \le \|x\|_\infty + \|y\|_\infty$.

25. Let X be the set of continuous, p-integrable signals x: $[a, b] \to \mathbb{K}$, where $a < b$ and $\mathbb{K}$ is either $\mathbb{R}$ or $\mathbb{C}$. Show that $(X, \|x\|_{p,\,[a,b]})$ is a normed linear space.

26. Let X be a normed linear space. Show that the norm is a continuous map: For any $x \in X$ and $\varepsilon > 0$ there is a $\delta > 0$ such that for any $y \in X$, if $\|y - x\| < \delta$, then $|\,\|y\| - \|x\|\,| < \varepsilon$. Show continuity for the algebraic operations on X: addition and scalar multiplication.

27. Show that the map $d(x, y) = \|x - y\|_p$ is a metric. How must the set of signals $x(t)$ be restricted in order to rigorously show this result? Explain how to remove the restrictions.

28. Suppose $a \sim b$ is an equivalence relation on a set S. Show that $\{[a]: a \in S\}$ partitions S.

29. Consider the analog signals having at most a finite number of discontinuities on $[a, b]$, where $a < b$, and let $1 \le p < \infty$. We restrict ourselves to the Riemann integral, suitably extended to handle piecewise continuous functions.

(a) Show that the set of all such signals does not constitute a normed linear space. In particular, exhibit a signal $x(t)$ which is nonzero and yet $\|x\|_p = 0$.

(b) Show that the relation $x \sim y$ if and only if $\|y\|_p = \|x\|_p$ is an equivalence relation.

(c) Let $[x] = \{y(t): \|y\|_p = \|x\|_p\}$. Show that $[x] = [y]$ if and only if $x(t)$ and $y(t)$ are identical except at a finite number of points.

(d) Define $\mathbf{\underline{L}}^p[a, b] = \{[x] \mid x \in L^p[a, b]\}$. Further define an addition operation on these equivalence classes by $[x] + [y] = [s(t)]$, where $s(t) = x(t) + y(t)$. Show that this addition operation makes $\mathbf{\underline{L}}^p[a, b]$ into an additive Abelian group: it is commutative, associative, has an identity element, and each element has an additive inverse. Explain the nature of the identity element for $\mathbf{\underline{L}}^p[a, b]$. For a given $[x] \in \mathbf{\underline{L}}^p[a, b]$, what is its additive inverse, $-[x]$? Explain.

(e) Define scalar multiplication for $\mathbf{\underline{L}}^p[a, b]$ by $c[x] = [cx(t)]$. Show that $\mathbf{\underline{L}}^p[a, b]$ thereby becomes a vector space.

(f) Define a norm on $\mathbf{\underline{L}}^p[a, b]$ by $\|[x]\|_p = \|x\|_p$. Show that this makes $\mathbf{\underline{L}}^p[a, b]$ into a normed linear space.

(g) Apply the same reasoning to $L^p(\mathbb{R})$.

30. Suppose X is a normed linear space and $x, y \in X$. Applying the triangle inequality to the expression $x_n = x_m + x_n - x_m$, show that $\left|\|x_n\| - \|x_m\|\right| \le \|x_n - x_m\|$.

31. Let $S = \{x_n\}$ and $T = \{y_n\}$ be Cauchy sequences in a normed linear space X. Define the relation $S \sim T$ to mean that S and T get arbitrarily close to one another, that is, $\lim_{n\to\infty}\|x_n - y_n\| = 0$.

(a) Show that $\sim$ is an equivalence relation.

(b) Let $[S] = \{T: T \sim S\}$; set $B = \{[S]: S = \{x_n\}$ is Cauchy in $X\}$. Define addition and scalar multiplication on B. Show that these operations are well-defined; and show that B is vector space.

(c) Define a norm for B. Show that it is well-defined, and verify each property.

(d) For $x \in X$, define $f(x) = [\{x_n\}]$, where $x_n = x$ for all n, and let $Y = \text{Range}(f)$. Show that $f: X \to Y$ is a normed linear space isometry.

(e) Show that if C is any other Banach space that contains X, then C contains a Banach subspace that is isometric to $Y = f(X)$, where f is given in (d).

32. If X and Y are normed spaces over $\mathbb{K}$ and $T: X \to Y$ is a linear operator, then show the following:

(a) Range(T) is a normed linear space.

(b) The null space of T is a normed linear space.

(c) The inverse map $T^{-1}: \text{Range}(T) \to X$ exists if and only $Tx = 0$ implies $x = 0$.

33. Prove the following alternative version of Schwarz's inequality: If $x, y \in L^2(\mathbb{R})$, then $|\langle x, y\rangle| \le \|x\|_2\|y\|_2$.

34. Suppose $F = \{f_n(t): n \in \mathbb{Z}\}$ is a frame in a Hilbert space H, and T is the frame operator T given by (3.88). Just to review the definition of the frame operator and inner product properties, please show us that T is linear. From its definition, show that the frame adjoint operator is also linear.

35. Suppose that a linear operator U is positive: $U \ge 0$. Show the following [26]:

(a) If I is the identity map, then $U + I$ is invertible.

(b) If V is a positive operator, then $U + V$ is positive.

36. Show that the following are σ-algebras:

(a) $\wp(\mathbb{R})$, the set of all subsets of $\mathbb{R}$.

(b) $\{\varnothing, \mathbb{R}\}$.

37. Let $S \subset \mathcal{P}(\mathbb{R})$. Show that there is a smallest σ-algebra that contains S. (The Borel sets is the class where S is the family of open sets in $\mathbb{R}$.)

38. Let Σ be a σ-algebra. Show Σ is closed under countable intersections.

39. Let $\Sigma = \wp(\mathbb{R})$. Show the following are measures:

(a) $\mu(\varnothing) = 0$ and $\mu(A) = \infty$ if $A \neq \varnothing$.

(b) For $A \in \Sigma$, define $\mu(A) = N$ if A contains exactly N elements and $\mu(A) = \infty$ otherwise.

(c) Show that if μ is a measure on Σ and $c > 0$, then $c\mu$ is also a measure on Σ.

40. Show the following properties of lim inf and lim sup:

(a) $\liminf\{a_n\}$ is the smallest limit point in the sequence $\{a_n\}$.

(b) $\limsup\{a_n\}$ is the largest limit point in the sequence $\{a_n\}$.

(c) $\liminf\{a_n\} = \limsup\{a_n\}$ if and only if the sequence $\{a_n\}$ converges to some limit value $a = \lim_{n\to\infty}\{a_n\}$; show that when this limit exists a $= \liminf\{a_n\} = \limsup\{a_n\}$.

(d) If $a_n \le b_n$, then $\liminf\{a_n\} \le \liminf\{b_n\}$.

(e) Provide an example of strict inequality in the above.

(f) $\limsup\{-a_n\} = -\liminf\{a_n\}$.

41. Show that the general differential equation governing the nth derivative of the Dirac delta is

$$f(t)\delta^{(n)}(t) = \sum_{m=0}^{n} (-1)^m \cdot \frac{n!}{(n-m)!} f^{(m)}(0)\delta^{(n-m)}(t). \tag{3.260}$$

42. Derive the following:

(a) The scaling law for an arbitrary distribution (3.157).

(b) Associativity (3.159).

43. (a) Calculate the amplitude for the unit-area of the Abel function (3.230).

(b) Calculate the amplitude for the unit-area of the decaying exponential (3.234).

(c) Graph each of these approximate identities for scales of $a = 1, 10, 100, 1000$.

(d) Verify that the behavior is consistent with the conditions governing an approximate identity.

44. Show that

(a) The rectangular pulse (3.235) is a weight function for any value of the parameter σ.

(b) Verify that this unit-area pulse acts as an approximate identity by explicitly showing

$$\lim_{a \to \infty} \int_{-\infty}^{\infty} f_a(t)\phi(t)\,dt = \phi(0). \tag{3.261}$$

45. Prove the differentiation theorem for convolution of distributions (3.199).

46. Demonstrate the validity (as approximating identities) of the alternative weight functions expressed by (3.224) and (3.225).

The following problems expand on the text's presentation, require some exploratory thinking, and are suitable for extended projects.

47. Let $1 \le p \le \infty$, $x \in L^p(\mathbb{R})$, and $h \in L^1(\mathbb{R})$.

(a) Show that $|y(t)| \le \int |x(t-s)||h(s)|\,ds$;

(b) $y = x * h \in L^p(\mathbb{R})$;

(c) $\|y\|_p \le \|x\|_p \|h\|_1$.

48. Consider a signal analysis matching application, where signal prototypes $P = \{p_1, p_2, ..., p_M\}$ are compared to candidates $C = \{c_1, c_2, ..., c_N\}$, using a distance measure $d(p, c)$. Explain why each of the following properties of $d(p, c)$ are useful to the application design.

(a) $d(p, c) \ge 0$ for all $p \in P$, $c \in C$.

(b) $d(p, c) = 0$ if and only if $p = c$.

(c) $d(p, c) = d(c, p)$ for all p, c.

(d) For any s, $d(p, c) \le d(p, s) + d(s, c)$.

Collectively, these properties make the measure $d(p, c)$ a metric. Consider matching applications where the $d(p,c)$ violates one metric property but obeys the other three. What combinations of of the three properties still suffice for a workable matching application? Explain how the deficiency might be overcome. Provide examples of such deficient match measures. Does deleting a particular metric property provide any benefit to the application—for instance, an ambiguity of analysis that could aid the application?

49. Develop the theory of half-infinite analog signals spaces. Provide formal definitions of the half-infinite L^p spaces: $L^p(-\infty, a]$ and $L^p[a, +\infty)$ and show that these are normed linear spaces. Are they complete? Are they inner product spaces? (You may use the Riemann integral to define L^p for this problem.)

50. Using Lebesgue measure to define it, show that L^∞ is complete [44].

51. Study the matched filtering technique of Section 3.3.2.3 for signal pattern detection. Assume that for computer experiments, we approximate analog convolution with discrete sums.

(a) Show that the the Schwarz inequality implies that the method gives a match measure of unit magnitude if and only if the candidate $x(t)$ and prototype $p(t)$ are constant multiples of one another.

(b) Show that we can generalize the match somewhat by subtracting the mean of each signal before computing the normalized cross-correlation, then the normalized cross-correlation has unit magnitude if and only if the signals are related by $x(t) = Ap(t) + B$, for some constants A, B.

(c) Consider what happens when we neglect to normalize the prototype signal. Show that the matching is still satisfactory but that the maximum match value must be the norm of the prototype pattern, $\|p\|$.

(d) Suppose further that we attempt to build a signal detector without normalizing the prototype. Show that this algorithm may fail because it finds false positives. Explain using examples how the match measure can be larger where $x(t)$ in fact is not a constant multiple of $p(t)$.

(e) What are the computational costs of matched filtering?

(f) How can the computational cost be reduced? What are the effects of various fast correlation methods on the matching performance? Justify your results with both theory and experimentation;

(g) Develop some algorithms and demonstrate with experiments how coarse-to-fine matching can be done using normalized cross-correlation [34].

52. Study the exponential and sinusoidal basis decompositions. Assume the exponential signals constitute an orthonormal basis for $L^2[-\pi, \pi]$.

(a) Show that any $x(t) \in L^2[-\pi, \pi]$ can be expressed as a sum of sinusoidal harmonics. From (3.76) set $a_n = c_n + c_{-n}$ and $jb_n = c_{-n} - c_n$. Show that

$$x(t) = \frac{a_0}{2} + \sum_{n=1}^{\infty} a_n \cos(nt) + \sum_{n=1} b_n \sin(nt). \tag{3.262}$$

(b) Give the spanning set for $L^2[-\pi, \pi]$ implied by (3.262).

(c) Show that the sinusoids are also orthogonal.

(d) By dividing up the real line into 2π-wide segments, $[-\pi + 2n\pi, \pi + 2n\pi]$, give an orthonormal basis for $L^2(\mathbb{R})$.

(e) Consider a square-integrable signal on $L^2(\mathbb{R})$, such as a Gaussian $\exp(-At^2)$ for some $A > 0$. Find a formula, based on the inner product on $L^2[-\pi, \pi]$ for the Fourier series coefficients that arise from the basis elements corresponding to this central interval.

(f) Consider the Fourier series expansion on adjacent intervals, say $[-\pi, \pi]$ and $[\pi, 3\pi]$. Show that the convergence of partial sums of the Fourier series to the signal $x(t)$ exhibits artifacts near the endpoints of the intervals. Explain these anomalies in terms of the periodicity of the exponentials (or sinusoids) on each interval.

(g) Does the selection of sinusoidal or exponential basis functions affect the partial convergence anomaly discovered in (e)? Explain.

(h) What happens if we widen the intervals used for fragmenting $\mathbb{R}$? Can this improve the behavior of the convergence at endpoints of the intervals?

(i) Summarize by explaining why identifying frequency components of general square-integrable signals based on sinusoidal and exponential Fourier series expansions can be problematic.

53. Section 3.3.4.4 argued that frame decompositions support basic signal analysis systems. One desirable property left open by that discussion is to precisely characterize the relation between two sets of coefficients that represent the same incoming signal. This exercise provides a partial solution to the uniqueness problem [42]. Let $F = \{f_n(t) : n \in \mathbb{Z}\}$ be a frame in a Hilbert space H, let $T = T_F$ be its associated frame operator (3.88), and let $S = T^*T$.

(a) If $x \in H$, define $a_n = \langle x, S^{-1}f_n \rangle$, and then show that $x = \Sigma a_n f_n$.

(b) If there are $c_n \in \mathbb{C}$ such that $x = \Sigma c_n f_n$, then

$$\sum_n |c_n|^2 = \sum_n |a_n|^2 + \sum_n |a_n - c_n|^2. \tag{3.263}$$

(c) Explain how the representation of x in terms of the dual frame for F is optimal in some sense.

(d) Develop an algorithm for deriving this optimal representation.

54. Let $F = \{f_n(t): n \in \mathbb{Z}\}$ be a frame in Hilbert space H. Prove the following:

(a) If an element of the frame is removed, then the reduced sequence is either a frame or not complete (closure of its linear span is everything) in H.

(b) Continuing, let $S = T^*T$, where T^* is the adjoint of the frame operator $T = T_F$. Show that if $\langle f_k, S^{-1}f_k \rangle \neq 1$, then $F \setminus \{f_k\}$ is still a frame.

(c) Finally, prove that if $\langle f_k, S^{-1}f_k \rangle = 1$, then $F \setminus \{f_k\}$ is not complete in H.

55. Let us define some variations on the notion of a basis. If $E = \{e_n\}$ is a sequence in a Hilbert space H, then E is a *basis* if for each $x \in H$ there are unique complex scalars a_n such that x is a series summation, $x = \Sigma_n a_n e_n$. The basis E is *bounded* if $0 \leq \inf\{e_n\} \leq \sup\{e_n\} < \infty$. The basis is *unconditional* if the series converges unconditionally for every element x in H. This last result shows that the uniqueness of the decomposition coefficients in a signal processing system can in fact be guaranteed when the frame is chosen to be *exact*. Let $F = \{f_n(t): n \in \mathbb{Z}\}$ be a sequence in a Hilbert space H. Show that F is an exact frame if and only if it is a bounded unconditional basis for H.

56. Repeat Problem 52, except allow for a frame-based signal decomposition. In what ways does the presence of redundancy positively and negatively affect the resulting decompositions?

CHAPTER 4

Time-Domain Signal Analysis

This chapter surveys methods for time-domain signal analysis. Signal analysis finds the significant structures, recognizable components, shapes, and features within a signal. By working with signals as functions of a discrete or analog time variable, we perform what is called time-domain signal analysis. We view signals as discrete or analog functions. We distinguish one signal from another by the significant differences in their magnitudes at points in time. Ultimately, then, time-domain methods rely in a central way on the level of the signal at an instant, over an interval, or over the entire domain of the signal. Our analysis tools will be both crude and sophisticated, and our achievements will be both problematic and successful. From the perspective of Hilbert space analysis, we will find that our time-domain signal analysis methods often involve the use of basis functions for signal subspaces that have irregular, blocky shapes.

Signal analysis does encompass many signal processing techniques, but it ultimately goes beyond the signal-in, signal-out framework: Signal analysis breaks an input signal down into a nonsignal form. The output could be, for example, an interpretation of the input, recognition results, or a structural description of the source signal. In time-domain analysis we study signals without first deriving their frequency content. Signals are viewed simply as functions of time (or of another independent spatial variable). This chapter's methods depend upon the foundation in determinate and indeterminate signals, systems, and Hilbert space laid previously, together with some calculus, differential geometry, and differential equations. In contrast, a frequency-domain analysis finds the frequencies within a signal by way of Fourier transformation—or, in more modern vein, using the Gabor or wavelet transforms—and uses the frequency information to interpret the signal.

After some philosophical motivation, this chapter considers some elementary signal segmentation examples. This discussion identifies problems of noise, magnitude, frequency, and scale in detecting special signal regions. Signal edges are the obvious boundaries between segmentable signal regions, but detecting them reliably and optimally proves to be harder that it would seem at first glance. Matched filtering is introduced, but the theoretical justifications are postponed. Scale space decomposition gives a complete and general method for the segmentation and structural description of a signal. Pattern recognition networks offer a hybrid scheme for signal detection.

Signal Analysis: Time, Frequency, Scale, and Structure, by Ronald L. Allen and Duncan W. Mills
ISBN: 0-471-23441-9

They combine ideas from matched filtering and decomposition by scale; provide for design rules, as in conventional pattern recognition systems; and have a network training scheme similar to neural networks. Hidden Markov models (HMMs) are a statistical recognition tool applied widely in speech and text interpretation. Later chapters will expand these techniques further by showing how to use these signal decomposition tools with frequency and mixed-domain signal processing.

We shall cover in this chapter, as promised, signal analysis techniques that are not commonly found in the standard signal processing texts. But we do more than validate the title of the whole book. By confronting the deep problems involved in finding the content and structure of signals, right after basic presentations on linear systems, we will find both a motivation and a conceptual foundation for the study of frequency and scale in signals. The first three chapters embellished the traditional presentation of signal processing—discrete and analog signals, random signals, periodicity, linear systems—with an introductory development of the theory of Hilbert spaces. The important Hilbert space notion of inner product gives us a theoretical tool for finding the similarity of two signals and therefore a point of departure for finding one signal shape within another. Now the exposition tilts toward techniques for breaking an input signal down into a nonsignal form.

Time-domain signal analysis will prove to have some serious limitations. Some types of signals will prove amenable to the time-domain methods. We can engineer signal analysis solutions for many process monitoring problems using time-domain techniques. Other application areas—especially those where signals contain periodic components—will cause trouble for our methods. Speech recognition and vibration analysis are examples. For such applications, the signal level itself is not so important as the regular assumption of some set of values by the signal. If we retreat again to the abstract vantage point of Hilbert space, we will see that our methods must now rely on basis functions for signals that are regular, textured, periodic, sinusoidal, exponential, and so on. Although this discovery qualifies our successes in preliminary signal analysis applications, it leads naturally to the study of the Fourier transforms of analog and discrete signals, and later it leads to the modern mixed-domain techniques: time-frequency and time-scale transforms.

This decomposition into structural features may be passed on to higher-level interpretation algorithms. Machine recognition of digitized speech is an example. So, by this view, signal analysis is the front-line discipline within artificial intelligence. While it makes extensive use of standard signal processing techniques, the analysis of a signal into some interpretable format is a distinct, challenging, and perhaps neglected area of engineering education and practice.

A small diagram, due to Professor Azriel Rosenfeld,[1] who first formulated it for images, illustrates the relationships between data processing, image synthesis (graphics), image processing, and image analysis (computer vision) (Figure 4.1a). Reducing the dimension of the data, we arrive at a similar diagram for signals (Figure 4.1b)

[1]For many years Professor Rosenfeld has used this diagram in his computer vision courses (personal communication). The authors have often heard—and, in the absence of a contrary reference in the scientific literature, do believe—that the diagram is attributable to Azriel Rosenfeld.

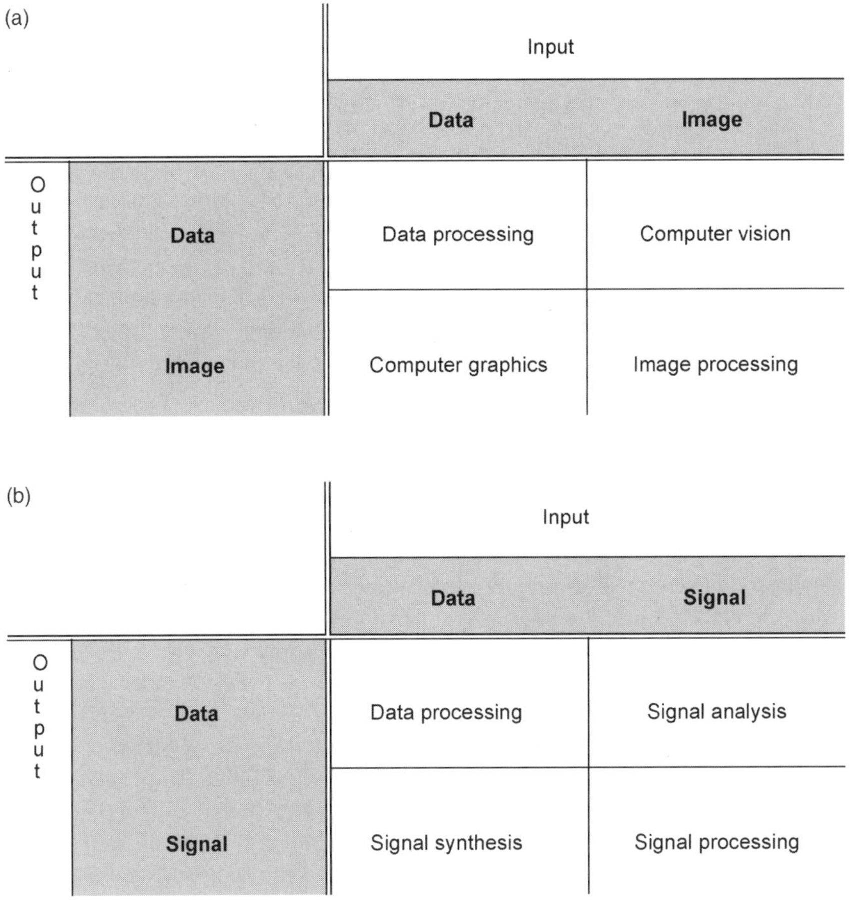

(a)

		Input	
		Data	**Image**
Output	**Data**	Data processing	Computer vision
	Image	Computer graphics	Image processing

(b)

		Input	
		Data	**Signal**
Output	**Data**	Data processing	Signal analysis
	Signal	Signal synthesis	Signal processing

Fig. 4.1. (a) An input–output diagram of the disciplines that work with two-dimensional signals (i.e., images). Conventional data processing began as soon as programmable machines became available. Image processing (for example, enhancement) has been a successful technology since the 1960s. With the advent of parallel processing and fast reduced instruction set computers, computer graphics has reached such a level of maturity that is both appreciated and expected by the public in general (example: morphing effects in cinema). Computer vision stands apart as the one technology of the four that remains problematic. We seem to be decades away from the development of autonomous, intelligent, vision-based machines. (b) An input–output diagram of the computer technologies that use signals. The rediscovery of the fast Fourier transform in the mid-1960s gave digital signal processing a tremendous boost. Computer music is now commonplace, with an example of signal synthesis possible even on desktop computers since the 1980s. Like computer vision, though, signal analysis is still an elusive technology. Where such systems are deployed, they are greatly constrained, lack generality, and adapt poorly, if at all, to variations in their signal diets.

and reveal the input–output relationships between data processing, signal synthesis, signal analysis, and signal processing. Most computing tasks fall into the data processing categories of Figure 4.1, which covers financial and accounting programs, database applications, scientific applications, numerical methods, control programs, and so on. Signal processing occupies the other diagonal square, where signal data are output from a signal data input. The other squares, converting data to signal information and converting signal information to data, generally do not correspond to classes in the university curriculum, except perhaps in some specialized programs or big departments that offer many courses. An application that accepts abstract data inputs and produces a signal output is an example of a signal synthesis system. Courses such as this have been quite rare. Recently, though, some computer science departments are offering multimedia courses that cover some aspects of producing digital music, speech, and sounds. A computer program that accepts a signal and from it generates a data output is in essence a signal analysis system. It is really the form of the data output that is critical, of course. We envision a description—a breakdown of content, or an *analysis*—of the signal that does not resemble the original at all in form.

The great advances in mixed-domain signal transforms over the last 10 years are the primary reason for reworking the conventional approach to signals. These time-frequency and time-scale transforms cast new light on the Fourier transform, expose its limitations, and point to algorithms and techniques that were either impossible or terribly awkward using the standard Fourier tools. Moreover, advances in computer hardware and software make technologies based on this new theory practically realizable. Applications—commercial products, too—are appearing for speech understanding, automatic translation, fingerprint identification, industrial process control, fault detection, vibration analysis, and so on. An unexpected influence on these emerging signal analysis techniques comes from biology. Research into hearing, vision, the brain, psychophysics, and neurology has been supplemented and stimulated by the investigations into the mathematics of the new signal transforms, neural networks, and artificial intelligence. This book will show that a fruitful interaction of artificial intelligence and signal analysis methods is possible. It will introduce students, engineers, and scientists to the fresh, rapidly advancing disciplines of time-frequency and time-scale signal analysis techniques.

In universities, signal processing is taught in both computer science and electrical engineering departments. (Its methods often arise in other disciplines such as mathematics, mechanical engineering, chemical engineering, and physics.) The courses are often almost identical by their course descriptions. There is controversy, though, about whether signal processing even belongs in the computer science curriculum, so similar in content is it to the electrical engineering course. One point of this book is to show that mainstream computer science methods are applicable to signal analysis, especially as concerns the computer implementation of algorithms and the higher-level interpretation methods necessary for complete signal analysis applications.

Signal analysis includes such technologies as speech recognition [1–4], seismic signal analysis [5, 6], interpretation of medical instrumentation outputs [7], fault detection [8], and online handwriting recognition [9]. Computer user interfaces may nowadays comprise all of the Rosenfeld diagram's technologies for both one-dimensional signals and two-dimensional signals (images): speech synthesis and recognition; image generation, processing, and interpretation; artificial intelligence; and conventional user shells and services [10, 11].

It is possible to slant a signal processing course to computer science course themes. Here, for example, we explore time-domain signal analysis and will see that tree structures can be found for signals using certain methods. One can also emphasize signal transforms in computer science signal processing instead of the design of digital filters which is taught with emphasis in the electrical engineering signal processing course. This too we will explore; however, this task we must postpone until we acquire an understanding of signal frequency through the various Fourier transforms. There are filter design packages nowadays that completely automate the design process, anyway. Understanding transforms is, however, essential for going forward into image processing and computer vision. Finally, one also notices that signal processing algorithms—with complex-valued functions, arrays having negative indices, dynamic tree structures, and so forth—can be elegantly implemented using modern computer languages known almost exclusively by students in the university's computer science department.

In the first sections of this chapter, we consider methods for finding basic features of signals: edges and textures. An edge occurs in a signal when its values change significantly in magnitude. The notion of texture seems to defy formal description; at an intuitive level, it can be understood as a pattern or repetition of edges. We develop several techniques for finding edges and textures, compare them, and discover some problems with their application. Edge and texture detection will not be complete in this chapter. Later, having worked out the details of the Fourier, short-time Fourier, and wavelet transforms, we shall return to edge and texture analysis to check whether these frequency-domain and mixed-domain methods shed light on the detection problems we uncover here.

4.1 SEGMENTATION

Segmentation is the process of breaking down a signal into disjoint regions. The union of the individual regions must be the entire domain of the signal. Each signal segment typically obeys some rule, satisfies some property, or has some numerical parameter associated with it, and so it can be distinguished from neighboring segments. In speech recognition, for example, there may be a segmentation step that finds those intervals which contain an utterance and accurately separates them from those that consist of nothing but noise or background sounds.

This section begins with an outline of the formal concept of segmentation. There are many approaches to the segmentation task, and research continues to add new techniques. We shall confine our discussion to three broad areas: methods based on signal levels, techniques for finding various textures within a signal, and region growing and merging strategies. Later sections of this chapter and later chapters in the book will add further to the segmentation drawer of the signal analysis toolbox.

4.1.1 Basic Concepts

Segmentation is a rudimentary type of signal analysis. Informally, segmentation breaks a signal's domain down into connected regions and assigns a type indication to each region. For signals, the regions are intervals: open, half-open, or closed. Perhaps a region is a single point. Some thinking in the design of a signal segmentation method usually goes into deciding how to handle the transition points between regions. Should the algorithm designer consider these boundaries to be separate regions or deem them part of one of the neighboring segments? This is not too difficult a question, and the next section's edge detection methods will prove useful for advanced segmentation strategies. However, for two-dimensional signals (images) the situation is extremely complex. The connected regions can assume quite complicated shapes[2]; and their discovery, description, and graphical representation by machine too often bring current computer vision systems to their practical limits.

Definition (Segmentation). A segmentation $\Sigma = (\Pi, L)$ of a signal f consists of a partition $\Pi = \{S_1, S_2, ...\}$ of Dom(f) into regions and a logical predicate L that applies to subsets of Dom(f). The predicate L identifies each S_i as a maximal region in which f is homogeneous. Precisely, then, segmentation requires the following [12]:

- $\mathrm{Dom}(f) = S_1 \cup S_2 \cup S_3 \cup ...$, where the S_i are disjoint.
- $L(S_i)$ = True for all i.
- $L(S_i \cup S_j)$ = False when S_i and S_j are adjacent in Dom(f) and $i \neq j$.

It is common to call the regions segments, but they are not necessarily intervals. Commonly, the segments are finite in number, only a specific region of interest within the signal domain is subject to segmentation, and some mathematical operation on the signal defines the logical predicate. It is also very much an application-specific predicate. For example, one elementary technique to segment the meaningful parts of a signal from background noise is to threshold the signal. Let $f(n)$ be a noisy signal, $T > 0$, $M = \{n: |f(n)| \geq T\}$, and $N = \{n: |f(n)| < T\}$. Let $\Pi = \{M, N\}$, and let L be the logical predicate "All signal values in this region exceed T or all signal values in this region do not exceed T." Then $S = (\Pi, L)$ is a segmentation of the signal into meaningful signal and background noise regions. Of course, different threshold values produce different segmentations, and it is possible that neither the M nor N is a connected subset of the integers. Most signal analysts, in point of fact, never formulate the logical predicate for segmentation; the predicate is implicit, and it is usually obvious from the computations that define the partition of the signal's domain.

[2]The famous Four-Color Problem is in fact a question of how many types are necessary to lable a map segmented into countries. For every map ever drawn, it had been shown to be possible to color the bounded regions with only four colors. Therefore, conjecture held that four-colorings of the plane were always possible, but for centuries this simple problem defined solution. Only recently, with a grand effort by both humans and computers, has the question been answered in the affirmative.

Definition (Labeling, Features, Pattern). Let $\Pi = \{S_1, S_2, \ldots\}$ be a partition of the domain of signal f. Then a labeling of f for Π is a map $\Lambda: \Pi \rightarrow \{\Lambda_1, \Lambda_2, \ldots\}$. Ran($\Lambda$) is the set of labels, which categorize subsets of Dom(f) in Π. If $\Lambda(S_i) = \Lambda_j$, then S_i is called a signal feature with label Λ_j. A pattern is set of features in a signal.

Labeling often follows segmentation in signal analysis systems. The nature of the signal analysis application determines the set of labels which apply to signal regions. The segmentation operation uses some logical predicate—which embodies an algorithm or computation—to decompose the signal into homogeneous regions. The regions are distinct from one another and suitable for labeling. Together, the segmentation and labeling operations assign distinct signal regions to predetermined categories, appropriate to the application. This results in a primitive description of the signal's content known as feature detection. The task of finding an assemblage of features in a signal is called pattern detection or pattern recognition.

By associating the regions with a label, the positions and time-domain extents of signal features are known. One problem is that the regions of a general segmentation may be disconnected, in which case the feature is located in multiple parts of the signal. This is a little awkward, but, in general, we shall not demand that our segmentations produce connected sets. But the partition elements do contain intervals, and a maximal interval within a segment does specify the location of a feature. Thus, finding the maximal intervals that have a given label accomplishes the signal analysis task of registration.

Definition (Registration). Let f be a signal, $\Sigma = (\Pi, L)$ be a segmentation of f, and let $\Lambda: \Pi \rightarrow \{\Lambda_1, \Lambda_2, \ldots\}$ be a labeling of f with respect to Σ. If $\Lambda(S_i) = \Lambda_j$, $I \subseteq S_i$ is an (open, closed, or half-open) interval, and I is contained in no other connected subset of S_i, then I registers the feature with label Λ_j.

Higher-level algorithms may process the labeling and revise it according to rules, prior knowledge of the signal's content, or some model of what form the signal should take. Such high-level algorithms may employ a very small set of rules. But many signal analysis applications use a full rule database or implement a complete expert system for interpreting labeled signals. They often use graph and tree structures. By this point, the design of a signal analysis system is well into the realm of artificial intelligence. Without delving deeply into artificial intelligence issues, Section 4.2.4 discusses some elementary statistical measures and algorithms for refining the segmentation and labeling steps. This is called region splitting and merging. Region merging can be a basis for classification. If the goal is to find a sufficiently large signal region or a region with a particular registration, then a region merging procedure applied to the results of a first-cut segmentation might be adequate. Later, in Section 4.7, we will introduce what are called consistent labeling methods for revising the region label assignments. This strategy uses constraints on both the regions and the labels applied to them. The constraints on signal features can be formulated as a directed graph, a familiar data structure from artificial intelligence. When a signal's features can be labeled so as to obey the constraints, which might be derived from a signal prototype or model, then the signal has been classified.

4.1.2 Examples

Some examples of segmentation, labeling, and registration should help to make these abstract concepts more concrete.

Example (Signal versus Noise Detection). A logical predicate, N, for "either noise or signal," may apply to subsets of the domain of a signal, f. (To be precise, we are using the exclusive "or" here: noise or signal, but not containing both characteristics.) A region $R_n \subseteq \mathrm{Dom}(f)$ for which $N(R_n) = \mathrm{True}$ may be a region of noise, and another region $S_n \subseteq \mathrm{Dom}(f)$ for which $N(S_n) = \mathrm{True}$ may be meaningful information. Many applications involve signals whose noise component are of a much lower magnitude than their information-bearing component. (This is not always the case; signals may follow an active-low policy, and the choice of labels for low- and high-magnitude segments must reflect this.) Reasonably, then, we might apply a threshold to the signal for distinguishing signal from noise. Figure 4.2 shows examples of

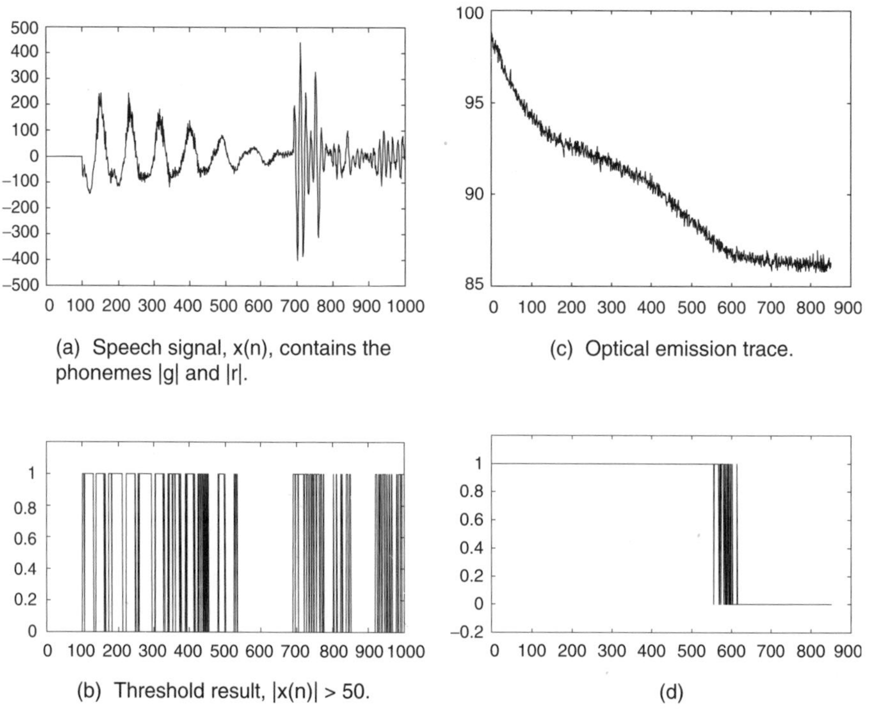

Fig. 4.2. Signal segmentation in speech recognition and industrial process control. Detecting a speaker's presence by thresholding works when the background noise level is sufficiently low (panels (a) and (b)). But background noises or artifacts of speech (tongue clicks, lip smacks, sniffs, sighs, loud breaths) will often be assigned to a speech region. Panels (c) and (d) show a representative optical trace from a plasma etching reactor. The controller monitors a carbon monoxide optical emission signal, whose high level indicates the presence of etching byproducts in the chamber.

signal segmentation based on thresholding. The goal is to separate the signal into a meaningful signal region, which has a high magnitude, and a background noise region, which has a relatively low magnitude. If the noise magnitude remains modest and the meaninful signal regions do not fade, then one-dimensional tasks like this are quite straightforward. Segmenting images into high-magnitude objects and low-magnitude background is called blob detection, an erstwhile colloquialism that has become standard technical parlance over the years. In two dimensions, because of the difficulties in tracking boundaries, blob detection is often problematic. The equivalent one-dimensional task—bump rather than blob detection—is much easier of course, because the topology of the line is simpler than the topology of the plane.

Application (Touch-Tone Telephone Pulse Detection). This example continues our discussion of digital telephony. Touch dialing telephones have largely supplanted the old rotary dialing units in the public service telephone network. Instead of a series of electromechanically generated pulses, modern telephones generate dual-tone multifrequency (DTMF) pulses for dialing [13]. The telephone company's central office (CO) equipment decodes these pulses; sets up the appropriate communication links; rings the far telephone; and sends a ringback signal, which simulates the sound a distant phone ring, to the near-end telephone. When someone lifts the far-end handset, the CO switching equipment sends a brief test tone through the circuit as a continuity check. The CO listens for facsimile (FAX) equipment tones and then for modem tones; in their absence, it initiates echo cancellation (see Chapter 2), and the call is complete. Table 4.1 shows the tone dialing scheme. Figure 4.3 illustrates a dual-tone sinusoidal pulse in noise and a sample segmentation. This segmentation produces two sets; however, due to the sinusoidal components in the pulse, there are numerous low-magnitude signal values interspersed in the time interval where we expect to detect the pulse. One strategy to improve the segmentation is to integrate the signal, and another strategy is to adjust the labeling to eliminate noise segments that are too brief.

Early telephones had two twin-tee feedback oscillators for producing DTMF signals. Later, and cheaper, designs were digital. Nowadays, digital signal processors

TABLE 4.1. DTMF Telephone Dialing Signal Specifications[a]

	1209 Hz	1336 Hz	1477 Hz	1633 Hz
697 Hz	1	2	3	A
770 Hz	4	5	6	B
852 Hz	7	8	9	C
941 Hz	*	0	#	D

[a] DTMF pulses consist of two tones from a high- and a low-frequency group. Telephones in public use do not sport the special keys on the right: A, B, C, D. The telephone company reserves them for test equipment and diagnostic signaling applications. DTMF tones pass at a 10-Hz rate. Pulse width is between 45 ms and 55 ms.

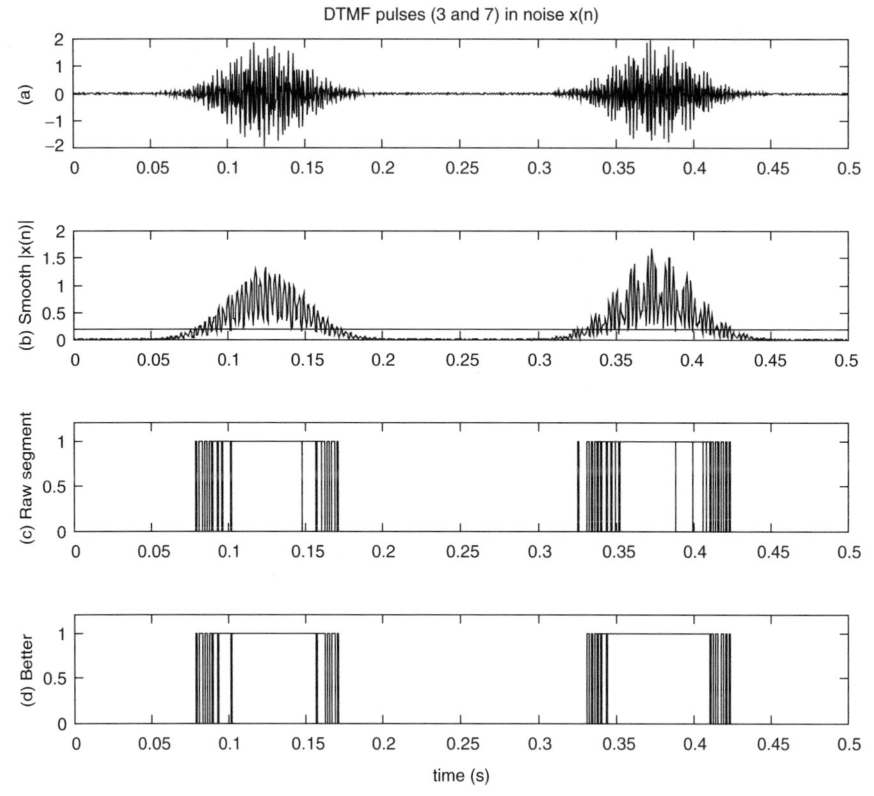

Fig. 4.3. An example of dual-tone multi-frequency (DTMF) pulse detection for telephone company dialing signals. Each number on the telephone dial is coded as a dual frequency pulse (a). Simple threshold (b) and labeling operations (c) mark tones (tone present = 1, background noise = 0). This naive, bottom-up procedure shows that no pulse segment is sufficiently long in duration to be a valid DTMF pulse; hence it finds no pulses present. The improved scheme (d) constrains the labeling operation so that short-duration noise segments are merged with signal regions, thus properly finding two 50-ms-wide pulses. It is possible—and more conventional for that matter—to filter the pulse (in this case its magnitude) for more consistent thresholding rather than rely on consistent labeling.

can be programmed to generate and decode DTMF signals [14, 15]. It is important to note that this example only considers the relatively easy problem of finding a blob-like pulse of the proper width for a DTMF pulse. A complete, practical solution must find the frequencies within the signal blob, verify that only two significant tones exist, and check that the pair of tones corresponds to a cell in Table 4.1. We will consider this more complex task in Chapter 7, after we develop methods for signal frequency analysis in Chapters 5 and 6.

4.1.3 Classification

Following segmentation and labeling, signal analysis may take a further step—classification of the entire signal. Classification is also called recognition. It is much like labeling, except that the term "labeling" in signal and image analysis parlance tends to mean a tentative classification. Labels are applied to small fragments of the signal, meant for later review, and facilitate the application of goal-driven rules. In contrast, the recognition step applies a broader signal region, which comprises several labeled, goal-directed fragments.

Definition (Classification). Let $F = \{f_1, f_2, ...\}$ be a family of signals. Then a classifier for F is a map C: $F \rightarrow \{C_1, C_2, ...\}$. The range of C is the set of classes, which categorize signals in F.

In view of our imprecise distinction between labeling and classifying, and with regard to the overlap of usage within the signal analysis community, we need to be casual about this definition. An application may break a long-term signal down into large regions and attempt to identify each chunk. While analyzing each large region, the remainder of the signal can be considered to be zero, and it is quite irrelevant about whether we deem the remaining chunks to be separate signals or just extended pieces of the original. The distinction, therefore, between labeling and classification revolves around which is the preliminary step and which is the ultimate step in the signal analysis.

The concepts of segmentation, labeling, and classification imply a scheme for constructing signal analysis systems (Figure 4.4). Most signal analysis systems work in the order: segmentation, labeling, classification. This strategy proceeds from low-level signal data, through assignment of labels to signal regions, to finally classify the entire signal. Therefore, this is a data-driven or bottom-up methodology. To impart a goal-driven or top-down aspect to the procedure, it is possible to constrain the labeling procedure. In this case, a high-level, rule-based algorithm reviews the initial labels and adjusts their assignments to signal features. The labeled regions may be too small in extent, as the narrow low-magnitude regions in the example of Fig. 4.3, and hence they are merged into the surrounding meaningful signal region.

Example (Voiced versus Unvoiced Speech Segmentation). Linguists typically classify speech events according to whether the vocal cords vibrate during the pronunciation of a speech sound, called a phone. Phones are speech fragments that represent the basic, abstract components of a natural language, called phonemes (Table 4.2). If the vocal cords do vibrate, then there is said to be a voiced speech event. If there is no vocal cord vibration, then the phoneme is unvoiced. It is also possible that a speech signal contains no speech sound; thus, it is simply background, or noise. One approach to segmenting speech classifies its portions as voiced (V), unvoiced (U), or noise (N). For example, a digital recording of the English phrase "linear fit" begins, divides the two words, and ends with noise

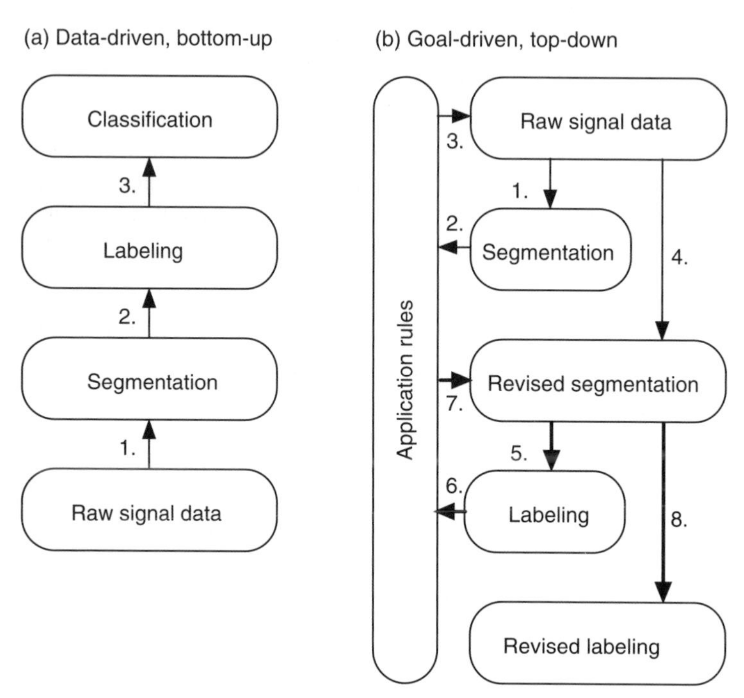

Fig. 4.4. Data-driven versus goal-driven signal analysis systems. In the data-driven, bottom-up scheme of (a), the processing proceeds from segmentation, to labeling, to classification without looking back. The goal-driven, top-down strategy in (b) revises earlier results according to rules specific to the application domain.

regions. The most apparent phonemes, then, are /l I n i ɜ f I t/, of which /f/ and /t/ are unvoiced. A preliminary segmentation, Σ_0, by voiced/unvoiced/noise classification is (N, V, N, U, V, U, N), respecting the stipulation that no adjacent regions have the same type. Actually, there are a number of unvoiced events that only become apparent when the speech signal is digitized and spread out over time. One may find, for example, momentary unvoiced aspirations and even periods of noise, surrounding the voiced segments. A refined segmentation, Σ_1, therefore supplies: (N, U, V, U, V, U, V, U, V, U, V, U, N, U, V, U, N). In practical speech recognition applications, surprisingly, this segmentation is too crude! Subtle periods of background noise, with no voiced or unvoiced sound present, intrude into spoken words. A modern, sophisticated segmentation algorithm finds that several *N* regions split the unvoiced regions U in the refined segmentation above. This means that several of the unvoiced intervals have much shorter time extents than Σ_1 would indicate. The benefit is that a higher-level interpretation algorithm may be better able to recognize the brief U boundaries of the V segments as trailing and leading aspirations instead

TABLE 4.2. Phonemes and Word Examples from American English [16][a]

Phoneme	Example	Class	Phoneme	Example	Class
/i/	even	Front vowel	/I/	signal	Front vowel
/e/	basis	Front vowel	/ε/	met	Front vowel
/ae/	at	Front vowel	/a/	father	Mid vowel
/Λ/	but	Mid vowel	/⊃/	all	Mid vowel
/schwa/	signal	Mid vowel	/u/	boot	Back vowel
/o/	boat	Back vowel	/U/	foot	Back vowel
/ɨ/	roses	Back vowel	/ɜ/	Hilbert	Mid vowel
/a^{w}/	down	Dipthong	/a^{y}/	cry	Dipthong
/⊃^{y}/	boy	Dipthong	/y/	yet	Semivowel glide
/w/	wit	Semivowel liquid	/r/	rent	Semivowel glide
/l/	linear	Semivowel liquid	/m/	segment	Nasal consonant
/n/	nose	Nasal consonant	/η/	Nguyen	Nasal consonant
/p/	partition	Unvoiced stop	/t/	fit	Unvoiced stop
/k/	kitten	Unvoiced stop	/b/	bet	Voiced stop
/d/	dog	Voiced stop	/g/	gain	Voiced stop
/h/	help	Aspiration	/f/	fit	Unvoiced fricative
/θ/	thanks	Unvoiced fricative	/s/	sample	Unvoiced fricative
/sh/	shape	Unvoiced fricative	/v/	vector	Voiced fricative
/∂/	that	Voiced fricative	/z/	zoo	Voiced fricative
/zh/	closure	Voiced fricative	/ch/	channel	Affricate
/j/	Jim	Affricate	/?/	no sound	Glottal stop

[a]Vowels are voiced, and they break down further into front, mid, and back classifications, according to the location of the tongue's hump in the mouth. Other essentially vowel-like sounds are the dipthongs and the semivowels. Nasal consonants are voiced, and the affricates are unvoiced. The glottal stop is a special symbol that indicates a momentary suspension of motion by the vocal cords (glottis). For example, without the /?/ symbol, the phoneme sequences for "I scream" and "ice cream" are identical. As big as it is, this table is far from panoramic; it offers but a glimpse of the natural language segmentation problem.

of, for example, unvoiced fricatives. Figure 4.5 illustrates such a speech segmentation example. We shall continue to expose the intricacies of natural language interpretation is this and the remaining chapters; there is much to cover.

Many problems involve a combination of segmentation procedures. For example, in digital telephony, it may be necessary to distinguish voice from DTMF pulses. Many commercial telephony applications that rely on DTMF detection require this capability. A pervasive—some would call it pernicious—application is the office voice mail system. The user presses telephone buttons to control the selection, playback, archival, and deletion of recorded messages. The voice mail application detects the DTMF and performs the appropriate function. Background office noise or talking confuses the DTMF classifier. Thus, a sophisticated DTMF detector sorts out the many possible natural language sounds from the dual-tone signaling pulses. Vowel sounds, such as /i/ and /u/, typically contain two sinusoidal components; and unvoiced fricatives, such as /s/ and /f/, are hard to discern from background

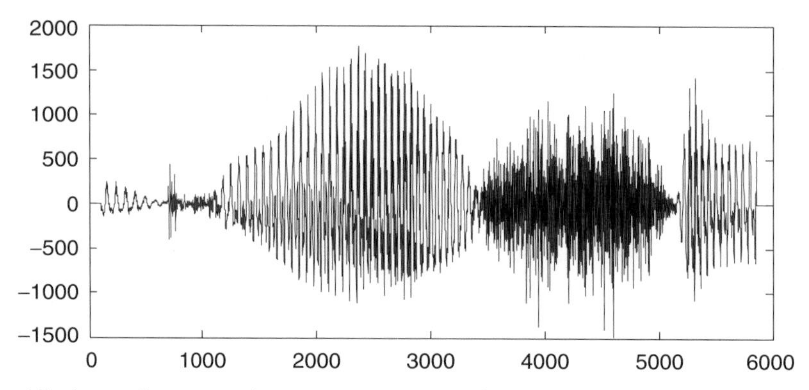

Fig. 4.5. A speech segmentation example. At a sampling rate of 16 kHz, the word "greasy" is shown. The corresponding phonemes are /g/, /r/, /i/, /s/, and /i/; of these /s/ is voiceless.

telephone line noise. Application success revolves around a careful design of segmentation, labeling, and classification algorithms. Sometimes a data-driven computational flow works, even for distinguishing DTMF signals from human voices. Where the bottom-up techniques break down, top-down constraints and goal-directed, artificial intelligence strategies become attractive.

4.1.4 Region Merging and Splitting

Once the domain of a signal f is broken down by a segmentation into a partition $\Pi = \{S_1, S_2, ...\}$, it may be necessary to combine some of the regions together. This reduction in the number of separate regions of the domain is called *region merging* (Figure 4.6). Region merging generally follows labeling. It is invoked because some goal-directed rules have discovered flaws in the labeling. We already considered this situation in the realm of speech analysis. A digitized speech fragment might receive an initial phoneme label, because of its high magnitude. However, subsequent analysis of its time-domain extent, and possibly its position at the beginning or ending of what appears to be a full-word utterance, might relegate the fragment to the category of a tongue click. Merging can also be useful when the signal segmentation must be made as simple as possible—for instance to facilitate signal compression or speed up subsequent analysis steps.

The choice of criteria for region merging depends heavily on the signal analysis application. For example, suppose that two segmentation regions have average signal values that are so close that they seem to be caused by the same physical process. Let μ_R be the mean within region R, and let be μ_S the mean within region S. If $|\mu_R - \mu_S| < \varepsilon$, where ε is an application-specific threshold, then we replace R and S in Π by $R' = R \cup S$. Merging continues by comparing the mean of R' against the mean of the remaining regions. Another statistic useful for merging two adjacent regions, especially when the segmentation is based on measures of signal texture (Section 4.3), is

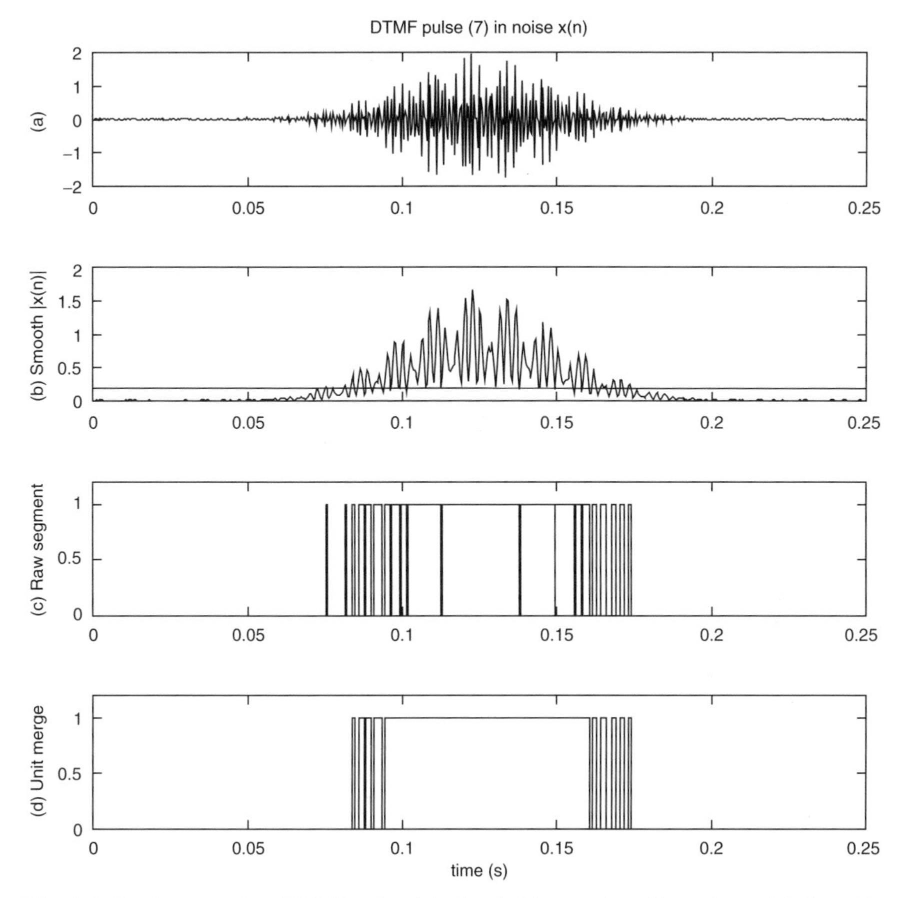

Fig. 4.6. Region merging: DTMF pulse (a), threshold operation (b), and raw labeling (c). The marked pulse segments are too small for a valid tone signal. Simply merging unit width segmented regions into their larger, differently marked neighbors isolates a valid DTMF pulse of 50-ms contiguous length (d).

a similarity in their variances. An edge detector could test the boundary between regions. If the edge detector fails to show a sufficiently distinct edge between the two areas, then they are candidates for merging. These are all low-level considerations for merging regions. If the application is so well characterized as to have a set of goal-directed rules that govern the labeling of signal regions, then the rules may be applied to the preliminary segmentation results. Regions can be joined, therefore, to better satisfy the top-down specifications. Finally, note that the results of the merging operations will supersede the original logical predicate that defined the segmentation.

Splitting a region is harder than merging. When the regions already exist, then the main problem is choosing the right criteria, out of many possibilities, for deciding whether to merge them. When a region is examined as a candidate for splitting, however, what partition of it begins the task? One possibility is to divide a region of interest, $S = [a, b] \subseteq \text{Dom}(f)$, into segments of equal length, $S = S_1 \cup S_2 \cup S_3 \cup \cdots \cup S_N$, where the S_i are disjoint. A good choice for segment length is the smallest time interval for meaningful features of signal f. Then a statistical measure, the mean or variance for example, applied to each segment in succession, can indicate whether the segment should be set apart from the rest. With the original S now subdivided, we can consider whether to unite the various S_i, just as in the above region merging algorithm. Splitting is then a kind of inverse of merging. The end result is to split the original region $S = R_1 \cup R_2 \cup \cdots \cup R_M$, where $M \le N$ and each R_i is a union of a different subset of $\{S_1, S_2, \ldots, S_N\}$. The obvious problem with this approach is that it directly depends upon the arbitrary partition of S into the segments, $S_1, S_2, \ldots, S_N$.

But there is a deeper problem with both the region merging and the "inverse" region splitting algorithms above. They both depend on the order in which the regions are selected for merging or splitting. This is due to the lack of transitivity in the "is similar to" relations we apply. Let's explore this. If we begin with the first segment, we may have $|\mu_1 - \mu_2| < \varepsilon$, and so we unite them, $R_1 = S_1 \cup S_2$. Now, the mean of R_1 is the statistical expectation, $E[\{f(n) \mid n \in R_1\}] = E[f(R_1)]$, and we compare it to μ_3. But it is quite possible that the criterion for merging S_3 into R_1 would be met, while at the same time we have $|\mu_1 - \mu_3| \ge \varepsilon$ and $|\mu_2 - \mu_3| \ge \varepsilon$. The merging of the subregions (and thus the splitting of the original region of interest) depends on the order in which the subregions are considered. To avoid this, we can try to cluster the subregions around some entity that does not depend on an individual region. Suppose we select M values $k_1 < k_2 < k_3 < \cdots < k_M$, where $k_i \in \text{Ran}(f)$. (Perhaps we specify $k_1 = \min\{f(n) \mid n \in [a, b]\}$, and $k_M = \max\{f(n) \mid n \in [a, b]\}$, but this is not essential.) Then we define the subregion S_i to belong to cluster j if $|\mu_i - k_j| \le |\mu_i - k_m|$, for all m, $1 \le m \le M$. That is, the clusters comprise those subregions whose mean falls closest to a target signal value among $\{k_1, k_2, \ldots, k_M\}$. We take $R_j = \cup\{S_i \mid S_i \text{ belongs to cluster } j\}$.

The above approach is called *nearest-neighbor clustering*. The same method can be followed using the variance as a clustering measure. The drawback for this elementary approach is its dependence upon the target values. Another possibility is to discover the target values. We consider all possible assignments of the S_i to clusters indicated by labels $\Lambda_1, \Lambda_2, \ldots, \Lambda_M$. For each permutation, we let $R_j = \cup\{S_i \mid S_i \text{ is assigned to cluster } j\}$ and $X_j = \{\mu_i \mid \mu_i = E[f(S_i)] \text{ and } S_i \text{ is assigned to } \Lambda_j\}$. A cluster is made up of similar subregions if the average values in each X_j are close to one another. Let the target values be $k_j = E[X_j]$. Thus, if we set $\text{Var}[X_j] = E[(X_j - E[X_j])^2] = E[(X_j - k_j)^2]$, then a reasonable measure of the inhomogeneity of our cluster assignment is $\text{Var}[X_1] + \text{Var}[X_2] + \cdots + \text{Var}[X_M]$. An assignment that minimizes this measure is an optimal (i.e., minimal variance) clustering for M labels.

4.2 THRESHOLDING

The threshold system is a nonlinear, but time-invariant operation on signals. In signal analysis applications, thresholding is used to separate the signal domain into regions according to the value that the signal assumes. Thresholding systems are a principal tool for signal segmentation. Thresholding can be applied globally or locally.

4.2.1 Global Methods

For purposes of segmentation, one of the easiest techniques is to select a single threshold value and apply it to the entire signal—global thresholding. For example, for discrete signals, $f(n)$, if $T > 0$, then $M = \{n: |f(n)| \geq T\}$ is the meaningful component, and $N = \{n: |f(n)| < T\}$ is the noise component of $f(n)$, respectively. Of course, with this division of the signal, the meaningful component will have some noise present. These jagged, low-magnitude artifacts are sometimes called "fuzz," "clutter," or "grass," and they are unavoidable for real-world signals. More precisely, the thresholding operation just picks out that part of the signal that contains something more—something relevant to the analysis application—than just the background noise.

The examples in the previous section showed that this technique works for signal features that are sufficiently distinct from the background noise. Difficulties arise, however, when the signal features of interest diminish over time, blend into the noise, or contain oscillatory components. Sometimes supplemental signal filtering helps. Sometimes a more sophisticated labeling procedure, which imposes constraints on the segmentation, corrects any flaws in the preliminary partitioning of the signal's domain. In any case, the main problem is to determine the appropriate threshold.

4.2.2 Histograms

There are several methods for finding global thresholds. Simple inspection of source signals may indeed suffice. Another method is to histogram the signal, producing a density map of the signal values. Histograms count the number of values that fall within selected ranges, or bins, over the signal domain. This is akin to segmentation, but it partitions the signal's range rather than its domain. The histogram is also an approximation of the signal level probability density function. This observation is the key idea in several of the optimal threshold selection techniques in the signal analysis literature.

Definition (Histogram). Let f be a discrete signal and let $\Pi = \{B_1, B_2, ...\}$ be a partition of the range of f. Then the B_k are *bins* for Ran(f). The histogram of f with respect to Π is defined by $h(k) = \#(f^{-1}(B_k))$. In other words, the value $h(k)$ is the cardinality of $f^{-1}(B_k)$—the number of $n \in$ Dom(f) with $f(n) \in B_k$.

Even though the definition confines the histogram idea to discrete signals, it is still a very general formulation. It allows for a countably infinite number of bins as well as for infinite histogram values. It is more practical to restrict the domain of discrete signals $f(n)$ to an interval and specify a finite number of bins. That is, suppose $f(n)$ is a digital signal with q-bit values and that f(n) is of finite support (i.e., $f(n) = 0$ outside some finite interval, I). Then $0 \le f(n) \le 2^q - 1$. Let $\Pi = \{\{k\} \mid 0 \le k \le 2^q - 1\}$. Then Π is a set of bins for Ran(f), and $h(k) = \#(\{n \in I \mid f(n) = k\}) = \#(f^{-1}(\{k\}))$ is a histogram for f. Since h counts domain elements only within I, $h(k)$ is a discrete signal with support in $[0, 2^q - 1]$.

Example (Digital Image Histograms). Many one-dimensional signal processing and analysis tasks arise in computer vision. Histograms, in particular, are fundamental to early image analysis tasks. Figure 4.7 shows a gray-scale text image and its histogram. Threshold selection is easy. But this is due to the close cropping of white background from the black text. Including larger areas of background in the image would result in an apparently unimodal histogram. A lesson in text image analysis is that successful segmentation of words into letters depends on a prior successful segmentation of the page into words or lines.

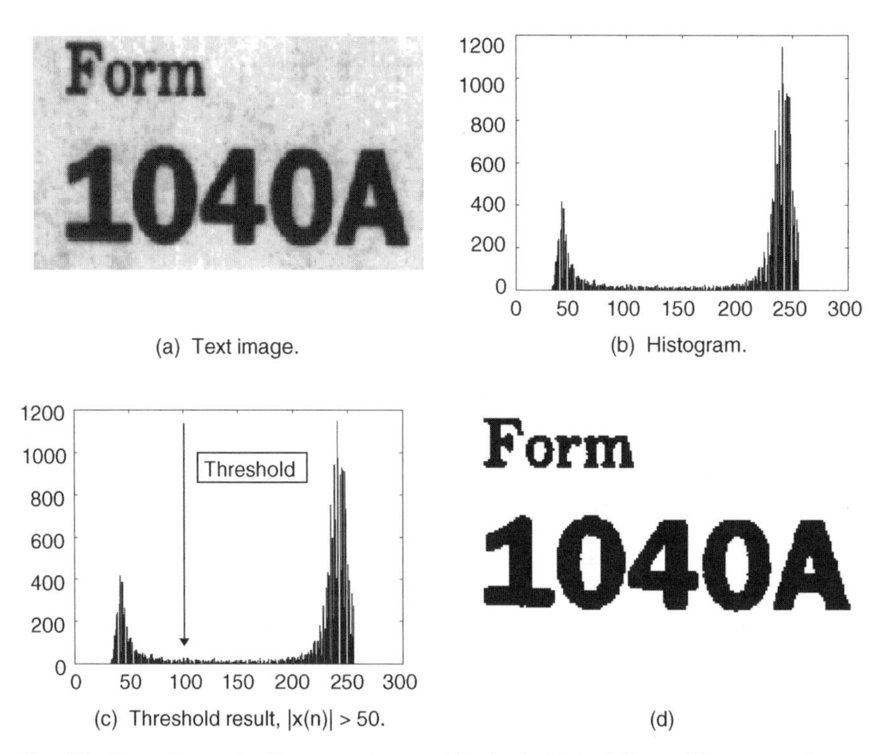

Fig. 4.7. A text image familiar to readers working in the United States. The gray-scale text image (a) has been cropped to exclude surrounding background areas. In the histogram (b), the modes correspond to the largely white and largely black picture elements. Inspection of the histogram reveals a threshold (c), and using this threshold segments the text into purely black letters and purely white background (d).

It is important, however, to automate the histogram threshold selection procedure for autonomous signal analysis systems. A valley-finding algorithm is sometimes effective. If the significant and background signal values are both large in number and different in magnitude, then two large modes appear in the histogram. Beginning from the histogram midpoint, a reverse-gradient search across the histogram should find the valley between the modes. Let $h(k)$ be the signal histogram, $T \in \mathrm{Dom}(h)$, and search to another k as follows:

- If $T-1 \in \mathrm{Dom}(h)$ and $h(T-1) \le h(T)$, then continue the search at $T-1$; otherwise.
- If $T+1 \in \mathrm{Dom}(h)$ and $h(T+1) \le h(T)$, then continue the search at $T+1$; otherwise.
- Accept the value of T last searched as the threshold.

There are several drawbacks to this procedure, and they are common to every type of reverse-gradient technique. The algorithm stops at the first local minimum it finds. The simple valley-finding technique has a direction preference, which may not accord well with the signal analysis application. And there is no guarantee that the final T the method selects does not lie in a local minimum interior to one of the histogram's modes. The next technique offers to improve the valley-finding algorithm by building a global cost function.

If we think of the histogram as a valley, possibly containing foothills, between two mountain ranges, then selecting a threshold amounts to deciding how much effort it takes to get over the mountains to one boundary or the other. At each point on the histogram, we associate a cost of leaving the valley; it is the minimum of the cost of reaching the first and the cost of reaching the last histogram bin. Of course, only uphill walking requires effort, so we increment the separate costs to the far bins only when the histogram is decreasing on the left and increasing on the right. A point with maximal cost lies in the valley and is a feasible threshold for separating the two modes. Less metaphorically, therefore, an algorithm for finding a threshold in a histogram $h(k)$ is as follows:

$$\mathrm{Cost}_L(t) = \sum_{\substack{k \le t \\ h(k-1) > h(k)}} [h(k-1) - h(k)], \tag{4.1}$$

$$\mathrm{Cost}_H(t) = \sum_{\substack{k \ge t \\ h(k) < h(k+1)}} [h(k+1) - h(k)], \tag{4.2}$$

$$\mathrm{Cost}(t) = \min\{\mathrm{Cost}_L(t), \mathrm{Cost}_H(t)\}, \tag{4.3}$$

$$T = \max_t \{\mathrm{Cost}(t)\} \tag{4.4}$$

We call the above technique a global valley-finding algorithm.

Techniques such as the above algorithm (4.1)–(4.4) furnish useful thresholds for many signal segmentation tasks. However, as the exercises show, it easy to find signals for which the algorithm fails. Furthermore, natural signals or those from engineered systems may have meaningful regions that become lost when the histogram procedure removes their time domain locality. The question therefore arises whether there is some way of inproving the histogram estimate, or, more optimistically, whether there is an optimal method for finding a threshold for segmentation of an image from its histogram.

4.2.3 Optimal Thresholding

In fact, there are several optimal methods for finding thresholds from signal histograms. Each such "optimal" method discovers the best threshold based on some particular assumption about the histogram's statistics. This is a natural thought, since the histogram approximates the probability density function of the signal values. Accordingly, let us consider some techniques for optimal threshold selection. We shall begin with work that is now classic and end with some quite recent improvements in this thread of investigation.

4.2.3.1 Parametric Approaches. Suppose that we know the some of the statistical behavior of signals that arrive as inputs to an analysis application. For example, we might have knowledge of the statistical distribution of signal values, along with knowledge of the likelihoods that signal features have certain labels; and, perhaps the most valuable information of all, we might even know the probabilities for the classification of signals. Such information on statistical parameters associated with a signal generation mechanism is the basis for the parametric approach to signal threshold determination.

Suppose that we know the probabilities that a discrete signal value is high-magnitude (meaningful), P_H, or low-magnitude (background, noise), P_L. These are called the a priori probabilities of meaningful and noise components, respectively. For example, in a digital telephony system, DTMF pulses might occur at 10 Hz with an average pulse width of 48 ms. Hence we assume a priori probabilities of $P_H = .48$ and $P_L = .52$. Some system designs rely on preliminary statistical studies to develop a priori probabilites. Other strategies are adaptive, and the probabilities change slowly while the system operates.

Segmentation errors are due to labeling a noise value as meaningful or a meaningful value as noise. Thus, if a threshold T for a signal histogram produces errors with probability $E(T)$, then

$$E(T) = P_H E_L(T) + P_L E_H(T), \tag{4.5}$$

where $E_L(T)$ and $E_H(T)$ are the probabilities of incorrectly labeling a signal value as noise and as meaningful, respectively. To find the minimum labeling error, we differentiate (4.5) and solve the equation $dE/dT = 0$ for the threshold T:

$$\frac{dE}{dT} = P_H \frac{dE_L}{dT} + P_L \frac{dE_H}{dT} = 0. \tag{4.6}$$

Now, this scheme does not work at all unless we can find estimates for $E_L(T)$ and $E_H(T)$. The idea is to approximate the distributions of signal values in the histograms by standard statistical distributions. In the histogram of Figure 4.7, for instance, the modes resemble normal (Gaussian) density functions. From the tutorial on probability theory in Section 1.8, the Central Limit Theorem shows that whatever the distributions we observe in a histogram, then (given their bounded variance) the average of a great many of these random variables always approaches a Gaussian distribution. Let us assume, therefore, Gaussian distributions of both noise and meaningful signal. Thus,

$$q_L(t) = \frac{1}{\sqrt{2\pi}\sigma_L} e^{-(t-\mu_L)^2/(2\sigma_L^2)} \tag{4.7}$$

is the probability density function for the signal noise values, where μ_L and σ_L are the mean and standard deviation, respectively. Similarly,

$$q_H(t) = \frac{1}{\sqrt{2\pi}\sigma_H} e^{-(t-\mu_H)^2/(2\sigma_H^2)}, \tag{4.8}$$

where μ_H and σ_H are the mean and standard deviation of the signal noise values, respectively. Since noise values are on average less than meaningful signal values, we know that $\mu_L < \mu_H$. From (4.7) and (4.8), it follows that

$$E_L(T) = \int_{-\infty}^{T} q_H(t)\, dt, \tag{4.9}$$

and

$$E_H(T) = \int_{T}^{\infty} q_L(t)\, dt. \tag{4.10}$$

Differentiating (4.9) and (4.10) with respect to T shows that $dE_L/dT = q_H(T)$ and $dE_H/dT = -q_L(T)$. We substitute expressions for these derivatives—(4.7) and (4.8)—into (4.6) to obtain

$$\frac{P_H}{\sigma_H} e^{-(T-\mu_H)^2/(2\sigma_H^2)} = \frac{P_L}{\sigma_L} e^{-(T-\mu_L)^2/(2\sigma_L^2)}. \tag{4.11}$$

We take the natural logarithm on both sides of (4.11), simplify, and reveal a quadratic equation in T:

$$\left(\sigma_L^2-\sigma_H^2\right)T^2+2\left(\sigma_H^2\mu_L-\sigma_L^2\mu_H\right)T+\left(\sigma_L^2\mu_H^2-\sigma_H^2\mu_L^2\right)-2\ln\left(\frac{P_H\sigma_L}{P_L\sigma_H}\right)=0. \tag{4.12}$$

The quadratic equation (4.12) may have two, one, or zero solutions, depending upon the statistics of the true signal and its noise component. It may be necessary to compare the performance of two possible thresholds using (4.5). And any solution T for (4.12) must be in the range of the signal in question. The exercises further explore these ideas.

This procedure is due to Chow and Kaneko [17]. They applied the technique for finding the left ventricle cardioangiograms—x-ray images acquired after injecting a contrast-producing dye into the heart. Some observations on the method are:

- The method requires *a priori* knowledge of the signal, namely the probabilities of the true signal, P_H, and of the background, P_L.
- It is a parametric method, in that it assumes a particular model of the signal histogram and then derives parameters that best describe the model.
- To discover the parameters of the sum of Gaussian distributions in Chow and Kaneko's approach, it is necessary to fit the model to the actual histogram data.
- Thus, parameter determination requires, for example, a least-squares fit of the model to the data, and an accurate or numerically well-behaved convergence is not guaranteed.
- Moreover, as our own examples show, the model (e.g., a sum of two normal distributions) may not be appropriate for the signal histogram.

These are difficulties with any parametric technique. The next section considers a nonparametric strategy. It is an alternative that does not presuppose statistical distributions for the signal values. Then, in Section 4.2.3.3 we will revisit parametric methods. There we will examine a method inspired by information theory that avoids the assumption of *a priori* probabilities.

4.2.3.2 Nonparametric Approaches. A nonparametric approach to threshold determination assumes no knowledge of the statistical parameters that derive from a signal's values. Thus, nonparametric strategies include the valley finding tactics covered earlier. Valley finding methods do not assume any statistical distribution of meaningful signal and noise values, even though these particular algorithms are rather primitive. Methods can combine, too, for better performance. We can use a global valley-finding algorithm to split the histogram with a preliminary threshold, T. We then determine the statistical parameters of the noise and true signal by separate least-squares fits to the histogram for $t < T$ and for $t > T$. And, finally, we apply the Chow and Kaneko technique to improve the estimate of T. If the segmentation that results from the threshold selection is unsatisfactory, goal-directed strategies are worth investigating. It may be possible to alter the preliminary threshold, recompute the mode statistics, or choose another statistical distribution for modeling the histogram.

Let us turn to a nonparametric approach for threshold determination and signal segmentation proposed by Otsu [18]. Otsu hoped to avoid some of the difficulties we noted above. The idea is to select a threshold to segment the signal into labeled regions of minimal variance in signal levels. Let $P_L(t)$ and $P_H(t)$ be the probability of background values and true signal values, respectively. In Chow and Kaneko's approach, these were needed *a priori*; in Otsu's algorithm, on the other hand, these are approximated from the histogram and are functions of the threshold value. We segment the signal values into two groups, according to the threshold. And let us suppose, again without loss of generality, that background values are low and meaningful signal values are high. Then, a measure of within group variance is

$$\sigma_w^2(t) = P_L(t)\sigma_L^2(t) + P_H(t)\sigma_H^2(t), \tag{4.13}$$

where $\sigma_L(t)$ and $\sigma_H(t)$ are the standard deviations of the noise and the meaningful signal, respectively. In order to find t so that (4.13) is minimized, we must also find the statistical distribuitons of the low-level and the high-level regions of the signal. Chow and Kaneko's parametric approach assumes that the histogram is a sum of two normal densities and that *a priori* probabilities are known.

Because it directly approximates the histogram statistics from threshold values and does not make assumptions about *a priori* noise and true signal probabilities, Otsu's method is essentially unsupervised. Let us see how the method works. Suppose that $\mathrm{Ran}(f) \subseteq [0, N-1]$, set $S_k = \{k\}$ for $0 \le k < N$, and suppose that $\mathrm{Dom}(f)$ is finite. Then $\{Sk \mid \}$ is a partition of $\mathrm{Ran}(f)$. Define

$$p_k = \frac{\#(f^{-1}(S_k))}{\#\mathrm{Dom}(f)}, \tag{4.14}$$

Then, p_k is a discrete probability density function for f. Hence,

$$P_L(t) = \sum_{k=0}^{t-1} p_k, \tag{4.15}$$

and

$$P_H(t) = \sum_{k=t}^{N-1} p_k. \tag{4.16}$$

Let $\{\Lambda_L, \Lambda_H\}$ be labels for the noise and meaningful regions of the signal. Then, the conditional probability that $f(n) = k$, given that n has label Λ_L and the threshold is t, is $P(k \mid \Lambda_L) = p_k/P_L(t)$. Similarly, $P(k \mid \Lambda_H) = p_k/P_H(t)$. This observation permits us to write down the following values for the parameters of the distributions comprising the histogram:

$$\mu_L(t) = \sum_{k=0}^{t-1} kP(k \mid \Lambda_L) = \sum_{k=0}^{t-1} k \frac{p_k}{P_L(t)}, \tag{4.17}$$

$$\mu_H(t) = \sum_{k=t}^{N-1} kP(k \mid \Lambda_H) = \sum_{k=t}^{N-1} k \frac{p_k}{P_H(t)}, \tag{4.18}$$

$$\sigma_L^2(t) = \sum_{k=0}^{t-1} (k - \mu_L(t))^2 P(k \mid \Lambda_L) = \sum_{k=0}^{t-1} (k - \mu_L(t))^2 \frac{p_k}{P_L(t)}, \tag{4.19}$$

and

$$\sigma_H^2(t) = \sum_{k=t}^{N-1} (k - \mu_H(t))^2 P(k \mid \Lambda_H) = \sum_{k=t}^{N-1} (k - \mu_H(t))^2 \frac{p_k}{P_H(t)}. \tag{4.20}$$

Having found the histogram statistics that follow from each possible threshold value, we are in a position to search over all threshold values for T which minimizes the within-group variance. Specifically, by a exhaustive search we find the optimal threshold T which satisfies

$$\sigma_w^2(T) = \min_{0 \le t < N} \sigma_w^2(t). \tag{4.21}$$

Otsu's method merits consideration in applications where human intervention in the signal analysis process must be minimized. It does not need *a priori* probability estimates. It does not make any assumptions about the distribution of histogram values. Two problems weigh on the approach, however:

- It requires a search and recomputation of the statistics (4.17)–(4.20) over all possible threshold values.
- There may not be a unique minimum in (4.21), and, unless some goal-directed are imposed, there is no criterion for selecting one variation-reducing threshold over another with the same effect.

The exercises explore some algorithm refinements that reduce the recomputation burden. But the second point is troublesome. If we knew the within-group variance to be unimodal, then the search would always identify an optimal threshold.

Experiments supported—and Otsu conjectured—variance unimodality within segmented regions, but it does not hold true in general. In summary, nonparametric strategies generally involve more computational tasks than their parametric cousins. Thus, a preliminary survey of the statistical parameters of input signals to an analysis application may be warranted. A study of the statistics within histograms, for example, may prove the feasibility of the simpler parametric strategy.

4.2.3.3 An Information-Theoretic Approach.

Attempting to avoid the modality problems inherent to Otsu's algorithm, Kittler and Illingworth [19] approached the problem by trying to find the mixture of Gaussian probability distributions that best matches the signal histogram for a given threshold value. For their optimality criterion, they employed relative entropy [20, 21], an information-theoretic tool, which we introduced in Section 1.8.4. They adopted a model of the histogram as a scaled sum of two normal distributions. Thus, theirs is a parametric approach akin to Chow and Kaneko's; however, it avoids the supposition of *a priori* probabilities for noise and meaningful signal segments, and therefore it represents a significant extension to the parametric method.

Following Kittler and Illingworth, let us suppose that p_k is given by (4.14), and q_k is an alternative distribution. Then the relative entropy, $I(p, q)$, of the distribution p_k with respect to q_k is

$$I(p,q) = \sum_{k=0}^{N-1} p_k \log_2 \frac{p_k}{q_k} = \sum_{k=0}^{N-1} p_k \log_2 p_k - \sum_{k=0}^{N-1} p_k \log_2 q_k$$
$$= -H(p) - \sum_{k=0}^{N-1} p_k \log_2 q_k, \tag{4.22}$$

where $H(p)$ is the entropy of p,

$$H(p) = \sum_{k=0}^{N-1} p_k \frac{1}{\log_2 p_k}. \tag{4.23}$$

It can be shown that $I(p, q) \geq 0$ for all distributions p and q. Furthermore, $I(p, q) = 0$ if and only if $p = q$. Finally, let us note that $\log_2(p_k/q_k)$ is the information increment, given that the signal $f(n) = k$, that supports the histogram $h(k)$ having distribution p_k instead of q_k. Thus, the average information in favor of $h(k)$ following distribution p_k instead of q_k is $I(p, q)$, from (4.22). If q_k is a scaled sum of two normal distributions, then

$$q_k = \frac{a_1}{\sqrt{2\pi}\sigma_1} e^{-(k-\mu_1)^2/(2\sigma_1^2)} + \frac{a_2}{\sqrt{2\pi}\sigma_2} e^{-(k-\mu_2)^2/(2\sigma_2^2)}, \tag{4.24}$$

where a_1 and a_2 are constants. Now, q_k represents the histogram $h(k)$ better when signal values $f(n) = k$ discriminate in its favor over p_k; in other words, (4.24) should be minimized for best representing $h(k)$ with a sum of two Gaussians scaled by a_1 and a_2. Since $H(p)$ depends only upon the given histogram, this means that

$$I(p,q)+H(p) = -\sum_{k=0}^{N-1} p_k \log_2 q_k \tag{4.25}$$

should be minimized.

We can minimize (4.25) by approximating the statistical parameters of the two components of the distribution q_k for each candidate threshold, $0 < t < N-1$. If t lies between well-separated means, μ_1 and μ_2, then we should expect reasonable estimates. Thus, for each such t, we take $\mu_{1,t} = \lfloor t/2 \rfloor$ (the floor, or integer part of $t/2$). Let $a_{1,t}$ and $\sigma_{1,t}$ be the mean and standard deviation of $\{p_1, p_2, \ldots p_{t-1}\}$, respectively. We set $\mu_{2,t} = \lfloor (N-t)/2 \rfloor$, and let $a_{2,t}$ and $\sigma_{2,t}$ be the mean and standard deviation of $\{p_t, p_{t+1}, \ldots, p_{N-1}\}$, respectively. Lastly, we substitute these values into (4.24) to obtain

$$q_k(t) = \frac{a_{1,t}}{\sqrt{2\pi}\sigma_{1,t}} e^{-(k-\mu_{1,t})^2/(2\sigma_{1,t}^2)} + \frac{a_{2,t}}{\sqrt{2\pi}\sigma_{2,t}} e^{-(k-\mu_{2,t})^2/(2\sigma_{2,t}^2)}. \tag{4.26}$$

Then, the optimal threshold is T where

$$-\sum_{k=0}^{N-1} p_k \log_2 q_k(T) \le -\sum_{k=0}^{N-1} p_k \log_2 q_k(t) \tag{4.27}$$

for all t, $0 < t < N-1$.

An extensive literature on thresholding testifies that no single technique guarantees correct signal segmentation for all applications. General surveys include [22, 23]. Others concentrate on thresholding text [24, 25] or map [26] images. Figure 4.8 shows a signal, its histogram, and the results of thresholding it using the Otsu algorithm. Simple threshold selection methods and bottom-up signal classification strategies very often work just as well as the more sophisticated techniques. When they fail, combining methods is fruitful. Many of the methods we cover can be extended, often in straightforward ways, to multiple threshold values. Such an approach looks for multiple thresholds and partitions the domain into several regions according to signal magnitude. This is also called signal quantization, since it maps signal values that spread over a wide range to a set of signal values that vary significantly less. The exercises explore this important problem. More challenging applications require some top-down, goal-directed mechanism for improving the thresholding, segmentation, and labeling of a signal.

Finally, let us confess that we have neglected an important technique for signal segmentation: local or adaptive thresholding.

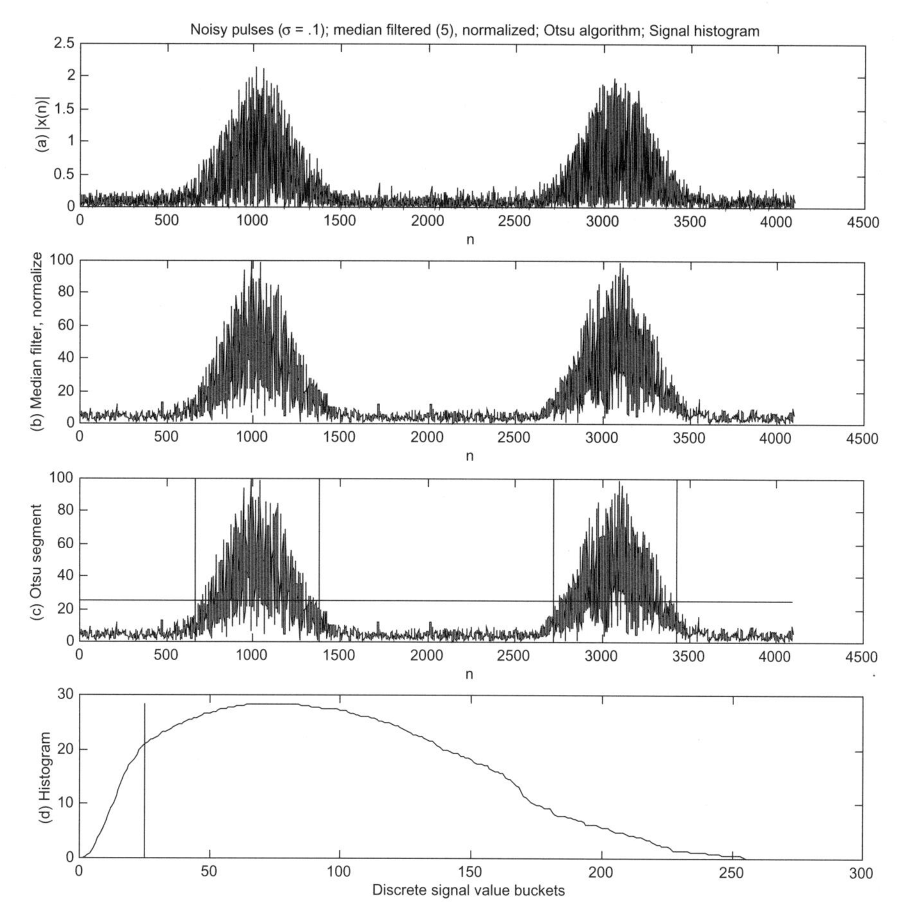

Fig. 4.8. A pulse signal (a), median filtered and normalized (b), threshold from Otsu's algorithm (c), and the signal histogram (d).

4.2.4 Local Thresholding

The methods covered above, by more or less sophisticated means, attempt to arrive at a global threshold value that will break down the signal. In many problems, the threshold that successfully segments a signal in one area fails to perform adequately in another area. But there is nothing in the global thresholding methods we have detailed above—based on histogram analysis—that prevents them from being applied to finite regions of the signal domain. When we find and apply thresholds locally within a signal's domain, the technique is called local thresholding or—since the threshold value adapts to the statistics or entropy of the signal locally—adaptive thresholding.

It may be possible, and indeed it may be essential, to vary the threshold at different places within the signal domain. This could be the case where the gain of the signal processing elements of the system varies. Falling signal strength could cause meaningful components to be misclassified as background noise. Thus, the threshold should adapt to the overall magnitude of local regions of the signal. Varying illumination, shading, and object reflectance make adaptive thresholding vital to many image analysis applications. Optical character recognition is but one example [27]. In digital telephony, echo cancellation should be suspended during episodes of double-talk, when both the near-end and far-end speakers are talking. But the threshold level for deciding that the near-end speaker is talking needs to adapt to the signal level of the far-end speaker; otherwise, cancellation may occur when both speakers are talking softly or might not take place when the far-end speaker is very loud. One approach is to declare double-talk when the the near-end signal level $s(n) > (1/2)\max\{x(n), x(n-1), ..., x(n-N)\}$, for some $N > 0$. By comparing levels over a whole range of recent far-end speaker voice magnitudes, the algorithm accomodates the unknown echo delay in the near-end circuit [28].

4.3 TEXTURE

Signal segmentation often requires that regions with certain regular patterns of signal values be distinguished from one another. It is not the absolute signal level—as with the previous section's principal concern, thresholding—but rather the repetitive transition of the signal through some value range that is of interest in such an application. This section covers methods for segmenting a signal into regions of different texture.

A threshold-based segmentation algorithm does not easily handle signals with periodic components. Examples covered in the previous sections, such as the speech fragments and the dual-tone multifrequency telephone system signaling pulses, are cases in point. It is possible to apply a preliminary smoothing filter to the values prior to thresholding. This blends the oscillation together and reduces the average value of the signal over a region, and then the thresholding operation properly isolates the region of interest. Regions found by one threshold, T_1, but not by another threshold $T_2 \neq T_1$ constitute a distinct texture field. Perhaps several filters and thresholds are necessary. As an alternative, we might consider segmenting the large-magnitude signal regions from the low-magnitude signal regions and then estimating an overall distance between peaks or between troughs. Areas that show different average intervals between crests or troughs represent different textured signal components. We are faced with a number of algorithm design issues, though, whatever approach we select. It is desirable to arrive at some indication of texture that produces a measure directly from the signal values, without preliminary filtering followed by thresholding.

Intuitively, the notion of texture incorporates the presence of smoothness or roughness and the overall character of repetition and continuity as opposed to rupture or discontinuity. One tool in wide use by computer vision researchers is a collection of photographs of textures, originally intended for use by artists, the Brodatz textures [29]. Many pioneering researchers in the field pursued the problem of texture characterization from psychological standpoint [30]. Some, in particular, worked with the principles of Gestalt psychology in mind [31, 32], which stipulates that pictorial features group together in such a way that the overall constellation achieves a form which cannot be expressed by descriptions of the individual elements. These pictorial features include proximity, similarity, continuation, symmetry, closure, and common fate.

Over the years, research efforts in texture analysis have been confined to three broad categories: statistical, spectral, and structural. Statistical approaches apply moment computations to the signal values: variance, skew, and kurtosis, for example [33]. Spectral approaches analyze the signal by comparing it to sinusoidal or Gabor elementary functions (Chapter 1) of varying frequencies [34]. And structural approaches endeavor to find a set of texture elements or primitives that obey a repeating pattern or a relational grammar in the signal [35].

Perhaps one of the most pressing problems is to define the type of repetitive structure, the texture regions in the signal that the application must find. Readers might well suppose that here we would introduce a formal definition of texture; this has, after all, been much the pattern of presentation so far. The U.S. industrial standard for surface metrology [36] defines it as repetitive or random deviations from the nominal surface within the three-dimensional topography of the surface. One might well object to this definition, since the terms "repetitive" and "random" contradict one another! Unfortunately, there is no suitable formal definition for texture, and there is no best approach to segmentation. The methods used are mostly ad hoc, and, if there can be anything close to a definition of texture, it is whatever the chosen method seems to find within the signal! The whole notion of what constitutes texture and what does not is quite uncertain. Let us consider the principal methods for segmenting signals according to their texture content in the next three sections. Though we lack a precise formal definition of texture, each approach constitutes a practical, working concept of what texture is—indeed it is whatever that method finds.

4.3.1 Statistical Measures

The statistical approach to texture segmentation and classification is the most widely used. However problematic the formal definition of texture is, methods to characterize and quantify it are quite important in areas such as materials science and remote sensing. In surface metrology, for example, many standard statistical measures of one- and two-dimensional textures are in use. There are also a number spectral measures of texture, and we will cover them, albeit briefly, in the next section. Some of these texture parameters are admittedly ad hoc, but they are nevertheless widely implemented and widely respected indications of the presence of texture within signals.

4.3.1.1 Basic Measures. One way to measure the amount of variation in a signal is to average the departure of the signal from its mean over a fixed interval. This statistical parameter of signal texture is the roughness average. It is the most widely used measure in the world, especially for manufactured surface characterization, and it has been standardized in the United States and internationally [33, 36–38].

Definition (Roughness Average). Let $x(t)$ be an integrable analog signal; let $[a, b]$ an interval, $a < b$; and let μ be the mean of $x(t)$ over $[a, b]$. Then the roughness average of $x(t)$ over $[a, b]$ is

$$R_a(x(t),[a,b]) = \frac{1}{b-a}\int_a^b |x(t)-\mu|\,dt. \tag{4.28a}$$

And if $x(n)$ is a discrete signal, $[a, b]$ is an interval, $a < b$, and μ is the mean of $x(n)$ over $[a, b]$, then the roughness average of $x(n)$ over $[a, b]$ is

$$R_a(x(n),[a,b]) = \frac{1}{b-a}\sum_{n=a}^{n=b} |x(n)-\mu|. \tag{4.28b}$$

When the interval over which we compute the roughness average is understood, it is common to write this parameter as $R_a(x)$. If the signal values are also unambiguous, it is simply R_a.

The roughness average indicates a change in signal variation, but it fails to explicate the distribution of crests, troughs, and transients in the signal. Somewhat more sophisticated statistical measures can be of help. We can, for example, better understand the distribution of signal values by applying a variance measure to the signal regions of interest. As a texture measure, the variance is also a standard.

Definition (Root Mean Square (RMS) of Roughness). If $x(t)$ is an analog signal, $[a, b]$, is an interval, $a < b$, and μ is the mean of $x(t)$ over $[a, b]$, then the RMS of roughness for $x(t)$ over $[a, b]$ is

$$R_q(x(t),[a,b]) = \left(\frac{1}{b-a}\int_a^b |x(t)-\mu|^2 dt\right)^{1/2}. \tag{4.29a}$$

For analog signals, the equivalent definition is

$$R_q(x(n),[a,b]) = \left(\frac{1}{b-a}\sum_{n=a}^{b-1} |x(n)-\mu|^2\right)^{1/2}. \tag{4.29b}$$

It is common to omit the signal and region from the specification of R_a and R_q, since the problem context often makes it understood.

Example (Surface Profilometers and Atomic Force Microscopes). Let us look at the segmentation of surface profiles by texture characterization. Instruments such as profilometers and atomic force microscopes (AFM) acquire one-dimensional (and in many commercial instruments, two-dimensional) height profiles of a surface. Surface texture measurements from such profiles are critical for the control and diagnosis of the fabrication processes. These measures include such aspects of surface texture as fine-scale roughness, more widely spaced regularities called waviness, and directional features called lay.

The roughness average is by far the most widely computed and reported in manufactured surface characterization. R_a detects general signal profile variations, and a significant change over one part of the signal domain indicates a fundamental change in the process that produces the signal. Nevertheless, it has limitations. To wit, the roughness average parameter fails to detect the presence or absence of widely separated signal transients. R_a completely overlooks subpatterns of texture. (Structural techniques are better suited to this type of problem; see Section 4.2.3.3.) The intervals between texture elements are also invisible to R_a. Some help in characterizing surface texture comes from the R_q parameter. It measures the distribution of deviations from the mean of the signal, so when it is large, it is an indication that the rough features of the signal have wide magnitude variability.

Other texture segmentation methods rely on peak-to-valley measurements within signal regions. Let us consider a few of these next and then proceed to some texture measures that arise from applying statistical moment ideas to a signal's values.

Definition (R_t, R_z). The total indicated reading over $[a, b]$, R_t, is the difference between the signal maximum and signal minimum over the interval. Thus, for an analog signal, $x(t)$, we define

$$R_t(x(t), [a, b]) = \max\{x(t): t \in [a, b]\} - \min\{x(t) \mid t \in [a, b]\}. \tag{4.30a}$$

The analog for a discrete signal, $x(n)$, is

$$R_t(x(n), [a, b]) = \max\{x(n): t \in [a, b]\} - \min\{x(n) \mid t \in [a, b]\}. \tag{4.30b}$$

The five-point peak parameter, R_z, is

$$R_z(x,[a,b]) = \frac{1}{5}\left(\sum_{k=1}^{5}(p_k - v_k)\right), \tag{4.31}$$

where $p_k \geq p_{k+1}$ are the five highest values x takes on $[a, b]$, and $v_{k+1} \geq v_k$ are the five lowest values x takes on $[a, b]$.

These parameters find application in diverse disciplines. Seismologists use the total indicated reading parameter to calculate earthquake magnitude according to the Richter scale.[3] Let $s(t)$ be the seismograph needle's deviation from the centerline of a paper strip chart at time t during an event. If the epicenter is 100 km

[3]Charles F. Richter (1900–1985), a seismologist at the California Institute of Technology, established the popular logarithmic scale for earthquake magnitude in 1935.

away from the instrument, then the seismologist computes the Richter magnitude, $M_L = \log_{10}[R_t(s)]$. The problem with the total indicated reading as a measure of texture is that it is sensitive to impulse noise in the signal. The R_z parameter is one of several that surface metrologists, for example, use to avoid this difficulty. An even larger average is tempting in (4.31); this provides roughness estimates with better immunity to occasional burrs and pits that blemish a generally good surface.

4.3.1.2 Higher-Order Moments. Further texture analysis measures proceed from an analysis of the signal using statistical moments. Following our histogramming discussion, suppose that we can obtain the discrete probability density function for the values of $x(n)$: $p_k = P[x(n) = k]$. For example, let $x(n)$ be a digital signal with $L \le x(n) \le M$ for $a \le n \le b$, where $a < b$. We form the histogram $h(k) = \#(\{n \in [a, b] \mid f(n) = k\})$ for each $k \in [L, M]$. If $M > L$, we then set $p_k = h(k)/(M-L)$; otherwise, $p_k = 1$. Then p_k is a discrete probability density function. The mean of x on $[L, M]$ is $\mu = (Lp_L + (L+1)p_{L+1} + \cdots + Mp_M)$. Then the variance, σ^2, skew, μ_3, and kurtosis, μ_4, are, respectively, as follows:

$$\sigma^2 = \sum_{k=L}^{M} (k-\mu)^2 p_k, \tag{4.32}$$

$$\mu_3 = \frac{1}{\sigma^3} \sum_{k=L}^{M} (k-\mu)^3 p_k, \tag{4.33}$$

$$\mu_4 = \frac{1}{4}\left(\sum_{k=L}^{M} (k-\mu)^4 p_k \right). \tag{4.34}$$

The R_q texture parameter is a variance measure. The skew measures the asymmetry of the values about the signal mean. The kurtosis measures the relative heaviness of the outlying signal values within the texture region. Some authors define kurtosis by subtracting three from (4.34); this ensures that a Gaussian distribution has zero kurtosis, but it does not affect the measure as a tool for signal segmentation.

These texture parameters supply some of the information missing from the roughness average figure. To use any of these parameters as a basis for signal segmentation, we compute the parameter over selected regions of the signal domain and label the regions according to the parameter's numerical value. A more informative segmentation of the signal is possible by calculating several of the parameters and assigning labels to the regions that represent combinations of significant texture parameters.

Notice that there is a distinct difference between the application of these texture measures and the previous section's thresholding algorithms to segmentation problems. When thresholding, the segmentation regions were discovered, signal value after signal value, by applying the thresholding criterion. On the other hand, to apply the texture measures, an interval or region of the signal domain must be used to calculate the statistical quantity, R_a or R_q, for example. Thus, a texture segmentation commonly starts with a preliminary partition of the signal domain into intervals, called frames, with texture measures assigned to the intervals. The bounds of these preliminary texture regions may be adjusted later with split and merge criteria, as we considered in Section 4.1.4. For precise registration of texture areas, a data-driven alternative exists. An application may apply texture measures to frames of a minimum size, say $N > 0$. After applying the measure to region $[a, b]$, where $b - a = N$, the statistics are compared to one or more threshold values. If $[a, b]$ contains texture, then the measures are applied to the larger region $[a, b + 1]$; otherwise the frame becomes $[a + 1, b + 1]$ and texture finding continues. Each time a region contains texture, the application attempts to expand it on the right, until at some iteration, say $[a, b + k + 1]$, the texture indication test fails. Then $[a, b + k]$ is declared to be texture-laden, and processing continues with the next minimal frame $[b + k + 1, b + k + N + 1]$. Albeit computationally expensive, this scheme avoids any top-down split and merge procedures.

4.3.1.3 Co-occurrence Matrices.

One significant drawback to the above methods for texture segmentation is that they utilize no distance measures between intensity features within the texture. The moment parameters, such as R_q, do incorporate a notion of breadth of variation. In manufactured surface characterization, R_q is often touted as an alternative to the widely quoted R_a value. Nevertheless, signal and image analysts reckon that the most powerful techniques are those that compute statistics for the distribution of signal values separated various time intervals [39]. The next definition [40] incorporates the distance between texture highlights into our statistical indicators of texture.

Definition (Co-occurence Matrix). Let $x(n)$ be a digital signal; let $L \le x(n) \le K$ for some integers L, K and for $a \le n \le b$; and let δ be a time interval. Then $M_\delta = [m_{i,j}]_{N \times N}$ is the $N \times N$ matrix defined by $m_{i,j} = \#\{(x(p), x(q)) \mid p, q \in [a, b], x(p) = i$ and $x(q) = j$, and $\delta = |p - q|\}$. The co-occurrence matrix for $x(n)$ and time interval δ is defined $P_\delta = [m_{i,j}/N_\delta]$, where $N_\delta = \#\{(x(p), x(q)) \mid p, q \in [a, b]$ and $\delta = |p - q|\}$.

Thus, $m_{i,j}$ contains a count of the number of pairs (p, q) for which $x(p) = i$, $x(q) = j$, and p and q are time δ apart. P_δ estimates the joint probability that two signal will take values i and j at a displacement δ apart. Also, it is not necessary to restrict $x(n)$ to be a digital signal; as long as its range is finite, the co-occurrence matrix can be defined.

Example (Co-occurrence Matrix). Let $x(n) = [..., 0, \underline{1}, 2, 1, 1, 2, 0, 0, 1, 0, 2, 2, 0, 1, 1, 0, ...]$ be a digital signal, and suppose we compute the co-occurrence matrices for $\delta = 1$, 2, and 3 within the interval $0 \le n \le 15$. This signal is roughly sawtooth in shape, with the ramps positioned three time instants apart. We compute the co-occurrence matrics, P_1, P_2, and P_3 as follows:

$$15 \times P_1 = M_1 = \begin{bmatrix} 1 & 3 & 1 \\ 2 & 2 & 2 \\ 2 & 1 & 1 \end{bmatrix}, \quad (4.35a)$$

$$14 \times P_2 = M_2 = \begin{bmatrix} 1 & 2 & 2 \\ 2 & 1 & 2 \\ 2 & 2 & 0 \end{bmatrix}, \quad (4.35b)$$

$$13 \times P_3 = M_3 = \begin{bmatrix} 3 & 1 & 1 \\ 2 & 1 & 1 \\ 0 & 3 & 1 \end{bmatrix}, \quad (4.35c)$$

where $N_1 = 15$, $N_2 = 14$, and $N_3 = 13$. Notice that the values are spread out in M_1 and M_2, the main diagonal values are relatively small, and there are few outliers. In contrast, M_3 contains two large values, several small values, and a maximal probability on the diagonal. If the structures within $x(n)$ are square pulses rather than ramps, the matrices M_1, M_2, and M_3 are even more distinct. The exercises explore these ideas further.

To use the co-occurrence matrix as a texture segmentation tool, the time interval δ must be selected according to the size and spacing of the signal regions. Suppose, for instance, that the signal $x(n)$ contains high-magnitude regions approximately Δ time instants apart and that we calculate the matrix P_δ, where δ is smaller than Δ. If $|p - q| < \delta$, then the values $x(p)$ and $x(q)$ are likely to fall into the same region. Hence P_δ entries on the diagonal should be large. Large values do not concentrate on P_δ's diagonal when δ is smaller than the typical region size. If a textured signal contains features of two sizes, δ and ε, then we should expect to find matrices P_δ and P_ε to be largely diagonal. In general, many values of δ are unnecessary for good texture segmentation results.

The idea of co-occurrence measures in texture analysis dates back to the early work of Julesz [41]. The co-occurrence matrix entry on row i and column j, $P_\delta(i, j)$, gives the probability that a sampling of a signal value and its neighbor δ time instants away will have values i and j. It is also possible to sample a point and two others, δ_1 and δ_2 time instants away, to generate third-order co-occurrence statistics. Similarly, fourth- and high-order co-occurrence measures are possible. Julesz conjectured that humans cannot distinguish textures that contain identical first- and second-order co-occurrence statistics; thus, visual texture fields may differ in their third- or higher-order statistics, but this effect is too subtle for the eye–brain system to detect. Julesz's thesis was tremendously attractive to computer vision researchers. It promised to bound the ever-growing number of texture measures by invoking a discriminability criterion based on human pattern detection performance. Possibly, too, researchers could anchor a definition of texture itself in the second-order co-occurrence statistics. But counterexamples were soon found. Julesz and other investigators were able to synthesize texture fields, which humans could distinguish, that had different third- or fourth-order co-occurrence statistics but identical first- and second-order statistics [42]. Under further scrutiny, the synthesized counterexamples themselves were shown to have different local and global co-occurrence statistics. The human visual system excels at perceiving distinct local image characteristics and can group local variations into global patterns over wide areas of the visual field. This causes many visual illusions. Thus, persons viewing the counter-example textures were able to discriminate regions within them, even though the global low-order co-occurrence statistics were identical. Ultimately, it appears that something very close to the Julesz thesis holds true and that humans cannot distinguish textures that locally have the same first- and second-order co-occurrence statistics [43].

However powerful the method of co-occurrence matrices, it evidently turns an analysis problem of a one-dimensional entity—the signal—into the two-dimensional analysis problem of analyzing the co-occurrence matrix. Hence, the key to achieving any analytical power from the method is to keep the co-occurrence matrices and their number small. For developing texture descriptors, researchers have suggested a wide variety of parameters obtained from the co-occurrence matrices [44, 45]. Briefly, let us review some of the most important ones.

Definition (Co-occurrence Matrix Texture Descriptors). Let $x(n)$ be a digital signal; $L \le x(n) \le K$ for some integers L, K and for $a \le n \le b$; let P_δ be the co-occurrence matrix for $x(n)$ with time interval δ; and denote the element at row i and column j of P_δ by $P_\delta(i, j)$. Then, the angular second moment, or energy uniformity, T_a; contrast, T_c; inverse difference moment, T_d; entropy, T_e; and maximum, T_m, descriptors are as follows:

$$T_a(\delta) = \sum_{i=L}^{K} \sum_{j=L}^{K} P_\delta^2(i, j), \tag{4.36}$$

$$T_c(\delta) = \sum_{i=L}^{K} \sum_{j=L}^{K} (i - j)^2 P_\delta(i, j), \tag{4.37}$$

$$T_d(\delta) = \sum_{i=L}^{K} \sum_{j=L}^{K} \frac{P_\delta(i, j)}{1 + (i - j)^2}, \tag{4.38}$$

$$T_e(\delta) = -\sum_{i=L}^{K} \sum_{j=L}^{K} P_\delta(i, j) \log_2 P_\delta(i, j), \tag{4.39}$$

$$T_m(\delta) = \max\{P_\delta(i, j) \mid L \le i, j \le K\}. \tag{4.40}$$

These values are easy to compute and their magnitudes shed light on the nature of the co-occurrence matrix. Note that T_a is smaller for uniform P_δ values and larger for widely varying co-occurrence matrix entries. Low-T_c and high-T_d descriptors indicate heavy groupings of values on P_δ's main diagonal. The entropy descriptor is large when P_δ values are relatively uniform. The maximum value can be thresholded, or compared to $(T_a(\delta))^{1/2}$, to detect extreme co-occurrence matrix entries. It may also be useful to study the behavior of the descriptors when the time interval δ varies [46].

Now let us turn to another method for texture segmentation. The descriptors it generates turn out be useful for extracting different sizes of repetitiveness in a signal.

4.3.2 Spectral Methods

The spectral approach to texture segmentation applies to textured signals that are very periodic in nature. The analysis tool used in spectral approaches to texture is consequently the sinusoids, or, more generally, the complex exponential, e^{x+jy}. Rudimentary statistical measures of texture, such as the roughness average, R_a, do not adequately account for the presence of different periodic trends in the signal. For instance, a broad undulation may be modulated by a more frequent periodic phenomenon—a ripple on top of a wave. In machined surface characterization, the term for the broad undulations is waviness. Metrologists distinguish waviness from roughness, which is a variation on a finer scale. But this signal phenomenon is hardly confined to the science of characterizing manufactured surfaces. The next example explores this type of textured signal in the context of biomedicine.

Example (Waviness). Figure 4.9 shows some biomedical signals taken from an anesthetized dog: a blood pressure trace (in millimeters of mercury) and an

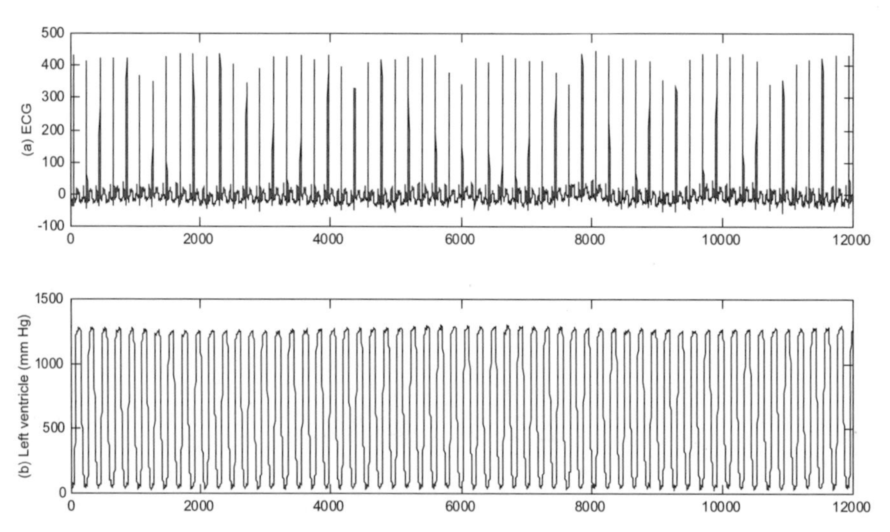

Fig. 4.9. Examples of short-term and long-term periodicities in signals. Panel (a) shows the electrocardiogram from an anesthetized dog. In panel (b) an anesthetized dog's left ventricular blood pressure indicates a short-term variation for the heart contraction and a long-term undulation due to the dog's breathing. This signal has a distinct waviness trend, a long-term undulation, which contrasts with the short-term pressure cycles.

electrocardiogram (dimensionless units). The left ventricle pressure is oscillatory, and—as expected—the histogram has two peaks, or modes, corresponding to the high- and low-pressure intervals within the heart beat. It is easy to identify a threshold that finds the individual pulses, but isolating the gentle waviness that underlies their progression is not as easy. Furthermore, the moment-based statistical methods we considered above do not help to identify the waviness. Such different periodicities in signals occur throughout signal processing and analysis. Problems in biomedicine, digitized speech, industrial control, vibration analysis, remote sensing, shape recognition, and, of course, surface characterization provide many other examples. We need to discover, for such applications, a method distinguish between the short-term periodic features from the long-term undulations in a signal. In surface characterization, the minute variations indicate the roughness of a signal profile, and large-scale, broad repetitions show the presence of waviness.

The sinusoidal functions naturally spring to mind in problems involving short-term and long-term periodicity. Not surprisingly, in order to segment signals containing both a waviness and a roughness character, signal analysts generally employ spectral measures. A comparison of the signal with sinusoids of varying frequency is the natural approach. This is the basis of the spectral method. To compare a signal to a sinusoid, we might proceed as with normalized cross-correlation, by taking the inner product of a portion of the signal with a sinusoidal function. In fact, a family of sinusoids of different frequencies can encompass the range of fine texture, medium texture, and waviness that a signal contains. This is called a spectral strategy, because in its fullest application it can expose an entire range, or spectrum, of periodicities in a signal's values. The foundation of this approach was laid, in fact, by our study of Hilbert spaces in Chapters 2 and 3. Our exploration of spectral approaches to signal texture will be introductory only. In fact, research indicates that statistical methods—and the co-occurrence matrix technique in particular—hold sway over spectral methods [39, 44, 47]. Notwithstanding the superiority of statistical approaches, there is one application for which spectral methods offer an intuitive and powerful approach to texture characterization: the identification of different periodicities—the roughness from the waviness—within a signal.

Let us contemplate the problem of discovering the various periodicities in a discrete signal $x(n)$ over an interval $[a, b] \subset \mathbb{Z}$. Let us assume that $a = 0$ and $b = N - 1$, thus translating our signal to align the beginning of the texture region of interest with the origin. Trigonometric functions, especially the sinusoids, are the natural tool with which to test $x(n)$ on $[0, N - 1]$ for periodic behavior. We know from the introduction to discrete signal frequency in Chapter 1 that the signals $\cos(2\pi nk/N)$, for $k = 0, 1, ..., \lfloor N/2 \rfloor$, range from the lowest ($k = 0$) to the highest (k = largest integer less than $N/2$) possible frequency on $[0, N - 1]$. The inner product $\langle x(n), \cos(2\pi nk/N)\rangle$ of the signals $x(n)$ and $\cos(2\pi nk/N)$ restricted to $[0, N - 1]$ measures of similarity of $x(n)$ to the sinusoid $\cos(2\pi nk/N)$. It is convenient to assume that $x(n) = 0$ outside $[0, N - 1]$. Thus, those values of k over the range $0, 1, ... , \lfloor N/2 \rfloor$, for which $\langle x(n), \cos(2\pi nk/N)\rangle$ has a relatively large magnitude, indicate the presence of a significant periodicity in $x(n)$ of frequency $\omega = 2\pi k/N$ radians per sample. Let's capture this concept by defining the periodicity descriptors $X_c(k)$ for a signal $x(n)$ defined on $[0, N-1]$:

$$X_c(k) = \left\langle x(n), \cos\left(\frac{2\pi nk}{N}\right)\right\rangle = \sum_{n=0}^{N-1} x(n)\cos\left(\frac{2\pi nk}{N}\right). \tag{4.41}$$

Note that for values of $k = \lfloor N/2 \rfloor + 1, \lfloor N/2 \rfloor + 2, ..., N-1$, the descriptors repeat in reverse order; specifically, $X_c(N-k) = X_c(k)$ for $k = 1, 2, ..., \lfloor N/2 \rfloor$. We can also define periodicity descriptors based on the sine function,

$$X_s(k) = \left\langle x(n), \sin\left(\frac{2\pi nk}{N}\right)\right\rangle = \sum_{n=0}^{N-1} x(n)\sin\left(\frac{2\pi nk}{N}\right). \tag{4.42}$$

In (4.42) we see that $X_s(N-k) = -X_s(k)$ for $k = 1, 2, ..., \lfloor N/2 \rfloor$, so that for $k > \lfloor N/2 \rfloor$, the $X_s(k)$ descriptors may be useful for detecting sign-inverted periodic signal components also.

One difficulty with the periodicity descriptors is that we do not in general know whether $x(n)$ has a maximal value at $n = 0$ as does $\cos(2\pi nk/N)$. In other words, the texture field may be shifted so that it does not align with the sinusoids used to compute the inner product periodicity measures. Suppose we shift $x(n)$ values in a circular fashion; thus, we let $y(n) = x((n+p) \bmod N)$. We desire that our texture descriptors respond equally well to $y(n)$ and $x(n)$. That is, the descriptors should support some kind of translation invariance. Equivalently, we can shift the sinusoids by amount p and compute new descriptors, $X_{c,p}(k) = \langle x(n), \cos(2\pi n(k-p)/N)\rangle$. This accounts for possible translation of the texture in the source signal $x(n)$, but now we have quite a large computation task. We must compute $X_{c,p}(k)$ for all possible offsets p and all possible frequencies $2\pi k/N$. Can we relate the shifted descriptor, $X_{c,p}(k)$, to the original $X_c(k)$ descriptor values in such a way that we avoid computation of $X_{c,p}(k)$ for multiple offsets p? We calculate,

$$\begin{aligned} X_{c,p}(k) &= \sum_{n=0}^{N-1} x(n)\cos\left(\frac{2\pi(n-p)k}{N}\right) = \sum_{n=0}^{N-1} x(n)\cos\left(\frac{2\pi nk}{N} - \frac{2\pi pk}{N}\right) \\ &= \sum_{n=0}^{N-1} x(n)\left[\cos\left(\frac{2\pi nk}{N}\right)\cos\left(\frac{2\pi pk}{N}\right) + \sin\left(\frac{2\pi nk}{N}\right)\sin\left(\frac{2\pi pk}{N}\right)\right] \\ &= X_c(k)\cos\left(\frac{2\pi pk}{N}\right) + X_s(k)\sin\left(\frac{2\pi pk}{N}\right), \end{aligned} \tag{4.43}$$

which shows that the descriptor $X_{c,p}(k)$ depends not only on $X_c(k)$ but on the sine-based descriptors as well. In other words, as the repetitive pattern of $x(n)$ shifts, both the cosine-based descriptor and the sine-based descriptor vary.

It turns out that a translation-invariant descriptor arises by combining the cosine- and sine-based descriptors into the exponential $\exp(j2\pi nk/N) = \cos(2\pi nk/N) +$

$j\sin(2\pi nk/N)$. Equation (4.43) already hints of this. We form the inner product of $x(n)$ with $\exp(j2pnk/N)$, as above, to get an exponential-based descriptor, $X(k)$:

$$X(k) = \left\langle x(n), \exp\left(\frac{2\pi jnk}{N}\right)\right\rangle = \sum_{n=0}^{N-1} x(n)\exp\left(\frac{-2\pi jnk}{N}\right)$$
$$= \sum_{n=0}^{N-1} x(n)\cos\left(\frac{2\pi nk}{N}\right) - j\sum_{n=0}^{N-1} x(n)\sin\left(\frac{2\pi nk}{N}\right) = X_c(k) - jX_s(k). \tag{4.44}$$

Now let's consider computing $X(k)$ values for a translated texture $y(n) = x((n+p) \bmod N)$, or, equivalently, by computing the inner product, $X_p(k) = \langle x(n), \exp(2\pi jn(k-p)/N)\rangle$:

$$X_p(k) = \left\langle x(n), \exp\left(\frac{2\pi j(n-p)k}{N}\right)\right\rangle = \sum_{n=0}^{N-1} x(n)\exp\left(\frac{-2\pi j(n-p)k}{N}\right)$$
$$= \exp\left(\frac{2\pi jpk}{N}\right)\sum_{n=0}^{N-1} x(n)\exp\left(\frac{-2\pi jnk}{N}\right) = \exp\left(\frac{2\pi jpk}{N}\right)X(k). \tag{4.45}$$

Now we have $|X_p(k)| = |\exp(2\pi jpk/N)X(k)| = |X(k)|$. In other words, we have shown the following:

Proposition (Translation Invariance of $|X(k)|$). The complex norm of the periodicity descriptor, $|X(k)|$, is invariant with respect to the modulo-N translation of the textured source signal, $x(n)$, defined by (4.44) on $[0, N-1]$.

This, then, is the tool we need to isolate periodic components within textures according to their different frequencies.

The values $X(k)$ in (4.44), for $k = 0, 1, ..., N-1$, represent the discrete Fourier transform of the signal $x(n)$ on $[0, N-1]$. We know the signals $X(k)$ already from Chapter 2. There we considered them, after a suitable normalization, as an orthonormal set on $[0, N-1]$, $\{e_k(n) \mid 0 \le k \le N-1\}$:

$$e_k(n) = \frac{\exp(j2\pi kn/N)}{\sqrt{N}}[u(n) - u(n-N)], \tag{4.46}$$

where $u(n)$ is the unit step signal. It is common to call the values $X(k) = \langle x(n), \exp(2\pi jnk/N)\rangle$ the Fourier coefficients of $x(n)$, even though, precisely speaking, we need to normalize them according to the inner product relation on an orthonormal set (4.46). Readers must pay close attention to the definitions that textbooks provide. When discussing Hilbert space decompositions, textbooks usually normalize the coefficients according to (4.46), whereas ordinarily they omit the $N^{-1/2}$ factor in the DFT's definition.

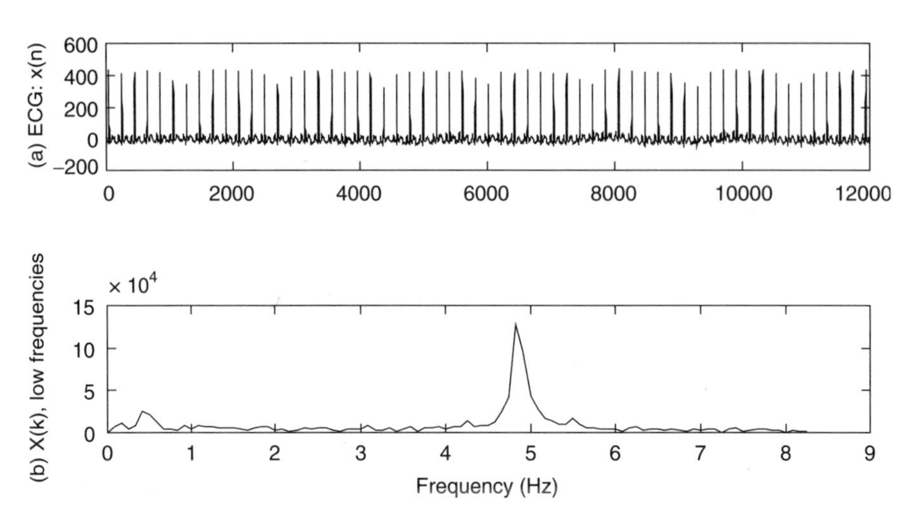

Fig. 4.10. Biomedical signal waviness extraction. The spectral method helps to identify the two sinusoidal components of the ECG signal (a). The main periodic trend is just below 5 Hz, and the waviness is the small peak below 0.5 Hz (b).

The theory and applications of the DFT will comprise much of Chapters 6 and 7. Let us content ourselves here with two applications of spectral texture analysis: first, to the biomedical signals of Fig. 4.9 and, second, to digitized speech segmentation and recognition. Figure 4.10 shows the results of applying the spectral texture descriptors $|X(k)|$ to biomedical signals.

Example (Biomedical Signal Frequency Analysis). Consider the ECG signal for the anesthesized dog.

Application (Voiced versus Unvoiced Speech Segmentation). This example continues the discussion of speech phoneme segmentation that we began in Section 4.2.2 with voiced and unvoiced consonants. Linguists typically classify speech events according to whether the vocal cords vibrate during the pronunciation of a speech sound, or phoneme (Table 4.3). If the vocal cords do vibrate, then there is said to be a voiced speech event. If there is no vocal cord vibration, then the phoneme is unvoiced. It is also possible that a speech signal contains no speech sound; thus, it is simply background, or noise. One approach to segmenting speech classifies its portions as voiced (V), unvoiced (U), or noise (N). For example, a digital recording of the English phrase "linear fit" begins, divides the two words, and ends with noise regions. The most apparent phonemes, then, are /l I n i ɜ f I t/, of which /f/ and /t/ are unvoiced. A preliminary segmentation, Σ_0, by voiced/unvoiced/noise classification is (N, V, N, U, V, U, N), respecting the stipulation that no adjacent regions have the same type. Actually, there are a number of unvoiced events that only become apparent when the speech signal is digitized and spread out over time. One may find, for example, momentary unvoiced aspirations and even periods

TABLE 4.3. More Complete Table of Phonemes and Examples from American English [12].[a]

Phoneme	Example	Class	Phoneme	Example	Class
/i/	even	Front vowel	/I/	signal	Front vowel
/e/	basis	Front vowel	/ε/	met	Front vowel
/ae/	at	Front vowel	/a/	father	Mid vowel
/Λ/	but	Mid vowel	/⊃/	all	Mid vowel
/schwa/	signal	Mid vowel	/u/	boot	Back vowel
/o/	boat	Back vowel	/U/	foot	Back vowel
/ɨ/	roses	Back vowel	/ɜ/	Hilbert	Mid vowel
/a^{W}/	down	Dipthong	/a^{y}/	cry	Dipthong
/⊃^{y}/	boy	Dipthong	/y/	yet	Semivowel glide
/w/	wit	Semivowel liquid	/r/	rent	Semivowel glide
/l/	linear	Semivowel liquid	/m/	segment	Nasal consonant
/n/	nose	Nasal consonant	/η/	Nguyen	Nasal consonant
/p/	partition	Unvoiced stop	/t/	fit	Unvoiced stop
/k/	kitten	Unvoiced stop	/b/	bet	Voiced stop
/d/	dog	Voiced stop	/g/	gain	Voiced stop
/h/	help	Aspiration	/f/	fit	Unvoiced fricati
/θ/	thanks	Unvoiced fricative	/s/	sample	Unvoiced fricati
/sh/	shape	Unvoiced fricative	/v/	vector	Voiced fricative
/∂/	that	Voiced fricative	/z/	zoo	Voiced fricative
/zh/	closure	Voiced fricative	/ch/	channel	Affricate
/j/	Jim	Affricate	/?/	no sound	Glottal stop

[a]Vowels are voiced, and they break down further into front, mid, and back classifications, according to t location of the tongue's hump in the mouth. Vowels typically consist of two oscillatory components, call formants. Other essentially vowel-like sounds are the dipthongs and the semivowels. The glottal stop is special symbol that indicates a momentary suspension of motion by the vocal cords (glottis). For examp without the /?/ symbol, the phoneme sequences for "I scream" and "ice cream" are identical. As big as it this table is far from panoramic; it offers but a glimpse of the natural language segmentation problem.

of noise, surrounding the voiced segments. A refined segmentation, Σ_1, therefore supplies: (N, U, V, U, V, U, V, U, V, U, V, U, N, U, V, U, N). In practical speech recognition applications, surprisingly, this segmentation is too crude! Subtle periods of background noise, with no voiced or unvoiced sound present, intrude into spoken words. A modern, sophisticated segmentation algorithm finds that several N regions split the unvoiced regions U in the refined segmentation above. This means that several of the unvoiced intervals have much shorter time extents than Σ_1 would indicate. The benefit is that a higher-level interpretation algorithm may be better able to recognize the brief U boundaries of the V segments as trailing and leading aspirations instead of, for example, unvoiced fricatives. Figure 4.5 illustrates such a speech segmentation example. We shall continue to expose the intricacies of natural language interpretation is this and the remaining chapters. Chapter 9, in particular, introduces several frequency-domain signal analysis tools for speech interpretation.

Vowels contain two frequency components called formants. It is hard to adapt the level-based approach of thresholding to these oscillatory parts of signals. Also, the statistical measures of texture do not account well for the time intervals between transitions, and yet this is the most important aspect of signal frequency. What we need to do is to discover some way to extract the degree of similarity between some raw, noisy signal and a pure sinusoid.

4.3.3 Structural Approaches

When the precise description of local texture elements is important, then the structual approach to texture segmentation comes to the fore. The statistical and spectral methods provide numerical estimates for the amount and distribution of variability within signal regions. Simple statistical measures lose the repeating parts of the signal in the sums and integrations of their mathematical implementation. The co-occurrence matrix method retains this information when co-occurrence matrices are computed for several values of the time offset, δ. The structural approach, on the other hand, relies on texture descriptors that provide a model of a local pattern that is replicated, in one or another degree, at one or another locations across the domain of the signal. Thus, the structural approach has a pattern recognition flavor. We need to develop some basic tools for structural pattern recognition it order to cover this approach adequately, and we will do this later in Section 4.7.

4.4 FILTERING AND ENHANCEMENT

This section considers some supplementary operations that can greatly improve the performance of low-level segmentation processes before the thresholding operation: convolutional smoothing, median filtering, morphological filtering, and histogram enhancement.

4.4.1 Convolutional Smoothing

Let us here expand upon the idea of filtering a signal before it passes to a threshold operation. The purpose of preliminary filtering is to prevent labeling mistakes before they happen. Three types of filters are in common use for this purpose: convolutional filters of the type we covered in Chapters 2 and 3, for discrete and analog signals, respectively; median filters; and morphological filters.

We noted by example in the previous section that oscillatory areas of signals may contain mixtures of low- and high-magnitude signal values. Such a region may be meaningful signal in its totality, and for many applications it is not correct to separately segment the low-magnitude regions and label them as background. A linear smoothing filter proved useful for this purpose. Chapter 2 introduced the *moving average* system given by

$$y(n) = \frac{1}{2N+1} \sum_{k=-N}^{N} x(k). \qquad (4.47)$$

It is also called a box filter, from the shape of its impluse response. This linear, translation-invariant (LTI) system averages the input signal values within a window of length $2N+1$ around each $x(n)$. Its effect on a oscillatory signal region, $[a, b]$, is to blur the high and low signal values together. Thus, for some threshold $T > 0$, the system's output, $y(n) = (Hx)(n)$, has magnitude $|y(n)| \geq T$ within $[a, b]$. It is also possible to smooth using an infinite impulse response filter. For example, $y(n) = ay(n-1) + x(n)$, with impulse response $h(0) = 1$, and $0 < a < 1$ is a possibility (Section 2.4.2). Smoothing with LTI systems blends oscillations together into more blob-like, easily thresholded regions. It also reduces background noise when the signal of interest is of relatively low magnitude.

The moving average system of (4.47) is not the only filter available for smoothing signals. One problem that readers may already note (see Chapter 2) is that this filter is not causal. So it cannot be applied to a data stream as values arrive in real time. It also has sharp edges where its support ends, which, for some types of input signals, causes its output to change abruptly. We will find in Chapter 7 some problems with how this filter treats frequencies inside the input signal; in fact, this motivates us to search for alternatives. A causal box filter averages only the current and some previous samples: $y(n) = (1/3)[x(n) + x(n-1) + x(n-2)]$, for example.

Noise in signals varies widely. It is usually a detriment to signal quality; but in some interesting cases, it is actually helpful. In digital telephony, for example, comfort noise is generated and automatically inserted into the line during episodes when both talkers are quiet. The reason for this is that silent periods can be used to carry other voices. The telephone company's central office multiplexing equipment routinely disconnects the channel, fills it with other voice traffic, and swiftly provides an alternative link when someone speaks. But it is disconcerting to almost everyone to hear a silent line. Realizing this, telecommunication engineers design the multiplexers to inject a small amount of white noise onto the two ends of the interrupted link. Since the synthetic noise level is matched to the background noise present while the conversants speak, the change is imperceptible. Another example of the value of noise synthesis is in active noise cancellation.

Now let us return to our discussion of noise removal. We consider a filter that has been widely used for noise mitigation—the discrete version of the Gaussian. This filter has the impulse response

$$h(n) = \frac{1}{\sqrt{2\pi}\sigma} e^{-\frac{1}{2}\left(\frac{\tau n}{\sigma}\right)^2}, \tag{4.48}$$

where $\tau > 0$ is some sampling interval. This filter emphasizes the signal values near the value, $x(n)$. Thus, it differs from the moving average filter, which considers all quantities equally valuable in a neighborhood $[x(n-N), \ldots, x(n+N)]$ around $x(n)$. We shall investigate this type of filter carefully in the last section of this chapter. For pattern recognition applications based on signal shape over multiple scales and convolutional filtering, there are strong theoretical reasons for using Gaussian filters above all others.

4.4.2 Optimal Filtering

It is also possible to derive filters that are, in some sense at least, optimal for removing the kind of noise that affects input signals to an application. Here, we will outline two approaches for designing such filters: One method uses least-squares principles [48–50], and another involves constrained minimization—a Lagrange[4] multipliers approach [51]. Time-domain methods for noise removal, such as these, generally depend upon statistical information about the signal values.

We can pose the search for an optimal noise removal filter as a least-squares search. Suppose that the noise removal filter H is linear and translation-invariant. Then the Convolution Theorem for LTI Systems (Chapter 2) instructs us that $y = Hx$ is given by convolution of $h(n) = (H\delta)(n)$. We assume a real-valued, noisy input signal $x(n)$. Let us further seek a finite impulse response (FIR) filter, of width $M = 2p + 1$, supported on $[m - p, m + p]$, and ask what filter coefficients $h(n)$ give the best estimate of x at time instant p: $y(p) = (Hx)(p)$. We view the FIR filtering operation as a matrix operation, $Ah = b$, where the rows of matrix A are noisy samples of the input signal $x(n)$ from $n = m-p$ to $n = m+p$; h is a column vector with entries $h(-p), h(-p+1), ..., h(p)$; and b is a row vector with FIR filtered values of the input. Suppose we filter N sets of noisy input windows of signal $x(n)$ around $n = p$, $(a_{1,1}, a_{1,2}, ..., a_{1,M})$, $(a_{2,1}, a_{2,2}, ..., a_{2,M})$, ..., $(a_{N,1}, a_{N,2}, ..., a_{N,M})$, with the filter $h(n)$. Suppose further that we know the true value (i.e., before corrupted by noise) of the input signal for each sample set at $n = p$; call them $b_1, b_2, ..., b_N$. Then we are in fact asking for the vector h for which the matrix product Ah is the best approximation to the ground truth vector b, representing uncorrupted input signal values $x(n)$ at $n = p$. Our quest for a best noise removal filter reduces to a problem of finding the column vector h, given a matrix A, which upon matrix multiplication Ah is closest to known signal values b. The theory of this problem is easiest if we consider the closest approximation to b in vector norm: we need to minimze $\|Ah - b\|^2$. In other words, the best filter problem reduces to a classic least squares minimization problem. The following theorem—mainly from linear algebra and numerical analysis [48, 50, 52], but drawn with a signal processing twist—summarizes this unconstrained search for the best FIR noise removal filter.

Theorem (Least-Squares Optimal Noise Removal Filters). If the vectors x, y, and b are real and the matrices A and M are real, then with the above premises and notation, the following statements are true:

1. If the columns of matrix A are linearly independent, then A^TA is symmetric and positive definite: $(A^TA)^T = A^TA$, and $x^T(A^TA)x \geq 0$ for all vectors $x \neq 0$.
2. If the square matrix M is positive definite, then the quadratic equation in a real vector x,

[4]Joseph Louis Lagrange (1736–1813), professor of mathematics at Turin, Berlin, and Paris, discovered the powerful technique of introducing new variables, subject to additional constraints, into an equation to be solved.

$$Q(x) = (1/2)x^T Mx - x^T c, \tag{4.49}$$

has a minimum of

$$Q(M^{-1}c) = (-1/2)c^T M^{-1} c, \tag{4.50}$$

when $Mx = c$.

3. The vector $h = (A^T A)^{-1} A^T b$ minimizes $\|Ah - b\|^2$.
4. The FIR filter $h(n) = [h(-p), h(-p+1), \ldots, \underline{h(0)}, \ldots, h(p)]$, given by the vector h of (3) is optimal for noise removal from signal $x(n)$, where matrix A in (3) is a set of noisy windows of $x(n)$, and b is the set of noise-free values of $x(n)$ corresponding to A's noisy sample sets.

Proof: To see statement 1, consider some vector x, and note that $x^T(A^T A)x = (Ax)^T(Ax) = \|Ax\|^2$. Since the columns of A are linearly independent, Ax, which is a linear combination of columns of A (the column space of A), cannot be 0 unless $x = 0$. Hence, $A^T A$ is positive definite. Symmetry follows from the basic fact that $(AB)^T = B^T A^T$. For statement 2, observe that if $Mx = c$, then

$$Q(y) - Q(x) = (1/2)(y - x)^T M(y - x) \geq 0 \tag{4.51}$$

by the fact that M is positive definite; so $Q(y) \geq Q(x)$, and the minimum (4.50) follows. Turning to statement 3, now note that $\|Ah - b\|^2 = (Ah - b)^T(Ah - b) = x^T A^T Ah - h^T A^T b - b^T Ah + b^T b$. So minimizing $\|Ah - b\|^2$ is equivalent to minimizing $h^T A^T Ah - h^T A^T b - b^T Ah$. Also, thinking of the 1×1 matrices here as real numbers, we find $h^T A^T b = b^T Ah$. So the minimization of $\|Ah - b\|^2$ reduces to a minimization of $Q(h) = (1/2)h^T(A^T A)h - h^T A^T b$. But this is precisely (4.49) with $M = A^T A$ and $c = A^T b$. Finally, statement 4 of the theorem simply expresses the equivalence of convolution using an FIR filter, $h(n)$, with vector dot products. Filter $h(n)$ is optimal in the sense that any other noise removal filter $g(n)$ will produce errors with respect to know values for $x(n)$ at $n = p$ that are larger in l^2 norm than those which $h(n)$ produces. ■

Example. Suppose that a signal $x(n)$ is corrupted with Gaussian noise, and we seek an optimal noise removal filter H with $(H\delta)(n) = h(n) = [h(-1), \underline{h(0)}, h(1)]$. We record a series of windows of width 3 of the input signal $x(n)$ around $n = 0$: $[a_{1,1}, a_{1,2}, a_{1,3}]$, $[a_{2,1}, a_{2,2}, a_{2,3}]$, ..., $[a_{m,1}, a_{m,2}, a_{m,3}]$. In this case, the normal equations become $Ah = b$, where, for example,

$$A^T = \begin{bmatrix} 1.12 & 0.94 & 0.77 & 0.96 & 1.83 & 0.52 & 0.91 \\ 0.86 & 1.31 & 0.94 & 1.19 & 2.15 & 0.70 & 1.03 \\ 1.19 & 2.72 & 0.32 & 0.65 & -0.04 & 0.89 & 1.21 \end{bmatrix}. \tag{4.52}$$

Upon solving the normal equations, we find $h(n) = [h(-1), \underline{h(0)}, h(1)] = [1.08, .92, 1.17]$, very close to the box or moving average filter to which our original intuition guided us. Now let us consider the possibilty that the noise on $x(n)$ is correlated; in particular, if we assume a correlation between the noise magnitude at $x(0)$ and that at $x(-1)$ and $x(1)$, then we arrive at normal equations with the following coefficient matrix

$$A^T = \begin{bmatrix} 1.12 & 0.98 & 0.85 & 0.90 & 0.73 & -0.11 & 1.13 \\ 0.86 & 1.24 & 1.32 & 1.23 & 1.07 & 0.89 & 1.20 \\ 1.19 & 1.69 & 0.79 & 0.97 & 1.01 & 0.35 & 1.03 \end{bmatrix}. \tag{4.53}$$

The solution to the normal equations provides an estimate $h(n) = [h(-1), \underline{h(0)}, h(1)] = [0.85, 1.12, 0.93]$, emphasizing the central filter value. Thus it is the correlation between successive noise values within the input signal $x(n)$ that governs the departure of the optimal filter from the moving average filter.

It is possible to adopt a different least-squares approach for optimal noise removal. For example, a popular approach [49] is to fit quadratic (or higher-order) polynomials to the signal values to obtain families of filters of varying support. There result filters that smooth signals, but do not resemble the simple box or Gaussian filters at all. Table 4.4 shows representative finitely supported smoothing filters that derive from this procedure.

Now let's consider a constrained approach to optimal noise removal filter design. Again, we seek an LTI system H with an impulse response $h = H\delta$, of width $M = 2p + 1$, supported on $[m-p, m+p]$. Again we ask what $h(n)$ must then be to optimally remove noise from known signal $x(n)$. Since we are concerned once more with an FIR filter and a noisy source signal $x(n)$, we may view both as row vectors. We view the system output $y = Hx$ as an estimator of the random vector x which has expected value vector $\mu b = E[x]$, where b is the M-dimensional row vector of all ones. In vector terms we seek a row vector h such that $\hat{\mu} = \langle h, x \rangle$ is an estimator of $y(m) = (Hx)(m)$. Here, $\langle\ ,\ \rangle$ is the vector inner (dot) product operation. We desire the random variable $\hat{\mu}$ to be unbiased, so that its mean on $[m-p, m+p]$ is the same as $x(n)$'s value on this interval. In other words, $E[\hat{\mu}] = \mu$. This condition implies the following condition on impulse response h:

TABLE 4.4. Smoothing Filters Derived from Polynomial Fits to Signal Values[a]

Points	Quadratic Polynomial Fit:	Quartic Polynomial Fit:
5	[..., $\underline{.4857}$, .3429, −.0857]	[..., $\underline{1.500}$, −.5000, .2500]
7	[..., $\underline{.3333}$, .2857, .1429, −.0952]	[..., $\underline{.5671}$, .3247, −.1299, .0216]
9	[..., $\underline{.2554}$, .2338, .1688, .0606, −.0909]	[..., $\underline{.4172}$, .3147, .0699, −.1282, .0350]
11	[..., $\underline{.2075}$, .1958, .1608, .1026, .0210, −.0839]	[..., $\underline{.3333}$, .2797, .1399, −.0233, −.1049, .0420]

[a]In the table, all of the impulse responses are even, and only the values for non-negative time instants are shown. Such noise removal filters are popular for smoothing data acquired from laboratory instruments [49].

Proposition (Unbiased Estimator). If the estimator $\hat{\mu}$ is unbiased, then

$$\sum_{n=m-p}^{m+p} h(n) = 1. \tag{4.54}$$

Proof: Since $E[\hat{\mu}] = \mu$, we have $E[\hat{\mu}] = E[\langle h, x\rangle] = \langle h, E[x]\rangle = \mu\langle h, b\rangle$, and therefore $\langle h, b\rangle = 1$. In terms of the impulse response of system H, this is (4.54); the proof is complete. ■

Let us continue this notational framework for the next two results. If we require the further condition that the estimator $\hat{\mu}$ have minimal variance, the following holds:

Proposition (Variance of Unbiased Estimator). If the estimator $\hat{\mu}$ is unbiased, then $\mathrm{Var}[\hat{\mu}] = \langle h, h\Sigma\rangle$, where $\Sigma = \mathrm{E}[(x - \mu b), (x - \mu b)]$ is the covariance matrix of the random vector x.

Proof: We calculate

$$\begin{aligned}\mathrm{Var}[\hat{\mu}] &= E[(\hat{\mu} - E[\hat{\mu}])^2] = E[(\langle h, x\rangle - \mu)^2] = E[(\langle h, x\rangle - \langle h, \mu b\rangle)^2] \\ &= E[\langle h, h\langle(x - \mu b), (x - \mu b)\rangle\rangle] = \langle h, hE[(x - \mu b), (x - \mu b)]\rangle = \langle h, h\Sigma\rangle. \end{aligned} \tag{4.55}$$

■

Now we wish to minimize $\mathrm{Var}[\hat{\mu}]$ subject to the constraint of the first Unbiased Estimator Proposition. The next theorem solves this is typical Lagrange multipliers problem.

Theorem (Unbiased Minimal Variance Estimator). If the estimator $\hat{\mu}$ is unbiased and has minimal variance, then

$$\mathbf{h} = \left(\frac{1}{\left\langle \mathbf{b}\Sigma^{-1}, \mathbf{b}\right\rangle}\right)\mathbf{b}\Sigma^{-1}. \tag{4.56}$$

Proof: To apply Lagrange multipliers to the optimization problem, let us introduce a function, $L(h)$, with an additional parameter, λ:

$$L(h) = \mathrm{Var}[\hat{\mu}] + \lambda(\langle h, b\rangle - 1) = \langle h, h\Sigma\rangle + \lambda(\langle h, b\rangle - 1). \tag{4.57}$$

Recall that the impulse response of the optimal filter we seek, $h(n)$, was supported on $[m-p, m+p]$. So the vector h is a $1 \times M = 1 \times (2p+1)$ row vector: $h = (h_1, h_2, \ldots, h_M)$. We next calculate the partial derivatives:

$$\left(\frac{\partial L}{\partial h_1}, \frac{\partial L}{\partial h_2} \cdots \frac{\partial L}{\partial h_M} \right) = 2\mathbf{h}\Sigma + \lambda\mathbf{b}. \tag{4.58}$$

Since the partial derivatives (4.58) are zero where $L(h)$ has a minimum, we solve $0 = 2h\Sigma + \lambda b$ for the vector h:

$$\mathbf{h} = -\left(\frac{\lambda}{2}\right)\mathbf{b}\Sigma^{-1}. \tag{4.59}$$

Taking the inner product of both sides of (4.59) with the all-ones vector b and applying the constraint $\langle h, b \rangle = 1$ gives

$$1 = \langle \mathbf{h}, \mathbf{b} \rangle = -\left(\frac{\lambda}{2}\right)\langle \mathbf{b}\Sigma^{-1}, \mathbf{b}. \rangle \tag{4.60}$$

and

$$\lambda = \frac{-2}{\langle \mathbf{b}\Sigma^{-1}, \mathbf{b} \rangle}. \tag{4.61}$$

Finally, substituting (4.61) into (4.59) gets us to (4.56), and the proof is complete. ■

The determination of the optimal filter $h(n)$ therefore depends entirely on the covariance matrix Σ.

Example. Suppose we seek a filter supported on $[-1, 1]$, and the random noise embedded in input signal $x(n)$ is uncorrelated. If the variances at $x(-1)$ and $x(1)$ are equal, say they are $\alpha\sigma^2$, where σ^2 is the variance of $x(0)$, then we have the following:

$$\Sigma = \sigma^2 \begin{bmatrix} \alpha & 0 & 0 \\ 0 & 1 & 0 \\ 0 & 0 & \alpha \end{bmatrix}, \tag{4.62}$$

$$h = \frac{1}{1+2\alpha^{-1}} \begin{bmatrix} \alpha^{-1} \\ 1 \\ \alpha^{-1} \end{bmatrix}. \tag{4.63}$$

Example. Suppose that a filter with support on [−1, 1] is again necessary, but that the random noise is correlated. Suppose that values one time instant apart have correlation ρ and values two time instants apart have correlation ρ^2. Then we have the following

$$\Sigma = \sigma^2 \begin{bmatrix} 1 & \rho & \rho^2 \\ \rho & 1 & \rho \\ \rho^2 & \rho & 1 \end{bmatrix}, \tag{4.64}$$

$$h = \frac{1}{(1-\rho)(3-\rho)} \begin{bmatrix} (1-\rho) \\ (1-\rho)^2 \\ (1-\rho) \end{bmatrix}. \tag{4.65}$$

The above development of optimal noise removal filters using Lagrange multipliers is usually presented for designing neighborhood smoothing operators in image processing [51]. We have adapted it here for one-dimensional filters. One consequence of this is that the covariance matrices (which are $M \times M$ rather than $M^2 \times M^2$, where M is the filter size) are much easier to invert.

There are some situations where convolutional filtering does not do a good job at enhancement before thresholding and segmentation. The next section considers such cases and develops some simple nonlinear smoothing filters which have proven to be very powerful and very popular.

4.4.3 Nonlinear Filters

Filtering with LTI systems may well result in an incorrect segmentation in those situations where signals contain transient phenomena, such as sharp, high-magnitude spikes. From simple faith in the principal theorems of probability, such as the Central Limit Theorem, we tend to dismiss the possibility of very high magnitude spikes in our data. After all, after repeated trials the sum of any distributions will tend toward a Gaussian distribution. Some 68.26% of the signal values should lie within one standard deviation, σ, of the signal mean, μ. Over 95% of the values should be less than 2σ from μ. And huge values should be extremely rare, using this logic. But this reasoning is flawed. Such transients may occur to severe, short-duration noise in the process that generates the source signal or result from imperfections in the signal acquisition apparatus.

The reason LTI filters become problematic is that for a very large impulse, the output of the smoothing signal $y(n)$ resembles the filter's impulse response. The linearity is the problem. Another problem with linear smoothing filters is their tendency to blur the sharp edges of a signal. Good noise removal requires a filter with a wide support. But this erodes sharp edges, too. Thus, when the thresholding operation is applied, the regions labeled as meaningful signal may be smaller in extent than they would be without the preliminary filtering. Perhaps separate regions blend together, rendering the final form of the signal unrecognizable by the classification step in the application. The problem here is the wide support of the filter impulse response.

Two types of nonlinear filters are widely used for impulse noise removal: median and morphological filters. Properly chosen, these filter a noisy signal without leaving a large-magnitude response that interferes with the subsequent thresholding operation. Like convolutional filters, median and morphological noise removal are handy before thresholding. By dint of their nonlinearity, however, they can preserve the binary nature of a thresholded signal; they can also be used to clean out the speckle noise that often persists in a signal after thresholding.

4.4.3.1 Median Filters.

A discrete median filter, H, accepts a signal $x(n)$ as input and produces $y(n)$, which is the median value of $x(n)$ in a neighborhood around $x(n)$.

Definition (Median Filter). If N is a positive integer, then the median filter $y = Hx$ for the neighborhood of width $2N+1$ is defined by $y(n) = \text{median}\{x(m) \mid n-M \leq m \leq n+N\}$.

To compute a median filter, the values $x(n-N)$, $x(n-N+1)$, ... , $x(n+N)$ must be sorted and the middle element selected. It is possible, and maybe useful, to have asymmetric median filters as well as to take $y(n)$ to be the median of a set that does not contain $x(n)$; nevertheless, the centered filter is the most common.

As an enhancement tool, the median filter is usually an early signal processing stage in an application, and most applications use it to improve segmentation operations that depend on some kind of thresholding. We have noted that convolutional filters tend to smear signal edges and fail to fully eradicate sharp transient noise. The main points to keep in mind about the median filter as an enhancement tool are as follows:

- It removes impulse noise.
- It smoothes the signal.
- The median can be taken over a wider interval so that the filter removes transients longer than a time instant in duration.
- Median filters preserve sharp edges.

In addition, since median filters involve no floating point operations, they are well-suited for implementation on fixed-point digital signal processors (which are much faster than their floating-point capable cousins) and in real-time applications [53]. An alternative to using wide median filters is to apply a small support median filter many times in succession. Eventually this reduces the original signal to the median root, which is no longer affected by further passes through the same filter [54].

4.4.3.2 Morphological Filters.

From the very beginning of this book, all of the signal operations have been algebraic—additive, multiplicative, or based upon some extension thereof. To conceptualize analog signals, we freely adopted real- and complex-valued functions of a real variable. To conceptualize discrete signals, we made them into vectors-without-end, as it were. Analog signal operations and analog system theory derived straight from the theory of functions of a real variable. And vector operations inspired discrete signal manipulations, too. Thus, analog and discrete systems added signals, multiplied signals by a scalar value, convolved them, and so on. This is not at all surprising. Signals are time-ordered numerical values, and nothing is more natural for numbers than arithmetic operations. But there is another, distinctly different, and altogether attractive viewpoint for all of signal processing and analysis: mathematical morphology.

Mathematical morphology, or simply morphology, relies on set-theoretic operations rather than algebraic operations on signals. Morphology first casts the concept of a signal into a set theoretic framework. Then, instead of addition and multiplication of signal values, it extends and contracts signals by union and intersection. The concept at the fore in morphology is shape, not magnitude.

It ought to be easier to understand the morphological perspective by thinking for a moment in terms of binary images in the plane, instead of signals. A binary image is a map I: $\mathbb{R} \times \mathbb{R} \to \{0, 1\}$. The binary image, I, and the subset of the plane, $A = I^{-1}(\{1\}) = \{(s, t) \in \mathbb{R} \times \mathbb{R} \mid I(s, t) = 1\}$ mutually define one another. (The image is precisely the characteristic function of the set A.) So we can think of a binary image as either a function or a set. Now suppose we have a small circle $C = \{(s, t) \mid s^2 + t^2 \le r\}$, with $r > 0$. Suppose we place the center of C on top of a boundary point of $(a, b) \in A$; we find all of the points, (r, s) covered by C, $B = \{(r, s) \mid (r - a)^2 + (s - b)^2 \le r^2\}$; and we set $A' = A \cup B$. We continue this placement, covering, and union operation for all of the boundary points of A. The result is the set A with a shape change effected for it by the morphological shaping element, B. This scheme works just as well for binary-valued signals: An analog binary signal is the characteristic function of a subset of the real line, and a discrete binary signal is the characteristic function of a subset of the integers.

Let's begin our brief tour of mathematical morphology with the definitions and basic properties for binary morphological operations. This will introduce the terminology and concepts as well as provide a theoretical foundation for our later extension to digital signals. Just as we can identify a binary image with a binary signal, we can also characterize a binary discrete signal with a subset of the integers or a binary analog signal with a subset of the real line. The signal is the characteristic function of the set. Although morphologists work their craft on either $\mathbb{Z}^n$ or $\mathbb{R}^n$ [55–57], we shall consider digital signals and restrict our morphological investigation to discrete n-space, $\mathbb{Z}^2$. Note that the development could proceed as well for n-tuples from an abstract normed linear space.

Definition (Translate of a Set). Let $A \subseteq \mathbb{Z}^2$ and $k \in \mathbb{Z}^2$. Then the translate of A by k is $A_k = A + k = \{a + k \mid a \in A\}$.

Definition (Dilation). Let $A, B \subseteq \mathbb{Z}^2$. Then the dilation of A by B is

$$A \oplus B = \bigcup_{b \in B} (A + \mathbf{b}). \tag{4.66}$$

The set B in (4.66) is called a structuring element; in this particular instance, it is a dilation kernel.

From this meager beginning, we see the development of diverse, interesting, useful properties.

Proposition (Dilation Properties). Let $A, B, C \subseteq \mathbb{Z}^2$ and $k \in \mathbb{Z}$. Then, $A \oplus B = \{a + b \mid a \in A \text{ and } b \in B\}$. Moreover, dilation is commutative, $A \oplus B = B \oplus A$; associative, $(A \oplus B) \oplus C = A \oplus (B \oplus C)$; translation invariant, $(A + k) \oplus B = (A \oplus B) + k$; increasing: if $A \subseteq B$, then $A \oplus D \subseteq B \oplus D$; distributive over unions: $A \oplus (B \cup C) = (A \oplus B) \cup (A \oplus C)$.

Proof: Let's show translation invariance. We need to show that the two sets, $(A + k) \oplus B$ and $(A \oplus B) + k$ are equal; thus, we need to show they contain the same elements, or, equivalently, that they are mutual subsets. If $x \in (A + k) \oplus B$, then there is $a \in A$ and $b \in B$ such that $x = (a + k) + b = (a + b) + k$. Since $a + b \in A \oplus B$, we know $(a + b) + k = x \in (A \oplus B) + k$. Thus, $(A + k) \oplus B \subseteq (A \oplus B) + k$. Now let's consider an element $x \in (A \oplus B) + k$. It must be of the form $y + k$, for some $y \in A \oplus B$, by the definition of the translate of $(A \oplus B)$. Again, there must be $a \in A$ and $b \in B$ such that $y = a + b$. This means $x = (a + b) + k = (a + k) + b \in (A + k) \oplus B$. Thus, $(A \oplus B) + k \subseteq (A + k) \oplus B$. Since these two are mutual subsets, the proof of translation invariance is complete. The other properties are similar, and we leave them as exercises. ■

Notice that the distributivity property implies that dilations by large structuring elements can be broken down into a union of dilations by smaller kernels.

How can dilation serve as a noise removal operation within a digital signal analysis application? To answer this question we adopt a set-theoretic description for discrete signals, $x(n)$, based on the graph of the signal, $\text{Graph}(x) = \{(n, x(n)) \mid n \in \mathbb{Z}\}$. Graph($x$) is a subset of the discrete plane, so it describes x(n) in set theory terms, but we need other tools to adequately define signal dilation. Our exposition draws the concepts of umbra and top surface from the tutorial of Haralick et al. [55].

Definition (Umbra). Let $x(n)$ be a digital signal, $x\colon \mathbb{Z} \to [0, N]$, for some natural number $N \geq 0$. Then the umbra of x is $\text{Umbra}(x) = \{(n, m) \in \mathbb{Z} \times [0, N] \mid m \leq x(n)\}$.

Definition (Top Surface). If $A \subseteq \mathbb{Z}^2$ and $B = \{b \in \mathbb{Z} \mid (b, k) \in A, \text{ for some } k \in \mathbb{Z}\}$, then the top surface for A is the function, $T[A]$, that has domain B and is defined $T[A](b) = \max\{k \in \mathbb{Z} \mid (b, k) \in A, \text{ for some } k \in \mathbb{Z}\}$.

Thus, the umbra of a digital signal $x(n)$ is the planar set consisting of Graph(x) and all the points beneath it. The umbra operation turns a signal into a set, and the top surface operation turns a set into a function. Clearly, if $x(n)$ is a digital signal, $x\colon \mathbb{Z} \to [0, N]$, for some natural number $N \geq 0$, then $x = T[\text{Umbra}(x)] = T[\text{Graph}(x)]$. The next definition formally captures the idea of digital signal dilation.

Definition (Signal Dilation). Let $x(n)$ and $h(n)$ be digital signals, $x: \mathbb{Z} \rightarrow [0, N]$ and $h: \mathbb{Z} \rightarrow [0, N]$ for some natural number $N \geq 0$. Then the dilation of f by h, is $f \oplus h = T[\text{Umbra}(f) \oplus \text{Umbra}(h)]$.

Dilation is useful for removing transient signal features, such as splintering, when a signal contains a sharp gap within the overall blob-like structure. This can occur due to noise in the signal acquisition equipment. It may even be a genuine aspect of the signal, but its presence disturbs subsequent analysis algorithms. From our earlier examples of electrocardiography, recall that one occasional characteristic of electrocardiograms (ECGs) is such splintering. Its presence should not only be detected, but for purposes of heart rate monitoring, the application should not misread the ECG's splinters as individual ventricular contractions. Linear smoothing spreads the splinter into a large canyon in the signal. Using a dilation kernel on the umbra of the signal erases the splintering, although it produces a signal with an overall higher mean. If the dilation kernel is not as wide as the gap, then it has an insignificant enhancement effect. For digital signals, this decreases the effective dynamic range of the signal. However, rudimentary dilation kernels do a good job of filling gaps in signal structures.

There is a dual operation to dilation—called erosion; the dilation of A by B is equivalent to erosion of the complement, $A^c = \mathbb{Z}^2 \setminus A = \{b \in \mathbb{Z}^2 \mid b \notin A\}$, by the reflection of B across the origin.

Definition (Erosion). Let $A, B \subseteq \mathbb{E}^n$. Then the erosion of A by B is

$$A \ominus B = \bigcap_{b \in B} (A - b). \tag{4.67}$$

The structuring element B is also called an erosion kernel. The next proposition collects some properties of erosion. Note the symmetry between these properties and those in the Dilation Properties Proposition.

Proposition (Erosion Properties). If $A, B, C \subseteq \mathbb{Z}^2$, and $k \in \mathbb{Z}$, then, $A \ominus B = \{d \in \mathbb{Z}^2 \mid d + b \in A \text{ for all } b \in B\} = \{d \in \mathbb{Z}^2 \mid B_d \subseteq A\}$; $(A + k) \ominus B = (A \ominus B) + k$; $A \ominus (B + k) = (A \ominus B) - k$; $A \subseteq B$ implies $A \ominus C \subseteq B \ominus C$; $(A \cap B) \ominus C = (A \ominus B) \cap (B \ominus C)$; and, finally, $(A \ominus B)^c = A^c \oplus (-B)$, where $-B = \{-k \mid k \in B\}$.

Proof: Exercises. ∎

The last of the Erosion Properties illustrates the dual nature of dilation and erosion. Note that erosion is neither commutative nor associative. Note also that translation invariance does not hold true for a translated erosion kernel. Let's continue with a definition of signal erosion and its application to enhancement.

Definition (Signal Erosion). Let $x(n)$ and $h(n)$ be digital signals, $x: \mathbb{Z} \rightarrow [0, N]$ and $h: \mathbb{Z} \rightarrow [0, N]$ for some natural number $N \geq 0$. Then the dilation of f by h, is $f \ominus h = T[\text{Umbra}(f) \ominus \text{Umbra}(h)]$.

Erosion can remove spike transients. Such transients present difficulties for linear enhancement filters. The convolution operation tends to blur the spike. Linear smoothing superimposes a replica of its kernel's impulse response on the signal and can obliterate small local features in the process. An erosion kernel removes the spike without spreading it into the rest of the signal's values. If the kernel is narrower than the transient, then it treats the transient as a blob, part of the main signal structure.

The following points summarize the behavior of dilation and erosion enhancement operators:

- (Splinter removal) Dilation blends signal structures separated by gaps narrower than the kernel width.
- (Spike removal) Erosion removes sharp, narrow, upward transients in the signal.
- Dilation has the undesirable effects of adding to the overall signal level and creating new, fine-scale signal features.
- Erosion has the undesirable effects of reducing the overall signal level and destroying existing, fine-scale signal features.

The drawbacks for both dilation and erosion as enhancement tools seem to be counterposed. They modify the signal mean in opposite ways. The have opposite effects on small signal features. These observations have led morphologists to compose dilation and erosion operations while using the same kernel. The operations are not inverses, and it turns out that this composition benefits later analysis steps, such as histogram derivation, thresholding, labeling, and region merging and splitting operations.

Definition (Opening). If $A, B \subseteq \mathbb{Z}^2$, then the opening of A by structuring element B is $A \circ B = (A \ominus B) \oplus B$. If $x(n)$ and $h(n)$ are digital signals, x: $\mathbb{Z} \rightarrow [0, N]$ and h: $\mathbb{Z} \rightarrow [0, N]$ for some natural number $N \geq 0$, then the opening of f by h, is $f \circ h = (f \ominus h) \oplus h$.

Definition (Closing). If $A, B \subseteq \mathbb{Z}^2$, then the closing of A by structuring element B is $A \bullet B = (A \oplus B) \ominus B$. If $x(n)$ and $h(n)$ are digital signals, x: $\mathbb{Z} \rightarrow [0, N]$ and h: $\mathbb{Z} \rightarrow [0, N]$ for some natural number $N \geq 0$, then the closing of f by h, is $f \bullet h = (f \oplus h) \ominus h$.

4.5 EDGE DETECTION

The edges of a signal mark the significant changes of its values, and edge detection is the process of determining the presence or absence of such significant changes. This is not a very satisfying definition, of course, since there is much room for disagreement over what makes a change in a signal significant or insignificant. It is most often the nature of the edge detection application that resolves such disputes. As we have already noted in segmentation problems, goal-directed considerations play a considerable role in designing edge detectors. Not all of the operations involved can proceed directly from the signal data to the edge detection result without some overall perspective on what the problem under study requires for a correct result. Thus, in edge detection there is again an interaction between bottom-up (or data-driven) approaches and top-down (or goal-driven) methods. Whatever the tools employed—edge detectors, texture analysis, local frequency components, scale-based procedures—this interplay between data-driven and goal-driven methods will continue to be at work in our signal interpretation efforts.

Despite the apparent simplicity of formulating the problem, edge detection is quite difficult. Among the first attempts to analyze signals and images were edge detection techniques. We will explore these methods in some depth, and this study will bring us up to an understanding of the current research directions, performance issues, and debates surrounding derivative-based edge finding. A second main approach is to fit members of a collection of edge models to a signal. Note that edge models are simply primitive shapes to be found in a signal. Since, as numerical analysts well know, finding an approximation to the derivative of a quantized function is a problematic undertaking, many researchers regard this approach as inherently more robust. Let us look a little further at these two approaches and their advocates.

Remembering that our signals are just one-dimensional functions of a discrete or continuous independent variable, we can take the magnitude of the signal derivative as a starting point for building an edge detector. Furthermore, since the second derivative changes sign over the extent of an edge, it is feasible to base an edge detector on approximations of the second derivative of a signal. Many of the early experiments in edge detection relied upon derivative-based approaches [58–62].

Another approach to edge detection is to fit selected edge-shaped masks or patterns to the signal. The edge detection patterns are signals themselves and (reasonably assuming that we are working with finite-energy signals) are elements of a Hilbert space. For a conceptualization of the edge detection problem, we can resort to our Hilbert space theory from Chapters 2 and 3. The edge detection operators can be orthonormalized. Thus, the edge detection masks become the basis elements $\{e_i: i \in I\}$ of a subspace of linear combinations of perfect edge-containing signals. The whole edge detection problem becomes one of finding the inner products of a signal $x(t)$ with edge detector basis elements $e_i(t)$. Notice, however, that the basis elements are not smoothly undulating functions such as the sinusoids or exponentials as used in the discrete Fourier transform (e.g., Section 4.3.2). Rather, the basis elements contain sharp discontinuities in the first- or higher-order derivatives. These sharp breaks in the basis functions match the shape of the edges to be detected. Thus, large values of $\langle x(t), e_i(t - t_0)\rangle$ indicate the presence of an edge of type i at location t_0 in $x(t)$. This too was a popular path of edge detection pioneers [61, 63–65]. Again, the edge basis elements are simply templates that contain elementary shapes, and this method, therefore, can be considered as a type of template matching. Section 4.6 of this chapter furnishes some basic template matching tools; the inner product operation lies at the heart of such methods.

These references are but a few of those that have appeared in the research literature over the last 30 years. Even at this writing, near the turn of the twenty-first century, investigators twist the interpretation of what is significant in a signal, adjust the mix of top-down and bottom-up goals, optimize in yet another way, and append a new variant to the vast edge detection literature [66–69]. Indeed, in this book we will continue to remark upon and deepen our understanding of edge detection with each new signal processing and analysis tool we develop.

Edge detection is a basic step in signal analysis. The output of an edge detector is a list of the location, quality, and type of edges in a signal. Interpretation algorithms could be designed to work on the list of edges. For an example from speech analysis, between certain pairs of edges, and based on the signal values therein, the interpretation could be that this portion of the signal represents irrelevant background noise. Other parts of the voice signal, bounded by edges, represent human speech. Further analysis may reveal that one significant section is a voiced consonant, another a fricative, and so on. In general, we find the edges in signals as a first step in analysis because the edges mark the transition between important and unimportant parts of a signal or between one and another significant part of a signal. Edge detection draws the boundaries within which later, more intricate algorithms must work. Edge guidelines allow interpretation processes to work in parallel on distinct parts of the signal domain. Finally, preliminary edge detection prevents the waste of processing time by higher-level algorithms on signal fragments that contain no useful information.

4.5.1 Edge Detection on a Simple Step Edge

To introduce the many problems that arise—even in the most elementary detection problem—let us consider a simple step edge in a signal. The unit step signal provides a natural example of a step edge. For discrete signal analysis, $u(n)$, the discrete unit step is a perfect step edge located at $n = 0$. What do we expect that a step edge detector should look like?

This much is obvious: the step edge we consider may not have unit height, and it may not be located at the origin. If the amplitude of the edge diminishes, then it is a less pronounced change in the signal, and our edge detector ought to produce a smaller response. Thus, a first consideration is that our edge detector should be linear: Its output is greater, given a larger amplitude in the edge and therefore a larger amplitude in the underlying signal. The second consideration is that an edge should be detected by the operator wherever it happens to be located within the signal. In other words, whether the step is at the origin or not is irrelevant: We should detect the edge at any time. The implication of our second consideration is that a first-cut edge detector is translation invariant. It is thus linear and translation invariant and, by the results of Chapters 2 and 3, must be convolutional.

Let us suppose that we are working with discrete signals. If the edge detector is $y = Hx$, then the output y is given by convolution with the impulse response, $h(n)$, of the detector system, H:

$$y(n) = (x * h)(n) = \sum_{k=-\infty}^{\infty} x(k)h(n-k). \tag{4.68}$$

There may be noise present in the signal $x(n)$ so that the sum in (4.68) does not readily simplify and the response is irregular. Noisy or not, the response of the system H to input $x(n)$ will be a signal $y(n)$ with discrete values:

$$\{..., y(-3), y(-2), y(-1), y(0), y(1), y(2), y(3), ...\}. \tag{4.69}$$

To find the edge, then, we have to threshold the output, supposing that an edge is present in $x(n)$ whenever $|y(n)| \geq T$. If we are very lucky, then there will be only one value that exceeds the threshold, along with a single detection result for a single step edge. If there is a range of high response values, then it may be necessary to select the maximum or a set of maximal responses.

In the case of an analog edge detection problem, the output of the convolution integral (4.70) will be continuous. There will be no hope of thresholding to discover a single detection result unless the threshold coincides with the maximum response value. We must be content with seeking a maximum or a set of maximal responses.

$$y(t) = (x * h)(t) = \int_{-\infty}^{+\infty} x(t-s)h(s)\,ds = \int_{-\infty}^{+\infty} h(t-s)x(s)\,ds. \tag{4.70}$$

It might seem that at this early stage in conceptualizing an edge detector that it is wrong to settle so quickly on a convolutional operator. For example, if the input signal is attenuated to $Ax(t)$, $|A| < 1$, then the response of the detector to the attenuation will be $y(t) = A(x*h)(t)$. The same edges—no matter how insignificant they become because of the attenuation—will still be detected as maximal responses due to the linearity of the operator H. So if the goals of our application change, and it becomes necessary to ignore sufficiently small changes in the input signal, then our edge detector will fail by falsely indicating significant transitions in $x(t)$. Sometimes top-down specifications demand non-linearity. Note, however, that we can accommodate such a top-down scheme by adjusting the threshold parameter. If certain jumps should be detected as edges and others should be passed, then the kind of nonlinearity we need can be implemented by thresholding all responses and then selecting the maximal response from those that exceed the threshold. In other words, if Edges[x] is a predicate or list of edges of input signal $x(t)$, then Edges[x] = $\{(t, x(t)) : |y(t)| \geq T$ and t is a local maximum of $y = x*h\}$.

There is a further difficulty with uncritical convolutional edge detection. If the input signal contains spike noise, then the output of the edge operator will have a large response to the spike. In fact, around an isolated impulse, the response will tend to look like the impulse response of the edge detector itself.

Example (Difference of Boxes Operator). One of the earliest edge detectors is the Difference of Boxes (DOB) filter [60], defined by $y = h_{\mathrm{DOB}}*x$, where, for some $L > 0$,

$$h_{\mathrm{DOB}}(t) = \begin{cases} -1, & -L \leq t < 0, \\ 1, & 0 \leq t \leq L. \end{cases} \tag{4.71}$$

Clearly, the DOB operator is designed to smooth the data on both sides of a possible edge and subtract the earlier values from the later ones. Any sudden transition at time $t = 0$ appears in the output as a large-magnitude value. The DOB operator emphasizes values near $t = 0$ just as much as values near $|t| = L$. It is also typical to zero the center value of the DOB operator, $h_{DOB}(0) = 0$.

Example (Derivative of Gaussian Operator). Another popular operator for convolutional edge detection is the first derivative of the Gaussian (dG). The impulse response of this system is given by $h_{dG}(t) = -t\exp[-(t/\sigma)^2/2]$. The dG operator is an alternative to the DOB convolution kernel; it emphasizes values near the edge over values further away. A disadvantage is that it has an infinite extent. Practically, of course, the kernel can be truncated, since the signal diminishes rapidly.

We can add some normally distributed noise to the analog unit step signal to produce noisy step edges. Convolution with DOB and dG operators produces satisfactory results (Figure 4.11).

Application (Plasma Etch Endpoint Detection). Consider an example from a problem of controlling an industrial process, Figure 4.2. A spectrometer monitors the carbon monoxide (CO) optical discharge from a plasma reactor that is etching an oxide layer on a silicon wafer [70]. This is a critical step in the production of integrated circuit chips. The CO is a chemical byproduct of the plasma reaction with the silicon dioxide on the wafer. When the CO spectral line falls abruptly, the layer of silicon dioxide that is exposed to the plasma has been cleared by the plasma reaction.

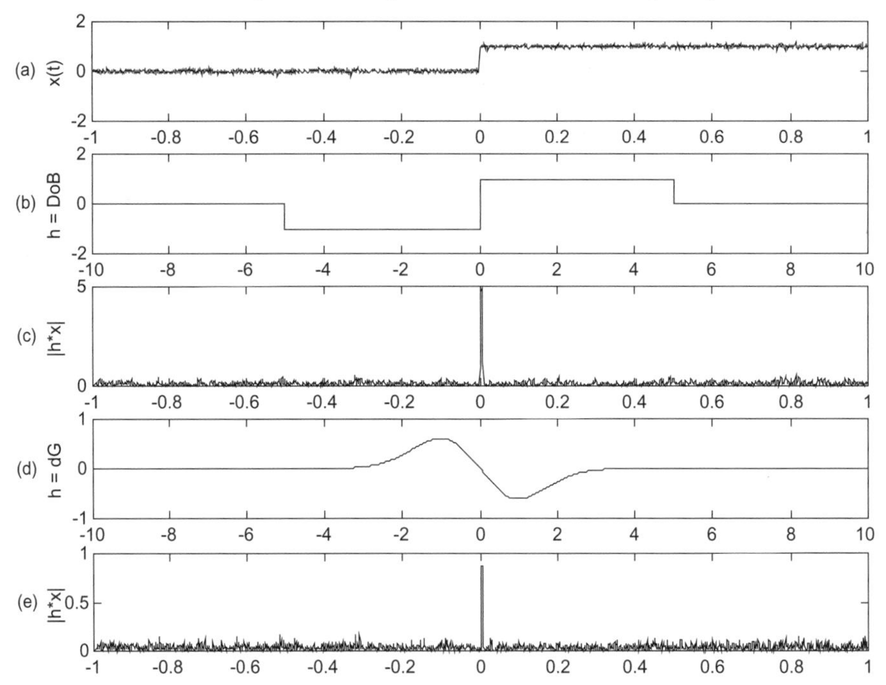

Fig. 4.11. An analog step edge in noise (a). The DOB and dG operators are shown in (b) and (d), respectively. Panels (c) and (e) show the result of convolving the signal with the DOB and dG operators, respectively.

It is essential to stop the plasma reaction as close as possible to the time when the CO intensity falls to avoid plasma damage to circuit layers underneath the target oxide. Thus, plasma etch endpoint detection can be largely viewed as a real-time edge detection problem. Plasma processes are very much subject to transients in the optical emissions. These can be caused by a reactor or acquisition fault or process instability. Transients can result in anomalous edge detection and an incorrect identification of etch process endpoint. As we found in Section 4.3.3, one the solution is to introduce a nonlinear filter that removes impulse noise prior to the linear filtering and application of a convolutional edge detection operator [e.g., the DOB operator (4.71)].

Most plasma etch endpoint detection algorithms rely on simple signal thresholding and first- and second-derivative estimates on the optical emission signal. When variability arises in the endpoint trace, perhaps due to imprecise control of previous thin film deposition steps or to the presence of different circuit configurations on the etched wafers, then strategies that consider the shape of the optical emission signal become attractive [71]. We will consider this aspect later in Section 4.7. In other processes, the signal edge is more difficult to discern, and endpoint detection methods that model the real-time optical emission and detect out-of-trend conditions become important [72]. Lastly, there is the top-down consideration that complete translation invariance is not appropriate for the edge detector. In fact, the process control application of Figure 4.2 is such a case. Typically, it is known in advance, by the process's nature, that the edge cannot occur except within a certain time range. In the case of plasma etching reactor control, it is common to specify an initiation of etch period during which no collapse in the emission spectral line should be detected. The upshot is that the convolutional operator is only applied within in the region known to contain valid edges. An alternative is to run the edge detector over the entire domain of the input signal and, in a further post-processing step, delete edges that occur outside the appropriate interval as spurious.

The next proposition sheds some further light on the nature of convolutional edge detectors [66]. We state and prove this proposition for analog signals and leave the discrete case for an exercise.

Proposition (Maximal Response Edge Detector). If a step edge detection operator locates edges from the maximal responses of a convolutional operator, H, and provides a unique maximum for $u(t)$ at $t = 0$, then

(i) $h(t)$ can only have one zero, at $t = 0$.
(ii) The impulse response of system H, $h(t)$, must be odd.

Proof: The first part of the proposition follows from the fundamental theorem of calculus. Since

$$y(t) = \int_{-\infty}^{+\infty} u(t-s)h(s)\,ds = \int_{-\infty}^{t} h(s)\,ds, \tag{4.72}$$

we know that $dy/dt = h(t)$. Since the response $y(t)$ has a maximum at $t = 0$, its derivative must be zero; therefore $h(0) = 0$. If the detector, H, responds correctly to the perfect step edge of $u(t)$ at $t = 0$, then this can be the only maximal point of the response, and it follows that $h(t)$ has precisely one zero.

For (ii) we can decompose $h(t)$ into its even and odd parts and then apply the fundamental theorem of calculus to get the result. Let $h(t) = h_e(t) + h_o(t)$ where $h_e(t)$ is even and $h_o(t)$ is odd. Applying the convolution theorem for analog LTI systems, let $t > 0$, and note that

$$y(t) = \int_{-\infty}^{+\infty} u(t-s)h(s)\,ds = \int_{-\infty}^{t} h(s)\,ds = \int_{-\infty}^{t} h_e(s)\,ds + \int_{-\infty}^{t} h_o(s)\,ds, \tag{4.73}$$

$$y(t) = \int_{-\infty}^{-t} h_o(s)\,ds + \int_{-t}^{0} h_o(s)\,ds + \int_{-\infty}^{-t} h_e(s)\,ds + \int_{-t}^{0} h_e(s)\,ds + \int_{0}^{t} h_o(s)\,ds + \int_{0}^{t} h_e(s)\,ds. \tag{4.74}$$

The finite integrals over $h_o(t)$ cancel, and the finite integrals over $h_e(t)$ are identical; thus,

$$y(t) = \int_{-\infty}^{-t} h_o(s)\,ds + \int_{-\infty}^{-t} h_e(s)\,ds + 2\int_{0}^{t} h_e(s)\,ds = \int_{-\infty}^{-t} h(s)\,ds + 2\int_{0}^{t} h_e(s)\,ds. \tag{4.75}$$

Now from (4.75) we may write

$$y(t) - y(-t) = 2\int_{0}^{t} h_e(s)\,ds. \tag{4.76}$$

Let us now differentiate (4.76), applying the chain rule to the left-hand side and the fundamental theorem to the right-hand side. This gives $2h_e(t) = 0$. Since t was arbitrary, $h_e(t) = 0$ for all $t > 0$. Moreover, since $h_e(t)$ is even, $h_e(t) = 0$ for all t. ■

4.5.2 Signal Derivatives and Edges

At first glance, finding the abrupt changes, or edges, in a signal amounts to finding the points at which the signal has a large derivative. Since we are familiar with analog differentiation from calculus and discrete differentiation from numerical analysis, it would seem that we should already possess the proper tools to begin a formal study of edge detection as a task within signal analysis. But the presence of even a minute amount of noise in a signal can lead to wild variations in its derivative at a time instant. Differentiation is known as a classical ill-posed problem or unstable process; in systems that perform differentiation, small differences in the input signal lead to large differences in the output signal [73]. The standard approach to such ill-posed problems is to convert them to well-posed problems by smoothing the input data. Thus, for signal analysis systems, some preliminary signal conditioning is appropriate.

We recall how to take the derivative of signals from calculus as a limit,

$$x'(t_0) = \left.\frac{dx}{dt}\right|_{t=t_0} = \lim_{t \to t_0} \frac{x(t) - x(t_0)}{t - t_0}. \tag{4.77}$$

This limit leads to standard formulas for differentiation of polynomials, algebraic functions, and trigonometic and exponential functions. Thus, we can obtain estimates for derivatives of discrete signals by finding an interpolating polynomial among a set of discrete vaues. The error estimates that this procedure gives are not as good as when the time-domain signal is approximated; this is a practical manifestation of the ill-posed nature of the differentiation process [74].

Deriving discrete derivatives for the analysis of digital signals by first developing an interpolating function, however, is not necessary for an approximate slope value near a point. We can derive discrete derivative formulas for a signal $x(n)$ by assuming that $x(n)$ arises from sampling an analog signal $x_a(t)$: $x(n) = x_a(nT)$, where T is the sampling interval. To get a formula for the discrete first and second derivatives around time instant $n = 0$, we expand $x_a(t)$ in a Taylor series and neglect the error terms:

$$x(1) = x(0) + x_a'(0) + \frac{1}{2} x_a''(0), \tag{4.78a}$$

$$x(-1) = x(0) - x_a'(0) + \frac{1}{2} x_a''(0). \tag{4.78b}$$

Subtracting (4.78b) from (4.78a) gives a simple formula for the first derivative:

$$x_a''(0) = \frac{1}{2}[x(1) - x(-1)]. \tag{4.79}$$

Adding (4.78a) from (4.78b) gives a simple formula for the second derivative:

$$x_a''(0) = x(1) - 2x(0) + x(-1). \tag{4.80}$$

Thus, the system $y = Hx$ with impulse response $h(n) = [1/2, \underline{0}, -1/2]$ approximates a discrete first derivative. And the system $y = Gx$ with impulse response $g(n) = [1, \underline{-2}, 1]$ approximates the second derivative. Recall that the system $y = Gx$ from Chapter 2 is called the discrete Laplacian operator. There we noted that it detected edges in signals, producing a zero response within flat and linearly sloped signal regions. Now we discover the discrete Laplacian's true nature: It is just a discrete version of the second derivative. The above analysis easily generalizes to approximate discrete derivatives with wider support. We find $x(2)$ and $x(-2)$ in terms of the Taylor series of $x_a(t)$, apply (4.79) and (4.80), and find that

$$x_a'(0) = \frac{1}{12}[x(-2) - 8x(-1) + 8x(1) - x(2)], \tag{4.81}$$

$$x_a''(0) = \frac{1}{12}[-x(-2) + 16x(-1) - 30x(0) + 16x(1) - x(2)]. \tag{4.82}$$

The above rudimentary edge detectors are intuitive, but they turn out to be less than ideal. To see this, we need an appropriate mathematical foundation. Thus, let's formally define signal edges and motivate the conditions on edge detectors that should constitute opimality. This will provide us with some initial results on optimal edge detectors.

Definition (Analog Signal Edge). Let $x(t)$ be an analog signal. Then $x(t)$ has an edge at $t = t_0$ if for some $n > 0$, there is a discontinuity in the nth derivative of $x(t)$. (We consider the 0th derivative of $x(t)$ to be the signal itself.)

It requires a little care to correctly define edges for a discrete signal. By their nature, discrete signals are composed entirely of discontinuities. So we adopt the following definition.

Definition (Discrete Signal Edge). Let $x(n)$ be a discrete signal. Then there is an edge in $x(n)$ at a point n_0 if there is a discontinuity in the derivative of the signal $x_a(t) = x(\lfloor t \rfloor) + (t - \lfloor t \rfloor)x(\lceil t \rceil) - x(\lfloor t \rfloor)$, where $\lfloor t \rfloor$ is the floor of t, and $\lceil t \rceil$ is the ceiling of t.

These definitions do encompass the variety of edge shapes that we find in signals. Of course, the step edge is a discontinuity in the signal itself. When the signal assumes a dihedral or "roof" shape, then the discontinuity lies in the first derivative. Thus, we shall concentrate on step edges in our theoretical development, since other edge shapes can be analyzed by computing the signal derivative and then applying our step edge analysis methods.

4.5.3 Conditions for Optimality

Let us now return to the simple step edge and complete an analysis of the edge detection problem. Specifically, three conditions for an edge detector H are as follows:

(i) The detector should have a high signal-to-noise ratio (SNR). This means that a better edge detector has a higher response to the edge within a signal than to the surrounding noise.

(ii) H should be well-localized about the true signal edge. The output maxima of H should cluster around the true edge, and better edge detectors show a tighter group of maximal responses.

(iii) There should be only one response to a single edge. If H is good, then it should present a number of apparently valid alternatives for the precise location, or registration, of the edge.

Canny chose the overall goals of (i)–(iii) for his detector design [75], which became the overwhelming favorite in research literature and signal analysis applications for many years. His strategy is to develop mathematical formulations for the goals and then to find convolution kernels that maximize the products of individual criteria.

Let's consider just the first two criteria above and see how this analysis unfolds. Let $h(t)$ be the convolution kernel that we seek and let $u(t)$ be the unit step. We will need to compute convolution integrals, $y(t) = (h * u)(t)$, so we must limit the time-domain extent of both $h(t)$ and $s(t)$, or assume that they belong to a signal space (Chapter 3) that supports closure under convolution. To achieve this, we assume that the edge detection filter has support on a finite interval $[-L, L]$. Now we define one form of the signal-to-noise ratio for analog systems and prove a property for the SNR under white Gaussian noise (Chapter 3).

Definition (Signal-to-Noise Ratio). Let $y = Hx$ be an analog system and let $n(t)$ be a noise signal. Then the signal-to-noise ratio (SNR) for H at $t = 0$ is

$$\mathrm{SNR}_{H,t=0} = \frac{|y(0)|}{\left(E\left[\int_{-\infty}^{+\infty} h(t)n(-t)\,dt\right]^2\right)^{\frac{1}{2}}}, \tag{4.83}$$

where $E[y]$ is the expectation operator.

That is, the SNR is a ratio of the system output without noise present to the system's expected output when the input is pure noise.

Proposition (SNR under White Gaussian Noise). Suppose H is an analog system and $n(t)$ is white Gaussian noise. Then $\mathrm{SNR}_{H,t=0} = |y(0)|/(n_0\|h\|_2)$, where n_0 is the standard deviation of $n(t)$.

Proof: From stochastic processes theory [76], we have that

$$E\left[\int_{-\infty}^{+\infty} h(t)n(-t)\,dt\right]^2 = E\left[\int_{-\infty}^{+\infty} h^2(t)n^2(-t)\,dt\right] = n_0^2\int_{-\infty}^{+\infty} h^2(t)\,dt; \tag{4.84}$$

hence,

$$\begin{aligned}\mathrm{SNR}_{H,t=0} &= \frac{|y(0)|}{\left(E\left[\int_{-\infty}^{+\infty} h(t)n(-t)\,dt\right]^2\right)^{\frac{1}{2}}} = \frac{|y(0)|}{\left(n_0^2\int_{-\infty}^{+\infty} h^2(t)\,dt\right)^{\frac{1}{2}}} \\ &= \frac{|y(0)|}{n_0\left(\int_{-\infty}^{+\infty} h^2(t)\,dt\right)^{\frac{1}{2}}} = \frac{|y(0)|}{n_0\|h\|_2}\end{aligned} \tag{4.85}$$

and the proof is complete. ■

This observation allows Canny [75] to specify a signal-to-noise performance measure for the edge detection system H which does not depend on the input signal $x(t)$; in particular, the optimal edge detector must maximize (4.85). It turns out the convolution kernel that provides a maximum SNR, given that the underlying step edge contains noise, is just the reversed step itself. This result is known as the Matched Filter Theorem. We introduce matched filtering in Section 4.6.2, but we need to have the tools of Fourier transform analysis at our disposal in order to complete its theoretical justification. Granting the Matched Filter Theorem, then, the convolutional kernel that is optimal for finding step edges is precisely the DOB operator. Canny attempted to prove that the DOB operator enjoyed an optimal localization property as well. The development proceeds as follows.

Let $y = Hx = H(u + n)$, where $u(t)$ is the (noiseless) unit step signal and $n(t)$ is white Gaussian noise. Then $y = Hu + Hn$ by linearity. Let $w = Hu$ and let $m = Hn$. If a maximum of the detector output appears at time $t = t_0$, then $y'(t_0) = w'(t_0) + m'(t_0) = 0$. Differentiation of the convolution integral for $w = h * u$ gives $w'(t_0) = h(-t_0) = -h(t_0)$, since h must be odd (by the Maximal Response Edge Detector Proposition, Section 4.5.1). If t_0 is close to the ideal edge at $t = 0$, then by a Taylor series expansion for $h(t)$, we know that $h(t_0) \approx h(0) + t_0 h'(0)$. Since $h(0)$ must be zero by the same Proposition, it follows that $t_0 h'(0) = m'(t_0)$. This implies that $h'(0)^2 E[(t_0)^2] = E[m'(t_0)^2]$. But, as in (4.84),

$$E\left[m'(t_0)^2\right] = n_0^2 \int_{-\infty}^{+\infty} h'(t)^2 \, dt. \tag{4.86}$$

Thus,

$$E\left[(t_0)^2\right] = \frac{E\left[m'(t_0)^2\right]}{h'(0)^2} = \frac{n_0^2 \int_{-\infty}^{+\infty} h'(t)^2 \, dt}{h'(0)^2}, \tag{4.87}$$

and the optimal localization criterion that Canny proposes is $(E[(t_0)^2])^{-1/2}$. Canny argued that the DOB operator remains optimal for both this localization criterion and the SNR criterion, for instance by seeking to maximize the product of (4.85) and (4.87). Canny adduced the third criterion, namely that there should be no multiple responses around a noisy edge in order to derive his optimal edge detection kernel. It turns out, after some lengthy calculus of variations, that the operator is very similar to the dG kernel [75].

Some time later, Tagare and deFigueiredo [66] pointed out some flaws in the reasoning above. The principal points are that

- There can be multiple time values $t = t_0$ for which a maximum occurs and $y'(t_0) = w'(t_0) + m'(t_0) = 0$. The performance measure must account for the distribution of all such maxima, and there is no reason to prefer one above the others.
- Equation (4.87) computes the variance of the maximum time $t = t_0$. But, by the previous point, the value of t_0 varies with every realization of the white noise process, and the substitution of (4.86) into $h'(0)^2 E[(t_0)^2] = E[m'(t_0)^2]$ to get (4.87) is invalid.

Tagare and deFigueiredo propose an alternative localization criterion. They propose that an optimal edge detector $y = h * x$ should maximize the quantity

$$J_{H,t=0} = \frac{\int_{-\infty}^{+\infty} t^2 h(t)^2 \, dt}{\int_{-\infty}^{+\infty} \omega^2 \left\| H(\omega) \right\|^2 \, d\omega}, \tag{4.88}$$

where $H(\omega)$ is the Fourier transform of the impulse response, $h(t)$, of the edge detector. Since we will develop Fourier transform theory in the next two chapters and outline its applications in Chapter 7, we defer the exposition of the new localization criterion. It turns out, interestingly enough, that the optimal kernel for convolutional edge detection under the criterion (4.88) is the dG function.

4.5.4 Retrospective

Edge detection has played a central role in the theoretical development of signal and image analysis since it was first applied, in the early 1950s, in attempts at image understanding with digital computers [58]. A large number of interdisciplinary researchers have worked on edge detection from the standpoint of building biologically motivated signal and image analysis systems. Edge detection at a variety of resolutions played a central role in Marr's account of animal and machine vision systems [77]. Marr[5] drew careful parallels between biological and computer vision mechanisms and inspired a generation of computer vision researchers to do the same. In particular, Marr conjectured that edge analysis across many scales would be sufficient for a characterization of a signal or image. Since biological signal analysis systems were capable of edge detection and multiscale analysis, it would appear that a major bond would then exist between electronic computers and their software on the one hand and animal brains and their experiences on the other. We will have occasion to revisit this conjecture later in the book; the theory of time-scale transforms, or wavelet theory, would shed light on the question a dozen or so years after Marr proposed it.

From a computational perspective, Canny's work [75]—which comprised his Master's thesis work [78] at the Massachusetts Institute of Technology—was thought to have essentially solved the edge detection problem. There was a period when just about everyone's research paper included a Canny edge detector, if it mentioned edge detection at all. Some practitioners used a derivative of the Gaussian operator as a close approximation to the Canny kernel, which was difficult to compute. Only after the passage of several years did minor flaws in the theory become apparent; fortuitously, those many papers of those many researchers that

[5]David Courtenay Marr (1945–1980). Spurred by Marr's critique ["Artificial intelligence: a personal view," *Artificial Intelligence*, vol. 9, pp. 37–48, 1977], computer vision rose from a collection of ad hoc techniques to a broad discipline unifying ideas from biology, psychology, computer science, and robotics. One of the prominent pioneers of computer vision, Marr died quite young, in the midst of a most productive carreer.

resorted to the dG shortcut were following the optimal road after all! There has begun another round of research papers, controversies, and algorithmic emendations [66, 67, 79, 80]. The most recent research, in addition to emphasizing the dG kernel, has found interconnections between diverse methods reaching back some 20 years, and a unifying framework has been elaborated [68, 69].

4.6 PATTERN DETECTION

Many of the most important signal analysis problems involve detecting more than a single edge in a signal. There may be several edges in a signal region of interest, and ascertaining the signal levels, slopes, and transitional shapes between the edges may be critical to correctly interpreting the signal. The basic signal pattern detection or recognition problem involves finding a region $a \le t \le b$ within a candidate signal $x(t)$ that closely matches a prototype signal $p(t)$ on $[0, b-a]$. For discrete signals, the detection problem reduces to comparing finite sets of values. It seems simple; but in the presence of noise or other distortions of the unknown signals, complications do arise. The difficulties worsen when the detection problem allows the size of the prototype within the candidate to vary. Indeed, the problem can be trivial, challenging, problematic, and frustrating and may even defy robust solution. Signal pattern detection remains a topic of research journals, experimental results, and tentative solutions. This section presents three basic for pattern detection approaches: correlation methods, structural methods, and statistical methods.

4.6.1 Signal Correlation

Correlating two signals seems to be the natural approach to pattern detection. Its ultimate foundation is a dot product relation of similarity between vectors, which generalizes to the inner product operation for Hilbert spaces. Some care in the formulation is essential, however. Thus, we first explore the method of normalized cross-correlation in Section 4.6.1.1. There is a subsequent result, called the Matched Filtering Theorem, which indicates that this approach is in fact optimal. We do not, however, yet possess the theoretical tools with which to prove the matched filtering result; so we simply state it herein and postpone the proof until Chapter 7.

4.6.1.1 Normalized Cross-Correlation. The Cauchy–Schwarz Inequality, covered in Chapters 2 and 3, provides a mathematically sound approach to the signal matching problem. Let's suppose that we are dealing with an unknown, candidate signal x and a prototype signal p, which may come from a collection of model signals. The application may require a comparison of x with each prototype in the collection. Prototype signals represent patterns that we expect to find in the input to the signal analysis application; thus, it is reasonable to stipulate that the patterns are finitely supported on an interval $I = [a, b]$.

For this discussion, let us assume that we are working in the discrete realm, so that the signal space tools from Chapter 2 apply. It is useful to have a measure of match, or distance, between two signals, $d(x, y)$, that is a metric; that is, it statisfies the following:

- The positive definite property: $d(x, y) \geq 0$ for all x, y.
- The identity property: $d(x, y) = 0$ if and only if $x = y$.
- The symmetry property: $d(x, y) = d(y, x)$ for all x, y.
- The triangle inequality: For any z, $d(x, y) \leq d(x, z) + d(z, y)$.

If we take the signals to be square-summable, then the l^2 norm is a comparison measure which is a metric, $d(x, y) = \|x - y\|_2$. Assuming that we work with l^2 signals, a measure of the mismatch between two square-summable signals is the l^2 norm; in other words, if x is an unknown signal, then we need to minimize $\|x - p\|_2$ among all prototype signals p from a family of models our application attempts to detect. If the registration of pattern $p(n)$ within candidate $x(n)$ must be found, then we seek the offset k that provides the smallest $\|x(n+k) - p(n)\|_2$ for all appropriate offsets k of $p(n)$ relative to $x(n)$.

Choosing $d(x, y) = \|x - y\|_2$ as a match measure on real-valued signals is equivalent to using the inner product $\langle x, y\rangle$. To see this, suppose we compare a square-summable candidate signal $x(n)$ to a square-summable prototype $y(n)$. Then,

$$\|x-y\|_2^2 = \sum_{n=-\infty}^{\infty} (x(n)-y(n))^2 = \sum_{n=-\infty}^{\infty} (x(n))^2 + \sum_{n=-\infty}^{\infty} (y(n))^2 - 2\sum_{n=-\infty}^{\infty} x(n)y(n)$$
$$= \|x\|_2^2 + \|y\|_2^2 - 2\langle x, y\rangle. \tag{4.89}$$

Thus, we see that $d(x, y)$ depends on both $\|y\|_2$ and the inner product $\langle x, y\rangle$. If we require that all prototype vectors have unit norm, $\|y\| = 1$, then the equivalence between minimizing $d(x, y)$ and maximizing $\langle x, y\rangle$ is clear.

The inner product relation inspires an important method of matching signals: normalized cross-correlation. We introduced cross-correlation for discrete signals in Chapter 2, where it was noted to be a kind of backwards convolution. Suppose that we have a real-valued pattern signal $p(n)$, supported on the interval $[0, N]$: $p(n) = [p_0, p_1, \ldots, p_N]$. We search for the best match of $p(n)$ at offset location k in real-valued signal $x(n)$. A measure of the match between the two at offset k is the inner product, $y(k) = \langle p(n-k), x(n)\rangle$. When $y(k)$ is at a maximum value, then we have found the offset k at which the prototype pattern $p(n)$ best matches the source signal $x(n)$. This operation, in effect, correlates the shifted pattern $p(n-k)$ with a window of the signal $x(n)$; this windowed function is just $x(n)[u(n-k)-u(n-k-N-1)]$, which is zero outside of the interval $[k, k+N]$. To evaluate $y(k)$, notice that

$$y(k) = \sum_{n=-\infty}^{\infty} p(n-k)x(n) = \sum_{n=k}^{n=N} p(n-k)x(n) \leq \left(\sum_{n=k}^{N} p^2(n-k)\right)^{\frac{1}{2}} \left(\sum_{n=k}^{N} x^2(n)\right)^{\frac{1}{2}}. \tag{4.90}$$

The inequality in (4.90) follows from the Cauchy–Schwarz Inequality for discrete signals applied to $p(n-k)$ and the windowed signal $x(n)[u(n-k)-u(n-k-N-1)]$.

Recall from Chapter 2 that the Cauchy–Schwarz Inequality states that if $x(n)$ and $p(n)$ are in l^2, then the product $x(n)p(n)$ is in l^1, and $\|xp\|_1 \le \|x\|_2 \, \|p\|_2$. Recall that equality holds if and only if $x = cp$ for some constant c. Thus, in (4.90) there is equality if and only if $p(n-k) = cx(n)[u(n-k)-u(n-k-N-1)]$, where c is a constant. The problem with using $y(k)$ as a measure of the match between $p(n-k)$ and $x(n)$ over the region $[k, k+N]$ is that the second term on the right in (4.90) depends on the registration value k. Thus, we take as the measure of match between the pattern $p(n-k)$ and the signal $x(n)$ on $[k, k+N]$ to be the normalized cross-correlation,

$$C_{p(n-k),x(n)} = \frac{y(k)}{\left(\sum\limits_{n=k}^{N} x^2(n)\right)^{\frac{1}{2}}} = \frac{\sum\limits_{n=k}^{N} p(n-k)x(n)}{\left(\sum\limits_{n=k}^{N} x^2(n)\right)^{\frac{1}{2}}}. \tag{4.91}$$

The match measure (4.91) assumes its maximum value $C_{p(n-k),x(n)} = 1$ when the pattern is an exact multiple of the windowed input signal $x(n)$. Thus as a pattern recognition measure, normalized cross-correlation finds patterns that match our prototypes up to an amplification or attenuation factor.

Matching by normalized cross-correlation, dependent as it is on the inner product notion, is quite intuitive. The inner product operation is closed in our Hilbert spaces of square-summable discrete and square-integrable analog signals; we thus have a rich variety of signals with which to work. The inner product generalizes the dot product relation of finite-dimensional vector spaces, which is a geometrically satisfying relation of similarity between two vectors. Without a doubt, these factors recommend normalized cross-correlation as one of the first strategies to be employed in a signal analysis application. It is not the only method, however. Depending upon the application, it may suffer from a couple of problems:

- Computing the sums of squares in to arrive at the l^2 norms is time-consuming.
- There may be a lot of near-misses to the optimal registration of the pattern $p(n)$ in the unknown candidate signal $x(n)$.

To avoid the first problem, it may be feasible to apply the l^∞ norm, taking $d(p(n-k), x(n)) = \|p(n-k) - x(n)[u(n-k)-u(n-k-N-1)]\|_\infty$ as a match measure on real-valued signals instead. Computing this distance measure involves only a comparison of the $N+1$ values of $p(n-k)$ with those of $x(n)$ on the window $[k, k+N]$. In real-time applications, using digital signal processors, for example, this tactic can often help to meet critical time constraints. The second arises when there are a number of structures present in $x(n)$ that resemble the model pattern $p(n)$. An example of this situation arises during electrocardiogram analysis. Suppose a model QRS complex pulse has been extracted. Normalized cross-correlation produces many close responses. Thus, it may be necessary to search for certain critical signal features—other than raw signal values—around which more robust pattern detection methods might be constructed. This leads to the structural techniqes we consider next.

It is also possible to develop normalized cross-correlation for analog signals. Following the derivation of discrete normalized cross-correlation very closely, along with using the Cauchy–Schwarz inequality from Chapter 3, we obtain

$$C_{p(t-s),x(t)} = \frac{y(s)}{\left(\int_I x^2(t)\,dt\right)^{\frac{1}{2}}} = \frac{\left(\int_I p(t-s)x(t)\,dt\right)^{\frac{1}{2}}}{\left(\int_I x^2(t)\,dt\right)^{\frac{1}{2}}}, \tag{4.92}$$

where s is the offset of prototype signal $p(t)$ into input signal $x(t)$, and I is the interval that contains the support of $p(t)$. Deriving this result is left as an exercise.

The next section sheds some light on the importance of cross-correlation for the detection of patterns that are embedded in noise.

4.6.1.2 Matched Filtering: A Glimpse. Understanding that real-life signals contain noise, it is natural to wonder what is the best way to match a signal against a pattern. If the detection process is a linear, translation-invariant system, H, and the input signal, $x(t)$, is corrupted by additive white noise, $s(t)$, then there is an answer to this intriguing question. It turns out that the impulse response of the system H, $h(t)$, is—up to a scaling (amplifying or attenuating) factor—none other than a reflected and translated version of the input signal $x(t)$. The impulse response $h(t)$ is called the matched filter for the signal $x(t)$. Let us state this as a theorem for analog signals, but postpone the proof until we have covered the Fourier transform theory.

Theorem (Matched Filter). Suppose $x(t) = f(t) + n(t)$, where $f(t)$ is a signal of known shape, and $n(t)$ is a zero mean white noise process. Then the optimal LTI system H for detecting $f(t)$ within $x(t)$ has impulse response $h(t) = cf(t_0 - t)$ for some constants c and t_0.

Proof: Chapter 5 prepares the foundation in analog Fourier theory. Chapter 7 proves the Matched Filter Theorem as an application of Fourier methods. ■

In developing the normalized cross-correlation method for signal matching, we found that the best possible match occurs when the pattern is identical to the source signal up to a constant scale factor on the shifted support interval of the prototype. Note that the convolution implied by the Matched Filter Theorem is the same as a cross-correlation with a reflected version of the source. Thus, matched filtering theory formulates an important converse to normalized cross-correlation, namely, that the best filter—for reasonable noise assumptions—is in fact just the target source pattern itself. Many communication theory texts cover matched filtering [81, 82], there are full reviews in the engineering literature [83], and Chapter 7 points out additional resources in print.

If matched filtering is both the obvious and optimal approach, why study anything else? One answer, as with "optimal" thresholding and "optimal" noise removal, is that the conditions that validate the optimality do not always hold. Furthermore, the optimal methods for such problems are often computationally burdensome, even untractable. Convolution, at the heart of matched filtering, can be expensive for signals, and on images may require special-purpose computing hardware. Thus, reasonably fast suboptimal techniques are desirable; a lot of these methods arise in conference presentations, appear in print, and become the basis for practical applications.

4.6.2 Structural Pattern Recognition

A structural pattern recognition application discovers the physical relationships between labeled signal regions and unites the regions into a graph structure. Then the application compares the graph structures that it derives from candidate signals against those from prototype patterns. The structural analysis application estimates the degree of match or mismatch between these graphs. If the match measure exceeds a theoretically or empirically derived threshold, then the candidate signal matches the prototype pattern; recognition is complete.

In signal analysis, structural descriptions derived from signals tend, like their source data, to be one-dimensional entities. Commonly the structure is just a vector, and the components of the vector are numerical weights that indicate the relative presence of some expected characteristic of the signal over a time interval. In computer vision, the structural descriptions are often two-dimensional, like the images they come from. Not unexpectedly, the feature relationships are more intriguing, more varied, and more difficult to analyze. Sometimes it is important to extract features at a range of scales of the signal's domain. Then there is a family of feature vectors, and a feature at a coarse scale may resolve into a set of finer scale characteristics. This forms a two-dimensional structural description of the signal: There are large-scale or low-resolution features at the top; features of intermediate size in the middle; and high-resolution, small-scale features at the bottom. For instance, we might find a texture region when segmenting a signal into wide regions and then resolve it into small curves, edges, and flats. At the finest scale, discrete signal features are just the signal values. The discrete world limits signal resolution. With analog signals, resolution refinment need never stop. This is called a pyramid representation. Pyramidal structural descriptions for signals are an important and widely studied analysis tool. However, with their introduction into the signal analysis application, the problems become as intricate as they are in computer vision.

This section first explains how to reduce the amount of signal data subject to analysis by the pervasive strategy of feature extraction. Section 4.6.2.2 follows with some basic measures for signal matching by comparing feature vectors. The vectors contain fewer numbers than the signals, so comparisons are quick—a happy contrast to correlation-based approaches. Next, we develop some formal theory for structural descriptions of signals. The last subsection explains a comparison measure for structural descriptions. This measure turns out to be a metric, one of our fundamental theoretical results in structural pattern recognition.

4.6.2.1 Feature Extraction. Extraction of feature vectors reduces the dimensionality of the matching problem. The expense of computing the normalized cross-correlation match measure is a drawback in its application. This is all the more the case when a candidate signal must be compared against many prototype signals.

Structural pattern recognition attempts to place input signals $\{f_1, f_2, \ldots, f_N\}$ into categories $\{C_1, C_2, \ldots, C_K\}$ by forming an intermediate, structural description of the signal and then using the structural description as the basis for classification. The simplest form of structural description is the feature vector, $v = (v_1, v_2, \ldots, v_M)$. Components of the feature vector are numerical, usually real but sometimes complex numbers. By thresholding the magnitude of the feature vector components, labels can be assigned to the features to build an abstract or symbolic description of the signal. Selecting the right feature vector is both quite open to design and quite critical to the recognition application's success.

Let's consider how we might extract a feature vector from a signal. First, suppose that we have a candidate signal, $x(n)$, and a set of model or prototype signals $\{e_i(n) \mid 1 \le i \le N\}$ defined on some interval $[a, b]$. The plan is to develop a vector $\mathbf{v} = (v_1, v_2, \ldots, v_N)$ such that its components, v_i, are a measure of the similarity between $x(n)$ and prototype e_i. Any of the norms we know from the theory of normed linear spaces in Chapter 2 are suitable; we may choose $v_i = \|x - e_i\|_p$, the l^p norm of the difference between the candidate signal and prototype e_i. This measures the mismatch between x and e_i. The l^p norm is a metric and is suitable for pattern detection applications.

But the results on normalized cross-correlation and matched filtering in Section 4.6 inspire us to use the inner product $v_i = \langle x, e_i \rangle$. Since the inner product $\langle x, e_i \rangle$ depends on the magnitude of e_i, the prototypes should all have the same magnitude; otherwise a large-magnitude prototype will skew all of the matches toward itself. Orthogonality of prototypes is important also. Suppose that $\langle e_1, e_2 \rangle \neq 0$. Without orthogonality, a signal which is a scalar multiple of e_1, say $y = ce_1$ on $[a, b]$, will have a similarity measure to e_2 of $\langle y, e_2 \rangle \neq 0$. Thus, although it is a perfect scaled (amplified or attenuated) replica of prototype e_1, y shows a similarity to e_2. The solution to this identification conundrum is to simply stipulate that the prototypes be orthogonal and of unit length. Hence, orthonomal bases—such as we developed in Chapters 2 and 3 for Hilbert spaces—play an important role in pattern recognition theory. An important problem is how to choose the basis for pattern recognition applications. Of course, we can orthogonalize any linearly independent set of signals on $[a, b]$, using, for instance, the Gram–Schmidt procedure sketched in our reflections on Hilbert space. However, the resulting basis set, $\{e_i\}$, may not resemble the original patterns. One way to ensure the construction of a pattern recognition basis set that preserves the features of input signals is to build the basis set according to candidate signal statistics. A classical technique, the Karhunen–Loève transform, can be applied to select the best possible basis [84, 85]—for instance, when the problem is a simple edge shape [65].

There are many other ways to extract feature vectors from signals. We can study a single signal region with basis functions, as above. The basis function inner products can also be applied to a series of intervals. But other processing steps are possible. The presence of a given signal level can be used as a feature vector component. The presence of a texture parameter—roughness or waviness—is a common technique in feature vector designs. Often the signal features have fixed registrations, assigned as part of the signal analysis application's design. In other situations, an input signal is subject to an initial segmentation, perhaps by signal level (thresholding), texture analysis, or edge detection. This determines the presence and registration of useful signal content. Then the feature vector is extracted from a set of intervals that derive from the preliminary segmentation. It is common, too, to generate a single large feature vector, $v = \{v_{i,j} \mid 1 \le i \le N, 1 \le j \le M\}$, where the components $v_{i,j}$ for $1 \le j \le M$, represent M different feature values, each applied to the same region of interest of the signal, S_i, $1 \le i \le N$. Statisitcal analysis of feature vectors may reveal correlations, and some of them can be deleted from the final application. Optimal selection of features is possible and explored fully in pattern recognition treatises [85]. Feature vector components can also be labeled. Components that exceed certain thresholds that exceed certain thersholds receive an appropriate label, and the vectors of labels are processed by higher-level, goal-directed artificial intelligence algorithms.

Let's consider some examples of feature extraction to see some of the alternatives.

Example (Plasma Etch Reactor Endpoint Detection). The semiconductor manufacturing industry uses plasma reactors to selectively remove materials from the surface of silcon or gallium arsenide wafers [70]. Plasma chemistry is notoriously complex. It is difficult to know how fast etching proceeds and when to halt the process. A popular technique for ending a plasma etch process is to monitor the optical emissions from reaction species for gross changes. The sudden disappearance of an etching byproduct emission indicates the endpoint of the etch cycle. Alternatively, the appearance of optical spectra characteristic of an underlying layer that should be preserved means that the reaction should halt. If a spectrograph monitors the appropriate emissions, its output is digitized, and a computer processes this digital signal, then this at first appears to be a simple real-time edge detection task. However, differences in plasma chemistry across runs, reactor chambers, and semiconductor wafer patterns combine to make this control strategy quite problematic. Moreover, as the semiconductor industry continues to reduce the scale of integrated circuits, etched areas get very small, targeted emission species diminish, and distinct endpoints become increasingly harder to identify. Feature extraction methods in common use for recognizing plasma etch endpoint include the following:

- Estimates of the first or second derivatives of the endpoint trace
- Approxmiations of the signal level within fixed regions of the trace
- Estimates of the departure of the endpoint signal from expected models of the optical emission trace
- Estimates of the endpoint trace curvature in fixed regions of the endpoint trace

Example (Internal Combustion Engine Knock Detection). In automobile engines there is an especially noticeable and harmful abnormal combustion situation known as knock. Knock occurs after the spark plug fires and results from a spontaneous combustion of unburned gasoline vapors in the cylinder. It produces a sharp, metallic, clanking sound, and is very harmful to engine components. By mounting an accelerometer on the engine block and recording the engine vibration signal, Molinaro and Castanié were able to digitally analyze engine knock [86]. From a vibration signal $x_a(t)$, the researchers acquired digital samples $x(n)$, $1 \le n \le N$, and extracted feature vectors $\boldsymbol{v} = (v_1, v_2, ..., v_p)$. The researchers studied a large variety of possible features, v_i, including the signal energy averaged over N samples,

$$v_1 = E_x = \frac{\sum_{n=1}^{N} |x(n)|^2}{N}. \tag{4.93}$$

Another set of features was derived from the histogram values of $|x(n)|$,

$$v_{2+i} = \#\{x(n) \mid k_i \le |x(n)| = k_{i+1}\}, \tag{4.94}$$

where $K = \{k_0, k_1, ..., k_{q-1}\}$, $0 \le i < q$, determines a set of intervals within the range of $|x(n)|$. Another set of parameters, designed to capture the significant periodicities in the $x(n)$ values, are given by $|X(k)|^2$, where, as in our study of texture segmentation in Section 4.3.2, we have

$$X(k) = \sum_{n=0}^{N-1} x(n) \exp\left(\frac{-2\pi jnk}{N}\right). \tag{4.95}$$

Molinaro and Castanié keep $|X(k)|^2$ for $k = 1, 2, ..., \lceil N/2 \rceil$ as feature vector elements. Two other families of feature vector components are more exotic: Prony and cepstral coefficients. The investigators model the trend of $x(n)$ values with a Prony model of order P:

$$x(k) \approx \sum_{k=0}^{p} A_k e^{j\theta_k} e^{n(\alpha_k + j2\pi f_k)}. \tag{4.96}$$

The construction of the model is outside the present scope, but related to the z-transform construction of Chapter 8 [87]. The cepstral coefficients, c_i, are Fourier coefficients of the Fourier transform of |the logarithm of $X(k)$|, given in (4.95):

$$\log|X(k)|^2 \approx \sum_{n=-Q}^{Q} c_i e\left(\frac{-2\pi jnk}{N}\right). \tag{4.97}$$

Both Prony model parameters and cepstral expansion coefficients are useful for compactly encoding the frequency and stability information within a signal. Prony parameters figure in time series analysis studies [88] and cepstral expansions have been widely used in speech detection applications [89]. The Prony model parameters of A_k, θ_k, α_k, and f_k and the cepstral coefficients c_k become feature vector elements. The point of this discussion is not to erode the reader's confidence is selecting feature vectors, but, rather, to show some of the possibilities and variety of feature vector components used in a modern application. In fact, Molinaro and Castanié reduce the number of feature vectors by statistical techniques before invoking their ultimate detection criteria.

4.6.2.2 Distance Measures for Feature Vectors. Let's consider two basic methods for comparing feature vectors. In a typical signal analysis application, feature vectors for a library of prototype signals, or models, may be archived in a database. Feature vectors are similarly extracted from input signals—candidates or unknowns—and compared to the library of prototype vectors.

Once the data reduction from time-series signal to finite feature vector is finished, it is possible to apply the Euclidean norm, from finite-dimensional vector spaces, to match prototype signals to candidates. This is called the *minimum distance classifier.*

Definition (Minimum Distance Classifier). Suppose $\boldsymbol{v}$ is a feature vector extracted from a application input signal, $x(n)$, and $\{\boldsymbol{e}_m \mid 1 \le m \le M\}$ is the set of feature vectors of prototype patterns. Then the minimum distance classifier recognizes $x(n)$ as being of type k if $\|\boldsymbol{v} - \boldsymbol{e}_k\| \le \|\boldsymbol{v} - \boldsymbol{e}_m\|$ for all m, $1 \le m \le M$.

Another popular classifier works on labeled feature vector components. Suppose that signal feature vector components are derived from a partition of the domain of signal $x(n)$; that is, each feature vector component represents a particular subset of Dom(x). Let $\Pi = \{S_1, S_2, ...\}$ be the partition of Dom(x). Suppose also that Λ is labeling of x for Π, Λ: $\Pi \to \{\Lambda_1, \Lambda_2, ...\ \}$. Typically, labels are applied to feature vector components if the component's magnitude exceeds a threshold associated with the feature. If feature vectors have been extracted from library prototypes as well, then candidate signals can be matched against library prototypes by comparing the labelings. One such method of comparing vectors of labels and labels applied to the vector components is the Hamming distance.

Definition (Hamming Distance Classifier). The Hamming distance between a candidate signal label vector $\boldsymbol{u} = (\alpha_1, \alpha_2, ..., \alpha_N\}$ and a prototype vector $\boldsymbol{w} = (\beta_1, \beta_2, ..., \beta_N\}$ is the number of positions in which $\boldsymbol{u}$ and $\boldsymbol{w}$ differ. We write the Hamming distance between label vectors $\boldsymbol{u}$ and $\boldsymbol{w}$ as $H(\boldsymbol{u}, \boldsymbol{w})$.

It is easy, and left as an exercise, to show that the Hamming distance is a metric.

4.6.3 Statistical Pattern Recognition

The third approach to pattern recognition that we will consider is statistical pattern recognition. It is possible to resort once again to a least-squares approach. The least-squares coefficient matrix derives from a large number of samples and represents the known knowledge of input signals to the analysis application [89, 90]. We considered this very general and very powerful approach to the problem of finding an optimal noise-removal filter in Section 4.3. The extension of the method to pattern recognition is straightforward. Of the many statistical approaches, we will examine one of the most important: the Bayes classifier.

4.6.3.1 Bayes Classifier. The Bayes classifier is a fundamental tool in pattern recognition. It is a parametric approach, in that statistical parameters associated with the source patterns (signals or images) are assumed or approximated by the application. As the first chapter already hinted, this pattern classification method relies on Bayes's formula for conditional probabilities.

Statistical pattern recognition, like correlation-based matching and structural recognition, attempts to associate a class or category, $\{C_1, C_2, ..., C_K\}$, with each input signal $\{f_1, f_2, ..., f_N\}$. In order to develop statistics for each input signal, the signal is decomposed into a feature vector, $\boldsymbol{v} = (v_1, v_2, ..., v_M)$. Each component of the feature vector is numerical, in order to develop statistics for the likelihood of features, classes, and for features within signal classes. Selecting the right feature vector is both quite open to design and quite critical to the recognition application's success.

Standard pattern recognition texts cover the Bayes classifier [91–93]. One of the earliest applications of Bayes classifiers to the design of a character recognition system was by Chow and dates from the 1950s [94].

4.6.3.2 Statistical Decision Rules. This section explains how statistical decision rules for deciding class membership can be made based on the likelihood of each class of signal occurring and on the probability distributions of feature vectors among classes. We examine the problem in several ways. First we frame the problem as a search for a set of discriminant functions that indicate the resemblance of a signal to members of a class. Then we consider the problem of finding risk functions. That is, we seek functions that measure the risk of misclassification, and our problem transposes to one of minimizing the risk of a mistaken classification of a signal. Finally, we pose the problem in terms of the Bayes rule for conditional probabilities. We find that this provides a reasonable statistical tool for building discriminant and risk functions, although there are a number of probability density functions that must be determined.

Discriminant and risk functions are closely related. Suppose that we assigning signals $\{f_1, f_2, ...\}$ to classes $C = \{C_1, C_2, ..., C_K\}$. For each signal, $f \in \{f_1, f_2, ...\}$, a feature vector, $\boldsymbol{v} = (v_1, v_2, ..., v_M)$, is generated. We desire a set of discriminant functions, $D_1, D_2, ..., D_K$, one for each signal class. The idea is that $D_k(\boldsymbol{v})$ tells us how strongly signals with features $\boldsymbol{v}$ resemble signals from class C_k. A discriminant-based classifier assigns signal f with feature vector $\boldsymbol{v}$ to class C_k if $D_k(\boldsymbol{v}) > D_i(\boldsymbol{v})$ for all $i \neq k$. The complementary idea is the risk function. Now we seek functions, R_1, $R_2, ..., R_K$, such that $R_k(\boldsymbol{v})$ tells us the risk of classifying f with features $\boldsymbol{v}$ as belonging to class C_k. How strongly do signals with features $\boldsymbol{v}$ resemble signals from class C_k? A risk-based classifier places signal f with feature vector $\boldsymbol{v}$ into class C_k if $R_k(\boldsymbol{v}) < R_i(\boldsymbol{v})$ for all $i \neq k$. Taking $R_k(\boldsymbol{v}) = -D_k(\boldsymbol{v})$ makes an easy transition from a discriminant-based classifier to a risk-based classifier.

Now let's consider how to use statistical information about feature vectors and classes to develop statistical discriminant functions. Suppose that we know the a priori probability of occurrence of each of the classes C_k, $P(C_k)$. Suppose further that for each class, C_k, we know the probability density function for the feature vector $\boldsymbol{v}$, $p(\boldsymbol{v}|C_k)$. The conditional probability, $P(C_k|\boldsymbol{v})$, provides the likelihood that class k is present, given that the input signal has feature vector $\boldsymbol{v}$. If we could compute $P(C_k|\boldsymbol{v})$ for each C_k and $\boldsymbol{v}$, then this would constitute a statistical basis for selecting one class over another for categorizing the input signal f. But the Bayes formula for conditional probabilities (Chapter 1) provides a tool for calculating this a posteriori probability:

$$P(C_k \mid v) = \frac{p(v \mid C_k)P(C_k)}{p(v)} = \frac{p(v \mid C_k)P(C_k)}{\sum_{i=1}^{K} p(v \mid C_i)P(C_i)}, \tag{4.98}$$

where $p(v)$ is the probability density function for feature vector v. This inspires the Bayes decision rule:

Definition (Bayes Decision Rule). Given a signal f and a feature vector v derived from f, the Bayes Decision Rule is to classify f as belonging to class $C_k \in C = \{C_1, C_2, ..., C_K\}$ if $P(C_k|v) > P(C_i|v)$ for all $i \neq k$; otherwise, classify f as belonging to some C_k where $P(C_k|v)$ is maximal.

We can also improve upon our concept of risk by incorporating the probabilistic ideas. Suppose that the cost of assigning signal f with feature vector v to class C_k, when it really belongs to class C_i is $r(k, i)$, where $r(k, i) = 0$ if $k = i$. Typically, $r(k, i)$ is a 1 or 0, for the cost of an error or the cost of no error, respectively. Then, the total cost simply counts the number of misclassifications. In a real application, it may be necessary to provide more informative risk estimates and cost counters. For example, in a character recognition system, the cost might be the time estimate for human assistance to the application. Then, incorporating the a posteriori probabilities from Bayes's formula, the risk of placing f in class k, $R(C_k, v)$, is the sum of the individual misclassification risks,

$$R(C_k, v) = \sum_{i=1}^{K} r(k,i)P(C_k \mid v). \tag{4.99}$$

Classifying signal f with feature vector v as belonging to class C_k according to whether $R(C_k, v) \leq R(C_i, v)$ for all $i \neq k$ implements the Bayes decision rule (4.99).

Now let us consider possible discriminant functions based on the Bayes formula (4.98). We may begin by assigning the discriminants $D_k(v) = P(C_k|v)$, but noting that, given feature vector v, for each such k, $1 \leq k \leq K$, the denominator in (4–6.1) is identical, we can simplify the discriminants to have

$$D_k(\mathbf{v}) = p(\mathbf{v} \mid C_k)P(C_k). \tag{4.100}$$

And it will be convenient in a moment to take the natural logarithm of (4.100) to obtain the alternative discriminant function

$$D_k(\mathbf{v}) = \log(p(\mathbf{v} \mid C_k)) + \log(P(C_k)). \tag{4.101}$$

Indeed, any constant can be added to each of a family of discriminant functions without changing their characterization capability. Each of the family of discriminants may be multiplied by any positive constant. Finally, a monotonic increasing function, such as the logarithm, may be be applied to all of the discriminants without affecting the classification results. These operations are often useful for simplifying the evaluation of discriminants.

4.6.3.3 Estimating Probabilities. Let us now turn to the practical problem of how to estimate the probabilities that are necessary for applying the Bayes rule for pattern recognition. Upon examinination of (4.98), we see that applying the Bayes formula assumes that we have knowledge of:

- The a priori probability of class occurrence C_k, $P(C_k)$
- The probability density function for the feature vector $\boldsymbol{v}$, given the presence of a signal of class C_k, $p(\boldsymbol{v}|C_k)$

How are we to estimate these probabilities?

It may be possible to know the class probabilities, $P(C_k)$, from the nature of the application. For example, in a character recognition application, the class probabilities come from known probabilities of the presence of characters in text. This could come from a broad scientific study of the problem area. To wit, the development of classifiers for the optical character recognition systems used by the United States Postal Service required the collection and analysis of hundreds of thousands of character samples [95–97]. And lacking the luxury of a broad scientific study, the character recognition classifier designer can perform a sampling of representative texts. In the case of speech phoneme recognition, these probabilities are known, but if the application concerns a special application area (i.e., spoken numbers for a telephone dialing application utilizing voice recognition), then more uniform probabilities apply. There are some subtleties, to be sure. In a spoken number recognition system, one must be alert to common alternative pronunciations for numbers. The system designer might assume that $P(C_1) = \cdots = P(C_9) = .1$, with C_n signifying the utterance of a nonzero number. But two common alternatives exist for saying "zero": "zero" and "oh." Thus, the design might posit two classes for this utterance; thus, $P(C_0) = P(C_z) = 0.05$, for saying "oh" or "zero," respectively.

The class conditional feature probabilities are a further problem. If we conduct a thorough statistical study of a large number of samples from each class, then it is possible to arrive at rough ideas of the distribution functions. But when a signal f arrives at the input to the system, and the front-end processing modules derive its feature vector v, how are we to compute the probability density functions $p(\boldsymbol{v}|C_k)$ for each class C_k? If we answer that we should once more adopt a parametric stance—that is, we assume a particular probability density function for the conditional densities required by (4.98)—then there is a elegant and practical resolution to this problem.

It should come as no surprise that the distribution of choice is the Gaussian. Not only is it justified by the Central Limit Theorem, but it provides the most tractable mathematical theory. Since we are dealing with feature vectors, we must consider the multivariate normal distribution. For the probability density function $p(\boldsymbol{v}|C_k)$, this parametric assumption is

$$p(v\,|\,C_k) = \frac{\exp[-\frac{1}{2}(\boldsymbol{v}-\mu_k)^T \Sigma_k^{-1}(\boldsymbol{v}-\mu_k)]}{\sqrt{\det(\Sigma_k)(2\pi)^M}}, \tag{4.102}$$

where $\boldsymbol{v} = (v_1, v_2, ..., v_M)$ is the feature vector of length M extracted from signal f; μ_k is the mean for feature vectors from signals of class C_k, $\mu_k = E[\boldsymbol{v}|C=C_k]$; $(\boldsymbol{v} - \mu_k)^T$ is the transpose of $\boldsymbol{v} - \mu_k$; and Σ_k is the $M \times M$ covariance matrix for feature vectors of C_k signals, $\det(\Sigma_k)$ is its determinant, and Σ_k^{-1} is its inverse.

The parametric assumption allows us to estimate the conditional probabilities, $p(\boldsymbol{v}|C_k)$. This requires a preliminary classifier training step in order to establish the statistical parameters that the analysis application uses to classify incoming signals. Data from a large number of signals for each of the different classes $\{C_1, C_2, ..., C_K\}$ is collected. For each such signal, its feature vector $\boldsymbol{v}$ is derived. The average of the feature vectors from class C_k is μ_k. Once we have computed the means for all of the feature vectors from signals in a class C_k, then we can compute the covariance matrices, $\Sigma_k = E[(\boldsymbol{v}-\mu_k)(\boldsymbol{v}-\mu_k)^T$. Once the covariance matrix is computed for a class C_k, its determinant and inverse can be calculated (this is problematic when the number of features is large). The feature vector averages, $\{\mu_1, \mu_2, \ldots, \mu_K\}$, and the covariance matrices, $\{\Sigma_1, \Sigma_2, \ldots, \Sigma_K\}$, are stored for analyzing the signals that the classifier accepts as inputs. This completes the training of the classifier.

The steps in running the classifier are as follows. For every signal f with feature vector $\boldsymbol{v}$, the Bayes classifier

- Computes (4.102) for each class C_k;
- Calculates the a posteriori probability $P(C_k|\boldsymbol{v})$ for each k, $1 \le k \le K$;
- Classifies signal f as belonging to class C_k where $P(C_k|\boldsymbol{v})$ is maximal.

4.6.3.4 Discriminants. Now let us consider different discriminant functions that support a Bayes classifier. Different discriminants arise from the statistics of the feature vectors of input signals to the classifier. In particular, special cases of the covariance matrix Σ_k in (4.102) result in significant simplifications to the classifier's computations.

Let us first look at an alternative discriminant function for (4.98), namely the natural logarithm. Since the natural logarithm is monotone increasing, taking $\log(P(C_k|\boldsymbol{v}))$ provides the same recognition decision as $P(C_k|\boldsymbol{v})$ itself. Thus, if we assume normally distributed feature vectors (4.102), then we have

$$\begin{aligned} D_k(\boldsymbol{v}) &= \log(p(\boldsymbol{v}\,|\,C_k)) + \log(P(C_k)) \\ &= -\frac{1}{2}(\boldsymbol{v}-\mu_k)^T \Sigma_k^{-1}(\boldsymbol{v}-\mu_k) - \frac{m\log(2\pi)}{2} - \frac{\log(\det(\Sigma_k))}{2} + \log(P(C_k)). \end{aligned} \tag{4.103}$$

The term $m\log(2\pi)/2$ does not depend on C_k, so it can be ignored; thus, we may set

$$D_k(\boldsymbol{v}) = -\frac{1}{2}(\boldsymbol{v}-\mu_k)^T \Sigma_k^{-1}(\boldsymbol{v}-\mu_k) - \frac{\log(\det(\Sigma_k))}{2} + \log(P(C_k)). \tag{4.104}$$

In (4.104), $\det(\Sigma_k)$ and $P(C_k)$ can be calculated from the training data, so only the vector product $(\boldsymbol{v}-\mu_k)\Sigma_k^{-1}(\boldsymbol{v}-\mu_k)^T$ needs to be calculated for the feature vector $\boldsymbol{v}$ of every input signal f.

Three simplifying assumptions make the discriminants (4.104) easier to compute:

- The a priori probabilities $P(C_k)$ of the signal classes are all equal.
- The features vectors are statistically independent and have the same variance.
- The convariance matrices are all the same.

In the first case, we may drop the $\log(P(C_k))$ term from (4.104). This helps, but the inversion and determinant of the covariance matrix pose a greater threat to computational tractability.

In the second of these special cases we find $\Sigma_k = \sigma^2 I$, where I is the $M \times M$ identity matrix. Then $\Sigma_k^{-1} = \sigma^{-2} I$, which is independent of class. This allows us to trim the discriminant functions further, so that now, having removed all class-independent terms, we arrive at

$$D_k(v) = -\frac{1}{2\sigma^2}(v - \mu_k)^T (v - \mu_k) + \log(P(C_k)). \tag{4.105}$$

Making the additional assumption that class membership likelihood is the same for all C_k, we find that maximizing $D_k(v)$ in (4.105) is the same as minimizing $(v - \mu_k)(v - \mu_k)^T = \|v - \mu_k\|^2$. This classifier we are already familiar with from Section 4.6.2.2. It is the minimum distance match between features drawn from the input signal f and the mean feature vectors. In other words, a Bayes classifier with statistically independent features of equal variance reduces to a minimum distance classifier. The class C_k for which feature vector v is closest to μ_k is the class to which we assign input signal f.

The Bayes classifier is optimal; and when the feature vectors obey a Gaussian distribution, the discriminant functions (4.103) are the proper tool for separating input signals into classes. There are nevertheless some important difficulties with the Bayes classifier, and we must note them. The feature vectors may not, in point of fact, be normally distributed. This makes the computation of the discriminant problematic. Sometimes, alternative features that more closely follow a Gaussian distribution can be selected. Furthermore, if the number of features is large, the computation of the determinant and inversion of the covariance matrix become intractable. There are, finally, some philosophical reasons for objecting to the Bayes classifier [98].

4.7 SCALE SPACE

This section studies a signal analysis technique known as *scale-space decomposition*. From our first studies of representative signal interpretation problems, we noted that the determination of the size of a signal component is a critical step in analyzing the signal. One task in automated electrocardiography, to recall an example from the first chapter, is to distinguish between splintered and normal contractions of the heart's left ventricle. It is the time-domain extent of the signal's jump that determines whether there is a normal contraction or an abnormal, spasmodic contraction. And in edge detection, we noted that understanding scale is a necessary part of simple edge detection. For an edge at a fine scale could just as well be considered an insignificant signal aberration at a coarse scale.

Soon after researchers began examining the magnitude of signal derivatives in study of edges, their interest extended to signal curvature, which is a local measure of signal shape. There are several quite natural reasons for this transition. Psychological experiments revealed that a few curved line fragments suffice to impart object shape information—Attneave's famous example is a sleeping cat—to a human subject [99]. Also, from calculus, the sign of the second derivative determines the curvature of a signal: Where the second derivative is positive defines a concave up region, where it is negative defines a concave down region, and where it is zero an inflection point exists. The problem is that the second derivative information is very noisy in real signals and images, resulting in erratic segmentation in regions of different curvatures. Researchers turned increasingly toward multiple resolution methods that would support the precious information content from curvature. Hierarchical methods for image processing and recognition within natural scenes were disclosed, for example [100, 101].

We have already noted Marr's contribution to multiscale edge detection, which formed the cornerstone of a very important trend in signal and image analysis [77]. Marr oriented the attention of the scientific and engineering communities to the links between the algorithms of engineered systems and the processes within biological systems—for example, animals [102]. Marr's strategy was to study animal sensory systems, especially vision, and from them derive the inspiration for machine vision system designs. Researchers in the new field of psychophysics, which studies the brain's sensory processes at a stage where they are still independent of consciousness, had found evidence of multiple resolution, orientation-sensitive processing channels in animal vision systems [103–105]. Aware of the biological vision system research, recognizing its link to the multiple resolution image analysis efforts, and building upon their earlier work in edge detection, Marr and his coworkers proposed a three-stage architecture for vision systems:

- The raw primal sketch, which segments the signal into edges and concavity regions;
- The extraction of geometric information relative to the observer;
- The determination of geometric information independent of the observer.

While Marr's scheme is a vision system formulation, we wish to focus on signal interpretation and shall reduce his architecture into one dimension. Let's replace Marr's visual terms with auditory terms. Thus, we ought to look not for a raw primal sketch so much, perhaps, as a raw primal listen. And we ought to think of signal or sound content rather than think of geometric information, with its planar and spatial connotations. An example should help. When you hear a whistle at a railroad crossing, the raw primal sketch consists of the auditory edge at the onset of train's whistle and the increasing intensity thereafter. The sudden realization that a train is coming closer to you constitutes the observer–relative signal content. The reflection that a train is about to cross through the intersection makes up the observer-independent information in the experience.

Computationally, the raw primal sketch begins by convolution of the signal with the second derivative of the Gaussian. If $f(t)$ is a signal and $g(t, \sigma)$ is the Gaussian of standard deviation σ and zero mean, then we let $F(t, \sigma)$ be defined by

$$F(t, \sigma) = f(t) * \frac{\partial^2}{\partial x^2} g(t, \sigma). \tag{4.106}$$

We recognize this as the edge detection operation of Marr and Hildreth [62]. The next raw primal step is to link the edges together and segment the signal into regions of concavity (concave down) and convexity (concave up). The remaining steps in Marr's schema, the development of signal information relative to the observer and signal information independent of the observer, then follow.

Note that there does not seem to be any clear direction within Marr's algorithm to decide at what scales σ the edge detection operation (4.106) should be developed. The second two steps in the algorithm are less clear than the first, and they contain a number of thresholds and parameters whose values are difficult determine without extensive experimentation. Finally, there is no clear relation between the primal sketch information in the different channels. How does the content of a large σ channel affect the processing of the conent of a small σ channel? Does processing occur from large to small or from small to large σ channels? These difficulties with the conception brought some researchers to critique Marr's architecture. Prominent among the skeptics, Pentland, deemed it to be too data-driven and lacking in its qualitative aspects [106]. Soon enough, however, in the early 1980s—beginning what would turn out to be a decade of signal analysis research breakthroughs—an important step in demonstrating qualitative computer vision strategies was taken by Witkin [107] and by Koenderink [108] with the concept of scale-space representation.

Like Marr's theory, the scale-space representation smoothes an image with a family of Gaussian filters. There are two substantive differences, however: A full set of smoothed images is maintained, and there is a critical interconnection between the regions of concavity at different scales, σ. We shall see that a complete description of the signal's shape results. The description proceeds from the smallest to the largest scale, and each concavity feature of the signal ranked according to its significance.[6]

Scale space decomposition thus furnishes a useful signal analysis paradigm. It identifies concavity as a critical feature of signals. It highlights the link that this feature shares with biological vision systems, as, for instance, a scattering of curved lines immediately suggests a shape. It shows how the features at one scale affect those at another scale. We shall soon see how to derive a complete graphical or

[6]Interestingly, the first exploration of scale-space decomposition was in the area of theoretical economics. James L. Stansfield, working at the Artificial Intelligence Laboratory at the Massachusetts Institute of Technology in the late 1970s, studied zero crossings under Gaussian smoothing while tracking commodity trends ["Conclusions from the commodity expert project," MIT AI Laboratory Memo, No. 601, November 1980].

structural description of the signal from this multiscale concavity information. Furthermore, this structural description enables us to readily build pattern recognition or object matching applications. We can nevertheless use it to perform a time-domain analysis of a signal, identify signal features of different scale, and derive a nonsignal structure that is useful for interpreting the signal. That is the goal of signal analysis, and scale space decomposition is the methodological exemplar.

We will first examine scale space as originally conceived—in analog form. We will consider the type of nonsignal structures that the scale space decomposition produces for a signal. We will state and prove (for simple, but important classes of signals) the theorems that give the method its power. And we shall highlight some of the drawbacks of the classic continuous-time form of scale-space decomposition. Interestingly enough, it was not until some years after the development of the analog scale-space theory that discrete versions of the theory were discovered. We will look at some approaches to discrete scale-space decomposition and close this section with some applications.

4.7.1 Signal Shape, Concavity, and Scale

Scale-space decomposition of a signal begins by smoothing the signal with a family of Gaussian filters. The smoothing operation is a convolution, so it is linear and translation invariant (Chapter 3). Furthermore, all of the smoothing kernels are the same, except for the standard deviation of the Gaussian, σ, which increases with the amount of smoothing performed. This procedure, quite familiar after our experiments with edge detection, produces a series of representations of the signal at different scales or resolutions. The highest resolution (and hence the smallest scale) occurs with the original signal, and we may derive coarser resolution representations of the signal by increasing σ. This collection of smoothed versions constitutes a scale space decomposition of the signal.

The next idea is to look for regions of curvature in the scale space decomposition's signals. The signal derivative remains the principal tool for recovering signal shape. Recall from calculus that the extrema of the nth derivative of a signal d^nf/dt^n are the zeros of its next higher derivative, $d^{(n+1)}f/dt^{(n+1)}$. Consider in particular the sign of the second derivative of a signal. (Withhold for a moment the objection, drawn from general intuition and practical edge detection endeavors, that this derivative is noisy and misleading.) Where $d^2f/dt^2 < 0$, the signal is concave down; where $d^2f/dt^2 = 0$ and $d^3f/dt^3 \neq 0$ there is a zero crossing of the second derivative, or a point of inflection, using calculus parlance; and regions where $d^2f/dt^2 > 0$ are concave up, or convex. Thus, the sign of the second derivative is the basis for signal segmentation (Section 4.1).

Definition (Curvature). Curvature is a measure of how rapidly the graph of a signal, $G = \{(t, f(t)): t \in \mathrm{Dom}(f)\}$ is turning in on itself. More formally, the osculating circle to the graph G at a point $(t_0, f(t_0)) \in G$ is the circle that is tangent to the curve at $(t_0, f(t_0))$. And the curvature κ equals $1/\rho$, where ρ is the radius of the osculating circle to the graph G.

Note that curvature can be obtained by fitting a polynomial to the signal values and computing the derivative of the polynomial model. Alternatively, a circle can be fitted to the signal data. A least-squares formulation of the problem exists and is widely used in the analysis of machined surfaces [33].

Proposition. The curvature has an analog signal representation as well; letting $y = f(t)$ and denoting differentiation by $y'(t)$, it is given by

$$\kappa(t) = \frac{y''(t)}{\left(1+(y'(t))^2\right)^{3/2}}. \tag{4.107}$$

Proof: Calculus; also Ref. 109. ∎

From the proposition, we can classify concave-up and concave-down regions of the signal not only by the sign of the curvature (positive and negative, respectively), but also by the magnitude of the curvature. For a candidate signal analysis method, this is an attractive concept for signal structure. It has considerable descriptive power. There is a link with animal vision systems. There is an important geometric connotation as well, via the idea of the osculating circle. The problem with curvature is that, despite its descriptive power and its link with biological vision systems, in the presence of signal noise it is quite problematic. This objection has likely occurred to many readers, and we need to address the issue now. Gaussian smoothing removes the local bumpiness of the signal. At larger scales, when the smoothing Gaussian kernel has a larger variance, the concave and convex regions of the signal that persist must be more significant.

Suppose $f(t)$ is an analog signal and $G(t, \sigma, \mu) = \sigma^{-1}(2\pi)^{-1/2}\exp(-(x-\mu)^2/(2\sigma^2))$ is the Gaussian with standard deviation σ and mean μ. Let $g(t, \sigma) = G(t, \sigma, 0)$. Convolution of $f(t)$ with $g(t, \sigma)$ gives $F(t, \sigma)$:

$$F(t,\sigma) = f(t) * g(t,\sigma) = \frac{1}{\sigma\sqrt{2\pi}} \int_{-\infty}^{+\infty} f(u)\exp\left(-\frac{(x-u)^2}{2\sigma^2}\right) du. \tag{4.108}$$

We are concerned with the behavior of concavity regions as we vary σ in (4.108). But the regions where the signal is concave down and concave up are separated by those where the second derivative of $F(t, \sigma)$ is zero:

$$\frac{\partial^2}{\partial t^2} F(t,\sigma) = F_{tt}(t,\sigma) = 0. \tag{4.109}$$

Thus, we can track the concavity regions of a signal by simply keeping track of the zero crossings of the second derivative (4.109). Notice that derivatives of $F(t, \sigma)$ can be computed by the convolution of $f(t)$ with the Gaussian's derivative of the same order:

Proposition. Let $f(t)$ be an analog signal and let $g(t, \sigma)$ be the zero mean Gaussian with variance σ^2. Then

$$\frac{\partial^n}{\partial t^n} F(t,\sigma) = f * \frac{\partial^n}{\partial t^n} g(t,\sigma). \tag{4.110}$$

Proof: Write out the convolution integral for (4.110) and interchange the order of the differentiation and integration. ■

Before this gets too abstract, let us consider an example. Consider a fourth-degree polynomial with two concave-up regions surrounding a narrow concave-down region. At each scale σ we determine the zero crossings of the second derivative (4.109), and over a range of scales we can draw a contour plot of the zero crossings. Notice that with sufficient smoothing the concave-down region disappears. As σ increases, the locations of the two zero crossings get closer together, eventually meet, and then there is only a convex region. For this example, once we mark the regions as concave or convex, this marking remains; a concave region does not merge with a convex region.

If this behavior is general, then we have a very great simplification in our task of segmenting the various smoothed versions of $f(t)$. All we have to do is follow the zero crossings. Where a pair meet, we know that smoothing has obliterated a convex (or concave) region and two surrounding concave (or convex, respectively) regions will merge. We can use this simplifying assumption if we know that Gaussian smoothing never creates new zero crossings, but may only destroy them as the scale of the smoothing increases. Let's pursue this line of thinking. Consider how the contour plot of zero crossings might look were the smoothing at some scale σ_0 to create a new zero crossing located at time t_0. In this case, we know that for coarser scales, $\sigma > \sigma_0$, there are regions of opposite concavity on either side of time t_0. On which side of t_0 does the concave-up region lie? We have to reexamine the signal smoothed at scale σ each time such a new zero crossing appears during the smoothing process. Depending on the complexity of the signal, this could be quite a chore! Could a large number of zero crossings abruptly appear at some scale? Could the number of regions we have to type according to concavity increase forever as we continue to convolve with Gaussians of ever-wider support? Indeed, what kind of "smoothing" do we have here that puts new wrinkles in our signal as we proceed?

Fortunately, there is a deep theoretical result for scale space decomposition that relieves us of all of these worries. The theorem is that Gaussian smoothing (4.108) never introduces additional structure as the scale parameter s increases. That is, new zero crossings of the second derivative of $F(t, \sigma)$ do not appear with increasing σ. There is a converse too: If a convolution kernel never introduces additional structure, then it must be the Gaussian. Together, these two results are the foundation of scale space theory.

4.7.2 Gaussian Smoothing

A variety of factors motivates the use of the Gaussian signal for smoothing a signal. Of course, tradition among signal analysts is one reason for using it. It was one of the smoothing operators used in some of the earliest edge detection efforts, and we found in Section 4.5.3 that the derivative of the Gaussian is an optimal step edge finder. But the Gaussian and its derivatives have a number of attractive properties, and these properties motivate its use for edge detection as well as for the more general scale space approach to signal analysis.

For scale-space decomposition the Gaussian is well-behaved. In particular, it has the following properties:

- (Symmetry Property) It is symmetric and strictly decreasing about its mean.
- As $\sigma \to 0$, $F(t, \sigma) \to f(t)$; that is, for small scales σ, the smoothed signal resembles the original.
- As $\sigma \to \infty$, $F(t, s) \to E(f(t))$; that is, for large scales σ, the smoothed signal approaches the mean of $f(t)$.
- The Gaussian is an $L^1(\mathbb{R})$ signal (absolutely integrable), and it is C^∞ (infinitely differentiable).
- (Causality Property) As σ increases, zero crossings of $F_{tt}(t, \sigma)$ may disappear, but new ones cannot arise.

While the first four of the above properties are quite nice, the Causality Property is so important that it elevates Gaussian smoothing, in a specific sense, to the status of the only possible choice for a scale-space smoothing kernel. And as indicated in the previous section, this property has an important converse, namely, that the Gaussian is unique in this regard.

The following Scale-Space Kernel Conditions formalize the above properties. We need to state these conditions for a general, candidate smoothing kernel $k(t, \sigma)$. We shall invoke these conditions later, in the course of proving our theorems. These are basic properties that we require for the filtering kernel of any scale-based signal decomposition, and so we state the conditions as a definition.

Definition (Scale-Space Kernel Conditions). A function $k(t, \sigma)$ is a scale-space kernel if it satisfies the following five conditions:

1. $k(t, \sigma)$ is the impulse of a linear, translation-invariant system: If $f(t)$ is an analog signal, then the smoothed version of $f(t)$ at scale σ is given by $F(t, \sigma) = f(t)*k(t, \sigma)$.
2. For different values of σ, $k(t, \sigma)$ should always maintain the same fundamental shape: $k(t, \sigma) = (1/\sigma^2)m(t/\sigma)$ for some one-dimensional signal $m(u)$.
3. As σ decreases, $k(t, \sigma)$ approaches the Dirac delta $\delta(t)$, so that $F(t, \sigma)$ approaches $f(t)$: as $\sigma \to 0$, we have $k(t, \sigma) \to \delta(t)$.

4. $k(t, \sigma)$ is an even signal; as $\sigma \to \infty$, or as $t \to \infty$, we have $k(t, \sigma) \to 0$.
5. The Causality Property holds for $k(t, \sigma)$.

While they might appear much too specific and technical upon first inspection, the Scale-Space Kernel Conditions are quite well-motivated. By requiring that the decomposition be linear and translation-invariant, the Convolution Theorem for Analog LTI Systems (Chapter 3) allows us to write the smoothing operation as a convolution.

4.7.2.1 *Sufficiency of the Gaussian.* Let us prove that the Gaussian is sufficient to produce a scale-space decomposition of a signal which does not create zero crossings in the smoothed second derivatives of $F(t, \sigma)$. Again, suppose $G(t, \sigma, \mu) = \sigma^{-1}(2\pi)^{-1/2}\exp(-(x-\mu)^2/(2\sigma^2))$ is the Gaussian with standard deviation σ and mean μ, and set $g(t, \sigma) = G(t, \sigma, 0)$. Suppose that the smoothed signal $F(t, \sigma)$ is given by the convolution of signal $f(t)$ with $g(t, \sigma)$:

$$F(t, \sigma) = f(t) * g(t, \sigma) = \int_{-\infty}^{+\infty} f(u) g(t-u, \sigma)\, du. \tag{4.111}$$

Our five-step proof relies on concepts from calculus:

- We consider the zero crossings of the second derivative $F_{tt}(t, \sigma)$ in (4.111) as curves in the (t, σ) plane.
- This allows us to invoke the implicit function theorem and second derivative conditions for a local maximum along the (t, σ) plane curves.
- We develop some straightforward characterizations of the Causality Property in a proposition.
- Any signal $k(t, \sigma)$ that is a solution of a particular form of the heat diffusion equation also satisfies one of the proposition's equivalent conditions for the Causality Property.
- Finally, since the Gaussian does solve the diffusion equation, we know that it provides a scale-based decomposition that eliminates structure as the scale of the signal smoothing increases.

To begin with, let us vary σ and observe the behavior of zero crossings of the second derivative of $F(t, \sigma)$. Let $E(t, \sigma) = F_{tt}(t, \sigma) = (\partial^2/\partial t^2)F(t, \sigma)$. Zero crossings are solutions of $E(t, \sigma) = 0$, and such pairs form curves in the (t, σ) plane. The curves may extend over the entire range of scales for which smoothing is performed, say from $0 \le \sigma \le \sigma_{max}$. Possibly, the curve has a local minimum or maximum at a certain time value, $t = t_0$. This situation allows us to write σ as a function of t along the curve: $\sigma = \sigma(t)$. Our structural description for $f(t)$ at scale σ is its segmentation into regions that are concave-up and concave-down, bounded by the zero crossings where $E(t, \sigma) = 0$. The desired Causality Property tells us that such zero crossings cannot increase with scale. Equivalently, this means that as smoothing proceeds, it continues to remove structure from the signal. This is the crucial idea: We can formulate the condition that zero crossings diminish as the scale of smoothing enlarges by examining the behavior of the curves $\sigma(t)$ where $E(t, \sigma) = 0$ is an extremum in (t, σ) space. The next result provides some elementary facts about zero crossing curves. The next proposition provides some basic results, useful for showing that when the scale of Gaussian filtering increases, the structural detail of a signal diminishes.

Proposition (Zero Crossing Conditions). Consider the curve $E(t, \sigma) = 0$ in the (t, σ) plane. Let the time variable t parameterize the curve $\sigma = \sigma(t)$, so that $E(t, \sigma) = F_{tt}(t, \sigma) = F_{tt}(t, \sigma(t))$ is parameterized by t as well. Then:

- The Causality Property holds if and only if each local extremum of $\sigma(t)$ is a local maximum: $\sigma'(t_0) = 0$ implies $\sigma''(t_0) < 0$.
- Along the curve E, $\sigma' = d\sigma/dt = -E_t/E_\sigma = -(\partial E/\partial t)/(\partial E/\partial \sigma)$.
- The Causality Property holds if and only if whenever $\sigma'(t_0) = 0$, then

$$\left.\frac{\partial^2 \sigma}{\partial t^2}\right|_{t=t_0} = -\left.\frac{E_{tt}}{E_\sigma}\right|_{t=t_0} < 0. \tag{4.112}$$

Proof: Condition (i) is a differential calculus formulation of our observation that an arch-shaped curve $E(t, \sigma) = 0$ is acceptable, while a trough-shaped curve is not.

To see the second condition, first note that we can parameterize the curve $E(t, \sigma) = 0$ with some parameter, say u. By the chain rule for vector-valued functions,

$$\frac{dE}{du} = \nabla E \cdot \left(\frac{dt}{du}, \frac{d\sigma}{du}\right) = \frac{\partial E}{\partial t}\frac{dt}{du} + \frac{\partial E}{\partial \sigma}\frac{d\sigma}{du}. \tag{4.113}$$

Since $E(t, \sigma) = 0$ along the curve, the derivative dE/du in (4.113) is also zero along the curve. Next, we may choose the parameterizing variable, $u = t$, the time variable. This little ruse produces

$$\frac{dE}{dt} = 0 = \frac{\partial E}{\partial t}\frac{dt}{dt} + \frac{\partial E}{\partial \sigma}\frac{d\sigma}{dt} = E_t + E_\sigma \frac{d\sigma}{dt}, \tag{4.114}$$

which gives condition (ii).

Condition (iii) follows from the first two. Indeed, applying the quotient rule for derivatives to condition (ii), we find

$$\frac{d^2\sigma}{dt^2} = -\frac{E_{tt}}{E_\sigma} + \frac{E_t}{(E_\sigma)^2}. \tag{4.115}$$

From (4.114), note that $\sigma'(t_0) = 0$ if and only if $E_t = 0$ at $t = t_0$. Thus, (4.115) ensures yet another equivalent condition: $\sigma''(t_0) = -(E_{tt}/E_\sigma)$ at $t = t_0$. The inequality in condition (iii) follows from condition (i), completing the proof. ■

This next theorem is a basic theoretical result in scale-space decomposition. It shows that a Gaussian filtering kernel can be the foundation for a scale-based signal decomposition method, which removes signal structure as the scale of the smoothing increases.

Theorem (Sufficiency). Suppose that convolution with the Gaussian, $g(t, \sigma) = G(t, \sigma, 0) = 2^{-1/2}\sigma^{-1}\exp(-t^2/(2\sigma^2))$, smoothes the signal $f(t)$ at scale σ to produce $F(t, \sigma)$:

$$F(t,\sigma) = f(t) * g(t,\sigma) = \int_{-\infty}^{+\infty} f(u)g(t-u,\sigma)\,du = \frac{1}{\sigma\sqrt{2}}\int_{-\infty}^{+\infty} f(u)e^{-\frac{(t-u)}{2\sigma^2}}\,du. \tag{4.116}$$

Then the Causality Property holds; that is, $F_{tt}(t, \sigma)$ zero crossings may disappear—but can never appear—as σ increases.

Proof: Consider the partial differential equation

$$\frac{\partial^2 U}{\partial t^2} = \frac{1}{\sigma}\frac{\partial U}{\partial \sigma}. \tag{4.117}$$

This is a form of the heat or diffusion equation from physics, introduced already in Chapter 1. We can easily check that the Gaussian $g(t, \sigma)$ solves (4.117) by calculating:

$$g_{tt}(t,\sigma) = \frac{\partial^2 g}{\partial^2 t} = \left(\frac{t^2}{\sigma^4} - \frac{1}{\sigma^2}\right) g(t,\sigma) \tag{4.118}$$

and

$$g_\sigma(t,\sigma) = \frac{\partial g}{\partial \sigma} = \left(\frac{t^2}{\sigma^3} - \frac{1}{\sigma}\right) g(t,\sigma). \tag{4.119}$$

so that (4.117) holds for $U(t, \sigma) = g(t, \sigma)$. However, we can show that $F(t, \sigma)$ and hence $E(t, \sigma)$ satisfy the diffusion equation as well. In fact, since

$$F_{tt}(t,\sigma) = \frac{\partial^2}{\partial t^2} \int_{-\infty}^{+\infty} f(u)g(t-u,\sigma)\,du = \int_{-\infty}^{+\infty} f(u)g_{tt}(t-u,\sigma)\,du = \int_{-\infty}^{+\infty} f(u)\frac{1}{\sigma}g_{\sigma}(t-u,\sigma)\,du$$

$$= \frac{1}{\sigma}\frac{\partial}{\partial\sigma}\int_{-\infty}^{+\infty} f(u)g(t-u,\sigma)\,du = \frac{1}{\sigma}F_{\sigma}(t,\sigma) \tag{4.120}$$

we have $E_{tt} = F_{tttt} = (1/\sigma)F_{\sigma tt} = (1/\sigma)F_{tt\sigma} = (1/\sigma)E_{\sigma}$. This shows that $E(t, \sigma)$ satisfies the diffusion equation (4.117) and the weaker inequality (4.112). A Gaussian kernel thereby guarantees that increasing the scale of smoothing creates no new signal structure. This completes the sufficiency proof. ■

4.7.2.2 Necessity of the Gaussian. It is quite a bit harder to show that the Gaussian is the only kernel that never allows new signal structure to arise as the scale of smoothing increases. The necessity proof involves several steps:

- We consider a candidate filtering kernel $k(t, \sigma)$ and represent the signal to be analyzed, $f(t)$, as a sum of Dirac delta functions.
- Because of the Sifting Property of the Dirac delta (Chapter 3), this transposes the convolution integral into a discrete sum, and, using the Zero Crossing Conditions Proposition above, there arises a set of simultaneous linear equations, $Ax = b$.
- If these simultaneous equations have a solution, then we can find a signal $f(t)$ for which the proposed kernel $k(t, \sigma)$ creates new structure as σ increases.
- We show that we can always solve the simultaneous equations unless the proposed kernel $k(t, \sigma)$ satisfies a special differential equation, which is a general form of the diffusion equation.
- We prove that the only kernel that satisfies this differential equation is the Gaussian.
- Thus, if the filtering kernel $k(t, \sigma)$ is not Gaussian, it cannot be a solution to the special differential equation; this implies that we can solve the simultaneous linear equations; and this solution at long last reveals a signal, $f(t)$, for which our proposed $k(t, \sigma)$ creates at least one new inflection point during smoothing.

Let's follow the above plan, beginning with some technical lemmas. We show how to derive a set of simultaneous equations by representing $f(t)$ as a sum of Dirac delta functions and using the Zero Crossing Conditions. From the discussion of the Dirac delta's Sifting Property (Chapter 3), $f(t)$ can be represented as the limit of such a sum.

Lemma (Zero Crossing Conditions for Dirac Sum Signals). Suppose that $k(t, \sigma)$ satisfies the Scale-Space Kernel Conditions and that the signal $f(t)$ has the following representation as a sum of Dirac delta functions:

$$f(t) = \sum_{i=1}^{n} c_i \delta(t - t_i). \tag{4.121}$$

As before we define

$$F(t,\sigma) = f(t) * k(t, \sigma) = \int_{-\infty}^{+\infty} f(u)k(t-u,\sigma)\,du, \tag{4.122}$$

and we set $E(t, \sigma) = F_{tt}(t, \sigma)$. To avoid excess subscript clutter, we also define $M(t, \sigma) = k_{tt}(t, \sigma) = (\partial^2/\partial t^2)k(t, \sigma)$. If $E(t_0, \sigma) = 0$ for some $t = t_0$, then

$$\sum_{i=1}^{n} c_i M(t_0 - t_i,\sigma) = 0, \tag{4.123}$$

and at an extremum (t_0, σ) of the curve $E(t, \sigma) = 0$ the following equations hold:

$$\sum_{i=1}^{n} c_i M_t(t_0 - t_i,\sigma) = 0 \tag{4.124}$$

and

$$\frac{\sum_{i=1}^{n} c_i M_{tt}(t_0 - t_i,\sigma) = 0}{\sum_{i=1}^{n} c_i M_{\sigma}(t_0 - t_i,\sigma) = 0} > 0. \tag{4.125}$$

Proof: Since $f(t)$ is a sum of Dirac delta functions (4.121), the Sifting Property of the delta function implies

$$E(t,\sigma) = \frac{\partial}{\partial t^2}\int_{-\infty}^{\infty} f(u)k(t-u,\sigma)\,du = \int_{-\infty}^{\infty}\left\{\sum_{i=1}^{n} c_i \delta(t - t_i)\right\}\frac{\partial}{\partial t^2}k(t-u,\sigma)\,du$$

$$= \sum_{i=1}^{n} c_i k_{tt}(t_i,\sigma) = \sum_{i=1}^{n} c_i M(t_i,\sigma). \tag{4.126}$$

At a point (t_0, σ) on a zero crossing curve, $E(t_0, \sigma) = 0$, so (4.123) follows from (4.126). Furthermore, the Zero Crossing Conditions Proposition showed that at an extreme point (t_0, σ) on the zero crossing curve we must have $E_t(t_0, \sigma) = 0$, and by

(4.126) this entails (4.124). Finally, note that (4.125) follows from equation (4.112) of the same proposition, and the proof is complete. ■

Corollary (Necessary Linear Equations). Let the lemma's assumptions still hold and let $t_0, t_1, t_2, \ldots, t_n$ be arbitrary. Then, for $P < 0$, the following four simultaneous equations,

$$\begin{bmatrix} M(t_0-t_{1,\sigma}) & M(t_0-t_{2,\sigma}) & \cdots & M(t_0-t_{n,\sigma}) \\ M_t(t_0-t_{1,\sigma}) & M_t(t_0-t_{2,\sigma}) & \cdots & M_t(t_0-t_{n,\sigma}) \\ M_{tt}(t_0-t_{1,\sigma}) & M_{tt}(t_0-t_{2,\sigma}) & \cdots & M_{tt}(t_0-t_{n,\sigma}) \\ M_\sigma(t_0-t_{1,\sigma}) & M_\sigma(t_0-t_{2,\sigma}) & \cdots & M_\sigma(t_0-t_{n,\sigma}) \end{bmatrix} \begin{bmatrix} c_1 \\ c_2 \\ \vdots \\ c_n \end{bmatrix} = \begin{bmatrix} 0 \\ 0 \\ P \\ 1 \end{bmatrix}, \tag{4.127}$$

have no solution $(c_1, c_2, \ldots, c_n)$.

Proof: The existence of $(c_1, c_2, \ldots, c_n)$ satisfying (4.127) with $P < 0$ violates (4.125). ■

Once we assume that $f(t)$ is a sum of Dirac delta functions, the lemma and its corollary show that the Causality Property imposes extremely strong conditions on the filtering kernel. In (4.127), finding $(c_1, c_2, \ldots, c_n)$ is equivalent to finding $f(t)$, and it would appear that solutions for such an underdetermined set of linear equations should be plentiful. If for some $t_0, t_1, t_2, \ldots, t_n, P$, with $P < 0$, we could discover a solution $(c_1, c_2, \ldots, c_n)$, then this would give us a signal $f(t)$ and a location in scale space (t_0, σ) at which the kernel $k(t, \sigma)$ fails the Causality Property: New structure unfolds when σ increases at (t_0, σ). The matrix of second-, third-, and fourth-order partial derivatives in (4.127) must be very special indeed if a candidate smoothing kernel is to support the Causality Property. Our goal must be to show that the Causality Property guarantees that for any $t_0, t_1, t_2, \ldots, t_n, P < 0$, (4.127) defies solution. The next proposition recalls a linear algebra result that helps bring the peculiarities of (4.127) to light.

Proposition. Let $\boldsymbol{M}$ be an $m \times n$ matrix, let $\boldsymbol{b}$ be an $m \times 1$ vector, and let $[\boldsymbol{M} \mid \boldsymbol{b}]$ be the $m \times (n+1)$ matrix whose first n columns are the same as $\boldsymbol{M}$ and whose last column is $\boldsymbol{b}$. Then the following are equivalent:

(i) The equation $\boldsymbol{Mx} = \boldsymbol{b}$ has a solution.

(ii) Rank $\boldsymbol{M}$ = rank $[\boldsymbol{M} \mid \boldsymbol{b}]$; that is,

$$\operatorname{rank}\begin{bmatrix} M_{1,1} & M_{1,2} & \cdots & M_{1,n} \\ M_{2,1} & \cdots & \cdots & M_{2,n} \\ \vdots & \cdots & \cdots & \vdots \\ M_{m,1} & M_{m,2} & \cdots & M_{m,n} \end{bmatrix} = \operatorname{rank}\begin{bmatrix} M_{1,1} & M_{1,2} & \cdots & M_{1,n} & b_1 \\ M_{2,1} & \cdots & \cdots & M_{2,n} & b_2 \\ \vdots & \cdots & \cdots & \vdots & \vdots \\ M_{m,1} & M_{m,2} & \cdots & M_{m,n} & b_m \end{bmatrix}. \tag{4.128}$$

(iii) For all vectors $\boldsymbol{y} = (y_1, y_2, \ldots, y_m)$: If $\langle \boldsymbol{y}, \boldsymbol{b} \rangle = 0$, then $y_1(M_{1,1}, M_{1,2}, \ldots, M_{1,n}) + y_2(M_{2,1}, M_{2,2}, \ldots, M_{2,n}) + \cdots + y_m(M_{m,1}, M_{m,2}, \ldots, M_{m,n}) = 0$.

Proof: Recall that the column space of the matrix $\boldsymbol{M}$ is the set of vectors spanned by the column vectors of $\boldsymbol{M}$, and the row space of is the set of vectors spanned by rows of $\boldsymbol{M}$. From linear algebra, the dimensions of these two spaces are the same—the rank of $\boldsymbol{M}$. Now, (i) is clearly equivalent to (ii), because (i) is true if and only if $\boldsymbol{b}$ is in $\boldsymbol{M}$'s column space. Also note that the row space of $\boldsymbol{M}$ must have the same dimension as the row space of $[\boldsymbol{M} \mid \boldsymbol{b}]$. So if vector $\boldsymbol{y} = (y_1, y_2, ..., y_m)$ is such that the linear combination of rows of $\boldsymbol{M}$, $y_1(M_{1,1}, M_{1,2}, ..., M_{1,n}) + y_2(M_{2,1}, M_{2,2}, ..., M_{2,n}) + \cdots + y_m(M_{m,1}, M_{m,2}, ..., M_{m,n}) = 0$, and also $y_1(M_{1,1}, M_{1,2}, ..., M_{1,n}, b_1) + y_2(M_{2,1}, M_{2,2}, ..., M_{2,n}, b_2) + ... + y_m(M_{m,1}, M_{m,2}, ..., M_{m,n}, b_m)$ is nonzero in the last component (that is, $\langle \boldsymbol{y}, \boldsymbol{b} \rangle \neq 0$), then the dimension of the row space of $[\boldsymbol{M} \mid \boldsymbol{b}]$ would exceed the dimension of the row space of $\boldsymbol{M}$, a contradiction. Thus, (ii) entails (iii). Finally, (iii) says that any vector $\boldsymbol{y} \perp \boldsymbol{b}$ must also be orthogonal to every column of $\boldsymbol{M}$. This means that $\boldsymbol{b}$ is in the column space of $\boldsymbol{M}$, $\boldsymbol{Mx} = \boldsymbol{b}$ is solvable, and the proof is complete. ■

The next proposition reveals a differential equation that scale space kernels must satisfy.

Proposition (Necessary Differential Equation). Suppose that $k(t, \sigma)$ satisfies the Scale-Space Kernel Conditions and that the signal $f(t)$ has a representation as a sum of Dirac delta functions (4.121), as in the lemma. Then there are constants, A, B, C, and D, with D/C > 0, such that

$$\frac{Ak(t,\sigma)}{\sigma^2} + \frac{Bk_t(t,\sigma)}{\sigma} + Ck_{tt}(t,\sigma) = \frac{Dk_\sigma(t,\sigma)}{\sigma}. \tag{4.129}$$

Proof: Consider the vector $\boldsymbol{b}$, where $\boldsymbol{b} = (0, 0, P, 1)$ and $P > 0$. Note that it is easy to find a vector $\boldsymbol{y} = (y_1, y_2, y_3, y_4)$ such that $\langle \boldsymbol{y}, \boldsymbol{b} \rangle = 0$. Applying the previous proposition and the Necessary Linear Equations Corollary, it follows that $y_1 M_{1,1}(t_0 - t_1) + y_2 M_t(t_0 - t_1) + y_3 M_{tt}(t_0 - t_1) + y_4 M_\sigma(t_0 - t_1) = 0$. Since (4.127) has a solution for any $t_0, t_1, t_2, \ldots, t_n$ we may write this as $y_1 M_{1,1}(t) + y_2 M_t(t) + y_3 M_{tt}(t) + y_4 M_\sigma(t) = 0$. Let $A = y_1\sigma^2$, $B = y_2\sigma$, $C = y_3$, and $D = -y_4\sigma$. Then,

$$\frac{AM(t,\sigma)}{\sigma^2} + \frac{BM_t(t,\sigma)}{\sigma} + CM_{tt}(t,\sigma) = \frac{DM_\sigma(t,\sigma)}{\sigma}. \tag{4.130}$$

Observe that $D/C = (-y_4\sigma)/y_3 = P\sigma > 0$. We require, however, that $k(t, \sigma)$, not just its second derivative, $M(t, \sigma)$, satisfy the differential equation. Any two filters with second derivatives satisfying (4.130) must differ by a function with zero second derivative. That is, their difference is a linear term. Since the Scale-Space Kernel Conditions require that as $t \to \infty$, we have $k(t, \sigma) \to 0$, this linear term must be identically zero. Thus, $k(t, \sigma)$ satisfies (4.129). ■

Equation (4.129) is a general form of the heat or diffusion equation. We have come a long way and have taken nearly all the steps toward proving that a filtering kernel which obeys the Scale-Space Kernel Conditions is necessarily Gaussian. The next theorem solves the generalized heat equation. To accomplish this, we must make use of the analog Fourier transform, the formal presentation of which will not be given until the next chapter. We fancy that many readers are already familiar with this technique for solving differential equations. We apologize to the rest for asking them to see into the future and all the more so for suggesting a premonition of the Fourier transform!

Some readers may wish to skip the proofs of the next two theorems, perhaps because the heat equation's solution through Fourier transformation is already familiar, or perhaps to return to them after assimilating Chapter 5's material. The first result solves the generalized heat equation. Because the solution involves an integration, it is less than satisfying, however. The second theorem remedies this. We apply the Scale-Space Kernel Conditions to our general solution and derive a result that clearly shows the Gaussian nature of all structure reducing scale-space filtering kernels.

Theorem (Generalized Heat Equation). If $k(t, \sigma)$ is a solution to the differential equation (4.129), then $k(t, \sigma)$ is the Gaussian

$$k(t,\sigma) = \sigma^{1+\frac{a}{d}} \sqrt{D/2\pi C} \int_{-\infty}^{+\infty} x(t-u) e^{-\frac{D}{2C}\left(\frac{u}{\sigma}+\frac{B}{D}\right)^2} du. \tag{4.131}$$

Proof: Let us first simplify (4.129) with the substitution $k(t, \sigma) = \sigma^{a/d} q(t, \sigma)$. This provides a heat equation in more familiar form,

$$\frac{Bq_t(t,\sigma)}{\sigma} + Cq_{tt}(t,\sigma) = \frac{Dq_\sigma(t,\sigma)}{\sigma}, \tag{4.132}$$

which we must solve for $q(t, \sigma)$. Further simplifications are possible, but we need to reach into the next chapter for the representation of a signal by the analog Fourier transform. The normalized radial Fourier transform of a signal $x(t)$ is given by

$$X(\omega) = \frac{1}{\sqrt{2\pi}} \int_{-\infty}^{+\infty} x(t) e^{-j\omega t}\, dt, \tag{4.133}$$

where $j^2 = -1$. Here we adopt a widely used convention that lowercase letters stand for time-domain signals and that uppercase letters stand for their Fourier transform counterpart. In the near future, we shall verify that if $x(t)$ is absolutely integrable or has finite energy, then its Fourier transform $X(\omega)$ exists (4.133). Likewise, if $X(\omega) \in l^1(\mathbb{R})$ or $X(\omega) \in l^2(\mathbb{R})$, then an inverse normalized radial Fourier transform exists, which is given by

$$x(t) = \frac{1}{\sqrt{2\pi}} \int_{-\infty}^{+\infty} X(\omega) e^{j\omega t}\, d\omega. \tag{4.134}$$

The key idea is to insert the representation of $q(t, \sigma)$ by its transform from $Q(\omega, \sigma)$ into the simplified heat equation (4.132). Then, after removing $(2\pi)^{-1}$ from each term, (4.132) becomes

$$\frac{B}{\sigma}\frac{\partial}{\partial t}\int_{-\infty}^{+\infty} Q(\omega,\sigma)e^{j\omega t}\,d\omega + C\frac{\partial^2}{\partial t^2}\int_{-\infty}^{+\infty} Q(\omega,\sigma)e^{j\omega t}\,d\omega = \frac{D}{\sigma}\frac{\partial}{\partial \sigma}\int_{-\infty}^{+\infty} Q(\omega,\sigma)e^{j\omega t}\,d\omega. \tag{4.135}$$

Interchanging the order of integration and differentiation in (4.135) gives

$$\frac{B}{\sigma}\int_{-\infty}^{+\infty} Q(\omega,\sigma)(j\omega)e^{j\omega t}\,d\omega + C\int_{-\infty}^{+\infty} Q(\omega,\sigma)(j\omega)^2 e^{j\omega t}\,d\omega = \frac{D}{\sigma}\int_{-\infty}^{+\infty} Q_\sigma(\omega,\sigma)e^{j\omega t}\,d\omega. \tag{4.136}$$

By the existence of the inverse Fourier transform, the integrands on either side of (4.136) must be equal. After a bit of algebra, the differential equation simplifies considerably:

$$\frac{B\omega j}{D}Q(\omega,\sigma) - \frac{C\sigma\omega^2}{D}Q(\omega,\sigma) = Q_\sigma(\omega,\sigma). \tag{4.137}$$

Let us now separate the variables in (4.137), to obtain

$$\left(\frac{B\omega j}{D} - \frac{C\sigma\omega^2}{D}\right)\partial\sigma = \frac{\partial Q}{Q}. \tag{4.138}$$

We integrate both sides from 0 to r,

$$\int_0^r \left(\frac{B\omega j}{D} - \frac{C\sigma\omega^2}{D}\right)\partial\sigma = \int_0^r \frac{\partial Q}{Q}. \tag{4.139}$$

where $r > 0$ is an integration limit, and remove the logarithms arising from the integration on the right-hand side of (4.139) by exponentiation. Letting $r = \sigma$ then gives

$$Q(\omega,\sigma) = Q(\omega,0)e^{\frac{jB\omega\sigma}{D}}e^{\frac{-C\omega^2\sigma^2}{2D}}. \tag{4.140}$$

Thus we have found the Fourier transform of $q(t, \sigma) = \sigma^{-a/d}k(t, \sigma)$. We can find $q(t, \sigma)$ by once again applying the inverse Fourier transform operation,

$$q(t,\sigma) = \frac{1}{\sqrt{2\pi}}\int_{-\infty}^{+\infty} Q(\omega,0)e^{\frac{jB\omega\sigma}{D}}e^{\frac{-C\omega^2\sigma^2}{2D}}e^{j\omega t}\,d\omega, \tag{4.141}$$

which is very close to the form we need. To continue simplifying, we utilize some further properties of the Fourier transform. Notice two things about the integrand in (4.141): It is a product of frequency-domain signals (i.e., their independent variable is ω), and one of the frequency-domain terms is a frequency-domain Gaussian, namely $\exp(-C\omega^2\sigma^2/2D)$. Therefore, one of the properties that we can now apply is the Convolution Theorem for Fourier Transforms. It says that the Fourier transform of a convolution, $s = x*y$, is the attenuated product of the Fourier Transforms; specifically, $S(\omega) = (2\pi)^{-1/2}X(\omega)Y(\omega)$. The second property that comes to mind from inspecting the integrand in (4.141) is the formula for the Fourier transform of the Gaussian signal. This states that if $\lambda > 0$, then the Fourier Transform of the Gaussian $g(t) = \exp(-\lambda t^2)$ is $G(\omega) = (2\lambda)^{-1/2}\exp(-\omega^2/4\lambda)$. We let $X(\omega) = Q(\omega, 0)$ and $Y(\omega, \sigma) = \exp(jB\omega\sigma/D)\exp(-C\omega^2\sigma^2/2D)$. Then $q(t, \sigma)$ is the inverse Fourier transform of $X(\omega)Y(\omega, \sigma)$, so we have $q(t, \sigma) = (2\pi)^{-1/2}x*y$. That is,

$$q(t,\sigma) = \frac{1}{\sqrt{2\pi}} \int_{-\infty}^{+\infty} x(t-u)y(u,\sigma)\, du, \tag{4.142}$$

where the Fourier transforms of $x(t)$ and $y(t, \sigma)$ are $X(\omega)$ and $Y(\omega, \sigma)$, respectively. Let us set aside the mysterious signal $x(t)$ for a moment and consider the $y(t, \sigma)$ factor in (4.142). Since the Fourier transform of $s(t, \sigma) = \sigma(D/C)^{1/2}\exp(-Dt^2/(2C\sigma^2))$ is $\exp(-C\sigma^2\omega^2/2D)$, it can easily be shown that the Fourier transform of $y(t) = s(t+B\sigma/D)$ is $Y(\omega, \sigma)$. Thus,

$$q(t,\sigma) = \sigma\sqrt{D/2\pi C} \int_{-\infty}^{+\infty} x(t-u)e^{-\frac{D}{2C}\left(\frac{u}{\sigma}+\frac{B}{D}\right)^2} du, \tag{4.143}$$

and, recalling that $k(t, \sigma) = \sigma^{a/d}q(t, \sigma)$, there follows (4.131), completing the proof of the theorem. ■

Now, the next theorem applies the Scale-Space Kernel Conditions to (4.131).

Theorem (Necessity of the Gaussian). Suppose that $k(t, \sigma)$ satisfies the Scale-Space Kernel Conditions and that the signal $f(t)$ has a representation as a sum of Dirac delta functions (4.121), as in the lemma. Then $k(t, \sigma)$ is a Gaussian of the form

$$k(t,\sigma) = \sigma^{1+\frac{a}{d}}\sqrt{D/2\pi C} \int_{-\infty}^{+\infty} x(t-u)e^{-\frac{D}{2C}\left(\frac{u}{\sigma}+\frac{B}{D}\right)^2} du. \tag{4.144}$$

Proof: Recall the time-delayed Gaussian $y(t) = s(t+B\sigma/D)$ from the previous proof. Since the smoothing filter must not permit zero crossings to shift as the scale of smoothing varies, this Gaussian must be centered at the origin; in other words, $B\sigma/D = 0$, and hence $B = 0$. Thus,

$$k(t,\sigma) = \sigma^{1+\frac{a}{d}}\sqrt{D/2\pi C}\int_{-\infty}^{+\infty} x(t-u)e^{-\frac{Du^2}{2C\sigma^2}}\,du. \tag{4.145}$$

As the scale of smoothing decreases, the smoothed signal must resemble the original, $f(t)$, and so in the limit, as $\sigma \to 0$, the Scale-Space Kernel Conditions demand that $k(t, \sigma) \to \delta(t)$, the Dirac delta. The consequence of this condition is

$$\lim_{\sigma\to 0} k(t,\sigma) = \lim_{\sigma\to 0} \sigma^{1+\frac{a}{d}}\sqrt{\frac{D}{2\pi C}}\int_{-\infty}^{+\infty} e^{-\frac{D(t-u)^2}{2C\sigma^2}} x(u)\,du$$

$$= \lim_{\sigma\to 0} \sigma^{\frac{a}{d}}\sqrt{\frac{D}{2\pi C}}\int_{-\infty}^{+\infty} \sigma e^{-\frac{D(t-u)^2}{2C\sigma^2}} x(u)\,du = \delta(t). \tag{4.146}$$

This completes the proof. ■

This proof contains a valuable lesson: A judicious application of the Fourier integral representation removes the differential equation's partial time derivatives. Higher powers of the frequency variable, ω, replace the derivatives, but this is tolerable because it leads to a simpler differential equation overall. This is a powerful—and quite typical—application of the Fourier transform. Applied mathematicians usually resort to the Fourier transformation for precisely this purpose and no others. For our own purposes, however, the Fourier transform does much more than expedite computations; it is the principal tool for studying the frequency content of analog signals. The next three chapters, no less, explore Fourier theory in detail: analog signal frequency in Chapter 5, discrete signal frequency in Chapter 7, and frequency-domain signal analysis in Chapter 9.

The efforts of many researchers have helped to elucidate the theory of scale-space decompositions. Our treatment here follows most closely the presentation of one-dimensional scale-space decomposition by Yuille and Poggio [110]. Other theoretical studies of scale-space decomposition include Ref. 111. Applications of scale space filtering include the recognition of two-dimensional shapes by extracting object boundaries, formulating the boundary as a curvature signal, and deriving the scale-space representation of the boundary curvature [112]. There results a graph structure, which can be matched against model graphs using artificial intelligence techniques, or matched level-by-level using the structural pattern detection techniques of Section 4.6.2.

Now, when an image is represented in its scale-space decomposition, no arbitrary preferred scale for the objects represented in the decomposed is used. Instead, the set of scales at which distinct regions of different curvature sign arise and disappear is found. Then the changes in the curvature are related to one another in a structure called an interval tree by Witkin. This further abstracts the structural description of the signal from its actual value, $f(x)$, by defining an interval of scales, and an interval of space, x, over which the salient curvature features of the signal are to be found. The signal is segmented in both scale and spatial coordinates.

4.8 SUMMARY

In the introduction to this chapter, we reflected on the distinction, sometimes subtle and sometimes profound, between signal processing and signal analysis. This chapter's methods work primarily with operations on the signal values in the time domain. In some cases, such as the problem of extracting periodicities from the signal values, statistics on signal levels proved to be inadequate. We resorted to comparisons, in the form of inner products, of the given signal to sinusoidal or exponential models. This makes the break with time-domain methods into the realm of frequency-domain methods. Now, we need to look deeper into the theory of analog signal frequency, discrete signal frequency, and the applications that arise from these studies. This task will occupy us for the next five chapters. Then we will consider the combination of the methods of both domains: Time-frequency transforms are the subject of Chapter 10, and time-scale transforms are covered in Chapter 11. This chapter is preparation, with a time-domain perspective, for Chapter 9 on frequency-domain signal analysis and for Chapter 12 on mixed-domain analysis.

REFERENCES

1. F. Jelinek, Continuous speech recognition by statistical methods, *Proceedings of the IEEE*, vol. 64, no. 4, pp. 532–556, April 1976.
2. S. Young, A review of large-vocabulary continuous-speech recognition, *IEEE Signal Processing Magazine*, pp. 45–57, September 1996.
3. K. Lee and R. Reddy, *Automatic Speech Recognition*, Boston: Kluwer Academic Publishers, 1989.
4. L. Rabiner and B.-H. Juang, *Fundamentals of Speech Recognition*, Englewood Cliffs, NJ: Prentice-Hall, 1993.
5. R. L. Wesson and R. E. Wallace, Predicting the next great earthquake in California, *Scientific American*, pp. 35–43, February 1985.
6. Ö. Yilmaz, *Seismic Data Processing*, Tulsa, OK: Society for Exploratory Geophysics, 1987.
7. M. Akay, *Biomedical Signal Processing*, San Diego, CA: Academic Press, 1994.
8. R. Isermann, Process fault detection based on modeling and estimation methods—a survey, *Automatica*, vol. 20, no. 4, pp. 387–404, 1984.
9. C. C. Tappert, C. Y. Suen, and T. Wakahara, The state of the art in on-line handwriting recognition, *IEEE Transactions on Pattern Analysis and Machine Intelligence*, vol. 12, no. 8, pp. 787–808, 1990.

10. A. Marcus and A. Van Dam, User interface developments in the nineties, *IEEE Computer,* pp. 49–57, September 1991.
11. J. Preece, *Human-Computer Interaction*, Reading, MA: Addison-Wesley, 1994.
12. S. L. Horowitz and T. Pavlidis, Picture segmentation in a directed split-and-merge procedure, *Proceedings of the Second International Joint Conference on Pattern Recognition,* Copenhagen, pp. 424–433, August 1974.
13. P. D. van der Puije, *Telecommunication Circuit Design*, New York: Wiley, 1992.
14. P. Mock, Add DTMF generation and decoding to DSP-μP designs, *Electronic Design News*, March 21, 1985; also in K.-S. Lin, ed., *Digital Signal Processing Applications with the TMS320 Family*, vol. 1, Dallas: Texas Instruments, 1989.
15. A. Mar, ed., *Digital Signal Processing Applications Using the ADSP-2100 Family*, Englewood Cliffs, NJ: Prentice-Hall, 1990.
16. J. L. Flanagan, *Speech Analysis, Synthesis, and Perception*, 2nd ed., New York: Springer-Verlag, 1972.
17. C. K. Chow and T. Kaneko, Automatic boundary detection of the left ventricle from cineangiograms, *Computers and Biomedical Research*, vol. 5, pp. 388–410, 1972.
18. N. Otsu, A threshold selection method from gray-level histograms, *IEEE Transactions on Systems, Man, and Cybernetics,* vol. SMC-9, no. 1, pp. 62–66, January 1979.
19. J. Kittler and J. Illingworth, On threshold selection using clustering criteria, *IEEE Transactions on Systems, Man, and Cybernetics,* vol. SMC-15, pp. 652–655, 1985.
20. S. Kullback and R. A. Leibler, On information and sufficiency, *Annals of Mathematical Statistics,* vol. 22, pp. 79–86, 1951.
21. S. Kullback, *Information Theory and Statistics*, Mineola, NY: Dover, 1997.
22. J. S. Weszka, A survey of threshold selection techniques, *Computer Graphics and Image Processing,* vol. 7, pp. 259–265, 1978.
23. P. K. Sahoo, S. Soltani, and A. K. C. Wong, A survey of thresholding techniques, *Computer Vision, Graphics, and Image Processing*, vol. 41, pp. 233–260, 1988.
24. P. W. Palumbo, P. Swaminathan, and S. N. Srihari, Document image binarization: evaluation of algorithms, *Applications of Digital Image Processing IX*, SPIE, vol. 697, pp. 278–285, 1986.
25. M. Kamel and A. Zhao, Extraction of binary character/graphics images from grayscale document images, *CVGIP: Graphical Models and Image Processing*, vol. 55, no. 3, pp. 203–217, 1993.
26. O. D. Trier and T. Taxt, Evaluation of binarization methods for document images, *IEEE Transactions on Pattern Analysis and Machine Intelligence,* vol. 17, no. 3, pp. 312–315, March 1995.
27. J. M. White and G. D. Rohrer, Image thresholding for character image extraction and other applications requiring character image extraction, *IBM Journal of Research and Development*, vol. 27, no. 4, pp. 400–411, July 1983.
28. D. L. Duttweiler, A twelve-channel digital voice echo canceller, *IEEE Transactions on Communications*, vol. COM-26, no. 5, pp. 647–653, May 1978.
29. P. Brodatz, *Textures: A Photographic Album for Artists and Designers,* New York: Dover, 1966.
30. B. Julesz, Experiments in the visual perception of texture, *Scientific American*, vol. 232, pp. 34–43, 1975.

31. J. Beck, Similarity grouping and peripheral discrimination under uncertainty, *American Journal of Psychology*, vol. 85, pp. 1–19, 1972.

32. M. Wertheimer, Principles of perceptual organization, in *Readings in Perception*, D. C. Beardslee and M. Wertheimer, eds., Princeton, NJ: Van Nostrand-Reinhold, 1958.

33. D. J. Whitehouse, *Handbook of Surface Metrology*, Bristol, UK: Institute of Physics, 1994.

34. J. Beck, A. Sutter, and R. Ivry, Spatial frequency channels and perceptual grouping in texture segmentation, *Computer Vision, Graphics, and Image Processing*, vol. 37, pp. 299–325, 1987.

35. F. Tomita, Y. Shirai, and S Tsuji, Description of texture by a structural analysis, *IEEE Transactions on Pattern Analysis and Machine Intelligence*, vol. PAMI-4, no. 2, pp. 183–191, 1982.

36. *Surface Texture: Surface Roughness, Waviness and Lay*, ANSI B46.1, American National Standards Institute, 1978.

37. *Surface Roughness—Terminology—Part 1: Surface and its Parameters*, ISO 4287/1, International Standards Organization, 1984.

38. *Surface Roughness—Terminology—Part 2: Measurement of Surface Roughness Parameters, ISO 4287/2*, International Standards Organization, 1984.

39. R. W. Conners and C. A. Harlow, A theoretical comparison of texture algorithms, *IEEE Transactions on Pattern Analysis and Machine Intelligence*, vol. PAMI-2, no. 3, pp. 204–222, May 1980.

40. A. Rosenfeld and A. C. Kak, *Digital Picture Processing*, vol. 2, Orlando, FL: Academic Press, 1982.

41. B. Julesz, Visual pattern discrimination, *IRE Transactions on Information Theory*, vol. IT-8, no. 2, pp. 84–92, February 1961.

42. B. Julesz, E. N. Gilbert, and J. D. Victor, Visual discrimination of textures with identical third-order statistics, *Biological Cybernetics*, vol. 31, no. 3, pp. 137–140, 1978.

43. A. Gagalowicz, A new method for texture field synthesis: Some applications to the study of human vision, *IEEE Transactions on Pattern Analysis and Machine Intelligence*, vol. PAMI-3, no. 5, pp. 520–533, September 1981.

44. R. M. Haralick, K. S. Shanmugam, and I. Dinstein, Textural features for image classification, *IEEE Transactions on Systems, Man, and Cybernetics*, vol. SMC-3, no. 6, pp. 610–621, November 1973.

45. S. W. Zucker and D. Terzopoulos, Finding structure in co-occurrence matrices for texture analysis, *Computer Graphics and Image Processing*, vol. 12, no. 3, pp. 286–308, March 1980.

46. D. Chetverikov, Measuring the degree of texture regularity, *Proceedings of the International Conference on Pattern Recognition*, Montreal, pp. 80–82, 1984.

47. J. S. Weszka, C. R. Dyer, and A. Rosenfeld, A comparative study of texture measures for terrain classification, *IEEE Transactions on Systems, Man, and Cybernetics*, vol. SMC-6, pp. 269–285, 1976.

48. G. Strang, *Introduction to Applied Mathematics*, Wellesley, MA: Wellesley-Cambridge, 1986.

49. A. Savitsky and M. J. E. Golay, Smoothing and differentiation of data by simplified least squares procedures, *Analytical Chemistry*, vol. 36, pp. 1627–1639, 1964.

50. C. L. Lawson and R. J. Hanson, *Solving Least Squares Problems*, Englewood Cliffs, NJ: Prentice-Hall, 1974.
51. R. M. Haralick and L. G. Shapiro, *Computer and Robot Vision*, vol. 1, New York: Addison-Wesley, 1992.
52. A. Jennings and J. J. McKeown, *Matrix Computation*, 2nd ed., Chichester, West Sussex, England: Wiley, 1992.
53. E. R. Dougherty and P. A. Laplante, *Introduction to Real-Time Imaging*, Bellingham, WA: SPIE, 1995.
54. N. C. Gallagher, Jr. and G. L. Wise, A theoretical analysis of the properties of median filters, *IEEE Transactions on Acoustics, Speech, and Signal Processing*, vol. ASSP-29, pp. 1136–1141, 1981.
55. R. M. Haralick, S. R. Sternberg, and X. Zhuang, Image analysis using mathematical morphology, *IEEE Transactions on Pattern Analysis and Machine Intelligence*, vol. PAMI-9, no. 4, pp. 532–550, July 1987.
56. M.-H. Chen and P.-F. Yan, A multiscaling approach based on morphological filtering, *IEEE Transactions on Pattern Analysis and Machine Intelligence*, vol. 11, no. 7, pp. 694–700, July 1989.
57. P. Maragos, Pattern spectrum and multiscale shape representation, *IEEE Transactions on Pattern Analysis and Machine Intelligence*, vol. 11, no. 7, pp. 701–716, July 1989.
58. L. S. Kovasznay and H. M. Joseph, Processing of two-dimensional patterns by scanning techniques, *Science,* vol. 118, no. 3069, pp. 475–477, 25 October 1953.
59. L. G. Roberts, Machine perception of three-dimensional solids, in *Optical and Electro-optical Information Processing*, J. T. Tippet, ed., Cambridge, MA: MIT Press, 1965.
60. A. Rosenfeld and M. Thurston, Edge and curve detection for visual scene analysis, *IEEE Transactions on Computers,* vol. 20, no. 5, pp. 562–569, May 1971.
61. J. M. S. Prewitt, Object enhancement and extraction, in *Picture Processing and Psychopictorics*, B. S. Lipkin and A. Rosenfeld, eds., New York: Academic Press, 1970.
62. D. Marr and E. Hildreth, Theory of Edge Detection, *Proceedings of the Royal Society of London B.* vol. 207, pp. 187–217, 1979.
63. M. F. Heukel, An operator which locates edges in digitized pictures, *Journal of the Association for Computing Machinery,* vol. 18, pp. 113–125, 1971.
64. F. O'Gorman, Edge detection using Walsh functions, *Artificial Intelligence*, vol. 10, pp. 215–223, 1978.
65. R. A. Hummel, Feature detection using basis functions, *Computer Graphics and Image Processing*, vol. 9, pp. 40–55, 1979.
66. H. D. Tagare and R. J. P. deFigueiredo, On the localization performance measure and optimal edge detection, *IEEE Transactions on Pattern Analysis and Machine Intelligence,* vol. 12, no. 12, pp. 1186–1190, 1990.
67. K. L. Boyer and S. Sarkar, Comments on "On the localization performance measure and optimal edge detection," *IEEE Transactions on Pattern Analysis and Machine Intelligence,* vol. 16, no. 1, pp. 106–108, January 1994.
68. R. J. Qian and T. S. Huang, Optimal edge detection in two-dimensional images, *IEEE Transactions on Image Processing,* vol. 5, no. 7, pp. 1215–1220, 1996.
69. M. Gökmen and A. K. Jain, $\lambda\tau$-space representation of images and generalized edge detector, *IEEE Transactions on Pattern Analysis and Machine Intelligence,* vol. 19, no. 6, pp. 545–563, June 1997.

70. D. M. Manos and D. L. Flamm, *Plasma Etching: An Introduction*, Boston: Academic Press, 1989.

71. R. L. Allen, R. Moore, and M. Whelan, Multiresolution pattern detector networks for controlling plasma etch reactors, in *Process, Equipment, and Materials Control in Integrated Circuit Manufacturing*, Proceedings SPIE 2637, pp. 19–30, 1995.

72. C. K. Hanish, J. W. Grizzle, H.-H. Chen, L. I. Kamlet, S. Thomas III, F. L. Terry, and S. W. Pang, Modeling and algorithm development for automated optical endpointing of an HBT emitter etch, *Journal of Electronic Materials*, vol. 26, no. 12, pp. 1401–1408, 1997.

73. V. Torre and T. A. Poggio, On edge detection, *IEEE Transactions on Pattern Analysis and Machine Intelligence,* vol. PAMI-8, no. 2, pp. 147–163, March 1986.

74. C. F. Gerald and P. O. Wheatley, *Applied Numerical Analysis*, 4th ed., Reading, MA: Addison-Wesley, 1990.

75. J. Canny, A computational approach to edge detection, *IEEE Transactions on Pattern Analysis and Machine Intelligence,* vol. PAMI-8, no. 6, pp. 679–698, November 1986.

76. A. Papoulis, *Probability, Random Variables, and Stochastic Processes*, New York: McGraw-Hill, 1965.

77. D. Marr, *Vision*, New York: W. H. Freeman, 1982.

78. J. F. Canny, *Finding Edges and Lines in Images*, Technical Report No. 720, MIT Artificial Intelligence Laboratory, June 1983.

79. H. D. Tagare and R. J. P. deFigueiredo, Reply to "On the localization performance measure and optimal edge detection," *IEEE Transactions on Pattern Analysis and Machine Intelligence,* vol. 16, no. 1, pp. 108–110, January 1994.

80. J. Koplowitz and V. Greco, On the edge location error for local maximum and zero-crossing edge detectors, *IEEE Transactions on Pattern Analysis and Machine Intelligence,* vol. 16, no. 12, pp. 1207–1212, December 1994.

81. L. W. Couch II, *Digital and Analog Communication Systems*, 4th ed., Upper Saddle River, NJ: Prentice-Hall, 1993.

82. S. Haykin, *Communication Systems,* 3rd ed., New York: Wiley, 1994.

83. G. L. Turin, An introduction to digital matched filters, *Proceedings of the IEEE*, vol. 64, pp. 1092–1112, 1976.

84. A. Rosenfeld and A. C. Kak, *Digital Picture Processing,* vol. 1, 2nd ed., New York: Academic, 1982.

85. J. R. Tou and R. C. Gonzalez, *Pattern Recognition Principles,* Reading, MA: Addison-Wesley, 1974.

86. F. Molinaro and F. Castanié, Signal processing pattern classification techniques to improve knock detection in spark ignition engines, *Mechanical Systems and Signal Processing*, vol. 9, no. 1, pp. 51–62, January 1995.

87. J. G. Proakis and D. G. Manolakis, *Digital Signal Processing: Principles, Algorithms, and Applications,* 2nd ed., New York: Macmillan, 1992.

88. A. V. Oppenheim and R. W. Schafer, *Discrete-Timel Signal Processing,* Englewood Cliffs, NJ: Prentice-Hall, 1989.

89. S. S. Yau and J. M. Garnett, least-mean-square approach to pattern classification, in *Frontiers of Pattern Recognition,* S. Wantanabe, ed., New York: Academic Press, pp. 575–588, 1972.

90. N. J. Nilsson, *Learning Machines*, New York: McGraw-Hill, 1965.

91. K. Fukunaga, *Introduction to Statistical Pattern Recognition*, 2nd ed., Boston: Academic Press, 1990.
92. P. R. Devijver and J. Kittler, *Pattern Recognition: A Statistical Approach*, Englewood Cliffs, NJ: Prentice-Hall, 1982.
93. R. O. Duda and P. E. Hart, *Pattern Classification and Scene Analysis*, New York: Wiley, 1973.
94. C. K. Chow, An optimum character recognition system using decision functions, *Transactions of the IRE on Electronic Computers,* vol. EC-6, pp. 247–254, 1957.
95. J. Schürmann, Multifont word recognition system with application to postal address reading, *Proceedings of the Third International Joint Conference on Pattern Recognition*, pp. 658–662, 1976.
96. W. Doster, Contextual postprocessing system for cooperation with a multiple-choice character-recognition system, *IEEE Transactions on Computers*, vol. C-26, no. 11, pp. 1090–1101, November 1977.
97. R. Ott, Construction of quadratic polynomial classifiers, *Proceedings of the Third International Joint Conference on Pattern Recognition*, 1976.
98. Y.-H. Pao, *Adaptive Pattern Recognition and Neural Networks*, Reading, MA: Addison-Wesley, 1989.
99. F. Attneave, Some informational aspects of visual perception, *Psychological Review*, vol. 61, no. 3, pp. 183–193, 1954.
100. S. Tanimoto and T. Pavlidis, A hierarchical data structure for picture processing, *Computer Graphics and Image Processing,* vol. 4, No. 2, pp. 104–119, June 1975.
101. .R. Y. Wong and E. L. Hall, Sequential hierarchical scene matching, *IEEE Transactions on Computers,* vol. C-27, No. 4, pp. 359–366, April 1978.
102. D. Marr, T. Poggio, and S. Ullman, Bandpass channels, zero-crossings, and early visual information processing, *Journal of the Optical Society of America,* vol. 69, pp. 914–916, 1979.
103. D. H. Hubel and T. N. Wiesel, Receptive fields, binocular interaction and functional architecture in the cat's visual cortex, *Journal of Physiology,* vol. 160, pp. 106–154, 1962.
104. F. W. C. Campbell and J. Robson, Application of Fourier analysis to the visibility of gratings, *Journal of Physiology*, vol. 197, pp. 551–566, 1968.
105. D. H. Hubel, *Eye, Brain, and Vision*, New York: W. H. Freeman, 1988.
106. A. P. Pentland, Models of image formation, in *From Pixels to Predicates*, A. P. Pentland, ed. Norwood, NJ: Ablex, 1986.
107. A. P. Witkin, Scale-space filtering, *Proceedings of the 8th International Joint Conference on Artificial Intelligence,* Karlsruhe, West Germany, 1983. See also A. P. Witkin, Scale-space filtering, in *From Pixels to Predicates*, A. P. Pentland, ed., Norwood, NJ: Ablex, 1986.
108. J. Koenderink, The structure of images, *Biological Cybernetics,* vol. 50, pp. 363–370, 1984.
109. M. P. do Carmo, *Differential Geometry of Curves and Surfaces*, Englewood Cliffs, NJ: Prentice-Hall, 1976.
110. A. L. Yuille and T. A. Poggio, Scaling theorems for zero crossings, *IEEE Transactions on Pattern Analysis and Machine Intelligence,* vol. PAMI-8, no. 1, pp. 15–25, January 1986.

111. Babaud, A. P. Witkin, M. Baudin, and R. O. Duda, Uniqueness of the Gaussian kernel for scale-space filtering, *IEEE Transactions on Pattern Analysis and Machine Intelligence,* vol. PAMI-8, no. 1, pp. 26–33, January 1986.

112. F. Mokhtarian and A. Mackworth, Scale-based description and recognition of planar curves and two-dimensional shapes, *IEEE Transactions on Pattern Analysis and Machine Intelligence,* vol. PAMI-8, no. 1, pp. 34-43, January 1986.

PROBLEMS

1. Consider the analog Gaussian kernel $h(t) = \exp(-t^2)$.

(a) Show that $h(t)$ contains a concave up portion for $t < 0$, a concave-down section symmetric about $t = 0$, a concave up portion for $t > 0$, and exactly two points of inflection where the second derivative of $h(t)$ is zero.

(b) Find a logical predicate, applied to subsets of Dom(h), that segments $h(t)$ into the regions described in (a).

(c) Suppose that a system designer elects to prefilter discrete signals by an approximate, discrete version of $h(t)$ and wishes to maintain the aspects of concavity of $h(t)$ in the filter $h(n)$. What is the minimal support of the filter $h(n)$?

(d) Show that if the analog signal $f(t)$ is segmented according to whether d^2f/dt^2 is negative, zero, or positive and d^3f/dt^3 exists, then segments with $d^2f/dt^2 < 0$ cannot be adjacent to segments with $d^2f/dt^2 > 0$.

2. Consider a system $y = Hx$ that should remove impulse noise from $x(n)$ to produce $y(n)$.

(a) If H is LTI, show that the moving average, or box filter $h(n) = [1/3, \underline{1/3}, 1/3]$ can fail to do an adequate job as the impulse response of H.

(b) Show that any LTI filter can fail to do an adequate job as $h = H\delta$.

3. Suppose we filter out impulse noise by a median filter, $y = Hx$.

(a) Show that H is translation invariant.

(b) Show that H is nonlinear.

(c) Show that the median filter $(Hx)(n) = \text{median}\{x(n-1), x(n), x(n+1)\}$ can remove impulse noise.

(d) Show that when performing preliminary smoothing of a signal before thresholding, a median filter $y(n) = \text{median}\{x(n-1), x(n), x(n+1)\}$ preserves signal edges better than box filters.

4. Consider a morphological filter, $y = Hx$.

(a) Show that if H is a dilation, then H is translation invariant but not linear.

(b) Show that if H is an erosion, then H is translation invariant but not linear.

5. Suppose f(n) is a discrete signal and $T > 0$. Let $M = \{n: |f(n)| > T\}$, $N = \{n: |f(n)| \leq T\}$, $\Pi = \{M, N\}$, and L be the logical predicate "(1) For all n, $|f(n)| > T$;

or (2) for all n, $|f(n)| < T$; or for all n, $|f(n)| = T$." Prove that $\Sigma = (\Pi, L)$ is a segmentation of $f(n)$.

6. Let $f(t)$ be an analog signal which is twice differentiable. Show that the sign of the second derivative is the basis for a segmentation of $f(t)$ into concave-down regions, concave-up regions, and zeros of the second derivative.

7. Consider the discrete signal $x(n) = [..., 0, 2, -1, 3, 1, 8, \underline{10}, 9, 11, 5, 2, -1, 1, 0, . . .]$.

(a) Suppose values $|x(n)| \geq T$ are labeled Object and values $|x(n)| < T$ are labeled Background. Consider the effect of thresholding this signal for $T = 5$ with and without preliminary filtering by the moving average filter of width $N = 3$:

$$y(n) = \frac{1}{2N+1} \sum_{k=-N}^{N} x(k).$$

(b) What happens to the segmentation as the filter width increases? Does changing the threshold help?

(c) What happens to the segmentation if a causal box filter performs the preliminary filtering operation? Does this affect the registration of the blob?

(d) Examine the effect of a high-magnitude noise impulse several time samples apart from the blob; suppose, therefore, that $x(-10) = 12$. Consider the effect of preliminary box filtering and threshold changes on segmentation and blob registration.

(e) Can a top-down rule be applied to correct the labeling of impulse noise? Does the width of the smoothing filter affect the success of the rule?

8. Prove that a discrete signal edge detector which locates edges from the maximal responses of a convolutional operator, H, satisfies the following:

(i) The impulse response of H, $h(n)$, must be odd.

(ii) $h(n)$ can only have one zero, at $n = 0$.

9. Consider the estimator $\hat{\mu}$ found in the Unbiased Minimal Variance Estimator Theorem (Section 4.3.2). Suppose that $\boldsymbol{x}$ is a random vector, representing a window $[m - p, m + p]$ of the noisy source signal $x(n)$, and that $\boldsymbol{b}$ is the unit vector of all ones of length $2p + 1$. If the estimator $\hat{\mu}$ is unbiased and has minimal variance, show that

$$\mathrm{Var}[\hat{\mu}] = \left(\frac{1}{\boldsymbol{b} \Sigma^{-1} \boldsymbol{b}^T} \right),$$

where $\Sigma = E[(\boldsymbol{x} - \mu 1), (\boldsymbol{x} - \mu 1)]$ is the covariance matrix of the random vector $\boldsymbol{x}$.

10. Develop the analog version of normalized cross-correlation. Consider an analog signal $x(t)$ input in which a pattern $p(t)$ must be found at an unknown offset s. Suppose that $p(t)$ has finite support contained in the real interval I. Show the following:

(a) The normalized cross-correlation,

$$C_{p(t-s),x(t)} = \frac{y(s)}{\left(\int_I x^2(t)\,dt\right)^{\frac{1}{2}}} = \frac{\left(\int_I p(t-s)x(t)\,dt\right)^{\frac{1}{2}}}{\left(\int_I x^2(t)\,dt\right)^{\frac{1}{2}}},$$

where s is the offset of prototype signal $p(t)$ into input signal $x(t)$ and where I is the interval that contains the support of $p(t)$, is a measure of match.

(b) The cross-correlation, $C_{p(t-s),x(t)}$, assumes a unity maximum when $x(t) = cp(t-s)$ on the interval I for some constant c.

11. Check that all signals in the family of exponentials $B = \{(2\pi)^{-1}\exp(2\pi jnt/T)$: $n \in \mathbb{Z}\}$ have $\|(2\pi)^{-1}\exp(2\pi jnt/T)\|_1 = 1$ in the Hilbert space $L^1[0, T]$.

12. Let $g(t, \sigma, \mu) = \sigma^{-1}(2\pi)^{-1/2}\exp(-(x-\mu)^2/(2\sigma^2))$ be the Gaussian with standard deviation σ and mean μ. Show that $g(t, \sigma, \mu)$ is symmetric and strictly decreasing about its mean, μ.

13. Let $F(t, \sigma)$ be defined as in Equation (4–5c) for analog signal $f(t)$. Show that for small scales σ, the smoothed signal $F(t, \sigma)$ resembles the original as $\sigma \to 0$; that is, we have $F(t, \sigma) \to f(t)$.

14. Show that as $\sigma \to \infty$, $F(t, s) \to E(f(t))$; that is, for large scales σ, the smoothed signal approaches the mean of $f(t)$.

15. Show that the Gaussian is an $L^1(\mathbb{R})$ signal (absolutely integrable), and it is C^∞ (infinitely differentiable).

16. Show that if $f(t)$ is an analog signal and $g(t, \sigma)$ is Gaussian with zero mean and variance σ^2, then

$$\frac{\partial^n}{\partial t^n}F(t,\sigma) = f * \frac{\partial^n}{\partial t^n}g(t,\sigma).$$

17. Consider the following region merging approach to real-time signal segmentation. The application receives a digitized signal, $f(n)$, and segments it into Noise and Signal regions, with labels Λ_N and Λ_S, respectively. Three thresholds are provided: T_N, T_S, and ε. If $x(n) < T_N$, then $x(n)$ is marked as a Λ_N value. If $x(n) > T_S$, then $x(n)$ is labeled Λ_S. If $T_N \le x(n) \le T_S$, then $x(n)$ is labeled the same as $x(n-1)$ if $|x(n) - x(n-1)| < \varepsilon$, and $x(n)$ is deemed the opposite of $x(n-1)$ otherwise.

(a) Show how this real-time segmentation works for $f(n) = [..., \underline{0}, 2, 4, 3, 1, 8, 6, 9, 11, 7, 3, 5, 1, 0, ...]$, with $T_N = 4$, $T_S = 7$, and $\varepsilon = 3$.

(b) What constraints should exist on T_N, T_S, and ε for a useful algorithm?

(c) What is the effect of impulse noise on the algorithm?

(d) Let N_n and S_n be the average of the Λ_N and the Λ_S values, respectively, prior to time instant n. Suppose that the algorithm is changed so that when $T_N \leq x(n) \leq T_S$, then $x(n)$ is labeled Λ_N when $|x(n) - N_n| < \varepsilon$, and then $x(n)$ is labeled Λ_S when $|x(n) - S_n| < \varepsilon$. How does this twist affect the operation of the algorithm?

(e) Critique this algorithm for real-time speech segmentation. Consider the presence of velar fricatives, noise, and voiced and unvoiced consonants in the digitized speech signal.

(f) Explain how a delay in making the labeling decision for a new signal value $x(n)$ might help improve the segmentation for the front-end of a speech recognition application.

(g) What characteristics of digitized speech signals should be taken into account in order to size the labeling decision delay in *(f)*? Explain.

18. Critique the following region splitting algorithm. We begin with a region of interest for a discrete signal $f(n)$, $S = [a, b]$, and a partition of S into (as closely as possible) equal-length subintervals, $S = S_1 \cup S_2 \cup \ldots \cup S_N$. We first compute the mean over the entire region S, μ. Then we begin with the region whose mean is closest to μ; by renumbering the regions, if necessary, we may suppose that it is S_1. Then we compute all of the differences, $d(1, i) = |\mu_1 - \mu_i|$, for $i > 1$, and put $R_1 = S_1 \cup \{S_i \mid i > 1 \text{ and } d(1, i) < \varepsilon\}$. We then perform this same operation with the remaining S_i, which are disjoint from R_1, to form R_2, and so on. The process eventually halts after M iterations, and S is split into M regions: $S = R_1 \cup R_2 \cup \cdots \cup R_M$.

19. Show that if $x(n)$ is a digital signal, $x\colon \mathbb{Z} \to [0, N]$, for some natural number $N \geq 0$, then $x = T[\mathrm{Umbra}(x)] = T[\mathrm{Graph}(x)]$.

20. Consider Chow and Kaneko's optimal threshold finding method, where the quadratic equation below must be solved for T:

$$(\sigma_L^2 - \sigma_H^2)T^2 + 2(\sigma_H^2\mu_L - \sigma_L^2\mu_H)T + (\sigma_L^2\mu_H^2 - \sigma_H^2\mu_L^2) - 2\ln\left(\frac{P_H\sigma_L}{P_L\sigma_H}\right) = 0.$$

(a) Suppose that the variances of the meaningful signal and the background noise are the same. Show that there can be at most one threshold, T.

(b) Show that if the variances of the meaningful signal and the background noise are the same and the *a priori* probabilities of signal and noise are the same, then $T = (mL + mH)/2$.

(c) Can there be two solutions to the quadratic equation? Explain.

(d) If there is only one solution to the quadratic equation, is it necessarily a valid threshold? Explain.

(e) Can there be no solution to the equation? Explain.

21. Suppose a signal $f(n)$ has regions of interest with high values above a relatively low background noise level. Suppose also that a threshold T on the signal histogram produces errors with probability $E(T) = P_H E_L(T) + P_L E_H(T)$, where $E_L(T)$ and $E_H(T)$ are the likelihoods of incorrectly labeling f(n) as noise and as meaningful, respectively.

(a) Assume that the histogram modes obey a log-normal distribution, and, following the Chow–Kaneko method of the text in Section 4.2, use

$$\frac{dE}{dT} = P_H \frac{dE_L}{dT} + P_L \frac{dE_H}{dT} = 0$$

to find T for a minimal labeling error.

(b) As in (a), find a T that minimizes labeling error if the histogram modes obey a Rayleigh distribution.

22. Consider the valley-finding algorithm for finding a threshold using signal histograms.

(a) Show that the search stops when the current threshold T is a local minimum.

(b) Show by example that the algorithm has a direction preference; that is, it tends to select a threshold toward the noise mode or toward the meaningful signal mode.

(c) Find and explicate a technique for resolving the direction preference of the algorithm.

(d) Show by example that the algorithm may halt its search well within the mode of the histogram's noise mode or the true signal mode.

(e) Show that a gradient calculation based on a wider interval may alleviate somewhat the problem exposed in part (d).

(f) For what types of signals does the algorithm find a threshold that is at the limit of the domain of the histogram? Explain.

23. Let f be a discrete signal with a finite domain with $\mathrm{Ran}(f) \subseteq [0, N-1]$. For $0 \leq k < N$, define

$$p_k = \frac{\#(f^{-1}(\{k\}))}{\#\mathrm{Dom}(f)}.$$

Show that p_k is a discrete probability density function for f.

24. Let f and p_k be as in the previous problem. Let μ_L, μ_H, and μ be the low-level, high-level, and total mean of $h(k)$, the histogram for f. Let σ_L, σ_H, and σ be the low-level, high-level, and total standard deviations of $h(k)$.

(a) Show

$$\sum_{k=0}^{t-1} p_k(k-\mu_L(t))(\mu_L(t)-\mu) + \sum_{k=t}^{N-1} p_k(k-\mu_H(t))(\mu_H(t)-\mu) = 0.$$

(b) Following Otsu [18] define the between-group variance to be

$$\sigma_b^2(t) = P_L(t)(\mu_L(t)-\mu)^2 + P_H(t)(\mu_H(t)-\mu)^2 mp$$

and prove that

$$\sigma^2(t) = \sigma_w^2(t) + \sigma_b^2(t).$$

(Thus, minimizing within-group variance is equivalent to maximizing between-group variance.)

(c) Show the following for $0 \le t < N-1$:

$$p_k = \frac{\#(f^{-1}(\{k\}))}{\#\mathrm{Dom}(f)},$$

(d) Explain how the relationships in (c) reduce the computational burden of finding an optimal threshold (R. M. Haralick and L. G. Shapiro, *Computer and Robot Vision*, vol. 1, New York: Addison-Wesley, 1992).

25. Consider the likelihood ratio of the distribution q_k with respect to p_k,

$$L(p,q) = \sum_{k=0}^{N-1} p_k \log_2 \frac{p_k}{q_k} = \sum_{k=0}^{N-1} p_k \log_2 p_k - \sum_{k=0}^{N-1} p_k \log_2 q_k.$$

(a) Show that $L(p, q) \ge 0$ for all discrete probability distributions p and q. (*Hint*: Note that $\ln(t) < t - 1$ for all $t > 0$, where $\ln(t)$ is the natural logarithm; $\ln(2)\log_2(t) = \ln(t)$; and $p_0 + p_1 + \cdots + p_{N-1} = 1$.)

(b) Show that $L(p, q) = 0$ if and only if $p = q$.

(c) Show that L is not symmetric; that is, find examples for p_k and q_k such that p_k and p_k are discrete probability density functions, but $L(p, q) \ne L(q, p)$.

(d) Does the triangle inequality hold for the likelihood ratio?

26. Prove the following properties of dilation. Let $A, B, C \subseteq \mathbb{Z}^2$ and $k \in \mathbb{Z}$.

(a) $A \oplus B = \{a + b \mid a \in A \text{ and } b \in B\}$.

(b) $A \oplus B = B \oplus A$.

(c) $(A \oplus B) \oplus C = A \oplus (B \oplus C)$.

(d) If $A \subseteq B$, then $A \oplus C \subseteq B \oplus C$.

(e) $A \oplus (B \cup C) = (A \oplus B) \cup (A \oplus C)$.

(f) $A \oplus (B + k) = (A \oplus B) + k$.

27. Prove the following properties of erosion. Suppose $A, B, C \subseteq \mathbb{Z}^2$ and $k \in \mathbb{Z}$.

(a) $A \ominus B = \{d \in \mathbb{Z}^2 \mid d + b \in A \text{ for all } b \in B\}$.

(b) $A \ominus B = \{d \in \mathbb{Z}^2 \mid B_d \subseteq A\}$.

(c) $(A + k) \ominus B = (A \ominus B) + k$.

(d) $A \ominus (B + k) = (A \ominus B) - k$.

(e) If $A \subseteq B$, then $A \ominus C \subseteq B \ominus C$.

(f) $(A \cap B) \ominus C = (A \ominus B) \cap (B \ominus C)$.

(g) If we define $-B$ to be the reflection of the set B, $-B = \{-k \mid k \in B\}$, then we have $(A \ominus B)^C = A^C \oplus (-B)$.

28. Consider the Bayes classifier discriminant that we developed in (4.104):

$$D_k(\boldsymbol{v}) = -\frac{1}{2}(\boldsymbol{v} - \mu_k)^T \Sigma_k^{-1} (\boldsymbol{v} - \mu_k) - \frac{\log(\det(\Sigma_k))}{2} + \log(P(C_k)).$$

The text considered the simplifications brought from assuming that the feature vectors $\boldsymbol{v}$ are statistically independent and have equal variances. What simplification, if any, results from assuming simple statistical independence? Explain.

29. Let $x(n) = [..., \underline{0}, 1, 1, 2, 1, 2, 1, 0, 1, 2, 2, 1, 0, 1, 1, 0, ...]$ be a digital signal.

(a) Compute the co-occurrence matrices, P_1, P_2, and P_3, within the interval $0 \le n \le 15$.

(b) Compute the energy of values on the main diagonal of P_1, P_2, and P_3, and compare it to the energy of off-diagonal values. How do the two compare? Explain.

(c) What should one expect from co-occurrence matrices P_4, P_5, and P_6?

(d) What differences exist between the co-occurrence matrices computed for a signal with sawtooth features versus a signal with square pulse features?

30. Suppose it is required that an analysis application detect signals containing large-scale, high-magnitude regions of width N separated by low-magnitude regions of width M.

(a) What co-occurrence matrices should be computed? Explain.

(b) Describe an algorithm using these co-occurrence matrices that meets the application's needs.

(c) How do different values of M and N affect the algorithm?

31. Suppose that $f(n) = A\cos(2\pi\omega n) + B\cos(2\pi\Omega n) + Cr(n)$ is a discrete signal with $A > B > C > 0$, $\Omega > \omega$, and $r(n)$ is uniformly distributed noise. Thus, $f(n)$ consists of a sinusoidal roughness component, $A\cos(2\pi\omega n)$, and a waviness component, $B\cos(2\pi\Omega n)$.

(a) Explore the feasibility of using a co-occurrence matrix method to detect waviness versus roughness in $f(n)$. Propose and explain such an algorithm. For what time intervals, δ, does the algorithm calculate P_δ? How are the entries of the P_δ used to discover the waviness and roughness within $f(n)$?

(b) How can the size of the P_δ matrices be kept small?

(c) Explain whatever assumptions you must make on ω, Ω, A, B, and C so that the proposed algorithm works.

32. Consider a discrete signal $x(n) = x_a(nT)$, where $T > 0$ is the sampling interval.

(a) Expand $x(1)$, $x(2)$, $x(-1)$, and $x(-2)$ in terms of the Taylor series of $x_a(t)$, and show that the system with impulse response $h(n) = [-1/12, 2/3, \underline{0}, -2/3, 1/12]$ is a first-derivative operation on signals.

(b) Similarly, show that $g(n) = [-1/12, 4/3, \underline{-5/2}, 4/3, -1/12]$ is a second-derivative operator on signals.

33. Consider an application that extracts and labels feature vectors of input signals, then compares them to labeled feature vectors derived from a library of prototype signals.

(a) Show that the Hamming distance, $H(u, v)$ between two vectors of labels $\boldsymbol{u} = (\alpha_1, \alpha_2, ..., \alpha_N\}$ and $\boldsymbol{w} = (\beta_1, \beta_2, ..., \beta_N\}$ is a metric.

(b) Consider the Levenshtein distance, $L(\boldsymbol{u}, \boldsymbol{v})$, defined as the total number of substitutions into $\boldsymbol{u}$ and transpositions of components of $\boldsymbol{u}$ necessary to convert $\boldsymbol{u}$ into $\boldsymbol{v}$. Prove or disprove: $L(\boldsymbol{u}, \boldsymbol{v})$ is a metric.

CHAPTER 5

Fourier Transforms of Analog Signals

This chapter furnishes a detailed introduction to the theory and application of the Fourier transform—the first of several transforms we shall encounter in this book. Many readers, including engineers, scientists, and mathematicians, may already be familiar with this widely used transform. The Fourier transform analyzes the frequency content of a signal, and it has four variations, according to whether the time-domain signal is analog or discrete, periodic or aperiodic. The present chapter covers the two analog transforms: the Fourier series, for periodic signals, and the Fourier transform proper, for aperiodic signals.

Technology involving filtering, modulation, and wave propagation all rely heavily upon frequency analysis accomplished by the Fourier transform operation. But biological systems execute spectral analysis as well. Our senses, especially hearing and sight, are living examples of signal processors based on signal frequency spectra. The color response of the human eye is nothing more than the end result of optical signal processing designed to convert solar electromagnetic waves into the various hues of the visible electromagnetic spectrum. On a daily basis, we are exposed to sounds which are easily classified according to high and low pitch as well as purity—we are all too aware of a tenor or soprano who wobbles into a note. All instances of frequency-domain analysis, these life experiences beg the question of how engineered systems might achieve like results.

This chapter develops the first of several practical frequency-domain analysis tools. Indeed we already have practical motivations:

- Experiments in finding the period of apparently periodic phenomena, such as example of sunspot counts in the first chapter
- Attempts to characterize texture patterns in the previous chapter

Our actual theoretical development relies heavily upon the general notions of Hilbert space and orthogonal functions developed in Chapter 3. For the mathematician, who may already have a thorough understanding of the Fourier series as a complete orthonormal expansion, Chapters 5 and 6 present an opportunity to get

Signal Analysis: Time, Frequency, Scale, and Structure, by Ronald L. Allen and Duncan W. Mills
ISBN: 0-471-23441-9

down to the business of calculating the coefficients and functions which shed so much information about the physical world.

The transform consists of two complementary operations. The first is the *analysis*—that is, the breaking down of the signal into constituent parts. In the case of Fourier analysis, this involves generation and interpretation of coefficients whose magnitude and phase contain vital information pertaining to the frequency content of a signal. In the case of the continuous Fourier transform studied in this chapter, these coefficients are a continuous function of frequency as represented by the Fourier transform $F(\omega)$. The Fourier series, which is applicable to periodic waveforms, is actually a special case of this continuous Fourier transform, and it represents spectral data as a discrete set of coefficients at selected frequencies.

The second operation involves *synthesis*, a mathematical picking up of pieces, to reconstruct the original signal from $F(\omega)$ (or from the set of discrete Fourier coefficients, if appropriate), as faithfully as possible. Not all waveforms readily submit to Fourier operations, but a large set of practical signals lends itself quite readily to Fourier analysis and synthesis. Information obtained via Fourier analysis and synthesis remains by far the most popular vehicle for storing, transmitting, and analyzing signals. In some cases the analysis itself cannot be performed, leaving synthesis out of the question, while in others the physically valid analysis is available, but a reconstruction via Fourier synthesis may not converge. We will consider these issues in some detail as Chapter 5 develops. Some waveforms amenable to Fourier analysis may be better suited to more advanced transform methods such as time-frequency (windowed) Fourier transforms or time-scale (wavelet) transforms considered in later chapters. However, the basic notion of 'frequency content' derived from Fourier analysis remains an important foundation for each of these more advanced transforms.

Communication and data storage systems have a finite capacity, so the storage of an entire spectrum represented by a continuous function $F(\omega)$ is impractical. To accommodate the combined requirements of efficiency, flexibility, and economy, a discrete form of the Fourier transform is almost always used in practice. This discrete Fourier transform (DFT) is best known in the widely used fast Fourier transform (FFT) algorithm, whose development revolutionized data storage and communication. These algorithms are discussed in Chapter 7, but their foundations lie in the concepts developed in Chapters 5 and 6.

Introductory signal processing [1–5] and specialized mathematics texts [6–9] cover continuous domain Fourier analysis. Advanced texts include Refs. [10–12]. Indeed, the topic is almost ubiquitous in applied mathematics. Fourier himself developed the Fourier series, for analog periodic signals, in connection with his study of heat conduction.[1] This chapter presupposes some knowledge of Riemann integrals, ideas of continuity, and limit operations [13]. Familiarity with Lebesgue integration, covered briefly in Chapter 3, remains handy, but definitely not essential [14].

[1]Jean-Baptiste Joseph Fourier (1768–1830). The French mathematical physicist developed the idea without rigorous justification and amid harsh criticism, to solve the equation for the flow of heat along a wire [J. Fourier, *The Analytical Theory of Heat*, New York: Dover, 1955].

Essential indeed are the fundamentals of analog L^p and abstract function spaces [15, 16]. We use a few unrigorous arguments with the Dirac delta. Chapter 6 covers the generalized Fourier transform and distribution theory [17, 18]. Hopefully this addresses any misgivings the reader might harbor about informally applying Diracs in this chapter.

5.1 FOURIER SERIES

Consider the problem of constructing a synthesis operation for periodic signals based on complete orthonormal expansions considered in Chapter 3. More precisely, we seek a series

$$x_n(t) = \sum_{k=1}^{n} c_k \phi_k(t) \tag{5.1}$$

which converges to $x(t)$, a function with period T, as n approaches infinity. Equation (5.1) is a statement of the synthesis problem: Given a set of coefficients c_k and an appropriate set of orthonormal basis functions $\{\phi_1(t), \phi_2((t), \ldots, \phi_n(t))\}$, we expect a good facsimile of $x(t)$ to emerge when we include a sufficient number of terms in the series. Since the linear superposition (5.1) will represent a periodic function, it is not unreasonable to stipulate that the $\phi_k(t)$ exhibit periodicity; we will use simple sinusoids of various frequencies, whose relative contributions to $x(t)$ are determined by the phase and amplitude of the c_k. We will stipulate that the basis functions be orthonormal over some fundamental interval $[a, b]$; intuitively one might consider the period T of the original waveform $x(t)$ to be sufficiently "fundamental," and thus one might think that the length of this fundamental interval is $b - a = T$. At this point, it is not obvious where the interval should lie relative to the origin $t = 0$ (or whether it really matters). But let us designate an arbitrary point $a = t_0$, requiring that the set of $\{\phi_k(t)\}$ is a complete orthonormal basis in $L^2[t_0, t_0 + T]$:

$$\langle \phi_i, \phi_l \rangle = \int_{t_0}^{t_0 + T} \phi_i(t)\phi_j(t)^* dt = \delta_{ij}, \tag{5.2}$$

where δ_{ij} is the Kronecker[2] delta.

We need to be more specific about the form of the basis functions. Since periodicity requires $x(t) = x(t + T)$, an examination of (5.1) suggests that it is desirable to select a basis with similar qualities: $\phi_k(t) = \phi_k(t + T)$. This affords us the prospect of a basis set which involves harmonics of the fundamental frequency $1/T$. Consider

$$\phi_k(t) = A_0 e^{jk2\pi Ft} = A_0 e^{jk\Omega t}, \tag{5.3}$$

[2]This simple δ function takes its name from Leopold Kronecker (1823–1891), mathematics professor at the University of Berlin. The German algebraist was an intransigent foe of infinitary mathematics—such as developed by his pupil, Georg Cantor—and is thus a precursor of the later *intuitionists* in mathematical philosophy.

where $F = 1/T$ cycles per second (the frequency common unit is the hertz, abbreviated Hz; one hertz is a single signal cycle per second). We select the constant A_0 so as to normalize the inner product as follows. Since

$$\langle \phi_l, \phi_m \rangle = A_0^2 \int_{t_0}^{(t_0+T)} e^{jl\Omega t} e^{-jm\Omega t} p dt = \delta_{lm}, \tag{5.4a}$$

if $m = l$, then

$$\langle \phi_m, \phi_m \rangle = A_0^2 \int_{t_0}^{t_0+T} dt = A_0^2 T. \tag{5.4b}$$

Setting $A_0 = 1/\sqrt{T}$ then establishes normalization. Orthogonality is easily verified for $m \neq 1$, since

$$\begin{aligned} \langle \phi_l | \phi_m \rangle &= \frac{1}{T} \int_{t_0}^{t_0+T} e^{j(l-m)\Omega t} dt \\ &= \frac{1}{T} \int_{t_0}^{t_0+T} (\cos[(l-m)\Omega t] + j\sin[(l-(m))\Omega t]) dt = 0. \end{aligned} \tag{5.5a}$$

This establishes orthonormality of the set

$$\left\{ \frac{1}{\sqrt{T}} e^{jk\Omega t} \right\} \tag{5.5b}$$

for integer k.

When the set of complex exponentials is used as a basis, all negative and positive integer k must be included in the orthonormal expansion to ensure completeness and convergence to $x(t)$. (We can readily see that restricting ourselves to just positive or negative integers in the basis, for example, would leave a countably infinite set of functions which are orthogonal to each function in the basis, in gross violation of the notion of completeness.)

Relabeling of the basis functions provides the desired partial series expansion for both negative and positive integers k:

$$x_{2N+1}(t) = \sum_{k=-N}^{N} c_k \frac{1}{\sqrt{T}} e^{jk\Omega t}. \tag{5.6}$$

Completeness will be assured in the limit as $N \rightarrow \infty$:

$$\lim_{N \rightarrow \infty} x_{2N+1}(t) = \sum_{k=-\infty}^{\infty} c_k \frac{1}{\sqrt{T}} e^{jk\Omega t} = x(t), \tag{5.7}$$

where the expansion coefficients are determined by the inner product,

$$c_k = \langle x(t), \phi_k(t) \rangle = \int_{t_0}^{t_0+T} x(t) \frac{1}{\sqrt{T}} e^{-jk\Omega t} dt. \tag{5.8}$$

Remark. The c_k in (5.8) are in fact independent of t_0, which can be shown by the following heuristic argument. Note that all the constituent functions in (5.8)—namely $x(t)$, as well as $\cos(k\Omega t)$ and $\sin(k\Omega t)$, which make up the complex exponential—are (at least) T-periodic. As an exercise, we suggest the reader draw an arbitrary function which has period T: $f(t + T) = f(t)$. First, assume that $t_0 = 0$ and note the area under $f(t)$ in the interval $t \in [0, T]$; this is, of course, the integral of $f(t)$. Next, do the same for some nonzero t_0, noting that the area under $f(t)$ in the interval $t \in [t_0, t_0 + T]$ is unchanged from the previous result; the area over $[0, t_0]$ which was lost in the limit shift is compensated for by an equivalent gain between $[t_0, t_0 + T]$. This holds true for any finite t_0, either positive or negative, but is clearly a direct consequence of the periodicity of $x(t)$ and the orthogonal harmonics constituting the integral (5.8). Unless otherwise noted, we will set $t_0 = 0$, although there are some instances where another choice is more appropriate.

5.1.1 Exponential Fourier Series

We can now formalize these concepts. There are two forms of the Fourier series:

- For exponential basis functions of the form $Ae^{jk\Omega t}$
- For sinusoidal basis functions of the form $A\cos(k\Omega t)$ or $A\sin(k\Omega t)$

The exponential expansion is easiest to use in signal theory, so with it we begin our treatment.

5.1.1.1 Definition and Examples. The Fourier series attempts to analyze a signal in terms of exponentials. In the sequel we shall show that broad classes of signals can be expanded in such a series. We have the following definition.

Definition (Exponential Fourier Series). The *exponential Fourier series* for $x(t)$ is the expansion

$$x(t) = \sum_{k=-\infty}^{\infty} c_k \phi_k(t), \tag{5.9}$$

whose basis functions are the complete orthonormal set,

$$\phi_k(t) = \frac{1}{\sqrt{T}} e^{jk\Omega t}, \tag{5.10}$$

and whose expansion coefficients take the form (5.8).

According to the principles governing complete orthonormal expansions, (5.9) predicts that the right-hand side converges to $x(t)$, provided that the infinite summation is performed. In practice, of course, an infinite expansion is a theoretical ideal, and a cutoff must be imposed after a selected number of terms. This results in a partial series defined thusly:

Definition (Partial Series Expansion). A *partial Fourier series* for $x(t)$ is the expansion

$$x(t) = \sum_{k=-N}^{N} c_k \phi_k(t) \tag{5.11}$$

for some integer $0 < N < \infty$.

The quality of a synthesis always boils down to how many terms (5.11) should include. Typically, this judgment is based upon how much error can be tolerated in a particular application. In practice, every synthesis is a partial series expansion, since it is impossible to implement (in a finite time) an infinite summation.

Example (Sine Wave). Consider the pure sine wave $x(t) = \sin(\omega t)$. The analysis calculates the coefficients

$$c_k = \int_0^T \left(\sin \Omega t \frac{1}{\sqrt{T}} [\cos k\Omega t - j \sin k\Omega t] \right) dt . \tag{5.12}$$

Orthogonality of the sine and cosine functions dictates that all c_k vanish except for $k = \pm 1$:

$$c_{\pm 1} = \frac{\mp j}{\sqrt{T}} \left\{ \int_0^T [\sin(\Omega t)]^2 dt \right\} = (\mp j) \frac{\sqrt{T}}{2} . \tag{5.13}$$

Synthesis follows straightforwardly:

$$x(t) = (-j) \frac{\sqrt{T}}{2} \left(\frac{e^{j\Omega t}}{\sqrt{T}} \right) + j \frac{\sqrt{T}}{2} \left(\frac{e^{(-j)\Omega t}}{\sqrt{T}} \right) = \sin(\Omega t) . \tag{5.14}$$

Example (Cosine Wave). For $x(t) = \cos(\omega t)$ there are two equal nonzero Fourier coefficients:

$$c_{\pm 1} = \frac{1}{\sqrt{T}} \int_0^T [\cos(\Omega t)]^2 dt = \frac{\sqrt{T}}{2} . \tag{5.15}$$

Remark. Fourier analysis predicts that each simple sinusoid is composed of frequencies of magnitude $|\Omega|$, which corresponds to the intuitive notion of a pure oscillation. In these examples, the analysis and synthesis were almost trivial, which stems from the fact that $x(t)$ was projected along the real (in the case of a cosine) or imaginary (in the case of a sine) part of the complex exponentials comprising the orthonormal basis. This property—namely a tendency toward large coefficients when the signal $x(t)$ and the analyzing basis match—is a general property of orthonormal expansions. When data pertaining to a given signal is stored or transmitted, it is often in the form of these coefficients, so both disk space and bandwidth can be reduced by a judicious choice of analyzing basis. In this simple example of Fourier analysis applied to sines and cosines, only two coefficients are required to perform an exact synthesis of $x(\ddot{t})$. But Fourier methods do not always yield such economies, particularly in the neighborhood of transients (spikes) or jump discontinuities. We will demonstrate this shortly. Finally, note that the two Fourier coefficients are equal (and real) in the case of the cosine, but of opposite sign (and purely imaginary) in the case of the sine wave. This results directly from symmetries present in the sinusoids, a point we now address in more detail.

5.1.1.2 Symmetry Properties. The Fourier coefficients acquire special properties if $x(t)$ exhibits even or odd symmetry. Recall that if $x(t)$ is odd, $x(-t) = -x(t)$ for all t, and by extension it follows that the integral of an odd periodic function, over any time interval equal to the period T, is identically zero. The sine and cosine harmonics constituting the Fourier series are odd and even, respectively. If we expand the complex exponential in the integral for c_k,

$$c_k = \int_0^T \frac{x(t))}{\sqrt{T}}[\cos(k\Omega t) - j\sin(k\Omega t)]\, dt, \tag{5.16}$$

then some special properties are apparent:

- If $x(t)$ is real and even, then the c_k are also real and even, respectively, in k-space; that is, $c_k = c_{-k}$.
- If $x(t)$ is real and odd, then the coefficients are purely imaginary and odd in k-space: $c_{-k} = -c_k$.

The first property above follows since the second term in (5.16) vanishes identically and since $\cos(k\Omega t)$ is an even function of the discrete index k. If even–odd symmetries are present in the signal, they can be exploited in numerically intensive applications, since the number of independent calculations is effectively halved. Most practical $x(t)$ are real-valued functions, but certain filtering operations may transform a real-valued input into a complex function. In the exercises, we explore the implications of symmetry involving complex waveforms.

Example (Rectangular Pulse Train). Consider a series of rectangular pulses, each of width t and amplitude A_0, spaced at intervals T, as shown in Figure 5.1. This waveform is piecewise continuous according to the definition of Chapter 3, and in due course it will become clear this has enormous implications for synthesis. The inner product of this waveform with the discrete set of basis functions leads to a straightforward integral for the expansion coefficients:

$$c_k = \langle x(t), \phi_k(t)\rangle = \frac{A_0}{\sqrt{T}}\int_0^{\frac{\tau}{2}}(\cos k\Omega t - j\sin k\Omega t)\, dt + \frac{A_0}{\sqrt{T}}\int_{T-\frac{\tau}{2}}^{T}(\cos k\Omega t - j\sin k\Omega t)\, dt \tag{5.17}$$

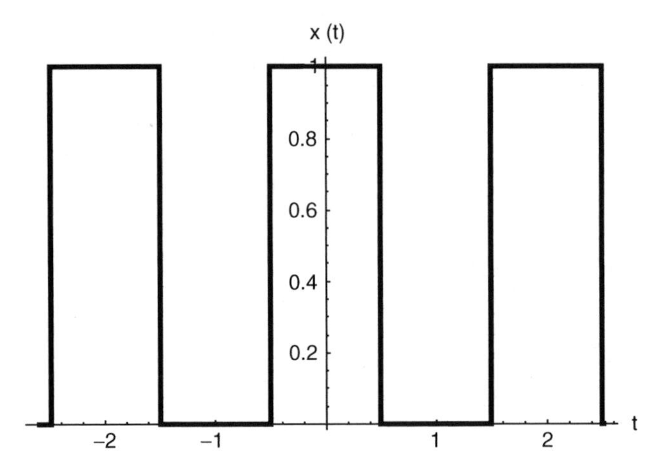

Fig. 5.1. A train of rectangular pulses. Shown for pulse width $\tau = 1$, amplitude $A_0 = 1$, and period $T = 2$.

Some algebra reduces this to the succinct expression

$$c_k = \frac{A_0}{\sqrt{T}} \cdot \tau \cdot \frac{\sin\left(\frac{k\Omega\tau}{2}\right)}{\left(\frac{k\Omega\tau}{2}\right)}. \tag{5.18}$$

Example (Synthesis of Rectangular Pulse). In Figure 5.2 we illustrate the synthesis of periodic rectangular pulses for several partial series, using (5.10) and (5.16).

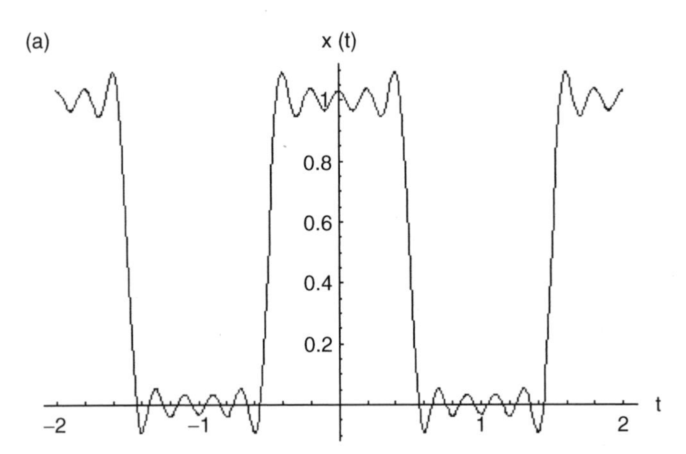

Fig. 5.2. Synthesis of the rectangular pulse train. (a) Partial series $N = 10$, (b) $N = 50$, (c) $N = 100$. The number of terms in the series is $2N + 1$.

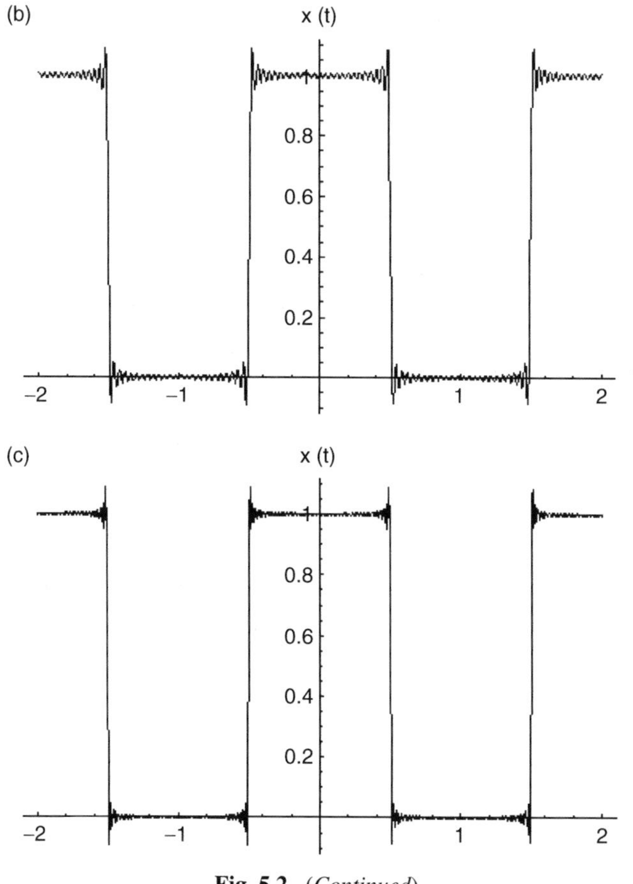

Fig. 5.2 *(Continued)*

5.1.2 Fourier Series Convergence

We are now in a position to prove the convergence of the exponential Fourier series for a signal $x(t)$. We shall consider two cases separately:

- At points where $x(t)$ is continuous;
- At points where $x(t)$ has a jump discontinuity.

5.1.2.1 Convergence at Points of Continuity. It turns out that the Fourier series does converge to the original signal at points of continuity. We have the following theorem.

Theorem (Fourier Series Convergence). Suppose $S_N(s)$ is a partial series summation of the form

$$S_N(s) = \sum_{k=-N}^{N} c_k \frac{1}{\sqrt{T}} e^{jk\Omega t}, \tag{5.19a}$$

where N is a positive integer. If $x(t)$ is continuous at s (including points of continuity within piecewise continuous functions), then

$$\lim_{N \to \infty} S_N(s) = x(s). \tag{5.19b}$$

Proof: Consider the partial series summation:

$$S_N(s) = \sum_{k=-N}^{N} \left\langle x(t), \frac{e^{jk\Omega t}}{\sqrt{T}} \right\rangle \frac{1}{\sqrt{T}} e^{jk\Omega t}. \tag{5.20}$$

Writing the inner product term (in brackets) as an explicit integral, we have

$$S_N(s) = \frac{1}{T} \sum_{k=-N} \int_0^T x(t) e^{jk\Omega(s-t)} dt = \frac{2}{T} \int_0^T x(t) \cdot K(s-t)\, dt, \tag{5.21}$$

where

$$K(s-t) = \frac{1}{2} + \sum_{k=1}^{N} \cos(k\Omega(s-t)). \tag{5.22}$$

The function $K(s-t)$ reduces—if we continue to exploit the algebraic properties of the exponential function for all they are worth—to the following:

$$K(s-t) = \mathrm{Re}\left[\left(1 - \frac{1}{2}\right) + \sum_{k=1}^{N} e^{jk\Omega(s-t)}\right]. \tag{5.23a}$$

This reduces to the more suggestive form,

$$K(s-t) = \frac{\sin\left[\left(N + \frac{1}{2}\right)(s-t)\right]}{2\sin\left[\frac{1}{2}(s-t)\right]}. \tag{5.23b}$$

Returning to the partial series expansion (5.21), the change of integration variable $u = s - t$ gives

$$S_N(s) = -\int_s^{s-T} x(s-u) \left[\frac{\sin\left(N + \frac{1}{2}\right)u}{T\sin\left(\frac{T}{2}\right)}\right] du. \tag{5.24}$$

The quantity in brackets is the *Dirichlet kernel*[3],

$$D_N(u) = \frac{\sin\left(N + \frac{1}{2}\right)u}{T\sin\left(\frac{u}{2}\right)}, \tag{5.25}$$

[3]P. G. Legeune Dirichlet (1805–1859) was Kronecker's professor at the University of Berlin and the first to rigorously justify the Fourier series expansion. His name is more properly pronounced "Dear-ah-klet."

whose integral exhibits the convenient property

$$-\int_{s}^{s-T} D_N(u)\, du = 1. \tag{5.26}$$

Equation (5.26) is easily demonstrated with a substitution of variables, $2v = u$, which brings the integral into a common tabular form:

$$-\int_{s}^{s-T} D_N(u)\, du = -\frac{1}{T}\int_{0}^{-\frac{T}{2}} 2\cdot\frac{\sin 2N+1)v}{\sin v}\, dv = \frac{-2}{T}\left[2\sum_{m=1}^{N}\frac{\sin(mv)}{2m}+v\right]\Bigg|_{0}^{-\frac{T}{2}} = 1. \tag{5.27}$$

The beauty of this result lies in the fact that we can construct the identity

$$x(s) = -\int_{s}^{s-T} x(s)D_N(u)\, du \tag{5.28}$$

so that the difference between the partial summation $S_N(s)$ and the original signal $x(s)$ is an integral of the form

$$S_N(s)-x(s) = \frac{-1}{T}\int_{s}^{s-T} g(s,u)\cdot\sin\left[\left(N+\frac{1}{2}\right)u\right] du, \tag{5.29}$$

where

$$g(s,u) = \frac{x(s-u)-x(s)}{\sin\left(\frac{u}{s}\right)}. \tag{5.30}$$

In the limit of large N, (5.29) predicts that the partial series summation converges pointwise to $x(s)$ by simple application of the Riemann–Lebesgue lemma (Chapter 3):

$$\lim_{N\to\infty}[S_N(s)-x(s)] = \lim_{N\to\infty}\int_{s}^{s-T}\frac{-g(s,u)}{T}\sin\left[\left(N+\frac{1}{2}\right)u\right] du = 0, \tag{5.31}$$

thus concluding the proof. ∎

The pointwise convergence of the Fourier series demonstrated in (5.31) is conceptually reassuring, but does not address the issue of how rapidly the partial series expansion actually approaches the original waveform. In practice, the Fourier series is slower to convergence in the vicinity of sharp peaks or spikes in $x(t)$. This aspect of the Fourier series summation is vividly illustrated in the vicinity of a step discontinuity—of the type exhibited by rectangular pulse trains and the family of sawtooth waves, for example. We now consider this problem in detail.

5.1.2.2 Convergence at a Step Discontinuity. It is possible to describe and quantify the quality of convergence at a jump discontinuity such as those exhibited by the class of piecewise continuous waveforms described in Chapter 3. We represent such an $x(t)$ as the sum of a continuous part $x_c(t)$ and a series of unit steps, each term of which represents a step discontinuity with amplitude $A_k = x(t_k^+) - x(t_k^-)$ located at $t = t_k$:

$$x(t) = x_c(t) + \sum_{k=1}^{M} A_k u(t - t_k). \tag{5.32}$$

In the previous section, convergence of the Fourier series was established for continuous waveforms and that result applies to the $x_c(t)$ constituting part of the piecewise continuous function in (5.32). Here we turn to the issue of convergence in the vicinity of the step discontinuities represented by the second term in that equation. We will demonstrate that

- The Fourier series converges pointwise at each t_k.
- The discontinuity imposes oscillations or ripples in the synthesis, which are most pronounced in the vicinity of each step. This artifact, known as the *Gibbs phenomenon*,[4] is present in all partial series syntheses of piecewise continuous $x(t)$; however, its effects can be minimized by taking a sufficiently large number of terms in the synthesis.

The issue of Gibbs oscillations might well be dismissed as a mere mathematical curiosity were it not for the fact that so many practical periodic waveforms are piecewise continuous. Furthermore, similar oscillations occur in other transforms as well as in filter design, where ripple or overshoot (which are typically detrimental) arise from similar mathematics.

Theorem (Fourier Series Convergence: Step Discontinuity). Suppose $x(t)$ exhibits a step discontinuity at some time t about which $x(t)$ and its derivative have well-behaved limits from the left and right, $t_{(l)}$ and $t_{(r)}$, respectively. Then $x(t)$ converges pointwise to the value

$$x(t) = \frac{[x(t_{(r)}) + x(t_{(l)})]}{2}. \tag{5.33}$$

Proof: For simplicity, we will consider a single-step discontinuity and examine the synthesis

$$x(t) = x_c(t) + A_s u(t - t_s), \tag{5.34}$$

[4]The Yale University chemist, Josiah Willard Gibbs (1839–1903), was the first American scientist of international renown.

where the step height is $A_s = x(t_s+) - x(t_s-)$. We begin by reconsidering (5.24):

$$S_N(s) = \int_0^T x(t)\left(\frac{\sin\left[\left(N+\frac{1}{2}\right)(s-t)\right]}{T\sin\left[\frac{1}{2}(s-t)\right]}\right) dt = \int_{-\frac{T}{2}}^{\frac{T}{2}} x(s-t)D_N(\Omega t)\, dt. \tag{5.35}$$

For convenience, we have elected to shift the limits of integration to a symmetric interval $[-T/2, T/2]$. Furthermore, let us assume that the discontinuity occurs at the point $t = t_s = 0$. (These assumptions simplify the calculations enormously and do not affect the final result. The general proof adds complexity which does not lead to any further insights into Fourier series convergence.) It is convenient to break up the integral into two segments along the t axis:

$$S_N(s) = \int_{-T/2}^{T/2} x_c(s-t)D_N(\Omega t)\Omega\, dt + A_0 \cdot \int_{-T/2}^{T/2} u(s-t)D_N(\Omega t)\Omega\, dt, \tag{5.36}$$

where $A_0 = x(0_{(r)}) - x(0_{(1)})$ is the magnitude of the jump at the origin. In the limit $N \to \infty$, the first term in (5.36) converges to $x_c(t)$, relegating the discontinuity's effect to the second integral, which we denote $e_N(s)$:

$$\varepsilon_N(s) = A_0 \cdot \int_{-\frac{T}{2}}^{\frac{T}{2}} u(t)D_N(\Omega(s-t))t\Omega\, dt = A_0 \cdot \int_0^{\frac{T}{2}} D_N(\Omega(s-t))\Omega\, dt. \tag{5.37}$$

The task at hand is to evaluate this integral. This can be done through several changes of variable. Substituting for the Dirichlet kernel provides

$$\varepsilon_N(s) = A_0 \cdot \int_0^{\frac{T}{2}} \left[\frac{\sin\left[\left(N+\frac{1}{2}\right)\Omega(s-t)\right]}{T\sin\left[\frac{\Omega(s-t)}{2}\right]}\right]\Omega\, dt. \tag{5.38}$$

Defining a new variable $u = \Omega(t-s)$ and expanding the sines brings the dependence on variable s into the upper limit of the integral:

$$\varepsilon_N(s) = A_0 \cdot \int_0^{-\Omega\left(s+\frac{T}{2}\right)} \left[\frac{\sin(nu)\cdot\cos\left(\frac{u}{2}\right)}{2\pi\sin\left[\frac{u}{2}\right]} + \cos(nu)\right] du. \tag{5.39}$$

Finally, we define a variable $2v = u$ which brings (5.39) into a more streamlined form

$$\varepsilon_N(s) = A_0 \cdot \int_0^{-\frac{\Omega}{2}\left(s+\frac{T}{2}\right)} \left[\frac{\sin(2nv)\cdot\cos(v)}{2\pi\sin(v)} + \cos(2nv)\right] dv. \tag{5.40}$$

This is as close as we can bring $e_N(s)$ to an analytic solution, but it contains a wealth of information. We emphasize that s appears explicitly as an upper limit in each integral. Tabulation of (5.40) produces an oscillatory function of s in the neighborhood of $s = 0$; this accounts for the ripple—Gibbs oscillations—in the partial series synthesis near the step discontinuity. As we approach the point of discontinuity at the origin, (5.40) can be evaluated analytically:

$$\varepsilon_N(s) = A_0 \cdot \int_0^{\frac{\pi}{2}} \left[\frac{\sin(2nv)\cdot\cos(v)}{2\pi\sin(v)} + \cos(2nv)\right] dv = A_0 \cdot \left[\left(\frac{1}{\pi}\cdot\frac{\pi}{2}\right) + 0\right] = \frac{1}{2}A_0. \tag{5.41}$$

(Note that in going from (5.40) to (5.41), a sign change can be made in the upper limit, since the integrand is an even function of the variable v.) Accounting for both the continuous portion $x_c(t)$—which approaches $x(0_{(l)})$ as $N \to \infty$ and as $t \to 0$—and the discontinuity's effects described in (5.41), we find

$$x(0) = x_c(0) + \frac{1}{2}[x(0_{(r)}) - x(0_{(l)})] = \frac{1}{2}[x(0_{(r)}) + x(0_{(l)})]. \tag{5.42}$$

A similar argument works for a step located at an arbitrary t; this provides the general result

$$x(0) = \frac{1}{2}[x(t_{(r)}) + x(t_{(l)})], \tag{5.43}$$

and the proof is complete. ■

Figure 5.3 illustrates the convergence of $e_N(s)$ near a step discontinuity in a rectangular pulse train. Note the smoothing of the Gibbs oscillations with increasing N.

As N approaches infinity and at points t where $x(t)$ is continuous, the Gibbs oscillations get infinitesimally small. At the point of discontinuity, they contribute an amount equal to one-half the difference between the left and right limits of $x(t)$, as dictated by (5.42).

When approaching this subject for the first time, it is easy to form misconceptions about the nature of the convergence of the Fourier series at step discontinuties, due to the manner in which the Gibbs oscillations (almost) literally cloud the issue. We complete this section by emphasizing the following points:

- The Gibbs oscillations do *not* imply a failure of the Fourier synthesis to converge. Rather, they describe how the convergence behaves.

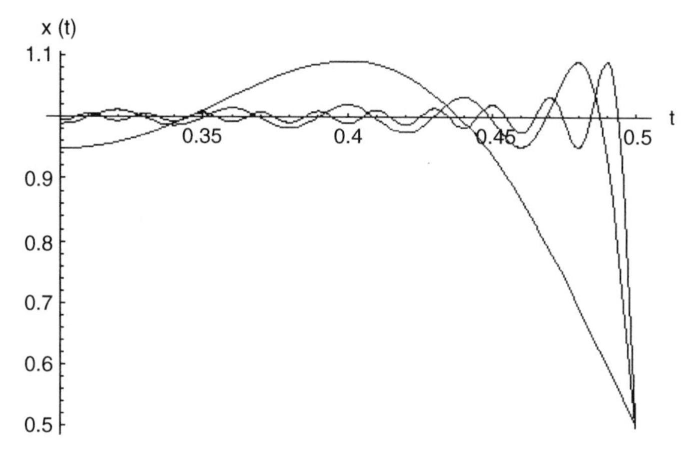

Fig. 5.3. Convergence of the Fourier series near a step, showing Gibbs oscillations for N = 10, 50, 100. For all N, the partial series expansion converges to 1/2 at the discontinuity.

- The Fourier series synthesis converges to an exact, predictable value at the point of discontinuity, namely the arithmetic mean of the left and right limits of $x(t)$, as dictated by (5.43).
- In the vicinity of the discontinuity, at points for which $x(t)$ is indeed continuous, the Gibbs oscillations disappear in the limit as N becomes infinite. That is, the synthesis converges to $x(t)$ with no residual error. It is exact.

5.1.3 Trigonometric Fourier Series

Calculations with the exponential basis functions make such liberal use of the orthogonality properties of the constituent sine and cosine waves that one is tempted to reformulate the entire Fourier series in a set of sine and cosine functions. Such a development results in the trigonometric Fourier series, an alternative to the exponential form considered in the previous section.

Expanding the complex exponential basis functions leads to a synthesis of the form

$$x(t) = \sum_{k=-\infty}^{\infty} c_k \phi_k(t) = \frac{1}{\sqrt{T}} \sum_{k=-\infty}^{\infty} c_k[\cos(k\Omega t) + j\sin(k\Omega t)]. \tag{5.44}$$

Since $\cos(k\Omega t)$ and $\sin(k\Omega t)$ are even and odd, respectively, in the variable k, and since for $k = 0$ there is no contribution from the sine wave, we can rearrange the summation and regroup the coefficients. Note that the summations now involve only the positive integers:

$$x(t) = \frac{1}{\sqrt{T}}\left\{c_0 + \sum_{k=1}^{\infty} (c_k + c_{-k})\cos(k\Omega t)\right\} + \sum_{k=1}^{\infty} (c_k - c_{-k})\sin(k\Omega t). \tag{5.45}$$

The zeroth coefficient has the particularly simple form:

$$c_0 = \langle x(t), 1\rangle = \int_0^T x(t)\,dt, \tag{5.46}$$

where 1 is the unit constant signal on $[0, T]$. Regrouping terms gives an expansion in sine and cosine:

$$x(t) = a_0 + \sum_{k=1}^{\infty} a_k \cos(k\Omega t) + \sum_{k=1}^{\infty} b_k \sin(k\Omega t), \tag{5.47}$$

where

$$a_k = \frac{c_k + c_{-k}}{\sqrt{T}}, \tag{5.48a}$$

$$b_k = j \cdot \frac{c_k - c_{-k}}{\sqrt{T}}, \tag{5.48b}$$

and

$$a_0 = \frac{1}{\sqrt{T}} \int_0^T x(t)\,dt. \tag{5.48c}$$

Under circumstances where we have a set of exponential Fourier series coefficients c_k at our disposal, (5.47) is a valid definition of the trigonometric Fourier series. In general, this luxury will not be available. Then a more general definition gives explicit intergrals for the expansion coefficients, a_k and b_k, based on the inner products $\langle x(t), f_m(t)\rangle$, where $f_m(t) = C_m \cos(m\Omega t)$ or $S_m \sin(m\Omega t)$ and C_m and S_m are normalization constants.

The C_m are determined by expanding the cosine inner product:

$$\begin{aligned}\langle x(t), C_m \cos(m\Omega t)\rangle = {} & \langle a_0, C_m \cos(m\Omega t)\rangle + \sum_{k=1}^{\infty} a_k \langle C_k \cos(k\Omega t), C_m \cos(m\Omega t)\rangle \\ & + \sum_{k=1}^{\infty} a_k \langle S_k \sin(k\Omega t), C_m \cos(m\Omega t)\rangle\end{aligned} \tag{5.49}$$

Consider each term above. The first one vanishes for all m, since integrating cosine over one period $[0, T]$ gives zero. The third term also vanishes for all k, due to the orthogonality of sine and cosine. The summands of the second term are zero, except for the bracket

$$\langle C_m \cos(m\Omega t), C_m \cos(m\Omega t)\rangle = \frac{T}{2} C_m^{\ 2}. \tag{5.50}$$

To normalize this inner product, we set

$$C_m = \sqrt{\frac{2}{T}} \tag{5.51}$$

for all m. Consequently, the inner product defining the cosine Fourier expansion coefficients a_k is

$$a_k = \left\langle x(t), \sqrt{\frac{2}{T}}\cos(k\Omega t)\right\rangle. \tag{5.52}$$

The sine-related coefficients are derived from a similar chain of reasoning:

$$b_k = \left\langle x(t), \sqrt{\frac{2}{T}}\sin(k\Omega t)\right\rangle. \tag{5.53}$$

Taking stock of the above leads us to define a Fourier series based on sinusoids:

Definition (Trigonometric Fourier Series). The *trigonometric Fourier series* for $x(t)$ is the orthonormal expansion

$$x(t) = a_0 + \sum_{k=1}^{\infty} a_k \phi_k(t) + \sum_{k=1}^{\infty} b_k \Psi_k(t), \tag{5.54}$$

where

$$\phi_k(t) = \sqrt{\frac{2}{T}}\cos(k\Omega t) \tag{5.55a}$$

and

$$\Psi_k(t) = \sqrt{\frac{2}{T}}\sin(k\Omega t). \tag{5.55b}$$

Remark. Having established both the exponential and trigonometric forms of the Fourier series, note that it is a simple matter to transform from one coefficient space to the other. Beginning in (5.48a), we derived expressions for the trigonometric series coefficients in terms of their exponential series counterparts. But these relations are easy to invert. For $k > 0$, we have

$$c_k = \frac{\sqrt{T}}{2}(a_k - jb_k) \tag{5.56a}$$

and

$$c_{-k} = \frac{\sqrt{T}}{2}(a_k + jb_k). \tag{5.56b}$$

Finally, for $k = 0$, we see

$$c_0 = a_0\sqrt{T}. \tag{5.57}$$

5.1.3.1 *Symmetry and the Fourier Coefficients.* As in the case of the exponential Fourier coefficients, the a_k and b_k acquire special properties if $x(t)$ exhibits even or odd symmetry in the time variable. These follow directly from (5.52) and (5.53), or by the application of the previously derived c_k symmetries to (5.45). Indeed, we see that

- If $x(t)$ is real and odd, then the a_k vanish identically, and the b_k are purely imaginary.
- On the other hand, if $x(t)$ is real and even, the b_k vanish and the a_k are real quantities.

The even/odd coefficient symmetry with respect to k is not an issue with the trigonometric Fourier series, since the index k is restricted to the positive integers.

5.1.3.2 *Example: Sawtooth Wave.* We conclude with a study of the trigonometric Fourier series for the case of a sawtooth signal. Consider the piecewise continuous function shown in Figure 5.4a. In the fundamental interval $[0, T]$, $x(t)$ consists of two segments, each of slope μ. For $t \in [0, T/2]$:

$$x(t) = \mu t, \tag{5.58a}$$

and for $t \in [T/2, T]$:

$$x(t) = \mu(t-T). \tag{5.58b}$$

The coefficients follow straightforwardly. We have

$$b_n = \frac{T}{2}\langle x(t), \sin(n\Omega t)\rangle = \frac{2\mu}{T}\int_0^T t\sin(n\Omega t)\, dt + \left(-\frac{4h}{T}\right)\int_{T/2}^{T} \sin(n\Omega t)\, dt. \tag{5.59}$$

The first integral on the right in (5.59) is evaluated through integration by parts:

$$\frac{2\mu}{T}\int_0^T t\sin(n\Omega t)\, dt = \frac{-2h}{\pi n}. \tag{5.60}$$

The second integral is nonzero only for $n = 1, 3, 5, \ldots$,

$$\frac{-4h}{T}\int_{T/2}^{T} \sin(n\Omega t)\, dt = \frac{4h}{\pi n}. \tag{5.61}$$

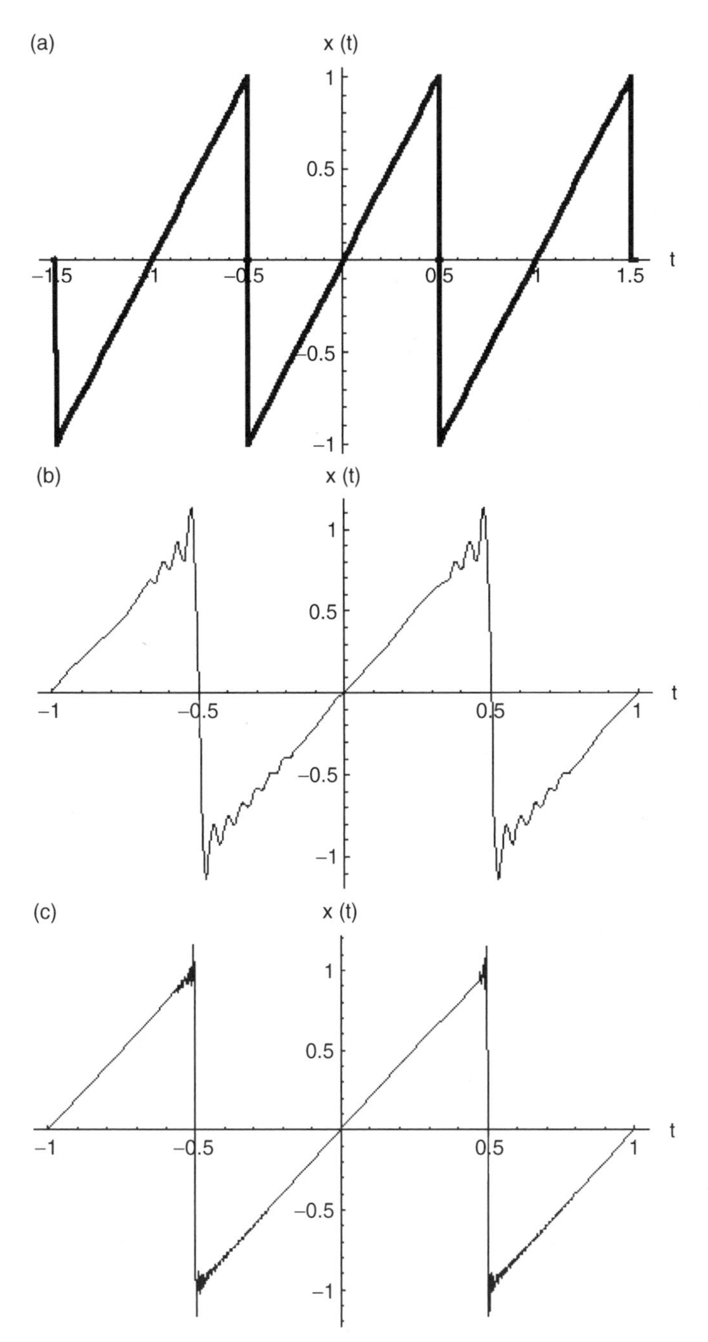

Fig. 5.4. Synthesis of the sawtooth wave using the trigonometric Fourier series. (a) The original waveform. (b) Partial series, $N = 20$. (c) For $N = 100$. (d) For $N = 200$. There are $N + 1$ terms in the partial series. (e) Details illustrating Gibbs oscillation near a discontinuity, for $N = 20$, 100, and 200. Note that all partial series converge to $x_N(t) = 0$ at the discontinuity.

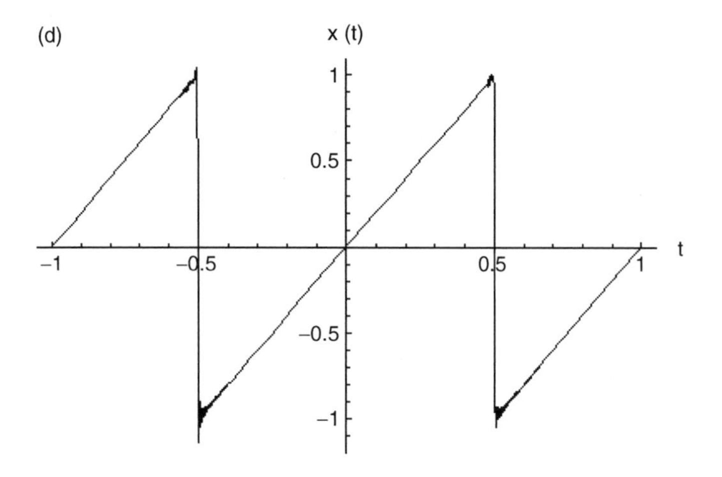

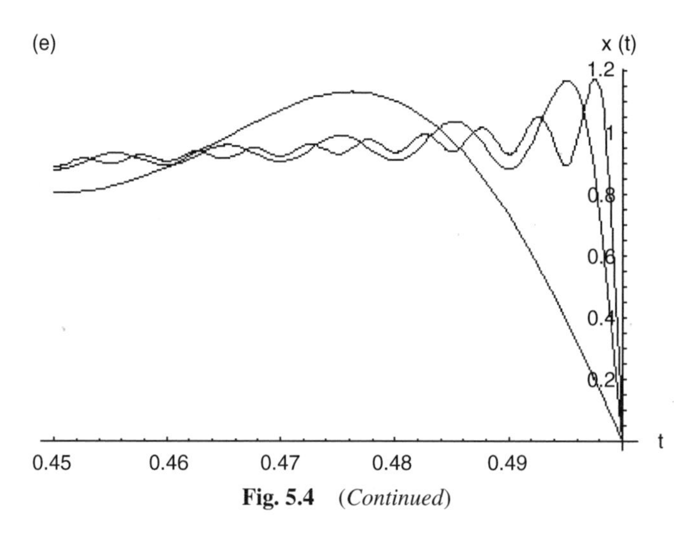

Fig. 5.4 (*Continued*)

Therefore, for $n = 1, 3, 5, \ldots,$

$$b_n = \frac{2h}{\pi n}, \tag{5.62a}$$

while for $n = 2, 4, 6, \ldots,$

$$b_n = \frac{-2h}{\pi n}. \tag{5.62b}$$

Since $x(t)$ exhibits odd symmetry in t, the coefficients for the cosine basis are identically zero for all n:

$$a_n = 0. \tag{5.63}$$

Example (Sawtooth Wave Synthesis). Figure 5.4 illustrates several partial series syntheses of this signal using the coefficients (5.62a). The Gibbs oscillations are clearly in evidence. The convergence properties follow the principles outlined earlier and illustrated in connection with the rectangular pulse train.

Remark. From the standpoint of physical correctness, the exponential and trigonometric series are equally valid. Even and odd symmetries—if they exist—are more easily visualized for the trigonometric series, but mathematically inclined analysts find appeal in the exponential Fourier series. The latter's formalism more closely relates to the Fourier transform operation considered in the next section, and it forms the basis for common numerical algorithms such as the fast Fourier transform (FFT) discussed in Chapter 7.

5.2 FOURIER TRANSFORM

In the case of periodic waveforms considered in the previous section, the notion of "frequency content" is relatively intuitive. However, many signals of practical importance exhibit no periodicity whatsoever. An isolated pulse or disturbance, or an exponentially damped sinusoid, such as that produced by a resistor–capacitor (RC) circuit, would defy analysis using the Fourier series expansion. In many practical systems, the waveform consists of a periodic sinusoidal carrier wave whose envelope is modulated in some manner; the result is a composite signal having an underlying sinusoidal structure, but without overall periodicity. Since the information content or the "message," which could range from a simple analog sound signal to a stream of digital pulses, is represented by the modulation, an effective means of signal analysis for such waves is of enormous practical value. Furthermore, all communications systems are subject to random fluctuations in the form of noise, which is rarely obliging enough to be periodic.

5.2.1 Motivation and Definition

In this section, we develop a form of Fourier analysis applicable to many practical aperiodic signals. In fact, we will eventually demonstrate that the Fourier series is a special case of the theory we are about to develop; we will need to equip ourselves, however, with a mathematical arsenal appropriate to the task. Many notions will carry over from the Fourier series. The transformation to frequency space—resulting in an analysis of the waveform in terms of its frequency content—will remain intact. Similarly, the synthesis, whereby the original signal is reconstructed based on the frequency spectrum, will be examined in detail. We will develop the criteria by which a given waveform will admit a transform to frequency space and by which the resulting spectra will admit a viable *synthesis*, or *inverse transform*.

Since our nascent Fourier transform involves integrals, analysis and synthesis relations lean heavily on the notion of absolute integrability. Not surprisingly, the analog L^p signal spaces—in particular, $L^1(\mathbb{R})$ and $L^2(\mathbb{R})$—will figure prominently.

We recall these abstract function spaces from Chapter 3: $L^p(\mathbb{K}) = \{x(t) \mid \|x\|_p < \infty\}$. Here

$$\|x\|_p = \left[\int_{-\infty}^{\infty} |x(t)|^p dt\right]^{\frac{1}{p}} \tag{5.64}$$

is the L^p norm of $x(t)$ and $\mathbb{K}$ is either the real numbers $\mathbb{R}$ or the complex numbers $\mathbb{C}$. We define the set of bounded signals to be L^∞. These signal classes turn out to be Banach spaces, since Cauchy sequences of signals in L^p converge to a limit signal also in L^p. L^1 is also called the space of *absolutely integrable* signals, and L^2 is called the space of *square-integrable* signals. The case of $p = 2$ is special: L^2 is a Hilbert space. That is, there is an inner product relation on square-integrable signals $\langle x, y\rangle \in \mathbb{K}$, which extends the idea of the vector space dot product to analog signals.

In the case of the Fourier series, the frequency content was represented by a set of discrete coefficients, culled from the signal by means of an inner product involving the signal and a discrete orthonormal basis:

$$c_k = \langle x(t), \phi_k(t)\rangle = \int_{t_0}^{t_0+T} x(t)\frac{1}{\sqrt{T}}e^{-jk\Omega t}dt. \tag{5.65}$$

One might well ask whether a similar integral can be constructed to handle nonperiodic signals $f(t)$. A few required modifications are readily apparent. Without the convenience of a fundamental frequency or period, let us replace the discrete harmonics $k\omega$ with a continuous angular frequency variable ω, in radians per second. Furthermore, all values of the time variable t potentially contain information regarding the frequency content; this suggests integrating over the entire time axis, $t \in [-\infty, \infty]$. The issue of multiplicative constants, such as the normalization constant $1/\sqrt{T}$, appears in a different guise as well. Taking all of these issues into account, we propose the following definition of the Fourier transform:

Definition (Radial Fourier Transform). The radial Fourier transform of a signal $f(t)$ is defined by the integral,

$$F(\omega) = \int_{-\infty}^{\infty} f(t)e^{-j\omega t}dt. \tag{5.66a}$$

It is common to write a signal with a lowercase letter and its Fourier transform with the corresponding uppercase letter. Where there may be confusion, we also write $F(\omega) = \mathcal{F}[f(t)](\omega)$, with a "fancy F" notation.

Remark. Note that the Fourier transform operation $\mathcal{F}$ is an analog system that accepts time domain signals $f(t)$ as inputs and produces frequency-domain signals

$F(\omega)$ as outputs. One must be cautious while reading the signal processing literature, because two other definitions for $\mathcal{F}$ frequently appear:

- The *normalized* radial Fourier transform;
- The *Hertz* Fourier transform.

Each one has its convenient aspects. Some authors express a strong preference for one form. Other signal analysts slip casually among them. We will mainly use the radial Fourier transform, but we want to provide clear definitions and introduce special names that distinguish the alternatives, even if our terminology is not standard. When we change definitional forms to suit some particular analytical endeavor, example, or application, we can then alert the reader to the switch.

Definition (Normalized Radial Fourier Transform). The *normalized radial Fourier transform* of a signal $f(t)$ is defined by the integral,

$$F(\omega) = \frac{1}{\sqrt{2\pi}} \int_{-\infty}^{\infty} f(t) e^{-j\omega t} dt. \tag{5.66b}$$

The $(2\pi)^{-1}$ factor plays the role of a normalization constant for the Fourier transform much as the factor $1/\sqrt{T}$ did for the Fourier series development. Finally, we have the Hertz Fourier transform:

Definition (Hertz Fourier Transform). The *Hertz Fourier transform* of a signal $x(t)$ is defined by the integral

$$x(f) = \int_{-\infty}^{\infty} x(t) e^{-j2\pi f t} dt. \tag{5.66c}$$

Remark. The units of ω in both the radial and normalized Fourier transforms are in radians per second, assuming that the time variable t is counted in seconds. The units of the Hertz Fourier transform are in hertz (units of inverse seconds or cycles per second). A laboratory spectrum analyzer displays the Hertz Fourier transform—or, at least, it shows a reasonably close approximation. So this form is most convenient when dealing with signal processing equipment. The other two forms are more convenient for analytical work. It is common practice to use ω (or Ω) as a radians per second frequency variable and use f (or F) for a Hertz frequency variable. But we dare to emphasize once again that Greek or Latin letters do no more than hint of the frequency measurement units; it is rather the particular form of the Fourier transform definition in use that tells us what the frequency units must be.

The value of the Fourier transform at $\omega = 0$, $F(0)$, is often called, in accord with electrical engineering parlance, the *direct current* or *DC* term. It represents that portion of the signal which contains no oscillatory, or *alternating current* (*AC*), component.

Now if we inspect the radial Fourier transform's definition (5.66a), it is tempting to write it as the inner product $\langle x(t), e^{j\omega t} \rangle$. Indeed, the Fourier integral has precisely this form. However, we have not indicated the signal space to which $x(t)$ may belong. Suppose we were to assume that $x(t) \in L^2(\mathbb{R})$. This space supports an inner product, but that will not guarantee the existence of the inner product, because, quite plainly, the exponential signal, $e^{j\omega t}$ is not square-integrable. Thus, we immediately confront a theoretical question of the Fourier transform's existence. Assuming that we can justify this integration for a wide class of analog signals, the Fourier transform does appear to provide a measure of the amount of radial frequency ω in signal $x(t)$. According to this definition, the frequency content of $x(t)$ is represented by a function $X(\omega)$ which is clearly analogous to the discrete set of Fourier series coefficients, but is—as we will show— a continuous function of angular frequency ω.

Example (Rectangular Pulse). We illustrate the radial Fourier transform with a rectangular pulse of width $2a > 0$, where

$$f(t) = 1 \tag{5.67}$$

for $-a \le t < a$, and vanishes elsewhere. This function has compact support on this interval and its properties under integration are straightforward when the Fourier transform is applied:

$$F(\omega) = \int_{-\infty}^{\infty} f(t) e^{-j\omega t} dt = \int_{-a}^{a} e^{-j\omega t} dt = 2a \left[\frac{\sin(\omega a)}{\omega a} \right]. \tag{5.68}$$

The most noteworthy feature of the illustrated frequency spectrum, Figure 5.5, is that the pulse width depends upon the parameter a.

Note that most of the spectrum concentrates in the region $\omega \in [-\pi/a, \pi/a]$. For small values of a, this region is relatively broad, and the maximum at $\omega = 0$ (i.e.,

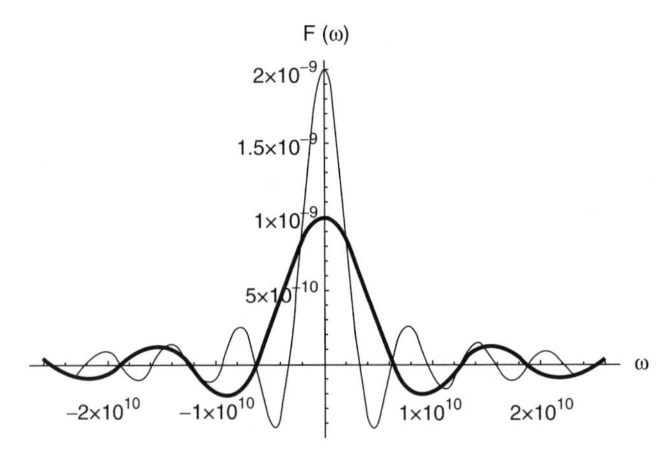

Fig. 5.5. The spectrum for a 1-ns rectangular pulse (solid line), and a 2-ns pulse. Note the inverse relationship between pulse width in time and the spread of the frequency spectrum.

the DC contribution) is, relatively speaking, low. This is an indication that a larger proportion of higher frequencies are needed to account for the relatively rapid jumps in the rectangular pulse. Conversely, as the pulse width increases, a larger proportion of the spectrum resides near the DC frequency. In fact, as the width of the pulse approaches infinity, its spectrum approaches the Dirac delta function $\delta(\omega)$, the generalized function introduced in Chapter 3. This scaling feature generalizes to all Fourier spectra, and the inverse relationship between the spread in time and the spread in frequency can be formalized in one of several uncertainty relations, the most famous of which is attributed to Heisenberg. This topic is covered in Chapter 10.

Example (Decaying Exponential). By their very nature, transient phenomena are short-lived and often associated with exponential decay. Let $\alpha > 0$ and consider

$$f(t) = e^{-\alpha t}u(t), \tag{5.69}$$

which represents a damped exponential for all $t > 0$. This signal is integrable, and the spectrum is easily calculated:

$$F(\omega) = \int_0^{\infty} e^{-\alpha t}e^{-j\omega t}dt = \frac{-1}{\alpha + j\omega}e^{-(\alpha - j\omega)t}\bigg|_0^{\infty} = \frac{1}{\alpha + j\omega}. \tag{5.70}$$

Remark. $F(\omega)$ is characterized by a singularity at $\omega = j\alpha$. This pole is purely imaginary—which is typical of an exponentially decaying (but nonoscillatory) response $f(t)$. In the event of decaying oscillations, the pole has both real and imaginary parts. This situation is discussed in Chapter 6 in connection with the modulation theorem. In the limit $\alpha \to 0$, note that $f(t) \to u(t)$, but (it turns out) $\mathcal{F}[u(t)](\omega)$ does *not* approach $1/j\omega$. In this limit, $f(t)$ is no longer integrable, and the Fourier transform as developed so far does not apply. We will rectify this situation with a generalized Fourier transform developed in Chapter 6.

TABLE 5.1. Radial Fourier Transforms of Elementary Signals

Signal Expression	Radial Fourier Transform
$f(t)$	$F(\omega) = \int_{-\infty}^{\infty} f(t)e^{-j\omega t}dt$
Square pulse: $u(t+a) - u(t-a)$	$2a\left[\frac{\sin(\omega a)}{\omega a}\right] = 2a\mathrm{sinc}(\omega a)$
Decaying exponential: $e^{-\alpha t}u(t)$, $\alpha > 0$	$\frac{1}{\alpha + j\omega}$
Gaussian: $e^{-\alpha t^2}$, $\alpha > 0$	$\sqrt{\frac{\pi}{\alpha}}e^{\frac{-\omega^2}{4\alpha}}$

5.2.2 Inverse Fourier Transform

The integral Fourier transform admits an inversion formula, analogous to the synthesis for Fourier series. One might propose a Fourier synthesis analogous to the discrete series (5.9):

$$f(t) \approx \int_{-\infty}^{\infty} F(\omega)e^{j\omega t}d\omega. \tag{5.71}$$

In fact, this is complete up to a factor, encapsulated in the following definition:

Definition (Inverse Radial Fourier Transform). The *inverse radial Fourier transform* of $F(\omega)$ is defined by the integral

$$f(t) = \frac{1}{2\pi}\int_{-\infty}^{\infty} F(\omega)e^{j\omega t}d\omega. \tag{5.72}$$

The Fourier transform and its inverse are referred to as a *Fourier transform pair.*

The inverses for the normalized and Hertz variants take slightly different forms.

Definition (Inverse Normalized Fourier Transform). If $F(\omega)$ is the normalized Fourier transform of $f(t)$, then the *inverse normalized Fourier transform* of $F(\omega)$ is the integral

$$f(t) = \frac{1}{\sqrt{2\pi}}\int_{-\infty}^{\infty} F(\omega)e^{j\omega t}d\omega. \tag{5.73a}$$

Definition (Inverse Hertz Fourier Transform). If $X(f)$ is the Hertz Fourier transform of $x(t)$, then the *inverse Hertz Fourier transform* of $X(f)$ is

$$x(t) = \int_{-\infty}^{\infty} X(f)e^{j2\pi ft}df. \tag{5.73b}$$

Naturally, the utility of this pair is constrained by our ability to carry out the integrals defining the forward and inverse transforms. At this point in the development one might consider the following:

- Does the radial Fourier transform $F(\omega)$ exist for all continuous or piecewise continuous functions?
- If $F(\omega)$ exists for some $f(t)$, is it always possible to invert the resulting spectrum to synthesize $f(t)$?

The answer to both these questions is *no*, but the set of functions which are suitable is vast enough to have made the Fourier transform the stock and trade of signal analysis. It should come as no surprise that the integrability of $f(t)$, and of its spectrum, can be a deciding factor. On the other hand, a small but very important set of common signals do not meet the integrability criteria we are about to develop, and for these we will have to extend the definition of the Fourier transform to include a class of generalized Fourier transform, treated in Chapter 6.

We state and prove the following theorem for the radial Fourier transforms; proofs for the normalized and Hertz cases are similar.

Theorem (Existence). If $f(t)$ is absolutely integrable—that is, if $f(t) \in L^1(\mathbb{R})$—then the Fourier transform $F(\omega)$ exists.

Proof: This follows directly from the transform's definition. Note that

$$|F(\omega)| = \left|\int_{-\infty}^{\infty} f(t)e^{-j\omega t}dt\right| \leq \int_{-\infty}^{\infty} |f(t)|\left|e^{-j\omega t}\right|dt = \int_{-\infty}^{\infty} |f(t)|\, dt. \tag{5.74}$$

So $F(\omega)$ exists if

$$\int_{-\infty}^{\infty} |f(t)|dt < \infty; \tag{5.75}$$

that is, $f(t) \in L^1(\mathbb{R})$. ■

Theorem (Existence of Inverse). If $F(\omega)$ is absolutely integrable, then the inverse Fourier transform $\mathcal{F}^{-1}[F(\omega)](t)$ exists.

Proof: The proof is similar and is left as an exercise. ■

Taken together, these existence theorems imply that if $f(t)$ and its Fourier spectrum $F(\omega)$ belong to $L^1(\mathbb{R})$, then both the analysis and synthesis of $f(t)$ can be performed. Unfortunately, if $f(t)$ is integrable, there is no guarantee that $F(\omega)$ follows suit. Of course, it *can* and often *does*. In those cases where synthesis (inversion) is impossible because $F(\omega)$ not integrable, the spectrum is still a physically valid representation of frequency content and can be subjected to many of the common operations (filtering, band-limiting, and frequency translation) employed in practical systems. In order to guarantee both analysis and synthesis, we need a stronger condition on $f(t)$, which we will explore in due course. For the time being, we will further investigate the convergence of the Fourier transform and its inverse, as applied to continuous and piecewise continuous functions.

Theorem (Convergence of Inverse). Suppose $f(t)$ and $F(\omega)$ are absolutely integrable and continuous. Then the inverse Fourier transform exists and converges to $f(t)$.

Proof: Define a band-limited inverse Fourier transform as follows:

$$f_{\Omega}(t) = \frac{1}{2\pi}\int_{-\Omega}^{\Omega} F(\omega)e^{j\omega t}\, d\omega. \tag{5.76}$$

In the limit $\Omega \to \infty$, (5.76) should approximate $f(t)$. (There is an obvious analogy between the band-limited Fourier transform and the partial Fourier series expansion.) Replacing $F(\omega)$ with its Fourier integral representation (5.66a) and interchanging the limits of integration, (5.76) becomes

$$f_{\Omega}(t) = \frac{1}{2\pi}\int_{-\Omega}^{\Omega}\left[\int_{-\infty}^{\infty} f(\tau)e^{-j\omega t}\,d\tau\right]e^{j\omega t}\,d\omega = \int_{-\infty}^{\infty} f(\tau)K_{\Omega}(t-\tau)\,d\tau, \tag{5.77a}$$

where

$$K_{\Omega}(t-\tau) = \frac{\sin\Omega(t-\tau)}{\pi(t-\tau)} \equiv \int_{-\Omega}^{\Omega} e^{j\omega(t-\tau)}\,d\omega. \tag{5.77b}$$

There are subtle aspects involved in the interchange of integration limits carried out in the preceding equations. We apply Fubini's theorem [13, 14] and the assumption that both $f(t)$ and $F(\omega)$ are in $L^1(\mathbb{R})$. This theorem, which we reviewed in Chapter 3, states that if a function of two variables is absolutely integrable over a region, then its iterated integrals and its double integral over the region are all equal. In other words, if $\|x(t, \omega)\|_1 < \infty$, then:

- For all $t \in \mathbb{R}$, the function $x_t(\omega) = x(t, \omega)$ is absolutely integrable (except—if we are stepping up to Lebesgue integration—on a set of measure zero).
- For all $\omega \in \mathbb{R}$, the function $x_{\omega}(t) = x(t, \omega) \in L^1(\mathbb{R})$ (again, except perhaps on a measure zero set).
- And we may freely interchange the order of integration:

$$\int_{-\infty}^{\infty}\left[\int_{-\infty}^{\infty} x(t, \omega)\,dt\right]d\omega = \int_{-\infty}^{\infty}\left[\int_{-\infty}^{\infty} x(t, \omega)\,d\omega\right]dt. \tag{5.78}$$

So we apply Fubini here to the function of two variables, $x(\tau, \omega) = f(\tau)e^{-j\omega\tau}e^{j\omega t}$, with t fixed, which appears in the first iterated integral in (5.77a). Now, the function

$$K_{\Omega}(x) = \frac{\sin\Omega x}{\pi x} \tag{5.79}$$

is the Fourier kernel. In Chapter 3 we showed that it is one of a class of generalized functions which approximates a Dirac delta function in the limit of large Ω. Thus,

$$\lim_{\Omega\to\infty} f_{\Omega}(t) = \lim_{\Omega\to\infty}\int_{-\Omega}^{\Omega} f(\tau)\frac{\sin\Omega(t-\tau)}{\pi(t-\tau)}\,d\tau = \int_{-\infty}^{\infty} f(\tau)\delta(t-\tau)\,d\tau = f(t). \tag{5.80}$$

completing the proof. ■

5.2.3 Properties

In this section we consider the convergence and algebraic properties of the Fourier transform. Many of these results correspond closely to those we developed for the Fourier series. We will apply these properties often—for instance, in developing analog *filters*, or frequency-selective convolutional systems, in Chapter 9.

5.2.3.1 Convergence and Discontinuities. Let us first investigate how well the Fourier transform's synthesis relation reproduces the original time-domain signal. Our first result concerns time-domain discontinuities, and the result is quite reminiscent of the case of the Fourier series.

Theorem (Convergence at Step Discontinuities). Suppose $f(t) \in L^1(\mathbb{R})$ has a step discontinuity at some time t. Let $F(\omega) = \mathcal{F}[f(t)](\omega)$ be the radial Fourier transform of $f(t)$ with $F(\omega) \in L^1(\mathbb{R})$. Assume that, in some neighborhood of t, $f(t)$ and its derivative have well-defined limits from the left and from the right: $f(t_{(l)})$ and $f(t_{(r)})$, respectively. Then the inverse Fourier transform, $\mathcal{F}^{-1}[F(\omega)](t)$, converges pointwise to the value,

$$\mathcal{F}^{-1}[F(\omega)](t) = \frac{[f(t_{(r)}) + f(t_{(l)})]}{2}. \tag{5.81}$$

Proof: The situation is clearly analogous to Fourier series convergence at a step discontinuity. We leave it as an exercise to show that the step discontinuity (assumed to lie at $t = 0$ for simplicity) gives a residual Gibbs oscillation described by

$$\varepsilon_N(t) = A_0\left[\int_{-\infty}^{0} \frac{\sin v}{\pi v}\,dv + \int_{0}^{\Omega t} \frac{\sin v}{\pi v}\,dv\right] = \frac{1}{2}A_0 + A_0 \int_{0}^{\Omega t} \frac{\sin v}{\pi v}\,dv, \tag{5.82a}$$

where the amplitude of the step is

$$A_0 = [f(0_{(r)}) - f(0_{(l)})]. \tag{5.82b}$$

Therefore in the limit as $\Omega \to \infty$,

$$\varepsilon_N(0) = \frac{1}{2}A_0. \tag{5.83}$$

Hence the inverse Fourier transform converges to the average of the left- and right-hand limits at the origin,

$$f(0) = f_c(0) + \frac{1}{2}[f(0_{(r)}) - f(0_{(l)})] = \frac{1}{2}[f(0_{(r)}) + f(0_{(l)})]. \tag{5.84}$$

For a step located at an arbitrary $t = t_s$ the result generalizes so that

$$f(0) = f_c(t_s) + \frac{1}{2}[f(t_{s(r)}) - f(t_{s(l)})] = \frac{1}{2}[f(t_{s(r)}) + f(t_{s(l)})], \tag{5.85}$$

and the proof is complete. ■

The Gibbs oscillations are an important consideration when evaluating Fourier transforms numerically, since numerical integration over an infinite interval always involves approximating infinity with a suitably large number. Effectively, they are band-limited Fourier transforms; and in analogy to the Fourier series, the Gibbs oscillations are an artifact of truncating the integration.

5.2.3.2 Continuity and High- and Low-Frequency Behavior of Fourier Spectra. The continuity of the Fourier spectrum is one of its most remarkable properties. While Fourier analysis can be applied to both uniform and piecewise continuous signals, the resulting spectrum is *always* uniformly continuous, as we now demonstrate.

Theorem (Continuity). Let $f(t) \in L^1(\mathbb{R})$. Then $F(\omega)$ is a uniformly continuous function of ω.

Proof: We need to show that for any $\varepsilon > 0$, there is a $\delta > 0$, such that $|\omega - \theta| < \delta|$ implies that $|F(\omega) - F(\theta)| < \varepsilon|$. This follows by noting

$$|F(\omega + \delta) - F(\omega)| = \int_{-\infty}^{\infty} f(t)(e^{-j\delta t} - 1)e^{-j\omega t}\, dt \le 2\|f\|_1 . \tag{5.86}$$

Since $|F(\omega + \delta) - F(\omega)|$ is bounded above by $2\|f\|_1$, we may apply the Lebesgu-dominated convergence theorem (Chapter 3). We take the limit, as $\delta \to 0$, of $|F(\omega + \delta) - F(\omega)|$ and the last integral in (5.86). But since $e^{-j\delta t} \to 1$ as $\delta \to 0$, this limit is zero:

$$\lim_{\delta \to 0} |F(\omega + \delta) - F(\omega)| = 0 \tag{5.87}$$

and $F(\omega)$ is continuous. Inspecting this argument carefully, we see that the limit of the last integrand of (5.86) does not depend on ω, establishing uniform continuity as well. ■

Remark. This theorem shows that absolutely integrable signals—which includes every practical signal available to a real-world processing and analysis system—can have no sudden jumps in their frequency content. That is, we cannot have $|F(\omega)|$ very near one value as ω increases toward ω_0, and $|F(\omega)|$ approaches a different value as ω decreases toward ω_0. If a signal is in $L^1(\mathbb{R})$, then its spectra are smooth. This is an interesting situation, given the abundance of piecewise continuous waveforms (such as the rectangular pulse) which are clearly in $L^1(\mathbb{R})$ and, according to this theorem, exhibit continuous (note, not piecewise continuous) spectra. Moreover, the uniform continuity assures us that we should find no cusps in our plots of $|F(\omega)|$ versus ω.

Now let us consider the high-frequency behavior of the Fourier spectrum. In the limit of infinite frequency, we shall show $F(\omega) \to 0$. This general result is easily demonstrated by the Riemann–Lebesgue lemma, a form of which was examined in Chapter 3 in connection with the high-frequency behavior of simple sinusoids (as distributions generated by the space of testing functions with compact support). Here the lemma assumes a form that suits the Fourier transform.

Proposition (Riemann–Lebesgue Lemma, Revisited). If $f(t)$ is integrable, then

$$\lim_{|\omega| \to \infty} |F(\omega)| = 0. \tag{5.88}$$

Proof: The proof follows easily from a convenient trick. Note that

$$e^{-j\omega t} = -e^{-j\omega t - j\pi} = -e^{-j\omega\left(t + \frac{\pi}{\omega}\right)}. \tag{5.89}$$

Thus, the Fourier integral can be written

$$F(\omega) = -\int_{-\infty}^{\infty} f(t) e^{-j\omega\left(t + \frac{\pi}{\omega}\right)} dt = -\int_{-\infty}^{\infty} f\left(t - \frac{\pi}{\omega}\right) e^{-j\omega t} dt. \tag{5.90}$$

Expressing the fact that $F(\omega) \equiv \frac{1}{2}[F(\omega) + F(\omega)]$ by utilizing the standard and revised representations (as given in (5.90)) of $F(\omega)$, we have

$$F(\omega) = \frac{1}{2} \int_{-\infty}^{\infty} \left[f(t) - f\left(t - \frac{\pi}{\omega}\right)\right] e^{-j\omega t} d\omega, \tag{5.91}$$

so that

$$|F(\omega)| \le \int_{-\infty}^{\infty} \left|\left[f(t) - f\left(t - \frac{\pi}{\omega}\right)\right]\right| d\omega. \tag{5.92}$$

Taking the high-frequency limit, we have

$$\lim_{\omega \to \infty} |F(\omega)| \le \lim_{\omega \to \infty} \int_{-\infty}^{\infty} \left|\left[f(t) - f\left(t - \frac{\pi}{\omega}\right)\right]\right| d\omega = 0, \tag{5.93}$$

and the lemma is proven. ■

Taken in conjunction with continuity, the Riemann–Lebesgue lemma indicates that the spectra associated with integrable functions are well-behaved across all frequencies. But we emphasize that despite the decay to zero indicated by (5.93), this does not guarantee that the spectra decay rapidly enough to be integrable. Note that the Fourier transform $F(\omega)$ of $f(t) \in L^1(\mathbb{R})$ is bounded. In fact, we can easily estimate that $\|F\|_\infty \le \|f\|_1$ (exercise).

5.2.3.3 Algebraic Properties. These properties concern the behavior of the Fourier transform integral under certain algebraic operations on the transformed signals. The Fourier transform is an analog system, mapping (some) time-domain signals to frequency-domain signals. Thus, these algebraic properties include such operations that we are familiar with from Chapter 3: scaling (amplification and attenuation), summation, time shifting, and time dilation.

Proposition (Linearity). The integral Fourier transform is linear; that is,

$$\mathcal{F}\left[\sum_{k=1}^{N} a_k f_k(t)\right](\omega) = \sum_{k=1}^{N} a_k F_k(\omega). \tag{5.94}$$

Proof: This follows from the linearity of the integral. ∎

From a practical standpoint, the result is of enormous value in analyzing composite signals and signals plus noise, indicating that the spectra of the individual components can be analyzed and (very often) processed separately.

Proposition (Time Shift). $\mathcal{F}[f(t-t_0)](\omega) = e^{-j\omega t_0}F(\omega)$.

Proof: A simple substitution of variables, $v = t - t_0$, applied to the definition of the Fourier transform leads to

$$\mathcal{F}[f(t-t_0)](\omega) = \int_{-\infty}^{\infty} f(t-t_0)e^{-j\omega t}dt = e^{-j\omega t_0}\int_{-\infty}^{\infty} f(v)e^{-j\omega v}dv = e^{-j\omega t_0}F(\omega) \tag{5.95}$$

completing the proof. ∎

Remark. Linear systems often impose a time shift of this type. Implicit in this property is the physically reasonable notion that a change in the time origin of a signal $f(t)$ does not affect the magnitude spectrum $|F(\omega)|$. If the same signal arises earlier or later, then the relative strengths of its frequency components remain the same since the energy $\|F(\omega)\|_2$ is invariant.

Proposition (Frequency Shift). $\mathcal{F}[f(t)e^{j\omega_0 t}](\omega) = F(\omega - \omega_0)$.

Proof: Writing out the Fourier transform explicitly, we find

$$\mathcal{F}[f(t)e^{j\omega t_0}](\omega) = \int_{-\infty}^{\infty} f(t)e^{-j(\omega - \omega_0)t}\, dt = F(\omega - \omega_0) \tag{5.96}$$

completing the proof. ■

Remark. This is another result which is central to spectral analysis and linear systems. Note the ease with which spectra can be translated throughout the frequency domain by simple multiplication with a complex sinusoidal phase factor in the time domain. Indeed, (5.96) illustrates exactly how radio communication and broadcast frequency bands can be established [19–21]. Note that the Fourier transform itself is not translation invariant. The effect of a frequency shift is shown in Figure 5.6. Note that ω_0 can be positive or negative.

The simplicity of the proof belies the enormous practical value of this result. Fundamentally, it implies that by multiplying a waveform $f(t)$ by a sinusoid of known frequency, the spectrum can be shifted to another frequency range. This idea makes multichannel communication and broadcasting possible and will be explored more fully in Chapter 6.

Proposition (Scaling). Suppose $a \neq 0$. Then

$$F[f(at)](\omega) = \frac{1}{|a|}F\left(\frac{\omega}{a}\right). \tag{5.97}$$

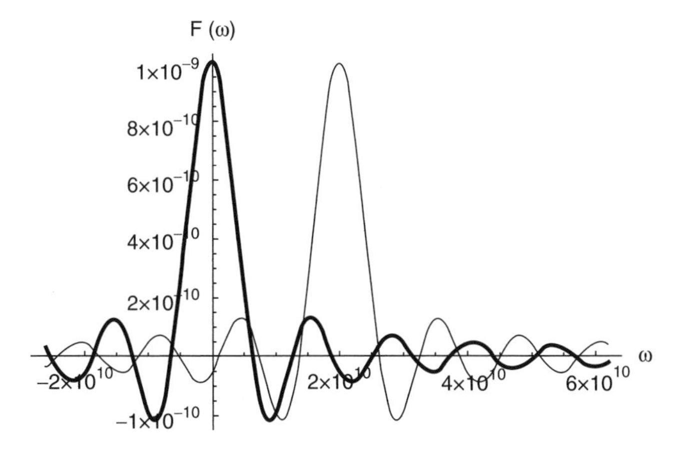

Fig. 5.6. Frequency translation. A "sinc" spectrum (solid line) and same spectrum shifted in frequency space by an increment $\omega_0 = 2 \times 10^{10}$.

Proof: Consider the cases of the scale parameter, $a > 0$ and $a < 0$, separately. First, suppose $a > 0$. With the substitution, $v = at$, it follows that

$$\mathcal{F}[f(at)](\omega) = \frac{1}{a}\int_{-\infty}^{\infty} f(v)e^{-j(\omega/a)v}dv = \frac{1}{a}F\left(\frac{\omega}{a}\right). \tag{5.98}$$

Following a similar argument for $a < 0$, carefully noting the signs on variables and the limits of integration, we find

$$\mathcal{F}[f(at)](\omega) = \frac{-1}{a}\int_{\infty}^{-\infty} f(v)e^{-j(\omega/a)v}dv = \frac{-1}{a}F\left(\frac{\omega}{a}\right). \tag{5.99}$$

In either case, the desired result (5.97) follows. ■

Scaling in the time domain is a central feature of the wavelet transform, which we develop in Chapter 11. For example, (5.97) can be used to describe the spectral properties of the crucial 'mother' wavelet, affording the proper normalization and calculation of wavelet coefficients (analogous to Fourier coefficients).

The qualitative properties previously observed in connection with the rectangular pulse and its spectrum are made manifest by these relations: The multiplicative scale a in the time-domain scales as a^{-1} in the spectrum. A Gaussian pulse serves as an excellent illustration of scaling.

Example (Gaussian). The Gaussian function

$$f(t) = e^{-\alpha t^2} \tag{5.100}$$

and its Fourier transform are shown in Figure 5.7. Of course, we assume $\alpha > 0$. Panel (a) shows Gaussian pulses in the time domain for $\alpha = 10^{11}$ and $\alpha = 11^{11}$. Panel (b) shows the corresponding Fourier transforms.

We will turn to the Gaussian quite frequently when the effects of noise on signal transmission are considered. Although noise is a nondeterministic process, its statistics—the spread of noise amplitude—often take the form of a Gaussian. Noise is considered a corruption whose effects are deleterious, so an understanding of its spectrum, and how to process noise so as to minimize its effects, plays an important role in signal analysis. Let us work out the calculations. In this example, we have built in a time scale, $a = \sqrt{\alpha}$, and we will trace its effects in frequency space:

$$F(\omega) = \int_{-\infty}^{\infty} e^{-\alpha t^2}[\cos(\omega t) - j\sin(\omega t)]\,dt. \tag{5.101}$$

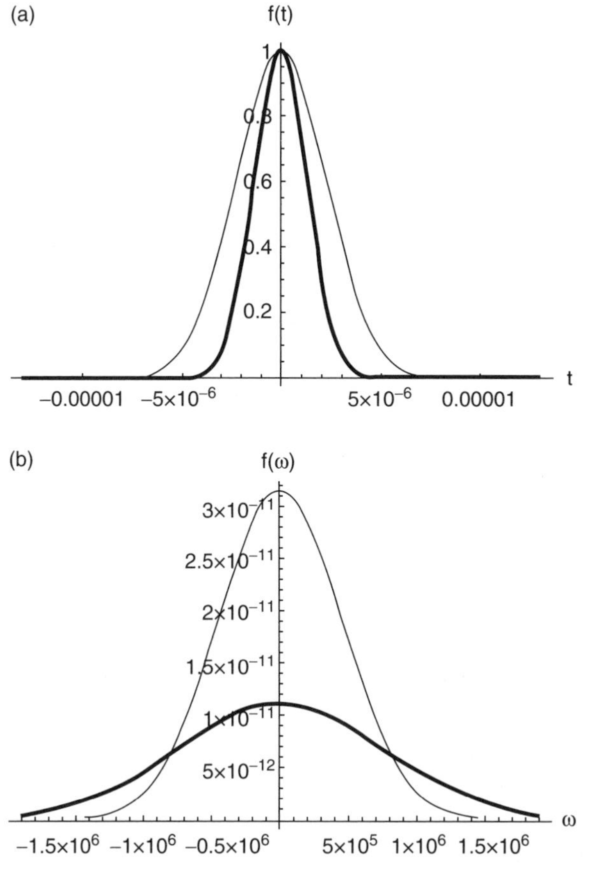

Fig. 5.7. (a) Gaussian pulses in the time domain, for $\alpha = 10^{11}$ and $\alpha = 11^{11}$ (solid lines). (b) Corresponding Fourier transforms.

If we informally assume that the integration limits in (5.101) approach infinity in perfect symmetry, then—since sine is an odd signal—we can argue that the contribution from $\sin(\omega t)$ vanishes identically. This leaves a common tabulated integral,

$$F(\omega) = \int_{-\infty}^{\infty} e^{-\alpha t^2} \cos(\omega t) dt = \sqrt{\frac{\pi}{\alpha}} e^{\frac{-\omega^2}{4\alpha}} . \tag{5.102}$$

Note several features:

- The Fourier transform of a Gaussian is again a Gaussian.
- Furthermore, the general effects of scaling, quantified in (5.97), are clearly in evidence in Figure 5.7 where the Gaussian spectra are illustrated.

Corollary (Time Reversal)

$$\mathcal{F}[f(-t)](\omega) = F(-\omega). \tag{5.103}$$

Proof: This is an instance of the scaling property. ■

Proposition (Symmetry)

$$\mathcal{F}[F(t)](\omega) = 2\pi f(-\omega). \tag{5.104}$$

Proof: A relationship of this sort is hardly surprising, given the symmetric nature of the Fourier transform pair. Since

$$\mathcal{F}^{-1}[F(\omega)](t) = \frac{1}{2\pi}\int_{-\infty}^{\infty} F(\omega)e^{j\omega t}\,d\omega, \tag{5.105a}$$

it follows that

$$f(-t) = \frac{1}{2\pi}\int_{-\infty}^{\infty} F(\omega)e^{-j\omega t}\,d\omega. \tag{5.105b}$$

With a simple change of variables, we obtain

$$f(-\omega) = \frac{1}{2\pi}\int_{-\infty}^{\infty} F(t)e^{-j\omega t}\,d\omega, \tag{5.106}$$

concluding the proof. ■

From a practical standpoint, this symmetry property is a convenient trick, allowing a list of Fourier transform pairs to be doubled in size without evaluating a single integral.

5.2.3.4 Calculus Properties. Several straightforward but useful properties are exhibited by the Fourier transform pair under differentiation. These are easily proven.

Proposition (Time Differentiation). Let $f(t)$ and $\frac{d^k f}{dt^k}$ be integrable functions, and suppose $\lim_{|t| \to \infty} \frac{d^k f}{dt^k} = 0$ for all $k = 0, 1, \ldots, n$. Then

$$\mathcal{F}\left[\frac{d^n f}{dt^n}\right](\omega) = (j\omega)^n F(\omega). \tag{5.107}$$

Proof: We establish this result for the first derivative; higher orders follow by induction. Representing the Fourier integral by parts gives

$$\int_{-\infty}^{\infty} \frac{d}{dt} f(t) e^{-j\omega t} dt = f(t) e^{-j\omega t} \Big|_{-\infty}^{\infty} - \int_{-\infty}^{\infty} f(t)(-j\omega) e^{-j\omega t} dt. \tag{5.108}$$

Under the expressed conditions, the result for the first derivative follows immediately:

$$\mathcal{F}\left[\frac{df}{dt}\right](\omega) = j\omega F(\omega). \tag{5.109}$$

Repeated application of this process lead to (5.107). ■

Proposition (Frequency Differentiation)

$$\mathcal{F}[(-jt)^n f(t)](\omega) = \frac{d^n}{d\omega^n} F(\omega). \tag{5.110}$$

Proof: The proof is similar to time differentiation. Note that the derivatives of the spectrum must exist in order to make sense of this propostion. ■

The differentiation theorems are useful when established $f(t)$ or spectra are multiplied by polynomials in their respective domains. For example, consider the case of a Gaussian time signal as in (5.100), multiplied by an arbitrary time-dependent polynomial. According to the frequency differentiation property,

$$\mathcal{F}\left[(a_0 + a_1 t + \cdots + a_k t^k) e^{-\alpha t^2}\right](\omega) = \left[a_0 + \frac{-a_1}{j} \frac{d}{d\omega} + \cdots + \frac{a_k}{(-j)^k} \frac{d^k}{d\omega^k}\right] \sqrt{\frac{\pi}{\alpha}} e^{-\omega^2/(4\alpha)} \tag{5.111}$$

so that the act of taking a Fourier transform has been reduced to the application of a simple differential operator. The treatment of spectra corresponding to pure polynomials defined over all time or activated at some time t_0 will be deferred until Chapter 6, where the generalized Fourier transform of unity and $u(t - t_0)$ are developed.

Now let us study the low-frequency behavior of Fourier spectra. The Riemann–Lebesgue lemma made some specific predictions about Fourier spectra in the limit of infinite frequency. At low frequencies, in the limit as $\omega \to 0$, we can formally expand $F(\omega)$ in a Maclaurin series,

$$F(\omega) = \sum_{k=0}^{\infty} \frac{\omega^k}{k!} \frac{d^k F(0)}{d\omega^k}. \tag{5.112}$$

The Fourier integral representation of $F(\omega)$ can be subjected to a Maclaurin series for the frequency-dependent exponential:

$$F(\omega) = \int_{-\infty}^{\infty} f(t)\left[\sum_{k=0}^{\infty} \frac{(-j\omega t)^k}{k!}\right] dt = \sum_{k=0}^{\infty} (-j)^k \frac{\omega^k}{k!} \int_{-\infty}^{\infty} t^k f(t)\, dt. \quad (5.113)$$

The last integral on the right is the kth moment of $f(t)$, defined as

$$m_k = \int_{-\infty}^{\infty} t^k f(t)\, dt \quad (5.114)$$

If the moments of a function are finite, we then have the following proposition.

Proposition (Moments)

$$\frac{d^k F(0)}{d\omega^k} = (-j)^k m_k, \quad (5.115)$$

which follows directly on comparing (5.112) and (5.113).

The moment theorem allows one to predict the low-frequency behavior of $f(t)$ from an integrability condition in time. This is often useful, particularly in the case of the wavelet transform. In order to qualify as a wavelet, there are necessary conditions on certain moments of a signal. This matter is taken up in Chapter 11.

Table 5.2 lists radial Fourier transformation properties. Some of these will be shown in the sequel.

5.2.4 Symmetry Properties

The even and odd symmetry of a signal $f(t)$ can have a profound effect on the nature of its frequency spectrum. Naturally, the impact of even–odd symmetry in transform analysis comes about through its effect on integrals. If $f(t)$ is odd, then its integral over symmetric limits $t \in [-L, L]$ vanishes identically; if $f(t)$ is even, this integral may be nonzero. In fact, this property was already put to use when discussing the Gaussian and its spectrum.

Not all functions $f(t)$ exhibit even or odd symmetry. But an arbitrary $f(t)$ may be expressed as the sum of even and odd parts: $f(t) = f_e(t) + f_o(t)$, where

$$f_e(t) = \frac{1}{2}[f(t) + f(-t))] \quad (5.116a)$$

and

$$f_o(t) = \frac{1}{2}[f(t) - f(-t)]. \quad (5.116b)$$

For example, in the case of the unit step,

$$f_e(t) = \frac{1}{2}[u(t) + u(-t))] = \frac{1}{2} \quad (5.117a)$$

and

$$f_o(t) = \frac{1}{2}[u(t) - u(-t)] = \frac{1}{2}\text{sgn } t. \tag{5.117b}$$

where sgn t is the signum function. These are illustrated in Figure 5.8 using the unit step as an example.

TABLE 5.2. Summary of Radial Fourier Transform Properties

Signal Expression	Radial Fourier Transform or Property
$f(t)$	$F(\omega) = \int_{-\infty}^{\infty} f(t)e^{-j\omega t}dt$ (Analysis equation)
$f(t) = \frac{1}{2\pi}\int_{-\infty}^{\infty} F(\omega)e^{j\omega t}d\omega$	$F(\omega)$ (Inverse, synthesis equation)
$af(t) + bg(t)$	$aF(\omega) + bG(\omega)$ (Linearity)
$f(t-a)$	$e^{-j\omega a}F(\omega)$ (Time shift)
$f(t)\exp(j\theta t)$	$F(\omega - \theta)$ (Frequency shift, modulation)
$f(at)$, $a \neq 0$	$\frac{1}{\lvert a\rvert}F\left(\frac{\omega}{a}\right)$ (Scaling, dilation)
$f(-t)$	$F(-\omega)$ (Time reversal)
$\left[\frac{d^n f(t)}{dt^n}\right]$	$(j\omega)^n F(\omega)$ (Time differentiation)
$(-jt)^n f(t)$	$\frac{d^n}{d\omega^n}F(\omega)$ (Frequency differentiation)
$\lVert x\rVert_2 = \frac{\lVert X(\omega)\rVert_2}{\sqrt{2\pi}}$	Plancherel's theorem
$\langle f, g\rangle = \frac{1}{2\pi}\langle F, G\rangle$ $f, g \in L^2(\mathbb{R})$	Parseval's theorem
$f * h$, where $f, h \in L^2(\mathbb{R})$	$F(\omega)H(\omega)$
$f(t)h(t)$	$(2\pi)^{-1}F(\omega) * H(\omega)$

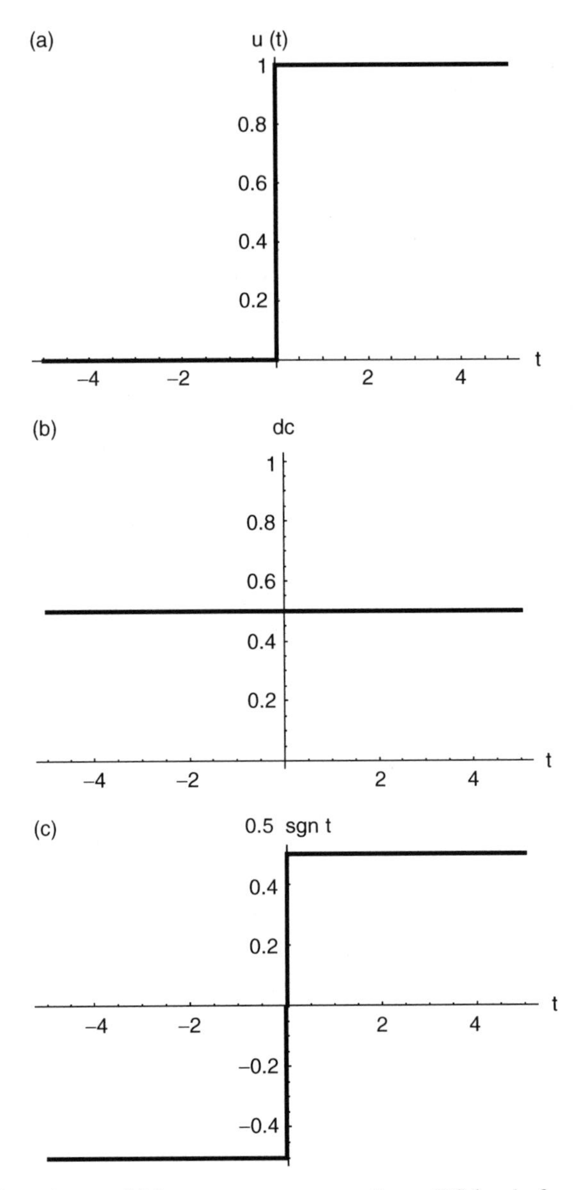

Fig. 5.8. (a) The unit step. (b) Its even-symmetry portion, a DC level of amplitude 1/2. (c) Its odd-symmetry portion, a signum of amplitude 1/2.

Whether the symmetry is endemic to the $f(t)$ at hand, or imposed by breaking it into even and odd parts, an awareness of its effects on the Fourier transform can often simplify calculations or serve as a check. Consider the spectrum of an arbitrary $f(t)$ written as the sum of even and odd constituents, $f(t) = f_e(t) + f_o(t)$:

$$F(\omega) = \int_{-\infty}^{\infty} [f_e(t) + f_o(t)] e^{-j\omega t} dt$$
$$= \int_{-\infty}^{\infty} [f_e(t) + f_o(t)] \cos(\omega t)\, dt - j \int_{-\infty}^{\infty} [f_e(t) + f_o(t)] \sin(\omega t)\, dt. \quad (5.118)$$

Elimination of the integrals with odd integrands reduces (5.110) to the elegant form

$$F(\omega) = \int_{-\infty}^{\infty} f_e(t) \cos(\omega t) dt - j \int_{-\infty}^{\infty} f_o(t) \sin(\omega t)\, dt. \quad (5.119)$$

This is a general result, applicable to an arbitrary $f(t)$ which may be real, complex, or purely imaginary. We consider these in turn.

5.2.4.1 Real f(t). A Fourier spectrum will, in general, have real and imaginary parts:

$$\mathrm{Re}[F(\omega)] = \int_{-\infty}^{\infty} f_e(t) \cos(\omega t)\, dt, \quad (5.120a)$$

which is an even function of ω (since $\cos(\omega t)$ is even in this variable) and

$$\mathrm{Im}[F(\omega)] = \int_{-\infty}^{\infty} f_o(t) \sin(\omega t)\, dt, \quad (5.120b)$$

which inherits the odd ω symmetry of $\sin(\omega t)$.

According to (5.120a), if $f(t)$ is even in addition to being real, then $F(\omega)$ is also real and even in ω. The Gaussian is a prime example, and many "mother wavelets" considered in Chapter 11 are real-valued, even functions of time. On the other hand, (5.120b) implies that if $f(t)$ is real but of odd symmetry, its spectrum is real and odd in ω.

5.2.4.2 Complex f(t). An arbitrary complex $f(t)$ can be broken into complex even and odd constituents $f_e(t)$ and $f_o(t)$ in a manner similar to the real case. When an expansion similar to (5.118) is carried out, it becomes apparent that $F(\omega)$ is, in general, complex, and it will consist of even and odd parts, which we denote $F_e(\omega)$ and $F_o(\omega)$. It is straightforward to show that the transforms break down as follows:

$$\mathrm{Re}[F_e(\omega)] = \int_{-\infty}^{\infty} \mathrm{Re}[f_e(t)] \cos(\omega t)\, dt, \quad (5.121a)$$

$$\mathrm{Re}[F_o(\omega)] = \int_{-\infty}^{\infty} \mathrm{Im}[f_o(t)] \sin(\omega t)\, dt, \quad (5.121b)$$

$$\mathrm{Im}[F_e(\omega)] = \int_{-\infty}^{\infty} \mathrm{Im}[f_e(t)]\cos(\omega t)\,dt, \tag{5.121c}$$

and

$$\mathrm{Im}[F_o(\omega)] = -\int_{-\infty}^{\infty} \mathrm{Re}[f_o(t)]\sin(\omega t)\,dt. \tag{5.121d}$$

The reader may easily verify that the earlier results for real $f(t)$ can be derived as special cases of (5.121a)–(5.121d).

5.2.4.3 *Imaginary f(t).* This is also a special case of the above, derived by setting $\mathrm{Re}[f_e(t)] = \mathrm{Re}[f_o(t)] = 0$. In particular, note that if $f(t)$ is imaginary and odd, then $F(\omega)$ is odd but real. If $f(t)$ is imaginary and even, then the spectrum is also even but imaginary.

Most signals $f(t)$ are real-valued, but there are notable cases where a signal may be modified, intentionally or as a by-product of transmission and processing, to become complex or even purely imaginary. Examples include exponential carrier modulation and filtering. Either operation may impose a phase shift that is not present in the original signal.

5.2.4.4 *Summary.* Familiarity with the symmetry properties of the Fourier transform can reduce unnecessary calculations and serve as a check of the final results. In the event that a waveform is not intrinsically odd or even, it is not always necessary, or even advisable, to break it into even and odd constituents, but doing so may be helpful when one is calculating transforms by hand and has access to a limited set of tabulated integrals. In most practical situations, numerical implementation of the Fourier transform, such as the fast Fourier transform (FFT) considered in Chapter 7, will handle the symmetries automatically.

5.3 EXTENSION TO $L^2(\mathbb{R})$

This section extends the Fourier transform to square-integrable signals. The formal definition of the Fourier transform resembles an inner-product integral of a signal $f(t)$ with the exponential $\exp(j\omega t)$. The inner product in $L^2(\mathbb{R})$ works as a measure of similarity between two signals, so $F(\omega)$, when it exists, indicates how much of radial frequency ω we find in $f(t)$. This is intuitive, simple, and attractive.

There are some fundamental difficulties, however, with this quite informal reasoning. We have shown that the Fourier transform of a signal $f(t)$ exists when $f(t)$ is absolutely integrable, but $L^1(\mathbb{R})$ signals are not the best realm for signal theorizing. In Chapter 3, for example, we found that they do not comprise an inner product space; that alone immediately breaks our intuitive concept of the Fourier integral as an inner product. Moreover, the Fourier transform of an $L^1(\mathbb{R})$ signal is not necessarily integrable, so we cannot assume to take its inverse transform. An example is the square pulse, $x(t) = u(t + 1) - u(t - 1)$ whose radial Fourier transform is a sinc function, $X(\omega) = 2a\text{sinc}(\omega a)$.

Several approaches exist for developing a Fourier transform for $f(t) \in L^2(\mathbb{R})$. These methods include:

(i) Defining $F(\omega)$ as an infinite series expansion using special Hermite functions [22];

(ii) Writing square-integrable signals as limits of elements of the Schwarz class S of infinitely differentiable, rapidly decreasing signals (Chapter 3) [10];

(iii) Using the familiar L^p spaces, in particular the intersection of L^1 and L^2, which is dense in L^2, as the seed for a general Fourier transform for square-integrable signals [23, 24].

We follow the last approach above.

5.3.1 Fourier Transforms in $L^1(\mathbb{R}) \cap L^2(\mathbb{R})$

Toward defining a Fourier transform for finite-energy signals, the main ideas are to:

- Show the validity of the forward transforms for a narrow signal class: $L^1 \cap L^2$
- Argue that this is a dense set within L^2, so we can write any general square-integrable $f(t)$ as a limit of integrable, finite-energy signals: $f(t) = \lim_{n\to\infty} f_n(t)$ where $\{f_n(t) \in L^1 \cap L^2 \mid n \in \mathbb{N}\}$;
- Then extend the transform to square-integrable by defining the transform as a limit $F(\omega) = \lim_{n\to\infty} F_n(\omega)$, where $F_n(\omega) = \mathcal{F}[f_n(t)]$.

Theorem. If $f(t) \in L^1 \cap L^2$, then its Fourier transform $F(\omega)$ is square-integrable.

Proof: Consider a rectangular pulse of width 2α in the frequency domain. For $-\alpha < \omega < \alpha$

$$P_\alpha(\omega) = 1 \tag{5.122}$$

and vanishes outside this interval. Our strategy is to incorporate one such pulse inside the spectral energy integral and consider the limit as the pulse width becomes infinite. We claim

$$\lim_{\alpha\to\infty} \int_{-\infty}^{\infty} P_\alpha(\omega)|F(\omega)|^2 d\omega = \int_{-\infty}^{\infty} |F(\omega)|^2 d\omega = \|F(\omega)\|_2. \tag{5.123}$$

Let us verify that this integral is finite: $F \in L^2(\mathbb{R})$. Inserting the explicit integrals for $F(\omega)$ and its complex conjugate into (5.123) gives

$$\int_{-\infty}^{\infty} P_\alpha(\omega)F(\omega)F^*(\omega)\, d\omega = \int_{-\infty}^{\infty} f(t) \int_{-\infty}^{\infty} f^*(\tau)\left\{ \int_{-\infty}^{\infty} e^{j\omega(\tau - t)} P_\alpha(\omega) d\omega \right\} d\tau\, dt. \tag{5.124}$$

The integral in curly brackets is proportional to the Fourier kernel,

$$\int_{-\infty}^{\infty} e^{j\omega(\tau - t)} P_\alpha(\omega) d\omega = 2\pi K_\alpha(\tau - t) = 2\pi \frac{\sin\alpha(\tau - t)}{\pi(\tau - t)}, \tag{5.125}$$

so thereby takes the streamlined form,

$$\int_{-\infty}^{\infty} P_\alpha(\omega)|F(\omega)|^2 d\omega = \int_{-\infty}^{\infty} f(t) \int_{-\infty}^{\infty} f^*(\tau) K_\alpha(\tau - t)\, d\tau\, dt. \tag{5.126}$$

A substitution of variables, $v = t - \tau$, provides

$$\int_{-\infty}^{\infty} P_\alpha(\omega)|F(\omega)|^2 d\omega = 2\pi \int_{-\infty}^{\infty} C_f(v) K_\alpha(v)\, dv, \tag{5.127}$$

where

$$C_f(v) = \int_{-\infty}^{\infty} f(v + \tau) f(\tau)\, d\tau. \tag{5.128}$$

Taking the limit as $\alpha \to \infty$, the kernel behaves like a Dirac:

$$\lim_{\alpha \to \infty} \int_{-\infty}^{\infty} P_\alpha(\omega)|F(\omega)|^2 d\omega = \int_{-\infty}^{\infty} |F(\omega)|^2 d\omega = 2\pi \int_{-\infty}^{\infty} C_f(v)\delta(v)\, dv$$
$$= 2\pi C_f(0) = 2\pi \int_{-\infty}^{\infty} |f(t)|^2 dt \tag{5.129}$$

Since $\|f\|_2 < \infty$, so too is $\|F\|_2 < \infty$. ■

An interesting and valuable corollary from the above proof is the following result showing the proportionality of the energy of a signal and its Fourier transform.

Corollary (Plancherel's Theorem). If $f(t) \in L^1(\mathbb{R}) \cap L^2(\mathbb{R})$, then $\|F\|_2 = (2\pi)^{1/2}$ $\|f\|_2$.

Proof: Evident from the theorem's proof (5.129). ■

Corollary. The radial Fourier transform $\mathcal{F}: L^1 \cap L^2 \to L^2$ is a bounded linear operator with $\|\mathcal{F}\| = (2\pi)^{1/2}$.

Proof: Recall from Chapter 3 that a bounded linear operator T is a linear map T: $N \to K$ of normed spaces when there is a constant $0 \le M$ such that $\|x\|_N \le M\|Tx\|_K$ for all $x \in N$. The Fourier transform is linear, which follows from the linearity of the integral. Also, the norm of T is $\|T\| = \sup\{\|x\|_N / \|Tx\|_K \mid x \in N, x \neq 0\}$; as long as T is bounded, this set is nonempty, and so it must have a least upper bound. From Plancherel's theorem, the ratio between the norms in N and K is always $(2\pi)^{1/2}$, giving the bound condition. ■

Corollary. If $\{f_n(t) \mid n \in \mathbb{N}\}$ is a Cauchy sequence in $L^1 \cap L^2$, then the sequence of Fourier transforms $\{F_n(\omega) \mid n \in \mathbb{N}\}$ is also Cauchy in L^2.

Proof: The Fourier transform is linear, so Plancherel's theorem implies $\|F_m - F_n\|_2 = (2\pi)^{1/2}\|f_m - f_n\|_2$. ■

5.3.2 Definition

Now we are in a position to define the Fourier transform for $L^2(\mathbb{R})$. We can write any general square-integrable f(t) as a limit of integrable, finite-energy signals: $f(t) = \lim_{n\to\infty} f_n(t)$. It is easy to find the requisite sequence by setting $f_n(t) \in L^1 \cap L^2$ to be $f(t)$ restricted to $[-n, n]$ and zero otherwise. In Chapter 3, we noted that $L^1 \cap L^2$ is dense in L^2, and by the last corollary the Fourier transforms $\{F_n(\omega) \mid n \in \mathbb{N}\}$ also comprise a Cauchy sequence in L^2.

Definition (Fourier Transform for L^2(R)). If $f(t) \in L^2(\mathbb{R})$, then we define the *Fourier transform* of $f(t)$ by $F(\omega) = \lim_{n\to\infty} F_n(\omega) = \mathcal{F}+[f(t)](\omega)$, where $\{f_n(t) \mid n \in \mathbb{N}\}$ is any Cauchy sequence in $L^1 \cap L^2$ that converges to $f(t)$, and $F_n(\omega) = \mathcal{F}[f_n(t)](\omega)$.

Remark. $F(\omega)$ must exist because $L^2(\mathbb{R})$ is complete and $\{F_n(\omega) \mid n \in \mathbb{N}\}$ is Cauchy. In order for the definition to make sense, we need to show the following:

- The designation of $\lim_{n\to\infty} F_n(\omega)$ to be the Fourier transform of $f(t)$ must be shown independent of what particular sequence $\{f_n(t) \mid n \in \mathbb{N}\}$ is taken as having $f(t)$ as its limit.
- The definition of $F(\omega)$ should match the conventional definition in terms of the Fourier transform analysis equation when $f(t) \in L^1(\mathbb{R})$ too.

We introduce a very temporary notation $\mathcal{F}$+ for the extension. Once we show that the extension of the Fourier transform from $L^1 \cap L^2$ to all of L^2 makes mathematical sense, then we can forget the superscript "+" sign. This next proposition shows that the Fourier transform on $L^2(\mathbb{R})$ is in fact well-defined and agrees with our previous definition for absolutely integrable signals.

Proposition (Well-Defined). The Fourier transform of $f(t) \in L^2(\mathbb{R})$, $F(\omega) = \lim_{n\to\infty} F_n(\omega)$, where $F_n(\omega) = \mathcal{F}[f_n(t)](\omega)$ and $\lim_{n\to\infty} f_n(t) = f(t)$. Then:

(i) $F(\omega)$ is well-defined; that is, it does not depend on the choise of limit sequence.

(ii) If $f(t) \in L^1(\mathbb{R})$, and $F(\omega)$ is given by the radial Fourier transform analysis equation (5.66a), then $F(\omega) = \lim_{n\to\infty} F_n(\omega)$.

(iii) $\mathcal{F}+$: $L^2 \to L^2$ is a norm-preserving extension of the map $\mathcal{F}$: $L^1 \cap L^2 \to L^2$ defined in Section 5.3.1.

Proof: That the limit $F(\omega)$ does not depend on the particular sequence whose limit is $f(t)$ follows from Plancherel's theorem. For the second claim, let $f_n(t)$ be $f(t)$ restricted to $[-n, n]$. Note that the $F_n(\omega) = \mathcal{F}[f_n(t)](\omega)$ in fact converge pointwise to $F(\omega)$ given by (5.66a), and any other Cauchy sequence $\{g_n(t) \mid n \in \mathbb{N}\}$ in $L^1 \cap L^2$ which converges to $f(t)$ must converge to $F(\omega)$ almost everywhere as well [24]. The third point follows immediately. ■

The next result shows inner products are preserved by $\mathcal{F}+$.

Corollary (Parseval's Theorem). If $f(t)$, $g(t) \in L^2(\mathbb{R})$ with radial Fourier transforms $F(\omega)$ and $G(\omega)$, respectively, then $\langle f, g\rangle = (2\pi)^{-1}\langle F, G\rangle$.

Proof: This follows because we can define the inner product in terms of the norm by the polarization identity (Chapter 2) for inner product spaces [15]:

$$4\langle f, g\rangle = \|f+g\|_2^2 - \|f-g\|_2^2 + \frac{\|f-jg\|_2^2}{j} - \frac{\|f+jg\|_2^2}{j}. \tag{5.130}$$

■

Corollary (Plancherel's Theorem). If $f(t) \in L^2(\mathbb{R})$ with radial Fourier transform $F(\omega)$, then $\|f\|_2 = (2\pi)^{-1/2}\|F\|_2$.

Proof: By Parseval's relation for $L^2(\mathbb{R})$ signals above. ■

Now that we have successfully extended the Fourier transform to all finite-energy signals, let us agree to drop the special notation $\mathcal{F}+$ for the extension and consider Domain($\mathcal{F}$) = $L^2(\mathbb{R})$. Now that we have enough machinery, we can build a theory of analog signal frequency quite rapidly. For example, signals with almost everywhere identical spectra must themselves be identical almost everywhere.

Corollary (Uniqueness). Let $f(t)$, $g(t) \in L^2(\mathbb{R})$ with radial Fourier transforms $F(\omega)$ and $G(\omega)$, respectively. Suppose $F(\omega) = G(\omega)$ for almost all $\omega \in \mathbb{R}$. Then $f(t) = g(t)$ for almost all $t \in \mathbb{R}$.

Proof: If $F(\omega) = G(\omega)$ for almost all $\omega \in \mathbb{R}$, then $\|F - G\|_2 = 0$. But by Plancherel's theorem for $L^2(\mathbb{R})$, we then know $\|f - g\|_2 = 0$, whence $f(t) = g(t)$ for almost all $t \in \mathbb{R}$ by the properties of the Lebesgue integral (Chapter 3). ■

Theorem (Convolution). Let $f(t)$, $h(t) \in L^2(\mathbb{R})$ with radial Fourier transforms $F(\omega)$ and $H(\omega)$, respectively, and let $g(t) = (f * h)(t)$ be the convolution of f and h. Then $G(\omega) = F(\omega)H(\omega)$.

Proof: By the Schwarz inequality, $g = (f * h) \in L^1(\mathbb{R})$, and it has a Fourier transform $G(\omega)$. Let us expand the convolution integral inside the Fourier transform analysis equation for $g(t)$:

$$\int_{-\infty}^{\infty} (f*h)(t)e^{-j\omega t}dt = \int_{-\infty}^{\infty}\int_{-\infty}^{\infty} f(s)h(t-s)ds e^{-j\omega t}dt. \tag{5.131}$$

Since $g \in L^1(\mathbb{R})$, we can apply Fubini's theorem (Section 3.4.2.4) to the integrand of the Fourier analysis equation for $g(t)$. Interchanging the order of integration gives

$$\int_{-\infty}^{\infty}\int_{-\infty}^{\infty} f(s)h(t-s)ds e^{-j\omega t}dt = \int_{-\infty}^{\infty} f(s)e^{-j\omega s}ds \int_{-\infty}^{\infty} h(t-s)e^{-j\omega(t-s)}dt. \tag{5.132}$$

The iterated integrals on the right we recognize as $F(\omega)H(\omega)$. ■

The convolution theorem lies at the heart of analog filter design, which we cover in Chapter 9.

Finally, we observe that the normalized and Hertz Fourier transforms, which are a scaling and a dilation of the radial transform, respectively, can also be extended precisely as above.

5.3.3 Isometry

The normalized radial Fourier transform, extended as above to finite enery signals, in fact constitutes an isometry of $L^2(\mathbb{R})$ with itself. We recall that an isometry T between Hilbert spaces, H and K, is a linear map that is one-to-one and onto and preserves inner products. Since $\langle x, y\rangle_H = \langle Tx, Ty\rangle_J$, T also preserves norms, and so it must be bounded; in fact, $\|T\| = 1$. Conceptually, if two Hilbert spaces are isometric, then they are essentially identical.We continue working out the special properties of the radial Fourier transform $\mathcal{F}$ and, as the last step, scale it to $(2\pi)^{-1/2}\mathcal{F}$, and thereby get the isometry.

The following result is a variant of our previous Plancherel and Parseval formulas.

Theorem. Let $f(t)$, $g(t) \in L^2(\mathbb{R})$ with radial Fourier transforms $F(\omega)$ and $G(\omega)$, respectively. Then,

$$\int_{-\infty}^{\infty} F(\omega)g(\omega)d\omega = \int_{-\infty}^{\infty} f(t)G(t)\, dt. \tag{5.133}$$

Proof: We prove the result in two steps:

(i) First for $f(t)$, $g(t) \in (L^1 \cap L^2)(\mathbb{R})$;

(ii) For all of $L^2(\mathbb{R})$, again using the density of $L^1 \cap L^2$ within L^2.

Let us make the assumption (i) and note that this stronger condition implies that $F(\omega)$ and $G(\omega)$ are bounded (exercises). Then, since $F(\omega) \in L^{\infty}(\mathbb{R})$, the Hölder inequality gives $\|Fg\|_1 \le \|F\|_{\infty}\|g\|_1$. Because the integrand $F(\omega)g(\omega)$ is absolutely integrable, Fubini's theorem allows us to interchange the order of integration:

$$\int_{-\infty}^{\infty} F(\omega)g(\omega)\, d\omega = \int_{-\infty}^{\infty}\left(\int_{-\infty}^{\infty} f(t)e^{-j\omega t}dt\right)g(\omega)\, d\omega = \int_{-\infty}^{\infty}\left(\int_{-\infty}^{\infty} g(\omega)e^{-j\omega t}d\omega\right)f(t)\, dt. \tag{5.134}$$

Notice that the integral in parentheses on the right-hand side of (5.134) is precisely $G(t)$, from which the result for the special case of $L^1 \cap L^2$ follows.

For (ii), let us assume that $\lim_{n\to\infty} f_n(t) = f(t)$ and $\lim_{n\to\infty} g_n(t) = g(t)$, where f_n, $g_n \in (L^1 \cap L^2)(\mathbb{R})$. Then $F(\omega) = \lim_{n\to\infty} F_n(\omega)$, where $F_n(\omega) = \mathcal{F}[f_n(t)](\omega)$ and $G(\omega) = \lim_{n\to\infty} G_n(\omega)$, where $G_n(\omega) = \mathcal{F}[g_n(t)](\omega)$. Then, f_n, g_n, F_n, and $G_n \in L^2(\mathbb{R})$, so that by the Schwarz inequality, $F_n g_n$ and $f_n G_n \in L^1(\mathbb{R})$. The Lebesgue Dominated Convergence theorem applies (Section 3.4.2.3). By part (i) of the proof, for all $n \in \mathbb{N}$, $\int F_n g_n = \int f_n G_n$. Taking limits of both sides gives

$$\lim_{n\to\infty}\int F_n g_n = \int \lim_{n\to\infty} F_n g_n = \int Fg = \lim_{n\to\infty}\int G_n f_n = \int \lim_{n\to\infty} G_n f_n = \int Gf. \tag{5.135}$$

as required. ■

We know that every $f \in L^2(\mathbb{R})$ has a radial Fourier transform $F \in L^2(\mathbb{R})$ and that signals with (almost everywhere) equal Fourier transforms are themselves (almost everywhere) equal. Now we can show another result—an essential condition for the isometry, in fact—that the Fourier transform is onto.

Theorem. If $G \in L^2(\mathbb{R})$, then there is a $g \in L^2(\mathbb{R})$ such that $F(g)(\omega) = G(\omega)$ for almost all $\omega \in \mathbb{R}$.

Proof: If $G(\omega) \in L^2(\mathbb{R})$, we might well guess that the synthesis formula for the case $G(\omega) \in (L^1 \cap L^2)(\mathbb{R})$ will give us a definition of $g(t)$:

$$g(t) = \frac{1}{2\pi}\int_{-\infty}^{\infty} G(\omega)e^{j\omega t}d\omega. \tag{5.136}$$

We need to show that the above integral is defined for a general $G(\omega)$, however. If we let $H(\omega) = G(-\omega)$ be the reflection of $G(\omega)$, then $H \in L^2(\mathbb{R})$. We can take its radial Fourier transform, $\mathcal{F}[H]$:

$$p\frac{1}{2\pi}\mathcal{F}[H](t) = \frac{1}{2\pi}\int_{-\infty}^{\infty} H(\omega)e^{-j\omega t}d\omega. \tag{5.137}$$

A change of integration variable in (5.137) shows that $(2\pi)^{-1}\mathcal{F}[H](t)$ has precisely the form of the radial Fourier synthesis equation (5.136). We therefore propose $g(t) = (2\pi)^{-1}\mathcal{F}[H](t) \in L^2(\mathbb{R})$. Now we need to show that the Fourier transform of $g(t)$ is equal to $G(\omega)$ almost everywhere. We calculate

$$\|\mathcal{F}g - G\|_2^2 = \langle \mathcal{F}g, \mathcal{F}g\rangle - 2\text{Real}\langle \mathcal{F}g, G\rangle + \langle G, G\rangle . \tag{5.138}$$

We manipulate the middle inner product in (5.138),

$$\text{Real}\langle \mathcal{F}g, G\rangle = \text{Real}\langle G, \mathcal{F}g\rangle = \text{Real}\int G(\overline{\mathcal{F}g}) = \text{Real}\int \mathcal{F}G(\overline{g(-t)}), \tag{5.139a}$$

applying the previous Parseval result to obtain the last equality above. Using the definition of $g(t)$, we find

$$\text{Real}\langle \mathcal{F}G, \frac{1}{2\pi}[\mathcal{F}H](-t)\rangle = \text{Real}\left\langle \mathcal{F}G, \frac{1}{2\pi}[\mathcal{F}H](-t)\right\rangle = \frac{1}{2\pi}\text{Real}\langle \mathcal{F}G, \mathcal{F}G\rangle. \tag{5.139b}$$

By Parseval's theorem, $\langle \mathcal{F}g, \mathcal{F}g\rangle = 2\pi\langle g, g\rangle$ and $\langle \mathcal{F}G, \mathcal{F}G\rangle = 2\pi\langle G, G\rangle$, which is real. Thus, putting (5.138), (5.139a), and together implies

$$\begin{aligned}\|\mathcal{F}g - G\|_2^2 &= 2\pi\langle g, g\rangle - \frac{2}{2\pi}\langle \mathcal{F}G, \mathcal{F}G\rangle + \frac{1}{2\pi}\langle \mathcal{F}G, \mathcal{F}G\rangle = 2\pi\langle g, g\rangle - \frac{1}{2\pi}\langle \mathcal{F}G, \mathcal{F}G\rangle \\ &= \frac{2\pi}{(2\pi)^2}\langle \mathcal{F}[G(-\omega)], \mathcal{F}[G(-\omega)]\rangle - \frac{1}{2\pi}\langle \mathcal{F}G, \mathcal{F}G\rangle.\end{aligned} \tag{5.140}$$

But $\|\mathcal{F}G\|_2 = \|\mathcal{F}[G(-\omega)]\|_2$, so the last term in (5.140) is zero: $\|Fg - G\|_2^2 = 0$ almost everywhere, and the theorem is proven. ■

Corollary (Isometry of Time and Frequency Domains). The normalized radial Fourier transform $(2\pi)^{-1/2}\mathcal{F}$, where $\mathcal{F}$ is the radial Fourier transform on $L^2(\mathbb{R})$, is an isometry from $L^2(\mathbb{R})$ onto $L^2(\mathbb{R})$.

Proof: Linearity follows from the properties of the integral. We have shown that $\mathcal{F}$ on $L^2(\mathbb{R})$ is one-to-one; this is a consequence of the Parseval relation. The map $\mathcal{F}$ is also onto, as shown in the previous theorem. Since $\langle Fx, Fy\rangle = 2\pi\langle x, y\rangle$, we now see clearly that $(2\pi)^{-1/2}\mathcal{F}$ preserves inner products and constitutes an isometry. ■

Remark. The uniqueness implied by this relationship assures that a given Fourier spectrum is a true signature of a given time-domain signal $f(t)$. This property is a valuable asset, but we emphasize that two signals are equivalent if their spectra are identical across the *entire* frequency spectrum. Deviations between spectra, even if they are small in magnitude or restricted to a small range of frequencies, can result in large discrepancies between the respective $f(t)$. This is a legacy of the complex exponentials which form a basis of the Fourier transform. They are defined across the entire line (both in frequency and time), so small perturbations run the risk of infecting the entire Fourier synthesis. In practical situations, where spectral information is stored and transmitted in the form of discrete Fourier coefficients, errors or glitches can wreak havoc on the reconstruction of $f(t)$. This inherent sensitivity to error is one of the least appealing attributes of Fourier analysis.

The time-domain and frequency-domain representations of a square-integrable signal are equivalent. Neither provides more information. And Fourier transformation, for all its complexities, serves only to reveal some aspects of a signal at the possible risk of concealing others.

5.4 SUMMARY

The Fourier series and transform apply to analog periodic and aperiodic signals, respectively.

The Fourier series finds a set of discrete coefficients associated with a periodic analog signal. These coefficients represent the expansion of the signal on the exponential or sinusoidal basis sets for the Hilbert space $L^2[0, T]$. We shall have more to say about the Fourier series in Chapter 7, which is on discrete Fourier transforms.

For absolutely integrable or square-integrable aperiodic signals, we can find a frequency-domain representation, but it is an analog, not discrete, signal. We have had to consider three different Banach spaces in our quest for a frequency-domain description of a such a signal: $L^1(\mathbb{R})$, $(L^1 \cap L^2)(\mathbb{R})$, and $L^2(\mathbb{R})$. We began by defining the Fourier transform over the space of absolutely integrable signals. Then we considered the restricted transform on $L^1 \cap L^2$, but noted that this transform's range is in L^2. Applying the limit theorems available with the modern (Lebesgue) integral to this restricted signal class, we were able to extend the transform to the full space of square-integrable signals. Ultimately, we found an isometry between the time and frequency domain representations of a finite energy signal.

5.4.1 Historical Notes

Prior to Fourier, there were a number of attempts by other mathematicians to formulate a decomposition of general waves into trigonometric functions. D'Alembert, Euler, Lagrange, and Daniel Bernoulli used sinusoidal expansions to account for the vibrations of a string [4]. Evidently, ancient Babylonian astronomers based their predictions on a rudimetary Fourier series [10]. Fourier applied trigonometric series to the heat equation, presented his results to the French Academy of Sciences, and published his result in a book [25]. Criticism was severe, however, and the method was regarded with suspicion until Poisson, Cauchy, and especially Dirichlet (1829) provided theoretical substantiation of the Fourier series.

Plancherel proved that L^2 signals have L^2 Fourier transforms in 1910. The basis for so doing, as we have seen in Section 5.3, is the modern Lebesgue integral and the powerful limit properties which it supports. The L^2 theory of the Fourier integral is often called the Plancherel theory.

5.4.2 Looking Forward

The next chapter generalizes the Fourier transform to include even signals that are neither absolutely integrable nor square-integrable. This so-called generalized Fourier transform encompasses the theory of the Dirac delta, which we introduced in Chapter 3. Chapters 7 and 8 consider the frequency-domain representation for discrete signals. Chapter 9 covers applications of analog and discrete Fourier transforms.

REFERENCES

1. H. Baher, *Analog and Digital Signal Processing*, New York: Wiley, 1990.
2. J. A. Cadzow and H. F. van Landingham, *Signals, Systems, and Transforms*, Englewood Cliffs, NJ: Prentice-Hall, 1989.
3. L. B. Jackson, *Signals, Systems, and Transforms*, Reading, MA: Addison-Wesley, 1991.
4. A. V. Oppenheim, A. S. Willsky, and S. H. Nawab, *Signals and Systems*, Englewood Cliffs, NJ: Prentice-Hall, 1989.
5. R. E. Ziemer, W. H. Tranter, and D. R. Fannin, *Signals and Systems: Continuous and Discrete*, New York: Macmillan, 1989.
6. J. S. Walker, *Fourier Analysis*, New York: Oxford University Press, 1988.
7. D. C. Champeney, *A Handbook of Fourier Theorems*, Cambridge: Cambridge University Press, 1987.
8. G. B. Folland, *Fourier Analysis and its Applications*, Pacific Grove, CA: Wadsworth and Brooks/Cole, 1992.
9. R. E. Edwards, *Fourier Series: A Modern Introduction*, vol. I, New York: Hold, Rinehart and Winston, 1967.
10. H. Dym and H. P. McKean, *Fourier Series and Integrals*, New York: Academic, 1972.
11. E. M. Stein and G. Weiss, *Introduction to Fourier Analysis on Euclidean Spaces*, Princeton, NJ: Princeton University Press, 1971.
12. A. Zygmund, *Trigonometric Series*, vols. I & II, 2nd ed., Cambridge: Cambridge University Press, 1977.
13. M. Rosenlicht, *Introduction to Analysis*, New York: Dover, 1978.
14. H. L. Royden, *Real Analysis*, 2nd. ed., Toronto: Macmillan, 1968.
15. E. Kreysig, *Introductory Functional Analysis with Applications*, New York: Wiley, 1989.
16. A. W. Naylor and G. R. Sell, *Linear Operator Theory in Engineering and Science*, New York: Springer-Verlag, 1982.
17. A. H. Zemanian, *Distribution Theory and Transform Analysis*, New York: Dover, 1987.
18. M. J. Lighthill, *Fourier Analysis and Generalized Functions*, New York: Cambridge University Press, 1958.
19. A. B. Carlson, *Communication Systems*, 3rd ed., New York: McGraw-Hill, 1986.
20. L. W. Couch III, *Digital and Analog Communication Systems*, 4th ed., Upper Saddle River, NJ: Prentice-Hall, 1993.
21. S. Haykin, *Communication Systems*, 3rd ed., New York: Wiley, 1994.
22. N. Wiener, *The Fourier Integral and Certain of Its Applications*, London: Cambridge University Press, 1933.
23. C. K. Chui, *An Introduction to Wavelets*, San Diego, CA: Academic, 1992.

24. W. Rudin, *Real and Complex Analysis*, 2nd ed., New York: McGraw-Hill, 1974.
25. I. Gratton-Guiness, *Joseph Fourier 1768–1830*, Cambridge, MA: MIT Press, 1972.

PROBLEMS

1. Find the exponential Fourier series coefficients (5.9) for the following signals.
  (a) $x(t) = \cos(2\pi t)$.
  (b) $y(t) = \sin(2\pi t)$.
  (c) $s(t) = \cos(2\pi t) + \sin(2\pi t)$.
  (d) $z(t) = x(t - \pi/4)$.
  (e) $w(t) = 5y(-2t)$.

2. Find the exponential Fourier series coefficients for the following signals:
  (a) Signal $b(t)$ has period $T = 4$ and for $0 \le t < 4$, $b(t) = u(t) - u(t - 2)$, where $u(t)$ is the analog unit step signal.
  (b) $r(t) = tb(t)$, where $b(t)$ is given in (a).

3. Let $x(t) = 7\sin(1600t - 300)$, where t is a (real) time value in seconds. Give:
  (a) The amplitude of x.
  (b) The phase of x.
  (c) The frequency of x in radians/second.
  (d) The frequency of x in Hz (cycles/second).
  (e) The period of x.
  (f) Find the exponential Fourier series coefficients for $x(t)$.

4. Suppose $x(t)$ has period $T = 1$ and $x(t) = t^2$ for $0 \le t < 1$.
  (a) Find the exponential Fourier series coefficients for $x(t)$.
  (b) Sketch and label the signal $y(t)$ to which x's Fourier series synthesis equation converges.

5. Find the exponential Fourier series coefficients for the periodic sawtooth signal $x(t)$ (Figure 5.9).

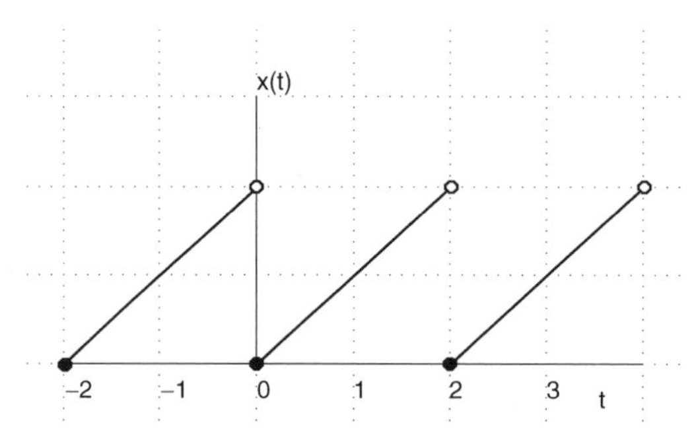

Fig. 5.9. Sawtooth signal $x(t)$.

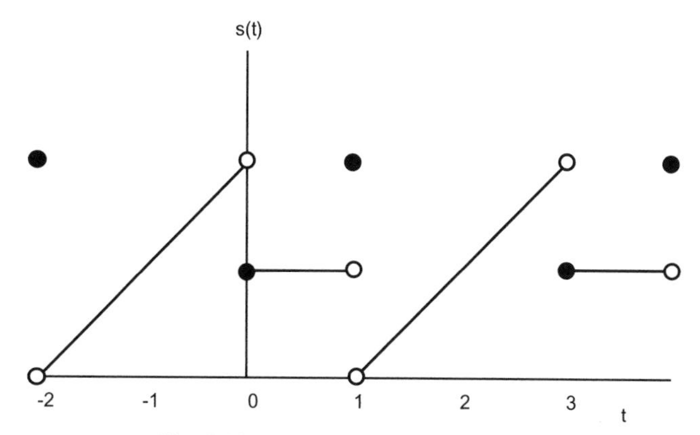

Fig. 5.10. Another sawtooth signal $s(t)$.

6. Consider the signal $s(t)$ shown in Figure 5.10.

(a) Find the exponential Fourier series for the signal $s(t)$.

(b) Consider the signal $y(t)$ to which the Fourier series synthesis equation for $s(t)$ converges. Sketch $y(t)$ and label the graph to show the exact values of $y(t)$ at places where it is the same and where it differs from $s(t)$.

7. The impulse response of an analog linear time invariant (LTI) system H is $h(t) = u(t + 50) - u(t)$, where $u(t)$ is the unit step signal.

(a) What is the response of the system to the signal $x(t) = u(t)e^{-t}$?

(b) Find the radial Fourier transform of $h(t)$.

(c) Does $x(t)$ have a radial Fourier transform? If so, find it; otherwise, give a reason why $X(\omega)$ does not exist; and, in any case, explain your answer.

8. Consider the analog signal $x(t) = [u(t + 1) - u(t - 1)]$, where $u(t)$ is the unit step signal.

(a) Find the radial Fourier transform of $x(t)$, $X(\omega) = \mathcal{F}[x(t)](\omega)$.

(b) Let $y(t) = x(t - 2)$; find $Y(\omega)$.

(c) Find $\mathcal{F}[x(2t)]$.

(d) Find $\mathcal{F}[x(t/5)]$.

(e) Find $\mathcal{F}[\sin(t)x(t)]$.

9. Let H be a linear time-invariant (LTI) analog system; $y = Hx$; $h = H\delta$; and $X(\omega)$, $Y(\omega)$, and $H(\omega)$ are their respective radial Fourier transforms. Which of the following are true? Explain.

(a) $Y(\omega)/X(\omega)$ is the Fourier transform of h.

(b) $y(t)/h(t) = x(t)$.

(c) $y(t) = x(t)*h(t)$, where $*$ is the analog convolution operation.

10. Prove or disprove the following statement: If an periodic analog signal $x(t)$ is represented by a Fourier series, but this series does not converge to $x(t)$ for all t, then $x(t)$ is not continuous.

11. Prove or disprove the following statement: If $x(t) \in L^2(\mathbb{R})$ is an odd signal (i.e., $x(t) = -x(-t)$), and $X(\omega)$ is the radial Fourier transform of x, then $X(0) = 0$.

12. Suppose the analog signals x(t) and h(t) have radial Fourier transforms $X(\omega) = u(\omega + 1) - u(\omega - 1)$ and $H(\omega) = \exp(-\omega^2)$, respectively. Let the signal $y = x * h$.

(a) Find $x(t)$.

(b) Find $h(t)$.

(c) Find $Y(\omega)$.

(d) Find $y(t)$.

13. Suppose that $X(\omega)$ and $Y(\omega)$ are the radial Fourier transforms of $x(t)$ and $y(t)$, respectively, and let $h(n)$ be a discrete signal with

$$x(t) = \sum_{n=-\infty}^{\infty} h(n)y(t-n). \tag{5.141}$$

(a) Find an expression for $X(\omega)$.

(b) What kind of conditions should be imposed upon the discrete signal $h(n)$ so that your answer in (a) is mathematically justifiable? Explain.

14. Show that the radial Fourier transform for analog signals is a linear operation. Is it also translation invariant? Explain.

15. Show that if $F(\omega)$ is absolutely integrable, then the inverse Fourier transform $\mathcal{F}^{-1}[F(\omega)](t)$ exists.

16. Suppose that analog periodic signal x(t) has exponential Fourier series coefficients c_k:

$$c_k = \langle x(t), \phi_k(t) \rangle = \int_{t_0}^{t_0+T} x(t)\frac{1}{\sqrt{T}}e^{-jk\Omega t}\,dt. \tag{5.142}$$

Prove the following symmetry properties:

(a) If $x(t)$ is real-valued and even, then the c_k are also real and even: $c_k = c_{-k}$.

(b) If $x(t)$ is real and odd, then the c_k are purely imaginary and odd: $c_{-k} = {}_{-}c_k$.

17. For analog signals $x(t)$ and $y(t) = x(t - a)$, show that the magnitudes of their radial Fourier transforms are equal, $|X(\omega)| = |Y(\omega)|$.

18. Prove or disprove: For all analog signals $x(t) \in L^2(\mathbb{R})$, if $x(t)$ is real-valued, then $X(\omega)$ is real-valued.

19. Let $x(t) \in L^1(\mathbb{R})$ be a real-valued analog signal and let $X(\omega)$ be its radial Fourier transform. Which of the following are true? Explain.

(a) $X(\omega)$ is bounded: $X(\omega) \in L^\infty(R)$.

(b) $|X(\omega)| \to 0$ as $|\omega| \to \infty$.

(c) $X(\omega)$ is unbounded.

(d) $X(0) = 0$.

(e) $X(\omega)$ has a Fourier transform.

(f) $X(\omega)$ has an inverse Fourier transform.

(g) $X(\omega)$ has an inverse Fourier transform and it is identical to x(t).

(h) $X(\omega) \in L^1(\mathbb{R})$ also.

20. Let $x(t) \in L^2(\mathbb{R})$ be a real-valued analog signal and let $X(\omega)$ be its radial Fourier transform. Which of the following are true? Explain.

(a) $X(\omega)$ is bounded: $X(\omega) \in L^\infty(R)$.

(b) $|X(\omega)| \to 0$ as $|\omega| \to \infty$.

(c) $X(\omega)$ is unbounded.

(d) $X(0) = 0$.

(e) $X(\omega)$ has a Fourier transform.

(f) $X(\omega)$ has an inverse Fourier transform.

(g) $X(\omega)$ has an inverse Fourier transform and it is identical to $x(t)$.

(h) $X(\omega) \in L^2(\mathbb{R})$ also.

21. Let $x(t) \in L^1(\mathbb{R})$. Show that:

(a) Fourier transform $X(\omega)$ of $x(t)$ is bounded.

(b) $\|X\|_\infty \le \|x\|_1$.

22. Loosely speaking, an analog *low-pass filter H* is a linear, translation-invariant system that passes low frequencies and suppresses high frequencies. We can specify such a system more precisely with the aid of the Fourier transform. Let $h(t)$ be the impulse response of H and let $H(\omega) = \mathcal{F}(h(t))(\omega)$ be its Fourier transform. For a low-pass filter we require $|H(0)| = 1$ and $|H(\omega)| \to 0$ as $|\omega| \to \infty$.

(a) Show that if $|H(0)| \ne 1$ but still $|H(0)| \ne 0$ and $|H(\omega)| \to 0$ as $|\omega| \to \infty$, then we can convert H into a low-pass filter by a simple normalization;

(b) Show that, with an appropriate normalization, the Gaussian signal is a low-pass filter (sometimes we say that any system whose impulse response can be so normalized is a low-pass filter).

23. An analog *high-pass filter H* is a linear, translation-invariant system that suppresses low frequencies and passes high frequencies. Again, if $h(t)$ is the impulse response of H and $H(\omega) = \mathcal{F}(h(t))(\omega)$, then we stipulate that $|H(0)| = 0$ and $|H(\omega)| \to 1$ as $|\omega| \to \infty$. Explain why our Fourier theory might not accept analog high-pass filters.

24. An analog *bandpass filter H* passes a range of frequencies, suppressing both low and high frequencies.

(a) Formalize the idea of a bandpass filter using the spectrum of the impulse response of H.

(b) Give an example of a finite-energy bandpass filter.

(c) Let $h(t)$ be an analog low-pass filter and let $g(t)$ be an analog bandpass filter. What kind of filters are $h * h$, $g * h$, and $g * g$? Explain your answer.

(d) Why would we not consider the case that $g(t)$ is an analog high-pass filter? Explain this too.

(e) Formulate and formalize the concept of an analog *bandstop* (also *banddreject* or *notch*) filter. Can these be absolutely integrable signals? Finite-energy signals? Explain.

25. Suppose $h(t)$ is the impulse response for an analog filter H that happens to be a *perfect low-pass filter*. That is, for some $\omega_c > 0$, $|H(\omega)| = 1$ for $|\omega| \leq \omega_c$, and $|H(\omega)| = 0$ for all $|\omega| > \omega_c$.

(a) Show that as a time-domain filter, H is noncausal.

(b) Explain why, in some sense, H is impossible to implement.

(c) Show that $h(t) \in L^2(\mathbb{R})$, but $h(t) \notin L^1(\mathbb{R})$.

26. Suppose an analog signal, $x(t)$, has radial Fourier transform, $X(\omega)$, given by Figure 5.11

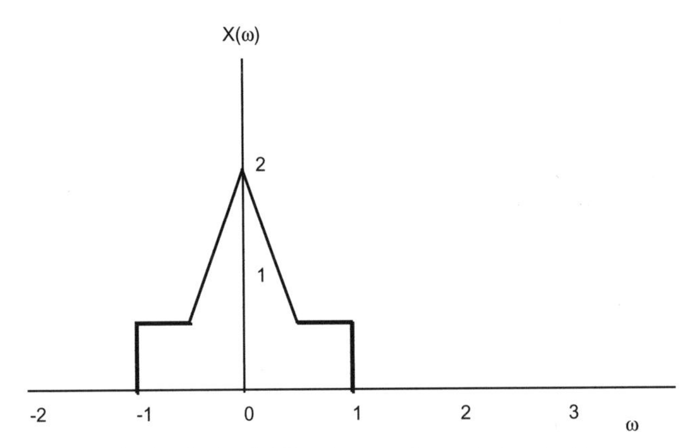

Fig. 5.11. Radial Fourier transform of signal $x(t)$.

Without attempting to compute $x(t)$, sketch $Y(\omega)$ for the following signals $y(t)$:

(a) $y(t) = x(2t)$.

(b) $y(t) = x(t/3)$.

(c) $y(t) = (x * g)(t)$, where $g(t) = \exp(-t^2)$.

(d) $y(t) = \cos(t)x(t)$.

(e) $y(t) = \sin(t)x(t)$.

(f) $y(t) = \exp(j5t)x(t)$.

27. Let $f(t)$, $h(t) \in L^2(\mathbb{R})$ and $F(\omega)$, $H(\omega)$ be their radial Fourier transforms.

(a) If $g(t) = f(t)h(t)$, show that $g(t) \in L^2(\mathbb{R})$.

(b) Show that $G(\omega) = (2\pi)^{-1}F(\omega) * H(\omega)$.

28. Develop an alternative approach to showing that signals with identical spectra must be the same. Let $f_1(t)$ and $f_2(t)$ have identical Fourier transforms; that is, $\mathcal{F}[f_1(t)] = \mathcal{F}[f_2(t)] = G(\omega)$.

(a) For an arbitrary $h(t)$ show that an alternative formulation of Parseval's relation holds:

$$2\pi \int_{-\infty}^{\infty} f_2(t)h(t)\, dt = \int_{-\infty}^{\infty} G(-\omega)H(\omega)\, d\omega. \tag{5.143}$$

(b) Show that the above leads to

$$2\pi \int_{-\infty}^{\infty} [f_1(t) - f_2(t)]h(t)\, dt = 0. \tag{5.144}$$

(c) Explain why it follows that $f_1(t) = f_2(t)$ for almost all t.

CHAPTER 6

Generalized Fourier Transforms of Analog Signals

This chapter extends Fourier analysis to common signals that lie outside of the spaces $L^1(\mathbb{R})$ and $L^2(\mathbb{R})$.

The theory of $L^1(\mathbb{R})$ and $L^2(\mathbb{R})$ Fourier transforms is broad enough to encompass a large body of signal processing and analysis. The foundation provided for the transform allows us to discover the frequency content of analog signals. One might be content with the situation as it stands, but several common and practical functions are neither absolutely integrable nor of finite energy. For example:

- The simple sinusoids $f(t) = \sin(\omega_0 t)$ and $f(t) = \cos(\omega_0 t)$. It is difficult to imagine functions for which the notion of frequency content is any more straightforward, yet the radial Fourier transform

$$F(\omega) = \int_{-\infty}^{\infty} f(t)e^{-j\omega t}dt = \mathcal{F}[f(t))](\omega) \tag{6.1}$$

 does not converge. Similar comments clearly apply to the complex exponential.
- The function $f(t) = c_0$, where c_0 is a positive or negative constant. Constant electrical signals are called direct current (DC) signals in engineering. Again, the notion of frequency content for this DC signal could hardly be more intuitive, but convergence of the Fourier integral fails.
- The unit step $u(t)$ and its close relative the signum, sgn t (see Figure 5.8), which clearly do not belong to the class of integrable or square-integrable functions.

Texts devoted to distributions and generalized Fourier transforms are Refs. 1–3. Mathematical analysis texts that also introduce the theory include Refs. 4–6.

6.1 DISTRIBUTION THEORY AND FOURIER TRANSFORMS

Our first encounter with useful integrals that defy solution using classical methods of calculus arose in Chapter 3, where integration of classically troublesome entities,

Signal Analysis: Time, Frequency, Scale, and Structure, by Ronald L. Allen and Duncan W. Mills
ISBN: 0-471-23441-9

such as the derivative of the unit step, were elegantly handled through distributions. The theoretical framework employed test functions of rapid descent which were classically well-behaved and generated a calculus of distributions simply because the classical notions of derivatives could be applied directly to the test functions themselves. These developments suggest that if the troublesome signals listed above are treated as distributions, and the test functions have traditional Fourier transforms, then a theory of generalized Fourier transforms, embracing the selected distributions, can be formulated.

Consider replacing the complex exponential with some function $\Phi(t)$ which is sufficiently well-behaved to allow the integral over time, namely

$$\int_{-\infty}^{\infty} f(t)e^{-j\omega t}\,dt \to \int_{-\infty}^{\infty} f(t)\Phi(t)\,dt \tag{6.2}$$

to converge. Intuitively, one needs a $\Phi(t)$ which decays rapidly enough to counter the lack of integrability inherent in $f(t)$. Two key points follow:

- Each of the nonintegrable signals $f(t)$ under consideration is a function of slow growth (Chapter 3) and therefore represents a regular distribution of slow growth when set in the context of generalized integrals.
- The class of testing functions is Fourier transformable in the "regular" sense of Section 5.2; this is our link to frequency space.

The study of distributions in the time domain was based on the classical concept of integration by parts. Similarly, the classically derived Parseval relations extend the theory of distributions into the frequency domain. We propose the following:

Definition (Generalized Fourier Transform). Let $f(t)$ be a distribution of slow growth. Note that if $\phi(\alpha)$ is a testing function of rapid descent, we can define a Fourier transform,

$$\Phi(\beta) = \int_{-\infty}^{\infty} \phi(\alpha)e^{-j\beta\alpha}\,d\alpha. \tag{6.3}$$

By Parseval's theorem

$$\int_{-\infty}^{\infty} F(\omega)\phi(\omega)\,d\omega = \int_{-\infty}^{\infty} f(t)\Phi(t)\,dt. \tag{6.4}$$

The function $F(\omega)$ is the *generalized Fourier transform* of $f(t)$.

Remark. In the event that $f(t)$ is integrable, the generalized Fourier transform is merely an expression of Parseval's theorem for such functions. Consequently, $F(\omega)$ is a bona fide generalized Fourier transform encompassing both the integrable functions (which are also covered by the Paresval relation) and the distributions of slow growth.

In the generalized Fourier transform, note that ω and t within the integrals (6.4) are merely continuous variables; the simple form of the generalized Fourier transform may be extended so that

$$\int_{-\infty}^{\infty} F(\omega)\phi(\omega)\,d\omega = \int_{-\infty}^{\infty} f(t)\Phi(t)\,dt = \int_{-\infty}^{\infty} F(t)\phi(t)\,dt = \int_{-\infty}^{\infty} f(\omega)\Phi(\omega)\,d\omega. \quad (6.5)$$

This is of more than academic interest and allows for greater dexterity when deriving the properties of the generalized Fourier transform.

6.1.1 Examples

How does our formulation of the generalized Fourier transform perform for the important, simple signals? Let us investigate the case of constant (DC) signals and impulses.

Example (DC Waveform). Let $f(t) = 1$ for all $t \in \mathbb{R}$. This signal represents a constant DC level for all values of t and is a function of slow growth. The generalized Fourier transform takes the form

$$\int_{-\infty}^{\infty} F(\omega)\phi(\omega)\,d\omega = \int_{-\infty}^{\infty} \Phi(t)\,dt = F[\Phi(t)](\omega)\big|_{\omega=0} = 2\pi\phi(-\omega)\big|_{\omega=0}. \quad (6.6)$$

The quantity following the last equality is simply $2\pi\phi(0)$, which can be written in terms of the Dirac delta:

$$\int_{-\infty}^{\infty} F(\omega)\phi(\omega)\,d\omega = 2\pi\int_{-\infty}^{\infty} \delta(\omega)\phi(\omega)\,d\omega. \quad (6.7)$$

Comparing both sides of (6.7), it is readily apparent that

$$F(\omega) = 2\pi\delta(\omega) \quad (6.8)$$

represents the spectrum of the constant DC signal. This result supports the intuitively appealing notion that a constant DC level represents clusters its entire frequency content at the origin. We have already hinted at this in connection with the Fourier transform of the rectangular pulse in the limit of large width; in a sense, (6.8) is the ultimate expression of the scaling law for a rectangular pulse.

The time-dependent Dirac delta represents the converse:

Example (Impulse Function). Consider a Dirac delta impulse, $f(t) = \delta(t)$. The generalized Fourier transform now reads

$$\int_{-\infty}^{\infty} F(\omega)\phi(\omega)\, d\omega = \int_{-\infty}^{\infty} \delta(t)\Phi(t)\, dt = \mathrm{F}[\phi(t)](\omega)\Big|_{\omega=0} = \int_{-\infty}^{\infty} \phi(t)\, dt$$
$$= \int_{-\infty}^{\infty} \phi(\omega)\, d\omega. \tag{6.9}$$

From a comparison both sides,

$$\mathcal{F}[\delta(t)](\omega) = 1 \tag{6.10}$$

for all $\omega \in \mathrm{R}$. The Dirac delta function's spectrum therefore contains equal contributions from all frequencies. Intuitively, this result is expected.

6.1.2 The Generalized Inverse Fourier Transform

The reciprocity in the time and frequency variables in (6.4) leads to a definition of a generalized inverse Fourier transform.

Definition (Generalized Inverse Fourier Transform). Let $F(\omega)$ be a distribution of slow growth. If $\Phi(\beta)$ is a testing function of rapid descent, then it generates an inverse Fourier transform:

$$\phi(\alpha) = \int_{-\infty}^{\infty} \Phi(\beta) e^{j\beta\alpha} d\beta. \tag{6.11}$$

Once again, by Parseval's theorem

$$\int_{-\infty}^{\infty} F(\omega)\phi(\omega)\, d\omega = \int_{-\infty}^{\infty} f(t)\Phi(t)\, dt, \tag{6.12}$$

and $f(t)$ is called the *generalized inverse Fourier transform* of $F(\omega)$. This definition is so intuitive it hardly needs to be written down.

No discussion of the generalized Fourier transform would be complete without tackling the remaining functions of slow growth which are central to many aspects of signal generation and analysis. These include the sinusoids and the appropriate piecewise continuous functions such as the unit step and signum functions. Their generalized spectra are most easily determined by judicious application of selected properties of the generalized Fourier transform. Prior to completing that discussion, it is useful to illustrate some general properties of the generalized Fourier transform.

6.1.3 Generalized Transform Properties

In the previous examples we have emphasized that the generalized Fourier transform is an umbrella which encompasses standard, classical Fourier-integrable as well as slow growth functions considered as regular distributions. The properties of the classically defined Fourier transform demonstrated in Chapter 5 apply with little or no modification to the generalized transform. Naturally, the methods for proving them involve the nuances specific to the use of generalized functions. A general strategy when considering properties of the generalized Fourier transform is to begin with integrals (6.5) and allow the desired parameter (scale, time shift) or operator (differential) to migrate to the classical Fourier transform of the test function, where its effects are easily quantified. The reader should study the following examples carefully. The generalized Fourier transform is elegant but seductive; a common pitfall is to rearrange the generalized transform so it resembles the familiar classical integral and then "declare" a transform when in fact the classical integral will not coverge because the integrand is not integrable.

Proposition (Linearity). Let $f(t)$ represent the linear combination of arbitrary distributions of slow growth,

$$f(t) = \sum_{k=1}^{N} a_k f_k(t). \tag{6.13}$$

Then,

$$\int_{-\infty}^{\infty} F(\omega)\phi(\omega)\, d\omega = \sum_{k=1}^{N} a_k \int_{-\infty}^{\infty} f_k(t)\Phi(t)\, dt = \int_{-\infty}^{\infty} \left\{ \sum_{k=1}^{N} a_k F_k(\omega) \right\} \phi(\omega)\, d\omega. \tag{6.14}$$

The expected result follows:

$$F(\omega) = \sum_{k=1}^{N} a_k F_k(\omega). \tag{6.15}$$

Proposition (Time Shift or Translation). Let $f(t)$ be a distribution of slow growth subjected to a time shift t_0 such that $f(t) \to f(t-t_0)$. Then,

$$\mathcal{F}[f(t-t_0)](\omega) = \mathcal{F}[f(t)]e^{-j\omega t_0}. \tag{6.16}$$

Use of the defining generalized Fourier transform relations leads to the following equalities (we reduce clutter in the integral by suppressing the (ω) suffix in $\mathcal{F}$):

$$\int_{-\infty}^{\infty} \mathcal{F}[f(t-t_0)]\phi(\omega)\, d\omega = \int_{-\infty}^{\infty} f(\alpha-t_0)\Phi(\alpha)\, d\alpha = \int_{-\infty}^{\infty} f(\gamma)\Phi(\gamma+t_0)\, d\gamma. \quad (6.17)$$

The change of variable, $\gamma \equiv \alpha - t_0$, in the last integral of (6.17) places the time shift conveniently within the classical Fourier transform of the test function. From here, matters are straightforward:

$$\begin{aligned}\int_{-\infty}^{\infty} f(\gamma)\Phi(\gamma+t_0)\, d\gamma &= \int_{-\infty}^{\infty} \mathcal{F}[f(t)]\mathcal{F}^{-1}[\Phi(\gamma+t_0)]\, d\omega \\ &= \int_{-\infty}^{\infty} \mathcal{F}[f(t)]\phi(\omega)e^{-j\omega t_0}\, d\omega \end{aligned} \quad (6.18)$$

so that

$$\int_{-\infty}^{\infty} \mathcal{F}[f(t-t_0)]\phi(\omega)\, d\omega = \int_{-\infty}^{\infty} \mathcal{F}[f(t)]\phi(\omega)e^{-j\omega t_0}\, d\omega, \quad (6.19)$$

and the property is proven:

$$\mathcal{F}[f(t-t_0)](\omega) = \mathcal{F}[f(t)]e^{-j\omega t_0}. \quad (6.20)$$

We leave the remaining significant properties of the generalized Fourier transform to the exercises. As we have noted, they are identical to the properties of the standard integral transform, and the proofs are straightforward.

Remark. In the case of regular functions $f(t)$ considered in Chapter 5, the validity of time differentiation property,

$$\mathcal{F}\left[\frac{d}{dt}f(t)\right](\omega) = j\omega F(\omega), \quad (6.21)$$

was conditioned upon $\lim_{|t| \to \infty} f(t) = 0$. No such restriction applies to distributions of slow growth, since the convergence of the generalized Fourier transform is assured by the decay of the testing functions of rapid descent.

Using the properties of the generalized transform, we can resume calculating spectra for the remaining functions of slow growth. These are central to much of signal analysis. We cover the signum function, the unit step, and the sinusoids.

Example (Signum). Consider the case $f(t) = \operatorname{sgn}(t)$. The differentiation property implies

$$\mathcal{F}\left[\frac{d}{dt}\operatorname{sgn}(t)\right](\omega) = j\omega F(\omega) = \mathcal{F}[2\delta(t)], \tag{6.22}$$

where we have used the recently derived transform of the Dirac delta function. From here, one is tempted to conclude that the desired spectrum $F(\omega) = \frac{2}{j\omega}$. However, a certain amount of care is required since in general $\omega F_1(\omega) = \omega F_2(\omega)$ does not imply $F_1(\omega) = F_2(\omega)$. This is another instance of unusual algebra resulting from the Dirac delta property, derived in Chapter 3: $\omega\delta(\omega) = 0$. Under the circumstances, this allows for the possibility of an additional term involving an impulse function, so that

$$\omega F_1(\omega) = \omega[F_2(\omega) + c_0\delta(\omega)], \tag{6.23}$$

where c_0 is a constant to be determined. Returning to the example, with $F_1(\omega) \equiv F(\omega)$ and $\omega F_2(\omega) \equiv 2$, we obtain a complete and correct solution:

$$F(\omega) = \frac{2}{j\omega} + c_0\delta(\omega). \tag{6.24}$$

A determination of c_0 can be made by appealing to symmetry. Since $\operatorname{sgn}(t)$ is a real, odd function of t, its transform must be purely imaginary and odd in the frequency variable ω. Hence, we conclude that $c_0 = 0$. The spectrum is shown in Figure 6.1.

Based on this result, we can proceed to the unit step.

Example (Unit Step). Let $f(t) = u(t)$. Then,

$$\mathcal{F}[u(t)](\omega) = \pi\delta(\omega) + \frac{1}{j\omega}. \tag{6.25}$$

This proof is left as an exercise. Note that the two terms above, $F_e(\omega)$ and $F_o(\omega)$, represent even and odd portions of the frequency spectrum. These may be obtained directly from even and odd components of $f(t)$, $f_e(t)$, and $f_o(t)$, respectively, in accordance with symmetries developed in Chapter 5. The Dirac delta impulse is a legacy of $f_e(t)$, which is not present in the signum function. The resulting spectrum Figure 6.1b is complex.

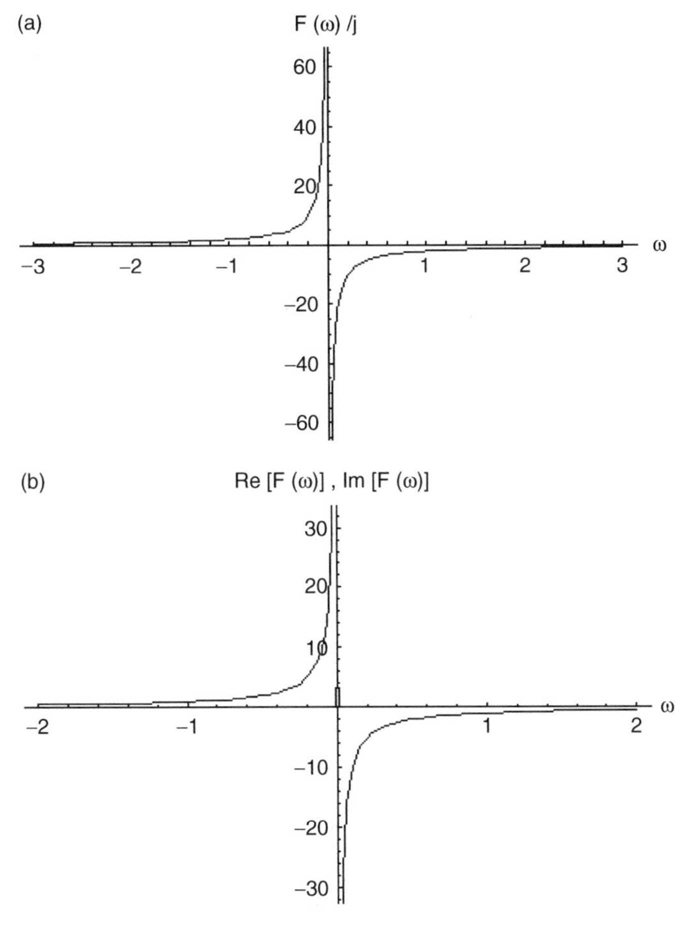

Fig. 6.1. (a) The Fourier transform of sgn(t) is purely imaginary and inversely proportional to ω. (b) The transform of the unit step consists of real Dirac delta function and an imaginary part, as shown.

Examples (Powers of t). The frequency differentiation property

$$\mathcal{F}[(-jt)^n f(t)](\omega) = \frac{d^n}{dt^n}F(\omega) \tag{6.26}$$

leads to several useful Fourier transforms involving powers of t and generalized functions. For integer $n \geq 0$

$$\mathcal{F}[t^n](\omega) = 2\pi j^n \cdot \frac{d^n\delta(\omega)}{d\omega^n}, \tag{6.27}$$

$$\mathcal{F}[t^n u(t)](\omega) = j^n \cdot \left[\pi \frac{d^n \delta(\omega)}{d\omega^n} + \frac{1}{j} \frac{(-1)^n n!}{\omega^{n+1}}\right], \tag{6.28}$$

$$\mathcal{F}[t^n \operatorname{sgn} t](\omega) = (-2) j^{n+1} \frac{(-1)^n n!}{\omega^{n+1}}. \tag{6.29}$$

The derivations are straightforward an left as an exercise. As expected, each of the above spectra contain singularities at $\omega = 0$ on account of the discontinuity in $f(t)$.

Integral powers of $|t|$ are easily handled. For even n,

$$\mathcal{F}[|t|^n](\omega) = \mathcal{F}[t^n](\omega), \tag{6.30}$$

so (6.27) applies. For odd n, note the convenient relation

$$\mathcal{F}[|t|^n](\omega) = \mathcal{F}[t^n \operatorname{sgn}(t)](\omega). \tag{6.31}$$

Remark. Treatment of fractional exponents $0 < n < 1$ and the theory of generalized Fourier transforms for $f(t)$ exhibiting logarithmic divergences is possible, but outside our scope.

Inverse integral powers of t, which are clearly neither integrable nor square integrable, readily yield generalized Fourier spectra. For example,

$$\mathcal{F}\left[\frac{1}{t}\right](\omega) = -j \cdot \pi \cdot \operatorname{sgn}(\omega) \tag{6.32}$$

follows from the application of the symmetry property to $\mathcal{F}[\operatorname{sgn}(t)](\omega)$. Repeated application of time differentiation leads to a more general result for integer $m > 0$:

$$\mathcal{F}\left[\frac{1}{t^m}\right](\omega) = \frac{-\pi}{(m-1)!} \cdot j^m \cdot \omega^{m-1} \cdot \operatorname{sgn}(\omega). \tag{6.33}$$

Example (Complex Exponential). Let $f(t) = e^{j\omega_0 t}$ represent a complex exponential with oscillations at a selected frequency ω_0. According to the frequency shift property of the generalized Fourier transform,

$$\mathcal{F}[g(t)e^{j\omega_0 t}](\omega) = G(\omega - \omega_0), \tag{6.34}$$

the simple substitution $g(t) = 1$— the constant DC signal—provides the desired result:

$$\mathcal{F}[e^{j\omega_0 t}](\omega) = 2\pi\delta(\omega - \omega_0). \tag{6.35}$$

An example of this transform is shown in Figure 6.2a. Not surprisingly, the spectrum consists of an oscillation at a single frequency. In the limit $\omega_0 \to 0$, the spectrum reverts to $2\pi\delta(\omega)$, as fully expected.

Proposition (General Periodic Signal). Let $f(t)$ represent a periodic distribution of slow growth with period T. Then

$$\mathcal{F}[f(t)](\omega) = \frac{2\pi}{\sqrt{T}} \sum_{n=-\infty}^{\infty} c_n \delta(\omega - n\omega_0), \tag{6.36}$$

where the c_n represent the exponential Fourier series coefficients for $f(t)$ and $\omega_0 = 2\pi T$.

This is almost trivial to prove using the linearity property as applied to an exponential Fourier series representation of the periodic signal:

$$f(t) = \frac{1}{\sqrt{T}} \sum_{n=-\infty}^{\infty} c_n e^{jn\omega_0 t}. \tag{6.37}$$

This leads immediately to

$$\mathcal{F}[f(t)](\omega) = \frac{1}{\sqrt{T}} \sum_{n=-\infty}^{\infty} c_n \mathcal{F}[e^{jn\omega_0 t}] \tag{6.38}$$

from which the desired result (6.36) follows.

This is an important conclusion, demonstrating that the Fourier series is nothing more than a special case of the generalized Fourier transform. Furthermore, upon application of the Fourier inversion, the sifting property of the Dirac delta readily provides the desired synthesis of $f(t)$:

$$\frac{1}{2\pi} \int_{-\infty}^{\infty} \mathcal{F}[f(t)] e^{j\omega t} d\omega = \sum_{n=-\infty}^{\infty} \frac{c_n}{\sqrt{T}} \int_{-\infty}^{\infty} \delta(\omega - n\omega_0) e^{j\omega t} d\omega \tag{6.39}$$

which trivially reduces to the series representation of $f(t)$ as given in (6.37).

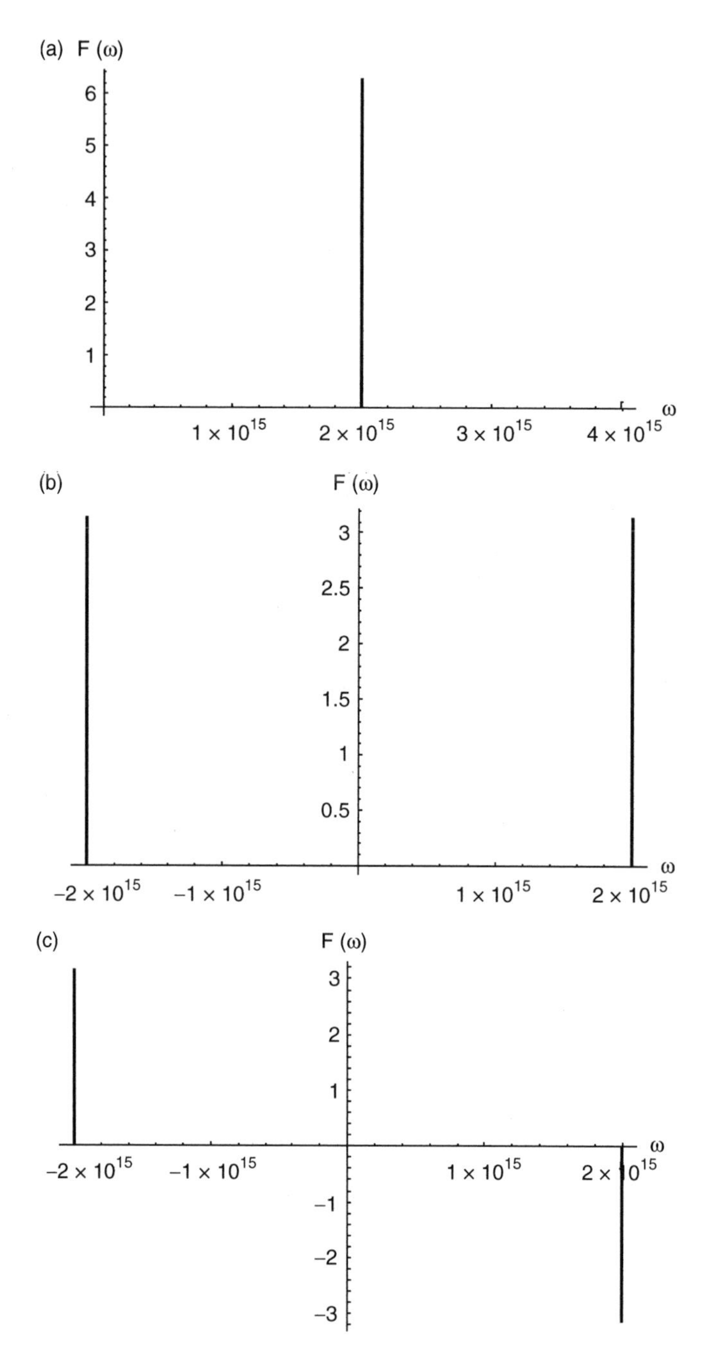

Fig. 6.2. Fourier transforms of (a) the complex exponential with fundamental frequency $\omega_0 = 2 \times 10^{15}$ rad/s, (b) a cosine of the same frequency, and (c) the corresponding sine wave.

Example (Cosine and Sine Oscillations). Let $f(t) = \cos(\omega_0 t)$. From the Euler relation—linking the sinusoids to the complex exponential—and the linearity property, we obtain

$$\mathcal{F}[\cos(\omega_0 t)](\omega) = \pi[\delta(\omega+\omega_0)+\delta(\omega-\omega_0)]. \tag{6.40}$$

Hence, the cosine exhibits a two-sided spectrum with contributions at $\omega = \pm\omega_0$, as Figure 6.2b illustrates. Note that in the process of forming a sinusoid, the spectral amplitude 2π inherent in the complex exponential has been redistributed equally amongst the positive and negative frequencies.

In the case of the sinusoid $f(t) = \sin(\omega_0 t)$, similar arguments demonstrate that

$$\mathcal{F}[\sin(\omega_0 t)](\omega) = j\pi(\delta(\omega+\omega_0)-\delta(\omega-\omega_0)). \tag{6.41}$$

The sine spectrum, shown in Figure 6.2c, is similar in form to the cosine but is, according to symmetry arguments, an odd function in frequency space. We will make use of both of the previous examples in the sequel. In particular, when we develop transforms that combine time- and frequency-domain information, the calculations of sinusoidal spectra will play an important role.

Generalized functions, particularly the Dirac delta function, arise repeatedly in applications and theoretical development of signal analysis tools. Far from being fringe elements in our mathematical lexicon, generalized functions provide the only mathematically consistent avenue for addressing the Fourier transform of several important waveforms. And, as we have just demonstrated, they link two analysis tools (the discrete Fourier series and the continuous Fourier transform) which initially appeared to be fundamentally distinct.

6.2 GENERALIZED FUNCTIONS AND FOURIER SERIES COEFFICIENTS

In this section, we apply generalized functions to develop an alternative technique for evaluating the Fourier coefficients of selected piecewise continuous periodic signals. We have encountered a number of such waveforms in earlier chapters, including the sawtooth wave and the train of rectangular pulses. In Chapter 5 we analyzed such waveforms by application of the the Fourier series expansion of periodic signals in terms of a sinusoidal orthonormal basis. There are no calculations performed in this section which could not, in principle, be performed using the well-established methods previously covered in this chapter and in Chapter 5, so casual readers can safely skip this section without a loss of the essentials. Nonetheless, readers who master this section will emerge with the following:

- A method that affords the determination of Fourier series coefficients—for a certain class of periodic functions—without the use of integrals;
- Further experience applying generalized functions to Fourier analysis, including the Fourier expansion of a periodic train of impulse functions known as the *Dirac comb*; and
- An introduction to linear differential equations as they apply to Fourier analysis.

The central theme of these developments is the Dirac delta function and its role as the derivative of a step discontinuity. This discontinuity may appear in one or more of the derivatives of $f(t)$ (including the zeroth-order derivative), and this is the tie-in to differential equations. Our discussion is heuristic and begins with the Fourier series expansion of an impulse train. This forms the analytical basis for the other piecewise continuous functions considered in this section.

6.2.1 Dirac Comb: A Fourier Series Expansion

The term "Dirac comb" is a picturesque moniker for a periodic train of Dirac delta functions (Figure 6.3). The Dirac comb is a periodic generalized function, and it is natural to inquire into its Fourier series representations. The discussion had been deliberately slanted to emphasize the role of differential equations in selected problems where the Dirac comb is applicable. We derive the trigonometric and exponential Fourier series representations of the Dirac comb prior to examining some practical problems in the next section.

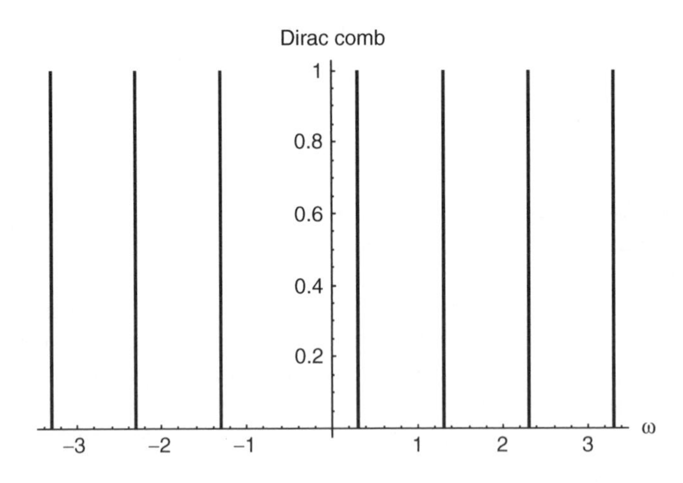

Fig. 6.3. A Dirac comb. By definition, the comb has unit amplitude. The version illustrated here has a unit period and is phase-shifted relative to the origin by an increment of 0.3.

6.2.1.1 Dirac Comb: Trigonometric Fourier Series. Let us revisit the periodic sawtooth wave discussed in Chapter 5. There is nothing sacred about our selection of the sawtooth wave to demonstrate the desired results other than the fact that its step transitions are such that a Dirac comb structure appear in the derivatives of the sawtooth wave. The Dirac comb is an odd function of period T with a sine Fourier series expansion:

$$x(t) = \sum_{k=1}^{\infty} b_k \sin(k\Omega t), \tag{6.42}$$

where $\Omega = (2\pi)/T$, and

$$b_k = -\frac{4h}{2\pi k}(-1)^k. \tag{6.43}$$

The signal $x(t)$ consists of a continuous portion $f(t)$ separated by periodically spaced steps of magnitude $-2h$:

$$x(t) = f(t) - 2h \sum_{m=-\infty}^{\infty} u\left(t - T\left(m + \frac{1}{2}\right)\right), \tag{6.44}$$

whose derivative is

$$x'(t) = \frac{2h}{T} - 2h \sum_{m=-\infty}^{\infty} \delta\left(t - \left(m + \frac{1}{2}\right)T\right). \tag{6.45}$$

Substituting the sine Fourier series representation for $x(t)$ into the left-hand side of (6.45) gives

$$\sum_{k=1}^{\infty} b_k \cdot k\Omega \cdot \cos(k\Omega t) = \frac{2h}{T} - 2h \sum_{m=-\infty}^{\infty} \delta\left(t - \left(m + \frac{1}{2}\right)T\right). \tag{6.46}$$

Therefore,

$$\sum_{m=-\infty}^{\infty} \delta\left(t - \left(m + \frac{1}{2}\right)T\right) = \frac{1}{T} + \frac{2}{T} \sum_{k=1}^{\infty} (-1)^k \cos(k\Omega t). \tag{6.47}$$

This is one form of the Dirac comb whose teeth are arranged along the t axis according to the sawtooth wave used in the derivation. A cleaner and more general form of the Dirac comb may be obtained through a time shift of T/2, giving the basic series representation for a canonical Dirac comb with impulses placed at integer values of T:

$$\sum_{m=-\infty}^{\infty} \delta(t-mT) = \frac{1}{T} + \frac{2}{T}\sum_{k=1}^{\infty}(-1)^k \cos(k\Omega(t+T/2))$$

$$= \frac{1}{T} + \frac{2}{T}\sum_{k=1}^{\infty}\cos(k\Omega t). \tag{6.48}$$

Note that the series representation for a Dirac comb of arbitrary phase shift relative to the origin can always be obtained from the canonical representation in (6.48) .

6.2.1.2 Dirac Comb: Exponential Fourier Series. The exponential Fourier series representation,

$$x(t) = \sum_{k=-\infty}^{\infty} c_k e^{jk\Omega t}, \tag{6.49a}$$

can be derived directly from first principles or from the trigonometric form using the conversion derived in Chapter 5. The result is elegant:

$$c_n = c_{-n} = \frac{1}{\sqrt{T}} \tag{6.49b}$$

for all integer n. Therefore,

$$\sum_{m=-\infty}^{\infty} \delta(t-mT) = \frac{1}{\sqrt{T}}\sum_{k=-\infty}^{\infty} e^{jk\Omega t}. \tag{6.50}$$

6.2.2 Evaluating the Fourier Coefficients: Examples

The problem of finding Fourier series expansion coefficients for piecewise continuous functions from first principles, using the Fourier basis and integration, can be tedious. The application of a Dirac comb (particularly its Fourier series representations), to this class of functions replaces the integration operation with simpler differentiation.

We will proceed by example, considering first the case of a rectified sine wave and selected classes of rectangular pulse waveforms. In each case, we develop a differential equation that can then be solved for the Fourier expansion coefficients. As we proceed, the convenience as well as the limitations of the method will become apparent. Mastery of these two examples will provide the reader with sufficient understanding to apply the method to other piecewise continuous waveforms.

6.2.2.1 Rectified Sine Wave.

Consider a signal

$$x(t) = |A_0 \sin \omega_0 t| , \tag{6.51}$$

where $\omega_0 = (2\pi)/T$, as in Figure 6.4. Now, $x(t)$ is piecewise continuous with discontinuities in its derivative at intervals of $T/2$ (*not* T). The derivative consists of continuous portions equal to the first derivative of the rectified sine wave, separated by step discontinuities of magnitude $\omega_0 A_0$:

$$x'(t) = \frac{d}{dt}|A_0 \sin \omega_0 t| + 2\omega_0 A_0 \sum_{n=-\infty}^{\infty} u(t - n\tau) , \tag{6.52}$$

where $\tau = T/2$. Taking a further derivative,

$$x''(t) = -\omega_0^2 x(t) + 2\omega_0 A_0 \sum_{n=-\infty}^{\infty} \delta(t - n\tau) , \tag{6.53}$$

brings in a train of impulses and—equally important—a term proportional to the original waveform. Substituting the trigonometric series representation of the impulse train and rearranging terms, we have the differential equation

$$x''(t) + \omega_0^2 x(t) = 2\omega_0 A_0 \left[\frac{1}{\tau} + \frac{2}{\tau} \sum_{n=1}^{\infty} \cos(n\Omega t) \right] , \tag{6.54}$$

where $\Omega = ((2\pi)/T)$.

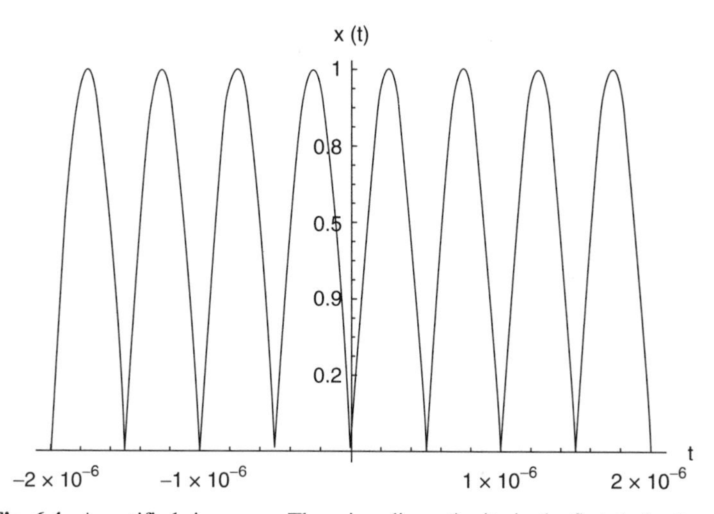

Fig. 6.4. A rectified sine wave. There is a discontinuity in the first derivative.

Remark. This is a second-order, linear differential equation for $x(t)$ whose presence in physical science is almost ubiquitous. The order refers to the highest derivative in the equation. Linearity implies no powers of $x(t)$ (or its derivative), greater than one. Due to the oscillatory nature of its solutions $x(t)$, it is termed the *wave equation*. In the form above, it contains a separate time-dependent term (in this case, representing the Dirac comb) on the right-hand side of (6.54). Depending on the physical context, this is referred to as the source term (in electromagnetic theory) or a forcing function (in circuit analysis and the study of dynamical systems). This equation is a major player in a number of disciplines. When no forcing function is present, the right-hand side vanishes, leaving the *homogeneous wave equation*

$$x''(t) + \omega_0^2 x(t) = 0 \tag{6.55}$$

whose solutions are the simple sinusoids of period T: $\sin(\Omega t)$ and $\cos(\Omega t)$, or linear combinations thereof.

Returning to the problem at hand, we can obtain expressions for the trigonometric Fourier series coefficients of $x(t)$ by substituting a trigonometric series representation,

$$x(t) = \frac{a_0}{2} + \sum_{n=1}^{\infty} [a_n \cos(n\Omega t)] + b_n \sin(n\Omega t), \tag{6.56}$$

and the second derivative

$$x''(t) = \sum_{n=1}^{\infty} -(n\Omega)^2 [a_n \cos(n\Omega t)] + b_n \sin(n\Omega t) \tag{6.57}$$

into (6.54). Then we solve the two resulting equations for a_n and b_n by equating the $\cos(n\Omega t)$ and $\sin(n\Omega t)$ components. So,

$$\omega_0^2 \cdot \frac{1}{2} a_0 = 2\omega_0 A_0 \frac{1}{\tau}, \tag{6.58}$$

giving

$$a_0 = \frac{4A_0}{\omega_0 \tau}, \tag{6.59}$$

and for $n \neq 0$,

$$-(n\Omega)^2 a_n + \omega_0^2 a_n = \frac{2}{\tau}. \tag{6.60}$$

Thus,

$$a_n = \frac{4A_0}{\pi}\left(\frac{1}{1-4n^2}\right). \tag{6.61}$$

The b_n vanish identically, as would be expected considering the even symmetry of $x(t)$, and this is confirmed by the governing equation:

$$(-(n\Omega)^2 + \omega_0^2) \cdot b_n = 0. \tag{6.62}$$

The exponential Fourier series representation can be obtained through these trigonometric coefficients or by direct substitution of the exponential Fourier series representation of the Dirac comb (6.50) into (6.54).

6.2.2.2 Periodic- "Shaped" Rectangular Pulse. Another problem we investigated in Section 5.1 is the periodic rectangular pulse train. Now let us consider a more general version of this waveform, consisting of a piecewise continuous portion denoted $p(t)$, with steps of magnitude A and B:

$$x(t) = p(t) + A_0 \sum_{n=-\infty}^{\infty} u[t-(p+nT)] - B_0 \sum_{n=-\infty}^{\infty} u[t-(q+nT)]. \tag{6.63}$$

In this notation, the pulse width is $d = q - p$. For the moment, we defer specification of a particular form for $p(t)$, but our experience with the previous example suggests that some restrictions will apply if we are to solve for the Fourier coefficients via a linear differential equation. Experience also suggests that the differential equation governing this situation will be of first order, since the Dirac comb appears when the first derivative is taken:

$$x'(t) = p'(t) + A_0 \sum_{n=-\infty}^{\infty} \delta[t-(p+nT)] - B_0 \sum_{n=-\infty}^{\infty} \delta[t-(q+nT)]. \tag{6.64}$$

Substituting the appropriate trigonometric Fourier series (6.48) for the impulse trains and expanding the cosines within the series leads to

$$x'(t) = p'(t) + \frac{(A_0 - B_0)}{T} + C(t) + S(t), \tag{6.65}$$

where

$$C(t) = \frac{2}{T}\sum_{n=1}^{\infty} \cos(n\omega_0 t)[A_0\cos(n\omega_0 p) - B_0\cos(n\omega_0 q)] \tag{6.66}$$

and

$$S(t) = \frac{2}{T}\sum_{n=1}^{\infty} \sin(n\omega_0 t)[A_0\sin(n\omega_0 p) - B_0\sin(n\omega_0 q)]. \tag{6.67}$$

Notice that for arbitrary choice of p and q, $x(t)$ is neither even nor odd; the expansion involves both sine and cosine components, as it should.

To complete this problem, $p(t)$ needs to be specified. If we are going to apply this technique successfully, we will have to restrict $p(t)$ so that the first-order differential equation governing $x(t)$ will be linear. One reasonable option is to specify that $p(t)$ is a linear function of t. Thus,

$$p'(t) = \frac{(B_0 - A_0)}{T}. \tag{6.68}$$

In this instance, the differential equation (6.65) is linear because $p(t)$ returns a constant upon differentiation. (From the previous example involving the rectified sine wave, it is obvious that $p(t)$ itself does not have to be a linear function of its independent variable in order to generate a linear governing differential equation for $x(t)$, but a recursion—as exhibited by the sinusoids—is necessary.) Returning to the problem at hand, we substitute the general Fourier series expansion for the derivative of $x(t)$ into (6.65) and solve for the Fourier coefficients,

$$a_n = \frac{-2}{n\omega_0 T}[A_0\sin(n\omega_0 p) - B_0\sin(n\omega_0 q)] \tag{6.69}$$

and

$$b_n = \frac{2}{n\omega_0 T}[A_0\cos(n\omega_0 p) - B_0\cos(n\omega_0 q)]. \tag{6.70}$$

As a check, notice that in the limit $A_0 = B_0$, $p = -q$ we generate the special case of a flat (zero-slope) rectangular pulse train of even symmetry, which was treated in Chapter 5. In this case, (6.69) predicts $b_n = 0$, as the even symmetry of $x(t)$ would dictate, and

$$a_n = \frac{4A_0}{n\omega_0 T}\sin[n\pi(d/T)]. \tag{6.71}$$

This is consistent with our previous derivation in Chapter 5 using the standard inner product with the Fourier basis.

Remarks. The problems associated with applying Fourier analysis to nonlinear differential equations can be appreciated in this example. Instead of equaling a constant, suppose that the derivative of $p(t)$ is proportional to some power, $p^n(t)$. Substituting the Fourier series for $p(t)$ would result in multiple powers of the Fourier coefficients, in various combinations, whose determination would be difficult, if not impossible. Furthermore, the example involving the rectified sine wave highlights the convenience associated with the fact that derivatives of the sinusoids are recursive: Up to multiplicative constants, one returns to the original function upon differentiating twice. These observations illustrate why the sinusoids (and their close relatives) figure so prominently in the solution of second-order linear differential equations.

6.3 LINEAR SYSTEMS IN THE FREQUENCY DOMAIN

Communication is the business of passing information from a source to a receiver as faithfully as possible. This entails preparation or encoding of the message, which is then impressed upon a waveform suitable for transmission across a channel to the receiver. At the receiver end, the signal must be decoded and distributed to the intended recipients. If all has gone well, they are provided with an accurate reproduction of the original information. The technical ramifications of each step are vast and involve the questions of analog versus digital encoding, the suitability of the transmission channel, and the design of suitable decoding apparatus—all of which are impacted in some way by the techniques described throughout this book.

This section is intended to provide a basic introduction to *filtering* and *modulation*, with an emphasis on the time and frequency domains implied by Fourier analysis. Filtering implies conditioning in the frequency domain; typically a given filter is designed to highlight or suppress portions of the spectrum. Filtering, in its ideal form, is conceptually simple, but in practice involves nuances and tradeoff due to restrictions imposed by the real world.

Modulation is an operation that inhabits the time domain; it is here that we connect the information-bearing message and a carrier signal, whose role is to aid in transporting the information across the designated channel. From the standpoint of our present knowledge base, the details of modulation are quite user-friendly, and we will present a somewhat detailed account of amplitude and frequency modulation—AM and FM—whose practical role in communications needs no introduction [7–9].

Since filtering and modulation involve the interaction of waveforms with linear systems, we rely extensively on the linear systems principles introduced in Chapter 3.

Two relatively simple theorems involving the Fourier transform establish the foundations of filtering and modulation. These are the *convolution theorem* and the *modulation theorem*, which we prove below. There are few electronic communication devices that do not, in some way, make use of the analytical mileage they provide.

6.3.1 Convolution Theorem

A *filter* is a linear system designed to suppress or enhance selected portions of a signal spectrum. In Chapter 3 we established an input–output relation for a linear, time-invariant system based on the system impulse response, denoted $h(t)$, and the input signal $f(t)$:

$$g(t)= \int_{-\infty}^{\infty} f(\tau)h(t-\tau)\, d\tau. \tag{6.72}$$

The convolution theorem relates the spectrum of the output $g(t)$ to those of the input and the system impulse response:

Theorem (Convolution). Let $f_1(t)$ and $f_2(t)$ be two functions for which radial Fourier transforms $F_1(\omega)$ and $F_1(\omega)$ exist and let $f(t) = (f_1 * f_2)(t)$. Then the Fourier spectrum of $f(t)$ is

$$F(\omega) = F_1(\omega)F_2(\omega). \tag{6.73}$$

Proof: By definition,

$$F(\omega) = \int_{-\infty}^{\infty} e^{-j\omega t}\left[\int_{-\infty}^{\infty} f_1(\tau)f_2(t-\tau)\, d\tau\right] dt. \tag{6.74}$$

Interchanging the order of integration gives

$$F(\omega) = \int_{-\infty}^{\infty} f_1(\tau)\left[\int_{-\infty}^{\infty} e^{-j\omega t} f2(t-\tau)\, dt\right] d\tau. \tag{6.75}$$

The time-shift property of the Fourier transform takes care of the integral with respect to t, so that

$$F(\omega) = \int_{-\infty}^{\infty} f_1(\tau)e^{-j\omega\tau}F_2(\omega)\, d\tau = F_1(\omega)F_2(\omega), \tag{6.76}$$

completing the proof. ■

It is hard to imagine a simpler relationship between spectra. Set in the context of linear systems, the input and output spectra are linked:

$$G(\omega) = F(\omega)H(\omega), \tag{6.77}$$

so that $G(\omega)$ can be shaped or modified by an appropriately designed and implemented system transfer function $H(\omega)$. This forms the backbone of filter design. It will be considered in more detail following a proof of the modulation theorem, which is effectively a converse to the convolution theorem.

6.3.2 Modulation Theorem

Modulation is an operation whereby two or more waveforms, typically an information-bearing modulating signal $m(t)$ and a sinusoidal carrier $c(t)$, are multiplied to form a composite. The termwise product signal $f(t)$ is appropriate for transmission across a communication channel:

$$f(t) = m(t)c(t). \tag{6.78}$$

The modulation theorem relates the spectrum of the composite to those of the constituent modulating wave and carrier:

Theorem (Modulation). Let $f_1(t)$ and $f_2(t)$ be two functions for which Fourier transforms $F_1(\omega)$ and $F_2(\omega)$ exist. Let $f(t) = f_1(t)f_2(t)$. Then the Fourier transform of $f(t)$ is a convolution in the frequency domain:

$$F(\omega) = \frac{1}{2\pi}\int_{-\infty}^{\infty} F_1(\alpha)F_2(\omega-\alpha)\,d\alpha. \tag{6.79}$$

Proof: The Fourier transform of the time product,

$$F(\omega) = \int_{-\infty}^{\infty} e^{-j\omega t} f_1(t)f_2(t)\,dt, \tag{6.80}$$

can be rearranged by substitution of the inverse Fourier transform of $f_2(t)$:

$$F(\omega) = \frac{1}{2\pi}\int_{-\infty}^{\infty} f_1(t)\left[\int_{-\infty}^{\infty} F_2(\gamma)e^{j(\gamma-\omega)t}d\gamma\right]dt. \tag{6.81}$$

A change of variables, $\alpha = \omega - \gamma$, gives (noting carefully the signs and integration limits),

$$F(\omega) = \frac{-1}{2\pi}\int_{-\infty}^{\infty} f_1(t)\left[\int_{\infty}^{(-\infty)} F_2(\omega-\alpha)e^{-j\alpha t}\,d\alpha\right] dt = \frac{1}{2\pi}\int_{-\infty}^{\infty} F_1(\alpha)F_2(\omega-\alpha)\,d\alpha, \tag{6.82}$$

and the proof is complete. ■

The exact form of this spectrum depends heavily upon the nature of $f_1(t)$ and $f_2(t)$ and, in the framework of a modulated carrier signal, gives

$$F(\omega) = \frac{1}{2\pi}\int_{-\infty}^{\infty} M(\alpha)C(\omega-\alpha)\,d\alpha = \frac{1}{2\pi}\int_{-\infty}^{\infty} C(\alpha)M(\omega-\alpha)\,d\alpha. \tag{6.83}$$

The *productive* ↔ *convolution* relationship is one of the most elegant and useful aspects of the Fourier transform. It forms the basis for the design and application of linear filters considered in the next section.

Example (Damped Oscillations). When pure sinusoids are applied to a damped exponential (whose spectrum contained a pole along the imaginary axis; see Chapter 5), the pole acquires a real part. Consider

$$f(t) = e^{-\alpha t}\sin(\omega_0 t)u(t), \tag{6.84}$$

where α is a positive definite constant. Designating $f_1(t) = \sin(\omega_0 t)$ and $f_2(t) = e^{-\alpha t}u(t)$ then (6.79) gives

$$\mathcal{F}[f_1(t)f_2(t)](\omega) = \frac{1}{2\pi}\int_{-\infty}^{\infty}\left[\frac{\pi}{j}\{\delta(\gamma-\omega_0)-\delta(\gamma+\omega_0)\}\right]\frac{1}{\alpha+j(\omega_0-\gamma)}\,d\gamma. \tag{6.85}$$

This reduces, after some algebra, to

$$\frac{1}{2j}\left[\frac{1}{\alpha+j(\omega-\omega_0)}-\frac{1}{\alpha+j(\omega+\omega_0)}\right] = \frac{\omega_0}{(\alpha+j\omega)^2+\omega_0^2}. \tag{6.86}$$

There are poles in the complex plane located at

$$\omega = \pm\,\omega_0 + j\alpha \tag{6.87}$$

whose real parts are proportional to the frequency of oscillation. The imaginary part remains proportional to the decay. For $\cos(\omega_0 t)$, the spectrum is similar,

$$F(\omega) = \frac{\alpha+j\omega}{(\alpha+j\omega)^2+\omega_0^2} \tag{6.88}$$

but exhibits a zero at $\omega = j\alpha$. Note that it is impossible to distinguish the spectra of the sine and cosine on the basis of the poles alone.

6.4 INTRODUCTION TO FILTERS

To design a filter, it is necessary to specify a system transfer function $H(\omega)$ that will pass frequencies in a selected range while suppressing other portions of the input spectrum. A *filter design* is a specification of $H(\omega)$, including the frequency bands to be *passed*, those to be *stopped*, and the nature of the transition between these regions. In general, $H(\omega)$ is a complex-valued function of frequency,

$$H_0(\omega) = He^{-j\omega\Theta(\omega)}, \tag{6.89}$$

composed of a real-valued *amplitude spectrum* $H_0(\omega)$ and a *phase spectrum* $\Theta(\omega)$. In this chapter, we will work with the class of so-called *ideal filters,* whose transitions between the *stop bands* and the *pass bands* are unit steps:

Definition (Ideal Filter). An *ideal filter* is a linear, translation-invariant system with a transfer function of the form

$$H_0(\omega) = \sum_{n=1}^{N} a_n u(\omega - \omega_n) \tag{6.90}$$

and a zero phase spectrum across all frequencies: $\Theta(\omega) = 0$.

The amplitude spectrum of an ideal filter is characterized by an integer number N transitions, each of which is a unit step of amplitude an at specified frequencies ω_n. The idealization is twofold:

- The unit step transitions are abrupt and perfectly clean. In practice, the transition exhibits rolloff—that is, it is gradual—and overshoot, which is signal processing parlance for oscillations or ripple near the corners of the step transitions, similar to Gibbs's oscillations.
- It is impossible to design and implement a filter whose phase spectrum is identically zero across all frequencies.

The nuisance imposed by a nonzero phase spectrum can be readily appreciated by the following simple illustration. Suppose an audio waveform $f(t)$ acts as an input to a linear system representing a filter with a transfer function $H(\omega)$; for the purposes of illustration we will assume that the amplitude spectrum is unity across all frequencies. The output signal $g(t)$ is characterized by a spectrum,

$$G(\omega) = e^{-j\omega\Theta(\omega)} F(\omega) \tag{6.91}$$

so that when $G(\omega)$ is inverted back to the time domain, the nonzero phase introduces time shifts in $g(t)$. If the input $f(t)$ were an audio signal, for example, $g(t)$ would sound distorted, because each nonzero phase would introduce a time shift that is a function of $\Theta(\omega)$. (Such *phasing* introduces a reverberation and was deliberately applied to audio entertainment during the so-called psychedelic era in the late 1960s. In more serious communication systems, such effects are not conducive to faithful and accurate data transmission.) In practice, there are ways to minimize phase distortion, but for the present discussion we will continue to inhabit the ideal world with zero phase.

Filter types are classified according to the frequency bands they pass, and the user makes a selection based upon the spectral characteristics of the signal he intends to modify via application of the filter. A signal $m(t)$ whose spectrum is clustered around $\omega = 0$ is termed *baseband.* In audio signals, for example, the power resides in the under-20-kHz range, and visual inspection of the spectrum shows a spread in the frequency domain whose nominal width is termed the *bandwidth*. The precise measure of bandwidth may depend upon the context, although a common measure of spectral spread is the 3-dB bandwidth:

Definition (3-dB Bandwidth). The *3-dB bandwidth* occupied by a spectrum $F(\omega)$ is the frequency range occupied by the signal as measured at the point at which the squared magnitude $|F(\omega)|^2$ is equal to one-half its maximum value.

The use of the squared magnitude allows the definition to encompass complex-valued spectra and eliminates any issues with +/– signs in the amplitude, which have no bearing on frequency spread. This definition of bandwidth applies equally well to baseband and bandpass spectra, but will be illustrated here with a Gaussian at baseband.

Example (3-dB Bandwidth of a Gaussian). In the previous chapter we noted that the spectrum of a Gaussian pulse $f(t) = e^{-\alpha t^2}$ was a Gaussian of the form

$$F(\omega) = \sqrt{\frac{\pi}{\alpha}} e^{-\omega^2/4\alpha}. \tag{6.92}$$

According to the definition, the 3-dB bandwidth is the spread in frequency between the points defined by the condition

$$\frac{\pi}{\alpha} e^{-\omega^2/2\alpha} = \frac{1}{2}\frac{\pi}{\alpha} \tag{6.93}$$

or

$$\omega^2 = -2\alpha \ln\frac{1}{2} = 2\alpha \ln 2. \tag{6.94}$$

These points are $\omega = \pm\sqrt{2\ln 2}\sqrt{\alpha}$ so that the total 3-dB bandwidth is

$$\Delta\omega = 2\sqrt{2\ln 2}\sqrt{\alpha}. \tag{6.95}$$

As expected, large values of α result in a greater spectral spread. In communications systems it is common to describe performance in Hz (cycles/s), which scales the bandwidth accordingly,

$$\Delta f = \Delta\omega/2\pi. \tag{6.96}$$

The typical baseband audio signal is not exactly Gaussian, but occupies approximately 40 kHz (i.e., 2 × 20 kHz), a relatively small increment in (Hertz) frequency space. Television picture signals carry more information—including audio and visual signals—and occupy approximately 9 MHz.

There are other definitions of frequency spread which will be introduced when appropriate.

Much of analog and digital communication involves translating baseband signals in frequency space and filtering to suit the needs of a given system. Frequency translation will be discussed in the next subsection. Prior to that, we turn to an illustration of three common filter types and their uses.

6.4.1 Ideal Low-pass Filter

A low-pass filter is characterized by transitions $\omega_1 = -\omega_t$ and $\omega_2 = \omega_t$ with associated transition amplitudes $a_{1,2} = \pm 1$, as illustrated in Figure 6.5a. The ideal filter has created a passband in the interval $[-\omega_t, \omega_t]$, while suppressing all other frequencies by creating a stopband in those regions. The effect of low-pass filtering is to rid a signal of unwanted high frequencies, which can occur in several contexts. If we plan to sample and digitize a baseband signal, for example, frequencies above a certain limit will end up contaminating the reconstructed waveform since information from the high frequencies will be spuriously thrown into the lower frequency range. This phenomenon is known as aliasing—high frequencies are falsely identified with the lower—and the best course of action is to rid the signal of the offending spectral content prior to sampling.

Low-pass filters are useful when a baseband signal needs to be isolated from other signals present in the received waveform. In selected modulation schemes, the process in which a baseband signal is recovered at the receiver introduces an additional waveform residing near a higher frequency. This waveform is useless and the baseband signal can be isolated from it with a suitable low-pass filter.

6.4.2 Ideal High-pass Filter

A high-pass filter passes all frequencies $|\omega| > \omega_t$. As with the low-pass filter, we locate transitions at $\omega_1 = -\omega_t$ and $\omega_2 = \omega_t$ but the associated transition amplitudes are $a_1 = -1$, $a_2 = -1$, as illustrated in Figure 6.5b. A primary application of high-pass filtering involves cutting out redundant portions of a signal spectrum to reduce overhead associated with bandwidth. In the forthcoming discussion on modulation, we will consider this in further detail.

6.4.3 Ideal Bandpass Filter

A bandpass filter is characterized by four transitions $\omega_1 = \omega_{t1}$, $\omega_1 = \omega_{t2}$, $\omega_1 = \omega_{t3}$, and $\omega_1 = \omega_{t4}$, with associated transition amplitudes $a_{1,2} = \pm 1$, $a_{3,4} = \pm 1$. As illustrated in Figure 6.5c, two passbands have been created which effectively isolate a band in the middle region of the spectrum.

Example (Shannon Function). The Shannon function

$$f(t) = \frac{\sin(2\pi t) - \sin(\pi t)}{\pi t} \tag{6.97}$$

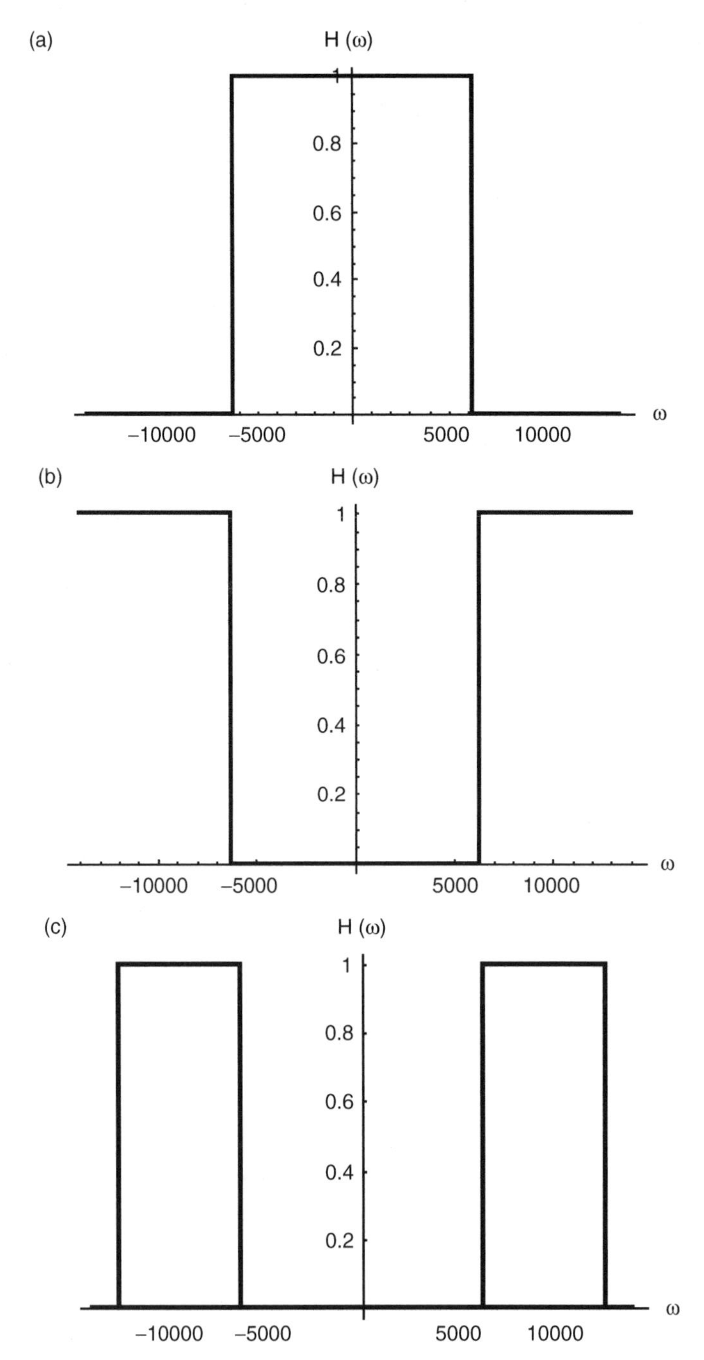

Fig. 6.5. (a) Ideal filter types. (a) Ideal low-pass filter, shown with a transition frequency f_t = 1 kHz. (b) Ideal high-pass filter. (c) Ideal bandpass filter, illustrated with passband widths of 1 kHz.

has the remarkable property of exhibiting and ideal two-sided bandpass spectrum. This is easily demonstrated. Note that the Shannon function is the difference of two sinc terms,

$$f_1(t) = \frac{\sin(2\pi t)}{\pi t} = 2\frac{\sin(2\pi t)}{2\pi t} \tag{6.98}$$

and

$$f_2(t) = \frac{\sin(\pi t)}{\pi t}, \tag{6.99}$$

each of which is integrable, so that one can evaluate the Fourier integral directly. Alternatively, we can apply the symmetry property

$$\mathcal{F}[F(t)](\omega) = 2\pi f(-\omega) \tag{6.100}$$

to the problem of the unit rectangular pulse supported on the interval $t \in [-a, a]$, whose spectrum was (see Chapter 5)

$$F(\omega) = 2a\frac{\sin(a\omega)}{a\omega}. \tag{6.101}$$

It follows immediately that the spectra of $f_1(t)$ and $f_2(t)$ are unit amplitude rectangular pulses of width 4π and 2π, respectively:

$$F_1(\omega) = u(\omega + 2\pi) - u(\omega - 2\pi), \tag{6.102}$$

$$F_2(\omega) = u(\omega + \pi) - u(\omega - \pi). \tag{6.103}$$

The composite spectrum of the Shannon function is the difference $F(\omega) = (F_2(\omega) - F_1(\omega))$ of two nested rectangular pulses, forming a perfect two-sided bandpass spectrum with transition frequencies $\omega_{t1} = -2\pi$, $\omega_{t2} = -\pi$, $\omega_{t3} = \pi$, $\omega_{t4} = 2\pi$. In terms of the unit step function,

$$F(\omega) = u(t + 2\pi) - u(t + \pi) + u(t - \pi) + u(t - 2\pi). \tag{6.104}$$

As expected given the properties of the Shannon function, the spectrum is a real function of even symmetry.

In general, bandpass filters are useful for isolating non-baseband spectra. For example, consider a multiuser communication link in which several operators are simultaneously transmitting information over several channels, each allocated to a given frequency range. Tuning in to a particular user typically involves some form of bandpass filter to isolate the desired channel.

Example (Derivative of a Gaussian). The Gaussian

$$g(t) = e^{-\alpha t^2} \tag{6.105}$$

exhibited a low-pass spectrum

$$\mathcal{F}[g(t)](\omega) = \sqrt{\frac{\pi}{\alpha}}e^{-\omega^2/\alpha}. \tag{6.106}$$

Multiplying the time-domain Gaussian by a pure sinusoid is one method of translating the bulk of the signal energy to higher frequencies to create a spectrum that approximates a bandpass filter (as we consider in the next section). Alternatively, one can induce the necessary waviness by taking the second derivative,

$$f(t) = -\frac{d^2}{dt^2}g(t) = 2\alpha[1 - 2\alpha t^2]e^{-\alpha t^2}. \tag{6.107}$$

Its spectrum,

$$F(\omega) = -(j\omega)(j\omega)\mathcal{F}[g(t)](\omega) = \omega^2\sqrt{\frac{\pi}{\alpha}}e^{-\omega^2/(4\alpha)} \tag{6.108}$$

demonstrates bandpass characteristics in the form of two quasi-Gaussian pass bands centered about

$$\omega = \pm 2\sqrt{\alpha}. \tag{6.109}$$

The characteristic is hardly ideal, because it passes portions of all finite frequencies except at DC ($\omega = 0$), but as such could be used to eliminate any unwanted DC portion of a waveform. Note that the lobes are not perfect Gaussians due to the effect of the quadratic factor; thus the use of the term "centered" in connection with (6.109) is only approximate. This also complicates the calculation of 3-dB bandwidth, a matter that is taken up in the exercises.

Remark. Both the Shannon function and the second derivative of the Gaussian are localized *atoms* in the time domain and make suitable *wavelets* (Chapter 11). In wavelet applications, their bandpass characteristics are used to advantage to select out features in the neighborhood of specific frequency.

6.5 MODULATION

The implications of the modulation theorem are far-reaching and quite useful. Most audio and video information begins as a baseband signal $m(t)$ whose frequency range is typically inappropriate for long-distance radio, TV, satellite, and optical fiber links. (Most often, a basic problem is attenuation in the channel, due to absorption in the transmission medium, at frequencies in the kHz regime.) There is also the question of multiple users. Whatever the medium, hundreds of audio and video programs are communicated simultaneously and must be so transferred without interference. Since the bandwidth of an individual audio or video signal is relatively small compared to the total bandwidth available in a given transmission medium, it is convenient to allocate a slot in frequency space for each baseband signal. This allocation is called *frequency division multiplexing* (FDM). The modulation theorem makes this multichannel scheme possible.

Modulation theory is treated in standard communications theory books [7–9].

6.5.1 Frequency Translation and Amplitude Modulation

Let us reconsider the notion of modulation, where by our baseband signal $m(t)$ is multiplied by an auxiliary signal $c(t)$, to form a composite waveform $f(t)$. The composite is intended for transmission and eventual demodulation at the receiver end. Thus,

$$f(t) = m(t)c(t), \tag{6.110}$$

where $c(t)$ is a sinusoidal *carrier* wave,

$$c(t) = A_c\cos(\omega_c t) = A_c\cos(2\pi f_c t). \tag{6.111}$$

As this scheme unfolds, we will find that the carrier effectively translates the baseband information to a spectral region centered around the carrier frequency f_c. Multiplication by a sinusoid is quite common in various technologies. In various parts of the literature, the carrier signal is also referred to as the *local oscillator* signal, *mixing* signal, or *heterodyning* signal, depending upon the context.

The modulation theorem describes the Fourier transform of the composite signal. Let $m(t) = f_1(t)$ and $c(t) = f_2(t)$. Then

$$\begin{aligned} F_2(\omega) &= \frac{2\pi}{\sqrt{T}}\left[\frac{\sqrt{T}}{2}\delta(\omega+\omega_c) + \frac{\sqrt{T}}{2}\delta(\omega+\omega_c)\right] \\ &= \pi[\delta(\omega+\omega_c) + \delta(\omega+\omega_c)], \end{aligned} \tag{6.112}$$

where we used the exponential Fourier series with $c_1 = c_{-1} = \sqrt{T}/2$. Designating the Fourier transform of $m(t)$ by $M(\omega)$, the spectrum of the composite signal is, according to the modulation theorem,

$$F(\omega) = \frac{1}{2}\int_{-\infty}^{\infty} M(\alpha)\delta(\omega+(\omega_c-\alpha))\,d\alpha + \frac{1}{2}\int_{-\infty}^{\infty} M(\alpha)\delta(\omega-(\omega_c-\alpha))\,d\alpha. \tag{6.113}$$

Using simple algebra to rearrange the arguments of the delta functionals and making use of their even symmetry, we can reduce the above to straightforward integrals:

$$F(\omega) = \frac{1}{2}\int_{-\infty}^{\infty} M(\alpha)\delta(\alpha-(\omega+\omega_c))\,d\alpha + \frac{1}{2}\int_{-\infty}^{\infty} M(\alpha)\delta(\alpha-(\omega-\omega_c))\,d\alpha \tag{6.114}$$

Equation (6.114) evaluates easily, resulting in

$$F(\omega) = \frac{1}{2}M(\omega+\omega_c) + \frac{1}{2}M(\omega-\omega_c). \tag{6.115}$$

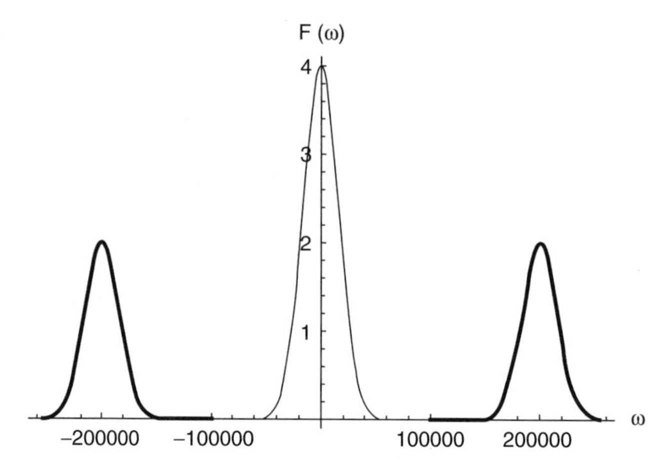

Fig. 6.6. Frequency translation resulting from carrier modulation of a baseband Gaussian spectrum of amplitude 4. For purposes of illustration, f_C was set at 100 kHz, but most broadcast systems utilize carrier frequencies up to several orders of magnitude higher.

The composite spectrum consists of two facsimiles of the baseband spectra, translated in frequency so that they are centered around $\omega = \pm\omega_c$. The power in the original signal has been equally split between the two portions of the frequency-translated spectrum. The situation is illustrated in Figure 6.6 for a hypothetical Gaussian baseband spectrum.

6.5.2 Baseband Signal Recovery

Our emerging picture of multiuser communications systems consists of multiple baseband signals. Each baseband signal centers around a given carrier frequency, which is broadcast and available to end users. Each end user, in turn, can recover the desired information-bearing baseband signal in a number of ways.

One method of recovery requires the receiver to multiply the incoming composite signal by a local oscillator with frequency ω_s, giving

$$s(t) = [m(t)\cos(\omega_c t)]\cos(\omega_s t)$$
$$= \frac{m(t)}{2}\{\cos((\omega_c+\omega_s)t + \cos(\omega_c-\omega_s)t)\}. \tag{6.116}$$

Let us assume that the carrier and local oscillator frequencies differ by some amount $2\pi\Delta f$:

$$\omega_s = \omega_c + 2\pi\Delta f. \tag{6.117}$$

Then,

$$s(t) = \frac{m(t)}{2}[\cos\{2\omega_c + 2\pi\Delta f\}t + \cos(2\pi\Delta f t)]. \tag{6.118}$$

If the frequency deviation Δf is identically zero, this reduces to

$$s(t) = \frac{m(t)}{2}\cos(2\omega_c t) + \frac{m(t)}{2}. \tag{6.119}$$

This is the sum of a half-amplitude baseband signal centered around $2\omega_c$ and a similar contribution residing at baseband.

Since systems are designed so that the carrier frequency is much larger than the highest frequency present in the baseband signal, these two contributions are well-separated in the frequency domain. The simple application of a low-pass filter to output described by (6.119) allows the user to eliminate the unwanted power near the frequency $2\omega_c$, leaving the half-amplitude baseband waveform intact. In the case of a multiuser channel, all of these double-frequency waves can be eliminated by a low-pass filter.

When $\Delta f = 0$, the local oscillator is said to be *synchornized* with the carrier. In practice, such precise tuning is not always possible, and a familiar problem with this technique is signal distortion and fading. This occurs, for instance, when the local oscillator drifts from the desired frequency. The source of this fading is evident in (6.118). A small amount of mistuning has little effect on the first term, since it is usually easy to maintain $\Delta f < f_c$; filtering removes this term. On the other hand, the second term is more sensitive to frequency adjustment. Any frequency deviation is going to cause distortion and undesired fading as $\cos(2\pi\Delta f t)$ periodically nears zero. Naturally, as Δf increases, the recovered baseband signal fades with greater frequency. Furthermore, the second term in (6.118) is effectively a baseband signal translated to an carrier frequency Δf. If this value gets too large—even a fraction of the typical baseband frequencies in $m(t)$—there is the possibility of translating a portion of the spectrum outside the passband of the low-pass filter used to retrieve $\frac{1}{2}m(t)$. This is not a trivial matter, because frequency deviations are usually specified as a percentage of the carrier frequency; so even a few percent can be a problem if the carrier frequency is relatively large.

6.5.3 Angle Modulation

Angle modulation is a method whereby the phase of the carrier wave is modulated by the baseband signal $m(t)$. That is,

$$f(t) = A_c \cos(\omega_c t + \phi_c(t)), \tag{6.120}$$

where the *phase deviation*

$$\phi_c(t) = \phi_c[m(t)] \tag{6.121}$$

is a function to be specified according to the application. The term angle modulation refers to the time-varying angle between the fixed carrier oscillation and the added phase $\phi_c(t)$. In practice, two functional relationships (6.121) are common. The first is a direct proportionality between the phase and the baseband modulation:

$$\phi_c(t) = \text{const} \times m(t), \tag{6.122}$$

which is referred to simply as *phase modulation* (PM). Another common arrangement makes the phase offset proportional to the integral of the baseband signal:

$$\phi_c(t) = \text{const} \times \int_{-\infty}^{t} m(\tau)\, d\tau, \tag{6.123}$$

which is called *frequency modulation* (FM), for reasons that will emerge as we proceed.

This classification scheme can seem confusing at first glance. Bear in mind that both phase modulation and frequency modulation do, in their respective ways, modulate the phase of the carrier signal. Furthermore, the astute reader has probably wondered how it is possible to distinguish between a PM and an FM waveform by inspection. And, as a matter of fact, you cannot distinguish between them visually. Indeed, for most purposes in this book, the distinction is academic, since in either case $\phi_c(t)$ is simply some function of t. The distinction between PM and FM arises in the implementation. Without explicit knowledge as to how the phase offset $\phi_c(t)$ was constructed, FM and PM are effectively identical. For this reason, it is often sufficient to lapse into the generic label *angle modulation* to describe these waveforms. Of course, the end user, whose task it is to extract information (i.e., the baseband signal $m(t)$) from a given signal will find it of inestimable value to know whether a PM- or FM-style implementation was actually used in the transmission.

Example (Angle Modulation). Much of this can be clarified by looking at a typical angle modulated signal. Consider a quadratic phase offset of the form

$$\phi_c(t) = \Omega t^2. \tag{6.124}$$

We illustrate the resulting angle modulated waveform (6.120) in Figure 6.7

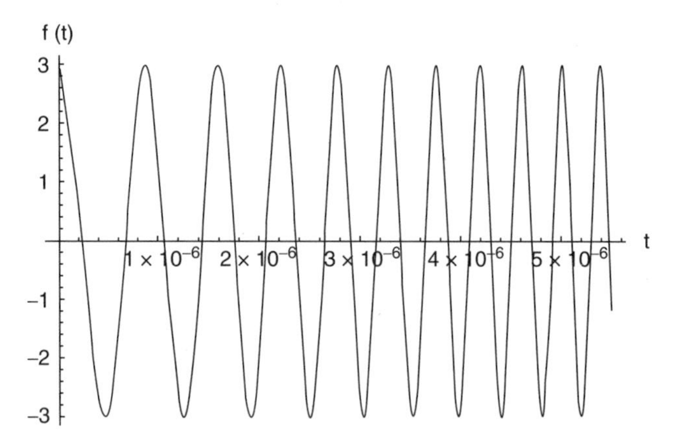

Fig. 6.7. Angle modulation. The illustrated waveform has a carrier frequency $f_c = 1$ MHz, amplitude $A_c = 3$, and $\Omega = 10^{12}$. The chirp induced by time-varying frequency is clearly in evidence.

Note the constant envelope (equal to A_c), which makes the class of *single-frequency* angle-modulated signals readily distinguishable from their amplitude-modulated counterparts (on the other hand, the superposition of multiple carriers can result in a time-varying envelope, as we will see). Furthermore, observe the apparent variation in frequency over time. This phenomenon, known as chirp, for its resemblance to the sound made by inhabitants of the avian world, is the most distinctive feature of angle modulated waveforms.

This motivates the following definition:

Definition (Instantaneous Frequency). Intuitively, the *instantaneous frequency* of a cosine-based angle modulated waveform (6.120) is defined

$$\omega(t) = \frac{d}{dt}[\omega_c t + \phi_c(t)] = \omega_c + \frac{d}{dt}\phi_c(t). \tag{6.125}$$

From this perspective, in which both carrier and time-varying phase effects are lumped into a general phase offset, the term *frequency* modulation makes sense. In the limit of vanishing or constant phase, the instantaneous frequency defaults to that of the carrier, as expected. According to (6.123), when employing an FM system, the baseband signal $m(t)$ is proportional to the second term in the instantaneous frequency defined in (6.125).

For the example in (6.124), the instantaneous frequency is a linear function of time, equal to $\omega_c + \Omega t$. This *linear chirp* is one of several common modulation schemes that involve higher-order polynomial or inverse powers of t and that are considered in the exercises.

More complicated signals, which may involve multiple frequency components, require the extra degree of freedom afforded by the complex exponential representation

$$f(t) = A(t)e^{j\Phi(t)}. \tag{6.126}$$

Taking the real part gives

$$f(t) = A(t)\cos\Phi(t), \tag{6.127}$$

leading to a general definition of instantaneous frequency:

$$\omega(t) \equiv \frac{d\Phi(t)}{dt}. \tag{6.128}$$

6.5.3.1 Multiple Frequencies. The time-varying amplitude $A(t)$ is a natural occurrence in signals that consist of multiple oscillations. For example, consider a simple composite signal consisting of two equal-amplitude pure oscillations represented by $\cos(\omega_1 t)$ and $\cos(\omega_2 t)$. Representing the composite as a sum of complex exponentials, it is easy to show that

$$f(t) = Ae^{j\omega_1 t} + Ae^{j\omega_2 t} = A\cos(\Delta t)e^{j\Sigma t}, \tag{6.129}$$

where Δ is the difference,

$$\Delta \equiv \frac{\omega_1 - \omega_2}{2}, \tag{6.130}$$

and the sum

$$\Sigma \equiv \frac{\omega_1 + \omega_2}{2} \tag{6.131}$$

is the average of the two pure oscillations. For this simple case, it is also the instantaneous frequency according to (6.128).

6.5.3.2 *Instantaneous Frequency versus the Fourier Spectrum.* The instantaneous frequency should not be confused with the Fourier spectrum. This is general, but we will illustrate the point by examining the instantaneous and Fourier spectra of FM signals with sinusoidal phase deviation,

$$\phi_c(t) = k \cdot \sin(\omega_m t). \tag{6.132}$$

This is a useful analysis, since arbitrary information $m(t)$ can be decomposed into a spectrum of sinusoids. The spread of instantaneous frequencies is quantified by the *frequency deviation* defined as the maximum difference between the carrier frequency and the instantaneous frequency (6.128),

$$\Delta\omega \equiv sup[|\omega_c - \omega(t)|] = k\omega_m. \tag{6.133}$$

The derivation of (6.133) is straightforward and left as an exercise. It is common to express the amplitude k as the ratio $\Delta\omega/\omega_m$ and call it the *modulation index*. Equation (6.133) implies that the range of instantaneous frequency occupies a range $\omega_c \pm \Delta\omega$ implying a nominal bandwidth of instantaneous frequencies $2\Delta\omega$. This is intuitively appealing since it is directly proportional to the amplitude and frequency of the phase deviation. In the limit that either of these vanish, the signal reverts to a pure carrier wave.

We now turn to the Fourier spectrum. It can be shown that

$$f(t) = A_c \cos(\omega_c t + k \cdot \sin(\omega_m t)) \tag{6.134}$$

can be elegantly expressed as a superposition a carrier wave and an infinite set of discrete oscillations in multiples (harmonics) of ω_m:

$$f(t) = J_0(k)\cos(\omega_c t) - \sigma_{\text{odd}} + \sigma_{\text{even}}, \tag{6.135}$$

where

$$\sigma_{\text{odd}} \equiv \sum_{n=1}^{\infty} J_{2n-1}(k)[\cos(\omega_c - (2n-1)\omega_m)t - \cos(\omega_c + (2n-1)\omega_m)t] \tag{6.136}$$

and

$$\sigma_{\text{even}} \equiv \sum_{n=1}^{\infty} J_{2n}(k)[\cos(\omega_c - 2n\omega_m)t - \cos(\omega_c + 2n\omega_m)t]. \tag{6.137}$$

In the exercises we lead the reader through the steps necessary to arrive at (6.135). The coefficients $J_p(k)$ are pth-order *Bessel functions of the first kind* and arise in a number of disciplines, notably the analysis of optical fibers, where good engineering treatments of the Bessel functions may be found [10]. Most scripted analysis tools make also them available as predefined functions. In general, they take the form of damped oscillations along the k axis [11]; by inspection of (6.135), they act as weights for the various discrete spectral components present in the signal. The carrier is weighted by the zeroth-order Bessel function, which is unity at the origin and, for k much smaller than unity, can be approximated by the polynomial,

$$J_0(k) \cong 1 - \frac{k^2}{4}. \tag{6.138}$$

The first *sideband* is represented by

$$J_1(k) \cong \frac{k}{2} - \frac{k^3}{16}. \tag{6.139}$$

The remaining sidebands, weighted by the higher-order Bessel functions for which $p > 1$, can be approximated as a single term,

$$J_p(k) \cong \frac{1}{p!}\left(\frac{k}{2}\right)^p \tag{6.140}$$

for small k. In the limit of zero phase deviation, the representation (6.135) reverts to a pure carrier wave $f(t) = \cos(\omega_c t)$, as expected. For k sufficiently small, but nonzero, the signal power consists mainly of the carrier wave and a single pair of sidebands oscillating at $\pm\omega_m$. Operation in this regime is termed *narrowband* FM.

In closing, we highlight the salient difference between Fourier spectral components and the instantaneous frequency: The spectrum of an FM signal with sinusoidal phase deviation is a superposition of Dirac impulses $\delta(\omega - (\omega_c \pm p\omega_m))$ for all integers p. On the other hand, the instantaneous frequency is continuous and oscillates through the range $\omega_c \pm \Delta\omega$.

The general case, in which information $m(t)$ is applied to the phase deviation, as given by (6.121), leads to analytically complex spectra that are beyond the scope of this discussion. But an appropriately behaved $m(t)$ can be naturally decomposed into pure Fourier oscillations similar to the sinusoids. So the simple case presented here is a foundation for the more general problem.

6.6 SUMMARY

This chapter has provided tools for studying the frequency content of important signals—sinusoids, constants, the unit step, and the like—for which the standard

Fourier analysis in Chapter 5 does not work. This generalized Fourier transform rests on the theory of distributions, which we covered in Chapter 3. The properties of the new transform are direct generalizations of those for the standard Fourier transform. An inverse transform exists as well. Further more, the generalized transform is equal to the standard transform for signals in the spaces $L^1(\mathbb{R})$ and $L^2(\mathbb{R})$.

The transform theory for generalized functions draws the links between the Fourier series and transform.

We applied the generalized transform to the study of communication systems. Understanding modulation schemes, for example, depends on spectral analysis of sinusoidal signals, and for this purpose the generalized transform makes the calculations particularly simple. We also considered the design of basic frequency selective linear, translation-invariant systems—filters. Chapter 9 will delve deeper into the theoretical and practical aspects of filter design using traditional Fourier analysis techniques.

Our next step is to develop the frequency theory task for the realm of discrete signals. Now, as we observed in the interplay between the ideas in Chapters 2 and 3, it is easier to justify a discrete summation (even if it has infinite limits) than an integration. Therefore, Chapter 7's mathematical work turns out to be much more concrete. With a discrete signal Fourier theory, computer implementations of frequency domain signal analysis becomes possible. We shall also build a link between analog and discrete signals through the famous Sampling Theorem, so our continuous domain results will appear once again.

REFERENCES

1. A. H. Zemanian, *Distribution Theory and Transform Analysis*, New York: Dover, 1987.
2. M. J. Lighthill, *Fourier Analysis and Generalized Functions*, New York: Cambridge University Press, 1958.
3. I. Halperin, *Introduction to the Theory of Distributions*, Toronto: University of Toronto Press, 1952.
4. R. Beals, *Advanced Mathematical Analysis*, New York: Springer-Verlag, 1987.
5. E. M. Stein and G. Weiss, *Introduction to Fourier Analysis on Euclidean Spaces*, Princeton, NJ: Princeton University Press, 1971.
6. L. Debnath and P. Mikusinski, *Introduction to Hilbert Spaces with Applications*, 2nd ed., San Diego, CA: Academic Press, 1999.
7. A. B. Carlson, *Communication Systems*, 3rd ed., New York: McGraw-Hill, 1986.
8. L. W. Couch III, *Digital and Analog Communication Systems*, 4th ed., Upper Saddle River, NJ: Prentice-Hall, 1993.
9. S. Haykin, *Communication Systems*, 3rd ed., New York: Wiley, 1994.
10. A. W. Snyder and J. D. Love, *Optical Waveguide Theory*, London: Chapman and Hall, 1983.
11. T. Okoshi, *Optical Fibers*, New York: Academic Press, 1982.
12. I. J. Schoenberg, Contribution to the problem of approximation of equidistant data by analytic functions, *Quarterly of Applied Mathematics*, vol. 4, pp. 45–99, 112–141, 1946.

13. M. Unser, Splines: A perfect fit for signal and image processing, *IEEE Signal Processing Magazine*, vol. 16, no. 6, pp. 22–38, November 1999.

PROBLEMS

1. Assume that $f(t)$ is a distribution of slow growth and prove the Fourier transform properties listed in Chapter 5.

2. Show that

(a)

$$F\left[\frac{d}{dt}\delta(t)\right](\omega) = j\omega, \tag{6.141}$$

(b)

$$F[t](\omega) = j2\pi \cdot \frac{d}{d\omega}(\delta(\omega)). \tag{6.142}$$

3. Show that

(a)

$$F[\cos(\omega_0 t)u(t)](\omega) = \frac{\pi}{2} \cdot [\delta(\omega - \omega_0) + \delta(\omega + \omega_0)] + j \cdot \frac{\omega_0}{(\omega_0^2 - \omega^2)}, \tag{6.143}$$

(b)

$$F[\sin(\omega_0 t)u(t)](\omega) = -j\frac{\pi}{2} \cdot [\delta(\omega - \omega_0) - \delta(\omega + \omega_0)] + \frac{\omega_0}{(\omega_0^2 - \omega^2)}. \tag{6.144}$$

4. Demonstrate the following generalized Fourier transforms, where $u(t)$ is the unit step.

(a)

$$F[u(t)](\omega) = \pi\delta(\omega) + \frac{1}{j\omega}, \tag{6.145}$$

(b)

$$F[(-jt)^n f(t)](\omega) = \frac{d^n}{dt^n}F(\omega), \tag{6.146}$$

(c)

$$F[t^n](\omega) = 2\pi j^n \cdot \frac{d^n\delta(\omega)}{d\omega^n}, \tag{6.147}$$

(d)

$$F[t^n u(t)](\omega) = j^n \cdot \left[\pi\frac{d^n\delta(\omega)}{d\omega^n} + \frac{1}{j}\frac{(-1)^n n!}{\omega^{n+1}}\right], \tag{6.148}$$

(e)

$$F[t^n \operatorname{sgn}(t)](\omega) = (-2)j^{n+1}\frac{(-1)^n n!}{\omega^{n+1}}. \quad (6.149)$$

5. A cosine-modulated signal $s(t) = m(t)\cos(\omega_c t)$ is recovered by multiplying it by $\cos(\omega_c t + \theta)$, where θ is a constant. (a) If this product is subjected to a low-pass filter designed to reject the contribution at $2\omega_c$, write the expression for the resulting waveform. (b) If the baseband signal $m(t)$ occupies a range $f \in [300, 3000]$ Hz, what is the minimum value of the carrier frequency ω_c for which it is possible to recover $m(t)$ according to the scheme in part a)? (c) What is the maximum value of the phase θ if the recovered signal is to be 95% of the maximum possible value?

6. The AM cosine-modulated signal $s(t) = m(t)\cos(\omega_c t)$ is recovered by multiplying by a periodic signal $\rho(t)$ with period k/f_c, where k is an integer. (a) Show that $m(t)$ may be recovered by suitably filtering the product $s(t)\rho(t)$. (b) What is the largest value of k for which it is possible to recover $m(t)$ if the baseband signal $m(t)$ occupies a range $f \in [0, 9000]$ Hz and the carrier frequency f_c is 1 MHz?

7. Consider two signals with *quadratic chirp*:

$$f_1(t) = a_1\cos(bt^2 + ct), \quad (6.150)$$

$$f_2(t) = a_2\cos(bt^2). \quad (6.151)$$

(a) Derive expressions for the instantaneous frequency of each. Comparing these, what purpose is served by the constant c?

(b) For convenience let $a_1 = b = c = 1$ and plot $f_1(t)$ over a reasonable interval (say, 10 to 30 s). Qualitatively, how is this plot consistent with the instantaneous frequency derived in part (a)?

(c) Let $a_2 = 1$ and plot the composite signal $f(t) = f_1(t) + f_2(t)$ over a 30-s interval. Are the effects of the instantaneous frequencies from part (a) still evident? Compared to the single waveform of part (b), what is the most noteworthy feature induced by superposing $f_1(t)$ and $f_2(t)$?

8. A signal exhibits *hyperbolic chirp*:

$$f(t) = a\cos\left(\frac{\alpha}{\beta - t}\right). \quad (6.152)$$

(a) Let $\alpha_1 = 1000$ and $\beta_1 = 900$, and plot $f(t)$ over a sufficiently large interval, say $t \in [0,3500]$.

(b) What qualitative effects are controlled by the parameters α and β? Let $\alpha_1 = 500$ and $\beta_1 = 740$ and replot $f(t)$.

(c) Derive an expression for the instantaneous frequency of $f(t)$ and plot your result using the signal parameters in part (a).

9. Consider an FM signal modulated by multiple cosines,

$$f(t) = \cos\left[\omega_c t + \sum_{n=1}^{N} k_n \cdot \cos(n \cdot \omega_m t)\right]. \tag{6.153}$$

For $N = 1, 2$, derive expressions for

(a) The instantaneous frequency of $f(t)$.

(b) The frequency deviation.

10. Derive the Bessel function expansion for the sinusoidally modulated FM signal, (6.135). *Hint*: Make use of the identities

$$\cos(k \cdot \sin\omega_m t) = J_0(k) + 2\sum_{n=1}^{\infty} J_{2n}(k)\cos(2n \cdot \omega_m t), \tag{6.154}$$

$$\sin(k \cdot \sin\omega_m t) = 2\sum_{n=1}^{\infty} J_{2n-1}(k)\sin((2n-1) \cdot \omega_m t) \tag{6.155}$$

and the trigonometric relations

$$\cos x \cdot \cos y = \frac{1}{2}\cos(x-y) + \frac{1}{2}\cos(x+y), \tag{6.156}$$

$$\sin x \cdot \sin y = \frac{1}{2}\cos(x-y) - \frac{1}{2}\cos(x+y). \tag{6.157}$$

11. A carrier wave is angle modulated sinusoidally.

(a) In principle, how many values of the modulation index k result in a completely suppressed (i.e., zero) carrier wave? List the first five of these values.

(b) On the basis of your answer in (a), is the zeroth-order Bessel function perfectly periodic?

(c) Let $k = 0.1$ and plot the ratio $J_p(k)/J_0(k)$ as a function of the order p, for $p \in [0, 10]$. Qualitatively, what is the effect of increasing p? (Of course, for our purposes, only integer values of p are relevant.)

(d) For $k \in [0, 0.1]$, plot the ratios $J_1(k)/J_0(k)$ and $J_2(k)/J_0(k)$. What is happening to the amplitudes of the sidebands relative to the carrier as k is increased? Is this true for the remaining sidebands as well? Outside of the narrowband FM operating regime, can such behavior be expected for arbitrary k?

12. Consider a unit-amplitude FM carrier signal which is modulated by two sinusoids,

$$f(t) = \cos(\omega_c t + k_1 \cdot \sin\omega_1 t + k_2 \cdot \sin\omega_2 t). \tag{6.158}$$

(a) Demonstrate the existence of sidebands at harmonics of the sum and difference frequencies $\omega_1 + \omega_2$ and $\omega_1 - \omega_2$ as well as at harmonics of ω_1 and ω_2.

(b) Show that in the limit of small k_1 and k_2 we may approximate $f(t)$ as a linear superposition of cosine and sine carrier waves,

$$f(t) \approx \cos\omega_c t - (k_1 \cdot \sin\omega_1 t + k_2 \cdot \sin\omega_2 t)\sin\omega_c t \tag{6.159}$$

Hint: Apply the approximations, valid for small x:

$$\cos x \cong 1\,, \tag{6.160}$$

$$\sin x \cong x\,. \tag{6.161}$$

13. As an application of Fourier transforms and their generalized extensions, this problem develops part of Schoenberg's Theorem on spline functions [12, 13]. We stated the theorem in Chapter 3: If $x(t)$ is a spline function of degree n having integral knots $K = \{m = k_m : m \in \mathbb{Z}\}$, then there are constants c_m such that

$$s(t) = \sum_{m=-\infty}^{\infty} c_m \beta_n(t-m)\,. \tag{6.162}$$

In (6.162) the B-spline of order zero is

$$\beta_0(t) = \begin{cases} 1 & \text{if } -\frac{1}{2} < t < \frac{1}{2} \\ \frac{1}{2} & \text{if } |t| = \frac{1}{2} \\ 0 & \text{if otherwise} \end{cases} \tag{6.163}$$

and higher-order B-splines are defined as

$$\beta_n(t) = \underbrace{\beta_0(t)*\beta_0(t)*\ldots\beta_0(t)}_{n+1 \text{ times}}\,. \tag{6.164}$$

(a) Explain why $\beta_n(t)$ has a Fourier transform.

(b) Let $B_n(\omega) = \mathcal{F}(\beta_n)(\omega)$ be the Fourier transform of $\beta_n(t)$. Following Ref. 13, show that

$$B_n(\omega) = \left[\frac{\sin\left(\frac{\omega}{2}\right)}{\frac{\omega}{2}}\right]^{n+1} = \left[\frac{e^{j\omega/2} - e^{-j\omega/2}}{j\omega}\right]^{n+1}. \tag{6.165}$$

(c) Let $y_n(t) = u(t)t^n$ be the one-sided polynomial of degree n. Show that

$$\frac{d^{n+1}}{dt^{n+1}} y_n(t) = n!\delta(t), \tag{6.166}$$

where $\delta(t)$ is the Dirac delta.

(d) Conclude that $Y_n(\omega) = n!/(j\omega)^{n+1}$.

(e) Next, show that

$$B_n(\omega) = \frac{Y_n(\omega)[e^{j\omega/2} - e^{-j\omega/2}]^{n+1}}{n!}. \tag{6.167}$$

(f) Use the binomial expansion from high-school algebra to get

$$B_n(\omega) = \frac{1}{n!} \sum_{k=0}^{n+1} \binom{n+1}{k} (-1)^k e^{-j\omega\left(k - \frac{n+1}{2}\right)} Y(\omega). \tag{6.168}$$

(g) Infer that

$$\beta_n(t) = \frac{1}{n!} \sum_{k=0}^{n+1} \binom{n+1}{k} (-1)^k y\left(t - k + \frac{n+1}{2}\right) \tag{6.169}$$

and that $\beta_n(t)$ is a piecewise polynomial of degree n.

(h) Show that $\frac{d^{n+1}}{dt^{n+1}} \beta_n(t) = n!\delta(t)$ is a sum of shifted Diracs.

(i) Show that an nth-order spline function on uniform knots is a sum of scaled, shifted B-splines.

CHAPTER 7

Discrete Fourier Transforms

We have already discovered a rich theory of frequency transforms for analog signals, and this chapter extends the theory to discrete signals. In fact, the discrete world presents fewer mathematical subtleties. Several reasons compel us to cover the difficult theory first. It was historically prior, for one thing. Fourier developed his techniques for the practical solution of the heat equation long before engineers worried about pen-and-pencil computations for the frequency content of a sampled signal. Beginning with the treatment of analog frequency transforms furnishes—especially in the case of the classic Fourier series—a very clear foundation for comprehending the idea of the frequency content of a signal. A general periodic signal becomes a sum of familiar sinusoids, each with its well-known frequency value. Finally, it is easy to relate the discrete theory to analog notions and thereby justify the claims that such and such a value does represent the discrete signal spectral content at some frequency.

We begin frequency transforms for discrete signals by covering the discrete Fourier transform (DFT), which Chapters 2 and 4 have already introduced. Chapter 2 covered the DFT only very briefly, in the context of providing an example of an orthonormal subset of the discrete Hilbert space l^2. In studying the analysis of signal textures in Chapter 4, our statistical methods proved inadequate for characterizing certain periodic trends within signals. An example is separating the fine-scale roughness from the coarse-scale waviness of a signal. Although statistical techniques tend to obscure the repetitive appearance of signal features, there is no such problem with spectral texture measures. We found that the magnitude of the inner product of a discrete signal with an exponential provided a translation invariant measure of the presence of a discrete frequency component. Thus, two very different motivations already exist for our pursuit of discrete frequency theory, and the DFT in particular: The complex exponentials upon which it is based are orthogonal on finite intervals $[0, N-1] \subset \mathbb{Z}$, and it is very useful for signal texture analysis.

The discrete world's DFT is analogous to the analog Fourier series. It works with discrete periodic signals. The discovery of a fast algorithm for computing the DFT, called the fast Fourier transform (FFT), coincided with the phenomenal development of computing technology in the middle part of the twentieth century. The

Signal Analysis: Time, Frequency, Scale, and Structure, by Ronald L. Allen and Duncan W. Mills
ISBN: 0-471-23441-9

FFT completely changed the nature of the signal processing discipline and the work habits—indeed the consciousness—of signal analysts [1]. By reducing the complexity of the computation from an $O(N^2)$ algorithm to an $O(N\log_2 N)$ problem, the FFT makes real-time frequency analysis of signals practical on digital computers.

Another discrete transform must be used when studying the frequencies within discrete aperiodic signals. It has a terrible name: the discrete-time Fourier transform (DTFT). Like the DFT, it extracts the frequency content of discrete signals. But unlike the DFT, the DTFT outputs a continuous range of frequencies from an aperiodic input signal. So similar are the acronyms that it is easily confused with the DFT, and, while its appellation boasts "discrete time," this is only a half-truth, because it in fact produces an analog result. Nevertheless, the awful moniker has stuck. We live with it. We repeat it. The DTFT is the discrete world's mirror image of the Fourier transform.

The next chapter covers a generalization of the DTFT, called the z-transform. It has applications in the stability analysis of discrete systems, solutions for discrete-time difference equations, and subsampling and upsampling operations.

This chapter's last sections develop the principles underlying the famous sampling theorem of Shannon[1] and Nyquist.[2] This result effectively builds a frequency analysis bridge between the world of analog signals and the world of discrete signals [2, 3].

7.1 DISCRETE FOURIER TRANSFORM

We have already made acquaintance with the discrete Fourier transform in Chapters 2 and 4. Now we seek a more formal foundation for the theory of the frequency content of discrete signals. All signal frequency analysis applications that rely on digital computers use the DFT, so we will have regular opportunities to refer back to this groundwork and even extend it in the later chapters of this book.

We first took note of the DFT in Chapter 2. The discrete complex exponentials, restricted to a finite interval, are an orthogonal set within the space of square-summable signals, l^2. Thus, if we consider the subspace of l^2 consisting of signals that are zero outside $[0, N-1]$, then the signals

$$u_k(n) = e^{\frac{-j2\pi kn}{N}}[u(n) - u(n-N)], \tag{7.1}$$

[1]Claude E. Shannon (1916–2001) first detailed the affinity between Boolean logic and electrical circuits in his 1937 Master's thesis at the Massachusetts Institute of Technology. Later, at Bell Laboratories, he developed much of theory of reliable communication, of which the sampling theorem is a cornerstone.

[2]Harry Nyquist (1887–1976) left Sweden at age 18 for the United States. As a Bell Laboratories scientist, he provided a mathematical explanation for thermal noise in an electrical resistance, discovered the relation between analog signal frequency and digital sampling rate that now bears his name, and acquired 138 patents.

form an orthogonal set on $[0, N-1]$. We can normalize the family $\{u_k(n) \mid k = 0, 1, 2, ..., N-1\}$ by dividing each signal in (7.1) by $N^{1/2}$. We don't need to consider $k > N-1$, because these signals repeat themselves; this is due to the 2π-periodicity of the exponentials: $\exp(-j2\pi kn/N) = \exp[-j2\pi(k+N)/N]$.

Chapter 4 further acquainted us to the discrete Fourier transform through our study of signal texture. In particular, signals may have different periodic components within them: short-term variations, called roughness (in the parlance of surface metrology), and the long-term variations, called waviness. One way to distinguish and measure the two degrees of repetitiveness is to take the inner product over $[0, N-1]$ of $x(n)$ with exponentials of the form (7.1). Any relatively large magnitude of the resulting inner product, $\langle x(n), \exp(2\pi nk/N)\rangle$ on $[0, N-1]$ indicates a correspondingly large presence of a periodic component of discrete frequency $\omega = 2\pi k/N$. This idea could be augmented by performing a normalized cross-correlation of $x(n)$ with prototype signals $\exp(2\pi nk/N)$ as in Section 4.6.1.

7.1.1 Introduction

Our interest in the discrete Fourier transform is twofold. From the viewpoint of Hilbert spaces—where it furnishes a particularly elegant example of an orthogonal basis—we have a theoretical interest in exploring the discrete Fourier transform. From texture interpretation—the rich, seemingly endless field from which so many research endeavors arise—we acquire a practical interest in better understanding and applying the DFT. Let us then formally define the DFT, prove that it forms a complete representation of discrete periodic signals, and consider some examples.

Definition (Discrete Fourier Transform). Suppose $x(n)$ is a discrete signal and $N > 0$. Then the discrete signal $X(k)$ defined by

$$X(k) = \sum_{n=0}^{N-1} x(n)\exp\left(\frac{-2\pi jnk}{N}\right), \tag{7.2}$$

where $0 \le k \le N-1$, is the *discrete Fourier transform* of $x(n)$ on the interval $[0, N-1]$. Equation (7.2) is called the *analysis equation* for the DFT. In general, $X(k)$ is complex; the complex norm, $|X(k)|$, and complex phase, $\arg(X(k))$, for $0 \le k < N$, are called the *discrete magnitude spectrum* and *discrete phase spectrum,* respectively, of $x(n)$.

DFT conventions vary widely. A popular notation is to use lowercase Latin letters for the time-domain discrete signal, $s(n)$, and the corresponding uppercase letter for the DFT, $S(k)$. Some authors like to use the hat mark: The DFT of $x(n)$ is $X(k)$ Also, the systems theory notation, $S = \mathcal{F}s$, is handy; $\mathcal{F}$ is the discrete system that accepts a signal $s(n)$ with period N and produces its DFT, $S(k)$. Note that the definition of the system $\mathcal{F}$ in this case depends on N. Different values for the period of the discrete input signals produce different discrete systems. We must register yet another warning about the varieties of DFT definitions in the literature. Equation

(7.2) is the most common definition of the DFT in the engineering research literature. There is also a discrete Fourier series (DFS) which multiplies each $X(k)$ in (7.2) by N^{-1}. We introduce the DFS in Section 7.1.2, and there it becomes clear that its particular formulation helps draw a link between analog and discrete frequency transforms. Scientists often define the DFT with a positive exponent. Some authors prefer to normalize the DFT coefficients by a factor of $N^{-1/2}$. Finally, when using mathematical software packages, one must always be alert to the possibility that the package indexes arrays beginning with one, not zero. In such a situation the DC term of the DFT is associated with $k = 1$, the smallest periodic component with $k = 2$ and $k = N$, and so on.

Example (Discrete Delta Signal). Consider the signal $x(n) = [\underline{1}, 0, 0, 0, 0, 0, 0, 0]$ on the interval [0, 7]. We have $X(k) = 1$, for $0 \le k < 8$. So the computation of the delta signal's transform is quite uncomplicated, unlike the case of the analog Fourier series.

Example (Discrete Square Pulse). Consider the signal $x(n) = [1, \underline{1}, 1, 1, 0, 0, 0, 0]$. Its DFT is $X(k) = [\underline{4}, 1-(1 + \sqrt{2})j, 0, 1 - (\sqrt{2} - 1)j, 0, 1 + (\sqrt{2} - 1)j, 0, 1 + (1 + \sqrt{2})j]$. Note the symmetry: $X(k)$ and $X(8 - k)$ are complex conjugates for $0 < k \le 7$. If we shift the square pulse so that $y(n) = [\underline{0}, 0, 1, 1, 1, 1, 0, 0]$, then $Y(k) = [\underline{4}, -(1 + \sqrt{2})-j, 0, (\sqrt{2} - 1) + j, 0, (\sqrt{2} - 1) - j, 0, -(1 + \sqrt{2}) + j]$. Although the time-domain pulse has translated, the zeros of the frequency-domain coefficients are in the same locations for both $X(k)$ and $Y(k)$. Indeed, there is a time shift property for the DFT, just as we found for the analog Fourier transform and Fourier series. Since $x(n) = y(n - m)$, where $m = 2$, we have $X(k) = Y(k)\exp(-2\pi jkm/8)$. The DFT can be visualized by plotting the coefficients as points in the complex plane, or as separate plots of the magnitude and the phase (Figure 7.1). Since the magnitude of the DFT coefficients do not change with translation of the time-domain signal, it is most convenient to plot the magnitude or the energy components of $X(k)$—$|X(k)|$ or $|X(k)|^2$, respectively.

Example (Ramp Pulse). Consider the signal $x(n) = [\underline{1}, 2, 3, 0, 0, 0]$ on [0, 5], with period $N = 6$. We find $X(k) = 1 + 2\exp(-\pi jk/3) + 3\exp(-\pi jk)$.

7.1.1.1 Inversion Formula. There is an inversion theorem for the DFT, and, as we found in studying analog transforms, it is the key to understanding the basic properties.

Theorem (DFT Inverse). Suppose $x(n)$ is a discrete signal and $X(k)$ is the DFT of $x(n)$ on $[0, N - 1]$. Then

$$x(n) = \frac{1}{N} \sum_{k=0}^{N-1} X(k) \exp\left(\frac{2\pi jnk}{N}\right). \tag{7.3}$$

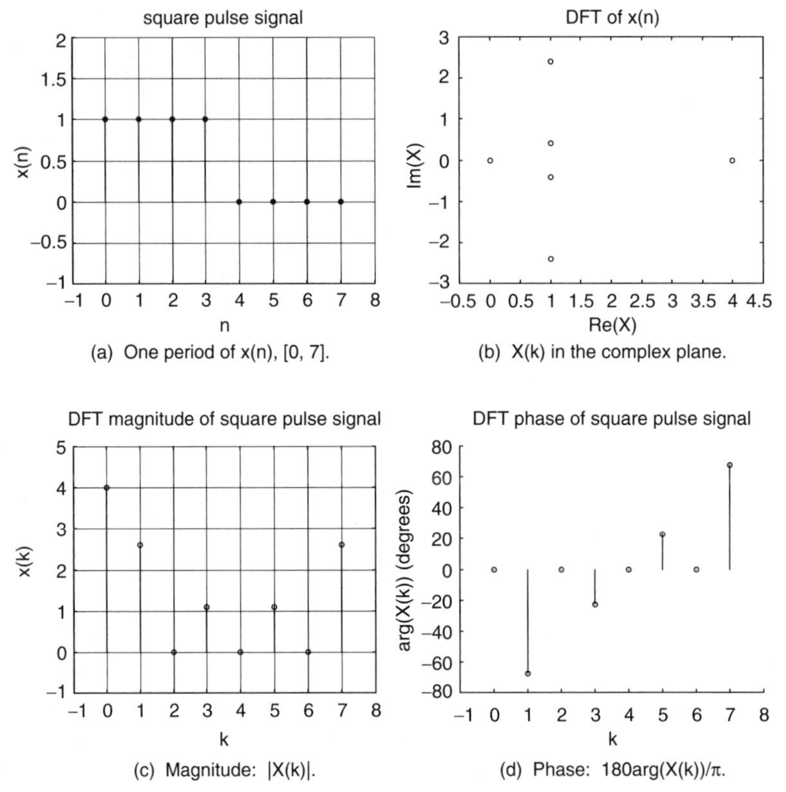

(a) One period of x(n), [0, 7].

(b) X(k) in the complex plane.

(c) Magnitude: |X(k)|.

(d) Phase: 180arg(X(k))/π.

Fig. 7.1. Some examples of the DFT computation. Panel (a) shows a single period of a simple square wave, $x(n)$. In panel (b), the DFT of $x(n)$, $X(k)$ is shown; note that there are only six distinct values, so the circled point at the origin represents three values: $X(2) = X(4) = X(6) = 0$. More revealing is the magnitude of the DFT, $|X(k)|$, shown in panel (c). The phase of $X(k)$ is shown in panel (d); note the linear progression of the phases of the nonzero $X(k)$ values. This clue indicates that DFT coefficients may be visualized as magnitudes associated with points on the unit circle of the complex plane.

Proof: We substitute the expression $X(k)$, given by (7.2) into the right-hand side of (7.3) and work through the exponential function algebra. This brute force attack gives

$$
\begin{aligned}
\frac{1}{N}\sum_{k=0}^{N-1} X(k)\exp\left(\frac{2\pi jnk}{N}\right) &= \frac{1}{N}\sum_{k=0}^{N-1}\left(\sum_{r=0}^{N-1} x(r)\exp\frac{-2\pi jrk}{N}\right)\exp\left(\frac{2\pi jnk}{N}\right) \\
&= \frac{1}{N}\sum_{r=0}^{N-1}\sum_{k=0}^{N-1} x(r)\exp\left(\frac{-2\pi jrk}{N}\right)\exp\left(\frac{2\pi jnk}{N}\right) \\
&= \sum_{r=0}^{N-1} x(r)\left[\frac{1}{N}\sum_{k=0}^{N-1}\exp\left(\frac{2\pi j(n-r)k}{N}\right)\right],
\end{aligned} \tag{7.4}
$$

and it appears that we have quite a mess on our hands! However, the following lemma shows that the bracketed expression in (7.4) has a gratifying, simple form.

Lemma. Let $N > 0$, and let p and k be integers. Then, if for some $m \in \mathbb{Z}$, we have $p = mN$, then

$$\frac{1}{N}\sum_{k=0}^{N-1} \exp\left(\frac{2\pi jpk}{N}\right) = 1; \tag{7.5a}$$

otherwise

$$\frac{1}{N}\sum_{k=0}^{N-1} \exp\left(\frac{2\pi jpk}{N}\right) = 0. \tag{7.5b}$$

Proof of lemma: Let $a = \exp(2\pi jp/N)$. Then, expanding the summation, for instance in (7.5a), gives

$$\frac{1}{N}\sum_{k=0}^{N-1} \exp\left(\frac{2\pi jpk}{N}\right) = \frac{1}{N}\left(1 + a + a^2 + \cdots + a^{N-1}\right). \tag{7.6}$$

But if $p = mN$, then $a = \exp(2\pi jmN/N) = \exp(2\pi jm) = 1$. In this case, the right-hand side of (7.6) is unity. If $p/N \notin \mathbb{Z}$, then $a \neq 1$ and $1 - a \neq 0$. This implies $1 + a + a^2 + \cdots + a^{N-1} = (1 - a^N)/(1 - a)$. However, $aN = 0$, and in this case the right-hand side of (7.6) is zero, proving the lemma. ∎

Let us then continue proving the theorem. The term $(n - r)$ in the bracketed expression in (7.4) is either an integral multiple of N or it is not. Suppose $p = (n - r) = mN$ for some $m \in \mathbb{Z}$; by the lemma, therefore, the bracketed expression is unity. But since $0 \leq n, r \leq N - 1$, we know that $p = n - r$ is a multiple of N only if $n - r = 0$, that is, $n = r$. So the bracketed sum in (7.4) is zero unless $n = r$:

$$\frac{1}{N}\sum_{k=0}^{N-1} \exp\left(\frac{2\pi j(n-r)k}{N}\right) = \begin{cases} 0 & n \neq r, \\ 1 & \text{otherwise.} \end{cases} \tag{7.7}$$

Thus, the whole expression in (7.4) reduces to $x(n)$, and the proof of the theorem is complete. ∎

Definition (Inverse DFT). Equation (7.3) is called the *synthesis equation* for the DFT. The expression (7.3) is also called the inverse discrete Fourier transform (IDFT).

The first term, $X(0)$, is often called the DC (direct current) component of the DFT for $x(n)$, since it contains no periodic (i.e., alternating current) component. This whispers of Fourier analysis's electrical engineering heritage, although nowadays everyone—engineers, scientists, even university business school professors running trend analyses—uses the term. Note that when $x(n)$ is reconstructed from its DFT, the synthesis equation summation begins with the factor $X(0)/N$, which is the average value of $x(n)$ on the interval $[0, N-1]$.

Corollary (Discrete Fourier Basis). Let K be the set of discrete signals supported on the interval $[0, N-1]$, and let $u(n)$ be the unit step signal. Then K is a Hilbert subspace of l^2, and the windowed exponential signals, $\{u_k(n) \mid 0 \le k \le N-1\}$, where

$$u_k(n) = \frac{\exp(2\pi jkn/N)}{\sqrt{N}}[u(n) - u(n-N)], \tag{7.8}$$

form an orthonormal basis for K.

Proof: Recall from Section 2.7.1 that a Hilbert subspace is an inner product subspace that is complete in the sense that every Cauchy sequence of elements converges to a subspace element. This is easy to show for K. So let us consider the windowed exponentials (7.8). Note that if $0 \le k, 1 \le N-1$, then

$$\langle u_k, u_l \rangle = \sum_{n=0}^{N-1} \frac{\exp(j2\pi kn/N)}{\sqrt{N}} \frac{\overline{\exp(j2\pi ln/N)}}{\sqrt{N}} = \frac{1}{N}\sum_{n=0}^{N-1} \exp[j2\pi(k-l)n/N]. \tag{7.9}$$

The theorem's lemma shows that the sum on the right-hand side of (7.9) is zero unless $k = 1$, in which case it is unity. But this is precisely the orthonormality condition. We must show too that the orthonormal set is complete; that is, we need to show the other sense of completeness, which specifies that every element of the subspace K is arbitrarily close to a linear combination of elements of $\{u_k(n) \mid 0 \le k \le N-1\}$. Let $X(k)$ be given by (7.2). For $0 \le n \le N-1$, the theorem implies

$$x(n) = \frac{1}{N}\sum_{k=0}^{N-1} X(k)\exp\left(\frac{2\pi jnk}{N}\right) = \frac{1}{\sqrt{N}}\sum_{k=0}^{N-1} X(k)u_k(n). \tag{7.10}$$

This shows that $x(n)$ is a linear combination of the $\{u_k(n)\}$. ■

Corollary (DFT for Discrete Periodic Signals). Suppose $x(n)$ is a discrete signal with period $N > 0$: $x(n) = x(n+N)$ for all n. Then,

$$x(n) = \frac{1}{N}\sum_{k=0}^{N-1} X(k)\exp\left(\frac{2\pi jnk}{N}\right). \tag{7.11}$$

for all n.

Proof: Note that on the finite interval $[0, N-1]$, the theorem implies (7.11). But $x(n) = x(n + N)$, and the right-hand side of (7.11) is also periodic with period N, so the corollary holds. ■

Corollary (DFT Uniqueness). Suppose $x(n)$ and $y(n)$ are discrete signals, and $X(k)$ and $Y(k)$ are their respective DFTs on $[0, N-1]$. If $X = Y$ on $[0, N-1]$, then $x(n) = y(n)$ on $[0, N-1]$.

Proof: If $X(k) = Y(k)$ for all $0 \le k \le N-1$, then both $x(n)$ and $y(n)$ are given by the same inversion formula (7.3). So $x(n) = y(n)$ for all $0 \le n \le N-1$. ■

So the transform's uniqueness on an interval follows from the inversion equation. We followed a similar roundabout route toward showing transform uniqueness with the analog Fourier series and Fourier transform. In the analog world, the integration of signals, possibly containing discontinuities, complicates the uniqueness propositions, of course. Supposing Riemann integration, we can claim uniqueness only up to a finite number of impulse discontinuities. And, allowing the more robust Lebesgue integration, we can only claim that signals that are equal, except perhaps on a set of measure zero, have identical transforms. However, with discrete signals, owing to the finite sums used in computing the DFT, the transform values are truly unique.

Corollary (DFT Uniqueness). Suppose $x(n)$ and $y(n)$ discrete signals, both with period equal to N. If their DFTs are equal, $X(k) = Y(k)$ for all $0 \le k \le N-1$, then $x(n) = y(n)$ for all n.

Proof: Combine the proofs of the previous two corollaries. ■

If $X(k) = Y(k)$ for all $0 \le k \le N-1$, then both $x(n)$ and $y(n)$ are given by the same inversion formula (7.3). So $x(n) = y(n)$ for all $0 \le n \le N-1$.

The DFT Uniqueness Corollary encourages us to characterize the DFT as the appropriate transform for periodic discrete signals. If a signal $x(n)$ has period $N > 0$, then, indeed, the synthesis equation provides a complete breakdown of $x(n)$ in terms of a finite number of exponential components. But, the DFT is also applied to the restriction of aperiodic discrete signals to an interval, say $s(n)$ on $[a, b]$, with $b - a = N - 1$. In this case, the analysis equation (7.2) is used with $x(n) = s(a - n)$. Of course, the synthesis equation does not give $s(n)$; rather, it produces the periodic extension of $s(n)$ on $[a, b]$.

7.1.1.2 Further Examples and Some Useful Visualization Techniques. Therefore, let us explore a few examples of the DFT's analysis and synthesis equations before proceeding to link the DFT to the frequency analysis of analog signals. These examples show that it is quite straightforward to compute the DFT analysis equations coefficients $X(k)$ for a trigonometric signal $x(n)$. It is not necessary, for instance, to explicitly perform the sum of products in the DFT analysis equation (7.2).

Example (Discrete Sinusoids). Consider the discrete sinusoid $x(n) = \cos(\omega n)$. Note that $x(n)$ is periodic if and only if ω is a rational multiple of 2π: $\omega = 2\pi p/q$ for some $p, q \in \mathbb{Z}$. If $p = 1$ and $q = N$, then $\omega = 2\pi/N$, and $x(n)$ is periodic on $[0, N-1]$ with fundamental period N. Signals of the form $\cos(2\pi kn/N)$ are also periodic on $[0, N-1]$, but since $\cos(2\pi kn/N) = \cos(2\pi(N-k)n/N)$, they are different only for $k = 1, 2, ..., \lfloor N/2 \rfloor$. We recall these facts from the first chapter and note that a like result holds for discrete signals $y(n) = \sin(\omega n)$, except that $\sin(2\pi kn/N) = -\sin(2\pi(N-k)n/N)$. We can explicitly write out the DFT synthesis equations for the discrete sinusoids by observing that

$$\cos\left(\frac{2\pi nk}{N}\right) = \frac{1}{2}\exp\left(\frac{2\pi jnk}{N}\right) + \frac{1}{2}\exp\left(\frac{2\pi jn(N-k)}{N}\right), \tag{7.12a}$$

$$\sin\left(\frac{2\pi nk}{N}\right) = \frac{1}{2j}\exp\left(\frac{2\pi jnk}{N}\right) - \frac{1}{2j}\exp\left(\frac{2\pi jn(N-k)}{N}\right), \tag{7.12b}$$

for $0 \le k \le \lfloor N/2 \rfloor$. Equations (7.12a) and (7.12b) thereby imply that $X(k) = X(N-k) = N/2$ with $X(m) = 0$, for $0 \le m \le N-1$, $m \ne k$; and $Y(k) = -Y(N-k) = (-jN/2)$, with $Y(m) = 0$, for $0 \le m \le N-1$, $m \ne k$. The factor of N in the expressions for $X(k)$ and $Y(k)$ ensures that the DFT synthesis equation (7.3) holds for $x(n)$ and $y(n)$, respectively.

Example (Linear Combinations of Discrete Sinusoids). If we multiply a discrete sinusoid by a constant scale factor, $v(n) = Ax(n) = A\cos(\omega n)$, then the DFT coefficients for $v(n)$ are $V(k) = AX(k)$. This is a *scaling* property of the DFT. This is clear from the analysis (7.2) and synthesis (7.3) equations. Furthermore, if $v(n) = x(n) + y(n)$, then $V(k) = X(k) + Y(k)$, where $V(k)$, $X(k)$, and $Y(k)$ are the DFT coefficients of $v(n)$, $x(n)$, and $y(n)$, respectively. This is a *superposition* property of the DFT. Thus, it is easy to find the DFT synthesis equation for a linear combinations of discrete sinusoids from this linearity property and the previous example.

Example (Phase of Discrete Sinusoids). If the sinusoid $x(n) = \cos(\omega n)$ has period $N > 0$, then so does $y(n) = \cos(\omega n + \phi)$. Since $y(n) = [\exp(j\omega n + j\phi) + \exp(-j\omega n - j\phi)]/2 = [\exp(j\phi)\exp(j\omega n) + \exp(-j\phi)\exp(-j\omega n)]/2$, we can use the scaling and superposition properties of the previous example to find the DFT coefficients of $Y(k)$ in terms of $X(k)$. Notice also that the sinusoidal signal's phase shift, ϕ, does not change the complex magnitude of the DFT coefficients. This property we noted in our study of textured signal periodicity in Chapter 4.

A common and useful notation allows us to write the DFT is a more compact form.

Definition (Phase Factor). For $N > 0$, the Nth root of unity, $W_N = e^{-2\pi j/N}$, is called the *phase factor* of order N.

The fast Fourier transform algorithms in Section 7.1.4 exploit the symmetries of the phase factor that appear in the DFT analysis and synthesis equations.

Now, the powers of W_N, $(W_N)^k = e^{-2\pi jk/N}$, $0 \le k < N$, all lie on the unit circle in the complex plane: $(W_N)^k = e^{-2\pi jk/N} = \cos(2\pi k/N) - j\sin(2\pi k/N)$. Then we can rewrite the DFT analysis and synthesis equations as follows:

$$X(k) = \sum_{n=0}^{N-1} x(n) W_N^{nk}, \tag{7.13}$$

$$x(n) = \frac{1}{N} \sum_{k=0}^{N-1} X(k) W_N^{-nk}. \tag{7.14}$$

Thus, $X(k)$ is a polynomial of degree $N - 1$ in $(W_N)^k$, and $x(n)$ is a polynomial of degree $N - 1$ in $(W_N)^{-n}$. That is, if we define the complex polynomials, $P(z) = x(0) + x(1)z + x(2)z^2 + \cdots + x(N-1)z^{N-1}$ and $p(z) = (1/N)[X(0) + X(1)z + X(2)z^2 + \cdots + X(N-1)z^{N-1}]$, then $X(k) = P((W_N)^k)$, and $x(n) = p((W_N)^{-n})$. The DFT of $x(n)$ is just the complex polynomial $P(z)$ evaluated at specific points on the unit circle, namely $(W_N)^k = e^{-2\pi jk/N}$, $0 \le k < N$. Similarly, the IDFT is the complex polynomial $p(z)$ evaluated at points $(W_N)^{-n}$ on the unit circle, $0 \le n < N$. In fact, these are the same points, just iterated in the opposite order. We will extend this idea of representing the DFT as a complex polynomial restricted to a set of discrete points in the next chapter; in fact, the concept of the z-transform carries it to the extreme. For now we just want to clarify that the DFT coefficients can be thought of as functions of an integer variable k or of points on the unit circle $z = \cos(2\pi k/N) - j\sin(2\pi k/N)$. This idea enables us to better visualize some of the behavior of the transform for specific signals. For example, we may study the transforms of square pulse signals for various periods N and various duty cycles (percent of non zero coefficients) as in Figure 7.2.

For square pulse signals, such as in Figure 7.2, there is a closed-form expression for the DFT coefficients. Suppose $x(n)$ has period $N > 0$, and $x(n)$ is zero except for the first M values, $x(0) = x(1) = \cdots = x(M-1) = 1$, with $0 < M < N - 1$. Then $X(k)$ is a partial geometric series in $(W_N)^k$: $X(k) = 1 + (W_N)^k + (W_N)^{2k} + \cdots + (W_N)^{k(M-1)}$. Assuming that $(W_N)^k \ne 1$, we calculate

$$X(k) = \frac{1-\left(W_N^k\right)^M}{1-W_N^k} = \frac{W_N^{kM/2}\left(W_N^{-kM/2} - W_N^{kM/2}\right)}{W_N^{k/2}\left(W_N^{-k/2} - W_N^{k/2}\right)} = \frac{W_N^{kM/2}\left(e^{\pi jkM/N} - e^{-\pi jkM/N}\right)}{W_N^{k/2}\left(e^{\pi jk/N} - e^{-\pi jk/N}\right)}$$

$$= \frac{W_N^{kM/2}\left(2j\sin(\pi kM/N)\right)}{W_N^{k/2}\left(2j\sin(\pi k/N)\right)} = W_N^{k(M-1)/2}\frac{\left(\sin(\pi kM/N)\right)}{\left(\sin(\pi k/N)\right)} \tag{7.15}$$

Thus, for $k = 0$, $X(k) = M$, and for $0 < k < M - 1$, the DFT coefficient $X(k)$ is given by the product of a complex $2N$th root of unity and a ratio of sinusoids: $\sin(\pi kM/N)$ and $\sin(\pi k/N)$. Since $\sin(\pi kM/N)$ oscillates M times faster than $\sin(\pi k/N)$, there are

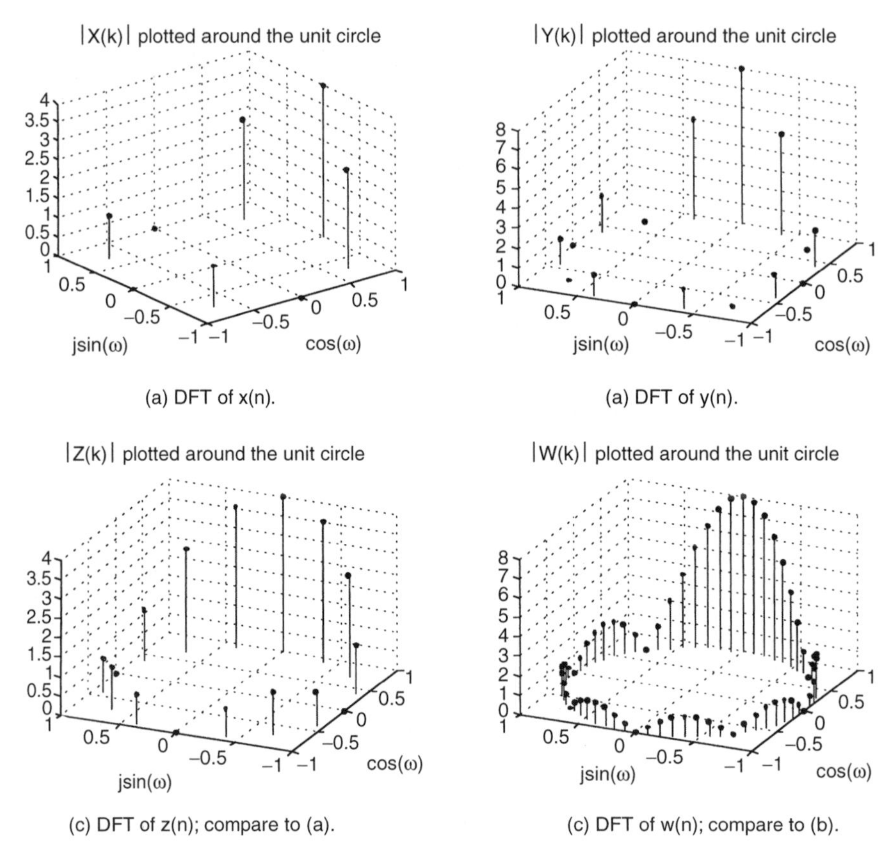

(a) DFT of x(n).

(a) DFT of y(n).

(c) DFT of z(n); compare to (a).

(c) DFT of w(n); compare to (b).

Fig. 7.2. Visualizing the DFT around the unit circle of the complex plane. We set $\omega = 2\pi k/N$ and plot the DFT magnitudes of some signals relative to $\cos(\omega)$ and $\sin(\omega)$ around the unit circle in the complex plane. In part (a), $|X(k)|$ for $x(n) = [\underline{1}, 1, 1, 1, 0, 0, 0, 0]$. In part (b), $|Y(k)|$ for $y(n) = [\underline{1}, 1, 1, 1, 1, 1, 1, 1, 0, 0, 0, 0, 0, 0, 0, 0]$. In part (c), $|Z(k)|$ for $z(n) = [\underline{1}, 1, 1, 1, 0, 0, 0, 0, 0, 0, 0, 0, 0, 0, 0, 0]$. In part (d), $w(n)$ has period $N = 64$, with $w(n) = 1$ for $0 \leq n \leq 7$, and $w(n) = 0$ otherwise. Note that the size of the square pulse within the signal's period determines the number of lobes, and the number of analysis equation summands determines the detail within the lobes.

M lobes in the discrete magnitude spectrum $|X(k)|$, if we count the big lobe around $k = 0$ twice (Figure 7.2).

7.1.1.3 Simple Applications. Let us briefly show how to use the DFT in applications on real sampled signal data. This section shows how the magnitudes of DFT coefficients indicate significant periodic trends in the original analog signal. As an example, we perform a discrete Fourier analysis of the solar disturbances associated with sunspots. Most people are aware that sunspot numbers increase dramatically during these solar disturbances, affect high-frequency radio communication on earth, and tend to occur in approximately 11-year cycles.

Since discrete signals arise from sampling analog signals, the first step is to suppose $x(n) = x_a(nT)$, where $x_a(t)$ is an analog signal, $T > 0$ is the sampling interval, and $x_a(t)$ has period NT. If we write $x(n)$ in terms of its DFT synthesis equation (7.3), then we see that the sinusoidal components of smallest frequency are for $k = 1$ and $k = N - 1$: $\cos(2\pi n/N)$, $\sin(2\pi n/N)$, $\cos(2\pi(N-1)n/N)$, and $\sin(2\pi(N-1)n/N)$. These discrete sinusoids come from sampling analog sinusoids with fundamental period NT, for example, $\sin(2\pi n/N) = \sin(2\pi t/(NT))|_{t=Tn}$. But this is precisely the analog sinusoid with fundamental frequency $1/NT$ Hz.

Application (Wolf Sunspot Numbers). The first chapter introduced the Wolf sunspot numbers as an example of a naturally occurring, somewhat irregular, but largely periodic signal. From simple time-domain probing of local signal extrema, we can estimate the period of the sunspots. The DFT is a more powerful tool for such analyses, however. Given the Wolf sunspot numbers over some 300 years, a good estimate of the period of the phenomenon is possible. Figure 7.3 shows the time-domain sunspot data from 1700 to 1996. The sampling interval T is 1 year ($T = 1$), and we declare 1700 to be year zero for Wolf sunspot number data. Thus, we perform a frequency analysis of the signal, using the DFT on $w(n)$ over the interval $[1700–1700, 1996–1700] = [0, N-1]$. Figure 7.3 shows the DFT, $W(k)$, and its magnitude; there is a huge peak, and for that the rest of the analysis is straightforward. Brute force search finds the maximum magnitude $|W(k)|$ for $0 < k < (1996-1700 + 1)/2$ at time instant $k = k_0$. We do not need to consider higher k values, since the terms above $k = 148 = \lfloor 297/2 \rfloor$ represent discrete frequencies already counted

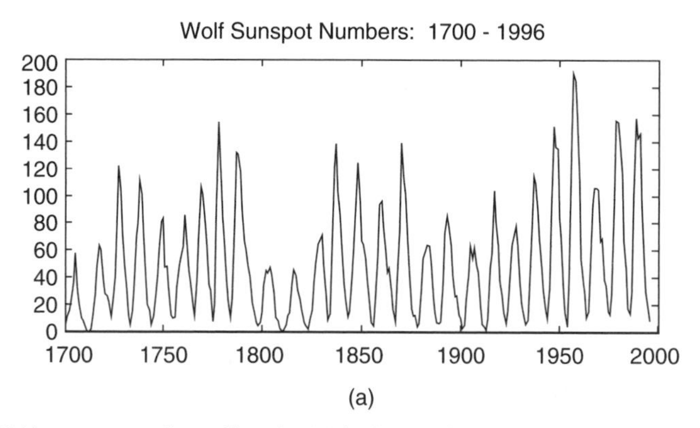

Fig. 7.3. Wolf sunspot numbers. Signal $w(n)$ is the Wolf sunspot number, a composite figure equal to $10G + S$, where G is the average number of sunspot groups and S is the average number of spots reported by a team of international observers. In panel (a), the time-domain plot of $w(n)$ from 1700 to 1996 is shown; note the periodicity. Panel (b) plots the DFT coefficients $W(k)$ in the complex plane but this representation does not aid interpretation. The power of the DFT signal, $P(k) = |W(k)|^2$, appears in panel (c), with the DC term zeroed. The maximum power value occurs at $k = 27$, which represents a sinusoidal component with a period of $297/27 = 11$ years.

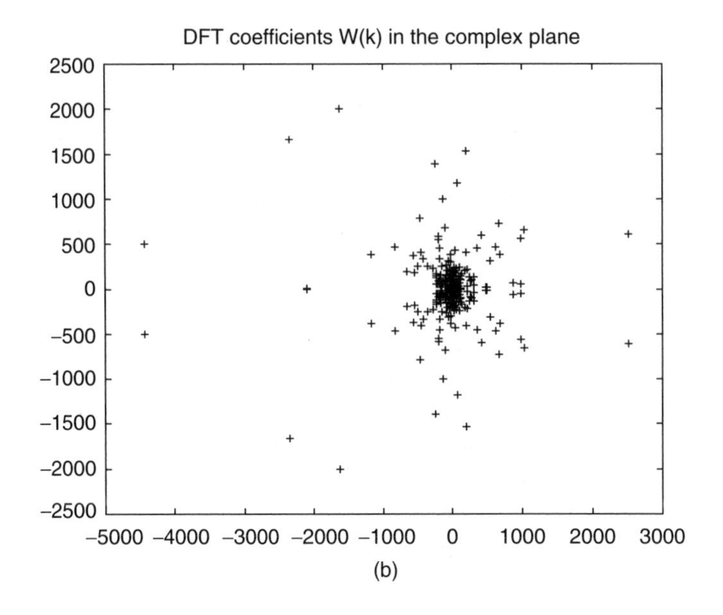

(b)

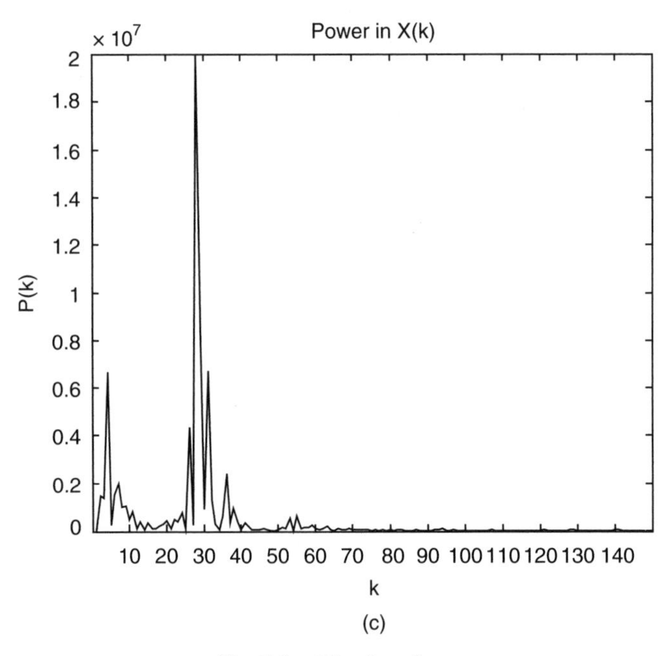

(c)

Fig. 7.3 (*Continued*)

among the lower k values. The frequency resolution of the DFT on $N = 297$ samples, each separated by $T = 1$ year, is $1/NT$. Therefore, the biggest periodic component in the sunspot cycle is k_0/NT cycles/year, and this corresponds to a sunspot period of NT/k_0 years.

This same analysis serves other oscillatory signal analysis applications considered already: electrocardiology, manufactured surface waviness analysis, and tone detection. We will consider its further practical uses, strengths, and weaknesses in Chapter 9.

Note that the theoretical presentation of the DFT proceeds in a completely formal manner. A modest amount of algebra, along with the nice properties of the complex exponential signal, are just enough to develop a complete theory. There are no subtleties concerning discontinuities such as we had to overcome with the analog Fourier series. The discrete nature of the signals and the decomposition of a discrete periodic signal into a finite set of discrete exponentials demand no exotic tools such as distributions or Dirac delta functions. The complex exponential signal's elegance and a little algebra are sufficient to develop the entire theory. We could now prove a bunch of theorems about properties of the DFT. But our goal is develop the tools, the understanding, and the fundamental insights of signal analysis; we should not think to just throw the theory at the reader. Before proceeding to a lengthy list of the DFT's properties, let's explore the links that the DFT shares with the tools we already know for the frequency analysis of analog signals. In particular, we shall show that a discrete signal's DFT coefficients approximate certain of the Fourier series coefficients for an analog periodic source signal.

7.1.2 The DFT's Analog Frequency-Domain Roots

Perhaps the clearest insight into how the discrete Fourier transform reveals the frequency content of a discrete signal is to explore its relationship to the Fourier series for analog periodic signals. We begin by defining a variant of the DFT that takes a form very similar to the analog Fourier series.

Definition (Discrete Fourier Series). Suppose $x(n)$ is a discrete signal and $N > 0$. Then the discrete signal $c(k) = c_k$, defined by

$$c(k) = \frac{1}{N}\sum_{n=0}^{N-1} x(n)\exp\left(\frac{-2\pi jnk}{N}\right) \tag{7.16}$$

where $0 \le k \le N-1$, is the *discrete Fourier series* (DFS) of $x(n)$ on the interval $[0, N-1]$. Equation (7.16) is called the DFS *analysis equation* for $x(n)$.

Note that if $x(n)$ has DFT coefficients $X(k)$ and DFS coefficients $c(k)$ on $[0, N-1]$, then $X(k) = c(k)/N$. Except for the factor N, the DFT and the DFS share an identical theory. Corresponding to the DFT's synthesis equation, there is a DFS *synthesis equation*:

$$x(n) = \sum_{k=0}^{N-1} c(k)\exp\left(\frac{2\pi jnk}{N}\right), \tag{7.17}$$

where $c(k)$ are the DFS analysis equation coefficients for $x(n)$ on $[0, N-1]$. Equation (7.17) also defines the inverse discrete Fourier series (IDFS). The two concepts are so similar that in the literature one must pay close attention to the particular form of the definitions of the DFT and DFS.

The next theorem clarifies the relationship between the DFS and the analog Fourier series. This shows that these discrete transforms are in fact simple approximations to the FS coefficients that we know from analog signal frequency theory. We are indeed justified in claiming that the DFS and DFT provide a frequency-domain description of discrete signals.

Theorem (Relation Between DFS and FS). Let $x_a(t)$ be an analog signal with period $T > 0$. Suppose $N > 0$, $\Delta t = T/N$, $F = 1/T$, and $x(n) = x_a(n\Delta t)$ is a discrete signal that samples $x_a(t)$ at intervals Δt. Finally, let $c(k)$ be the kth DFS coefficient (7.16) for $x(n)$, and let $c_a(k)$ is the kth analog Fourier series coefficient for $x_a(t)$ on $[0, T]$. That is,

$$c_a(k) = \frac{1}{T}\int_0^T x(t)e^{-j2\pi kFt}\,dt, \tag{7.18}$$

where $0 < k < N-1$. Then, $c(k)$ is the trapezoidal rule approximation to the FS integral (7.18) for $c_a(k)$, using the intervals $[0, \Delta t]$, $[\Delta t, 2\Delta t]$, ..., $[(N{-}1)\Delta t, N\Delta t]$.

Proof: The integrand in (7.18) is complex, but the trapezoidal rule can be applied to both its real and imaginary parts. Suppose we let $y(t) = x_a(t)\exp(-j2\pi kFt)$ be the integrand. Recalling the trapezoidal rule from calculus (Figure 7.4), we can see that a typical trapezoid has a base of width Δt and average height $[y(n\Delta t) + y((n+1)\Delta t)]/2$. In other words, an approximation to $c_a(k)$ is

$$\hat{c}_a(k) = \frac{1}{T}\left\{\begin{aligned}&\left(\frac{y(0\cdot\Delta t)+y(1\cdot\Delta t)}{2}\right)\Delta t + \left(\frac{y(1\cdot\Delta t)+y(2\cdot\Delta t)}{2}\right)\Delta t + \cdots\\ &+\left(\frac{y((N-1)\cdot\Delta t)+y(N\cdot\Delta t)}{2}\right)\Delta t\end{aligned}\right\}. \tag{7.19}$$

Collecting terms inside the braces of (7.18) and observing that $y(0) = y(N\Delta t) = y(T)$ gives

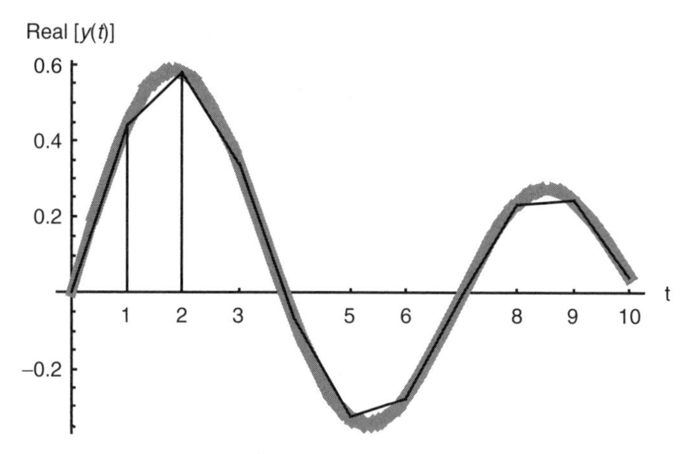

Fig. 7.4. Approximating the Fourier series integral by the trapezoidal rule. The signal $x_a(t)$ has period $T > 0$. The trapezoidal rule applies to the real and imaginary parts of the integrand in the Fourier series analysis equation, $y(t) = x_a(t)\exp(-j2\pi kFt)$. It is necessary to include sufficient trapezoids to span an entire period of the analog signal $x(t)$; in this illustration $T = 10$.

$$\hat{c}_a(k) = \frac{\Delta t}{T}\sum_{n=0}^{N-1} x_a(n\Delta t)\exp(-j2\pi kFn\Delta t) = \frac{1}{N}\sum_{n=0}^{N-1} x(n)\exp\left(\frac{-j2\pi kn}{N}\right) = c(k). \quad (7.20)$$

This shows that the trapezoidal rule approximation to the FS integral is precisely the DFS coefficient $c(k)$ and completes the proof. ■

Thus, the DFS is a straightforward approximation of the analog Fourier series components. And the DFT is just the DFS scaled by the period of the discrete signal. Again, these transforms are appropriate for discrete periodic signals. Of course, one may take any discrete signal, $x(n)$, restrict it to a finite interval, $[0, N-1]$, where $N > 0$, and then perform the DFT analysis equation computation for the $x(n)$ values for $0 \le n < N$. The result is N complex numbers, $X(0), X(1), \ldots, X(N-1)$. This is like computing the DFT for the periodic extension of $x(n)$. The result of computing the IDFT on $X(0), X(1), \ldots, X(N-1)$, is not the original signal $x(n)$; it is $x_p(n) = (1/N)[X(0) + X(1)e^{2\pi jkn/N} + \cdots + X(N-1)e^{2\pi j(N-1)n/N}$, which is periodic with period N. Another transform is necessary for the study of frequency in aperiodic signals—the discrete-time Fourier transform (DTFT). As the DFT is the discrete world's replica of the Fourier series, so the counterpart to Fourier transform for discrete signals is the DTFT. Before considering the aperiodic case, however, let us explain some of the properties of the DFT and how these lead to efficient algorithms for its computation.

7.1.3 Properties

This section explains the many elegant properties of the discrete Fourier transform. Generally speaking, these theorems on linearity, symmetry, and conservation of signal energy result from the special algebraic characteristics of the complex exponential function. Interesting in themselves, they are useful as well in analyzing signals. As an example, the energy in DFT coefficients does not change as the underlying periodic signal is translated. This property we noted in Chapter 4, and its motivation was our search for a translation-invariant, spatial-frequency-based texture indicator. Our further studies of the properties of DFT, especially its computational symmetries, will lead to the fast Fourier transform (FFT) algorithm. For signals with certain periods—especially powers of two—the FFT offers a dramatic reduction in the computational burden of computing discrete frequency components with the DFT.

Let us begin with some basic properties of the DFT. We have already noted that the DFT of a signal $x(n)$ on $[0, N-1]$, $N > 0$, has period N.

Proposition (Linearity, Time Shift, and Frequency Shift). Let $x(n)$ and $y(n)$ be periodic signals with period $N > 0$, let a and b be constants, and let $X(k)$ and $Y(k)$ be their DFTs, respectively. Then:

(a) (Linearity) The DFT of $ax(n) + by(n)$ is $aX(k) + bY(k)$.
(b) (Time Shift) The DFT of $x(n-m)$ is $(W_N)^{km}X(k)$.
(c) (Frequency Shift) The IDFT of $X(k-m)$ is $(W_N)^{-nm}x(n)$.

In other words, delaying a time-domain signal by m samples is equivalent to multiplying each DFT coefficient in the frequency domain, $X(k)$, by the factor $(W_N)^{km} = e^{-2\pi jkm/N}$. And delaying the frequency-domain signal $X(k)$ by m samples reflects a time-domain multiplication of each value $x(n)$ by $(W_N)^{-nm}$.

Proof: Linearity is easy and left as an exercise. Let $z(n) = x(n-m)$ and let $r = n - m$. Let's apply the DFT analysis equation directly to $z(n)$:

$$
\begin{aligned}
Z(k) &= \sum_{n=0}^{N-1} z(n)\exp\left(\frac{-j2\pi kn}{N}\right) = \exp\left(\frac{-j2\pi km}{N}\right)\sum_{n=0}^{N-1} x(n-m)\exp\left(\frac{-j2\pi k(n-m)}{N}\right) \\
&= W_N^{kM}\sum_{n=0}^{N-1} x(n-m)\exp\left(\frac{-j2\pi k(n-m)}{N}\right) = W_N^{kM}\sum_{r=-m}^{N-1-m} x(r)\exp\left(\frac{-j2\pi kr}{N}\right) \\
&= W_N^{kM}\sum_{r=0}^{N-1} x(r)\exp\left(\frac{-j2\pi kr}{N}\right) = W_N^{kM}X(k).
\end{aligned}
\tag{7.21}
$$

Since $x(r)$ and $\exp(-2\pi jkr/N)$ both have period N, the summation over r in (7.21) may start at any index; in particular, if we start the summation at $r = 0$, we have precisely the DFT analysis formula for $X(k)$. The proof of the frequency shift property is similar and is left as an exercise. ■

Definition (Discrete Convolution). Let $x(n)$ and $y(n)$ be periodic signals with period $N > 0$. Then the signal $z(n)$ defined by

$$z(n) = \sum_{k=0}^{N-1} x(k)y(n-k) \tag{7.22}$$

is called the *discrete convolution* of x and y: $z = x * y$.

Note that the definition of discrete convolution can be extended to the case where one of the signals (or both) is not periodic. The expression (7.22) is computed for the periodic extension of the signals over $[0, N-1]$. We then have the following theorem that relates convolutions of discrete signals to the termwise products of their DFTs. Although we are working with frequency transforms of a greatly different formal nature, the comparison to the analog Fourier transform's Convolution Theorem is striking.

Theorem (Convolution in Time). Let $x(n)$ and $y(n)$ be periodic signals with period $N > 0$, $X(k)$ and $Y(k)$ their DFTs, and $z(n) = x * y$. Then the DFT of $z(n)$ is $Z(k) = X(k)Y(k)$.

Proof: A direct attack is fruitful. We write out the expression for $Z(k)$ according to the DFT analysis equation, insert the convolution formula for z in terms of x and y, and then separate the terms.

$$Z(k) = \sum_{n=0}^{N-1} z(n)W_N^{kn} = \sum_{n=0}^{N-1} (x*y)(n)W_N^{kn} = \sum_{n=0}^{N-1}\left(\sum_{m=0}^{N-1} x(m)y(n-m)\right)W_N^{kn}. \tag{7.23}$$

Interchanging the order of summation in (7.23) is the key step. This permits us to collect m-summation terms associated with x and n-summation terms associated with y together and expose the product of their DFTs.

$$\begin{aligned} Z(k) &= \sum_{m=0}^{N-1}\sum_{n=0}^{N-1} x(m)y(n-m)W_N^{kn} = \sum_{m=0}^{N-1} x(m)\sum_{n=0}^{N-1} y(n-m)W_N^{kn} \\ &= \sum_{m=0}^{N-1} x(m)W_N^{km}\sum_{n=0}^{N-1} y(n-m)W_N^{k(n-m)} = \sum_{m=0}^{N-1} x(m)W_N^{km}\sum_{r=0}^{N-1} y(r)W_N^{kr} = X(k)Y(k). \end{aligned} \tag{7.24}$$

We let $r = n - m$ for a change of summation variable in the penultimate summation of (7.24). Since $y(n)$ is periodic with period N, the summation from $r = -m$ to $N - 1 - m$ is the same as the summation from $r = 0$ to $N - 1$. ■

Theorem (Convolution in Frequency). That is, let $x(n)$ and $y(n)$ be periodic signals with period $N > 0$; let $X(k)$ and $Y(k)$ be their DFTs, and let $z(n) = x(n)y(n)$ be the termwise product of x and y. Then the DFT of $z(n)$ is $Z(k) = (1/N)X(k)*Y(k)$, where $X(k)*Y(k)$ is the discrete convolution of $X(k)$ and $Y(k)$.

Proof: Similar to the Convolution in Time Theorem. ■

Proposition (Symmetry). Let signal $x(n)$ have period $N>0$ and $X(k)$ be its DFT. Then:

(a) The DFT of $x^*(n)$, the complex conjugate of $x(n)$, is $X^*(N-\mathrm{k})$, and the DFT of $x^*(N-n) = x^*(-n)$ is $X^*(k)$.
(b) The DFT of $x_e(n) = (1/2)(x(n) + x^*(N-n))$, the even part of $x(n)$, is $\mathrm{Re}[X(k)]$, the real part of $X(k)$.
(c) The DFT of $x_o(n) = (1/2)(x(n) - x^*(N-n))$, the odd part of $x(n)$, is $j\mathrm{Im}[X(k)]$, where $\mathrm{Im}[\mathrm{X(k)}]$ is the imaginary part of $X(k)$.
(d) The DFT of $\mathrm{Re}[x(n)]$ is $X_e(k) = (1/2)(X(k) + X^*(N-k))$, the even part of $X(k)$.
(e) The DFT of $j\mathrm{Im}[x(n)]$ is $X_o(k) = (1/2)(X(k) - X^*(N-\mathrm{k}))$, the odd part of $X(k)$.

Proof: Easy. ■

Proposition (Real Signals). Let signal $x(n)$ be real-valued with period $N > 0$, and let $X(k)$ be its DFT. Then:

(a) $X(k) = X^*(N-k)$.
(b) The DFT of $x_e(n)$ is $\mathrm{Re}[X(k)]$.
(c) The DFT of $x_o(n)$ is $j\mathrm{Im}[X(k)]$.

Proof: Also easy. ■

Theorem (Parseval's). Let $x(n)$ have period $N > 0$ and let $X(k)$ be its DFT. Then

$$\sum_{m=0}^{N-1} x(n)\bar{x}(n) = \frac{1}{N}\sum_{k=0}^{N-1} X(k)\bar{X}(k). \tag{7.25}$$

Proof: Although it seems to lead into a messy triple summation, here again a stubborn computation of the frequency-domain energy for $X(k)$ on the interval $[0, N-1]$ bears fruit. Indeed,

$$\begin{aligned}
\sum_{k=0}^{N-1} X(k)\bar{X}(k) &= \sum_{k=0}^{N-1}\left(\sum_{m=0}^{N-1} x(m)W_N^{km}\right)\overline{\left(\sum_{n=0}^{N-1} x(n)W_N^{kn}\right)} \\
&= \sum_{k=0}^{N-1}\left(\sum_{m=0}^{N-1} x(m)W_N^{km}\right)\left(\sum_{n=0}^{N-1} \overline{x(n)}\,\overline{W_N^{kn}}\right) \\
&= \sum_{k=0}^{N-1}\sum_{m=0}^{N-1}\sum_{n=0}^{N-1} x(m)\overline{x(n)}W_N^{k(m-n)} \\
&= \sum_{m=0}^{N-1}\sum_{n=0}^{N-1} x(m)\overline{x(n)}\sum_{k=0}^{N-1} W_N^{k(m-n)}.
\end{aligned} \tag{7.26}$$

The last summation at the bottom in (7.26) contains a familiar expression: the partial geometric series in $(W_N)^{m-n}$ of length N. Recall from the lemma within the proof of the DFT Inverse Theorem (Section 7.1.1.1) that this term is either N or 0, according to whether $m = n$ or not, respectively. Thus, the only products $x(m)x^*(n)$ that will contribute to the triple sum in (7.26) are those where $m = n$, and these are scaled by the factor N. Therefore,

$$\sum_{k=0}^{N-1} X(k)\bar{X}(k) = \sum_{m=0}^{N-1}\sum_{n=0}^{N-1} x(m)\overline{x(n)} \sum_{k=0}^{N-1} W_N^{k(m-n)} = N \sum_{n=0}^{N-1} x(n)\overline{x(n)}, \tag{7.27}$$

and the proof is complete. ■

We have explored some of the theory of the DFT, noted its specific relation to the analog Fourier series, and considered its application for finding the significant periodicities in naturally occurring signals. In particular, we presented an example that uncovered the period of the sunspot cycle by taking the DFT of the discrete signal giving Wolf sunspot numbers over several centuries. We know from Chapter 4's attempts to analyze signals containing significant periodicities (textures, speech, tone detection, and the like) that pure time-domain methods—such as statistical approaches—can prove quite awkward. We do need the DFT for computer implementation, and the next section explores, therefore, the efficient computation of the DFT on digital computers.

7.1.4 Fast Fourier Transform

The fast Fourier transform (FFT) has been known, it turns out, since the time of Gauss.[3] Only recently, however, has it been widely recognized and utilized in signal processing and analysis. Indeed, its original rediscovery in the 1960s marks the beginning of an era in which digital methods supplanted analog methods in signal theory and applications.

7.1.4.1 Computational Cost. Let us begin by studying the computational costs incurred in the DFT analysis and synthesis equations. Clearly, if the time-domain signal, $x(n)$ on $[0, N-1]$, is complex-valued, then the operations are nearly identical. The IDFT computation requires an additional multiplication of a complex value by the factor $(1/N)$, as an inspection of the equations shows:

$$X(k) = \sum_{n=0}^{N-1} x(n)\exp\left(\frac{-2\pi jnk}{N}\right) = \sum_{n=0}^{N-1} x(n)W_N^{nk}, \tag{7.28a}$$

$$x(n) = \frac{1}{N}\sum_{k=0}^{N-1} X(k)\exp\left(\frac{2\pi jnk}{N}\right) = \frac{1}{N}\sum_{k=0}^{N-1} X(k)W_N^{-nk}. \tag{7.28b}$$

[3]Gauss, writing in a medieval form of Latin, made progress toward the algorithm in his notebooks of 1805. [M. T. Heideman, D. H. Johnson, and C. S. Burrus, 'Gauss and the history of the fast Fourier transform,' *IEEE ASSP Magazine*, vol. 1, no. 4, pp. 14–21, October 1984.]

But the principal computational burden lies in computing the complex sum of complex products in (7.28a) and (7.28b).

Consider the computation of $X(k)$ in (7.28a). For $0 \le k \le N-1$, the calculation of $X(k)$ in (7.28a) requires N complex multiplications and $N-1$ complex sums. Computing all N of the $X(k)$ values demands N^2 complex multiplications and $N^2 - N$ complex additions. Digital computers implement complex arithmetic as floating point operations on pairs of floating point values. Each complex multiplication, therefore, requires four floating point multiplications and two floating point additions; and each complex addition requires two floating point additions. So the total floating point computation of an N-point FFT computation costs $4N^2$ multiplications and $2(N^2 - N) + 2N^2$ additions. Other factors to consider in an FFT implementation are:

- Storage space and memory access time for the $x(n)$ and $X(k)$ coefficients;
- Storage space and memory access time for the $(W_N)^{kn}$ values;
- Loop counting and termination checking overheads.

Ultimately, as N becomes large, however, the number of floating point additions and multiplications weighs most significantly on the time to finish the analysis equation. Since the number of such operations—whether they are complex operations or floating point (real) operations—is proportional to N^2, we deem the DFT an order-N^2, or $O(N^2)$ operation.

FFT algorithms economize on floating point operations by eliminating duplicate steps in the DFT and IDFT computations. Two properties of the phase factor, W_N, reveal the redundancies in the complex sums of complex products (7.28a) and (7.28b) and make this reduction in steps possible:

- Phase factor periodicity: $(W_N)^{kn} = (W_N)^{k(n+N)}$;
- *Phase factor symmetry:* $[(W_N)^{kn}]^* = (W_N)^{k(N-n)}$.

Two fundamental approaches are decimation-in-time and decimation-in-frequency.

7.1.4.2 Decimation-in-Time. Decimation-in-time FFT algorithms reduce the DFT into a succession of smaller and smaller DFT analysis equation calculations. This works best when $N = 2^p$ for some $p \in \mathbb{Z}$. The N-point DFT computation resolves into two $(N/2)$-point, each of which resolves into two $(N/4)$-point DFTs, and so on.

Consider, then, separating the computation of $X(k)$ in (7.28a) into even and odd n within $[0, N-1]$:

$$\begin{aligned} X(k) &= \sum_{m=0}^{(N/2)-1} x(2m)W_N^{2km} + \sum_{m=0}^{(N/2)-1} x(2m+1)W_N^{(2m+1)k} \\ &= \sum_{m=0}^{(N/2)-1} x(2m)W_N^{2km} + W_N^k \sum_{m=0}^{(N/2)-1} x(2m+1)W_N^{2km} \end{aligned} \tag{7.29}$$

Note the common $(W_N)^{2km}$ phase factor in both terms on the bottom of (7.29). This leads to the key idea for the DFT's time-domain decimation:

$$W_N^2 = \exp\left(\frac{-2\pi j}{N}\right)^2 = \exp\left(\frac{-4\pi j}{N}\right) = \exp\left(\frac{-2\pi j}{N/2}\right) = W_{\left(\frac{N}{2}\right)}. \tag{7.30}$$

We set $y(m) = x(2m)$ and $z(m) = x(2m + 1)$. Then $y(m)$ and $z(m)$ both have period $N/2$. Also, (7.30) allows us to write (7.29) as a sum of the $N/2$-point DFTs of $y(m)$ and $z(m)$:

$$X(k) = \sum_{m=0}^{(N/2)-1} y(m) W_{\left(\frac{N}{2}\right)}^{km} + W_N^k \sum_{m=0}^{(N/2)-1} z(m) W_{\left(\frac{N}{2}\right)}^{km} = Y(k) + W_N^k Z(k). \tag{7.31}$$

From (7.31) it is clear that an N-point DFT is the computational equivalent of two $N/2$-point DFTs, plus $N/2$ complex multiplications, plus $N/2$ complex additions. Figure 7.5 illustrates the process of decomposing an 8-point DFT into two 4-point DFTs.

Does this constitute a reduction in computational complexity? The total cost in complex operations is therefore $2(N/2)^2 + 2(N/2) = N + N^2/2$ complex operations. For large N, the $N^2/2$ term, representing the DFT calculations, dominates. But the division by two is important! Splitting the $Y(k)$ and $Z(k)$ computations in the same way reduces the computation of the two DFTs to four $N/4$-point DFTs, plus $2(N/4)$ complex multiplications, plus $2(N/4)$ complex additions. The grand total cost

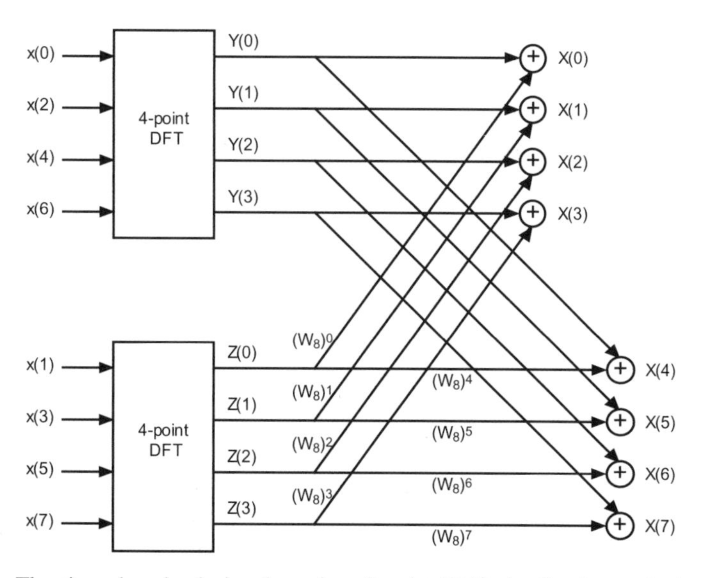

Fig. 7.5. The time domain decimation of an 8-point DFT. An 8-point analysis problem decomposes into two preliminary 4-point problems, followed by a scalar multiplication and a summation. This is only the first stage, but it effectively halves the number of complex operations necessary for computing the DFT of $x(n)$ on $[0, 7]$.

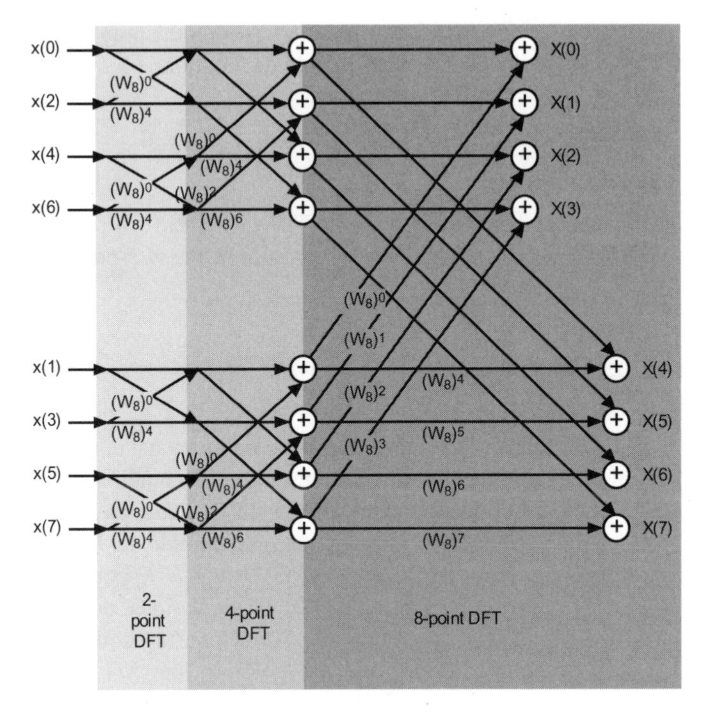

Fig. 7.6. Fully decimated 8-point DFT. Further decomposition steps expose an elementary operation that recurs throughout the computation. Notice also that the original data elements must be sorted in bit-reversed order at the beginning of the algorithm. Binary index numbers are used; that is, $x(000)$ is followed by $x(100)$, instead of $x(001)$, with which it swaps places. Next comes $x(010)$, since its bit-reversed index stays the same. But the next signal value must be $x(110)$, which swaps places with $x(011)$. This computational trick allows the in-place computation of DFT coefficients.

$N + [N + 4(N/4)^2] = N + N + N^2/4 = 2N + N^2/4$. The next iteration trims the cost to $3N + N^2/8$. And this can continue as long as N contains factors of two—$\log_2 N$ times. Thus, the fully executed decimation-in-time reduces the computational burden of an N-point DFT from $O(N^2)$ to $O(N\log_2 N)$.

Figure 7.6 illustrates the process of decomposing an 8-point DFT down to the final 2-point problem.

It turns out that a single basic operation underlies the entire FFT algorithm. Consider the 8-point problem in Figure 7.6. It contains three DFT computation stages. Four two-point problems comprise the first stage. Let us examine the structure of the first stage's operation. Pairs of the original signal elements are multiplied by either $(W_8)^0 = 1$ or $(W_8)^4 = -1$, as shown in Figure 7.7a. Thereafter, pairs of the two-point DFT coefficients are similarly multiplied by either of two possible powers of W_8: $(W_8)^0 = 1$ or $(W_8)^4 = -1$ again, or, $(W_8)^2$ and $(W_8)^6$.

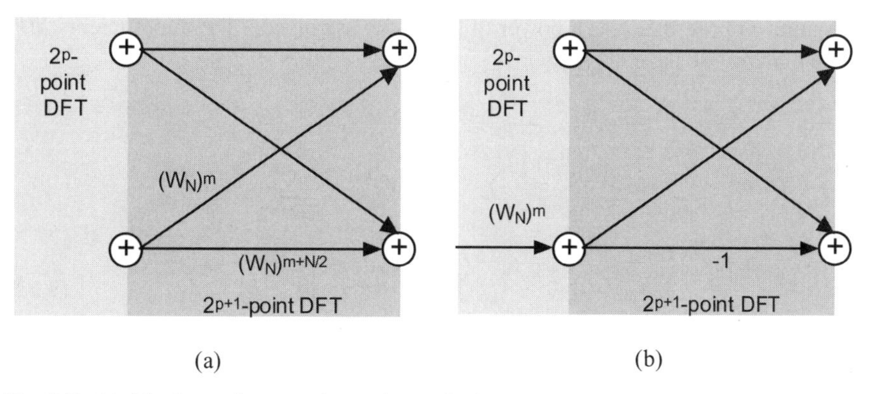

(a) (b)

Fig. 7.7. (a), The butterfly operation or its equivalent, simpler form (b), occurs throughout the FFT computation.

Indeed, at any of the later stages of computation, the multiplying phase factors always assume the form $(W_8)^m$ and $(W_8)^{m+4}$. Because their ratio is $(W_8)^{m+4-m} = (W_8)^4 = -1$, we may further simplify the rudimentary operation (Figure 7.7b), eliminating one complex multiplication. The resulting crisscross operation is called a butterfly, which the flow graph vaguely resembles. Some aesthetically pleasing term does seem appropriate: The butterfly operation reveals an elegant structure to the DFT operation, and the algorithms in Section 7.1.4.4 make efficient use of this elegance.

Now let us consider the rearrangement of the original signal values $x(0)$, $x(1)$, $x(2)$, $x(3)$, $x(4)$, $x(5)$, $x(6)$, $x(7)$ into the proper order for the four initial butterflies: $x(0)$, $x(4)$, $x(2)$, $x(6)$, $x(1)$, $x(5)$, $x(3)$, $x(7)$. This represents a bit-reversed reading of the indices of the data, because we observe that if the indices of the signal values are written in binary form, that is, $x(0) = x(000)$, $x(1) = x(001)$, and so on, then the necessary rearrangement comes from reading the binary indices backwards: $x(000)$, $x(100)$, $x(010)$, $x(110)$, $x(001)$, $x(101)$, $x(011)$, $x(111)$. At each of the three stages of the computation, we maintain eight complex values, beginning with the original signal data in bit-reversed order. Then we perform butterfly operations, with the index difference between value pairs doubling at each stage.

Now let us consider another approach to the efficient computation of the DFT by splitting the frequency domain values $X(k)$ into smaller and smaller groups.

7.1.4.3 Decimation-in-Frequency.

Suppose again that we are faced with the problem of computing the DFT coefficients $X(k)$ for a signal $x(n)$ on $[0, N-1]$. In the previous section, we split the DFT analysis equation sum over $0 \le n \le N-1$ into two sums: for n even and for n odd. Now we divide the frequency-domain values $X(k)$ for $0 \le k \le N-1$ into even k and odd k. The result is an alternative, efficient algorithm for computing DFT coefficients called the decimation-in-frequency FFT.

Again, suppose that N is a power of 2, and let us consider DFT coefficients $X(k)$ where $k = 2m$. We have

$$X(2m) = \sum_{n=0}^{N-1} x(n)W_N^{2mn} = \sum_{n=0}^{(N/2)-1} x(n)W_N^{2mn} + \sum_{n=N/2}^{N-1} x(n)W_N^{2mn}$$

$$= \sum_{n=0}^{(N/2)-1} x(n)W_N^{2mn} + \sum_{n=0}^{(N/2)-1} x\left(n+\frac{N}{2}\right)W_N^{2m(n+\frac{N}{2})} \tag{7.32}$$

Not unexpectedly, phase factor properties fortuitously apply. We observe that $(W_N)^{2m(n+N/2)} = (W_N)^{2mn}(W_N)^{mN} = (W_N)^{2mn}(W_{N/2})^{mn}$. Hence, for $0 \le m < N/2$, we have

$$X(2m) = \sum_{n=0}^{(N/2)-1} x(n)W_N^{2mn} + \sum_{n=0}^{(N/2)-1} x\left(n+\frac{N}{2}\right)W_N^{2m(n+\frac{N}{2})}$$

$$= \sum_{n=0}^{(N/2)-1} x(n)W_{N/2}^{mn} + \sum_{n=0}^{(N/2)-1} x\left(n+\frac{N}{2}\right)W_{N/2}^{2mn}$$

$$= \sum_{n=0}^{(N/2)-1} \left[x(n)+x\left(n+\frac{N}{2}\right)\right]W_{N/2}^{2mn}. \tag{7.33}$$

This last result shows that the $X(k)$ coefficients, for k even, can be calculated by an $(N/2)$-point DFT. Turning to the remaining $X(k)$, for $k = 2m + 1$ odd, we find

$$X(2m+1) = \sum_{n=0}^{(N/2)-1} x(n)W_N^{(2m+1)n} + \sum_{n=N/2}^{N-1} x(n)W_N^{(2m+1)n}$$

$$= \sum_{n=0}^{(N/2)-1} x(n)W_N^{(2m+1)n} + \sum_{n=0}^{(N/2)-1} x\left(n+\frac{N}{2}\right)W_N^{(2m+1)(n+N/2)}$$

$$= \sum_{n=0}^{(N/2)-1} x(n)W_N^{(2m+1)n} + W_N^{(2m+1)(N/2)} \sum_{n=0}^{(N/2)-1} x\left(n+\frac{N}{2}\right)W_N^{(2m+1)n} \tag{7.34}$$

Now it is time to invoke the phase factor properties: $(W_N)^{(2m+1)(n+N/2)} = (W_N)^{2mn}(W_N)^{m(N/2)} = -1$. Therefore, (7.34) simplifies to

$$
\begin{aligned}
X(2m+1) &= \sum_{n=0}^{(N/2)-1} x(n)W_N^{(2m+1)n} - \sum_{n=0}^{(N/2)-1} x\left(n+\frac{N}{2}\right)W_N^{(2m+1)n} \\
&= \sum_{n=0}^{(N/2)-1} \left[x(n)-x\left(n+\frac{N}{2}\right)\right]W_N^{(2m+1)n} \\
&= \sum_{n=0}^{(N/2)-1} \left[x(n)-x\left(n+\frac{N}{2}\right)\right]W_N^{2mn}W_N^{n} \\
&= \sum_{n=0}^{(N/2)-1} \left[x(n)-x\left(n+\frac{N}{2}\right)\right]W_{N/2}^{mn}W_N^{n} \\
&= \sum_{n=0}^{(N/2)-1} \left[W_N^n\left\{x(n)-x\left(n+\frac{N}{2}\right)\right\}\right]W_{N/2}^{mn}.
\end{aligned}
\tag{7.35}
$$

This shows that we can calculate $X(k)$ for k odd by an $(N/2)$-point DFT of the complex signal $y(n) = (W_N)^n[x(n) - x(n + (N/2))]$. Together (7.33) and (7.35) demonstrate that an N-point DFT computation can be replaced by two $(N/2)$-point DFT computations. As in the decimation-in-time strategy, we can iterate this divide-and-compute strategy as many times as N is divisible by two. The decimation-in-frequency result is an $O(N\log_2 N)$ algorithm too.

7.1.4.4 Implementation. This section considers FFT implementation in a modern high-level language, C++, and in assembly language on a representative digital signal processor, the Motorola 56000/56001.

Let's examine a C++ implementation of a radix-2 decimation-in-time algorithm, drawn from a classic FORTRAN coding of the FFT [4]. This algorithm uses the new standard template library contained in the header file <complex>. It replaces the traditional C++ complex arithmentic library, <complex.h>, which defines complex numbers as instances of a class whose member variables are two double-precision floating point numbers. The new template library allows us to construct complex numbers using the C++ float data type, for example, by declaring:

```
complex<float> x;
```

Specifying the `float` data type in the template conserves memory space.

Figure 7.8 shows the C++ implementation of the FFT.

Most references present FFT algorithms in FORTRAN [6–8], but also in C [9, 10]. The FFT can be implemented on a special computer architecture—called a shuffle-exchange network—that interconnects multiple processors in a manner similar to the FFT's butterfly flow diagram. This facilitates either the butterfly operation or the bit-reversed ordering of data, but not both at once, because interprocessor communication bottlenecks occur. Nevertheless, it is still possible to improve the FFT algorithm by an additional $O(N^{1/2})$ [11].

```
#include <math.h>
#include <stdlib.h>
#define PI 3.14159265358979
#include <use_ansi.h>
#include <complex>     //ISO/ANSI std template library
using namespace std;
int fft(complex<double> *x, int nu)
{ // sanity check to begin with:
  if (x == NULL || nu <= 0)
    return 0;
  int N = 1 << nu;              //N=2**nu
  int halfN = N >> 1;           //N/2
  complex<double> temp, u, v;
  int i, j, k;
  for (i = 1, j = 1; i < N; I++){  //bit-reversing data:
    if (i < j){ temp = x[j-1]; x[j-1] = x[i-1]; x[i-1] = temp;}
    k = halfN;
    while (k < j){j -= k; k >>= 1;}
    j += k;
  }
  int mu, M, halfM, ip;
  double omega;
  for (mu = 1; mu <= nu; mu++) {
    M = 1 << mu;        // M = 2**mu
    halfM = M >> 1;     // M/2
    u = complex<double>(1.0, 0.0);
    omega = PI/(double)halfM;
    w = complex<double>(cos(omega), -sin(omega));
    for (j = 0; j < halfM;j++){
      for (i = j; i < N; i += M){
        ip = i + halfM;
        temp = x[ip]*u;
        x[ip] = x[i] - temp;
        x[i] += temp;
      }
      u  *= w;
    }
    u  *= w;
  }
  return 1;
}
```

Fig. 7.8. A C++ implementation of the FFT. This algorithm uses the complex data type in the ISO/ANSI standard template library [5].

```
;Radix-2 decimation in time FFT macro call
fftr2a    macro points, data, coeff
          move #points/2, n0
          move #1,n2
          move #points/4, n6
          move #-1,m0
          move m0,m1
          move m0,m4
          move m0,m5
          move #0,m6              ; addr mod for bit-rev addr
          do   #@cvi (@log (points)/@log(2)+0.5),_end_pass
          move #data,r0
          move r0,r4
          lua  (r0)+n0,r1
          move #coef,r6
          lua  (r1)-,r5
          move n0,n1
          move n0,n4
          move n0,n5
          do   n2,_end_grp
          move x:(r1),x1     y:(r6),y0   ;sin and cos tables
          move x:(r5),a      y:(r6),b    ;load data values
          move x:(r6)+n6,x0
          do   n0,_end_bfy
          mac  x1,y0,b       y:(r1)+,y1  ;decim in time begins
          macr -x0,y1,b      a:x:(r5)+   y:(r0),a
          subl b,a           x:(r0),b    b,y: (r4)
          mac  -x1,x0,b      x:(r0)+,a   a,y: (r5)
          macr -y1,y0,b      x:(r1),x1
          subl b,a           b,x:(r4)+   y:(r0),b
_end_bfy
          move a,x:(r5)+n5        y:(r1)+n1,y1
          move x:(r0)+n0,x1       y:(r1)+n1,y1
_end_grp
          move n0,b1              ; div bfys/group by 2
          lsr  b             n2,a1
          lsl  a             b1,n0
          move a1,n2
_end_pass
          endm
```

Fig. 7.9. An Assembly language implementation of the FFT on the Motorala 56001 DSP chip [12]. The X and Y memory banks contain the input data's real and imaginary components, respectively. The X and Y memory banks also hold the cosine and sine tables for the exponential function implementation. The algorithm bit-reverses the output data stream.

It is interesting to examine the implementation of the FFT on a DSP processor (Figure 7.9). Let us now briefly describe some architectural features of this processor that support the implementation of digital signal processing algorithms. These features include:

- Multiple memory areas (X and Y) and data paths for two sets of signal data.
- No-overhead loop counting.
- A multiply-accumulate operation that runs in parallel with X and Y data movements.
- Built-in sine and cosine tables (in the case of the 56001) to avoid trigonometric function library calls.
- Addressing modes that support memory indexing needed for convolution operations.

The FFT code in Figure 7.9 explois these processing features for a compact implementation of decimation-in-time algorithm.

There are also FFT algorithms for $N \neq 2^n$. These general radix algorithms are more complicated, consume more memory space, and are less speedy than the dyadic decimation-in-time and decimation-in-frequency computations. Nevertheless, they are superior to the DFT and useful when the success of an application depends on fast frequency component computations. Another method, Goertzal's algorithm, uses a convolutional representation of the DFT's analysis equation as the basis for a fast computation. This algorithm is attractive for digital signal processors, because of their special hardware. Some of the exercises explore these FFT algorithms.

7.2 DISCRETE-TIME FOURIER TRANSFORM

In our study of signals and their frequency-domain representations, we have yet to cover one possibility: the class of discrete aperiodic signals. The frequency transform tool that applies to this case is the discrete-time Fourier transform (DTFT). It turns out that not all aperiodic discrete signals have DTFTs. We demonstrate that the transform exists for the important classes of absolutely summable and square-summable signals.

There is a generalization of the DTFT—the z-transform—which provides additional analytical capabilities suitable for those signals for which there is no DTFT. We cover the z-transform in Chapter 8. Chapter 9 explains a variety of applications of both the DTFT and the z-transform.

7.2.1 Introduction

We start the discussion with a formal definition—the mathematician's style at work once again. The definition involves an infinite sum, and so, as with the Fourier transform's infinite integral, for a given signal there are questions of the validity of the transform operation. We momentarily set these concerns aside to review a few examples. Then we turn to the important question of when the DTFT sum converges and prove an inversion result. The DTFT is an important theoretical tool, since it provides the frequency domain characterization of discrete signals from the most important signal spaces we know from Chapter 2: the absolutely summable (l^1) signals and the square summable (l^2) signals.

7.2.1.1 Definition and Examples. Let us begin with the abstract definition of the DTFT and follow it with some simple examples. These provide some indication of the transform's nature and make the subsequent theorems more intuitive. The DTFT's definition does involve an infinite sum; for a given signal, therefore, we must eventually provide answers to existential questions about this operation.

Definition (Discrete-Time Fourier Transform). If $x(n)$ is a discrete signal, then the analog signal discrete signal $X(\omega)$ defined by

$$X(\omega) = \sum_{n=-\infty}^{+\infty} x(n)\exp(-jn\omega), \tag{7.36}$$

where $\omega \in \mathbb{R}$, is the *radial discrete-time Fourier transform* (DTFT) of $x(n)$. The units of ω are radians/second. We often refer to (7.36) as the DTFT analysis equation for $x(n)$. If we take $\omega = 2\pi f$, then we can define the (Hertz) DTFT also:

$$X(f) = \sum_{n=-\infty}^{+\infty} x(n)\exp(-j2\pi nf). \tag{7.37}$$

Generally, here, and in most texts, the DTFT takes the form (7.36) and, unless otherwise specified, we use the radial form of the transform.

Like the analog Fourier series, the DTFT is a transform that knows not what world it belongs in. The Fourier series, we recall from Chapter 4, transforms a periodic analog signal, $x(t)$, into a discrete signal: the Fourier coefficients $c(k) = c_k$. And the DTFT maps a discrete signal, $x(n)$, to an analog signal, $X(\omega)$. This prevents the FS and the DTFT from being considered bona fide systems. Both the Fourier transform and the DFT, however, can be viewed as systems. The FT maps analog signals to analog signals, and it may be considered a partial function on the class of analog signals. We know from Chapter 4's study of the FT that the absolutely integrable (L^1) signals, for example, are in the domain of the the FT system. Also, the DFT maps discrete signals of period N to discrete signals of period N. We surmise (rightly, it turns out) that, owing to the doubly infinite sum in the analysis equation, the study of the DTFT will involve much more mathematical subtlety than the DFT.

Clearly, the DTFT sum exists whenever the signal $x(n)$ is finitely supported; that is, it is zero outside some finite interval. So for a wide—and important—class of signals, the DTFT exists. Without worrying right now about the convergence of the analysis equation (7.36) for general signals, we proceed to some examples of the radial DTFT.

Example (Discrete Delta Signal). Consider the signal $\delta(n)$, the discrete delta function. This signal is unity at $n = 0$, and it is zero for $n \neq 0$. For any $\omega \in \mathbb{R}$, all of the summands in (7.36) are zero, save the $n = 0$ term, and so we have $X(\omega) = 1$ for all $\omega \in \mathbb{R}$.

Example (Discrete Square Pulse). Consider the signal $h(n) = [1, 1, \underline{1}, 1, 1]$, which is unity for $-2 \leq n \leq 2$ and zero otherwise. We recognize this signal as the impulse response of a Moving Average System. If $y = Hx = h * x$, then the system H averages the five values around $x(n)$ to produce $y(n)$. We calculate

$$H(\omega) = \sum_{n=-\infty}^{+\infty} h(n)\exp(-jn\omega) = \sum_{n=-2}^{+2} \exp(-jn\omega)$$

$$= e^{2j\omega} + e^{j\omega} + 1 + e^{-j\omega} + e^{-2j\omega} = 1 + 2\cos(\omega) + 2\cos(2\omega). \tag{7.38}$$

The notation might cause confusion, but we will use $H(\omega)$ for the DTFT of discrete signal $h(n)$ and simply H for the system with impulse response $h(n)$. Figure 7.10 plots $H(\omega)$.

Example (Exponential Signal). Consider the signal $x(n) = 2^{-n}u(n)$, and its DTFT analysis equation,

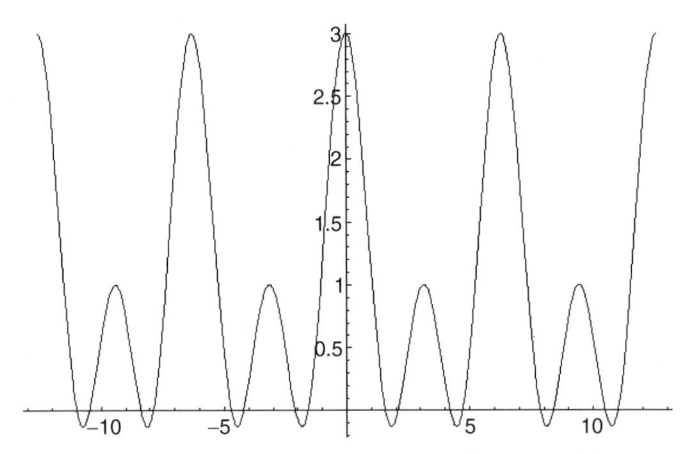

Fig. 7.10. Graph of $H(\omega)$ where $h(n) = [1, 1, \underline{1}, 1, 1]$. $H(\omega)$ is a 2π-periodic analog signal, and it assumes a maximum value at $\omega = 0$. In general, $H(\omega)$ is complex, and in such cases we prefer to plot $|H(\omega)|$ and $\arg(H(\omega))$ on one period: $[0, 2\pi]$ or $[-\pi, \pi]$.

$$X(\omega) = \sum_{n=-\infty}^{+\infty} x(n)\exp(-jn\omega) = \sum_{n=0}^{+\infty}\left(\frac{1}{2}\right)^n \exp(-jn\omega) = \sum_{n=0}^{+\infty}\left(\frac{e^{-j\omega}}{2}\right)^n$$

$$= \frac{1}{1-\left(\frac{e^{-j\omega}}{2}\right)} = \frac{1}{1-\frac{\cos(\omega)}{2}+j\frac{\sin(\omega)}{2}} = \frac{1-\frac{\cos(\omega)}{2}-j\frac{\sin(\omega)}{2}}{\left(1-\frac{\cos(\omega)}{2}\right)^2+\left(\frac{\sin(\omega)}{2}\right)^2}. \quad (7.39)$$

Signal $x(n)$ is a geometric series of the form $1 + a + a^2 + \cdots$, where $a = (1/2)\exp(-j\omega)$. Since $|a| = 1/2 < 1$, the step from an infinite sum to the simple quotient $1/(1 - a)$ on the bottom left side of (7.39) is justified. Figure 7.11 shows the magnitude and phase of the DTFT of $x(n)$.

Notice that the DTFT (7.36) appears to be in the form of an l^2 inner product. Given $x(n)$, we can informally write $X(\omega) = \langle x(n), \exp(j\omega n)\rangle$, because the l^2 inner

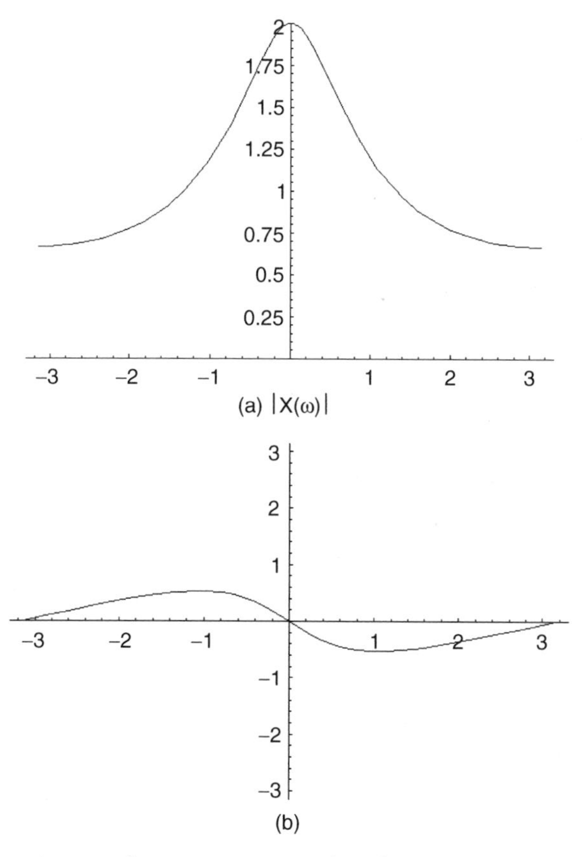

Fig. 7.11. DTFT of $x(n) = 2^{-n}u(n)$. Panel (a) shows $|X(\omega)|$ on $[-\pi, \pi]$. Panel (b) plots $\arg(X(\omega))$.

product, like the DTFT analysis equation, is precisely the sum of all products of terms $x(n)\exp(-j\omega n)$. Conceptually (and still informally), then, for each ω the DTFT $X(\omega)$ measures the similarity between $x(n)$ and $\exp(j\omega n)$. So $X(\omega)$ is the spectral content of the discrete signal $x(n)$ for radial frequency ω. We cannot formalize this intuition unless we know more about the convergence of the analysis equation sum. The signal $\exp(j\omega n)$ is not square-summable, so we do not know, without more investigation, whether the inner product-like sum converges. The convergence of the DTFT analysis equation is a question that we must therefore address, and fortunately, there are satisfactory answers.

7.2.1.2 Existence. Let us consider two of the most important classes of discrete signals: absolutely-summable signals and square-summable signals. Do representatives of these function spaces always have discrete-time Fourier transforms? Guaranteeing a transform for l^2 signals, like the case of the analog L^2 Fourier transforms in Chapter 5, requires some care. There is, however, a very easy existence proof for the DTFT of l^1 signals.

Theorem (DTFT Existence for Absolutely Summable Signals). Suppose $x(n) \in l^1$. Then the DTFT of $x(n)$, $X(\omega)$, exists and converges absolutely and uniformly for all $\omega \in \mathbb{R}$.

Proof: To show absolute convergence of the analysis equation, we need to show that

$$\sum_{n=-\infty}^{+\infty} \left|x(n)\exp(-jn\omega)\right| = \sum_{n=-\infty}^{+\infty} \left|x(n)\right|\left|\exp(-jn\omega)\right| = \sum_{n=-\infty}^{+\infty} \left|x(n)\right| < \infty. \tag{7.40}$$

But the last term in (7.40) is precisely the l^1 norm of $x(n)$, and this must be finite, since $x(n) \in l^1$. So the analysis equation formula for $X(\omega)$ does indeed converge to a limit for any ω. We recall from real and complex analysis [13, 14] that the uniform convergence of the series (7.36), means that for every $\varepsilon > 0$ there is an $N > 0$ such that if $m, n > N$, then for all ω:

$$\left|\sum_{k=m}^{k=n} x(k)\exp(-jk\omega)\right| < \varepsilon. \tag{7.41}$$

The main point of uniform convergence is that the Cauchy criterion (7.41) applies to all ω, independent of N, which depends only on the choice of ε. If N must vary with ε, then ordinary, but not uniform, convergence exists. Now, (7.40) shows convergence for all ω; the interval $[-\pi, \pi]$ is closed and bounded, hence it is *compact*; and a convergent series on a compact subset of the real line is uniformly convergent on that subset. Therefore, since we know that $X(\omega)$ is periodic with period $[-\pi, \pi]$, the DTFT analysis equation series converges uniformly to $X(\omega)$ for all ω. ■

Corollary (Continuity of DTFT). Suppose $x(n) \in l^1$. Then the DTFT of $x(n)$, $X(\omega)$, is continuous.

Proof: It is the sum of a uniformly convergent series, and each partial sum in the series,

$$S_N(\omega) = \sum_{n=-N}^{n=N} x(n)\exp(-jn\omega), \tag{7.42}$$

is a continuous function of ω. ■

Now consider a signal $x(n) \in l^2$. We are familiar with the signal space l^2 from Chapter 2: It is the Hilbert space of square-summable discrete signals. It has an inner product, $\langle x, y\rangle$, which measures how alike are two l^2 signals; a norm, $\|x\|_2$, which derives from the inner product, $\|x\|_2 = (\langle x, x\rangle)^{1/2}$; and orthonormal bases, $\{e_i : i \in \mathbb{N}\}$, which allow us to decompose signals according to their similarity to the basis elements, $\langle e_i, x\rangle$. Thus, Hilbert spaces extend the convenient analytical tools (which one finds, for example, in finite-dimensional vector spaces), to the doubly infinite "vectors" of signal processing and analysis—discrete signals. Let us next show that for square-summable signals the DTFT exists.

Theorem (DTFT Existence for Square-Summable Signals). Suppose $x(n) \in l^2$. Then the DTFT of $x(n)$, $X(\omega)$, exists for all $\omega \in \mathbb{R}$.

Proof: Let $\omega \in \mathbb{R}$ and consider the partial DTFT analysis equations sums,

$$X_N(\omega) = \sum_{n=-N}^{+N} x(n)\exp(-jn\omega), \tag{7.43}$$

where $N \in \mathbb{N}$, the natural numbers. (We shall use the notation $X_N(\omega)$ for DTFT partial sums quite a bit.) Let H be the Hilbert space of square-integrable analog signals (cf. Chapter 3) on $[-\pi, \pi]$: $H = L^2[-\pi, \pi]$. For each N, $X_N(\omega)$ has period 2π and is square-integrable on $[-\pi, \pi]$:

$$\|X_N(\omega)\|_H^2 = \int_{-\pi}^{\pi} \left| \sum_{n=-N}^{N} x(n)\exp(-jn\omega) \right|^2 d\omega < \infty; \tag{7.44}$$

hence, $X_N(\omega) \in L^2[-\pi, \pi]$. Let us denote the L^2-norm on H of a square-integrable signal, $s(t)$, by $\|s(t)\|_H$. We wish to show that $(X_N(\omega): N \in \mathbb{N})$ is an $L^2[-\pi, \pi]$ Cauchy sequence; that is, given $\varepsilon > 0$, we can choose $N > M$ both sufficiently large so that $\|X_N - X_M\|_H < \varepsilon$. This would imply that the limit of the sequence $(X_N(\omega))$, which is the series sum of the analysis equation for $x(n)$, does in fact converge to an element of H: $X(\omega)$. Note that

$$X_N(\omega) - X_M(\omega) = \sum_{n=M+1}^{N} x(n)\exp(-jn\omega) + \sum_{n=-M-1}^{-N} x(n)\exp(-jn\omega). \tag{7.45}$$

Thus, using the orthogonality of signals $\exp(-jwn)$ on $[-\pi, \pi]$ and the properties of the inner product on H, we find

$$\|X_N(\omega)-X_M(\omega)\|_H^2$$

$$=\left\langle \sum_{n=M+1}^{N} x(n)e^{-j\omega n}+\sum_{n=-M-1}^{-N} x(n)e^{-j\omega n},\ \sum_{m=M+1}^{N} x(m)e^{-j\omega m}+\sum_{m=-M-1}^{-N} x(m)e^{-j\omega m}\right\rangle$$

$$=\left\langle \sum_{n=M+1}^{N} x(n)e^{-j\omega n},\ \sum_{m=M+1}^{N} x(m)e^{-j\omega m}+\sum_{m=-M-1}^{-N} x(m)e^{-j\omega m}\right\rangle$$

$$+\left\langle \sum_{n=-M-1}^{-N} x(n)e^{-j\omega n},\ \sum_{m=M+1}^{N} x(m)e^{-j\omega m}+\sum_{m=-M-1}^{-N} x(m)e^{-j\omega m}\right\rangle$$

$$=\left\langle \sum_{n=M+1}^{N} x(n)e^{-j\omega n},\ \sum_{m=M+1}^{N} x(m)e^{-j\omega m}\right\rangle+\left\langle \sum_{n=M+1}^{N} x(n)e^{-j\omega n},\ \sum_{m=-M-1}^{-N} x(m)e^{-j\omega m}\right\rangle$$

$$+\left\langle \sum_{n=-M-1}^{-N} x(n)e^{-j\omega n},\ \sum_{m=M+1}^{N} x(m)e^{-j\omega m}\right\rangle+\left\langle \sum_{n=-M-1}^{-N} x(n)e^{-j\omega n},\ \sum_{m=-M-1}^{-N} x(m)e^{-j\omega m}\right\rangle$$

$$=\sum_{n=M+1}^{N}\sum_{m=M+1}^{N} x(n)\bar{x}(m)\left\langle e^{-j\omega n},e^{-j\omega m}\right\rangle+\sum_{n=M+1}^{N}\sum_{m=-M-1}^{-N} x(n)\bar{x}(m)\left\langle e^{-j\omega n},e^{-j\omega m}\right\rangle$$

$$+\sum_{n=-M-1}^{-N}\sum_{m=M+1}^{N} x(n)\bar{x}(m)\left\langle e^{-j\omega n},e^{-j\omega m}\right\rangle+\sum_{n=-M-1}^{-N}\sum_{m=-M-1}^{-N} x(n)\bar{x}(m)\left\langle e^{-j\omega n},e^{-j\omega m}\right\rangle$$

$$=2\pi\sum_{n=M+1}^{N} x(n)\bar{x}(n)+0+0+2\pi\sum_{n=-M-1}^{-N} x(n)\bar{x}(n). \tag{7.46}$$

We have used $\langle \exp(-j\omega n), \exp(-j\omega m)\rangle = 2\pi\delta_{n,m}$, where $\delta_{n,m}$ is the Kronecker delta, to simplify the first and last double summations and to discard the two middle double summations in (7.46). Now, since $x(n) \in l^2$, it has a finite l^2-norm; in other words,

$$\|x(n)\|^2=\sum_{n=-\infty}^{\infty} x(n)\bar{x}(n)<\infty. \tag{7.47}$$

This means that for $\varepsilon > 0$, we can find $N > M$ sufficiently large so that

$$\sum_{n=M+1}^{N} x(n)\bar{x}(n)<\varepsilon \tag{7.48a}$$

and

$$\sum_{n=-M-1}^{-N} x(n)\bar{x}(n)<\varepsilon. \tag{7.48b}$$

Together, (7.48a) and (7.48b) imply that (7.46) can be made arbitrarily small. Thus the sequence $(X_N(\omega): N \in \mathbb{N})$ is Cauchy in $H = L^2[-\pi, \pi]$. Since H is a Hilbert space, this Cauchy sequence converges to a signal, the DTFT of $x(n)$, $X(\omega) \in H$. ■

7.2.1.3 Inversion. This section studies the problem of finding an inverse for the DTFT. Those frequency transforms covered so far—the Fourier series, the Fourier transform, and the discrete Fourier transform—all have inverses, assuming, in the case of the FS and FT that our signals belong to certain function spaces. So we expect no exceptions from the frequency transform for discrete aperiodic signals. Now, the DTFT of a discrete signal $x(n)$ is a periodic function $X:\mathbb{R} \to \mathbb{C}$, so the inverse must transform a periodic analog signal into a discrete signal. One transform, familiar from Chapter 5, does precisely this—the Fourier series. For an analog periodic signal, the FS finds a discrete set of frequency coefficients. We shall see that there is in fact a very close relationship between the inverse relation for the DTFT and the analog FS.

Our first theorem provides a simple criterion for the existence of an inverse.

Theorem (Inverse DTFT). Suppose that $x(n)$ has a DTFT, $X(\omega)$, and that the analysis equation for $X(\omega)$ converges uniformly on $[-\pi, \pi]$. Then, for all $n \in \mathbb{Z}$,

$$x(n) = \frac{1}{2\pi} \int_{-\pi}^{+\pi} X(\omega) \exp(j\omega n)\, d\omega \cdot \tag{7.49}$$

Proof: The stipulation that the analysis equation's convergence be uniform is critical to the proof. The DTFT analysis equation for $x(n)$ is a limit of partial sums:

$$X(\omega) = \lim_{N\to\infty} X_N(\omega) = \lim_{N\to\infty} \sum_{n=-N}^{+N} x(n) \exp(-jn\omega) = \sum_{n=-\infty}^{+\infty} x(n) \exp(-jn\omega). \tag{7.50}$$

After changing the dummy summation variable, we insert (7.50) directly into the integrand of (7.49):

$$\frac{1}{2\pi} \int_{-\pi}^{+\pi} X(\omega) \exp(j\omega n)\, d\omega = \frac{1}{2\pi} \int_{-\pi}^{+\pi} \left(\lim_{N\to\infty} \sum_{m=-N}^{+N} x(m) \exp(-jm\omega) \right) \exp(j\omega n)\, d\omega. \tag{7.51}$$

The uniform convergence of the limit in (7.51) permits us to interchange the integration and summation operations [13]:

$$\begin{aligned} \frac{1}{2\pi} \int_{-\pi}^{+\pi} X(\omega) \exp(j\omega n)\, d\omega &= \frac{1}{2\pi} \lim_{N\to\infty} \sum_{m=-N}^{+N} x(m) \int_{-\pi}^{+\pi} \exp[j\omega(n-m)]\, d\omega \\ &= \frac{1}{2\pi} \sum_{m=-\infty}^{+\infty} x(m) \delta_{m,n} = x(n), \end{aligned} \tag{7.52}$$

where $\delta_{n,m}$ is the Kronecker delta. ■

Corollary (Inverse DTFT for Absolutely Summable Signals). If $x(n) \in l^1$, then for all $n \in \mathbb{Z}$,

$$x(n) = \frac{1}{2\pi} \int_{-\pi}^{+\pi} X(\omega) \exp(j\omega n)\, d\omega, \tag{7.53}$$

where $X(\omega)$ is the DTFT of $x(n)$.

Proof: The analysis equation sum for $X(\omega)$ converges uniformly by the DTFT Existence Theorem for Absolutely Summable Signals in the previous section. Hence the Inverse Theorem above implies that the formula (7.53) is valid. ■

Definition (Inverse Discrete-Time Fourier Transform). If $X(\omega)$ is a 2π-periodic analog signal and $x(n)$, as defined by (7.53), exists, then $x(n)$ is the *inverse discrete-time Fourier transform* (IDTFT) of $X(\omega)$. Equation (7.53) is also called the DTFT *synthesis equation.*

This last result (7.53) highlights an intriguing aspect of the DTFT. Equation (7.53) says that if $x(n)$ is absolutely summable, then $x(n)$ is the nth Fourier series coefficient for $X(\omega)$. To understand this, recall that if the analog signal $y(s) \in L^1[0, T]$ has period $T > 0$, then the Fourier series analysis equation gives (unnormalized) Fourier coefficients,

$$c_k = \frac{1}{T} \int_{-T/2}^{+T/2} y(s) \exp(-2\pi jkFs)\, ds, \tag{7.54}$$

where $F = 1/T$. The companion synthesis equation reconstructs $y(s)$ from the c_k:

$$y(s) = \sum_{k=-\infty}^{+\infty} c_k \exp(2\pi jkFs). \tag{7.55}$$

The reconstruction is not perfect, however. When $y(s)$ contains a discontinuity at $s = s_0$, then the synthesis equation converges to a value midway between the left- and right-hand limits of $y(s)$ at $s = s_0$. We can easily transpose (7.55) to the form of the DTFT analysis equation. We need only set $x(k) = c_k$, $\omega = -2\pi Fs$, and $X(\omega) = y(s) = y(-\omega/(2\pi F))$. Therefore, analog periodic signals have discrete spectral components given by the Fourier series coefficients. And discrete aperiodic signals have continuous periodic spectra given by their 2π-periodic DTFTs. There is more than an affinity between the discrete aperiodic signals and the analog periodic signals; there is, as we shall see in a moment, a Hilbert space isomorphism.

Now, we have found an inversion relation for the DTFT and defined the synthesis equation for absolutely summable signals, but what about l^2 signals? This is our most important l^p space, since it supports an inner product relation. Is there an inverse relation similar to (7.53) for square-summable discrete signals? The following theorem helps to answer this question in the affirmative, and it is one step toward showing the essential sameness of l^2 and $L^2[-\pi, \pi]$.

Theorem (DTFT Existence for Square-Summable Signals). If $x(n) \in l^2$, then there is an analog signal $y(s) \in L^2[-\pi, \pi]$ such that

$$x(n) = \frac{1}{2\pi} \int_{-\pi}^{+\pi} y(s) \exp(jsn)\, ds, \tag{7.56a}$$

for all $n \in \mathbb{Z}$, and

$$y(s) = \sum_{n=-\infty}^{+\infty} x(n) \exp(-jsn), \tag{7.56b}$$

Proof: Recall the Riesz–Fischer Theorem from Chapter 2. Suppose we take $H = L^2[-\pi, \pi]$ as the Hilbert space and $\{e_n(s): n \in \mathbb{Z}\} = \{(2\pi)^{-1/2}\exp(-jsn): n \in \mathbb{Z}\}$ as the orthonormal set in H which Riesz–Fischer presupposes. Then the Riesz–Fischer result states that if $x(n) \in l^2$, then there is a $w(s) \in H$ such that $\langle w, e_n(s)\rangle = x(n)$, and $w = \Sigma x(n)e_n(s)$. Furthermore, w is unique in the following sense: Any other $h \in H$ for which this holds can differ from w only on a set of measure zero; in other words, $\|w - h\|_2 = 0$, where $\|\cdot\|_2$ is the norm on $L^2[-\pi, \pi]$:

$$\|y\|_2 = \left(\int_{-\pi}^{+\pi} |y(s)|^2 ds \right)^{\frac{1}{2}}. \tag{7.57}$$

Continuing to apply this previous abstract theorem to our present concrete problem, we must have a $w(s) \in H$ such that

$$x(n) = \langle w, e_n \rangle = \langle w, \exp(-jsn) \rangle = \int_{-\pi}^{+\pi} w(s) \frac{\exp(jsn)}{\sqrt{2\pi}}\, ds \tag{7.58}$$

and

$$w(s) = \sum_{n=-\infty}^{+\infty} x(n) e_n(s) = \sum_{n=-\infty}^{+\infty} x(n) \frac{\exp(-jsn)}{\sqrt{2\pi}}. \tag{7.59}$$

Setting $y(s) = (2\pi)^{1/2} w(s)$ completes the proof. ■

How exactly does this help us answer the question of whether square-summable discrete signals have DTFTs? Briefly, $x(n) \in l^2$ does have a DTFT: We take $X(\omega) = y(\omega)$, where y is the $L^2[-\pi, \pi]$ signal guaranteed by the theorem. The problem is that $X(\omega)$ need not be continuous; therefore, there are many possible choices for $X(\omega)$ in

$L^2[-\pi, \pi]$ that obey the DTFT synthesis equation. The various choices may differ on a set of measure zero, so that the norm of their difference, computed with an integral of adequate power (such as the Lebesgue integral) is zero. This should be no surprise. We recall Chapter 5's lesson, for example, that the Fourier series sum of a signal converges to the midpoint of a signal discontinuity. The FS imperfectly recovers the original periodic analog signal. If $x(n) \in l^1$, on the other hand, then the convergence of the DTFT analysis equation (or, alternatively, the convergence of the Fourier series sum) is uniform, so that $X(\omega)$ is continuous and pointwise unique on $[-\pi, \pi]$.

Corollary (Embedding of l^2 into $L^2[-\pi, \pi]$). For $x(n) \in l^2$, then set $\mathcal{F}(x) = (2\pi)^{-1/2}X(\omega) \in L^2[-\pi, \pi]$. Then $\mathcal{F}$ is a Hilbert space isomorphism between l^2 and (the equivalence classes of signals that are equal almost everywhere) its image $\mathcal{F}[l^2]$.

Proof: The theorem guarantees that a unique (up to a set of measure zero) $X(\omega)$ exists, so $\mathcal{F}$ is well-defined. It is also clear that the DTFT is a linear mapping from l^2 to $L^2[-\pi, \pi]$, and so too is $\mathcal{F}$ (exercise). We need to show as well that $\langle x, y\rangle = \langle \mathcal{F}x, \mathcal{F}y\rangle$, for all $x, y \in l^2$. Let $Y(\omega)$ be the DTFT of $y(n)$. Working from within the realm of $L^2[-\pi, \pi]$, we find

$$\langle X,Y\rangle = \int_{-\pi}^{+\pi} X(\omega)\overline{Y}(\omega)\,d\omega = \int_{-\pi}^{+\pi}\left[\sum_{n=-\infty}^{+\infty} x(n)\exp(-j\omega n)\right]\left[\sum_{k=-\infty}^{+\infty}\overline{y}(k)\exp(j\omega k)\right]d\omega$$

$$= \int_{-\pi}^{+\pi}\left[\sum_{n=-\infty}^{+\infty}\sum_{k=-\infty}^{+\infty} x(n)\overline{y}(k)\exp^{-j\omega(n-k)}\right]d\omega. \tag{7.60}$$

Since $x, y \in l^2$, the partial sums $X_N(\omega)$ and $Y_N(\omega)$ converge absolutely; for example, we have $|x(n)\exp(-j\omega n)| = |x(n)|$, and $\Sigma|x(n)|^2 = (\|x\|_2)^2 < \infty$. This justifies the step to a double summation of products [13]. And, because the double sum on the bottom of (7.60) converges on the closed set $[-\pi, \pi]$, it converges uniformly. This allows us to interchange the summation and integration operations, obtaining

$$\langle X,Y\rangle = \sum_{n=-\infty}^{+\infty}\sum_{k=-\infty}^{+\infty}\int_{-\pi}^{+\pi} x(n)\overline{y}(k)\exp^{-j\omega(n-k)}\,d\omega = 2\pi\sum_{n=-\infty}^{+\infty} x(n)\overline{y}(n)$$

$$= 2\pi\langle x(n), y(n)\rangle. \tag{7.61}$$

Only the terms with $n = k$ in the integral of (7.61) are nonzero. Finally, since $\mathcal{F}(x) = (2\pi)^{-1/2}X(\omega)$, $\langle \mathcal{F}x, \mathcal{F}y\rangle = \langle x, y\rangle$. ■

An embedding therefore exists from the discrete Hilbert space l^2 into the continuous Hilbert space $L^2[-\pi, \pi]$. This Hilbert subspace of $L^2[-\pi, \pi]$, the image of l^2 under $\mathcal{F}$, $\mathcal{F}[l^2]$, is essentially just like l^2. Is it the case, perhaps owing to the intricacies of its analog signal elements, that the full Hilbert space, $L^2[-\pi, \pi]$, is fundamentally more complex than l^2? The requisite tools of formal analog theory—Dirac delta functions, Lebesgue integration, questions of separability, and so on—make it tempting to conclude that $L^2[-\pi, \pi]$ ought to be a richer mathematical object than the drab, discrete l^2. Moreover, the embedding that we have given is a straightforward application of the abstract Riesz–Fischer theorem; no technical arguments using the specific characteristics of $L^2[-\pi, \pi]$ signals are necessary. So it might well be concluded that the orthogonal complement, $\mathcal{F}[l^2]^{\perp}$, is indeed nontrivial.

No, the truth is quite the opposite: The mapping $\mathcal{F}(x(n)) = (2\pi)^{-1/2}X(\omega)$ from l^2 into $L^2[-\pi, \pi]$ is indeed a Hilbert space isomorphism. We can show this if we can find a set of signals in the image of l^2 under the embedding relation, $\mathcal{F}$, that is dense in $L^2[-\pi, \pi]$. In general, questions of orthogonality and finding embeddings (also called injections) of one Hilbert space into another tend to admit easier answers. But showing that one or another set of orthogonal elements spans the entire Hilbert space—the question of completeness—is quite often a daunting problem. Fortunately, we already have the crucial tool in hand, and the next corollary explains the result.

Corollary (Isomorphism of l^2 and $L^2(-\pi, \pi)$). Let $x(n) \in l^2$, let $X(\omega)$ be the DTFT of $x(n)$, and set $\mathcal{F}(x) = (2\pi)^{-1/2}X(\omega) \in L^2[-\pi, \pi]$. Then $\mathcal{F}$ is a Hilbert space isomorphism.

Proof: Consider some $Y(\omega) \in L^2[-\pi, \pi]$. We need to show that Y is arbitrarily close to some element of the image of $\mathcal{F}$, $\mathcal{F}[l^2]$. From Chapter 5, Y has a Fourier series representation,

$$Y(\omega) = \sum_{k=-\infty}^{+\infty} c_k \exp(2\pi jkF\omega) = \sum_{k=-\infty}^{+\infty} c_k \exp(jk\omega)$$

$$= \lim_{K\to\infty} \sum_{k=-K}^{+K} c_k \exp(jk\omega) = \lim_{K\to\infty} Y_K(\omega), \tag{7.62}$$

where

$$c_k = \frac{1}{T}\int_{-T/2}^{+T/2} Y(\omega)\exp(-2\pi jkF\omega)\,d\omega = \frac{1}{2\pi}\int_{-\pi/2}^{+\pi/2} Y(\omega)\exp(-jk\omega)\,d\omega. \tag{7.63}$$

Since $T = \pi - (-\pi) = 2\pi$ and $F = 1/T = 1/(2\pi)$, (7.62) shows that see that $Y(\omega)$ is really the limit of the $Y_K(\omega)$. But each $Y_K(\omega)$ is a linear combination of exponentials, $\exp(jk\omega)$, which are in the image, $\mathcal{F}[l^2]$. Since Y was arbitrary, this implies that the span of the exponentials is dense in $L^2[-\pi, \pi]$, or, equivalently, that its closure is all of $L^2[-\pi, \pi]$. ■

The next section contains an extended study of a single example that illustrates some of the convergence problems that arise when taking the DTFT of a signal that is not absolutely summable.

7.2.1.4 Vagaries of $X(\omega)$ convergence in $L^2[-\pi, \pi]$. The search for those classes of discrete signals, $x(n)$, that have a DTFT leads to difficulties with the convergence of the analysis equation sum. We have just shown that absolutely summable and square-summable discrete signals have DTFTs; if $x(n) \in l^2$, however, we cannot guarantee that the DTFT, $X(\omega)$, is unique. Now we explore in some detail an example of a square-summable signal that is not in l^1. Its DTFT is not unique, due to the presence of discontinuities in $X(\omega)$. Moreover, its convergence is tainted by spikes near the points of discontinuity that persist even as the partial analysis equation sums converge (in the l^2 norm) to $X(\omega)$.

Example (Discrete Sinc Signal). Consider the signal $x(n)$ defined as follows:

$$x(n) = \frac{\operatorname{sinc}(n)}{\pi} = \begin{cases} \dfrac{\sin(n)}{\pi n} & \text{if } n \neq 0, \\[2ex] \dfrac{1}{\pi} & \text{if } n = 0. \end{cases} \tag{7.64}$$

Although $x(n) \notin l^1$, since its absolute value decays like n^{-1}, we do find $x(n) \in l^2$, because $|x(n)|^2$ is dominated by $(\pi n)^{-2}$, which does converge. If we let $\tilde{X}(\omega) = 1$ for $|\omega| \leq 1$ and $\tilde{X}(\omega) = 0$ otherwise, then

$$x(n) = \frac{1}{2\pi} \int_{-\pi}^{+\pi} \tilde{X}(\omega) \exp(j\omega n)\, d\omega \cdot \tag{7.65}$$

Thus, the IDTFT of $\tilde{X}(\omega)$ is $x(n)$. Figure 7.12 shows $\tilde{X}(\omega)$ and the discrete signal that results from its IDTFT.

Is $\tilde{X}(\omega)$ the DTFT of $x(n)$? Not exactly, but let us study the situation further. Since (7.65) is a Fourier series analysis equation, and $x(n)$ is a FS coefficient for $\tilde{X}(\omega)$, we can consider the limit of the corresponding Fourier series sum, $X(\omega)$. Then $X(\omega)$ converges for $\pm\omega = 1$ to $X(\omega) = 1/2$, the midpoint between the discontinuities:

$$\begin{aligned} X(\omega) &= \lim_{N \to \infty} \sum_{n=-N}^{n=+N} x(n) \exp(-j\omega n) \\ &= \sum_{n=-\infty}^{n=+\infty} x(n) \exp(-j\omega n) = \begin{cases} 1 & \text{if } |\omega| < 1, \\ 0 & \text{if } |\omega| > 1, \\ 1/2 & \text{if } |\omega| = 1. \end{cases} \end{aligned} \tag{7.66}$$

This is an unlucky result: for each $\omega \in [-\pi, \pi]$ the partial sums in (7.66) have a limit, but it is $X(\omega)$, not $\tilde{X}(\omega)$. The convergence of the partial sums in (7.66) is not uniform,

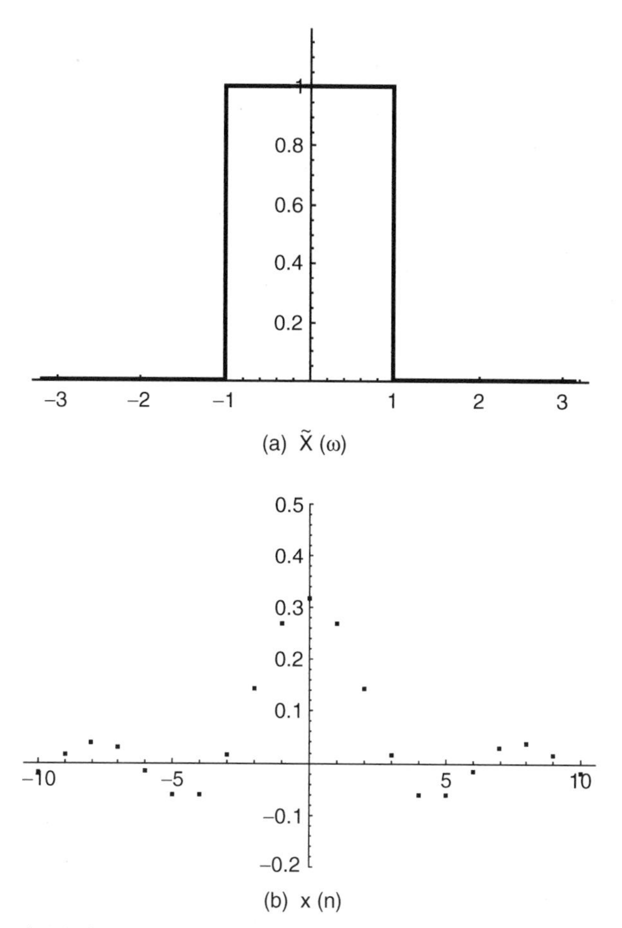

Fig. 7.12. Panel (a) shows a square pulse signal on $[-\pi, \pi]$. The result of applying the IDTFT to this pulse is shown in panel (b).

as it would be if $x(n)$ were absolutely summable. Instead, the convergence is in the $L^2[-\pi, \pi]$ norm, $\|\cdot\|_2$, and it allows the limit of the DTFT analysis equation, $X(\omega)$, to differ from the signal from which we derived $x(n)$ in the first place, $\tilde{X}(\omega)$.

Example (Gibbs Phenomenon). Examining the partial sums of (7.62) exposes a further feature of convergence under $\|\cdot\|_2$, namely the Gibbs[4] phenomenon. Spikes appear near the step edges in $\tilde{X}(\omega)$ and do not diminish with increasingly long partial sums (Figure 7.13).

[4]Josiah W Gibbs (1839–1903), an American chemist, physicist, and professor at yale University. Gibbs devised the vector dot product, $v \cdot w$, and the cross product, $v \times w$, and investigated the famous spike in Fourier series convergence. Although he was the first scientist of international stature from the United States, Yale neither appreciated his capabilities nor remunerated him for his service. Gibbs supported himself on an inheritance over the cource of a decade at Yale.

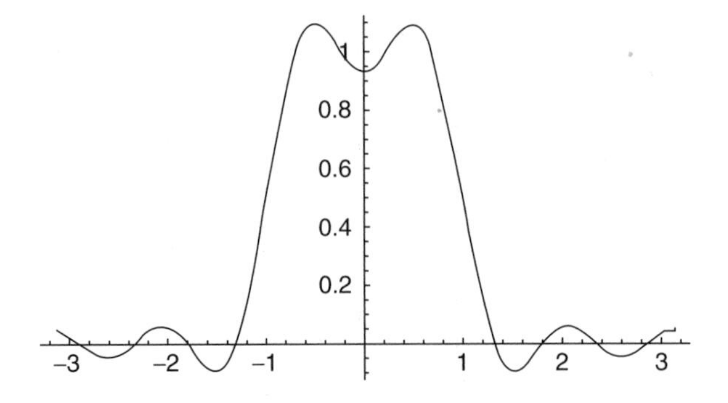

(a) Partial DTFT sum, $X_N(\omega)$, N = 5;

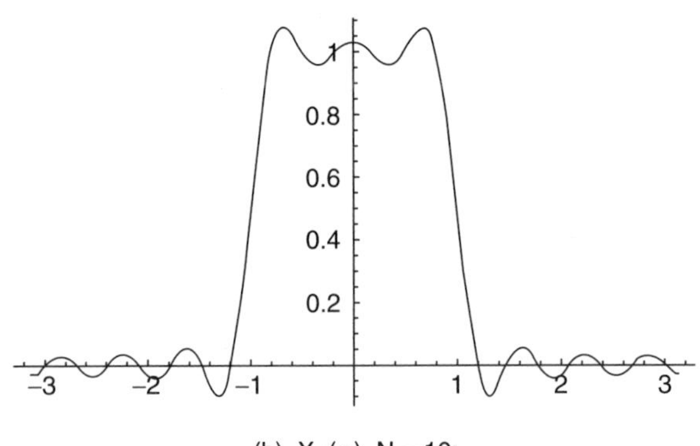

(b) $X_N(\omega)$, N = 10;

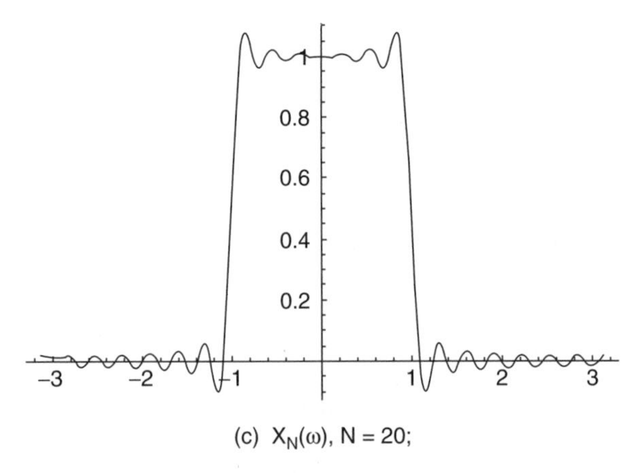

(c) $X_N(\omega)$, N = 20;

Fig. 7.13. A study of partial DTFT analysis equation sums for the square-summable signal $x(n) = \pi^{-1}\sin(n)$. Panel (a) shows $X_5(\omega)$; panel (b) shows $X_{10}(\omega)$; panel (c) shows $X_{20}(\omega)$; and panel (d), $X_{50}(\omega)$, shows a persistent ringing effect at the discontinuities.

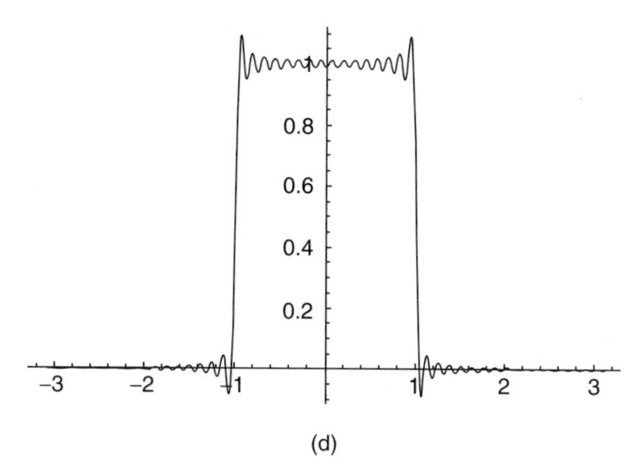

Fig. 7.13 (*Continued*)

Although its name has a supernatural resonance, there are specific, clearly definable reasons for the Gibbs phenomenon. In fact, it occurs whenever there is a discontinuity in a square integrable signal on $[-\pi, \pi]$. To more clearly grasp the reasons for the phenomenon, let us consider the partial DTFT sums, $X_N(\omega)$:

$$X_N(\omega) = \sum_{n=-N}^{n=+N} x(n)\exp(-j\omega n) = \sum_{n=-N}^{n=+N} \left\{ \frac{1}{2\pi} \int_{-\pi}^{+\pi} X(\Omega)\exp(j\Omega n)\, d\Omega \right\} \exp(-j\omega n)$$

$$= \frac{1}{2\pi} \sum_{n=-N}^{n=+N} \left\{ \int_{-\pi}^{+\pi} X(\Omega)\exp(j(\Omega-\omega)n)\, d\Omega \right\}. \tag{7.67}$$

The IDTFT of $X(\omega)$ replaces $x(n)$ in the finite summation of (7.67). $X(\omega)$ itself is the limit of these partial DTFT sums as $N \to \infty$. (We already know the definition of $X(\omega)$: It is the step function equal to unity on $[-1, 1]$ and zero elsewhere on $[-\pi, +\pi]$. It is thus possible to simplify the integral in (7.67), but we resist in order to show how the following development does not depend on the specific nature of $X(\omega)$.) Next, we set $\theta = \Omega - \omega$ toward a change of the integration variable, thereby giving

$$X_N(\omega) = \frac{1}{2\pi} \sum_{n=-N}^{n=+N} \left\{ \int_{-\pi-\omega}^{\pi-\omega} X(\theta+\omega)\exp(j\theta n)\, d\theta \right\}$$

$$= \frac{1}{2\pi} \sum_{n=-N}^{n=+N} \left\{ \int_{-\pi}^{\pi} X(\theta+\omega)\exp(j\theta n)\, d\theta \right\} \tag{7.68}$$

Note that the integrand is 2π-periodic, which permits us to take the limits of integration from $[-\pi, +\pi]$ instead of $[-\pi - \omega, \pi - \omega]$. Let us now interchange the finite summation and integration operations in (7.68) to obtain

$$X_N(\omega) = \int_{-\pi}^{\pi} X(\theta+\omega)\left\{\frac{1}{2\pi}\sum_{n=-N}^{n=+N}\exp(j\theta n)\right\}d\theta = \int_{-\pi}^{\pi} X(\theta+\omega)D_N(\theta)d\theta. \tag{7.69}$$

Therefore, the partial DTFT synthesis equation sums, $X_N(\omega)$, are given by the cross-correlation on $[-\pi, +\pi]$ of X and the Dirichlet[5] kernel of order N, $D_N(\theta)$. Chapter 5 introduced the Dirichlet kernel in connection with the problem of the Fourier series sum's convergence . It is an algebraic exercise to show that

$$D_N(\theta) = \frac{1}{2\pi}\sum_{n=-N}^{n=+N}\exp(j\theta n) = \frac{1}{2\pi}\frac{\sin\left(N\theta+\frac{\theta}{2}\right)}{\sin\left(\frac{\theta}{2}\right)} = \frac{1}{2\pi}+\frac{1}{\pi}\sum_{n=1}^{n=N}\cos(\theta n), \tag{7.70}$$

and therefore, for any $N > 0$,

$$\int_{0}^{\pi} D_N(\theta)\ d\theta = \frac{1}{2} = \int_{-\pi}^{0} D_N(\theta)\ d\theta\ \cdot \tag{7.71}$$

Now, from (7.70) the Dirichlet kernel is an even function. So changing the variable of integration in (7.69) shows $X_N(\omega)$ to be a convolution of X and $D_N(\theta)$:

$$X_N(\omega) = \int_{-\pi}^{\pi} X(\theta+\omega)D_N(\theta)\ d\theta = \int_{-\pi}^{\pi} X(\theta)D_N(\omega-\theta)\ d\theta\ \cdot \tag{7.72}$$

Now we understand the root cause of the Gibbs phenomenon. Because $X(\theta)$ in (7.72) is zero for $|\theta| > 1$, has discontinuities at $\theta = \pm 1$, and is unity for $|\theta| < 1$, the convolution integral produces a response that has a spike near the discontinuity. The spike's height is roughly the sum of the area of $D_N(\theta)$ under its main hump (Figure 7.14) plus the tail of the Dirichlet kernel that overlaps with the nonzero part of X.

The crux of the problem is how the first peak in the evaluation of the convolution integral behaves as $N \to \infty$. Empirically, as evidenced by Figure 7.13, the convolution generates a spike that shrinks in width but does not diminish in height. Clearly, the height of $D_N(\theta)$ does increase as $\theta \to 0$. Hence, it would appear that the area under the Dirichlet kernel between the first two zero crossings, $\pm\theta_N$, where $\theta_N = \pi/(N + 1/2)$, does not fall below some positive value. In fact, this area

[5]Peter Gustav Lejeune Dirichlet (1805–1859) studied convergence kernels for the Fourier series and problems in potential theory. He provided a proof of Fermat's Last Theorem for the case $n = 5$.

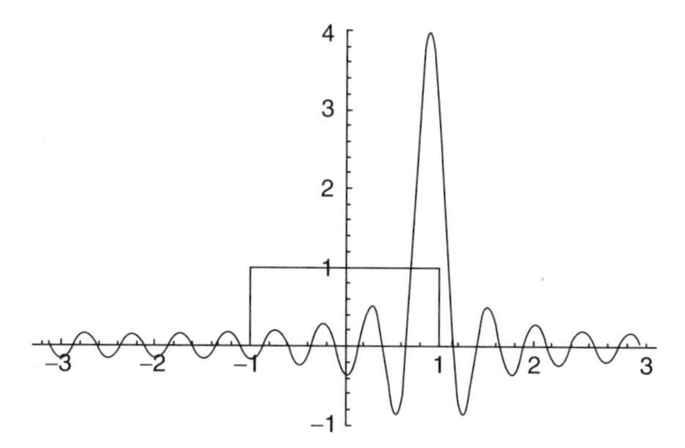

Fig. 7.14. Convolution with $D_N(\theta)$ at a discontinuity. A convolution integral of the analog signal X with the Dirichlet kernel of order N (here, $N = 12$) gives the partial DTFT summation for X, $X_N(\omega)$. The output is equal to the area under $D_N(\theta)$ that overlaps the nonzero area of the square pulse signal X.

remains constant! As $N \to \infty$, its decreasing width is perfectly balanced by the increasing height. Let us investigate:

$$\int_{-\theta_N}^{+\theta_N} D_N(\theta)\, d\theta = 2\int_{0}^{\theta_N} D_N(\theta)\, d\theta = \frac{1}{\pi}\int_{0}^{\theta_N} \frac{\sin\left(N\theta + \frac{\theta}{2}\right)}{\sin\left(\frac{\theta}{2}\right)}\, d\theta. \tag{7.73}$$

Making the change of integration variable $\Omega = N\theta + \theta/2$, we find

$$\int_{-\theta_N}^{+\theta_N} D_N(\theta)\, d\theta = \frac{1}{\pi}\int_{0}^{\pi} \frac{\sin(\Omega)}{\sin\left(\frac{\Omega}{2N+1}\right)} \frac{2}{2N+1}\, d\Omega = \frac{2}{\pi}\int_{0}^{\pi} \frac{\sin(\Omega)}{\Omega} \frac{\left(\frac{\Omega}{2N+1}\right)}{\sin\left(\frac{\Omega}{2N+1}\right)}\, d\Omega. \tag{7.74}$$

As $N \to \infty$, $\Omega/(2N + 1) \to 0$; taking the limit of (7.74) as $N \to \infty$ and interchanging the limit and integration operations on the right-hand side of (7.74) gives

$$\lim_{N\to\infty} \int_{-\theta_N}^{+\theta_N} D_N(\theta)\, d\theta = \frac{2}{\pi}\int_{0}^{\pi} \frac{\sin(\Omega)}{\Omega} \lim_{N\to\infty} \frac{\left(\frac{\Omega}{2N+1}\right)}{\sin\left(\frac{\Omega}{2N+1}\right)}\, d\Omega = \frac{2}{\pi}\int_{0}^{\pi} \frac{\sin(\Omega)}{\Omega}\, d\Omega \tag{7.75}$$

showing that the between the first zero crossings of $D_N(\theta)$ is constant. The area under $\mathrm{sinc}(\Omega)$ from 0 to π is approximately 1.85194, so the convolution integral (7.72) for the partial DTFT summation, $X_N(\omega)$, evaluates to approximately

$(2/\pi)(1.85194) = 1.17898$. To find the approximate value of the convolution integral (7.72) at one of the Gibbs spikes, we must add the main hump's contribution to the area under the small oscillations in the tail of $D_N(\theta)$. What is this latter value? Figure. 7.14 shows the situation; these small osciallations overlap with the nonzero part of $X(\omega)$ and affect the convolution result. We know from (7.71) that the entire integral of $D_N(\theta)$ over $[-\pi, \pi]$ is unity. Thus, the integral over $[-1, 1]$ is also near unity, since (exercise) the areas of the oscillations decrease like $1/n$. So, we have

$$1 = \int_{-\theta_N}^{+\theta_N} D_N(\theta)\, d\theta + 2\int_{+\theta_N}^{\pi} D_N(\theta)\, d\theta \approx 1.17898 + 2\int_{+\theta_N}^{\pi} D_N(\theta)\, d\theta, \tag{7.76}$$

and therefore the maximum of the convolution, $X_N(\omega_{max})$, approximates to

$$X_N(\omega_{max}) = \int_{-\pi}^{\pi} X(\theta) D_N(\omega_{max} - \theta)\, d\theta \approx \int_{-\theta_N}^{+\theta_N} D_N(\theta)\, d\theta + \int_{+\theta_N}^{\pi} D_N(\theta)\; d\theta$$

$$\approx 1.17898 - \frac{0.17898}{2} = 1.08949. \tag{7.77}$$

A careful review of this analysis should convince the reader that

- Any step discontinuity in $X(\omega)$ will produce the same result.
- The amount of overshoot around the discontinuity is approximately 9% of the step height of the discontinuity.

The first point follows from the fact that for sufficiently high N values, the Dirichlet kernel's main peak will be so narrow that $X(\omega)$ will appear flat before and after the jump. That is, $D_N(\theta)$ localizes the discontinuity of $X(\omega)$. The second point is perhaps easier to witness from the present example. The size of the Gibbs phenomenon's spike is given by an integral, which is a linear operation, and so scaling the step height causes the convolution's output to scale accordingly. The rich Fourier analysis literature provides further explanations and generalizations of the Gibbs phenomenon [15–17].

7.2.1.5 *Some Final Points.* This section concludes our preliminary study of the DTFT by stating two theoretical results.

Not all of our abstract l^p signal spaces support a DTFT; however, we have the following result [18]:

Theorem (DTFT Existence for p-Summable Signals). Suppose $x(n) \in l^p$, for $1 < p < 2$. Then there exists an $X(\omega) \in L^q[-\pi, \pi]$, where p and q are conjugate exponents (that is, $p^{-1} + q^{-1} = 1$), such that the values $x(n)$ are the Fourier series coefficients for $X(\omega)$.

Thus, we have a rich variety of signal spaces whose members have DTFTs. In signal analysis, the limiting cases—l^1 and l^2—are the most important. For $x(n) \in l^1$, the partial sums, $X_N(\omega)$, converge uniformly on $[-\pi, \pi]$ to a continuous signal, $X(\omega)$. Because $X(\omega)$ is continuous on a closed interval, $[-\pi, \pi]$, it is also bounded: $X(\omega) \in L^\infty$. The partial DTFT sums, $X_N(\omega)$, converge pointwise when $x(n)$ is absolutely summable. This is not the case for square-summable signals; we have seen that pointwise convergence does not necessarily occur for $x(n) \in l^2$. A long-standing problem has been to characterize the set of points for which the Fourier series of an $L^2[-\pi, \pi]$ signal converges. Many mathematicians supposed that convergence occurs almost everywhere, and in 1966 Carleson finally proved [19] the following theorem:

Theorem (Carleson's). If $X(\omega) \in L^2[-\pi, \pi]$, then the Fourier series for X converges almost everywhere to $X(\omega)$.

This result was soon generalized by Hunt to any $L^p[-\pi, \pi]$, $1 < p < \infty$ [20]. The exceptional case is $L^1[-\pi, \pi]$. And what a exception it is: In 1926, Kolmogorov[6] found an unbounded, discontinuous function, $f \in L^1[-\pi, \pi]$, whose Fourier series diverges everywhere from f [21, 22].

7.2.2 Properties

Let us enumerate the properties of the discrete-time Fourier transform. We assume throughout that discrete signals, say $x(n)$, belong to some signal space that supports a DTFT operation. Many of these are similar in flavor to the properties of previously covered transforms: the Fourier series, Fourier transform, and discrete Fourier transform. Consequently, we leave several of the proofs as exercises.

Proposition (Linearity, Time-Shift, and Frequency Shift). Let $x(n)$ and $y(n)$ be discrete signals and let $X(\omega)$ and $Y(\omega)$ be their DTFTs, respectively. Then

(a) (Linearity) The DTFT of $ax(n) + \text{by}(n)$ is $aX(\omega) + bY(\omega)$.
(b) (Time Shift) The DTFT of $x(n - m)$ is $\exp(-j\omega m)X(\omega)$.
(c) (Frequency Shift) The IDTFT of $X(\omega - \theta)$ is $\exp(-j\theta n)x(n)$.

Proof: Exercise. ∎

[6] A. N. Kolmogorov (1903–1987), professor of mathematics at Moscow State University, investigated problems of topology and analysis and established the axiomatic approach to probability theory.

Proposition (Frequency Differentiation). Suppose $x(n)$ is a discrete signal, $X(\omega)$ is its DTFT, and the partial sums $X_N(\omega)$ converge uniformly on $[-\pi, \pi]$ to $X(\omega)$. Then the DTFT of $nx(n)$ is $jdX(\omega)/d\omega$.

Proof: To prove transform properties, one may work from either end of a proposed equality; in this case, it is easier to manipulate the derivative of the DTFT. To wit,

$$j\frac{d}{d\omega}X(\omega) = j\frac{d}{d\omega}\sum_{n=-\infty}^{\infty} x(n)\exp(-jn\omega) = j\sum_{n=-\infty}^{\infty} x(n)\frac{d}{d\omega}\exp(-jn\omega). \tag{7.78}$$

The interchange of the differentiation and infinite summation operations is valid because the DTFT analysis equation is uniformly convergent. Taking the derivative of the summand in (7.78) and pulling the constants through the summation gives

$$j\frac{d}{d\omega}X(\omega) = j\sum_{n=-\infty}^{\infty} x(n)(-jn)\exp(-jn\omega) = \sum_{n=-\infty}^{\infty} nx(n)\exp(-jn\omega). \tag{7.79}$$

This is precisely the DTFT analysis equation for the signal $nx(n)$. ■

Without a doubt, the most important property of the DTFT is the Convolution-in-Time Theorem. This result shows that convolving two signals in the time domain is equivalent to multiplying their frequency domain representations. Since we are dealing with aperiodic signals, there is no need to redefine convolution for a finite interval, as we did with the DFT in (7.22). The convolution-in-time property is the key to understanding signal filtering—the selective suppression of frequency bands within a signal. We shall resort to this theorem many times in the chapters that follow.

Theorem (Convolution in Time). Let $x(n)$ and $y(n)$ be signals, let $X(\omega)$ and $Y(\omega)$ be their DTFTs, and let $z = x * y$. If the convolution sum for $z(n)$ converges absolutely for each integer n, then the DTFT of $z(n)$ is $Z(\omega) = X(\omega)Y(\omega)$.

Proof: Among all of the theoretical investigations into all of the transforms studied so far, we should note a distinct computational compatibility between the transform integral (or summation) and the convolution operation. The DTFT is no exception. We substitute the expression for the convolution, $z = x * y$, directly into the DTFT analysis equation for $Z(\omega)$:

$$
\begin{aligned}
Z(\omega) &= \sum_{n=-\infty}^{\infty} z(n)\exp(-j\omega n) = \sum_{n=-\infty}^{\infty} (x*y)(n)\exp(-j\omega n) \\
&= \sum_{n=-\infty}^{\infty}\left(\sum_{k=-\infty}^{\infty} x(k)y(n-k)\right)\exp(-j\omega n) \\
&= \sum_{n=-\infty}^{\infty}\sum_{k=-\infty}^{\infty} x(k)y(n-k)\exp(-j\omega(n-k))\exp(-j\omega k) \\
&= \sum_{k=-\infty}^{\infty}\sum_{n=-\infty}^{\infty} x(k)y(n-k)\exp(-j\omega(n-k))\exp(-j\omega k) \\
&= \sum_{k=-\infty}^{\infty} x(k)\exp(-j\omega k)\sum_{n=-\infty}^{\infty} y(n-k)\exp(-j\omega(n-k)) \\
&= X(\omega)\sum_{n=-\infty}^{\infty} y(n-k)\exp(-j\omega(n-k)) \\
&= X(\omega)\sum_{m=-\infty}^{\infty} y(m)\exp(-j\omega m) = X(\omega)Y(\omega).
\end{aligned}
\tag{7.80}
$$

We use the absolute convergence of the convolution sum to justify writing the iterated summation as a double summation and to subsequently switch the order to the summation. A routine change of summation variable, $m = n - k$, occurs in the last line of (7.80). ∎

There is an important link back to our results on linear, translation-invariant systems. Recall from Chapter 2 that the convolution relation characterizes LTI systems. If H is LTI, then the output, $y = Hx$, is the convolution of $h = H\delta$ with x: $y = h * x$. Thus we have the following corollary.

Corollary (Convolution in Time). If H is an LTI system, h is the impulse response of H, and $y = Hx$, then $Y(\omega) = X(\omega)H(\omega)$, assuming their DTFTs exist.

Proof: Note that $y = h * x$ and apply the theorem. ∎

We shall establish yet another transform Convolution Theorem when we study the z-transform in Chapter 8. The next theorem is a companion result. It establishes for the DTFT a familiar link: Multiplication in the time domain equates with convolution in the frequency domain. This theorem has a z-transform variant, too.

Theorem (Convolution in Frequency). Suppose $x(n)$ and $y(n)$ are discrete signals; $X(\omega)$, $Y(\omega) \in L^2[-\pi, \pi]$ are their respective DTFTs; and $z(n) = x(n)y(n)$ is their termwise product. Then the DTFT of $z(n)$, $Z(\omega)$, is given by the scaled convolution of $X(\omega)$ and $Y(\omega)$ in $L^2[-\pi, \pi]$:

$$Z(\omega) = \frac{1}{2\pi} \int_{-\pi}^{+\pi} X(\theta) Y(\omega - \theta)\, d\theta. \tag{7.81}$$

Proof: The right-hand side of (7.81) is the integral of the product of infinite summations whose terms contain the complex exponential—for instance, $\exp(-j\theta n)$. We have already witnessed numerous cases where the summands cancel owing to the 2π-periodicity of the exponential. Therefore, let us work from the $L^2[-\pi, \pi]$ side of (7.81). Indeed, we compute,

$$\begin{aligned}
\frac{1}{2\pi} \int_{-\pi}^{+\pi} X(\theta) Y(\omega - \theta)\, d\theta &= \frac{1}{2\pi} \int_{-\pi}^{+\pi} \sum_{n=-\infty}^{\infty} x(n) \exp(-j\theta n) \sum_{k=-\infty}^{\infty} y(k) \exp[-j(\omega - \theta)k]\; d\theta \\
&= \frac{1}{2\pi} \sum_{n=-\infty}^{\infty} \sum_{k=-\infty}^{\infty} x(n) y(k) \exp(-j\omega k) \int_{-\pi}^{+\pi} \exp(-j\theta n) \exp(j\theta k)\, d\theta \\
&= \frac{1}{2\pi} \sum_{n=-\infty}^{\infty} \sum_{k=-\infty}^{\infty} x(n) y(k) \exp(-j\omega k) \int_{-\pi}^{+\pi} \exp[-j\theta(n-k)]\; d\theta.
\end{aligned} \tag{7.82}$$

Once again the last integral is zero, unless $n = k$; in this case it evaluates to 2π. Thus, all of the terms of the double summation on the bottom of (7.82) are zero, save those where $n = k$. Our strategy works, and we find

$$\begin{aligned}
\frac{1}{2\pi} \int_{-\pi}^{+\pi} X(\theta) Y(\omega - \theta)\, d\theta &= \frac{1}{2\pi} \sum_{n=-\infty}^{\infty} \sum_{k=-\infty}^{\infty} x(n) y(k) \exp(-j\omega k) \int_{-\pi}^{+\pi} \exp[-j\theta(n-k)]\, d\theta \\
&= \frac{2\pi}{2\pi} \sum_{n=-\infty}^{\infty} x(n) y(n) \exp(-j\omega n) = \sum_{n=-\infty}^{\infty} z(n) \exp(-j\omega n) = Z(\omega)
\end{aligned} \tag{7.83}$$

which completes the proof. ■

Note that the above proof allows for the case that the signal $y(n)$ may be complex-valued. This observation gives us the following corollary.

Corollary. Again let $x(n)$ and $y(n)$ be discrete signals, and let $X(\omega), Y(\omega) \in L^2[-\pi, \pi]$ be their respective DTFTs. Then,

$$\sum_{n=-\infty}^{\infty} x(n) \overline{y}(n) = \frac{1}{2\pi} \int_{-\pi}^{+\pi} X(\theta) \overline{Y}(\theta)\, d\theta. \tag{7.84}$$

Proof: Before commencing with the proof, let us observe that the left-hand side of (7.84) is the l^2 inner product of $x(n)$ and $y(n)$, and the right-hand side is just the inner product of $X(\omega)$ and $Y(\omega)$ scaled by the factor $(2\pi)^{-1}$. Well, in the course of establishing the embedding isomorphism from l^2 into $L^2[-\pi, \pi]$ as Hilbert spaces, we already proved this result. Nonetheless, we can convey some of the symmetry properties and computational mechanics of the DTFT by offering another argument. So set $w(n) = \bar{y}(n)$ and $z(n) = x(n)w(n) = x(n)\bar{y}(n)$. By the theorem, then, the DTFT of $z(n)$ is

$$Z(\omega) = \frac{1}{2\pi} \int_{-\pi}^{+\pi} X(\theta) W(\omega - \theta)\, d\theta. \tag{7.85}$$

Therefore,

$$Z(0) = \left. \left(\sum_{n=-\infty}^{+\infty} z(n) e(-j\omega n) \right) \right|_{\omega=0} = \sum_{n=-\infty}^{+\infty} x(n)\bar{y}(n) = \frac{1}{2\pi} \int_{-\pi}^{+\pi} X(\theta) W(-\theta)\, d\theta. \tag{7.86}$$

What is $W(-\theta)$ in the integral on the right-hand side of (7.86)? By the algebra of complex conjugates, however, we find

$$\bar{Y}(\theta) = \overline{\left[\sum_{n=-\infty}^{+\infty} y(n) e(-j\theta n) \right]} = \sum_{n=-\infty}^{+\infty} \bar{y}(n) e(j\theta n) = \sum_{n=-\infty}^{+\infty} \bar{y}(n) e[-j(-\theta)n] = W(-\theta). \tag{7.87}$$

Putting (7.86) and (7.87) together establishes the theorem's result. ■

Corollary (Parseval's Theorem). If $x(n) \in l^2$, then $\|x(n)\|_2 = (2\pi)^{-1} \|X(\omega)\|_2$.

Proof: Take $x(n) = y(n)$ in the previous corollary. (Observe that the two norms in the statement of Parseval's theorem are taken in two different Hilbert spaces: l^2 and $L^2[-\pi, \pi]$.) ■

Note that Parseval's theorem too follows from our earlier Hilbert space isomorphism. Many frequency transform theorems at first glance appear to be almost miraculous consequences of the properties of the exponential function, or its sinusoidal parts, or the definition of the particular discrete or analog transform. But in fact they are mere instances of Hilbert space results. That very general, very abstract, partly algebraic, partly geometric theory that we studied in Chapters 2 and 3 provides us with many of the basic tools for the frequency domain processing and analysis of signals.

Parseval's theorem shows that signal energy in the time domain is proportional to signal energy in the frequency domain. This has some practical applications. In Chapter 9 we shall consider frequency-domain analysis of signals. We sought methods for discovering the periodicities of signals in Chapter 4, and to some extent we were successful in applying statistical and structural methods toward this end. Using discrete Fourier transforms, such as the DFT or the DTFT, we can obtain a description of the signal in terms of its frequency content. Then, in order to decide whether one or another frequency is present in the time-domain signal, we examine the frequency-domain representation for significant values at certain frequencies. But what constitutes a significant value? We can threshold signals in the frequency domain, just as we did in the time domain in Chapter 4. But, again, how do we set the threshold for what constitutes a significant frequency component? Parseval's theorem tells us that we can look for a sufficient portion of the signal's energy within a frequency range. We know that the overall frequency-domain energy is proportional to the overall time-domain energy, and the time-domain energy is computable from the signal values. Thus, we can select a threshold for the frequency domain based on some percentage of time-domain energy. Since we know that the total frequency domain energy is proportional to time-domain energy, we do not even have to examine other bands once the threshold is exceeded in some range of frequencies.

As with the DFT, there are a variety of DTFT symmetry properties. At this stage in our exposition, these are routine, and we leave them as exercises. The next section covers a property of the DTFT that applies to linear, translation-invariant systems. It turns out that with the DTFT, we can show that LTI systems have a very benign effect on exponential signals.

7.2.3 LTI Systems and the DTFT

Let us return to the idea of a discrete, linear, translation-invariant system. We introduced discrete LTI systems in the second chapter, and there we showed that LTI systems are characterized by the convolution relation. The system output, $y = Hx$, is the convolution of the input with a fixed signal, $y = h * x$, where $h = H\delta$. The discrete signal $h(n)$ is called the impulse response of the LTI system H. There is a close, important relationship between LTI systems and the DTFT.

Recall that an eigenvector for a finite-dimensional linear map, T, is a vector $\boldsymbol{v}$ for which $\boldsymbol{Tv} = \boldsymbol{av}$, for some constant a. Similarly, we can define an eigenfunction for a system, to be a signal for which $y = Hx = ax$, for some constant value a.

Theorem (Eigenfunctions of LTI Systems). If H is an LTI system, $y = Hx$, where $x(n) = \exp(j\omega n)$, then $x(n)$ is an eigenfunction of the system H.

Proof: By the Convolution Theorem for LTI systems, we have

$$y(n) = (h * x)(n) = \sum_{k=-\infty}^{+\infty} h(k)x(n-k) = \sum_{k=-\infty}^{+\infty} h(k)\exp[j\omega(n-k)]$$

$$= \exp(j\omega n) \sum_{n=-\infty}^{+\infty} h(k)\exp(-j\omega k) = \exp(j\omega n)H(\omega), \tag{7.88}$$

where $H(\omega)$ is the DTFT of $h(n)$. ∎

The theorem inspires the following definition.

Definition (Frequency Response of LTI Systems). If H is an LTI system, and $h = H\delta$, is the impulse response of H, then $H(\omega)$, the DTFT of $h(n)$, is called the *frequency response* of H.

Note that we have a notational conflict, in that we are using the uppercase H to denote both the LTI system and its frequency response. Of course, writing the frequency response as a function of ω helps to distinguish the two. The context usually makes clear which is the system and which is the frequency-domain representation of the impulse response. So we persist in using the notation, which is a widespread signal theory convention.

It is the behavior of $H(\omega)$ as a function of ω that determines how an LTI system, H, affects the frequencies within a discrete signal. The Eigenfunctions Theorem showed that if $x(n) = \exp(j\omega n)$ is an exponential signal, and $y = Hx$, then $y(n) = x(n)H(\omega)$. So the magnitude of the output, $|y(n)|$, is proportional to the magnitude of the input, $|x(n)|$, and the constant of proportionality is $|H(\omega)|$. Thus, if $|H(\omega)|$ is small, then the system suppresses exponential signals of radial frequency w. And if $|H(\omega)|$ is large, then H passes exponentials $\exp(j\omega n)$. What is meant, however, by the "frequencies within a discrete signal?" If the input signal consists of a pure exponential, $x(n) = \exp(j\omega n) = \cos(\omega n) + j\sin(\omega n)$, of frequency ω radians/second, then the frequency component within the signal is the exponential. It consists of a real and an imaginary sinusoidal component. And, by the theorem, the system's frequency response determines how the system affects the frequency components of $x(n)$. Furthermore, suppose the input signal consists of a sum of scaled exponentials,

$$x(n) = \sum_{k=M}^{N} c_k \exp(j\omega_k n). \tag{7.89}$$

By linearity we have

$$\begin{aligned} y(n) = H[x(n)] = H\left[\sum_{k=M}^{N} c_k \exp(j\omega_k n)\right] &= \sum_{k=M}^{N} c_k H\left[\exp(j\omega_k n)\right] \\ = \sum_{k=M}^{N} c_k H\left[\exp(j\omega_k n)\right] = \sum_{k=M}^{N} c_k \exp(j\omega_k n) H(\omega_k) &= \sum_{k=M}^{N} H(\omega_k) c_k \exp(j\omega_k n). \end{aligned} \tag{7.90}$$

Thus, the output consists of the sum of $x(n)$'s frequency components, each further attenuated or amplified by its corresponding value, $H(\omega_k)$. But is it still correct to refer to a general signal's frequency components?

Within certain classes of aperiodic signals, it is indeed possible to approximate them arbitrarily well with sums of exponentials, such as in (7.89). If we can show this, then this justifies the above characterization of the frequency response as admitting and suppressing the various frequency components within a signal (7.90). Let us state our desired result as a theorem.

Theorem (Frequency Components of Aperiodic Signals). Let the signal $x(n)$ have DTFT $X(\omega)$. If $x(n)$ is absolutely summable or square-summable, then it can be approximated arbitrarily well by linear combinations of exponential signals of the form $\{\exp(jn\omega) : \omega = \pi(2m - M)/M$ for some m, $0 < m < M - 1$, $1 < M\}$.

Proof: The key idea is to approximate the synthesis equation integral representation for $x(n)$ in terms of $X(\omega)$:

$$x(n) = \frac{1}{2\pi}\int_{-\pi}^{+\pi} X(\omega)\exp(j\omega n)\,d\omega. \tag{7.91}$$

The trapezoidal rule approximates $x(n)$ by dividing the interval $[-\pi, \pi]$ into $N > 0$ segments of equal width, $2\pi/N$, and summing the areas of the trapezoidal regions. Let $y(n, \omega) = X(\omega)\exp(j\omega n)$. Then, $y(n, -\pi) = y(n, \pi)$, and after some simplification, we get

$$\begin{aligned} x(n) &\approx \frac{1}{M}\left[y(n,-\pi) + y(n,-\pi+2\pi/M) + y(n,-\pi+4\pi/M) + \cdots + y(n,\pi-2\pi/M)\right] \\ &= \frac{1}{M}\sum_{m=0}^{M-1} y(n,-\pi+2\pi m/M) = \frac{1}{M}\sum_{m=0}^{M-1} X(-\pi+2\pi m/M)\exp[jn(-\pi+2\pi m/M)] \\ &= \frac{1}{M}\sum_{m=0}^{M-1} X(-\pi+2\pi m/M)\exp\left[jn\pi\left(\frac{2m-M}{M}\right)\right]. \end{aligned} \tag{7.92}$$

Since (7.92) is a linear combination of terms of the form $A_m\exp(jn\omega)$, where A_m is a constant, and $\omega = \pi(2m - M)/M$, the proof is complete. ■

Now, it is possible to investigate the effect an LTI system H, where $h = H\delta$, has on an aperiodic input signal $x(n)$. We first closely approximate $x(n)$ by a linear combination of exponential terms, $\exp(jn\omega)$, $\omega = \pi(2m - M)/M$, as given by the Frequency Components Theorem. By the Eigenfunctions Theorem, the various component exponential terms scale according to the value of the DTFT of $h(n)$, $H(\omega)$:

$$\begin{aligned} y(n) = (h*x)(n) &= \left(h*\sum_{m=0}^{M-1} A_m\exp\left[jn\frac{\pi(2m-M)}{M}\right]\right)(n) \\ &= \sum_{m=0}^{M-1} A_m\exp\left[jn\frac{\pi(2m-M)}{M}\right]H\left(\frac{\pi(2m-M)}{M}\right) = \sum_{m=0}^{M-1} A_m\exp\left(jn\omega_m\right)H(\omega_m). \end{aligned} \tag{7.93}$$

Depending on the value of $H(\omega_m)$ at various values of ω_m, then, the frequency components of $x(n)$ are attenuated or amplified by the LTI system H.

Example (Perfect High-Frequency Attenuation System). Let us explore a system H that removes all frequency components above a certain fixed radial frequency, ω_c, from a discrete aperiodic signal, $x(n)$. There is a practical need for such systems. They remove noise from signals prior to segmentation and classification, as we noted in Chapter 4. If the time-domain filter $h(n)$ removes all frequency components above ω_c, the DTFT of $h(n)$, $H(\omega)$, must be zero for $|\omega| > |\omega_c|$, as Figure 7.15 shows. We can compute $h(n)$ for the square pulse in Fig. 7.15(a) as follows.

$$h(n) = \frac{1}{2\pi}\int_{-\pi}^{+\pi} X(\omega)exp(jn\omega)\, d\omega = \frac{1}{2\pi}\int_{-\omega_c}^{+\omega_c} \exp(jn\omega)\, d\omega$$

$$= \frac{1}{2\pi}\frac{\exp(jn\omega)}{jn}\Bigg|_{-\omega_c}^{+\omega_c} = \frac{\sin(n\omega_c)}{\pi n} = \frac{\omega_c}{\pi}\operatorname{sinc}(n\omega_c). \tag{7.94}$$

However flawless it may be in the frequency domain, as a time-domain noise removal filter, $h(n)$ is quite imperfect. Two fundamental problems render it physically impossible to implement:

- It has infinite support.
- The system H is non causal: $h(n) = 0$ for $n < 0$.

The first point means that we can never finish the convolution sum necessary to calculate the output $y = Hx$. Of course, we can come very close to approximating the

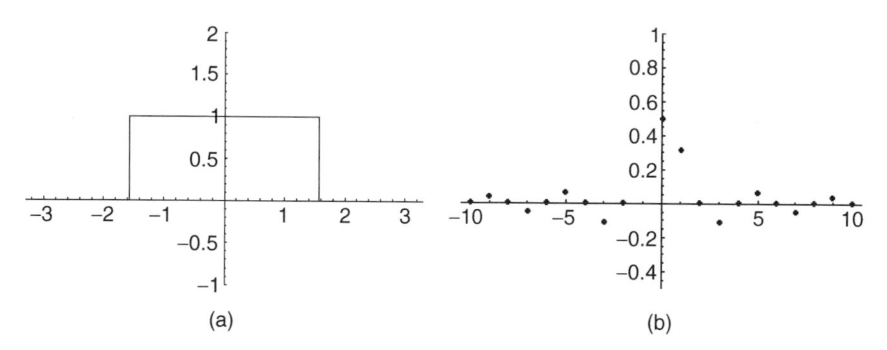

Fig. 7.15. Perfect high-frequency removal. The system with impulse response $h(n)$ and DTFT $H(\omega)$ will remove all frequencies above ω_c if $H(\omega) = 0$ for $|\omega| > \omega_c$. We also expect that H will perfectly preserve frequencies within the range $[-\omega_c, \omega_c]$; in other words, if $|\omega| \le \omega_c$, then $H(\omega) = 1$. Thus, $H(\omega)$ resembles a square pulse centered in $[-\pi, \pi]$, as in panel (a). Here, $\omega_c = \pi/2$. The time-domain sinc signal in (b) gives rise to such a frequency-domain representation.

perfect filter output by prolonging the summation until further terms are negligible. But worse is the non causality. This implies that computing the convolution $y = h * x$ requires future values of the input signal $x(n)$. In the next chapter we shall uncover filter design techniques that avoid these problems.

7.3 THE SAMPLING THEOREM

Presentations of signal theory throughout the text have alternated between analog developments and discrete developments. We have worked either in the analog world or in the discrete world. Sampling an analog signal at regular intervals produces a discrete signal, but so far nothing has been proffered as an interconnection between the analog source and the discrete result. However, now have the theoretical tools in hand to effect a unification of the two realms of signal theory. The unification takes place using not time domain methods, but rather frequency-domain methods. Perhaps this is not so unexpected. After all, we noted that the discrete-time Fourier transform very much resembles the analog Fourier series and that there is a Hilbert space isomorphism between the analog space $L^2[-\pi, \pi]$ (or, more precisely, its equivalence classes of signals equal almost everywhere) and the discrete space l^2.

7.3.1 Band-Limited Signals

One of the key ideas in linking the analog and discrete worlds is the notion of a bandlimited analog signal. This means, informally, that the frequency-domain representation of the signal has finite support. Physically, this is a realistic assumption, as no physical signal can have frequency components that go arbitrarily high in frequency. Nature can only shake so fast.

Definition (Band-Limited Signal). An analog signal $x(t)$ is *band-limited* if its Fourier transform, $X(\omega)$, exists and has finite support.

To discover the connection to band-limited signals, let us consider anew the operation of sampling an analog signal. If $x(t)$ is a continuous, absolutely integrable analog signal and $T > 0$, then it may be sampled at intervals T to produce the discrete signal $s(n) = x(nT)$. Let us suppose that $s(n) \in l^1$ so that the DTFT sum converges uniformly to $S(\omega)$,

$$S(\omega) = \sum_{n=-\infty}^{+\infty} s(n)\exp(-jn\omega), \tag{7.95}$$

and, hence, that $s(n)$ is represented by the DTFT synthesis equation,

$$s(n) = \frac{1}{2\pi}\int_{-\pi}^{+\pi} S(\omega)\exp(j\omega n)\,d\omega \tag{7.96}$$

Because $x(t) \in L^1$, $x(t)$ is represented by the FT's synthesis equation,

$$x(t) = \frac{1}{2\pi} \int_{-\infty}^{+\infty} X(\omega) \exp(j\omega t)\, d\omega \tag{7.97}$$

where $X(\omega)$ is the radial Fourier transform of $x(t)$. Since $s(n) = x(nT)$, we get another representation of $s(n)$:

$$s(n) = x(nT) = \frac{1}{2\pi} \int_{-\infty}^{+\infty} X(\omega) \exp(j\omega nT)\, d\omega \tag{7.98}$$

Now we have a derivation of the discrete signal values from both a discrete frequency representation (7.96) and from an analog frequency representation (7.98). The observation is both easy and important. (The reason that we did not take note of this earlier is that only now do we have Fourier and inverse Fourier transforms for both analog and discrete signals!) The key to discovering the hidden bond between the analog and discrete signal domains lies in finding mathematical similarities in the integrands of the DTFT and FT synthesis equations.

Connecting the two integrands in (7.96) and (7.98) involves both the sampling interval, T, and the bandwidth of the original analog signal, $x(t)$. Since the IFT integral (7.98) has infinite limits, and the IDTFT integral has finite limits, the prospects for relating the two integrands seem dim. Note, however, that if the FT of $x(t)$ is band-limited, then the nonzero values of $X(\omega)$ are confined to an interval, $[-b, +b]$, where $b > 0$. The FT integral representation of $s(n)$ becomes

$$s(n) = x(nT) = \frac{1}{2\pi} \int_{-b}^{+b} X(\omega) \exp(j\omega nT)\, d\omega \tag{7.99}$$

A change of integration variable, $\theta = \omega T$, converts (7.99) into the form

$$s(n) = x(nT) = \frac{1}{2\pi T} \int_{-bT}^{+bT} X\left(\frac{\theta}{T}\right) \exp[jn\theta]\, d\theta \tag{7.100}$$

Now, if the interval $[-bT, bT] \subseteq [-\pi, \pi]$, then the integrals (7.96) and (7.100) are comparable. Suppose we choose T small enough so that this is true; we space the discrete samples of $x(t)$ so close together that $bT < \pi$. Since $X(\theta/T)$ is zero outside $[-bT, bT]$,

$$s(n) = x(nT) = \frac{1}{2\pi} \int_{-\pi}^{+\pi} \frac{1}{T} X\left(\frac{\theta}{T}\right) \exp[jn\theta]\, d\theta \tag{7.101}$$

(Observe carefully that none of this analysis would be valid without $x(t)$ being band-limited.) Now there are two different representations of the discrete signal $x(n)$: one (7.96) dependent on $S(\omega)$, the DTFT of $s(n)$, and another (7.101) dependent on a scaled, amplified portion of the FT of $x(t)$: $T^{-1}X(\theta/T)$. We know that the DTFT is unique, since it is invertible, and therefore we have $S(\omega) = T^{-1}X(\omega/T)$.

Let us summarize. If the analog signal, $x(t)$, is band-limited, then a sufficiently high sampling rate, T, guarantees that the DTFT of $s(n) = x(nT)$ and the (scaled, amplified) FT of $x(t)$ are equal.

In principle, this fact allows us to reconstruct $x(t)$ from its discrete samples. Suppose, again, that $s(n) = x(nT)$ and that T is so small that $bT < \pi$, where $X(\omega)$ is zero outside $[-b, b]$. From the samples, $s(n)$, we compute the DTFT, $S(\omega)$. Now, by the preceding considerations, $S(\omega) = T^{-1}X(\omega/T)$; in other words, $TS(\omega T) = X(\omega)$. Now we can compute $x(t)$ as the IFT of $X(\omega)$. So, indeed, the samples of a band-limited analog signal can be chosen close enough together that the original signal can be recovered from the samples.

The next section we will give this abstract observation some practical value.

7.3.2 Recovering Analog Signals from Their Samples

Now let us work toward a precise characterization of the conditions under which an analog signal is recoverable by discrete samples. One outcome of this will be an elucidation of the concept of aliasing, which occurs when the conditions for ideal reconstruction are not completely met.

In the previous section we studied the relationship between the DTFT and the FT for band-limited signals $x(t)$ and discrete signals $s(n) = x(nT)$, $T > 0$. The next theorem relaxes the assumption that $x(t)$ is band-limited.

Theorem (DTFT and FT). Suppose that $x(t) \in L^1$ is an analog signal, $T > 0$, and $s(n) = x(nT)$. If $s(n) \in l^1$ so that the DTFT sum converges uniformly to $S(\omega)$, then

$$S(\omega) = \frac{1}{T}\sum_{k=-\infty}^{+\infty} X\left(\frac{\omega + 2\pi k}{T}\right). \tag{7.102}$$

Proof: Continuing the development of the previous section, we set $\theta = \omega T$ for a change of integration variable:

$$s(n) = x(nT) = \frac{1}{2\pi}\int_{-\infty}^{+\infty} X(\omega)\exp(j\omega nT)\,d\omega = \frac{1}{2\pi T}\int_{-\infty}^{+\infty} X\left(\frac{\theta}{T}\right)\exp(j\theta n)\,d\theta. \tag{7.103}$$

Let $Y(\theta) = T^{-1}X(\theta/T)$. Then,

$$s(n) = x(nT) = \frac{1}{2\pi}\int_{-\infty}^{+\infty} X(\omega)\exp(j\omega nT)\,d\omega = \frac{1}{2\pi}\int_{-\infty}^{+\infty} Y(\theta)\exp(j\theta n)\,d\theta. \tag{7.104}$$

If we assume that $x(t)$ is band-limited, then X—and hence Y—have finite support; this reduces (7.104) to a finite integral as in the previous section. Let us not assume here that $x(t)$ is band-limited and instead investigate how (7.104) can be written as a sum of finite integrals. Indeed, we can break (7.104) up into 2π-wide chunks as follows:

$$s(n) = \frac{1}{2\pi}\int_{-\infty}^{+\infty} Y(\theta)\exp(j\theta n)\,d\theta = \frac{1}{2\pi}\sum_{k=-\infty}^{\infty}\int_{-\pi+2\pi k}^{\pi+2\pi k} Y(\theta)\exp(j\theta n)\,d\theta \tag{7.105}$$

The insight behind this is that the chunk of $Y(\theta)$ corresponding to $k = 0$ should look like $S(\omega)$ on $[-\pi, \pi]$, and the others, corresponding to $k = 0$, should be negligible if T is small and $x(t)$ is approximately band-limited. Now set $\phi = \theta - 2\pi k$ to get

$$\begin{aligned} s(n) &= \frac{1}{2\pi}\sum_{k=-\infty}^{\infty}\int_{-\pi}^{\pi} Y(\phi+2\pi k)\exp[j(\phi+2\pi k)n]\,d\phi \\ &= \frac{1}{2\pi}\sum_{k=-\infty}^{\infty}\int_{-\pi}^{\pi} Y(\phi+2\pi k)\exp(j\phi n)\,d\phi = \frac{1}{2\pi}\int_{-\pi}^{\pi}\sum_{k=-\infty}^{\infty} Y(\phi+2\pi k)\exp(j\phi n)\,d\phi \end{aligned} \tag{7.106}$$

The interchange of the order of the summation and the integration is allowable, because the sum converges uniformly to $Y(\theta)$ on $\mathbb{R}$. Now we have (7.106) in the form of the DTFT synthesis equation (7.96) for $s(n)$:

$$s(n) = \frac{1}{2\pi}\int_{-\pi}^{\pi} S(\omega)\exp(j\omega n)\,d\omega = \frac{1}{2\pi}\int_{-\pi}^{\pi}\sum_{k=-\infty}^{\infty} Y(\omega+2\pi k)\exp(j\omega n)\,d\omega \tag{7.107}$$

Since the DTFT is invertible, together (7.107) entails

$$S(\omega) = \sum_{k=-\infty}^{\infty} Y(\omega+2\pi k) = \frac{1}{T}\sum_{k=-\infty}^{\infty} X\left(\frac{\omega+2\pi k}{T}\right) \tag{7.108}$$

as desired. ■

Equation (7.108) shows that $S(\omega)$, the DTFT of $s(n)$, is the sum of an infinite number of copies of $Y(\omega)$, each translated by $2\pi k$ (Figure 7.16). Note that the sum of shifted copies of $Y(\omega)$ is 2π-periodic. The situation of interest is when there is no overlap in the shifted $Y(\omega)$ components in Figure 7.16. In this case, $Y(\omega)$ resembles a single period of the DTFT of $s(n)$, $S(\omega)$. We may recover $x(t)$ from the discrete samples, because we then know that $X(\omega) = TY(T\omega)$, and $x(t)$ derives from $X(\omega)$ via the FT synthesis equation. What criteria are necessary for there to be no overlap of the shifted versions of $Y(\omega)$? The famous Shannon–Nyquist theorem answers this question.

Theorem (Shannon–Nyquist Sampling Theorem). Suppose that $x(t) \in L^1$ is an analog signal, $T > 0$, and $s(n) = x(nT)$. If $s(n) \in l^1$, so that the DTFT sum converges uniformly to $S(\omega)$, then $x(t)$ may be recovered from the samples $s(n)$ if

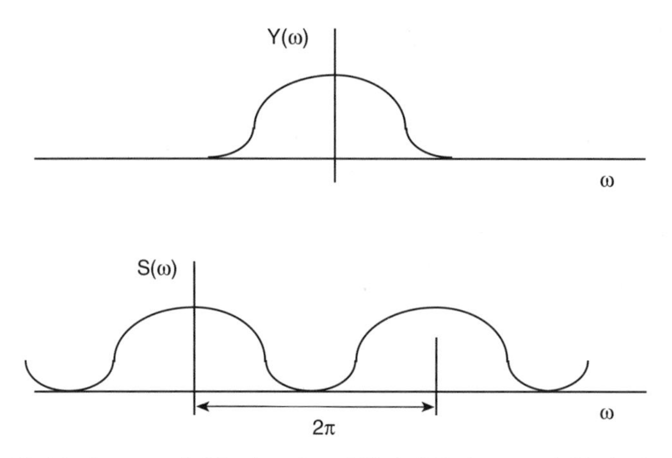

Fig. 7.16. S(ω) is the sum of shifted copies of $Y(\omega)$. If Y decays quickly in the frequency domain, then its translated copies overlap only slightly or not at all.

- $x(t)$ is band-limited.
- The sampling frequency $F = T^{-1} > 2F_{\max}$, where $|X(\omega)| = 0$ for $\omega > 2\pi F_{\max}$.

Proof: We deduce a series of equivalent criteria that prevent overlap. Now, since the difference between two translates is 2π, no overlap occurs when all of the nonzero values of $Y(\omega)$ lie within $[-\pi, \pi]$. Thus, $Y(\omega) = 0$ for $|\omega| > b > 0$, for some $b < \pi$. This means that $X(\omega) = 0$ for $|\omega| > |b/T|$, since $Y(\omega) = T^{-1}X(\omega/T)$. Equivalently, for no overlap, it must be the case that $x(t)$ is band-limited, and its nonzero spectral values within $[-\pi/T, \pi/T]$. Let $0 < B$ be the least upper bound of $\{\omega : X(\omega) > 0 \text{ or } X(-\omega) > 0\}$. Then $B < \pi/T$. But, B is measured in radians per second, and to give it in hertz we need to use the radians-to-hertz conversion formula, $\omega = 2\pi f$. Thus, the criterion for no overlap becomes $2\pi F_{\max} < \pi/T$, where $F_{\max}$ is the maximum frequency component of $x(t)$ in hertz, and T is the sampling period $s(n) = x(nT)$. Finally, this means precisely that $2F_{\max} < 1/T = F$. ■

When the analog signal $x(t)$ is not band-limited, aliasing is inevitable, because the shifted versions of $Y(\omega)$ must overlap. When $x(t)$ is band-limited and the sampling interval is too large, the shifted versions of $Y(\omega)$ overlap with one another and in their summation produce artifacts that are not part of the shape of the true analog spectrum of $x(t)$. Figure 7.17 illustrates a situation with aliasing.

These results motivate the following definition [3].

Definition (Nyquist Rate). If a signal $x(t)$ is band-limited, then its *Nyquist rate* is $F = 2F_{\max}$, where $F_{\max}$ is the least upper bound of values ω, where $|X(\omega)| \neq 0$.

Thus, sampling at intervals T such that $1/T$ is above the Nyquist rate permits perfect reconstruction of the analog signal $x(t)$ from its discrete samples, $s(n) = x(nT)$.

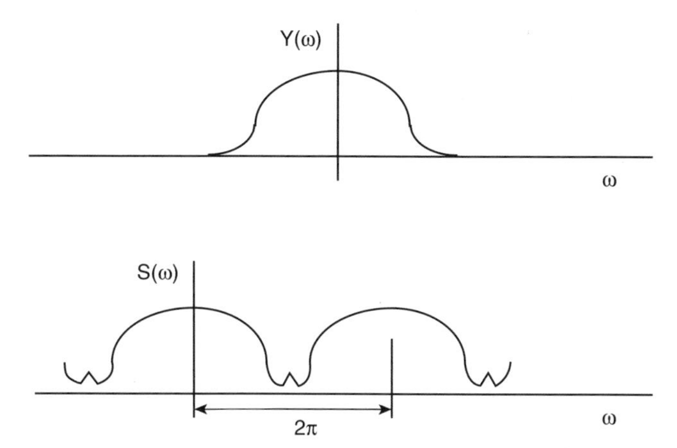

Fig. 7.17. Aliasing occurs when the shifted copies of $Y(\omega)$ overlap. In this situation, a single period of the spectrum of $s(n)$, $S(\omega)$, is *not* an exact replica of $X(\omega)$. High frequency components are added into some of the low frequency components. The result is that the reconstructed analog signal does not equal the original time domain signal, $x(t)$.

We now know, in principle, the conditions under which an analog signal may be reconstructed from its discrete samples. The next two sections limn out the theory of sampling. Section 7.3.3 provides an elegant reconstruction formula; it shows how to rebuild the analog signal from a simple set of interpolating signals. Section 7.3.4 casts a shadow on all these proceedings, however. A central result, the Uncertainty Principle, informs us that a signal with good frequency-domain behavior (as regards sampling and reconstruction) must have poor time-domain characteristics.

7.3.3 Reconstruction

The next theorem gives a formula for reconstructing, or interpolating, an analog signal from its samples. The conditions discovered above for ideal reconstruction must apply, of course. And there are some qualifications to this result that should be kept in mind:

- It assumes that perfect discrete samples are obtained in the sampling operation.
- The interpolating signals are not finitely supported.
- There are an infinite number of signals that must be summed to achieve perfect reconstruction.

Clearly, there are practical concerns with implementing analog signal reconstruction using this method. The reconstruction derives from evaluating the Fourier transform synthesis equation integral over a single period of the DTFT of $s(n)$, the signal samples.

Theorem (Shannon–Nyquist Interpolation Formula). Suppose that $x(t) \in L^1$ is an analog signal, $T > 0$, and $s(n) = x(nT)$. Let $s(n) \in l^1$ so that the DTFT sum converges uniformly to $S(\omega)$. Also, let $2F_{\max} < F = 1/T$, where $F_{\max}$ is the maximum frequency component of $x(t)$ in hertz. Then $x(t)$ may be recovered from $s(n)$ by the following sum:

$$x(t) = \sum_{n=-\infty}^{+\infty} s(n)\,\mathrm{sinc}\left(\frac{\pi t}{T} - n\pi\right) \tag{7.109}$$

Proof: The criterion for reconstruction applies, $x(t)$ is band-limited, and we find $x(t)$ from the IFT integral:

$$x(t) = \frac{1}{2\pi}\int_{-\infty}^{+\infty} X(\omega)\exp(j\omega t)\,d\omega = \frac{1}{2\pi}\int_{-\pi/T}^{+\pi/T} X(\omega)\exp(j\omega t)\,d\omega \tag{7.110}$$

Now, the DTFT of $s(n)$ is given by (7.108) for all $\omega \in \mathbb{R}$, and (because there is no overlap of the shifted versions of the Fourier transform) for $\omega \in [-\pi/T, \pi/T]$, we have $TS(\omega) = X(\omega/T)$, whence

$$x(t) = \frac{T}{2\pi}\int_{-\pi/T}^{+\pi/T} S(T\omega)\exp(j\omega t)\,d\omega \tag{7.111}$$

Inserting the DTFT analysis equation sum for $S(T\omega)$ in (7.111) and interchanging integration and summation gives

$$\begin{aligned} x(t) &= \frac{T}{2\pi}\int_{-\pi/T}^{+\pi/T}\left[\sum_{n=-\infty}^{\infty} s(n)\exp(-jT\omega n)\right]\exp(j\omega t)\,d\omega \\ &= \frac{T}{2\pi}\sum_{n=-\infty}^{\infty} s(n)\int_{-\pi/T}^{+\pi/T}\exp(-jT\omega n)\exp(j\omega t)\,d\omega \\ &= \frac{T}{2\pi}\sum_{n=-\infty}^{\infty} s(n)\int_{-\pi/T}^{+\pi/T}\exp[j\omega(t-Tn)]\,d\omega \end{aligned} \tag{7.112}$$

The last definite integral evaluates to a sinc signal:

$$\begin{aligned} x(t) &= \frac{T}{2\pi}\sum_{n=-\infty}^{\infty} s(n)\int_{-\pi/T}^{+\pi/T}\exp[j\omega(t-Tn)]\,d\omega = \sum_{n=-\infty}^{\infty} s(n)\frac{\sin\left[\frac{\pi}{T}(t-Tn)\right]}{\left[\frac{\pi}{T}(t-Tn)\right]} \\ &= \sum_{n=-\infty}^{\infty} s(n)\,\mathrm{sinc}\left(\frac{\pi t}{T} - \pi n\right) \end{aligned} \tag{7.113}$$

and the proof is complete. ■

Another way to interpret this result is to note that, for a given sampling interval T, the set of the analog sinc functions, $\mathrm{sinc}(\pi t/T - \pi n)$, span the space of band-limited signals.

7.3.4 Uncertainty Principle

The concept of a band-limited signal demands further scrutiny. If it was not at the beginning of this discussion, it is certainly clear now how important band-limited signals are for signal processing. Whenever the motivation is to cast an analog signal in digital form and reproduce it—perhaps with some intermediate processing steps applied—then a major consideration is how well discrete samples can represent the original signal. Technologies such as compact disc players succeed or fail based on whether they can sample signals fast enough to beat the Nyquist rate. In analyzing a signal, we also begin with a real-world—analog—signal; then we sample it, process it, and load it into a computer. In the computer, software algorithms build a structural description of the signal and then attempt to classify, identify, and recognize the signal or its fragments. Admittedly, the algorithms may destroy the original form of the signal. But the representation by the interpolation formula is useful, since the coefficients of the expansion indicate a certain signal similarity to the interpolating sinc functions. These may be a basis for classification. And this whole classification procedure gets started with an observation that the source analog signals enjoy a strict limitation on the extent of their frequency content.

This does beg the question, How common are band-limited signals? A signal, $x(t)$, is band-limited when $X(\omega) = \mathcal{F}[x(t)]$ has finite support. If the signal is band-limited, but still has high-frequency components, then a proportionately higher sampling frequency is necessary for ideal signal reconstruction. So, in general, we seek signals whose spectral values are concentrated, or localized, about the origin, $\omega = 0$. We confess that real signals—be they analog or discrete—do not continue unabated forever in the time domain; they must eventually die out. And for practical reasons, such as available memory in a signal analysis system, this time-domain locality is an important consideration. But can we also expect good frequency-domain behavior from finitely supported analog signals?

It is easy to see that a nonzero signal cannot be finitely supported in both domains, because if $x(t)$ and $X(\omega)$ have finite support, then $x(t) = x(t)[u(t + a) - u(t - a)]$ for some $a > 0$. The FT of $x(t)$ is therefore the convolution of the FT of $x(t)$ and the analog boxcar signal $b(t) = u(t + a) - u(t - a)$. But $B(\omega)$ is a sinc-type function, and since $X(\omega)$ is nonzero, the convolution of the two in the frequency domain does not have finite support, a contradiction.

Let us state and prove another famous result, the Uncertainty Principle, which shows that there is an insurmountable tradeoff between frequency-domain locality and time-domain locality. First, however, we need to define the concept of the locality of a signal in the time and frequency domains. We invoke concepts from statistics; namely, the locality of a signal is associated with the second moment, or the variance, of its values.

Definition (Time- and Frequency-Domain Locality). The *time-domain locality* of a signal $x(t)$ is

$$\Delta_t^2(x) = \int_{-\infty}^{+\infty} |x(t)|^2 t^2 dt, \tag{7.114a}$$

and its *frequency-domain locality* is

$$\Delta_\omega^2(X) = \int_{-\infty}^{+\infty} |X(\omega)|^2 \omega^2 d\omega \tag{7.114b}$$

The uncertainty principle holds for signals that decay faster than the reciprocal square root signal. This is necessary for the convergence of the second-order moment integral.

Theorem (Uncertainty Principle). Suppose that $x(t)$ is an analog signal, $\|x\|_2 = 1$, and $x^2(t)t \to 0$ as $t \to \infty$. Then

$$\sqrt{\frac{\pi}{2}} \le \Delta_t(x)\Delta_\omega(X). \tag{7.115}$$

Proof: The idea is to apply the analog Cauchy–Schwarz inequality to $tx(t)x'(t)$:

$$\left| \int_{-\infty}^{+\infty} tx(t)x'(t)\, dt \right|^2 \le \int_{-\infty}^{+\infty} |tx(t)|^2 dt \int_{-\infty}^{+\infty} |x'(t)|^2 dt = \Delta_t^2(x) \int_{-\infty}^{+\infty} |x'(t)|^2 \tag{7.116}$$

Now, $x'(t)$ has radial FT $j\omega X(\omega)$, so the analog version of Parseval's formula (Chapter 5) implies that

$$\int_{-\infty}^{+\infty} |x'(t)|^2 = \frac{1}{2\pi}\Delta_\omega^2(X). \tag{7.117}$$

Hence,

$$\left| \int_{-\infty}^{+\infty} tx(t)x'(t)\, dt \right|^2 \le \Delta_t^2(x)\frac{1}{2\pi}\Delta_\omega^2(X). \tag{7.118}$$

The integral in (7.118) is our focus; using the chain rule on its integrand and then integrating it by parts gives

$$\int_{-\infty}^{+\infty} tx(t)x'(t)\,dt = \frac{1}{2}\int_{-\infty}^{+\infty} t\frac{\partial x^2(t)}{\partial t}x'(t)\,dt = \frac{1}{2}tx^2(t)\Big|_{-\infty}^{\infty} - \frac{1}{2}\int_{-\infty}^{+\infty} x^2(t)\,dt \tag{7.119}$$

Now, $x^2(t)t \to 0$ as $t \to \infty$ and $\|x\|_2 = 1$ imply

$$\int_{-\infty}^{+\infty} tx(t)x'(t)\,dt = \frac{1}{2}\int_{-\infty}^{+\infty} t\frac{\partial x^2(t)}{\partial t}x'(t)\,dt = 0 - \frac{1}{2}\int_{-\infty}^{+\infty} x^2(t)\,dt = -\frac{1}{2}\|x\|^2 = -\frac{1}{2}. \tag{7.120}$$

Hence, from (7.118),

$$\left|-\frac{1}{2}\right|^2 = \frac{1}{4} \le \Delta_t^2(x)\frac{1}{2\pi}\Delta_\omega^2(X), \tag{7.121}$$

from which (7.115) follows. ■

In the exercises, it is shown that the Gaussian signal achieves this lower bound in the product of joint time- and frequency- domain locality.

Thinking about the Fourier transform and the Uncertainty Principle, we can understand how poor is its joint locality. Allowing that we may Fourier transform signals of slow decay (Chapter 6) using the generalized FT, the FT of a sinusoid is a pair of pulses in the frequency domain. Also, the FT of a pulse $\delta(t)$ is the constant $\omega = 1$. Thus, signals with extreme locality in one domain transform into signals with no locality whatsoever in the other domain. We will discover the problems that this lack of joint locality causes when we work through frequency-domain applications in Chapter 9. Chapters 10 and 11 develop transformation theories—very modern theories, it turns out—that furnish good local time and frequency decompositions of signals. Finally, in the last chapter, we apply these short-time Fourier and wavelet transforms to signal analysis problems.

7.4 SUMMARY

The detailed investigation and intense interest in discrete frequency transforms is a relatively recent phenomenon, and this is an altogether curious circumstance in view of the very tractable nature of the mathematical underpinnings. Analog theory—as some readers now just catching their breath would doubtlessly urge—is much more difficult. Historically, discrete frequency transforms have been explored since the time of Gauss, but it is only with the development of digital computers that the fast computational methods have attracted wide interest and investigation.

Most of the exercises are basic problems that reinforce the concepts developed in the text. The next chapter considers an extension of the DTFT, called the z-transform. Chapter 9 considers applications of frequency-domain analysis to signal interpretation problems.

REFERENCES

1. J. W. Cooley and J. W. Tukey, An algorithm for the machine calculation of complex Fourier series, *Mathematics of Computation,* vol. 19, pp. 297–301, 1965.
2. C. E. Shannon, A mathematical theory of communication, *Bell Systems Technical Journal,* vol. 27, pp. 379–423 and pp. 623–656, 1948.
3. H. Nyquist, Certain topics in telegraph transmission theory, *Transactions of the AIEE,* vol. 47, pp. 617-644, 1928.
4. J. W. Cooley, P. A. Lewis, and P. D. Welch, Historical notes on the fast Fourier transform, *IEEE Transactions on Audio and Electroacoustics,* vol. AU-15, pp. 76–79, June 1967.
5. P. J. Plauger, *The C++ Standard Library*, PTR Publishers, 1997.
6. DSP Committee of the IEEE Society for Acoustics, Speech, and Signal Processing, *Programs for Digital Signal Processing*, New York: IEEE Press, 1979.
7. W. H. Press, B. P. Flannery, S. A. Teukolsky, and W. T. Vetterling, *Numerical Recipes*, Cambridge: Cambridge University Press, 1986.
8. C. S. Burrus and T. W. Parks, *DFT/FFT and Convolution Algorithms: Theory and Implementation*, New York: Wiley, 1985.
9. S. D. Stearns and R. A. David, *Signal Processing Algorithms in FORTRAN and C*, Englewood Cliffs, NJ: Prentice-Hall, 1988.
10. W. H. Press, B. P. Flannery, S. A. Teukolsky, and W. T. Vetterling, *Numerical Recipes in C*, Cambridge: Cambridge University Press, 1988.
11. H. S. Stone, *High-Performance Computer Architecture*, Reading, MA: Addison-Wesley, 1987.
12. *DSP56000/56001 Digital Signal Processor User's Manual*, Motorola, Inc., 1990.
13. M. Rosenlicht, *Introduction to Analysis*, New York: Dover, 1986.
14. E. Hille, *Analytic Function Theory*, vol. I, Waltham, MA: Blaisdell, 1959.
15. H. S. Carslaw, *An Introduction to the Theory of Fourier's Series and Integrals*, 3rd ed., New York: Dover, 1950.
16. H. F. Davis, *Fourier Series and Orthogonal Functions*, New York: Dover, 1963.
17. J. S. Walker, *Fourier Analysis*, New York: Oxford University Press, 1988.
18. D. C. Champeney, *A Handbook of Fourier Theorems*, Cambridge: Cambridge University Press, 1987.
19. L. Carleson, On convergence and growth of partial sums of Fourier series, *Acta Mathematica,* vol. 116, pp. 135–157, 1966.
20. R. A. Hunt, in *Orthogonal Expansions and Their Continuous Analogues*, Carbondale, IL: Southern Illinois University Press, pp. 235–255, 1968.
21. A. Zygmund, *Trigonometric Series*, vols. 1–2, Cambridge: Cambridge University Press, 1959.
22. Y. Katznelson, *An Introduction to Harmonic Analysis,* New York: Dover, 1976.

PROBLEMS

1. 1.For each of the following signals, $x(n)$, and intervals, $[0, N-1]$, find the discrete Fourier transform (DFT), $X(k)$:
   (a) $x(n) = \cos(\pi n/3)$ on $[0, 5]$
   (b) $x(n) = \sin(\pi n/3)$ on $[0, 5]$
   (c) $x(n) = \cos(\pi n/3)$ on $[0, 11]$
   (d) $x(n) = 3\sin(\pi n/3) + \cos(4\pi n/3)$ on $[0, 11]$
   (e) $x(n) = \exp(4\pi n/5)$ on $[0, 9]$
   (f) $x(n) = \cos(2\pi n/5 + \pi/4)$ on $[0, 4]$

2. Let $X = \mathcal{F}x$ be the system that accepts a signal of period $N > 0$ at its input and outputs the DFT of x.
   (a) Show that the system $\mathcal{F}$ is linear. That is, suppose that discrete signals $x(n)$ and $y(n)$ both have period $N > 0$. Show that the DFT of $s(n) = x(n) + y(n)$ is $S(k) = X(k) + Y(k)$, where $X(k)$ and $Y(k)$ are the DFTs of $x(n)$ and $y(n)$, respectively.
   (b) Show that the system $\mathcal{F}$ is not translation-invariant.

3. We may apply either the DFT or IDFT equation to transform a signal $x(n)$. Suppose that $x(n)$ has period $N > 0$.
   (a) Show that the DFT and the IDFT of $x(n)$ both have period N.
   (b) Show that if $X(k)$ is the DFT of $x(n)$, then the DFT of $X(k)$ is $Nx(-k)$.

4. Suppose the discrete signal, $x(n)$, has support on the finite interval, $[0, N-1]$, where $N > 0$.
   (a) Show that the signal $x_p(n)$ defined by

$$x_p(n) = \sum_{k=-\infty}^{\infty} x(n-kN) \tag{7.122}$$

   is periodic with period N and is identical to $x(n)$ on $[0, N-1]$.
   (b) Suppose we perform the DFT analysis equation calculation for $x(n)$'s values on $[0, N-1]$, giving $X(0), X(1), \ldots, X(N-1)$. Then define $y(n) = (1/N)[X(0) + X(1)e^{2\pi jkn/N} + \cdots + X(N-1)e^{2\pi j(N-1)n/N}$. Show that $y(n) = x_p(n)$ for all n.

5. Some of the first examples in this chapter showed that the delta signal $\delta(n)$ has DFT $\Delta(k) = 1$, and the signal signal $x(n) = [\underline{1}, 1, 1, 1, 0, 0, 0, 0]$ has DFT $X(k) = [\underline{4}, 1-(1+\sqrt{2})j, 0, 1-(\sqrt{2}-1)j, 0, 1+(\sqrt{2}-1)j, 0, 1+(1+\sqrt{2})j]$ on the interval $[0, 7]$. Find the DFT of the following signals, using only the properties of the DFT and without explicitly computing the DFT analysis equation's summation of products.
   (a) $y(n) = x(n-1) = [\underline{0}, 1, 1, 1, 1, 0, 0, 0]$
   (b) $y(n-k)$ for some $0 < k < 8$

(c) $y(n) = x(n) + \delta(n) = [\underline{2}, 1, 1, 1, 0, 0, 0, 0]$

(d) $y(n) = x(n) + \delta(n - 3)$

6. Prove the Convolution in Frequency Theorem. That is, let $x(n)$ and $y(n)$ be periodic signals with period $N > 0$; let $X(k)$ and $Y(k)$ be their DFTs; and let $z(n) = x(n)y(n)$, the termwise product of x and y. Show that the DFT of $z(n)$ is $Z(k) = (1/N)X(k)*Y(k)$, where $X(k)*Y(k)$ is the discrete convolution of $X(k)$ and $Y(k)$.

7. Let signal $x(n)$ be real-valued with period $N > 0$, and let $X(k)$ be its DFT. Prove the following symmetry properties:

(a) $\mathrm{Re}[X(k)] = \mathrm{Re}[X^*(N - k)] = \mathrm{Re}[X^*(-k)]$

(b) $\mathrm{Im}[X(k)] = -\mathrm{Im}[X^*(N - k)]$

(c) $|X(k)| = |X(N - k)|$

(d) $\arg(X(k)) = -\arg(X(n - k))$.

8. Suppose the DTFT, $X(\omega)$, of the signal $x(n)$ exists. Show that $X(\omega)$ is periodic with period 2π.

9. Suppose that H is a linear, translation-invariant (LTI) system, and let $h(n)$ be its impulse response.

(a) Suppose H is a finite impulse response (FIR) system. Show that the DTFT of $h(n)$, $H(\omega)$, exists.

(b) Suppose H is stable: if $x(n)$ is bounded, then $y = Hx$ is also bounded. Show that $H(\omega)$ exists.

10. Consider the two Hilbert spaces, l^2 and $L^2[a, b]$, where $a < b$. (Consider two signals in $L^2[a, b]$ to be the same if they are equal except on a set of Lebesgue measure zero.)

(a) Show that there is an isomorphism between the discrete Hilbert space l^2 and the analog Hilbert space $L^2[a, b]$.

(b) Give an explicit definition of a mapping, $\mathcal{G}$, between them that effects the isomorphism.

(c) The shifted impulses $\{u(n - k): k \in \mathbb{Z}\}$ constitute an orthogonal basis set for l^2; find, therefore, their image under G and show that it is an orthogonal basis as well.

(d) Are the exponential signals $\{\exp(j\omega n): k \in \mathbb{Z}\}$ an orthogonal basis set for $L^2[a, b]$? Explain.

11. Derive the following properties of the Dirichlet kernel, $D_N(\theta)$.

(a) Use the properties of the exponential function $\exp(j\theta n)$ to show

$$D_N(\theta) = \frac{1}{2\pi} \sum_{n=-N}^{n=+N} \exp(j\theta n) = \frac{1}{2\pi} + \frac{1}{\pi} \sum_{n=1}^{n=N} \cos(\theta n).$$

(b) Use the closed-form expression for the partial geometric series summation to show

$$D_N(\theta) = \frac{1}{2\pi} \frac{\sin\left(N\theta + \frac{\theta}{2}\right)}{\sin\left(\frac{\theta}{2}\right)}.$$

(c) Use (a) to prove

$$\int_{0}^{\pi} D_N(\theta)\, d\theta = \frac{1}{2} = \int_{-\pi}^{0} D_N(\theta)\, d\theta.$$

(d) From (b), show that $D_N(\theta)$ has its first two zero crossings at the points $\theta_N = \pm\pi/(N + 1/2)$. What is $D_N(0)$?

(e) Use (b) to sketch $D_N(\theta)$ for various values of N. Explain how $D_N(\theta)$ may be considered as the high-frequency sinusoid $\sin(N\theta + \theta/2)$ bounded above by the cosecant envelope $[\csc(\theta/2)]/(2\pi)$ and below by $-[\csc(\theta/2)]/(2\pi)$.

12. Consider the mapping $\mathcal{F}$ that takes $x(n) \in l^2$ to $X(\omega) \in L^2[-\pi, \pi]$, where $X(\omega)$ is the DTFT of $x(n)$. Show that $\mathcal{F}$ is linear, but not quite an isomorphism. Explain how to modify $\mathcal{F}$ so that it becomes a Hilbert space isomorphism.

13. Find the DTFT of the following signals.

(a) $e_k(n) = \delta(n - k)$

(b) $b(n) = [1, 1, 1, \underline{1}, 1, 1, 1]$

(c) $a(n) = b(n) + e_2(n) + 4e_{-3}(n)$

(d) $s(n) = (1/3)^n u(n)$

(e) $x(n) = (5)^n u(2 - n)$

(f) $h(n) = s(n) + 4b(n)$

(g) $y(n) = (x*h)(n)$

14. Find the IDTFT of the following signals.

(a) $E_k(\omega) = \exp(j\omega k)$

(b) $S(\omega) = 3\sin(-7j\omega)$

(c) $C(\omega) = 2\cos(3j\omega)$

(d) $P(\omega) = S(\omega)C(\omega)$

15. Let $x(n)$ and $y(n)$ be discrete signals and $X(\omega)$ and $Y(\omega)$ be their respective DTFTs. Then show the following linearity, time shift, frequency shift, and time reverse properties:

(a) The DTFT of $ax(n) + by(n)$ is $aX(\omega) + bY(\omega)$.

(b) The DTFT of $x(n - m)$ is $\exp(-j\omega m)X(\omega)$.

(c) The IDTFT of $X(\omega - \theta)$ is $\exp(-j\theta n)x(n)$.

(d) The DTFT of $x(-n)$ is $X(-\omega)$.

16. Let the signal $x(n)$ be real-valued and let $X(\omega)$ be its DTFT. If $z \in \mathbb{C}$, then let z^* be the complex conjugate of z, let Real(z) be its real part, let Imag(z) be its imaginary part, and let $\arg(z) = \tan^{-1}[\mathrm{Imag}(z)/\mathrm{Real}(z)]$ be the argument of z. Prove the following symmetry properties:

(a) $\mathrm{Real}(X(\omega)) = \mathrm{Real}(X(-\omega))$

(b) $-\mathrm{Imag}(X(\omega)) = \mathrm{Imag}(X(-\omega))$

(c) $X(\omega) = X^*(-\omega)$

(d) $|X(\omega)| = |X(-\omega)|$

(e) $\arg(X(\omega)) = -\arg(X(-\omega))$

17. Let the signal $x(n)$ be real-valued and $X(\omega)$ be its DTFT. If $x_e(n) = [x(n) + x(-n)]/2$ is the even part of $x(n)$, and $x_o(n) = [x(n) - x(-n)]/2$ is the odd part of $x(n)$, then find

(a) The DTFT of $x_e(n)$

(b) The DTFT of $x_o(n)$

18. Let the signal $x(n)$ be real-valued and let $X(\omega)$ be its DTFT. Show the following symmetry properties, which use the notation of the previous two problems:

(a) The DTFT of $x^*(n)$ is $X^*(-\omega)$, and the DTFT of $x^*(-n)$ is $X^*(\omega)$.

(b) The DTFT of $x_e(n)$ is Real($X(\omega)$), and the DTFT of $x_o(n)$ is j[Imag($X(\omega)$)].

(c) The DTFT of Real($X(n)$) is $X_e(\omega)$, and the DTFT of j[Imag($X(n)$)] is $X_o(\omega)$.

19. We know that the perfect high-frequency removal (low-pass) filter has impulse response

$$h(n) = \frac{\omega_c}{\pi}\mathrm{sinc}(n\omega_c) = \frac{\omega_c}{\pi}\frac{\sin(n\omega_c)}{n\omega_c} = \frac{\sin(n\omega_c)}{n\pi}.$$

(a) Consider the discrete system whose frequency response, $G(\omega)$, is unity for $|\omega| > \omega_c$ and zero otherwise. Explain why G may be considered a perfect high-pass filter. Find the time-domain filter, $g(n)$, corresponding to $G(\omega)$. Is $g(n)$ physically implementable? Explain.

(b) Consider the discrete system whose frequency response, $P(\omega)$, is unity for $\omega_h \geq |\omega| \geq \omega_l$ and zero otherwise. Explain the description of P as being an ideal bandpass filter. Find the time-domain filter, $g(n)$, corresponding to $G(\omega)$. Is $g(n)$ physically implementable? Explain.

(c) If $h(n)$ is an ideal time-domain low-pass filter, it is possible to approximate it by a finitely supported filter by zeroing terms above $n = N > 0$. In signal analysis applications, such as Chapter 4 considered, explain the possible uses of such a filter. For what types of applications is this filter useful? What applications are not served by this filter?

(d) Consider the questions in part (c) for perfect high-pass and bandpass filters.

20. Analog signal $x(t)$ has radial FT $X(\omega)$ shown below.

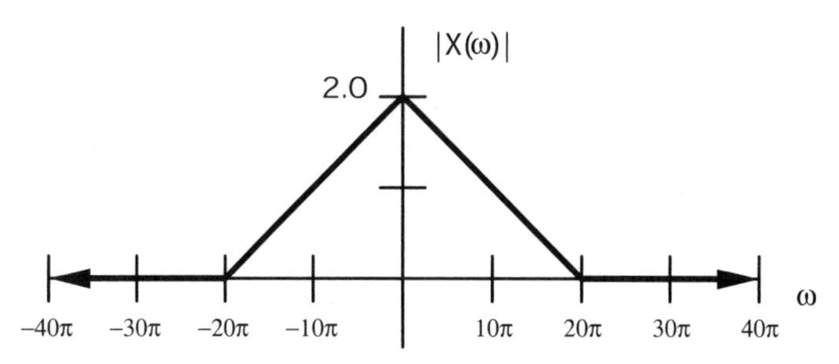

(a) What is the Nyquist rate for this signal in hertz?

(b) If $s(n) = x(nT)$, where $1/T = F = 15$ hertz, sketch the DTFT of s, $S(\omega)$.

(c) Sketch the radial FT of an ideal low-pass filter H such $y = Hs$ is not aliased when sampled at $F_S = 15$ hertz.

The remaining problems extend some of the ideas in the text.

21. Suppose we are given a formula for the DTFT of a discrete signal $h(n)$: $H(\omega) = P(\omega)/Q(\omega)$, where P and Q are both polynomials in ω. Develop two methods to find $h(n)$.

22. Show that a nonzero signal $x(n)$ cannot be finitely supported in both the time and frequency domains.

(a) Show that there is a $k > 0$ such that $x(n) = x(n)[u(n+k) - u(n-k)] = x(n)b(n)$, where $u(n)$ is the discrete unit step signal.

(b) Find the discrete-time Fourier transform of $b(n)$: $B(\omega)$.

(c) Apply the Convolution-in-Frequency Theorem to the product $x(n)b(n)$ to find an equivalent expression for $X(\omega)$.

(d) Derive a contradiction from the two expressions for $X(\omega)$.

23. Let $g(t) = A\exp(-\sigma t^2)$. Use the conditions for equality in the analog Schwarz inequality, and find constants A and σ so that

$$\sqrt{\frac{\pi}{2}} \le \Delta_t(g)\Delta_\omega(G)$$

CHAPTER 8

The *z*-Transform

The z-transform generalizes the discrete-time Fourier transform. It extends the domain of the DTFT of $x(n)$, the periodic analog signal $X(\omega)$, which is defined for $\omega \in \mathbb{R}$, to a function defined on $z \in \mathbb{C}$, the complex plane.

The motivations for introducing the z-transform are diverse:

- It puts the powerful tools of complex analysis at our disposal.
- There are possibilities for analyzing signals for which the DTFT analysis equation does not converge.
- It allows us to study linear, translation-invariant systems for which the frequency response does not exist.
- Some specialized operations such as signal subsampling and upsampling are amenable to z-transform techniques.
- It provides a compact notation, convenient for describing a variety of systems and their properties.

Having listed these z-transform benefits, we hasten to add that very often a series of z-transform manipulations concludes with a simple restriction of the transform to the DTFT. So we will not be forgetting the DTFT; on the contrary, it is the basic tool that we will be using for the spectral analysis of aperiodic discrete signals for the remainder of the book.

The development of z-transform theory proceeds along lines similar to those used in Chapter 7 for the DTFT. Many of the proofs of z-transform properties, for example, are very like the corresponding derivations for the DTFT. We often leave these results as exercises, and by now the reader should find them straightforward. This is a short chapter. It serves as a bridge between the previous chapter's theoretical treatment of discrete Fourier transforms and the diverse applications—especially filter design techniques—covered in Chapter 9.

Texts on systems theory introduce the z-transform and its analog world cousin, the Laplace transform [1, 2]. Books on digital signal processing [3–7] cover the z-transform in more detail. Explanations of the transform as a discrete filter design tool may also be found in treatments oriented to specific applications [8–10]. The

Signal Analysis: Time, Frequency, Scale, and Structure, by Ronald L. Allen and Duncan W. Mills
ISBN: 0-471-23441-9

z-transform was applied in control theory [11–13] long before it was considered for the design of digital filters [14]. Specialized treatises include Refs. 15 and 16.

8.1 CONCEPTUAL FOUNDATIONS

The z-transform has a very elegant, abstract definition as a power series in a complex variable. This power series is two-sided; it has both positive and negative powers of z, in general. Furthermore, there may be an infinite number of terms in the series in either the positive or negative direction. Like the DTFT, the theory of the z-transform begins with an investigation of when this doubly-infinite summation converges. Unlike the DTFT, however, the z-transform enlists a number of concepts from complex analysis in order to develop its existence and inversion results. This section introduces the z-transform, beginning with its abstract definition and then considering some simple examples.

Readers may find it helpful to review the complex variables tutorial in Chapter 1 (Section 1.7) before proceeding with the z-transform.

8.1.1 Definition and Basic Examples

The z-transform generalizes the DTFT to a function defined on complex numbers. To do this, we replace the complex exponential in the DTFT's definition with $z \in \mathbb{C}$. A simple change it is, but we shall nevertheless face some interesting convergence issues. For our effort, we will find that many of the properties of the DTFT carry through to the extended transform, and they provide us with tools for analyzing signals and systems for which the DTFT is not well-suited.

Definition (z-Transform). If $x(n)$ is a discrete signal and $z \in \mathbb{C}$, then its *z-transform*, $X(z)$, is defined by

$$X(z) = \sum_{n=-\infty}^{+\infty} x(n)z^{-n}. \tag{8.1}$$

To avoid some notation conflicts, the fancy-z notation, $X = \mathcal{Z}(x)$, is often convenient for writing the z-transform of $x(n)$. The signal $x(n)$ and the complex function $X(z)$ are called a z-transform pair. We also call (8.1) the z-transform *analysis equation*. Associated with a z-transform pair is a *region of convergence* (the standard acronym is ROC): $\mathrm{ROC}_X = \{z \in \mathbb{C}: X(z) \text{ exists}\}$. Sometimes as $|z|$ gets large, the value $X(z)$ approaches a limit. In this case, it is convenient to indicate that $\infty \in \mathrm{ROC}_X$. The notation $\mathbb{C}^+$ is useful for the so-called *extended complex plane*: $\mathbb{C}$ augmented with a special element, ∞.

Let us postpone, for a moment, convergence questions pertaining to the z-transform sum. Note that taking the restriction of complex variable z to the unit circle, $z = \exp(j\omega)$, and inserting this in (8.1), gives the DTFT. The DTFT is the restriction of the z-transform to the unit circle of the complex plane, $|z| = 1$: $X[\exp(j\omega)] = X(\omega)$, where the first "$X$" is the z-transform, and the second "X" is the DTFT of $x(n)$, respectively.

There is another form of the z-transform that uses only the causal portion of a signal.

Definition (One-sided z-Transform). If $x(n)$ is a discrete signal and $z \in \mathbb{C}$, then its *one sided z-transform*, $X^{+}(z)$, is defined by

$$X^{+}(z) = \sum_{n=0}^{+\infty} x(n) z^{-n}, \tag{8.2}$$

The one-sided, or unilateral, z-transform is important for the specialized problem of solving linear, constant-coefficient difference equations. Typically, one is given difference equations and initial conditions at certain time instants. The task is to find all the discrete signal solutions. The one-sided z-transform agrees with the standard two-sided transform on signals $x(n) = 0$ for $n < 0$. The linearity property is the same, but the shifting property differs. These ideas and an application are considered in the problems at the end of the chapter.

As with the DTFT, the infinite sum in the z-transform summation (8.1) poses convergence questions. Of course, the sum exists whenever the signal $x(n)$ has finite support; the corresponding z-transform $X(z)$ is a sum of powers (positive, zero, and negative) of the complex variable z. Let us consider some elementary examples of finding the z-transforms of discrete signals. Finding such z-transform pairs, $x(n)$ and $X(z)$, is typically a matter of finding the z-transform of a signal $y(n)$ which is similar to $x(n)$ and then applying the z-transform properties to arrive at $X(z)$ from $Y(z)$.

Example (Discrete Delta). Let us start simple by considering the discrete delta signal, $\delta(n)$. For any $z \in \mathbb{C}$, only the summand corresponding to $n = 0$ is nonzero in (8.1), and thus $\Delta(z) = \mathcal{Z}(\delta)(z) = 1$ for all $z \in \mathbb{C}$.

Example (Square Pulse). Again, let us consider the impulse response of the moving average system, H. It has impulse response $h(n) = [1, 1, \underline{1}, 1, 1]$; in other words, $h(n) = 1$ for $-2 \le n \le 2$, and $h(n) = 0$ otherwise. We write immediately

$$H(z) = \sum_{n=-\infty}^{+\infty} h(n) z^{-n} = \sum_{n=-2}^{+2} z^{-n} = z^{2n} + z^{n} + 1 + z^{-n} + z^{-2n}. \tag{8.3}$$

Note that $H(z)$ exists for all $z \in \mathbb{C}$, $z \ne 0$. Thus, a signal, $x(n)$, whose DTFT converges for all $\omega \in \mathbb{R}$ may have a z-transform, $X(z)$, which does not converge for all $z \in \mathbb{C}$. In general, a finitely supported signal, $x(n)$, that is nonzero for positive time instants will not have a z-transform, $X(z)$, which exists for $z = 0$.

Example (Causal Exponential Signal). Consider the signal $x(n) = a^{n}u(n)$. We calculate

$$X(z) = \sum_{n=-\infty}^{+\infty} x(n) z^{-n} = \sum_{n=0}^{\infty} a^{n} z^{-n} = \sum_{n=0}^{\infty} \left(\frac{a}{z}\right)^{n} = \frac{1}{1-\frac{a}{z}} = \frac{z}{z-a}, \tag{8.4}$$

where the geometric series sums in (8.4) to $z/(z-a)$ provided that $|a/z| < 1$. Thus, we have $\mathrm{ROC}_X = \{z \in \mathbb{C}: |a/z| < 1\} = \{z \in \mathbb{C}: |a| < |z|\}$. In other words, the region of convergence of the z-transform of $x(n) = a^n u(n)$ is all complex numbers lying outside the circle $|z| = a$. In particular, the unit step signal, $u(n)$, has a z-transform, $U(z) = 1/(1 - z^{-1})$. We may take $a = 1$ above and find thereby that $\text{ROC}_U = \{z \in \mathbb{C}: 1 < |z|\}$.

This example shows that a z-transform can exist for a signal that has no DTFT. If $|a| > 1$ in the previous example, for instance, then the analysis equation for $x(n)$ does not converge. But the z-transform, $X(z)$, does exist, as long as z lies outside the circle $|z| = a$ in the complex plane. Also, the ROC for this example was easy to discover, thanks to the geometric series form taken by the z-transform sum. There is a companion example, which we need to cover next. It illustrates the very important point that the ROC can be the only distinguishing feature between the z-transforms of two completely different signals.

Example (Anti-causal Exponential Signal). Consider the signal $y(n) = -a^n u(-n - 1)$. Now we find

$$Y(z) = \sum_{n=-\infty}^{+\infty} y(n)z^{-n} = \sum_{n=-\infty}^{-1} -a^n z^{-n} = -\sum_{n=1}^{\infty}\left(\frac{z}{a}\right)^n = -\frac{z}{a}\sum_{n=0}^{\infty}\left(\frac{z}{a}\right)^n$$
$$= -\left(\frac{\frac{z}{a}}{1-\frac{z}{a}}\right) = \frac{z}{z-a}, \tag{8.5}$$

with the convergence criterion $|z/a| < 1$. In this case, we have $\text{ROC}_Y = \{z \in \mathbb{C}: |z/a| < 1\} = \{z \in \mathbb{C}: |z| < |a|\}$. The region of convergence of the z-transform of $y(n) = -a^n u(-n - 1)$ is all complex numbers lying inside the circle $|z| = a$.

The upshot is that we must always be careful to specify the region of convergence of a signal's z-transform. In other words, given one algebraic expression for $X(z)$, there may be multiple signals for which it is the z-transform; the deciding factor then becomes the region of convergence.

8.1.2 Existence

Demonstrating the convergence of the z-transform for a particular signal makes use of complex variable theory. In particular, the z-transform is a Laurent series [17, 18].

Definition (Power and Laurent Series). A complex power series is a sum of scaled powers of the complex variable z:

$$\sum_{n=0}^{+\infty} a_n z^n. \tag{8.6}$$

A Laurent series is a two-sided series of the form

$$\sum_{n=-\infty}^{+\infty} a_n z^{-n} = \sum_{n=1}^{+\infty} a_n z^{-n} + \sum_{n=-\infty}^{0} a_n z^{-n}, \tag{8.7}$$

where the a_n are (possibly complex) coefficients. (Most mathematics texts do not use the negative exponent in the definition—their term z^n has coefficient a_n; our definition goes against the tradition so that its form more closely follows the definition of the z-transform.) One portion of the series consists of negative powers of z, and the other part consists of non-negative powers of z. We say the Laurent series (8.7) converges for some $z \in \mathbb{C}$ if both parts of the series converge.

Obviously, we are interested in the situation were the Laurent series coefficients are the values of a discrete signal, $x(n) = a_n$. From complex variable theory, which we introduced in the first chapter, the following results are relevant to Laurent series convergence. We prove the first result for the special case where the z-transform of $x(n)$ contains only non-negative powers of z; that is, it is a conventional power series in z. This will happen when $x(n) = 0$ for $n > 0$.

The next definition identifies upper and lower limit points within a sequence.

Definition (lim sup and lim inf). Let $A = \{a_n: 0 \le n < \infty\}$. Define $A_N = A \setminus \{a_n: 0 \le n < N\}$, which is the set A after removing the first N elements. Let κ_N be the least upper bound of A_N. Then the limit, κ, of the sequence $\{\kappa_N: N > 0\}$ is called the *lim sup* of A, written

$$\kappa = \lim_{N \to \infty} \{\kappa_N: N > 0\} = \lim_{n \to \infty} \sup A = \lim_{n \to \infty} \sup\{a_n: 0 \le n < \infty\}. \tag{8.8}$$

Similarly, if we let λ_N be the greatest lower bound of A_N, then the limit of the sequence $\{\lambda_N: N > 0\}$ is called the *lim inf*[1] of A:

$$\lambda = \lim_{N \to \infty} \{\lambda_N: N > 0\} = \lim_{n \to \infty} \inf A = \lim_{n \to \infty} \inf\{a_n: 0 \le n < \infty\}. \tag{8.9}$$

Sometimes a sequence of numbers does not have a limit, but there are convergent subsequences within it. The lim sup is the largest limit of a convergent subsequence, and the lim inf is the smallest limit of a convergent subsequence, respectively. The sequence has a limit if the lim sup and the lim inf are equal. Readers with advanced calculus and mathematical analysis background will find these ideas familiar [19–21]. We need the concept of the lim sup to state the next theorem. Offering a convergence criterion for a power series, it is a step toward finding the ROC of a z-transform.

Theorem (Power Series Absolute Convergence). Suppose $x(n)$ is a discrete signal; its z-transform, $X(z)$, has only non-negative powers of z,

$$X(z) = \sum_{n=0}^{+\infty} a_n z^n; \tag{8.10}$$

[1]These are indeed the mathematical community's standard terms. The lim sup of a sequence is pronounced "lim soup," and lim inf sounds just like its spelling.

and

$$\kappa = \lim_{n\to\infty} \sup\{|a_n|^{1/n} : 1 \le n < \infty\}. \tag{8.11}$$

Then $X(z)$ converges absolutely for all z with $|z| < \kappa^{-1}$ and diverges for all $|z| > \kappa^{-1}$. (This allows A to be unbounded, in which case $\kappa = \infty$ and, loosely speaking, $\kappa^{-1} = 0$.)

Proof: Consider some z such that $|z| < \kappa^{-1}$, and choose $\lambda > \kappa$ so that λ^{-1} lies between these two values: $|z| < \lambda^{-1} < \kappa^{-1}$. Because $\kappa = \lim\sup A$, there is an N such that $|a_n|^{1/n} < \lambda$ for all $n > N$. But this implies that $|a_n z^n| < |z\lambda|^n$ for $n > N$. Since $|z\lambda| < 1$ by the choice of λ, the power series (8.10) is bounded above by a convergent geometric series. The series must therefore converge absolutely. We leave the divergence case as an exercise. ■

Definition (Radius of Convergence). Let $\rho = \kappa^{-1}$, where κ is given by (8.11) in the Power Series Absolute Convergence Theorem. Then ρ is called the *radius of convergence* of the complex power series (8.10).

Corollary (Power Series Uniform Convergence). Suppose $x(n)$ is a discrete signal; its z-transform, $X(z)$, has only non-negative powers of z as in the theorem (8.10); and κ is defined as in (8.11). Then for any $0 < R < \rho = \kappa^{-1}$, $X(z)$ converges uniformly in the complex disk $\{z \in \mathbb{C}: |z| < R < \rho = \kappa^{-1}\}$.

Proof: For any disk of radius R, $0 < R < \rho$, the proof of the theorem implies that there is a convergent geometric series that bounds the power series (8.10). Since the convergence of the dominating geometric series does not depend on z, the sum of the series in z (8.10) can be made arbitrarily close to its limit independent of z. The convergence is therefore uniform. ■

Corollary (Analyticity of the Power Series Limit). Again, if $x(n)$ is a discrete signal; its z-transform, $X(z)$, has the form (8.10); and $\rho = \kappa^{-1}$ is given by (8.11), then $X(z)$ is an analytic function, the derivative $X'(z) = dX(z)/dz$ can be obtained by termwise differentiation of the power series (8.10), and $\mathrm{ROC}_X = \mathrm{ROC}_{dX/dz}$.

Proof: This proof was given already in Section 1.7, where we assumed the uniform convergence of the power series. From the Uniform Convergence Corollary, we know this to be the case within the radius of convergence, ρ; the result follows. ■

Finally, we consider the situation that most interests us, the Laurent series. The z-transform assumes the form of a Laurent series. We have, in fact, already developed the machinery we need to discover the region of convergence of a z-transform. We apply the Power Series Convergence Theorems above for both parts of the Laurent series: the negative and non-negative powers of z.

Theorem (z-Transform Region of Convergence). Let $x(n)$ be a discrete signal and let $X(z) = X_1(z) + X_2(z)$ be its z-transform (which may be two-sided). Suppose

$X(z) = X_1(z) + X_2(z)$, where $X_2(z)$ consists on non-negative powers of z, and $X_1(z)$ contains only negative powers of z. Then $X(z)$ converges absolutely within an annulus of the complex plane, $ROC_X = \{z \in \mathbb{C}: \rho_1 < |z| < \rho_2\}$, where ρ_1 and ρ_2 are the radii of convergence of $X_1(z)$ and $X_2(z)$, respectively. The convergence of the z-transform Laurent series is uniform within any closed annulus contained in ROC_X, and its limit, $X(z)$ is analytic within this same closed annulus.

Proof: $X(z)$'s true power series portion, $X_2(z)$, converges inside some circle $|z| = \rho_2$, where ρ_2 is the radius of convergence. The $X_1(z)$ portion of $X(z)$ converts to power series form by setting $w = z^{-1}$. Then the radius of convergence may be found for the power series $Y(w) = X_1(w^{-1}) = X_1(z)$. $Y(w)$ converges inside some circle of radius R_1, say, which means $X_1(z)$ converges outside the circle $\rho_1 = 1/R_1$. The region formed by intersecting the exterior of the circle $|z| = \rho_1$ and the interior of the circle $|z| = \rho_2$ is the annulus we seek. ■

Example. Suppose $x(n)$ is given as follows:

$$x(n) = \begin{cases} 3^{-n} & \text{for } n \geq 0, \\ -4^{n} & \text{for } n<0. \end{cases} \tag{8.12}$$

Let $x_1(n) = a^n u(n)$, where $a = 1/3$, and $x_2(n) = -b^n u(-n-1)$, where $b = 4$. Then $x(n) = x_1(n) + x_2(n)$. We have computed the z-transforms of signals of this form in earlier examples. We have $X(z) = X_1(z) + X_2(z)$, where $X_1(z)$ converges outside the circle $|z| = 1/3$, and $X_2(z)$ converges inside the circle $|z| = 4$ (Figure 8.1).

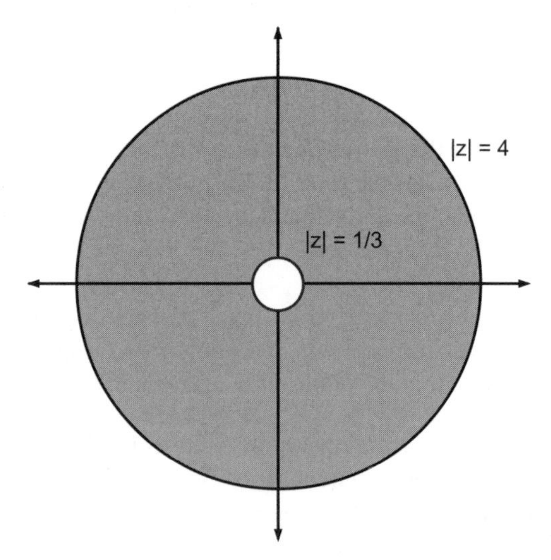

Fig. 8.1. The region of convergence of the z-transform of $x(n) = 3^{-n}u(n) - 4^n u(-n-1)$ is an open annulus in the complex plane. The causal portion of $x(n)$ produces a power series in z^{-1}, which converges outside the circle $|z| = 1/3$. The anti-causal part of $x(n)$ produces a power series in z, which converges inside the circle $|z| = 4$.

Let us now turn to the basic properties of the z-transform. The above example made tacit use of the linearity property. The z-transform properties we cover resemble very closely those of our previously studied transforms, especially the DTFT. One subtlety is the region of convergence, which we must account for during algebraic and analytic operations on the transformed signals.

8.1.3 Properties

The properties of the z-transform closely resemble those of the discrete transform that it generalizes—the discrete-time Fourier transform. With the z-transform, there is a twist, however; now the region of convergence of the transform figures prominently in validating the properties. We divide this section's results into two categories: basic properties and those that rely on the principles of contour integration in the complex plane.

8.1.3.1 Basic Properties. This theorems herein are much like those we developed for the DTFT. Their proofs are also similar, involve familiar methods, and we leave them as exercises for the most part. One caveat in dealing with the z-transform is the region of convergence; one must always be careful to specify this annulus and to consider the special cases of $z = 0$ and $z = \infty$.

Proposition (Linearity, Time Shift, Frequency Shift, and Time Reversal). Let $x(n)$ and $y(n)$ be discrete signals and let $X(z)$ and $Y(z)$ be their z-transforms, respectively. Then

(a) (Linearity) The z-transform of $ax(n) + by(n)$ is $aX(z) + bY(z)$, and its region of convergence contains $\text{ROC}_X \cap \text{ROC}_Y$.

(b) (Time Shift) The z-transform of $x(n - m)$ is $z^{-m}X(z)$, and its region of convergence is ROC_X, except, perhaps, that it may include or exclude the origin, $z = 0$, or the point at infinity, $z = \infty$;

(c) (Frequency Shift, or Time Modulation) Let $a \in \mathbb{C}$. Then the z-transform of $a^n x(n)$ is $Y(z) = X(z/a)$ with $\text{ROC}_Y = |a|\text{ROC}_X$.

(d) (Time Reversal) If $Z[x(n)] = X(z)$, and $y(n) = x(-n)$, then $Z[y(n)] = Y(z) = X(z^{-1})$, with $\text{ROC}_Y = \{z \in \mathbb{C}\colon z^{-1} \in \text{ROC}_X\}$.

Proof: In (a), the formal linearity is clear, but it is only valid where both transforms exist. If $a \neq 0$, then the ROC of $aX(z)$ is ROC_X, and a similar condition applies to $Y(z)$. However, when the two transforms are added, cancellations of their respective terms may occur. This expands the sum's ROC beyond the simple intersection of ROC_X and ROC_Y. Also in (b), multiplication of $X(z)$ by z^k for $k > 0$ may remove from ROC_X. However, if $X(z)$ contains only powers $z^{-|n|}$, for $n > k$, then multiplication by z^k will have no effect on ROC_X. Similarly, multiplication of $X(z)$ by z^k for $k < 0$ may remove 0 from ROC_X, and so on. The other cases are just as straightforward to list. Toward proving (c), let us remark that the power series expansion for $X(z/a)$ will have a new region of convergence that is scaled by $|a|$ as follows: If $\text{ROC}_X = \{z\colon r_1 < |z| < r_2\}$, then $\text{ROC}_Y = \{z\colon |a|r_1 < |z| < |a|r_2\}$. Time reversal is an exercise. ■

Proposition (Frequency Differentiation). Suppose $x(n)$ is a discrete signal, and $X(z)$ is its z-transform. Then the z-transform of $nx(n)$ is $-zdX(z)/dz$. The region of convergence remains the same, except that ∞ may be deleted or 0 may be inserted.

Proof: Similar to the Frequency Differentiation Property of the DTFT. Note that within the region of convergence, the z-transform Laurent series is differentiable. Since we multiply by a positive power of z, will be deleted from the ROC if the highest power of z in $X(z)$ is z^0. Similarly, 0 may be inserted into the ROC if the lowest power of z in $X(z)$ is z^{-1}. ■

Example. Suppose $x(n) = -na^n u(-n-1)$. Find $X(z)$. We already know from the example of the anti-causal exponential signal in Section 8.1.1 that the z-transform of $-a^n u(-n-1)$ is $(1 - a/z)^{-1}$, with ROC = $\{z \in \mathbb{C}: |z| < |a|\}$. Thus, the frequency differentiation property applies and we have

$$X(z) = -z\frac{d}{dz}\left(1-\frac{a}{z}\right)^{-1} = z\left(1-\frac{a}{z}\right)^{-2}\frac{d}{dz}\left(1-\frac{a}{z}\right) = \frac{az}{(z-a)(z-a)}. \tag{8.13}$$

Also, $\mathrm{ROC}_X = \{z \in \mathbb{C}: |z| < |a|\}$. The properties are useful in finding the z-transforms of new signals.

The z-transform is not without a convolution theorem. Sometimes signal processing systems must deal with signals or system impulse responses for which the DTFT does not converge. A useful tool in this instance is the z-transform. And (not unexpectedly by this point!) there is a convolution result for the z-transform; it finds use in studying LTI systems, questions of stability, subsampling, and interpolation operations for discrete signals.

Theorem (Convolution in Time). Let $x(n)$ and $y(n)$ be signals; let $X(z)$ and $Y(z)$ their z-transforms, respectively; and let $w = x * y$. Then the z-transform of $w(n)$ is $W(z) = X(z)Y(z)$, and $\mathrm{ROC}_W \supseteq \mathrm{ROC}_X \cap \mathrm{ROC}_Y$.

Proof: The proof of the DTFT Convolution-in-Time Theorem extends readily to the z-transform:

$$\begin{aligned} W(z) &= \sum_{n=-\infty}^{\infty} w(n)z^{-n} = \sum_{n=-\infty}^{\infty} (x*y)(n)z^{-n} = \sum_{n=-\infty}^{\infty}\left(\sum_{k=-\infty}^{\infty} x(k)y(n-k)\right)z^{-n} \\ &= \sum_{n=-\infty}^{\infty}\sum_{k=-\infty}^{\infty} x(k)y(n-k)z^{-(n-k)}z^{-k} = \sum_{k=-\infty}^{\infty}\sum_{n=-\infty}^{\infty} x(k)y(n-k)z^{-(n-k)}z^{-k} \\ &= \sum_{k=-\infty}^{\infty} x(k)z^{-k}\sum_{n=-\infty}^{\infty} y(n-k)z^{-(n-k)} = X(z)Y(z). \end{aligned} \tag{8.14}$$

If $x \in \mathrm{ROC}_X \cap \mathrm{ROC}_Y$, then the z-transform's Laurent series converges absolutely. This justifies the step from an iterated summation to a double summation as well as the interchange in the order of summation in (8.14). Hence, $\mathrm{ROC}_W \supseteq \mathrm{ROC}_X \cap \mathrm{ROC}_Y$. ■

Corollary (LTI System Function). Suppose H is a discrete LTI system $y = Hx$; its impulse response is $h = H\delta$; and $X(x)$, $Y(z)$, and $H(z)$ are the z-transforms of $x(n)$, $y(n)$, and $h(n)$, respectively. Then, $Y(z) = H(z)X(z)$.

Proof: The output signal $y = h * x$ by the Convolution Theorem for LTI Systems, and the result follows from the theorem. ■

Definition (System or Transfer Function). Let H be a discrete LTI system $y = Hx$ and $h = H\delta$ its impulse response. Then the z-transform of $h(n)$, $H(z)$ is called the system function or the transfer function for the system H.

Remark. For our multiple uses of the uppercase "H," we once again ask the reader's indulgence. Here, both the system itself and its impulse response, a complex-valued function of a complex variable, are both denoted "H."

8.1.3.2 Properties Involving Contour Integration. With the DTFT synthesis equation, we can identify a time-domain signal $x(n)$ with the analog Fourier Series coefficients of its DTFT, $X(\omega)$. The interconnection is at once elegant and revealing. The z-transform is like the DTFT, in that it is a discrete transform with an inversion relation which involves a continuous domain integration operation. However, because the domain of definition of the z-transform is a region in the complex plane, the inversion formula becomes quite exotic: It depends on a complex contour integral.

Readers who skimmed the material in Section 1.7 may wish to review it more carefully before proceeding with the next several theorems.

Theorem (Inversion). Suppose $x(n)$ is a discrete signal and $X(z)$ is its z-transform. If C is any simple, counterclockwise, closed contour of the complex plane; the origin is in the interior of C; and $C \subseteq \mathrm{ROC}_X$, then

$$x(n) = \frac{1}{2\pi j} \oint_C X(z) z^{n-1} dz. \tag{8.15}$$

Proof: The integrand in (8.15) contains a power of z, and the contour is closed; this suggests the Cauchy integral theorem from Section 1.7.3.

$$\frac{1}{2\pi j} \oint_C X(z) z^{n-1}\, dz = \frac{1}{2\pi j} \oint_C \left[\sum_{k=-\infty}^{k=+\infty} x(k) z^{-k} \right] z^{n-1} dz = \frac{1}{2\pi j} \sum_{k=-\infty}^{k=+\infty} x(k) \oint_C z^{n-k-1}\, dz. \tag{8.16}$$

Once again, inserting the analysis equation directly into the integral and then interchanging the order of summation and integration pays off. Since the z-transform is absolutely and uniformly convergent within its ROC, the order of summation and integration is unimportant in (8.16). Recall, from Section 1.7.3, the Cauchy integral theorem:

$$\frac{1}{2\pi j}\oint_C z^m dz = \begin{cases} 0 & \text{if} \quad m \neq -1, \\ 1 & \text{if} \quad m = -1. \end{cases} \tag{8.17}$$

As a consequence, all terms in the summation of are zero except for the one where $n = k$. This implies that

$$x(n) = \frac{1}{2\pi j}\oint_C X(z) z^{n-1} dz, \tag{8.18}$$

as desired. ■

Equation (8.18) is the z-transform synthesis equation.

Example (Unit Circle Contour). Suppose now that $Z(x(n)) = X(z)$ and that ROC_X contains the unit circle: $C = \{z: |z| = 1\} \subseteq \mathrm{ROC}_X$. Then, $z = \exp(j\omega)$ on C, $\omega \in [-\pi, +\pi]$; $dz = j\exp(j\omega)\, d\omega$; and evaluating the inverse z-transform contour integral gives

$$\frac{1}{2\pi j}\oint_C X(z) z^{n-1} dz = \frac{1}{2\pi j}\oint_{|z|=1} X(z) z^{n-1} dz = \frac{1}{2\pi j}\int_{-\pi}^{+\pi} X(e^{j\omega})(e^{j\omega})^{n-1}\, je^{j\omega} d\omega$$

$$= \frac{1}{2\pi}\int_{-\pi}^{+\pi} X(e^{j\omega})(e^{j\omega})^n\, d\omega = \frac{1}{2\pi}\int_{-\pi}^{+\pi} X(\omega)\exp(jn\omega)\, d\omega. \tag{8.19}$$

In (8.19), $X(e^{j\omega})$ is the z-transform evaluated at $z = \exp(j\omega)$, and $X(\omega)$ is the DTFT of $x(n)$ evaluated at ω, $-\pi \leq \omega \leq \pi$. This example thus shows that if C is the unit circle and it lies within ROC_X, then the inverse z-transform relation reduces to the IDTFT.

Prominent among the z-transform's basic properties in the previous section, the Convolution-in-Time Theorem linked convolution in time to simple multiplication in the z-domain. The next theorem is its counterpart for z-domain convolutions. Although this results lacks the aesthetic appeal of the DTFT's Convolution-in-Frequency Theorem, it will nevertheless prove useful for discrete filter design applications in the next chapter.

Theorem (Convolution in the z-Domain). Let $s(n) = x(n)y(n)$ be the termwise product of $x(n)$ and $y(n)$; $Z(x(n)) = X(z)$, with $\mathrm{ROC}_X = \{z \in \mathbb{C}: r_X < |z| < R_X\}$; and $Z(y(n)) = Y(z)$, where $\mathrm{ROC}_Y = \{z \in \mathbb{C}: r_Y < |z| < R_Y\}$. Then, $\mathrm{ROC}_S \supseteq \{z \in \mathbb{C}: r_X r_Y < |z| < R_X R_Y\}$. Furthermore, let C be a simple, closed contour of the complex plane whose interior contains the origin. If $\mathrm{ROC}_{X(w)} \cap \mathrm{ROC}_{Y(z/w)}$ contains C, then

$$S(z) = \frac{1}{2\pi j}\oint_C X(w) Y\left(\frac{z}{w}\right) w^{-1} dw. \tag{8.20}$$

Proof: The z-transform analysis equation for $S(z)$ is

$$S(z) = \sum_{n=-\infty}^{+\infty} s(n)z^{-n} = \sum_{n=-\infty}^{+\infty} x(n)y(n)z^{-n} = \sum_{n=-\infty}^{+\infty} \left[\frac{1}{2\pi j} \oint_C X(w)w^{n-1} \right] y(n)z^{-n}, \tag{8.21}$$

where the z-transform synthesis equation (8.18), with dummy variable of integration w, replaces $x(n)$ inside the sum. This substitution is valid for any simple, closed path C, when $C \subseteq \mathrm{ROC}_X$. Summation and integration may change order in , so long as $z \in \mathrm{ROC}_S$, where uniform convergence reigns:

$$S(z) = \frac{1}{2\pi j} \oint_C X(w) \sum_{n=-\infty}^{+\infty} y(n) \left(\frac{z}{w} \right)^{-n} w^{-1} dw = \frac{1}{2\pi j} \oint_C X(w) Y\left(\frac{z}{w} \right) w^{-1} dw. \tag{8.22}$$

When does $S(z)$ exist? We need $C \subseteq \mathrm{ROC}_X = \{w \in \mathbb{C}\colon r_X < |w| < R_X\}$ and $z/w \in \mathrm{ROC}_Y = \{z \in \mathbb{C}\colon r_Y < |z| < R_Y\}$. The latter occurs if and only if $r_Y < |z/w| < R_Y\}$; this will be the case if $|w|r_Y < |z| < |w|R_Y$ for $w \in \mathrm{ROC}_X$. Hence ROC_S includes $\{z \in \mathbb{C}\colon r_X r_Y < |z| < R_X R_Y\}$. The contour integral in (8.22) will exist whenever $w \in C$ and $z/w \in \mathrm{ROC}_Y$; in other words, $C \subseteq \mathrm{ROC}_X \cap \mathrm{ROC}_{Y(z/w)}$, as stated. ■

Corollary (Parseval's Theorem). Suppose that $x(n), y(n) \in l^2$, $Z(x(n)) = X(z)$, and $Z(y(n)) = Y(z)$. If C is a simple, closed contour whose interior contains the origin, and $C \subseteq \mathrm{ROC}_X \cap \mathrm{ROC}_{Y^*(1/w^*)}$, then

$$\langle x, y \rangle = \sum_{n=-\infty}^{+\infty} x(n)y^*(n) = \frac{1}{2\pi j} \oint_C X(w) Y^*\left(\frac{1}{w^*} \right) w^{-1} dw. \tag{8.23}$$

Proof: The inner product $\langle x, y \rangle$ exists, since x and y are square-summable, and if $s(n) = x(n)y^*(n)$, then $s(n) \in l^1$ (Cauchy–Schwarz). If $Z(s(n)) = S(z)$, then

$$S(z) = \sum_{n=-\infty}^{+\infty} x(n)y^*(n)z^{-n}, \tag{8.24}$$

so that

$$\langle x, y \rangle = \sum_{n=-\infty}^{+\infty} x(n)y^*(n) = S(1). \tag{8.25}$$

It is an easy exercise to show that $Z(y^*(n)) = Y^*(z^*)$; and together with the Convolution in z-Domain Theorem (8.22), this entails

$$\langle x, y \rangle = S(1) = \frac{1}{2\pi j} \oint_C X(w) Y^*\left(\frac{1}{w^*} \right) w^{-1} dw, \tag{8.26}$$

completing the proof. ■

8.2 INVERSION METHODS

Given a complex function of a complex variable, $X(z)$, there are three methods for finding the time domain signal, $x(n)$, such that $Z[x(n)] = X(z)$. These approaches are:

- Via the inverse z-transform relation, given by the contour integral (8.15).
- Through an expansion of $X(z)$ into a Laurent series; then the $x(n)$ values read directly from the expansion's coefficient of z^{-n}.
- By way of algebraic manipulation (especially using partial fractions) of $X(z)$ into a form in which its various parts are readily identified as the z-transforms of known discrete signals. This approach relies heavily on the z-transform's basic properties (Section 8.2.1).

Actually, the second two methods are the most useful, because the contour integrals prove to be analytically awkward. This section considers some examples that illustrate each of these z-transform inversion tactics.

8.2.1 Contour Integration

Let us look first at the easiest nontrivial example using contour integration in the complex plane to discover the discrete signal $x(n)$ whose z-transform is the given complex function $X(z)$. On first reading, this section can be reviewed casually. But those readers who accepted our invitation—several chapters back—to skip the complex variables tutorial should note that those ideas are key to this approach for z-transform inversion.

Example (Inversion by Contour Integration). Suppose $X(z) = z/(z - a)$, $a \neq 0$, with $\mathrm{ROC}_X = \{z \in \mathbb{C}: |a| < |z|\}$. Of course, we already know the signal whose z-transform is $X(z)$; it is the causal exponential signal $a^n u(n)$ from the example in Section 8.1.1. But, in order to learn the technique, let us proceed with pretenses toward discovery. From the z-transform synthesis equation (8.15), we may choose the contour C to be a circle outside $z = |a|$ and immediately write

$$x(n) = \frac{1}{2\pi j}\oint_C X(z)z^{n-1}dz = \frac{1}{2\pi j}\oint_C \frac{z^n}{(z-a)}\,dz. \tag{8.27}$$

Is the contour integral (8.27) easy to evaluate? To the fore, from complex analysis, comes a powerful tool: the Cauchy residue theorem (Section 1.7.3). Assume that C is a simple, closed curve; $a_m \notin C$, $1 < m < M$; $f(z)$ is a complex function, which is analytic (has a derivative df/dz) on and within C, except for poles ($|f(z)| \to \infty$ near a pole) at each of the a_m. The residue theorem then states that

$$\frac{1}{2\pi j}\oint_C f(z)\,dz = \sum_{m=1}^{M} \mathrm{Res}(f(z), a_m). \tag{8.28a}$$

Recall that the residue of $f(z)$ at the pole $z = p$ is given by

$$\operatorname{Res}(f(z), p) = \begin{cases} \dfrac{1}{(k-1)!} g^{(k-1)}(p) & \text{if } p \in \text{Interior}(C), \\ 0 & \text{otherwise,} \end{cases} \tag{8.28b}$$

where k is the order of the pole, $g(z)$ is the nonsingular part of $f(z)$ near p, and $g^{(k-1)}(p)$ is the $(k-1)$th derivative of $g(z)$ evaluated at $z = p$. (Complex function $f(z)$ has a pole of order $(k-1)$ at $z = 0$ if there exists $g(z)$ such that $f(z) = g(z)/(z-p)^k$, $g(z)$ is analytic near $z = p$, and $g(p) \neq 0$.) Let us continue the example. To find $x(n)$, $n \geq 0$, we set $f(z) = z^n/(z-a)$, as in (8.27). Note that $f(z)$ is analytic within C, except for a first-order pole at $z = a$. Therefore, $g(z) = z^n$, and we have $\operatorname{Res}(f(z), a) = g^{(0)}(a) = x(n) = a^n$. For non-negative values of n, computing the contour integral (8.27) is generally quite tractable. Elated by this result, we might hope to deduce values $x(n)$, $n < 0$, so easily. When $n < 0$, however, there are multiple poles inside C: a pole at $z = a$ and a pole of order n at $z = 0$. Consider the case $n = -1$. We set $f(z) = z^{-1}/(z-a)$. Thus, the pole at $z = p = 0$ is of order $k = 1$, since $g(z) = (z-a)^{-1}$ is analytic around the origin. Therefore, $\operatorname{Res}(f(z), z = 0) = g^{(0)}(p) = g^{(0)}(0) = (-a)^{-1}$. There is another residue, and we must sum the two, according to (8.28b). We must select a different analytic part of $f(z)$ near the pole at $z = a$; we thus choose $g(z) = z^{-1}$, so that $g(z)$ is analytic near $z = a$ with $f(z) = g(z)/(z-a)^1$. Consequently, this pole is also first order. Since $g^{(0)}(a) = a^{-1} = \operatorname{Res}(f(z), z = a)$, we have $x(-1) = \operatorname{Res}(f(z), z = 0) + \operatorname{Res}(f(z), z = a) = (-a)^{-1} + a^{-1} = 0$. Now let us turn our attention to the case $n = -2$. Now $f(z) = z^{-2}/(z-a)$, whence the pole at $z = 0$ is of order 2. Still, by (8.28b), $x(-2) = \operatorname{Res}(f(z), z = 0) + \operatorname{Res}(f(z), z = a)$, but now $f(z)$ has a pole of order 2 at $z = 0$. First, we set $g(z) = (z-a)^{-1}$ as before, but now we find $\operatorname{Res}(f(z), z = 0) = g^{(1)}(0) = -1(0-a)^{-2} = -a^{-2}$. For the pole at $z = a$, we put $g(z) = z^{-2}$ and verify that $\operatorname{Res}(f(z), z = a) = a^{-2}$. Thus, coincidentally, $x(-2) = -a^{-2} + a^{-2} = 0$. It is possible to show that, indeed, $x(n) = 0$ for $n < 0$. Therefore, $x(n) = a^n u(n)$ for all n.

The lesson of the example is that z-transform inversion by complex contour integration is sophisticated, easy, and fun for $n > 0$ where the integrand in (8.27) has first order poles, but tedious when there are higher-order poles. We seek simpler methods.

8.2.2 Direct Laurent Series Computation

Let us now try to exploit the idea that the z-transform analysis equation is a two-sided power, or Laurent, series in the complex variable z. Given by the z-transform inversion problem are a complex function of z, $X(z)$, and a region of convergence, ROC_X. The solution is to find $x(n)$ so that $\mathcal{Z}(x(n)) = X(z)$. Direct Laurent series computation solves the inversion problem by algebraically manipulating $X(z)$ into a form that resembles the z-transform analysis equation (8.1). Then the $x(n)$ values read off directly as the coefficients of the term z^{-n}. Not just any algebraic fiddling will do; the method can go awry if the algebraic manipulations do not stay in consonance with the information furnished by ROC_X.

Once again, we consider a well-known example to learn the technique.

Example (Laurent Series Computation for Causal Discrete Signal). Suppose $X(z) = z/(z - a)$, $a \neq 0$, with $\mathrm{ROC}_X = \{z \in \mathbb{C}: |a| < |z|\}$. Of course, we already know the signal whose z-transform is $X(z)$; it is the causal exponential signal $a^n u(n)$. Performing long division on $X(z)$ produces a Laurent series:

$$X(z) = \frac{z}{z-a} = z - a\,\overline{)\,z\quad}^{\,1 + az^{-1} + a^2 z^{-2} + a^2 z^{-2} + \cdots}. \tag{8.29}$$

Using the z term in $z - a$ as the principal divisor produces a quotient that is a Laurent expansion. Since ROC_X is the region outside the circle $|z| = a$, we see by inspection that $x(n) = u(n)a^n$.

Now we consider the same $X(z)$, but allow that ROC_X is inside the circle $|z| = a$.

Example (Laurent Series Computation for Anti-causal Discrete Signal). Suppose $X(z) = z/(z - a)$, $a \neq 0$, with $\mathrm{ROC}_X = \{z \in \mathbb{C}: |z| < |a|\}$. Performing long division again, but this time using the $-a$ term in $z - a$ as the principal divisor, produces a different Laurent series:

$$X(z) = \frac{z}{z-a} = -a + z\,\overline{)\,z\quad}^{\,-a^{-1}z - a^{-2}z^2 - a^{-3}z^3 - a^{-4}z^4 - \cdots}. \tag{8.30}$$

The algebraic manipulation takes into account the fact that ROC_X is the region inside the circle $|z| = a$, and the expansion is in positive powers of z. This means that $x(n)$ is anti-causal: $x(n) = -u(-n - 1)a^n$.

The next example complicates matters a bit more.

Example (Quadratic Denominator). Suppose $X(z) = z(z - 2)^{-1}(z - 1)^{-1}$, with $\mathrm{ROC}_X = \{z \in \mathbb{C}: |2| < |z|\}$. Attacking the problem directly with long division gives

$$X(z) = \frac{z}{(z-2)(z-1)} = z^2 - 3z + 2\,\overline{)\,z\quad}^{\,z^{-1} + 3z^{-2} + 7z^{-3} + 15z^{-4} - \cdots}. \tag{8.31}$$

We observe that $x(n) = u(n)(2^n - 1)$ from the derived form of the Laurent series (8.31). We can check this result by using linearity. Note that $X(z) = z(z-2)^{-1} - z(z-1)^{-1}$. The first term is the z-transform of $u(n)2^n$, and its radius of convergence is $\{z \in \mathbb{C}: 2 < |z|\}$. The second term is the z-transform of $u(n)1^n$, with $\mathrm{ROC} = \{z \in \mathbb{C}: 1 < |z|\}$. Therefore, their difference, $u(n)(2^n - 1^n)$, has z-transform $z(z-2)^{-1} - z(z-1)^{-1}$, whose radius of convergence equals $\{z \in \mathbb{C}: 2 < |z|\} \cap \{z \in \mathbb{C}: 1 < |z|\} = \{z \in \mathbb{C}: 2 < |z|\} = \mathrm{ROC}_X$.

Our method of checking this last example leads to the table lookup technique of the section.

8.2.3 Properties and z-Transform Table Lookup

The last method for computing the inverse z-transform is perhaps the most common. We tabulate a variety of z-transforms for standard signals and use the various properties of the transform to manipulate a given $X(z)$ into a form whose components are z-transforms of the known signals. Typically, then, $x(n)$ is a linear combination of these component signals. One standard trick that is useful here is to break up a complicated rational function in z, $X(x) = P(z)/Q(z)$, where P and Q are polynomials, into a sum of simpler fractions that allows table lookup. This is called the partial fractions method, and we will consider some examples of its use as well.

Example. Suppose $X(z) = (1 - az)^{-1}$, with $\mathrm{ROC}_X = \{z \in \mathbb{C}: |z| < |a|^{-1}\}$. From a direct computation of the z-transform, we know that $Z[a^n u(n)] = z/(z - a)$, with $\mathrm{ROC} = \{z \in \mathbb{C}: |z| > |a|\}$. Let $y(n) = a^n u(n)$ and $x(n) = y(-n)$. The time-reversal property implies

$$X(z) = \frac{z^{-1}}{(z^{-1} - a)} = \frac{1}{(1 - az)}, \tag{8.32}$$

with $\mathrm{ROC}_X = \{z \in \mathbb{C}: z^{-1} \in \mathrm{ROC}_Y\} = \{z \in \mathbb{C}: |z| < |a|^{-1}\}$, as desired.

Table 8.1 provides a list of common signals, their z-transforms, and the associated regions of convergence. These pairs derive from

TABLE 8.1. Signals, Their z-Transforms, and the Region of Convergence of the z-Transform

$x(n)$	$X(z)$	ROC_X
$\delta(n - k)$	z^{-k}	$k > 0: \{z \in \mathbb{C}^+: z \neq 0\}$ $k < 0: \{z \in \mathbb{C}^+: z \neq \infty\}$
$a^n u(n)$	$z/(z - a)$	$\{z \in \mathbb{C}: \lvert a\rvert < \lvert z\rvert\}$
$-a^n u(-n-1)$	$z/(z - a)$	$\{z \in \mathbb{C}: \lvert z\rvert < \lvert a\rvert\}$
$a^{-n} u(-n)$	$\dfrac{1}{(1-az)}$	$\{z \in \mathbb{C}: \lvert z\rvert < \lvert a\rvert^{-1}\}$
$-a^{-n} u(n-1)$	$\dfrac{1}{(1-az)}$	$\{z \in \mathbb{C}^+: \lvert z\rvert > \lvert a\rvert^{-1}\}$
$na^n u(n)$	$az/(z^2 - 2az + a^2)$	$\{z \in \mathbb{C}^+: \lvert a\rvert < \lvert z\rvert\}$
$-na^n u(-n - 1)$	$az/(z^2 - 2az + a^2)$	$\{z \in \mathbb{C}: \lvert z\rvert < \lvert a\rvert\}$
$\cos(an)u(n)$	$\dfrac{z^2 - \cos(a)z}{z^2 - 2\cos(a)z + 1}$	$\{z \in \mathbb{C}: 1 < \lvert z\rvert\}$
$\sin(an)u(n)$	$\dfrac{\sin(a)z}{z^2 - 2\cos(a)z + 1}$	$\{z \in \mathbb{C}: 1 < \lvert z\rvert\}$
$u(n)/(n!)$	$\exp(z)$	$\{z \in \mathbb{C}\}$
$n^{-1}u(n-1)(-1)^{n+1}a^n$	$\log(1 + az^{-1})$	$\{z \in \mathbb{C}^+: \lvert a\rvert < \lvert z\rvert\}$

- Basic computation using the z-transform analysis equation;
- Application of transform properties;
- Standard power series expansion from complex analysis.

Example. Suppose that

$$X(z) = \frac{z^2 - \left(\frac{\sqrt{2}}{2}\right)z}{(z^2 - \sqrt{2}z + 1)}, \tag{8.33}$$

with $\mathrm{ROC}_X = \{z \in \mathbb{C}: |z| > 1\}$. The table entry, $\cos(an)u(n)$, applies immediately. Taking $a = \pi/4$ gives $x(n) = \cos(\pi n/4)u(n)$. Variants of a z-transform pair from Table 8.1 can be handled using the transform properties. Thus, if

$$Y(z) = \frac{1 - \left(\frac{\sqrt{2}}{2}\right)z^{-1}}{(z^2 - \sqrt{2}z + 1)}, \tag{8.34}$$

then $Y(z) = z^{-2}X(z)$, so that $y(n) = x(n + 2) = \cos[\pi(n + 2)/4]u(n + 2)$ by the time shift property.

Example (Partial Fractions Method). Suppose that we are given a rational function in the complex variable z,

$$X(z) = \frac{2z^2}{2z^2 - 3z + 1}, \tag{8.35}$$

where $\mathrm{ROC}_X = \{z \in \mathbb{C}: |z| > 1\}$. The partial fraction technique factors the denominator of (8.35), $2z^2 - 3z + 1 = (2z - 1)(z - 1)$, with an eye toward expressing $X(z)$ in the form

$$X(z) = 2z\frac{z}{2z^2 - 3z + 1} = 2z\left[\frac{A}{2z - 1} + \frac{B}{z - 1}\right], \tag{8.36}$$

where A and B are constants. Let us concentrate on finding the inverse z-transform of $Y(z) = X(z)/(2z)$, the bracketed expression in (8.36). Table 8.1 covers both of its terms: they are of the form $(1 - az)^{-1}$. The sum of these fractions must equal $z(2z^2 - 3z + 1)^{-1}$, so $A(z - 1) + B(2z - 1) = z$. Grouping terms involving like powers of z produces two equations: $A + 2B = 1$, $A + B = 0$. Hence,

$$\frac{X(z)}{2z} = \frac{z}{2z^2 - 3z + 1} = \left[\frac{-1}{2z - 1} + \frac{1}{z - 1}\right] = \left[\frac{1}{1 - 2z} - \frac{1}{1 - z}\right] = Y(z). \tag{8.37}$$

Now, $y(n) = -2^{-n}u(n - 1) + u(n - 1) = (1 - 2^{-n})u(n - 1)$ by linearity and Table 8.1. Also, $\mathrm{ROC}_Y = \{z \in \mathbb{C}^+: |z| > 2^{-1}\} \cap \{z \in \mathbb{C}^+: |z| > 1\} = \{z \in \mathbb{C}^+: |z| > 1\}$.

Therefore, the z-transform of $x(n) = 2y(n+1) = (2 - 2^{-n})u(n)$ is $2zY(z)$ by the time shift property.

Example (Partial Fractions, Multiple Roots in Denominator). Now suppose that the denominator of $X(z)$ has multiple roots:

$$X(z) = \frac{z}{(z-1)(z+2)^2}. \tag{8.38}$$

It turns out that a partial fractions expansion of $X(z)/z$ into, say,

$$\frac{X(z)}{z} = \frac{1}{(z-1)(z+2)^2} = \frac{A}{(z-1)} + \frac{B}{(z+2)^2} \tag{8.39}$$

does not work, in general. Rather, the partial fractions arithmetic is satisfactory when $X(z)/z$ is broken down as follows:

$$\frac{X(z)}{z} = \frac{1}{(z-1)(z+2)^2} = \frac{A}{(z-1)} + \frac{B}{(z+2)^2} + \frac{C}{(z+2)}. \tag{8.40}$$

The solutions are $A = 1/7$, $B = 2/7$, and $C = -1/7$. Applying linearity, time shift, and Table 8.1 completes the example. This is left as an exercise.

8.2.4 Application: Systems Governed by Difference Equations

The above theory applies directly to the study of linear, translation-invariant systems where a difference equation defines the input–output relation. Chapter 2 introduced this kind of system (Sections 2.4.2 and 2.10). We shall see here and in Chapter 9 that:

- For such systems, the transfer function $H(z)$ is a quotient of complex polynomials.
- Difference equations govern a wide variety of important signal processing systems.
- Recursive algorithms very efficiently implement these systems on digital computers.
- The filters that arise from difference equations can be derived straightforwardly from equivalent analog systems.
- For almost any type of filter—low-pass, high-pass, bandpass, or band-reject—a difference equation governed system can be devised that very well approximates the required frequency selection behavior.

In fact, the filters within almost all signal analysis systems derive from difference equations, and we describe them by the z-transform of their impulse response.

Suppose that the difference equation for a system H is

$$\begin{aligned} y(n) &= b_0 x(n) + b_1 x(n-1) + \cdots + b_M x(n-M) \\ &\quad - a_1 y(n-1) - a_2 y(n-2) - \cdots - a_N y(n-N). \end{aligned} \tag{8.41}$$

We see that $y(n)$ can be computed from its past N values, $\{y(n) \mid 1 \le n \le N\}$, the current input value $x(n)$, and the past M values of input signal $\{x(n) \mid 1 \le n \le M\}$. Collecting output terms on the left-hand side and input terms on the right-hand side of (8.41), taking the z-transform of both sides, and finally applying the shift property, we have

$$Y(z)\left[1 + \sum_{k=1}^{N} a_k z^{-k}\right] = X(z)\left[\sum_{m=0}^{M} b_m z^{-m}\right]. \tag{8.42}$$

Hence,

$$H(z) = \frac{Y(z)}{X(z)} = \frac{\left[\sum_{m=0}^{M} b_m z^{-m}\right]}{\left[1 + \sum_{k=1}^{N} a_k z^{-k}\right]}, \tag{8.43}$$

confirming that the system function for H is a rational function of a single complex variable.

Now the methods of z-transform inversion come into play. The partial fractions technique converts the rational function (8.43) into a sum of simpler terms to which table lookup applies. Thus, we can find the impulse response $h(n)$ of the LTI system H. Finally, we can compute the response of H to an input signal $x(n)$ by convolution $y(n) = (h * x)(n)$.

Example (Smoothing System). The system H smoothes input signals by weighting the previous output value and adding it to the weighted input value as follows:

$$y(n) = Ay(n-1) + Bx(n). \tag{8.44}$$

By z-transforming both sides of (8.44), we get

$$Y(z) = Az^{-1}Y(z) + BX(z), \tag{8.45}$$

so that

$$H(z) = \frac{Y(z)}{X(z)} = \frac{B}{1 - Az^{-1}}. \tag{8.46}$$

Assuming that the system is causal, so that $h(n) = 0$ for $n < 0$, we have

$$h(n) = BA^n u(n), \tag{8.47}$$

by Table 8.1.

8.3 RELATED TRANSFORMS

This section introduces two other transforms: the chirp z-transform (CZT) and the Zak transform (ZT). A short introduction to them gives the reader insight into recent research efforts using combined analog and discrete signal transformation tools.

8.3.1 Chirp *z*-Transform

The CZT samples the z-transform on a spiral contour of the complex plane [7, 22]. The CZT transform has a number of applications [23]:

- It can efficiently compute the discrete Fourier transform (DFT) for a prime number of points.
- It can be used to increase the frequency resolution of the DFT, zooming in on frequency components (Chapter 9).
- It has been applied in speech [8, 24], sonar, and radar signal analysis [6, 25], where chirp signals prevail and estimations of their parameters—starting frequency, stopping frequency, and rate of frequency change—are crucial.

8.3.1.1 Definition. Recall that evaluating the z-transform $X(z)$ of $x(n)$ on the unit circle $z = \exp(j\omega)$ gives the discrete-time Fourier transform: $X(\omega) = [Z(x)](e^{j\omega})$. If $N > 0$ and $x(n)$ is finitely supported on the discrete interval $[0, N - 1]$, then $X(z)$ becomes

$$X(z) = \sum_{n=0}^{N-1} x(n)z^{-n}. \tag{8.48}$$

Furthermore, if $\omega = 2\pi k/N$, $0 \le k < N$, so that $z = \exp(2\pi jk/N)$, then the DTFT analysis equation (8.48) becomes a discrete Fourier transform of $x(n)$. So we are evaluating the z-transform $X(z)$ on a discrete circular contour of the complex plane. The idea behind the CZT is evaluate the z-transform on a discrete spiral—as distinct from purely circular—contour. We use the notation and generally follow the presentation of Ref. 7.

Definition (Chirp *z*-Transform). Let $A = A_0\exp(2\pi j\theta_0)$; $W = W_0\exp(2\pi j\phi_0)$; M, $N > 0$ be natural numbers; $x(n) = 0$ outside $[0, N - 1]$; and set $z_k = AW^{-k}$ for $0 \le k < M$. The chirp z-transform of $x(n)$ with respect to A and W is

$$X_{A,W}(k) = \sum_{n=0}^{N-1} x(n){z_k}^{-n} = \sum_{n=0}^{N-1} x(n)A^{-n}W^{nk}. \tag{8.49}$$

If $A = 1$, $M = N$, and $W = \exp(-2\pi j/N)$, then the CZT gives the DFT of order N for the signal $x(n)$ (exercise). Figure 8.2 shows a typical discrete spiral contour for a CZT.

Further note that if $W_0 > 1$, then the contour spirals inward, whereas $W_0 < 1$ means the contour winds outward (Figure 8.2).

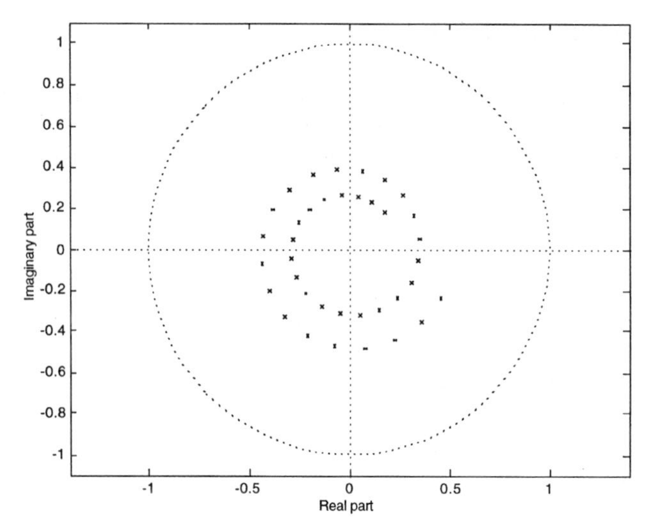

Fig. 8.2. Discrete spiral path for a CZT. Innermost point is $A = (.25)\exp(j\pi/4)$. Ratio between the M = 37 contour samples is $W = (0.98)\exp(-j\pi/10)$. Unit circle $|z| = 1$ is shown by dots.

8.3.1.2 Algorithm. An efficient implementation of the CZT, as a weighted convolution of two special functions, is possible as follows. Using our earlier notation, define the discrete signals $v(n)$ and $y(n)$ by

$$v(n) = W^{\frac{n^2}{2}}, \tag{8.50a}$$

$$y(n) = x(n)v(n)A^{-n}. \tag{8.50b}$$

Since

$$nk = \frac{n^2 + k^2 - (k-n)^2}{2}, \tag{8.51}$$

we calculate

$$X_{A,W}(k) = \sum_{n=0}^{N-1} x(n)A^{-n}W^{\frac{n^2+k^2-(k-n)^2}{2}} = W^{\frac{k^2}{2}} \sum_{n=0}^{N-1} y(n)v(k-n). \tag{8.52}$$

Equation (8.52) gives $X_{A,W}(k)$ as the convolution of $y(n)$ and $v(n)$, but weighted by the factor $v(k)$. Now, thanks to the convolution theorem for the DFT, we can compute discrete convolutions by Fourier transforming both signals, taking the frequency-domain product term-by-term, and then inverse transforming the result. Hence, if we have an efficient fast Fourier transform algorithm available, then a CZT may much more efficiently compute the DFT for a problematic—even prime—order N.

Here are the steps in the CZT algorithm [7]:

(i) First, we define $y(n)$:

$$y(n) = \begin{cases} A^{-n} W^{2/2} x(n) & \text{if } n \in [0, N-1], \\ 0 & \text{otherwise.} \end{cases} \tag{8.53}$$

(ii) Next, we determine the size of the FFT operation to perform. Inspecting the convolution equation (8.52), where k ranges between 0 and $M - 1$, we see that we need $v(n)$ values for $-N + 1 \le n \le M - N$, for a total of $M - N - (-N + 1) + 1 = M$ samples. Since $y(n)$ is supported on $[0, N - 1]$, the convolution result will be supported on $[0, (M - 1) + (N - 1)] = [0, M + N - 2]$. So the full $y{*}v$ requires $(M + N - 2) + 1 = M + N - 1$ samples (of which, for the CZT, we only care about M of them). Thus, we pick a power 2^P (or another FFT-suitable composite integer L), so that $M + N - 1 \le L$. This will be the order of the fast forward transforms and of the inverse transform after pointwise multiplication in the frequency domain.

(iii) We set $v(n)$ to be L-periodic such that

$$v(n) = \begin{cases} W^{-n^2/2} & \text{if } n \in [0, M-1], \\ W^{-(L-n)^2/2} & \text{if } n \in [L-N+1, L-1], \\ 0 & \text{otherwise.} \end{cases} \tag{8.54}$$

(iv) Compute the FFTs, $Y(k)$ and $V(k)$, of $y(n)$ and $v(n)$, respectively.

(v) Compute $G(k) = Y(k)V(k)$, for $0 \le k \le L - 1$.

(vi) Compute the inverse FFT of $G(k)$: $g(n)$.

(vii) Set

$$X_{A,W}(k) = W^{\frac{k^2}{2}} g(k) \qquad \text{for} \qquad 0 \le k \le M-1. \tag{8.55}$$

Evidently, the computational burden within the algorithm remains the three fast transforms [7]. Each of these requires on the order of $L\log_2(L)$ operations, depending, of course, on the particular FFT available. So we favor the CZT when $L\log_2(L)$ is much less than the cost of a full-convolution MN operation.

8.3.2 Zak Transform

The Zak transform (ZT) is an important tool in Chapter 10 (time-frequency analysis). The transform and its applications to signal theory are covered in Refs. 26 and 27.

8.3.2.1 Definition and Basic Properties. The Zak transform maps an analog signal $x(t)$ to a two-dimensional function having independent variables in both time and frequency. We know that restricting the z-transform to the unit circle $|z| = 1$

gives the discrete-time Fourier transform. The idea behind the Zak transform is that discrete signals generally come from sampling analog signals $x(n) = x_a(nT)$, for some $T > 0$, and that we can compute a DTFT for a continuum of such sampled analog signals.

Definition (Zak Transform). Let $a > 0$ and $x(t)$ be an analog signal. Then the Zak transform with parameter a of $x(t)$ is

$$X_a(s,\omega) = \sqrt{a} \sum_{k=-\infty}^{\infty} x(as - ak)\exp(2\pi j\omega k). \tag{8.56}$$

Remark. We use a fancy $\mathcal{Z}_a$ for the map taking an analog signal to its Zak transform: $(\mathcal{Z}_a x)(s, \omega) = X_a(s, \omega)$. Our transform notation uses the sign notation $x(s - k)$ following [28]. Generally, we take $a = 1$ and omit it from the notation: $(\mathcal{Z}x)(s, \omega) = X(s, \omega)$; this is the form of the definition we use later in several parts of Chapter 10.

Proposition (Periodicity). If $X_a(s, \omega)$ is the ZT of $x(t)$, then

$$X_a(s+1,\omega) = \exp(2\pi j\omega)X_a(s,\omega), \tag{8.57a}$$

$$X_a(s,\omega+1) = X_a(s,\omega). \tag{8.57b}$$

Proof: Exercise. ■

Observe that with parameter s fixed $y(-k) = x(s - k)$ is a discrete signal. If we further set $z = \exp(2\pi j\omega)$, then the ZT summation with $a = 1$ becomes

$$X(s,\omega) = \sum_{k=-\infty}^{\infty} y(-k)\exp(2\pi j\omega k) = \sum_{k=-\infty}^{\infty} y(k)\exp(-2\pi j\omega k) = \sum_{k=-\infty}^{\infty} y(k)z^{-k}. \tag{8.58}$$

Equation (8.58) is the z-transform of $y(k)$ evaluated on the unit circle. More precisely, it is the DTFT of $y(k)$ with 2π frequency dilation.

8.3.2.2 An Isomorphism. We now show an interesting isomorphism between the Hilbert space of finite-energy analog signals $L^2(\mathbb{R})$ and the square-integrable two-dimensional analog signals on the unit square $S = [0, 1] \times [0, 1]$.

The Zak transform is in fact a unitary map from $L^2(\mathbb{R})$ to $L^2(S)$; that is, $\mathcal{Z}$ is a Hilbert space map that takes an $L^2(\mathbb{R})$ orthonormal basis to an $L^2(S)$ orthonormal basis in a one-to-one and onto fashion [27, 28].

Lemma. Let $b(t) = u(t) - u(t - 1)$ be the unit square pulse, where $u(t)$ is the analog step function. Then $\{b_{m,n}(t) = \exp(2\pi jmt)b(t - n) \mid m, n \in \mathbb{Z}\}$ is a basis for $L^2(\mathbb{R})$. Moreover, $\{e_{m,n}(s, t) = \exp(2\pi jms)\exp(2\pi jnt) \mid m, n \in \mathbb{Z}\}$ is a basis for $L^2(S)$.

Proof: Apply Fourier series arguments to the unit intervals on $\mathbb{R}$. The extension of Fourier series to functions of two variables is outside our one-dimensional perspective, but is straightforward, and can be found in advanced Fourier analysis texts (e.g., Ref. 29.) ∎

Theorem (Zak Isomorphism). The Zak transform $\mathcal{Z}$: $L^2(\mathbb{R}) \to L^2(S)$ is unitary.

Proof: Let's apply the ZT to the Fourier basis on $L^2(\mathbb{R})$, $\{b_{m,n}(t) \mid m, n \in \mathbb{Z}\}$ of the lemma:

$$
\begin{aligned}
(\mathcal{Z}b_{m,n})(s,\omega) &= \sum_{k=-\infty}^{\infty} b_{m,n}(s-k)\exp(2\pi j\omega k) \\
&= \exp(2\pi jms)\exp(-2\pi jn\omega) \sum_{k=-\infty}^{\infty} b(s-n-k)\exp(2\pi j\omega(n-k)) \\
&= \exp(2\pi jms)\exp(-2\pi jn\omega)(\mathcal{Z}b)(s,\omega) = e_{m,-n}(s,\omega)(\mathcal{Z}b)(s,\omega). \qquad (8.59)
\end{aligned}
$$

Let us reflect on the last term, $(\mathcal{Z}b)(s, \omega)$ for $(s, \omega) \in S$. We know that

$$
(\mathcal{Z}b)(s,\omega) = \sum_{k=-\infty}^{\infty} b(s-k)\exp(2\pi j\omega k). \qquad (8.60)
$$

On the interior of S, $(0, 1) \times (0, 1)$, we have $b(s - k) = 0$ for all $k \neq 0$. So only one term counts in the infinite sum (8.60), namely $k = 0$, and this means $(\mathcal{Z}b)(s, \omega) = 1$ on the unit square's interior. On the boundary of S, we do not care what happens to the ZT sum, because the boundary has (two-dimensional) Lebesgue measure zero; it does not affect the $L^2(S)$ norm. Thus, $\mathcal{Z}$ sends $L^2(\mathbb{R})$ basis elements $b_{m,n}$ to $L^2(S)$ basis elements $e_{m,-n}$, and is thus unitary. ∎

8.4 SUMMARY

The z-transform extends the discrete-time Fourier transform from the unit circle to annular regions complex plane, called regions of convergence. For signal frequency, the DTFT is the right inspection tool, but system properties such as stability can be investigated with the z-transform. Readers may recall the Laplace transform from system theory and differential equations work; it bears precisely such a relationship to the analog Fourier transform (Chapter 5). The Laplace transform extends the definition of the Fourier transform, whose domain is the real numbers, to regions of the complex plane.

The next chapter covers frequency-domain signal analysis, including both analog and digital filter design. It most assuredly explores further z-transform techniques.

This chapter closed with an introduction to two related tools: the chirp z-transform and the Zak transform. The CZT is a discretized z-transform computed on a custom contour. If the contour follows the unit circle, then the CZT can be used to save some computational steps that we would ordinarily suffer when computing a DFT of difficult (prime) order. Or, careful contour selection with the CZT gives more frequency coefficients in a narrow application range than the Fourier transform. The Zak transform's isomorphism property effectively converts questions about $L^2(\mathbb{R})$ analog signals into questions about finite-energy signals on the unit square. Analytically, the unit square, even though it is two-dimensional, is often easier to deal with. This benefit of the ZT makes it especially powerful when we study frames based on windowed Fourier atoms in Chapter 10.

8.4.1 Historical Notes

Kaiser [14] introduced the z-transform into the signal processing discipline from control theory only in the early 1960s. At the time, digital computer applications had stimulated interest in discrete transforms, filtering, and speech processing. Filters are systems that pass some frequency ranges while suppressing others, and they are common at the front end of a signal analysis system that must interpret oscillatory data streams. It turns out—as we shall see in the next chapter—that very good filters can be built out of simple recursive structures based on difference equations. The z-transform readily gives the system function for such difference equations as a rational function of a single complex variable: $H(z) = B(z)/A(z)$. We have developed straightforward algebraic methods for inverting such rational functions, which in turn reveals the system impulse response and allows us to calculate the system response to various inputs.

In the late 1960s, Bluestein [30] first showed how to compute the DFT using a chirped linear filtering operation. The formalization of CZT algorithm and many of its original applications are due to Rabiner, Schafer, and Rader [22, 23].

The ZT arrives relatively late to signal theory from physics [31], where Zak developed it independently for solid-state applications. Janssen introduced it into the mainstream signal analysis literature [26]. The transform has been many places—indeed, Gauss himself may have known of it [28].

8.4.2 Guide to Problems

Readers should find most problems straightforward. Problems 2 and 3 explore some of the limit ideas and radius of converge concepts used in the chapter. There is a z-transform characterization of stable systems, which is developed in the later problems. Finally, some computer programming tasks are suggested.

REFERENCES

1. A. V. Oppenheim, A. S. Willsky, and S. H. Nawab, *Signals and Systems*, 2nd ed., Upper Saddle River, NJ: Prentice-Hall, 1997.
2. J. A. Cadzow and H. F. Van Landingham, *Signals, Systems, and Transforms*. Englewood Cliffs, NJ: Prentice-Hall, 1983.

3. H. Baher, *Analog and Digital Signal Processing.* New York: Wiley, 1990.
4. J. G. Proakis and D. G. Manolakis, *Digital Signal Processing*, 2nd ed., New York: Macmillan, 1992.
5. R. A. Roberts and C. T. Mullis, *Digital Signal Processing*. Reading, MA: Addison-Wesley, 1987.
6. A. V. Oppenheim and R. W. Shafer, *Discrete-Time Signal Processing*, Englewood Cliffs, NJ: Prentice-Hall, 1989.
7. L. R. Rabiner and B. Gold, *Theory and Application of Digital Signal Processing*, Englewood Cliffs, NJ: Prentice-Hall, 1975.
8. T. Parsons, *Voice and Speech Processing*, New York: McGraw-Hill, 1987.
9. M. Akay, *Biomedical Signal Processing*, San Diego, CA: Academic Press, 1994.
10. F. Scherbaum, *Of Poles and Zeros: Fundamentals of Digital Seismology*, Dordrecht, The Netherlands: Kluwer, 1996.
11. C. H. Wilts, *Principles of Feedback Control*, Reading, MA: Addison-Wesley, 1960.
12. C. L. Phillips and H. T. Nagle, Jr., *Digital Control System Analysis and Design*, Englewood Cliffs, NJ: Prentice-Hall, 1984.
13. K. Ogata, *Discrete-Time Control Systems*, Englewood Cliffs, NJ: Prentice-Hall, 1987.
14. J. F. Kaiser, Design methods for sampled data systems, *Proceedings of the First Annual Allerton Conference on Circuits and System Theory*, pp. 221–236, 1963. Also in L. R. Rabiner and C. M. Rader, eds., *Digital Signal Processing*, New York: IEEE Press, 1972.
15. E. I. Jury, *Theory and Applications of the Z-Transform Method*, Malabar, FL: Krieger, 1982.
16. R. Vich, *Z-Transform Theory and Applications*, Boston: D. Reidel, 1987.
17. E. Hille, *Analytic Function Theory*, vol. I, Waltham, MA: Blaisdell, 1959.
18. L. V. Ahlfors, *Complex Analysis,* 2nd ed., New York: McGraw-Hill, 1966.
19. M. Rosenlicht, *An Introduction to Analysis*, New York: Dover, 1978.
20. W. F. Rudin, *Real and Complex Analysis*, 2nd ed., New York: McGraw-Hill, 1974.
21. R. Beals, *Advanced Mathematical Analysis*, New York: Springer-Verlag, 1987.
22. L. R. Rabiner, R. W. Shafer, and C. M. Rader, The chirp z-transform algorithm, in L. R. Rabiner and C. M. Rader, eds., *Digital Signal Processing*, New York: IEEE Press, pp. 322–328, 1972.
23. L. R. Rabiner, R. W. Shafer, and C. M. Rader, The chirp z-transform algorithm and its applications, *Bell System Technical Journal*, vol. 48, pp. 1249–1292, May 1969.
24. R. W. Shafer and L. R. Rabiner, System for automatic formant analysis of voiced speech, *Journal of the Acoustical Society of America*, vol. 47, no. 2, pp. 634–648, February 1970.
25. M.I. Skolnik, *Introduction to Radar Systems*, 2nd ed., New York: McGraw-Hill, 1986.
26. A. J. E. M. Janssen, The Zak transform: A signal transform for sampled time-continuous signals, *Philips Journal of Research*, vol. 43, no. 1, pp. 23–69, 1988.
27. C. E. Heil and D. F. Walnut, Continuous and discrete wavelet transforms, *SIAM Review*, vol. 31, pp. 628–666, December 1989.
28. I. Daubechies, *Ten Lectures on Wavelets*, Philadelphia: SIAM, 1992.
29. A. Zygmund, *Trigonometric Series*, vols. I & II, 2nd ed., Cambridge: Cambridge University Press, 1985.
30. L. I. Bluestein, A linear filtering approach to the computation of the discrete Fourier transform, *IEEE Transactions on Audio and Electroacoustics*, vol. AU-18, pp. 451–455,

December 1970. Reprinted in L. R. Rabiner and C. M. Rader, eds., *Digital Signal Processing*, New York: IEEE Press, 1972.

31. J. Zak, Finite translations in solid state physics, *Physics Review Letters*, vol. 19, pp. 1385–1397, 1967.

PROBLEMS

1. Find the z-transform and ROC for each of the following signals:

(a) $x(n) = u(n-5) - u(n+2)$, where $u(n)$ is the discrete unit step signal. Can one simply apply the linearity and shift properties to $x(n)$ for the right answer?

(b) $\delta(n-4) + u(n)$, where $\delta(n)$ is the discrete impulse.

(c) $x(n) = 3^n u(-n) + n2^{-n}u(n)$.

(d) $x(n) = u(n)[n2^{n-1}]$.

(e) $x(n) = u(n)[n2^{n-1} + n]$.

(f) $x(n) = 1/n!$

(g) $x(n) = u(-n-1)(1/3)^n$.

(h) $x(n) = u(n)(-1/5)^n + u(-n-1)(1/2)^n$.

2. Consider the lim sup and lim inf of a sequence, $A = \{a_n: 0 \le n < \infty\}$. Suppose we have defined elements of the sequence as follows: $a_0 = 0$; $a_n = 1 + 1/n$, if n is even; and $a_n = -1 - 1/n$, if n is odd.

(a) Show that the sequence A has no limit.

(b) Show that the lim sup A is 1.

(c) Show that the lim inf of A is –1.

(d) Let $A_N = A \setminus \{a_n: 0 \le n < N\}$ and κ_N be the least upper bound of A_N. Show that $\kappa_N \le \kappa_M$ if $M < N$.

(e) Show that a sequence $B = \{b_n: 0 \le n < \infty\}$ has a limit if and only if its lim inf and its lim sup are equal. What about the cases where the limit is $\pm\infty$?

(f) Show that

$$\lim_{n\to\infty} \inf\{b_n: 0 \le n < \infty\} = -\lim_{n\to\infty} \sup\{-b_n: 0 \le n < \infty\}. \tag{8.61}$$

3. Suppose $Z(x(n)) = X(z)$ has only non-negative powers of z:

$$X(z) = \sum_{n=0}^{+\infty} a_n z^n. \tag{8.62}$$

Let

$$\kappa = \lim_{n\to\infty} \sup\{|a_n|^{1/n}: 1 \le n < \infty\}, \tag{8.63}$$

so that $\rho = \kappa^{-1}$ is the radius of convergence of $X(z)$. Show that the radius of convergence for the derivative,

$$X'(z) = \frac{dX(z)}{dz} = \sum_{n=0}^{+\infty} n a_n z^{n-1}, \tag{8.64}$$

is also ρ.

4. Let $Z(x(n)) = X(z)$. Show the following z-transform symmetry properties:
   (a) $Z[x^*(n)] = X^*(z^*)$, where z^* is the complex conjugate of z.
   (b) (Time Reversal) If $y(n) = x(-n)$, then $Z[y(n)] = Y(z) = X(z^{-1})$, and $\mathrm{ROC}_Y = \{z \in \mathbb{C}: z^{-1} \in \mathrm{ROC}_X\}$.
   (c) If $y(n) = \mathrm{Real}[x(n)]$, then $Y(z) = [X(z) + X^*(z^*)]/2$.
   (d) If $y(n) = \mathrm{Imag}[x(n)]$, then $Y(z) = j[X^*(z^*) - X(z)]/2$.
   (e) Find the z-transform of $x_e(n)$, the even part of $x(n)$.
   (f) Find the z-transform of $x_o(n)$, the odd part of $x(n)$.

5. Suppose $X(z) = z/(z - a)$, $a \neq 0$, with $\mathrm{ROC}_X = \{z \in \mathbb{C}: |a| < |z|\}$. In the first example of Section 8.1.1, we found that $x(-1) = x(-2) = 0$ and claimed that $x(n) = 0$ for $n < -2$. For the last case, $n < -2$, verify that
   (a) $\mathrm{Res}(f(z), z = 0) = -a^{-n}$.
   (b) $\mathrm{Res}(f(z), z = a) = a^{-n}$.
   (c) $x(n) = 0$ for $n < -2$.

6. Suppose $X(z) = z/(z - a)$, $a \neq 0$, with $\mathrm{ROC}_X = \{z \in \mathbb{C}: |z| < |a|\}$. Using the method of contour integration, find $x(n)$ for all $n \in \mathbb{Z}$.

7. Suppose $X(z) = z(z - 2)^{-1}(z - 1)^{-1}$.
   (a) Let $\mathrm{ROC}_X = \{z \in \mathbb{C}: |z| < 1\}$. With the method of inverse z-transformation by computation of the Laurent series, find $x(n)$.
   (b) Suppose now that $\mathrm{ROC}_X = \{z \in C: 2 > |z| > 1\}$. Is it possible to use the long division method to find the Laurent series form of $X(z)$ and thence find $x(n)$? Explain.

8. Suppose that

$$X(z) = \frac{z}{(z-1)(z+2)^2}, \tag{8.65}$$

   and $\mathrm{ROC}_X = \{z \in \mathbb{C}: 2 < |z|\}$.
   (a) Find A, B, and C to derive the expansion of $z^{-1}X(z)$ into partial fractions:

$$\frac{X(z)}{z} = \frac{1}{(z-1)(z+2)^2} = \frac{A}{(z-1)} + \frac{B}{(z+2)^2} + \frac{C}{(z+2)}. \tag{8.66}$$

   (b) Find the discrete signal whose z-transform is $z^{-1}X(z)$.
   (c) Find the discrete signal whose z-transform is $X(z)$.

9. Again suppose

$$X(z) = \frac{z}{(z-1)(z+2)^2}. \tag{8.67}$$

Find all of the discrete signals whose z-transforms are equal to $X(z)$. For each such signal,

(a) State the region of convergence.

(b) Sketch the region of convergence.

(c) State whether the signal is causal, anticausal, or neither.

10. Signal $x(n)$ has z-transform $X(z)/z = 1 / (z^2 - 3z/2 - 1)$. Find three different possibilities for $x(n)$ and give the ROC of $X(z)$ for each.

11. If $x(n)$ has z-transform $X(z) = z / (z^2 - 5z - 14)$, then find three different possibilities for $x(n)$ and give the ROC of $X(z)$ for each.

12. Let $X^+(z)$ be the one-sided z-transform for $x(n)$.

(a) Show that the one-sided z-transform is linear.

(b) Show that the one-sided z-transform is not invertible by giving examples of different signals that have the same transform.

(c) Show that if $x(n) = 0$ for $n < 0$, then $X^+(z) = X(z)$.

(d) Let $y(n) = x(n - k)$. If $k > 0$, show that the shift property becomes

$$Y^+(z) = x(-k) + x(-k+1)z^{-1} + \cdots + x(-1)z^{-m+1} + z^{-m}X^+(z). \tag{8.68}$$

13. A simple difference equation,

$$y(n) = ay(n-1) + x(n), \tag{8.69}$$

describes a signal processing system. Some signed fraction $0 < |a| < 1$ of the last filtered value is added to the current input value $x(n)$. One application of the one-sided z-transform is to solve the difference equation associated with this system [4, 6]. Find the unit step response of this system, given the initial condition $y(-1) = 1$, as follows.

(a) Take the one-sided z-transforms of both sides of (8.69):

$$Y^+(z) = a\left[Y^+(z)z^{-1} + y(-1)\right] + X^+(z). \tag{8.70}$$

(b) Use the initial condition to get

$$Y^+(z) = \frac{a}{(1-az^{-1})} + \frac{1}{(1-z^{-1})(1-az^{-1})}. \tag{8.71}$$

(c) Apply the partial fractions method to get the inverse z-transform:

$$y(n) = \frac{(1-a^{n+2})}{(1-a)}u(n). \tag{8.72}$$

14. The Fibonacci[2] sequence is defined by $f(-2) = 1, f(-1) = 0$, and

$$f(n) = f(n-1) + f(n-2). \tag{8.73}$$

(a) Show that $f(0) = f(1) = 1$.

(b) Using the one-sided z-transform [4], show

$$y(n) = \frac{u(n)}{2^{n+1}\sqrt{5}}\left[(1+\sqrt{5})^{n+1} - (1-\sqrt{5})^{n+1}\right]. \tag{8.74}$$

15. Consider the system H given by the following difference equation:

$$y(n) = 0.25 * y(n-2) + 0.25 * y(n-1) + 0.5 * x(n). \tag{8.75}$$

(a) Find the system function $H(z)$.

(b) Assuming that H is a causal system, find $h(n)$.

(c) Give ROC_H for the causal system.

(d) What are the poles for the system function?

(e) Is the system stable? Explain.

16. Show that a discrete LTI system H is causal, $h(n) = 0$ for $n < 0$, if and only if ROC_H is $\{z: |z| > r\}$ for some $r > 0$.

17. Show that a discrete LTI system H is stable (bounded input implies bounded output signal) if and only if its z-transform ROC includes the unit circle $|z| = 1$.

18. Show that a causal LTI system H is stable if and only if all of the poles of $H(z)$ lie inside the unit circle $|z| = 1$.

19. Consider the causal system H given by the following difference equation:

$$y(n) = Ay(n-1) + Bx(n). \tag{8.76}$$

(a) Find necessary and sufficient conditions on constants A and B so that H is stable.

(b) Find the unit step response $y = Hu$, where $u(n)$ is the unit step signal.

(c) Show that if A and B satisfy the stability criteria in (a), then the unit step response in (b) is bounded.

(d) Find the poles and zeros of the system function $H(z)$.

20. Assume the notation for chirp z-transform of Section (8.31).

(a) Show that if $A = 1$, $M = N$, and $W = \exp(-2\pi j/N)$ in (8.49), then $X_{A,W}(k) = X(k)$, where $X(k)$ is the DFT of order N for the signal $x(n)$.

[2]Leonardo of Pisa (c. 1175–1250) is known from his father's name. The algebraist and number theorist asked a question about rabbits: If an adult pair produces a pair of offspring, which mature in one month, reproduce just as their parents, and so on, then how many adult rabbits are there after N months? The answer is F_N, the Nth Fibonacci number.

(b) Show that $W_0 > 1$ implies an inward spiral and $W_0 < 1$ produces an outward spiral path.

21. Derive the periodicity relations for the Zak transform (8.57a, 8.57b).

The next few problems are computer programming projects.

22. As a programming project, implement the CZT algorithm of Section 8.3.1.2. Compare the fast CZT algorithm performance to the brute-force convolution in (8.52). Use the algorithm to compute some DFTs for large prime orders. Compare the CZT-based algorithm to straight DFT computations.

23. Implement the z-transform for finitely supported discrete signals in a computer program. Verify the convolution property of the z-transform by calculating the z-transforms, $X(z)$ and $Y(z)$, of two nontrivial signals, $x(n)$ and $y(n)$, respectively; their convolution $z(n)$; and the z-transform $Z(z)$. Finally, confirm that $Z(z) = X(z)Y(z)$ with negligible numerical error.

24. Consider the implementation of the inverse z-transform on a digital computer. Which approach might be easiest to implement? Which is the most general? Develop an application that handles some of possible forms of $X(z)$. Explain the strengths and limitations of the application.

CHAPTER 9

Frequency-Domain Signal Analysis

Frequency-domain signal analysis covers a wide variety of techniques involving the Fourier transformation of the signal. The signal's frequency domain representation is then manipulated, decomposed, segmented, classified, and interpreted. One central idea is that of a *filter*: a linear, translation-invariant system that allows one band of frequencies to appear in the output and suppresses the others. Where signal elements of interest occupy a restricted spectrum, filters invariably enter into the early processing of candidate signals. In other ways—often purely theoretical—frequency-domain analysis is important. For example, in this chapter we substantiate the methods of matched filtering and scale-space decomposition, and the Fourier transform plays a crucial role.

The main tools for frequency-domain analysis are of course the discrete signal transforms: the discrete-time Fourier transform (DTFT); its generalization, the z-transform; and especially the discrete Fourier transform (DFT). Many of the introductory applications proceed from Chapter 1 examples. There are extensions of techniques already broached in Chapters 4 and 6. Modern spectral analysis applications have a digital computer at their heart, and they rely on the either the DFT or one of its many fast versions. Some signal filtering applications use infinite impulse response (IIR) filtering methods, implemented using feedback, as discussed in Chapter 2. The DTFT is convenient for obtaining a theoretical understanding of how such filters suppress and enhance signal frequencies. We often begin with an analog filter and convert it to a discrete filter. Thus, we shall have occasion to use the continuous-domain Fourier transform. We also briefly explain how the Laplace transform, a generalization of the Fourier transform, can be used in certain analog filter constructions. The z-transform figures prominently in this conversion process.

We are generally working with complex-valued signals, but the thresholding, segmentation, and structural decomposition methods that Chapter 4 developed for time-domain signal analyis are just as useful in the frequency domain. For example, to find the spectral region where a source signal contains significant energy, we threshold the signal's magnitude or squared magnitude (power) spectrum. We know that thresholding is often improved by filtering the data, so we are inclined to filter the frequency-domain magnitudes too. This leads directly to the technique of windowing

Signal Analysis: Time, Frequency, Scale, and Structure, by Ronald L. Allen and Duncan W. Mills
ISBN: 0-471-23441-9

time-domain signal slices before Fourier transformation. The meaning of the analytical results can be quite different, of course; but as long as we understand the transform relation clearly and capture that in the application design, then the time-domain and frequency-domain procedures are remarkably similar. In some applications, the results of this analysis convey the signal content. For other tasks, an inverse transformation back to the time domain is required. In any case, the principal tools are the Fourier transform and its inverse, in both their analog and discrete formulations.

Our theoretical resources include Chapters 5, 7, and 8. This chapter introduces some further theory, appropriate to the particular applications upon which we focus. Specific applications include tone detection, speech recognition and enhancement, and chirp analysis. Some of these experiments show that the Fourier transform is precisely the tool we need to make the application work. Further reflection reveals problems in applying the Fourier transforms. This motivates a search for frequency transforms that incorporate a time-domain element: time-frequency and time-scale transforms, which are topics for the final three chapters.

Fourier-domain techniques also offer many insights into our earlier material. Scale space and random signals are considered once more, this time from the vantage point of the new frequency-domain methods. The last two sections detail the construction of special filters for frequency analysis and possible structures for their application. Later chapters will draw the link between this approach to signal processing and the notion of a multiresolution analysis of the L^2 Hilbert space.

References on Fourier transforms include Refs. 1–5. Popular signal processing texts that introduce Fourier methods and construct filters from the theory are Refs. 6–10. An older book that concludes its thorough coverage of signal theory with detailed application studies in speech and radar signal analysis is Ref. 11.

Notation. The present discussion covers both analog and discrete filters. We use the following notations for clarity:

(i) *Analog filters*: The impulse response is $h(t)$, the radial Fourier transform is $H(\Omega)$, and the Laplace transform is $H_L(s)$; in some contexts, we insert the subscript a: $h_a(t)$, $H_a(\Omega)$, and $H_{L,a}(s)$.

(ii) *Discrete filters*: The impulse response is $h(n)$, or any FORTRAN-like integer independent variable such as $h(i)$, $h(k)$, or $h(m)$; the discrete time Fourier transform is $H(\omega)$; and the discrete Fourier transform is $H(k)$.

(iii) We continue to use $j^2 = -1$.

9.1 NARROWBAND SIGNAL ANALYSIS

The most basic frequency-domain analysis task involves detecting and interpreting more or less isolated periodic components of signals. For some signals, or at least for some part of their domain, a few oscillatory components contain the bulk of the energy. It is such *narrowband* signals—sinusoids (tones) and dual-tones, mainly, but we could also allow complex-valued exponentials into this category—that we begin our study of Fourier transform applications. We distinguish narrowband signals from wideband signals, where the energy is spread over many frequencies. A signal that contains sharp edges, for example, will generally have frequency-domain energy dispersed across a wide spectral range.

Although basic, a tone detection application leads to important practical concepts: noise removal, filtering, phase delay, group delay, and windowing. *Filters* are frequency-selective linear translation invariant systems. Filtering a signal can change the time location of frequency components. For instance, a filter might retard a sinusoidal pulse and the signal envelope itself, depending on their frequency. These time lags define the *phase* and *group delay*, respectively. Knowing them is crucial for application designs that must compensate for filter delays. Finally, we often need to analyze signals in chunks, but so doing invariably corrupts the signal spectrum. Only by looking at the signal through a well-constructed *window* can we mitigate this effect. This concept leads directly to the modern theory of the windowed Fourier transform (Chapter 10) and eventually to wavelets (Chapter 11).

Theory from earlier chapters now becomes practice. For designing filters, we employ the discrete-time Fourier transform (DTFT). For implementing filters on a computer, we use the discrete Fourier transform (DFT) or one of its fast Fourier transform (FFT) schemes. Some infinite impulse response filter implementations are particularly powerful, and we visualize their possible recursive implementation on a computer through the z-transform (Chapter 8). We definitely do not need the continuous-domain Fourier transform (FT), right? Quite wrong: We can obtain very good discrete filters by first designing an analog filter and then converting it into a discrete filter. It is perhaps a surprising fact, but this is the preferred method for constructing high-performance discrete filters. We even use the Fourier series (FS); after a bit of contemplation, we realize that the FS converts an analog $L^2[0, 2\pi]$ signal into a discrete l^2 signal—just like the inverse DTFT. Indeed, as mathematical objects, they are one and the same.

9.1.1 Single Oscillatory Component: Sinusoidal Signals

Let the real-valued signal $x(n)$ contain a single oscillatory component and perhaps some corrupting noise. We have seen such examples already in the first chapter—$x(n)$ is the Wolf sunspot count, for instance. Now, in earlier chapters, introductory applications explored the discrete Fourier transform as a tool for detecting such periodicities. The source signal's oscillatory component manifests itself as an isolated cluster of large magnitude spectral values. We can usually segment the frequency components with a simple threshold around the maximum value; this is a straightforward application of the DFT.

A tone is a time-domain region of a signal that consists of only a few sinusoidal components. Briefly, detection steps are as follows:

(i) Select signal regions that may contain tones.

(ii) For noise removal and frequency selection, processing the signal through various filters may benefit the analysis steps that follow.

(iii) Fourier transform the signal over such regions.

(iv) For each such region, examine the spectrum for large concentrations of signal energy in a few frequency coefficients.

(v) Optionally, once a possible tone has been identified through frequency-domain analysis, return to the time domain to more precisely localize the tone.

The discrete signal sometimes arises from sampling an analog signal $x_a(t)$: $x(n) = x_a(nT)$, where $T > 0$ is the sampling period. But perhaps—as in the case of sunspot estimates—the signal is naturally discrete. If we take select a window of signal values, $0 \le n < N$, we can compute the DFT over these N samples:

$$X(k) = \sum_{n=0}^{N-1} x(n) e^{\frac{-2\pi jkn}{N}} . \tag{9.1}$$

In (9.1), $X(0)$ represents the direct current (DC), or constant, or zero frequency component. The signal average over the interval $[0, N-1]$ is $X(0)/N$, and it represents zero cycles per sample (hertz) in the transform. If $x(n)$ is real, then $X(k) = X(N-k)$ for all $1 \le k \le N-1$. The transform is invertible:

$$x(n) = \frac{1}{N} \sum_{k=0}^{N-1} X(k) e^{\frac{2\pi jkn}{N}} . \tag{9.2}$$

Equations (9.1) and (9.2) are the analysis and synthesis equations, respectively, for the DFT. If $x_a(t) = \cos(2\pi t/NT)$ is a real-valued analog sinusoid with frequency $(NT)^{-1}$ hertz, then upon sampling it becomes $x(n) = x_a(nT) = \cos(2\pi n/N)$. We can expand $2x(n) = [\exp(2\pi jn/N) + \exp(2\pi jn(N-1)/N)]$, which is a synthesis equation (9.2) for the sinusoid. Thus, the smallest frequency represented by the transform coefficients, the frequency resolution of the DFT, is $(NT)^{-1}$. This means that energy from source signal periodicities appears in the transform coefficients in at least two different places. If N is even, then the highest frequency represented by the transform values is $1/(2T) = (NT)^{-1} \times (N/2)$ hertz. If N is odd, then the two middle coefficients share the Nyquist frequency energy.

This section explains the basic methods for applying the discrete Fourier transform (DFT) and discrete-time Fourier transform (DTFT) for detecting oscillatory components of signals. Strictly speaking, the DFT applies to periodic signals and the DTFT to aperiodic signals.

9.1.2 Application: Digital Telephony DTMF

Let us consider the problem of recognizing multiple oscillatory components in a source signal. If the source contains multiple periodic components, then the transform exhibits multiple clusters of large values. To spot significant oscillatory signal features, we might invoke the more powerful threshold selection methods covered in Chapter 4, applying them to the magnitude spectrum instead of the signal amplitude values. Thresholding the magnitude spectrum does work, but it does not take us very far.

This is the limitation of peak finding in the magnitude spectrum: These magnitudes represent frequencies over the entire time domain of the source signal. If oscillations at different times have the same frequency, or those that occur at the same time overlap with others of different wavelengths, then this potentially crucial information for signal understanding is lost in the Fourier transformation. Many applications involve signals with localized frequency components. What complicates such applications is getting the Fourier transform—an inherently global transformation—to work for us in a time localized fashion.

This application—however humble—inspires three general approaches for identifying and localizing signal frequency components:

- Preliminary time-domain analysis, arriving at a segmentation of the signal's values, and subsequent frequency-domain analysis on the segments of promise (Section 9.1.2.2);
- An important tool—the *time-frequency map* or *plane*—which generally decomposes the signal into pieces defined by the time interval over which they occur and the frequencies over which their oscillatory components range (Section 9.1.2.3);
- Another significant tool—the *filter bank*—which directs the signal values into a parallel array of frequency selective linear, translation-invariant (LTI) systems (filters) and analyzes the outputs jointly (Section 9.1.2.4).

It will become clear that these three alternatives couple their time- and frequency-domain analyses ever more closely. Thus, in the first case, algorithms finish the time-domain segmentation and hand the results over to spectral analysis. Using the second alternative's time-frequency plane, in contrast, we decompose the signal into pieces that represent a particular time interval and a particular frequency span. The analyses within the two domains, instead of working in strict cascade, operate simultaneously, albeit through restricted time and frequency windows. Finally, in a filter bank, the signal's values are streamed into the filters in parallel, and application logic interprets the output of the filters. Since this can occur with each signal sample, the output of the intepretive logic can be associated with the particular time instant at which the frequency-domain logic makes a decision. So the filter bank, at least as we sketch it here, comprises a very intimate merging of both time- and frequency-domain signal analysis.

9.1.2.1 Dual-Tone Pulses in Noise. Let us review the discrete dual-tone multifrequency (DTMF) pulses that modern digital telephone systems use for signaling [12]. Table 9.1 shows the standard pairs.

True DTMF decoders—such as in actual use at telephone company central offices—must stop decoding DTMF pulses when there is speech on the line. One frequency-domain trait that allows an application to detect the presence of human voices

TABLE 9.1. DTMF Frequency Pairs[a]

Low (Hz) \ High (Hz):	1209	1336	1477	1633
697	1	2	3	A
770	4	5	6	B
852	7	8	9	C
941	*	0	#	D

[a]The letter tones are generally reserved for the telephone company's signaling, testing, and diagnostic uses.

voices is that speech contains second and third harmonics [13], which the DTMF tones by design do not [12]. For example, a vowel sound could contain significant energy at 300 Hz, 600 Hz, and 900 Hz. An upcoming speech analysis application confirms this. But note in Table 9.1 that the second harmonic of the low tone at 697 Hz (that would be approximately 1.4 kHz) lies equidistant from the high tones at 1336 Hz and 1477 Hz. Later in this chapter, we will consider mixed speech and DTMF tones and see how to discriminate between control tones and voice. For now, let us return to the basic tone detection problem.

Suppose we represent a DTMF telephony pulse by a sum of two sinusoids chosen from the above table enclosed within a Gaussian envelope. If we sample such an analog signal at $F_s = 8192$ Hz, then the sampling period is $T = 8192^{-1}$ s.

Let us briefly cover the synthesis of the dual-tone multifrequency signal used in this example.

The analog sinusoidal signals for a "5" and "9" tone are, respectively,

$$s_5(t) = \sin(2\pi t F_{5,a}) + \sin(2\pi t F_{5,b}), \tag{9.3a}$$

$$s_9(t) = \sin(2\pi t F_{9,a}) + \sin(2\pi t F_{9,b}), \tag{9.3b}$$

where $F_{5,a} = 770$, $F_{5,b} = 1336$, $F_{9,a} = 852$, and $F_{5,b} = 1477$ as in Table 9.1. We need to window these infinite duration signals with Gaussians that—effectively—die to zero outside a small time interval. We take the window width $L = 50$ ms, let $\sigma = L/2$, and use the window functions

$$g_5(t) = e^{-\frac{(t-t_5)^2}{2\sigma^2}}, \tag{9.4a}$$

$$g_9(t) = e^{-\frac{(t-t_9)^2}{2\sigma^2}}, \tag{9.4b}$$

where $t_5 = 0.125$ s and $t_9 = 0.375$ s are the centers of the "5" and "9" pulse windows, respectively. Let $x(t) = s_5(t)g_5(t) + s_9(t)g_9(t) + n(t)$, where $n(t)$ is a noise term.

The noise term $n(t)$ could be genuine noise arduously derived from a real system, such as a radioactive source or galactic background radiation. Or it could be pseudo-random noise conveniently synthesized on a digital computer.[1] In order to control the noise values for the experiments in this section, let us make noise by assuming some realistic distribution functions and using some standard algorithms for the synthesis.

A variety of methods exist for synthesizing noise. For example, an early algorithm for generating a uniform random variable [14] called the *congruential* generator is

$$x(n) = [Ax(n-1)](\mathrm{mod}\ M), \tag{9.5}$$

where A is the *multiplier*, M is the *modulus*, and the iteration typically starts with a choice for $x(0)$, the *seed* value. The method produces values in the range $[0, M-1)$ and works better for large M. For uniform noise on $[0, 1)$ divide (9.5) by M.

A better algorithm is the *linear congruential* generator:

$$x(n) = [Ax(n-1)+C](\mathrm{mod}\ M), \tag{9.6}$$

where C is the *increment*. If $C = 0$, then (9.6) reduces to (9.5). The following values make the congruential method work well [15]: $A = 16{,}807$; $M = 2^{31} - 1$; and $C = 0$. For the linear congruential iteration, nice choices are $A = 8{,}121$; $M = 134{,}456$; and $C = 28{,}411$ [16].

There is a standard algorithm for generating pseudo-random normally (Gaussian) distributed sequences [17]. Let $x_1(n)$ and $x_2(n)$ be two uniformly distributed random variables on (0, 1) and define

$$y_1(n) = \cos(2\pi x_2(n))\sqrt{-2\ln(x_1(n))}, \tag{9.7a}$$

$$y_2(n) = \sin(2\pi x_2(n))\sqrt{-2\ln(x_1(n))}. \tag{9.7b}$$

Then $y_1(n)$ and $y_2(n)$ are zero-mean normally distributed random variables. References on random number generation include Refs. 18 and 19.

The chapter exercises invite readers to change signal to noise ratios and explore the impact on detection algorithms.

We begin with an analysis of the DTMF signal using the discrete Fourier transform on the entire time interval of $N = 4096$ points. Let $x(n) = x_a(nT)$ be the discretized input, where $T = 8192^{-1}$s, $x_a(t)$ is the real-world analog representation, and $n = 0, 1, ..., N-1 = 4095$. Figure 9.1 (second from top) shows the magnitude of the DFT coefficients (9.1). Even though the presence of DTMF tones in the signal is clear, we cannot be sure when the tones occurred. For example, we detect two low tones, 770 and 852 Hz, and we can see two high tones, 1336 and 1477 Hz, but this global frequency-domain representation does not reveal whether their presence

[1]"Anyone who considers arithmetical methods of producing random digits is, of course, in a state of sin" (John von Neumann). And having repeated the maxim that everyone else quotes at this point, let us also confess that "Wise men make proverbs, but fools repeat them" (Samuel Palmer).

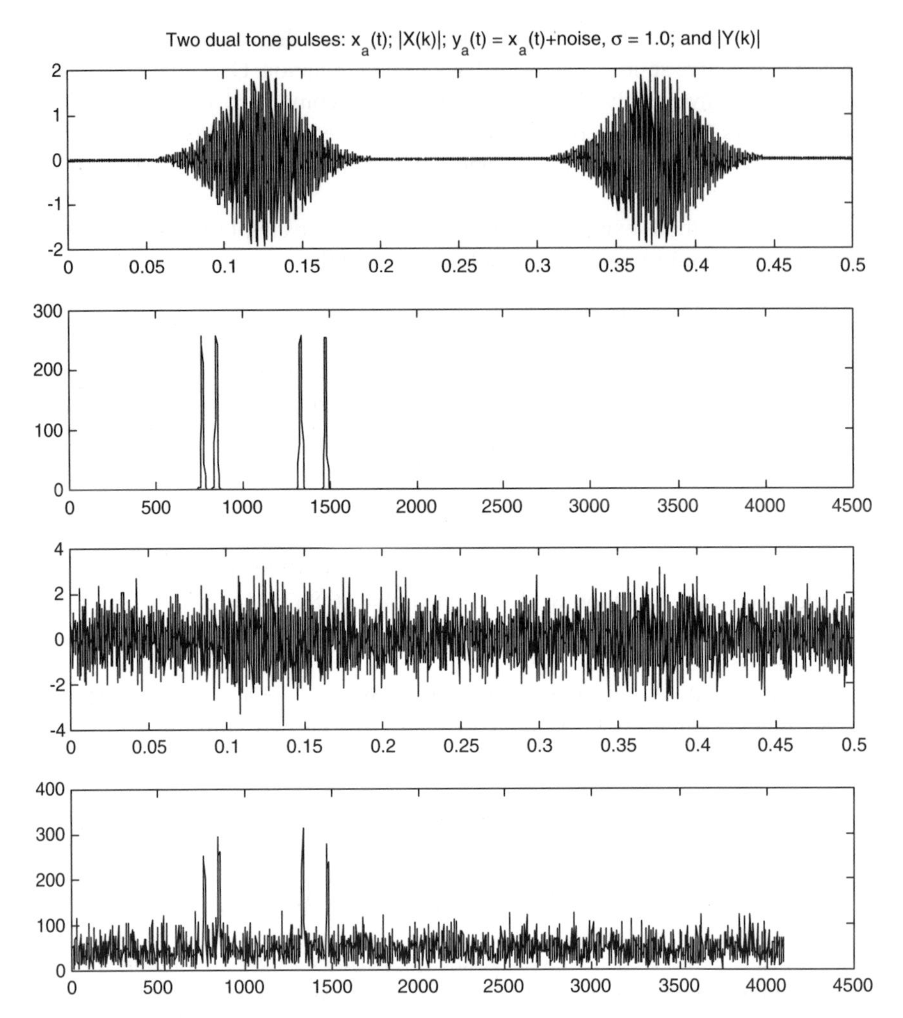

Fig. 9.1. DTMF numbers "5" and "9" tones (top). Sampling produces $x(n) = x_a(nT)$, where $T = 8192^{-1}$ s. Discrete Fourier transformation gives $|X(k)|$ (second). Adding noise of zero mean, normal distribution, and standard deviation $\sigma = 1$ effectively buries the signal (third). Yet the characteristic peaks remain in the magnitude spectrum of the noisy signal (bottom).

indicates that the time-domain signal contains a "5" pulse, a "9" pulse, a "6" pulse, an "8" pulse, or some invalid combination of tones.

The DFT efficiently detects signal periodicity. Adding a considerable amount of noise to the pure tone signals used above demonstrates this, as can be seen in the lower two panels of Figure 9.1. Should the application need to ascertain the mere existence of a periodity, obscuring noise is no problem; there is just a question of seeing the spike in $|X(k)|$. But it does become a problem when we need to find the time-domain extent of the tone—when one tone occurs earlier than another. Indeed, high noise levels can confound as simple an application as DTMF detection.

9.1.2.2 Preliminary Time-Domain Segmentation. A straightforward approach is to preface frequency-domain interpretation with time-domain segmentation. Chapter 4's diverse thresholding methods, for example, can decide whether a piece of time-domain signal $x(n)$ merits Fourier analysis. For dual-tone detection, the appropriate steps are as follows:

(i) Segment the time-domain signal $x(n)$ into background regions and possible DTMF tone regions.
(ii) Compute the DFT of $x(n)$ on possible tone segments.
(iii) Apply the DTMF specifications and check for proper tone combinations in the candidate regions.

Time-domain signal segmentation methods are familiar from Chapter 4. If we know the background noise levels beforehand, we can assume a fixed threshold T_x. Of course, the source $x(n)$ is oscillatory, so we need to threshold against $|x(n)|$ and merge nearby regions where $|x(n)| \geq T_x$. Median filtering may help to remove narrow gaps and small splinters at the edge of high magnitude regions. If we know the probability of DTMF tones, then a parametric method such as the Chow and Kaneko algorithm [20] may be useful. However, if $x(n)$ contains other oscillatory sounds, such as speech, or the tones vary in length and temporal separation, then a nonparametric algorithm such as Otsu's [21] or Kittler and Illingworth's [22] may work better. In any case, the first step is to produce a preliminary time-domain segmentation into possible tone signal versus noise (Figure 9.2).

Nevertheless, segmentation by Otsu's method fails to provide two complete candidate pulse regions for noise levels only slightly higher than considered in Figure 9.2. It is possible to apply a split and merge procedure to fragmented regions, such as considered in the exercises. However, these repairs themselves fail for high levels of noise such as we considered in the previous section.

Since the high level of noise is the immediate source of our time-domain segmentation woes, let us try to reduce the noise. Preliminary noise removal filtering comes to mind. Thus, we can apply some frequency-selective signal processing to the input DTMF signal before attempting the partition of the source into meaning signal and background noise regions.

Let us try a moving average filter. The motivation is that the zero mean noise locally wiggles around more than the sinusoidal pulses that comprise the DTMF information. So we anticipate that filter averaging ought to cancel out the noise but leave the sinusoidal DTMF pulses largely intact.

The *moving average filter* of order $N > 0$ is $h = H\delta$, where

$$h(n) = \begin{cases} \dfrac{1}{N} & \text{if } 0 \leq n \leq N-1, \\ 0 & \text{if otherwise,} \end{cases} \tag{9.8}$$

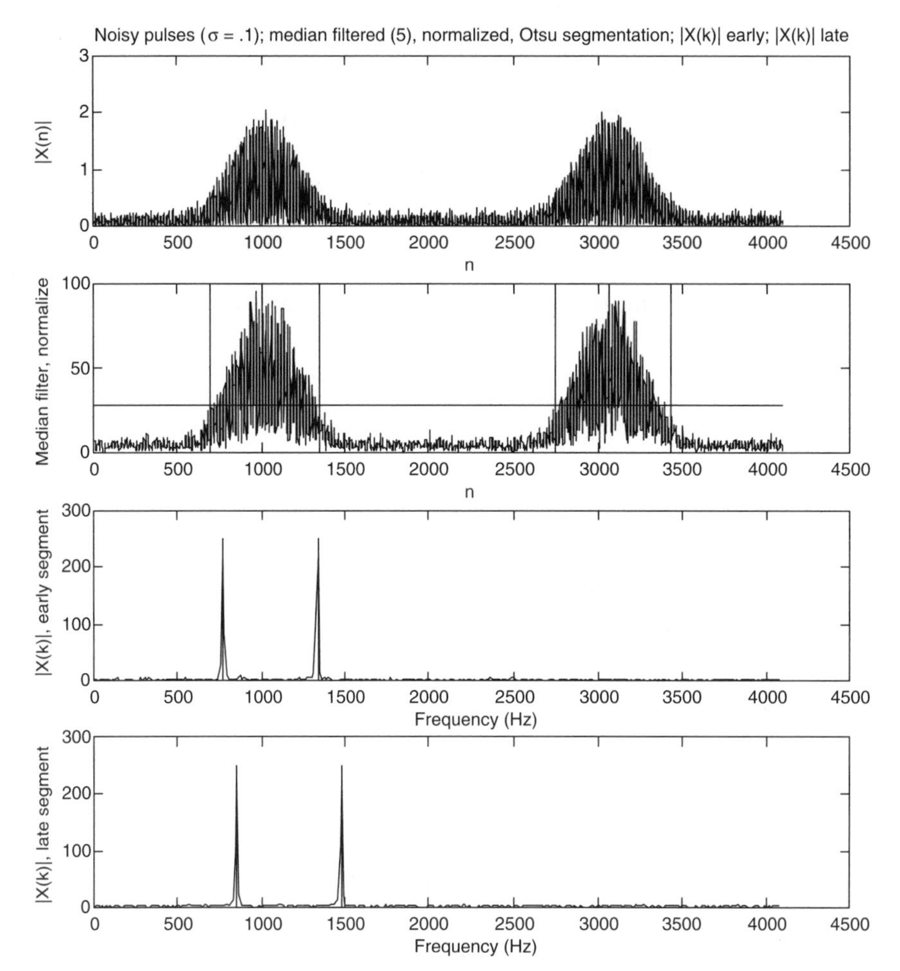

Fig. 9.2. Magnitudes of DTMF "5" and "9" tones within noise of zero mean, normal distribution, and moderate $\sigma = 0.1$ standard deviation. The second panel shows time-domain segmentation via the Otsu algorithm. Here, the magnitude spectra are median-filtered and normalized to a maximum of 100% before histogram construction and segmentation. The horizontal line is the Otsu threshold. The vertical lines are boundaries of the likely pulse regions. The lower panels show the magnitude spectra of DFTs on the two candidate regions. Note that the spikes correspond well to "5" and "9" dual-tone frequencies.

and $N > 0$ is the *order* of the filter. Let $x(n)$ be the pure DTMF signal and let us add normally distributed noise of mean $\mu_x = 0$ and standard deviation $\sigma_x = 0.8$ (Figure 9.3).

Why the time domain does not seem to aid the segmentation process is evident from examining the various frequency-domain representations. Figure 9.3 shows the spectral effect of filtering this signal with moving average filters of orders 3, 7, and 21. We can see that the smallest order is beneficial in terms of aiding a time-domain

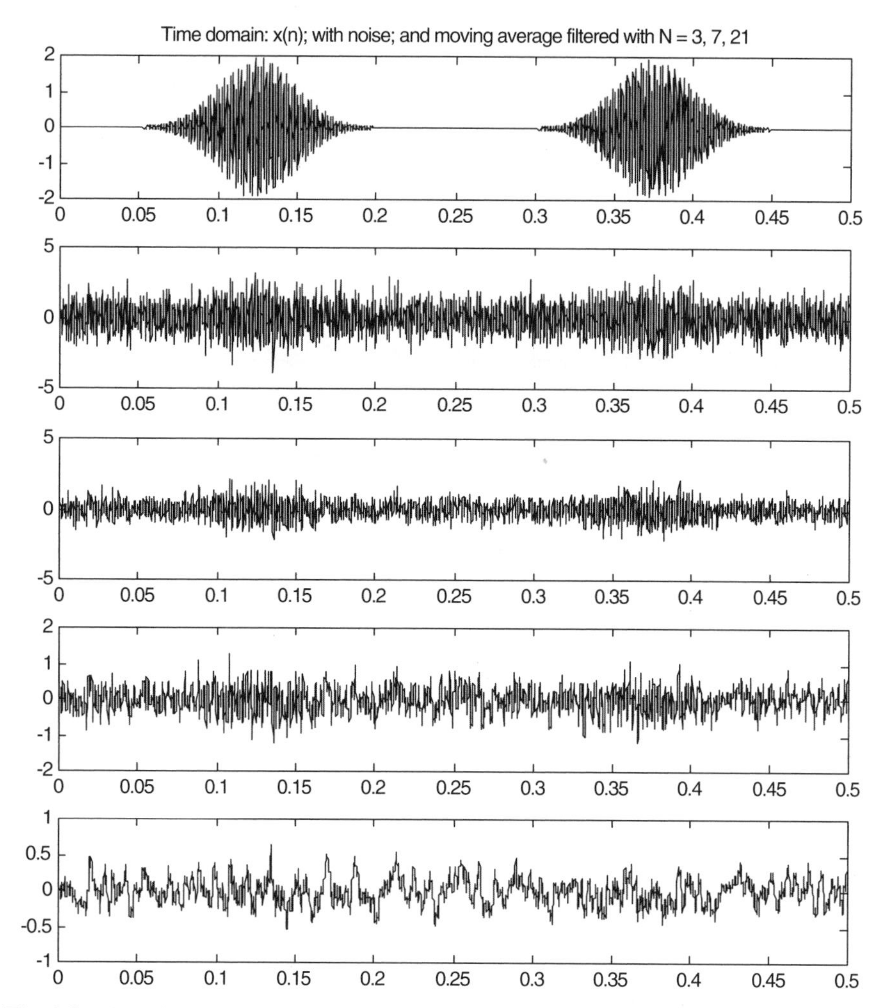

Fig. 9.3. Time-domain plots of the pure DTMF signal (top) and with Gaussian noise added, $\mu_x = 0$ and $\sigma_x = 0.8$ (second from top). The next three panels show $y = h*x$, with H a moving average filter of order $N = 3$, $N = 7$, and $N = 21$.

segmentation, but only slightly so. The higher-order filters are—if anything—a hindrance.

We can see that the moving average filter magnitude spectrum consists of a series of slowly decaying humps (Figure 9.4). Frequencies between the humps are suppressed, and in some cases the frequency buckets that correspond to our DTMF pulses are attenuated by the filter. The filter will pass and suppress frequencies in accordance with the DFT convolution theorem: $Y(k) = H(k)X(k)$.

This explains in part why the moving average filter failed to clarify the signal for time-domain segmentation. Although it is intuitive and easy, its frequency

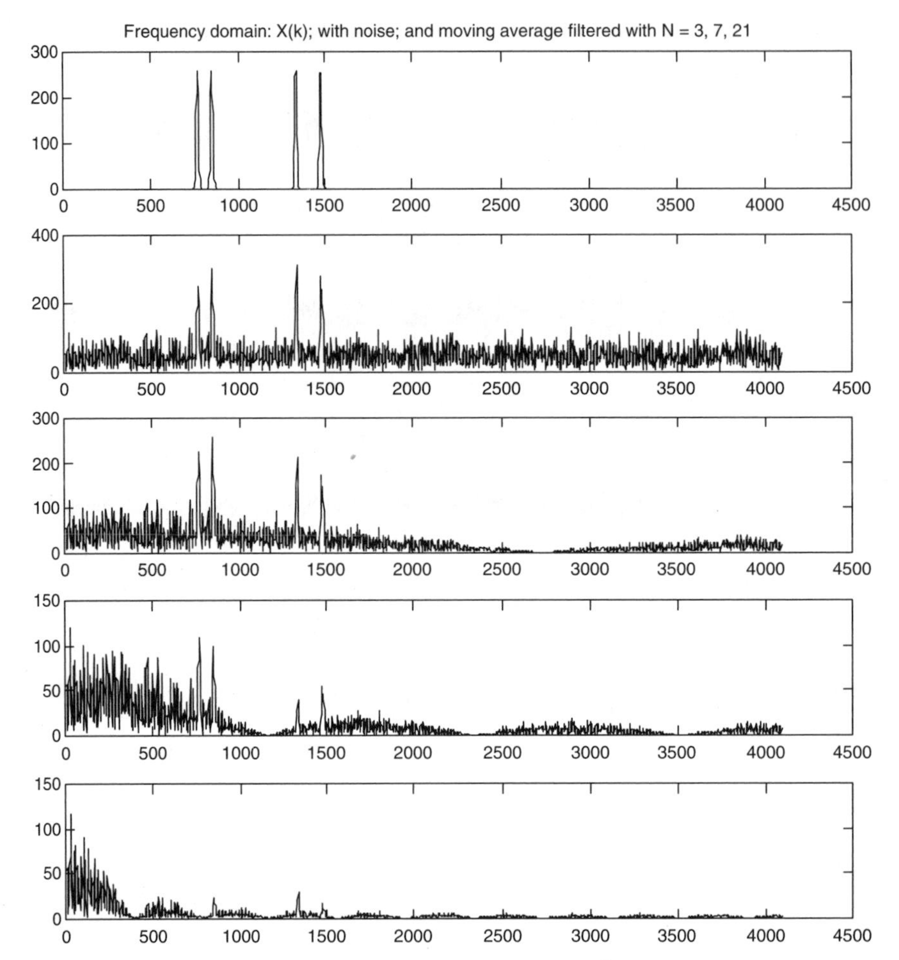

Fig. 9.4. Frequency-domain plots of the magnitude spectrum $|X(k)|$ of DTMF signal $x(n)$ (top); with Gaussian noise added, $\mu_x = 0$ and $\sigma_x = 0.8$ (b); and the final three panels are $|Y(k)| = |H(k)||X(k)|$, with H of order $N = 3$, $N = 7$, and $N = 21$.

suppression capabilities do not focus well for narrowband tones. We seem to find ourselves designing a raw signal that the moving average filter can improve. Later in this chapter, we shall discover better filters and learn how to build them in accord with the requirements of an application.

In fact, the second method, which applies the time-frequency map to the dual-tone signals, offers some help with the high noise problem.

9.1.2.3 Analysis in the Time-Frequency Plane. A time-frequency map is a two-dimensional array of signal frequencies plotted on one axis and their time location plotted on the other axis. This is a useful tool for signal interpretation problems where the preliminary segmentation is problematic or when there is scant a priori information on frequency ranges and their expected time spans. To decompose a signal into a time-frequency plane representation, we chop up its time domain into intervals of equal size and perform a Fourier analysis on each piece.

Let us explain the new tool using the DTMF application as an example. Application necessity drives much of the design of time-frequency maps. A method appropriate for the DTMF detection problem is as follows.

(i) Select a fixed window width, say $N = 256$. This corresponds to a frequency resolution of 32 Hz at $F_s = 8192$ Hz and a time-domain width of 31 ms.
(ii) This DFT length supports efficient calculation of the transform: the Fast Fourier Transform (FFT) algorithm (Chapter 7). Since we may have quite a few time windows, Candidate segments that are too small can be padded with zeros at the end to make, say, 256 points.
(iii) We can cover longer segments with 256-point windows, overlapping them if necessary.
(iv) We have to run the FFT computation over a whole set of time-domain windows; hence it may be crucial to use a fast transform and limit overlapping.
(v) A genuine DTMF application must account for proper pulse time width (23 ms, minimum, for decoding), frequency (within 3.5% of specification), and energy ratio (DTMF tone energy must exceed that of other frequencies present by 30 dB).
(vi) Once the application checks these details, it can then invoke the logic implied by Table 9.1 and make a final tone decision.

Let us form an array of dual tone energies plotted against time window location (Figure 9.5) and thereby interpret the signal. Sixteen disjoint 256-point windows cover the signal's time domain. Over each window, we compute the FFT. For each of 16 DTMF tones, we calculate the frequency-domain signal energy in the tone frequency range, the energy outside the tone frequency range, and the ratio of the two energies in dB.

Recall that there are two formulas for expressing a gain or ratio R_{dB} between signals Y_1 and Y_2 in decibels (dB). We use either magnitude or power (which is proportional to energy, the magnitude squared):

$$R_{\mathrm{dB}} = 20\log_{10}\left(\frac{M_1}{M_2}\right) = 10\log_{10}\left(\frac{P_1}{P_2}\right), \tag{9.9}$$

where M_i and P_i are the magnitude and power, respectively, of signal Y_i. For each time window we set $y(n) = x(n)$ restricted to the window. Let $Y(k)$ be the 256-point-FFT of $y(n)$ over the window. Then, for each DTMF tone frequency range in Figure 9.5, we take P_1 to be the sum of squared magnitudes of the transform values that lie within the tone frequency range, considering only discrete frequency values $0 \le k < 128$ that lie below the Nyquist value. We set P_2 to be the sum of squared magnitudes that remain; these represent other tones or noise. For example, for DTMF dual tone

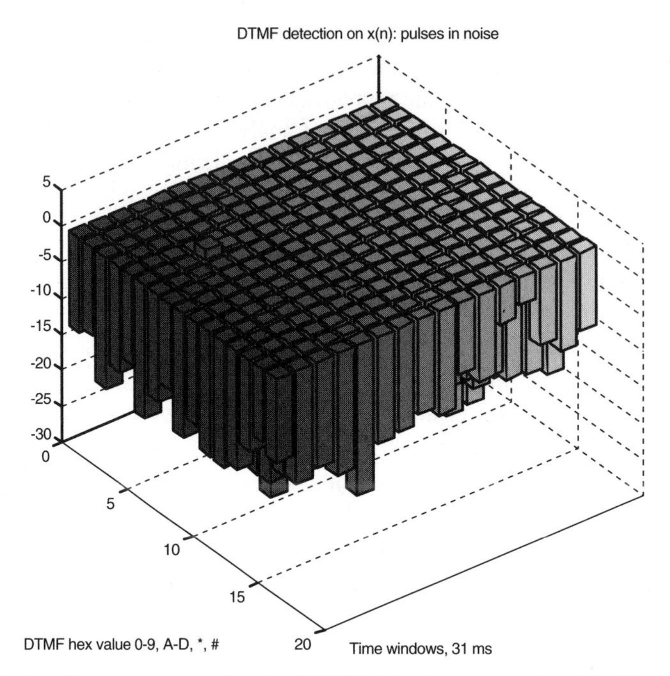

Fig. 9.5. A time-frequency array showing DTMF detection on a noisy $x(n)$. The DTMF "5" and "9" tones appear as tall blocks, representing high ratios of DTMF tone power to overall signal power (dB). Note, however, that the tones are just barely detectable, by a threshold above 0 dB.

"9," the energy lies in $Y(k)$ coefficients $26 \le k \le 27$ (which represent frequencies f (Hz) of $26 \times 32 = 832 \le f \le 864$, for the lower tone) and in $45 \le k \le 47$ (which represent frequencies $1440 \le f \le 1504 = 47 \times 32$) for the upper tone. Thus, we construct a 16×16 array of power ratios, DTMF versus non-DTMF.

Note that the joint frequency and time domain computations involve a tradeoff between frequency resolution and time resolution. When we attempt to refine the time location of a tone segment, we use a lower-order DFT, and the frequency resolution $(NT)^{-1}$ suffers. Since the sampling rate has been fixed at 8 kHz, on the other hand, improving the frequency resolution—that is, making $(NT)^{-1}$ *smaller*—requires a DFT over a larger set of signal samples and thus more imprecision in temporal location.

Let us consider the effect of noise on the time-frequency plane analysis. We have observed that preliminary time domain segmentation works well under benign noise. Heavier noise demands some additional time domain segmentation effort. Noise whose variation approaches the magnitude of the tone oscillations causes problems for segmentation, even though the Fourier analysis can still reveal the underlying tones.

The noisy signal in Figure 9.1 resists segmentation via the Otsu algorithm, for example. Without a time domain segmentation, we can try using small time domain windows and computing an array of coarse frequency resolution magnitude spectra,

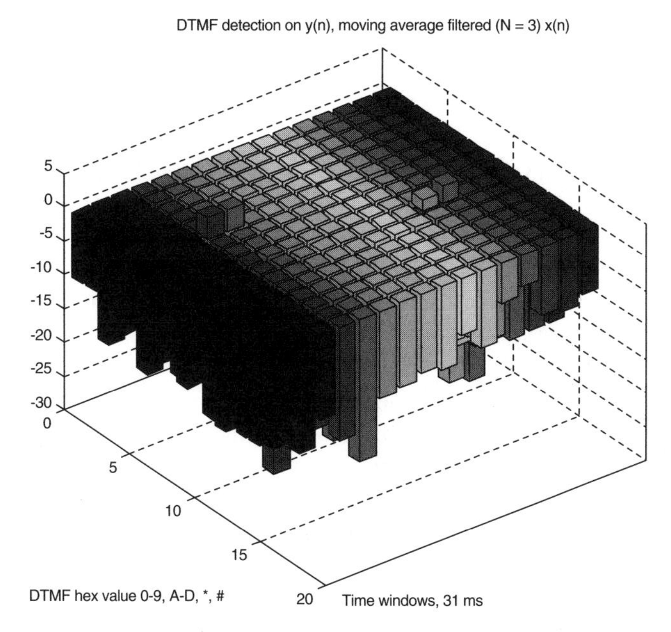

Fig. 9.6. A time-frequency array showing DTMF detection on a very noisy $x(n)$ subject to a moving average noise removal filter. The plot shows the ratio between frequency-domain DTMF power (dB) and non-DTMF power. The time location of the tones is clear, but the frequency discrimination shows little improvement.

such as in Figure 9.5. The problem is that the noisiness of the signal can obscure the tone peaks in the time-frequency array. Let us apply a moving average filter to the signal prior to time-frequency decomposition (Figure 9.6).

The moving average filter's poor performance is not so surprising. We have already empirically shown that its effectiveness is limited by our lack of control over the lobes that appear in its magnitude spectrum (Figure 9.4). We apparently require a filter that passes precisely the range of our DTMF signals, say from 600 to 1700 Hz, and stops the rest. Such a *bandpass filter* cuts down signal components whose frequencies lie outside the DTMF band.

We can construct such a filter H by specifying its frequency domain $H(k)$ as being unity over the discrete DTMF frequencies and zero otherwise. Let f_{LO} = 600 Hz, f_{HI} = 1700 Hz, the sampling rate F_s = 8192 Hz, and suppose N = 256 is the DFT order. The sampling interval is $T = F_s^{-1}$, so that the frequency resolution is $f_{res} = 1/(N \times T)$. The Nyquist rate is $f_{max} = (N/2) \times f_{res} = F_s/2$ = 4096 Hz. Hence, let us define $k_{LO} = (N/2) \times (f_{LO}/f_{max})$ and $k_{HI} = (N/2) \times (f_{HI}/f_{max})$. An ideal bandpass filter for this Fourier order is given by

$$H(k) = \begin{cases} 1 & \text{if } k_{LO} \le k \le k_{HI}, \\ 1 & \text{if } N - k_{HI} \le k \le N - k_{LO}, \\ 0 & \text{if otherwise.} \end{cases} \tag{9.10}$$

Then we find

$$h(n) = \frac{1}{N}\sum_{k=0}^{N-1} H(k)e^{\frac{2\pi jkn}{N}} . \tag{9.11}$$

This creates an N-point finite impulse response (FIR) filter (Figure 9.7). Except for the DC term $n = 0$, $h(n)$ is symmetric about $n = 128$.

We can filter the noisy $x(n)$ in either of two ways:

(i) Set up the filter as a difference equation, for example, using the Direct Form II architecture that we will cover later. This method is appropriate for on-line processing of the signal data.

(ii) Translate the filter $g(n) = h(n - 128)$ and perform the convolution $y(n) = (g*x)(n)$. This method is suitable for off-line applications, where all of the signal data is available and the noncausal filter $g(n)$ can be applied to it.

The results of the filtering are shown in the two lower panels of Figure 9.7. Note that the difference equation implementation produces a significant delay in the output

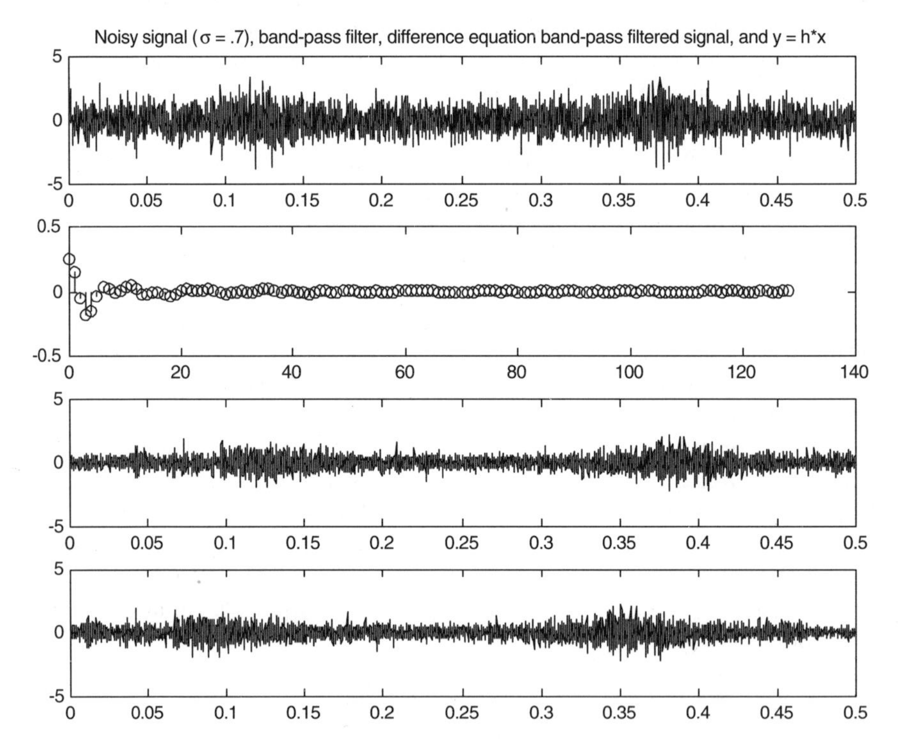

Fig. 9.7. Noisy signal $x(n)$, $\sigma = 0.7$ (top). Bandpass filter $h(n)$ for $0 \le n \le N/2$ (second from top). Alternative filtering results (lower panels).

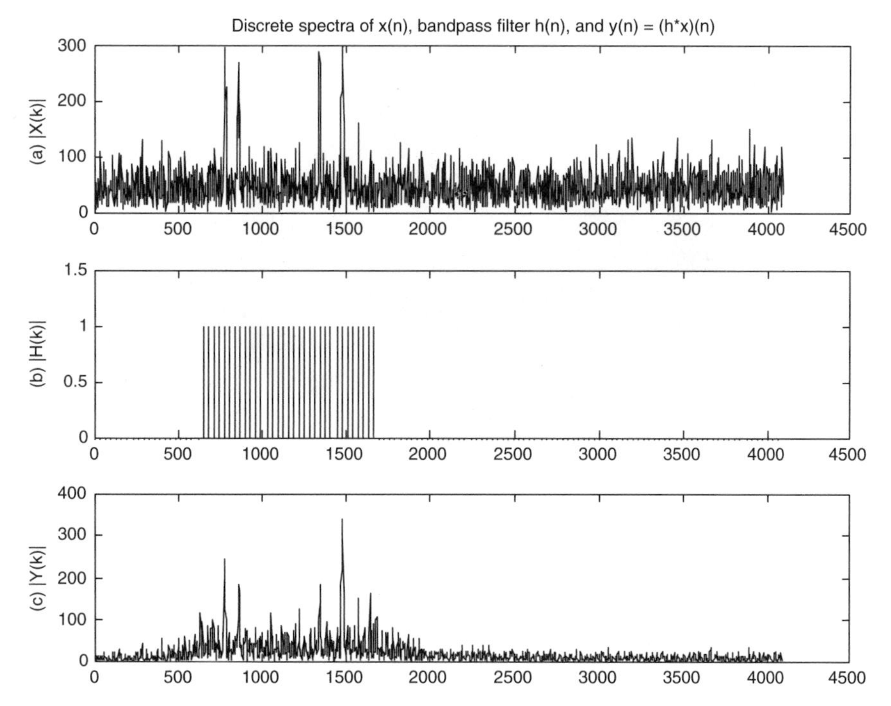

Fig. 9.8. Magnitude spectra from the bandpass filtering operation.

signal. Figure 9.8 shows the frequency-domain magnitude spectra of $X(k)$, $H(k)$, and $Y(k)$, where $y(n) = (Hx)(n)$.

An analysis of the noisy DTMF signal using a time-frequency map is shown in Figure 9.9. Note that the bandpass filter raises the peaks in the time-frequency plane, which potentially helps with detection under severe noise. The drawback is that a few false positive frequency markers appear as well. Why does the bandpass filter not do a clearly superior job compared to the simple moving average filter and analysis without prefiltering? Unfortunately, this bandpass filter is still forgiving to all noise in its pass band—that is, from 600 to 1700 Hz. For example, when the signal of interest is a DTMF "1" dual-tone (697 Hz and 1209 Hz), filtering with the above H will allow noise from 1.25 to 1.7 kHz into the output.

So bandpass filtering is a promising idea, but cleaning all DTMF tones with a single bandpass filter gives only modest results.

9.1.2.4 Filter Bank Decomposition and Analysis.

A third approach to dual-tone detection employs a frequency selective filter for each tone in the DTMF ensemble. Constructing, implementing, and applying so many filters seems onerous, but the humble results in the previous two sections encourage alternatives. The appropriate frequency domain tool for this approach is called a *filter bank*. Indeed, this is the conventional approach for DTMF detection, which often calls for online implementation and real-time detection [12].

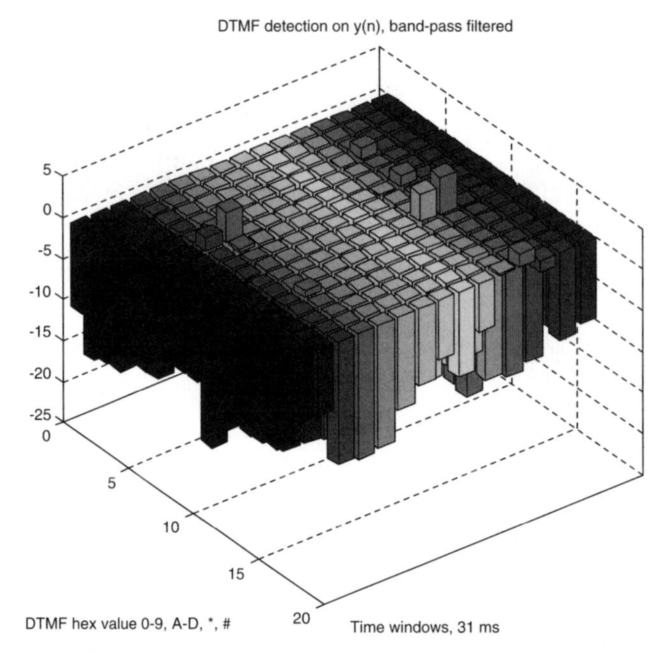

Fig. 9.9. Time-frequency map: A DTMF signal after bandpass filtering.

Filter banks have been carefully studied by signal processing researchers for applications involving compression and efficient signal transmission. More recently they have been subject to intense scrutiny because, when combined with subsampling operations, they are related to the theory of wavelets [23–25]. We will consider these ideas at the end of the chapter and follow up on them in the last two chapters especially. However, for now, our purposes are elementary.

We just want to assemble an array of filters whose parallel output might be read out to interpret a signal containing digital telephony dual-tones. Such simple filter banks are suitable for analysis applications where the input signal frequency ranges are generally known in advance, but the time at which they might occur is not known. If the several filters in the bank are implemented as causal filters, $h(n) = 0$ for $n < 0$, then the filter bank can process data as it arrives in real time.

To build a filter bank for the DTMF application, we set up bandpass filters with unit gain passbands centered about the eight (four low and four high) tones of Table 9.1. Each bandpass filter is designed exactly as in (9.10), except that the frequency range is narrowed to within 3.5% of the tone center frequencies. All filters have the same passband width, and the order of the DFT is $N = 200$ samples (Figure 9.10).

The result of filtering the input signal x(n), which contains DTMF tones "5" and "9" as well as normally distributed noise of zero mean and standard deviation $\sigma =$ 0.8, is shown in Figure 9.11.

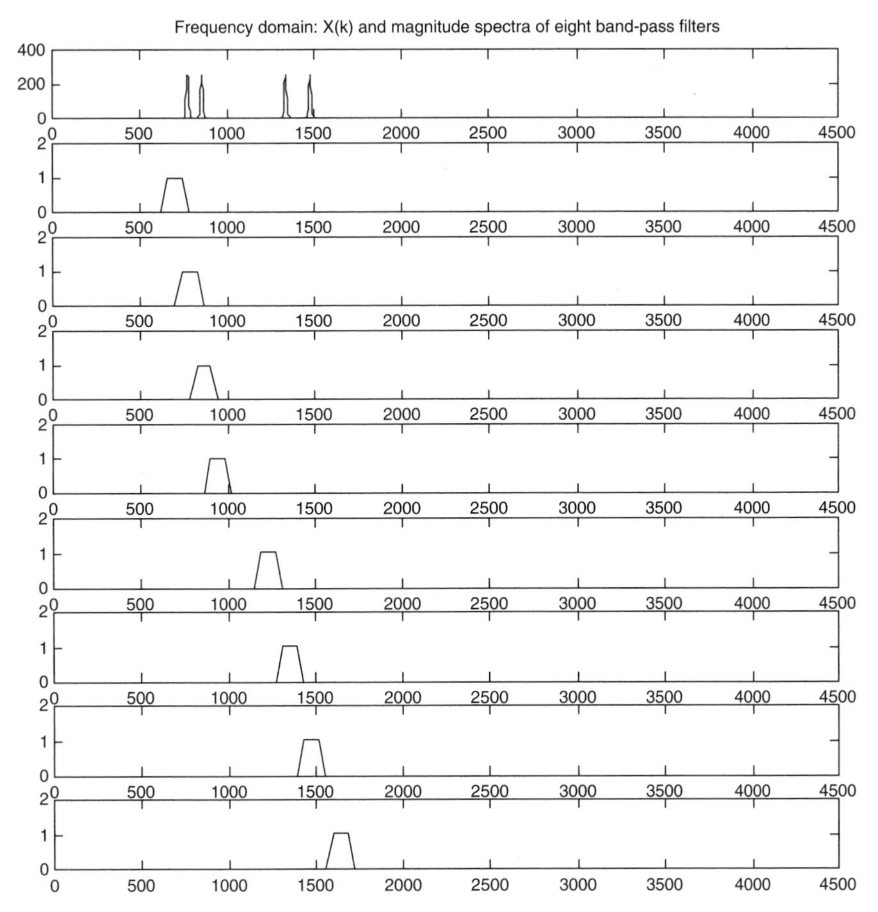

Fig. 9.10. Magnitude spectra of pure DTMF tones "5" and "9" (top) and bank of eight bandpass filters.

To complete the application, one could calculate the energy in a certain window of the last M samples. The dual-tone standard calls for 23 ms for decoding, so at the sampling rate of the example application, $M = 188$. If the energy exceeds a threshold, then the tone is detected. A valid combination of tones, one low tone and one high tone, constitutes a dual-tone detection.

The main problem with the filter bank as we have developed it is the delay imposed by the bandpass filters. The shifting of pulses after filtering must be compensated for in the later analysis stages, if there is a need to know exactly when the tones occurred. For example, do we know that the tones are delayed the same amount? If so, then the detection logic will be correct, albeit a little late, depending on the length of the filters. But if different frequencies are delayed differnet amounts, then we either need to uniformize the delay or compensate for it on a filter-by-filter basis.

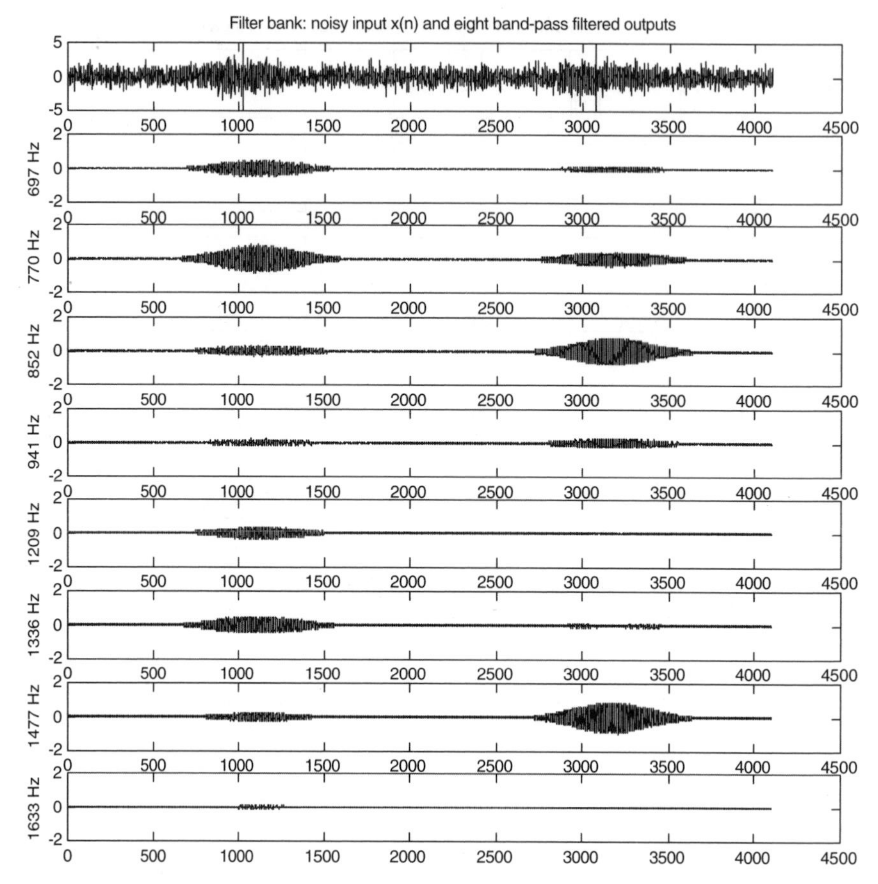

Fig. 9.11. Filter bank output, causally implemented. Note the large delay between the center of the input and output pulses. This is a consequence of the length of the filter, $N = 200$, and the causal implementation.

9.1.3 Filter Frequency Response

When input signals contain oscillatory components that are hidden within noise, the discrete Fourier transform reveals the periodicity as high magnitude spikes in the magnitude spectrum. Even when the signal is so immersed in noise that the time-domain representation refuses to betray the presence of sinusoidal components, this signal transformation is still effective. Though its power is evident for this purpose, the DFT nonetheless loses the time location of oscillations. And for this some time-domain analysis remains. But the noise obscures the time-domain location and extent of the oscillations. This can be a crucial factor in interpreting the signal. By noise removal filtering, however, we can improve visibility into the time-domain and better know the places where the periodicity hides. All of this suggests a theoretical study of the effect of filters on periodic trends in signals.

Consider an exponential signal $x(n)$ input into a linear, translation-invariant system H, producing output $y(n)$: $y = Hx$. If $\delta(n)$ is the discrete impulse and $h = H\delta$ is the *impulse response* of H, then $y(n) = (x * h)(n)$ is the convolution of $x(n)$ and $h(n)$:

$$y(n) = (x*h)(n) = \sum_{k=-\infty}^{\infty} x(k)h(n-k) = \sum_{k=-\infty}^{\infty} h(k)x(n-k). \tag{9.12}$$

Suppose $x(n) = e^{j\omega n}$ is the discrete exponential signal with radial frequency ω radians per sample. Then

$$y(n) = \sum_{k=-\infty}^{\infty} h(k)e^{j\omega(n-k)} = e^{j\omega n}\sum_{k=-\infty}^{\infty} h(k)e^{j\omega k} = e^{j\omega n}H(\omega), \tag{9.13}$$

where $H(\omega)$ is the *frequency response* of $h(n)$. An exponential input to an LTI system produces an exponential output of the same frequency, except amplified (or attenuated) by the factor $H(\omega)$. This basic Chapter 7 result tells us that LTI systems pass exponential signals directly from input to output, multiplied by a complex constant which depends on the signal frequency.

9.1.4 Delay

We have observed empirically that noise removal filtering—and by implication, convolutional filtering in general—imposes a delay on input signals. This section explains the theory of two types of signal delay caused by filtering: *phase delay* and *group delay.*

9.1.4.1 Phase Delay. Suppose $x(n)$ is a discrete signal and $y = Hx$ is a linear, translation-invariant (LTI) discrete system. If $x(n) = \exp(j\omega n)$ is a pure, complex-valued exponential signal, then $y(n) = H(\omega)\exp(j\omega n) = H(\omega)x(n)$, where $H(\omega)$ is the discrete-time Fourier transform (DTFT) of $h(n) = (H\delta)(n)$.

Consider a sinusoidal input signal $x(n) = \cos(n\omega) = [e^{j\omega n} + e^{-j\omega n}]/2$. Then

$$y(n) = \frac{H(\omega)e^{j\omega n}}{2} + \frac{H(-\omega)e^{-j\omega n}}{2} = \frac{H(\omega)e^{j\omega n}}{2} + \frac{\overline{\overline{H(\omega)}e^{j\omega n}}}{2}$$
$$= 2\mathrm{Real}\left[\frac{H(\omega)e^{j\omega n}}{2}\right] = \mathrm{Real}[H(\omega)e^{j\omega n}]. \tag{9.14}$$

But

$$\mathrm{Real}[H(\omega)e^{j\omega n}] = |H(\omega)|\cos[\mathrm{Arg}(H(\omega))e^{j\omega n}]. \tag{9.15}$$

If we set $\theta(\omega) = \mathrm{Arg}(H(\omega))$, then

$$y(n) = \mathrm{Real}[H(\omega)e^{j\omega n}] = |H(\omega)|\cos[\omega n + \theta(\omega)] = |H(\omega)|\cos\left[\omega\left(n + \frac{\theta(\omega)}{\omega}\right)\right]. \tag{9.16}$$

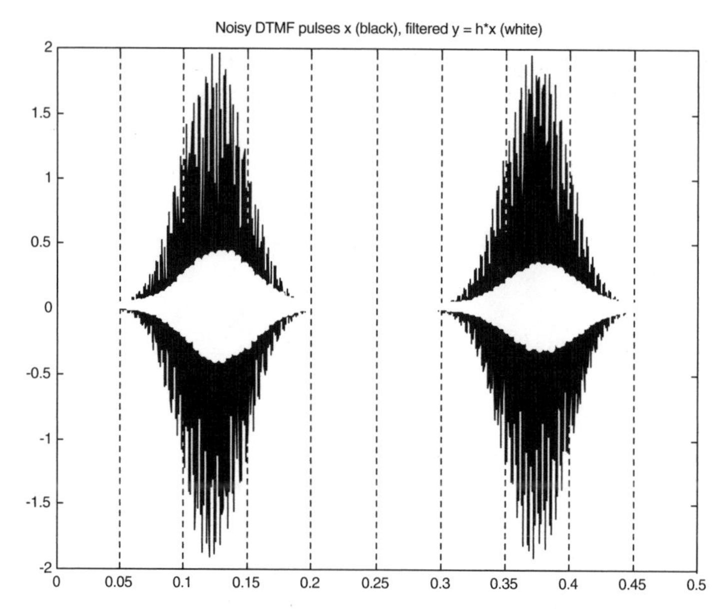

Fig. 9.12. Filter phase delay. Noisy DTMF pulses x (black), filtered $y = h*x$ (white).

So if the input $x(n)$ to H is a sinusoid, then the output $y(n)$ is a sinusoid too. Signals $x(n)$ and $y(n)$ have the same frequency, but $y(n)$ is scaled by $|H(\omega)|$ and phase shifted by $T_H = -\theta(\omega)/\omega$, which is called the *phase delay* of H [26].

If we apply a moving average filter of length $N = 101$ to the noisy DTMF pulses (), then the phase delay imposed by the filter is clearly evident (Figure 9.12).

So sinusoids too, subject to a complex scaling, pass directly through LTI systems. This helps explain the clarity with which the sinusoidal pulses of the DTMF application appear in the frequency domain. Also, we now have a tool, namely the phase delay, $T_H = -\theta(\omega)/\omega$ in (9.16) for comparing the delays induced by various filters.

9.1.4.2 Group Delay. Another type of filter delay occurs when source signals contain sinusoids of nearby frequencies that form an envelope. The superposition of the two sinusoids

$$x(n) = \cos(\omega_1 n) + \cos(\omega_2 n), \tag{9.17}$$

with $\omega_1 \approx \omega_2$, creates a long-term oscillation, called a *beat*. This holds as long as the filter H does not suppress the individual sinusoids; this means that ω_1 and ω_2 are in the *passband* of H.

By trigonometry, we can write $x(n)$ as a product of cosine functions, one of which gives the envelope, of frequency $\omega_1 - \omega_2$, and the other is a sinusoid whose frequency is the mean. Thus,

$$x(n) = 2\cos\left(n\frac{(\omega_1 - \omega_2)}{2}\right)\cos\left(n\frac{(\omega_1 + \omega_2)}{2}\right), \tag{9.18}$$

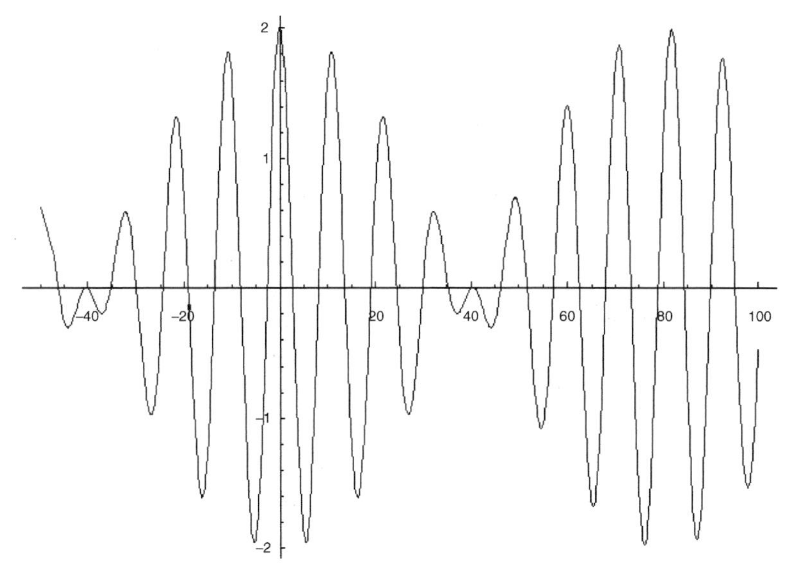

Fig. 9.13. Signal envelope formed by two sinusoids of approximately the same frequency.

which explains the amplitude modulated oscillation of Figure 9.13. Now impose a filter $y(n) = (h^*x)(n)$, with $h = H\delta$. Using (9.16) and (9.17) we have

$$y(n) = |H(\omega_1)|\cos(\omega_1 n + \theta(\omega_1)) + |H(\omega_2)|\cos(\omega_2 n + \theta(\omega_2)), \tag{9.19}$$

where $\theta(\omega) = \text{Arg}(H(\omega))$. Assume $\omega_1 \approx \omega_2$ and that these lie in the passband of H, which is to say $|H(\omega_1)| \approx |H(\omega_2)| \neq 0$. Thus,

$$y(n) = |H(\omega_1)|\{\cos(\omega_1 n + \theta(\omega_1)) + \cos(\omega_2 n + \theta(\omega_2))\}. \tag{9.20}$$

From trigonometry once again,

$$\begin{aligned} y(n) &= 2|H(\omega_1)|\left\{\cos\frac{\omega_1 n + \theta(\omega_1) - \omega_2 n - \theta(\omega_2)}{2}\right. \\ &\quad \left.\cdot \cos\left(\frac{\omega_1 n + \theta(\omega_1) + \omega_2 n + \theta(\omega_2)}{2}\right)\right\}. \end{aligned} \tag{9.21}$$

Rearranging the cosine arguments gives

$$\begin{aligned} y(n) &= 2|H(\omega_1)|\left\{\cos\left\{\frac{\omega_1 - \omega_2}{2}\left[n + \frac{\theta(\omega_1) - \theta(\omega_2)}{\omega_1 - \omega_2}\right]\right\}\right. \\ &\quad \left.\cdot \cos\left(\frac{\omega_1 + \omega_2}{2}\left[n + \frac{\theta(\omega_1) + \theta(\omega_2)}{\omega_1 + \omega_2}\right]\right)\right\}, \end{aligned} \tag{9.22}$$

where the first cosine defines the envelope of $y(n)$. This envelope is delayed by a factor $\frac{\theta(\omega_1)-\theta(\omega_2)}{\omega_1-\omega_2}$. As $\omega_1 \to \omega_2$, this delay factor becomes a derivative, which is called the *group delay* of the filter H: $T_G = -d\theta/d\omega$ [11].

9.1.4.3 *Implications.* Applications that require significant signal filtering must consider the phase and group delay inherent in the system. In our own humble DTMF example above, we noted the phase delay caused by the filter bank. In many scientific and engineering applications, delay considerations affect the actual choice of filters. We shall see later that certain types of finite impulse response (FIR) filters have *linear phase*, so that their group delay is constant. Such filters support signal processing without distortion, an important consideration in communications systems [27].

9.2 FREQUENCY AND PHASE ESTIMATION

The dual-tone multifrequency (DTMF) detection problem in the previous section required Fourier transformation of local signal slices in order to find coded tones. With many slices and many frequency bins, we built time-frequency maps. And thus, we were able to ascertain the presence of signal frequency components over the time span of the signal slice by thresholding for large-magnitude Fourier-domain values. In this section, we study the effectiveness of such techniques. Our methods will be limited and introductory, only a small part of the broad and involved theory of *spectral estimation*. In what appear to be obviously correct and quite straightforward approaches to the problem, we shall see that there are some surprising limitations.

This section introduces an important tool: *window functions*. These are special analog or discrete signals that are used to weight a local signal slice. This technique, called *windowing*, helps to suppress artifacts caused by Fourier transformation on a segment of time-domain signal values. Thus, windowing improves the estimation of local signal frequencies. The signal slice itself is called a *window* or a *region of interest*. Sometimes window functions are loosely called "windows" as well. The ideas are easy, and in context the terms are usually clear. In Chapter 10, we consider analog window functions as an instrument with which to generalize the Fourier transform. Here, we pursue signal analysis applications, computerized implementation, and our emphasis is consequently on discrete windowing.

The DTMF tutorial application did not weight the signal values before performing the discrete Fourier transform (DFT). The results were satisfactory, but we shall see later that applying a window function to the values produces a cleaner, easier to analyze time-frequency map. Moreover, we shall see that the window functions and the windowing method provide a straightforward method for designing discrete finite impulse response (FIR) filters.

In the present context, we can hardly to do justice to the vast research and engineering literature on spectral estimation [28–31].

9.2.1 Windowing

Let us experiment with a simple discrete sinusoid $x(n)$ and the task of computing its discrete Fourier transform (DFT) on a *window*—that is, over a restricted set of values. Three problematic cases emerge:

(i) Alignment of the DFT samples with the signal's spectrally significant portion;

(ii) Signal features that appear to an initial interpretation as frequency characteristics, but in fact arise from wholly different reasons—for example, the presence of an edge;

(iii) Proper sizing of the DFT for the samples.

The first two points affect one another.

9.2.1.1 Alignment and Edges. Let us consider a sinusoidal pulse and its frequency analysis on slices of varying alignments with the pulse event (Figure 9.14).

Windowing involves weighting the samples from the signal slice by window function values before computing the spectrum. We might speculate that Fourier magnitude spectra would be better represented by weighting the central values more

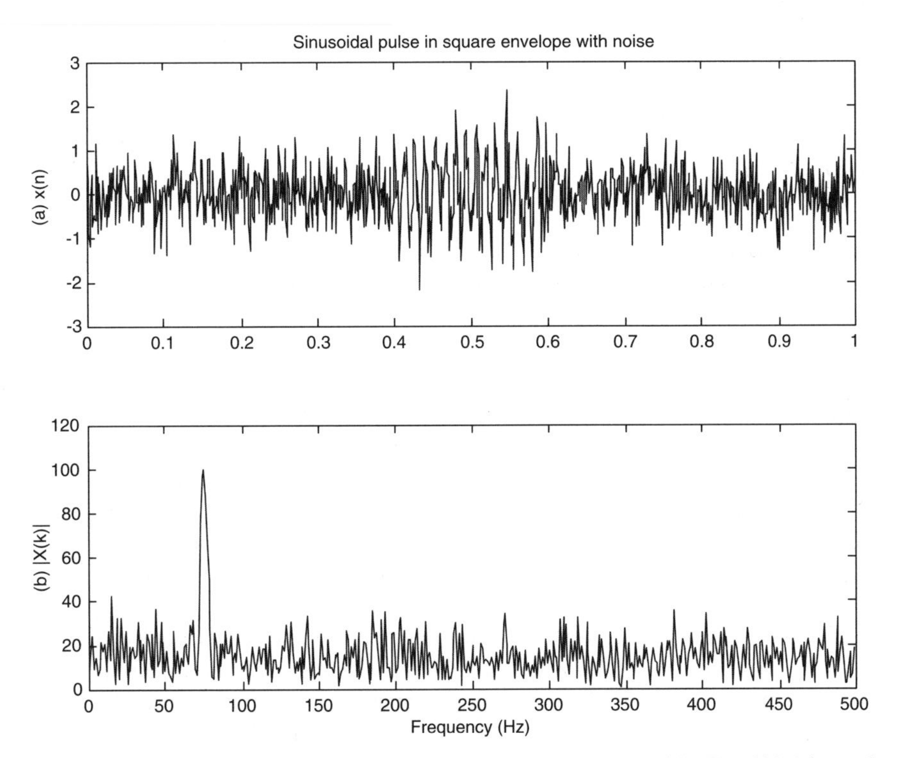

Fig. 9.14. Sinusoidal pulse in square envelope (75 Hz, 200 samples wide, T = .001 s) in moderate noise (top) and its magnitude spectrum (bottom).

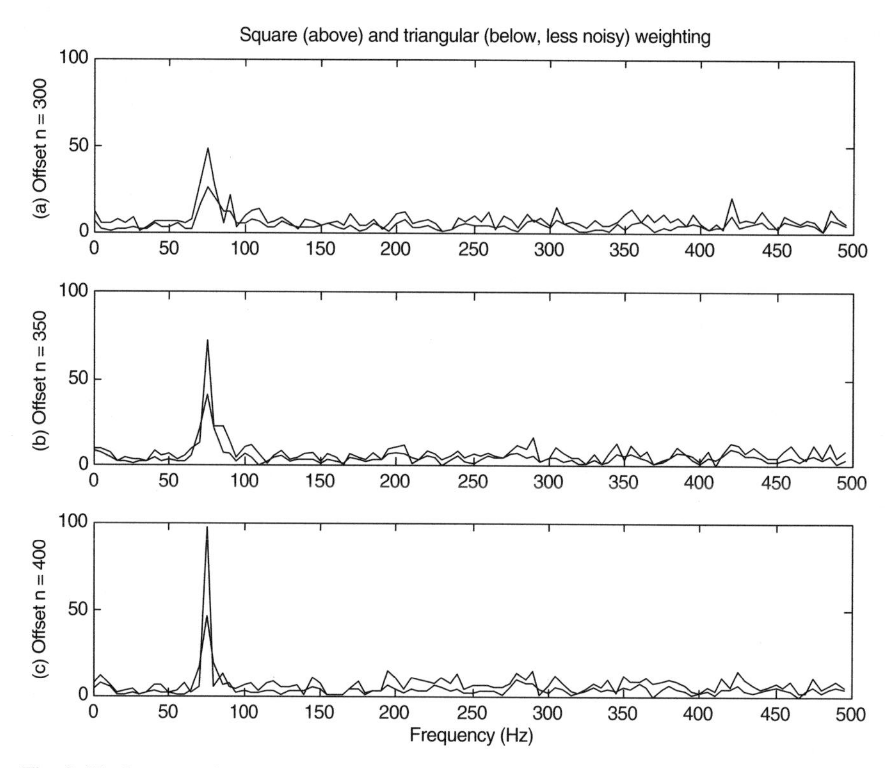

Fig. 9.15. Square pulse magnitude spectra, same DFT order ($N = 200$) at three different offsets: $n = 300$, $n = 350$, and $n = 400$ (full alignment).

than the peripheral values with a time slice from an input signal $x(n)$. The next experiment (Figure 9.15) shows the result of computing magnitude spectra for square and for triangular-weighted window functions.

The main effect of misalignment of the DFT window with the signal oscillations is a blurring of the magnitude spectrum spike. Improving the alignment—clearly—produces a more distinct spike, and invoking a weighting function (a triangular window) in this case offers only modest improvement.

In many applications, preliminary time-domain segmentation helps avoid this problem. Applications can detect signal edges early and use them to align spectral analysis windows. Sometimes edge detection can be based on signal level changes, but in other cases what constitutes an edge is a change in frequency content.

9.2.1.2 Window Size. Now we turn to another anticipated difficulty. Suppose that the signal slice for DFT computation aligns with the oscillation-bearing part of the signal, but the order of the Fourier transformation is a poor choice for the underlying frequency component. We know that a DFT of order N on data $x(n)$ sampled at $F_s = 1/T$ Hz will have frequency resolution $(NT)^{-1}$ and cover discrete frequencies $(NT)^{-1}$, $2(NT)^{-1}$, ..., $(2T)^{-1}$ as long as $x(n)$ is real-valued. Adding a pure sinusoid of one these frequencies—say $\omega_k = k(NT)^{-1}$, for $1 \le k \le N/2$—to $x(n)$ will alter only $X(k)$ and $X(N-k)$ [32]. Superimposing onto $x(n)$ a sinusoid of frequency $\omega \ne \omega_k$, for any $1 \le k \le N/2$, will perturb all of the $X(k)$. The difference caused by adding the sinusoid diminishes in magnitude like $1/|\omega - \omega_k|$ as $|\omega - \omega_k|$ increases (exercise).

9.2.2 Windowing Methods

Windows are weighting functions that attenuate signals at their discontinuities. When we cut out a piece of a signal and use it to compute its DFT, this effectively periodicizes the signal. The problem is that the signal's early values might differ greatly from the later values in the excised portion. So the effective periodic signal has a huge discontinuity, and this creates large spectral components that are due to the time slicing rather than the trend of the original signal.

The remedy is to suppress the signal slice at its boundaries with a window function. Window functions also serve as a tool for constructing FIR filters. There are a variety of window functions [7–9, 26, 33]:

- The *rectangular* window takes raw signal values without shaping them.
- The *Bartlett*[2] or *triangular* window weights them linearly away from the center.
- The *Hann*[3] window, sometimes called a "Hanning" window, is a modified cosine weighting function.
- The *Hamming*[4] window is also a modified cosine window.
- The *Blackman*[5] window is another modified cosine window.
- The *Kaiser*[6] window uses a Bessel function for shaping the signal slice.

It seems that throughout our experimentation in Section 9.1, we employed the rectangular window. For comparison, Table 9.2 lists the window functions. Note that the window domains $|n| \le \frac{N-1}{2}$ of Table 9.2 are convenient for applications not needing causal filters, such as off-line signal analysis tasks. It is also the form that we will use for the analog windows in the next chapter. Since the windows are zero outside this interval, as linear, translation-invariant system impulse responses, the window functions are all weighted moving average filters. They remove high frequencies and preserve low frequencies when convolved with other discrete signals.

[2]After M. S. Bartlett, who used this window to estimate spectra as early as 1950.

[3]Austrian meteorologist Julius von Hann introduced this window. At some point, perhaps due to confusion with Hamming's similar window or to the use of the term "hann" for the cosine windowing technique in general (as in Ref. 32), the name "Hanning" seems to have stuck.

[4]Richard.W. Hamming (1915–1998) used this window for improved signal spectral analysis, but the American mathematician is more widely known for having invented error correcting codes (*The Bell System Technical Journal*, April 1950).

[5]After Hamming's collaborator at Bell Telephone Laboratories, Ralph B. Blackman (1904–).

[6]Introduced by J. F. Kaiser of Bell Laboratories in 1974.

TABLE 9.2. Window Functions for $N > 0$ Samples[a]

Name	Definition
Rectangular	$w(n) = \begin{cases} 1 & \text{if } \lvert n\rvert \le \dfrac{N-1}{2} \\ 0 & \text{otherwise} \end{cases}$
Bartlett (triangular)	$w(n) = 1 - \dfrac{2\lvert n\rvert}{N-1} \quad \text{if } \lvert n\rvert \le \dfrac{N-1}{2}$
Hann	$w(n) = \dfrac{1}{2}\left[1 - \cos\dfrac{2\pi n}{N-1}\right] \quad \text{if } \lvert n\rvert \le \dfrac{N-1}{2}$
Hamming	$w(n) = 0.54 - 0.46\cos\dfrac{2\pi n}{N-1} \quad \text{if } \lvert n\rvert \le \dfrac{N-1}{2}$
Blackman	$w(n) = 0.42 + 0.5\cos\dfrac{2\pi n}{N-1} + 0.08\cos\dfrac{4\pi n}{N-1} \quad \text{if } \lvert n\rvert \le \dfrac{N-1}{2}$
Kaiser	$w(n) = \dfrac{I_0\left(\alpha\sqrt{1-\left(\dfrac{2n}{N-1}\right)^2}\right)}{I_0(\alpha)} \quad \text{if } \lvert n\rvert \le \dfrac{N-1}{2}$

[a]The table defines the windows as centered about $n = 0$. Outside the specified ranges, the windows are zero. It is straightforward to shift them so that they are causal [7]. The Kaiser window is defined in terms of the zeroth-order Bessel[7] function of the first kind (9.23) and a parameter α given below (9.24).

The summation

$$I_0(t) = 1 + \sum_{n=1}^{\infty}\left[\frac{1}{n!}\left(\frac{t}{2}\right)^n\right]^2 \tag{9.23}$$

defines the Bessel function. There is a standard formula [33] for the Kaiser window parameter α. To ensure a Kaiser window whose Fourier transform suppresses high-frequency components to more than $-\Delta$ dB, set

$$\alpha = \begin{cases} 0.1102(\Delta - 8.7) & \text{if } \Delta > 50, \\ 0.5842(\Delta - 21)^{0.4} + 0.07886(\Delta - 21) & \text{if } 50 \ge \Delta \ge 21, \\ 0 & \text{if } 21 > \Delta. \end{cases} \tag{9.24}$$

[7]Professor of astronomy, mathematician, and lifelong director of the Königsberg Observatory, Friedrich Wilhelm Bessel (1784–1846) devised the functions bearing his name for analyzing the motions of three bodies under mutual gravitation.

9.2.3 Power Spectrum Estimation

The Fourier transform magnitude spectrum has some important drawbacks. Our introductory digital telephony control tones application showed that—under moderate noise—measuring the contributions of frequency components by the relative magnitude of Fourier transform coefficients is effective.

To understand how this comes about, we have to consider signal noise in a mathematically tractable form.

9.2.3.1 *Power Spectral Density.* Let $\mathbf{x}$ be a discrete random signal; that is, $\mathbf{x} = \{x_n: n \in \mathbb{Z}\}$ is a family or *ensemble* of *random variables* (Chapter 1). This is an abstract formulation. What it means is that if a signal $x(n)$ has a random nature, then we do not know exactly what value it may take at any particular time instant $n \in \mathbb{Z}$. But we at least know that the values $x(n)$ might assume at $n = k$, for example, are given by a random variable, namely $x_k \in \mathbf{x}$. So by a random signal, we understand a signal that is random at all of its measured time instants; it is indeed an ensemble of random variables.

But that is not to say that we know nothing about the random signal $\mathbf{x}$. Associated with each random variable $r = x_n \in \mathbf{x}$ for some $n \in \mathbb{Z}$ is a *probability distribution function* F_r and a *probability density function* f_r such that $F_r(s) = P(r \le s)$, the probability that r does not exceed $s \in \mathbb{R}$. Moreover,

$$F_r(s) = \int_{-\infty}^{s} f_r(t)\, dt, \tag{9.25}$$

which is to say that $\frac{\partial}{\partial s}F_r(s) = f_r(s)$. To the skeptically inclined individual, these are almost incredible conditions, but they do approximate naturally occurring random signals fairly well. In any case, we need them for the theoretical development.

The distribution and density functions allow us to describe random variables with averages. If $r = x_n \in \mathbf{x}$ again, then we define its *mean*

$$\mu_r = \int_{-\infty}^{\infty} t f_r(t)\, dt = E[r] \tag{9.26}$$

and *standard deviation* σ_r, the square root of the *variance*: $\sigma_r^2 = E[r^2] - \mu_r^2$.

Generalizing for two random variables, $u = x_n$ and $v = x_m$ in $\mathbf{x}$, we assume a *joint distribution function* $F_{u,v}(s, t) = P(u \le s \text{ and } v \le t)$ and *joint density function* $\frac{\partial^2}{\partial s \partial t}F_{u,v}(s, t) = f_{u,v}(s, t)$. If $E[uv] = E[u]E[v]$, then random variables u and v are *uncorrelated* or *linearly independent*.

Power spectrum estimation studies how signal power distributes among frequencies. For finite-energy deterministic signals $x(t)$, the power of $x(t)$ in the (unsigned) band $0 < \Omega_0 < \Omega_1 < \infty$ comes from integrating $|X(\Omega)|^2$ over $[-\Omega_1, -\Omega_0] \cup [\Omega_0, \Omega_1]$.

But for random signals, the mathematical analysis depends on a special class of signals $x(t)$ that obey the following two conditions:

(i) $E[x(t)]$ does not depend on the process variable $t \in \mathbb{R}$.
(ii) $E[x(t)x(t+\tau)]$ is a function of τ and does not depend on $t \in \mathbb{R}$.

Such signals are called *wide-sense stationary* (WSS) [34]. We define the *autocorrelation* for a WSS random signal $x(t)$ to be $r_{xx}(\tau) = E[x(t)x(t+\tau)]$. It is easy to show that $E[x(t)x(s)] = r_{xx}(t-s)$ and that $r_{xx}(\tau)$ is an even signal. A special type of WSS random signal $x(t)$ has an autocorrelation function that is an impulse: $r_{xx}(\tau) = A\delta(t)$ for some constant $A \in \mathbb{R}$. This means that signal values are completely uncorrelated with their neighbors. Such random signals are called *white noise* processes; we shall explain this colorful terminology in a moment.

In order to study the spectrum of a noisy signal, we have to limit its time-domain extent. So for $L > 0$ let us define the *localization* of random signal $x(t)$ to $[-L, L]$:

$$x_L(t) = \begin{cases} x(t) & \text{if } -L \le t \le L, \\ 0 & \text{if otherwise.} \end{cases} \tag{9.27}$$

so that

$$X_L(\Omega) = \int_{-\infty}^{\infty} x_L(t) e^{-j\Omega t} dt = \int_{-L}^{L} x(t) e^{-j\Omega t} dt. \tag{9.28}$$

The energy of x_L is $\|x_L\|_2^2 = \frac{1}{2\pi}\|X_L\|_2^2 = \frac{1}{2\pi}\int_{-\infty}^{\infty}|X_L(\Omega)|^2 d\Omega$ by Parseval's identity.

The approximate energy of x_L in a narrow signed frequency band, $\Delta(\Omega) = \Omega_1 - \Omega_0$, is thus $|X_L(\Omega)|^2\Delta(\Omega)$. Since frequency is the reciprocal of time, $|X_L(\Omega)|^2/(2L)$ has units of energy, which is the product of power and time, or power divided by frequency. Therefore, we may define the *power spectral density* (PSD) for $x_L(t)$ to be $\frac{|X_L(\Omega)|^2}{2L}$. This is a random variable, and its expectation is $E\left[\frac{|X_L(\Omega)|^2}{2L}\right]$. It is tempting to define the PSD of $x(t)$ as the large time window $[-L, L]$ limit of such expectations:

$$X_{\mathrm{PSD}}(\Omega) = \lim_{L \to \infty} E\left[\frac{|X_L(\Omega)|^2}{2L}\right]. \tag{9.29}$$

But some caution is in order. We need to know that the limit (9.29) exists. A famous result shows that the desired limit operation is valid and moreover provides a way to compute it. The Wiener[8]–Khinchin[9] theorem, says that if $x(t)$ is a real-valued, WSS random signal with autocorrelation $r_{xx}(t) \in L^1(\mathbb{R})$, then

$$X_{\mathrm{PSD}}(\Omega) = R_{xx}(\Omega) = \int_{-\infty}^{\infty} r_{xx}(t) e^{-j\Omega t} dt. \tag{9.30}$$

While the exact values of a random signal are not known, it is a reasonable assumption that the autocorrelation of the signal is available. Indeed, the autocorrelation will tend to resemble a narrow pulse when local signal values $x(t + \tau)$ correlate poorly with a particular $x(t)$, and it will look like a broad pulse when $x(t + \tau)$ as a trend repeats $x(t)$. In any case, for τ large, $r_{xx}(\tau)$ diminishes, and we can often assume a mathematically tractable model for the autocorrelation. For example, from basic physical considerations, we can derive a model for the thermal noise across a resistor in an electric circuit. A purely theoretical example is the aforementioned white noise process. The Wiener–Khinchin theorem implies that the white noise process $r_{xx}(\tau) = A\delta(t)$ has $X_{\mathrm{PSD}}(\Omega) = A$, for $A \in \mathbb{R}$. Thus, its frequency spectrum is flat; it contains all "colors," as it were, and is therefore "white." It turns out that white noise models the thermal noise across a resistor and that its autocorrelation scales according to the absolute temperature of the circuit elements.

Similar ideas work for discrete random signals. If $x(n) \in l^2$ is a discrete deterministic signal with DTFT $X(\omega)$, then integrating $|X(\omega)|^2$ over $[-\omega_1, -\omega_0] \cup [\omega_0, \omega_1]$ gives the power in the band $0 < \omega_0 < \omega_1 < \pi$. A *wide-sense stationary* (*WSS*) *discrete* random signal satisfies the following:

(i) $E[x(n)]$ does not depend on the process variable $n \in \mathbb{Z}$.

(ii) $E[x(n)x(n + \nu)]$ is a function of ν and does not depend on $n \in \mathbb{Z}$.

The autocorrelation for a WSS random signal $x(n)$ is $r_{xx}(\nu) = E[x(n)x(n + \nu)]$. Again, $E[x(n)x(m)] = r_{xx}(n - m)$ and $r_{xx}(\nu)$ is symmetric about $\nu = 0$. Toward analyzing the power spectrum, for $L > 0$ we define

$$x_L(n) = \begin{cases} x(n) & \text{if } -L \le n \le L, \\ 0 & \text{if otherwise.} \end{cases} \tag{9.31}$$

[8]First-generation American mathematician Norbert Wiener (1894–1964) finished the doctoral program at Harvard at age 18, concentrating on philosophy of mathematics and logic. In the tradition of Plato, the great English scholar Bertrand Russell hoped to improve Wiener's philosophical insights by having him study more mathematics. But later encounters with G. H. Hardy, D. Hilbert, and E. G. H. Landau nudged the prodigy toward mathematical analysis. After some peregrination, Wiener took a ground-floor job as a mathematics instructor at the Massachussetts Institute of Technology. He eventually arose to full Professor, contributed substantially to statistical communication and control theory, and remained at MIT for the rest of his career.

[9]Soviet mathematician Aleksandr Yakovlevich Khinchin (1894–1959) established much of the early theory of stationary stochastic processes. The author of some 150 papers, he took a mathematics professorship at Moscow State University in 1927. He was a patron of the arts and theater. Election to the Soviet Academy of Sciences (1939) recognized Khinchin's contributions to ranging from probability, number theory, information theory, statistical physics, and quantum mechanics.

The Fourier spectrum of the localized random signal is

$$X_L(\omega) = \sum_{n=-L}^{L} x(n)e^{-j\omega n} = \sum_{n=-\infty}^{\infty} x_L(n)e^{-j\omega n}. \tag{9.32}$$

We define the PSD for $x_L(n)$ to be

$$X_{\mathrm{PSD}}(\omega) = \lim_{L\to\infty} E\left[\frac{|X_L(\omega)|^2}{2L+1}\right]. \tag{9.33}$$

There is a discrete version of the Wiener–Khinchin theorem. If $x(n)$ is a real-valued, WSS random signal with an absolutely summable autocorrelation function $r_{xx}(n)$, then

$$X_{\mathrm{PSD}}(\omega) = R_{xx}(\omega) = \sum_{n=-\infty}^{\infty} r_{xx}(n)e^{-j\omega n}. \tag{9.34}$$

Thus, for both analog and discrete random variables we are justified in defining the power spectral density, and it can be computed as long as the associated autocorrelation function is respectively L^1 or l^1. The exercises outline the proofs of both the analog and discrete Wiener–Khinchin theorems.

9.2.3.2 Periodogram. Now we consider approximating the power spectral density. The oldest and most straightforward approach is to compute the discrete time Fourier transform on a local time window $[-L, L]$ of sampled data points $x(n)$. Thus, we have

$$\tilde{X}_{L,\,\mathrm{PSD}}(\omega) = \frac{1}{2L+1}|X_L(\omega)|^2 = \frac{1}{2L+1}\left|\sum_{n=-L}^{L} x(n)e^{-j\omega n}\right|^2. \tag{9.35}$$

Generally, we would take $\omega = 2\pi k/T$ for $-L \le k \le L$ and compute (9.35) on a discrete set of frequencies. After all, although we used the discrete Fourier transform magnitude spectrum in the application examples of Section 9.1, we could have equally well used the squared magnitude spectrum. Also, due to the periodicity of the discrete Fourier transforms, we could equally well shift the local window of $x(n)$ values. In practice, a window of width $2M$ is chosen to enable a fast Fourier transform computation. In any event, (9.35) is a statistical estimator for the random variable $X_{\mathrm{PSD}}(\omega)$. The question before us is how well—for a particular frequency of interest, $-\pi < \omega \le \pi$—the estimate of $|X(\omega)|^2$ over $N = 2L + 1$ samples of noisy $x(n)$ compares to the actual power at that frequency.

Briefly, the problem with the estimated power spectrum $\tilde{X}_{\mathrm{PSD}}(\omega)$ is twofold:

(i) As the number of samples is increased, the mean of the estimate does not approach the actual mean; it is a *biased estimator.*

(ii) As the number of samples is increased, the variance of the estimate does not approach zero; it is an *inconsistent estimator.*

Signal processing [7] and spectrum estimation [29–31] texts explain this theory. Unfortunately, the development would drag us away from our signal analysis focus. We would instead like to emphasize that the problems of the periodogram as an estimator of the PSD can be addressed by applying the window functions we developed earlier in this section along with some straightforward averaging techniques.

9.2.3.3 Periodogram Improvement. Fortunately, there are some easy ways to improve the periodogram estimate $\tilde{X}_{L,\,\mathrm{PSD}}(\omega)$ of (9.35). We cover some classic methods that use no model of the signal, its spectrum, or its autocorrelation. These *nonparametric* techniques include:

- Bartlett's method smoothes the time-domain data by breaking the interval into smaller, equally sized segments and averaging the periodograms computed for each segment [35].
- Welch's algorithm smoothes the time-domain data by breaking the interval into smaller, equally sized segments, applying a window function to each segment, and allowing the windows to overlap [36].
- Another technique, due to Blackman and Tukey [37], relies directly on the Wiener–Khinchin theorem's identification of the PSD with the Fourier transform of the autocorrelation function.

Bartlett's method divides a set of $N = K \times M$ data points of $x(n)$ into K subwindows of length M. Thus, the signal values on subwindow k are $x_k(m) = x(kM + m)$, where $0 \le k \le K - 1$ and $0 \le m \le M - 1$. For each such subwindow we set

$$\tilde{X}_{k,\,\mathrm{PSD}}(\omega) = \frac{1}{M}\left|\sum_{m=0}^{M-1} x_k(m)e^{-j\omega m}\right|^2, \tag{9.36a}$$

and then average them all to get the estimate over $[0, N-1]$:

$$\tilde{X}_{N,\,\mathrm{PSD}}(\omega) = \frac{1}{K}\sum_{k=0}^{K-1} \tilde{X}_{k,\,\mathrm{PSD}}(\omega), \tag{9.36b}$$

The Welch algorithm improves upon the Bartlett method by

(i) Allowing the subwindows to overlap.

(ii) Applying a window function to the individual PSD estimates on the subwindows. The window function can be any of those described in Table 9.2.

The steps in the Blackman–Tukey algorithm are as follows:

(i) From a body of measured noisy signal data, the autocorrelation function for the random process is estimated.

(ii) One of the typical window functions—for example, the Hann window—is applied to the autocorrelation estimate.

(iii) The discrete Fourier transform is applied to windowed autocorrelation values.

Another class of periodogram improvement algorithms—called *parametric* methods—make a model for the noisy signal data. The idea is to assume that the signal arises from a linear system excited by white noise. The exercises cover the concept of noisy inputs to linear systems.

9.2.4 Application: Interferometry

An application that involves the precise estimation of signal frequency and phase is interferometry, which is based on the wave nature of electromagnetic radiation [38]. In interferometry, an input optical signal contains light combined from two different sources—for example, reflected from two different surfaces. If the original source of both reflecting beams is coherent (that is, the light waves are in phase with one another, such as from a laser), then the resulting interferogram will contain peaks and valleys of intensity, depending on the path distance of the component light waves. Of course, moving one reflecting surface by a wavelength amount produces the same light combination, and so the intensity only indicates relative changes in position between the two reflecting surfaces.

The technique enables us to measure minute differences in distance. Peaks in the interferogram correspond to when the peak of one sinusoidal wave matches up with the peak of the other. This is the length of the wave; and in the case of light, this value is quite small, from about 400 nm (violet) to 700 nm (red). Thus, optical interferometry is used in precision measurement and manufacture, such as semiconductor integrated circuit fabrication.

We consider a semiconductor manufacturing and control application of interferometry involving chemical mechanical planarization (CMP) of silicon wafers [39]. CMP has become an important process for ensuring the planarity of the wafer surface. A high degree of flatness eliminates flaws in later deposition steps. More importantly for modern integrated circuit manufacture, CMP is used for selectively removing thin layers of material on wafers that do not etch well in plasmas, such as copper.

Evident in the signal trace (Figure 9.16) at the top are:

(i) The long-term undulations in reflected intensity due to the combination of beams reflected from the surface and Si/SiO_2 interface;

(ii) Short-term vibrations around $f = 0.1$ Hz;

(iii) At the end of the trace, a change in process conditions causing wild instability of the sensor's measured reflectance.

Our interferometric interest is to isolate these three trends. The wavelength of the long-term oscillation, on the order of 0.0085 Hz, will be used to estimate the silicon oxide removal rate (Figure 9.17). The short-term 0.1 Hz oscillation can be removed with a notch filter in order to enhance the estimation of the removal rate. Also, with the short-term oscillation removed by a notch filter, it becomes possible to design a simple algorithm to compute the phase of the long-term reflectance oscillation and use this

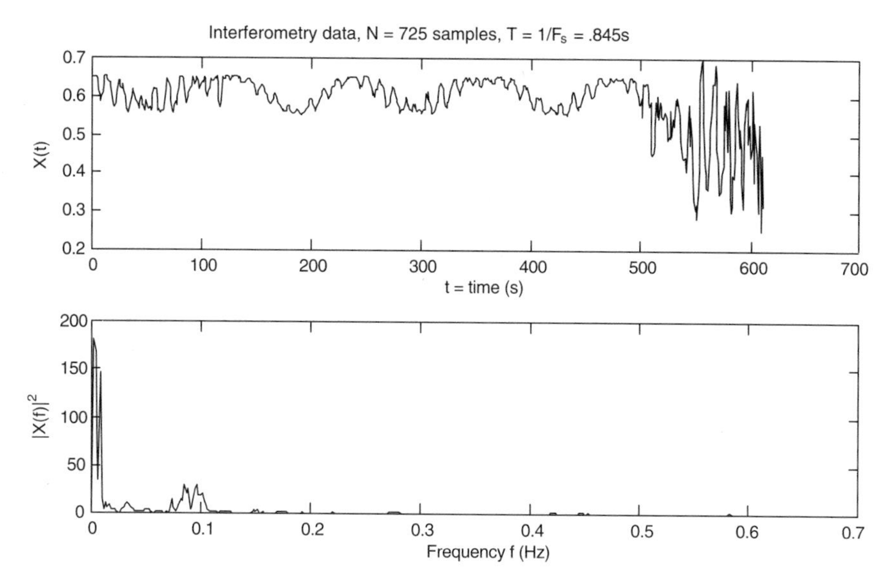

Fig. 9.16. Reflectance data from a CMP removal process on a silicon wafer with a surface silicon dioxide film. The upper panel contains the signal trace and the lower panel contains the magnitude spectrum.

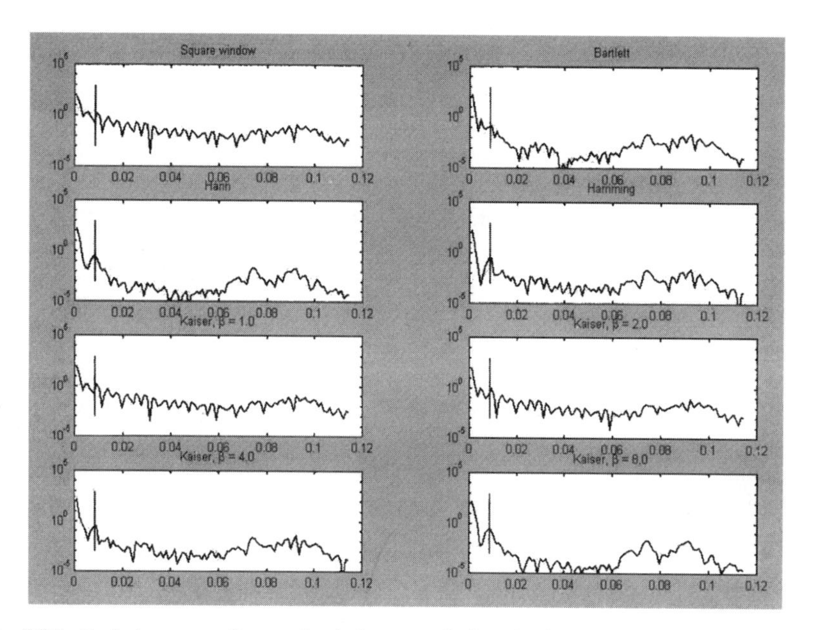

Fig. 9.17. Periodograms of several windows applied to the interferometic data from CMP processing of a silicon wafer. Top four periodograms: Square, Bartlett, Hann, and Hamming windows applied to signal before periodogram calculation. The bottom four plots show the efficacy of various Kaiser window parameters. The vertical line marks the frequency of the interference fringes.

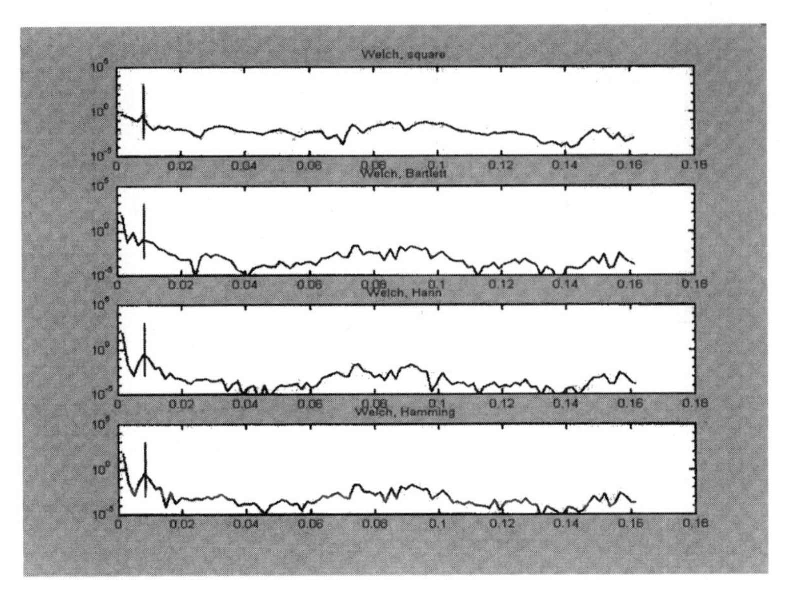

Fig. 9.18. Welch's method for periodogram computation.

as a measurement of the amount of top-level film removed. Finally, we are interested in detecting the signal instability at the end of the trace, which is an indication that the oxide has begun to disappear from the wafer, exposing the silicon substrate. This represents polish endpoint.

Let us also investigate one of the spectrogram improvement methods—in this case Welch's method with windows overlapped 50% and a variety of window functions applied before computing the local periodograms. The results are shown in Figure 9.18.

As a practical matter, some goal-directed information needs to be invoked in this example. A peak detector needs to be provided with a limited range in hertz for its search. Once the periodicity of the interferogram fringes is determined from the periodogram algorithm, it is possible to calculate the phase of individual points along the fringe trace. To do this, we would perform a discrete Fourier transform of the proper window size according to the wavelength of the long-term undulations. For the above sampling interval $T = 0.8451$ s, this would be $N = 140$ samples. The phase of points along the interference fringe trace could be computed by the complex argument of the first DFT coefficient, for example.

9.3 DISCRETE FILTER DESIGN AND IMPLEMENTATION

Discrete filters are suitable for computer implementation, and signal analysis applications depend on them directly. This section reviews some elements of filter theory, the z-transform, and ideal filters. It also covers the more practical aspects such as (a) filter approximation and (b) the steps in designing a discrete filter, and it explains basic methods on how to implement discrete filters.

The principal tools for filter design are the discrete Fourier transform (DFT), the discrete-time Fourier transform (DTFT), and the z-transform. One important design method is to derive a discrete filter from an analog filter. So we shall also use the continuous-domain Fourier transform as well as introduce briefly the Laplace transform. The Fourier transform properties allow us to convert one type of filter into another, considerably simplifying the mechanics of filter construction. For example, we generally design a low-pass filter and then convert it into the required bandpass filter.

9.3.1 Ideal Filters

Some applications, such as in the dual-tone multifrequency application above, require fine separation in the frequency domain. Some periodicities we need to pass through for further analysis, such as the DTMF, range from 697 to 1633 Hz. Others, such as high-frequency background noise and low-frequency interference, we prefer to suppress. Offline applications can Fourier transform large time-slices of data and select spectral components according to the frequency resolution of the transform. Online applications, though, must achieve frequency selection as data enters the system, in real time, and then pass the output to interpretation algorithms. Here the filters have to be causal and efficiently implemented. Since no high-resolution Fourier transform of the input data is possible in this situation, it becomes all the more important to design filters that distinguish between nearby frequencies.

9.3.1.1 Low Pass. We can eliminate all frequencies above the range of interest to an application by processing input signals through an ideal low-pass filter H. It is easy to describe such a filter using the discrete-time Fourier transform $H(\omega)$:

$$H(\omega) = \begin{cases} 1 & \text{if } |\omega| \le \omega_c, \\ 0 & \text{if otherwise,} \end{cases} \tag{9.37}$$

where ω_c is the *cutoff frequency* of the filter. This filter's perfect cutoff is ideal for separating one frequency from another. The only caveat is that a sufficiently high sampling rate must be chosen so that ω_c is close to the frequency of a discrete component.

The inverse DTFT serves as a tool for building discrete filters from such frequency-domain descriptions. The impulse response of the ideal low-pass system (Figure 9.19) is

$$h(n) = \frac{1}{2\pi}\int_{-\pi}^{\pi} H(\omega)e^{j\omega n}d\omega = \frac{1}{2\pi}\int_{-\omega_c}^{\omega_c} e^{j\omega n}d\omega = \frac{\sin(\omega_c n)}{\pi n}. \tag{9.38}$$

Using the inverse DTFT to generate a filter impulse response from description of its ideal frequency-domain representation is elegant and straightforward, but the resulting filters are often—and most particularly in the present case—quite impractical.

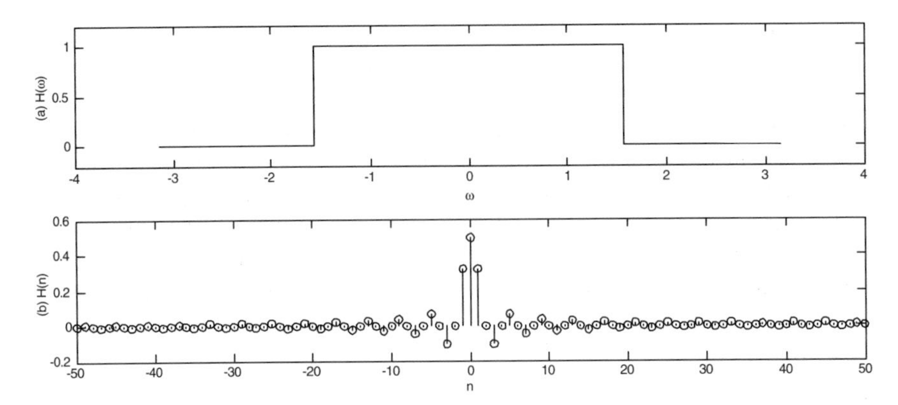

Fig. 9.19. Ideal low-pass filter in the frequency-domain (a) and truncation of impulse response in the time-domain (b).

The problems with $h(n)$ as given by (9.38) are as follows:

(i) It is an infinite impulse response (IIR) filter.

(ii) This means that it cannot be implemented by a straightforward convolution in applications.

(iii) Some IIR filters permit a recursive computation for their implementation, but in this case $h(n) \neq 0$ for arbitrarily large magnitude $n < 0$, so it is in fact unrealizable for applications.

(iv) The filter is not causal, so its realization requires future signal values and cannot work on real-time data streams.

(v) The filter is not stable, since its impulse response is not absolutely summable (Chapter 2).

But these problems are not catastrophic. We can truncate $h(n)$ so that it is supported within some reasonable interval $[-N, N]$, say. This is often satisfactory for applications where the raw signals contain high-frequency background noise. The perfect frequency-domain characteristic is lost (Figure 9.19b). We will see how to overcome these problems in a moment.

Next, however, let us look at ideal filters of other types and see how to build them out of low-pass filters.

9.3.1.2 Other Filters.

Applications require filters with varying frequency-domain characteristics. For example, the DTMF application of Section 9.1 uses a low-pass filter, a simple moving averager, to suppress background noise. In the filter output, this highlights the relatively pure dual tones and improves thresholding results. But an online version of the same application would need filters that select frequency ranges and reject oscillations below and above certain bands. Here we are speaking of a bandpass filter, and we have noted that an array of them in parallel operation can pass filtered signals to the application logic for tone classification.

The mechanics of converting a lowpass filter into one of an alternative frequency-domain behavior are fortunately quite straightforward. We require only a few DTFT properties (Chapter 7).

From a low-pass filter $h(n) = (H\delta)(n)$, it is easy to make a high-pass filter $f(n) = (F\delta)(n)$. An *all-pass* filter is the discrete impulse $\delta(n)$ or one of its shifted versions $\delta(n - n_0)$. We subtract $f(n) = \delta(n) - h(n)$ for a high-pass filter $f(n) = (G\delta)(n)$.

Two elementary methods exist for building a bandpass filter F. One way to do this, starting from a low-pass filter H with cutoff ω_c, is to use the frequency shifting property of the DTFT. If we set $f(n) = e^{j\omega_L n}h(n)$, then $F(\omega) = H(\omega - \omega_L)$. F is thus a bandpass filter with lower cutoff ω_L and upper cutoff $\omega_H = \omega_L + \omega_c$. The blemish on this otherwise elegant modification of $h(n)$ is the fact that $f(n)$ becomes a complex-valued discrete signal. A second basic technique is to combine a low-pass filter H, with cutoff ω_c, and a high-pass filter G, with cutoffs ω_L and ω_H. Typically, we get G itself by subtracting $h(n)$ from an all-pass system's impulse response. If our filters satisfy $\omega_c > \omega_L$, then the system composition $F(n) = G(H(n))$ will pass precisely the frequencies that lie in the passband overlap. We know that $f = F\delta = g*h$, the convolution of $g(n)$ and $h(n)$. So, if both $g(n)$ and $h(n)$ are real-valued, then $f(n) \in \mathbb{R}$ too. By the convolution theorem for discrete linear, translation-invariant systems, the DTFT of $f(n)$ is $F(\omega) = H(\omega)G(\omega)$. So, if H and G are ideal, then F will be a perfect bandpass filter with lower cutoff ω_L and upper cutoff ω_c.

We form band-reject or notch filters by subtracting a bandpass filter from an all-pass filter.

9.3.2 Design Using Window Functions

A useful discrete FIR filter design method uses the windowing concept from Section 9.2. This is a straightforward way to improve truncated perfect low-pass filters (Section 9.3.1.1). Again, let ω_c be the cutoff frequency so that $h(n)$ is given by

$$h(n) = \left. \frac{1}{2\pi}\int_{-\omega_c}^{\omega_c} e^{j\omega n}\, d\omega \right]_{-N \le n \le N} = \left. \frac{\sin(\omega_c n)}{\pi n} \right]_{-N \le n \le N} \tag{9.39}$$

as shown in Figure 9.20(a).

Let $w(n)$ be a window function, for example the Hann window, of Table 9.2. Set $g(n) = h(n)w(n)$ as in Figure 9.20c. Using signal multiplication in the time domain is equivalent to convolution in the frequency domain:

$$G(\omega) = \mathcal{F}[h(n)w(n)] = \frac{1}{2\pi}\int_{-\pi}^{\pi} H(\theta)W(\theta - \omega)\, d\theta. \tag{9.40}$$

Since discrete time Fourier transform of $w(n)$ has the shape of a weighted averaging function in the frequency domain, convolving it with $H(\omega)$ effectively smoothes the spectrum of $h(n)$. This blends away the Gibbs phenomenon ripples caused by the truncation of the perfect low-pass filter Figure 9.20b. The result is a magnitude spectrum almost completely devoid of the problematic ringing Figure 9.20d.

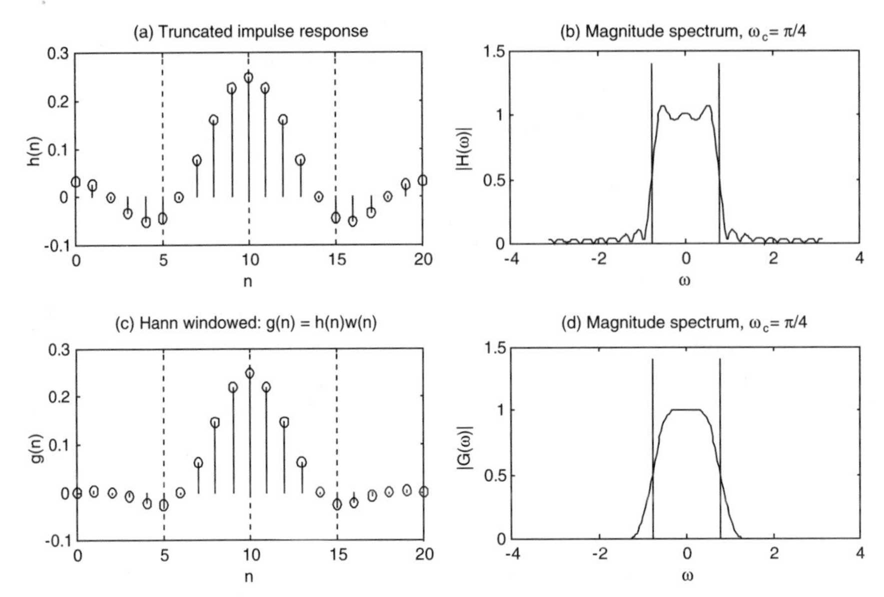

Fig. 9.20. FIR filter design using Hann window. Formerly perfect, but now truncated impulse response $h(n)$ (a). Vertical lines in the magnitude spectrum plot (b) mark the cutoff frequency. Panel (c) shows $g(n) = h(n)w(n)$, where $w(n)$ is a Hann window function having the same support as $h(n)$. The final panel (d) shows $|G(\omega)|$ with the Gibbs phenomenon ringing virtually eliminated.

9.3.3 Approximation

This section covers the approximation of ideal filters, the initial step among filter design tasks.

9.3.3.1 Design Criteria. Ideal filters are impossible to implement on a digital computer, but a filter that comes close to ideal performance is usually adequate for a signal processing and analysis application. Let us look at the magnitude spectrum of a typical discrete filter—in this case a third-order low-pass elliptic filter, which we will study in Section 9.5. A plot of the magnitude response illustrates design criteria (Figure 9.21).

Referring to Figure 9.21, we assume that the filter H has been normalized so that it has at most unit gain.

- The *passband* is the region in which the filter in which the magnitude response is near unity. For a low-pass filter, this is an interval around the origin: $\omega \in [0, \omega_p)$.
- The *stopband* is the region in which $H(\omega)$ is near zero: $\omega \in (\omega_s, \pi]$ for the low-pass filter in the figure.
- Between the passband and the stopband lies the *transition band*: $\omega \in [\omega_p, \omega_s]$.

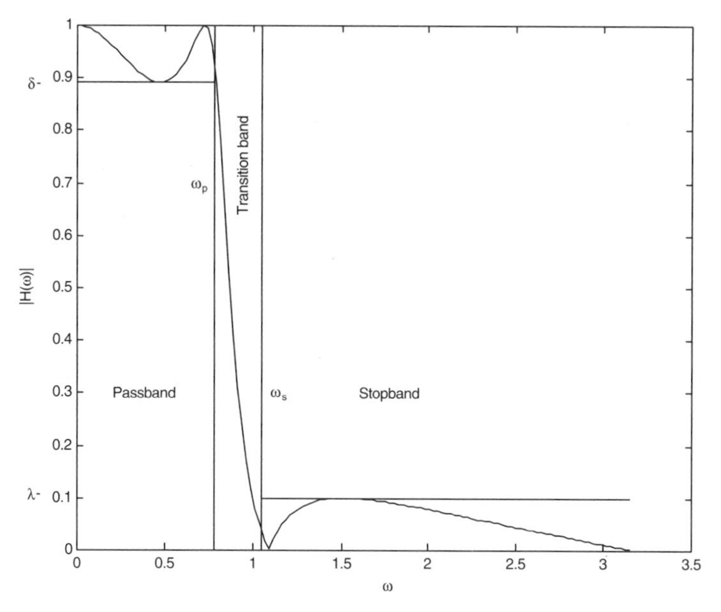

Fig. 9.21. Design characteristics of a low-pass filter (third-order low-pass elliptic), showing the positive frequencies $0 \le \omega \le \pi$.

- The *cutoff frequency* ω_c is a transition band value that is somewhat arbitrary, depending on the particular filter, but represents the frequency at which filter suppression begins. For our low-pass example, $\omega_p \le \omega_c \le \omega_s$.
- *Passband ripple* measures the variation of $|H(\omega)|$ within the passband. We typically specify a maximum value $\delta > 0$, such that $|H(\omega) - 1| \le \delta$ for $\omega \in [0, \omega_p)$. An equivalent, and sometimes convenient, mode of specification is to express tolerances in decibels (dB). Thus, $-\Delta \le 20\log_{10}|H(\omega)| \le \Delta$, for some $\Delta > 0$.
- *Stopband ripple* measures the variation of $|H(\omega)|$ within the stopband: There is a $\lambda > 0$, such that $|H(\omega)| \le \lambda$ for $\omega \in (\omega_s, \pi]$. Equivalently, in decibels: $20\log_{10}|H(\omega)| \le \Lambda$, for some $\Lambda < 0$.
- *Sharpness* indicates how narrow the transition band is. Applications that require fine frequency discrimination use filters with correspondingly sharp transition bands. One measure of sharpness is the average slope of $|H(\omega)|$ in the transition band. Another way to specify sharpness is to stipulate a derivative at a representative transition band value, such as ω_c or $(\omega_p + \omega_s)/2$.

The design criteria for other filters are similar. Thus, in a high-pass filter, the passband is at the upper end of the spectrum: $\omega_p \le \omega \le \pi$. In a bandpass filter, there are two stop bands (in the neighborhoods of 0 and π), and there is a single passband. Likewise, the notch filter has two outer passbands and a single inner stopband.

9.3.3.2 Design Steps. In signal analysis, our design steps for discrete filters are as follows:

(i) Determine the type (low-pass, high-pass, bandpass, or band-reject) of filter G appropriate to the application.
(ii) Select a low-pass filter that approximates an ideal filter's frequency-domain characteristics.
(iii) Design a suitable low-pass filter $h(n) = (H\delta)(n)$.
(iv) Convert the low-pass filter H to the application filter $g(n) = (G\delta)(n)$ using Fourier transform properties, ensuring that the approximation constraints on G are not compromised by the conversion.
(v) For an online application, maintain causality.
(vi) For IIR filters, maintain stability.
(vii) Achieve an efficient computer implementation.

It is sometimes feasible to design G directly, skipping the low-pass starter filter H and conversion. Once the filter type and approximation questions have been resolved in accord with application constraints—points (i) and (ii), above—the next step is to design a low-pass filter. The z-transform is the basic tool for working out these remaining design steps.

9.3.4 *Z*-Transform Design Techniques

We recall from the previous chapter that the *z-transform* $X(z)$ of a discrete signal $x(n)$ is given by

$$X(z) = \sum_{n=-\infty}^{\infty} x(n)z^{-n}, \tag{9.41}$$

where $z \in \mathbb{C}$. We recognize that if $z = e^{j\omega}$, then (9.41) is the discrete-time Fourier transform of $x(n)$. Evaluating the z-transform on the complex unit circle, $|z| = 1$, produces the DTFT. To avoid confusion, we may occasionally use the function notations $(\mathcal{Z}h)(z)$ or $H_z(z)$ for the z-transform and continue to use $H(\omega)$ for the DTFT. The z-transform came to signal processing from control theory in the early 1960s [40]. It is a staple topic of discrete signal processing books (for example, Refs. 7–11), and there are specialized texts covering the z-transform [41, 42].

The sum (9.41) converges absolutely on annular regions of the extended complex plane $\mathbb{C}^+ = \mathbb{C} \cup \{\infty\}$, called the *region of convergence*, ROC_X. The z-transform expression $X(z)$ for a signal is not unique; one must also specify ROC_X. Both the causal signal $x(n) = a^n u(n)$ and the anti-causal signal $y(n) = -a^n u(-n-1)$, for example, have the same z–transforms: $X(z) = Y(z) = z/(z-a)$. The difference is that $\mathrm{ROC}_X = \{z \in \mathbb{C}\colon |a| < |z|\}$ while $\mathrm{ROC}_Y = \{z \in \mathbb{C}\colon |z| < |a|\}$.

A theoretical subtlety on the region of convergence concerns square-summable signals. We know that discrete signals $h \in l^2$ have DTFTs, but it is possible that $h \notin l^1$. In fact, the discrete Hilbert space l^2 is isomorphic to the continuous Hilbert space $L^2[-\pi, \pi]$, understanding that $x, y \in L^2[-\pi, \pi]$ are considered identical if they are equal except on a set of Lebesgue measure zero (Chapter 3). Thus, such signals $h(n)$ possess a discrete-time Fourier transform, even though ROC_H does not include the unit circle $|z| = 1$.

9.3.4.1 System Function, Stability, Causality. If H is a discrete linear, translation-invariant system, such as we might use for filtering a signal, then the z-transform $H(z)$ of its impulse response $h = H\delta$ is called the *system* or *transfer function* of H. The convolution theorem for the z-transform tells us that if $y = Hx = h*x$, then $Y(z) = H(z)X(z)$, where $X(z)$ is the z-transform of input $x(n)$, $Y(z)$ is the z-transform of output $y(n)$, and $H(z)$ is the system function.

Recall that a discrete LTI system H is *stable* if and only if the impulse response $h = H\delta$ is absolutely summable: $h \in l^1$. But if H is stable, then $h(n)$ has a DTFT $H(\omega)$. Since $H(\omega) = H_Z(e^{j\omega})$, ROC_H evidently contains the unit circle, $|z| = 1$. The converse is also true: If $H_Z(z)$ converges absolutely on a region that contains the unit circle, then H is stable (exercise).

A signal $x(n)$ is *right-sided* means that $x(n) = 0$ for $n < N \in \mathbb{Z}$. In this case, its z-transform $X(z)$ (9.41) contains at most a finite number of positive powers of z, and ROC_X is the exterior of a circle. If $N \geq 0$, then $\infty \in \mathrm{ROC}_X$, and $x(n)$ is *causal*; that is, $x(n) = 0$ for $n < 0$. Similarly, if $x(n) = 0$ for $n > N \in \mathbb{Z}$, then we say that $x(n)$ is a *left-sided* sequence. The ROC of a left-sided signal is the interior of a circle, omitting perhaps the origin $z = 0$.

An LTI system H is *causal* if its impulse response $h = H\delta$ is causal: $h(n) = 0$ for $n \leq 0$. Thus, $h(n)$ is right-sided, and ROC_X is the exterior of a circle. If H is causal and $y = Hx$, then

$$y(n) = \sum_{k=0}^{\infty} h(k)x(n-k); \tag{9.42}$$

$y(n)$ can be computed without using future values of the input signal.

9.3.4.2 Systems Governed by Difference Equations. A wide variety of discrete systems are defined by a difference equation:

$$y(n) + \sum_{k=1}^{N} a_k y(n-k) = \sum_{m=0}^{M} b_m x(n-m). \tag{9.43}$$

Note that (9.43) allows us to compute a new output $y(n)$ if we know the previous N output values, the previous M input values, and the current input value. Thus, for real-time applications, signal filtering prior to frequency domain analysis is often implemented using filters governed by a difference equation. Offline applications, of course, do not worry about this detail, and they commonly employ noncausal filters.

If an LTI system H is governed by a difference equation, then its transfer function is rational. That is, $H(z) = P(z^{-1})/Q(z^{-1})$, where P and Q are complex polynomials. We can see this by taking the z-transform of both sides of (9.43):

$$Y(z) + \sum_{k=1}^{N} a_k Y(z) z^{-k} = \sum_{m=0}^{M} b_m X(z) z^{-m}. \tag{9.44}$$

Consequently,

$$Y(z)\left[1 + \sum_{k=1}^{N} a_k z^{-k}\right] = X(z) \sum_{m=0}^{M} b_m z^{-m}. \tag{9.45}$$

This becomes a rational function in z^{-1} by computing,

$$\frac{Y(z)}{X(z)} = \frac{\sum_{m=0}^{M} b_m z^{-m}}{\left[1 + \sum_{k=1}^{N} a_k z^{-k}\right]} = H(z). \tag{9.46}$$

Although the system function is given by the z-transform convolution theorem as a rational function in the complex variable z, it is often easier to work in terms of the variable z^{-1}, which is the z-transform of a unit delay.

9.3.4.3 Poles and Zeros Analysis. Let us continue to consider an LTI system H, defined by a difference equation. The rational system function $H(z) = Y(z)/X(z) = P(z^{-1})/Q(z^{-1})$ may also be characterized by its poles and zeros. The *poles* are the zeros of $Q(z^{-1})$, and—assuming that common factors are removed from $H(z)$—the *zeros* of $P(z^{-1})$ are those of $H(z)$ too.

To find the poles and zeros, we must factor $P(z^{-1})$ and $Q(z^{-1})$ into products of linear terms. The fundamental theorem of algebra guarantees that every complex polynomial factors into linear terms, unique except for their order. Many readers know theorem, but if we recount some complex variable theory from Chapter 1, an argument of Liouville[10] proves it fairly easily.

We may assume that some polynomial $P(z)$ has been reduced to its lowest terms and that it still has degree exceeding unity. If $P(c) = 0$ for some $c \in \mathbb{C}$, then $(z - c)$ evenly divides $P(z)$, so we must have $P(z) \neq 0$ for all $z \in \mathbb{C}$. This means that the reciprocal function $R(z) = 1/P(z)$ is defined for all $z \in \mathbb{C}$. But then $R(z)$ is everywhere differentiable, since its denominator is differentiable and has no zeros. Further as $|z|$ gets large, $|R(z)|$ gets small. So $R(z)$ is bounded and everywhere differentiable. But, by Liouville's theorem, a bounded, everywhere differentiable (analytic) function is constant (exercise). This means $R(z)$ is a constant, and we have a

[10]French mathematician Joseph Liouville (1809–1882) authored some 400 papers on number theory, integral equations, and differential geometry.

contradiction. It must be the case that we can always extract another root from a complex polynomial of degree two or more.

Now consider the case of a discrete causal LTI system H whose transfer function $H(z)$ is a rational function. Since H is causal, $h(n)$ is right-sided: $h(n) = 0$ for $n < 0$. ROC_H is the exterior of a circle. Since $H(z)$ is rational, its denominator is a complex polynomial $Q(z^{-1})$. The only values $z \in \mathbb{C}$ for which $H(z)$ does not exist are the zeros of $Q(z^{-1})$, which are the poles of $H(z)$. The number of poles is finite; it is at most the degree of $Q(z^{-1})$. Hence, there is a pole of largest modulus $|p|$, where $Q(p) = 0$. Finally, we conclude that ROC_H consists of the exterior of the circle defined by $|z| = |p|$. If all the poles of $H(z)$ are contained within the complex unit circle $|z| = 1$, then ROC_H contains the unit circle, and H is a stable system.

How a signal processing application implements a discrete filter on a digital computer depends on the support of its *impulse response*: *finite* (FIR) or *infinite* (IIR).

A simple convolution calculation suffices for an FIR filter. If $y = Hx$, and $h = H\delta$, then the filtered output $y(n)$ can be calculated from the input $x(n)$ via a convolution operation:

$$y(n) = \sum_{k=M}^{N} h(k)x(n-k). \tag{9.47}$$

Finite impulse response filters are particularly easy to implement. The weighting function $h(k)$ is the impulse response of the linear, translation invariant discrete filter. The application must store at least $N - M + 1$ values for this calculation.

The system function for an FIR filter is given by

$$\frac{Y(z)}{X(z)} = \sum_{k=M}^{N} h_k z^{-k} = H(z), \tag{9.48}$$

where we have written $h_k = h(k)$. Since $H(z)$ has no denominator polynomial, it is an all-zero filter. Thus, another way to characterize FIR and IIR filters is as follows:

- The system function for an FIR filter is an all-zero rational function of z^{-1}.
- The system function for an IIR filter has at least one pole.

FIR filters tolerate transient inputs. A transient in past $x(n)$ values eventually falls outside the window $[M, N]$ over which the sum is computed in (9.47), so an FIR filter's output is never permanently affected by input spikes.

FIR filters behave well. They are clearly stable; that is, a bounded output signal results from filtering a bounded input signal. If $x(n)$ represents time sampled data with a transient, then eventually the effect of the transient on the output $y = Hx$ will disappear.

This means that the delay of a sinusoid at the output is proportional to its frequency. This can be important for processing and analysis applications where it is important that tones not be scrambled by the filter. A preliminary filter for a speech recognition system, for example, should not turn a simple spoken word into a grunt followed by a squeak. Later we shall see that some FIR filters have linear phase.

Examples of FIR filters that are implemented as in (9.47) include moving average filters and weighted averaging filters. We explored moving average filters for noise reduction in the dual tone multifrequency detection application. There also we considered windowing the signal in order to obtain a time localized snapshot of its spectral characteristics.

The convolution calculation for an IIR filter is impossible to directly implement. However, a large, powerful, efficient, and therefore imporant class of IIR filters admits recursive implemenatation. Recursive filters save prior output values and combine them with current and past input values to produce a current output value. That is, these filters obey a difference equation (Chapters 2 and 8) of the form

$$y(n) + \sum_{k=1}^{N} a_k y(n-k) = \sum_{m=0}^{M} b_m x(n-m). \tag{9.49}$$

9.3.4.4 *Implementation.* Recursive implementation can be very efficient. In fact, certain IIR filters, defined by a difference equation, require fewer computations than equally powerful FIR filters. Moreover, these filters have almost linear phase.

To see schematically how we might implement a filter defined by a difference equation on a digital computer, let us consider the system defined by (9.49). We can see that if the blocks symbolized by z^{-1} store a value for a unit time, then the equation is implemented by the *Direct Form I* architecture [8] shown in Figure 9.22, where

$$y(n) = \sum_{m=0}^{M} b_m x(n-m) - \sum_{k=1}^{N} a_k y(n-k) = v(n) - \sum_{k=1}^{N} a_k y(n-k). \tag{9.50}$$

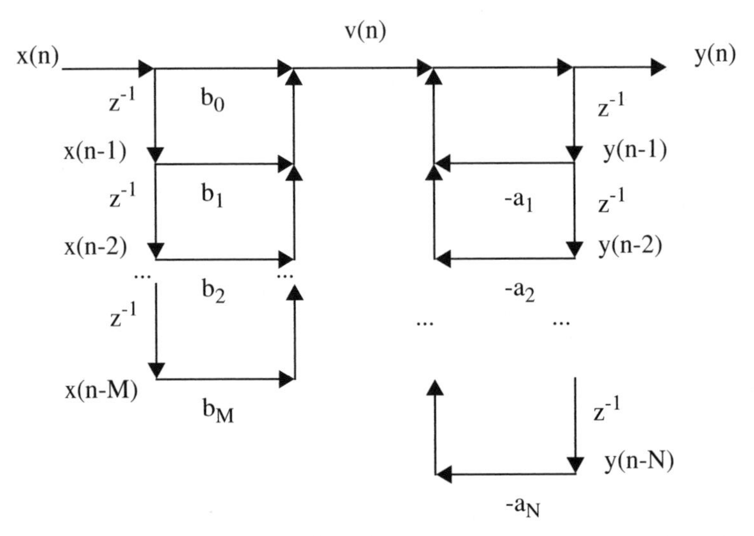

Fig. 9.22. Direct Form I implementation of a recursive system, governed by a difference equation.

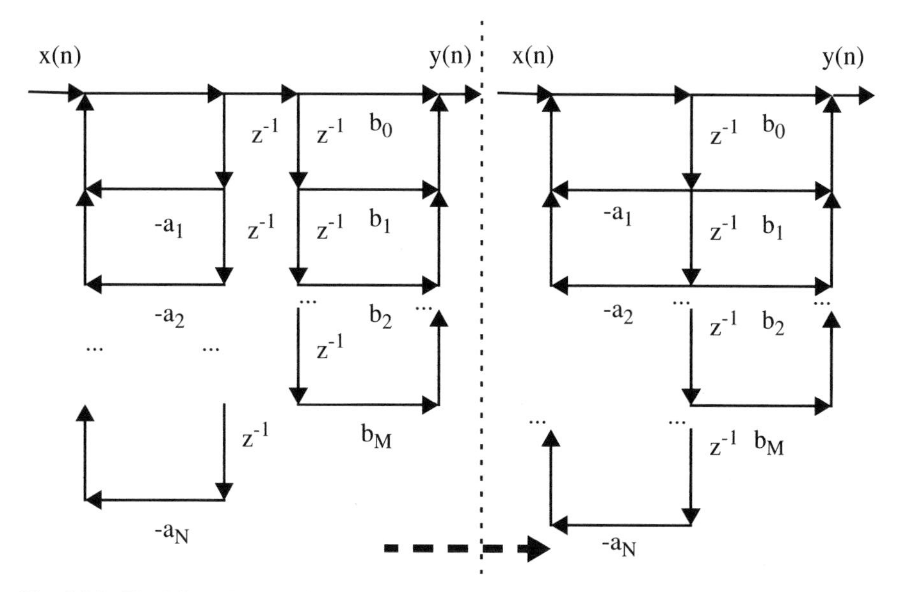

Fig. 9.23. The Direct Form II implementation of a recursive system is formed by (i) reversing the order of composition of the subsystems in the Direct Form I arrangement (as on the left) and (ii) then merging the duplicated delay operations (right).

The Direct Form I structure is seldom implemented, because a much more efficient scheme is possible. To see this, note that the lattice structures of Figure 9.22 cascades two LTI systems. One system produces $v(n)$ from input $x(n)$, and the other accepts $v(n)$ as an input and generates $y(n)$. We know from Chapter 2 that the composition of systems is characterized by the convolution of their respective impulse responses. Since convolution is commutative, so is composition. Thus, we can swap the order of the cascaded systems in Figure 9.22. This is shown on the left-hand side of Figure 9.23.

Reversing the order of these two subsystems leaves a network with two identical sets of sequential delays in the middle (Figure 9.23). The insight of the *Direct Form II* architecture is that the nodes across from one another contain the same mathematical result; the two halves can be joined, cutting the number of delay operations and economizing on memory locations. The Direct Form II, on the right-hand side of Figure 9.23, is a commonplace in computer implementations of signal processing systems.

When the difference equation coefficients (9.49) are all real, another efficient filter structure is possible. In this case, the poles and zeros of the system function $H(z)$ that have nonzero imaginary parts must consist of conjugate pairs. Then $H(z)$ is the product of terms of the form

$$G_p(z) = b_0\frac{1 + b_{1p}z^{-1}}{1 + a_{1p}z^{-1}} \tag{9.51a}$$

and

$$K_q(z) = \frac{1 + b_{1q}z^{-1} + b_{2q}z^{-2}}{1 + a_{1q}z^{-1} + a_{2q}z^{-2}}, \tag{9.51b}$$

where all of the coefficients are real. Some coefficients may be zero to account for unequal numbers of poles and zeros. Terms of the form (9.51a) are called *first-order sections*. Terms of the form (9.51b) are called *second-order sections*. We may implement each first- or second-order section as a Direct Form I or Direct Form II network. When we serialize all of them to implement $H(z)$, we have the *cascade* architecture for a system defined by a difference equation. Strategies exist for pairing poles and zeros in the cascade architecture so as to minimize round-off error in finite precision arithmetic [11].

9.3.5 Low-Pass Filter Design

This section covers several methods for designing discrete low-pass filters. The mechanics of converting an approximate low-pass filter to one of the other types (high-pass, bandpass, or band-reject) are the same as for ideal filters. This section also covers two particularly important procedures for converting continuous domain to discrete filters: the impulse invariance technique and the bilinear transformation.

We have already constructed and applied several ad hoc discrete filters. One obvious method is to use a frequency-domain mask:

(i) Fourier transform the signal. Thus, from $x(n)$, we find $X(k)$ according to the DFT analysis equation (9.1).

(ii) Mark out all but the frequencies of interest to the application. In other words, the application determines a binary mask $H(k) \in \{0, 1\}$ and we set $Y(k) = H(k)X(k)$. Note that this is equivalent to convolving $y(n) = (h*x)(n)$, where $h(n)$ is the inverse DFT of $H(k)$.

(iii) Analyze the result $Y(k)$ by thresholding, segmentation, and classification procedures in the spirit of ordinary time-domain analysis (Chapter 4); or, alternatively, inverse transform $Y(k)$ to continue time-domain analysis on $y(n)$.

(iv) Finally, interpret the signal.

There are advantages and disadvantages to the obvious approach. It offers perfect control of the filtering, assuming that the sampling rate is high enough. Also, it allows the application designer to work in both domains: time and frequency. This could be pertinent. But one drawback is that it is only feasible for offline applications. Where decisions about the signal must be made while the data arrive, this strategy does not work very well. Even for offline applications, the method could be expensive; the discrete Fourier transform requires a lot of arithmetical operations, and if the data do not naturally come in fast transform-sized chunks, this presents another problem of data windowing and interpretation across the window boundaries.

Another method is to construct low-pass filters intuitively, using weighted averages or recursive, sample-and-hold algorithms. Although it chagrins a signal theorist, probably most of the filters used in analysis applications are of this type. Such adhoc filters are useful for salt-and-pepper noise removal, blending away transients, and preparing signals for derivative or edge finding operations. One can investigate the spectral characteristics of such filters using the discrete Fourier transform, such as we carried out in Section 9.1.

9.3.5.1 *Impulse Invariance.* It is possible to begin with an analog low-pass filter and convert it to a discrete filter. Often, the useful characteristics for an application are known from their analog frequency-domain description. An example is the Gaussian; it is analytically attractive, it decays quickly, and filtering with wider kernels does not create additional time domain structure, as shown by scale-space theory (Chapter 4).

The *impulse invariance* technique is as follows:

(i) From the application, a specification of the filter's continuous-domain Fourier transform $H_a(\Omega)$ is generated.

(ii) The inverse Fourier transform is applied to find $h_a(t)$, the impulse response of the analog filter.

(iii) The analog impulse respose is sampled $h(n) = h_a(nT)$, where $T > 0$ is the sampling interval.

Now, impulse invariance is simple and commonly invoked in signal analysis applications. One question, however, is whether the discrete sampling in step (iii) above undoes the frequency-domain behavior that motivated the analog filter's selection in the first place.

The sampling theorem (Chapter 7) answers this question. We can write $H(\omega)$, the discrete-time Fourier transform of $h(n)$, in terms of $H_a(\Omega)$, the analog Fourier transform of $h_a(t)$:

$$H(\omega) = \frac{1}{T}\sum_{k=-\infty}^{\infty} H_a\left(\frac{j}{T}[\omega + 2k\pi]\right). \tag{9.52}$$

We recognize the form of (9.52). The DTFT of $h(n)$ consists of a superposition of scaled (amplified by T^{-1}) versions of H_a, shifted by $2\pi/T$ and dilated by T.

Note also that if $|H_a(\Omega)| \approx 1$ for $|\Omega| \approx 0$ in (9.52), then $H(\omega)$ will provide a gain of approximately $1/T$. With the sampling rate close to 1 Hz, the effect will be negligible. However, if T is small—that is, the sampling rate is high—then the discrete filter will have a very high gain. For this reason, it is customary to set $h(n) = Th_a(nT)$ [7, 40].

9.3.5.2 *Sampling Rational Analog Filters.* An important discrete filter design method is to sample an analog filter whose Fourier transform is a rational function. It turns out that the impulse invariance method applied to such an analog filter produces a discrete filter governed by a difference equation. Such analog filters are well-known from analog filter theory, and we will develop several of them later. This becomes a very powerful method for designing discrete filters with excellent design characteristics and efficient computer implementations.

Suppose that an analog low-pass filter H_a has impulse response $h_a(t)$. Let the radial Fourier transform of $h_a(t)$ be a quotient of polynomials in Ω:

$$H_a(\Omega) = \int_{-\infty}^{\infty} h_a(t) e^{-j\Omega t} dt = \frac{P(\Omega)}{Q(\Omega)}. \tag{9.53}$$

We can find $h_a(t)$ using a partial fractions expansion of $H_a(\Omega)$. This can be a messy manual computation, but it works just like partial fraction expansions for rational z-transforms. These are the steps to expand $H_a(\Omega)$ in partial fractions [10]:

(i) Normalize the fraction so that the denominator $Q(\Omega)$ has a leading coefficient of unity.

(ii) Since H_a is low-pass, as $|\Omega| \to \infty$, it must also be the case that $|H_a(\Omega)| \to 0$; the degree of $Q(\Omega)$ must exceed the degree of $P(\Omega)$:

$$H_a(\Omega) = \frac{P(\Omega)}{Q(\Omega)} = \frac{p_{M-1}\Omega^{M-1} + p_{M-2}\Omega^{M-2} + \dots + p_0}{\Omega^M + q_{M-1}\Omega^{M-1} + \dots + q_0}. \tag{9.54}$$

(iii) Factor the denominator $Q(\Omega)$ into its roots, ω_m, $1 \le m \le M$,

$$Q(\Omega) = \prod_{m=1}^{M} (\Omega - \Omega_m), \tag{9.55}$$

where we assume for now that the Ω_m are distinct. Possibly $Q(\Omega)$ has a form that allows us to easily derive its roots. In other cases, a computational method of root finding, such as the Traub–Jenkins algorithm may be employed [43].

(iv) Then $H_a(\Omega)$ has a partial fractions expansion of the form:

$$H_a(\Omega) = \sum_{m=1}^{M} \frac{c_m}{\Omega - \Omega_m}, \tag{9.56}$$

where c_m, $1 \le m \le M$, are constants.

(v) To find c_m, $1 \le m \le M$, note that—having assumed that the denominator's roots Ω_m are distinct—we see

$$c_m = [(\Omega - \Omega_m) H_a(\Omega)]\big|_{\Omega = \Omega_m}. \tag{9.57}$$

The partial fraction expansion enables us to write out the impulse response $h_a(t)$ of the analog low-pass filter H_a. Indeed, the inverse radial Fourier transform of $\frac{c_m}{\Omega - \Omega_m}$ is $jc_m e^{j\Omega_m t} u(t)$, where $u(t)$ is the unit step signal. This transformation is valid so long as the imaginary part of Ω_m is positive: $\mathrm{Imag}(\Omega_m) > 0$. By Fourier transform linearity, we have

$$h_a(t) = ju(t) \sum_{m=1}^{M} c_m e^{j\Omega_m t}. \tag{9.58}$$

Now we can discretize the filter by the impulse invariance technique, for instance. Let $T > 0$ be the sampling interval. Then,

$$h(n) = h_a(nT) = ju(n) \sum_{m=1}^{M} c_m e^{j\Omega_m nT}. \tag{9.59}$$

Taking the z-transform of (9.59) gives

$$H(z) = j \sum_{m=1}^{M} c_m Z[u(n)e^{j\Omega_m nT}](z) = \sum_{m=1}^{M} \frac{jc_m}{1 - e^{jT\Omega_m} z^{-1}}. \tag{9.60}$$

The important points about this derivation are as follows:

- Note that (9.60) is already in the form of a partial fractions expansion.
- It has in fact the same partial fractions expansion coefficients as given by (9.56) except for the factor of $j \in \mathbb{C}$.
- The poles of $H(z)$ are at $\exp(jT\Omega_m)$ for $m = 1, 2, \ldots, M$.
- The pole at $\exp(jT\Omega_m)$ will be inside the unit circle if and only if $\mathrm{Real}(jT\Omega_m) < 0$ and thus if and only if $\mathrm{Imag}(\Omega_m) > 0$.
- If we consider $H(z)$ to be the z-transform of a causal filter, then $\mathrm{Imag}(\Omega_m) > 0$ for all $m = 1, 2, \ldots, M$ implies that the region of convergence will be $\{z \in \mathbb{C}: |z| > a\}$, for some $1 > a > 0$, and the discrete filter H will therefore be *stable*.
- All of the partial fractions (9.60) are of the form $\frac{C}{1 - Az^{-1}} = \frac{Cz}{z - A}$, which is the z-transform of the LTI system governed by the difference equation $y(n) = Ay(n-1) + Cx(n)$.
- Finally, if the sampling rate $1/T$ differs substantially from unity, then we choose

$$H(z) = \sum_{m=1}^{M} T \frac{jc_m}{1 - e^{jT\Omega_m} z^{-1}} \tag{9.61}$$

in accord with (9.52) and the remarks thereafter.

Now let us turn to the special case of (9.55) where the denominator $Q(\Omega)$ has multiple roots. Suppose the root Ω_1 has multiplicity R and the remaining Ω_m, $2 \le m \le M$, are distinct. Then $H_a(\Omega)$ has a partial fractions expansion of the form

$$H_a(\Omega) = \sum_{r=1}^{R} \frac{c_{1,r}}{(\Omega - \Omega_1)^r} + \sum_{m=2}^{M} \frac{c_m}{\Omega - \Omega_m}. \tag{9.62}$$

where $c_{1,r}$, $1 \le r \le R$, and c_m, $2 \le m \le M$, are constants. The formula for calculating the $c_{1,r}$ is as follows:

$$c_{1,r} = \frac{\left[\frac{d^{R-r}}{d\Omega^{R-r}}(\Omega - \Omega_1)^R H_a(\Omega)\right]\Bigg|_{\Omega = \Omega_1}}{(R-r)!}. \tag{9.63}$$

If there are several multiple roots, then we follow the above procedure, inserting supplemental terms in the partial fractions expansion (9.62) and computing the coefficients with repeated derivatives (9.63).

Notice again that if the analog $H_a(\Omega)$ has at least one pole, then so will the z-transform (9.61), and the discrete filter will be IIR.

9.3.5.3 Laplace Transform Techniques. Analog filters and discrete filter designs from them are usually approached using the Laplace transform [6–11]. Readers are probably familiar with this tool from continuous-domain systems theory. The Laplace transform plays the same role in analog systems theory that the z-trasnform plays in discrete system theory. Let us briefly review how the transform is used with analog filters whose Fourier transform is a rational function. Specialized texts include Refs. 44 and 45.

The *Laplace transform* $X_L(s)$ of the analog signal $x(t)$ is defined by

$$X_L(s) = \int_{-\infty}^{\infty} x(t)e^{-st}\,dt, \tag{9.64}$$

where $s \in \mathbb{C}$. If H is an LTI system, then the Laplace transform $H_L(s)$ of its impulse response, $h = H\delta$, is also called the *system* or *transfer function* of H.

Note that if $x(t)$ has a Fourier transform $X(\Omega)$, then $X_L(j\Omega) = X(\Omega)$. The Fourier transform is the Laplace transform evaluated on the imaginary axis of the complex plane. If $s = \sigma + j\omega$, where $\sigma \in \mathbb{R}$, then $X_L(s)$ is the Fourier transform of $x(t)e^{-\sigma t}$. Transform convergence depends on the relation of $x(t)$ to the exponential factor $\exp(-\sigma t)$, and it does not depend on the imaginary part of $s = \sigma + j\omega$. Hence, the Laplace transform converges on vertical strips in the complex plane.

A couple of basic examples show that the Laplace transform must be associated with a *region of convergence* (ROC). If $x(t) = e^{-At}u(t)$, then $X_L(s) = 1/(s + a)$ and the $\mathrm{ROC}_X = \{s \in \mathbb{C}: \mathrm{Real}(s) > -a\}$. If $y(t) = -e^{-At}u(-t)$, then $Y_L(s) = 1/(s + a)$ and the

$\mathrm{ROC}_Y = \{s \in \mathbb{C}: \mathrm{Real}(s) < -a\}$. Recall that an analog LTI system H is stable if and only if the impulse response $h = H\delta$ is absolutely integrable: $h \in L^1$. But this means $h(t)$ has a Fourier transform $H(\Omega)$. Since $H(\Omega) = H_L(j\Omega)$, ROC_H must contain the imaginary axis.

Now suppose $x(t)$ is *right-sided*: $x(t) = 0$ for $t < a \in \mathbb{R}$. If ROC_X contains the vertical line $\mathrm{Real}(s) = b \in \mathbb{R}$, then ROC_X contains $\{s \in \mathbb{C} : \mathrm{Real}(s) \geq \mathrm{Real}(b)\}$. This is fairly easy to see, because, for such $s \in \mathbb{C}$, $\exp(-\mathrm{Real}(s)t) \leq \exp(-\mathrm{Real}(b)t)$ for $t > 0$ and the transform integral (9.64) will still exist. The ROC of a right-sided signal is a right half-plane. Similarly, if $x(t)$ is *left-sided* ($x(t) = 0$ for $t > a \in \mathbb{R}$), then ROC_X is a left half-plane. Now consider the case of a causal LTI system H whose transfer function $H_L(s)$ is a rational function. Since H is causal, $h(t) = 0$ for $t < 0$. In other words, $h(t)$ is right-sided, and ROC_H is a right half-plane. Since $H_L(s)$ is rational, its denominator is a complex polynomial $Q(s)$. The only values $s \in \mathbb{C}$ for which $H_L(s)$ does not exist are the zeros of $Q(s)$, which are the rational function's *poles*. As there are only a finite number of poles of $H_L(s)$, ROC_H must be the half-plane to the right of the zero of $Q(s)$ with the largest real part.

We invert the Laplace transform using much the same methods as z-transform inversion. Rational functions $X_L(s)$ can be decomposed into a partial-fractions representation and the linearity propert applied to elementary transforms to arrive at $x(t)$. Table 9.3 lists basic Laplace transform properties.

Let us turn now to the use of the Laplace transform in designing discrete filters from rational analog filters [7]. Let the Laplace transform of $h_a(t)$ be a quotient of complex polynomials

$$H_L(s) = \int_{-\infty}^{\infty} h_a(t)e^{-st}dt = \frac{P(s)}{Q(s)} \tag{9.65}$$

TABLE 9.3. Summary of Laplace Transform Properties

Signal Expression	Laplace Transform or Property
$x(t)$	$X_L(s) = \int_{-\infty}^{\infty} x(t)e^{-st}dt$
$z(t) = ax(t) + by(t)$	$aX_L(\omega) + bY_L(\omega)$ (Linearity, $\mathrm{ROC}_X \cap \mathrm{ROC}_Y \subseteq \mathrm{ROC}_Z$)
$y(t) = x(t-a)$	$e^{-sa}X_L(s)$ (Time shift, $\mathrm{ROC}_X = \mathrm{ROC}_Y$)
$y(t) = x(t)\exp(at)$	$X_L(s-a)$ (Frequency shift, modulation, $\mathrm{ROC}_Y = \{s: s-a \in \mathrm{ROC}_X\}$)
$y(t) = x(at)$, $a \neq 0$	$\frac{1}{\lvert a\rvert}XL_L\left(\frac{s}{a}\right)$ (Scaling, dilation, $\mathrm{ROC}_Y = \{s: s/a \in \mathrm{ROC}_X\}$)
$y(t) = (x * h)(t)$	$F(s)H(s)$ (Convolution, $\mathrm{ROC}_X \cap \mathrm{ROC}_H \subseteq \mathrm{ROC}_Y$)

and let its partial fractions expansion be

$$H_L(s) = \sum_{m=1}^{M} \frac{d_m}{s - s_m}, \tag{9.66}$$

where d_m, $1 \le m \le M$, are constants, and the poles s_m are distinct. Laplace transform linearity and inversion of the summands in (9.66) implies

$$h_a(t) = u(t) \sum_{m=1}^{M} d_m e^{s_m t}. \tag{9.67}$$

Impulse invariance applies exactly as above. We set

$$h(n) = h_a(nT) = u(n) \sum_{m=1}^{M} d_m e^{s_m nT}, \tag{9.68}$$

and the z-transform of $h(n)$ is

$$H(z) = \sum_{m=1}^{M} d_m \mathcal{Z}[u(n) e^{d_m nT}](z) = \sum_{m=1}^{M} \frac{d_m}{1 - e^{Ts_m} z^{-1}}. \tag{9.69}$$

Notice that if $d_m = jc_m$ and $s_m = j\Omega_m$, then the two expressions for $H(z)$, (9.69) and (9.60), are identical. The conditions for stability and causality are similar too. From a Fourier transform perspective, we need the poles of the rational function $H_a(\Omega) = \frac{P(\Omega)}{Q(\Omega)}$ to have positive imaginary parts. From the Laplace transform standpoint, however, we require the poles of $H_L(s) = \frac{P(s)}{Q(s)}$ to have negative real parts. Of course, if Ω_0 is a pole of $H_a(\Omega)$, then $j\Omega_0$ is a pole of $H_L(s)$.

9.3.5.4 Bilinear Transformation. The bilinear transformation obtains a discrete filter from the frequency domain representation of an analog filter by directly mapping the analog frequency values to discrete frequency values. What sort of operation performs such a mapping? Note that analog frequencies can be arbitrarily large, $-\infty < \Omega < +\infty$, whereas discrete frequencies are limited to a 2π-wide interval: $-\pi < \omega \le \pi$. So we seek a function that maps the real line to the circumference of a circle. The arctangent, $\tan^{-1}$, maps $\mathbb{R}$ to the interval $(-\pi/2, \pi/2)$. Let T be the sampling interval. Then the following relation maps continuous to discrete frequency values:

$$\omega = 2\tan^{-1}\left(\frac{\Omega T}{2}\right). \tag{9.70}$$

Observe that as $\Omega \to \pm\infty$, the maximum analog frequency values, then $\omega \to \pm\pi$, respectively, the maximum discrete frequency values.

Suppose that $H_a(\Omega)$ is an analog lowpass filter with cutoff frequency Ω_c. Then the bilinear transformation (9.70) allows us to define a discrete low-pass filter as follows:

$$H(\omega) = H_a\left(\frac{2}{T}\tan\left(\frac{\omega}{2}\right)\right). \tag{9.71}$$

The cutoff frequency for $H(\omega)$ is $\omega_c = 2\tan^{-1}\left(\frac{\Omega_c T}{2}\right)$.

How does the scaling factor T come to appear in (9.70)? One practical reason is that for small frequencies, T controls the rate of change of ω with respect to Ω: $d\omega/d\Omega \approx T$. Some authors (e.g., Ref. 46) set $\omega = \frac{2}{T}\tan^{-1}\left(\frac{\Omega T}{2}\right)$ to ensure that for low frequencies $\omega \approx \Omega$. Another way of justifying the frequency mapping (9.70) is to consider the relation between poles of rational analog filters and the discrete filters obtained from them. If $s = s_m$ is a pole of the Laplace transform $H_L(s)$ of a filter (9.66), then corresponding to it is a pole $z = \exp(s_m T)$ of the z-transform $H(z)$ (9.69). This suggests a mapping from the Laplace s-plane to the z-plane: $z = e^{sT}$. Thus,

$$z = e^{sT} = \frac{e^{s(T/2)}}{e^{-s(T/2)}} \approx \frac{1 + s\frac{T}{2}}{1 - s\frac{T}{2}} = \frac{2 + sT}{2 - sT}, \tag{9.72}$$

where we approximate the quotient on the right in (9.72) using the first two Taylor series terms for the exponential function. This implies

$$s \approx \frac{2}{T}\left(\frac{z-1}{z+1}\right), \tag{9.73}$$

which relates the Laplace transform variable to the z-transform variable. To relate continuous and discrete frequency responses, we use $z = e^{j\omega}$ and $s = j\Omega$ in (9.73), treating it as an equality. After a little algebra, (9.70) results (exercise).

9.3.6 Frequency Transformations

There are convenient mappings of the independent variable of a filter's system function that convert a low-pass filter to a high-pass, bandpass, bandstop, or even another low-pass filter [7, 11, 33, 47].

9.3.6.1 Analog. Consider an analog low-pass filter with Laplace transform $H_L(s)$ and cutoff frequency $\Omega_c = 1$. The transformations are as follows:

(i) Let $\phi(s) = s/\Omega_H$. Then $H_L(\phi(s))$ is a low-pass filter with cutoff frequency Ω_H.
(ii) Let $\phi(s) = \Omega_H/s$. Then $H_L(\phi(s))$ is a high-pass filter with cutoff frequency Ω_H.

(iii) Let $\phi(s) = \dfrac{s^2 + \Omega_L\Omega_H}{s(\Omega_H - \Omega_L)}$. Then $H_L(\phi(s))$ is a bandpass filter with lower cutoff frequency Ω_L and upper cutoff frequency Ω_H.

(iv) Let $\phi(s) = \dfrac{s(\Omega_H - \Omega_L)}{s^2 + \Omega_L\Omega_H}$. Then $H_L(\phi(s))$ is a bandstop filter with lower cutoff frequency Ω_L and upper cutoff frequency Ω_H.

9.3.6.2 Discrete. Consider a discrete low-pass filter with z-transform $H(z)$ and cutoff frequency ω_c. The transformations are as follows:

(i) Let $\phi(z^{-1}) = \dfrac{z^{-1} - \alpha}{1 - \alpha z^{-1}}$. If we set $\alpha = \sin\left(\dfrac{\omega_c - \omega_H}{2}\right) / \sin\left(\dfrac{\omega_c + \omega_H}{2}\right)$, then $H(\phi(z))$ is a low-pass filter with cutoff frequency ω_H.

(ii) Let $\phi(z^{-1}) = -\dfrac{z^{-1} + \alpha}{1 + \alpha z^{-1}}$. If $\alpha = -\cos\left(\dfrac{\omega_c + \omega_H}{2}\right) / \cos\left(\dfrac{\omega_c - \omega_H}{2}\right)$, then $H(\phi(z))$ is a high-pass filter with cutoff frequency ω_H.

(iii) Let $\phi(z^{-1}) = -\dfrac{z^{-2} - 2\alpha\beta z^{-1} + \gamma}{\gamma z^{-2} - 2\alpha\beta z^{-1} + 1}$. If $\alpha = \cos\left(\dfrac{\omega_H + \omega_L}{2}\right) / \cos\left(\dfrac{\omega_H - \omega_L}{2}\right)$,

$\kappa = \cot\left(\dfrac{\omega_H - \omega_L}{2}\right)\tan\left(\dfrac{\omega_c}{2}\right)$, $\beta = \kappa/(\kappa + 1)$, and $\gamma = (\kappa - 1)/(\kappa + 1)$,

then H(ϕ(z)) is a bandpass filter with upper cutoff frequency ω_H and lower cutoff frequency ω_L.

(iv) Let $\phi(z^{-1}) = \dfrac{z^{-2} - 2\alpha\beta z^{-1} + \gamma}{\gamma z^{-2} - 2\alpha\beta z^{-1} + 1}$. If $\alpha = \cos\left(\dfrac{\omega_H + \omega_L}{2}\right) / \cos\left(\dfrac{\omega_H - \omega_L}{2}\right)$,

$\kappa = \tan\left(\dfrac{\omega_H - \omega_L}{2}\right)\tan\left(\dfrac{\omega_c}{2}\right)$, $\beta = 1/(\kappa + 1)$, and $\gamma = (1 - \kappa)/(\kappa + 1)$,

then $H(\phi(z))$ is a bandstop filter with upper cutoff frequency ω_H and lower cutoff frequency ω_L.

Example (Low-Pass to High-Pass). Suppose that $\omega_c = \pi/4$ and $\omega_H = 3\pi/4$ in (ii) above. Then $\alpha = -\dfrac{\cos(\pi/2)}{\cos(-\pi/4)} = 0$ and $H(\phi(z)) = H(-z)$.

9.3.7 Linear Phase

It is possible to construct causal finite impulse response filters with *linear phase*. Where the analysis steps in an application depend on linear phase during signal processing steps, this can be factor in favor of using finite impulse response (FIR) filters. Note that it is possible to have infinite impulse response (IIR) filters with linear phase. For example, if $h = H\delta$ and $h(n)$ is symmetric about $n = 0$, then H will have zero phase. However, we are interested in filters that are practically implementable, and therefore we require right-sided impuse responses: $h(n) = 0$ for $n < N$ for some N.

9.3.7.1 FIR Characterization. Let H be a linear, translation invariant system, let $h(n)$ be its impulse response, and let $H(\omega)$ be its discrete time Fourier transform. If $H(\omega) = e^{j\phi(\omega)}H_R(\omega)$, with $\phi(\omega)$ a linear function of ω, and $H_R(\omega) \in \mathbb{R}$, then we say H has *linear phase*.

The theoretical result we are going to prove is as follows. If H is a discrete causal FIR filter with impulse response $h = H\delta$ such that $h(n) = 0$ for $n > N - 1$, then H has linear phase if and only if for some $c \in \mathbb{C}$, with $|c| = 1$, $h(n) = ch^*(N - 1 - n)$, where h^* is the complex conjugate of h.

To begin, let us assume that H is causal, FIR, and has linear phase. Let $H(\omega) = e^{j\phi(\omega)}H_R(\omega)$, with $\phi(\omega) = a + b\omega$, for some $a, b \in \mathbb{R}$, and $H_R(\omega) \in \mathbb{R}$. Assume that Support$(h) = [0, N - 1]$ with $N > 0$, so that $h(0) \neq 0$ and $h(N - 1) \neq 0$. Let $A_H(\omega)$ be the amplitude function for $H(\omega)$ the DTFT of $h(n)$: $H(\omega) = e^{j\phi(\omega)}A_H(\omega)$. Then

$$H(\omega) = \sum_{-\infty}^{\infty} h(n)e^{-j\omega n} = \sum_{n=0}^{N-1} h(n)e^{-j\omega n}, \tag{9.74}$$

$$H(\omega) = e^{j[2a + 2b\omega - \phi(\omega)]}A_H(\omega) = e^{2ja}e^{2jb\omega}e^{-j\phi(\omega)}A_H(\omega) = e^{2ja}e^{2jb\omega}H^*(\omega). \tag{9.75}$$

Note that $H(\omega)$ and its complex conjugate $H^*(\omega)$ are both 2π–periodic; we must have $2b = K \in \mathbb{Z}$ on the right-hand side of (9.75). Let $c = e^{2ja}$ and $g(n) = h^*(-n)$, so that $G(\omega) = H^*(\omega)$. Then the discrete signal $s(n) = cg(n + K)$ has DTFT $S(\omega) = ce^{jK\omega}G(\omega) = ce^{jK\omega}H^*(\omega) = H(\omega)$. Because the DTFT is invertible, we must have $h(n) = s(n) = cg(n + K) = ch^*(-K - n)$. We know that for $n < 0$, $h(n) = h^*(n) = 0$. Also, if $n > -K$, we have $ch^*(-K - n) = h(n) = 0$. Thus, $-K = N-1$ because that is the upper limit of the support of $h(n)$, and so $h(n) = ch^*(N - 1 - n)$, as claimed.

Conversely, suppose Support$(h) = [0, N - 1]$ and $h(n) = ch^*(N - 1 - n)$ for some $c \in \mathbb{C}$ with $|c| = 1$. Applying the DTFT properties gives $H(\omega) = ce^{-j(N-1)\omega}H^*(\omega)$. Let $c = e^{j\theta}$. If $H(\omega) = e^{j\phi(\omega)}A_H(\omega)$, then $H^*(\omega) = e^{-j\phi(\omega)}A_H(\omega)$. Putting these together, we have

$$e^{j\phi(\omega)} = e^{j\theta}e^{-j\phi(\omega)}e^{-j(N-1)\omega}, \tag{9.76}$$

and thus for some $K \in \mathbb{Z}$,

$$\phi(\omega) = \frac{(1 - N)\omega}{2} + \frac{\theta}{2} + \pi K. \tag{9.77}$$

Clearly, $\phi(\omega)$ is a linear function of ω, and we are done.

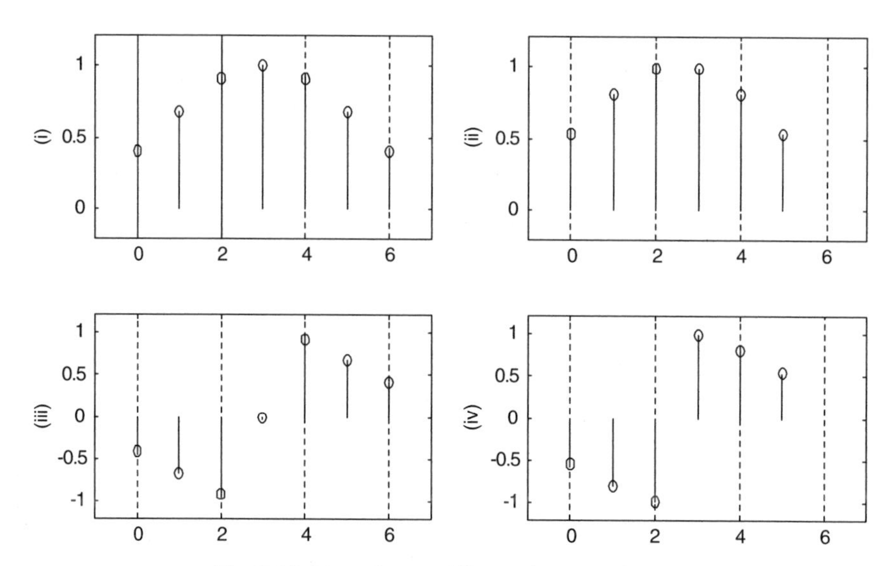

Fig. 9.24. Four classes of linear phase FIR filters.

9.3.7.2 Linear Phase Filter Classes. For real-valued filters $h = H\delta$, where $h(n)$ has support $[0, N-1]$, with $N > 0$, there are four categories of linear phase filters. We showed that a finite impulse response filter H has linear phase if and only if for some $c \in \mathbb{C}$, $|c| = 1$, $h(n) = ch^*(N-1-n)$. If $h(n) \in \mathbb{R}$, then $c = 1, -1$.

We can thus put the filter into one of four classes (Figure 9.24):

(i) $c = 1$ and $N = 2M + 1$ is odd.
(ii) $c = 1$ and $N = 2M$ is even.
(iii) $c = -1$ and $N = 2M + 1$ is odd.
(iv) $c = -1$ and $N = 2M$ is even.

9.3.7.3 Examples. We review two cases where the need for linear phase motivates the specific use of FIR filters: electroencephalogram (EEG) interpretation and seismogram intrpretation.

The first chapter explained the multichannel EEG [48], a biomedical signal that is often employed in studying brain functions and diagnosing injuries and illnesses. Electrodes attached to the scalp record the minute voltages produced by the interactions of large numbers of neurons. The signals are often quite noisy, and successive averaging is often employed to improve the quality of EEG traces. In studying auditory potentials—EEG traces of the part of the brain that is involved in the front-end processing of auditory nerve impulses—linear filtering has been investigated in order to improve upon successive averaging, the efficacy of which diminishes after a large number of sampling epochs. Frequencies above 2 kHz are removed by lowpass filtering, since they cannot be due to neuronal changes, which take place on the order of 1 ms. The remaining noise is usually at low frequencies, DC to about 150 Hz [49]. The most popular filtering methods for such an application use IIR filters derived from difference equations (Section 9.3.4.2), since they are efficient and generally have better sharpness than FIR filters. But the nonlinear phase response of the causal IIR filters distorts the EEG trace, making linear phase FIR filtering preferable [49].

In seismic processing, the signals are generally quite noisy, composed of many frequency components, which are filtered by their propagation through the earth. These different sinusoidal pulses arrive at the sensing unit—the seismometer—at different times [50], a kind of phase delay. Perhaps the most basic task of earthquake seismology is to estimate the arrival time of an event, so that different seismometer stations can compare seismograms and locate the epicenter. Against the background noise of minor earth movements, impacts from construction equipment, and vehicle traffic vibrations, the seismic station determines the edge of a significant transient. At what time this transient occurs for a seismic station depends on the group delay of Mother Earth acting as a filter. Thus, for automated seismogram interpretation, a signal processing filter that introduces a nonlinear phase delay into the system might distort the signal and cause an error in pinpointing the onset time of a seismic shock. In order to facilitate the comparison of arrival times among different stations with different equipment, it is essential that their diverse noise removal filters not introduce any frequency dependent delays at all. The filtering requirement is even more stringent; successful analysis of the seismogram usually demands *zero phase* filtering [51].

9.4 WIDEBAND SIGNAL ANALYSIS

This section considers signals that contain diverse spectral components. These signals include chirps, which consist of rising or falling tones; transient signals, such as seismic pulses; signals with sharp edges; and irregularly shaped signals, such as image object boundaries. Our earlier methods of periodicity detection are successful only with much simpler waveforms.

9.4.1 Chirp Detection

A chirp is a signal segment where the frequency rises or falls over time. Strictly speaking, of course, a chirp is not a narrowband signal. But locally, at least, the signal energy is contained in a narrow spectral range. If this is indeed the case, then the task of chirp analysis becomes based upon a series of pure tone detection problems where the detected tones regulary rise or fall. This section considers the case where locally, at least, a signal contains mainly one frequency component, but that the frequency itself is changing over time.

9.4.1.1 Synthetic Chirps. Section 6.5 presented the theory of signal modulation, which is the theoretical foundation of chirp signal analysis. Let us first consider

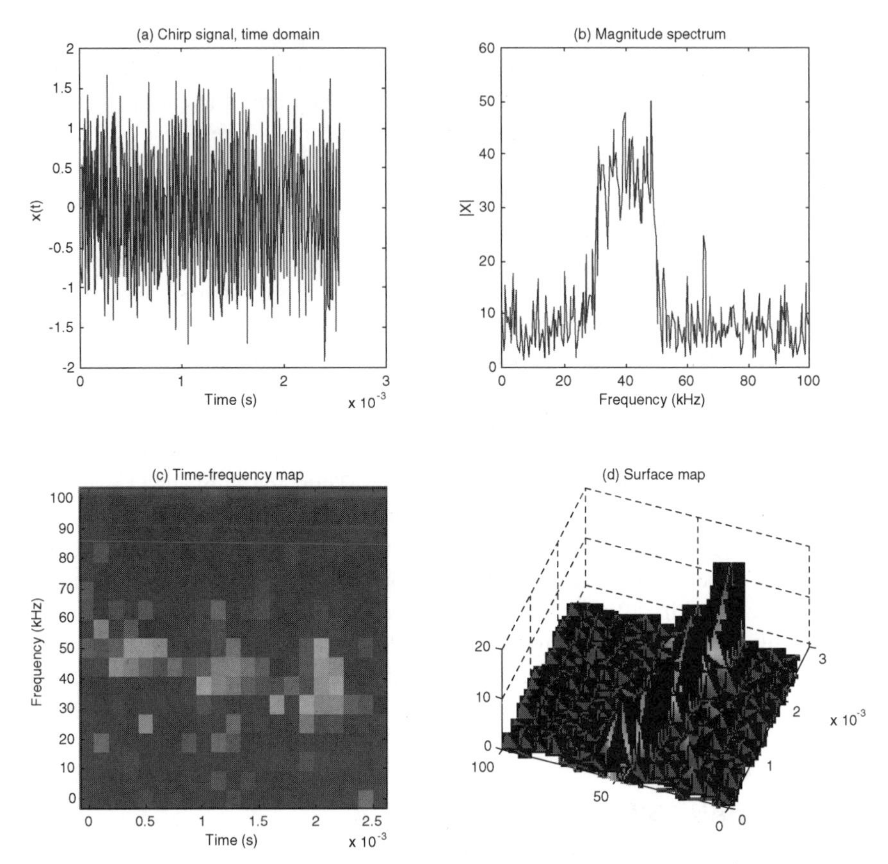

Fig. 9.25. Synthetic chirp with noise added (a). Panel (b) shows the magnitude spectrum. A time-frequency map (c) shows how the frequencies change over time. Surface maps (d) are useful aids for visualizing signal structure.

chirp signals for which the change in frequency over time is fairly simple—linear, for instance. Thus, we have $\omega(t) = t\omega_1 + \omega_0$ as shown in Figure 9.25.

From its time-domain plot Figure 9.25a, it is hard to understand the signal. The magnitude spectrum Figure 9.25b shows that a concentrated range of tones is present, in the range from 30 to 50 kHz, amidst some moderate background noise. But the time evolution of these frequencies and therefore the structure of the signal-itself remains unclear. We compute a sequence of 32-point discrete Fourier transforms over the time span of the signal. Overlapping the windows by eight samples helps smooth the time-frequency representation. The resulting map, shown in Figure 9.25c, reveals a linear chirp beginning at about 50 kHz and decreasing in frequency in a linear fashion down to about 30 kHz.

As long as its frequency content is, over a certain time interval, basically a tone, a chirp signal can be analyzed using tone detection techiques and a state machine.

Local discrete Fourier transforms carry out the frequency analysis, and some straightforward intepretive logic carry out the chirp discrimination. We can surmise the following steps:

(i) An order N for the number of samples for discrete Fourier transformation (DFT) is selected, depending on the sampling rate of the source signal and the range of frequencies expected in the chirp signal.
(ii) A DFT window overlap is selected.
(iii) DFT computations are conducted on the source signal within overlapping windows as chosen in (i) and (ii).
(iv) Where a relatively pure tone in the expected frequency range is found using local Fourier analysis, the state machine enters a tone detected state.
(v) Step (iv) is carried out again, and if there is a tone in the acceptable spectral range, then the tone is checked for purity, and the machine enters a state of increasing or decreasing tone frequency.
(vi) Continued local frequency analysis extends the time-domain support of the tone, breaks out of the tone detected state based on an invalid frequency response, or decides that the tone has the proper quality and range to continue the chirp defined in the current machine state.
(vii) This process continues until the chirp ends or the input signal is exhausted.

The main difficulties with this analysis is that it requires—for the most part—offline data analysis. That is, the DFT windows are applied around a time center value in a noncausal fashion. This could be an expensive operation, and for real-time processing, it may be impossible. One alternative might be to employ a bank of causal filters and seek significant outputs from the banks tuned to increasing or decreasing frequency bands. To achieve this, however, we need to devise filtering methods that are causal and sufficently efficient for online implementation.

9.4.1.2 Biological Signals: Bat Echolocation Chirp. Now let us study the echolocation chirp recorded from a large brown bat (*Eptesicus fuscus*).[11] The sampling period is $T = 7$ μs, and there are $N = 400$ samples in the data set. The time-domain signal oscillates and rises in amplitude, but the plotted values evince few other clues as to its structure (Figure 9.26a). The magnitude spectrum explains a little more. There are frequencies between 20 kHz and 50 kHz, centered more or less strongly around a spectral peak at some 35 kHz. The spectrum appears to be bimodal (Figure 9.26b). From a Fourier domain perspective, we cannot tell whether the modes are frequency components that appear at certain times, one after the other, or whether they substantially overlap and the bimodality is an artifact of relatively weaker middle frequencies.

[11]This data set is available from the signal processing information base (SPIB): `http://spib.rice.edu/spib.html`. The authors wish to thank Curtis Condon, Ken White, and Al Feng of the Beckman Center at the University of Illinois for the bat data and for permission to use it in this book.

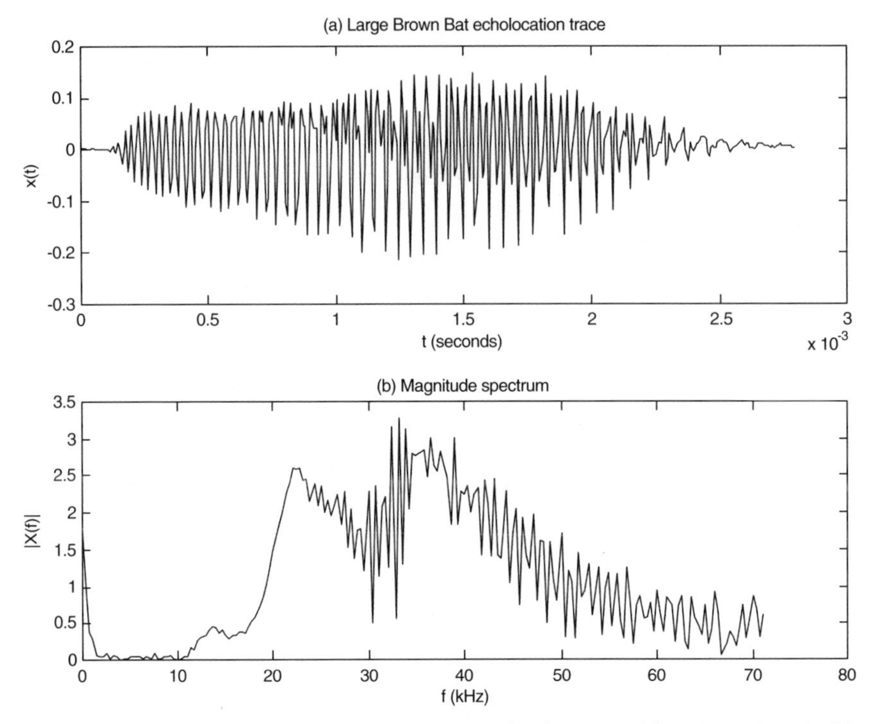

Fig. 9.26. Large brown bat echolocation pulse time-domian (a) and frequency-domain (b).

Our knowledge of the echolocation pulse's content changes dramatically when we develop the time-frequency map (Figure 9.27).

We use a boxcar window of width $N = 128$ to generate the local spectral values, overlapping successvie windows by $M = 120$ samples. This reveals three descending chirps (Figure 9.27a) and shows that the time-frequency plot is at least tri-modal. The bar-shaped artifacts most visible in the lower frequencies appear to correlate with window alignment. We try a Hann window function of length N and overlap M to improve the local frequency estimates, as shown in Figure 9.27b. This reduces the time-frequency artifacts, as one might expect. However, Hann windowing has the added benefit of resolving the time-frequency mode of highest initial frequency into two chirps; the echolocation pulse in fact contains four modes.

9.4.2 Speech Analysis

Let us now consider some low-level speech signal analysis problems. There is, to be sure, a large research literature on natural language processing from initial filtering methods, detection of utterances in noise, phoneme recognition, word recognition, contextual analysis, and artificial intelligence techniques for computerized speech understanding [13, 52–54].

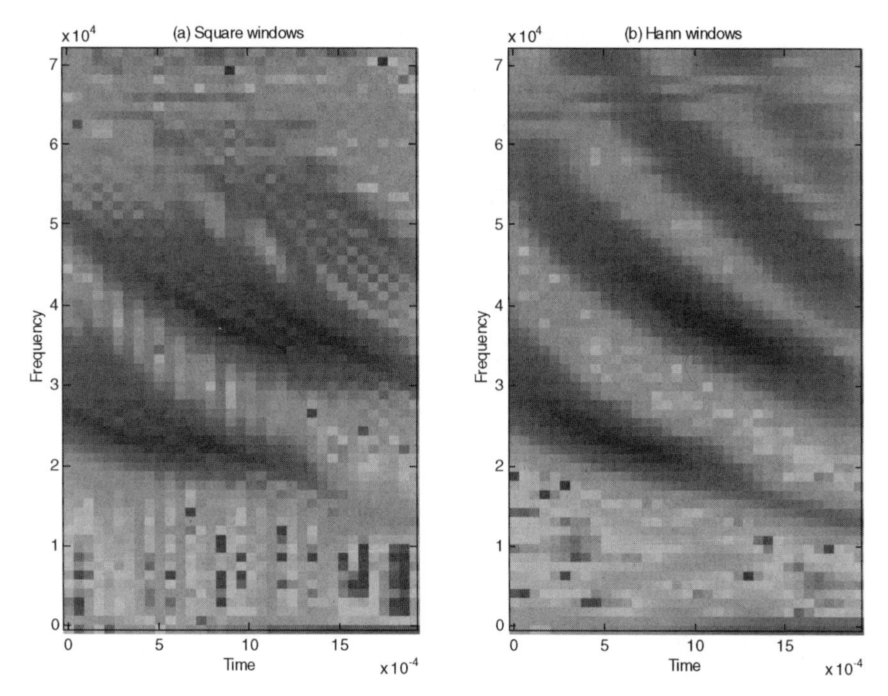

Fig. 9.27. The time-frequency map of the large brown bat echolocation pulse evidently has three components (a), each a descending chirp. Instead of a 128-point square window, panel (b) employs a 128-point Hann window to generate the local frequency information.

9.4.2.1 Formant Detection. *Formants* are relatively high-energy, tone-like components within a spoken word. They appear in the speech spectrum as isolated peaks. Consider, for example, a digitized voice fragment, consisting of a single word (Figure 9.28).

It is possible to discover formants using peak detection in the Fourier magnitude spectrum. The vowel phoneme /a/ in Figure 9.28c exhibits three strong peaks at approximately 300 Hz, 600 Hz, and 900 Hz. Such harmonics are characteristic of sounds produced in a tube, with a source at one end and open at the other. This crudely models the vocal tract, with the vocal cords at one end and the open mouth at the other. The vowel phoneme /i/ in in Figure 9.28e shows three resonant components as well as significant energy in many higher frequencies. Parts of the speech signal that do not contain high-energy tones, such as the /k/ in Figure 9.28b, cannot have formant structures.

Another formant detection task is to identify a *pitch* or *fundamental harmonic frequency* among the significant tones. This is the frequency of vibration of the vocal cords. In real speech recognition systems, this frequency must be identified and tracked, as it varies with the utterance as well as with the gender and emotional

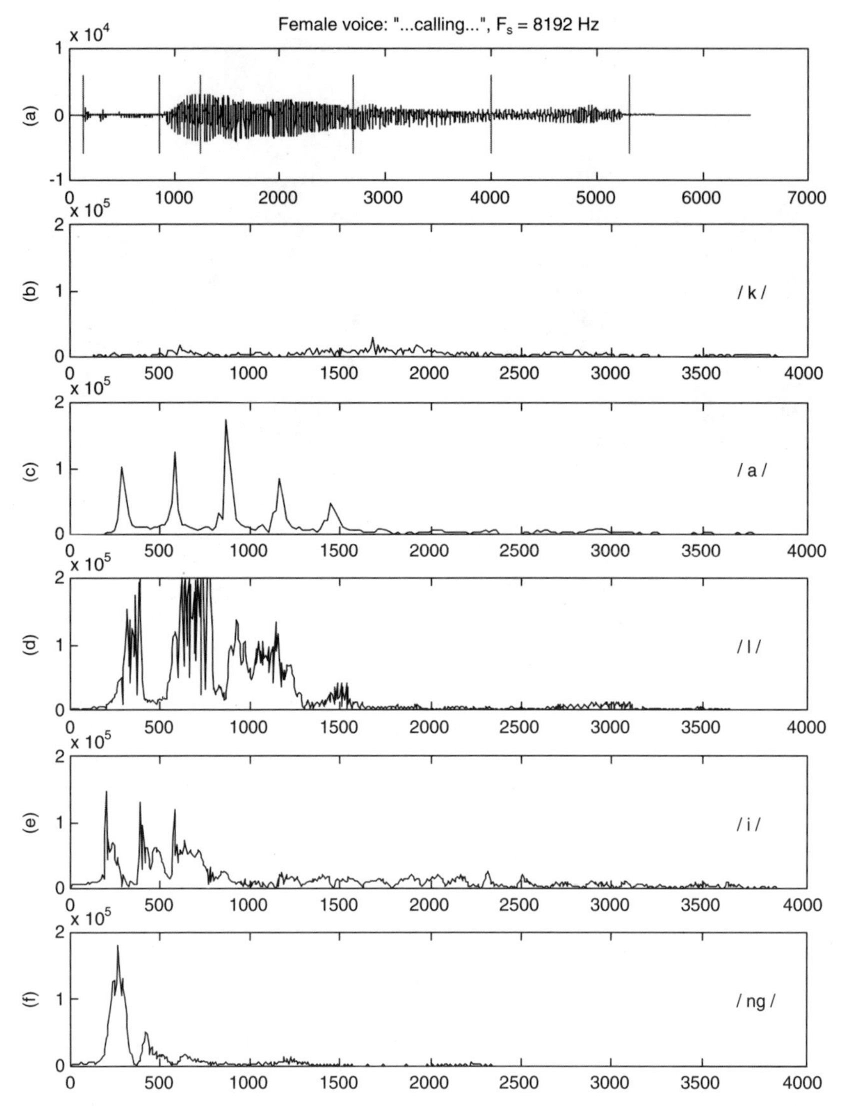

Fig. 9.28. Digitized voice (a) of a woman: "...calling...." Vertical lines mark the approximate locations of five phonemes. Panels (b)–(f) show the magnitude spectra for phonemes /k/, /a/, /l/, /i/, and /ng/.

state of the speaker. Some goal-directed information applies here. Pitch ranges from about 60 Hz to 300 Hz in adult males and up to 600 Hz in adult females.

A third formant detection task is to compare the center frequencies of significant tones. If such peaks represent formants, they must be integral multiples of the pitch frequency. Thus, in addition to ordinary peak finding, a formant detection

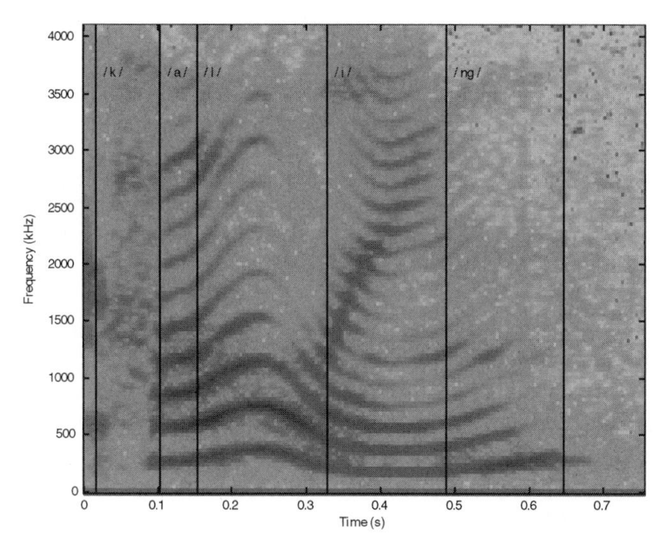

Fig. 9.29. Mixed-domain analysis of "calling" speech fragment. Using $N = 256$ point Hann windows for local frequency estimations centered $K = 32$ samples apart along the time axis produces the above time-frequency map. In many cases, voiced versus unvoiced speech segments can be found using the time-frequency map.

algorithm must include an assessment of the relative energy contributed by resonant frequencies.

9.4.2.2 Voice and Unvoiced Speech Determination. Speech sounds can be divided into *voiced* and *unvoiced* sounds, according to whether the vocal cords vibrate or do not vibrate, respectively (Figure 9.29). Unvoiced sounds split roughly into two categories: fricatives, such as /s/ or /f/, and aspirates, such as /k/.

9.4.2.3 Endpoint Detection. An important early speech analysis step involves automatic discrimination between background noise and speech signals. This segmentation procedure is important for automated compilation of speech databases and for detecting word boundaries in speech recognition systems. A fairly old but reliable method uses local energy and zero crossing rate parameters to isolate speech [55]. From a background noise sample, upper- and lower-energy parameters E_u and E_l, respectively, and a zero crossing threshold Z_c are determined. The algorithm refines the boundaries of a speech fragment in three stages as follows:

(i) The initial energy-based segmentation, say speech exists within $[M_u, N_u]$, is given by where local signal energy exceeds E_u.

(ii) The refined energy-based segmentation widens this interval to $[M_l, N_l]$ by searching outside $[M_u, N_u]$ for the points where energy diminishes to E_l.

(iii) Finally, a measure of local signal frequency—the zero crossing rate—is checked outside of $[M_l, N_l]$. If the rate exceeds Z_c three or more times in the 250-ms intervals on either side of $[M_l, N_l]$, then the speech fragment boundaries grow again to $[M_z, N_z]$, where $[M_u, N_u] \subseteq [M_l, N_l] \subseteq [M_z, N_z]$.

Noise versus speech discrimination problems continue to attract experimental researchers [56, 57].

9.4.3 Problematic Examples

Let us mention a couple of signal analysis problem domains where Fourier transform-based interpretation techniques begin to break down.

Seismic signals contain oscillations, but these oscillations are of unusually short time domain and are interspersed with transient artifacts that often thwart analysis. The Fourier transform is adequate for long-term periodic signal trends, but its efficiency as a signal descriptor diminishes with the duration of the oscillations. Geophysicists resorted to the short-time or windowed Fourier transform, which we cover in the next chapter, with some success. This theory constructs a time- and frequency-domain transform using the Fourier transform and the windowing techniques of Section 9.2. However, transient components are still problematic. Finally, efforts to get around the difficulties of windowed Fourier methods let to the development of the wavelet transform, which is the subject of Chapter 11.

Another problem area for Fourier methods is shape analysis. It seems that object shapes, however they might be described mathematically, are comprised of parts. One-dimensional methods, such as we develop here, can be applied to object boundaries, and we attempt such applications in the last chapter. There are Fourier transform-based approaches:

(i) An early method, called Fourier descriptors, approximates image object boundaries with varying numbers of Fourier series components.

(ii) Another method, the Fourier–Mellin transform, incorporates a temporal parameter into the transform.

These strategies work, but when the bounded object resolves into separate parts, the overall analytic techniques collapse. As a result, automated object recognition systems tend to retreat into structural methods of pattern recognition and cordon off, perhaps giving up on frequency-domain interpretation.

9.5 ANALOG FILTERS

Although our present goal is discrete time signal analysis and the frequency selective systems that support it, we have ample reasons for developing a respectable theory of analog filtering.

(i) While signal analysis is computerized interpretation of signals and therefore assumes a digital implementation, it relies upon discrete theory.

(ii) However, our ultimate source signals come from analog world.

(iii) We need to filter incoming analog so as to remove high frequencies that would otherwise cause aliasing during sampling, according to the Nyquist criterion.

(iv) We may be able to implement cheaper analog filters.

(v) A lot of discrete filter designs are based on analog filters.

(vi) We have shown that it is easy to derive discrete filters from analog versions—and especially if the analog filters have a rational transfer function.

(vii) Thus, if we develop a sound theory of analog filters, then we shall have a correspondingly sound theory of discrete filters.

(viii) Analog filters are historically prior.

(ix) The analog theory of continuous theory involves continuous sinusoidal signals, and hence the analog Fourier transform is a convenient theoretical tool. Anyway, this is consonant with our analog-first treatment of the frequency transforms.

Just as the contiuous Fourier transform is the natural setting for studying signal frequency, so the class of continuous domain, or analog, filters constitute the right beginning place for our study of frequency-domain signal analysis. Signal analysis usually takes place on a digital computer, but the frequency-selective algorithms that operate on digital signal representations often derive from continuous-domain filters. That is one reason for studying analog filters. But even before signals are digitized and fed to the computer, the signal digitization must often be filtered by analog means so that aliasing (Chapter 7) is minimized. This section presents some basic analog filters, introduces their characterisitic descriptors, and outlines the mechanics of converting one filter into another.

Conventional analog filter designs begin by examining the frequency-domain behavior $X(\omega)$ or the s-domain behavior $X_L(s)$ of analog signals $x(t)$. System theory texts covering analog filter theory and the Laplace transform include Refs. 6, 58, and 59. Texts that concentrate on the subsequent conversion to discrete time processing are [7–11, 26].

Classical electronics studies networks of electrical components—resistors, capacitors, and inductors—which implement the analog filtering operation in hardware. Circuit design texts cover the electronic circuit designs [60–63]. More modern electrical and computer engineering texts cover the design of hardware for digital filtering [64]. For signal analysis using digital computers, we need digital filters that selectively enhance and suppress signal frequency components of interest to the application. We derive the digital filters from their analog equivalents using some classic methods.

9.5.1 Introduction

A *filter* is a frequency-selective linear, translation-invariant system. Analog filtering takes place by virtue of the convolution property: $Y(\Omega) = H(\Omega)X(\Omega)$. So, if $|H(\Omega)|$ is small for values of Ω where it is desirable to suppress frequencies in the input $x(t)$

and $|H(\Omega)|$ is near unity where it is desirable to preserve frequencies in $x(t)$, then convolution with $h(t)$, $y(t) = (h*x)(t)$ performs the requisite frequency selection operation. To describe system filtering we specify the magnitude spectrum, $|H(\Omega)|$, or, equivalently, $|H(\Omega)|^2$. We are mainly interested in real-valued filters: $h(t) \in \mathbb{R}$. Since $H(-\Omega) = H^*(\Omega)$ in this case and since $|H(\Omega)| = |H^*(\Omega)|$, the filters herein have both positive and negative frequency components. Thus, the magnitude spectra are symmetric about the frequency-domain origin, $\Omega = 0$.

This section explains how to construct bandpass and high-pass filters from low-pass filters. The discussion begins with the Gaussian; its Fourier transform is also Gaussian, so it is a natural choice for a low-pass filter. Filter constructions depend on the Fourier transform properties from Chapter 5.

The analog convolution operation is once again denoted by the * operator: $y = x*h$. We define

$$y(t) = (x*h)(t) = \int_{-\infty}^{\infty} x(s)h(t-s)\,ds\,. \tag{9.78}$$

Section 6.4.1 introduced ideal analog filter types. One filter missing there is the notch or band-reject filter. It is a like a reverse bandpass, and it suppresses rather than preserves a range of frequencies.

9.5.2 Basic Low-Pass Filters

Since we can move easily from low-pass filters to any of the other three basic types—high-pass, bandpass, or band-reject—let us consider some examples.

9.5.2.1 Perfect. The ideal low-pass filter completely removes all frequencies higher than some given frequency and preserves the rest without amplifying or attenuating them. This is the familiar analog *moving average* or *boxcar* filter.

9.5.2.2 Gaussian. The Gaussian

$$g_{\mu,\sigma}(t) = \frac{1}{\sigma\sqrt{2\pi}}e^{-\frac{(t-\mu)^2}{2\sigma^2}} = g(t) \tag{9.79}$$

with mean μ and standard deviation σ has Fourier transform

$$G(\Omega) = \int_{-\infty}^{\infty} g(t)e^{-j\Omega t}dt = \exp\left(-\left[\frac{\sigma^2\Omega^2}{2} + j\Omega\mu\right]\right). \tag{9.80}$$

Its magnitude spectrum is also a Gaussian, centered at $\Omega = 0$. Thus, $g(t)$ is the impulse response of an analog low-pass filter. If $x(t)$ is an analog signal and $y = Gx = (g * x)(t)$ is the convolution with the Gaussian (9.79), then $Y(\Omega) = X(\Omega)G(\Omega)$, where $G(\Omega) = (\mathcal{F}g)(\Omega)$ is given by (9.80). Gaussians decay rapidly, faster than the inverse of any polynomial. Thus, the Fourier transform of the system response $Y(\Omega) = (\mathcal{F}y)(\Omega)$ will contain the $x(t)$ frequency components near $\Omega = 0$, but they will be suppressed by the product with the Gaussian $G(\Omega)$. This idea is basic to all low-pass filtering.

Now we also gain a deeper understanding of scale space analysis, introduced in Chapter 4. Recall that we showed that smoothing signals by ever broader Gaussian kernels had the unique property that no structure was created in the process. We understand signal structure to be determined by the concavity regions within the signal. Now (9.80) shows that the wider kernels are actually low-pass filters with smaller passbands. That is, the wider kernels progressively remove the high-frequency components, leaving relatively lower frequency undulations, and—more importantly—not creating additional changes in signal curvature.

A particularly important area of signal analysis is the detection, classification, and recognition of signal features that vary according to the size of their features—according to their *scale*, for instance.

9.5.2.3 Rational Functions. We can find other examples based on rational functions. For example, suppose $H(\Omega) = (1 + \Omega^2)^{-1}$. The inverse Fourier transform is

$$h(t) = \frac{1}{2\pi}\int_{-\infty}^{\infty} H(\Omega)e^{j\Omega t}\,d\Omega = \frac{e^{-|t|}}{2}, \tag{9.81}$$

which is easy to see from the forward radial transform $\mathcal{F}(\exp(-a|t|)) = 2a/(a + \Omega^2)$. Evidently, convolution with $h(t)$ performs a weighted averaging on input signal data. In signal analysis applications, the pulse $h(t)$ is often called a Lorentzian, and it is used to find peaks, valleys, and transients in general by the method of template matching.

9.5.2.4 Better Low-Pass Filters. It turns out that very good filters can be built by pursuing the idea of rational functions introduced in the previous example. The important features of such filters are as follows:

- They have superior cutoffs—sharper and more like perfect low-pass filters.
- Upon discrete conversion, they will be causal.
- They are stable.
- They have efficient implementations, relying on difference equations.

For signal analysis applications involving real-time data, such as in speech recognition or industrial control applications, causality is important. Of course, many signal analysis applications do not involve real-time data; that is, they are offline applications, and so causal systems are less critical. Nonetheless, very good filters can be developed for data whose discrete values are only known for the present and past.

9.5.3 Butterworth

Butterworth[12] *filters* have maximally flat pass and stop bands. Thus, the filter designer that values, above all else, reduced pass- and stop-band ripple inclines toward this filter.

9.5.3.1 Conditions for Optimally Flat Filters. We describe these filters via their frequency-domain representation. Their Fourier transforms are based on rational functions—quotients of continuous domain polynomials. The Butterworth filter specializes the rational function by looking at the Taylor series representations of its numerator and denominator.

Indeed, we have already constructed lowpass filters out of rational functions. The idea is to look at analog filters $h = H\delta$, such that their Fourier transform power spectra $|H(\Omega)|^2$ are rational functions $B(\Omega)/A(\Omega)$, where $A(\Omega)$ and $B(\Omega)$ are polynomials. We impose conditions on the rational functions so that we achieve our design criteria: passband performance, stopband performance, cutoff frequency, allowable ripple, and required transition band sharpness. Significantly, for signal analysis on digital computers, when the power spectrum of an analog filter is a rational function, then it can be used to derive a discrete filter.

A simple but useful fact is that if $h(t) \in \mathbb{R}$, then the squared magnitude spectrum $|H(\Omega)|^2$ is an even function of Ω. To see this, note that Fourier transform symmetry properties imply $H(-\Omega) = H^*(\Omega)$. So $|H(\Omega)|^2 = H(\Omega)H^*(\Omega) = H(\Omega)H(-\Omega)$. But then $|H(-\Omega)|^2 = H(-\Omega)H^*(-\Omega) = H(-\Omega)H(\Omega)$ too.

We thus consider $P(\Omega) = |H(\Omega)|^2 = B(\Omega)/A(\Omega)$, such that $P(\Omega)$ is symmetric about $\Omega = 0$. This means $A(\Omega)$ and $B(\Omega)$ are polynomials in Ω^2 (exercise). Thus,

$$A(\Omega) = a_0 + a_2\Omega^2 + a_4\Omega^4 \ldots + a_{2N}\Omega^{2N}, \tag{9.82a}$$

$$B(\Omega) = b_0 + b_2\Omega^2 + b_4\Omega^4 \ldots + b_{2M}\Omega^{2M}, \tag{9.82b}$$

with $a_0 = b_0$. We may assume that $a_0 = b_0 = 1$. A low-pass filter implies $|H(\Omega)| \to 0$ as $|\Omega| \to \infty$, so $N > M$. For the filter stopband to be maximally flat as $\Omega \to \infty$, the maximal number of numerator terms in (9.82b) should be zero. Thus, $b_2 = b_4 = \cdots = b_{2M} = 0$, and we see

$$P(\Omega) = |H(\Omega)|^2 = \frac{1}{A(\Omega)} = \frac{1}{1 + a_2\Omega^2 + a_4\Omega^4 \ldots + a_{2N}\Omega^{2N}}. \tag{9.83}$$

Butterworth criteria also require the filter's passband to be maximally flat at $\Omega = 0$, which entails $a_2 = a_4 = \cdots = a_{2n-2} = 0$. We define the cutoff frequency Ω_c of the

[12]After S. Butterworth, a British engineer who first analyzed this response profile ["On the theory of filter amplifiers," *Wireless Engineer*, vol. 7, pp. 536–554, 1930]. V. D. Landon later described this same filter as *maximally flat* [Cascade amplifiers with maximal flatness, *RCA Review*, vol. 5, pp. 347–362, 1941].

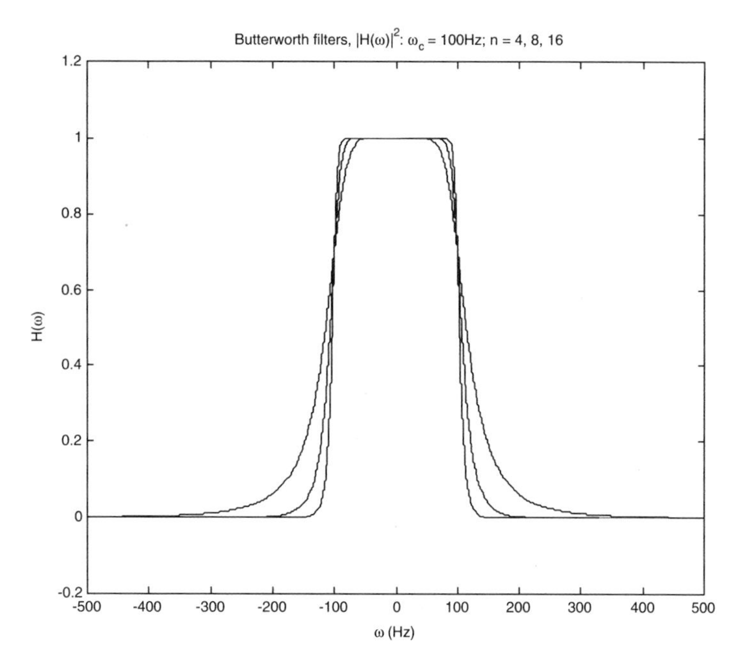

Fig. 9.30. Analog Butterworth filters for a few orders.

Butterworth filter by $(\Omega_c)^{-2n} = a_{2n}$. Thus, the *Butterworth filter* of order $N > 0$ is defined by its Fourier transform $H(\omega)$ (Figure 9.30):

$$H(\Omega) = \sqrt{\frac{1}{1 + \left(\frac{\Omega}{\Omega_c}\right)^{2N}}}; \tag{9.84}$$

an important next step in practical implementation is to decide which square roots to choose for $H(\Omega)$ in (9.84). Note that $H(\Omega_c) = 2^{-1/2} \approx 0.707$, for any filter order.

It is possible to invert the Butterworth filter Fourier transform $H(\Omega)$, truncate the impulse response $h(t)$, and then sample the result. As we have observed, this would induce some errors due to aliasing into the final filtering result. The filter designer might reduce these errors by preserving a large number of discrete samples. But this necessitates a time-consuming convolution operation. It turns out, however, that because a Butterworth filter has a rational Fourier transform, an efficient discrete implementation is possible using difference equations.

Now let us turn to the approximation problem for the Butterworth filter. The procedure differs depending on how the designer performs conversion from the analog to discrete filter form: impulse invariance or bilinear transformation.

9.5.3.2 Butterworth Approximation: Impulse Invariance. Given bounds on how far the filter's magnitude spectrum can stray from the ideal passband and stopband, the Butterworth filter approximation finds two parameters: the filter order $N > 0$ and the radial cutoff frequency Ω_c in (9.84). Suppose that we require a discrete low-pass Butterworth filter with unit DC gain: $|H(\omega)| = 1$ for $\omega = 0$, where $H(\omega)$ is the DTMF of the filter impulse response $h(n)$. We assume that the sampling frequency is high enough so that aliasing of $H_a(\Omega)$, the filter's (radial) Fourier transform, is not an issue. *This allows us to use the analog filter's magnitude response in the design approximation.*

Suppose that the passband is $|\Omega| < \Omega_p$ and the application requires that $|H_a(\Omega)|$ is within $\delta > 0$ of unity. Suppose the stopband begins at $\Omega_s > \Omega_p$, and we need $|H_a(\Omega)|$ to be within $\lambda > 0$ of zero. The Butterworth magnitude response is monotone; it suffices to consider the dual constraints:

$$|H(\Omega_p)| = \sqrt{\frac{1}{1 + \left(\frac{\Omega_p}{\Omega_c}\right)^{2N}}} \geq 1 - \delta \approx 1, \tag{9.85a}$$

$$|H(\Omega_s)| = \sqrt{\frac{1}{1 + \left(\frac{\Omega_s}{\Omega_c}\right)^{2N}}} \leq \lambda \approx 0. \tag{9.85b}$$

Filter designers often prefer approximation by differences in decibels. Thus, we say the passband magnitude (dB) is greater than a small negative value (Δ) and the stopband magnitude (dB) is less than a large negative value (Λ):

$$0 > 10\log_{10}|H(\Omega_p)|^2 \geq \Delta, \tag{9.86a}$$

$$10\log_{10}|H(\Omega_s)|^2 \leq \Lambda < 0. \tag{9.86b}$$

The above constraints reduce to

$$1 + \left[\frac{\Omega_p}{\Omega_c}\right]^{2N} \leq R_\Delta = 10^{-\frac{\Delta}{10}} \approx 1, \tag{9.87a}$$

$$1 + \left[\frac{\Omega_s}{\Omega_c}\right]^{2N} \geq R_\Lambda = 10^{-\frac{\Lambda}{10}} >> 1, \tag{9.87b}$$

but (9.85a) and (9.85b) give similar relations too. Let $R = \frac{\log(R_\Delta - 1)}{\log(R_\Lambda - 1)}$ so that solving for Ω_c gives

$$\log\Omega_c = \frac{R\log\Omega_s - \log\Omega_p}{R - 1}. \tag{9.88}$$

At this point, the filter designer makes choices. We can use (9.87a) to find the Butterworth filter order

$$\nu = \frac{\log(R_\Delta - 1)}{2\log\left(\frac{\Omega_p}{\Omega_c}\right)}, \tag{9.89}$$

which will generally not be a whole number. Designers usually round the order upward, taking the Butterworth order N to be the *integral ceiling* of ν in (9.89)—the integer greater than ν, or $Ceil(\nu)$. Using equality in either (9.87a) or (9.87b), the cutoff frequency must be recomputed. For instance, setting

$$\Omega_{c,N} = \frac{\Omega_p}{\sqrt[2N]{R_\Delta - 1}} \tag{9.90}$$

establishes a new cutoff frequency based on the upwardly rounded order and the passband constraint. The resulting low-pass Butterworth filter satisfies the passband condition determined by $\Omega_{c,N}$ and improves upon the stopband condition. This is a good approach for low sampling rates, when the aliasing caused by filter conversion using impulse invariance is a concern [7].

Alternatively, tight application timing contraints might force the designer to round down and choose a smaller N, the *floor* of ν in (9.89). Perhaps $N = Floor(\nu) \approx \nu$, or deviation from the passband and stopband specifications is acceptable. The designer might also opt for a cutoff frequency that favors an exactly met stopband. Assuming that the filter order is rounded upward, the filter then meets a more stringent passband specification. The cost of this option is that it does not counteract the aliasing arising from impulse invariance. The exercises explore these design alternatives.

We determine the ω_p and ω_s (radians/sample) from the sampling rate $F_s = 1/T$ and the application's specified analog passband and stopband frequencies (Hz).

Example (DTMF Passband and Stopband Calculations). Suppose we require a lowpass filter that removes high-frequency noise for the dual-tone multifrequency application of Section 9.1. Let the digital sampling rate be $F_s = 8192$ Hz. Then discrete frequency $\omega = \pm\pi$ corresponds to the Nyquist frequency of $F_s/2 = 4096$ Hz. DTMF tones range up to 1633 Hz. Thus, the lowpass filter bands could be specified by setting $\omega_p = (1633/4096) \times \pi \approx 0.3989\pi$, and, depending upon the desired filter sharpness, $\omega_s = (1800/4096) \times \pi \approx 0.4395\pi$.

Example (Butterworth impulse invariance approximation). We require a lowpass filter with a passband within 1 dB of unity up to $\omega_p = \pi/4$ and at least 5 dB below unity beyond $\omega_s = \pi/3$. Assume the sampling interval is $T = 1$. Thus, we require $\Delta = -1$, and $\Lambda = -5$. The approximation steps above give $R_\Delta = 1.2589$, $R_\Lambda = 3.1623$, and $R = -1.7522$. The exact cutoff frequency for this example is $\Omega_c = 0.9433$, but we elect to round $\nu = 3.6888$ upward to $N = 4$. Revising the cutoff frequency, we find $\Omega_{c,N} = 0.9299$.

9.5.3.3 Poles and Zeros Analysis. Let us continue the design of the Butterworth low-pass filter of order $N > 0$ by factoring the denominator of the squared magnitude response. Indeed, many analog filters satisfy

$$P_H(\Omega) = |H(\Omega)|^2 = \frac{1}{1+\left(\frac{\Omega}{\Omega_c}\right)^{2N}} = H(\Omega)H(-\Omega). \tag{9.91}$$

$P_H(\Omega)$ has $2N$ poles in the complex plane. Let the roots of the denominator in (9.91) be $\Omega_1, \Omega_2, ..., \Omega_{2N}$. In the case of the Butterworth squared magnitude, the roots lie on a circle of radius Ω_c in the complex plane. They are in fact the order-$2N$ roots of unity scaled by the cutoff frequency Ω_c. All we have to do to find $H(\Omega)$ that satisfies (9.91) is to select one pole from each pair $\{\Omega_i, -\Omega_i\} \subset \{\Omega_1, \Omega_2, ..., \Omega_{2N}\}$. But a judicious root selection allows us to construct a causal discrete filter governed by a difference equation and therefore having an efficient computer implementation. We can obtain a discrete difference equation from an analog filter that has a rational Fourier transform $H(\Omega)$ only if its poles have positive imaginary parts (Section 9.2.5.5). (Equivalently, if we are working with the Laplace transform, because $H(\Omega) = H_L(s)\big|_{s=j\Omega}$, this means that the poles of $H_L(s)$ must have negative real parts.) With causality and difference equation implementation in mind, we retain those roots, $\Omega_1, \Omega_2, ..., \Omega_N$, such that

$$H(\Omega) = \frac{1}{(\Omega-\Omega_1)(\Omega-\Omega_2)\ldots(\Omega-\Omega_N)}. \tag{9.92}$$

with $\mathrm{Imag}(\Omega_i) > 0$.

Let the partial fractions expansion of (9.92) be

$$H(\Omega) = \sum_{i=1}^{N} \frac{c_i}{\Omega-\Omega_i}, \tag{9.93}$$

where c_i, $1 \le i \le N$, are constants. The results of Section 9.2.5.5 give the corresponding discrete filter's transfer function:

$$H(z) = \sum_{i=1}^{N} T\frac{jc_i}{1-e^{jT\Omega_i}z^{-1}}, \tag{9.94}$$

where T is the sampling interval. We generally compute ω_p and ω_s using the sampling interval, as explained in the previous section. Then, we derive the order N and the cutoff frequency Ω_c from the Butterworth approximation. The poles and zeros analysis gives a Butterworth filter of order N with cutoff frequency Ω_c, no matter what sampling interval we choose; it is conventient to assume $T = 1$. It is usually necessary to scale the coefficients of the discrete filter's impulse response for unit DC gain. Let us see how this works by following through on the previous example.

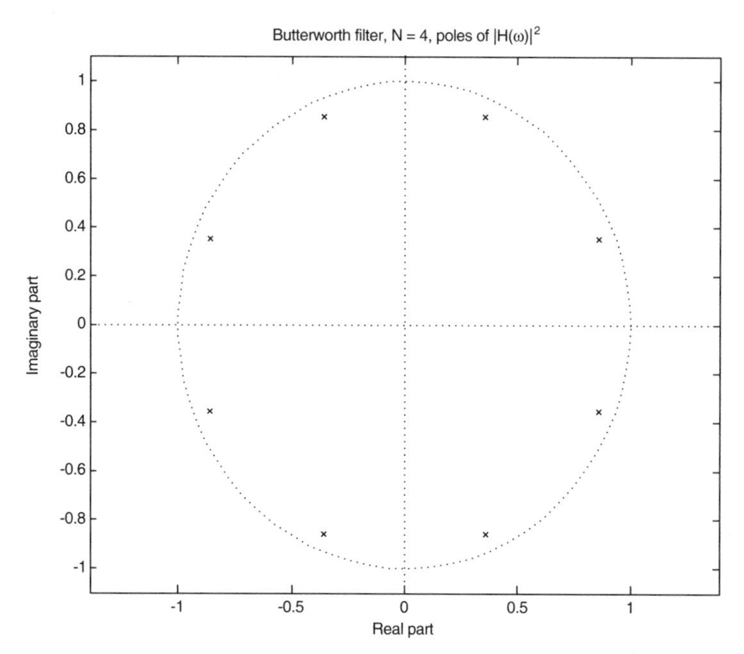

Fig. 9.31. Butterworth filter, $N = 4$, eight poles of $|H(\Omega)|^2$.

Example (Butterworth impulse invariance design, $N = 4$). Suppose we require a lowpass filter with order $N = 4$ and cutoff frequency $\Omega_c = 0.9299$, as in the previous example. Figure 9.31 shows the eight roots of the Butterworth squared magnitude function denominator. To find $H(\Omega)$, we select poles having a positive imaginary part. This selection corresponds to the poles of $H_L(s)$ whose real parts are negative, because if Ω_0 is a pole of $H(\Omega)$, then $H_L(j\Omega_0)$ is a pole of $H_L(s)$ (Table 9.4).

TABLE 9.4. Pole Selection in Butterworth Filter Design, $N = 4$, $\Omega_c = 0.9299$

Poles of $\lvert H(\Omega)\rvert^2$	Fourier-Selected Poles	Poles of $\lvert H_L(s)\rvert^2$	Laplace-Selected Poles
$-0.8591 + 0.3559j$	$-0.8591 + 0.3559j$	$-0.3559 - 0.8591j$	$-0.3559 - 0.8591j$
$-0.8591 - 0.3559j$		$0.3559 - 0.8591j$	
$-0.3559 + 0.8591j$	$-0.3559 + 0.8591j$	$-0.8591 - 0.3559j$	$-0.8591 - 0.3559j$
$-0.3559 - 0.8591j$		$0.8591 - 0.3559j$	
$0.3559 + 0.8591j$	$0.3559 + 0.8591j$	$-0.8591 + 0.3559j$	$-0.8591 + 0.3559j$
$0.3559 - 0.8591j$		$0.8591 + 0.3559j$	
$0.8591 + 0.3559j$	$0.8591 + 0.3559j$	$-0.3559 + 0.8591j$	$-0.3559 + 0.8591j$
$0.8591 - 0.3559j$		$0.3559 + 0.8591j$	

Thus, performing the partial fractions calculations for $H(z)$ gives

$$H(z) = T\left[\frac{j(-0.2379+0.5745j)}{1-e^{jT(-0.8591+0.3559j)}z^{-1}} + \frac{j(1.3869-0.5745j)}{1-e^{jT(-0.3559+0.8591j)}z^{-1}} + \frac{j(-1.3869-0.5745j)}{1-e^{jT(0.3559+0.8591j)}z^{-1}} + \frac{j(0.2379+0.5745j)}{1-e^{jT(0.8591+0.3559j)}z^{-1}}\right] \quad (9.95)$$

where T is the sampling interval. Writing $H(z)$ as a quotient of polynomials in z^{-1}, we see that (9.46) becomes

$$\frac{\sum_{m=0}^{M} b_m z^{-m}}{\left[1+\sum_{k=1}^{N} a_k z^{-k}\right]} = \frac{0.0873z^{-1}+0.1837z^{-2}+0.0260z^{-3}}{1+(-1.7091)z^{-1}+1.3967z^{-2}+(-0.5538)z^{-3}+0.0880z^{-4}}. \quad (9.96)$$

Now, finally, we can implement (9.95) with a cascade architecture and (9.96) with a Direct Form II. We can compute the impulse response $h(n)$ as in Figure 9.32 by feeding a discrete impulse through either of the difference equation architectures or by using (9.68).

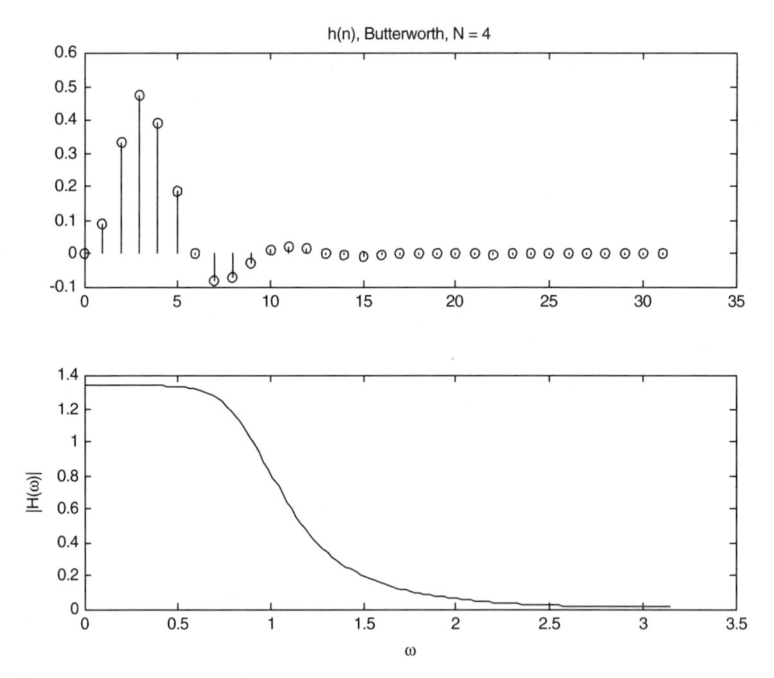

Fig. 9.32. Impulse response $h(n)$ of Butterworth low-pass filter, $N = 4$, $\Omega_c = 0.9299$ (top), and magnitude response (bottom) $|H(\omega)|$. Note that $H(0) \approx 1.3386$, so for unit DC gain, we scale $h(n)$ by $(1.3386)^{-1}$.

9.5.3.4 Butterworth Approximation: Bilinear Transformation. This section considers the Butterworth low-pass filter approximation using the bilinear transformation. The crucial difference in finding the filter order N and the cutoff frequency Ω_c is the frequency mapping $H(\omega) = H_a\left(\frac{2}{T}\tan\left(\frac{\omega}{2}\right)\right)$. Here, $H(\omega)$ is the DTMF of the desired discrete low-pass filter, $T > 0$ is the sampling interval, and $H_a(\Omega)$ is an analog Butterworth low-pass filter whose order N and cutoff frequency Ω_c remain to be found.

Let us suppose specifications on the passband and stopband similar to those with which we began the impulse invariance approximation. Thus,

$$0 > 10\log_{10}|H(\omega_p)|^2 = 10\log_{10}\left|H_a\left(\frac{2}{T}\tan\left(\frac{\omega_p}{2}\right)\right)\right|^2 \geq \Delta, \tag{9.97a}$$

$$10\log_{10}|H(\omega_s)|^2 = 10\log_{10}\left|H_a\left(\frac{2}{T}\tan\left(\frac{\omega_s}{2}\right)\right)\right|^2 \leq \Lambda < 0, \tag{9.97b}$$

where ω_p is the discrete passband frequency, $\omega_s > \omega_p$ is the stopband frequency, and $\Lambda < \Delta < 0$. Applying the Butterworth filter condition as before, we calculate

$$\left[\frac{2}{T\Omega_c}\tan\left(\frac{\omega_p}{2}\right)\right]^{2N} \leq 10^{-\frac{\Delta}{10}} - 1 = R_\Delta - 1, \tag{9.98a}$$

$$\left[\frac{2}{T\Omega_c}\tan\left(\frac{\omega_s}{2}\right)\right]^{2N} \geq 10^{-\frac{\Lambda}{10}} - 1 = R_\Lambda - 1. \tag{9.98b}$$

We treat these as equalities and solve for the filter order:

$$\nu = \frac{\log\left(\dfrac{R_\Delta - 1}{R_\Lambda - 1}\right)}{2\log\left(\dfrac{\tan(\omega_p/2)}{\tan(\omega_s/2)}\right)} \tag{9.99}$$

where, in general, ν is not an integer. Typically, we round ν upward: $N = \text{Ceil}(\nu)$. We can use equality in either (9.98a) or (9.98b) to find Ω_c. An advantage of bilinear transformation over impulse invariance is that there is no stopband aliasing. Thus, computing Ω_c in terms of ω_s gives

$$\Omega_c = \frac{2\tan(\omega_s/2)}{T}(R_\Lambda - 1)^{-\frac{1}{2N}}. \tag{9.100}$$

Example (Butterworth Bilinear Transformation Approximation). Suppose we require the same low-pass filter: $\omega_p = \pi/4$, $\omega_s = \pi/3$, $\Delta = -1$, and $\Lambda = -5$. Again, $R_\Delta = 1.2589$, $R_\Lambda = 3.1623$, but we calculate $\nu = 3.1957$ and choose $N = 4$. The exact analog cutoff frequency with $T = 1$ is $\Omega_c = 1.048589$, corresponding to a discrete filter cutoff of $\omega_c = 0.965788$. Note that as ν nears its integral floor—in this case $\nu \approx 3$—it might be feasible to round to the *lower* integral order. This could be useful when computing time is a consideration and some cushion exists for passband and stopband specifications.

Having derived the filter order N and the analog cutoff frequency Ω_c, the rest of the bilinear filter design steps are the same as for impulse invariance.

Example (Butterworth Bilinear Transformation Design, $N = 4$). For the low-pass filter with order $N = 4$ and cutoff frequency $\omega_c = 0.9658$, as in the previous example, the analog cutoff frequency is $\Omega_c = 1.0486$, since $\Omega_c = \frac{2}{T}\tan\left(\frac{\omega_c}{2}\right)$. Of the eight poles of

$$|H_a(\Omega)|^2 = \frac{1}{1+\left(\frac{\Omega}{\Omega_c}\right)^{2N}} = H_a(\Omega)_a H(-\Omega), \tag{9.101}$$

we select those having a positive imaginary part for $H_a(\Omega)$. The poles of the Laplace transform $H_{L,a}(s)$ will thus have negative real parts (Table 9.5).

Thus, using the chosen poles (Table 9.5) the rational Laplace transform is

$$H(s) = \frac{1}{s^4 + 2.7401s^3 + 3.7541s^2 + 3.0128s + 1.2090}. \tag{9.102}$$

Substituting the bilinear transform relation $s = \frac{2}{T}\left(\frac{z-1}{z+1}\right)$ with $T = 1$ into (9.102) gives the discrete transfer function,

$$H(z) = \frac{0.0166z^4 + 0.0665z^3 + 0.0997z^2 + 0.0665z + 0.0166}{z^4 - 1.5116z^3 + 1.2169z^2 - 0.4549z^2 + 0.0711}. \tag{9.103}$$

TABLE 9.5. Pole Selection in Butterworth Filter Design Using Bilinear Transformation, $N = 4$, $\Omega_c = 0.9658$.

$\|H_a(\Omega)\|^2$ Poles	Fourier Poles Selected	$\|H_{L,a}(s)\|^2$ Poles	Laplace Poles Selected
$-0.9688 + 0.4013j$	$-0.9688 + 0.4013j$	$-0.4013 - 0.9688j$	$-0.4013 - 0.9688j$
$-0.9688 - 0.4013j$		$0.4013 - 0.9688j$	
$-0.4013 + 0.9688j$	$-0.4013 + 0.9688j$	$-0.9688 - 0.4013j$	$-0.9688 - 0.4013j$
$-0.4013 - 0.9688j$		$0.9688 - 0.4013j$	
$0.4013 + 0.9688j$	$0.4013 + 0.9688j$	$-0.9688 + 0.4013j$	$-0.9688 + 0.4013j$
$0.4013 - 0.9688j$		$0.9688 + 0.4013j$	
$0.9688 + 0.4013j$	$0.9688 + 0.4013j$	$-0.4013 + 0.9688j$	$-0.4013 + 0.9688j$
$0.9688 - 0.4013j$		$0.4013 + 0.9688j$	

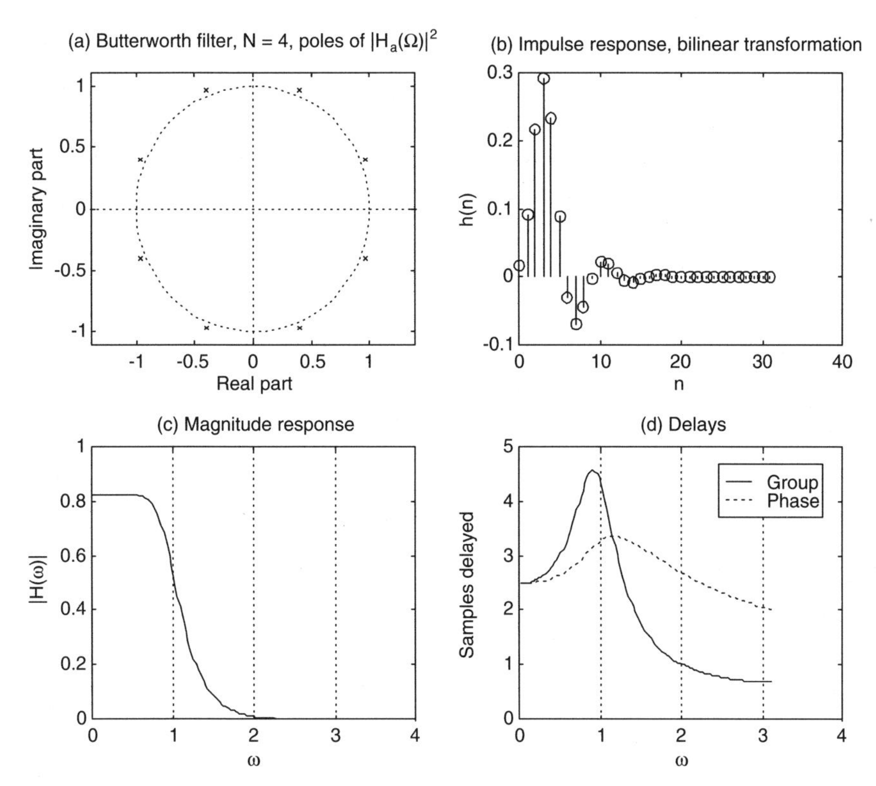

Fig. 9.33. Butterworth filter design, bilinear transformation. Pole locations for analog squared magnitude response (a). Impulse response $h(n)$ of Butterworth low-pass filter, $N = 4$, $\omega_c = 0.9658$ (b) and (c) magnitude response $|H(\omega)|$. Note that $H(0) \approx 0.8271$, so for unit DC gain, we scale $h(n)$ by $(0.8271)^{-1} \approx 1.2090$. Panel (d) shows the phase and group delay for this filter.

The partial fraction expansion of $H(z) = B(z)/A(z)$ is

$$\frac{B(z)}{A(z)} = \frac{-0.3048 + 0.1885j}{1-(0.4326+0.5780j)z^{-1}} + \frac{-0.3048 - 0.1885j}{1-(0.4326-0.5780j)z^{-1}} + \frac{0.1962 - 1.2480j}{1-(0.3232+0.1789j)z^{-1}} + \frac{0.1962 + 1.2480j}{1-(0.3232-0.1789j)z^{-1}} + 0.2337 \tag{9.104}$$

The filter has a difference equation implementation (9.49), from which the impulse response follows (Figure 9.33b).

Note that it is possible to skip the approximation step, instead stipulating a discrete cutoff frequency ω_c and guessing a filter order $N > 0$. After computing the associated analog cutoff Ω_c, the above poles and zeros analysis follows. This step produces an impulse response $h(n)$, from which one can derive a Fourier magnitude response $|H(\omega)|$. Should the passband or stopband not meet the anticipated constraints, the filter order is incremented and trial-and-error goes on. This is a good use case for a computerized filter design package.

Consider the group and phase delays of the above Butterworth low-pass filter. If $H(\omega) = e^{j\phi(\omega)}H_R(\omega)$ is a filter's DTFT, where $H_R(\omega) \in \mathbb{R}$, and $\phi(\omega)$ is its phase response, then its group delay is $-d\phi(\omega)/d\omega$. Since FIR filters enjoy linear phase (Section 9.3.6), their group delay is constant. Butterworth filters are IIR, so the group delay varies. In fact, Figure 9.33d illustrates that this system's group delay can change as much as four samples over the discrete frequency domain.

9.5.4 Chebyshev

Suppose a signal analysis application needs a low-pass filter with an especially sharp cutoff frequency, but tolerates some passband ripple. The Butterworth condition began by hypothesizing flatness in both the pass- and stopbands, so we need to relax one of these constraints. The first type of *Chebyshev*[13] filter approximation achieves a sharper transition than the Butterworth, but it does so at the cost of allowing passband ripple. On the other hand, the stopband is flat, and the designer can easily reduce the ripple to any positive value.

9.5.4.1 Chebyshev Polynomials and Equiripple Conditions. For the filter stopband to be maximally flat as $\Omega \to \infty$, any (rational) analog low-pass filter will have a squared Fourier magnitude response with a constant numerator (9.83). If we hope to improve upon the Butterworth filter's sharpness, we have to relax the Butterworth contraints. By the logic of the argument for maximally flat filters in Section 9.5.3.1, we must allow more a_{2k} to be nonzero. Let $\varepsilon^2 T(\Omega) = A(\Omega) - 1$ be the nonconstant part of the denominator in (9.83), where $\varepsilon > 0$ is a parameter controlling the passband ripple height. A low-pass filter with unity gain requires a denominator near unity when $\Omega = 0$, so let us stipulate that $|T(\Omega)| \le 1$ for $|\Omega| \le 1$. Then a suitable ε can always make $\varepsilon^2 T(\Omega)$ small for Ω near zero. Since $T(\Omega)$ is still a polynomial, its magnitude will get arbitrarily large as $|\Omega| \to \infty$, and so away from the origin, $P(\Omega) \to 0$. Are there such polynomials?

Indeed approximation theory provides us with precisely such polynomials. We set $T(\Omega) = T_N(\Omega)$, where $T_N(\Omega)$ is the *Chebyshev polynomial* of order $N \ge 0$ [43, 65]. These are defined recursively as follows: $T_0(\Omega) = 1$, $T_1(\Omega) = \Omega$, and

$$T_{N+1}(\Omega) = 2\Omega T_N(\Omega) - T_{N-1}(\Omega). \tag{9.105}$$

The Chebyshev polynomials have nice properties, which are explored in the exercises and exploited by our filter designs. It can be shown that

$$T_N(\Omega) = \cos(N\cos^{-1}(\Omega)), \tag{9.106a}$$

[13]In addition to his work on orthogonal functions, Russian mathematician Pafnuty Lvovich Chebyshev (1821–1894) proved Bertrand's conjecture: For $n > 3$, there is at least one prime number between n and $2n - 2$.

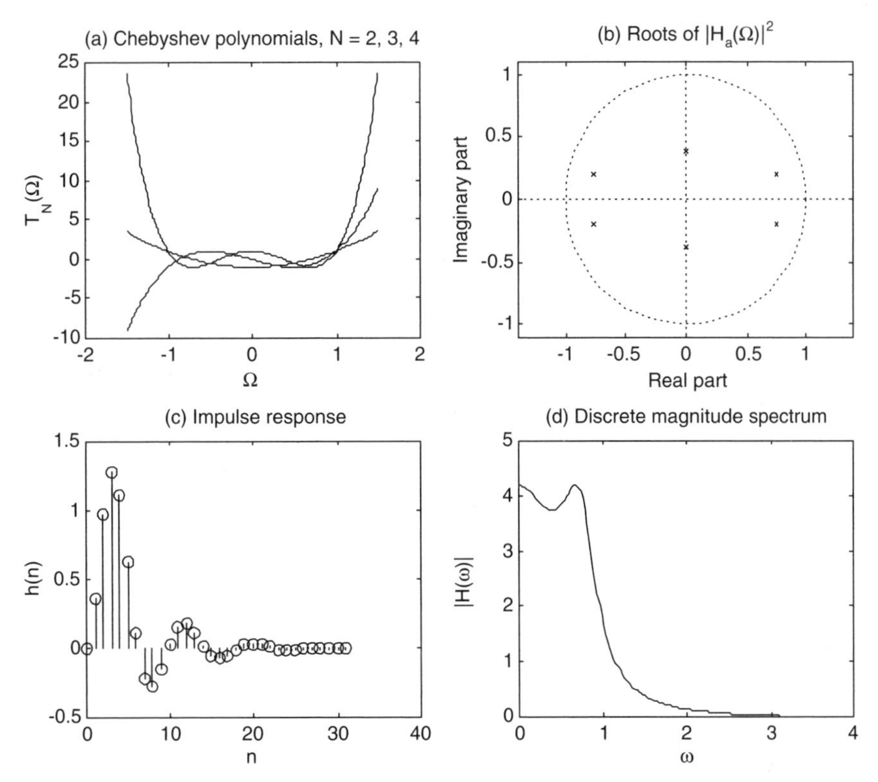

Fig. 9.34. Chebyshev polynomials for a few orders (a). Roots of the Fourier squared magnitude response for a third-order low-pass filter (b). Corresponding discrete impulse response (c) and magnitude spectrum (d).

so that indeed $|T_N(\Omega)| \le 1$ for $|\Omega| \le 1$. Furthermore, the polynomials are orthogonal on $[-1, 1]$. Chebyshev polynomials are also given for $|\Omega| > 1$ by the relation

$$T_N(\Omega) = \cosh(N\cosh^{-1}(\Omega)), \tag{9.106b}$$

as shown in Ref. 65 (see Figure 9.34).

Importantly for the filter approximation problem:

- If $1 \le N$ and $\Omega \in [-1, 1]$, then $T_N(\Omega)$ oscillates between -1 and $+1$.
- $T_N(\Omega)$ always achieves the minimum of -1 and the maximum of $+1$ on $[-1, 1]$ (for this reason it is called the *equal ripple approximation*).
- $T_N(1) = 1$ for all N.
- For $|\Omega| > 1$, $T_N(\Omega)$ is strictly increasing or strictly decreasing.

Thus, the *Chebyshev squared magnitude response* is defined by

$$|H_a(\Omega)|^2 = \frac{1}{A(\Omega)} = \frac{1}{1 + \varepsilon^2 T_N^2(\Omega/\Omega_c)}. \tag{9.107}$$

9.5.4.2 Impulse Invariance Approximation. For the Chebyshev low-pass filter approximation using the impulse invariance transformation, we are given a discrete passband frequency ω_p, a stopband frequency ω_s, an allowable passband ripple, and a required stopband attenuation value. We seek the analog cutoff frequency Ω_c, the filter order N, and the ripple parameter ε (9.107).

The Chebyshev polynomials' properties simplify the approximation. $|T_N(\Omega/\Omega_c)|$ is strictly increasing for $\Omega > \Omega_c$, so we can set $\Omega_c = \omega_p$. If the passband constraint is $10\log_{10}|H(\omega_p)|^2 \geq \Delta$, for some $\Delta < 0$ as in (9.86a), then the maximum departure from unity will occur for some $-\Omega_c \leq \Omega \leq \Omega_c$. Thus, we need $\varepsilon \geq \sqrt{10^{-\Delta/10} - 1}$. Here our design assumes that *ripple* is defined by the passband peak-to-valley difference. But some treatments assume that the ripple is half of this value—how far the passband magnitude strays from its mean [26]. So readers should be aware of the differences this assumption can make in the final design specifications. Finally, suppose the stopband specification (9.86b) is $10\log_{10}|H(\omega_s)|^2 \leq \Lambda$, for some $\Lambda < 0$, and the sampling rate is sufficiently high so that in the stopband, $|\omega| > \omega_s$,

$$|H(\omega)|^2 \approx |H_a(\Omega)|^2 = \frac{1}{1 + [\varepsilon T_N(\Omega/\Omega_c)]^2}. \tag{9.108}$$

Thus, we seek $N > 0$ such that

$$|H_a(\omega_s)|^2 = 10\log_{10}[1 + [\varepsilon T_N(\omega_s/\Omega_c)]^2]^{-1} \leq \Lambda. \tag{9.109}$$

We solve (9.109) as an equality for N,

$$N = \frac{\cosh^{-1}\left(\frac{\sqrt{10^{-\Lambda/10} - 1}}{\varepsilon}\right)}{\cosh\left(\frac{\omega_s}{\Omega_c}\right)}, \tag{9.110}$$

and round upward to the nearest integer.

Example (Chebyshev Impulse Invariance Approximation). Suppose we try the Chebyshev approximation on a filter with the same specifications as in the Butterworth impulse invariance design (Section 9.5.3.2). We need a passband within 1 dB of unity for $\omega < \omega_p = \pi/4$. This means $\Delta = -1$, $\Omega_c = \pi/4$, and $\varepsilon = 0.5088$. We need a stopband that is 5 dB or more below unity for $\omega > \omega_s = \pi/3$. We have $\Lambda = -5$, and (9.110) gives $N = 2.1662$, which we round up to $N = 3$. Thus, the Chebyshev approximation gives a third-order IIR filter, whereas the Butterworth approximation needed a fourth-order system—a benefit from allowing passband ripple.

Example (Chebyshev Impulse Invariance design, $N = 3$). Let us continue the previous example: $N = 3$, $\Omega_c = \pi/4$, and $\varepsilon = 0.5088$. The poles of $|H_a(\Omega)|^2$ are $-0.7587 \pm 0.1941j$, $0.7587 \pm 0.1941j$, and $\pm 0.3881j$. They lie on an ellipse in the complex plane [26] as shown in (Figure 9.34b). To find $H_a(\Omega)$, we select the three poles with positive imaginary parts. (Equivalently, for the Laplace transform-based filter derivation, these are $-0.1941 \pm 0.7587j$ and -0.3881.) Figure 9.34c shows the resulting impulse response. Figure 9.34d shows the discrete magnitude spectrum.

9.5.4.3 Bilinear Approximation. In a bilinear transformation, $\Omega_c = 2\tan\left(\frac{\omega_p}{2}\right)$ gives the cutoff frequency. The frequency mapping does not alter the passband ripple, so we can calculate ε just as with impulse invariance. For the filter order, the stopband condition says

$$10\log_{10}\left[\left(1 + \left[\varepsilon T_N\left(\frac{2}{\Omega_c}\tan\left(\frac{\omega_s}{2}\right)\right)\right]^2\right)^{-1}\right] \le \Lambda, \tag{9.111}$$

which implies

$$N \ge \frac{\cosh^{-1}\left(\frac{\sqrt{10^{-\Lambda/10} - 1}}{\varepsilon}\right)}{\cosh^{-1}\left(\frac{2}{\Omega_c}\tan\left(\frac{\omega_s}{2}\right)\right)}. \tag{9.112}$$

Example (Chebyshev Bilinear Approximation). Let us turn to the bilinear transformation method for the same filter design problem as above: $\omega_p = \pi/4$, $\omega_s = \pi/3$, $\Delta = -1$, $\Lambda = -5$, with sampling interval $T = 1$. Thus, $\Omega_c = 2\tan\left(\frac{\omega_p}{2}\right) = 0.8284$, using the bilinear frequency mapping. The ripple factor is $\varepsilon = 0.5088$. Solving (9.112) as an equality gives $N \approx 2.0018$, but we opt for the integral ceiling, setting $N = 3$.

Note that for the present design criteria, the Chebyshev filter comes very close to reducing the required filter order to $N = 2$. In fact, unless the application parameters are unusually rigid, this is an attractive possibility. The lesson is twofold:

- For the same filter order the Chebyshev filter has faster (sharper) rolloff than the equivalent Butterworth filter.
- It is possible to achieve the same rolloff as the equivalent Butterworth filter using a Chebyshev filter with a smaller order.

For a quicker rolloff, the Chebyshev suffers some passband ripple. Its transient response is also worse than the Butterworth, as shown below (Section 9.5.4.5).

Example (Chebyshev Bilinear Design, $N = 3$). Let us continue the above approximation: $N = 3$, $\varepsilon = 0.5088$, and $\Omega_c = 0.8284$. The poles of $|H_a(\Omega)|^2$ are $-0.8003 \pm 0.2047j$, $0.8003 \pm 0.2047j$, and $\pm 0.4094j$ (Figure 9.35a). The Laplace transform poles of choice are therefore $-0.2047 \pm 0.8003j$ and -0.4094. The analog system function is

$$H_{L,a}(s) = \frac{1}{s^3 + 0.8188s^2 + 0.8499s + 0.2793}. \tag{9.113}$$

Inserting the bilinear map $s = \frac{2}{T}\left(\frac{z-1}{z+1}\right)$ with $T = 1$ into (9.113) gives

$$H(z) = \frac{0.0754z^3 + 0.2263z^2 + 0.2263z + 0.0754}{z^3 - 1.8664z^2 + 1.4986z - 0.4637}. \tag{9.114}$$

$H(z)$ has partial fractions expansion

$$\frac{B(z)}{A(z)} = \frac{-0.6458 - 0.1170j}{1 - (0.6031 + 0.5819j)z^{-1}} + \frac{-0.6458 + 0.1170j}{1 - (0.6031 - 0.5819j)z^{-1}} + \frac{1.5297}{1 - (0.6602)z^{-1}} - 0.1627. \tag{9.115}$$

Figure 9.35 shows the impulse (b) and magnitude responses (c).

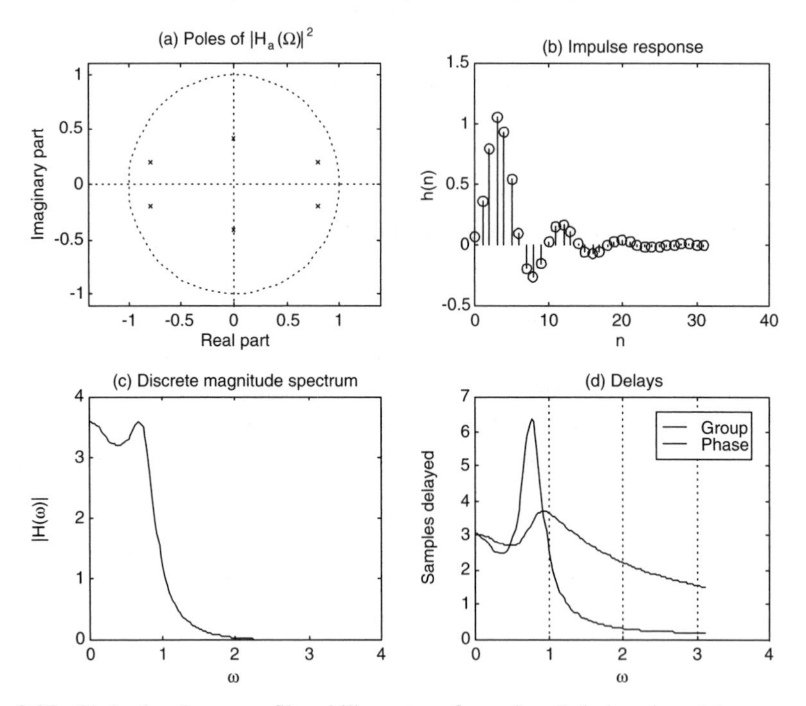

Fig. 9.35. Chebyshev low-pass filter, bilinear transformation. Pole locations (a), unnormalized discrete impulse response (b), magnitude response (c) and phase and group delay (d).

9.5.4.4 Phase and Group Delay of IIR Filters. Let us compare the group delay of Butterworth and Chebyshev filters. Figure 9.35d shows that the group delay of the Chebyshev filter can be as much as six samples for $\omega \in [0, \pi]$—worse than the equivalent Butterworth design Figure 9.33d. This is one thing the Chebyshev filter gives up in order to improve its rolloff performance.

9.5.4.5 Application: Transient Response Comparison. The Chebyshev filter's better rolloff, compared to the maximally flat filter, also costs it some transient response performance. To see this, let us consider a basic transient filtering application. Observe first that the Chebyshev filter exhibits a sharper rolloff Figure 9.36a.

We apply the Butterworth and Chebyshev low-pass filters (N = 3) developed in previous sections to the transient Figure 9.36b. The Chebyshev filter produces a longer delay and the ringing induced by the step edge persists for a longer time interval than with the Butterworth. On the other hand, the pulse's later sloped edge provokes only a little bad behavior from the Chebyshev filter Figure 9.36c.

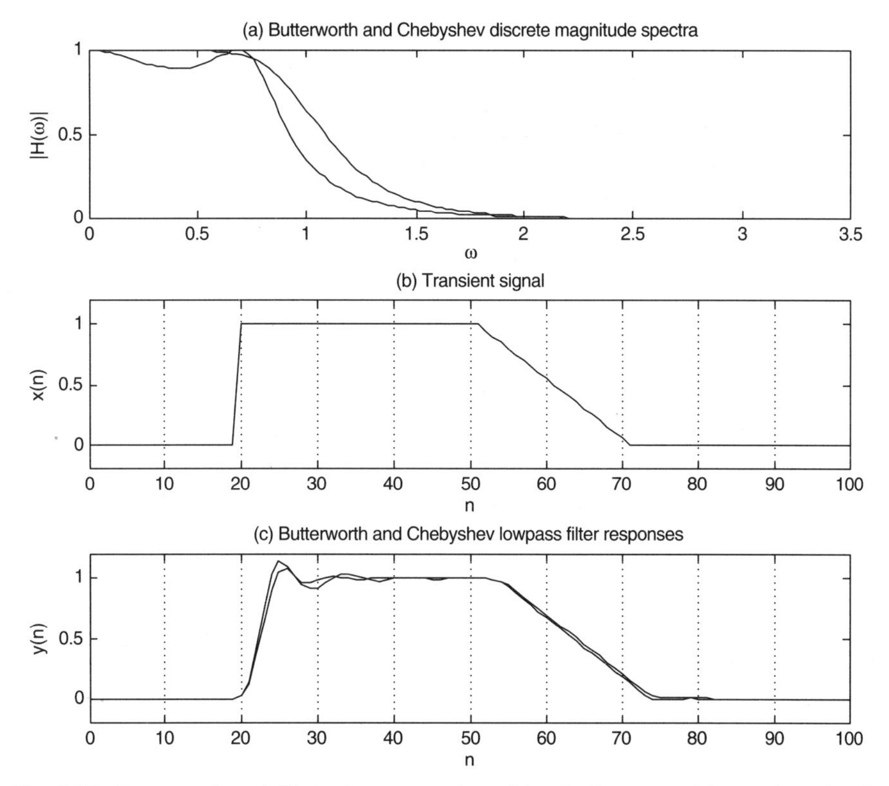

Fig. 9.36. Butterworth and Chebyshev comparison. Magnitude spectra (a), transient signal (b), and Butterworth and Chebyshev responses (c).

The following points should be taken into consideration when choosing the Chebyshev over the Butterworth filter for a signal analysis application:

- For less computational overhead with the same frequency discrimination performance, prefer the Chebyshev.
- If the frequencies in the filter passband will be further characterized by their relative strength, then the ripple in the Chebyshev becomes a detriment.
- If the application does not further analyze passband frequencies and is concerned with their mere presence or absence (such as in the DTMF application in Section 9.1) then the Chebyshev should be better.
- This is moreover the case when the application needs to segment relatively close bands of spectral information and sharp rolloff becomes a priority.
- Finally, if the time location of edges is important, and the application needs a crisp response from a crisp input edge to satisfactorily identify and locate the transition, then the Butterworth filter is superior.

9.5.5 Inverse Chebyshev

The inverse Chebyshev filter provides a flat passband and an equiripple stopband. This filter is also called the *Chebyshev Type II* filter.

9.5.5.1 Stopband Equiripple Conditions. The specification of the filter's squared magnitude response is based on the Chebyshev filter of Section 9.5.4 [26, 65]. The steps (Figure 9.37a) are as follows.

(i) We begin with the Chebyshev squared magnitude response function

$$|H_a(\Omega)|^2 = \frac{1}{1+\varepsilon^2 T_N^2(\Omega/\Omega_c)} = P(\Omega), \tag{9.116}$$

where $T_N(\Omega)$ is the order-N Chebyshev polynomial.

(ii) Subtract this response from unity, $1 - P(\Omega)$, to form the squared magnitude response of a high-pass filter having a flat passband and equiripple in the stopband.

(iii) Reverse the frequency axis to find $Q(\Omega) = 1 - P(\Omega^{-1})$:

$$Q(\Omega) = 1 - \frac{1}{1+\varepsilon^2 T_N^2(\Omega_c/\Omega)} = \frac{\varepsilon^2 T_N^2(\Omega_c/\Omega)}{1+\varepsilon^2 T_N^2(\Omega_c/\Omega)}, \tag{9.117}$$

which is the squared magnitude response of a low-pass filter. It has a maximally flat passband and puts the Chebyshev equiripple characteristic in the stopband.

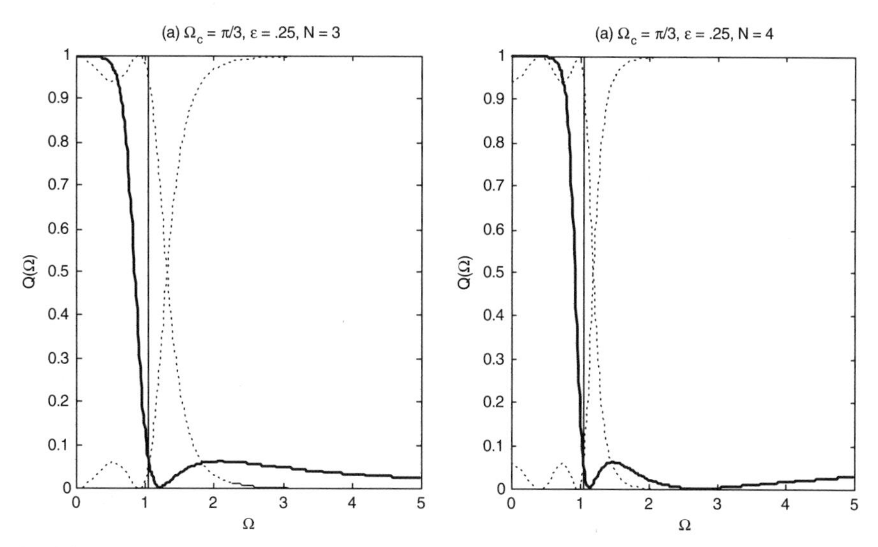

Fig. 9.37. Conversion from a Chebyshev to a third-order Inverse Chebyshev squared magnitude response function (a). The dotted traces are intermediate steps. The vertical line is Ω_c from (9.117) and represents the stopband frequency. For comparison, panel (b) shows an $N = 4$ squared magnitude response.

Note that $Q(\Omega)$ is a rational function in Ω^{-1}, but we can write it equivalently as a rational function of Ω. The poles of $Q(\Omega)$ are the reciprocals of the poles of $P(\Omega)$ (exercise).

9.5.5.2 *Impulse Invariance Approximation.* Consider the impulse invariance approximation for the inverse Chebyshev low-pass filter. Suppose the discrete passband frequency is ω_p, the stopband frequency is ω_s, the passband is within $\Delta < 0$ (dB) of unity, and the stopband ripple (dB) does not exceed $\Lambda < \Delta < 0$. We need the analog stopband frequency Ω_c, the filter order N, and the ripple parameter ε (9.117).

As remarked above, the parameter Ω_c in (9.117) specifies the analog stopband. So, although for an ordinary Chebyshev filter we took $\Omega_c = \omega_p$, now we set $\Omega_c = \omega_s$. The stopband condition is $10\log_{10}Q(\Omega) \le \Lambda$ for $\Omega \ge \Omega_c$. But as long as $\Omega \ge \Omega_c$ we have $\Omega_c/\Omega \le 1$ and so $1 + \varepsilon^2 T_N^2(\Omega_c/\Omega) \le 1 + \varepsilon^2$. Hence, we can determine the stopband ripple factor ε from the stopband condition with $\Omega = \Omega_c$: $10\log_{10}Q(\Omega_c) \le \Lambda$. Using the elementary Chebyshev polynomial property, $T_N(1) = 1$ for all $N \ge 0$, this reduces to

$$\varepsilon \le \sqrt{\frac{10^{\Lambda/10}}{1 - 10^{\Lambda/10}}}. \tag{9.118}$$

Observe that the ripple parameter does not depend on the filter order. Typically, we solve (9.118) as an equality to obtain ε.

The passband condition gives the analog filter order. If the discrete sampling rate is high enough, this means that for $\Omega < \omega_p < \omega_s = \Omega_c$ we can assume $10\log_{10}Q(\Omega) \geq \Delta$. for $\Omega < \omega_p$. $Q(\Omega)$ is strictly increasing as $\Omega \to 0$. Thus, we know that the passband condition applied to $\Omega = \omega_p$ is a worst case. The usual algebra boils this down to

$$\cosh\left[N\cosh^{-1}\left(\frac{\Omega_c}{\omega_p}\right)\right] = T_N\left(\frac{\Omega_c}{\omega_p}\right) \geq \frac{1}{\varepsilon}\sqrt{\frac{10^{\Delta/10}}{1-10^{\Delta/10}}}. \tag{9.119}$$

The familiar steps of taking (9.119) as an equality, solving for N, and rounding upward give the filter order.

Example (Inverse Chebyshev, Impulse Invariance, $N = 3$). We require a low-pass filter with a passband within 1 dB of unity up to $\omega_p = \pi/4$ and at least 5 dB below unity beyond $\omega_s = \pi/3$. Again, we see $\Delta = -1$ and $\Lambda = -5$ and assume $T = 1$. As above, we set $\Omega_c = \omega_s = \pi/3$. From (9.118) we find $\varepsilon = 0.680055$. Solving (9.119) as an equality produces $N = 2.1662$. This value we round up to $N = 3$. Since $T_3^2(\theta) = 16\theta^6 - 24\theta^4 + 9\theta^2$, it must be the case that

$$Q(\Omega) = \frac{16\Omega_c^{\ 6} - 24\Omega_c^{\ 4}\Omega^2 + 9\Omega_c^{\ 2}\Omega^4}{16\Omega_c^{\ 6} - 24\Omega_c^{\ 4}\Omega^2 + 9\Omega_c^{\ 2}\Omega^4 + \dfrac{\Omega^6}{\varepsilon^2}}, \tag{9.120}$$

where we have expressed $Q(\Omega)$ in positive powers of Ω (Figure 9.38a).

9.5.5.3 Poles and Zeros Analysis. Unlike the others we have already considered, this filter has finite zeros in the extended complex plane. For $H_a(\Omega)$ we select the poles of $Q(\Omega)$ which have positive imaginary parts. In general, among the zeros of a squared magnitude response function, we select one each of the numerator's conjugate roots.

Example (Inverse Chebyshev, Impulse Invariance, $N = 3$). Let us continue with the poles and zeros analysis of the previous example, for which the squared magnitude response is given by (9.117). The poles of $Q(\Omega)$ are $\pm 2.5848j$, $-1.0581 \pm 0.2477j$, and $1.0581 \pm 0.2477j$. The selected Laplace transform poles are thus -2.5848, $-0.2477 - 1.0581j$, and $-0.2477 + 1.0581j$. The zeros of $Q(\Omega)$ are $-1.1885 \pm 0.1276j$ and $1.1885 \pm 0.1276j$. They all have the same magnitude, so we can choose Laplace transform zeros to be $-0.1276 - 1.1885j$, and $-0.1276 + 1.1885j$. The poles and zeros plot for $Q(\Omega)$ is shown in Figure 9.38b.

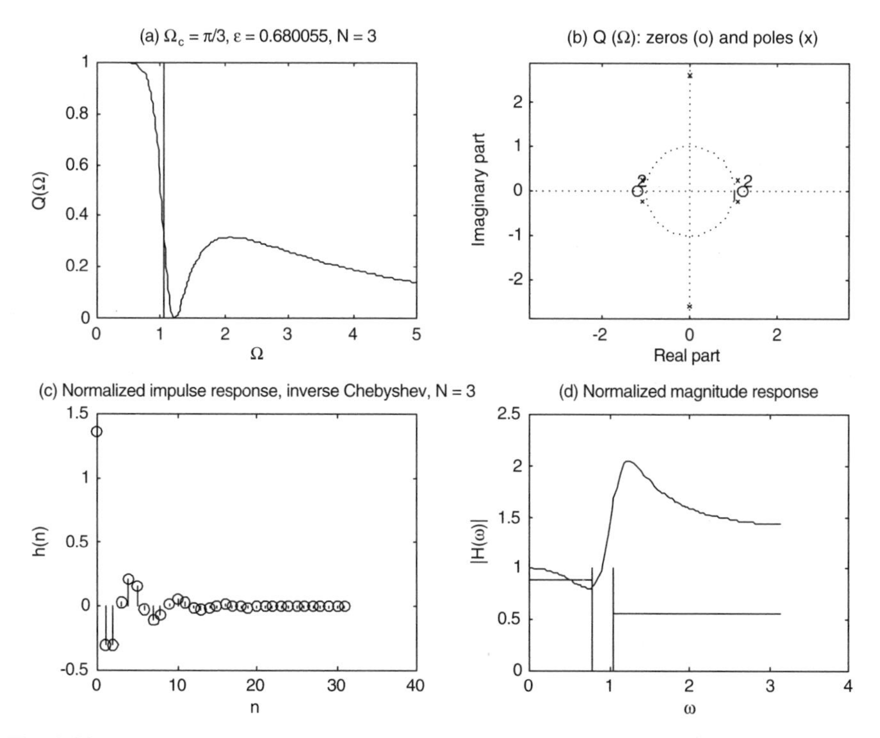

Fig. 9.38. The squared magnitude response $Q(\Omega)$ for an $N = 3$ analog filter approximated using the impulse invariance method (a). Associated poles and zeros (b), extracted discrete impulse response (c), and magnitude response (d). In panel (d), vertical lines mark the discrete passband and stopband, π/4 and π/3, respectively. The horizontal lines are the level criteria associated with the Δ and Λ parameters.

As a quotient of polynomials in z^{-1}, the discrete system function $H(z)$ is thus given by

$$\frac{\sum_{m=0}^{M} b_m z^{-m}}{\left[1 + \sum_{k=1}^{N} a_k z^{-k}\right]} = \frac{1 - 0.7624 z^{-1} + 0.6826 z^{-2}}{1 + (-0.8412) z^{-1} + 0.6671 z^{-2} + (-0.0460) z^{-3}}. \tag{9.121}$$

Figure 9.38(c) shows the impulse response. Note that it consists of an impulse and some low magnitude correction terms.

The magnitude response Figure 9.38d shows the effect of the rather loose stopband ripple constraint for this example. The filter sharpness is adequate, as can be seen from the bounds in the figure. However, there are high frequencies present in the analog filter, because of the allowed stopband ripple. Because of impulse invariance sampling, aliasing occurs and the derived discrete magnitude response does not satisfy the required stopband criterion. It might appear that increasing the sampling rate should reduce the aliasing. But unfortunately this causes the cutoff frequency of the analog filter to increase as well [7]. The best choice is to adjust the design parameters so as to meet the discrete filter's stopband criterion. The filter order may be increased, the ripple parameter may be reduced, or the bilinear approximation may be worthwhile.

9.5.5.4 Bilinear Approximation. Let us consider the bilinear approximation for the inverse Chebyshev low-pass filter. Suppose the discrete passband frequency is ω_p, the stopband frequency is ω_s, the passband is within $\Delta < 0$ (dB) of unity, and the stopband ripple (dB) does not exceed $\Lambda < \Delta < 0$. We need the analog stopband frequency Ω_c, the filter order N, and the ripple parameter ε (9.117) in order to specify the analog filter.

We know from our study of the inverse Chebyshev squared magnitude resonse that the Ω_c parameter governs the analog stopband frequency. Thus, $\Omega_c = \frac{2}{T}\tan\left(\frac{\omega_s}{2}\right)$ using the bilinear transformation. The passband and stopband conditions on the desired discrete filter are

$$0 > 10\log_{10}|H(\omega)|^2 = 10\log_{10}\left|Q\left(\frac{2}{T}\tan\left(\frac{\omega}{2}\right)\right)\right|^2 \geq \Delta, \tag{9.122a}$$

for $\omega < \omega_p$, and

$$10\log_{10}|H(\omega)|^2 = 10\log_{10}\left|Q\left(\frac{2}{T}\tan\left(\frac{\omega}{2}\right)\right)\right|^2 \leq \Lambda < 0, \tag{9.122b}$$

for $\omega > \omega_s$, where $Q(\Omega)$ is the analog squared magnitude response (9.117). By Chebyshev polynomial properties, $Q(\Omega)$ achieves its stopband maximum at the analog stopband value $\Omega = \Omega_c$. From (9.122b) we have $\varepsilon \leq \sqrt{\frac{10^{\Lambda/10}}{1-10^{\Lambda/10}}}$, as in the impulse invariance approximation (9.118). We assume an equality to compute the ripple parameter. In the passband constraint (9.122a) we use $\Omega = \Omega_p = \frac{2}{T}\tan\left(\frac{\omega_p}{2}\right)$ as a worst case. This entails

$$T_N(\Omega_c/\Omega_p) = \cosh^{-1}[N\cosh(\Omega_c/\Omega_p)] \geq \frac{1}{\varepsilon}\sqrt{\frac{10^{\Lambda/10}}{1-10^{\Lambda/10}}}. \tag{9.123}$$

Changing this relation to an equality, solving for N, and rounding to the integral ceiling gives the filter order.

Example (Inverse Chebyshev, Bilinear Approximation, $N = 3$). For a low-pass filter with a passband within 1 dB of unity for $\omega < \omega_p = \pi/4$ and at least 5 dB of attenuation for $\omega > \omega_s = \pi/3$, we again have $\Delta = -1$, $\Lambda = -5$. Let $T = 1$ be the sample

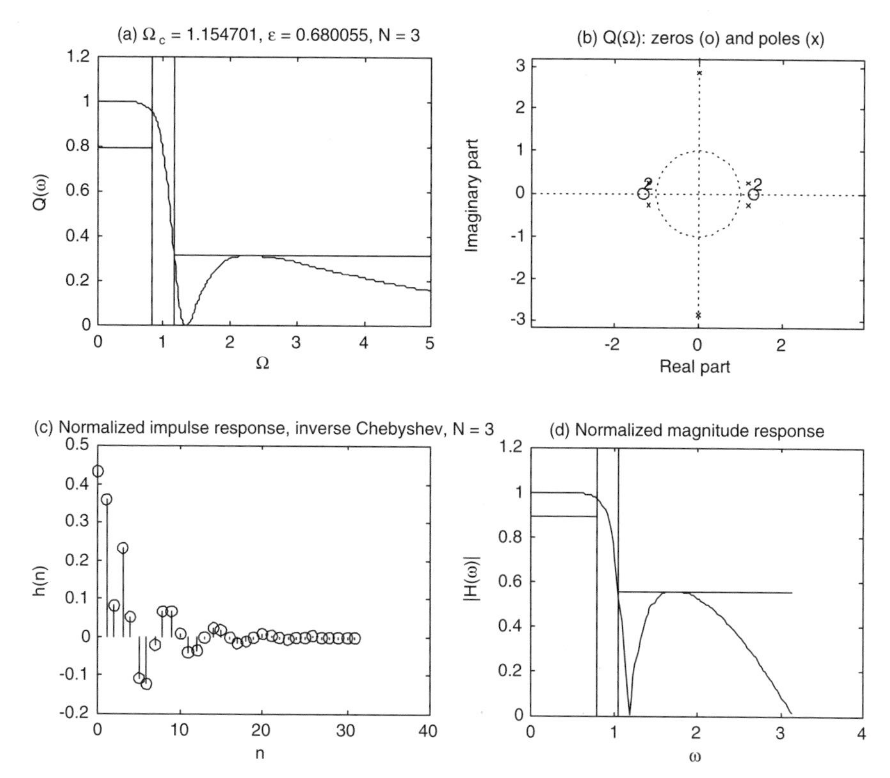

Fig. 9.39. The squared magnitude response $Q(\Omega)$ for an $N = 3$ analog filter approximated using the bilinear transformation method (a). Panel (b) shows the associated poles and zeros. Note that the zeros are second order. In (c), after normalization, is the impulse response. Finally, there is the normalized magnitude response (d). In panel (d), vertical lines mark the discrete passband and stopband, Ω_p and Ω_c, respectively. The horizontal lines are the level criteria associated with the Δ and Λ parameters.

distance. We find $\Omega_c = 2\tan(\omega_s/2) \approx 1.154701$ and $\Omega_p \approx 2\tan(\omega_p/2) \approx 0.8284$. Again, from (9.118) we get $\varepsilon = 0.680055$. Solving (9.123) as an equality produces $N = 2.0018$, and although we are breathtakingly close to a second order filter, we prudently round up to $N = 3$. The three estimates, Ω_c, ε, and N, give the squared magnitude response $Q(\Omega)$ (Figure 9.39a).

9.5.5.5 Poles and Zeros Analysis.

Among the poles of $Q(\Omega)$, we select those having positive imaginary parts to form $H_a(\Omega)$. After bilinear transformation, this filter too has finite zeros in the extended complex plane.

Example (Inverse Chebyshev, Bilinear Transformation, $N = 3$). Let us wrap up the previous example. In the squared magnitude response (9.117). The poles of $Q(\Omega)$ are $\pm 2.8657j$, $-1.1817 \pm 0.2550j$, and $1.1817 \pm 0.2550j$. The good Laplace transform poles are thus -2.8657, $-0.2550 - 1.1817j$, and $-0.2550 + 1.1817j$. The zeros of $Q(\Omega)$ are ± 1.3333 and each has order two. The Laplace transform zeros must be $\pm 1.3333j$. Figure 9.39b shows the poles and zeros plot for $Q(\Omega)$. Thus,

$$H_{L,a}(s) = \frac{s^2 + 1.7778}{s^3 + 3.3757s^2 + 2.9228s + 4.1881}. \tag{9.124}$$

Applying the bilinear map $s = \frac{2}{T}[(z-1)/(z+1)]$ with $T = 1$ to (9.124) gives

$$H(z) = \frac{0.1832z^3 + 0.0423z^2 + 0.0423z + 0.1832}{z^3 - 0.6054z^2 + 0.5459z + 0.1219}. \tag{9.125}$$

The z-transform $H(z)$ has partial fractions expansion

$$H(z) = \frac{-0.0821 - 0.0082j}{1-(0.3917+0.7293j)z^{-1}} + \frac{-0.0821 + 0.0082j}{1-(0.3917-0.7293j)z^{-1}} + \frac{-1.1551}{1-(-0.1779)z^{-1}} + 1.5026 = \frac{B(z)}{A(z)}. \tag{9.126}$$

Feeding a discrete impulse $\delta(n)$ through the difference equation implementation for (9.126) gives the unnormalized impulse response. This we scale (Figure 9.39c) by a factor of $(0.4245)^{-1}$ so that the magnitude response has unit DC value, as shown in Figure 9.39d.

9.5.6 Elliptic Filters

The fourth common transfer function—the *elliptical*, or *Cauer*[14] (1958), filter—has ripple in both the passband and stopband, nonlinear phase response, and the fastest rolloff from passband to stopband for a given IIR filter order [26, 66, 67].

The squared magnitude response function for the elliptic or Cauer filter[14] is

$$|H_a(\Omega)|^2 = \frac{1}{1+\varepsilon^2 R^2(\Omega)} = \frac{B(\Omega)}{A(\Omega)}, \tag{9.127}$$

where $R(\Omega) = U(\Omega)/V(\Omega)$ is a rational function, $\varepsilon > 0$ is a parameter, $R(0) = 0$, and the Degree(U) > Degree(V).

The rational function approximations (9.127) for the elliptic filter response are derived from the analytic properties of the Jacobi elliptic functions, which are encountered in the study of nonlinear oscillations. This oscillatory behavior gives rise to the passband and stopband ripple associated with the elliptic filter transfer function. Under certain conditions, the elliptic functions are qualitatively similar to

[14]German circuit theorist Wilhelm Cauer (1900–1945) invented and patented elliptic filters in the mid-1930s. While on his way to his office to get some papers, Cauer was arrested and executed by troops taking control of Berlin at the end of World War II. A short note on Cauer's life and accomplishments is given by A. Fettweis, Fifty years since Wilhelm Cauer's death, *IEEE Transactions on Circuits and Systems—I: Fundamental Theory and Applications*, vol. 42, no. 4, pp. 193–194, April 1995.

the trigonometric sine and cosine. But they have advanced general features which, when properly manipulated, give rise to better rolloff characteristics—for a given filter order—than the Chebyshev filter. Since the reader may have only minimal exposure to the elliptic functions, we will describe the analytical background for Cauer's elliptic filter prescription prior to developing the rational function approximations implied by (9.127).

9.5.6.1 Elliptic Functions and Integrals. Just as the Chebyshev filter response had an analytic description in terms of the circular functions (e.g., (9.106a)), the general elliptic filter response,

$$|H_a(\Omega)|^2 = \frac{1}{1+\varepsilon^2 R_N^{\ 2}(\Omega)}, \tag{9.128}$$

can be described in terms of a class of analytic functions known as the *Jacobi elliptic sine*, designated $sn(z, m)$, where the parameter $0 \le m < 1$ is the modulus, and $z \equiv u + jv$ is the argument, which may be complex-valued. The Cauer design calls for a function of the form

$$R_N(\Omega) = sn(f \cdot z + c, m), \tag{9.129}$$

where the argument consists of a factor *f*; a constant additve offset *c*—both of which can be specified to give the desired filter response; and a variable z that is the inverse of a Jacobi elliptic sine,

$$z = sn^{-1}(\Omega/\Omega_c, m). \tag{9.130}$$

The *Jacobi*[15] *elliptic sine* of modulus m is defined by

$$sn(z, m) = \sin(\phi(z, m)). \tag{9.131}$$

The argument u is described by the *elliptic integral of the first kind*:

$$u(\phi, m) = \int_0^{\phi} \frac{1}{\sqrt{(1-m^2\sin^2\theta)}}\, d\theta, \tag{9.132}$$

where the modulus is restricted to the interval $0 \le m < 1$. The function $\phi(u, m)$ the inverse of the $u(\phi, m)$, for fixed modulus m, and when we refer to a specific value we denote ϕ as the amplitude of the elliptic integral [33]. For the special case of amplitude $\phi = \pi/2$, the elliptic integral of the first kind is a function only of the modulus and reduces to the *complete elliptic integral of the first kind* (Figure 9.40),

$$K(m) = \int_0^{\pi/2} \frac{1}{\sqrt{(1-m^2\sin^2\theta)}}\, d\theta. \tag{9.133}$$

[15]Carl Gustav Jacobi (1804–1851), along with Gauss and Legendre, contributed to the early theory.

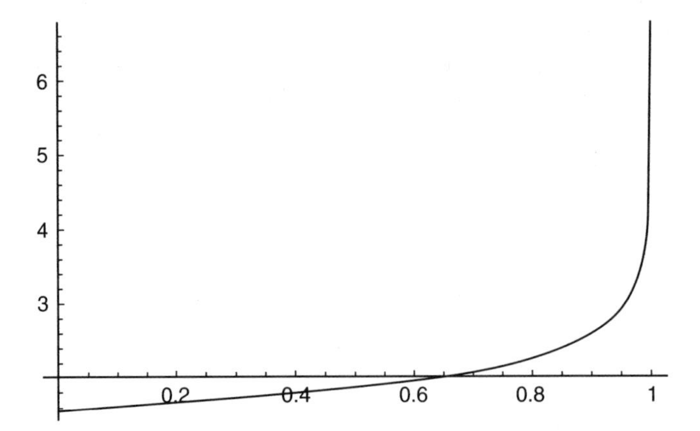

Fig. 9.40. The complete elliptic integral as a function of the modulus m. It is real-valued, but becomes singular as $m \to \infty$.

The complement to this integral is defined

$$K'(m) = \int_0^{\pi/2} \frac{1}{\sqrt{(1 - m_1^{\,2} \sin^2\theta)}} \, d\theta, \tag{9.134}$$

where $m_1 = 1 - m$. (We follow the usual convention and denote the complement by a prime, but emphasize that it has nothing to do with differentiation.) From (9.134) it is obvious that

$$K(m_1) = K'(m) \tag{9.135}$$

and we will use these interchangeably.

In the real-valued interval $x \in [0, \infty]$, the Jacobi elliptic sine is qualitatively similar to a sine wave: It is real-valued, restricted in amplitude to the interval $[-1, 1]$, and exhibits oscillations which qualitatively resemble a pure sinusoid, as illustrated in Figure 9.41. As this illustration suggests, $K(m)$ is one-fourth of a full period of the elliptic sine. For this reason it is also known as the *real quarter period.*

The Jacobi elliptic sine exhibits a richness that surpasses the simpler pure sinusoid. The most important new property is double periodicity,

$$sn(z + r \cdot 4K + s \cdot 4K', m) = sn(z, m), \tag{9.136}$$

where r and s are arbitrary integers. In Figure 9.42 and Figure 9.43 we illustrate the elliptic sine along other important intervals of the complex plane.

The validity of the Cauer's construction of the elliptic filter response is dependent upon the value of the Jacobi elliptic sine at several strategic points in the complex plane. For convenience, these points are listed in table Table 9.6. This table applies to both even- and odd-order elliptic filters.

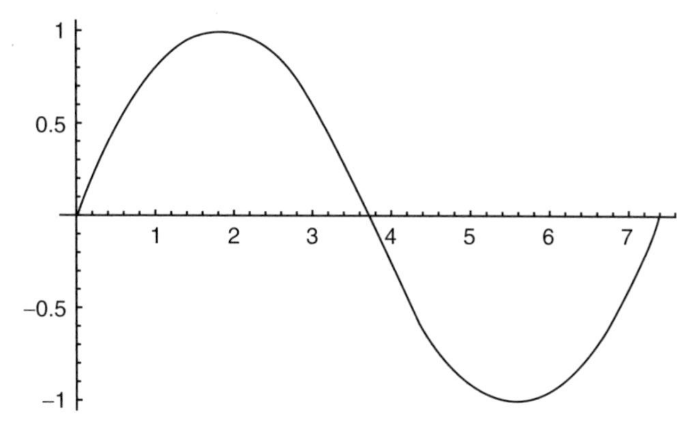

Fig. 9.41. The Jacobi elliptic sine $sn(x, m)$ on the interval $x \in [0, 4K]$. The case $m = 0.5$ is shown.

The foregoing illustrations were selected because they will aid in the understanding of Cauer's design of the elliptic filter response. Note that along these chosen intervals, $sn(z; m)$ is real-valued, although the argument itself may acquire nonzero real and imaginary parts. Although excursions from the intervals selected here may result in generally complex values for $sn(z; m)$, Cauer's design conveniently limits us to these intervals in which the elliptic sine remains real-valued. With suitable manipulation of the free constants f and c in (9.129), we can ensure continuity of the response at the transition points between the pass- and stopbands. The third column in Table 9.6 gives the points in the frequency plane at which the conditions in the first two columns are applied. These issues are considered in the next section.

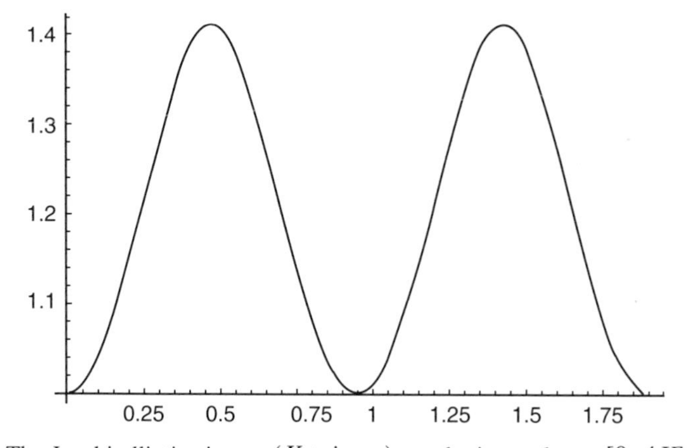

Fig. 9.42. The Jacobi elliptic sine $sn(K + jy, m)$ on the interval $y \in [0, 4K]$. The case $m = 0.5$ is shown.

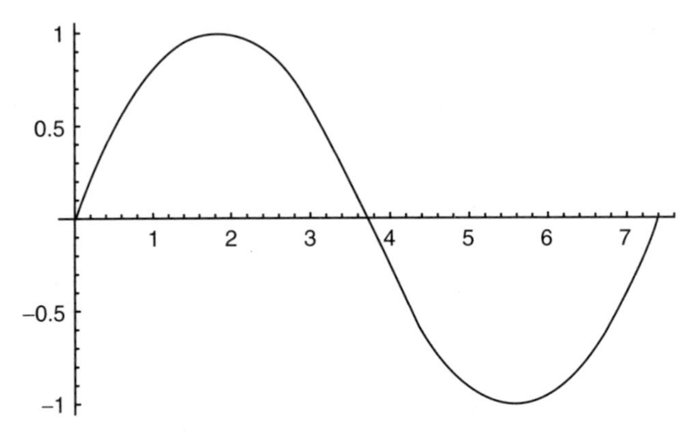

Fig. 9.43. The Jacobi elliptic sine $sn(x + 4jK', m)$ on the interval $x \in [0, 4K]$, shown for $m = 0.5$.

9.5.6.2 Elliptic Filter Response. For the elliptic low-pass filter approximation using the impulse invariance transformation, we are given a discrete passband frequency Ω_p, a stopband frequency Ω_s, an allowable passband ripple, and a required stopband attenuation value. We seek the filter order N and the ripple parameter ε. As in the case of the Chebyshev response, it is not possible to specify a specific set of parameters $\{\Omega_p, \Omega_s, N, \varepsilon\}$ that are identically satisfied by a low-pass elliptic filter response, but an allowable upper bound on the passband ripple and a lower bound on the stopband attenuation can be achieved, with both pass- and stopbands exhibiting a flat response.

The design of an elliptic filter response involves the specification of parameters m, p, and integer N such that these specific acceptable bounds can be achieved. At the same time, the filter response must be continuous at $\Omega = \Omega_p$ and $\Omega = \Omega_s$ while also satisfying acceptable bounds at $\Omega = 0$ and in the limit $\Omega \to \infty$. Continuity can be achieved by proper specification of the constants f and c (see (9.129)), while the proper bounds arise naturally from the behavior of the elliptic sine. The following relations are imposed by the design process and hold for arbitrary positive integer N. First,

$$m = \Omega_p^2 / \Omega_s^2 \tag{9.137}$$

TABLE 9.6. The Argument z and the associated value of the Jacobi Elliptic Sine as the Frequency Ω Traverses the Passband, Transition Band, and stopband of a Cauer Elliptic Filter[a]

z	$sn(z;m)$	Ω
0	0	0
$\beta K(m))$	1	Ω_p
$\beta K(m) \pm j\gamma K'(m)$	$m^{-1/2}$	Ω_s
$\alpha K(m) \pm j\gamma K'(m)$	∞	∞

[a]The integer α is even, the integers β and γ are odd.

which ensures that $m \in [0, 1]$, as required by the elliptic sine. Second, we define

$$f \equiv N\frac{K(p)}{K(m)} = \frac{K(p_1)}{K(m_1)}. \tag{9.138}$$

The second equality in (9.138) is imposed (it is not an identity) and gives the *flatness condition*,

$$N = \frac{K(m)}{K(p)}\frac{K(p_1)}{K(m_1)}. \tag{9.139}$$

Since parameter m is specified by the filter rolloff in (9.137), and the order N of the filter is typically specified in advance, relation (9.139) amounts to a default specification of the unknown parameter p. We note that (9.139) must be satisfied to give identically flat pass- and stopband ripple. In practice, deviations from an integer lead to good flatness provided they are small, and we will find it necessary to finesse the pass- and stopband ripple levels a_1 and a_2 to achieve something close to (9.139).

The value of remaining unknown, namely the additive offset c, depends on whether the filter is of odd or even order. We will now consider these cases in turn. For *odd* filter order N, the design process will result in a filter response having the following general characteristics $|H(\Omega/\Omega_p)|^2$ at selected critical frequencies:

$$|H(0)|^2 = 1, \tag{9.140}$$

$$|H(1)|^2 = \frac{1}{1+a_1^2}, \tag{9.141}$$

$$|H(\Omega_s/\Omega_p)|^2 = \frac{1}{1+a_2^2}, \tag{9.142}$$

$$|H(\infty)|^2 \to 0. \tag{9.143}$$

For filters of an *even* order N, the DC value of the response will be

$$|H(0)|^2 = \frac{1}{1+a_1^2} \tag{9.144}$$

but the other three points are the same as given in (9.141)–(9.143). In an actual design problem, the real-valued parameters a_1 and a_2 will specified according to the desired acceptable ripple and will be adjusted so as to leave the ripple within specified bounds while also providing a flat response in the pass- and stopbands.

Remark. Readers consulting further references on elliptic filter design may encounter alternative design procedures which result in a filter response with asymptotic behavior which deviates from that specified in (9.140) and (9.143), especially when consulting prepackaged tables or routines. Unless otherwise noted, we confine ourselves to the limits defined here.

Case of N Odd. For an elliptic filter of odd order we stipulate

$$c = 0. \tag{9.145}$$

Consider the conditions at the stopband $\Omega = \Omega_s$, as laid out in Table 9.6. Expressing (9.29), we have

$$R_N(\Omega_s/\Omega_p) = sn\left(N\frac{K(p)}{K(m)} \cdot (K(m) + jK(m_1)), m\right), \tag{9.146}$$

where we have also used expression (9.135). After straightforward algebra this reduces to

$$R_N(\Omega_s/\Omega_p) = sn(NK(p) + jK(p_1), p) = 1/\sqrt{p}. \tag{9.147}$$

Similarly, the passband edge at $\Omega = \Omega_p$ leads to

$$R_N(1) = sn(NK(p), p) = 1. \tag{9.148}$$

Substitution of (9.147) and (9.148) into the expression for the filter characteristics leads to the relations,

$$|H(\Omega_s/\Omega_p)|^2 = \frac{1}{1+\left(\frac{\varepsilon}{p}\right)} = \frac{1}{1+a_2^2} \tag{9.149}$$

and

$$|H(1)|^2 = \frac{1}{1+\varepsilon} = \frac{1}{1+a_1^2}. \tag{9.150}$$

Combining these lead to expressions for ε and the parameter p in terms of the pass- and stopband ripple:

$$\varepsilon = a_1^2, \tag{9.151}$$

$$p = \frac{a_1^2}{a_2^2}. \tag{9.152}$$

Remark. In the design of an elliptic characteristic for specified Ω_p and Ω_s, the essential relations are (9.139), (9.151), and (9.152). A successful design will involve juggling of a_1 and a_2 (within acceptable bounds) such that p from (9.152) will lead to close agreement with the flatness criterion (9.139). This can be done graphically utilizing packaged math routines, as we will do in the following example.

Example (Cauer Elliptic Filter, $N = 3$). Consider a specification calling for the design of an $N = 3$ elliptic filter with $\Omega_p = 0.9$, $\Omega_s = 1.39$, a maximum passband ripple of –2.0 dB, and a stopband ripple not exceeding –25.0 dB. As a first pass, one can barely make the ripple tolerances by setting $a_1 = 0.763$ and $a_2 = 17.9$, which lead to

$$10 \cdot \log[1/(1 + a_1^2)] = -1.99253, \tag{9.153}$$

$$10 \cdot \log[1/(1 + a_2^2)] = -25.0706. \tag{9.154}$$

However, the "integer" is given by

$$N = \frac{K(p_1)}{K(m_1)} \cdot \frac{K(m)}{K(p)} = 2.68387, \tag{9.155}$$

The effect of the deviation from the ideal value of 3 is to cause a departure from flatness, which is particularly notable in the stopband, as illustrated in Figure 9.44.

The problem can be alleviated by reducing the passband ripple such that $a_1 = 0.445$ and

$$N = \frac{K(p_1)}{K(m_1)} \cdot \frac{K(m)}{K(p)} = 3.00271, \tag{9.156}$$

which is close to the ideal target of 3. The reduction in passband ripple required to achieve this exceeds 1 dB, since

$$10 \cdot \log[1/(1 + a_1^2)] = -0.784659. \tag{9.157}$$

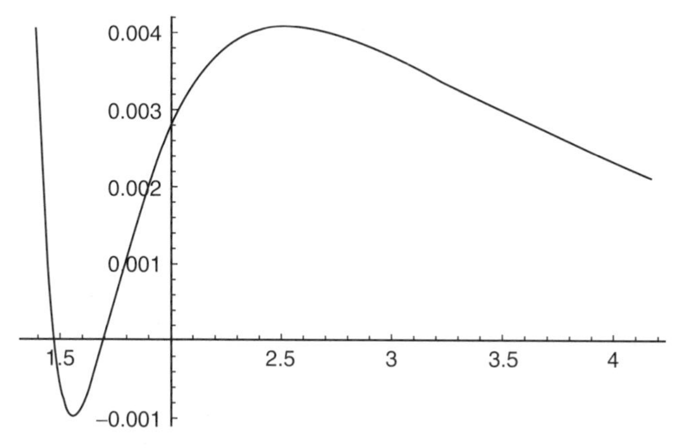

Fig. 9.44. Detail of the stopband response when the integer flatness condition is not met. Note that the characteristic becomes negative in a small region of the spectrum.

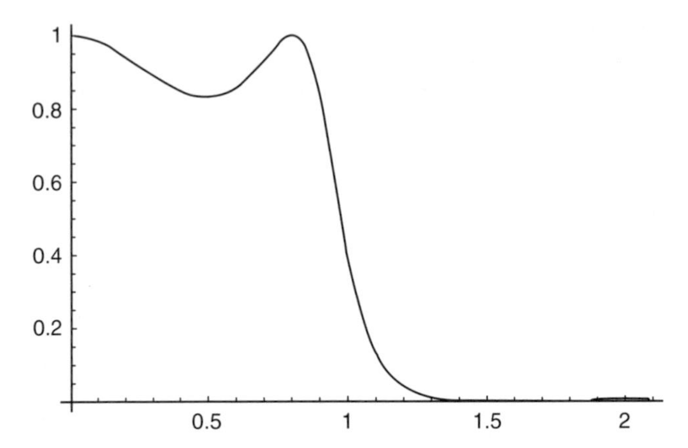

Fig. 9.45. The $N = 3$ elliptic filter response after the parameters have been adjusted to obtain pass- and stopband flatness.

However, the result is the desired flat elliptic filter response representing (9.128), as shown in Figure 9.45.

Remark. In practice, deviations from the integer ideal N have the greatest effect on the stopband response. This includes the flatness, as noted, but even small deviations can induce a small imaginary part to $H(\Omega)$. When setting up a plot, it is useful to specify the real part to eliminate the small but unwanted imaginary component.

Case of *N* Even. For an elliptic filter of even order we apply a nonzero offset to the argument,

$$c = K(p). \tag{9.158}$$

At the passband edge $\Omega = \Omega_p$ it is easy to show, since $N+1$ is an odd number,

$$R_N(1) = sn((N+1)K(p), p) = 1. \tag{9.159}$$

Likewise, at the stopband edge,

$$R_N(\Omega_s/\Omega_p) = sn((N+1)K(p) + iK(p_1), p) = 1/\sqrt{p}. \tag{9.160}$$

The effect of the nonzero offset (9.158) is to give edge conditions (9.159) and (9.160) identical to their counterparts in the odd-order case. The offset will have the effect of changing the elliptic filter response at zero frequency, but otherwise the even order characteristic resembles that of the odd-order case. These points are illustrated in the following example.

Example (Cauer Elliptic Filter, $N = 4$). Consider the specifications laid out in the example for $N = 3$, but suppose we require faster rolloff by specifying $\Omega_s = 1.12$. By increasing the order of the filter to 4, and setting $a_1 = 0.39$ and $a_2 = 23.9$, we obtain

$$10 \cdot \log[1/(1 + a_1^2)] = -0.614902, \tag{9.161}$$

$$10 \cdot \log[1/(1 + a_2^2)] = -27.5756, \tag{9.162}$$

and

$$N = \frac{K(p_1)}{K(m_1)} \cdot \frac{K(m)}{K(p)} = 4.01417, \tag{9.163}$$

which leads to a nominally flat response shown in Figure 9.46. Note that by increasing the order of the filter we have achieved the desired faster rolloff and brought the stopband ripple under the specification by more than 1.5 dB.

Remark. Note that the even-order elliptic response has the low-frequency limit,

$$|H(0)|^2 = \frac{1}{1 + a_1^2} = 0.86798. \tag{9.164}$$

This outcome is the legacy of the nonzero offset specified in (9.158): At $\Omega = 0$, the argument of the Jacobi elliptic sine is no longer zero, as it was in the odd-order construction.

9.5.7 Application: Optimal Filters

Finally, let us consider the problem of designing a filter that conditions a signal so that later processing preserves just the desired features of the input. To be more precise, suppose that an analog signal $x(t)$ contains an original trend $s(t)$ and an

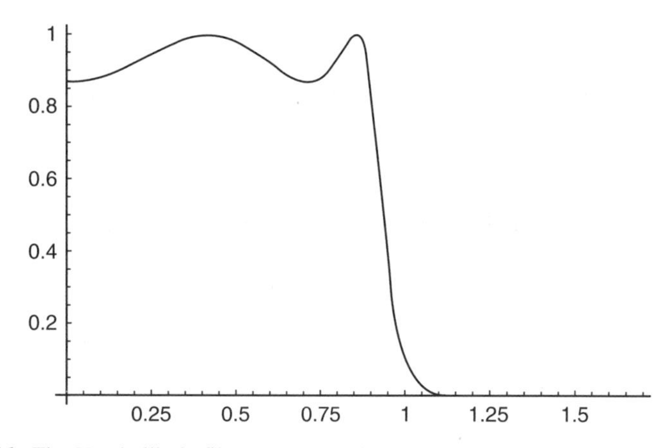

Fig. 9.46. The $N = 4$ elliptic filter response. The result is flat in the pass- and stopbands.

unknown corrupting noise component $n(t)$: $x(t) = s(t) + n(t)$. The signal $x(t)$ passes through a linear, translation-invariant system H, $y(t) = (Hx)(t)$. Assume all signals are real-valued. We seek the best H with real impulse response $h(t)$ so that $(x*h)(t)$ is optimally close to $s(t)$.

Frequency-domain methods provide a solution: the *optimal* or *Wiener*[16] filter [68]. Arguing informally, let us use the L^2 norm as a measure of how close $y(t)$ is to $s(t)$. For an optimal noise removal filter, then, we seek to minimize

$$\|y-s\|^2 = \frac{1}{2\pi}\|Y-S\|^2 = \frac{1}{2\pi}\int_{-\infty}^{\infty} |S(\omega) - S(\omega)H(\omega) - H(\omega)N(\omega)|^2 \, d\omega, \quad (9.165)$$

where we have used Parseval's result and the convolution theorem for the radial Fourier transform (Chapter 5). The integrand on the right-hand side of (9.165) is

$$S(\omega)\overline{S(\omega)}[1-H(\omega)]\overline{[1-H(\omega)]} - \overline{H(\omega)N(\omega)}[1-H(\omega)]S(\omega)$$
$$- H(\omega)N(\omega)\overline{[1-H(\omega)]S(\omega)} + N(\omega)\overline{N(\omega)}H(\omega)\overline{H(\omega)}. \quad (9.166)$$

The noise $n(t)$ is random, so it is uncorrelated with $s(t)$, and integrating products involving their respective Fourier transforms gives zero:

$$\|y-s\|^2 = \frac{1}{2\pi}\int_{-\infty}^{\infty} (|S(\omega)|^2|1-H(\omega)|^2 - |H(\omega)N(\omega)|^2)\, d\omega, \quad (9.167)$$

To find the minimum of $\|y-s\|^2$, we must minimize the integral in (9.167). Thus, we must minimize its integrand, and—arguing informally—the criterion for this is that the function $f(H) = |S|^2|1-H|^2 - |HN|^2$ has zero derivative. Taking the derivative with respect to H gives $\frac{\partial f}{\partial H} = H[|S|^2 + |N|^2] - |S|^2$. Setting $\frac{\partial f}{\partial H} = 0$ and solving gives

$$H = \frac{|S|^2}{|S|^2 + |N|^2}, \quad (9.168)$$

which is the Fourier transform of the optimal or Wiener filter for removing noise from the signal $x(t)$.

9.6 SPECIALIZED FREQUENCY-DOMAIN TECHNIQUES

This section introduces and applies some signal analysis methods arising from Fourier transform theory.

[16]Although it had been developed in 1942, Wiener's optimal filter was made public only in 1949, when the first edition of Ref. 68 was published. The theory had remained classified during World War II, because of its application to radar.

9.6.1 Chirp-*z* Transform Application

In many applications, the frequencies of interest within candidate signals are known in advance of their processing. Such a priori information can simplify the design of the analysis system. For instance, instead of computing a broad range of spectral values using the discrete Fourier transform, the engineer may elect to compute only a small portion of the spectrum, namely that part that might contain useful signal information. One way to focus in on a spectral interval without computing large numbers of useless coefficients is the use the *chirp-z transform* (CZT), introduced in the previous chapter (Section 8.3.1).

Let us first recall the basic ideas of the CZT. The CZT computes z-transform on a spiral contour in the complex plane [11]. It is determined by two parameters: A and W—the spiral starting point and arc step, respectively. Via an example, we shall see how to apply it to zoom in on DFT frequency components. As in Chapter 8, we take the notation of Rabiner and Gold [11]. Suppose that $A = A_0\exp(2\pi j\theta_0)$; $W = W_0\exp(2\pi j\phi_0)$; M, N are positive natural numbers; $x(n) = 0$ outside $[0, N-1]$; and $z_k = AW^{-k}$ for $0 \le k < M$. The chirp z-transform of $x(n)$ with respect to A and W is

$$X_{A,W}(k) = \sum_{n=0}^{N-1} x(n)z_k^{-n} = \sum_{n=0}^{N-1} x(n)A^{-n}W^{nk}. \tag{9.169}$$

The exercises of Chapter 8 explained that the CZT reduces to the DFT of order N when $A = 1$, $M = N$, and $W = \exp^{(-2\pi j/N)}$.

Let us return to the speech fragment considered earlier, "calling" in a female voice. The fundamental frequency range of the /a/ phoneme for a woman is from about $F_{lo} = 100$ Hz to $F_{hi} = 400$ Hz. We would like to design a CZT detector for this spectral range and apply it to the digitized speech sample. From a manual segmentation of the speech sample in question (Figure 9.28), we know that the /a/ phoneme occurs from samples $n = 800$ to $n = 1200$. It also degrades off into the /l/ sound immediately following.

To set up the CZT for this formant detection application, we set the sliding disjoint windows to consist of $N = 400$ samples. The sampling frequency $F_s = 8192$ Hz. Also,

$$A = \exp\left[\frac{2\pi j F_{lo}}{F_s}\right], \tag{9.170a}$$

$$W = \exp\left[-2\pi j\left(\frac{F_{hi} - F_{lo}}{NF_s}\right)\right]. \tag{9.170b}$$

Applying the algorithms of Chapter 8, we find the first six detection windows (Figure 9.47).

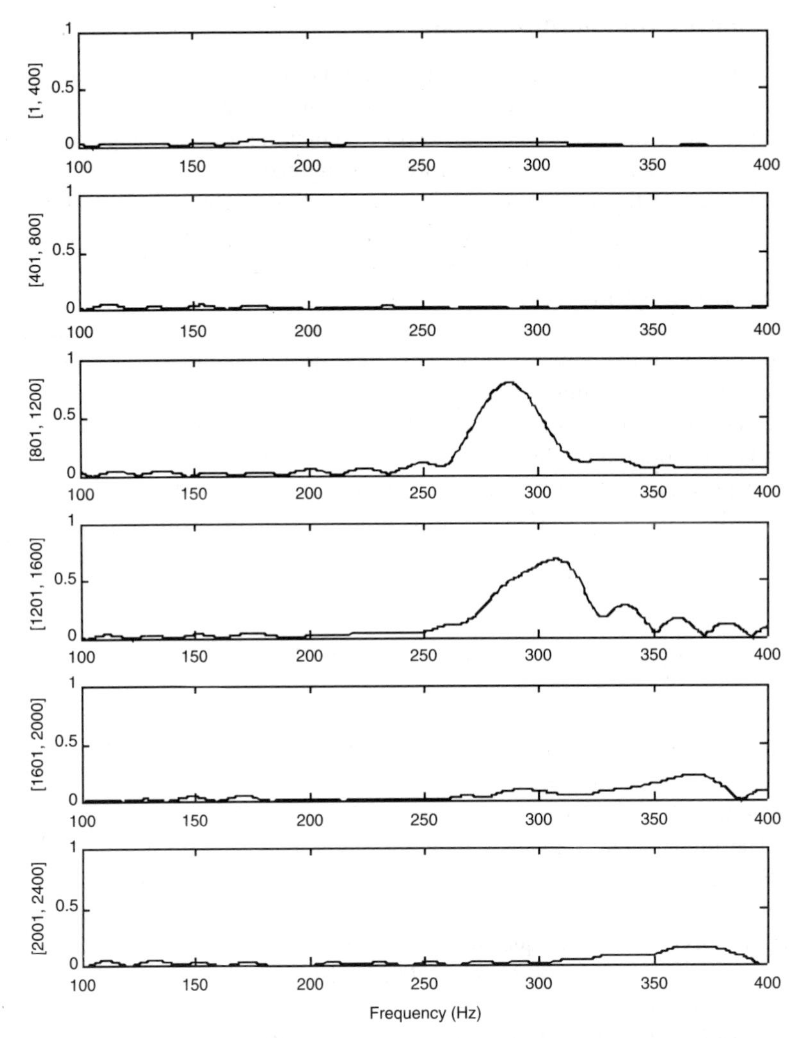

Fig. 9.47. Fundamental frequency detector for the /a/ phoneme based on the chirp-z transform. A lone peak rises in the third window, which concurs with a manual segmentation of the speech sample.

9.6.2 Hilbert Transform

The Hilbert transform[17] is a powerful technique that can be used to:

(i) Find the envelope of a signal.

(ii) Find the instantaneous phase of a signal.

[17]David Hilbert originated the idea in his papers on integral equations, reprinted in the book *Grundzuge einer allgemeinen Theorie der linearen Integralgleichungen*, Leipzig and Berlin: Teubner, 1912.

(iii) Find the instantaneous frequency of a signal.

(iv) Suppress one of the sidebands in order to create a single sideband (SSB) modulation of a signal.

The Hilbert transform has rich theory and many interesting properties [7, 69, 70]. We shall skim the theoretical material and show how the transform works in another speech analysis application.

9.6.2.1 Definition and Properties. There are analog and discrete Hilbert transforms. The *analog Hilbert transform* of a signal $x(t)$ is defined to be

$$x_H(t) = \frac{1}{\pi}\int_{-\infty}^{\infty} \frac{x(s)}{t-s}\,ds = (\mathcal{H}x)(t). \tag{9.171}$$

The integral, due to the singularity of its integrand at $t = s$, must be interpreted in a special way in order to make sense. The standard way to define the integral is by the Cauchy principal value [71]:

$$PV\int_{-\infty}^{\infty} \frac{x(s)}{t-s}\,ds = \lim_{\varepsilon \to 0^+}\left[\int_{\varepsilon}^{\infty} \frac{x(s)}{t-s}\,ds + \int_{-\infty}^{-\varepsilon} \frac{x(s)}{t-s}\,ds\right], \tag{9.172}$$

which is valid as long as the limit of the sum of the two partial integrals exists. The principal value is written with a PV before the integral sign to signify that a special, augmented form of the Lebesgue integral is supposed. Note that the individual limits of the integrals inside the square brackets of (9.172) may not exist. It is in general not permissable to move the limit operation inside the brackets when using the principal value of the integral.

Example (Square Pulse). Consider the signal $x(t) = u(t+1) - u(t-1)$. The function $h(t) = t^{-1}$ defies integration on $[-1, 1]$, because of the singularity at the origin. But using the Cauchy principal value, we can still compute $x_H(0)$:

$$x_H(0) = PV\int_{-\infty}^{\infty} \frac{x(s)}{0-s}\,ds = -\lim_{\varepsilon \to 0^+}\left[\int_{\varepsilon}^{1} \frac{1}{s}\,ds + \int_{-1}^{-\varepsilon} \frac{1}{s}\,ds\right] = 0. \tag{9.173}$$

Consequently, we can interpret the transform integral as a special kind of convolution. Let us investigate how the Hilbert transform system affects an analog signal. Let $h(t) = (\pi t)^{-1}$. Then $x_H(t) = (x*h)(t)$. The generalized Fourier transform of $h(t)$ is

$$H(\Omega) = \int_{-\infty}^{\infty} \frac{1}{\pi t} e^{-j\Omega t}\,dt = -j\,\mathrm{sgn}(\Omega) = \begin{cases} -j & \text{for } \Omega > 0, \\ 0 & \text{for } \Omega = 0, \\ j & \text{for } \Omega < 0, \end{cases} \tag{9.174}$$

The *duality principle* of the Fourier transform explains (9.174). If $x(t)$ has radial Fourier transform $y(\Omega)$, then $y(t)$ will have Fourier transform $2\pi x(-\Omega)$. From Chapter 6, we know that the Fourier transform of $(j/2)\mathrm{sgn}(t)$ is Ω^{-1}. Hence by duality, t^{-1} transforms to $(j/2)(2\pi)\mathrm{sgn}(-\Omega)$, and (9.174) follows.

The Hilbert transform system $x(t) \to x_H(t)$ is also called a *90° phase shift* or *quadrature filter* [50]. To see why, we look at the system's frequency-domain effect. The Fourier transform of $x_H(t)$ is $X_H(\Omega) = X(\Omega)H(\Omega) = -jX(\Omega)\mathrm{sgn}(\Omega)$. This operation multiplies positive spectral components by the factor $-j = \exp(-j\pi/2)$ and negative spectral components by $j = \exp(j\pi/2)$. These correspond to phase shifts of $-\pi/2$ and $\pi/2$, respectively. Thus, the Hilbert transform converts sines to cosines and vice versa. Let us examine at these basic transformations.

Example (Sinusoids). Consider $x(t) = \cos(\Omega_0 t)$ and $y(t) = \sin(\Omega_0 t)$. The generalized Fourier transforms of $x(t)$ and $y(t)$ are $X(\Omega) = \pi[\delta(\Omega - \Omega_0) + \delta(\Omega + \Omega_0)]$ and $Y(\Omega) = (\pi/j)[\delta(\Omega - \Omega_0) - \delta(\Omega + \Omega_0)]$. Note that $\delta(\Omega - \Omega_0)$ is a positive frequency impulse, whereas $\delta(\Omega + \Omega_0)$ lives across the origin, in $\Omega < 0$ land. The Fourier transform of $x_H(t)$ is $X_H(\Omega) = F(x_H)(\Omega) = -j\pi\mathrm{sgn}(\Omega)[\delta(\Omega - \Omega_0) + \delta(\Omega + \Omega_0)]$. But this is $(\pi/j)[\delta(\Omega - \Omega_0)] - (\pi/j)[\delta(\Omega + \Omega_0)] = Y(\Omega)$. Evidently, $x_H(t) = y(t)$. As an exercise, we leave the other relation $y_H(t) = -x(t)$ to the reader.

Typically, then, we compute the Hilbert transform of a signal $x(t)$ by examining the frequency domain product $X(\Omega)H(\Omega)$. This is usually much simpler than evaluating the Cauchy principal value integral (9.172), although the results can be counter-intuitive. The generalized Fourier transform of the signal $x(t) = 1$ is the Dirac $2\pi\delta(\Omega)$, for instance. Multiplication by $-j\mathrm{sgn}(\Omega)$ therefore gives zero.

We summarize analog Hilbert transform properties in Table 9.7 and leave the derivations as exercises. Note that if $X(0) = 0$, then the inverse transform is $\mathcal{H}^{-1} = -\mathcal{H}$. Also, many algebraic properties of the Fourier transform carry through to the Hilbert transform.

9.6.2.2 Discretization.

Moving toward computer applications, let us now consider how to define a discrete Hilbert transform. Again, the frequency-domain behavior is the key; we seek a discrete 90° phase shift system. Such a system would turn each cosine component $\cos(\omega n)$ in a signal $x(n)$ into a $\sin(\omega n)$ term and each $\sin(\omega n)$ into a $-\cos(\omega n)$.

First, we consider the case of aperiodic discrete signals $x(n)$. The appropriate 90° phase shift system should have a frequency response $H(\omega)$ given by

$$H(\omega) = \sum_{n=-\infty}^{\infty} h(n)e^{-j\omega n} = -j\,\mathrm{sgn}(\omega) = \begin{cases} -j & \text{for } \omega > 0, \\ 0 & \text{for } \omega = 0, \\ j & \text{for } \omega < 0. \end{cases} \tag{9.175}$$

TABLE 9.7. Some Analog Hilbert Transform Properties

Signal Expression	Hilbert Transform or Property
$x(t)$	$x_H(t) = \frac{1}{\pi} PV \int_{-\infty}^{\infty} \frac{x(s)}{t-s} ds = (\mathcal{H}x)(t)$ (Analysis equation)
$x_H(t)$	$(\mathcal{H}x_H)(t) = -x$ (Inverse, synthesis equation)
$ax(t) + by(t)$	$ax_H(t) + by_H(t)$ (Linearity)
dx/dt	dx_H/dt (Derivative)
$\langle x, x_H \rangle = \int_{-\infty}^{\infty} x(t)\overline{x_H(t)}\, dt = 0$ $x \in L^2(\mathbb{R})$	Orthogonality
$\|x\|_2 = \|x_H\|_2$ $x \in L^2(\mathbb{R})$	Energy conservation

The inverse DTFT computation gives

$$h(n) = \frac{1}{2\pi}\int_{-\pi}^{\pi} H(\omega)e^{j\omega n} d\omega = \frac{j}{2\pi}\int_{-\pi}^{0} e^{j\omega n} d\omega + \frac{-j}{2\pi}\int_{0}^{\pi} e^{j\omega n} d\omega = \begin{cases} 0 & \text{if } n \text{ is even,} \\ \dfrac{2}{n\pi} & \text{if } n \text{ is odd.} \end{cases} \tag{9.176}$$

So the above discrete Hilbert transform system is neither causal nor FIR. Now let us consider a discrete $x(n)$ with period $N > 0$.

Let $X(k)$ be the DFT of a real-valued signal $x(n)$ defined on $[0, N-1]$. So corresponding to each positive discrete frequency $k \in [1, N/2)$ there is a negative frequency $N-k \in (N/2, N-1]$. The DFT coefficients $X(0)$ and $X(N/2)$—corresponding the DC and Nyquist frequency values—are both real. Mimicking the analog Hilbert transform, let us therefore define the system function of the discrete Hilbert transform to be

$$H(k) = \begin{cases} 0 & \text{if } k = 0, \\ -j & \text{if } 1 \le k < \frac{N}{2}, \\ 0 & \text{if } k = N/2, \\ j & \text{if } \frac{N}{2} < k \le N-1. \end{cases} \tag{9.177}$$

We claim this works. For if $x(n) = A\cos(2\pi k_0 n/N)$, then its representation in terms of the inverse DFT is $x(n) = (A/2)\exp[2\pi j k_0 n/N] + (A/2)\exp[2\pi j(N-k_0)n/N]$. That is,

$X(k) = (A/2)\delta(k - k_0) + (A/2)\delta(k - (N - k_0))$. Multiplying by $H(k)$ gives $X(k)H(k) = (-jA/2)\delta(k - k_0) + (jA/2)\delta(k - (N - k_0))$. Applying the inverse DFT, this becomes $y(n) = (-jA/2)\exp[2\pi jk_0n/N] + (jA/2)\exp[2\pi j(N - k_0)n/N] = A\sin(2\pi k_0n/N)$. So $y(n) = x_H(n)$, as claimed. Similarly, discrete Hilbert transformation of $A\sin(2\pi k_0n/N)$ gives $-A\cos(2\pi k_0n/N)$.

Note that the Hilbert transform $x_H(n)$ of a discrete signal $x(n)$ on $[0, N - 1]$ loses the energy of both the DC term and the Nyquist frequency term. To find the impulse response of the discrete Hilbert transform system, we calculate the inverse DFT of (9.177) to get $h(n)$. This allows us to implement discrete Hilbert transforms on a digital computer. However, the value of Hilbert transform applications revolves around the related concept of the analytic signal, which the next section covers.

9.6.2.3 Analytic Signal.

Given an analog signal $x(t)$ and its Hilbert transform $x_H(t)$, the associated *analytic signal* [72, 73] is

$$x_A(t) = x(t) + jx_H(t). \tag{9.178a}$$

Although replacing a real-valued with a complex-valued signal may make things seem needlessly complicated, it does allow us to define the following related—and quite valuable—concepts. The *signal envelope* is

$$|x_A(t)| = \sqrt{x^2(t) + x_H^2(t)}. \tag{9.178b}$$

Thus, we can write the analytic signal as

$$x_A(t) = |x_A(t)|e^{j\phi(t)}, \tag{9.178c}$$

where the *instantaneous phase* $\phi(t)$ is

$$\phi(t) = \tan^{-1}\left[\frac{x_H(t)}{x(t)}\right]. \tag{9.178d}$$

In the first chapter, we argued that the derivative of the phase with respect to time is a reasonable way to define the *instantaneous radial frequency.* Hence, we set

$$\omega(t) = \frac{d}{dt}\phi(t). \tag{9.178e}$$

We may also define discrete versions of these notions. Notice that the signal envelope for a sinusoid is precisely its amplitude. Thus, the definition of signal envelope (9.178b) gives us a definition that applies to aperiodic signals, but reduces to what we should expect for the case of sinusoids. The imaginary part of the analytic signal (9.178a) fills in the gaps, as it were, left in the signal by its fine scale oscillations (Figure 9.48).

In many ways, the analytic signal is more important than the Hilbert transform itself. It is possible to show that the analytic signal satisfies the Cauchy–Riemann

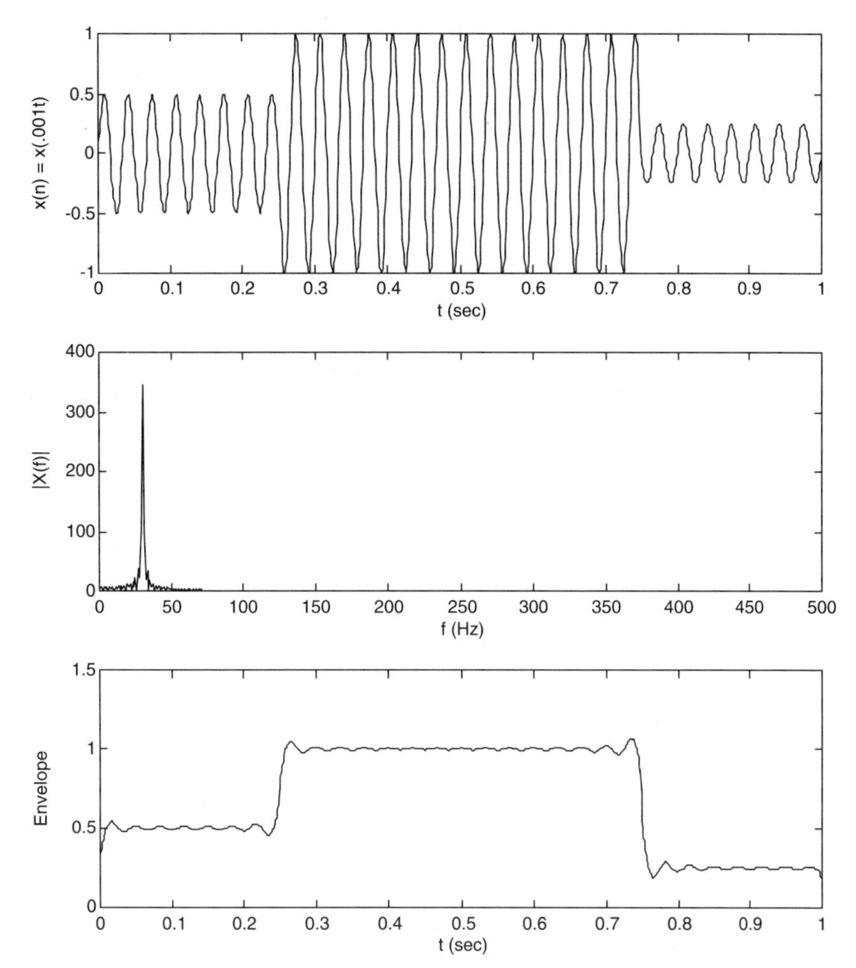

Fig. 9.48. Signal $x(t)$ contains a unit amplitude 30-Hz sinusoid that has been attenuated in the first and last 250 ms of its domain (top). Magnitude spectrum (middle panel). Signal envelope (bottom).

equations , so that it can be extended to an analytic function of a complex variable $x(z)$ [71, 74]. The next section contains an example of envelope computation on a speech signal.

9.6.2.4 Application: Envelope Detection. An important early task in speech analysis is to segment the input signal into regions containing utterances and those holding only background noise. The utterance portions can be further broken up into separate words, although this is by no means a simple task. One tool in either segmentation procedure is the signal envelope. Here, as an example, we compute the envelope of the speech signal considered earlier, namely, the "calling" clip.

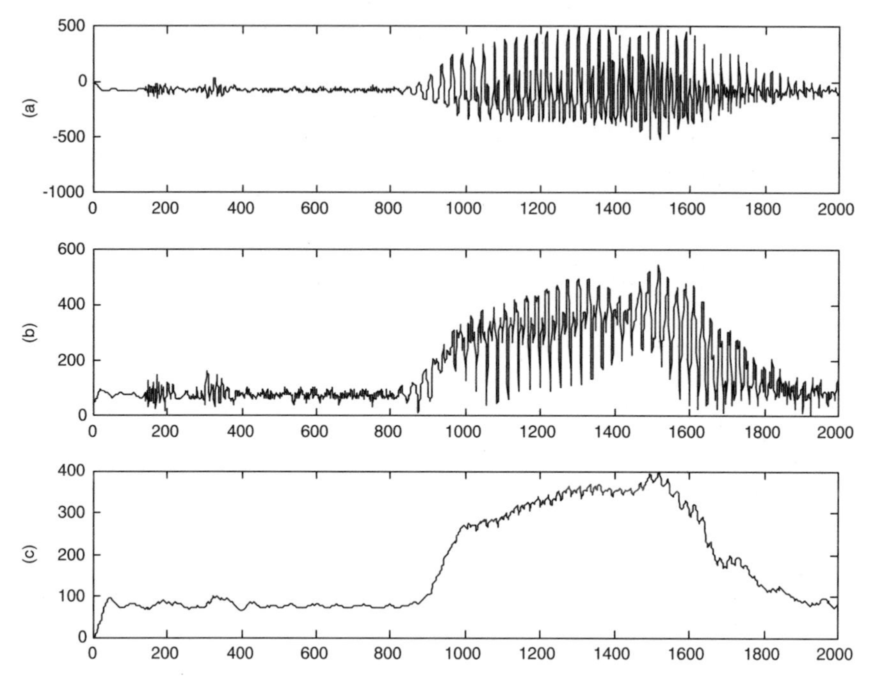

Fig. 9.49. Speech signal (a), its envelope (b), and its filtered envelope (c).

Working on the first 2000 speech samples (Figure 9.49a), we can compute the Hilbert transform, analytic signal, and envelope (Figure 9.49b). The problem is that the envelope remains rather jagged and therefore problematic for segmentation purposes.

One way to improve the envelope is to apply low-pass filter. In Figure 9.49c we created a third order Chebyshev Type II low-pass filter with stopband attenuation of 20 dB. The speech sampling rate is 8 kHz for this example, so the Nyquist rate is 4 kHz, and the cutoff frequency for the low-pass filter is 200 Hz.

In general, envelope detection problems can be improved with such smoothing filters. Even simple sample-and-hold filters with $H(z) = z/(z - a)$ may prove adequate for envelope amelioration.

Signal analysis applications requiring instantaneous phase or frequency computations—such as interferometry, for instance—may demand more refined filtering. A typical strategy is to use filters based on fitting methods, such as the Savitzky–Golay filters to the raw signal envelope before computing the phase and its derivative.

9.6.3 Perfect Reconstruction Filter Banks

In this chapter's first section we considered simple signal analysis problems using an array of filters selective of different frequency ranges. By examining the energy outputs of the separate filters, the frequency content according to time location of signals could be ascertained. This section investigates filter banks more deeply, and, in particular, takes up the problem of reconstructing the original signal from its separately filtered versions.

Why should this matter? There are two basic reasons:

- If a signal can be broken down into separate components and perfectly (or approximately) reconstructed, then this provides a basis for an efficient signal transmission and compression technology.
- There is also the possibility of constructing signal libraries for detection and interpretation purposes that support a coarse-to-fine recognition methodology but provide a compact library of signal prototypes.

One more involved reason is that a perfect reconstruction filter bank is closely related to a type of time-scale transform, the orthogonal wavelet transformation, which we shall cover in Chapter 11.

9.6.3.1 Laplacian Pyramid. An early and innovative approach combining signal scale and frequency-domain analysis is the Laplacian pyramid decomposition. Constructing hierarchical image decompositions was employed by Ref. 75 in their development of the Laplacian pyramid. They approached the problem of managing the sheer volume of information in a pixel image by making two points: First, the gray-scale pixel values are highly correlated in natural scenes; second, it is possible to decompose the original image into both a coarse representation which contains the gross features of the image and a difference image which contains sufficient information to reconstruct the original image from the coarse representation. Their objective was to remove the correlations that typically exist between neighboring pixels in natural scenes. This is a primary goal of image compression.

Burt and Adelson used a discrete filter, which in certain instances closely resembles a Gaussian, to derive the coarse images. The filtered representations are subsampled at twice the unit distance of the previous image to obtain new levels in the pyramid. The authors call this the Gaussian pyramid. This process for one-dimensional signals, passes the original signal at resolution level 0, $f(n)$, given by the digitizer, to the first coarser level of the Gaussian pyramid. The filter coefficients, $w(n)$, are chosen to have an approximately Gaussian shape by Burt and Adelson, although the technical conditions the authors impose on the $w(n)$ allow some quite different filters to arise [75].

To extract a difference signal from two successive layers of the Gaussian pyramid, Burt and Adelson began by inserting zeros between the values of the coarse pyramid level. This is necessary because the coarser level contains pixels whose unit of size is twice that of the finer level. The addition of zero elements causes extra high frequency components to be added to the signal when this up-sampling operation is performed. This requires a second smoothing operation. The new smoothed signal, the values of which are now taken at unit intervals, can be subtracted from the original signal. Figure 9.50 illustrates these operations on the signals. The coarse images are obtained by Gaussian-like filtering and the difference images are

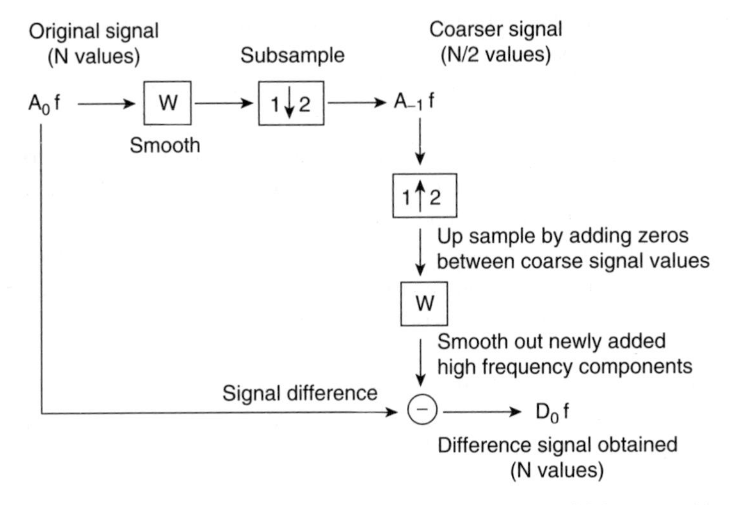

Fig. 9.50. Signal processing diagram of the Laplacian pyramid decomposition.

obtained by subtracting such filtered representations. Since the difference of two Gaussians so closely resembles the Laplacian of a Gaussian operation studied by Marr [77], Burt and Adelson called their construction the Laplacian pyramid.

The signal operations of Figure 9.50 may be repeated. Successive difference signals are produced together with a final coarse, or approximate, signal. The Laplacian pyramid, then, consists of $D_0, D_{-1}, D_{-2}, ..., D_{-J}, A_{-J}$. As is evident from the simple decomposition procedure shown in Figure 9.50, the finer resolution layers of the pyramid may be recovered from the appropriate difference and coarse signals.

When a one-dimensional signal with N samples is hierarchically analyzed with a Laplacian pyramid, the number of coefficients required increases to approximately $2N$. This is evident from the diagram (Figure 9.50).

The Laplacian pyramid provides a scale-based signal recognition strategy. Notice that quasi-Gaussian filters of identical shape applied at differing scales and basic arithmetic operations between the levels of the pyramid to decompose and

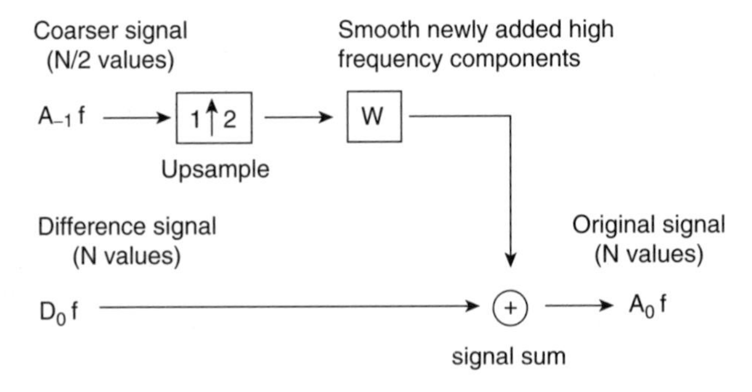

Fig. 9.51. Laplacian pyramid reconstruction of signal.

reconstruct the original image. The computational strategies of the Laplacian pyramid have been implemented in a series of special-purpose vision machines [76].

9.6.3.2 Exact Reconstruction Filter Banks. There is a type of pyramid decomposition that allows perfect signal reconstruction but does not increase the amount of data required for storing the decomposition components. Of course, this is an ideal feature for signal compression applications. But it can also be applied for progressive signal transmission as well as for coarse-to-fine recognition applications. Here we introduce the theory. We shall have occasion to refer back to it when we cover wavelet transforms in the last two chapters.

Consider a real-valued signal $a_0(n)$. The decomposition and reconstruction scheme resembles the Laplacian pyramid's signal flow (Figure 9.52). For the present and the sequel (Chapter 11, in particular) it is useful to set up the following notations. Given the filter impulse responses $h(n)$ and $g(n)$, we set $\overline{h}(n) = h(-n)$ and $\overline{g}(n) = g(-n)$ to be their reflections in time. Observe that $\overline{H}(\omega) = H^*(\omega)$.

Also, we have $a_1(n) = (a_0*\overline{h})(2n)$, which is the original signal convolved with $h(-n)$ and subsampled. Similarly, we have $d_1(n) = (a_0*\overline{g})(2n)$. Typically, we shall select $h(n)$ to be a low-pass filter and $g(n)$ to be a high-pass filter. In terms of the classic paper [78], $a_1(n)$ and $d_1(n)$ are the first-level approximate and detail signals, respectively, in the decomposition of source signal $a_0(n)$. Furthermore, we obtain $a_1'(n)$ by upsampling $a_1(n)$ and then $b_1'(n)$ by filtering with $h'(n)$. Similarly, $c_1'(n)$ comes from filtering $d_1'(n)$ with $g'(n)$, where $d_1'(n)$ is an upsampled version of $d_1(n)$. Finally, $a_0'(n) = (a_1' * h')(n) + (d_1' * g')(n)$. We seek conditions that will guarantee $a_0(n) = a_0'(n)$.

Let us note some properties of the upsampling and downsampling operations. First, let $x(n)$ be a discrete signal and $y(n) = x(2n)$. Then

$$Y(2\omega) = \sum_{n=-\infty}^{\infty} x(2n)e^{-2jn\omega} = \frac{X(\omega) + X(\omega+\pi)}{2}. \tag{9.179}$$

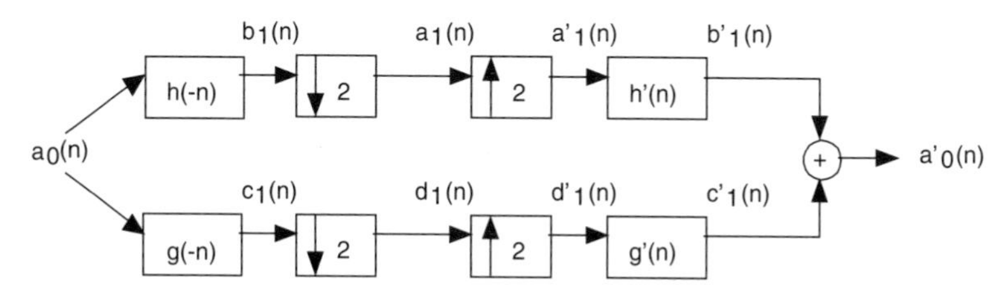

Fig. 9.52. Decomposition and reconstruction signal paths. With the proper choice of filters, the original signal can be recovered and yet its decomposition does not increase the amount of data to be stored.

The second term on the right in (9.179) is called the *folding* or *aliasing* term. This is what wrecks the desired exact reconstruction. Next, if $y(n)$ comes from $x(n)$ by upsampling and inserting zeros,

$$y(n) = \begin{cases} x(m) & \text{if } n = 2m, \\ 0 & \text{if otherwise.} \end{cases} \tag{9.180}$$

then $Y(\omega) = X(2\omega)$. We leave the straightforward proofs as exercises.
The next result is due to Ref. 79.

Theorem. Let $h(n)$ and $g(n)$ be real-valued impulse responses for discrete filters. Then the decomposition and reconstruction filter bank (Figure 9.52) performs perfect reconstruction if and only if

(i) $H^*(\omega+\pi)H'(\omega) + G^*(\omega+\pi)G'(\omega) = 0$ and
(ii) $H^*(\omega)H'(\omega) + G^*(\omega)G'(\omega) = 2$.

Proof: Referring to Figure 9.52, let us calculate the frequency response of the reconstructed signal $a_0'(n)$. We know that $B_1(\omega) = A_0(\omega)\overline{H}(\omega)$ and $C_1(\omega) = A_0(\omega)\overline{G}(\omega)$. Thus,

$$\begin{aligned} 2A_1(2\omega) &= B_1(\omega) + B_1(\omega+\pi) = A_0(\omega)\overline{H}(\omega) + A_0(\omega+\pi)\overline{H}(\omega+\pi) \\ &= A_0(\omega)H^*(\omega) + A_0(\omega+\pi)H^*(\omega+\pi) \end{aligned} \tag{9.181a}$$

and similarly,

$$2D_1(2\omega) = A_0(\omega)G^*(\omega) + A_0(\omega+\pi)G^*(\omega+\pi). \tag{9.181b}$$

On the reconstruction side of the diagram, we have

$$A_0'(\omega) = B_1'(\omega) + C_1'(\omega) = A_1(2\omega)H'(\omega) + D_1(2\omega)G'(\omega). \tag{9.182}$$

Substituting (9.181a) and (9.181b) into (9.182) and simplifying leads to

$$\begin{aligned} A_0'(\omega) &= \frac{1}{2}A_0(\omega)[H^*(\omega)H'(\omega) + G^*(\omega)G'(\omega)] \\ &\quad + \frac{1}{2}A_0(\omega+\pi)[H^*(\omega+\pi)H'(\omega) + G^*(\omega+\pi)G'\omega]. \end{aligned} \tag{9.183}$$

Inspecting (9.183), we see that the only way that $A_0'(\omega)$ can equal $A_0(\omega)$ is if the nonaliased term is doubled, and the aliased term is zero. These are precisely the conditions (i) and (ii) of the theorem's statement. ■

The theorem gives a necessary and sufficient condition on the reconstruction filters $h'(n)$ and $g'(n)$ so that the decomposition scheme provides exact reconstruction.

Theorem. Let $h(n)$ and $g(n)$ be as in the previous theorem. Then the decomposition and reconstruction scheme (Figure 9.52) performs perfect reconstruction only if

$$H^*(\omega)H'(\omega) + H^*(\omega+\pi)H'(\omega+\pi) = 2. \tag{9.184}$$

Proof: After taking complex conjugates, we may write the result of the previous theorem

(iii) $H(\omega+\pi)H'^*(\omega) + G(\omega+\pi)G'^*(\omega) = 0$ and
(iv) $H(\omega)H'^*(\omega) + G(\omega)G'^*(\omega) = 2$.

Let us rewrite these relations in matrix form as follows:

$$\begin{bmatrix} H(\omega) & G(\omega) \\ H(\omega+\pi) & G(\omega+\pi) \end{bmatrix} \begin{bmatrix} H'^*(\omega) \\ G'^*(\omega) \end{bmatrix} = \begin{bmatrix} 2 \\ 0 \end{bmatrix}. \tag{9.185}$$

If the determinant is nonzero, (9.185) can be solved,

$$\begin{bmatrix} H'^*(\omega) \\ G'^*(\omega) \end{bmatrix} = \frac{2}{\Delta(\omega)} \begin{bmatrix} G(\omega+\pi) \\ -H(\omega+\pi) \end{bmatrix}, \tag{9.186}$$

where $\Delta(\omega) = H(\omega)G(\omega+\pi) - G(\omega)H(\omega+\pi)$. Note too that $\Delta(\omega+\pi) = -\Delta(\omega)$. From (9.186) we get

$$H'^*(\omega+\pi) = \frac{2G(\omega+2\pi)}{\Delta(\omega+\pi)} = \frac{-2G(\omega)}{\Delta(\omega)}, \tag{9.187}$$

implying

$$G(\omega)G'^*(\omega) = \frac{\Delta(\omega)H'^*(\omega+\pi)}{-2}\frac{(-2)H(\omega+\pi)}{\Delta(\omega)} = H'^*(\omega+\pi)H(\omega+\pi). \tag{9.188}$$

Using the complex conjugate of (9.188) with (ii) of the previous theorem gives (9.184) above. ■

Corollary. Under the theorem's assumptions,

$$G^*(\omega)G'(\omega) + G^*(\omega+\pi)G'(\omega+\pi) = 2. \tag{9.189}$$

Proof: Exercise. ■

Definition (Quadrature Mirror Filters). If the decomposition filter is the same as the reconstruction filter $h(n) = h'(n)$, then it is called a *quadrature mirror filter* (QMF) or *conjugate mirror filter* (CMF).

Corollary. If $h(n)$ is QMF, then

$$|H(\omega)|^2 + |H(\omega+\pi)|^2 = 2. \tag{9.190}$$

Proof: From (9.184). ■

9.7 SUMMARY

This chapter explored a variety of frequency domain signal analysis applications, developed several related tools, explained how to construct and implement filters for frequency selective input processing, and studied the theoretical issues that arose in the course of experimentation.

We can divide analysis applications according to whether the source signals are basically narrowband or wideband:

(i) Narrowband signals are interpretable—at least locally—via pulse detection in the Fourier magnitude spectrum.
(ii) Wideband signals, on the other hand, demand more complicated procedures, multicomponent, and shape-based techniques.

For narrow band signals we developed three basic approaches:

(i) Time-domain segmentation, based on the histogram for example, prior to frequency-domain analysis.
(ii) Local windowing using a short (say $N = 256$) DFT in many locations, thus forming a time-frequency map.
(iii) The filter bank.

Either of the first two methods work well under moderate noise. Under heavy noise, or faced with real-time processing and interpretation constraints, a bank of carefully crafted filters, operating in parallel and equipped with decision logic at their output, can provide satisfactory results.

We significantly extended filter bank theory at the end of the chapter. We showed that subsampling combined with low- and high-pass filtering can be used to decompose signals for analysis. Moreover, under certain conditions on the filters, the filter bank supports an exact reconstruction algorithm using upsampling and filtering. We will revisit these ideas in Chapters 11 and 12, showing that there is a link between filter banks and the theory of orthogonal wavelets—a time-scale signal analysis technique.

The concepts of phase and group delay arose in frequency-domain signal analysis applications where discrete filters were necessary. Applications that filter incoming signals for noise removal, frequency selection, or signal shaping and then analyze the output must take into account the delay characteristics of the filter. We observed the phase delay of sinusoidal tones in the DTMF filter bank application, for example. Other applications involve group delay, such as speech analysis and edge detection.

Many applications require that the filters provide linear, or nearly linear, phase. In communication systems, for example, the information is carried on the envelope and the carrier has a constant frequency. Thus, nonlinear phase delay could well stretch and compress the frequencies in the signal so that the distortions render the signal unintelligible. Important aspects of seismic signal processing are to properly measure the time delays between different signal components. Thus, linear phase is crucial when filtering; and for this reason, finite impulse response (FIR) filters, which we proved to have linear phase, are preferred.

REFERENCES

1. J. S. Walker, *Fourier Analysis*, New York: Oxford University Press, 1988.
2. G. B. Folland, *Fourier Analysis and Its Applications*, Pacific Grove, CA: Wadsworth and Brooks/Cole, 1992.
3. R. E. Edwards, *Fourier Series: A Modern Introduction*, vol. I, New York: Holt, Rinehart and Winston, 1967.
4. T. W. Körner, *Fourier Analysis*, Cambridge: Cambridge University Press, 1988.
5. D. C. Champeney, *A Handbook of Fourier Theorems*, Cambridge: Cambridge University Press, 1987.
6. H. Baher, *Analog and Digital Signal Processing*, New York: Wiley, 1990.
7. A. V. Oppenheim and R. W. Schafer, *Discrete-Time Signal Processing*, Englewood Cliffs, NJ: Prentice-Hall, 1989.
8. J. G. Proakis and D. G. Manolakis, *Digital Signal Processing: Principles, Algorithms, and Applications*, 2nd ed., New York: Macmillan, 1992.
9. R. A. Roberts and C. T. Mullis, *Digital Signal Processing*, Reading, MA: Addison-Wesley, 1987.
10. A. V. Oppenheim, A. S. Willsky, and S. H. Nawab, *Signals and Systems*, 2nd ed., Upper Saddle River, NJ: Prentice-Hall, 1997.
11. L. R. Rabiner and B. Gold, *Theory and Application of Digital Signal Processing*, Englewood Cliffs, NJ: Prentice-Hall, 1975.
12. P. Mock, Add DTMF generation and decoding to DSP-μP designs, *Electronic Design News*, March 21, 1985; also in K.-S. Lin, ed., *Digital Signal Processing Applications with the TMS320 Family*, vol. 1, Dallas: Texas Instruments, 1989.
13. J. L. Flanagan, *Speech Analysis, Synthesis, and Perception*, 2nd ed., New York: Springer-Verlag, 1972.
14. D. H. Lehmer, Mathematical methods in large-scale computing units, *Proceeding of the 2nd Symposium on Large-Scale Digital Calculating Machines*, Harvard University Press, pp. 141–146, 1951.
15. S. K. Park and K. W. Miller, Random number generators: Good ones are hard to find, *Communications of the ACM*, vol. 31, no. 10, pp. 1192–1201, 1988.
16. W. H. Press, B. P. Flannery, S. A. Teukolsky, and W. T. Vetterling, *Numerical Recipes in C*, Cambridge: Cambridge University Press, 1988.
17. G. E. Box and M. E. Muller, A note on the generation of random normal deviates, *Annals of Mathematical Statistics*, vol. 29, pp. 610–611, 1958.
18. D. E. Knuth, *Seminumerical Algorithms*, 2nd ed., Reading, MA: Addison-Wesley, 1981.
19. L. Devroye, *Non-Uniform Random Variate Generation*, New York: Springer-Verlag, 1986.
20. C. K. Chow and T. Kaneko, Automatic boundary detection of the left ventricle from cineangiograms, *Computers and Biomedical Research*, vol. 5, pp. 388–410, 1972.

21. N. Otsu, A threshold selection method from gray-level histograms, *IEEE Transactions on Systems, Man, and Cybernetics*, vol. SMC-9, no. 1, pp. 62–66, January 1979.
22. J. Kittler and J. Illingworth, On threshold selection using clustering criteria, *IEEE Transactions on Systems, Man, and Cybernetics*, vol. SMC-15, pp. 652–655, 1985.
23. R. E. Crochiere and L. R. Rabiner, *Multirate Digital Signal Processing*, Englewood Cliffs, NJ: Prentice-Hall, 1983.
24. A. N. Akansu and R. A. Haddad, *Multiresolution Signal Decomposition: Transforms, Subband, and Wavelets*, Boston: Academic, 1992.
25. M. Vetterli and J. Kovacevic, *Wavelets and Subband Coding*, Upper Saddle River, NJ: Prentice-Hall, 1995.
26. T. W. Parks and C. S. Burrus, *Dignal Filter Design*, New York: Wiley, 1987.
27. L. W. Couch II, *Digital and Analog Communication Systems*, 4th ed., Upper Saddle River, NJ: Prentice-Hall, 1993.
28. R. B. Blackman and J. W. Tukey, *The Measurement of Power Spectra*, New York: Dover, 1958.
29. S. M. Kay, *Modern Spectral Estimation: Theory and Application*, Englewood Cliffs, NJ: Prentice-Hall, 1988.
30. L. L. Scharf, *Statistical Signal Processing: Detection, Estimation, and Time Series Analysis*, Reading, MA: Addison-Wesley, 1991.
31. D. G. Childers, ed., *Modern Spectral Analysis*, New York: IEEE Press, 1978.
32. C. Bingham, M. D. Godfrey, and J. W. Tukey, Modern techniques of power spectrum estimation, *IEEE Transactions on Audio and Electroacoustics*, vol. AU-15, pp. 56–66, June 1967. Also in Refs. 31, pp. 6–16.
33. A. Antoniou, *Digital Filters: Analysis and Design*, New York, McGraw-Hill, 1979.
34. A. Papoulis, *Probability, Random Variables, and Stochastic Processes*, 2nd ed., New York: McGraw-Hill, 1984.
35. M. S. Bartlett, Smoothing periodograms from time series with continuous spectra, *Nature (London)*, vol. 161, pp. 686–687, 1948.
36. P. D. Welch, The use of fast Fourier transform for the estimation of power spectra: A method based on time averaging over short, modified periodograms, *IEEE Transactions on Audio and Electroacoustics*, vol. AU-15, no. 2, pp. 70–73, 1967. Also in Refs. 31, p. 17–20.
37. R. B. Blackman and J. W. Tukey, *Measurement of Power Spectra from the Point of View of Communications Engineering*, New York: Dover, 1959.
38. D. Malacara, *Optical Shop Testing*, New York: Wiley-Interscience, 1992.
39. J. M. Steigerwald, S. P. Murarka, and R. J. Gutmann, *Chemical Mechanical Planarization of Microelectronic Materials*, New York: Wiley, 1997
40. J. F. Kaiser, Design methods for sampled data filters, *Proceedings of the 1st Allerton Conference on Circuit and System Theory*, pp. 221–236, 1963. Also in L. R. Rabiner and C. M. Rader, eds., *Dignal Signal Processing*, New York: IEEE Press, 1972.
41. E. I. Jury, *Theory and Applications of the Z-Transform Method*, Malabar, FL: Krieger, 1982.
42. R. Vich, *Z Transform Theory and Applications*, Boston: D. Reidel, 1987.
43. A. Ralston and P. Rabinowitz, *A First Course in Numerical Analysis*, New York: McGraw-Hill, 1978.

44. R. Beals, *Advanced Mathematical Analysis*, New York: Springer-Verlag, 1983.

45. E. D. Rainville, *The Laplace Transform: An Introduction*, New York: Macmillan, 1963.

46. F. J. Taylor, *Digital Filter Design Handbook*, New York: Marcel Dekker, 1983.

47. A. G. Constantinides, Spectral transformations for digital filters, *Proceedings of the IEE*, vol. 117, no. 8, pp. 1585–1590, August 1970.

48. M. Akay, *Biomedical Signal Processing*, San Diego, CA: Academic Press, 1994.

49. M. Granzow, H. Riedel, and B. Kollmeier, Linear phase filtering of early auditory evoked potentials, in A. Schick and M. Klatte, eds., *Seventh Oldenburg Symposium on Psychological Acoustics*, pp. 39–48, 1997.

50. J. F. Claerbout, *Fundamentals of Geophysical Data Processing with Applications to Petroleum Prospecting*, New York: McGraw-Hill, 1976.

51. F. Scherbaum, *Of Poles and Zeros: Fundamentals of Digital Seismology*, Dordrecht, The Netherlands: Kluwer, 1996.

52. T. Parsons, *Voice and Speech Processing*, New York: McGraw-Hill, 1987.

53. F. J. Owens, *Signal Processing of Speech*, New York: McGraw-Hill, 1993.

54. L. Rabiner and B.-H. Juang, *Fundamentals of Speech Recognition*, Englewood Cliffs, NJ: Prentice-Hall, 1993.

55. L. R. Rabiner and M. R. Sambur, An algorithm for determining the endpoints of isolated utterances, *Bell System Technical Journal*, vol. 54, no. 2, pp. 297–315, 1975.

56. L. Lamel et al., An improved endpoint detector for isolated word recognition, *IEEE Transactions on Acoustics, Speech, and Signal Processing*, vol. 29, pp. 777–785, Aug. 1981.

57. J. Junqua et al., A robust algorithm for word boundary detection in the presence of noise, *IEEE Transactions on Speech and Audio Processing*, vol. 2, no. 3, pp. 406–412, July 1994.

58. R. E. Ziemer, W. H. Tranter, and D. R. Fannin, *Signals and Systems: Continuous and Discrete*, 3rd ed., New York: Macmillan, 1993.

59. L. B. Jackson, Signals, *Systems, and Transforms*, Reading, MA: Addison-Wesley, 1991.

60. E. A. Guillemin, *Synthesis of Passive Networks*, New York: Wiley, 1957.

61. J. E. Storer, *Passive Network Synthesis*, New York: McGraw-Hill, 1957.

62. M. E. Van Valkenburg, *Analog Filter Design*, New York: Holt, Rinehart and Winston, 1982.

63. L. Weinberg, *Network Analysis and Synthesis*, Huntington, NY: Kreiger, 1975.

64. V. K. Madiseti, *VLSI Digital Signal Processors*, Piscataway, NJ: IEEE Press, 1995.

65. R. W. Daniels, *Approximation Methods for Electronic Filter Design*, New York: McGraw-Hill, 1974.

66. W. Cauer, *Synthesis of Linear Communication Networks*, vols. I and II, New York: McGraw-Hill, 1958.

67. D. A. Calahan, *Modern Network Synthesis*, vol. 1, New York: Hayden, 1964.

68. N. Wiener, *Extrapolation, Interpolation, and Smoothing of Stationary Time Series*, New York: Technology Press of MIT and Wiley, 1957.

69. R. N. Bracewell, *The Fourier Transform and Its Applications*, New York: McGraw-Hill, 1978.

70. S. L. Hahn, *Hilbert Transforms in Signal Processing*, Norwood, MA: Artech House, 1996.

71. L. V. Ahlfors, *Complex Analysis*, 2nd ed., New York: McGraw-Hill, 1966.

72. S. L. Marple, Computing the discrete-time “analytic signal” via FFT, *IEEE Transactions on Signal Processing*, vol. 47, no. 9, pp. 2600–2603, September 1999.

73. D. Gabor, Theory of communication, *Journal of the IEE*, vol. 93, pp. 429–457, 1946.

74. L. Cohen, *Time-Frequency Analysis*, Englewood Cliffs, NJ: Prentice-Hall, 1995.

75. P. J. Burt and E. H. Adelson, The Laplacian pyramid as a compact image code, *IEEE Transactions on Communication*, vol. 31, no. 4, pp. 532–540, April 1983.

76. P. J. Burt, Smart sensing within a pyramid vision machine, *Proceedings of the IEEE*, vol. 76, no, 8, pp. 1006–1015, August 1988.

77. D. Marr, *Vision*, New York: W. H. Freeman, 1982.

78. S. Mallat, A theory for multiresolution signal decomposition: The wavelet representation, *IEEE Transactions on Pattern Analysis and Machine Intelligence*, vol. 11, no. 7, pp. 674–693, July 1989.

79. M. Vetterli, Filter banks allowing perfect reconstruction, *Signal Processing*, vol. 10, no. 3, pp. 219–244, April 1986.

PROBLEMS

1. Suppose an analog signal is sampled $x(n) = xa(nT)$, where T is the sampling period. A discrete Fourier transform of order N follows: $X(k) = (\mathcal{F}x)(k)$. Find the frequency resolution, the Nyquist frequency, and the highest frequency represented by the DFT coefficients:

(a) $N = 80$, $T = 0.02$;

(b) $N = 10$, sampling frequency $F = 1$ kHz;

(c) $N = 21$, $T = 1$;

(d) $N = 256$, sampling frequency = 8192 Hz.

2. Which of the following finite impulse response (FIR) filters have linear phase? zero phase?

(a) $x(n) = u(n + 2) - u(n - 2)$, where $u(n)$ is the discrete unit step signal.

(b) $y(n) = u(n + 2) - u(n - 3)$.

(c) $v(n) = (-1)^n x(n)$.

(d) $w(n) = (-1)^n y(n)$.

(e) Can an IIR filter have linear phase?

3. Provide sketches of all four types of linear phase filters H with $h(n) \in \mathbb{R}$ [26].

4. Let $r_{xx}(\tau) = E[x(t)x(t + \tau)]$ be the autocorrelation for a wide-sense stationary (WSS) analog random signal $x(t)$. Prove:

(a) $E[x(t)x(s)] = r_{xx}(t - s)$.

(b) $r_{xx}(\tau) = r_{xx}(-\tau)$.

(c) If $x = Ay$ for some constant A, find $r_{yy}(\tau)$.

(d) State and the corresponding properties for the autocorrelation $r_{ss}(\kappa) = E[s(n)s(n + \kappa)]$ of a discrete WSS random signal $s(n)$.

5. Suppose data from a noisy signal is collected for two seconds at a sampling rate of 8 kHz.

(a) If a periodogram is calculated for the entire data set, what is its frequency resolution? What is the Nyquist frequency?

(b) Suppose that Bartlett's method is tried for the purpose of improving the periodogram. The data is partitioned into sixteen disjoint windows. Now what is the frequency resolution? What is the Nyquist frequency?

(c) It is decided to try Welch's method using windows of a larger size than in part (b), but to overlap them by fifty percent. How many windows are needed to cut the frequency resolution in half? Sketch your window layout.

6. Let H be a discrete LTI system, let $H(z)$ be its system function, and let ROC_H be the region of convergence of $H(z)$.

(a) Show that if H is *stable* (if the input $x(n)$ is bounded, then the output $y(n) = (Hx)(n)$ is also bounded), then this implies that ROC_H contains the unit circle $|z| = 1$ of the complex plane.

(b) Suppose that $\{z \in \mathbb{C}: |z| = 1\} \subset ROC_H$. Show that H is stable.

(c) Given: an example of a discrete signal $h(n)$ that has a discrete-time Fourier transform $H(\omega)$, but ROC_H does not include the unit circle. Under this circumstance, can the system $y = Hx = h*x$ be causal, anti-causal, or both?

7. Let H be a discrete LTI system, $h(n)$ its impulse response, $H(z)$ its transfer function, and ROC_H the region of convergence of $H(z)$.

(a) Show that if H is stable, then ROC_H contains the unit circle $|z| = 1$.

(b) Show the converse: If $ROC_H \supset \{z \in \mathbb{C}: |z| = 1\}$, then H is stable.

8. Let $x(n)$ be a discrete signal, $X(z)$ its z-transform, and ROC_X the region of convergence.

(a) Show that $x(n)$ is right-sided implies that ROC_X is the exterior of a circle in the complex plane.

(b) More particularly, if $x(n)$ is causal, show that $\infty \in ROC_X$.

(c) If $x(n)$ is left-sided, show that ROC_X is the interior of a circle and may or may not include 0.

(d) Give a condition on a left-sided signal $x(n)$ so that $0 \in ROC_X$.

(e) Give a characterization in terms of ROC_H for causal, stable LTI systems $y = Hx$.

9. Let $x(t)$ have Fourier transform $X(\Omega)$ and Laplace transform $X_L(s)$.

(a) Show that $X_L(j\Omega) = X(\Omega)$, for $\Omega \in \mathbb{R}$.

(b) Let $s = \Sigma + j\Omega$, where $\Sigma \in \mathbb{R}$. Show that $X_L(s)$ is the Fourier transform of $x(t)e^{-\Sigma t}$.

(c) Show that Laplace transform convergence does not depend on the imaginary part of $s = \Sigma + j\Omega$.

(d) Conclude that $X_L(s)$ converges on vertical strips in the complex plane.

10. Let $x(t) = \exp(-a|t|)$ be the Lorentzian function, where $a > 0$.

(a) Show that $X(\Omega) = 2a/(a + \Omega^2)$.

(b) Sketch $x(t)$ for $a = 1, 2, 3$.

(c) Explain how convolution with $x(t)$ performs a weighted averaging of input signal data.

11. Suppose we map the Laplace s-plane to the z-plane via $z = e^{sT}$, where we shall assume equality in (9.72).

(a) Show that this implies $s = \frac{2}{T}\left(\frac{z-1}{z+1}\right)$, the bilinear relation between the Laplace and z-transformations.

(b) Let $z = e^{j\omega}$ and $s = j\Omega$ in (a) and derive the bilinear mapping (9.70).

12. Derive the Laplace transform properties given in Table 9.3.

13. Let $T_N(\Omega)$ be the Chebyshev polynomial of order $N \geq 0$.

(a) Show that $T_N(1) = 1$ for all N.

(b) Show $T_N(\Omega)$ is even if N is odd, and $T_N(\Omega)$ is odd if N is even.

(c) Show all the zeros of $T_N(\Omega)$ lie on the open interval $(-1, 1)$.

(d) Show that $|T_N(\Omega)| \leq 1$ on $[-1, 1]$.

14. Let $P(\Omega)$ and $Q(\Omega)$ be the squared magnitude response for the Chebyshev and inverse Chebyshev low-pass filters, respectively.

(a) Show that the poles of $Q(\Omega)$ are the reciprocals of the poles of $P(\Omega)$.

(b) Find $P(\Omega)$ for IIR filters of order 2 and 3.

(c) Find $Q(\Omega)$ for IIR filters of order 2 and 3.

(d) Under what conditions is $Q(\Omega)$ an all-pass filter? Explain.

(e) Find the zeros of $Q(\Omega)$.

(f) Show that the zeros of $Q(\Omega)$ do not depend on the stopband ripple function.

15. Compute the Hilbert transform $x_H(t)$ for the following signals:

(a) $x(t) = \cos(2\pi t)$.

(b) $x(t) = \sin(-3\pi t)$.

(c) $x(t) = \cos(5\pi t) + 2\sin(2\pi t)$.

(d) $x(t) = \delta(t)$, the Dirac delta.

(e) $x(t) = C_0$, a constant signal.

(f) $x(t) = u(t)$, the unit step signal.

(g) $x(t) = u(t + 1) - u(t - 1)$.

16. Let $x(t) = \cos(\Omega_0 t)$ and $y(t) = \sin(\Omega_0 t)$.

(a) Show that $y_H(t) = -x(t)$, where $y_H(t)$ is the Hilbert transform of $y(t)$.

(b) Compute the analytic signal $x_A(t)$.

(c) Compute the signal envelope for $x(t)$.

(d) Compute the instantaneous phase $\phi(t)$.

(e) Compute the instantaneous frequency $\omega(t)$.

17. Show that the Hilbert transform is linear: If $x(t)$ and $y(t)$ are analog signals and $z(t) = Ax(t) + By(t)$, then $z_H(t) = Ax_H(t) + By_H(t)$.

18. Let $x(n)$ be discrete with period $N > 0$. Show that the discrete Hilbert transformation of $A\sin(2\pi k_0 n/N)$ is $-A\cos(2\pi k_0 n/N)$.

19. Suppose the discrete filter H has impulse responses $h(n)$. Suppose $\overline{h}(n) = h(-n)$ and let $\overline{H}(\omega)$ be its DTFT.

(a) Show that $\overline{H}(\omega) = H^*(\omega)$, the complex conjugate of $H(\omega)$.

(b) Let $y(n) = x(2n)$. Show that $Y(2\omega) = [X(\omega) + X(\omega + \pi)]/2$.

(c) Let $y(n) = x(n/2)$. Show that $Y(\omega) = X(2\omega)$.

20. Let $h(n)$ and $g(n)$ be discrete filters that provide an exact reconstruction scheme as in Figure 9.52. Show that $G^*(\omega)G'(\omega) + G^*(\omega + \pi)G'(\omega + \pi) = 2$.

Advanced problems and projects:

21. This problem outlines a proof of the Wiener–Khinchin theorem for discrete random signals. Assume the notation from Section 9.2.3.1.

(a) First show that

$$\sum_{n=-L}^{L} \sum_{m=-L}^{L} x(n)x(m)e^{-j\omega(n-m)} = |X_L(\omega)|^2, \tag{9.191}$$

where $X_L(\omega)$ is the local discrete time Fourier transform of real-valued WSS random signal $x(n)$, and we assume that the autocorrelation function $r_{xx}(\nu)$ is absolutely summable: $r_{xx}(\nu) \in l^1$.

(b) Apply the expectation operator to both sides of (9.191) to get

$$\begin{aligned} E[|X_L(\omega)|^2] &= \sum_{n=-L}^{L} \sum_{m=-L}^{L} E[x(n)x(m)]e^{-j\omega(n-m)} \\ &= \sum_{n=-L}^{L} \sum_{m=-L}^{L} r_{xx}(n-m)e^{-j\omega(n-m)}. \end{aligned} \tag{9.192}$$

(c) Divide by $2L + 1$ and take limits on both sides of (9.192). From the absolute summability of $r_{xx}(\nu)$, argue that the equal limits of the double summation can be replaced with independent limits to get

$$\begin{aligned} \lim_{L\to\infty} \frac{1}{2L+1} E[|X_L(\omega)|^2] &= \lim_{L\to\infty} \frac{1}{2L+1} \sum_{n=-L}^{L} \sum_{m=-L}^{L} r_{xx}(n-m)e^{-j\omega(n-m)} \\ &= \lim_{L\to\infty} \frac{1}{2L+1} \sum_{n=-L}^{L} \lim_{K\to\infty} \sum_{m=-K}^{K} r_{xx}(n-m)e^{-j\omega(n-m)}. \end{aligned} \tag{9.193}$$

(d) Conclude as follows:

$$X_{\text{PSD}}(\omega) = \lim_{L\to\infty}\frac{1}{2L+1}\sum_{n=-L}^{L} R_{xx}(\omega)$$
$$= \lim_{L\to\infty}\frac{2L+1}{2L+1}R_{xx}(\omega) = R_{xx}(\omega). \qquad (9.194)$$

22. This problem outlines a proof of the Wiener–Khinchin theorem (9.30) for WSS analog random signals $x(t)$. Assume the notation from Section 9.2.3.1.

(a) By interchanging the order of integration, show that

$$E\left[\int_{-\infty}^{\infty}\int_{-\infty}^{\infty} x_L(t)x_L(s)e^{-j\omega(t-s)}dsdt\right] = E[|X_L(\omega)|^2], \qquad (9.195)$$

where $X_L(\omega)$ is the local radial Fourier transform of $x(t)$.

(b) Use the results of probability theory: $E[ax + by] = aE[x] + bE[y]$ and $r_{xx}(t - s) = E[x(s)x(t)]$. Show that the expectation operation may be moved inside the integral in (9.195) and therefore that

$$\int_{-\infty}^{\infty}\int_{-\infty}^{\infty} E[x_L(t)x_L(s)e^{-j\omega(t-s)}]\,dsdt = \int_{-L}^{L}\int_{-L}^{L} r_{xx}(t-s)e^{-j\omega(t-s)}dsdt. \qquad (9.196)$$

(c) Let $u = t - s$ for a change of integration variable in (9.196) and show that the iterated integral becomes $\int_{-2L}^{2L} r_{xx}(u)e^{-j\omega u}[2L-|u|]\,du$.

(d) Put the above expressions together and take limits to get

$$\lim_{L\to\infty}\frac{1}{2L}E[|X_L(\omega)|^2] = \lim_{L\to\infty}\int_{-2L}^{2L} r_{xx}(u)e^{-j\omega u}\left[\frac{2L-|u|}{2L}\right]du. \qquad (9.197)$$

(e) Show that the integrand on the right-hand side of (9.197) is bounded by $|r_{xx}(u)|$ so that Lebesgue's dominated convergence theorem (Chapter 3) applies.

(f) Since the the limit and integration operations can be interchanged, show

$$X_{\text{PSD}}(\omega) = \int_{-\infty}^{\infty} r_{xx}(u)e^{-j\omega u}\left\{\lim_{L\to\infty}\left[\frac{2L-|u|}{2L}\right]\right\}du = R_{xx}(\omega), \qquad (9.198)$$

where $R_{xx}(\omega)$ is the radial Fourier transform of $r_{xx}(t)$.

23. Consider a linear system $y = Hx$, where the WSS random signal $x(t)$ is the input and $y(t)$ is the corresponding output. Let $r_{xx}(\tau) = E[x(t)x(t + \tau)]$ and $r_{yy}(\tau) = E[y(t)y(t + \tau)]$.

(a) Show that the cross correlation $r_{xy}(\tau) = r_{xx}(\tau) * \overline{h(-\tau)}$.

(b) Show that $r_{yy}(\tau) = r_{xy}(\tau) * h(\tau)$.

(c) Show that $Y_{PSD}(\Omega) = X_{PSD}(\Omega)|H(\Omega)|^2$, where $H(\Omega)$ is the radial Fourier transform of $h(t)$.

24. Suppose that real-valued discrete signal $x(n)$ is sampled at $F_s = 1/T$ Hz. We select a window of $N > 0$ values and compute the DFT of order N. Thus, the DFT $X(k)$ has frequency resolution $(NT)^{-1}$ and its coefficients represent discrete frequencies $(NT)^{-1}$, $2(NT)^{-1}$, ..., $(2T)^{-1}$.

(a) Show that adding a pure sinusoid of one these frequencies—say $\omega_k = kF_s/N$, for $1 \le k \le N/2$—to $x(n)$ will alter only $X(k)$ and $X(N-k)$.

(b) Verify experimentally using that $y(n) = x(n) + \cos(2\pi nk/N)$ a sinusoid of frequency $\omega \ne \omega_k$, for any $1 \le k \le N/2$, will perturb all of the $X(k)$.

(c) Show that the difference caused by adding the sinusoid diminishes in magnitude like $1/|\omega - \omega_k|$ as $|\omega - \omega_k|$ increases [32].

25. This problem motivates use of the sampling interval T in the bilinear transformation (9.70). The derivation closely follows [33]. Consider the analog integrator system with impulse response

$$h(t) = \begin{cases} 0 & \text{if } t < 0, \\ 1 & \text{if } t \ge 0. \end{cases} \tag{9.199}$$

(a) Let $y = Hx$, so that $y(t) = (h*x)(t)$ and show that for $0 < a < b$,

$$y(b) - y(a) = \int_a^b x(t)\, dt. \tag{9.200}$$

(b) Argue that as $a \to b$,

$$y(b) - y(a) \approx \frac{b-a}{2}[x(a) + x(b)]. \tag{9.201}$$

(c) Let $a = nT - T$ and $b = nT$ so that (9.201) becomes a discrete integrator and show that

$$Y(z) - z^{-1}Y(z) = \frac{T}{2}[z^{-1}X(z) + X(z)]. \tag{9.202}$$

(d) Show that the discrete integrator has z-transform

$$H(z) = \frac{T(z+1)}{2(z-1)}. \tag{9.203}$$

(e) Show that the Laplace transform of the analog integrator is $H(s) = s^{-1}$.

(f) Argue that an analog system defined by a difference equation (9.65) can be implemented using adders, amplifiers, and integrators.

(g) Replace every analog element by its corresponding discrete element and conclude that the discrete transfer function corresponding to (9.65) is given by

$$s = \frac{2(z-1)}{T(z+1)}. \tag{9.204}$$

26. Prove Liouville's theorem, which we used to justify the fundamental theorem of algebra. Let $|f(z)| \le B$ on $\mathbb{C}$ be everywhere differentiable, or *analytic*.

(a) Show that the Cauchy Residue Theorem (Chapter 1) implies

$$f(z) = \frac{1}{2\pi j}\oint_C \frac{f(s)}{s-z}\,ds, \tag{9.205}$$

where C is a large circular contour centered about $z \in \mathbb{C}$.

(b) Show that

$$f'(z) = \frac{1}{2\pi j}\oint_C \frac{f(s)}{(s-z)^2}\,ds. \tag{9.206}$$

(c) If C has radius $R > 0$, show that

$$|f'(z)| \le \frac{B}{R}. \tag{9.207}$$

(d) Since the radius of C may be arbitrarily large, conclude that $f'(z) = 0$ and $f(z) \equiv$ constant.

27. Suppose a rational function $P(\Omega) = B(\Omega)/A(\Omega)$ is even: $P(\Omega) = P(-\Omega)$ for all $\Omega \in$ R. Show that $A(\Omega)$ and $B(\Omega)$ have no terms of odd degree.

28. Consider the analog Lorentzian signal $h(t)$ with rational (radial) Fourier transform. $H(\Omega) = (1 + \Omega^2)^{-1}$.

(a) Show that $h(t)$ is given by

$$h(t) = \frac{1}{2\pi}\int_{-\infty}^{\infty} H(\Omega)e^{j\Omega t}\,d\Omega = \frac{e^{-|t|}}{2}. \tag{9.208}$$

(b) Using synthetic signals, evaluate the performace of $h(t)$ as a peak, valley, and transient detector.

(c) Evaluate the signal analysis performance of $h(t)$ for detecting the same signal features in the presence of synthetic additive noise.

(d) Identify suitable real data sets and report on the Lorentzian's capabilities for detecting these same features.

29. Develop software for a Butterworth filter design tool.

(a) Given a discrete cutoff frequency ω_c and filter order $N > 0$, compute the analog cutoff Ω_c.

(b) Find the above poles of the analog squared Fourier magnitude response function.

(c) Select poles in order to ensure a stable filter.

(d) Derive the discrete impulse response $h(n)$.

(e) Compute the discrete magnitude response $|H(\omega)|$.

(f) Compute the margin by which the filter H passes or fails the passband and stopband constraints.

(g) Provide an approximation feature that computes the filter order given ω_c as well as passband and stopband criteria.

CHAPTER 10

Time-Frequency Signal Transforms

Time-frequency signal transforms combine traditional Fourier transform signal spectrum information with a time location variable. There results a two-dimensional transformed signal having an independent frequency variable and an independent time variable. Such a signal operation consititutes the first example of a *mixed-domain signal transform.*

Earlier discussions, many applications, and indeed our entire theoretical approach considered signal analysis strategies based upon time, frequency, or scale. *Time-domain* methods are adequate for tasks such as edge detection, elementary segmentation, correlation-based shape recognition, and some texture analysis problems. But in many situations, the inherent periodicity within signal regions pushes us toward a decomposition of the signal according to its *frequency content.*

Frequency—or *spectral*—analysis enters the picture as a tool for discovering a signal's sinusoidal behavior. But this is an inherently global approach. Standard spectral analysis methods, which the Fourier transform in both its analog and discrete guises completely typifies, suffer signal interpretation difficulties when oscillations of interest exist only within a limited signal region. As we discovered in the previous chapter, windowing the signal improves local spectral estimates.

Another approach takes a standard signal shape element, shrinks or expands it into a library of local signal forms, and then considers how well different regions of the signal match one or another such local prototypes. This is an analysis based on *signal scale*. The first chapter provided tutorial sketches of all three approaches. Later chapters built up theoretical tools, demonstrated their applications, and discovered some limitations. So far we have worked out much theory and many applications involving time- and frequency-domain techniques, but we have not formalized the notion of a scale-based analysis.

We now have a sufficient theoretical foundation and practical motivation to explore combined methods. The idea is to mix time-domain methods with either the frequency- or the scale-domain approach. Both combinations provide avenues for structural signal decomposition. The theory is rich and powerful. It has developed rapidly in the last few years. We elect to start with the methods that are most intuitive and, in fact, historically prior: the time-frequency transform techniques.

Signal Analysis: Time, Frequency, Scale, and Structure, by Ronald L. Allen and Duncan W. Mills
ISBN: 0-471-23441-9

The Fourier transform is the fundamental tool for frequency-domain signal analysis. It does allow us to solve some problems that confound time-domain techniques. The mapping $X(\omega) = \mathcal{F}[x(t)]$ lays out the frequency content of a signal $x(t)$, albeit in complex values, and its large magnitude $|X(\omega)|$ indicates the presence of a strong sinusoidal component of frequency ω radians per second in $x(t)$. We can construct filters and assemble them into filter banks in order to search for spectral components in frequency ranges of interest. All such strategies stem from the convolution theorem, which identifies time-domain convolution—and hence linear, time-invariant processing—with frequency-domain multiplication. The caveat is that standard Fourier techniques depend on a knowledge of the entire time-domain extent of a signal. Even the filter bank highlights ranges of frequencies that existed in the signal for all time: past, present, and future.

But many signals have salient periodic features only over limited time intervals. Although a global analysis is theoretically possible, it may not be practical or efficient. Consider, for example, an orchestra that must play a two-hour symphony, and let us fancy that the composer employs a Fourier transform music style that assigns each instrument just one tone for the entire duration of the performance. The superposition of the various tones, each constantly emitted for two hours by the musicians, does indeed produce the composer's envisioned piece. Of course, the orchestra has but a finite number of musicians, so what is in effect here is really a Fourier series music synthesis. The conductor's job is greatly simplified, perhaps reducing to a few minor pre-concert modifications to the chosen tones. Concert hall owners could well be drawn to encourage such an artform; it would allow them to hire low-paid, unskilled musicians and cut the conductor's hours. The problem of course is that it would be nearly impossible to get the right tonal mix to compose a Fourier symphony. A few hertz too far in this or that direction generates not a symphony but a cacophany instead. Localizing the tones works much better. The composer uses a local frequency synthesis, assigning tones to moments in time; the musicians—they must be artists of supreme skill and dedication—read the musical notation and effect the appropriate, time-limited tones; and the conductor orchestrates the entire ensemble, settting the tempo and issuing direction as the performance proceeds. The composition of the signal in terms of time-localized tones is far easier to understand, communicate, replicate, and modify.[1]

10.1 GABOR TRANSFORMS

The previous chapter studied the strategy of time-limiting, or windowing, a signal before calculating its spectrum. This technique—of which there are many variants—furnishes better estimates of the signal's spectrum, because it restricts the signal

[1]Interestingly enough, there is a musical composition style that combines long-term tones to produce desired harmonies: "spectral music." Its resonances evolve slowly, retain a distinctly synthetic character, and thereby differ greatly from traditional 12-tone music. French composer Gérard Grisey (1946–1998), winner of the Rome prize, pioneered this form.

values to those over which the relevant oscillatory signal features should appear. The spectrogram of the signal $x(t)$ relative to the window function $w(t)$ is the squared magnitude of the Fourier transform of the product $s(t) = x(t)w(t)$: $|\mathcal{F}[s(t)]|^2 = |\mathcal{F}[x(t)w(t)]|^2 \geq 0$. Applications can therefore compare or threshold spectrogram values in order to decide whether one frequency is more significant than another or whether an individual frequency is significant, respectively. With the spectrogram, of course, the application design may need to search through possible time locations as well as through possible frequency ranges when seeking local spectral components. That is, Fourier applications tend to be one-dimensional, in contrast to short-time Fourier applications, which are inherently two-dimensional in nature.

We first explore the basic ideas of the transform, working with its original analog formulation. Section 10.1.2 develops the idea of the time-frequency plane. The Gabor transform partitions the (t, ω)-plane into equally sized regions, which Gabor dubbed "logons," from the ancient Greek word *logos*, meaning *word* or *account*. Logons are now generally called *time-frequency windows* or *atoms*. These time-frequency cells contain the signal's local frequency information, and their derivation provides a structural interpretation. Time-frequency windows with smaller t-dimensions provide higher signal time resolution, and those with tighter ω-dimensions have better signal frequency resolution. So small time-frequency cells are good, but we will eventually discover that a lower limit on cell size exists.

We generalize the Gabor transform further in Section 10.2 to include general window functions. It is proven, however, that among the many short-time Fourier techniques, the Gabor transform has smallest time-frequency windows. A Gaussian window, therefore, provides a joint time and frequency resolution superior to all other window functions: Hanning, Hamming, Kaiser, Bartlett, and so on. Finally, we derive the discretization of the Gabor transform in Section 10.3.

The Gabor transform uses a Gaussian window to create a window of time from which the spectrum of the local signal values are computed. Gaussian signals possess a magic property: Their Fourier transform is also a Gaussian. And this fact imparts an utterly elegant mathematical development to the study of the transform. But elegance is not the only reason for starting our study of short-time Fourier methods with the Gabor transform. In a sense that we will make precise momentarily, the Gabor transform is in fact the optimal short-time Fourier transform.

Carefully note that the width of the Gaussian window, as measured by its variance, is fixed throughout the transformation. Allowing it to vary has proven useful in many applications, but doing so undermines the essence of the transform as a time-frequency tool. It could indeed be argued that varying the window width makes it more like a time-scale transform. The location of the time-domain window, on the other hand, does change and becomes a variable of the two-dimensional, complex valued, Gabor transform function.

After Gabor's original paper [1], occasional research contributions related to Gabor transforms appeared sporadically in the scientific and engineering literature over the next 30 years. Interest in mixed-domain transforms accelerated with the discovery of the wavelet transform in the mid-1980s. There are now a variety of tutorial articles [2–4] on time-frequency transforms. Books devoted to Gabor analysis and the broader category of time-frequency transforms include Refs. 5–9. Their introductory chapters and those concentrating on the short-time Fourier transforms—of which the Gabor transform, by using a Gaussian window, is a particular case—are the most accessible. Treatments of time-scale transforms, or wavelets, often contain material introducing time-frequency transforms; we have found the material in Refs. 10–13 to be particularly useful.

10.1.1 Introduction

The Gabor transform picks a particular time-limiting window—the Gaussian—and generalizes the windowed spectrum computation into a full signal transform. The goal is to capture both the frequency components of a signal and their time locality in the transform equation. Of course, a Gaussian window is not truly finite in extent; its decay is so fast, however, that as a practical computation matter it serves the purpose of localizing signal values. Finite windows are possible with species [14–18].

Definition (Gabor Transform). Let $g(t)$ be some Gaussian of zero mean:

$$g(t) = Ae^{-Bt^2}, \tag{10.1}$$

where $A, B > 0$. If $x(t) \in L^2(\mathbb{R})$ is an analog signal, then its *Gabor transform*, written $X_g(\mu, \omega)$, is the radial Fourier transform of the product $x(t)g(t-\mu)$:

$$X_g(\mu, \omega) = \int_{-\infty}^{\infty} x(t) e^{-\frac{(t-\mu)^2}{2\sigma^2}} e^{-j\omega t}\, dt. \tag{10.2}$$

We will occasionally use the "fancy G" notation for the Gabor transform: $X_g(\mu, \omega) = \mathcal{G}_g[x(t)](\mu, \omega)$. The windowing function $g(t)$ in (10.1) remains fixed for the transform. If its parameters are understood—for instance, it may be the Gaussian of zero mean and standard deviation $\sigma > 0$—then we may drop the subscript g for the windowing function.

No reader can have overlooked the fact that we define the Gabor transform for $L^2(\mathbb{R})$ signals. Analog Fourier analysis (Chapter 5) shows that square-integrable signals have Fourier transforms which are also in $L^2(\mathbb{R})$. Thus, if $x(t) \in L^2(\mathbb{R})$ and $g(t)$ is a Gaussian, then $x(t)g(t-\mu) \in L^2(\mathbb{R})$ also, and the Fourier transform integral (10.2) therefore exists. Now for each μ, $\mathcal{F}[x(t)g(t-\mu)](\omega) \in L^2(\mathbb{R})$, and this will therefore enable us to find a Gabor inverse transform, or synthesis formula.

It is possible to specify a particular normalization for the Gaussian window used in the Gabor transform. For example, we might choose $\|g(t)\|_1 = 1$ or $\|g(t)\|_2 = 1$, where $\|\cdot\|_p$ is the norm in the Banach space $L^p(\mathbb{R})$ of Chapter 3. Gaussian signals belong to both spaces. Either choice makes some Gabor transform properties look nice but not others. We generally normalize the window with respect to the $L^1(\mathbb{R})$ norm, so that our windowing functions are zero-mean Gaussians of standard deviation $\sigma > 0$, $g_{0,\sigma}(t)$:

$$X_g(\mu, \omega) = \frac{1}{\sigma\sqrt{2\pi}} \int_{-\infty}^{\infty} x(t) e^{-\frac{(t-\mu)^2}{2\sigma^2}} e^{-j\omega t} dt. \tag{10.3}$$

The exercises explore how these alternative Gabor transform normalizations affect various transform properties.

Observe that the Gabor transform, unlike the analog Fourier transform, is a function of two variables. There is a time-domain variable μ, which is the center or mean of the window function, and a frequency-domain variable, ω. Since a time-domain variable—namely the location of the window's center, μ—is a parameter of the transform, the inverse Gabor transform involves a two-dimensional, or iterated integral. Figure 10.1 shows the Gabor transform scheme.

It is also possible to vary the width of the window, which is determined by σ, the standard deviation of the Gaussian. However, this changes the fundamental analytical nature of the transform operation, and our theory endeavors to avoid this. If σ changes while ω remains fixed, then the effect of the transform is to find oscillatory components of radial frequency ω over signal regions of varying width. But this is the defining characteristic of a scale-based signal analysis. The size of the prototype signal changes. When ω and σ both vary, we lapse into a hybrid scale and frequency approach. This does aid some applications. But our present purpose is to reveal the strengths and weaknesses of pure time-frequency methods, and therefore we fix σ for each particular Gabor transform formulation.

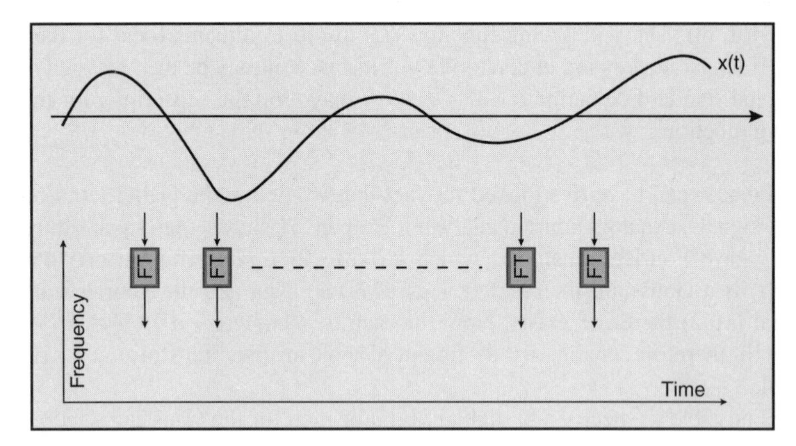

Fig. 10.1. The Gabor transform finds the spectral content of $x(t)$ within a Gaussian window $g(t - \mu)$. The two-dimensional transform function takes parameters μ, the window's center, and ω, the frequency of the exponential $\exp(j\omega t)$.

10.1.2 Interpretations

There are several fruitful ways to interpret the resulting Gabor transform:

- The most immediate way to visualize $X_g(\mu, \omega)$ is to see it as the Fourier transform of the windowed—and therefore essentially time-limited—signal $x(t)g(t-\mu)$.
- Secondly, in resonance with our physical concept of the Fourier transform, we can think of $X_g(\mu, \omega)$ as an inner product relation that measures the similarity of $x(t)$ to the pulse $g(t)\exp(j\omega t)$, a Gabor elementary function (GEF).
- Parseval's theorem provides yet a third interpretation: an inner product measure of the similarity of the Fourier transforms of $x(t)$ and $g(t)\exp(j\omega t)$.
- Finally, the Fourier transform's convolution theorem shows that the Gabor transform is a filtering operation on $X(\omega)$, the Fourier transform of $x(t)$.

So the idea of windowing a signal $x(t)$ with a Gaussian and making the location of the window a parameter of the transform richly interconnects concepts in signal spaces and transforms. In fact, we encounter two more interpretations of this many-faceted transform later in the chapter! But these first four carry us a long ways, so let us further investigate them.

The most immediate observation is that the Gaussian function $g(t)$ windows signal values of $x(t)$ in a neighborhood of around the point $t = \mu$. This springs right out of the definiton. The windowing effect suppresses oscillatory components of $x(t)$ distant from $t = \mu$. The Gabor transform of $x(t)$, $X_g(\mu, \omega)$, is thus the frequency content of $x(t)$ in this Gaussian-trimmed region.

Another perspective on the Gabor transform follows, if we recall that the product $g(t)\exp(-j\omega t)$ in the integrand (10.2) is the complex conjugate of a Gabor elementary function,[2] introduced in Chapter 1. Thus, if $x(t) \in L^2(\mathbb{R})$, then for each $\omega \in \mathbb{R}$, the Gabor transform integral is an inner product: $\langle x(t), g(t-\mu)\exp(j\omega t)\rangle$. Or, if the Gaussian has zero mean and standard deviation $\sigma > 0$, then $X_g(\mu, \omega) = \langle x(t), g_{\mu,\sigma}(t)\exp(j\omega t)\rangle$. Beginning with elementary vector spaces, through abstract inner product spaces, and finally with Hilbert spaces, the inner product relation continues to serve as our yardstick for establishing signal similarity. Hence, the Gabor transform $X_g(\mu, \omega)$ measures the similarity between $x(t)$ and the Gabor elementary function $g(t-\mu)\exp(j\omega t)$—an important idea which leads directly to the next point and figures prominently in the sequel.

Our third view of the Gabor transform follows from applying Parseval's formula to inner product relation:

$$\langle x(t), g(t-\mu)e^{j\omega t}\rangle = \frac{1}{2\pi}\langle X(\theta), \mathcal{F}[g(t)e^{j\omega t}]\rangle = X_g(\mu, \omega). \tag{10.4}$$

[2]Gabor actually used the Hertz formulation of the Fourier transform in his landmark 1946 paper. He applied the results to human hearing, observing that, up to about 1 kHz and independent of pulse width, we can distinguish some 50% of audible GEFs. Above that frequency, our sense rapidly deteriorates; Gabor concluded that cheaper means of transmission—although perhaps hampered by a poorer frequency, response—might replace more faithful and expensive systems [19].

Note that we are fixing ω so that the Gabor elementary function $g(t)\exp(j\omega t)$ is a pure function of t. The dummy variable for $\mathcal{F}[x(t)](\theta) = X(\theta)$ in (10.4) changes from the usual ω to avoid a conflict. Thus, the Gabor transform is a (scaled) similarity measure between the Fourier transforms of $x(t)$ and the GEF $g(t-\mu)\exp(j\omega t)$.

The convolution theorem for the radial Fourier transform provides a fourth insight into the Gabor transform. Convolution in time is equivalent to multiplication in frequency. And, reversing the transform direction, convolution in frequency corresponds to termwise multiplication in time. Therefore, if $y(t) = g(t)\exp(j\omega t)$, it follows that

$$X_g(\mu, \omega) = \langle x(t), g(t-\mu)e^{j\omega t}\rangle = \frac{1}{2\pi}(X*Y)(\theta) = \frac{1}{2\pi}\int_{-\infty}^{\infty} X(\zeta)Y(\theta-\zeta)\, d\zeta. \tag{10.5}$$

Gabor transforming a signal $x(t)$ is the same as filtering $X(\theta)$ with the Fourier transform of the Gabor elementary function $y(t) = g(t-\mu)\exp(j\omega t)$.

These several interpretations lead us to further study the Gaussian, the GEFs, inner products, and convolution operations in both the time and frequency domains.

10.1.3 Gabor Elementary Functions

By now, Gabor elementary functions $y(t) = g(t-\mu)\exp(j\omega t)$ are quite familiar. We introduced them as early as Chapter 1, noted their applicability to spectral analysis of signal texture in Chapter 4, and considered them as amplitude-modulated sinusoidal carrier signals in Chapter 5. They have other names, too: Gabor atoms or windowed Fourier atoms. Now we have seen that $y(t)$ is a signal *model*—or *prototype*—to which the Gabor transform compares $x(t)$. This section continues our investigation of these important signal prototypes.

In the time domain, $y(t) = g(t)\exp(j\omega t)$ is a complex exponential that is amplitude modulated by a Gaussian $g(t)$. The Gaussian envelope—let us say it has mean μ and standard deviation σ—$g_{\mu,\sigma}(t)$ amplitude modulates the real and imaginary parts of $\exp(j\omega t)$. From a communications theory standpoint, the latter are sinusoidal carrier signals. The real part of $y(t)$ is $\cos(\omega t)$-modulated by $g_{\mu,\sigma}(t)$; hence, $\mathrm{Real}[g_{\mu,\sigma}(t)\exp(j\omega t)]$ is even. And its imaginary part is $\sin(\omega t)$ inside the same envelope, making $\mathrm{Imag}[g_{\mu,\sigma}(t)\exp(j\omega t)]$ an odd signal. The GEFs exhibit more or less oscillations as the frequency of the exponential carrier signal increases or decreases, respectively, under a modulating Gaussian pulse of constant width. This changes the shape of the model signal as in Figure 10.2.

Altering the spread of the Gaussian envelope (given by its width parameter σ) while leaving ω constant also produces prototype signals of different shapes. The large sinusoidal oscillations persist over a wider time-domain region (Figure 10.3). This behavior typifies time-scale analysis methods, which depend upon comparing source signals with models of variable time-domain extent, but similar shape. Unless we tie the frequency ω to the Gaussian's standard deviation σ, then the Gabor elementary functions will exhibit different basic shapes as σ changes. This

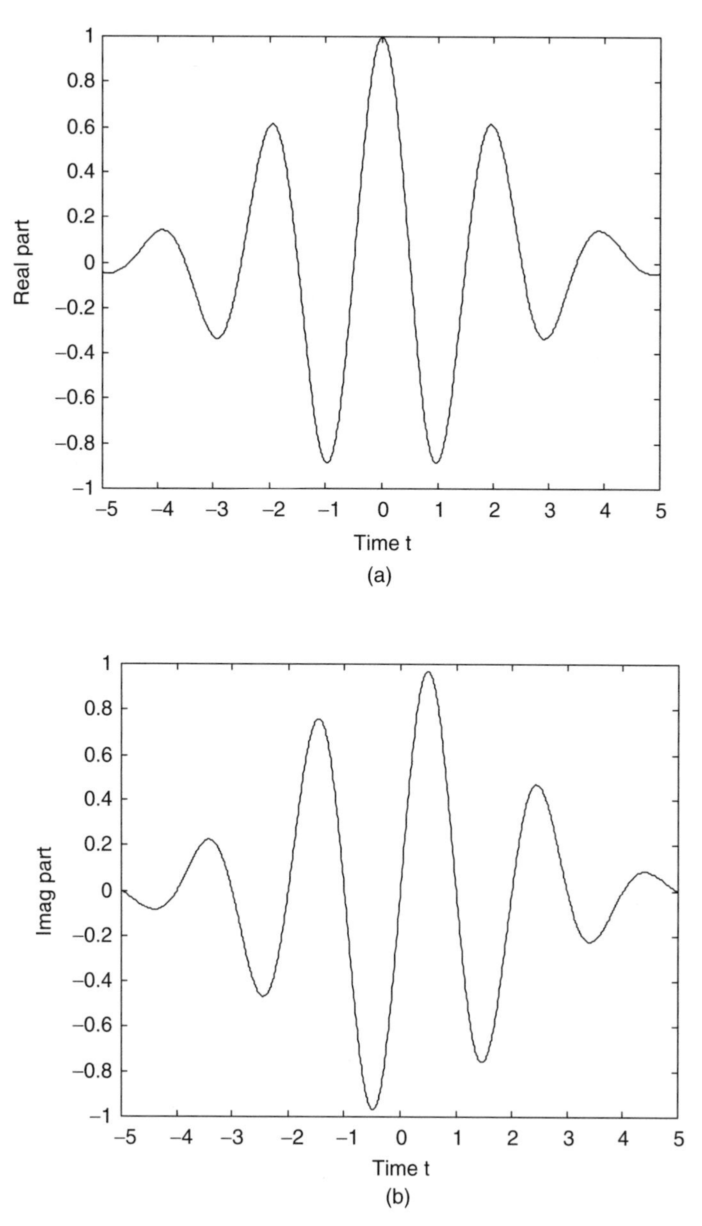

Fig. 10.2. Gabor elementary functions. The cosine term (a) represents the real part of a GEF and is an even signal. The sine term represents the imaginary part and is an odd signal (b). Panels (c) and (d) show the effect of frequency doubling.

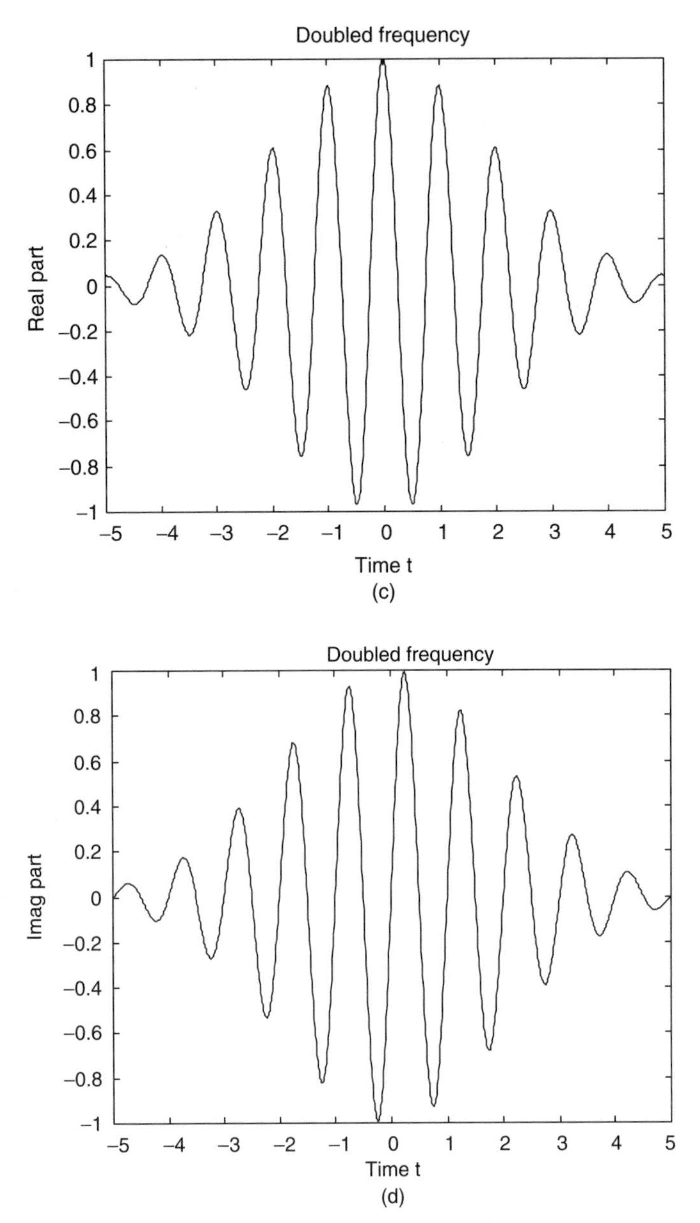

Fig. 10.2 (*Continued*)

technique is often called the *adaptive Gabor transform*. But a bona fide time-frequency transform should be able to reveal all frequencies within a local signal region; we cannot mathematically link ω and σ. Consequently, to preserve the time-frequency nature of the Gabor transform and short-time Fourier tools in general, we avoid flexible windows. Time-scale transforms (Chapter 11) use dilation to

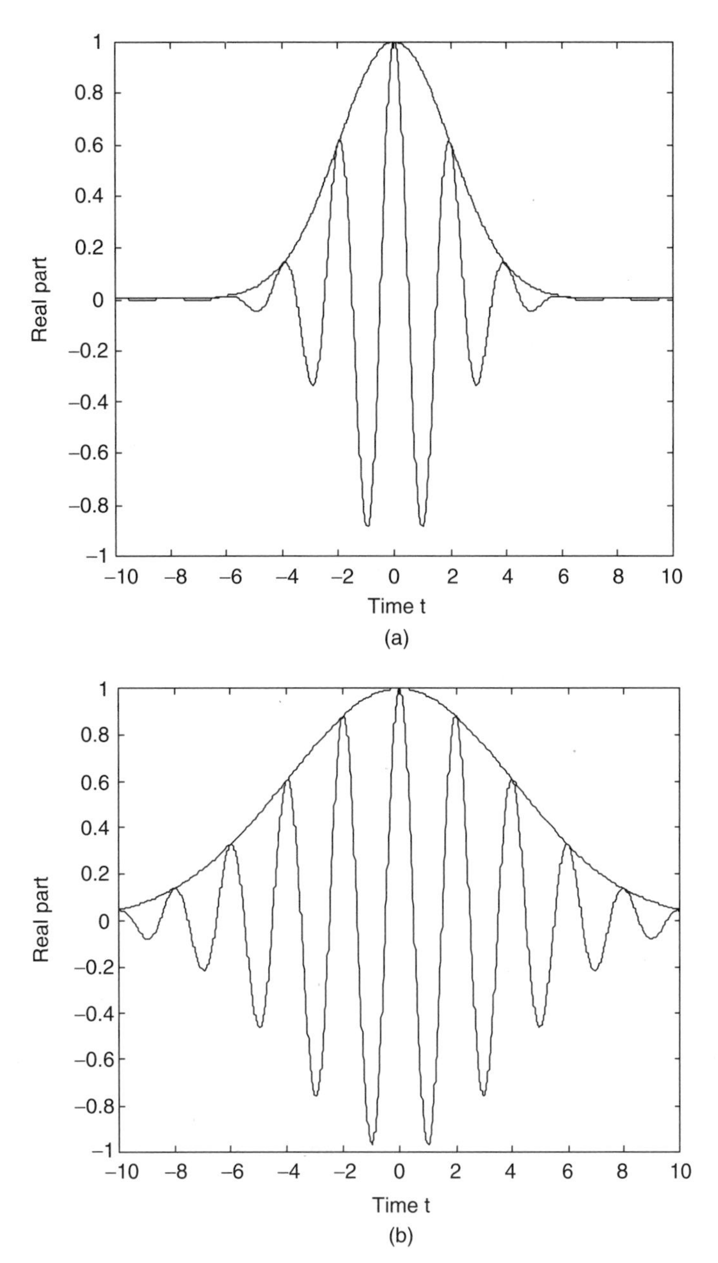

Fig. 10.3. Window size variation. With the radial frequency ω fixed, the shape of a Gabor elementary function signal (*a*) changes as the window expands (*b*).

maintain a basic shape while changing the size of a prototype signal. This chapter's exercises preview this idea.

Now let us consider the frequency-domain representation of Gabor elementary functions. The formula for the Fourier transform of $y(t) = g_{\mu,\sigma}(t)\exp(j\omega_0 t)$ derives from the radial Fourier transform's properties (Chapter 5).

Proposition (Fourier Transform of Gabor Elementary Function). Let $\sigma > 0$ and suppose $g(t) = g_{\mu,\sigma}(t)$ is the Gaussian with mean μ and standard deviation σ. Let $y(t) = g(t)\exp(j\omega_0 t)$ be a Gabor elementary function with envelope $g(t)$ and radial frequency ω_0. Then

$$Y(\omega) = \frac{1}{2\pi}\exp\left[-\frac{\sigma^2}{2}(\omega-\omega_0)^2 + j(\omega-\omega_0)\mu\right]. \tag{10.6}$$

Proof: In Chapter 5 we calculated the radial Fourier transform of the Gaussian; there results a Gaussian once more: $\mathcal{F}[\exp(-t^2)](\omega) = \pi^{1/2}\exp(-\omega^2/4)$. The Fourier transform properties allow us to write out the Fourier transform for $g(t) = g_{\mu,\sigma}(t)$:

$$G(\omega) = \int_{-\infty}^{\infty} g(t)e^{-j\omega t}dt = \exp\left(-\left[\frac{\sigma^2\omega^2}{2} + j\omega\mu\right]\right) \tag{10.7}$$

whose magnitude $|G(\omega)|$ is a Gaussian centered at $\omega = 0$. Applying the generalized Fourier transform handles the exponential factor: $\mathcal{F}[\exp(j\omega_0 t)](\omega) = \delta(\omega - \omega_0)$. A termwise multiplication $y(t) = x_1(t)x_2(t)$ in time has Fourier transform $Y(\omega) = X_1(\omega)*X_2(\omega)/(2\pi)$. This implies $\mathcal{F}[g(t)\exp(j\omega_0 t)] = G(\omega)*\delta(\omega - \omega_0)/(2\pi)$, the convolution of a Gaussian with a shifted Dirac delta. Making θ the integration variable for continuous-domain convolution, we compute:

$$\begin{aligned} Y(\omega) &= \mathcal{F}[g(t)e^{j\omega_0 t}] = \frac{1}{2\pi}\int_{-\infty}^{\infty}\delta(\theta-\omega_0)e^{-[\sigma^2(\omega-\theta)^2 + j(\omega-\theta)\mu]}d\theta \\ &= \frac{1}{2\pi}e^{-(\sigma^2(\omega-\omega_0)^2 + j(\omega-\omega_0)\mu)}. \end{aligned} \tag{10.8}$$

We use the Sifting Property of the Dirac delta in (10.8), the last expression of which is precisely the value (10.6). ■

Remarks. In (10.8) observe that $|Y(\omega)|$ is a scaled (amplified or attenuated) Gaussian pulse centered at $\omega = \omega_0$ in the frequency domain (Figure 10.4). To find the Fourier transform of Real[$y(t)$], we write $\cos(\omega_0 t) = [\exp(j\omega_0 t) + \exp(-j\omega_0 t)]/2$. Its spectrum is a pair of impulses at $|\omega| = \omega_0$; hence, a convolution like (10.8) produces a sum of two Gaussians. A similar procedure (exercises) works for the imaginary part of $\exp(j\omega_0 t)$ and gives us the Fourier transform of Imag[$y(t)$].

Simple experiments demonstrate that for the Gabor elementary function a reciprocal relationship apparently exists between time- and frequency-domain window widths (Figure 10.4). Further elucidation requires us to formalize the concept of window width, which is a topic covered in Section 10.2.4.

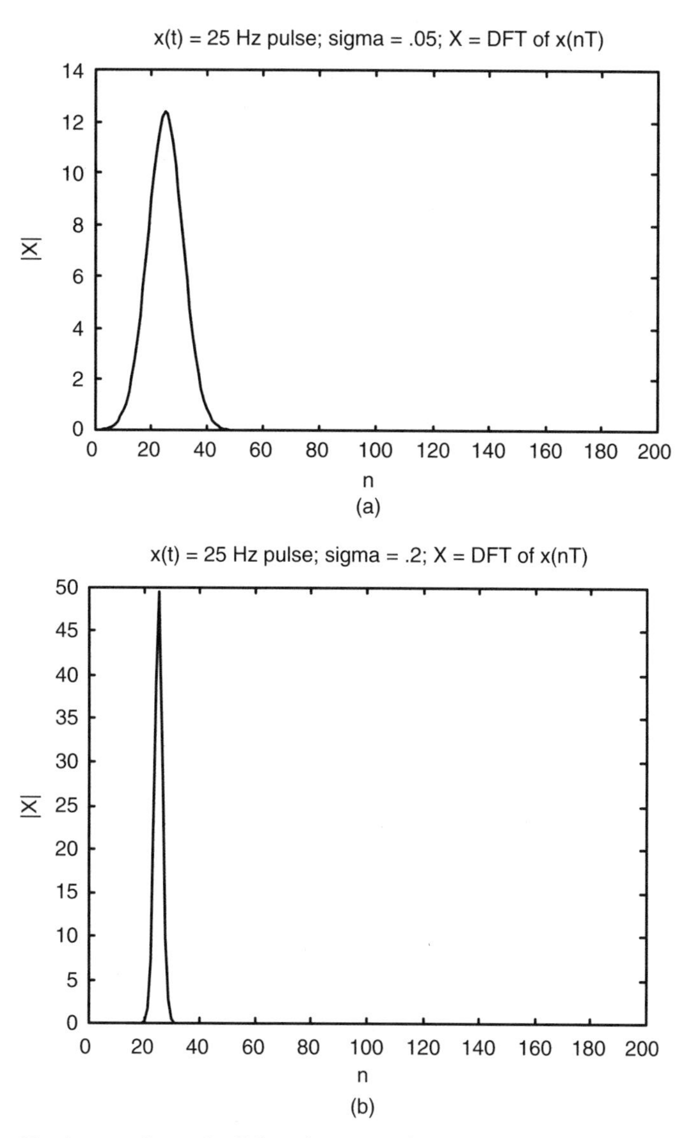

Fig. 10.4. Fourier transform of a Gabor elementary function. A narrow time-domain signal, $y(t) = g_{\mu,\sigma}(t)\exp(j\omega_0 t)$, has a wide magnitude spectrum (a). As the time-domain width of the Gaussian envelope grows, the frequency-domain window width shrinks (b).

10.1.4 Inversion

Recalling that all of the analog and discrete versions of the Fourier transform of Chapters 5 and 7 have inverse relations, let us consider the same problem for the Gabor transform. Suppose we transform with respect to a Gaussian window

$g(t) = g_{\mu,\sigma}(t)$. We call the forward transform relation, $X(\omega) = \mathcal{F}[x(t)]$, the *analysis equation*, and we call the inverse Fourier transform relation, $x(t) = \mathcal{F}^{-1}[X(\omega)]$, the *synthesis equation*. Of course, $X_g(\mu, \omega)$ is the radial Fourier transform of the windowed signal $x(t)g_{\mu,\sigma}(t)$, so its synthesis equation produces not $x(t)$, but $x(t)g_{\mu,s}(t)$ back again. How can we get $x(t)$ from $X_g(\mu, \omega)$? Because $X_g(\mu, \omega)$ is a function of two variables, we can integrate a second time, letting the Gaussian's location μ vary and using it as an integration variable.

We divide our efforts according to whether the Gabor transform of $x(t)$ is integrable. As with the Fourier transform development in Chapter 5, if $X_g(\mu, \omega) \in L^1(\mathbb{R})$, then we can freely interchange integration limits. If we assume $x(t) \in L^2(\mathbb{R})$ and nothing more, then a limiting argument is once again necessary.

10.1.4.1 Assuming Transform Integrability. The following lemma is a direct consequence of our choice of $g_{\mu,\sigma}(t)$ for the Gabor transform windowing signal. It shows that, for each μ, the Gabor transform $X_g(\mu, \omega)$ represents a local piece of $\mathcal{F}[x(t)]$. Independent variable ω corresponds to the spectral frequency, and μ represents the spectral fragment's time location. Indeed, when we integrate all of these pieces together (10.9), the result is the full signal spectrum $X(\omega)$.

Lemma. Suppose $x(t) \in L^2(\mathbb{R})$; $\mu, \sigma \in \mathbb{R}$, $\sigma > 0$; $g(t) = g_{\mu,\sigma}(t)$ is the Gaussian window with mean μ and standard deviation σ; and let $X_g(\mu, \omega) \in L^1(\mathbb{R})$ be the Gabor transform of $x(t)$. Then,

$$X(\omega) = \int_{-\infty}^{\infty} X_g(\mu, \sigma)\, d\mu. \tag{10.9}$$

Proof: Let us expand the integrand in (10.9):

$$\int_{-\infty}^{\infty} X_g(\mu, \sigma)\, d\mu = \int_{-\infty}^{\infty}\left(\int_{-\infty}^{\infty} x(s)g(s)e^{-j\omega s}\, ds\right) d\mu = \int_{-\infty}^{\infty} x(s)e^{-j\omega s}\left(\int_{-\infty}^{\infty} g_{\mu,\sigma}(s)\, d\mu\right) ds. \tag{10.10a}$$

Fubini's theorem [11, 12] states that if an iterated integral's integrand is absolutely integrable, then it is identical to the double integral, and the order of iteration is irrelevant. The interchange of limits (10.10a) is possible by applying Fubini's theorem, which is possible because the integrand is absolutely integrable. The inner integrand is unity, since $\|g\|_1 = 1$. Consequently,

$$\int_{-\infty}^{\infty} X_g(\mu, \sigma)\, d\mu = \int_{-\infty}^{\infty} x(s)e^{-j\omega s}\, ds = X(\omega). \tag{10.10b}$$

■

Now we can prove an initial inverse Gabor transform relationship for the situation where $X_g(\mu, \omega)$ is integrable.

Theorem (Inverse Gabor Transform or Synthesis Equation). Suppose $\sigma > 0$; $x(t) \in L^2(\mathbb{R})$; $g(t) = g_{\mu,\sigma}(t)$ is the Gaussian window with mean μ and standard deviation σ; and let $X_g(\mu, \omega) \in L^1(\mathbb{R})$ be the Gabor transform of $x(t)$. Then,

$$x(t) = \frac{1}{2\pi}\int_{-\infty}^{\infty}\left(\int_{-\infty}^{\infty} X_g(\mu, \omega)e^{j\omega t}d\omega\right)d\mu. \tag{10.11}$$

Proof: Using the definition of the Gabor transform for $x(t)$, we have

$$\frac{1}{2\pi}\int_{-\infty}^{\infty}\left(\int_{-\infty}^{\infty} X_g(\mu, \omega)e^{j\omega t}d\omega\right)d\mu = \frac{1}{2\pi}\int_{-\infty}^{\infty} e^{j\omega t}\left(\int_{-\infty}^{\infty} X_g(\mu, \omega)\,d\mu\right)d\omega. \tag{10.12}$$

We use the assumption that $X_g(\mu, \omega) \in L^1(\mathbb{R})$ to infer $X_g(\mu, \omega)\exp(j\omega t) \in L^1(\mathbb{R})$ as well; Fubini's theorem then implies (10.12). Using the lemma to evaluate the parenthesized integral on the right-hand side of (10.12) gives

$$\frac{1}{2\pi}\int_{-\infty}^{\infty} e^{j\omega t}\left(\int_{-\infty}^{\infty} X_g(\mu, \omega)d\mu\right)d\omega = \frac{1}{2\pi}\int_{-\infty}^{\infty} e^{j\omega t}X(\omega)\,d\omega = x(t) \tag{10.13}$$

as desired. ∎

The next result is a time-frequency version of Plancherel's theorem. It shows that the Gabor transform preserves signal energy. We do not have a perfect proportion, since the equation depends on the $L^2(\mathbb{R})$ norm of the window function.

We interpose a lemma that shows how to compute the Fourier transform of a Gabor transform.

Lemma (Fourier Transform of Gabor Transform). Suppose $\sigma > 0$; $x(t) \in L^2(\mathbb{R})$; $g(t) = g_{\mu,\sigma}(t)$ is the Gaussian window with mean μ and standard deviation σ; and let $X_g(\mu, \omega) \in L^1(\mathbb{R})$ be the Gabor transform of $x(t)$. Then, for each $\omega \in \mathbb{R}$ we can Fourier transform the signal $X_g(\mu, \omega)$, viewing it as a function of μ. So,

$$\int_{-\infty}^{\infty} X_g(\mu, \omega)e^{-j\mu\theta}d\mu = \frac{1}{2\pi}X(\omega+\theta)G(\theta) = \frac{1}{2\pi}X(\omega+\theta)e^{-\frac{\theta^2\sigma^2}{2}}, \tag{10.14}$$

where $G(\theta)$ is the radial Fourier transform of $g(t)$.

Proof: Expanding $X_g(\mu, \omega)$ in the integrand (10.14) gives

$$\int_{-\infty}^{\infty}\left[\int_{-\infty}^{\infty} x(t)g(t)e^{-j\omega t}dt\right]e^{-j\mu\theta}d\mu = \int_{-\infty}^{\infty}\left(\left[\int_{-\infty}^{\infty} x(t)g_{0,\sigma}(t-\mu)e^{-j\omega t}dt\right]\right)e^{-j\mu\theta}\,d\mu$$

$$= \int_{-\infty}^{\infty}\left(\left[\int_{-\infty}^{\infty} x(t)g_{0,\sigma}(t-\mu)e^{-j\omega t}e^{j\omega\mu}dt\right]\right)e^{-j\mu\theta}e^{-j\omega\mu}d\mu. \tag{10.15}$$

The algebraic manipulations in (10.15) aim to change the expression's form into a convolution of $x(t)$ with the Gabor elementary function $y(s) = g_{0,\sigma}(s)\exp(j\omega s)$. Because $g_{0,\sigma}(t-\mu) = g_{0,\sigma}(\mu - t)$,

$$\int_{-\infty}^{\infty}\left[\int_{-\infty}^{\infty} x(t)g(t)e^{-j\omega t}dt\right]e^{-j\mu\theta}d\mu = \int_{-\infty}^{\infty}\left[\int_{-\infty}^{\infty} x(t)g_{0,\sigma}(\mu-t)e^{j\omega(\mu-t)}dt\right]e^{-j\mu\theta}e^{-j\omega\mu}d\mu$$

$$= \int_{-\infty}^{\infty}(x*y)(\mu)e^{-j\mu(\theta+\omega)}d\mu, \qquad (10.16)$$

which exposes a convolution integral, $(x * y)(\mu)$. But now the outer integral in is evidently a radial Fourier transform; invoking the convolution theorem,

$$\int_{-\infty}^{\infty}\left[\int_{-\infty}^{\infty} x(t)g(t)e^{-j\omega t}dt\right]e^{-j\mu\theta}\,d\mu = \int_{-\infty}^{\infty}(x*y)(\mu)e^{-j\mu(\theta+\omega)}\,d\mu$$

$$= X(\omega+\theta)Y(\omega+\theta). \qquad (10.17)$$

Now, $X(\omega + \theta)$ is the Fourier transform of $x(t)$ evaluated at $\omega + \theta$, as the lemma requires. $Y(\phi) = \mathcal{F}[y(s)](\phi) = \mathcal{F}[g_{0,\sigma}(s)\exp(j\omega s)](\phi)$ is the Fourier transform of a Gabor elementary function. In (10.8) we found $Y(\phi) = (2\pi)^{-1}\mathcal{F}[g_{0,\sigma}(s)](\phi - \omega)$. If $\mu = 0$, then $\mathcal{F}[g_{\mu,\sigma}(s)](\phi) = \exp(-\sigma^2\phi^2/2)$. So $Y(\omega + \theta) = (2\pi)^{-1}\mathcal{F}[g_{0,\sigma}(s)]([\omega + \theta] - \omega) = (2\pi)^{-1}[g_{0,\sigma}(s)](\theta)$. Finally,

$$X(\omega+\theta)Y(\omega+\theta) = X(\omega+\theta)\,\mathcal{F}[g_{0,\sigma}(t)e^{j\omega t}](\omega+\theta) = \frac{X(\omega+\theta)}{2\pi}e^{-\frac{\sigma^2\theta^2}{2}}, \qquad (10.18)$$

and the proof is complete. ■

10.1.4.2 Two-Dimensional Hilbert Spaces in Brief. Our next result is the time-frequency version of the Plancherel formula. Now, this theorem depends on the two-dimensional $L^2(\mathbb{R})$ norm. "You are so unaccustomed to speak in images," Adeimantus ironically remarks to Socrates in the Republic,[3] and we too have been—intentionally—so unaccustomed! The $L^2(\mathbb{R})$ norm applies to analog *images*, and up until now we have been deliberately partial to one-dimensional signal theory. Nonetheless, time-frequency methods, and mixed-domain techniques in general, often transgress into multidimensional or *image* analysis. This is to be expected, since the transforms do encode both time and frequency information as independent variables in the transformed signal. The theoretical extensions are gratefully straightforward. In fact, $L^2(\mathbb{R} \times \mathbb{R})$ is a Hilbert space also, and its theoretical development follows from our one-dimensional endeavors in Chapter 3. We do not need to spend a lot of time developing that multidimensional theory here; nevertheless, the concepts of the $L^2(\mathbb{R}^2)$ space and its norm are worth reviewing.

[3]*Republic*, vol. II, B. Jowett, translator, Oxford: Clarendon Press, 1964.

Definition ($\mathbf{L^2(\mathbb{R}^2)}$). A two-dimensional signal $x(s, t)$ is *square-integrable* or has *finite energy* if

$$\int_{-\infty}^{\infty}\int_{-\infty}^{-\infty} |x(s, t)|^2 \, ds dt < \infty. \tag{10.19}$$

We denote the set of all such signals by $L^2(\mathbb{R}^2)$ or $L^2(\mathbb{R} \times \mathbb{R})$. If $x(s, t) \in L^2(\mathbb{R}^2)$, then

$$\left[\int_{-\infty}^{\infty}\int_{-\infty}^{-\infty} |x(s, t)|^2 ds dt\right]^{\frac{1}{2}} = \|x\|_{2, L^2(d\mathbb{R}^2)} \tag{10.20}$$

is its $L^2(\mathbb{R}^2)$ norm. If the context is clear, then we omit the subscripted $L^2(\mathbb{R}^2)$ in (10.20). If $x(s, t)$ and $y(s, t)$ are in $L^2(\mathbb{R}^2)$, then we define their inner product by

$$\langle x, y\rangle_{L^2(\mathbb{R}^2)} = \int_{-\infty}^{\infty}\int_{-\infty}^{-\infty} x(s, t)\overline{y(s, t)} \, ds dt. \tag{10.21}$$

In a clear context, we drop the subscript and write (10.21) as $\langle x, y\rangle$. The exercises further cover the ideas of two-dimensional signal spaces.

Theorem (Gabor Transform Plancherel's). Suppose $\sigma > 0$; $x(t) \in L^2(\mathbb{R})$; $g(t) = g_{\mu,\sigma}(t)$ is the Gaussian with mean μ and standard deviation σ; and let $X_g(\mu, \omega) \in L^1(\mathbb{R})$ be the Gabor transform of $x(t)$. Then

$$\|x\|_2 = \sqrt{2\pi}\frac{\|X_g(\mu, \omega)\|_{2, L^2(\mathbb{R}^2)}}{\|g\|_2}. \tag{10.22}$$

Proof: Fubini's theorem applies to the double integral that defines the $L^2(\mathbb{R}^2)$ norm:

$$\|X_g(\mu, \omega)\|^2_{2, L^2(\mathrm{R}^2)} = \int_{-\infty}^{\infty}\int_{-\infty}^{\infty} |X_g(\mu, \omega)|^2 d\mu \, d\omega = \int_{-\infty}^{\infty}\left[\int_{-\infty}^{\infty} |X_g(\mu, \omega)|^2 d\mu\right] d\omega. \tag{10.23}$$

Since the inner integral is a function of the time domain variable μ, we can Fourier transform its integrand with respect to μ. An application of the Fourier transform Plancherel formula is then feasible:

$$\|X_g(\mu, \omega)\|^2_{2, L^2(\mathbb{R}^2)} = \int_{-\infty}^{\infty} \frac{1}{2\pi}\left[\int_{-\infty}^{\infty} |\mathcal{F}[X_g(\mu, \omega)](\theta)|^2 d\theta\right] d\omega. \tag{10.24}$$

Let $H(\theta) = \mathcal{F}[g_{0,\sigma}(t)](\theta)$, so that by the Lemma we find

$$\|X_g(\mu, \omega)\|^2_{2, L^2(\mathbb{R}^2)} = \frac{1}{2\pi}\int_{-\infty}^{\infty}\left[\int_{-\infty}^{\infty}\left|\frac{1}{2\pi}X(\omega+\theta)H(\theta)\right|^2 d\theta\right] d\omega$$

$$= \frac{1}{2\pi}\int_{-\infty}^{\infty}\frac{1}{(2\pi)^2}\left[\int_{-\infty}^{\infty}|X(\omega+\theta)|^2|H(\theta)|^2 d\theta\right] d\omega. \tag{10.25}$$

To evaluate the iterated integral (10.25), we swap the order of integration and use Plancherel two more times. The last expression above becomes

$$\|X_g(\mu, \omega)\|^2_{2, L^2(\mathbb{R}^2)} = \frac{1}{2\pi}\int_{-\infty}^{\infty}\frac{|H(\theta)|^2}{2\pi}\left[\int_{-\infty}^{\infty}\frac{|X(\omega+\theta)|^2}{2\pi}d\omega\right] d\theta$$

$$= \frac{\|x\|_2^2}{2\pi}\int_{-\infty}^{\infty}\frac{|H(\theta)^2|}{2\pi}d\theta = \frac{\|x\|_2^2}{2\pi}\|g_{0,\sigma}\|_2^2. \tag{10.26}$$

■

10.1.4.3 For General Square-Integrable Signals. This section develops the Plancherel and inverse results for square-integrable signals, dropping the assumption of integrability on the Gabor transform. We begin with a form of the Parseval theorem.

Theorem (Gabor Transform Parseval's). Suppose $\sigma > 0$; $x(t), y(t) \in L^2(\mathbb{R})$; $g(t) = g_{\mu,\sigma}(t)$ is the Gaussian window with mean μ and standard deviation σ; and let $X_g(\mu, \omega)$ and $Y_g(\mu, \omega)$ be the Gabor transforms of $x(t)$ and $y(t)$, respectively. Then

$$2\pi\|g\|_2^2\langle x, y\rangle = \int_{-\infty}^{\infty}\int_{-\infty}^{\infty} X_g(\mu, \omega)\overline{Y_g(\mu, \omega)}\, d\omega d\mu = \langle X_g, Y_g\rangle_{L^2(\mathbb{R}^2)}. \tag{10.27}$$

Proof: For fixed μ, we can apply the Parseval theorem to find

$$\int_{-\infty}^{\infty} X_g(\mu, \omega)\overline{Y_g(\mu, \omega)}\, d\omega = 2\pi\int_{-\infty}^{\infty}\mathcal{F}^{-1}X_g(\mu, \omega)\mathcal{F}^{-1}\overline{Y_g(\mu, \omega)}\, d\omega, \tag{10.28}$$

where $\mathcal{F}^{-1}$ is the inverse radial Fourier transform. Since the inverse Fourier transform of the Gabor transform $X_g(\mu, \omega)$ is the windowed signal $x(t)g_{\mu,\sigma}(t)$, we continue (10.28) as follows:

$$2\pi\int_{-\infty}^{\infty} x(t)g_{\mu,\sigma}(t)\overline{y(t)g_{\mu,\sigma}(t)}\, dt = 2\pi\int_{-\infty}^{\infty} x(t)\overline{y(t)}g^2{}_{\mu,\sigma}(t)\, dt. \tag{10.29}$$

Integrating (10.29) with respect to μ produces

$$\int_{-\infty}^{\infty}\int_{-\infty}^{\infty} X_g(\mu,\omega)\overline{Y_g(\mu,\omega)}\,d\omega d\mu = 2\pi\int_{-\infty}^{\infty}\int_{-\infty}^{\infty} x(t)\overline{y(t)}g^2_{\mu,\sigma}(t)\,dtd\mu$$

$$= 2\pi\int_{-\infty}^{\infty} x(t)\overline{y(t)}\int_{-\infty}^{\infty} g^2_{\mu,\sigma}(t)\,d\mu dt$$

$$= 2\pi\|g_{\mu,\sigma}\|_2^2\langle x, y\rangle. \tag{10.30}$$

Fubini's theorem and the Schwarz inequality (applied to the inner integral on the top right-hand side of (10.30), which is itself an inner product) allow us to interchange the integration order. ∎

The next result shows how to retrieve the original signal $x(t) \in L^2(\mathbb{R})$ from its Gabor transform $X_g(\mu, \omega)$. This is the Gabor transform inverse relation, but it has other names, too. It is sometimes called the *resolution of the identity* or simply the *synthesis equation* for the Gabor transform.

Theorem (Inverse Gabor Transform). Suppose $\sigma > 0$; $x(t) \in L^2(\mathbb{R})$; $g(t) = g_{\mu,\sigma}(t)$ is the Gaussian window with mean μ and standard deviation σ; and let $X_g(\mu, \omega)$ be the Gabor transform of $x(t)$. Then for all $s \in \mathbb{R}$, if $x(t)$ is continuous at s, then

$$x(a) = \frac{1}{(2\pi\|g\|_2^2)}\int_{-\infty}^{\infty} X_g(\mu,\omega)g_{\mu,\sigma}(a)e^{j\omega a}\,d\omega d\mu. \tag{10.31}$$

Proof: Consider a family of Gaussians $h_{a,s}(t)$, where $s > 0$. As $s \to 0$, they approximate a delta function, and, informally, from the sifting property we expect that

$$\lim_{s\to 0}\langle x(t), h_{a,s}(t)\rangle = x(a) \tag{10.32}$$

when $x(t)$ is continuous at $t = a$. If we set $y(t) = h_{a,s}(t)$, then we can apply the prior Plancherel theorem to obtain

$$\langle x, h_{a,s}\rangle = \frac{1}{2\pi\|g\|_2^2}\int_{-\infty}^{\infty}\int_{-\infty}^{\infty} X_g(\mu,\omega)\overline{\left[\int_{-\infty}^{\infty} h_{a,s}(t)g_{\mu,\sigma}(t)e^{-j\omega t}dt\right]}\,d\omega d\mu. \tag{10.33}$$

We calculate the limit

$$\lim_{s\to 0}\overline{\int_{-\infty}^{\infty} h_{a,s}(t)g_{\mu,\sigma}(t)e^{-j\omega t}dt} = \lim_{s\to 0}\int_{-\infty}^{\infty} h_{a,s}(t)g_{\mu,\sigma}(t)e^{j\omega t}dt$$

$$= \int_{-\infty}^{\infty}\lim_{s\to 0} h_{a,s}(t)g_{\mu,\sigma}(t)e^{j\omega t}dt$$

$$= \int_{-\infty}^{\infty}\delta(t-a)g_{\mu,\sigma}(t)e^{j\omega t}dt = g_{\mu,\sigma}(a)e^{j\omega a}. \tag{10.34}$$

Taking the same limit $s \to 0$ on both sides of (10.33) and interchanging limit and integration operations gives

$$\lim_{s \to 0} \langle x, h_{a,s} \rangle = \frac{1}{2\pi\|g\|_2^2} \int_{-\infty}^{\infty} \int_{-\infty}^{\infty} X_g(\mu, \omega) g_{\mu,\sigma}(a) e^{j\omega a} d\omega d\mu = x(a). \quad (10.35)$$

■

10.1.5 Applications

Let us pause the theoretical development for a moment to explore two basic Gabor transform applications: a linear chirp and a pulsed tone. These illustrate the use and behavior of the transform on practical signals. Studying the transform coefficients as dependent on the width of the transform window function will also lead us to important ideas about the relation between the transform's time and frequency resolutions.

10.1.5.1 Linear Chirp. This section discusses the Gabor transform for a linear chirp signal. A linear chirp is a sinusoidal function of a squared time variable At^2, where A is constant. Thus, as $|t|$ increases, the signal oscillations bunch up. Signal frequency varies with time in a linear fashion, and we anticipate that the Gabor transform will expose this behavior.

Let us consider the analog signal

$$X_a(t) = \begin{cases} \cos(At^2), & t \in [0, L] \\ 0 & \text{otherwise.} \end{cases} \quad (10.36)$$

The Gabor transform of $x_a(t)$ is

$$\mathcal{G}[x_a(t)](\mu, \omega) = (X_a)_g(\mu, \omega) = \int_0^L x_a(t) g_{\mu,\sigma}(t) e^{-j\omega t} dt. \quad (10.37)$$

We need to decide upon an appropriate value for the spread of the Gaussian, which is given by its standard deviation σ. Also, $(X_a)_g$ is a two-dimensional function, so we seek an image representation of the Gabor transform for a range of values, μ and ω.

We apply Section 7.1.2's ideas for approximating an analog transform with discrete samples. Recall that the discrete Fourier series (DFS) coefficients

$$c(k) = \frac{1}{N} \sum_{n=0}^{N-1} x(n) e^{\frac{-2\pi jnk}{N}}, \quad (10.38)$$

where $0 \le k \le N-1$, are a trapezoidal rule approximation to the Fourier series integral using the intervals $[0, T/N]$, $[T/N, 2T/N]$, ..., $[(N-1)T/N, T]$. Recall as well from

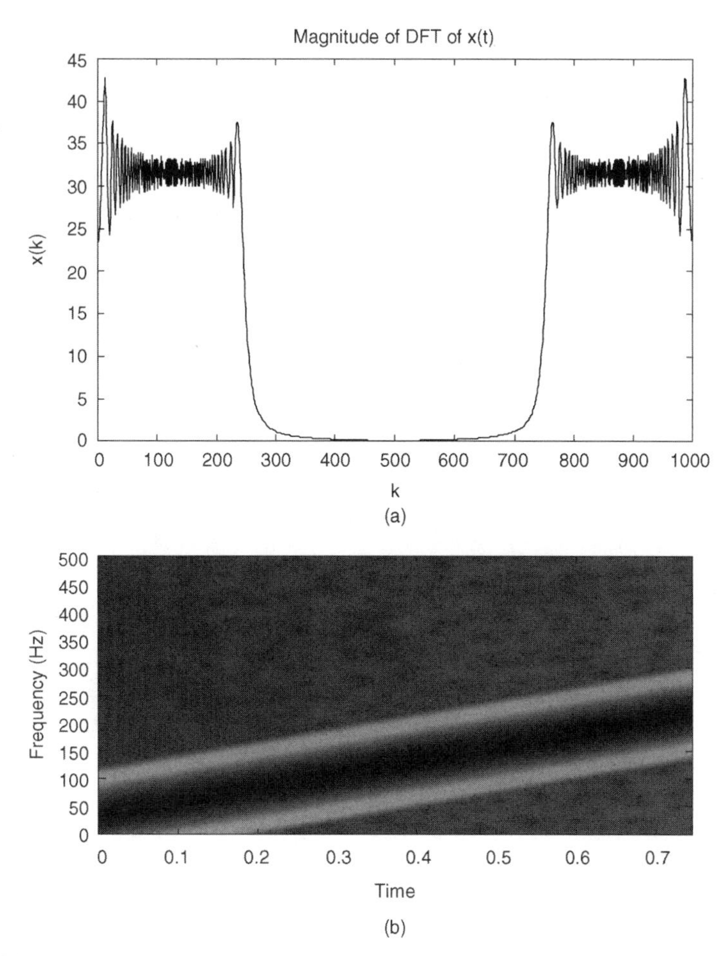

Fig. 10.5. Gabor transform of linear chirp, windowing with a Gaussian of $\sigma = 16$. The frequency of $x_a(t) = \cos(At^2)$ rises from 0 to 250 Hz over a 1–s time interval. Its magnitude spectrum $X(k)$ is shown in panel (a). Note the apparent presence of frequencies between 0 and 250 Hz, but that the time of a particular frequency is lost by the discrete Fourier transform (DFT). The Gabor transform reveals the time evolution of frequencies in $x(t)$, as shown in panel (b). Time values are shown along the bottom over the interval [0, 1], which represents samples n from 0 to 255. Image intensities represent Gabor magnitude spectral values $|\mathcal{G}[x_a](\mu, \omega)|$; darker values indicate larger magnitudes.

Chapter 7 that if $x(n)$ has discrete Fourier transform (DFT) coefficients $X(k)$ and DFS coefficients $c(k)$ on $[0, N-1]$, then $X(k) = Nc(k)$. Since we have to perform a discrete transform on N samples over an array of points, we choose that $N = 2^m$, for some m, so that the efficient fast Fourier transform (FFT) algorithm applies. We transform the windowed signal $x_a(t)g_{\mu,\sigma}(t)$ sampled at $t = 0, T/N, 2T/N, \ldots, (N-1)T/N$. Finally,

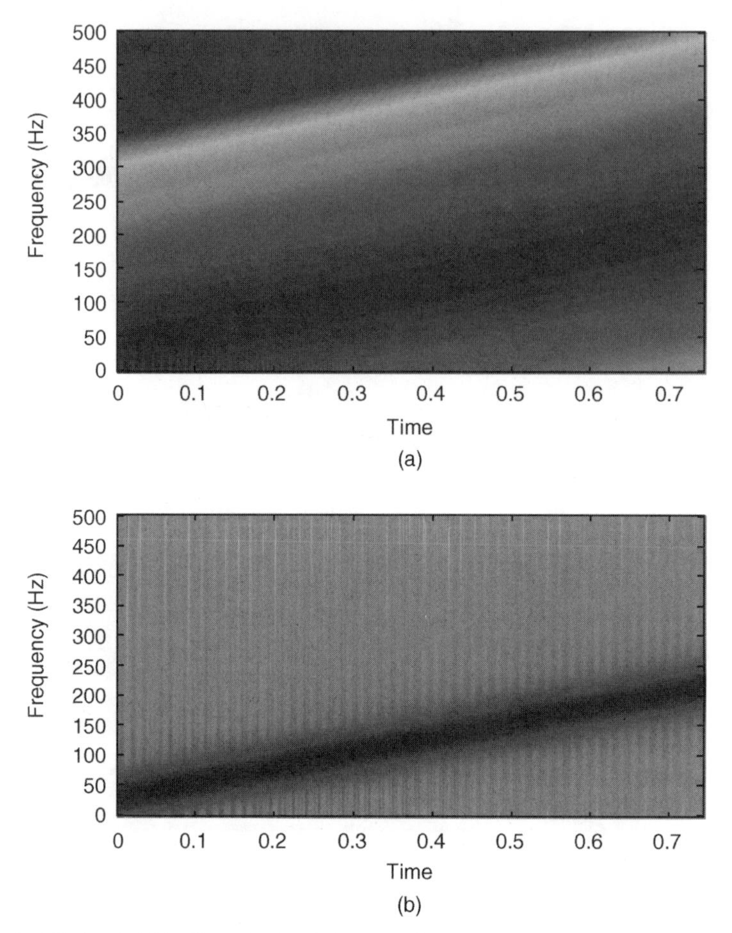

Fig. 10.6. Window width effects in the Gabor transform of a linear chirp. Windowing with a Gaussian of $\sigma = 4$ is shown in panel (a). Panel (b) shows the case of $\sigma = 64$.

we select $\sigma = 16$ as the standard deviation of the Gaussian window function for the transform (Figure 10.5).

What effect does the decision $\sigma = 16$ play for the transform? If σ increases, then the band of large transform coefficients shrinks. And decreasing the width of the transform's window function causes the sloping region of large magnitude coefficients to expand.

Carefully note in Figure 10.6 that broadening the time-domain window functions narrows the region of large magnitude values in the transformed signal. Indeed a reciprocal relation is manifest. This is an important characteristic. The next section further explores the link between time- and frequency-domain resolution under Gabor signal transformation.

10.1.5.2 *Pulsed Tone.* Now suppose that we begin with a time-domain pulse:

$$x_a(t) = \exp(-Bt^2)\cos(At). \tag{10.39}$$

We shall suppose that the pulse frequency is 50 Hz and consider different time-domain durations of $x_a(t)$, which are governed by stretching the Gaussian envelope, $\exp(-Bt^2)$. (Figures 10.7 and 10.8).

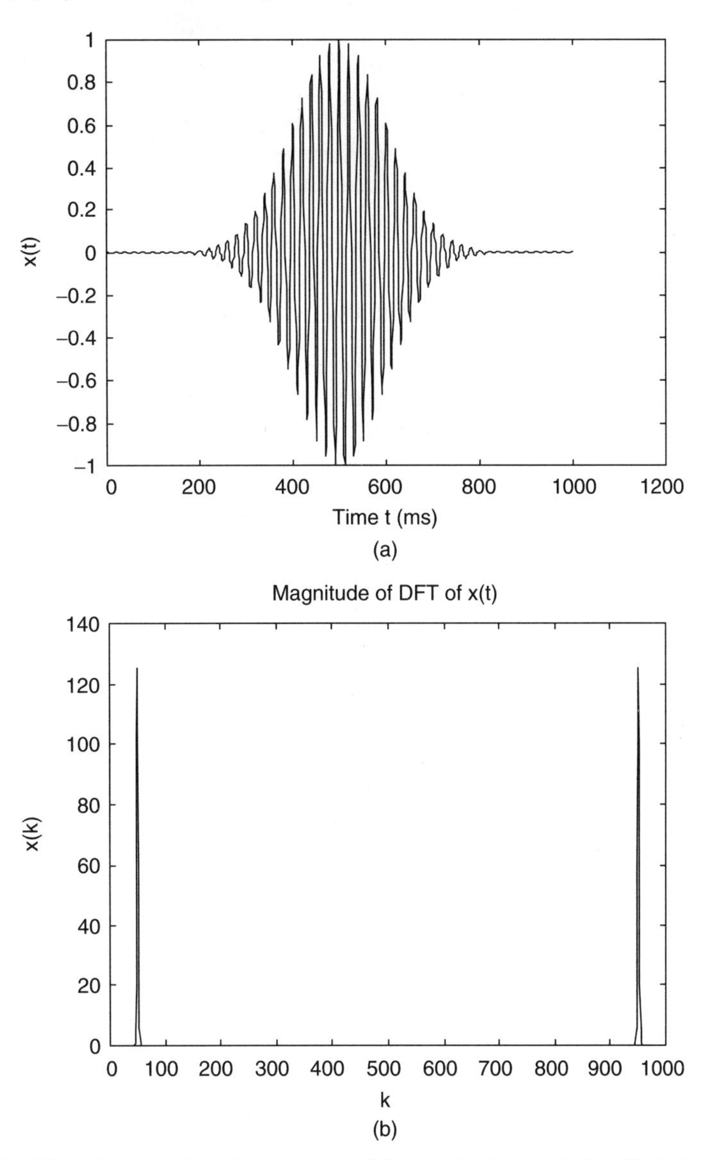

Fig. 10.7. Time-frequency localization tradeoff for a pulse tone. (a) The 50-Hz tone pulse rising and decaying in a 600-ms interval about $t = 0.5$ *s*. (b) Its Fourier spectrum shows the frequencies present but provides no time information. (c) The Gabor transform.

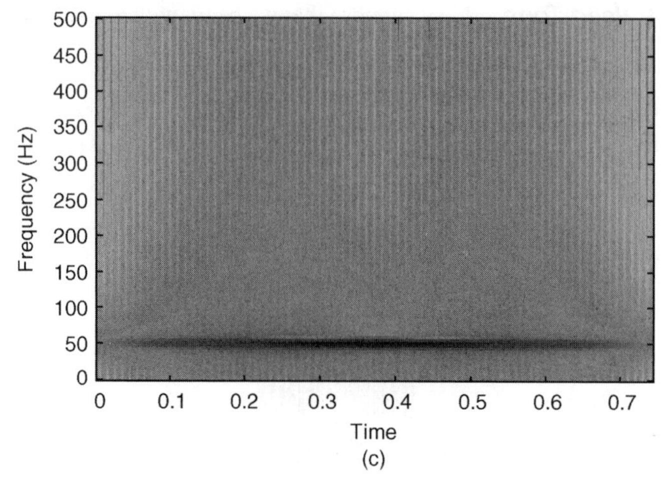

Fig. 10.7 (*Continued*)

This elementary experiment reveals that as the time-domain locality of the pulse increases, the frequency-domain locality decreases. In other words, it seems that as we gain a better knowledge of the time span a signal's frequencies occupy, then we lose knowledge of the specific frequencies it contains. This points to a fundamental tradeoff

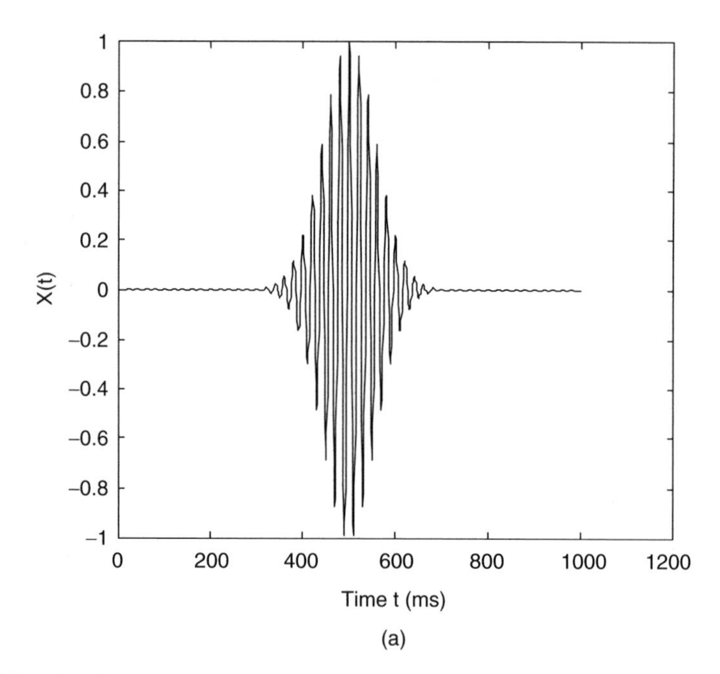

Fig. 10.8. Time-frequency localization tradeoff for a pulse tone. Panel (a) shows the 50-Hz tone pulse rising and decaying in a 300-ms interval about $t = 0.5$ s; (b) the more tightly localized pulse has a Gabor transform that is correspondingly dispersed.

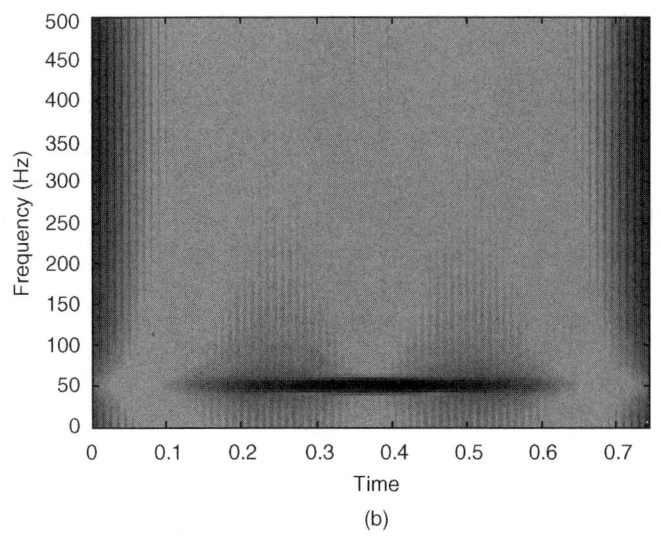

Fig. 10.8 *(Continued)*

in time-frequency analysis. It is also closely related to the Heisenberg Uncertaintly Principle, which we touched upon in our treatment of frequency-domain analysis.

10.1.6 Properties

Table 10.1 summarizes properties of the Gabor transformation, some of which are left as exercises.

TABLE 10.1. Gabor Transform Properties[a]

Signal Expression	Gabor Transform or Property
$x(t)$	$X_g(\mu, \omega)$
$ax(t) + by(t)$	$aX_g(\mu, \omega) + bY_g(\mu, \omega)$
$x(t-a)$	$e^{-j\omega a}X_g(\mu - a, \omega)$
$x(t)\exp(j\theta t)$	$X_g(\mu, \omega - \theta)$
$\|x\|_2 = \sqrt{2\pi}\dfrac{\|X_g(\mu, \omega)\|_{2, L^2(R^2)}}{\|g\|_2}$	Plancherel's theorem
$\langle x, y\rangle = \dfrac{1}{2\pi\|g\|_2^2}\langle X_g, Y_g\rangle$	Parseval's theorem
$x(t) = \dfrac{1}{(2\pi\|g\|_2^2)}\int\limits_{-\infty}^{\infty} X_g(\mu, \omega)g_{\mu, \sigma}(t)e^{j\omega t}d\omega d\mu$	Inverse, resolution of the identity, or synthesis equation

[a]In the table, $x(t)$ is square-integrable, and $g(t)$ is a Gaussian of mean μ and standard deviation σ.

10.2 SHORT-TIME FOURIER TRANSFORMS

A *short-time Fourier transform* (STFT) generalizes the Gabor transform by allowing a general window function. For the supporting mathematics to work, the theory requires constraints on the window functions. These we will elaborate in a moment. Once these theoretical details are taken care of, though, the general transform enjoys many of the same properties as the Gabor transform.

One might well ask whether a window shape other than the Gaussian can provide a better time-frequency transform. The answer is affirmative, but qualified. If the window shape matches the shape of signal regions to be analyzed, then an alternative window function offers somewhat better numerical results in signal detection applications. Thus, choosing the window to have roughly the same shape as the signals to be analyzed improves detection performance. These benefits are usually slight, however.

We know from the experiments with pulses and chirps at the end of the previous section that there is a tradeoff between time and frequency localization when using the Gabor transform. How does the selection of a transform window affect this behavior? It turns out that there is a hard lower limit on the joint time-frequency resolution of windowed Fourier transforms. Constricting the time-domain window so as to sharpen the time domain resolution results in a proportionately broader, more imprecise frequency-domain localization. This is a fundamental limitation on windowed Fourier methods. Its practical import is that signals with both low and high frequencies or with abrupt transients are difficult to analyze with this transform family. In fact, this limitation—which is a manifestation of the famous Heisenberg Uncertainty Principle—stimulated the search for alternative mixed domain transforms and was an impetus behind the discovery of the wavelet transform (Chapter 11).

Among all possible window functions, there is one signal in particular that shows the best performance in this regard: the Gaussian. Thus, the Gabor transform is the short-time Fourier transform with the best joint time-frequency resolution. So despite the benefits a special window may have, the Gabor transform prevails in all but certain specialized STFT-based signal analysis applications.

10.2.1 Window Functions

This section specifies those functions that may serve as the basis for a windowed transform. We formally define window functions and the resulting general window transform. We also develop some window function properties. This leads to a criterion for measuring joint time-frequency resolution. We prove the uncertainty principle, and the optimality of the Gabor transform follows as a corollary.

We should note right away that exponential signals modulated by window functions will play the role of structuring elements for signal analysis purposes. The short-time Fourier transform applies this structuring element at different time locations to obtain a set of time-ordered snapshots of the signal at a given frequency. When we later discretize the STFT, this idea will become clearer.

Definition (Window Function). If $x(t) \in L^2(\mathbb{R})$, $\|x(t)\|_2 \neq 0$, and $tx(t) \in L^2(\mathbb{R})$, then $x(t)$ is called a *window function.*

So, $x(t)$ is a window function when its squared magnitude, $|x(t)|^2$, has a second order moment. This technical condition is necessary for many of the properties of the windowed transform. Of course, the familiar functions we have used to improve signal spectra in Chapter 9 satisfy this definition.

Example (Gaussian). The Gaussian $g(t) = A\exp(-Bt^2)$, where $A \neq 0$ and $B > 0$, is a window function. The Gaussian has moments of all orders, as we can check by integrating by parts:

$$\int_{-\infty}^{\infty} |tg(t)|^2 dt = A\int_{-\infty}^{\infty} t^2 e^{-2Bt^2} dt = \left(\left. \frac{Ate^{-2Bt^2}}{-4B} \right|_{-\infty}^{\infty} + \left(\frac{A}{4B}\right) \int_{-\infty}^{\infty} e^{-2Bt^2} dt \right)$$

$$= \left(\frac{A}{4B}\right) \int_{-\infty}^{\infty} e^{-2Bt^2} dt. \tag{10.40}$$

The Fourier transform of a Gaussian is still a Gaussian, and therefore $G(\omega)$ is a window function in the frequency domain too. But many window functions have Fourier transforms that are not window functions by our definition.

Example (Square Pulse). The square pulse of width $2T > 0$, $w(t) = u(t + T) - u(t - T)$, is a window function. Indeed any non-trivial compactly supported $L^2(\mathbb{R})$ signal is a window function. The Fourier transform of $w(t)$, $W(\omega) = 2T\text{sinc}(T\omega)$, decays like ω^{-1} in the frequency domain. Thus, $\omega W(\omega) = 2\sin(T\omega) \notin L^2(\mathbb{R})$. So a window function does not necessarily have a Fourier transform that is a window function as well. Indeed, this occurs rarely, as the next result shows.

Lemma (Integrability). If $x(t)$ is a windowing function, then $x(t) \in L^1(\mathbb{R})$.

Proof: (A Schwarz Inequality exercise). ■

Proposition. If $x(t)$ is a discontinuous window function, then $X(\omega)$ cannot also be a window function.

Proof: If $X(\omega)$ is a window function, then it is absolutely integrable by the lemma, and its inverse Fourier transform $\mathcal{F}^{-1}[X(\omega)] = x(t)$ is continuous. But this is a contradiciton. ■

So constructing double pane windows requires some care. To do so we must find window functions that are continuous, decay quickly, and have Fourier transforms which are continuous with fast decay. Later, we will define the concept of the center and size of a window function. These definitions will lead to the uncertainty principle and the result that, among the short-time Fourier transforms, the Gabor transform alone possesses a window function with optimal time- and frequency-domain resolution.

10.2.2 Transforming with a General Window

It is not hard to generalize the Gabor transform to work with a general window, now that we have introduced the moment condition that a window function must satisfy. We will define the windowed transform for window functions and make the additional assuption that the Fourier transform of the window is also a window function for some of the properties. Drawing inspiration from the Gabor transform formalizations, we can easily draft a definition for a general windowed transform.

Definition (Short-Time Fourier Transform). Let $w(t)$ be a window function and $x(t) \in L^2(\mathbb{R})$. The *short-time Fourier transform (STFT) with respect to* $w(t)$, written $X_w(\mu, \omega)$, is the radial Fourier transform of the product $x(t)w(t-\mu)$:

$$X_w(\mu, \omega) = \int_{-\infty}^{\infty} x(t)w(t-\mu)e^{-j\omega t}\,dt. \tag{10.41}$$

The STFT is also known as the *windowed Fourier transform*. There is a "fancy W" notation for the short-time Fourier transform: $X_w(\mu, \omega) = (\mathcal{W}_w)[x(t)](\mu, \omega)$.

Remarks. The windowing function $w(t)$ in (10.41) remains fixed for the transform, as does the Gaussian in a Gabor transform. Indeed, our definition generalizes the Gabor transform: If $w(t)$ is a Gaussian, then the short-time Fourier transform with respect to $w(t)$ of $x(t)$ is precisely the Gabor transform of $x(t)$ using the Gaussian $w(t)$. We do not demand that the Fourier transform $(\mathcal{F}w)(\omega) = W(\omega)$ must also be a window function; when we turn to study time-frequency localization using the transform, however, we make this qualification.

10.2.2.1 Standard Windows. We can define an STFT for any of the windowing functions used to improve local spectra estimates in Chapter 9. We recall that windowing a signal $x(t)$ with a tapered window function reduces the size of Gibbs phenomenon sidelobes. Table 10.2 summarizes possible standard analog windows: rectangle, Bartlett (triangle), Hamming, Hanning, and Blackman functions.

Each of the standard window functions above has a discontinuity in a time-domain derivative of some order. We can develop the STFT using B-splines, however, and achieve smooth time-domain derivatives of arbitrarily high orders.

10.2.2.2 B-spline Windows. Another window function appropriate for the STFT involves B-splines, which we introduced in Section 3.2.5. Splines are popular in applied mathematics [14], computer graphics [15], and signal processing and analysis [16–18]. We recall the definition.

TABLE 10.2. Short-Time Fourier Transform Window Functions[a]

Name	Definition
Rectangle	$w(t) = \begin{cases} b & \text{if } (\lvert t\rvert \le a) \\ 0 & \text{otherwise.} \end{cases}$
Bartlett (triangle)	$w(t) = \begin{cases} \frac{b}{a}t + b & \text{if } -a \le t \le 0, \\ -\frac{b}{a}t + b & \text{if } 0 \le t \le a, \\ 0 & \text{otherwise.} \end{cases}$
Hanning (von Hann)	$w(t) = \begin{cases} b\cos^2\left(\frac{\pi t}{2a}\right) & \text{if } \lvert t\rvert \le a \\ 0 & \text{otherwise.} \end{cases}$
Hamming	$w(t) = \begin{cases} 0.54\text{b} + 0.46\text{b}\cos\left(\frac{\pi t}{a}\right) & \text{if } \lvert t\rvert \le a \\ 0 & \text{otherwise.} \end{cases}$
Blackman	$w(t) = \begin{cases} 0.42b + 0.5b\cos\left(\frac{\pi t}{a}\right) + 0.08b\cos\left(\frac{2\pi t}{a}\right) & \text{if } \lvert t\rvert \le a \\ 0 & \text{otherwise.} \end{cases}$

[a]Adjust parameter $a > 0$ for a window width appropirate to the signal features of interest. Adjust parameter $b > 0$ in order to normalize the window function.

Definition (B-spline). The B-spline of order zero is

$$\beta_0(t) = \begin{cases} 1 & \text{if } -\frac{1}{2} < t < \frac{1}{2} \\ \frac{1}{2} & \text{if } \lvert t\rvert = \frac{1}{2} \\ 0 & \text{if otherwise.} \end{cases} \tag{10.42}$$

and higher-order B-splines are found by successive convolution:

$$\beta_n(t) = \underbrace{\beta_0(t)*\beta_0(t)*\ldots\beta_0(t)}_{n\,+\,1\text{ times}}. \tag{10.43}$$

The B-splines are clearly window functions; $\beta_n(t)$ has compact support. Now let us examine the Fourier transform of $\beta_n(t)$. Let $B_n(\omega) = \mathcal{F}(\beta_n)(\omega)$. The Fourier transform convolution theorem implies

$$B_n(\omega) = \left[\frac{\sin\left(\frac{\omega}{2}\right)}{\frac{\omega}{2}}\right]^{n+1}. \tag{10.44}$$

So the denominator of (10.44) is ω^{n+1}; in case $n \geq 1$, we see $\omega B_n(\omega) \in L^2(\mathbb{R})$, so that $B_n(\omega)$ is indeed a window function.

In Section 10.2.4 we formulate, refine, and quantify the concept of the a windowed Fourier transform's time-frequency localization. A crucial precondition for frequency-domain locality is that the window function's Fourier transform must also be a window function. Note that both the Gabor transform and the B-spline windowed STFT enjoy this condition. Before addressing the idea of joint localization, however, let us cover some STFT properties.

10.2.3 Properties

Many of the properties of the Gabor transform carry over directly to the short-time Fourier transform. Like the specialized Gabor transform, the STFT obeys basic properties of linearity, time shift, and frequency shift. We state and leave as exercises the STFT Plancherel, Parseval, and inverse results.

Theorem (Short-Time Fourier Transform Parseval's). Suppose $x(t), y(t) \in L^2(\mathbb{R})$; $w(t)$ is a window function; and let $X_w(\mu, \omega)$ and $Y_w(\mu, \omega)$ be the STFTs of $x(t)$ and $y(t)$, respectively, based on windowing with $w(t)$. Then

$$2\pi\|w\|_2^2\langle x, y\rangle = \int_{-\infty}^{\infty}\int_{-\infty}^{\infty} X_w(\mu, \omega)\overline{Y_w(\mu, \omega)}\, d\omega d\mu = \langle X_w, Y_w\rangle_{L^2(\mathbb{R}^2)}. \tag{10.45}$$

Proof: Similar to Gabor transform (exercise). ■

Theorem (Short-Time Fourier Transform Plancherel's). Suppose $\sigma > 0$; $x(t) \in L^2(\mathbb{R})$; $w(t)$ is a window function; and let $X_w(\mu, \omega)$ be the STFT of $x(t)$. Then

$$\|x\|_2 = \sqrt{2\pi}\frac{\|X_g(\mu, \omega)\|_{2, L^2(\mathbb{R}^2)}}{\|g\|_2}. \tag{10.46}$$

Proof: Exercise. ■

Theorem (Inverse Short-Time Fourier Transform). Suppose $x(t) \in L^2(\mathbb{R})$, $w(t)$ is a window function, and let $X_w(\mu, \omega)$ be the STFT of $x(t)$. Then for all $a \in \mathbb{R}$, if $x(t)$ is continuous at a, then

$$x(a) = \frac{1}{(2\pi\|w\|_2^2)} \int_{-\infty}^{\infty} X_w(\mu, \omega) w(a) e^{j\omega a} d\omega d\mu . \tag{10.47}$$

Proof: Apply a limit argument to the Parseval formula, as with the Gabor transform (exercise). ■

10.2.4 Time-Frequency Localization

How precisely we can locate the frequency values within a signal using the short-time Fourier transform? Section 10.1.5 showed how the Gaussian window width dramatically affects the transform coefficients. Indeed, an improperly chosen window width—determined by the standard deviation σ—can render the transform information useless for intepreting signal evolution through time. The reason is not too hard to grasp. By narrowing the window, we obtain a more precise time frame in which frequencies of interest occur. But if we calculate the transform from discrete samples, then we cannot shrink σ too far; eventually the number of samples within the window are too few to compute the discrete signal frequencies. This is, of course, the threshold governed by the Nyquist rate. As σ decreases, then, the Gabor transform gains time-domain resolution, but it loses frequency-domain resolution at the same time.

10.2.4.1 Window Location and Size. To study the tradeoffs between time and frequency-domain resolution requires first of all a standard for measuring a signal's width or extent. The standard deviation of the enclosing Gaussian is a natural choice for the Gabor elementary function, $y(t) = g_{\mu,\sigma}(t)\exp(j\omega t)$. Recalling the Gaussian or normal distribution from the probability theory tutorial in Section 1.8, the probability that a normally distributed random variable has a value within one standard deviation of the mean μ is approximately 68%. That is, the area under the bell curve from $\mu - \sigma$ to $\mu + \sigma$ is about 0.68, whereas the total underlying area is unity. Thus, we propose 2σ for the "width" of $y(t)$, rather than a single standard deviation. Now, the standard deviation for a normally distributed random variable with density function $g_{\mu,\sigma}(t)$ is

$$\sigma = \left[\int_{-\infty}^{\infty} (t-\mu)^2 g_{\mu,\sigma}(t) dt \right]^{\frac{1}{2}} . \tag{10.48}$$

Can we extend this scheme to a general $x(t) \in L^2(\mathbb{R})$ which we propose to Gabor transform? The answer is, unfortunately, no; we do know that there are signals that have finite energy without being integrable. The canonical example in signal processing is $\mathrm{sinc}(t) = \sin(t)/t$. It is square-integrable, because $\mathrm{sinc}^2(t)$ decays like t^{-2} at infinity. However, $\mathrm{sinc}(t) \notin L^1(\mathbb{R})$, because, for instance, its Fourier transform is a square pulse, which is not continuous. Another problem is that the second moment integral (10.48) must also be valid. The following definition accounts for both difficulties, but we need a preliminary lemma.

Lemma. If $x(t)$ is a window function, then $t^{1/2}x(t) \in L^2(\mathbb{R})$.

Proof: This turns out to be a consequence—through the Schwarz inequality—of the square integrability of $x(t)$ and $tx(t)$. We leave this as an exercise. ■

Definition (Center and Radius). If $x(t)$ is a window function, then the *center* C_x and the *radius* ρ_x for $x(t)$ are given by

$$C_x = \frac{1}{\|x\|_2^2}\int_{-\infty}^{\infty} t|x(t)|^2\,dt \tag{10.49a}$$

and

$$\rho_x = \left[\frac{1}{\|x\|_2^2}\int_{-\infty}^{\infty}(t-C_x)^2|x(t)|^2\,dt\right]^{\frac{1}{2}}, \tag{10.49b}$$

respectively. The *diameter* or *width* of a windowing function $x(t)$ is $\Delta_x = 2\rho_x$.

Remark. The lemma assures us that the integral (10.49a) exists.

The more highly concentrated a signal $x(t)$ is about its center C_x, the smaller is its radius ρ_x. Let us consider a few examples of window functions before stating some of their basic properties.

Examples (Window Functions). Any Gaussian, $g(t) = A\exp(-Bt^2)$ is a window function as we already showed. All of the standard window functions of Table 10.2 are also window functions. The B-spline functions $\beta_n(t)$ are also window functions, and, for $n \geq 1$, $B_n(\omega)$ is a window function.

Now let us work out a few basic properties of window center and radius.

Lemma (Window Translation and Modulation). Suppose $x(t)$ is a window function and $y(t) = x(t + t_0)$. Then:

(a) $C_y = C_x - t_0$.

(b) If $X = \mathcal{F}x$ and $Y = \mathcal{F}y$ are the Fourier transforms of x and y, respectively, and X and Y are window functions, then $C_X = C_Y$.

(c) $\rho_y = \rho_x$.

(d) If $y(t) = \exp(-jC_X t)x(t + C_x)$ and $X = \mathcal{F}x$ and $Y = \mathcal{F}y$ are window functions, then $C_y = C_Y = 0$ and $\rho_y = \rho_x$.

Proof: By the Shifting and Modulation Properties of the Fourier transform (exercises). ■

Lemma (Radius of Derivative). Suppose $x(t) \in L^2(\mathbb{R})$ and is differentiable. If $x'(t) \notin L^2(\mathbb{R})$, then $\rho_X = \infty$.

Proof: Use Parseval's theorem for the radial Fourier transform and the formula for $\mathcal{F}[x'(t)](\omega)$. (exercise). ■

10.2.4.2 Uncertainty Principle. The next theorem is the classic Heisenberg Uncertainty Principle[4] for the Fourier transform [20, 21]. The theorem says as a signal becomes more concentrated about its time-domain center, it becomes more dispersed about its frequency domain center. Recent tutorials on the Uncertainty Principle include [22, 23].

Theorem (Heisenberg Uncertainty). Suppose $x(t) \in L^2(\mathbb{R})$, $X(\omega) = \mathcal{F}[x](\omega)$ is the radial Fourier transform of $x(t)$. Then $\rho_x \rho_X \geq \frac{1}{2}$.

Proof: We prove the Uncertainty Principle in two steps:

- First, for the happy circumstance that $x(t)$ obeys a special limit condition at infinity:

$$\lim_{t \to \infty} \sqrt{|t|}\,|x(t)| = 0\,; \tag{10.50}$$

  this condition does not necessarily hold for a square-integrable signal, of course; we could have $x(t) > \varepsilon > 0$ on some set S of measure zero, for example.
- Then, for the general case by writing $x(t)$ as a limit of such continuous, piecewise smooth signals.

Note that we may assume that $x(t)$ is a window function; otherwise, $\rho_x = \infty$, so that $\rho_x \rho_X \geq 1/2$. We assume $X(\omega)$ is a window function as well, since otherwise $\rho_X = \infty$ with the same consequence. In either exceptional case, we are done. The Window Translation and Modulation Lemma allows the further simplifying assumption that $C_x = C_X = 0$. Therefore,

$$\begin{aligned} \rho_x^2 \rho_{\mathcal{F}x}^2 &= \left[\frac{1}{\|x\|_2^2}\int_{-\infty}^{\infty} t^2 |x(t)|^2 dt\right]\left[\frac{1}{\|X\|_2^2}\int_{-\infty}^{\infty} \omega^2 |X(\omega)|^2 d\omega\right] \\ &= \frac{1}{\|X\|_2^2 \|x\|_2^2}\left[\int_{-\infty}^{\infty} |tx(t)|^2 dt\right]\left[\int_{-\infty}^{\infty} |\omega X(\omega)|^2 d\omega\right]. \end{aligned} \tag{10.51}$$

[4] Werner Heisenberg (1901–1976) discovered that the probable location of a particle trades off against its probable momentum. In 1927, Heisenberg showed that $\Delta p \Delta x \geq 2h$, where Δp represents the width of a particle's momentum distribution, Δx is the width of its position distribution, and h is Planck's constant [W. Heisenberg, *Physical Properties of the Quantum Theory*, New York: Dover, 1949].

Using Plancherel's theorem and the radial Fourier transform derivative formula gives

$$\rho_x^2 \rho_{\mathcal{F}x}^2 = \frac{\|x\|_2^{-4}}{2\pi}\left[\int_{-\infty}^{\infty} |tx(t)|^2 dt\right]\left[\int_{-\infty}^{\infty} |\mathcal{F}[x'(t)](\omega)|^2 d\omega\right]$$
$$= \frac{\|x\|_2^{-4}}{2\pi}\|tx(t)\|_2^2 \|\mathcal{F}[x'(t)](\omega)\|_2^2. \tag{10.52}$$

That is,

$$\|x\|_2^4 \rho_x^2 \rho_{\mathcal{F}x}^2 = \frac{1}{2\pi}\|tx(t)\|_2^2 \|x'(t)\|_2^2 2\pi = \|tx(t)\|_2^2 \|x'(t)\|_2^2. \tag{10.53}$$

Invoking the Schwarz inequality, $\|x\|_2 \|y\|_2 \geq \|xy\|_1$, on (10.53) gives

$$\|x\|_2^4 \rho_x^2 \rho_{\mathcal{F}x}^2 \geq \|tx(t)x'(t)\|_1^2 = \left[\int_{-\infty}^{\infty} |tx(t)||x'(t)|\, dt\right]^2 = \left[\int_{-\infty}^{\infty} |\overline{tx(t)}||x'(t)|\, dt\right]^2 \tag{10.54}$$

and, continuing our algebraic duties, we find that

$$\|x\|_2^4 \rho_x^2 \rho_{\mathcal{F}x}^2 \geq \left|\int_{-\infty}^{\infty} \overline{tx(t)}x'(t)dt\right|^2 = (|\langle x'(t), tx(t)\rangle|^2 \geq (\mathrm{Re}\langle x'(t), tx(t)\rangle)^2). \tag{10.55}$$

Now, we claim the following:

$$(\mathrm{Re}\langle x'(t), tx(t)\rangle) = -\frac{1}{2}\int_{-\infty}^{\infty} |x(t)|^2 dt = -\frac{1}{2}\|x(t)\|^2. \tag{10.56}$$

The trick behind the strange looking (10.56) is integration by parts on the inner product integral:

$$\int_{-\infty}^{\infty} \overline{tx(t)}x'(t)\, dt = \overline{tx(t)}x(t)\Big|_{-\infty}^{\infty} - \int_{-\infty}^{\infty} x(t)[\overline{tx'(t)} + \overline{x(t)}]\, dt$$
$$= 0 - \int_{-\infty}^{\infty} x(t)[\overline{tx'(t)} + \overline{x(t)}]\, dt. \tag{10.57}$$

Note that we have invoked (10.50) to conclude that $t|x(t)|^2 \to 0$ as $|t| \to \infty$. Separating the bottom of (10.57) into two integrals gives

$$\int_{-\infty}^{\infty} \overline{tx(t)}x'(t)\, dt = -\int_{-\infty}^{\infty} |x(t)|^2 dt - \overline{\int_{-\infty}^{\infty} \overline{tx(t)}x'(t)\, dt}. \tag{10.58}$$

After rearrangment, the claim (10.56) easily follows. We insert the result into the inequality (10.55), thereby finding

$$\|x\|_2^4 \rho_x^2 \rho_{\mathcal{F}x}^2 \geq \frac{\|x\|_2^4}{4} \tag{10.59}$$

and hence $\rho_x \rho_X \geq \frac{1}{2}$.

Let us proceed to the second step in the proof: removing the limit assumption (10.50) on the finite energy signal $x(t)$. We write $x(t)$ as the limit of a sequence of signals in the Schwarz space S of infinitely continuously differentiable, rapidly decreasing signals [21]:

$$x(t) = \lim_{n \to \infty} x_n(t). \tag{10.60}$$

We introduced the Schwarz space in Chapter 3 and know it to be dense in both $L^2(\mathbb{R})$ and $L^1(\mathbb{R})$. Since for all $x \in S$, we have $t|x(t)|^2 \to 0$ as $|t| \to \infty$, we have

$$\lim_{n \to \infty} \int_{-\infty}^{\infty} t\overline{x_n(t)} x_n'(t) dt = \lim_{n \to \infty} t\overline{x_n(t)} x_n(t) \Big|_{-\infty}^{\infty} - \lim_{n \to \infty} \int_{-\infty}^{\infty} x_n(t)[t\overline{x_n'(t)} + \overline{x_n(t)}]\, dt. \tag{10.61}$$

Because $x_n \in S$, which decreases faster than any polynomial, the integrands in (10.61) are absolutely integrable and we may interchange the integration and limit operations. Schwarz space elements are also continuous, so the first limit on the right-hand side of (10.61) is still zero. Thus,

$$\int_{-\infty}^{\infty} \lim_{n \to \infty} t\overline{x_n(t)} x_n'(t)\, dt = -\int_{-\infty}^{\infty} \lim_{n \to \infty} x(t)[t\overline{x_n'(t)} + \overline{x_n(t)}]\, dt. \tag{10.62}$$

But these limits are precisely (10.57). ■

The above proof follows Weyl's derivation [24], which he published in 1931.[5]

Thus, every windowed Fourier transform has a lower limit on its joint time-frequency resolution. If we work with a transform based on a window function $w(t)$ whose Fourier transform $W(\omega)$ is also a window function, then it makes sense to define the time-frequency resolution as the product $\rho_w \rho_W$. If we use a standard window function—a Hamming window, for example—whose Fourier transform is not itself a window function, then ρ_W is infinite. The Uncertainty Principle tells us that this is a hard lower bound: $\rho_w \rho_W \geq 1/2$. As a practical consequence, smaller time-domain window sizes result in proportionally large frequency-domain window

[5]The interests of Hilbert's student, Hermann Weyl (1885–1955), ranged from quantum mechanics to number theory. He showed, for instance, that given an irrational number r, the fractional parts of r, $2r$, $3r$, ..., etc., lie uniformly distributed on the interval (0, 1).

sizes. As we attempt to better locate a signal oscillation, we suffer a corresponding loss of accuracy in estimating the precise frequency of the oscillation.

There are window functions that achieve the lower bound on time-frequency localization given by the Heisenberg Uncertainty Principle. The next section shows that the optimally localizing window is none other than the Gaussian.

10.2.4.3 Optimally Localized Signals.

The Gabor transform is the short-time Fourier transform with the smallest time-frequency resolution. We identify *time-frequency resolution* with the joint product of the time-domain and frequency-domain radius: $\rho_x\rho_X$. To derive this optimality claim, we review the Uncertainty Principle's proof. Our scrutiny shows that inequality arises with its use of the Schwarz inequality [21].

Corollary (Optimal Time-Frequency Locality). We have $\rho_x\rho_X = \frac{1}{2}$ if and only if $x(t) = ae^{-bt^2}$ for some $a \in \mathbb{C}$ and $b \geq 0$.

Proof: We recall that $\|x\|_2\|y\|_2 \geq \|xy\|_1$ always, and equality occurs if and only if $x = cy$ for some constant $c \in \mathbb{C}$. In the context of the proof, then, optimally small time-frequency locality coincides with the condition $x'(t) = ct\overline{x(t)}$. Are there any $L^2(\mathbb{R})$ signals satisfying the above differential equation? It is a basic first-order differential equation, but before we note the solution, let us address two problems:

- Since the Uncertainty Principle deals with square-integrable signals $x(t)$, we understand this equality as occurring almost everywhere; that is, we require $x'(t) = ct\overline{x(t)}$ except on some set of measure zero.
- Furthermore, the proof depends on the fact that we can represent a general $x(t) \in L^2(\mathbb{R})$ as the limit of a sequence $x(t) = \lim_{n \to \infty} x_n(t)$, where $x_n(t) \in S$, the Schwarz space of infinitely continuously differentiable, rapidly decreasing signals. We must therefore show that for any $x(t)$ satisfying the differential equation, which is an $L^2(\mathbb{R})$ limit of $x_n(t) \in S$, that

$$x'(t) = \lim_{n \to \infty} x_n{}'(t). \tag{10.63}$$

In the convenient Schwarz space, the second point is straightforward. Indeed, we recall that $x(t) = y'(t)$ in Lebesgue integration theory means

$$y(t) = \int_0^t x(s)ds + y(0) \tag{10.64}$$

almost everywhere. We have $x(t) - x(0) = \lim_{n \to \infty} [x_n(t) - x_n(0)]$, and the $x_n(t) \in S$ are infinitely continuously differentiable; thus,

$$x(t) - x(0) = \lim_{n \to \infty} \int_0^t x_n{}'(s)\, ds = \int_0^t x'(s)\, ds. \tag{10.65}$$

To solve the differential equation, note that $t^{-1}x'(t) = c\overline{x(t)}$, whereby

$$\{t^{-1}x'(t)\}' = \{c\overline{x(t)}\}' = c\overline{x'(t)} = c\bar{c}tx(t) = |c|^2 tx(t). \tag{10.66}$$

If we let $b = |c|^2$, then the solutions to this second-order differential equation are of the form $x(t) = ae^{-bt^2}$, where $a \in \mathbb{C}$ is a constant. ■

Example (STFT based on a B-Spline Window). Suppose we use a B-spline function $\beta(t) = \beta_n(t)$, where $n \geq 1$, to define a short-time Fourier transform. We know that $\omega B_n(\omega) \in L^2(\mathbb{R})$, so that $B_n(\omega)$ is indeed a window function. The Uncertainty Principle applies. The Gaussian is not a B-spline, however, and we know therefore that $\rho_\beta \rho_B > 1/2$.

10.3 DISCRETIZATION

The short-time Fourier transform can also be discretized. There are two possible approaches:

- To compose discrete sums from values of a discrete signal $x(n)$, which is covered in Section 10.3.1.
- To sample the ordinary analog STFT analysis equation of an analog signal $x_a(t)$—the far more interesting and challenging problem—introduced in Section 10.3.2 and further explored in the sequel.

The second approach is our main emphasis. Its successful development leads to a new structural decomposition of finite-energy analog signals. It was also a focus of Gabor's original paper 1, a preoccupation of a number of later signal analysts, and the wellspring of much of our later insight into the nature of mixed-domain signal interpretation. We shall in fact pursue this idea for the remainder of this chapter.

10.3.1 Transforming Discrete Signals

Working with discrete signals, we can forumulate a purely discrete theory of windowed Fourier transforms. The results are not difficult to develop, and it turns out that they follow directly from discrete Fourier theorems. We are thus content to explicate only the discrete STFT synthesis and energy conservation equations.

We begin with a discrete signal $x(n)$ having period $N > 0$, $x(n) = x(n + N)$. Alternatively, we may select N samples $\{s(n): 0 \leq n < N\}$ from an arbitrary discrete signal $s(n)$ and consider the periodic extension $x(n) = s(n \bmod N)$. We require the discrete window function to be nonzero and have the same period as the signal to be transformed.

Definition (Discrete Short-Time Fourier Transform). Let $x(n)$ and $w(n)$ be discrete signals of period $N > 0$. Further suppose $w(n)$ is real and not identically zero on $[0, N-1]$. Then the *discrete short-time Fourier transform* (or *discrete windowed fourier transform*) of $x(n)$ with respect to $w(n)$ is

$$X_w(m,k) = \sum_{n=0}^{N-1} x(n)w(n-m)e^{-2\pi jk\frac{n}{N}}. \tag{10.67}$$

The signal $w(n)$ is called the *windowing function* for the transform.

Definition (Discrete Gabor Elementary Function). Let $w(n)$ be a discrete signal of period $N > 0$, with $w(n)$ is not identically zero on $[0, N-1]$. Then the *discrete Gabor elementary function* or *discrete Gabor atom* of discrete frequency $k \in [0, N-1]$ and location $m \in [0, N-1]$ is $w_{m,k}(n) = w(n-m)\exp(2\pi jkn/N)$.

As with its analog world counterpart, the discrete STFT can be viewed in several ways. In particular, we may think of (10.67) as giving

- For each $m \in [0, N-1]$, the discrete Fourier transform (DFT) of $x(n)w(n-m)$;
- For each $k \in [0, N-1]$, the inner product on $[0, N-1]$ of $x(n)$ with the discrete GEF $w_{m,k}(n)$.

The following theorem gives the synthesis equation for the discrete STFT.

Theorem (Inverse Discrete STFT). Let $x(n)$ and be a discrete signal with period $N > 0$; let $X_w(m, k)$ be its discrete STFT with respect to the windowing function $w(n)$; and, finally, let $\|w\|_2$ be the l^2-norm of $w(n)$ restricted to the interval $[0, N-1]$: $\|w\|_2 = [w^2(0) + w^2(1) + \cdots + w^2(N-1)]^{1/2}$. Then,

$$x(n) = \frac{1}{N\|w\|_2^2} \sum_{m=0}^{N-1} \sum_{k=0}^{N-1} X_w(m,k)w(n-m)e^{2\pi jk\frac{n}{N}}. \tag{10.68}$$

Proof: Substituting the definition of $X_w(m, k)$ into the double summation on the right-hand side of (10.68) gives

$$\sum_{m=0}^{N-1} \sum_{k=0}^{N-1} \sum_{p=0}^{N-1} x(p)w(p-m)e^{-2\pi jk\frac{p}{N}} w(n-m)e^{2\pi jk\frac{n}{N}}. \tag{10.69}$$

Rearrangment of the sums produces

$$\sum_{p=0}^{N-1} x(p) \sum_{m=0}^{N-1} w(p-m)w(n-m) \sum_{k=0}^{N-1} e^{2\pi jk\frac{(n-p)}{N}}. \tag{10.70}$$

Reciting what has become a familiar and fun argument, we note that the final sum in is zero unless $n = p$, in which case it is N. Therefore the entire triple summation is simply

$$Nx(n)\sum_{m=0}^{N-1} w(n-m)w(n-m) = Nx(n)\|w\|_2^2, \tag{10.71}$$

and the theorem follows. ■

Theorem (Discrete STFT Parseval's). Let $x(n)$ and be a discrete signal with period $N > 0$; let $X_w(m, k)$ be its discrete STFT with respect to the windowing function $w(n)$; and, finally, let $\|w\|_2$ be as in the previous theorem. Then,

$$\sum_{n=0}^{N-1} |x(n)|^2 = \frac{1}{N\|w\|_2^2}\sum_{m=0}^{N-1}\sum_{k=0}^{N-1} |X_w(m, k)|^2. \tag{10.72}$$

Proof: Let us expand the double summation on the right-hand side of (10.72):

$$\sum_{m=0}^{N-1}\sum_{k=0}^{N-1}\left[\sum_{p=0}^{N-1} x(p)w(p-m)e^{-2\pi jk\frac{p}{N}}\right]\left[\sum_{q=0}^{N-1} \overline{x(q)}w(q-m)e^{2\pi jk\frac{q}{N}}\right]. \tag{10.73}$$

Interchanging the sums we find that (10.73) becomes

$$\sum_{p=0}^{N-1}\sum_{q=0}^{N-1} x(p)\overline{x(q)}\left[\sum_{m=0}^{N-1} w(p-m)w(q-m)\right]\left[\sum_{k=0}^{N-1} e^{2\pi jk\frac{(q-p)}{N}}\right]. \tag{10.74}$$

The final bracketed sum is either N or 0, depending on whether $p = q$ or not, respectively. Since only the case $p = q$ contributes to the sum, we let $n = p = q$ and reduce the double summation on the left-hand side of (10.74) to a single sum over n:

$$N\sum_{n=0}^{N-1} x(n)\overline{x(n)}\left[\sum_{m=0}^{N-1} w(n-m)w(n-m)\right]. \tag{10.75}$$

Finally we see

$$N\|w\|_2^2\sum_{n=0}^{N-1} x(n)\overline{x(n)} = \sum_{m=0}^{N-1}\sum_{k=0}^{N-1} |X_w(m, k)|^2, \tag{10.76}$$

using the periodicity of $w(n)$. ■

10.3.2 Sampling the Short-Time Fourier Transform

Now let us turn to the deeper question of what happens we attempt to sample the STFT. We select a time-domain sampling interval $T > 0$ and a frequency-domain sampling interval $\Omega > 0$. These remain fixed for the discrete transform, and a complex-valued function on pairs of integers results. It might appear that our endeavors here will not differ radically in method and results from the work we did earlier in discretizing the Fourier transform. Quite the opposite turns out to be the case: Discretization of the windowed Fourier transform opens the door to a wealth of intriguing problems in signal analysis.

For one thing, discretizing the transform provides us with a ready breakdown of the signal into time localized frequency components, or time-frequency atoms. Each atom represents a spot in time. Each atom represents a possible frequency component. And—depending on the nature of our atomic signal building blocks—there is a way to measure the quantity of that frequency resident in the signal in the vicinity a discrete time instant. This is a structural decomposition of the signal. Chapters 4 and 9 explored time- and frequency- domain signal analysis, respectively. Among their lessons is the usefulness of a structural decomposition of the signal for purposes of classification, recognition, and interpretation. Time-frequency transforms benefit signal analysis by providing an elegant, formal mathematical theory as well as a relational description of the signal.

Discretization places Gabor's original problem on the agenda [1]. He proposed to model communication signals using families of discretely indexed signal elements, which he called *logons*, but which nowadays are known by various other monikers—*Gabor elementary functions*, *Gabor atoms*, *windowed Fourier atoms*, and so on. Can families of the form $\{\exp(2\pi jnt)g(t-m): m, n \in \mathbb{Z}\}$ provide an orthonormal basis for $L^2(\mathbb{R})$ signals? Their optimal joint time-frequency localization does recommend them, but neither Gabor nor any other signal analyst for decades after his suggestive 1946 paper could substantiate in theory what seemed so tantalizing for practice.

It was a negative answer to Gabor's insightful proposal that began to emerge in the 1980s, a decade marking a watershed of results in time-frequency and time-scale signal analysis. The rest of the chapter elaborates some of these apparently discouraging results for short-time Fourier methods. The next chapter suggests an alternative approach, motivated in part our understanding of the limitations inherent in atomic time-frequency signal decompositions. Chapter 11 does show that transformations that rely on signal scale instead—the wavelet transform in particular—may avoid the weaknesses of short-time Fourier techniques.

Definition (Discretized Short-Time Fourier Transform). Suppose that $X_w(\mu, \omega)$ is the STFT of $x(t) \in L^2(\mathbb{R})$ with respect to the window function $w(t)$. Given $T > 0$ and $\Omega > 0$, the *discretized short-time Fourier transform* is

$$X_w(m, n) = (X_w)_a(m\Omega, nT) = \int_{-\infty}^{\infty} x(t)w(t-mT)e^{-jn\Omega t}\,dt. \tag{10.77}$$

If distinguishing between the discrete and analog transform signals becomes a problem, then we can append a subscript a to the analog form, as in (10.77). Note that we are using the first discrete independent variable of $X_w(m, n)$ as the time index and are using the second variable as the frequency index.

10.3.3 Extracting Signal Structure

If we can find a sufficiently strong mathematical representation, then discretized short-time Fourier transforms provide an attractive means of describing signal structure. We have already covered the broad qualifications for such a represenation. It must be able to represent any candidate signal, for otherwise some inputs will avoid our decomposition method. The representation must also be stable, which, informally, means that changing the signal a little bit only perturbs the representation a little bit.

So, the question is, Can windowed Fourier atoms of the form

$$w_{m,n}(t) = e^{jn\Omega t}w(t-mT), \tag{10.78}$$

where $T > 0$ and $T\Omega = 2\pi$, serve as a complete signal representation? The two practical alternatives are that the family $\{w_{m,n}(t): m, n \in \mathbb{Z}\}$ constitutes either

- An orthonormal basis or
- A frame.

It is hoped that we can discover $\{w_{m,n}(t): m, n \in \mathbb{Z}\}$ that make up an orthonormal basis. Then every square-integrable signal $x(t)$ has a expansion in terms of Fourier coefficients, easily calcuated as the inner products of $x(t)$ with the $w_{m,n}(t)$:

$$x(t) = \sum_{m,n \in Z} \langle x(t), w_{m,n}(t)\rangle w_{m,n}(t). \tag{10.79}$$

If we fail to find such a basis, then computing the expansion coefficients (10.79) becomes problematic. Lack of a basis encumbers our signal analysis too. While we might be able to decompose a candidate signal $x(t)$ into a linear combination of atoms, $x(t) = \Sigma c_{m,n}w_{m,n}(t)$, we do not necessarily know the uniqueness of the expansion coefficients $c_{m,n}$ for representing $x(t)$. So the utility of the expansion coefficients as indicators of some signal component's presence or the lack thereof is very much compromised.

Should a basis not be available, we could search for a frame representation of $L^2(\mathbb{R})$ signals using the Gabor atoms (10.118). After all, we know from Chapter 3 that frame coefficients can characterize the source signal $x(t)$, and they support numerically stable reconstructions. This may be a good redoubt.

10.3.3.1 Discrete Time-Frequency Plane. Toward building a structural interpretation of a signal, we can place the expansion coefficients $c_{m,n}$ into an array. Thus, for a fixed frequency $n\Omega$, the rows of the array, $\{c_{m,n}: m \in \mathbb{Z}\}$, indicate the relative weight of frequency $n\Omega$ inside signal $x(t)$ at all time instants mT. Similarly, the columns record the frequencies at a given time instant. Refer to Figure 10.9.

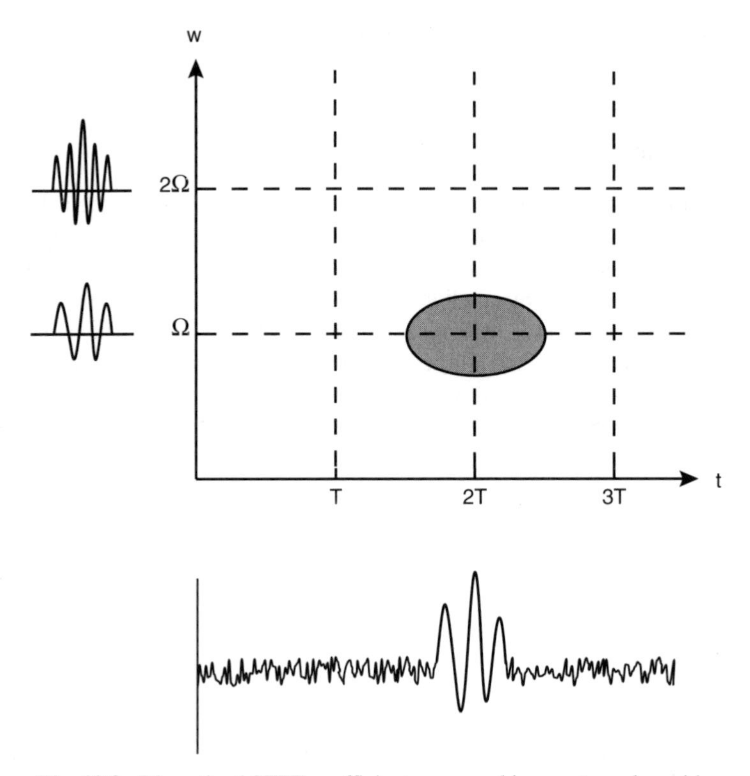

Fig. 10.9. Discretized STFT coefficients arranged in a rectangular grid.

Notice that upon discretizing the STFT we have a mathmatical signal transform that resolves a signal into finite regions of the time-frequency plane. The Fourier series of Chapter 5, in contrast, can only furnish time-frequency regions of infinite time- domain extent. When we covered discrete Fourier theory in Chapter 7, we studied the sampling theorem by which a band-limited analog signal can be reconstructed from sufficiently dense discrete samples. The sampling theorem too implies a partition of the time-frequency plane, except that its regions have an infinite frequency-domain extent. The STFT therefore marks a theoretical advance within our signal analytic understanding.

As a relational structure, this partition of the time-frequency plane is quite simple. Each region has the same size as its neighbors. We can, however, adjust the size of the regions to be smaller or larger in time or frequency by dilating our windowing function. The Uncertainty Principle imposes the constraint that the area of the STFT regions be no smaller than that given by the Gabor transform. Signal analysis applications based on STFT methods generally search the corresponding time-frequency mesh in order to understand signal content.

Let us consider some examples of how the time-frequency decomposition structure presents itself in applications.

Figure 10.10 illustrates a time-frequency mesh that contains a linear chirp and what is apparently a tone. Chirp signal energy concentration is fairly constant and can

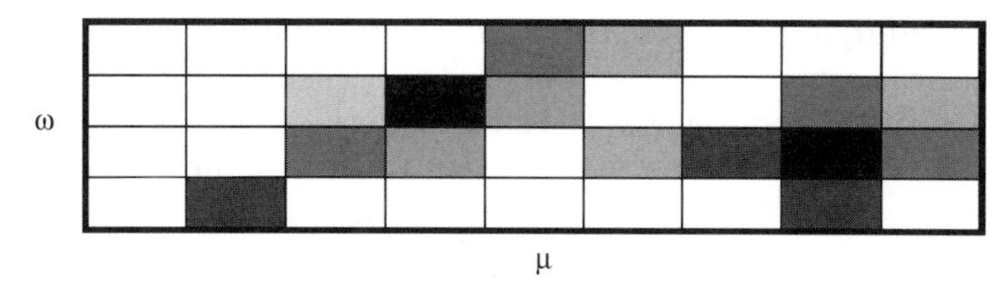

Fig. 10.10. Schematic representation of a signal with two components: a linear chirp and an isolated tone.

be tracked as a rising line over time. Tone signal energy concentration remains at a fixed frequency. Clicks or pops are characterized by a narrow time-domain extent and an extended, more or less uniform distribution of signal energy over a wide range of frequencies—a vertical linear structure. A sinusoidal tone is a horizontal linear structure. These ideas are only schematic, but they convey some of the signal varieties that are amenable to time-frequency analysis.

Let us now consider a speech analysis application. We have considered speech signal interpretation already in Chapters 4 and 9. In fact, in Chapter 9 we saw that many speech recognition systems have been developed using the basic technique of windowing the Fourier transform. If $x(t)$ is a speech signal, for example, then looking for a large percentage of signal energy in a pair of frequencies might indicate the presence of a vowel phoneme. Or, a broad dispersion of signal energy in a range of high frequencies could mark a velar fricative. Time-frequency signal decomposition offers a complete picture of the speech waveform. Figure 10.11 shows a contour diagram of the energies in a speech fragment.

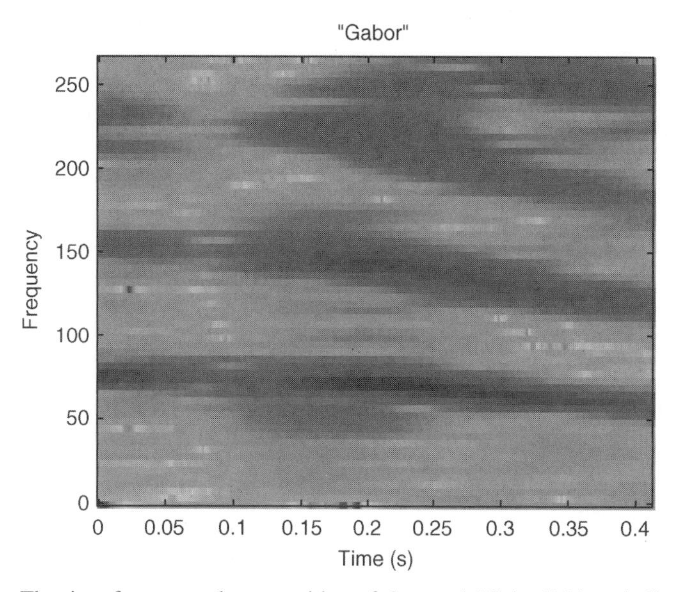

Fig. 10.11. The time-frequency decomposition of the word "Gabor." Lines indicate signal energy contours. Note that the two syllables can be segmented in time according to the distribution of signal energy along the frequency axis.

10.3.3.2 Identifying Significant Local Frequency Components. Many of the filtering, enhancement, and thresholding techniques we applied in time-domain signal analysis can help us find localized signal frequency components. For instance, we might use a threshold to indicate a significant frequency component. Furthermore, we might calculate the total amount of energy among all time-frequency cells at a certain time instant and count the number that contain say a majority of the energy. Such quantities can be assembled into feature vectors, for example. All of the methods we employed in Chapter 4 for thresholding signals apply equally well to thresholding values in the two-dimensional time-frequency plane. Indeed, this is an elementary instance of image analysis, and with it, our work in interpreting signal content begins to take advantage of techniques in image processing and computer vision.

10.3.4 A Fundamental Limitation

We have observed that the windowed Fourier transform provides an elegant and natural description of signal structure—a two-dimensional array, easily searched along time or frequency axes. With other decomposition techniques, especially those revolving around signal scale, structures may assume the form of a tree or some more general graph. Traversing graph structures consumes computer time. So even though our derived structures may be far simpler than the time-domain signal, we are still concerned to make the graphs sparse and conclude our exhaustive search as quickly as possible. Hence the problem before us is, How large can we choose the time and frequency sampling intervals, T and Ω, so that we still build a family of windowed Fourier atoms that provide an orthonormal basis or exact frame structure?

10.3.4.1 Nyquist Density. Our question directly concerns the power of short-time Fourier transforms for signal interpretation. Gabor studied the case $T\Omega = 2\pi$, suggesting that the Fourier expansion coefficients could be used to encode a signal for efficient transmission. Note too that for this case the time-domain sampling interval is $T = 2\pi/\Omega$. If a signal $x(t)$ has bandwidth Ω, then its highest frequency component is $\Omega/2$ radians per second, or $F_{max} = (\Omega/2)/(2\pi) = \Omega/(4\pi)$ hertz. By the Shannon–Nyquist sampling theorem, it can be reconstructed from discrete samples taken at intervals sampled at a rate not less than $F = T^{-1} = 2F_{max} = \Omega/(2\pi)$ hertz. We offer the formal definition.

Definition (Time-Frequency, Nyquist Densities). Let $x(t)$ have bandwidth Ω and be sampled at intervals $T > 0$. Then we define its *time-frequency density* to be $(T\Omega)^{-1}$. The *Nyquist density* is $(2\pi)^{-1}$.

In other words, Gabor's concern was to set the time- and frequency-domain sampling intervals so that $T\Omega = 2\pi$—that is, to sample at the Nyquist density. This is equivalent to time sampling at the largest interval allowable, by the sampling theorem, for analog signal reconstruction from discrete samples. Gabor proposed families of windowed Fourier atoms, separated from one another at the Nyquist limit $T = 2\pi\Omega^{-1}$. Our problem is to characterize Gabor's proposal for each of the three cases:

(i) $T\Omega < 2\pi$, when the sampling interval is less than the Nyquist frequency or, equivalently, when the time-frequency density exceeds the Nyquist density.
(ii) $T\Omega = 2\pi$, the original proposal of Gabor.
(iii) $T\Omega > 2\pi$.

10.3.4.2 Too Sparse: $T\Omega > 2\pi$. Gabor's proposal to represent signals using sums of windowed Fourier atoms does not in fact succeed for the sparse case, $T\Omega > 2\pi$. This case occurs when the time- domain sampling interval exceeds the maximum allowable for reconstructing an analog signal of bandwidth Ω by its discrete samples at times mT, $m \in \mathbb{Z}$. That is, families of Gabor atoms $\{w_{m,n}(t) = \exp(jn\Omega)w(t - mT)$: $m, n \in \mathbb{Z}\}$ cannot be a frame when $(T\Omega)^{-1}$ is less than the Nyquist density.

Using the Zak transform, introduced in Chapter 8, this result can be shown for the case where $(2\pi)^{-1}T\Omega$ is rational and exceeds unity. We will only consider a far simpler case: $\Omega = 2\pi$ and $T = 2$. We recount the following results from Ref. 12.

Lemma. Let $w(t) \in L^2(\mathbb{R})$; $w_{m,n}(t) = e^{jn\Omega t}w(t - mT)$, for $m, n \in \mathbb{Z}$; $\Omega = 2\pi$; and $T = 2$. Then there is an $x(t) \in L^2(\mathbb{R})$ such that $\|x\|_2 \neq 0$ and $\langle x(t), w_{m,n}(t)\rangle = 0$ for all $m, n \in \mathbb{Z}$.

Proof: Let $t \in [0, 1)$ and define

$$x(t+k) = (-1)^k\overline{w(t-k-1)}, \tag{10.80}$$

where $k \in \mathbb{Z}$. It is easily shown (exercise) that

(i) $x \in L^2(\mathbb{R})$.
(ii) $\|x\|_2 = \|w\|_2$.
(iii) $\|x\|_2 \neq 0$.

We contend that $\langle x(t), w_{m,n}(t)\rangle .= 0$ for all $m, n \in \mathbb{Z}$. Breaking up the inner product integral reveals

$$\begin{aligned}\langle x(t), w_{m,n}(t)\rangle &= \int_{-\infty}^{\infty} x(t)\overline{e^{jn\Omega t}w(t-2m)}\,dt \\ &= \int_0^1 e^{-jn\Omega t}\sum_{k=-\infty}^{\infty} x(t+k)\overline{w(k+t-2m)}\,dt \end{aligned} \tag{10.81}$$

But inserting (10.80), we observe that

$$\sum_{k=-\infty}^{\infty} x(t+k)\overline{w(k+t-2m)} = \sum_{k=-\infty}^{\infty} (-1)^k\overline{w(t-k-1)}\,\overline{w(t+k-2m)}. \tag{10.82}$$

On the right-hand side of (10.82) consider a summand,

$$(-1)^k \overline{w(t-k-1)}\,\overline{w(t+k-2m)}, \tag{10.83a}$$

for some $k \in \mathbb{Z}$. Let $i = 2m - k - 1$ and compare the term

$$(-1)^i \overline{w(t-i-m)}\,\overline{w(t+i-2m)}. \tag{10.83b}$$

The trick is that

$$(-1)^i \overline{w(t-i-m)}\,\overline{w(t+i-2m)} = (-1)^{-k-1} \overline{w(t-2m+k)}\,\overline{w(t-k-1)}, \tag{10.84}$$

which is the additive inverse of (10.83a). The upshot is that every summand in is complemented by its additive inverse also inside the summation; the sum is precisely zero! All inner products (10.81) are zero, and we have constructed a nontrivial $x(t)$ in the orthogonal complement of $\{w_{m,n}(t): m, n \in \mathbb{Z}\}$. ■

Now we can prove the theorem. Recall that a frame generalizes the notion of an orthogonal basis, yet provides stable signal reconstruction and complete signal representation. We introduced frame theory in Section 3.3.4.

Theorem. Let $w(t) \in L^2(\mathbb{R})$; $w_{m,n}(t) = e^{jn\Omega t}w(t - mT)$, for $m, n \in \mathbb{Z}$; $\Omega = 2\pi$; and $T = 2$. Then $\{w_{m,n}(t): m, n \in \mathbb{Z}\}$ cannot be a frame.

Proof: Let $x(t)$ be given by the lemma: nontrivial and orthogonal to all of the $w_{m,n}(t)$. If the $\{w_{m,n}(t): m, n \in \mathbb{Z}\}$ were a frame, then by the definition of frame, there exist $A > 0$ and $B > 0$ such that

$$A\|y\|^2 \le \sum_{m,n=-\infty}^{\infty} |\langle y, w_n\rangle|^2 \le B\|y\| \tag{10.85}$$

for all $y(t) \in L^2(\mathbb{R})$. The frame condition must hold for the lemma's $x(t)$ as well, but since $\langle x(t), w_{m,n}(t)\rangle = 0$ for all $m, n \in \mathbb{Z}$, we have a contradiction. ■

Remark. So there are no frames of windowed Fourier atoms, $w_{m,n}(t) = e^{jn\Omega t}w(t - mT)$, when the frequency- and time-domain sampling intervals are $\Omega = 2\pi$; and $T = 2$, respectively. This is perhaps not too surprising a result, given the Shannon–Nyquist sampling theorem.

We have shown our result for only a particular instance, $\Omega = 2\pi$ and $T = 2$, of the case $T\Omega > 2\pi$. An interesting, but somewhat technical, Zak transform application extends this same argument whenever $T\Omega > 2\pi$ and $T\Omega$ is a rational multiple of 2π [25]. Using advanced mathematical methods well beyond our present scope, it has been shown that whenever the time-frequency sampling is too sparse—whether either T and Ω are rational or irrational—then there are no frames of windowed Fourier atoms [26].

10.3.5 Frames of Windowed Fourier Atoms

Now let us consider another possibility: $T\Omega < 2\pi$. This is the dense time-frequency sampling case. Now, from a classic construction [27], it can be shown that we can build frames from windowed Fourier atoms when $T\Omega < 2\pi$. Here, we adapt the presentation in Ref. 12 to our own notation and show that collections of Gabor atoms $\{w_{m,n}(t) = \exp(jn\Omega)w(t - mT): m, n \in \mathbb{Z}\}$ can be a frame when $(T\Omega)^{-1}$ exceeds the Nyquist density, $(2\pi)^{-1}$.

Theorem. Let $w(t) \in L^2(\mathbb{R})$; let $w_{m,n}(t) = e^{jn\Omega t}w(t - mT)$, for $m, n \in \mathbb{Z}$; $T\Omega < 2\pi$; and suppose that $[-\pi/\Omega, \pi/\Omega] \supset \mathrm{Support}(w)$. Then for any $x(t) \in L^2(\mathbb{R})$,

$$\sum_{m,n=-\infty}^{\infty} |\langle x, w_{m,n}\rangle|^2 = \frac{2\pi}{\Omega}\int_{-\infty}^{\infty} |x(t)|^2\left(\sum_{k=-\infty}^{\infty} |w(t-kT)|^2\right) dt. \tag{10.86}$$

Proof: Let us expand the sum on the left-hand side of (10.86) as sums over $2\pi/\Omega$-wide intervals:

$$\sum_{m,n=-\infty}^{\infty} |\langle x, w_{m,n}\rangle|^2 = \sum_{m,n=-\infty}^{\infty} \left|\int_0^{\frac{2\pi}{\Omega}} e^{jn\Omega t} \sum_{k=-\infty}^{\infty} x\left(t+\frac{2\pi k}{\Omega}\right)\overline{w\left(t+\frac{2\pi k}{\Omega}-mT\right)}\, dt\right|^2. \tag{10.87}$$

Notice that the integral in (10.87) is a constant multiple of a Fourier series coefficient. The functions $\sqrt{\frac{\Omega}{2\pi}}e^{jn\Omega t} = e_n(t)$ are an orthonormal basis for the Hilbert space $H = L^2[0, 2\pi/\Omega]$, and we know therefore that $\|y\|_2^2 = \sum_n |\langle y, e_n\rangle|^2$ for any square-integrable $y(t)$ in H. (This is in fact a Parseval result for H, and its roots extend back to our very early algebraic result from abstract Banach spaces—Bessel's inequality.) Thus, for each $m \in \mathbb{Z}$ we are able to replace the sum over $n \in \mathbb{Z}$ in (10.87) with the square of the $L^2[0, 2\pi/\Omega]$ norm of the sum in the integrand:

$$\sum_{m,n=-\infty}^{\infty} |\langle x, w_{m,n}\rangle|^2 = \frac{2\pi}{\Omega}\sum_{m=-\infty}^{\infty}\int_0^{\frac{2\pi}{\Omega}} \left|\sum_{k=-\infty}^{\infty} x\left(t+\frac{2\pi k}{\Omega}\right)\overline{w\left(t+\frac{2\pi k}{\Omega}-mT\right)}\right|^2 dt. \tag{10.88}$$

Next, observe that for any m all of the summands over k inside the integral are zero except for possibly one. This is due to the choice of support for the window function $w(t)$. The right-hand side of simplifies, and we see

$$\sum_{m,n=-\infty}^{\infty} |\langle x, w_{m,n}\rangle|^2 = \frac{2\pi}{\Omega}\sum_{m,k=-\infty}^{\infty}\int_0^{\frac{2\pi}{\Omega}} \left|x\left(t+\frac{2\pi k}{\Omega}\right)\overline{w\left(t+\frac{2\pi k}{\Omega}-mT\right)}\right|^2 dt. \tag{10.89}$$

We can now reassemble the separate finite integrals to one over the entire real line:

$$\sum_{m,n=-\infty}^{\infty} |\langle x, w_{m,n}\rangle|^2 = \frac{2\pi}{\Omega}\int_{-\infty}^{\infty} |x(t)|^2 \left(\sum_{k=-\infty}^{\infty} |w(t-kT)|^2\right) dt. \tag{10.90}$$

■

The term in parentheses inside the integral (10.90) is crucial. If we can show that there are constants A, $B > 0$ such that A bounds this term below and B bounds this term above, then we will have found frame bounds and shown that the windowed Fourier atoms $\{w_{m,n}(t) = \exp(jn\Omega)w(t-mT)\colon m, n \in \mathbb{Z}\}$ do comprise a frame. The following corollary imposes a reasonable technical condition on the window function $w(t)$ [27], namely that the window function be continuous and positive on some interval about $t = 0$.

Corollary. Let $w(t) \in L^2(\mathbb{R})$ be as in the theorem. Further suppose that $w(t)$ is continuous and that there are $\varepsilon > 0$ and $1 > \delta > 0$ such that $|w(t)| > \varepsilon$ on $I = [-\delta\pi/\Omega, \delta\pi/\Omega]$. Then $\{w_{m,n}(t) = e^{jn\Omega t}w(t-mT)\colon m, n \in \mathbb{Z}\}$ are a frame.

$$\sum_{m,n=-\infty}^{\infty} |\langle x, w_{m,n}\rangle|^2 = \frac{2\pi}{\Omega}\int_{-\infty}^{\infty} |x(t)|^2 \left(\sum_{k=-\infty}^{\infty} |w(t-kT)|^2\right) dt. \tag{10.91}$$

Proof: Since

$$\sum_{m,n=-\infty}^{\infty} |\langle x, w_{m,n}\rangle|^2 = \frac{2\pi}{\Omega}\int_{-\infty}^{\infty} |x(t)|^2 \left(\sum_{k=-\infty}^{\infty} |w(t-kT)|^2\right) dt, \tag{10.92}$$

by the theorem, we seek positive constants α and β such that $\alpha < \Sigma|w(t-kT)|^2$ and $\Sigma|w(t-kT)|^2 < \beta$ for all t. Then we have

$$\frac{2\pi\alpha}{\Omega}\int_{-\infty}^{\infty} |x(t)|^2 dt \le \sum_{m,n=-\infty}^{\infty} |\langle x, w_{m,n}\rangle|^2 = \frac{2\pi\beta}{\Omega}\int_{-\infty}^{\infty} |x(t)|^2 dt, \tag{10.93}$$

so that $A = (2\pi\alpha)/\Omega$ and $B = (2\pi\beta)/\Omega$ constitute lower and upper frame bounds, respectively, for $\{w_{m,n}(t)\}$. By the assumption that $w(t)$ exceeds $\varepsilon > 0$ on the proper subinterval I, we can set $\alpha = \inf\{|w(t)|^2\colon t \in I\}$. Since α is the greatest lower bound of $|w(t)|^2$ on I, and $|w(t)| > \varepsilon$ on I, we know $\alpha \ge \varepsilon > 0$. The lower frame condition follows easily with bound $A = (2\pi\alpha)/\Omega$. To find the upper frame bound, we note that because the support of $w(t)$ is contained within the interval $[-\pi/\Omega, \pi/\Omega]$, only a finite number K of terms in the sum $\Sigma|w(t-kT)|^2$ will be nonzero. Since $w(t)$ is continuous and supported on $[-\pi/\Omega, \pi/\Omega]$, we may let M be its least upper bound; that is, $M = \|w\|_\infty$. We can then set $\beta = \sup\{\Sigma|w(t-kT)|^2 : t \in \mathbb{R}\} \le \text{KM}$, and with $B = (2\pi\beta)/\Omega$ we can verify the upper frame bound property. ■

Remarks. Recall from our general discussion of frames in Section 3.3.4 that the frame reconstruction algorithm is more efficient when the frame is tight. We can see in the theorem that finding $w(t)$ so that $\Sigma|w(t-kT)|^2$ is constant does provide us with a tight frame: $A = B = 2\pi(\Omega T)^{-1}$. In fact, it is fairly straightforward to concoct window functions $w(t)$ so that this expression is constant. Moreover, the construction method gives windows with compact support arbitrarily good smoothness. We refer the reader to the literature for details [12, 25, 27].

The result of the theorem (10.86) can be used to find examples of windowed Fourier frames from special window functions.

Example. Suppose $w(t) = (1 + t^2)^{-1}$. Then $w(t)$ is bounded and absolutely integrable. If $T > 0$, then $\Sigma|w(t-kT)|^2$ has an upper and lower bound. One can show (exercise) that $\{w_{m,n}(t) = e^{jn\Omega t}w(t - mT)\colon m, n \in \mathbb{Z}\}$ are a frame if we take Ω to be sufficiently small.

Example. Now let $w(t) = g_{\mu,\sigma}(t)$, the Gaussian with mean μ and standard deviation σ. Again, $\Sigma\left|g_{\mu,\,\sigma}(t-kT)\right|^2$ is bounded above and below when $T > 0$, and we can use the theorem's criterion for showing that Gabor frames exist for a sufficiently small frequency sampling interval.

Before summarizing our results in pursuit of Gabor's problem, let us note an important necessary condition of windowed Fourier frames [12].

Theorem. Suppose $w(t) \in L^2(\mathbb{R})$; Ω, $T > 0$; and $\{w_{m,n}(t) = e^{jn\Omega t}w(t - mT)\colon m, n \in \mathbb{Z}\}$ constitute a frame with lower and upper bounds A and B, respectively. Then

$$A \le \frac{2\pi}{\Omega T}\|w\|_2^2 \le B. \tag{10.94}$$

Proof: Exercise. ■

10.3.6 Status of Gabor's Problem

We can briefly summarize the status of our search for frames of windowed Fourier atoms. There are three cases, which depend on the time- and frequency-domain sampling intervals, T and Ω, respectively. Our present understanding is as follows:

(i) When $T\Omega < 2\pi$ the time-frequency density is higher than the Nyquist density, and we have just constructed frames of windowed Fourier atoms in this case.

(ii) When $T\Omega = 2\pi$ the atom are at Nyquist density exactly; this is the alternative proposed by Gabor, and our analysis of it is not yet complete.

(iii) Finally, when $T\Omega > 2\pi$ we have noted that windowed Fourier frames do not exist in this situation; we proved a simple instance, and the research literature—portions of which rely on advanced analysis—completely covers the remaining cases.

We will in fact devote a considerable portion of the remainder of this chapter to Gabor's dividing line case. The applicability of the short-time Fourier transform (STFT) when time-frequency localization is of paramount importance hangs on this question. This question also vexed signal processing investigators for a number of years; we are especially interested in fully understanding the impact of windowed Fourier transform discretization when $T\Omega = 2\pi$.

Before turning to this question, however, let us consider another approach to time-frequency signal decompositions.

10.4 QUADRATIC TIME-FREQUENCY TRANSFORMS

There are classes of time-frequency transforms that do not depend on a windowing function. Instead, the transform relation emerges out of the properties of the analyzed signal. The signal $x(t)$ enters the transform integral as a quadratic rather than as linear term, as it does in the windowed Fourier transform. This transform family is therefore generally known as the *quadratic* time-frequency transformations. Its principal members are the *Wigner–Ville transform* (WVT) and the closely related *ambiguity function.*

Now, transforming without a window function appears to be quite advantageous, since the resulting procedure eliminates the effect window selection imposes on the transform's behavior. The short-time Fourier transform mixes spectral properties of the analyzed signal $x(t)$ together with those of the window function $w(t)$. Blindly perusing coefficients, we do not know whether their large magnitude results from signal or window properties. On the other hand, we do not often blindly process transform coefficients. Rather, the window function is typically chosen to isolate signal features of expected frequency content and time- domain extent; in the more typical application then, choosing a window function may well be the best first step.

Although avoiding window effects may recommend quadratic transforms, there are some more important considerations. We shall explore three significant properties of these transforms. This transform family:

- More precisely resolves certain standard cases of time-varying frequencies than does the STFT;
- Enjoys special properties called *marginal conditions* that allow them to act as distribution functions for a signal's spectral content;
- Has the significant drawback that transformed signals exhibit certain artifacts called *cross-terms* that hamper higher-level interpretation.

This is in fact a very rich transform family. An entire book could be written about these transforms, and many treatments devote considerable space to these transforms [2, 6, 9]. By our brief sketch we hope that the reader will acquire a more balanced opinion of the windowed Fourier transforms and an interest in further exploring the theory and application of quadratic transforms.

10.4.1 Spectrogram

We can base a quadratic time-frequency transform on the STFT. This is in fact just the spectrogram, which we define as follows.

Definition (Spectrogram). Let $x(t) \in L^2(\mathbb{R})$ and let $w(t)$ be a window function. The *spectrogram with respect to* $w(t)$, written $X_{S,w}(\mu, \omega)$, is

$$X_{S,w}(\mu, \omega) = |X(\mu, \omega)|^2 = \left| \int_{-\infty}^{\infty} x(t) w(t-\mu) e^{-j\omega t} dt \right|^2 , \tag{10.95}$$

where $X_w(\mu, \omega)$, is the STFT of $x(t)$ with respect to $w(t)$.

Thus, the spectrogram of $x(t)$ is the squared magnitude of the STFT of $x(t)$ with respect to $w(t)$. The spectrogram is thus a natural generalization of the windowed Fourier methods we have been comfortable in using. However, despite the more intuitive feel, spectrograms are far from being the most popular quadratic time-frequency transforms. For one thing, $X_{S,w}$ relies on a window function. But it also has some other undesirable traits that have motivated signal theorists to search out other transform techniques. Among these better transforms is the the classic transform of Wigner and Ville which we introduce next; we shall assess the merits of the spectrogram in this context.

10.4.2 Wigner–Ville Distribution

The Wigner–Ville distribution (WVD) is the oldest time-frequency transform and the preeminent quadratic signal representation. In fact it dates to the early 1930s when E. Wigner[6] applied it in quantum mechanics [28]. The communication theorist J. Ville[7] introduced the transform to the signal processing community some 16 years later [29].

The transform has been widely studied for signal analysis applications [30, 31]. It has also found use as an important tool in computer vision [32]. The WVD has some distinct advantages over the more intuitive spectrogram. But it is not without its faults.

One difficulty in applying the WVD is the presence of so-called *cross-* or *interference terms* among the transform coefficients. Indeed, many research efforts in time-frequency theory have concentrated on avoiding or ameliorating the effects of cross-terms when using this type of tool. This problem is covered in the Section 10.4.3.

10.4.2.1 Definition and Motivation. The Wigner–Ville distribution takes the Fourier transform of a product of the signal with its complex conjugate. Thus, it resembles the computation of the power spectral density.

[6]The Hungarian chemical engineer Eugene P. Wigner (1902–1996) immigrated to the United States to teach mathematics at Princeton University in 1930. He received the Nobel prize in 1963 for discoveries in atomic and elementary particle research.

[7]French communication researcher J. Ville developed the same transform as Wigner, but for the purposes of clarifying the concept of instantaneous frequency.

Definition (Wigner–Ville Distribution). If $x(t) \in L^2(\mathbb{R})$ is an analog signal, then its *Wigner–Ville distribution*, written $X_{WV}(\mu, \omega)$, is the radial Fourier transform of the product $x(\mu + t/2)x^*(\mu - t/2)$:

$$X_{WV}(\mu, \omega) = \int_{-\infty}^{\infty} x\left(\mu + \frac{t}{2}\right)\overline{x\left(\mu - \frac{t}{2}\right)}e^{-j\omega t}\,dt. \tag{10.96}$$

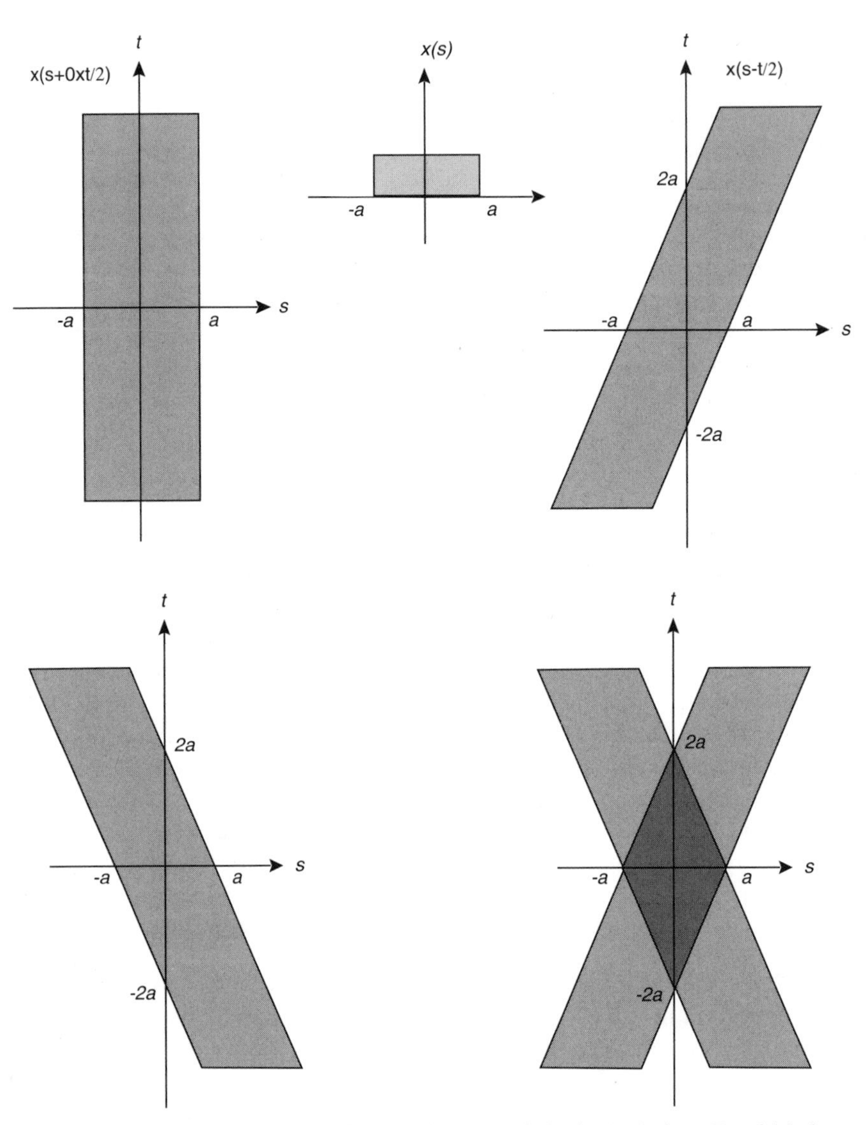

Fig. 10.12. Illustrating the support of $x(s + t/2)x^*(s - t/2)$ in the (s, t) plane. Panel (a) shows the support of $x(s)$. We move to two dimensions in Panel (b), but show the support of the degenerate function $x(s + 0 \times (t/2))$. The support of $x(s - t/2)$ is the parallelpiped region (c) and the product $x(s + t/2)x^*(s - t/2)$ has support mainly within the diamond (d).

Remark. We observe that the integral (10.96) is well-defined. It is the Fourier transform of $x(\mu + t/2)x^*(\mu - t/2)$, which must be absolutely integrable. This follows because both factors are in $L^2(\mathbb{R})$, and the Schwarz inequality ensures that $\|x(\mu + t/2)\|_2 \, \|x^*(\mu - t/2)\|_2 \geq \|x(\mu + t/2)x^*(\mu - t/2)\|_1$. Of course, $L^1(\mathbb{R})$ signals have Fourier transforms.

No window function appears in the definition (10.96), but there is an easy way to understand how folding the source signal $x(t)$ over on itself accomplishes the required time localization. Imagining that a signal's support lies mainly within the interval $[-a, a]$, the local region implied by the WVD transform is a diamond, as shown in Figure 10.12.

10.4.2.2 Properties. The properties of this transform are quite remarkable. To begin with, we can show that the Wigner–Ville distribution is *real-valued.*

Proposition (Real-Valued). Let $x(t) \in L^2(\mathbb{R})$ and $X_{WV}(\mu, \omega)$ be its WVD. Then $X_{WV}(\mu, \omega) \in \mathbb{R}$.

Proof: We calculate the complex conjugate of (10.96) and make the change of integration variable $s = -t$.

$$\overline{X_{WV}(\mu, \omega)} = \int_{-\infty}^{\infty} \overline{x\left(\mu + \frac{t}{2}\right)x\left(\mu - \frac{t}{2}\right)} e^{j\omega t} dt$$

$$= -\int_{\infty}^{-\infty} \overline{x\left(\mu - \frac{s}{2}\right)x\left(\mu + \frac{s}{2}\right)} e^{-j\omega s} ds = X_{WV}(\mu, \omega). \qquad (10.97)$$

Because $X^*_{WV}(\mu, \omega) = X_{WV}(\mu, \omega)$, it must be real. ■

Already we see that the WVD's properties are quite unlike those of the Fourier transform or its time-limited versions. However, the time- and frequency-domain shifting properties are familiar, as the next proposition shows. Symmetry properties are covered in the exercises.

Proposition (Time and Frequency Shift). Let $x(t) \in L^2(\mathbb{R})$ and $X_{WV}(\mu, \omega)$ be its WVD. Then

(a) If $s(t) = x(t - a)$, then $S_{WV}(\mu, \omega) = X_{WV}(\mu - a, \omega)$.
(b) If $y(t) = e^{j\theta t}x(t)$, then $Y_{WV}(\mu, \omega) = X_{WV}(\mu, \omega - \theta)$.

Proof: Exercise. ■

The double product of $x(t)$ terms in the WVD integral, which is the "quadratic" factor, spoils the transform's linearity. This is easy to see for the scaling property of the linearity. We let $y(t) = ax(t)$, where a is a constant. Then $Y_{WV}(\mu, \omega) = |a|^2 X_{WV}(\mu, \omega)$.

Superposition also fails for the WVD. We use an auxiliary transform as part of the argument.

Definition (Cross Wigner–Ville Distribution). If $x(t), y(t) \in L^2(\mathbb{R})$ are analog signals, then the *Cross Wigner–Ville Distribution*, written $X_{WV,y}(\mu, \omega)$, is the radial Fourier transform of the product $x(\mu + t/2)y^*(\mu - t/2)$:

$$X_{WV,y}(\mu, \omega) = \int_{-\infty}^{\infty} x\left(\mu + \frac{t}{2}\right)\overline{y\left(\mu - \frac{t}{2}\right)}e^{-j\omega t}\,dt. \tag{10.98}$$

One can easily show that $X_{WV,y}(\mu, \omega)$ is the complex conjugate of $Y_{WV,x}(\mu, \omega)$. If we set $s(t) = x(t) + y(t)$, then $S_{WV}(\mu, \omega) = X_{WV}(\mu, \omega) + Y_{WV}(\mu, \omega) + 2\mathrm{Real}[X_{WV,y}(\mu, \omega)]$.

Thus, both component properties of linearity fail for the WVD. The failure of superposition is the more serious deficiency. This defect causes artifacts, called *cross-terms*, in the WVD transform coefficients. The presence of cross-terms leads to difficulties of automatic interpretation, and removing them by various alternative transformations has been a major research goal of the last several years.

The next theorem reveals an interesting symmetry between time and frequency domain representations for the WVD. Besides having an eerie similarity to the inverse Fourier relation, it is also useful in calculations involving the WVD.

Theorem (Frequency-Domain Representation). Let $x(t) \in L^2(\mathbb{R})$, let $X(\omega)$ be its Fourier transform, and $X_{WV}(\mu, \omega)$ be its WVD. Then

$$X_{WV}(\mu, \omega) = \frac{1}{2\pi}\int_{-\infty}^{\infty} X\left(\omega + \frac{\theta}{2}\right)\overline{X\left(\omega - \frac{\theta}{2}\right)}e^{j\theta\mu}\,d\theta. \tag{10.99}$$

Proof: The key idea is to write $X_{WV}(\mu, \omega)$ as an inner product,

$$X_{WV}(\mu, \omega) = \left\langle x\left(\mu + \frac{t}{2}\right)e^{\frac{-j\omega t}{2}}, \quad x\left(\mu - \frac{t}{2}\right)e^{\frac{j\omega t}{2}}\right\rangle, \tag{10.100}$$

splitting the exponential between its two terms. We can then apply Parseval's theorem:

$$\begin{aligned} X_{WV}(\mu, \omega) &= \frac{1}{2\pi}\left\langle \mathcal{F}x\left(\mu + \frac{t}{2}\right)e^{\frac{-j\omega t}{2}}, \mathcal{F}x\left(\mu - \frac{t}{2}\right)e^{\frac{j\omega t}{2}}\right\rangle \\ &= \frac{1}{2\pi}\langle 2X(\omega + 2\phi)e^{j\mu(\omega + 2\phi)}, 2X(\omega - 2\phi)e^{j\mu(\omega - 2\phi)}\rangle. \end{aligned} \tag{10.101}$$

The final inner product in (10.101) simplifies to the integral

$$X_{WV}(\mu, \omega) = \frac{4}{2\pi} \int_{-\infty}^{\infty} X(\omega + 2\phi)\overline{X(\omega - 2\phi)} e^{4j\phi\mu} d\phi, \quad (10.102)$$

whereupon the substitution $\theta = 4\phi$ gives (10.99). ∎

Corollary (Fourier Transform of WVD). Let $x(t) \in L^2(\mathbb{R})$, let $X(\omega)$ be its Fourier transform, and $X_{WV}(\mu, \omega)$ be its WVD. Then, with ω fixed and viewing $X_{WV}(\mu, \omega)$ as a signal with independent time variable μ, we have

$$\mathcal{F}[X_{WV}(\mu, \omega)](\theta) = \int_{-\infty}^{\infty} X_{WV}(\mu, \omega) e^{-j\mu\theta} d\mu = X\left(\omega + \frac{\theta}{2}\right)\overline{X\left(\omega - \frac{\theta}{2}\right)}. \quad (10.103)$$

Proof: Apply the theorem to the WVD term in the integral (exercise).

Table 10.3 summarizes WVD properties. Some of the table's properties are left as exercises.

10.4.2.3 Examples. Let us look at some WVD calculations on standard example signals. These examples are revealing, because they show how the WVD improves upon the frequency resolving capability of the STFT.

TABLE 10.3. Wigner–Ville Distribution Properties[a]

Signal Expression	WVD or Property
$x(t)$	$X_{WV}(\mu, \omega)$
$ax(t)$	$\lvert a\rvert^2 X_{WV}(\mu, \omega)$
$x(t) + y(t)$	$X_{WV}(\mu, \omega) + Y_{WV}(\mu, \omega) + 2\text{Real}[X_{WV,y}(\mu, \omega)]$
$x(t-a)$	$X_{WV}(\mu - a, \omega)$
$x(t)\exp(j\theta t)$	$X_{WV}(\mu, \omega - \theta)$
$x(t)\exp(j\theta t^2)$	$X_{WV}(\mu, \omega - 2\theta\mu)$
$x(t/a)$, with a > 0	$aX_{WV}(\mu/a, a\omega)$
$X_{WV}(\mu, \omega) = \frac{1}{2\pi} \int_{-\infty}^{\infty} X\left(\omega + \frac{\theta}{2}\right)\overline{X\left(\omega - \frac{\theta}{2}\right)} e^{j\theta\mu} d\theta$	Frequency-domain representation
$\int_{-\infty}^{\infty} X_{WV}(\mu, \omega) e^{-j\mu\theta} d\mu = X\left(\omega + \frac{\theta}{2}\right)\overline{X\left(\omega - \frac{\theta}{2}\right)}$	Fourier transform of WVD

[a]In this table, $x(t)$ and $y(t)$ are square-integrable.

Example (Dirac). Let $x(t) = \delta(t - a)$. Then $X_{WV}(\mu, \omega) = \delta(\mu - a)$. To verify this formula, we utilize the Frequency-Domain Representation Theorem. Then.

$$X_{WV}(\mu, \omega) = \frac{1}{2\pi}\int_{-\infty}^{\infty} e^{-j\left(\omega+\frac{\theta}{2}\right)a} e^{j\left(\omega-\frac{\theta}{2}\right)a} e^{j\theta\mu} d\theta = \frac{1}{2\pi}\int_{-\infty}^{\infty} e^{j(\mu-a)} d\theta = \delta(\mu - a). \tag{10.104}$$

The interesting aspect of this example is that the WVD maps an impulse to an impulse. The time-frequency representation that the WVD provides is just as temporally localized as the original time-domain signal. In the (μ, ω) plane, realm of the WVD, the signal $\delta(\mu - a)$ is a Dirac knife edge, infinitely high and infinitely narrow, parallel to the ω-axis and passing through the point $\mu = a$. This stands in stark contrast to the STFT. The windowed Fourier transformation of the same Dirac impulse $x(t) = \delta(t - a)$ is $X_w(\mu, \omega) = w(a - \mu)e^{-j\omega a}$, an exponential modulated by the transform window $w(t)$ situated over the point $\mu = a$.

Example (Sinusoid). Let $x(t) = e^{jat}$. Then $X_{WV}(\mu, \omega) = (2\pi)^{-1}\delta(\omega - a)$. This can be shown using the WVD properties (Table 10.3) or by direct computation as above.

Example (Gaussian Pulse). Now let $g(t) = g_{\alpha,\sigma}(t)$, the Gaussian of mean α and standard deviation σ. Then

$$G_{WV}(\mu, \omega) = \frac{e^{-(\sigma\omega)^2}}{2\pi^{3/2}\sigma} e^{-\left(\frac{\mu-\alpha}{\sigma}\right)^2}. \tag{10.105}$$

Notice in this example that the WVD of a Gaussian pulse is always positive. The only signals $x(t)$ for which $X_{WV}(\mu, \omega)$ is positive are linear chirps, $\exp(jbt^2 + jat)$, modulated by a Gaussian envelope Ref. [7].

Example (Square Pulse). Let $s(t) = u(t + 1) - u(t - 1)$, the square pulse supported on the interval $[-1, 1]$. Then

$$S_{WV}(\mu, \omega) = \frac{2s(\mu)}{\omega}\sin(2\omega(1 - |\mu|)). \tag{10.106}$$

Thus, although the WVD is real-valued, its values can be negative.

These examples illustrate the trade offs between the windowed Fourier transforms and the WVD. There are still other time-frequency transforms, of course; Ref. 33 compares the frequency resolution efficiencies of several of them.

10.4.2.4 Densities and Marginals. Now we turn to an important feature of the Wigner–Ville distribution, a set of properties that distinguish it from the short-time Fourier transform, namely its density function-like character. What does this mean? In the case of the Fourier transform, we have been content to use fractions of signal energy as an indication that a frequency range is significant within a signal $x(t)$. Thus, the energy of $x(t)$ is E_x:

$$E_x^2 = \int_{-\infty}^{\infty} |x(t)|^2 \, dt = \|x\|_2^2 = \frac{1}{2\pi} \int_{-\infty}^{\infty} |X(\omega)|^2 \, d\omega = \frac{\|X\|_2^2}{2\pi}. \tag{10.107}$$

We may normalize $x(t)$ or $X(\omega)$ so that the have unit energy, or we may elect to use the normalized radial Fourier transform and equate time and frequency-domain energies. The exercises explore the use of the Hertz Fourier transform for quantifying energy distributions in both time and frequency-domains. In any case then, like a probability density function, the fraction of signal energy between ω_0 and ω_1 is given by

$$E_x^2[\omega_0, \omega_1] = \int_{\omega_0}^{\omega_1} |X(\omega)|^2 d\omega, \tag{10.108}$$

where we have normalized so that $\int_{-\infty}^{\infty} |X(\omega)|^2 \, d\omega = 1$.

Now, we are interested in transform representations of signals that have both a time- and a frequency-domain independent variable. Our question is whether such transforms can have joint density function behavior as well. For this to be the case, we should require that the signal transform assumes non-negative values and obey certain marginal integral conditions.

Definition (Marginals). The time-frequency transform $P(\mu, \omega)$ of $x(t) \in L^2(\mathbb{R})$ obeys the *marginal conditions* if

$$P(\mu, \omega) \geq 0, \tag{10.109a}$$

$$\frac{1}{2\pi} \int_{-\infty}^{\infty} P(\mu, \omega) \, d\omega = |x(t)|^2, \tag{10.109b}$$

$$\int_{-\infty}^{\infty} P(\mu, \omega) \, d\mu = |X(\omega)|^2, \tag{10.109c}$$

where $X(\omega)$ is the radial Fourier transform of $x(t)$.

(These conditions are somewhat imperfect, due to the scaling factor in (10.109b). We can escape the scaling by using a Hertz Fourier transform. All we really require is that the marginal integral with respect to one variable be proportional to the signal energy with respect to the other variable; see the exercises.) The idea behind the definition is that $P(\mu, \omega)$ represents a relative amount of the signal per unit time and per unit frequency. Summing the distribution over frequency values should produce a relative amount of signal per unit time. Finally, summing over time should produce a signal strength per unit frequency.

The interpretation of the WVD as a kind of probability density function seems to gain steam from the fact that its values are real; we have already seen from the example of the square pulse, however, that the values can be negative. In contrast, the spectrogram, because it is a squared norm, is always non-negative. However, the WVD does satisfy marginal conditions, which the spectrogram does not.

Theorem (Marginals). Let $x(t) \in L^2(\mathbb{R})$, let $X(\omega)$ be its Fourier transform, and let $X_{WV}(\mu, \omega)$ be its WVD. Then

$$\int_{-\infty}^{\infty} X_{WV}(\mu, \omega)\, d\omega = 2\pi|x(\mu)|^2, \tag{10.110a}$$

$$\int_{-\infty}^{\infty} X_{WV}(\mu, \omega)\, d\mu = |X(\omega)|^2. \tag{10.110b}$$

Proof: We can directly evaluate the integral (10.110a) as follows:

$$\int_{-\infty}^{\infty} X_{WV}(\mu, \omega)\, d\omega = \int_{-\infty}^{\infty}\int_{-\infty}^{\infty} x\left(\mu + \frac{t}{2}\right)\overline{x\left(\mu - \frac{t}{2}\right)} e^{-j\omega t}\, dt d\omega. \tag{10.111}$$

Interchanging the order of integration on the right-hand side of (10.111) gives

$$\begin{aligned}
\int_{-\infty}^{\infty} x\left(\mu + \frac{t}{2}\right)\overline{x\left(\mu - \frac{t}{2}\right)} \int_{-\infty}^{\infty} e^{-j\omega t}\, d\omega dt &= 2\pi \int_{-\infty}^{\infty} x\left(\mu + \frac{t}{2}\right)\overline{x\left(\mu - \frac{t}{2}\right)}\delta(t)\, d\omega dt \\
&= 2\pi x\left(\mu + \frac{t}{2}\right)\overline{x\left(\mu - \frac{t}{2}\right)}\Bigg|_{t=0} \\
&= 2\pi x(\mu)\overline{x(\mu)} = 2\pi|x(\mu)|^2.
\end{aligned} \tag{10.112}$$

We leave the second marginal as an exercise. ■

Thus, we have shown that the WVD obeys a marginal condition akin to that of a joint probability density function. It is possible to show that employing a Hertz formulation of the WVD produces perfect marginal conditions for the transform. Unfortunately, the spectrogram fails the marginals, precisely because of the window function (exercises).

10.4.3 Ambiguity Function

Another quadratic time-frequency signal representation is the ambiguity function. Its formulation is much like the WVD, except that it swaps the time and integration variables in the defining Fourier integral.

Definition (Ambiguity Function). If $x(t) \in L^2(\mathbb{R})$ is an analog signal, then its *ambiguity function*, written $X_{AF}(\mu, \omega)$, is the radial Fourier transform of the product $x(t + \mu/2)x^*(t - \mu/2)$:

$$X_{AF}(\mu, \omega) = \int_{-\infty}^{\infty} x\left(t + \frac{\mu}{2}\right)\overline{x\left(t - \frac{\mu}{2}\right)}e^{-j\omega t}dt. \tag{10.113}$$

The following result relates the ambiguity function to the WVD.

Theorem (Ambiguity Function Characterization). Let $x(t) \in L^2(R)$ be an analog signal, $X_{AF}(\mu, \omega)$ its ambiguity function, and $X_{WV}(\mu, \omega)$ its Wigner–Ville distribution. Then,

$$X_{AF}(\mu, \omega) = \int_{-\infty}^{\infty}\int_{-\infty}^{\infty} X_{WV}(\nu, \theta)e^{-j(\nu\omega + \mu\theta)}d\nu d\theta. \tag{10.114}$$

Proof: Let us evaluate the integral on the right-hand side of (10.114) by splitting up the exponential into two one-dimensional Fourier transforms. Then the corollary to the Frequency-Domain Representation Theorem [(10.103) applies.

$$\int_{-\infty}^{\infty}\int_{-\infty}^{\infty} X_{WV}(\nu, \theta)e^{-j(\nu\omega + \mu\theta)}d\nu d\theta = \int_{-\infty}^{\infty} X\left(\omega + \frac{\theta}{2}\right)\overline{X\left(\omega - \frac{\theta}{2}\right)}e^{-j\mu\theta}d\theta. \tag{10.115}$$

Writing the integral on the right-hand side of (10.115) as an inner product and invoking Parseval's formula, we find

$$\int_{-\infty}^{\infty} X\left(\omega + \frac{\theta}{2}\right)\overline{X\left(\omega - \frac{\theta}{2}\right)}e^{-j\mu\theta}d\theta = \int_{-\infty}^{\infty} x\left(t + \frac{\mu}{2}\right)\overline{x\left(t - \frac{\mu}{2}\right)}e^{-j\omega t}dt. \tag{10.116}$$

But the last integral above is $X_{AF}(\mu, \omega)$. ■

Remark. Notice that the ambiguity function characterization (10.114) shows that $X_{AF}(\mu, \omega)$ is the two-dimensional Fourier transform of $X_{WV}(\mu, \omega)$.

10.4.4 Cross-Term Problems

While the WVD does have several advantages over the spectrogram—among them superior frequency-domain resolution, satisfaction of the marginals, and independence of a windowing function—it does have the misfortune of interference terms.

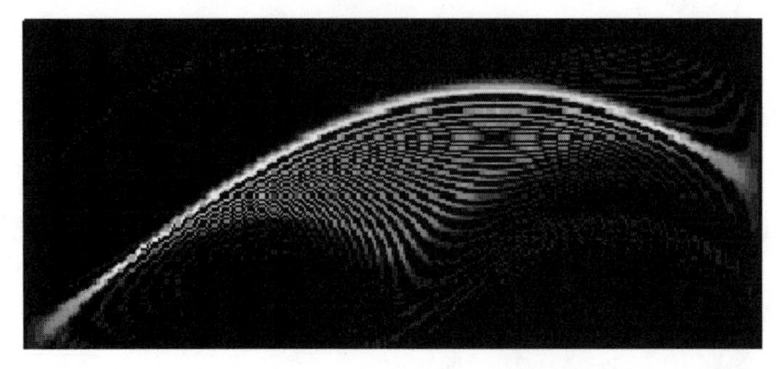

Fig. 10.13. An example of the presence of cross-terms in the WVD. The original signal is a chirp, first rising in frequency and then falling. High-energy coefficients appear beneath the main arc of energy in the transform plane, yet the time-domain signal contains no such tones. Higher-level interpretation routines are beset with the problem of separating such artifacts from genuine features of the signal's time evolution.

The cross-terms represent time-frequency domain energy in locations where it is not present in the original signal (Figure 10.13).

Cross-terms in WVD coefficients are, generally speaking, oscillatory in nature. As such, their effect on applications can be mitigated if not removed by filtering in the (μ, ω) plane. The *quid pro quo* is some loss of frequency resolution [9]. The principal approach to removing these oscillatory components is through frequency domain filtering. Thus, one manipulates the Fourier transform of $X_{WV}(\mu, \omega)$—the ambiguity function $X_{AF}(\mu, \omega)$. Several such methods are compared in Ref. 34. Modification of the ambiguity plane image, a two-dimensional undertaking, is outside our present scope, however.

Although they are in general tolerable, WVD interferences can be extreme in some cases. For example, for each pair of energy concentrations in the time-frequency plane, a possible cross-term region is created. Thus, if there are N significant temporal-spectral components of $x(t)$, then $X_{WV}(\mu, \omega)$ will have $N \times (N-1)$ interference term regions. This combinatorial explosion presents nasty problems for higher-level interpretation algorithms. Furthermore, by the algebraic nature of their origin, cross-term amplitudes can conceivably be double the magnitudes of their source pair of time-frequency energy modes. So how then can a high-level algorithm distinguish meaningful events from meaningless interferences?

10.4.5 Kernel Construction Method

Many of the latest time-frequency signal analysis research efforts have revolved around the problem of finding density-like distributions that obey the marginal conditions, but avoid cross-term effects.

The principal strategy is to introduce a third term into the defining transform integral [35].

Definition (Kernel-Based or Cohen's Class of Transforms). If $x(t) \in L^2(\mathbb{R})$ is an analog signal, then its *Cohen Class Transform* with respect to K, written $X_K(\mu, \omega)$, is

$$X_K(\mu, \omega) = \int_{-\infty}^{\infty}\int_{-\infty}^{\infty}\int_{-\infty}^{\infty} K(\theta, t)e^{j\theta(s-\mu)}x\left(s+\frac{t}{2}\right)\overline{x\left(s-\frac{t}{2}\right)}e^{-j\omega t}dtdsd\theta\,. \quad (10.117)$$

The kernel term can be thought of as smoothing the interferences. It is also possible to show that almost all time-frequency transforms, assuming a suitable choice of the kernel function, belong to the Cohen Class [7, 9].

Why resort to such triple integrals for basic one-dimensional signal representation? The spectrogram does not obey the marginals, so signal theorists sought a solution amongst transforms such as the WVD. Hope for a quadratic transform, however, was dashed by Wigner's theorem [36]. This result states that a quadratic time-frequency transform cannot obey the marginal conditions (10.109a)–(10.109c). Pursuing these very interesting ideas would require quite a bit of multidimensional transform development and take us far afield, however.

We return to the STFT family to answer conclusively the question Gabor posed: Can a critically sampled set of windowed Fourier atoms fully support representation and analysis?

10.5 THE BALIAN–LOW THEOREM

This section concludes the chapter by proving a famous result in time-frequency transform theory: the Balian–Low theorem. The theorem applies to the entire class of time-frequency (or windowed-Fourier) transforms. Balian–Low answers a question posed by discretizing the short-time Fourier transform: Can windowed Fourier atoms of the form

$$w_{m,n}(t) = e^{jn\Omega t}w(t-mT)\,, \quad (10.118)$$

where $T > 0$ and $T\Omega = 2\pi$ serve as a complete signal representation? We desire good time and frequency localization; we stipulate, therefore, that both $w(t)$ and its Fourier transform $W(\omega)$ are window functions. The two practical alternatives are that $\{w_{m,n}(t): m, n \in \mathbb{Z}\}$ constitutes either

- An orthonormal basis or
- A frame.

It is hoped that we can discover $\{w_{m,n}(t): m, n \in \mathbb{Z}\}$ that make up an orthonormal basis. Then every square-integrable signal $x(t)$ has a expansion in terms of Fourier coefficients, easily calcuated as the inner products of $x(t)$ with the $w_{m,n}(t)$:

$$x(t) = \sum_{m,n \in Z} \langle x(t), w_{m,n}(t)\rangle w_{m,n}(t)\,. \quad (10.119)$$

If we fail to find such a basis, then computing the expansion coefficients (10.119) becomes problematic. Lack of a basis encumbers our signal analysis too. While we might be able to decompose a candidate signal $x(t)$ into a linear combination of atoms, $x(t) = \Sigma c_{m,n} w_{m,n}(t)$, we do not necessarily know the uniqueness of the expansion coefficients $c_{m,n}$ for representing $x(t)$. So the utility of the expansion coefficients as indicators of some signal component's presence or the lack thereof is very much compromised.

Should a basis not be available, we could search for a frame representation of $L^2(\mathbb{R})$ signals using the Gabor atoms (10.118). After all, frame coefficients can characterize the source signal $x(t)$, and they support numerically stable reconstructions. This may be a good redoubt.

In either case, orthonormal basis or frame, we can build a structural interpretation of finite-energy signals x(t). The Balian–Low theorem dashes our hopes—both of them. We cover the theorem for the easier-to-prove case of orthonormal bases, first. Then we turn to Balian–Low's rejection of frames. To prove that no such frames exist, we need a special tool, namely the Zak transform. Frames were covered in Section 3.3.5, and the Zak transform was introduced at the end of Chapter 8.

10.5.1 Orthonormal Basis Decomposition

The Balian–Low theorem answers Gabor's original problem of finding well-localized signal representations using time-frequency atoms [25, 37–40]. It is also a negative result, for it shows the impossibility of finding well-localized, orthonormal decompositions based on windowed Fourier atoms when $T\Omega = 2\pi$.

We begin with a lemma. It is simple, but it allows us to reduce the proof of the theorem for all possible samplings $T\Omega = 2\pi$, to the specific case of $T = 1$ and $\Omega = 2\pi$.

Lemma. If $T > 0$, then the map $H(x(t)) = T^{1/2}x(Tt)$ is unitary on $L^2(\mathbb{R})$; that is,

(i) H is onto;

(ii) H preserves inner products, $\langle Hx, Hy\rangle = \langle x, y\rangle$.

Proof: For (i), let $y \in L^2(\mathbb{R})$ and choose $x(t) = T^{-1/2}y(t/T)$. Then $(Hx)(t) = y(t)$. For (ii) we change variables, $s = Tt$, in the inner product integral:

$$\langle Hx, Hy\rangle = \int_{-\infty}^{\infty} (Hx)(t)\overline{(Hy)(t)}\,dt = T\int_{-\infty}^{\infty} x(Tt)\overline{y(Tt)}\,dt = T\int_{-\infty}^{\infty} \frac{x(s)\overline{y(s)}}{T}\,ds = \langle x, y\rangle \tag{10.120}$$

This completes the proof. ■

Now we can prove the result of Balian and Low for the orthonormal basis situation.

Theorem (Balian–Low for Orthonormal Bases). Let $T\Omega = 2\pi$, let $w(t) \in L^2(\mathbb{R})$, and let the collection of windowed Fourier atoms $\{w_{m,n}(t): m, n \in \mathbb{Z}\}$ be given by (10.118). If $\{w_{m,n}(t): m, n \in \mathbb{Z}\}$ is an orthonormal basis for $L^2(\mathbb{R})$, then either

(i) $w(t)$ is not a window function:

$$\int_{-\infty}^{\infty} t^2 |w(t)|^2 dt = \|tw(t)\|_2^2 = \infty, \qquad \text{or} \tag{10.121}$$

(ii) $W(\omega)$ is not a window function:

$$\int_{-\infty}^{\infty} \omega^2 |W(\omega)|^2 d\omega = \|\omega W(\omega)\|_2^2 = \infty. \tag{10.122}$$

Proof: It is sufficient to prove the theorem for the special case $T = 1$ and $\Omega = 2\pi$. We know from the lemma that the scaling map $H(x(t)) = T^{1/2}x(Tt)$ is unitary on $L^2(\mathbb{R})$. If we examine its effect on atoms, we find

$$\begin{aligned} H(w_{m,n}(t)) &= \sqrt{T} w_{m,n}(Tt) = \sqrt{T} e^{jn\Omega Tt} w(Tt - mT) \\ &= e^{2\pi jnt} \sqrt{T} w(T(t-m)) = e^{2\pi jnt} (Hw)(t-m). \end{aligned} \tag{10.123}$$

The map H takes basis elements $w_{m,n}(t)$ with time and frequency sampling intervals T and Ω, respectively, to basis elements with corresponding sampling intervals 1 and 2π. So it suffices to prove the theorem for this special case—that is, for the image of the set $\{w_{m,n}(t): m, n \in \mathbb{Z}\}$ under H.

We note that the derivative $w'(t) \in L^2(\mathbb{R})$ if and only if $W(\omega)$ is a window function. We can check this relationship of differentiation to the window condition on $W(\omega)$ as follows:

$$\|w'(t)\|_2^2 = \frac{1}{2\pi}\|\mathcal{F}w'\|_2^2 = \frac{1}{2\pi}\int_{-\infty}^{\infty} |j\omega W(\omega)|^2 d\omega = \frac{1}{2\pi}\int_{-\infty}^{\infty} \omega^2 |W(\omega)|^2 d\omega. \tag{10.124}$$

Let us assume that both $w(t)$ and $W(\omega)$ are window functions. From (10.124) this is equivalent to assuming that $tw(t)$ and $w'(t)$ are square-integrable. Our goal is to show that this leads to a contradiction.

From the assumption, we can calculate the pair of inner products $\langle tw(t), -jw'(t)\rangle$ and $\langle -jw'(t), tw(t)\rangle$. Since $\{w_{m,n}(t): m, n \in \mathbb{Z}\}$ is an orthonormal basis for $L^2(\mathbb{R})$, we can expand both inner products as follows:

$$\langle tw(t), -jw'(t)\rangle = \sum_{m,n \in \mathbb{Z}} \langle tw(t), w_{m,n}(t)\rangle \langle w_{m,n}(t), -jw'(t)\rangle, \tag{10.125a}$$

and

$$\langle -jw'(t), tw(t)\rangle = \sum_{m,\,n\in Z} \langle -jw'(t), w_{m,\,n}(t)\rangle \langle w_{m,\,n}(t), tw(t)\rangle. \quad (10.125b)$$

Let us attend to the first inner product in the summations of (10.125a). We compute

$$\begin{aligned}\langle tw(t), w_{m,\,n}(t)\rangle &= \int_{-\infty}^{\infty} tw(t)\overline{e^{2\pi jnt}w(t-m)}\,dt\\ &= \int_{-\infty}^{\infty} (s+m)w(s+m)e^{-2\pi jns}\overline{w(s)}\,ds,\end{aligned} \quad (10.126)$$

where $s = t - m$. Since $w_{-m,-n}(t) = \exp(-2\pi jnt)w(t + m)$, we can write the final integral in (10.126) as a sum,

$$\langle tw(t), w_{m,\,n}(t)\rangle = \int_{-\infty}^{\infty} sw(s+m)e^{-2\pi jns}\overline{w(s)}\,ds + m\int_{-\infty}^{\infty} w_{-m,\,-n}(s)\overline{w(s)}\,ds. \quad (10.127)$$

The final term in (10.127) is zero. It is clearly so if $m = 0$, and if $m \neq 0$, then it is $m\langle w_{-m,-n}(t), w(t)\rangle$. But by the orthogonality of the $\{w_{m,n}(t)\}$, of which $w_{0,0}(t) = w(t)$, we get $\langle w_{-m,-n}(t), w(t)\rangle = 0$. Thus,

$$\langle tw(t), w_{m,\,n}(t)\rangle = \int_{-\infty}^{\infty} sw(s+m)e^{-2\pi jns}\overline{w(s)}\,ds = \langle w_{-m,\,-n}(s), sw(s)\rangle. \quad (10.128)$$

Now we can rewrite the inner product (10.125a) like this:

$$\langle tw(t), -jw'(t)\rangle = \sum_{m,\,n\in Z} \langle w_{-m,\,-n}(t), tw(t)\rangle \langle w_{m,\,n}(t), -jw'(t)\rangle. \quad (10.129)$$

Let us now divert our attention to the second inner product expansion (10.125b). We try the same strategy, expand the first inner product in the summation, and use the orthogonality of the $\{w_{m,n}(t)\}$ to simplify.

$$\langle -jw'(t), w_{m,\,n}(t)\rangle = -j\int_{-\infty}^{\infty} w'(t)e^{-2\pi jnt}\overline{w(t-m)}\,dt. \quad (10.130)$$

Integration by parts gives

$$j\langle -jw'(t), w_{m,\,n}(t)\rangle = w(t)\overline{w(t-m)}e^{-2\pi jnt}\Big|_{-\infty}^{\infty} - I_1(m,n) + I_2(m,n), \quad (10.131)$$

where $I_1(m, n)$ and $I_2(m, n)$ are the integrals

$$I_1(m, n) = -\int_{-\infty}^{\infty} (e^{-2\pi jnt} w(t)\overline{w'(t-m)}\, dt) \tag{10.132a}$$

and

$$I_2(m, n) = 2\pi jn \int_{-\infty}^{\infty} (e^{-2\pi jnt} w(t)\overline{w(t-m)}\, dt). \tag{10.132b}$$

Since $w(t) \in L^2(\mathbb{R})$, the first term on the right-hand side of (10.131) must be zero. The second integral (10.132b) is zero also. To see this, note that

$$I_2(m, n) = 2\pi jn \int_{-\infty}^{\infty} w(t)\overline{w(t-m)e^{j2\pi nt}}\, dt = 2\pi n \langle w(t), w_{m,-n}(t)\rangle . \tag{10.133}$$

But the final term in (10.133) is zero; either $n = 0$ or $\langle w(t), w_{m,-n}(t)\rangle = 0$ by orthogonality of the $\{w_{m,n}(t)\}$. Thus,

$$\langle -jw'(t), w_{m,n}(t)\rangle = jI_1(m, n) = j\int_{-\infty}^{\infty} w(t)e^{-2\pi jnt}\overline{w'(t-m)}\, dt . \tag{10.134}$$

Letting $s = t - m$ to change integration variables in (10.134) gives

$$\langle -jw'(t), w_{m,n}(t)\rangle = \int_{-\infty}^{\infty} w(s+m)e^{-2\pi jnt}\overline{[-jw'(s)]}\, ds = \langle w_{-m,-n}(t), -jw'(t)\rangle . \tag{10.135}$$

Thanks to this last result (10.135) we can rewrite the inner product (10.125b) as follows:

$$\langle -jw'(t), tw(t)\rangle = \sum_{m,n\in Z} \langle w_{-m,-n}(t), -jw'(t)\rangle \langle w_{m,n}(t), tw(t)\rangle . \tag{10.136}$$

Reversing the order of summation in (10.136) produces

$$\langle -jw'(t), tw(t)\rangle = \sum_{m,n\in Z} \langle w_{m,n}(t), -jw'(t)\rangle \langle w_{-m,-n}(t), tw(t)\rangle . \tag{10.137}$$

This is nothing else but the summation in (10.129); in other words, we have shown that $tw(t)$ and $-jw'(t)$ commute under the inner product relation:

$$\langle tw(t), -jw'(t)\rangle = \langle -jw'(t), tw(t)\rangle . \tag{10.138}$$

Contradiction looms. Indeed, computing the integral on the left-hand side of (10.138), we can integrate by parts to find

$$\int_{-\infty}^{\infty} tw(t)\overline{[-jw'(t)]}dt = jtw(t)\overline{w(t)}\Big|_{-\infty}^{\infty} - j\int_{-\infty}^{\infty} \overline{w(t)}\{w(t) + tw'(t)\}\, dt. \quad (10.139)$$

Again, $|w(t)|^2 \to 0$ as $t \to \infty$, because w(t) $\in L^2(\mathbb{R})$. This means

$$j\langle tw(t), -jw'(t)\rangle = \int_{-\infty}^{\infty} \overline{w(t)}w(t)\, dt - \int_{-\infty}^{\infty} \overline{tw(t)}w'(t)\, dt = \|w\|_2^2 + \langle -jw'(t), w(t)\rangle. \quad (10.140)$$

Of course, $\|w\| = \|w_{0,0}\| = 1$ by orthogonality of $\{w_{m,n}\}$, and this contradicts (10.138). ■

To illustrate the Balian–Low theorem, we consider two examples of orthonormal bases for $L^2(\mathbb{R})$.

Example (Fourier Basis). For the first example, we try $w(t) = u(t) - u(t-1)$, where $u(t)$ is the unit step signal, as the window function. Then the family $\{w_{m,n}(t) = \exp(2\pi jnt)w(t-m)\colon m, n \in \mathbb{Z}\}$ is an orthonormal basis for $L^2(\mathbb{R})$. In fact, for any fixed $m_0 \in \mathbb{Z}$, we can represent a general signal $x(t)$ restricted to $[m_0, m_0+1]$ by its Fourier series. And the collection of all such Fourier series suffices to construct $x(t)$. Orthogonality follows, of course, from the orthogonality of the exponentials on unit intervals. What does Balian–Low say? Either $w(t)$ or its Fourier transform $W(\omega)$ must not be a window function. Indeed the Fourier transform of $w(t)$ is a sinc function, its Fourier transform decays like ω^{-1}, and $\|\omega W(\omega)\|_2 = \infty$. So although $w(t)$ is well-localized in time, it is poorly localized in frequency.

Example (Shannon Basis). This example takes the other extreme. We now let $w(t) = \mathrm{sinc}(\pi t) = (\pi t)^{-1}\sin(\pi t)$. Then, once again, $\{w_{m,n}(t) = \exp(2\pi jnt)w(t-m)\colon m, n \in \mathbb{Z}\}$ is an orthonormal basis for $L^2(\mathbb{R})$. Although the Fourier transform of this signal is a window function, a square pulse, we now find $tw(t) \notin L^2(\mathbb{R})$.

We might yet hope to find an exact frame representation, having shown now that no orthonomal basis of Gabor elementary functions, or more general windowed Fourier atoms, can exist of the case of $T\Omega = 2\pi$. Orthonormality is, after all, an extremely strict constraint. And we understand that exact frames can support complete and stable signal representation for analysis purposes. Unfortunately, the looser requirement—that the windowed Fourier decomposition set should form an exact frame—is also impossible. The next section covers this more involved proof.

10.5.2 Frame Decomposition

The most elegant and accesible proof of the Balian–Low theorem for frames relies uses the notion of a signal's Zak transform.

10.5.2.1 Zak Transform Preliminaries. Recall from Chapter 8 that the Zak transform maps an analog signal $x(t)$ to a two-dimensional function having independent variables in both time and frequency. In this sense, it resembles the Gabor transform; however, it looks through no window function, and, much like the discrete-time Fourier transform, it is the limit of an infinite sum involving discrete samples of $x(t)$. With parameter $a = 1$, we define

$$(Zx)(s, \omega) = \sum_{k=-\infty}^{\infty} x(s-k)e^{2\pi j\omega k}. \tag{10.141}$$

The Zak transform's properties make it a particularly effective tool for studying frames based on windowed Fourier atoms. Reviews concerning the Zak transform, frames, and windowed Fourier expansions include Refs. 41 and 42.

The Zak transform is a unitary map from $L^2(\mathbb{R})$ to $L^2(S)$, where S is the unit square $[0, 1] \times [0, 1]$. The set of two-dimensional exponentials $\{e_{m,n}(t) = \exp(2\pi jmt)\exp(2\pi jnt): m, n \in \mathbb{Z}\}$ is a basis for $L^2(S)$. Because of this unitary map, the Zak transform converts questions about frames in $L^2(\mathbb{R})$ into questions about frames in $L^2(S)$, where the answers are generally easier to find. The next proposition shows that Zak transforming a Gabor atom is equivalent to a modulation operation.

Proposition (Modulation). Let $w(t) \in L^2(\mathbb{R})$; let $w_{m,n}(t) = \exp(2\pi jnt)w(t-m)$ for $m, n \in \mathbb{Z}$; and let $Z: L^2(\mathbb{R}) \to L^2(S)$ be the Zak transform, where S is the unit square $[0, 1] \times [0, 1]$. Then

$$(Zw_{m,n})(s, \omega) = e^{2\pi jns}e^{-2\pi j\omega m}(Zw)(s, \omega). \tag{10.142}$$

Proof: We compute

$$\begin{aligned}(Zw_{m,n})(s, \omega) &= \sum_{k=-\infty}^{\infty} e^{2\pi j\omega k}e^{2\pi jns}e^{-2\pi jnk}w(s-k-m)\\ &= e^{2\pi jns}\sum_{k=-\infty}^{\infty} e^{2\pi j\omega k}w(s-k-m)\end{aligned} \tag{10.143}$$

and find

$$\begin{aligned}(Zw_{m,n})(s, \omega) &= e^{2\pi jns}e^{-2\pi j\omega m}\sum_{k=-\infty}^{\infty} e^{2\pi j\omega(k+m)}w(s-(k+m))\\ &= (Zw)(s, \omega)e^{2\pi jns}e^{-2\pi j\omega m}\end{aligned} \tag{10.144}$$

as required. ■

Proposition (Norms). If $w(t) \in L^2(\mathbb{R})$; $w_{m,n}(t) = \exp(2\pi jnt)w(t-m)$ for $m, n \in \mathbb{Z}$; and Z: $L^2(\mathbb{R}) \to L^2(S)$ is the Zak transform, where S is the unit square $[0, 1] \times [0, 1]$, then

$$\sum_{m, n \in \mathbb{Z}} \left|\langle x, w_{m,n}\rangle\right|^2 = \|(Zx)(s, \omega)(Zw)(s, \omega)\|^2 . \tag{10.145}$$

Proof: From the transform's unitary property and the previous proposition, the sum in (10.145) expands as follows:

$$\begin{aligned}\sum_{m, n \in \mathbb{Z}} \left|\langle x, w_{m,n}\rangle\right|^2 &= \sum_{m, n \in Z} \left|\langle Zx, Zw_{m,n}\rangle\right|^2 \\ &= \sum_{m, n \in \mathbb{Z}} \left|\int_0^1\int_0^1 e^{-2\pi jns} e^{2\pi j\omega m} (Zx)(s, \omega)\overline{(Zw)(s, \omega)}\, ds d\omega\right|^2 .\end{aligned} \tag{10.146}$$

The two-dimensional exponentials are an orthonormal basis for $L^2(S)$. Hence, the Bessel relation for Hilbert spaces implies that the final sum in is in fact

$$\sum_{m, n \in \mathbb{Z}} \left|\langle x, w_{m,n}\rangle\right|^2 = \int_0^1\int_0^1 |(Zx)(s, \omega)|^2 |(Zw)(s, \omega)|^2\, ds d\omega. \tag{10.147}$$

as desired. ■

The theorem below uses the ideas on Lebesgue measure and integration from Chapter 3.

Theorem (Bounds). Let $w(t) \in L^2(\mathbb{R})$ and suppose $\{w_{m,n}(t) = e^{2\pi jnt}w(t-m)$: $m, n \in \mathbb{Z}\}$ is a frame, with lower and upper bounds A and B, respectively. Then for almost all s and ω we have

$$A \le |(Zw)(s, \omega)|^2 \le B . \tag{10.148}$$

Proof: The frame property implies

$$0 < A\|x\|^2 \le \sum_{m, n \in Z} \left|\langle x, w_{m,n}\rangle\right|^2 \le B\|x\|^2 < \infty \tag{10.149}$$

for any $x(t) \in L^2(\mathbb{R})$. By the previous proposition,

$$0 < A\|x\|^2 \le \int_0^1\int_0^1 |(Zx)(s, \omega)|^2 |(Zw)(s, \omega)|^2\, ds d\omega \le B\|x\|^2 < \infty \tag{10.150}$$

must hold for all finite energy $x(t)$ as well. We argue that this entails

$$0 < A \le \int_0^1 \int_0^1 |(Zw)(s, \omega)|^2 \, ds d\omega \le B < \infty . \tag{10.151}$$

To see this, let us suppose that, for example, $|(Zw)(s, \omega)|^2 < A$ on some subset $R \subseteq S$ with positive measure: $\mu(R) > 0$. Let $v(s, \omega)$ be the characteristic function on S and set $y = Z^{-1}v$; since Z is unitary, $\|y(t)\|^2 = \|v(s, \omega)\|^2 = \mu(R)$. Consequently,

$$\begin{aligned} A\mu(R) &= A\|y\|^2 < \int_0^1 \int_0^1 |(Zy)(s, \omega)|^2 |(Zw)(s, \omega)|^2 \, ds d\omega \\ &= \int_0^1 \int_0^1 |v(s, \omega)|^2 |(Zw)(s, \omega)|^2 ds d\omega. \end{aligned} \tag{10.152}$$

Now, because $v(s, \omega) = 1$ on $R \subseteq S$ and $v(s, \omega) = 0$ otherwise, this last integral becomes

$$\int_0^1 \int_0^1 |v(s, \omega)|^2 |(Zw)(s, \omega)|^2 \, ds d\omega = \iint_R |(Zw)(s, \omega)|^2 \, ds d\omega < \iint_R A \, ds d\omega = A\mu(R). \tag{10.153}$$

Together, (10.152) and (10.153) produce a contradiction. By a similar argument, so does the assumption $B < |(Zw)(s, \omega)|^2$. Showing this last step is left as an exercise, which finishes the proof. ∎

The next two propositions characterize time- and frequency-domain window functions as having differentiable derivatives of their Zak transforms.

Lemma (Window Function). If $x(t) \in L^2(\mathbb{R})$, and $x(t)$ is a window function, then

$$Z(tx(t))(s, \omega) = s(Zx)(s, \omega) + \frac{j}{2\pi}\left[\frac{\partial}{\partial \omega}(Zx)(s, \omega)\right]. \tag{10.154}$$

Proof: Applying the Zak transform (10.141) to $y(t) = tx(t)$, this is straightforward:

$$\begin{aligned} (Zy)(s, \omega) &= \sum_{k=-\infty}^{\infty} (s-k)x(s-k)e^{2\pi j\omega k} \\ &= s\sum_{k=-\infty}^{\infty} x(s-k)e^{2\pi j\omega k} - \sum_{k=-\infty}^{\infty} kx(s-k)e^{2\pi j\omega k}. \end{aligned} \tag{10.155}$$

The first summation on the bottom of (10.155) is $s(\mathcal{Z}x)(s, \omega)$. Partial differentiation of $(\mathcal{Z}x)(s, \omega)$ with respect to ω gives

$$\frac{\partial}{\partial\omega}(\mathcal{Z}x)(s, \omega) = \sum_{k=-\infty}^{\infty} 2\pi jke^{2\pi jk\omega}x(s-k), \tag{10.156}$$

and algebraic manipulation accounts for the second summation in (10.155). ■

Proposition (Window Function Characterization). Let $x(t) \in L^2(\mathbb{R})$ and $\mathcal{Z}$ be the Zak transform $\mathcal{Z}: L^2(\mathbb{R}) \to L^2(S)$, where S is the unit square $[0, 1] \times [0, 1]$. Then $x(t)$ is a window function if and only if

$$\frac{\partial}{\partial\omega}(\mathcal{Z}x)(s, \omega) \in L^2[S]. \tag{10.157}$$

Proof: If $x(t)$ is a window function, then the Window Function Lemma applies and (10.154) holds. Since $s(\mathcal{Z}x)(s, \omega) \in L^2(\mathbb{R})$, necessarily (10.157) holds. Conversely, suppose (10.157). Because both of the final two sums in (10.155) are in $L^2(S)$, so is their sum. Following equalities backwards in (10.155), we thereby find that

$$\sum_{k=-\infty}^{\infty} (s-k)x(s-k)e^{2\pi j\omega k} \in L^2(S). \tag{10.158}$$

But (10.158) is none other than the Zak transform expansion for $tx(t)$. Thus, $x(t)$ is a window function. ■

Lemma (Derivative). If $x(t), x'(t) \in L^2(\mathbb{R})$, then

$$\mathcal{Z}(x'(t))(s, \omega) = \frac{\partial}{\partial s}(\mathcal{Z}x)(s, \omega). \tag{10.159}$$

Proof: By differentiating the Zak transform sum (exercise). ■

Proposition (Derivative Characterization). Let $x(t) \in L^2(\mathbb{R})$, let $X(\omega)$ be its radial Fourier transform, and let $\mathcal{Z}$ be the Zak transform $\mathcal{Z}: L^2(\mathbb{R}) \to L^2(S)$, where S is the unit square $[0, 1] \times [0, 1]$. Then $X(\omega)$ is a window function if and only if

$$\frac{\partial}{\partial s}(\mathcal{Z}x)(s, \omega) \in L^2[S]. \tag{10.160}$$

Proof: The Fourier transform of $x'(t)$ is $j\omega X(\omega)$, so $x'(t) \in L^2(\mathbb{R})$ if and only if $\omega X(\omega) \in L^2(\mathbb{R})$; that is, $X(\omega)$ is a window function. Invoking the Derivative Lemma completes the proof (exercise). ■

The proof of the general Balian–Low theorem for frames uses the above Zak transform properties. The bounds theorem, however, implies a weaker version of the theorem, recapitulated from Ref. 13, where it is attributed to Yves Meyer.

Theorem (Meyer). Let w: $\mathbb{R} \to \mathbb{R}$ be continuous and suppose there are $\varepsilon > 0$ and $C > 0$ such that

$$|w(t)| \le \frac{C}{(1+|t|)^{1+\varepsilon}}. \tag{10.161}$$

Then $\{w_{m,n}(t) = e^{2\pi jnt}w(t-m)\colon m, n \in \mathbb{Z}\}$ cannot be a frame.

Proof: Since $|w(t)|$ is dominated by $C(1 + |t|)^{-1-\varepsilon}$, its Zak transform sum converges. Moreover, it must converge to a continuous function in $L^2(S)$, where S is the unit square $[0, 1] \times [0, 1]$. For the sake of contradiction, now suppose that $\{w_{m,n}(t) = e^{2\pi jnt}w(t-m)\colon m, n \in \mathbb{Z}\}$ constitute an $L^2(\mathbb{R})$ frame with lower and upper bounds A and B, respectively. Since $w(t)$ has a Zak transform, the bounds theorem (10.148) entails $0 < A \le |(Zw)(s, \omega)|$ for almost all $(s, \omega) \in S$. But $(Zw)(s, \omega)$ is continuous, and therefore $|(Zw)(s, \omega)| \ne 0$ for all $(s, \omega) \in S$.

The trick is to define, for each $s \in [0, 1]$, the curve, ζ_s: $[0, 1] \to \mathbb{C}$:

$$\zeta_s(\omega) = \frac{(Zw)(s,\omega)}{(Zw)(0,\omega)}. \tag{10.162}$$

Note that $\zeta_0(\omega) = 1$. Since $(Zw)(s + 1, \omega) = e^{2\pi j\omega}(Zw)(s, \omega)$, it follows as well that $\zeta_1(\omega) = e^{2\pi j\omega}$ for all $\omega \in [0, 1]$. But we cannot continuously map the horizontal line segment defined by $\zeta_0(\omega)$ to the unit circle defined by $\zeta_1(\omega)$ unless at some $r \in (0, 1)$ and some $\omega_0 \in [0, 1]$ we have $\zeta_r(\omega_0) = 0$. But then $(Zw)(r, \omega_0) = 0$, which contradicts the fact that $(Zw)(s, \omega) \ne 0$ for all $(s, \omega) \in S$. Indeed, $\{w_{m,n}(t)\colon m, n \in \mathbb{Z}\}$ cannot be a frame. ■

The next section proves the general Balian–Low theorem. We have already shown the result for orthonormal bases and for frames deriving from continuous window functions with a sufficient decay rate. In the general theorem, the window function assumption is much weaker: the windowing function $w(t)$ of $\{w_{m,n}(t)\colon m, n \in \mathbb{Z}\}$ need only have finite energy.

10.5.2.2 General Balian–Low Theorem. The idea of a frame generalizes the notion of an orthonormal basis. We introduced frames along with the theory of Hilbert spaces of analog signals in Section 3.3.4. Signal analysis using atomic signal models is possible with frames in the sense that such a decomposition:

- Uniquely represents candidate signals;
- Reconstructs a candidate signal in a numerically stable way from its decomposition coefficients.

Frame theory has classic beginnings—Ref. 43–45, for example—and numerous recent texts and papers cover its relationship to mixed-domain signal analysis [27, 42, 46].

We begin with two lemmas on applying the Zak transform to dual frame elements. Recall from basic frame theory (Section 3.3.4.3) that if $F = \{f_n(t): m, n \in \mathbb{Z}\}$ is a frame in a Hilbert space H, then the associated *frame operator* $\mathcal{T}: H \to l^2(\mathbb{Z})$ is defined by

$$\mathcal{T}(x)(n) = \langle x, f_n \rangle . \tag{10.163}$$

Frame operator $\mathcal{T}$ is linear; it is bounded; and, in particular, if B is the upper frame bound, then $\|\mathcal{T}(x)\|^2 \le B\|x\|^2$. Associated to $\mathcal{T}$ is the operator $\mathcal{S}: l^2(\mathbb{Z}) \to H$ defined for $y(n) \in l^2(\mathbb{Z})$ by

$$\mathcal{S}(y) = \sum_{n=-\infty}^{\infty} y(n) f_n . \tag{10.164}$$

In fact, we showed that S is the Hilbert space *adjoint operator* of $\mathcal{T}$: $S = \mathcal{T}^*$. The composition $\mathcal{T}^*\mathcal{T}$ happens to be an invertible map $\mathcal{T}^*\mathcal{T}: L^2(\mathbb{R}) \to L^2(\mathbb{R})$ given by

$$(\mathcal{T}^*\mathcal{T})(x) = \sum_{n=-\infty}^{\infty} \langle x, f_n \rangle f_n . \tag{10.165}$$

We can thus define the *dual frame* to F by applying the inverse of $\mathcal{T}^*\mathcal{T}$ to frame elements:

$$\tilde{F} = \left\{ (\mathcal{T}^*\mathcal{T})^{-1}(f_n) \right\}_{n \in Z} . \tag{10.166}$$

The dual frame idea is key in signal analysis applications. If the dual frame elements are given by $\tilde{f}_n = (\mathcal{T}^*\mathcal{T})^{-1} f_n$, then we have a reconstruction formula for $x(t)$ from both frame and dual frame elements:

$$x(t) = \sum_{n=-\infty}^{\infty} \langle x(t), f_n(t) \rangle \tilde{f}_n(t) \tag{10.167a}$$

and

$$x(t) = \sum_{n=-\infty}^{\infty} \langle x(t), \tilde{f}_n(t) \rangle f_n(t) . \tag{10.167b}$$

If $x(t), y(t) \in H$, then these formulas imply

$$\langle x(t), y(t) \rangle = \sum_{n=-\infty}^{\infty} \langle x(t), f_n(t) \rangle \langle \tilde{f}_n(t), y(t) \rangle \tag{10.168a}$$

and

$$\langle x(t), y(t)\rangle = \sum_{n=-\infty}^{\infty} \langle x(t), \tilde{f}_n(t)\rangle \langle f_n(t), y(t)\rangle . \tag{10.168b}$$

Our theory now combines frame and Zak transform concepts. We are also exploiting several different Hilbert spaces: $L^2(\mathbb{R})$, $L^2([0, 1] \times [0, 1])$, and $l^2(\mathbb{Z})$. Note that although we have formulated these properties for frame elements indexed by a single integral variable, we can specify a one-to-one correspondence between integers and their pairs, $k \leftrightarrow (m, n)$. Our families of windowed Fourier atoms $\{w_{m,n}(t): m, n \in \mathbb{Z}\}$ are doubly indexed, and we thus rewrite (10.163) through (10.166) accordingly. The next lemma shows the relationship between the Zak transform and the frame operator.

Lemma. Let $w(t) \in L^2(\mathbb{R})$; let $F = \{w_{m,n}(t) = \exp(2\pi jnt)w(t - m): m, n \in \mathbb{Z}\}$ be a frame; let $\mathcal{T}$ be the associated frame operator; and let $\mathcal{Z}: L^2(\mathbb{R}) \to L^2(S)$ be the Zak transform, where S is the unit square $[0, 1] \times [0, 1]$. If $x \in L^2(\mathbb{R})$ and $\mathcal{Z}x = y \in L^2(S)$, then $[\mathcal{Z}(\mathcal{T}^*\mathcal{T})\mathcal{Z}^{-1}]y = |\mathcal{Z}w|^2 y$.

Proof: We have $[\mathcal{Z}(\mathcal{T}^*\mathcal{T})\mathcal{Z}^{-1}]y = [\mathcal{Z}(\mathcal{T}^*\mathcal{T})]x$. Since $(\mathcal{T}^*\mathcal{T})x = \Sigma\langle x, w_{m,n}\rangle w_{m,n}$, we have

$$[\mathcal{Z}(\mathcal{T}^*\mathcal{T})\mathcal{Z}^{-1}]y = \sum_{m,n\in Z} \langle x, w_{m,n}\rangle \mathcal{Z}w_{m,n} = \sum_{m,n\in Z} \langle \mathcal{Z}x, \mathcal{Z}w_{m,n}\rangle \mathcal{Z}w_{m,n}. \tag{10.169}$$

But

$$\sum\langle \mathcal{Z}x, \mathcal{Z}w_{m,n}\rangle \mathcal{Z}w_{m,n} = \sum\langle y, e^{2\pi jns}e^{-2\pi j\omega m}\mathcal{Z}w\rangle e^{2\pi jns}e^{-2\pi j\omega m}\mathcal{Z}w, \tag{10.170}$$

by the Modulation Proposition in the previous section. Manipulating the inner product on the right-hand side of (10.170) gives

$$\sum\langle \mathcal{Z}x, \mathcal{Z}w_{m,n}\rangle \mathcal{Z}w_{m,n} = \sum\langle y\overline{\mathcal{Z}w}\mathcal{Z}w, e^{2\pi jns}e^{-2\pi j\omega m}\rangle e^{2\pi jns}e^{-2\pi j\omega m}. \tag{10.171}$$

Now, $\overline{\mathcal{Z}w}\mathcal{Z}w = |\mathcal{Z}w|^2$, and since the two-dimensional exponentials $e^{2\pi jns}e^{-2\pi j\omega m}$ are an orthonormal basis for $L^2(S)$, the last expression is precisely $|\mathcal{Z}w|^2 y$. ■

Lemma (Dual Frame). Let $w(t) \in L^2(\mathbb{R})$, let $F = \{w_{m,n}(t) = e^{2\pi jnt}w(t - m): m, n \in \mathbb{Z}\}$ be a frame, and let $\mathcal{T}$ be the frame operator on F. Further, let

$$\tilde{w}_{m,n} = (\mathcal{T}^*\mathcal{T})^{-1} w_{m,n} \tag{10.172}$$

be the dual frame elements for F and $\mathcal{Z}: L^2(\mathbb{R}) \to L^2(S)$ be the Zak transform, where S is the unit square $[0, 1] \times [0, 1]$. Then

$$(Z\tilde{w}_{m,n})(s,\omega) = \frac{(Zw_{m,n})(s,\omega)}{|(Zw)(s,\omega)|^2} = \frac{e^{2\pi jns}e^{-2\pi j\omega m}}{\overline{(Zw)(s,\omega)}}. \tag{10.173}$$

Proof: By the previous lemma, if $Zw_{m,n} = y \in L^2(S)$, then $[Z(\mathcal{T}^*\mathcal{T})Z^{-1}]y = |Zw|^2 y$. By definition of the dual frame, $\tilde{w}_{m,n} = (\mathcal{T}^*\mathcal{T})^{-1}w_{m,n}$. Fiddling with operators shows

$$Z\tilde{w}_{m,n} = Z(\mathcal{T}^*\mathcal{T})^{-1}w_{m,n} = Z(\mathcal{T}^*\mathcal{T})^{-1}Z^{-1}y = |Zw|^2 y = |Zw|^2 Zw_{m,n}. \tag{10.174}$$

The Bounds Theorem justifies division by $|Zw|^2$ in (10.174), and (10.173) follows from the Modulation Proposition. ■

These tools allow us to extend the Balian–Low theorem to frames.

Theorem (Balian–Low for Frames). Let $T\Omega = 2\pi$, $w(t) \in L^2(\mathbb{R})$, and let $W = \mathcal{F}w$ be the Fourier transform of w. If the collection of windowed Fourier atoms $F = \{w_{m,n}(t) = e^{\Omega jnt}w(t - mT): m, n \in \mathbb{Z}\}$ is a frame, then—once again—either

(i) $w(t)$ is not a window function:

$$\int_{-\infty}^{\infty} t^2|w(t)|^2\,dt = \|tw(t)\|_2^2 = \infty, \quad \text{or} \tag{10.175}$$

(ii) $W(\omega)$ is not a window function:

$$\int_{-\infty}^{\infty} \omega^2|W(\omega)|^2\,d\omega = \|\omega W(\omega)\|_2^2 = \infty. \tag{10.176}$$

Proof: By a scaling argument, such as we used in proving the theorem for orthonormal bases in Section 10.6.1, it suffices to prove the theorem for $T = 1$ and $\Omega = 2\pi$. Let us suppose that both $tw(t)$ and $\omega W(\omega)$ are square-integrable and seek a contradiction.

Let $Z : L^2(\mathbb{R}) \to L^2(S)$ be the Zak transform, where $S = [0, 1] \times [0, 1]$, and $\tilde{w} = (\mathcal{T}^*\mathcal{T})^{-1}w$, where $\mathcal{T}$ is the frame operator for F. We claim that

$$t\tilde{w}(t) \in L^2(\mathbb{R}) \tag{10.177a}$$

and

$$\frac{d}{dt}\tilde{w}(t) \in L^2(\mathbb{R}). \tag{10.177b}$$

Indeed, from the assumption that $tw(t) \in L^2(\mathbb{R})$, the Window Function Characterization Proposition (10.157) implies $\frac{\partial}{\partial\omega}(Zw)(s,\omega) \in L^2[S]$. Likewise, supposing $\omega W(\omega) \in L^2(\mathbb{R})$ gives $\frac{\partial}{\partial s}(Zw)(s,\omega) \in L^2[S]$ via the Derivative Characterization Proposition. But extracting derivatives is straightforward using the Dual Frame Lemma (10.173) with $m = n = 0$. Thus, $w_{m,n}(t) = w(t)$, and we calculate

$$\frac{\partial}{\partial s}(Z\tilde{w})(s,\omega) = \frac{\partial}{\partial s}\frac{(Zw)(s,\omega)}{|(Zw)(s,\omega)|^2} = \frac{\partial}{\partial s}\frac{1}{\overline{(Zw)(s,\omega)}} = \frac{-\frac{\partial}{\partial s}\overline{(Zw)(s,\omega)}}{\overline{(Zw)(s,\omega)}^2} \tag{10.178a}$$

and

$$\frac{\partial}{\partial\omega}(Z\tilde{w})(s,\omega) = \frac{\partial}{\partial\omega}\frac{(Zw)(s,\omega)}{|(Zw)(s,\omega)|^2} = \frac{\partial}{\partial\omega}\frac{1}{\overline{(Zw)(s,\omega)}} = \frac{-\frac{\partial}{\partial s}\overline{(Zw)(s,\omega)}}{\overline{(Zw)(s,\omega)}^2}. \tag{10.178b}$$

Denominators are nonzero almost everywhere in (10.178a) and (10.178b) by the Bounds Theorem on the Zak transform. That is, the expressions on the right-hand sides of (10.178a) and (10.178b) are in $L^2(S)$. Hence the partial derivatives $\frac{\partial}{\partial s}(Z\tilde{w})(s,\omega)$ and $\frac{\partial}{\partial\omega}(Z\tilde{w})(s,\omega)$ are in $L^2(S)$. Hence $t\tilde{w}(t) \in L^2(\mathbb{R})$ and $\tilde{w}'(t) \in L^2(\mathbb{R})$ by the Window Function and Derivative Characterization Propositions, respectively.

Now—working toward a contradiction along the same lines we used for the case of orthonormal bases—we claim that

$$\langle \tilde{w}, w_{m,n}\rangle = \langle w, \tilde{w}_{m,n}\rangle = \begin{cases} 1 & \text{if } m = n = 0 \\ 0 & \text{otherwise.} \end{cases} \tag{10.179}$$

To justify the claim, we apply the unitary Zak transform to the inner products of (10.179) and use (10.173) once more:

$$\begin{aligned}\langle w, \tilde{w}_{m,n}\rangle &= \langle Zw, Z\tilde{w}_{m,n}\rangle = \int_0^1\int_0^1 (Zw)(s,\omega)\overline{(Z\tilde{w}_{m,n})(s,\omega)}\, ds\, d\omega \\ &= \int_0^1\int_0^1 (Zw)(s,\omega)\frac{e^{-2\pi jns}e^{2\pi j\omega m}}{(Zw)(s,\omega)}\, ds\, d\omega.\end{aligned} \tag{10.180}$$

Consequently,

$$\langle w, \tilde{w}_{m,n}\rangle = \int_0^1\int_0^1 e^{-2\pi jns}e^{2\pi j\omega m}\, ds\, d\omega, \tag{10.181}$$

which will be unity when $m = n = 0$ and zero otherwise by the orthogonality of the dual exponentials on the unit square. Showing the same result for $\langle \tilde{w}, w_{m,n} \rangle$ is left as an exercise.

Our next claim is that

$$\langle tw(t), \tilde{w}'(t) \rangle = -\langle w'(t), t\tilde{w}(t) \rangle . \tag{10.182}$$

Recalling the reconstruction formula for frames (10.167a) and (10.167b), we write

$$\langle tw(t), \tilde{w}'(t) \rangle = \sum_{m,n=-\infty}^{\infty} \langle tw(t), \tilde{w}_{m,n}(t) \rangle \langle w_{m,n}(t), \tilde{w}'(t) \rangle \tag{10.183}$$

and work on inner products within the sum. We compute the first by expanding the inner product integral and using (10.179) (exercise):

$$\langle tw(t), \tilde{w}_{m,n}(t) \rangle = \langle w_{-m,-n}(t), t\tilde{w}(t) \rangle . \tag{10.184}$$

The second inner product in the sum of (10.183) involves integration by parts, and here the proof has much of the uncertainty principle's flavor.

$$\langle w_{m,n}(t), \tilde{w}'(t) \rangle = \int_{-\infty}^{\infty} e^{2\pi jnt} w(t-m) \overline{\tilde{w}'(t)}\, dt . \tag{10.185}$$

It follows, upon integrating by parts, that

$$\langle w_{m,n}(t), \tilde{w}'(t) \rangle = e^{2\pi jnt} w(t-m) \overline{\tilde{w}(t)} \Big|_{-\infty}^{\infty} - \int_{-\infty}^{\infty} e^{2\pi jnt} w'(t-m) \overline{\tilde{w}(t)}\, dt$$

$$- \int_{-\infty}^{\infty} 2\pi jn e^{2\pi jnt} w(t-m) \overline{\tilde{w}(t)}\, dt. \tag{10.186}$$

The first term on the right-hand side of (10.186) is zero. The inner product in the third term, $2\pi jn \langle w_{m,n}(t), \tilde{w}(t) \rangle$, is zero unless $m = n = 0$, as we proved above, and, thanks to the $2\pi jn$ factor, the entire term is necessarily zero. Changing the integration variable in the remaining term produces

$$\langle w_{m,n}(t), \tilde{w}'(t) \rangle = -\int_{-\infty}^{\infty} e^{2\pi jnt} w'(t-m) \overline{\tilde{w}(t)}\, dt = -\langle w'(t), \tilde{w}_{-m,-n}(t) \rangle . \tag{10.187}$$

Substituting (10.184) and (10.187) into (10.183) and reversing the summation, we discover

$$\langle tw(t), \tilde{w}'(t) \rangle = -\sum_{m,n=-\infty}^{\infty} \langle w'(t), \tilde{w}_{m,n}(t) \rangle \langle w_{m,n}(t), t\tilde{w}(t) \rangle = -\langle w'(t), t\tilde{w}(t) \rangle . \tag{10.188}$$

Is this last equality plausible? We claim that it is not, that it leads to a contradiction, and that, therefore, our assumption that $w(t)$ and $W(\omega)$ are both window functions is false. To verify this final claim, let us work the integration by parts on the inner product on the left-hand side of (10.188):

$$\langle tw(t), \tilde{w}'(t)\rangle = \int_{-\infty}^{\infty} tw(t)\overline{\tilde{w}'(t)}\,dt = -\int_{-\infty}^{\infty} w(t)\overline{\tilde{w}(t)}\,dt - \int_{-\infty}^{\infty} w'(t)\overline{t\tilde{w}(t)}\,dt. \tag{10.189}$$

The integrals on the right-hand side of (10.189) are inner products, so that

$$\langle tw(t), \tilde{w}'(t)\rangle = -\langle w(t), \tilde{w}(t)\rangle - \langle w'(t), t\tilde{w}(t)\rangle. \tag{10.190}$$

But we know $\langle tw(t), \tilde{w}'(t)\rangle = -\langle w'(t), t\tilde{w}(t)\rangle$ by (10.188) and $\langle w(t), \tilde{w}(t)\rangle = 1$ by (10.179). This exposes the contradiction and finishes the proof. ■

Remark. Integration by parts for $x, y \in L^2(\mathbb{R})$, $\int x'y = xy| - \int xy'$, generally presupposes that $x'y, xy' \in L^1(\mathbb{R})$. Using a limit argument, however, we can make only the additional assumptions that the derivatives x' and y' are square-integrable [12, 13]. We can specify $x_n \to x$ and $y_n \to y$, where $\{x_n\}$ and $\{y_n\}$ are Schwarz space elements, for example.

10.5.3 Avoiding the Balian–Low Trap

Let us add a final footnote to the saga of Gabor's problem. It is possible to escape the negative conclusion of the Balian–Low theorem only by giving up on some of its suppositions. One of these suppositions is the exponential term in the windowed Fourier atoms. This is something that probably seems quite natural given all the work we have done with signal transforms with the complex exponential at their heart. The idea is to extract orthonormal bases with good time-frequency localization by using *sines and cosines* instead [12].

10.6 SUMMARY

The Gabor transform is the most accessible mixed-domain transform tool. It is a representative of a broader class of short-time (or windowed) Fourier transforms (STFT). These transforms invoke a tiling of the time-frequency plane with regions of equal size. The Gabor transform tiles have the minimal area. Tilings of the time-frequency plane are a powerful technique for discovering signal structure, and, developing wavelet theory in the next chapter, we will explore the concept further. In point of fact, equally sized tiles can be a difficulty when analyzing signals that contain feature of different extents and transients; the wavelet transform has time-frequency tiles of varying size, and it was first developed as a transient-capable alternative to the Gabor transform.

Thus, there are some practical and theoretical limitations to the Gabor transform and the STFT in general:

- The size of the signal structures that must be analyzed becomes problematic when the time-domain window has already been fixed for the transform analysis.
- Computationally, the method does not support frames based on windowed Fourier atoms unless the time-frequency density is sufficiently dense.

Natural and synthetic signals have time-limited frequency components. One of the important observations from Chapter 9's study of frequency-domain signal analysis is that standard Fourier transform tools sometimes do a poor job of identifying these localized oscillations. Our principal approach was to trim the source signal $x(t)$ with a symmetric, decaying window function. This technique time-limits, or windows, a signal before calculating its spectrum. There are many variants, depending on the window's shape. Windowing furnishes better estimates of a signal's spectrum, because it restricts the signal values to those over which the relevant oscillatory waveform features should appear.

The short-time Fourier transform extends this idea of signal windowing to a full transform. It lays out the frequency content of a signal according to the time that the oscillatory components appear. Discretized, the windowed Fourier transform presents a complete structural description of a signal. In Chapters 4 and 9, such constructions were at best ad hoc. Now we can produce a full graphical representation of signal frequency components and the time of their occurrence. Moreover, a rich mathematical theory supports the application of this structural tool.

Of the many feasible window shapes upon which we can found a time-limited Fourier transformation, the one which uses a Gaussian window is the Gabor transform. It is the most natural of the various STFTs, and we introduced it to lead off the chapter. The Gabor transform's Gaussian window function has optimal time and frequency locality—a result of the classic Heisenberg Uncertainty Principle.

Applications generally use the squared norm of the transformed signal, called the spectrogram. We discovered, moreover, that spectrogram performance is satisfactory for many signal analysis tasks. In particular, it has seen wide and largely successful application in speech recognition. It has been the basis for many applications that need time-limited descriptions of signal spectra. The spectrogram of the signal $x(t)$ relative to the window function $w(t)$ is the squared magnitude of the Fourier transform of the product: $|\mathcal{F}[s(t)]|^2 = |\mathcal{F}[x(t)w(t)]|^2$. This is a nonnegative real value. Applications can therefore compare or threshold spectrogram values in order to decide whether one frequency is more significant than another or whether an individual frequency is significant, respectively. With the spectrogram, of course, the application design may need to search through possible time locations as well as through possible frequency ranges when seeking local spectral components. That is, Fourier applications tend to be one-dimensional, in contrast to short-time Fourier applications, which are inherently two-dimensional in nature.

Virtuous in their locality, short-time Fourier transform methods improve upon the necessarily global Fourier transform, but they are not without their problems. The previous chapter found the spectrogram adequate for many important signal analysis tasks. It can perform poorly, however, when signals contain sharply varying frequencies, such as chirps, transients, or unconstrained frequency modulation. This behavior is mitigated in the short-time Fourier transform, but not completely removed. There is still the problem of selecting a window for the transform operation. And there is the fundamental limitation for discrete methods that the Balian-Low theorem enforces. Since their critically sampled collections cannot be frames, and hence cannot provide stable signal reconstruction, we are led to hope that the signal universe is made up of more than just windowed Fourier atoms.

10.6.1 Historical Notes

A recent history of time-frequency analysis by one of the principal contributors to the discipline, L. Cohen, is Ref. 47. It includes an extensive bibliography.

The original time-frequency signal analysis technique is the Wigner–Ville distribution. E. Wigner proposed it for application to quantum mechanics [28]. J. Ville used it to explicate the notion of instantaneous frequency for communication theory purposes [29]. The WVD does not rely on a separate window function for generating the transform, using instead a bilinear term involving the original signal. This independence from window selection is at once its strength and weakness. The WVD and its more modern variants have been widely studied and are quite powerful; under certain conditions these transforms are optimal detectors for frequency-modulated signals [48]. It has been used as the cornerstone of a complete approach to biological and computer vision, for example [32].

The WVD has come under critical scrutiny because of the problematic cross-terms that the transform produces [7]. Some conferences have witnessed spirited debates over this transform's strengths and weaknesses. There are a variety of approaches for reducing cross-term effects, and a number of researchers were already investigating them in the early 1980s. The main line of attack was given by P. Flandrin in 1984 [49].

The general theory of quadratic kernel-based transforms is due to L. Cohen [35]. He introduced the *Cohen class* of distributions for applications in quantum mechanics—an area that has stimulated many original contributions to time-frequency signal theory. Later, Wigner published the result that quadratic time-frequency representations, such as his namesake distribution, cannot be simultaneously non-negative and obey the Marginal Conditions.

The Gabor transform is the most easily accessible time-frequency transform, and this is due to the analytic tractability of the Gaussian window function. Gabor's 1946 paper studied sets of signal atoms—Gabor elementary functions—with optimal joint resolution in the time and frequency domains. Gabor applied the theory to acoustics [19] and communication theory [1]. Gabor's conjecture—that optimally localized time-frequency atoms of spatial and spectral sampling intervals satisfying $T\Omega = 2\pi$ could be a foundation for signal analysis—was seriously undermined by the Balian–Low theorem. Such windowed Fourier atoms cannot comprise a frame; their signal reconstruction behavior is therefore unstable. Nevertheless, Gabor methods remain the mostly widely applied time-frequency transform.

Researchers in diverse areas—communications theory, speech recognition, seismic signal interpretation, image analysis, and vision research—have had a keen interest in the Gabor transform for many years. One surprise, in fact, issued from investigations into the behavior of neurons in the visual cortex of animals. Research showed that individual neurons respond to certain visual stimuli in ways that resemble the shapes of the real and imaginary parts of GEFs. Chapter 12 outlines these discoveries and provide references to the literature.

We now know from frame theory, largely developed by I. Daubechies and her coworkers [12, 25, 27] that frames of windowed Fourier atoms are possible for sufficiently dense time-frequency samplings. For many years, investigators pondered how to expand a signal with elementary functions based on a particular window function, such as the Gaussian. Sparse samplings preclude windowed Fourier frames, and at the Nyquist density they are only possible given poor time-frequency localization. Therefore, decomposing signals with windowed Fourier atoms was a major problem. The first solution was in fact given many years after Gabor's paper, by M. Bastianns [50, 51] using the Zak transform. Another strategy for finding expansion coefficients relied upon a neural network for their approximation [52]. Only recently have efficient algorithms for calculating the decomposition coefficients been disclosed [53].

The correct proof of the Balian–Low theorem eluded researchers for a number of years. The result was given independently by Balian [37] and Low [38]. Their proofs both contained the same technical gap, which was corrected several years later for the specific case of orthonormal bases [39] and later extended to frames [25, 40]. It is a hard-won result. Further research in this area produced a workaround for the Balian–Low theorem: Use sinusoids instead of exponentials for the atomic decomposition! Some examples of this approach are [54, 55].

10.6.2 Resources

Readers will find the following resources handy for working with time-frequency transforms:

- The Matlab and Mathematica commercial software packages, which we have used to generate many of the figures.
- The Time-Frequency ToolBox (TFTB), available over the web from CNRS in France; this public-domain software package, based on Matlab, contains a variety of tools for performing STFT, WVD, and other time-frequency signal transforms. We have used it for replicating the WVD analysis of the speech sample "Gabor."
- The small, but very educational, demonstration tool bundled with the treatise [8], the Joint Time-Frequency Analysis (JTFA) package. We have used JTFA to illustrate a number of STFT and WVD concepts in this chapter.

10.6.3 Looking Forward

Time-frequency transforms are problematic in certain applications, especially those with transients or local frequency information that defies any *a priori* demarcation of its frequency- and time-domain boundaries. Quadratic methods have better spectral resolution, but interference terms are sometimes hard to overcome. Finally, the Balian–Low theorem enforces a fundamental limitation on the joint time-frequency resolution capability of the windowed Fourier transforms.

This situation led to the discovery of another mixed-domain signal analysis tool—the wavelet transform, one of the great discoveries of mathematical analysis in the twentieth century. As we have already indicated, the wavelet transform uses a signal scale variable instead of a frequency variable in its tranform relation. This renders it better able to handle transient signal behavior, without completely giving up frequency selectivity. Readers seeking a popular introduction to wavelet theory, a review of the basic equations, and fascinating historical background will find Ref. 56 useful. A more mathematical treatment focusing on applications is Ref. 57. The next chapter introduces wavelets, and the final chapter covers both time-frequency and time-scale applications.

REFERENCES

1. D3. Gabor, Theory of communication, *Journal of the Institute of Electrical Engineers*, vol. 93, pp. 429–457, 1946.
2. F. Hlawatsch and G. F. Boudreaux-Bartels, Linear and quadratic time-frequency signal representations, *IEEE SP Magazine*, pp. 21–67, April 1992.
3. L. Cohen, Introduction: A primer on time-frequency analysis, in *Time-Frequency Signal Analysis*, B. Boashash, ed., Melbourne: Longman Cheshire, pp. 3–42, 1992.
4. S. Qian and D. Chen, Joint time-frequency analysis, *IEEE Signal Processing Magazine*, pp. 52–67, March 1999.
5. H. G. Feichtinger and T. Strohmer, eds., *Gabor Analysis and Algorithms: Theory and Applications*, Boston: Birkhäuser, 1998.
6. B. Boashash, ed., *Time-Frequency Signal Analysis*, Melbourne: Longman Cheshire, 1992.
7. L. Cohen, *Time-Frequency Analysis*, Englewood Cliffs, NJ: Prentice-Hall, 1995.
8. S. Qian and D. Chen, *Joint Time-Frequency Analysis: Methods and Applications*, Englewood Cliffs, NJ: Prentice-Hall, 1996.
9. P. Flandrin, *Time-Frequency/Time-Scale Analysis*, San Diego, CA: Academic Press, 1999.
10. C. K. Chui, *An Introduction to Wavelets*, San Diego, CA: Academic Press, 1992.
11. S. Mallat, *A Wavelet Tour of Signal Processing*, San Diego, CA: Academic Press, 1998.
12. I. Daubechies, *Ten Lectures on Wavelets*, Philadelphia: Society for Industrial and Applied Mathematics, 1992.
13. E. Hernandez and G. Weiss, *A First Course on Wavelets*, Boca Raton, FL: CRC Press, 1996.
14. J. H. Ahlberg, E. N. Nilson, and J. L. Walsh, *The Theory of Splines and Their Applications*, New York: Academic Press, 1967.

15. R. H. Bartels, J. C. Beatty, and B. A. Barsky, *Splines for Use in Computer Graphics*, Los Altos, CA: Morgan Kaufmann, 1987.

16. M. Unser, A. Aldroubi, and M. Eden, B-spline signal processing: Part I—theory, *IEEE Transactions on Signal Processing*, vol. 41, no. 2, pp. 821–833, February 1993.

17. M. Unser, A. Aldroubi, and M. Eden, B-spline signal processing: Part II—efficient design and applications, *IEEE Transactions on Signal Processing*, vol. 41, no. 2, pp. 834–848, February 1993.

18. M. Unser, Splines: A perfect fit for signal and image processing, *IEEE Signal Processing Magazine*, vol. 16, no. 6, pp. 22–38, November 1999.

19. D. Gabor, Acoustical quanta and the theory of hearing, *Nature*, vol. 159, no. 4044, pp. 591–594, May 1947.

20. G. B. Folland, *Fourier Analysis and Its Applications*, Pacific Grove, CA: Wadsworth, 1992.

21. H. Dym and H. P. McKean, *Fourier Series and Integrals*, New York: Academic, 1972.

22. J. J. Benedetto, Uncertainty principle inequalities and spectrum estimation, in *Recent Advances in Fourier Analysis and Its Applications*, J. S. Byrnes and J. F. Byrnes, eds., Dordrecht: Kluwer, pp. 143–182, 1990.

23. G. B. Folland and A. Sitaram, The uncertainty principle: A mathematical survey, *Journal of Fourier Analysis and Applications*, vol. 3, pp. 207–238, 1997.

24. H. Weyl, *The Theory of Groups and Quantum Mechanics*, New York: Dutton, 1931; also, New York: Dover, 1950.

25. I. Daubechies, The wavelet transform, time-frequency localization and signal analysis, *IEEE Transactions on Information Theory*, vol. 36, no. 5, pp. 961–1005, September 1990.

26. M. A. Rieffel, Von Neumann algebras associated with pairs of lattices in Lie groups, *Mathematische Annalen*, vol. 257, pp. 403–418, 1981.

27. I. Daubechies, A. Grossmann, and Y. Meyer, Painless nonorthogonal expansions, *Journal of Mathematical Physics*, vol. 27, no. 5, pp. 1271–1283, May 1986.

28. E. Wigner, On the quantum correction for thermodynamic equilibrium, *Physical Review*, vol. 40, pp. 749–759, 1932.

29. J. Ville, Théorie et applications de la notion de signal analytique, *Cables et Transmission*, vol. 2A, pp. 61–74, 1948.

30. F. Hlawatsch and P. Flandrin, The interference structure of the Wigner distribution and related time-frequency signal representations, in *The Wigner Distribution—Theory and Applications in Signal Processing*, W. F. G. Mecklenbrauker, ed., Amsterdam: Elsevier, 1993.

31. T. A. C. M. Claasen and W. F. G. Mecklenbrauker, The Wigner distribution—a tool for time-frequency signal analysis; part I: Continuous-time signals, *Philips Journal of Research*, vol. 35, no. 3, pp. 217–250, 1980.

32. H. Wechsler, *Computational Vision*, San Diego, CA: Academic Press, 1990.

33. B. Boashash, Time-frequency signal analysis, in S. Haykin, ed., *Advances in Spectrum Estimation*, Prentice-Hall, 1990.

34. F. Hlawatsch, T. G. Manickam, R. L. Urbanke, and W. Jones, Smoothed pseudo-Wigner distribution, Choi–Williams distribution, and cone-kernel representation: ambiguity-domain analysis and experimental comparison, *Signal Processing*, vol. 43, pp. 149–168, 1995.

35. L. Cohen, Generalized phase-space distribution functions, *Journal of Mathematical Physics*, vol. 7. pp. 781–786, 1966.

36. E. Wigner, Quantum-mechanical distribution functions revisited, in *Perspectives in Quantum Theory*, W. Yourgrau and A. van der Merwe, eds., New York: Dover, 1971.

37. R. Balian, Un principe d'incertitude fort en theorie du signal ou en mecanique quantique, *C.R. Acad. Sci. Paris*, vol. 292, series 2, pp. 1357–1362, 1981.

38. F. Low, Complete sets of wave packets, in *A Passion for Physics—Essays in Honor of Geoffrey Chew*, C. DeTar, ed., Singapore: World Scientific, pp. 17–22, 1985.

39. G. Battle, Heisenberg proof of the Balian–Low theorem, *Letters on Mathematical Physics*, vol. 15, pp. 175–177, 1988.

40. I. Daubechies and A. J. E. M. Jannsen, Two theorems on lattice expansions, *IEEE Transactions on Information Theory*, vol. 39, no. 1, pp. 3–6, January 1993.

41. A. J. E. M. Janssen, The Zak transform: A signal transform for sampled time-continuous signals, *Philips Journal of Research*, vol. 43, no. 1, pp. 23–69, 1988.

42. C. E. Heil and D. F. Walnut, Continuous and discrete wavelet transforms, *SIAM Review*, vol. 31, pp. 628–666, December 1989.

43. B. Sz-Nagy, Expansion theorems of Paley–Wiener type, *Duke Mathematical Journal*, vol. 14, pp. 975–978, 1947.

44. R. J. Duffin and A. C. Schaeffer, A class of nonharmonic Fourier series, *Transactions of the American Mathematical Society*, vol. 72, pp. 341–366, 1952.

45. R. M. Young, *An Introduction to Nonharmonic Fourier Series*, New York: Academic Press, 1980.

46. Y. Meyer, *Wavelets and Operators*, Cambridge: Cambridge University Press, 1992.

47. L. Cohen, Time-frequency analysis, in L. Atlas and P. Duhamel, Recent developments in the core of digital signal processing, *IEEE Signal Processing Magazine*, vol. 16, no. 1, pp. 22–28, January 1999.

48. B. Barkat and B. Boashash, Adaptive window in the PWVD for IF estimation of FM signals in additive Gaussian noise, *Proceedings of the ICASSP*, vol. 3, pp. 1317–1320, 1999.

49. P. Flandrin, Some features of time-frequency representations of multi-component signals, *IEEE International Conference on Acoustics, Speech and Signal Processing*, ICASSP-84, San Diego, CA, pp. 41.B.4.1–41B.4.4, 1984.

50. M. Bastiaans, Gabor's expansion of a signal into gaussian elementary signals, *Optical Engineering*, vol. 20, no. 4, pp. 594–598, July 1981.

51. M. Bastiaans, On the sliding-window representation in digital signal processing, *IEEE Transactions on Acoustics, Speech, and Signal Processing*, vol. ASSP-33, pp. 868–873, August 1985.

52. J. Daugman, Complete discrete 2-D Gabor transforms by neural networks for image analysis and compression, *IEEE Transactions on Acoustics, Speech, and Signal Processing*, vol. 36, July 1988.

53. J. Yao, Complete Gabor transformation for signal representation, *IEEE Transactions on Image Processing*, vol. 2, no. 2, pp. 152–159, April 1993.

54. I. Daubechies, A simple Wilson orthonormal basis with exponential decay, *SIAM Journal of Mathematical Analysis*, vol. 22, pp. 554–572, 1991.

55. H. Malvar, Lapped transforms for efficient transform/subband coding, *IEEE Transactions on Acoustics, Speech, and Signal Processing*, vol. 38, pp. 969–978, 1990.

56. B. B. Hubbard, *The World According to Wavelets*, Wellesley, MA: A. K. Peters, 1996.
57. Y. Meyer, *Wavelets: Algorithms and Applications*, Philadelphia: SIAM, 1993.

PROBLEMS

1. Suppose that $\mu, \sigma \in \mathbb{R}$, $\sigma > 0$; and $g(t) = g_{\mu,\sigma}(t)$ is the Gaussian signal with mean μ and standard deviation σ. Find the Gabor transform with respect to $g_{0,4}(t)$ of the following signals:
(a) $x(t) = \exp(6\pi jt)$
(b) $y(t) = \exp(-5\pi jt)$
(c) $z(t) = x(t) + y(t)$
(d) $\exp(j\omega_0 t)$
(e) $\sin(6\pi t)$
(f) $\cos(5\pi t)$

2. Using the notation of the first problem, find the Gabor transform with respect to $g_{0,1}(t)$ of the following signals:
(a) $x(t) = \delta(t)$, the Dirac delta
(b) $y(t) = \delta(t - 5)$
(c) $s(t) = x(t) + y(t)$
(d) $z(t) = \delta(t - r)$, where $r \in \mathbb{R}$

3. Using the notation of the first problem, find the Gabor transform with respect to $g_{0,1}(t)$ of the following signals:
(a) $x(t) = g_{2,4}(t)$
(b) $y(t) = g_{2,4}(t)\exp(6\pi jt)$
(c) $z(t) = g_{-2,4}(t)\cos(6\pi t)$
(d) $s(t) = g_{-2,4}(t)\sin(6\pi t)$

4. Let $x(t) = \exp(j\Omega t^2)$ be a linear chirp signal and $g(t) = g_{0,\sigma}(t)$ be the zero-mean Gaussian with standard deviation $\sigma > 0$.
(a) Find the Gabor transform $X_g(\mu, \omega)$ of $x(t)$.
(b) Show that the frequency at which $|X_g(\mu, \omega)|$ reaches a maximum value for $\mu = T$ is $\Omega_{max} = 2T\Omega$.
(c) Let $\phi(t) = \Omega t^2$, so that $\phi(t)$ is the phase of the signal $x(t)$; show that the instaneous frequency of $x(t)$, $d\phi/dt$ evaluated at $t = T$, is precisely Ω_{max} of part (b).

5. Suppose that $x(t)$ is an analog signal; $\mu, \sigma \in \mathbb{R}$, $\sigma > 0$; and $g(t) = g_{\mu,\sigma}(t)$ is the Gaussian signal with mean μ and standard deviation σ. Show the following:
(a) $g(t) \in L^2(\mathbb{R})$.
(b) If $x(t) \in L^2(\mathbb{R})$, then $x(t)g(t) \in L^2(\mathbb{R})$ and the Gabor transform of $x(t)$, $X_g(\mu, \omega)$ exists.

(c) Show that if $x(t) \in L^2(\mathbb{R})$, then its Gabor transform with respect to the window function $g(t)$ is the inner product of $x(t)$ and the Gabor elementary function $g(t)\exp(j\omega t)$.

(d) Find the norm of $g(t)$ in $L^2(\mathbb{R})$: $\|g(t)\|_2$.

6. With the notation of Problem 1, show the following:

(a) If $x(t) \in L^1(\mathbb{R})$, then $x(t)g(t) \in L^1(\mathbb{R})$ also.

(b) If $x(t) \in L^1(\mathbb{R})$, then $X_g(\mu, \omega)$ exists.

(c) Find an upper bound for $\|X_g(\mu, \omega)\|_1$, the $L^1(\mathbb{R})$ norm of $X_g(\mu, \omega)$.

7. Let $y(t) = g(t)\exp(j\omega_0 t)$, using the notation of Problem 1.

(a) Write the sinusoidal signal, $\sin(\omega_0 t)$, as a sum of exponentials and sketch its spectrum using Dirac delta functions.

(b) Using the Sifting Property of the Dirac delta, calculate $Y(\omega)$.

(c) Sketch the magnitude spectrum $|Y(\omega)|$ and the phase spectrum $\arg[Y(\omega)]$.

8. If $w(t)$ is a window function and $v(t) = w(t + t_0)$. Define the center C_w and radius of ρ_w of $w(t)$ by

$$C_w = \frac{1}{\|w\|_2^2} \int_{-\infty}^{\infty} t|w(t)|^2 dt, \tag{10.191}$$

$$\rho_w = \left[\frac{1}{\|w\|_2^2} \int_{-\infty}^{\infty} (t - C_w)^2 |w(t)|^2 dt\right]^{\frac{1}{2}}. \tag{10.192}$$

(a) Show that $v(t)$ is a window function also.

(b) Show $C_v = C_w - t_0$.

(c) If $W = \mathcal{F}w$ and $V = \mathcal{F}v$ are the Fourier transforms of w and v, respectively, and W and V are window functions, then $C_W = C_V$.

(d) $\rho_v = \rho_w$.

(e) If $x(t) = \exp(-jC_W t)w(t + C_w)$, then $C_x = C_X = 0$ and $\rho_x = \rho_w$.

9. Let $w(t)$ be a window function. Show the following:

(a) $s(t) = (1 + |t|)^{-1} \in L^2(\mathbb{R})$.

(b) $v(t) = (1 + |t|)w(t) \in L^2(\mathbb{R})$.

(c) Use the Schwarz inequality for analog Hilbert spaces (Chapter 3) and the previous two results to show that $w(t) \in L^1(\mathbb{R})$.

(d) Again using the Schwarz inequality, show that $t^{1/2}w(t) \in L^2(\mathbb{R})$.

10. This problem explores how our definitons of center and radius accord with the mean and standard deviation of a Gaussian pulse.

(a) Calculate the center and radius of $g(t) = g_{\mu,\sigma}(t)$, the Gaussian with mean μ and standard deviation σ. For instance, for ρ_x we find

$$\int_{-\infty}^{\infty} t|g(t)|^2 dt = \int_{-\infty}^{\infty} \frac{t}{\sigma^2 2\pi} e^{-\frac{(t-\mu)^2}{\sigma^2}} dt. \tag{10.193}$$

(b) Calculate the center and radius of the other standard window functions: the rectangular, triangular, Hanning, Hamming, and Blackman windows.

(c) Calculate the center and radius of the *B*-spline windows, $\beta_n(t)$.

11. Let $w(t) \in L^2(\mathbb{R})$ and define the signal $x(t)$ by

$$x(t+k) = (-1)^k \overline{w(t-k-1)}, \tag{10.194}$$

where $t \in [0, 1)$ and $k \in \mathbb{Z}$. Show the following:

(a) $x \in L^2(\mathbb{R})$

(b) $\|x\|_2 = \|w\|_2$

(c) $\|x\|_2 \neq 0$

12. Let $x(t) \in L^2(\mathbb{R})$, and $x'(t) \notin L^2(\mathbb{R})$. Show that $\rho_X = \infty$. [*Hint*: Use Parseval's theorem for the radial Fourier transform and the formula for $\mathcal{F}[x'(t)](\omega)$.]

13. Consider the two-dimensional $L^2(\mathbb{R}^2)$ signals $x(s, t)$, or *images*, that satisfy

$$\int_{-\infty}^{\infty} \int_{-\infty}^{-\infty} |x(s,t)|^2 \, ds dt < \infty. \tag{10.195}$$

(a) Show that $L^2(\mathbb{R}^2)$ is a vector space: it is closed under sums and scalar multiplication, and each element has an additive inverse.

(b) What is the zero element of $L^2(\mathbb{R}^2)$? Is it unique? Explain how to rectify this difficulty by establishing equivalence classes of images $[x] = \{y \in L^2(\mathbb{R}^2)$: $x(s, t) = y(s, t)$ for almost all $(s, t) \in \mathbb{R}^2\}$. Define vector addition of equivalence classes by $[x] + [y] = [z]$, where $z = x + y$. Define scalar multiplication analogously. Show that this definition makes sense.

(c) Define a norm on $L^2(\mathbb{R}^2)$ by

$$\left[\int_{-\infty}^{\infty} \int_{-\infty}^{-\infty} |x(s,t)|^2 \, ds dt\right]^{\frac{1}{2}} = \|x\|_{2, L^2(\mathbb{R}^2)}. \tag{10.196}$$

Show that $\|x\|$ in (10.196) is indeed a norm: $\|x\| > 0$, unless $x(t)$ is zero almost everywhere; $\|ax\| = |a| \cdot \|x\|$; and $\|x\| + \|y\| \geq \|x + y\|$.

(d) Show that $L^2(\mathbb{R}^2)$ with norm (10.196) is a Banach space; that is, every Cauchy sequence of finite-energy images converges to a finite energy image.

(e) If $x(s, t)$ and $y(s, t)$ are in $L^2(\mathbb{R}^2)$, then we define their inner product by

$$\langle x, y \rangle_{L^2(\mathbb{R}^2)} = \int_{-\infty}^{\infty} \int_{-\infty}^{-\infty} x(s, t)\overline{y(s, t)}\, ds dt\,. \tag{10.197}$$

Show that $\langle x, y \rangle$ is an inner product space: $\langle x + y, z \rangle = \langle x, y \rangle + \langle x, z \rangle$; $\langle ax, y \rangle = a\langle x, y \rangle$; $\langle x, x \rangle \geq 0$; $\langle x, x \rangle = 0$ if and only if $x(t) = 0$ almost everywhere; and $\langle x, y \rangle = \overline{\langle y, x \rangle}$;

(f) Show that $L^2(\mathbb{R}^2)$ is a Hilbert space.

14. Prove the Parseval theorem for the short-time Fourier transform. Suppose $x(t)$, $y(t) \in L^2(\mathbb{R})$; suppose $w(t)$ is a window function; and let $X_w(\mu, \omega)$ and $Y_w(\mu, \omega)$ be the STFTs of $x(t)$ and $y(t)$, respectively, based on windowing with $w(t)$. Then

$$2\pi \|w\|_2^2 \langle x, y \rangle = \int_{-\infty}^{\infty} \int_{-\infty}^{\infty} X_w(\mu, \omega)\overline{Y_w(\mu, \omega)}\, d\omega d\mu = \langle X_w, Y_w \rangle_{L^2(\mathbb{R}^2)}\,. \tag{10.198}$$

15. Prove the Plancherel formula for the short-time Fourier transform. Let $x(t) \in L^2(\mathbb{R})$, let $w(t)$ be a window function, and let $X_w(\mu, \omega)$ be the STFT of $x(t)$. Then

$$\|x\|_2 = \sqrt{2\pi} \frac{\|X_g(\mu, \omega)\|_{2, L^2(\mathrm{R}^2)}}{\|g\|_2}\,. \tag{10.199}$$

16. Prove the inversion formula for the short-time Fourier transform. Suppose $x(t) \in L^2(\mathbb{R})$, suppose $w(t)$ is a window function, and let $X_w(\mu, \omega)$ be the STFT of $x(t)$. Then for all $a \in \mathbb{R}$, if $x(t)$ is continuous at a, then

$$x(a) = \frac{1}{(2\pi \|w\|_2^2)} \int_{-\infty}^{\infty} X_w(\mu, \omega) w(a) e^{j\omega a}\, d\omega d\mu\,. \tag{10.200}$$

17. Provide an example of a signal $x(t) \in L^2(\mathbb{R})$ that fails to satisfy the special condition we assumed in the first part of the Uncertainty Principle's proof:

$$\lim_{t \to \infty} \sqrt{|t|}\, |x(t)| = 0\,; \tag{10.201}$$

[*Hint*: Define $x(t)$ so that $x(n) = \varepsilon > 0$ on the integers $\mathbb{Z}$.]

18. Restate and prove the Uncertainty Principle for the Hertz Fourier transform:

$$X(\omega) = \int_{-\infty}^{\infty} x(t) e^{-2\pi j \omega t}\, dt\,. \tag{10.202}$$

19. Let us derive a one-dimensional form of Heisenberg's Uncertainty Principle. Following the quantum mechanical viewpoint, we assume that the position and momentum describe the state of an electron, and a probability density function $|\phi(x)|^2$ governs its position. The probability that the particle is on the closed real interval $a \le x \le b$ is

$$\int_a^b |\phi(x)|^2 dx, \tag{10.203}$$

where we must have $\|\phi(x)\|_2 = 1$ so that ϕ is indeed a density. Define a momentum state function $\psi(\omega)$ as follows:

$$\psi(\omega) = \frac{\Phi\left(\frac{\omega}{h}\right)}{\sqrt{2\pi h}}, \tag{10.204}$$

where $\Phi(\omega)$ is the radial Fourier transform of $\phi(x)$, and h is a constant (Planck's).

(a) Show that $\|\psi(\omega)\|_2 = 1$, so that ψ is a density also.

(b) Let $\Delta_\phi = 2\rho_\phi$ and $\Delta_\psi = 2\rho_\psi$ be the diameters of ϕ and ψ, respectively, where ρ is the radius (10.192). Show that $\Delta_\phi \Delta_\psi \ge 2h$.

20. Suppose we are interested in time-frequency localization and thus require a short-time Fourier transform based on a window function $w(t)$ such that $W(\omega)$ is also a window function.

(a) Which of the standard window functions, if any, in Table 10.2 supports this requirement?

(b) Show that a Gaussian works.

(c) Show that any B-spline $\beta_n(t)$ of order $n \ge 1$ works too.

21. This problem explores the idea of changing the window width normalization for the Gabor transform. We defined the Gabor transform for an arbitrary Gaussian window $g_{0,\sigma}(t)$ of zero mean and arbitrary standard deviation $\sigma > 0$:

$$X_g(\mu, \omega) = \frac{1}{\sigma\sqrt{2\pi}} \int_{-\infty}^{\infty} x(t) e^{-\frac{(t-\mu)^2}{2\sigma^2}} e^{-j\omega t} dt. \tag{10.205}$$

(a) Suppose we are Gabor transforming with a window function $g(t)$ with $\|g(t)\|_1 = 1$, where $\|\cdot\|_1$ is the norm in the Banach space of absolutely integrable signals $L^1(\mathbb{R})$. What form do the following Gabor tranform properties take in this case: the inverse theorem, the Plancherel theorem, and the Parseval theorem?

(b) Suppose instead that we have used a Gaussian $g(t)$ with $\|g(t)\|_2 = 1$. Now what form do these same properties take?

22. Try to prove the bounds theorem for windowed Fourier frames without resorting to Zak transform results. Let $w(t) \in L^2(\mathbb{R})$ and its windowed Fourier atoms $\{w_{m,n}(t) = e^{jn\Omega t}w(t - mT)\colon m, n \in \mathbb{Z}\}$, constitute a frame. Show that $A \le \frac{2\pi}{\Omega T}\|w\|_2^2 \le B$, where A and B are the lower and upper frame bounds, respectively.

23. Suppose we select the following window function: $w(t) = (1 + t^2)^{-1}$.

- **(a)** Show that $w(t)$ is bounded (in $L^\infty(\mathbb{R})$) and absolutely integrable (in $L^1(\mathbb{R})$).
- **(b)** Let $T > 0$ be the time-domain sampling interval for discretizing the STFT with respect to $w(t)$. Show that $\Sigma|w(t-kT)|^2$ has an upper and lower bound.
- **(c)** Show that we can find a frequency-domain sampling interval $\Omega > 0$ such that $\{w_{m,n}(t) = e^{jn\Omega t}w(t - mT)\colon m, n \in \mathbb{Z}\}$ are a frame. How small must Ω be?
- **(d)** Repeat the above steps for the Gaussian window $w(t) = g_{\mu,\sigma}(t)$, the Gaussian with mean μ, and standard deviation σ.

24. Let $x(t) \in L^2(\mathbb{R})$ and $X_{WV}(\mu, \omega)$ be its Wigner–Ville distribution. Show the following:

- **(a)** If $s(t) = x(t - a)$, then $S_{WV}(\mu, \omega) = X_{WV}(\mu - a, \omega)$.
- **(b)** If $y(t) = e^{j\theta t}x(t)$, then $Y_{WV}(\mu, \omega) = X_{WV}(\mu, \omega - \theta)$.
- **(c)** If $y(t) = ax(t)$, then $Y_{WV}(\mu, \omega) = |a|^2X_{WV}(\mu, \omega)$.
- **(d)** If $y(t) = x(t/a)$ and $a > 0$, then $Y_{WV}(\mu, \omega) = aX_{WV}(\mu/a, a\omega)$.
- **(e)** If $y(t) = \exp(j\theta t^2)x(t)$, then $Y_{WV}(\mu, \omega) = X_{WV}(\mu, \omega - 2\theta\mu)$.

25. Let $x(t) \in L^2(\mathbb{R})$, $X(\omega)$ be its Fourier transform, and $X_{WV}(\mu, \omega)$ be its WVD. Show the following symmetry properties:

- **(a)** If $x(t)$ is real-valued and $X_{WV}(\mu, \omega) = X_{WV}(\mu, -\omega)$, then $X(\omega)$ is even: $X(\omega) = X(-\omega)$.
- **(b)** If $X_{WV}(-\mu, \omega) = X_{WV}(\mu, \omega)$ and $X(\omega)$ is real-valued, then $x(t)$ is even.

26. Let $x(t) \in L^2(\mathbb{R})$, $w(t)$ be a window function, and let $X_{S,w}(\mu, \omega)$ be the spectrogram of $x(t)$ with respect to $w(t)$. Develop a table of properties for $X_{S,w}(\mu, \omega)$ analogous to Table 10.3.

27. Let $x(t)$ and $y(t)$ be finite energy analog signals and let $X_{WV,y}(\mu, \omega)$ be the cross Wigner–Ville distribution of $x(t)$ with respect to y:

$$X_{WV,y}(\mu, \omega) = \int_{-\infty}^{\infty} x\left(\mu + \frac{t}{2}\right)\overline{y\left(\mu - \frac{t}{2}\right)}e^{-j\omega t}\,dt. \tag{10.206}$$

- **(a)** Show that $[X_{WV,y}(\mu, \omega)]^* = Y_{WV,x}(\mu, \omega)$.
- **(b)** If $s(t) = x(t) + y(t)$, show then that $S_{WV}(\mu, \omega) = X_{WV}(\mu, \omega) + Y_{WV}(\mu, \omega) + 2\text{Real}[X_{WV,y}(\mu, \omega)]$.
- **(c)** What is the relation between the cross Wigner–Ville distribution and the short-time Fourier transform?

28. Suppose $x(t) = e^{jat}$. Show that $X_{WV}(\mu, \omega) = (2\pi)^{-1}\delta(\omega - a)$.

29. Let $g(t) = g_{\alpha,\sigma}(t)$, the Gaussian of mean α and standard deviation $\sigma > 0$. Show that

$$X_{WV}(\mu, \omega) = \frac{e^{-(\sigma\omega)^2}}{2\pi^{3/2}\sigma} e^{-\left(\frac{\mu - \alpha}{\sigma}\right)^2}. \quad (10.207)$$

30. Let $s(t) = u(t + 1) - u(t - 1)$. Show that

$$S_{WV}(\mu, \omega) = \frac{2s(\mu)}{\omega} \sin(2\omega(1 - |\mu|)). \quad (10.208)$$

31. Let $x(t) \in L^2(\mathbb{R})$, let $X(\omega)$ be its Fourier transform, and let $X_{WV}(\mu, \omega)$ be its WVD. As a function of μ, show that $X_{WV}(\mu, \omega)$ has the following Fourier transform:

$$\mathcal{F}[X_{WV}(\mu, \omega)](\theta) = \int_{-\infty}^{\infty} X_{WV}(\mu, \omega) e^{-j\mu\theta} d\mu = X\left(\omega + \frac{\theta}{2}\right)\overline{X\left(\omega - \frac{\theta}{2}\right)}. \quad (10.209)$$

32. Complete the proof of the bounds theorem for windowed Fourier frames. Let $w(t) \in L^2(\mathbb{R})$ and suppose $\{w_{m,n}(t) = e^{2\pi jnt} w(t - m): m, n \in \mathbb{Z}\}$ is a frame, with lower and upper bounds A and B, respectively. Then for almost all s and ω we have $|(\mathcal{Z}w)(s, \omega)|^2 \le B$, where $\mathcal{Z}w$ is the Zak transform (parameter $a = 1$) of w:

$$(\mathcal{Z}x)(s, \omega) = \sum_{k = -\infty}^{\infty} x(s - k) e^{2\pi j\omega k}. \quad (10.210)$$

33. Prove the Zak transform derivative lemma: if $x(t), x'(t) \in L^2(\mathbb{R})$, then

$$\mathcal{Z}(x'(t))(s, \omega) = \frac{\partial}{\partial s}(\mathcal{Z}x)(s, \omega). \quad (10.211)$$

Justify interchanging the summation and differentiation operations when differentiating the Zak transform sum with respect to s.

34. Complete the proof of the Zak transform derivative characterization. Let $x(t) \in L^2(\mathbb{R})$; $X(\omega) = \mathcal{F}[x(t)]$ be its Fourier transform; and $\mathcal{Z}$ be the Zak transform $\mathcal{Z}$: $L^2(\mathbb{R}) \to L^2(S)$, where S is the unit square $[0, 1] \times [0, 1]$. Then the following are equivalent:

(a) $X(\omega)$ is a window function.

(b) $x'(t) \in L^2(\mathbb{R})$.

(c) $\frac{\partial}{\partial s}(\mathcal{Z}x)(s, \omega) \in L^2[S]$.

35. Let $w(t) \in L^2(\mathbb{R})$; $F = \{w_{m,n}(t) = e^{\Omega jnt} w(t - mT): m, n \in \mathbb{Z}\}$ be a frame; let $\mathcal{Z}: L^2(\mathbb{R}) \to L^2(S)$ be the Zak transform, where $S = [0, 1] \times [0, 1]$; and let $\tilde{w} = (\mathcal{T}^*\mathcal{T})^{-1} w = \mathrm{S}^{-1} w$, where $\mathcal{T}$ is the frame operator for F. If $k \in \mathbb{Z}$, show that

(a) Translations by k and the operator $\mathcal{S}$ commute:

$$(\mathcal{S}w)(t-k) = \mathcal{S}(w(t-k))\,. \tag{10.212}$$

(b) Modulations are also commutative under S transformation:

$$e^{2\pi jkt}(\mathcal{S}w(t)) = \mathcal{S}(e^{2\pi jkt}w(t))\,. \tag{10.213}$$

(c) $((\mathcal{T}^*\mathcal{T})^{-1}w)_{m,n} = (\mathcal{T}^*\mathcal{T})^{-1}(w_{m,n})$.

(d) $\langle \tilde{w}, w_{m,n}\rangle = \begin{cases} 1 & \text{if } m = n = 0, \\ 0 & \text{otherwise.} \end{cases}$

(e) Show that

$$\langle tw(t), \tilde{w}_{m,n}(t)\rangle = \langle w_{-m,-n}(t), t\tilde{w}(t)\rangle\,. \tag{10.214}$$

by expanding the inner product integral.

36. Develop an experiment with either real or synthetic data showing that an improperly chosen STFT window width can render the transform information useless for intepreting signal evolution through time.

37. Develop an experiment with either real or synthetic data showing the presence of cross-terms in the WVD of a signal. Consider the analysis of a linear chirp signal. Devise an algorithm to estimate the rate of change in frequency over time. How do the WVD's cross terms affect this algorithm? Suppose that a quadratic chirp is given, and explore the same issues.

38. Obtain or generate signals have significant transient phenomena in addition to localized frequency components. Develop experiments comparing the STFT and the WVD for the purposes of analyzing such signals.

39. Define the following Hertz version of the spectrogram:

$$X_{S,w}(\mu, f) = \left| \int_{-\infty}^{\infty} x(t)w(t-\mu)e^{-2\pi jft}dt \right|^2 . \tag{10.215}$$

(a) Show that

$$\int_{-\infty}^{\infty} X_{S,w}(\mu, f)df = \int_{-\infty}^{\infty} |s(t)w(t-\mu)|^2 dt. \tag{10.216}$$

(b) Also show

$$\int_{-\infty}^{\infty} X_{S,w}(\mu, f)d\mu = \int_{-\infty}^{\infty} |X(u)W(u-f)|^2 du. \tag{10.217}$$

(c) Show that the Hertz spectrogram does not satisfy the either the time or frequency marginal conditions.

(d) Define a Hertz version of the WVD.

(e) Show that the Hertz WVD satisfies the ideal Marginal Conditions.

CHAPTER 11

Time-Scale Signal Transforms

Petroleum seismologists discovered the modern form of the continuous wavelet transform in the mid-1980s. For some time, researchers had been using time-frequency transforms—such as the Gabor transform and its broader family of short-time Fourier transforms—for analyzing signals containing localized frequency components. Speech and seismic waveforms are representative examples. Windowed Fourier analysis becomes problematic, however, when the structure of the signal involves transients of varying scale. Then the short-time tools behave more like the global Fourier transform, and their approximations converge poorly. One idea put forward to improve decomposition convergence was to replace the frequency variable with a scale parameter in the transform relation. The basis functions for this new method were shifted and dilated versions of each other. So they looked like little waves: *wavelets*.

This research caught the eye of mathematicians who found that the new technique held a wealth of special properties. It could be discretized. Wavelets were close kin to theoretical tools used in the study of singular integral operators (mathematical physics), the frame signal decomposition structure (harmonic analysis), quadrature mirror filters (communication theory), and the scale space representation (signal and image analysis). And against the intuition of all theoreticians of the time, there were found orthonormal bases for the L^2 Hilbert space that consisted of smooth, rapidly decaying, similarly shaped elements: *orthonormal wavelets*.

This chapter develops both continuous and discrete scale-based transforms. The topics include the continuous wavelet transform; further development of the idea of frames, which we covered in Chapters 3 and 10; the concept of multiresolution analysis; orthogonal wavelets; discrete wavelet transforms; and, finally, the construction of multiresolution analyses and orthogonal wavelets. Wavelet decomposition furnishes an alternative approach for describing signal structure.

There is a rich research literature on wavelet transforms, including a history of the discipline [1] and many excellent introductory treatments [2–12].

Signal Analysis: Time, Frequency, Scale, and Structure, by Ronald L. Allen and Duncan W. Mills
ISBN: 0-471-23441-9

11.1 SIGNAL SCALE

In a variety of signal analysis applications we have taken note of the problems that arise due to the scale of signal features. The idea is that of the extent of recognizable portions of the signal or the width of regions of interest within the signal may vary over time. The scale of signal features affects the behavior of such signal analysis elements as edge detectors, shape detectors, and local frequency identification algorithms.

From several standpoints we have attempted to interpret signals, and from each of them we had to deal with the issue of scale in a rather informal manner. For instance, in Chapter 4 we pursued time domain techniques for understanding signals. Scale issues affect edge and peak detection, obviously, and even the decision about how wide noise removal filters should be must take into account the size of objects sought within the signal. In Chapter 9 we designed filters to find periodicities within signals. But when such oscillatory components are localized—and so they often are in natural signals—then the extent of the oscillatory phenomenon affects the outcome of the analysis.

The dilation of a function has the same basic shape. For scale-based signal analysis we generally use translations and *dilations* or *scalings* of a basic signal $\psi(t)$:

$$\psi_{a,b}(t) = \frac{1}{\sqrt{a}}\psi\left(\frac{t-b}{a}\right). \tag{11.1}$$

In the next section, we shall show, following Grossmann and Morlet [13], that dilations (variations in parameter a) and translations (variations in b) support a decomposition of a general function $\psi(t)$.

In the previous chapter we covered time-frequency transforms, which combine time and frequency information in the transformed signal. These time-frequency transforms achieve local frequency estimation, but the windowed Fourier transforms suffer from a fixed window size. It turns out, as a consequence, that they do not effectively handle signals with transients and components whose pitch changes rapidly. Making the time-domain window more localized (narrower) makes the frequency-domain window less localized (wider) and vice versa. Time-scale transforms can deal with these last problems; indeed we can mark this insight by petroleum geologists as the grand opening of modern wavelet theory. But time-scale transforms too have deficiencies. One such is the lack of translation-invariance. The final chapter explores some signal analysis applications and examines the tradeoffs between pure time-domain, time-frequency, and time-scale methods.

11.2 CONTINUOUS WAVELET TRANSFORMS

This section presents the continuous wavelet transform. The wavelet representation for one-dimensional signals was developed by Grossmann and Morlet to overcome the deficiencies of the Gabor transform for seismic applications [13]. Wavelets are special functions whose translations and dilations can be used for expansions of

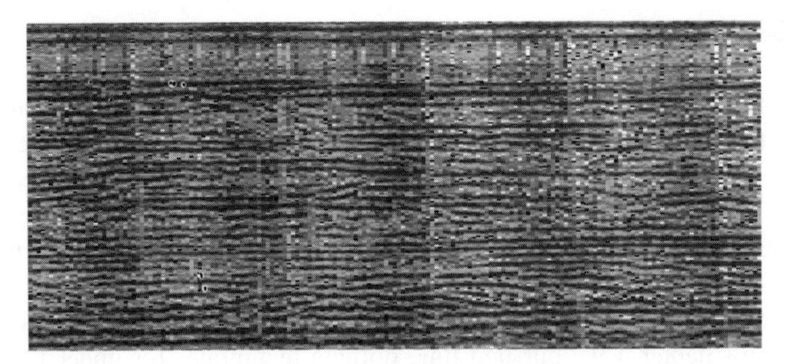

Fig. 11.1. Typical seismic section.

square-integrable functions. In the discussion of the fixed window size implicit in the Gabor representation in Section 2.4, it was noted that the Gabor representation is burdened by the problem of high-magnitude, high-frequency coefficients that is so typical of the Fourier transform.

11.2.1 An Unlikely Discovery

Seismic signals contain many irregular and isolated transients (Figure 11.1). The drawback of the Fourier transform is that it represents signal frequencies as present for all time, when in many situations, and in seismic signal interpretation in particular, the frequencies are localized. The Gabor transform and its more general variant, the short-time Fourier transform (STFT), provide local frequency analysis. One feature of the short-time transforms is that the window size remain fixed. This is acceptable as long as the signal frequency bursts are confined to regions approximating the size of the transform window.

However, in seismic applications, even the STFT becomes problematic. The problem is that seismic signals have many transients, and Grossmann and Morlet found the windowed Fourier algorithms to be numerically unstable. That is, a slight change in the input seismic trace results in a quite pronounced change in the decomposition coefficients. Grossmann and Morlet identified the fixed window size as contributing to the difficulty. Their solution was to keep the same basic filter shape, but to shrink its time-domain extent. That is, they resorted to a transform based on *signal scale*.

11.2.2 Basic Theory

This section introduces the fundamental ideas behind continuous-domain wavelet transforms.

11.2.2.1 Definition and Motivation. Let us begin with a formal definition of a wavelet. The idea is rather recent, and this special signal type passes through the scientific and engineering literature by means of a variety of monikers.

Definition (Analyzing Wavelet). The square-integrable signal $\psi(t)$ is an *analyzing wavelet* if it satisfies the *admissibility condition*

$$C_\psi = \int_{-\infty}^{\infty} \frac{|\Psi(\omega)|^2}{|\omega|} d\omega < \infty, \tag{11.2}$$

where $\Psi(\omega)$ is the radial Fourier transform of $\psi(t)$. The quantity in (11.2) is called the *admissibility factor.* Other names for analyzing wavelets are *basic wavelet, continuous wavelet, admissible wavelet,* and *mother wavelet.* In some analyses it is convenient to normalize the analyzing wavelet,

$$\sqrt{\langle \psi(t), \psi(t) \rangle} = 1, \tag{11.3}$$

but normalization is not a necessary condition for generating useful time-scale transforms or performing the inverse wavelet transform.

The admissibility condition makes possible the inversion relation for the transform. There are some further consequences, however: Wavelets are bandpass filters with a quick frequency cutoff characteristic and have zero mean in the time domain.

The *wavelet transform* is a time-scale transform that uses a scaled and translated version of the analyzing wavelet in a Hilbert space inner product to convert one-dimensional time-varying signals to a two-dimensional scale and translations space:

Definition (Wavelet Transform). Let

$$\psi_{a,b}(t) = \frac{1}{\sqrt{|a|}} \psi\left(\frac{t-b}{a}\right) \tag{11.4}$$

Let $f(t)$ be square-integrable. The wavelet transform of $f(t)$ is defined as the inner product

$$F_\psi(a, b) = \mathcal{W}[f(t)](a, b) = \int_{-\infty}^{\infty} f(t)\overline{\psi_{a,b}(t)} dt \equiv (f(t), \psi_{a,b}(t)). \tag{11.5}$$

The wavelet transform is a mapping from the one-dimensional time domain to a two-dimensional space consisting of a scale a and a translation b (Figure 11.2).

An inverse wavelet transform synthesizes $f(t)$ from the two-dimensional $W[f(t)]$ (a, b):

Definition (Inverse Wavelet Transform). The inverse wavelet transform is the two-dimensional integral,

$$f(t) = \frac{1}{C_\psi} \int_{-\infty}^{\infty} \int_{-\infty}^{\infty} W[f(t)](a, b)\psi(t)\, d\mu, \tag{11.6}$$

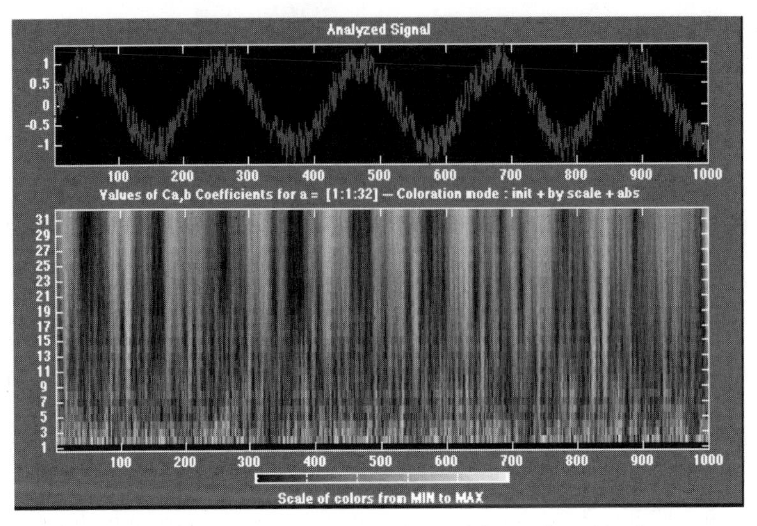

Fig. 11.2. Example of continuous wavelet transform of a noisy sine wave.

where

$$d\mu = \frac{dadb}{a^2} \tag{11.7}$$

and

$$C_\psi = \int_{-\infty}^{\infty} \frac{|\Psi(\omega)|^2}{|\omega|} d\omega. \tag{11.8}$$

Remarks. The definition of a wavelet is fundamentally quite simple, consisting of a *time* criterion (square integrability) and a *frequency* criterion expressed by the admissibility condition. Some references include normalization in the definition of a wavelet, but we emphasize that unit energy is an option, not a necessity. At the present level of development, (11.6) suggests that the admissibility condition (11.2) allows the inverse wavelet transform to be carried out. On cursory inspection the admissibility condition would suggest that the spectrum of an analyzing wavelet should decay rapidly for large $|\omega|$, and since wavelets are defined to be square-integrable, this is automatically fulfilled. On the other hand, the presence of $|\omega|$ in the denominator of (11.8) imposes two further requirements on the time domain behavior of the analyzing wavelet. One is obvious, the other is a bit more subtle, but both are relatively easy to satisfy, as the following discussion demonstrates.

Proposition. Let $\psi(t)$ be an analyzing wavelet as previously defined. Then the admissibility criterion is satisfied if

(i) the analyzing wavelet is of zero mean, that is,

$$\int_{-\infty}^{\infty} \psi(t)\, dt = 0, \tag{11.9}$$

and

(ii)

$$t\psi(t) \in L^1(\mathrm{R}). \tag{11.10}$$

Proof: The first condition is obvious: We require $\Psi(0) = 0$ to ensure that the integrand in (11.2) remains finite at $\omega = 0$. Equation (11.9) simply restates this in terms of the Fourier transform,

$$\lim_{\omega \to 0} \int_{-\infty}^{\infty} \psi(t) e^{-j\omega t} dt = \int_{-\infty}^{\infty} \psi(t)\, dt = 0. \tag{11.11}$$

The significance of the second criterion is best demonstrated by dividing the real line into three segments and examining the integral (11.8),

$$C_\psi = \int_{-\infty}^{-1} \frac{|\Psi(\omega)|^2}{|\omega|}\, d\omega + \int_{-1}^{1} \frac{|\Psi(\omega)|^2}{|\omega|}\, d\omega + \int_{1}^{\infty} \frac{|\Psi(\omega)|^2}{|\omega|}\, d\omega. \tag{11.12}$$

Our primary interest is the integral over the interval $t \in [-1, 1]$. According to the moment theorem developed in Chapter 5, if $t\psi(t) \in L^1(\mathbb{R})$, then the first derivative of $\Psi(\omega)$ exists and is bounded. Designate the maximum value of this derivative in the interval $|\omega| \le 1$:

$$\frac{d}{d\omega}\Psi(\omega) \le M. \tag{11.13}$$

According to the mean value theorem of differential calculus, if a function $g(\omega)$ is bounded and continuous on an interval $[a, b]$, then

$$\int_{a}^{b} g(\omega)\, d\omega \le M(b-a). \tag{11.14}$$

Designating $\frac{d}{d\omega}|\Psi(\omega)| = g(\omega)$, (11.14) implies

$$|\Psi(\omega)| \le M \cdot 2|\omega| \tag{11.15}$$

for $|\omega| \le 1$. This bound is actually tighter than implied by (11.15). Since $\Psi(0) = 0$, the relevant interval is effectively halved so that

$$|\Psi(\omega)| \le M \cdot |\omega|. \tag{11.16}$$

Returning to the admissibility condition, we now have

$$\int_{-1}^{1} \frac{|\Psi(\omega)|^2}{|\omega|}\, d\omega \le \int_{-1}^{1} M^2|\omega|\, d\omega \le M, \tag{11.17}$$

thus bounding one portion of the admissibility condition. The remaining two integrals along their respective semi-infinite intervals are easily handled. Since $\frac{|\Psi(\omega)|^2}{|\omega|} \ge 0$ over $|\omega| \le 1$, it follows that

$$\int_{-\infty}^{-1} \frac{|\Psi(\omega)|^2}{|\omega|}\, d\omega + \int_{1}^{\infty} \frac{|\Psi(\omega)|^2}{|\omega|}\, d\omega \le \int_{-\infty}^{\infty} \frac{|\Psi(\omega)|^2}{|\omega|}\, d\omega < \int_{-\infty}^{\infty} |\Psi(\omega)|^2 d\omega. \tag{11.18}$$

This is bounded by virtue of the L^2 Fourier transform. In summary, the overall proposition is proved by virtue of (11.11), (11.17), and (11.18). ■

Remark. The conditions (11.9) and (11.10) are not difficult to satisfy. The first criterion,

$$\int_{-\infty}^{\infty} \psi(t)\, dt = 0 \tag{11.19}$$

implies that a wavelet must oscillate about the time axis—it puts the wave into a *wave*let. The stipulation $t\psi(t) \in L^1(d\mathbb{R})$ can be met if (for example) $\psi(t) \in L^1(\mathbb{R})$ and *has compact support*. By definition, this would imply that any function $\psi(t) \in L^1(\mathbb{R}) \cap L^2(\mathbb{R})$ with zero mean is a wavelet.

Proposition (Fourier Representation of Wavelet Transform). Let $f(t) \in L^2(\mathbb{R})$. Then for a given scale a, the wavelet transform is proportional to a Fourier transform into the space of translations:

$$W[f(t)](a, b) = \frac{1}{\sqrt{2\pi}} \mathcal{F}[F(\gamma)](-b), \tag{11.20}$$

where

$$F(\gamma) = \sqrt{|a|} \cdot \mathcal{F}[f(t)](\gamma) \cdot \overline{\mathcal{F}[\psi(t)](a\gamma)}. \tag{11.21}$$

By definition and by Parseval's relation, it readily follows that

$$W[f(t)](a, b) = \langle f(t), \psi_{a,b}(t)\rangle = \frac{1}{\sqrt{2\pi}} \langle \mathcal{F}[f(t)](\gamma), \overline{\mathcal{F}[\psi_{a,b}(t)](\gamma)}\rangle. \tag{11.22}$$

However,

$$\mathcal{F}[\psi_{a,b}(t)](\gamma) = e^{j\gamma b} \mathcal{F}[\psi_{a,0}(t)](\gamma) = e^{j\gamma b} \sqrt{|a|}\, \mathcal{F}[\psi(t)](\gamma a) \tag{11.23}$$

so that (11.22) can be expressed in the desired form,

$$W[f(t)](a,b) = \frac{1}{\sqrt{2\pi}} \int_{-\infty}^{\infty} \sqrt{|a|} \cdot \mathcal{F}[f(t)](\gamma) \cdot \mathcal{F}[\psi_{a,b}(t)](\gamma) \cdot e^{-j\gamma b} d\gamma \tag{11.24}$$

and the proposition is proven. ∎

This intermediate result is useful for establishing the more important Parseval relation for the wavelet transform.

Theorem (Wavelet Transform Parseval Relations). Let $f(t) \in L^2(\mathbb{R})$ and $g(t) \in L^2(\mathbb{R})$, and let C_ψ be the admissibility coefficient as previously defined. Then

$$\int_{-\infty}^{\infty} \int_{-\infty}^{\infty} W[f(t)](a,b)\overline{W[g(t)](a,b)}\, d\mu = C_\psi \langle f(t), g(t) \rangle. \tag{11.25}$$

Proof: Let $F(\gamma)$ be d efined as in (11.21) and define

$$G(\gamma) \equiv \sqrt{|a|} \cdot \mathcal{F}[g(t)](\gamma) \cdot \overline{\mathcal{F}[\psi(t)](a\gamma)}. \tag{11.26}$$

Then according to the previous proposition,

$$\int_{-\infty}^{\infty} \int_{-\infty}^{\infty} W[f(t)](a,b)\overline{W[g(t)](a,b)}\, d\mu$$

$$= \int_{-\infty}^{\infty} \int_{-\infty}^{\infty} \frac{1}{\sqrt{2\pi}} \mathcal{F}[F(\gamma)](-b) \overline{\frac{1}{\sqrt{2\pi}} \mathcal{F}[G(\gamma)](-b)}\, d\mu \tag{11.27}$$

Using the Parseval relation to convert the b-space Fourier transforms back to γ space, the above integral takes the form

$$\frac{1}{\sqrt{2\pi}} \int_{-\infty}^{\infty} \int_{-\infty}^{\infty} \mathcal{F}[f(t)](\gamma) \overline{\mathcal{F}[g(t)](\gamma)} \cdot |\mathcal{F}[\psi(t)](a\gamma)|^2 d\gamma\, d\alpha, \tag{11.28}$$

where $d\alpha \equiv \frac{da}{a}$.

The integrals over a and γ can be separated so that (11.28) becomes

$$\frac{1}{\sqrt{2\pi}} \int_{-\infty}^{\infty} |\mathcal{F}[\psi(t)](a\gamma)|^2 d\alpha \int_{-\infty}^{\infty} \mathcal{F}[f(t)](\gamma) \overline{\mathcal{F}[g(t)](\gamma)}\, d\gamma. \tag{11.29}$$

Applying Parseval's relation to the second of these integrals gives a time-domain inner product:

$$\sqrt{2\pi} \langle f(t), g(t) \rangle. \tag{11.30}$$

The substitution of variables $\omega = a\gamma$ into the first integral implies $\frac{da}{a} = \frac{d\omega}{|\omega|}$ so that (11.29) takes the desired form,

$$C_{\psi}\langle f(t), g(t)\rangle, \tag{11.31}$$

completing the proof. ■

Theorem (Inverse Wavelet Transform). Let $f(t)$ be square-integrable. The synthesis problem for the continuous wavelet transform takes the form

$$f(t) = \frac{1}{C_{\psi}}\int_{-\infty}^{\infty}\int_{-\infty}^{\infty} W[f(t)](a, b)\psi_{a, b}(t)\, d\mu. \tag{11.32}$$

Proof: This inversion formula follows directly from the Parseval relation (11.25), which can be written

$$\int_{-\infty}^{\infty}\int_{-\infty}^{\infty} W[f(t)](a, b)\overline{\int_{-\infty}^{\infty} g(t)\overline{\psi_{a, b}(t)}\, dt}d\mu = C_{\psi}\langle f(t), g(t)\rangle. \tag{11.33}$$

This can be rearranged in the more suggestive form:

$$\int_{-\infty}^{\infty}\left[\int_{-\infty}^{\infty}\int_{-\infty}^{\infty} W[f(t)](a, b)\psi_{a, b}(t)\, d\mu\right]\overline{g(t)}\, dt = C_{\psi}\langle f(t), g(t)\rangle. \tag{11.34}$$

Since $g(t)$ is an arbitrary function in $L^2(\mathbb{R})$, (11.34) implies (11.25), and the proposition is proven. ■

Remark. Note that the wavelet $\psi_{a, b}(t)$ is not conjugated when taking the inverse transform (11.25), in contrast to the forward wavelet transform (11.5).

Since $\Psi(\omega) \in L^2(\mathbb{R})$, we must obtain $\Psi(\omega) \to 0$ as $\omega \to \infty$; hence $y(t)$ is a band-pass filter. The details are left as an exercise.

11.2.2.2 Algebraic Properties. As in the case of the Fourier transform, operations such as scaling, translation, and linear combination can be applied to both the analyzing wavelet and the signal waveform. The proofs are straightforward, some are given explicitly in the text, and others are left as exercises. In the following discussion, we assume all signals are square-integrable.

Let us first cover operations on the analyzing wavelet.

Proposition. Let α, β be complex constants and $\psi(t)$, $\phi(t)$ are wavelets. If we define $\theta(t) = \alpha\psi(t) + \beta\phi(t)$, then

$$W_{\theta}[f(t)](a, b) = \bar{\alpha}W_{\psi}[f(t)](a, b) + \bar{\beta}W_{\phi}[f(t)](a, b). \tag{11.35}$$

Proof: Follows trivially from the linearity of the integral (exercise). ■

Proposition (Translation of Analyzing Wavelet). Let γ be a real constant and $\psi(t)$ be a wavelet. If we define $\theta(t) = \psi(t-\gamma)$, then

$$W_\theta[f(t)](a,b) = W_\psi[f(t)](a,b+\gamma a). \tag{11.36}$$

Proof: By definition,

$$W_\theta[f(t)](a,b) = \int_{-\infty}^{\infty} f(t)\frac{1}{\sqrt{|a|}}\overline{\psi\left(\frac{t-b}{a}-\gamma\right)}\,dt = \int_{-\infty}^{\infty} f(t)\frac{1}{\sqrt{|a|}}\overline{\psi\left(\frac{t-(b+\gamma a)}{a}\right)}\,dt, \tag{11.37}$$

which proves the theorem. ■

Proposition (Scaling of Analyzing Wavelet). Let $\eta > 0$ and $\psi(t)$ be a wavelet. If $\theta(t) = \frac{1}{\eta}\psi\left(\frac{t}{\eta}\right)$, then

$$W_\theta[f(t)](a,b) = \frac{1}{\sqrt{\eta}}W_\psi[f(t)](a\eta,b). \tag{11.38}$$

Proof: Exercise. ■

Now let us turn to signal operations and the resulting wavelet transformations.

Proposition (Linearity). Let α, β be complex constants. If we define $\theta(t) = \alpha\psi(t) + \beta\phi(t)$, then

$$W_\theta[\alpha f(t) + \beta g(t)](a,b) = \alpha W_\psi[f(t)](a,b) + \beta W_\phi[f(t)](a,b). \tag{11.39}$$

Proof: The proof is straightforward and left as an exercise. Note the similarity to, and subtle difference between, this case and the similar operation on the analyzing wavelet. ■

Proposition (Translation). Let γ be a real constant. Then

$$W[f(t-\gamma)](a,b) = W[f(t)](a,b-\gamma). \tag{11.40}$$

Proof: Exercise. ■

Proposition (Scaling of Signal). Let $\eta > 0$. Then

$$W\left[\frac{1}{\eta}f\left(\frac{1}{\eta}\right)\right](a,b) = \frac{1}{\sqrt{\eta}}W[f(t)]\left(\frac{a}{\eta},\frac{b}{\eta}\right). \tag{11.41}$$

Proof: By change of variables $\tau = t/\eta$, it follows that

$$W\left[\frac{1}{\eta}f\left(\frac{1}{\eta}\right)\right](a, b) = \int_{-\infty}^{\infty} f(\tau)\frac{1}{\sqrt{|a|}}\overline{\psi\left(\frac{\eta\tau - b}{a}\right)}\,d\tau = \int_{-\infty}^{\infty} f(\tau)\frac{1}{\sqrt{|a|}}\overline{\psi\left(\frac{\tau-(b/\eta)}{(a/\eta)}\right)}\,d\tau. \tag{11.42}$$

Since

$$W[f(t)]\left(\frac{a}{\eta}, \frac{b}{\eta}\right) = \int_{-\infty}^{\infty} f(\tau)\sqrt{\frac{|\eta|}{|a|}}\overline{\psi\left(\frac{\tau-(b/\eta)}{(a/\eta)}\right)}\,d\tau \tag{11.43}$$

the desired relation (11.41) follows. ■

11.2.2.3 Synthesis with Positive Scale. One final set of properties follows when we restrict the dilation parameter a to positive values.

Practical signal analysis and synthesis algorithms benefit from the elimination of redundant data. We now demonstrate a condition under which the reconstruction (11.32) (and by inference, the forward wavelet transform) requires only positive values of the dilation. We show that this condition is met by *all real-valued wavelets*, which comprise the vast majority of continuous and discrete wavelets.

Proposition (Positive Dilation Values). If

$$\int_{0}^{\infty} \frac{|\Psi(\omega)|^2}{|\omega|}\,d\omega = \int_{-\infty}^{0} \frac{|\Psi(\omega)|^2}{|\omega|}\,d\omega, \tag{11.44}$$

then (note the limit on the domain of a)

$$f(t) = \frac{1}{C_\psi}\int_{0}^{\infty}\int_{-\infty}^{\infty} W[f(t)](a, b)\psi_{a,b}(t)\,d\mu, \tag{11.45}$$

where

$$C_\psi = \int_{0}^{\infty} \frac{|\Psi(\omega)|^2}{|\omega|}\,d\omega = \int_{-\infty}^{0} \frac{|\Psi(\omega)|^2}{|\omega|}\,d\omega. \tag{11.46}$$

Proof: First, note

$$\int_{-\infty}^{\infty} \frac{|\Psi(\omega)|^2}{|\omega|}\,d\omega = 2\int_{0}^{\infty} \frac{|\Psi(\omega)|^2}{|\omega|}\,d\omega = 2\int_{-\infty}^{0} \frac{|\Psi(\omega)|^2}{|\omega|}\,d\omega. \tag{11.47}$$

The overall proof is best carried out by reconsidering the steps leading up to the Parseval relation (11.25). Note that if (11.44) holds, the two auxiliary functions,

$$G(\gamma) \equiv \sqrt{|a|}\cdot\mathcal{F}[g(t)](\gamma)\cdot\overline{\mathcal{F}[\psi(t)](a\gamma)} \tag{11.48}$$

and

$$F(\gamma) \equiv \sqrt{|a|} \cdot \mathcal{F}[f(t)](\gamma) \cdot \overline{\mathcal{F}[\psi(t)](a\gamma)}, \tag{11.49}$$

display the necessary symmetry in a so that (11.27) can be reformulated as an integral over positive dilations only:

$$2\int_0^{\infty}\int_{-\infty}^{\infty} W[f(t)](a,b)\overline{W[g(t)](a,b)}\,d\mu = 2\left(\int_0^{\infty}\frac{|\Psi(\omega)|^2}{|\omega|}\,d\omega\right)\langle f(t), g(t)\rangle. \tag{11.50}$$

The factors of 2 cancel and, starting from (11.50), it is straightforward to reproduce the wavelet inversion formula, leading to the desired result (11.45). ■

This proposition is of more than passing interest, as indicated by our next observation.

Theorem (Real Wavelets). If $\psi(t)$ is a real-valued function, then (11.44) is satisfied.

Proof: This is easily established from the Fourier transform of $\psi(t)$,

$$\Psi(\omega) = \int_{-\infty}^{\infty} \psi(t)e^{-j\omega t}dt. \tag{11.51}$$

If $\psi(t) \in \mathbb{R}$, then

$$\Psi(-\omega) = \int_{-\infty}^{\infty} \psi(t)e^{j\omega t}dt = \overline{\Psi(\omega)}. \tag{11.52}$$

From here is easy to establish condition (11.44), since

$$\int_{-\infty}^{0} \frac{|\Psi(\omega)|^2}{|\omega|}\,d\omega = -\int_{0}^{-\infty} \frac{|\Psi(\omega)|^2}{|\omega|}\,d\omega. \tag{11.53}$$

With a simple substitution of variables $\eta = -\omega$, this can be rearranged to the desired result,

$$-\int_{0}^{-\infty} \frac{|\Psi(\omega)|^2}{|\omega|}\,d\omega = -\int_{\infty}^{0} \frac{|\Psi(\gamma)|^2}{|\gamma|}\,d\gamma = \int_{0}^{\infty} \frac{|\Psi(\gamma)|^2}{|\gamma|}\,d\gamma. \tag{11.54}$$

■

Remarks. Note how the condition (11.52) is explicitly used to establish the first equality in (11.54). Also, the importance of this theorem lies in the implication that all real-valued wavelets can lead to reconstruction on the half-plane $a \in [0, \infty]$.

Note that some authors *define* synthesis to occur over this restricted domain, but they are often tacitly restricting the discussion to real-valued $\psi(t)$, which form the overwhelming majority of practical wavelets. Selected complex-valued wavelets (to be considered later) may also satisfy (11.45) with a suitable redefinition of C_ψ, but whenever complex-valued wavelets are under consideration, the reader should exercise caution when performing reconstruction.

Table 11.1 summarizes our results so far.

11.2.2.4 Wavelets by Convolution. Convolution is a smoothing operation which preserves any existing localized properties of the functions involved. It is simple to show that under certain reasonable conditions, wavelets generate other wavelets through the convolution operation.

Theorem (Wavelets Through Convolution). If $\psi(t)$ is a wavelet and $\lambda(t) \in L^1$, then

$$\phi \equiv \psi * \lambda \tag{11.55}$$

is a wavelet.

Proof: We first need to establish that $\phi \in L^2$. This can be carried out in the time domain, but it is simpler to consider the frequency domain where

$$\mathcal{F}[\phi(t)](\omega) = \Psi(\omega)\Lambda(\omega). \tag{11.56}$$

TABLE 11.1. Wavelet Transform Properties[a]

Signal Expression	Wavelet Transform or Property
$\psi_{a,b}(t) = \frac{1}{\sqrt{\lvert a\rvert}}\psi\left(\frac{t-b}{a}\right)$	Dilation and translation of $\psi(t)$
$C_\psi = \int_{-\infty}^{\infty} \frac{\lvert\Psi(\omega)\rvert^2}{\lvert\omega\rvert} d\omega$	Admissibility factor
$f(t)$	$W[f(t)](a,b) = \int_{-\infty}^{\infty} f(t)\overline{\psi_{a,b}(t)}\, dt$
$W[f(t)](a,b) = \frac{1}{\sqrt{2\pi}}\mathcal{F}[F(\gamma)](-b)$	Fourier transform representation
$f(t) = \frac{1}{C_\psi}\int_{-\infty}^{\infty}\int_{-\infty}^{\infty} W[f(t)](a,b)\psi(t)\frac{da\,db}{a^2}$	Inverse
$\theta(t) = \alpha\psi(t) + \beta\phi(t)$	$W_\theta[f(t)] = \bar{\alpha}W_\psi[f(t)] + \bar{\beta}W_\phi[f(t)]$

[a] In the table, $\psi(t)$ is square-integrable.

It is easy to establish that this spectrum is L^2. While we cannot assert that $\Lambda(\omega)$ is integrable, it is certainly bounded and

$$\int_{-\infty}^{\infty} |\Psi(\omega)\Lambda(\omega)|^2 d\omega = \int_{-\infty}^{\infty} |\Psi(\omega)|^2 |\Lambda(\omega)|^2 d\omega < |\Lambda(\omega)|_{\max}{}^2 \int_{-\infty}^{\infty} |\Psi(\omega)|^2 d\omega < \infty, \tag{11.57}$$

which proves $\mathcal{F}[\phi(t)](\omega) \in L^2$. The inverse Fourier transform maps L^2 to L^2 so that

$$\phi(t) \in L^2. \tag{11.58}$$

The admissibility condition on $\phi(t)$ follows in a similar manner:

$$\int_{-\infty}^{\infty} \frac{|\Phi(\omega)|^2}{|\omega|} d\omega = \int_{-\infty}^{\infty} \frac{|\Psi(\omega)|^2}{|\omega|} |\Lambda(\omega)|^2 d\omega < |\Lambda(\omega)|_{\max}{}^2 \int_{-\infty}^{\infty} \frac{|\Psi(\omega)|^2}{|\omega|} d\omega < \infty. \tag{11.59}$$

Conditions (11.58) and (11.59) establish that $\phi(t)$ is a wavelet. ■

Now let us turn to some examples of analyzing wavelets.

11.2.3 Examples

Continuous analyzing wavelets are atomic functions with imposed oscillations. For example, we have seen that the Gaussian time-scale atom is not a wavelet, but operating on a Gaussian by taking one or more derivatives can impose the necessary waviness to ensure that the zero-mean condition (11.19) is satisfied.

11.2.3.1 First derivative of a Gaussian. Let us first consider the analyzing wavelet. A bona fide wavelet is created by applying the first derivative to a Gaussian,

$$\psi(t) = A_0 \cdot \left[-\frac{d}{dt} e^{-t^2}\right] = 2A_0 t e^{-t^2}. \tag{11.60}$$

The normalization constant A_0 can be determined by solving a straightforward Gaussian integral,

$$\int_{-\infty}^{\infty} |\psi(t)|^2 dt = 4A_0^2 \int_{-\infty}^{\infty} t^2 e^{-2t^2} dt \equiv 1, \tag{11.61}$$

which leads to

$$A_0 = \sqrt[4]{2/\pi}. \tag{11.62}$$

The normalization verifies that $\psi(t)$ is in fact square-integrable and by inspection, due to the odd symmetry of (11.60), the zero-mean condition

$$\int_{-\infty}^{\infty} \psi(t)\, dt = 0 \tag{11.63}$$

is assured.

Next, we check the admissibility criteria. The Fourier domain is easily handled by applying the time differentiation property,

$$\mathcal{F}[\psi(t)](\omega) = j\omega A_o \sqrt{\pi} e^{-\omega^2/4}. \tag{11.64}$$

Then for $\omega < 0$

$$\frac{|\psi(\omega)|^2}{|\omega|} = -A_o^2 \pi \omega e^{-\omega^2/2} \tag{11.65}$$

and for positive frequencies

$$\frac{|\psi(\omega)|^2}{|\omega|} = A_o^2 \pi \omega e^{-\omega^2/2}. \tag{11.66}$$

The coefficient (11.8) takes the form

$$C_\psi = -A_o^2 \pi \int_{-\infty}^{0} \omega e^{-\omega^2/2} d\omega + A_o^2 \pi \int_{0}^{\infty} \omega e^{-\omega^2/2} d\omega. \tag{11.67}$$

These integrals defined along the half-line can be evaluated by noting that each integrand can be represented as a derivative, so (11.67) now reads

$$C_\psi = A_o^2 \pi \int_{-\infty}^{0} \frac{d}{d\omega} e^{-\omega^2/2} d\omega - A_o^2 \pi \int_{0}^{\infty} \frac{d}{d\omega} e^{-\omega^2/2} d\omega \tag{11.68}$$

so

$$C_\psi = 2\pi A_0^2. \tag{11.69}$$

Remark. Note that (11.54) holds for this real-valued analyzing wavelet, as expected. If reconstruction uses only positive values of scale (as per (11.45)), then

$$C_\psi = \pi A_0^2 \tag{11.70}$$

should be used in place of (11.69).

Example (Gaussian Transient). We will generate and discuss the wavelet transform of a Gaussian pulse

$$f(t) = e^{-\alpha t^2}, \tag{11.71}$$

where α is a positive factor. The analyzing wavelet (11.60) with an applied scale a and translation b reads

$$\psi\left(\frac{t-b}{a}\right) = 2A_0 e^{-b^2/a^2} \cdot \left(\frac{t-b}{a}\right) e^{-\left(\frac{t^2 - 2bt}{a^2}\right)} \tag{11.72}$$

and the wavelet transform integral breaks down conveniently,

$$W[f(t)](a, b) = \frac{1}{\sqrt{a}} \int_{-\infty}^{\infty} f(t)\psi\left(\frac{t-b}{a}\right) dt = C(a, b)[I_1 - bI_2], \tag{11.73}$$

where

$$C(a, b) = \frac{2A_0}{a\sqrt{a}} e^{-b^2/a^2} \tag{11.74}$$

and the integrals

$$I_1 \equiv \int_{-\infty}^{\infty} te^{-(\alpha + (1/a^2))t^2 + (2b/a^2)t} dt = \frac{a\sqrt{\pi}}{\sqrt{a^2\alpha + 1}} \left(\frac{b}{a^2\alpha + 1}\right) e^{\left(\frac{b^2}{a^2}\right)\left(\frac{1}{a^2\alpha + 1}\right)} \tag{11.75}$$

and

$$I_2 \equiv \int_{-\infty}^{\infty} e^{-(\alpha + (1/a^2))t^2 + (2b/a^2)t} dt = \frac{a\sqrt{\pi}}{\sqrt{a^2\alpha + 1}} e^{\left(\frac{b^2}{a^2}\right)\left(\frac{1}{a^2\alpha + 1}\right)} \tag{11.76}$$

are evaluated using standard Gaussian integration. The result

$$W[f(t)](a, b) = \frac{-a^2 b\alpha\sqrt{\alpha}}{\sqrt{a(a^2\alpha + 1)}} e^{\left(\frac{b^2}{a^2}\right)\left(\frac{1}{a^2\alpha + 1}\right)} \tag{11.77}$$

is a two-dimensional function of scale and translation shown in Figure 11.3.

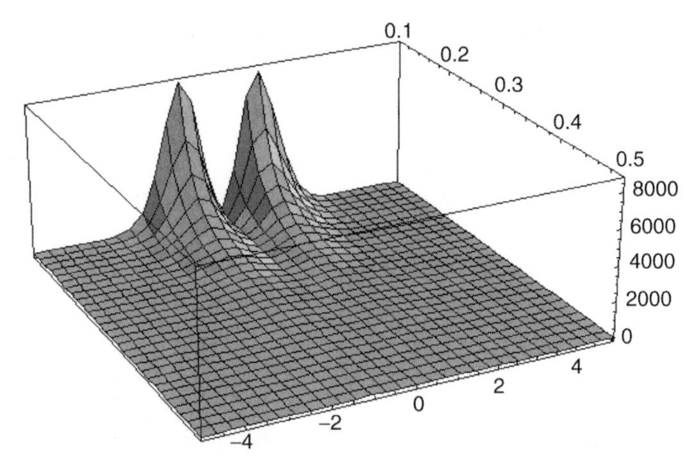

Fig. 11.3. Wavelet transform of Gaussian pulse.

Example (Rectangular Pulse). Consider a rectangular pulse of width D analyzed by the same wavelet as above. The wavelet transform again takes the form

$$W[f(t)](a,b) = \frac{1}{\sqrt{a}}\int_{-\infty}^{\infty} f(t)\psi\left(\frac{t-b}{a}\right)dt = C(a,b)[I_1 - bI_2], \tag{11.78}$$

where as before

$$C(a,b) = \frac{2A_0}{a\sqrt{a}}e^{-b^2/a^2}. \tag{11.79}$$

The integrals now read

$$I_1 \equiv \int_{-D/2}^{D/2} te^{-(1/a^2)t^2 + (2b/a^2)t}dt \tag{11.80}$$

and

$$I_2 \equiv \int_{-D/2}^{D/2} e^{-(1/a^2)t^2 + (2b/a^2)t}dt. \tag{11.81}$$

The central feature of each integral is the exponential

$$e^{-[f(a)t^2 + g(b)t]}, \tag{11.82}$$

where $f(a) \equiv 1/a^2$ and $g(a) \equiv -2b^2/a^2$. With proper manipulation, (11.81) and (11.82) can be handled analytically. It is a simple matter to complete the square on the argument of the above exponential, transforming it:

$$f(a)t^2 + g(b)t \to (f(a)t^2 + g(b)t + x) - x, \tag{11.83}$$

where $x = \frac{1}{4}\frac{g^2(b)}{f(a)}$. If we let $y \equiv \sqrt{f(a)}t + \frac{1}{2}\frac{g(b)}{\sqrt{f(a)}}$ and make a substitution of variables, then

$$I_1 \to \frac{1}{\sqrt{f(a)}}\int_{L_1}^{L_2}\left(y - \frac{1}{2}\frac{g(b)}{\sqrt{f(a)}}\right)e^{-y^2}e^{x} \cdot \frac{1}{\sqrt{f(a)}}dy, \tag{11.84}$$

where the limits $L_1 \equiv \frac{-d\sqrt{f(a)}}{2} + \frac{1}{2}\frac{g(b)}{\sqrt{f(a)}}$ and $L_2 \equiv \frac{d\sqrt{f(a)}}{2} + \frac{1}{2}\frac{g(b)}{\sqrt{f(a)}}$. This conveniently breaks into two terms,

$$I_1 = \frac{e^x}{f(a)}\left[\int_{L_1}^{L_2} ye^{-y^2}dy - \frac{1}{2}\frac{g(b)}{\sqrt{f(a)}}\int_{L_1}^{L_2} e^{-y^2}dy\right]. \tag{11.85}$$

Now $ye^{-y^2} = \left(-\frac{1}{2}\right)\frac{d}{dy}\left[e^{-y^2}\right]$ so

$$I_1 = \frac{e^x}{f(a)}\left[\left(-\frac{1}{2}\right)\left(e^{L_2^2} - e^{L_1^2}\right) - \frac{1}{2}\frac{g(b)}{\sqrt{f(a)}}[erf(L_2) + erf(L_1)]\right]. \tag{11.86}$$

With similar operations, it is easy to show

$$I_2 = \frac{e^x}{\sqrt{f(a)}}[erf(L_2) + erf(L_1)]. \tag{11.87}$$

11.2.3.2 Second Derivative of a Gaussian ("Mexican Hat"). Taking a further derivative provides an analyzing wavelet

$$\psi(t) = B_0 \cdot \left[\frac{de^{-t^2}}{dt^2}\right] = -2B_0[1 - 2t^2]e^{-t^2}. \tag{11.88}$$

It is readily shown that

$$\int_{-\infty}^{\infty} |\psi(t)|^2 dt = 3B_0^2\sqrt{\frac{\pi}{2}}, \tag{11.89}$$

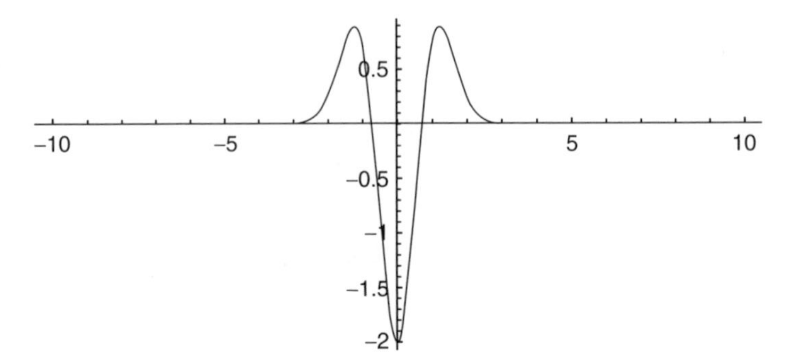

Fig. 11.4. Mexican hat wavelet.

so the normalization constant is

$$B_0 = \frac{1}{\sqrt{3}}\sqrt[4]{2/\pi}. \tag{11.90}$$

This wavelet is shown in Figure 11.4. It has an even symmetry, but equal area above and below the axis, so that (11.9) is satisfied.[1] These details are left as an exercise.

In the Fourier domain, the Mexican hat provides a spectrum

$$\mathcal{F}[\psi(t)](\omega) = -B_0\omega^2\sqrt{\pi}e^{-\frac{\omega^2}{4}}, \tag{11.91}$$

and the admissibility coefficient is

$$C_\psi \equiv \int_{-\infty}^{\infty}\frac{|\psi(\omega)|^2}{|\omega|}\,d\omega = \frac{2}{3}\sqrt{2\pi}. \tag{11.92}$$

Example (Gaussian Transient). Consider the Mexican hat applied to the Gaussian transient of (11.71). The scaled and translated analyzing wavelet is easily found:

$$\psi\left(\frac{t-b}{a}\right) = [d_0 - d_1 t + d_2 t^2]e^{-\frac{b^2}{a^2}}e^{-\left(\frac{t^2-2bt}{a^2}\right)}, \tag{11.93}$$

where

$$d_0 = -2B_0\left[1 - 2\frac{b^2}{a^2}\right], \tag{11.94}$$

$$d_1 = \frac{8B_0}{a^2}, \tag{11.95}$$

[1]It might resemble a traditional Mexican *sombrero* in cross section—hence the name.

and

$$d_2 = \frac{4B_0}{a^2}. \tag{11.96}$$

It is left as an exercise to show that the wavelet transform takes the form

$$\mathcal{W}[f(t)](a,b) = \frac{e^{-b^2/a^2}}{\sqrt{a}}[d_0 I_0 - d_1 I_1 + d_2 I_2] \tag{11.97}$$

with

$$I_0 \equiv \int_{-\infty}^{\infty} e^{-(\alpha + (1/a^2))t^2 + (2b/a^2)t} dt = \frac{a\sqrt{\pi}}{\sqrt{a^2\alpha + 1}} e^{\left(\frac{b^2}{a^2}\right)\left(\frac{1}{a^2\alpha + 1}\right)} \tag{11.98}$$

and I_1 as in (11.75), and

$$I_2 \equiv \int_{-\infty}^{\infty} t^2 e^{-(\alpha + (1/a^2))t^2 + (2b/a^2)t} dt$$

$$= \frac{a\sqrt{\pi}}{\sqrt{a^2\alpha + 1}} \left(\frac{a^2}{2(a^2\alpha + 1)}\right)\left(1 + \frac{4(b^2/a^2)}{a^2\alpha + 1}\right) e^{\left(\frac{b^2}{a^2}\right)\left(\frac{1}{a^2\alpha + 1}\right)}. \tag{11.99}$$

11.3 FRAMES

It is much easier to construct frames based upon wavelets than upon the short-time Fourier transform. Building computer applications requires us to work with discrete rather than continuous signal representations. One requirement for signal analysis is that our discrete representation be capable of representing any signal; this is a *completeness* or *spanning* condition. If we also ask that our discrete representation also be *numerically stable*—that is, a small change in a signal results in a small change in its decomposition coefficients—then we must use a *frame* representation.

As a generalization of orthonormal bases, Chapter 3 introduced frames. We remember that $F = \{f_n : n \in \mathbb{Z}\}$ from a Hilbert space H is a *frame* if there are $A, B \in \mathbb{R}$ such that $A > 0$, $B > 0$, and for all $x \in H$,

$$A\|x\|^2 \le \sum_{n=-\infty}^{\infty} |\langle x, f_n \rangle|^2 \le B\|x\|^2. \tag{11.100}$$

The frame F is *tight* if its lower and upper bounds—A and B, respectively—are equal. Any frame $F \subset H$ spans H. A frame F is *exact* if, when an element is removed from it, it ceases to be a frame. If F is orthonormal, then F is tight; in fact, $A = B = 1$, and F is exact.

Recall from Chapter 10 that the Balian–Low theorem imposes strict constraints on the time- and frequency-domain sampling intervals for a frame of windowed Fourier atoms. The time- and frequency-domain sampling intervals, T and Ω, respectively are critical:

(i) If $T\Omega < 2\pi$, then the time-frequency density $(T\Omega)^{-1}$ exceeds the Nyquist density $(2\pi)^{-1}$, and frames of windowed Fourier atoms are possible.

(ii) If $T\Omega > 2\pi$, then we are sampling below the Nyquist density and there are no windowed Fourier frames.

(iii) If we sample at precisely the Nyquist density, $T\Omega = 2\pi$, and $F = \{w_{m,n}(t) = e^{\Omega jnt}w(t - mT)\colon m, n \in \mathbb{Z}\}$ is a frame, then either $w(t)$ or its Fourier transform $W(\omega)$ is not well-localized (i.e., not a window function).

In this section we shall see that the wavelet transform is not so restrictive; one can find wavelets $\psi(t)$ that allow tight frames aslong as $\Omega \neq 0, 1$ and $T \neq 0$ [14, 15].

11.3.1 Discretization

The wavelet discretization procedure is analogous to discretization of time-frequency transforms. Instead of applying a time and frequency increment, we use a time and scale increment on a signal model. The signal model is an admissible wavelet $\psi(t)$.

We have noted that the continuous wavelet transform is an inner product. It measures similarity of $x(t)$ and $\psi_{a,\,b}(t) = \frac{1}{\sqrt{a}}\psi\left(\frac{t-b}{a}\right)$ as follows:

$$X_{\psi}(a, b) = \langle x, \psi_{a,\,b}\rangle = \int_{-\infty}^{\infty} x(t)\overline{\psi_{a,\,b}(t)}\,dt\,, \tag{11.101}$$

where $a, b \in \mathbb{R}$. The wavelet $\psi(t)$ must satisfy the admissibility condition (11.2). For simplicity, let us consider only the case $a > 0$ and assume $\psi(t) \in \mathbb{R}$. The inner product (11.101) measures the similarity of $x(t)$ and $a^{-1/2}\psi_{a,b}(t)$, which is a dilated version of $\psi(t)$, shifted so that it centers at time $t = b$.

Suppose we are searching a candidate signal for a prototype shape $\psi(t)$. This is a typical signal analysis problem. Perhaps the shape $\psi(t)$ resembles the signal trace we are trying to detect, or it maybe it responds significantly to some feature—such as an edge—that we can use in a structural description to identify the candidate. If we know the exact location and time-domain extent, we can fix $a, b \in \mathbb{R}$ and perform the inner product computation. If $x(t)$ happens to be a scalar multiple (an attenuated or amplified replica) of $\psi_{a,b}(t)$, then the Schwarz inequality

$$|\langle x(t), \psi_{a,b}(t)\rangle| \le \|x\|\,\|\psi_{a,b}\| \tag{11.102}$$

will be an equality. Thus, we threshold the inner product (11.102) as a percentage of $\|x\| \times \|\psi_{a,b}\|$ to obtain a measure of the match between prototype and candidate signals.

One the other hand, if we do not know the location and time extent—and this is the more common and daunting signal recognition problem—then the task of performing many inner products in (11.102) becomes a computational burden. We can correlate $\psi_{a,b}(t)$ with local values of $x(t)$, say restricted to $[b - c, b + c]$, for some $c > 0$. But then our inner product varies with the L^2 norm of $x(t)$ restricted to $[b - c, b + c]$. This is conventional normalized cross-correlation, where we divide the inner product by the norm of the candidate signal in a region of interest. Nevertheless, there is in principle a continuous range of scale factors, offsets, and (perhaps) window widths—a, b, and c, respectively. To make the analysis practical, we must choose a discrete set of locations and signal prototype sizes against which we compare the candidate waveform.

Let us start discretization with scale increment $a_0 > 0$. Our discussion closely follows [3]. Dyadic decomposition remains the most common. In this case $a_0 = 2$, and we have dilation steps $\psi(t/2)$, $\psi(t)$, $\psi(2t)$, $\psi(4t)$, and so on. These signal models are, respectively, twice as large, exactly the same, half as large, and one quarter as large in time-domain extent as the root scale element $\psi(t)$. If we let $a = a_0{}^m$, then $\psi(ta_0{}^{-m})$ is $a_0{}^m$ times wider than $\psi(t)$.

Now let us decide how to discretize the time domain. A moment's thought shows that we cannot just take $b = nb_0$ for some $b_0 > 0$ and $n \in \mathbb{Z}$. Note that if $a_0 = 2$, then signal prototypes at the scale $a = a_0{}^1$ have the shape of $\psi(t/2)$ and occupy twice the time-domain extent as at unit scale $a = 1$. Thus, we should cover the time-domain with step increments that are twice as far apart as at unit scale. That way, the time-domain coverage and overlap between prototypes at unit and double scale is proportional. Similarly, if $a = a_0{}^{-1}$, then models at this scale look like $\psi(2t)$ and take only half the time-domain width as at unit scale. We could repeat this logic at quadruple and quarter scales, but the point is that time-domain steps for scale $a = a_0{}^m$ should be in increments of the product $b_0 a_0{}^m$. For wavelet transform discretization, we employ wavelet atoms of the form

$$\psi_{m,n}(t) = a_0^{-\frac{m}{2}}\psi\left(\frac{t - nb_0 a_0^m}{a_0^m}\right) = a_0^{-\frac{m}{2}}\psi(a_0^{-m}t - nb_0). \tag{11.103}$$

Note that—in accord with other established notations [3, 10]—we use the first discrete index for the scale variable and use the second for the time variable.

As with the short-time Fourier transform, discretization implies a structural description of a signal. Windowed Fourier transforms produce a tiling of the time-frequency plane by signal atoms that occupy equally sized regions. In contrast, time-scale discretizations, as with wavelets, tile the plane with regions of varying size. Signal atoms tuned to higher frequencies have a more restricted time-domain support (Figure 11.5).

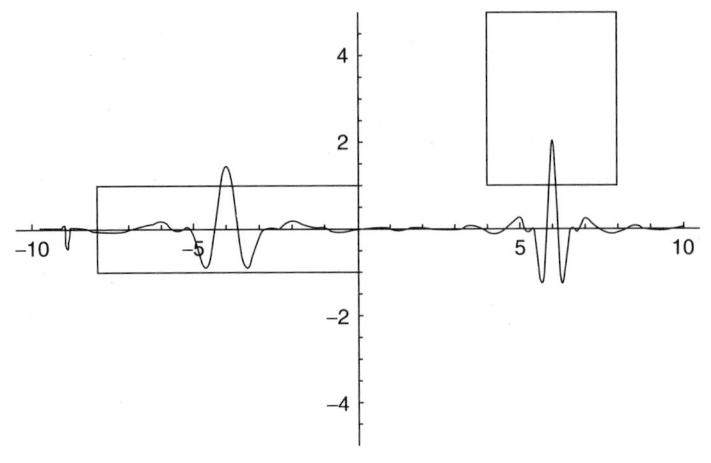

Fig. 11.5. Tiling of the time-frequency plane by a discretized wavelet transform.

11.3.2 Conditions on Wavelet Frames

In order for a discretization based on translations and dilations to constitute a frame, certain necessary conditions must obtain. We quote the following theorem.

Theorem (Necessity of Admissible Wavelet). Suppose $\psi(t) \in L^2(\mathbb{R})$, $a_0 > 0$, and

$$F = \left\{ \psi_{m,n}(t) = a_0^{-\frac{m}{2}} \psi(a_0^{-m} t - nb_0) \middle| m, n \in \mathbb{Z} \right\} \tag{11.104}$$

constitutes a frame with lower and upper bounds A and B, respectively. Then

$$Ab_0 \ln a_0 \le \int_0^{\infty} \frac{|\Psi(\omega)|^2}{\omega} d\omega \le Bb_0 \ln a_0 \tag{11.105a}$$

and

$$Ab_0 \ln a_0 \le \int_0^{\infty} \frac{|\Psi(\omega)|^2}{\omega} d\omega \le Bb_0 \ln a_0 , \tag{11.105b}$$

where $\Psi(\omega)$ is the (radial) Fourier transform of $\psi(t)$.

Proof: Due to Daubechies [3, 15]. ■

Remark. Interestingly, for a family of translates and dilates to be a frame, $\psi(t)$ must be admissible. One might think that the admissibility condition (11.2) is a technicality, concocted just to make the wavelet transform inversion work. We see now that it is essential for signal analysis using families of scaled, translated atoms—that is, for wavelet frames.

11.3.3 Constructing Wavelet Frames

This section covers one method for constructing tight wavelet frames [3, 14, 15]. Let $v(t)$ be real-valued, k times continuously differentiable, and approximate the unit step as follows:

$$v(t) = \begin{cases} 0 & \text{if } t \le 0, \\ 1 & \text{if } t \ge 1. \end{cases} \tag{11.106}$$

An example (Figure 11.6) of $v \in C^1$ is the following

$$v(t) = \begin{cases} 0 & \text{if } t \le 0, \\ \sin^2\left(\frac{\pi t}{2}\right) & \text{if } t \le 0 \le 1, \\ 1 & \text{if } t \ge 1. \end{cases} \tag{11.107}$$

Now let $a_0 > 1$ and $b_0 > 0$. We will specify two square-integrable signals, $\psi^+(t)$ and $\psi^+(t)$, by their *normalized* radial Fourier transforms, $\Psi^+(\omega)$ and $\Psi^-(\omega)$, respectively. Let $L = 2\pi[b_0(a_0^2 - 1)]^{-1}$, define

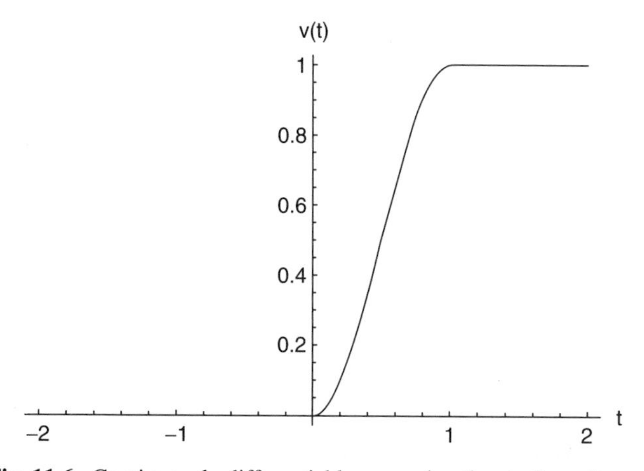

Fig. 11.6. Continuously differentiable approximation to the unit step.

$$\Psi^{+}(\omega) = (\ln a_0)^{-\frac{1}{2}} \begin{cases} 0 & \text{if } \omega \le L \text{ or } \omega \ge La_0^2, \\ \sin\left(\frac{\pi}{2}v\left(\frac{\omega - L}{L(a_0 - 1)}\right)\right) & \text{if } L \le \omega \le La_0, \\ \cos\left(\frac{\pi}{2}v\left(\frac{\omega - La_0}{La_0(a_0 - 1)}\right)\right) & \text{if } La_0 \le \omega \le La_0^2 \end{cases} \tag{11.108}$$

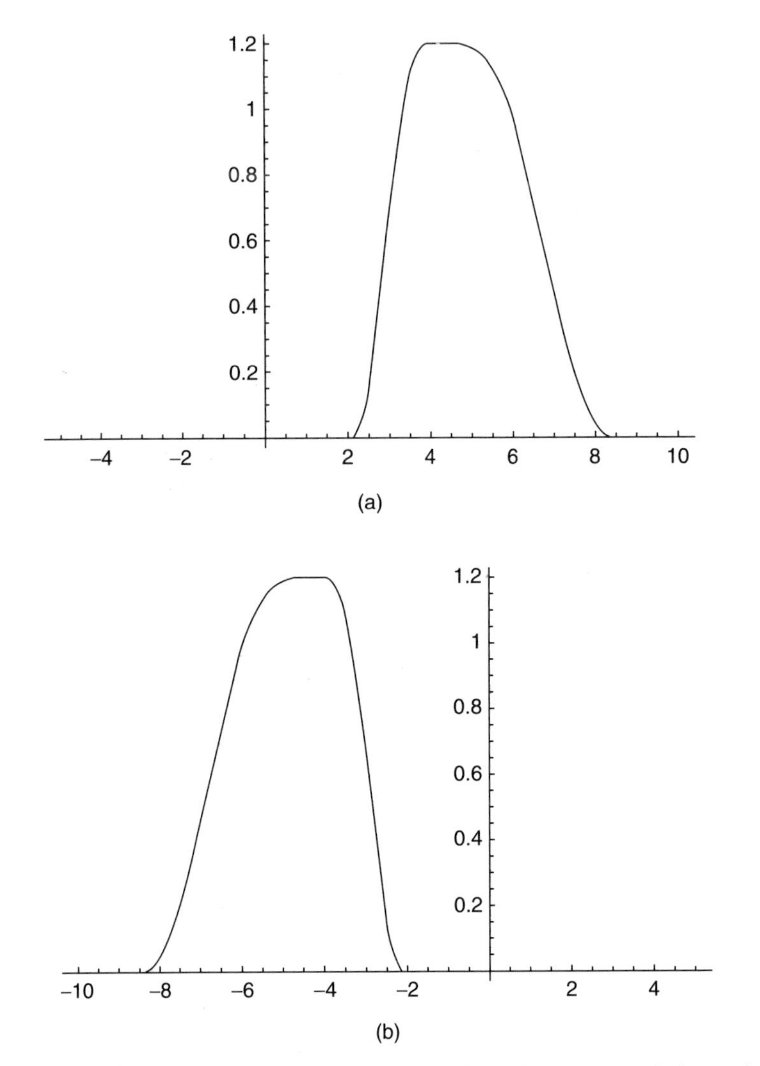

Fig. 11.7. Fourier transforms of atoms used for a tight frame based on translation and dilation: (a) $\Psi^{+}(\omega)$ and (b) $\Psi^{-}(\omega)$.

and set $\Psi^{-}(\omega) = \Psi^{+}(-\omega)$. Then—as Figure 11.7 illustrates for the choices $a_0 = 2$, $b_0 = 1$, and $v(t)$ given by (11.107)—$\Psi^{+}(\omega)$ is finitely supported on $[L, La_0^2]$.

The special construction (11.108) guarantees

$$\sum_{m=-\infty}^{\infty} \left|\Psi^{+}(a_0^m \omega)\right|^2 = \begin{cases} \dfrac{1}{\ln(a_0)} & \text{if } 0 < \omega, \\ 0 & \text{if } \omega \le 0. \end{cases} \tag{11.109}$$

To see this, note that the sequence $\{a_0{}^m \mid m \in \mathbb{Z}\}$ decreases toward zero as $m \to -\infty$ and increases toward ∞ as $m \to \infty$. If $\omega > 0$, then there must be exactly one $m \in \mathbb{Z}$ such that $\omega a_0{}^m \in [L, La_0]$, the interval of the sin() term in (11.108). Then the next summand's argument $\omega a_0{}^{m+1}$ falls in the interval supporting the cos() term, $[La_0, La_0{}^2]$. The consequence is that exactly two summands from (11.109) are non-zero: one sin() term for $\omega a_0{}^m$ and one cos() term for $\omega a_0{}^{m+1}$. The sum of their squares is unity justifying (11.109).

We turn to the frame condition. If $x(t) \in L^2(\mathbb{R})$ and $X(\omega)$ is its normalized radial Fourier transform, $X(\omega) = (2\pi)^{-(1/2)}\int x(t)\exp(-j\omega t)\,dt$, then following Ref. 3 we find

$$\begin{aligned}
&\sum_{m,n=-\infty}^{\infty} \left|\langle x, \psi^{+}{}_{m,n}\rangle\right|^2 \\
&= \sum_{m,n=-\infty}^{\infty} \left|\langle X, \Psi^{+}{}_{m,n}\rangle\right|^2 \\
&= \sum_{m,n=-\infty}^{\infty} a_0^m \left|\int_{-\infty}^{\infty} e^{j\omega n b_0 a_0^m} X(\omega)\Psi^{+}(a_0^m\omega)\,d\omega\right|^2 \\
&= \sum_m a_0^m \sum_n \left|\sum_k \int_{k\frac{2\pi}{\Omega}}^{(k+1)\frac{2\pi}{\Omega}} X(\omega)\Psi^{+}(a_0^m\omega)e^{j\omega n b_0 a_0^m}\,d\omega\right|^2 .
\end{aligned} \tag{11.110}$$

Our strategy has been to break down the integral over $\mathbb{R}$ in (11.110) into an infinite sum of integrals over a finite interval. Note that we have used the fact that the normalized radial Fourier transform is an isometry: $\langle x, y\rangle = \langle X, Y\rangle$. By resorting to the normalized frequency transform, we economize on 2π factors. We recall from basic Hilbert space theory that $\{e_n(t) = (\Omega/2\pi)^{1/2}\exp(jn\Omega t) \mid n \in \mathbb{Z}\}$ is an orthonormal basis for the Hilbert space $H = L^2[0, 2\pi/\Omega]$. Thus, by Chapter 2's abstract version of the Pythagorean theorem, $\|x\|_2^2 = \Sigma_n|\langle x, e_n\rangle|^2$. Above, we set $\Omega = b_0 a_0^m$, break up the integral over $\mathbb{R}$ into sections $2\pi/\Omega$ wide, and interchange summation and integral. With the substitution $\omega = \theta + 2\pi\frac{k}{\Omega}$, (11.110) continues as follows:

$$\sum_m a_0^m \sum_n \left| \int_0^{\frac{2\pi}{\Omega}} e^{j\theta n\Omega} \sum_k X\left(\theta + \frac{2\pi k}{\Omega}\right) \Psi^+\left(a_0^m \theta + \frac{2\pi k}{b_0}\right) d\theta \right|^2$$
$$= \sum_m a_0^m \sum_n \left| \int_0^{\frac{2\pi}{\Omega}} e^{j\theta n\Omega} Y(\theta) d\theta \right|^2 = \sum_m a_0^m \frac{2\pi}{\Omega} \|Y\|_2^2, \tag{11.111}$$

where $Y(\theta)$ is the summation in the integrand at the top of (11.111). Having disposed of one summation, we now backtrack, writing the $\|Y\|_2^2$ term (which is the L^2 norm over H) as an integral, first as a sum over a finite interval, and then as over all of $\mathbb{R}$:

$$\sum_m a_0^m \frac{2\pi}{\Omega} \|Y\|_2^2 = \frac{2\pi}{b_0} \sum_m \int_0^{\frac{2\pi}{\Omega}} Y(\theta)\overline{Y(\theta)}\, d\theta$$
$$= \frac{2\pi}{b_0} \sum_m \int_{-\infty}^{\infty} |X(\omega)|^2 \left|\Psi^+(a_0^m \omega)\right|^2 d\omega = \frac{2\pi}{b_0 \ln a_0} \int_0^{\infty} |X(\omega)|^2 d\omega. \tag{11.112}$$

A similar argument—with a little analytical discomfort but still no pain—gives

$$\sum_{m,\, n\, =\, -\infty}^{\infty} \left|\langle x, \psi^-{}_{m,\, n}\rangle\right|^2 = \frac{1}{2\pi b_0 \ln a_0} \sum_m \int_{-\infty}^{0} |X(\omega)|^2 d\omega, \tag{11.113}$$

where $\psi^-(t)$ has normalized Fourier transform $\Psi^-(\omega)$. Now we claim that $F = \{\psi_{m,n}{}^+(t)\} \cup \{\psi_{m,n}{}^-(t)\}$ is a frame. Indeed, (11.110) through (11.112) and (11.113) imply

$$\sum_{\substack{m,\, n\, =\, -\infty \\ \varepsilon\, =\, +,-}}^{\infty} \left|\langle x, \psi^{\varepsilon}{}_{m,\, n}\rangle\right|^2 = \frac{2\pi}{b_0 \ln a_0} \int_{-\infty}^{\infty} |X(\omega)|^2 d\omega = \frac{2\pi}{b_0 \ln a_0} \int_{-\infty}^{\infty} |x(t)|^2 dt; \tag{11.114}$$

we see that (11.114) is a tight frame with bounds $\dfrac{2\pi}{b_0 \ln a_0}$.

This construction—albeit clearly contrived—shows how readily we can construct frames for signal analysis based on translations and dilations. We can actually loosen the provisions $a_0 > 1$ and $b_0 > 0$ by reviewing the argument. We see that positive and negative integer powers of a_0 were used to justify (11.109). Hence, as long as $0 < a_0$ and $a_0 \neq 1$, the same argument applies (exercise).

Now, it turns out that this frame is not a particularly stellar choice for signal analysis [3]. For one thing, its frame elements do not consist entirely of translations and dilations of a single element; rather, there are two prototype patterns from which the others derive. Worse, however, is the fact that the elements in our special tight frame have poor time-domain decay. Their spectrum has finite support. The consequence is that inner products will have nonzero responses even when the analyzing frame elements are offset some distance away from the candidate signals. The exercises invite the reader to explore these ideas further.

11.3.4 Better Localization

We can construct frames based on translations and dilations of a Gaussian root signal. Such frames offer satisfactory time-domain decay. The drawback, however, is that the precise explication of the frame condition is not as elegant as in the previous section's special construction. There is a sufficient condition for wavelet frames [3], which, despite its ponderous formulation, allows one to estimate frame bounds for $\psi(t)$ having good time and frequency-domain decay.

11.3.4.1 Sufficient Conditions for Wavelet Frames.

Let us state some theorems due to Daubechies [3].

Theorem (Sufficiency). Suppose $\psi(t) \in L^2(\mathbb{R})$, $a_0 > 1$, $\Psi(\omega)$ is the (radial) Fourier transform of $\psi(t)$, and

$$\beta(s) = \sup_{1 \le |\omega| \le a_0} \left(\sum_{m=-\infty}^{\infty} \left|\Psi(a_0^m \omega)\right| \left|\Psi(a_0^m \omega + s)\right| \right). \tag{11.115}$$

Further assume that

$$\inf_{1 \le |\omega| \le a_0} \left(\sum_{m=-\infty}^{\infty} \left|\Psi(a_0^m \omega)\right|^2 \right) > 0; \tag{11.116a}$$

$$\sup_{1 \le |\omega| \le a_0} \left(\sum_{m=-\infty}^{\infty} \left|\Psi(a_0^m \omega)\right|^2 \right) < \infty; \tag{11.116b}$$

and, for some $\varepsilon > 0$, $\beta(s)$ decays as fast as $(1 + |s|)^{-1-\varepsilon}$. Then there is $B_0 > 0$ such that for any $b_0 < B_0$, $F = \{\psi_{m,n}(t) = a_0^{-(m/2)}\psi(a_0^{-m}t - nb_0) | m, n \in \mathbb{Z}\}$ is a frame.

Corollary (Bounds). With the theorem's assumptions and notation, let

$$C = \sum_{\substack{k=-\infty \\ k \neq 0}}^{\infty} \left[\beta\left(\frac{2\pi k}{b_0}\right)\beta\left(-\frac{2\pi k}{b_0}\right)\right]^{\frac{1}{2}}, \tag{11.117}$$

and suppose $b_0 < B_0$. Then $F = \{\psi_{m,n}(t) | m, n \in \mathbb{Z}\}$ has lower and upper frame bounds, A and B, respectively:

$$A = \frac{1}{b_0}\left[\inf_{1 \le |\omega| \le a_0}\left(\sum_{m=-\infty}^{\infty}\left|\Psi(a_0^m\omega)\right|^2\right) - C\right] \tag{11.118a}$$

and

$$B = \frac{1}{b_0}\left[\sup_{1 \le |\omega| \le a_0}\left(\sum_{m=-\infty}^{\infty}\left|\Psi(a_0^m\omega)\right|^2\right) + C\right]. \tag{11.118b}$$

Proofs: Again due to Daubechies [3, 15].

Remark. These technical conditions will be met, for example, if $\psi(t)$ obeys the following:

- Its time- and frequency-domain decay rates are not too slow.
- Its spectrum is zero for $\omega = 0$: $\Psi(0) = \int \psi(t)\, dt = 0$.

The conditions do imply that $\psi(t)$ is admissible (11.2). Moreover, under these mild constraints, there will be many combinations of scale and time steps for which F comprises a frame [3].

11.3.4.2 Example: Mexican Hat. Let us consider the Mexican hat function, introduced in Section 11.2.3.2 (Figure 11.4). This signal is the second derivative of the Gaussian: $\psi(t) = \exp(-t^2/2)$. Normalizing, $\|\psi\|_2 = 1$, gives

$$\psi(t) = \frac{2\pi^{-\frac{1}{4}}}{\sqrt{3}}(1 - t^2)\exp\left(-\frac{t^2}{2}\right). \tag{11.119}$$

Table 11.2 repeats some estimates for the lower and upper frame bounds, A and B, respectively, for frames based on translations and dilations of the Mexican hat [3].

Notice that as the time domain increment b_0 increases, then frame lower bound A decreases much faster toward zero; we might interpret this as indicating that there are finite energy signals that are more and more orthogonal to the frame elements. Decreasing the scale domain increment $a = a_0^{1/k}$, $k = 1, 2, 3, 4$, etc., mitigates this tendency.

TABLE 11.2. Lower and Upper Bound Estimates for Frames Based on the Mexican Hat

b_0	A	B	$a = 2^1$
0.25	13.091	14.183	
0.50	6.546	7.092	
0.75	4.364	4.728	
1.0	3.223	3.596	
1.25	2.001	3.454	
1.50	0.325	4.221	
1.75	—	—	No frame
b_0	A	B	$a = 2^{1/2}$
0.25	27.273	27.278	Nearly exact
0.50	13.673	13.676	
0.75	9.091	9.093	
1.0	6.768	6.870	
1.25	4.834	6.077	
1.50	2.609	6.483	
1.75	0.517	7.276	
b_0	A	B	$a = 2^{1/3}$
0.25	40.914	40.914	Nearly exact
0.50	20.457	20.457	
0.75	13.638	13.638	
1.0	10.178	10.279	
1.25	7.530	8.835	
1.50	4.629	9.009	
1.75	1.747	9.942	
b_0	A	B	$a = 2^{1/4}$
0.25	55.552	55.552	Nearly exact
0.50	27.276	27.276	
0.75	18.184	18.184	
1.0	13.586	13.690	
1.25	10.205	11.616	
1.50	6.594	11.590	
1.75	2.928	12.659	

What time- and scale-domain increments make the best choices? To answer this question, we recall the formula for reconstructing a signal $x(t)$ from frame elements (Section 3.3.4). Let $\{\psi_k(t): k \in \mathbb{Z}\}$ enumerate the doubly indexed frame F of translations and dilations of $y(t)$, $F = \{\psi_{m,n}(t)\}$. Then,

$$x = \sum_k \langle x, S^{-1}\psi_k\rangle \psi_k = \sum_k \langle x, \psi_k\rangle S^{-1}\psi_k, \tag{11.120}$$

where $S = T^*T$; T is the frame operator, $T_F(x)(k) = \langle x, \psi_k\rangle$; and T^* is the frame operator adjoint, $T^*(s) = \sum_{k=-\infty}^{\infty} s(k)\psi_k$, where $s(k) \in l^2(\mathbb{Z})$. Now by the Frame Characterization Theorem of Section 3.3.4.3, we can write the frame condition as $AI \le S \le BI$, where I is the identity operator on $L^2(\mathbb{R})$.

Suppose that the lower and upper frame bounds are almost equal, a condition that several of the alternatives in Table 11.2 allow [3]. As $B \to A$, $\varepsilon = B/A - 1 \to 0$, and the operator $S = T^*T$ is close to a midpoint operator between AI and BI: $S \approx \frac{A+B}{2}I$. Thus, $S^{-1} \approx \frac{2}{A+B}I$, and (11.120) becomes

$$x = \sum_k \langle x, S^{-1}\psi_k\rangle \psi_k \approx \frac{2}{A+B}\sum_k \langle x, \psi_k\rangle \psi_k. \tag{11.121}$$

Equation (11.121) is a simple, approximate reconstruction formula for $x(t)$ that is valid when the frame F is almost exact. Thus, choosing time and scale dilation factors that provide an almost exact frame facilitates reconstruction of $x(t)$ from its frame coefficients.

The next section develops wavelet theory that provides not just frames, but orthonormal bases for finite energy signals based on translations and dilations of a single prototype signal.

11.4 MULTIRESOLUTION ANALYSIS AND ORTHOGONAL WAVELETS

After the publication of Grossmann and Morlet's paper in 1984, wavelet methods attracted researchers—including the present authors—from a broad range of scientific and engineering disciplines.[2] The new scale-based transform posed an alternative to short-time Fourier techniques for seismic applications [13, 16]. It facilitated the construction of frames (Section 11.3), which are necessary for numerically stable signal modeling [14, 15]. Wavelets were used for analyzing sound waves [17] and adapted to multiscale edge detection [18]. Academic meetings were exciting. It was still unclear how powerful this tool could become.

[2]"Such a portentous and mysterious monster roused all my curiosity" (Melville).

The wavelet transform was but one of several mixed domain signal transforms known in the mid-1980s. Among the others were time-frequency techniques such as the short-time Fourier transform and the Wigner distribution (Chapter 10). Wavelet analysis, in contrast, represents a scale-based transform.

Recall that if we set $\psi_{a,b}(t) = \frac{1}{\sqrt{a}}\psi\left(\frac{t-b}{a}\right)$, then the wavelet transform

$$X_w(a,b) = \langle x, \psi_{a,b}\rangle = \int_{-\infty}^{\infty} x(t)\overline{\psi_{a,b}(t)}\, dt \tag{11.122}$$

measures the similarity of $x(t)$ and the scaled, shifted wavelet $\psi_{a,b}(t)$. This makes it a multiscale shape detection technique.

Because Grossmann and Morlet's wavelets are also bandpass filters, the convolution (11.122) effects a frequency selection from the source signal $x(t)$. Assuming that $\psi(t) \in L^2(\mathbb{R})$ is an analyzing wavelet and $\Psi(\omega)$ is its radial Fourier transform, the inverse wavelet transform is given by

$$x(t) = \frac{1}{C_\Psi}\int_0^{\infty}\int_{-\infty}^{\infty} X_w(a,b)\frac{\psi_{a,b}(t)}{a^2}\, db\, da\,, \tag{11.123}$$

where C_ψ is the admissibility factor, $C_\psi = \int_{-\infty}^{\infty}\frac{|\Psi(\omega)|^2}{|\omega|}\,d\omega < \infty$. Thus, the transform (11.122) characterizes the signal $x(t)$ and can be the basis for signal comparisons, matching, and interpretation.

To accomplish wavelet-based signal analysis on a computer requires, of course, that the transform be discretized. For example, we might study transform coefficients of the form $x_w(m,n) = X_w(m\Delta, nT) = \langle x(t), \psi_{m\Delta,nT}(t)\rangle$. This leads to the construction of wavelet frames, which support signal characterization and numerically stable representation. These benefits would be all the stronger if a wavelet frame could be somehow refined into an orthonormal basis.

Nonetheless, it was the intuition of pioneering researchers that—just as they had shown for windowed Fourier expansions—a Balian–Low type of result would hold for wavelets, precluding orthonormal bases. As exciting as the developments of the 1980s had been, the prospects for well-localized short-time Fourier bases appeared quite bleak. The critical time-frequency sampling density, $T\Omega = 2\pi$, does not permit frames let alone orthogonal windowed Fourier bases, unless either the windowing function $w(t)$ or its Fourier transform $W(\omega)$ fails to be well-localized: $\|tw(t)\|_2 = \infty$ or $\|\omega W(\omega)\|_2 = \infty$ (Section 10.5.1). Anticipating equally negative results for the new scale-based transforms too, Meyer [19] tried to prove a version of Balian–Low for wavelets. To his own and everyone else's surprise, he failed and instead found an orthonormal wavelet basis!

Meyer's basis [4,19] proceeds from a wavelet $\psi(t)$ whose normalized Fourier transform is given by

$$\Psi(\omega) = \begin{cases} \dfrac{e^{\frac{j\omega}{2}}}{\sqrt{2\pi}} \sin\left[\dfrac{\pi}{2}\upsilon\left(\dfrac{3|\omega|}{2\pi} - 1\right)\right] & \text{for} \quad \dfrac{2\pi}{3} \le |\omega| \le \dfrac{4\pi}{3}, \\ \dfrac{e^{\frac{j\omega}{2}}}{\sqrt{2\pi}} \cos\left[\dfrac{\pi}{2}\upsilon\left(\dfrac{3|\omega|}{2\pi} - 1\right)\right] & \text{for} \quad \dfrac{4\pi}{3} \le |\omega| \le \dfrac{8\pi}{3}, \\ 0 & \text{otherwise}. \end{cases} \tag{11.124}$$

In (11.124) $\upsilon(t)$ is a C^k signal, where $v(t) \approx u(t)$, except on (0, 1). It specializes the $v(t)$ used earlier; the extra proviso is $v(t) + v(1 - t) = 1$. Figure 11.8 shows Meyer's wavelet $\psi(t) = \mathcal{F}^{-1}[\Psi(\omega)]](t)$.

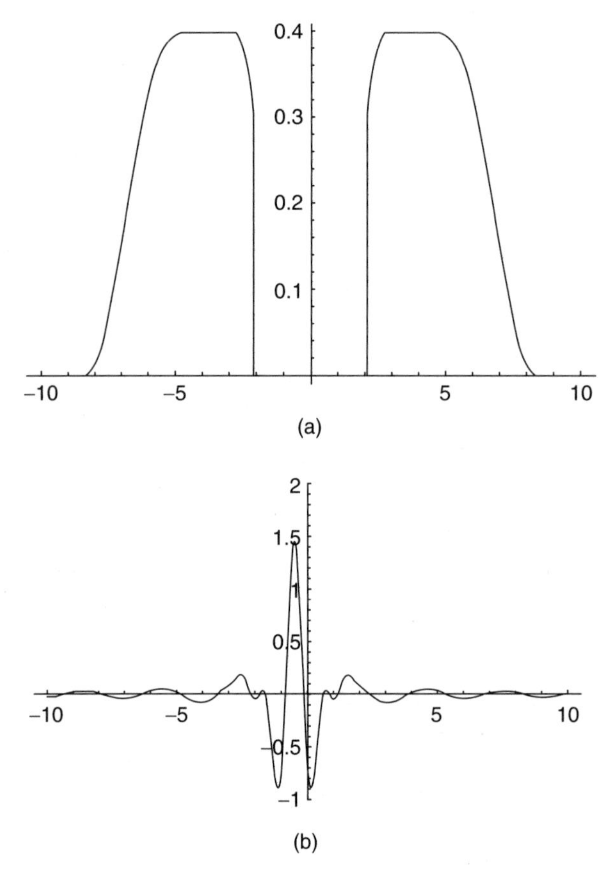

Fig. 11.8. Its normalized Fourier transform (a) and the Meyer wavelet (b). Its translations and dilations form an orthonormal basis for L^2 signals.

Until this discovery, mathematicians had more or less given up on finding orthonormal expansions for $L^2(\mathbb{R})$ using smooth basis elements. It was assumed that there had to be discontinuities in the time domain such as with Haar's basis, which was discovered some 75 years earlier [20], or in the frequency domain, such as with Shannon's $\mathrm{sinc}(t) = \sin(t)/t$ basis of Ref. 21 (Section 3.3.3). A later construction of Strömberg provides another orthonormal basis of continuous functions [22]; it predated Meyer's function by a few years and had been largely overlooked.

From the desks of several other mathematicians, more examples of orthonormal wavelet bases soon issued. But—except for the intricate calculations, carefully concocted estimations, and miraculous cancellations—there seemed to be no connection between these diverse constructions. Could there be no rules for building them? It was a fascinating mess.

11.4.1 Multiresolution Analysis

The unifying breakthrough came when Mallat [23] and Meyer elaborated the concept of a *multiresolution analysis* (MRA) for square-integrable signals. A computer vision researcher, Mallat was especially inspired by the similarities between some of the recent wavelet basis developments and work in pyramid decompositions for signal and image analysis such as the Laplacian pyramid [24], quadrature mirror filter banks employed in communication engineering [25], and scale space decompositions [26–28].

The MRA concept leads to a rich theory of the scale-based structure of signals. As a bonus, the MRA establishes a set of rules for constructing a wide range of orthonormal wavelet bases. Mallat and Meyer found the rules for building orthonormal wavelet bases in a quite unexpected place: the very applied areas of multiscale signal decomposition, image analysis, and efficient communication engineering. The discovery of important theoretical concepts out of utterly practical problems seems to be a distinct characteristic within the new discipline of wavelet analysis.

11.4.1.1 Definition. A multiresolution analysis of $L^2(\mathbb{R})$ is an abstract structure, but it has close links to several signal analysis ideas that we have already covered. We present the formal definition and develop some of the theory of bases made up of translations of a single root signal. The presentation follows closely and may be considered a tutorial on the classic papers [23, 29].

Definition (Multiresolution Analysis). A *multiresolution analysis* (or *multiresolution approximation*, MRA) is a chain of closed subspaces $\{V_i : i \in \mathbb{Z}\}$ in $L^2(\mathbb{R})$ such that the following conditions hold:

(i) The V_i are nested within one another: $\ldots \subset V_{-1} \subset V_0 \subset V_1 \subset V_2 \subset \ldots$.

(ii) The union of the V_i is dense in $L^2(\mathbb{R})$: $\overline{\bigcup_{n=-\infty}^{\infty} V_i} = L^2(\mathbb{R})$.

(iii) The intersection of the V_i is the signal of zero norm (zero almost everywhere), which we write $\bigcap_{i=-\infty}^{\infty} V_i = 0$.

(iv) Elements of the spaces are dyadically scaled (more precisely, *dilated*) versions of one another: $x(t) \in V_i \Leftrightarrow x(2t) \in V_{i+1}$.

(v) For any $x(t) \in V_0$ and any $k \in \mathbb{Z}$, $x(t-k) \in V_0$.

(vi) There is an isomorphism from V_0 onto the Hilbert space of square-summable discrete signals $I: V_0 \to l^2$ such that for any $k \in \mathbb{Z}$, if $I(x(t)) = s(n) \in l^2$, then $I(x(t-k)) = s(n-k)$.

Remark. We give the classic definition and notation for the MRA [23]. It has become common to index the V_i in the other direction: $V_i \supset V_{i+1}$. So readers must pay close attention to an author's V_i indexing.

Nowadays, many treatments (for instance, Refs. 3, 8, and 9) replace (vi) with the requirement that V_0 has an orthonormal basis of translates of a single finite-energy signal: $\{\phi(t-n) \mid n \in \mathbb{Z}\}$. This works. But so early on, it also seems incredible; we prefer to proceed from the apparently weaker criterion. In Section 11.4.2 we demonstrate that there is indeed a special function in V_0, called a *scaling function*, whose translates comprise an orthonormal basis of V_0.

Finally, note that by an *isomorphism* in (vi) we mean only a bounded, one-to-one, linear map, with a bounded inverse. Some mathematics texts, for example [30], define the term to mean also $\langle Ix, Iy\rangle = \langle x, y\rangle$, which implies an *isometry*; this we do not assume herein. The last MRA property (vi) is very strong, though. The isomorphism is a bounded linear map: There is an M such that $\|Ix\| \le M\|x\|$ for all $x \in V_0$. For linear maps this is equivalent to continuity. Range(I) is all of l^2. If it were an isometry, then we would be easily able to show that V_0 has a scaling function; but with our weaker assumption, this requires quite a bit more work.

11.4.1.2 Examples. Although it is rich with mathematical conditions, which might appear difficult to satisfy, we can offer some fairly straightforward instances of multiresolution analyses. Here are three examples where the root spaces consist of:

- Step functions;
- Piecewise linear functions;
- Cubic splines.

Note that the root spaces in these examples contain increasingly smooth signals.

Example (Step Functions). It is easiest to begin with a root space V_0 comprised of step functions and define the spaces of non-unit scale by dilation of V_0 elements. Let $u(t)$ be the analog unit step signal and set $V_0 = \{x(t) \in L^2(\mathbb{R}) \mid$ for all $n \in \mathbb{Z}$, there is a $c_n \in \mathbb{R}$ such that $x(t) = c_n[u(t-n) - u(t-n-1)]$ for $t \in (n, n+1)\}$. So elements of V_0 are constant on the open unit intervals $(n, n+1)$. The boundary values of the signals do not matter, since $\mathbb{Z}$ is a countable set and thus has Lebesgue measure zero. We define $V_i = \{y(t) \in L^2(\mathbb{R}) \mid \text{for some } x(t) \in V_0, y(t) = x(2^i t)\}$. Thus, V_i signals are constant on intervals $(n2^{-i}, (n+1)2^{-i})$, where $n \in \mathbb{Z}$. Let us show that each of the MRA properties holds.

(i) Signals that are constant on $(n2^{-i}, (n+1)2^{-i})$ for all $n \in \mathbb{Z}$ will also be constant on subintervals $(n2^{-i-1}, (n+1)2^{-i-1})$, so the first property holds.

(ii) From Chapter 3, we know that the step functions are dense in $L^2(\mathbb{R})$; since arbitrarily narrow steps are contained in V_i for i sufficiently large, we know that the V_i are dense in $L^2(\mathbb{R})$.

(iii) For a nonzero signal to be in all of the V_i, it would have to have arbitrarily wide steps, so the intersection property must be satisfied.

(iv) This is how we define the V_i for $i \neq 0$.

(v) Integral translates of signals in V_0 are obviously in V_0, since an integral translate is still constant on unit intervals.

(vi) If $x(t) \in V_0$, and $x(t) = c_n(u(t-n) - u(t-n-1)]$ for $t \in (n, n+1)$, then we set $I(x(t)) = s(n)$, where $s(n) = c_n$ for all $n \in \mathbb{Z}$; then $I(x(t-k)) = s(n-k)$, and I is an isomorphism. This is left as an exercise.

This MRA is the orthonormal basis of Haar in modern guise [20]. For analyzing blocky signals, this simple MRA is appropriate (Figure 11.9a). But when studying smoother signals, decompositions on the Haar set require quite a few coefficients in

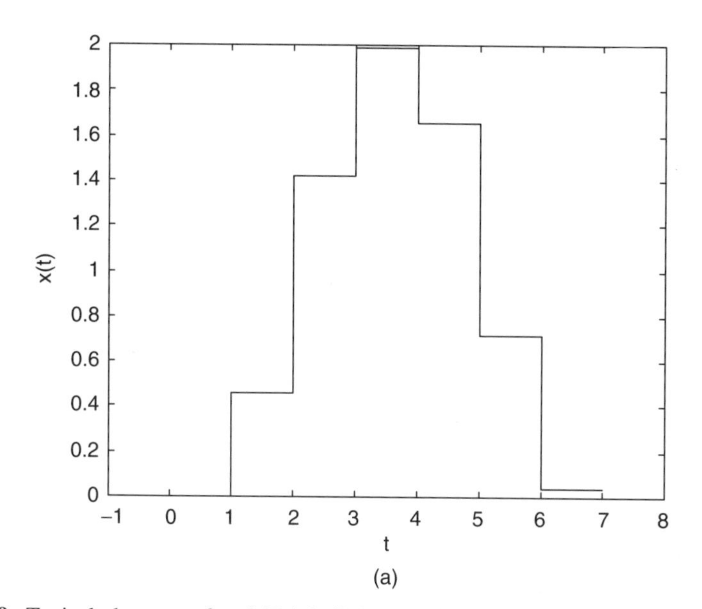

Fig. 11.9. Typical elements of an MRA built by step functions (a), piecewise linear functions (b), and cubic splines (c).

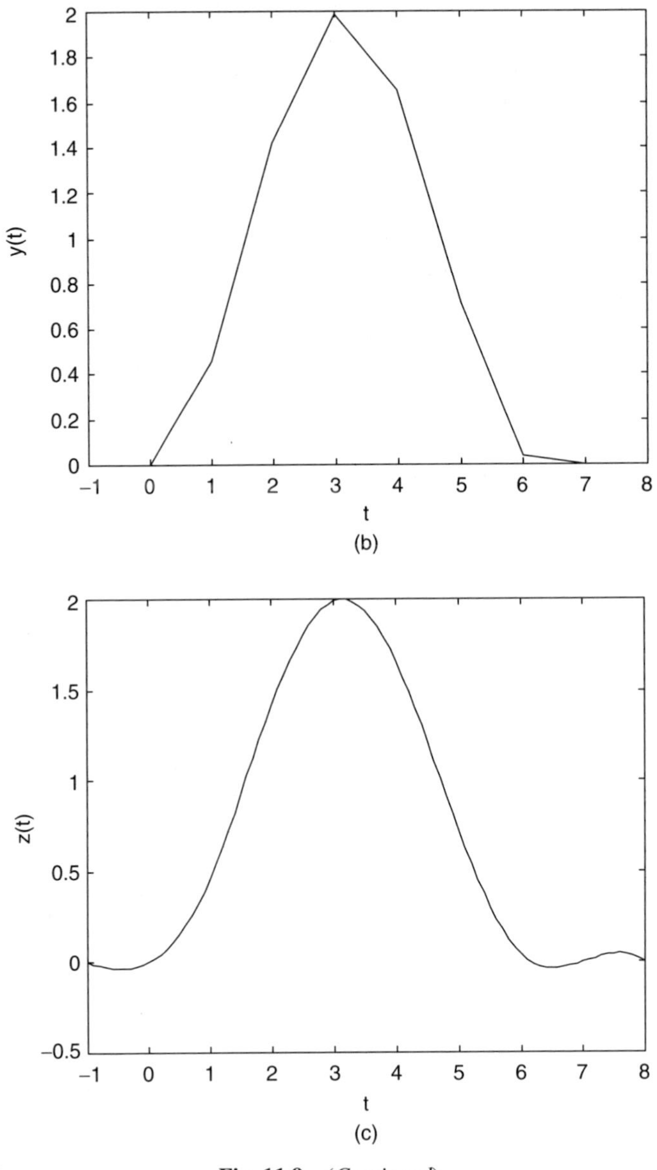

Fig. 11.9 *(Continued)*

order to smooth out the discontinuities present in the basis elements. So the approximations are often quite inefficient.

Example (Piecewise Linear Functions). Let us refine the multiscale structure so that it employs piecewise continuous signals. MRAs using such functions are better for signal analysis purposes when we have to interpret signals that do not contain abrupt jumps in value (Figure 11.9b). Let us define V_0 to be the $L^2(\mathbb{R})$ continuous functions that are piecewise linear on integral intervals $[n, n+1]$. We define the remaining spaces via the MRA's dilation property (iv): $x(t) \in V_i \Leftrightarrow x(2t) \in V_{i+1}$. The six MRA properties are clear, except perhaps for density (ii) and the isomorphism (vi). Note, however, that a piecewise continuous $x(t) \in V_0$ is determined uniquely by its values on integers. We define $I(x(t)) = s(n)$, where $s(n) = x(n)$ for all $n \in \mathbb{Z}$. We can approximate a step function to arbitrary precision with piecewise linear functions, and the step functions are dense in $L^2(\mathbb{R})$, so the second MRA property holds. Strömberg [22] elaborated this MRA's theory.

Example (Cubic Splines). A more interesting example relies on cubic splines. Here, the root space V_0 consists of all functions that are twice continuously differentiable and equal to a cubic polynomial on integral intervals $[n, n+1]$. Again, dilation defines the other V_i: $x(t) \in V_i \Leftrightarrow x(2t) \in V_{i+1}$. Numerical analysis texts, (e.g., Ref. 31, show that a continuous function can be approximated to any precision with a cubic spline, so the V_i are dense in $L^2(\mathbb{R})$.

As we continue to develop MRA theory, we shall return to these examples.

11.4.1.3 Links to Signal Analysis Legacy. Before expounding more theory, let us reflect on how the multiresolution analysis concept echoes many ideas from previous multiscale signal analysis techniques. In fact, stripped of their mathematical formalism, several of the multiresolution analysis properties (i)–(vi) have conceptual precedents in prior multiscale representations. Others embody ideas that were only implicit in the intuitive constructs of earlier methods of interpretation.

For example, the nested sequence of subspaces in (i) embodies the concept of the representations becoming ever finer in resolution. The inclusion property indicates that every coarse representation of a signal may also be considered to be a fine resolution version of some other waveform that has an even coarser shape. The subspaces are closed; each V_i contains the limit of its convergent function sequences. Coarse resolution representations are useful because:

- Using them can reduce the time required for pattern searches, such as in elementary edge and feature detection applications [32–34].
- Some signal features appear only at certain scales [35].

We need closure to guarantee that given a finite-energy signal $x(t)$ and an approximation error, there is some V_i that approximates $x(t)$ to within the tolerance.

What does the union property (ii) mean? If one looks in a sufficiently fine resolution space, then there is a finite energy signal that is arbitrarily close to any given signal. V_i signals are only approximations of real signals, but we can choose them to be very good approximations. Notice that (i) and (ii) encapsulate the intuitive notion of earlier researchers that scale is critical for structural decompositions of signals.

To interpret a signal we have to determine either the specific time extents of signal shapes, or we must search for shapes across all scales. This insight is the motivation behind scale space analysis [26–28], which we first considered in Chapter 4.

The intersection property (iii) tells us that, from a scale-based signal analysis standpoint, any meaningful signal must be visible to the MRA at some scale. That is, if a signal is composed of structures that have such a fine scale that it must appear in all the subspaces, then this signal must be the almost everywhere zero signal.

The next property concern resolution. Next, if a function is in space V_i, then its dilation by a factor of 2 is in the next higher resolution space V_{i+1}. Furthermore, its dilation by a factor of 1/2 is in the lower resolution space V_{i-1}. Thus, the implication of (iv) is that one ascends and descends in resolution by means of dilations, such as in the classic Laplacian pyramid construction [24].

The subspace V_0 contains signals that resolve features to the unit of measure. Signals in V_0 may be translated by integral amounts, and we are assured that the result remains in the root space (v).

Lastly, discrete samples characterize the MRA functions that model the signals. Property (vi) formalizes this by requiring an isomorphism between V_0 and the square-summable sequences of real numbers. This discretization has the further property that the discrete representations of functions are translated when the function is translated. Without this invariance, the discrete samples associated with a waveform $x(t)$ in V_0 might change with the translation of $x(t)$ by integral steps. For the lower resolution spaces V_i, $i < 0$, translation invariance does not hold. It will become apparent in applications that the overall situation for translation invariance is far from satisfactory; indeed, it is problematic.

11.4.1.4 Bases of Translates: Theory. Our first theoretical result on MRAs comes right out of the sixth criterion. It shows that the root space V_0 has a basis, consisting of integral translates of a single square-integrable signal. Multiresolution analysis exploits the special properties of such bases.

Proposition (Basis). If $\{V_i \mid i \in \mathbb{Z}\}$ is an MRA in $L^2(\mathbb{R})$, then there is $e(t) \in V_0$ such that $\{e(t-k) \mid k \in \mathbb{Z}\}$ is a basis for V_0.

Proof: Let $\delta(n-k)$ be the discrete delta signal delayed by $k \in \mathbb{Z}$. $\{\delta(n-k) \mid k \in \mathbb{Z}\}$ is an orthonormal basis for l^2. Let $I: V_0 \to l^2$ be the isomorphism of MRA property (vi). Since I is onto, we may set $e_k(t) = I^{-1}(\delta(n-k))$. Then $\{e_k(t) \mid k \in \mathbb{Z}\}$ is a basis for V_0. However, $I(e_0(t)) = \delta(n)$, so that the translation-invariance provision of (vi) also implies $I(e(t-k)) = \delta(n-k)$, whence $e_k(t) = e(t-k)$. ■

The proposition guarantees a basis that is useful for those pattern matching applications where we expect candidate signals containing the shape of the root element. The basis elements comprise our model or prototype signals. For computational purposes, we prefer such bases to be orthonormal, since that simplifies expansion coefficient computations. But, again, orthogonality requires more work. Let us explore some of the theory of such bases.

Definition (Riesz Basis). $E = \{e_n \mid n \in \mathbb{Z}\}$ is an *unconditional* or *Riesz basis* in a Hilbert space H if

(i) E spans H: For any $x \in H$, there is $s \in l^2$ such that

$$x = \sum_{n=-\infty}^{\infty} s(n)e_n. \tag{11.125a}$$

(ii) There are $0 < A \le B < \infty$ such that for any $s \in l^2$,

$$A\|s\| \le \left\| \sum_{n=-\infty}^{\infty} s(n)e_n \right\| \le B\|s\|. \tag{11.125b}$$

The constants A and B are called the *lower* and *upper Riesz basis bounds*, respectively.

Notice how (11.125b) cross-couples the norms of H and l^2. An orthonormal basis is also a Riesz basis. Note too that the lower bound condition in (11.125b) implies that a Riesz basis is linearly independent (exercises). The next result shows that property (vi) of an MRA is equivalent to V_0 having a Riesz basis.

Theorem (Riesz Basis Characterization). $E = \{e_k \mid k \in \mathbb{Z}\}$ is a Riesz basis in a Hilbert space H if and only if:

(i) There is an isomorphism I from H onto l^2 such that $I(e_k) = \delta(n - k)$, where $\delta(n)$ is the discrete delta.
(ii) I^{-1} is bounded.

Proof: Let $E = \{e_k \mid k \in \mathbb{Z}\}$ be a Riesz basis in H. Since for any $x \in H$, there is $s \in l^2$ such that (11.125a) holds, we may set $Ix = s$. This map is well-defined by the linear independence of E; that is, for each x, the coefficient sequence $\{s(n)\}$ is unique. The linearity follows from the properties of the square-summable sequences, and clearly $I(e_k) = \delta(n - k)$. The map is also onto, because any $s \in l^2$ defines an element of H as in (11.125a). This sum converges in H by the upper bound inequality in (11.125b). The boundedness of I follows as well: $\|Ix\| = \|s\| \le A^{-1}\|x\|$. The inverse $J = I^{-1}$ exists because I is one-to-one and onto (a *bijection*). $\|Js\| = \|x\| = \|\Sigma s(k)e_k\| \le B\|s\|$ by (11.125b), so J is bounded too.

Conversely, let $I: H \to l^2$ be an isomorphism obeying (i) and (ii). Let $x \in H$. We need to show that x is in the closure of Span(E). Let $Ix = s \in l^2$. In general, it does not necessarily follow that if an isomorphism is bounded, then its inverse is bounded. However, we are assuming $J = I^{-1}$ is bounded, so it is also continuous [30]. Thus,

$$\lim_{N\to\infty} J\left(\sum_{k=-N}^{N} s(k)\delta(n-k)\right) = \sum_{k=-\infty}^{\infty} s(k)J(\delta(n-k)) = \sum_{k=-\infty}^{\infty} s(k)e_k, \tag{11.126}$$

and x is a limit of elements in Span(E). Let $\|I\| = A^{-1}$ and $\|J\| = B$. Then A and B are the lower and upper Riesz bounds for E (11.125b). ■

Corollary. Property (vi) of an MRA is equivalent to V_0 having a Riesz basis of translates.

Proof: Since the isomorphism I in (vi) has a bounded inverse and I is onto, we may find $e_k(t)$ such that $I(e_k(t)) = \delta(n-k)$, where $\delta(n)$ is the discrete delta signal. The theorem tells us that $\{e_k(t) \mid k \in \mathbb{Z}\}$ is a Riesz basis. The translation invariance for V_0 implies that $e_k(t) = e(t-k)$. ■

The conditions (11.125a) and (11.125b) for a Riesz basis resemble the criterion for a frame, which we studied in Chapters 3 and 10. Indeed, the following corollary shows that a Riesz basis is a frame. Of course, the converse is not true; a Riesz basis must be linearly independent, while frames can be overcomplete.

Corollary (Frame). If $E = \{e_k \mid k \in \mathbb{Z}\}$ is a Riesz basis in a Hilbert space H, then E is a frame.

Proof: Let let $I: H \to l^2$ be the isomorphism promised in the Riesz basis characterization: $I(e_k) = \delta(n-k)$. Let I^* be the Hilbert space adjoint operator for I. We introduced the adjoint operator in Chapter 3 and therein applied it to the study of frames (Section 3.3.4). The adjoint cross-couples the inner product relations of H and l^2 so that if $s(n)$ is square-summable, then $\langle Ix, s\rangle = \langle x, I^*s\rangle$. Note that $I^*: l^2 \to H$ is an isomorphism, bounded, and in fact $\|I^*\| = \|I\|$. For example, to show I^* is one-to-one, let $I^*v = I^*w$ for some $v(n), w(n) \in l^2$. Then for any $h \in H$, $\langle h, I^*v\rangle = \langle h, I^*w\rangle$. But this implies $\langle Ih, v\rangle = \langle Ih, w\rangle$. Since I is onto and h is arbitrary, Ih could be any finite-energy discrete signal. In other words, $\langle s, v\rangle = \langle s, w\rangle$ for all $s \in l^2$. But this means $v = w$ too. We leave the remaining I^* details as exercises.

Now, let $f_k = I^*(\delta(n-k))$. Since $\delta(n-k) = (I^*)^{-1}(f_k)$ and $((I^*)^{-1})^{-1} = I^*$ is bounded, the proposition says that $F = \{f_k \mid k \in \mathbb{Z}\}$ is a Riesz basis in H. If $x \in H$, then by (11.125a) there is $s \in l^2$ such that $x = \sum_{k=-\infty}^{\infty} s(k) f_k$. We see that $\langle x, e_k\rangle = s(k)$ by calculating $\langle f_i, e_k\rangle = \langle I^*(\delta(n-i)), e_k\rangle = \langle \delta(n-i), I(e_k)\rangle = \langle \delta(n-i), \delta(n-k)\rangle$. So

$$\sum_{k=-\infty}^{\infty} \left|\langle x, e_k\rangle\right|^2 = \sum_{k=-\infty}^{\infty} |s(k)|^2 = \|s\|_2^2. \tag{11.127}$$

Since I^* and $(I^*)^{-1}$ are bounded and $x = I^*s$, we have

$$\frac{\|x\|^2}{\|I^*\|^2} \le \sum_{k=-\infty}^{\infty} \left|\langle x, e_k\rangle\right|^2 = \sum_{k=-\infty}^{\infty} |s(k)|^2 \le \|x\|^2 \left\|(I^*)^{-1}\right\|^2, \tag{11.128}$$

which is precisely a frame condition on E. ■

Theorem (Orthonormal Translates). Let $\phi(t) \in L^2(\mathbb{R})$ and $\Phi(\omega) = \mathcal{F}[\phi(t)](\omega)$ be its (radial) Fourier transform. The family $F = \{\phi(t-k) \mid k \in \mathbb{Z}\}$ is orthonormal if and only if

$$\sum_{k=-\infty}^{\infty} |\Phi(\omega + 2\pi k)|^2 = 1 \tag{11.129}$$

for almost all $\omega \in \mathbb{R}$.

Proof: An interesting application of Fourier transform properties makes this proof work. Let us define $a_k = \langle \phi(t), \phi(t-k) \rangle$. Note that—by a simple change of variable in the inner product integral—F is orthonormal if and only if a_k is zero when $k \neq 0$ and unity when $k = 0$. We calculate

$$a_k = \langle \phi(t), \phi(t-k) \rangle = \frac{1}{2\pi}\langle \Phi(\omega), \Phi(\omega)e^{-jk\omega} \rangle = \frac{1}{2\pi}\int_{-\infty}^{\infty} \Phi(\omega)\overline{\Phi(\omega)}e^{jk\omega}\, d\omega. \tag{11.130}$$

by the Parseval and shift properties. The right-hand integrand in is $|\Phi(\omega)|^2 e^{jk\omega}$. We break up the integral into 2π-wide pieces, invoke the exponential's periodicity, and swap the order of summation and integration to get

$$a_k = \frac{1}{2\pi}\sum_{n=-\infty}^{\infty}\int_0^{2\pi} |\Phi(\omega + 2\pi n)|^2 e^{jk\omega}\, d\omega = \frac{1}{2\pi}\int_0^{2\pi} e^{jk\omega} \sum_{n=-\infty}^{\infty} |\Phi(\omega + 2\pi n)|^2\, d\omega. \tag{11.131}$$

We move the sum into the integral, since $\Phi = \mathcal{F}\phi \in L^2(\mathbb{R})$, so that $|\Phi(\omega)|^2 \in L^1(\mathbb{R})$. Let us define

$$P_\Phi(\omega) = \sum_{n=-\infty}^{\infty} |\Phi(\omega + 2\pi n)|^2. \tag{11.132}$$

Now observe

$$\int_0^{2\pi} \sum_{n=-\infty}^{\infty} |\Phi(\omega + 2\pi n)|^2\, d\omega = \int_0^{2\pi} P_\Phi(\omega)\, d\omega = \int_{-\infty}^{\infty} |\Phi(\omega)|^2\, d\omega = \|\Phi(\omega)\|_2^2, \tag{11.133}$$

so that $P_\Phi(\omega)$ is finite for almost all $\omega \in \mathbb{R}$. This allows us to interchange summation and integration with the Lebesgue integral (Chapter 3) in (11.131). We can say more about $P_\Phi(\omega)$: It is 2π-periodic, and the right-hand side of (11.131) is precisely the expression for its Fourier series coefficient, which is a_k. Within the inner product there hides nothing less than a Fourier series analysis equation for the special periodic function $P_\Phi(\omega)$! We use this periodization argument a lot.

Let us check our claim that $P_\Phi(\omega) = 1$ almost everywhere if and only if the family of translates F is orthonormal. First, if $P_\Phi(\omega) = 1$, then (11.131) becomes

$$a_k = \frac{1}{2\pi}\int_0^{2\pi} e^{jk\omega}\,d\omega = \langle \phi(t), \phi(t-k)\rangle. \tag{11.134}$$

The integral in (11.134) is 2π if $k = 0$ and zero if $k \neq 0$. So $F = \{\phi(t-k)\}$ must be orthonormal. Conversely, suppose F is orthonormal so that $a_k = 1$ if $k = 0$ and $a_k = 0$ if $k \neq 0$. Because (11.131) gives the Fourier series coefficients for the 2π-periodic function $P_\Phi(\omega)$, we know that $P_\Phi(\omega)$ has all zero Fourier series coefficients except for its DC term, which is one. In other words,

$$P_\Phi(\omega) = \sum_{k=-\infty}^{\infty} a_k e^{jk\omega} = 1e^0 = 1. \tag{11.135}$$

■

The following corollary shows that when the translates of $\phi(t)$ are orthonormal, then its spectrum, as given by the support of $\Phi(\omega)$, cannot be too narrow [8]. Scaling functions cannot have simple frequency components. This result uses the Lebesgue measure of a set, an idea introduced in Section 3.4.1.

Corollary (Spectral Support). Let $\phi(t) \in L^2(\mathbb{R})$, $\Phi(\omega) = \mathcal{F}[\phi(t)](\omega)$ be its (radial) Fourier transform, let $\text{Support}(\Phi) = \{\omega \in \mathbb{R} \mid \Phi(\omega) \neq 0\}$, and let $\mu(A)$ be the Lebesgue measure of a measurable set A. If the family $F = \{\phi(t-k) \mid k \in \mathbb{Z}\}$ is orthonormal, then $\mu(\text{Support}(\Phi)) \geq 2\pi$. Under these assumptions, moreover, $\mu(\text{Support}(\Phi)) = 2\pi$ if and only if $|\Phi(\omega)| = \chi_A$, for some Lebesgue measurable $A \subset \mathbb{R}$ with $\mu(A) = 2\pi$.

Proof: Since $\|\phi\|_2 = 1$, we know $\|\Phi\|_2 = (2\pi)^{1/2}$, by Plancherel's formula. The theorem then implies $|\Phi(\omega)| \leq 1$ for almost all $\omega \in \mathbb{R}$. Consequently,

$$\mu(\text{Support}(\Phi)) = \int_{\text{Support}(\Phi)} 1\,d\omega \geq \int_{-\infty}^{\infty} |\Phi(\omega)|^2 d\omega = 2\pi. \tag{11.136}$$

Now suppose (11.136) is an equality, but $0 < |\Phi(\omega)| < 1$ on some set $B \subset \mathbb{R}$ with $\mu(B) > 0$. Then

$$\int_B |\Phi(\omega)|^2\,d\omega < \int_B 1\,d\omega = \mu(B) \tag{11.137}$$

and

$$\begin{aligned}\|\Phi\|_2^2 = 2\pi &= \int_{\text{Support}(\Phi)} |\Phi(\omega)|^2 d\omega < \mu(\text{Support}(\Phi)\backslash B) + \mu(B) \\ &= \mu(\text{Support}(\Phi)) = 2\pi\end{aligned} \tag{11.138}$$

a contradiction. Conversely, assume F is orthonormal and $|\Phi(\omega)| = \chi_A$, for some Lebesgue measurable $A \subset \mathbb{R}$ with $\mu(A) = 2\pi$. Then we quickly see that

$$\mu(A) = \mu(\text{Support}(\Phi)) = \|\Phi\|_2^2 = 2\pi. \tag{11.139}$$

■

Here is a second characterization of unconditional bases. In the next section, we use this result to find the scaling function for an MRA. We begin with a lemma [9].

Lemma. Suppose $\phi(t) \in L^2(\mathbb{R})$, $\Phi(\omega) = \mathcal{F}[\phi(t)](\omega)$ is its (radial) Fourier transform, $F = \{\phi(t-k) \mid k \in \mathbb{Z}\}$, and we define $P_\Phi(\omega)$ as above (11.132). If $s \in l^2$, then

$$\left\| \sum_{k=-\infty}^{\infty} s(k)\phi(t-k) \right\|_2^2 = \frac{1}{2\pi} \int_0^{2\pi} |S(\omega)|^2 P_\Phi(\omega)\, d\omega, \tag{11.140}$$

where $S(\omega)$ is the discrete-time Fourier transform of $s(k)$.

Proof: Let us consider a linear combination of the $\phi(t-k)$, $\sum_{k=p}^{q} s(k)\phi(t-k)$, where $s \in l^2$. Using the above periodization technique, we compute

$$\left\| \sum_{k=p}^{q} s(k)\phi(t-k) \right\|_2^2 = \frac{1}{2\pi} \left\| \sum_{k=p}^{q} s(k) e^{-jk\omega} \Phi(\omega) \right\|_2^2$$

$$= \frac{1}{2\pi} \int_{-\infty}^{\infty} \left| \sum_{k=p}^{q} s(k) e^{-jk\omega} \right|^2 |\Phi(\omega)|^2 d\omega = \frac{1}{2\pi} \int_0^{2\pi} \left| \sum_{k=p}^{q} s(k) e^{-jk\omega} \right|^2 P_\Phi(\omega)\, d\omega \tag{11.141}$$

By assumption, $s(k)$ is square-summable. Hence its discrete-time Fourier transform exists (Chapter 7), and we may pass to the double summation limit in (11.141). Indeed, as $p, q \to \infty$, the last integrand in (11.141) becomes $|S(\omega)|^2 P_\Phi(\omega)$, where $S(\omega)$ is the DTFT of $s(k)$. ■

Theorem (Riesz Translates Basis Characterization). Suppose $\phi(t) \in L^2(\mathbb{R})$, $\Phi(\omega) = \mathcal{F}[\phi(t)](\omega)$ is its (radial) Fourier transform, $F = \{\phi(t-k) \mid k \in \mathbb{Z}\}$, $0 < A \le B < \infty$, and we define $P_\Phi(\omega)$ as above (11.132). Then the following are equivalent:

(i) F is a Riesz basis with lower and upper bounds $\sqrt{A}$ and $\sqrt{B}$, respectively.

(ii) $A \le P_\Phi(\omega) \le B$ for almost all $\omega \in \mathbb{R}$.

Proof: Suppose (ii), and let $s \in l^2$. Then $A|S(\omega)|^2 \le |S(\omega)|^2 P_\Phi(\omega) \le B|S(\omega)|^2 < \infty$ almost everywhere, too, where $S(\omega)$ is the DTFT of $s(k)$. Integrating on $[0, 2\pi]$, we see

$$\frac{A}{2\pi} \int_0^{2\pi} |S(\omega)|^2 d\omega \le \frac{1}{2\pi} \int_0^{2\pi} |S(\omega)|^2 P_\Phi(\omega)\, d\omega \le \frac{B}{2\pi} \int_0^{2\pi} |S(\omega)|^2 d\omega. \tag{11.142}$$

We know that $2\pi\|s\|^2 = \|S\|^2$, and so the lemma implies

$$A\|s\|_2^2 \le \left\| \sum_{k=-\infty}^{\infty} s(k)\phi(t-k) \right\|_2^2 \le B\|s\|_2^2 ; \tag{11.143}$$

this is precisely the Riesz basis condition for lower and upper bounds $\sqrt{A}$ and $\sqrt{B}$, respectively.

Now let us assume (i) and try to show $A \le P_\Phi(\omega)$ almost everywhere on $[0, 2\pi]$. Following [9], we set $Q_{\Phi,a} = \{\omega \in [0, 2\pi]: P_\Phi(\omega) < a\}$. If the Lebesgue measure of $Q_{\Phi,a}$, $\mu(Q_{\Phi,a})$, is zero for almost all $a \in \mathbb{R}$, then $P_\Phi(\omega)$ diverges almost everywhere, and, in particular, $A \le P_\Phi(\omega)$. We can thus suppose that there is some $a \in \mathbb{R}$ such that $\mu(Q_{\Phi,a}) > 0$. Let χ_a be the characteristic function on $Q_{\Phi,a}$:

$$\chi_a(\omega) = \begin{cases} 1 & \text{if } \omega \in Q_{\Phi,a}, \\ 0 & \text{if } \omega \notin Q_{\Phi,a}. \end{cases} \tag{11.144}$$

By the theory of Lebesgue measure, if a set is measurable, then so is its characteristic function. This entitles us to compute the inverse discrete-time Fourier transform of $\chi_a(\omega)$:

$$x_a(n) = \frac{1}{2\pi}\int_0^{2\pi} \chi_a(\omega)e^{j\omega n}\,d\omega, \tag{11.145}$$

where $x_a \in l^2$. From (i) and the lemma we see

$$A\|x_a\|_2^2 \le \left\| \sum_{k=-\infty}^{\infty} x_a(k)\phi(t-k) \right\|_2^2 = \frac{1}{2\pi}\int_0^{2\pi} |\chi_a(\omega)|^2 P_\Phi(\omega)\,d\omega$$

$$= \frac{1}{2\pi}\int_0^{2\pi} \chi_a(\omega)P_\Phi(\omega)d\omega = \frac{1}{2\pi}\int_{Q_a} P_\Phi(\omega)d\omega. \tag{11.146}$$

By our choice of $Q_{\Phi,a}$, $P_\Phi(\omega) < a$ for $\omega \in Q_{\Phi,a}$, and (11.146) entails

$$A\|x_a\|_2^2 \le \frac{a}{2\pi}\mu(Q_{\Phi,a}). \tag{11.147}$$

But $\|x_a\|^2 = (2\pi)^{-1}\|\chi_a\|^2 = (2\pi)^{-1}\mu(Q_{\Phi,a})$, and, by (11.147), $A \le a$. This gives us a contradiction by the following observation. If $A \le P_\Phi(\omega)$ almost everywhere on $[0, 2\pi]$, then we are done. Otherwise, there must be some $U \subseteq \mathbb{R}$ such that $\mu(U) > 0$ and $P_\Phi(\omega) < A$. But then there must also be some $a > 0$ such that $P_\Phi(\omega) < a < A$ and $\mu(Q_{\Phi,a}) > 0$. But our argument above has just proven that $A \le a$, a contradiction.

Let us continue to assume (i) and try to show $P_\Phi(\omega) \le B$ for almost all $\omega \in [0, 2\pi]$. Define $P_{\Phi,a} = \{\omega \in [0, 2\pi]: P_\Phi(\omega) > a\}$. Much like before, if $\mu(P_{\Phi,a}) = 0$ almost everywhere, then $P_\Phi(\omega) = 0$ for almost all $\omega \in [0, 2\pi]$, and thus $P_\Phi(\omega) \le B$. Assume that some $a > 0$ gives $\mu(P_{\Phi,a}) > 0$. Now the argument parallels the one just given and is left as an exercise [9]. For an alternative proof, see [8]. ■

11.4.2 Scaling Function

From our study of bases of translates and Riesz expansions, we can show that every multiresolution analysis $V = \{V_i\}$ has a special *scaling function*, whose translates form an orthonormal basis for V_0. The MRA structure is appropriately named; the scaling function property further implies that signals in every V_i look like combinations of dilated versions of V_0 elements.

11.4.2.1 Existence. In the following result, the equalities in (11.148a) and (11.148b) are assumed to hold almost everywhere.

Proposition (Spanning Translates). If $x(t), y(t) \in L^2(\mathbb{R})$, $X(\omega)$ and $Y(\omega)$ are their respective radial Fourier transforms, and $s(k) \in l^2$, then the following are equivalent:

$$y(t) = \sum_{k=-\infty}^{\infty} s(k)x(t-k), \tag{11.148a}$$

$$Y(\omega) = S(\omega)X(\omega), \tag{11.148b}$$

where $S(\omega)$ is the discrete-time Fourier transform of $s(k)$.

Proof: Now, assuming (11.148b), we have

$$y(t) = \frac{1}{2\pi}\int_{-\infty}^{\infty} S(\omega)X(\omega)e^{j\omega t}d\omega = \frac{1}{2\pi}\int_{-\infty}^{\infty}\left(\sum_{k=-\infty}^{\infty} s(k)e^{-jk\omega}\right)X(\omega)e^{j\omega t}d\omega. \tag{11.149}$$

Hence,

$$y(t) = \sum_{k=-\infty}^{\infty} s(k)\frac{1}{2\pi}\int_{-\infty}^{\infty} X(\omega)e^{j\omega(t-k)}d\omega = \sum_{k=-\infty}^{\infty} s(k)x(t-k). \tag{11.150}$$

To show the converse, we work backwards through the equalities in (11.150) to the front of (11.149), a Fourier transform synthesis equation for $y(t)$. We must have (11.148b) except on a set of Lebesgue measure zero. ■

This section's main result comes from the classic source papers on wavelets and multiresolution analysis [23, 29].

Theorem (Scaling Function). If $\{V_i: i \in \mathbb{Z}\}$ is an MRA, then there is some $\phi(t) \in V_0$ such that $\{\phi(t-k): k \in \mathbb{Z}\}$ is an orthonormal basis of V_0.

Proof: By the Riesz Basis Characterization (Section 11.4.1.3), there is some $g(t) \in V_0$ such that $F = \{g(t-k) \mid k \in \mathbb{Z}\}$ is a Riesz basis for V_0. Let us say it has lower and upper bounds $\sqrt{A}$ and $\sqrt{B}$, respectively. The Riesz Translates Basis Characterization implies

$$A \le \sum_{k=-\infty}^{\infty} |G(\omega + 2\pi k)|^2 \le B \tag{11.151}$$

for almost all $\omega \in \mathbb{R}$, where $G(\omega)$ is the (radial) Fourier transform of $g(t)$. Note that the sum in (11.151) is the 2π-periodic function $P_G(\omega)$, defined in (11.132). The Riesz bounds on $P_G(\omega)$ allow us to define (almost everywhere) the $L^2(\mathbb{R})$ function

$$\Phi(\omega) = \frac{G(\omega)}{\sqrt{P_G(\omega)}}. \tag{11.152}$$

$\Phi(\omega)$ is the Fourier transform of $\phi(t) \in L^2(\mathbb{R})$, and our claim is that $\phi(t)$ works. The previous proposition implies $\phi(t) \in \overline{\text{Span}\{g(t-k)\}}$. Since V_0 is closed, $\phi(t) \in V_0$, and so $\phi(t-k) \in V_0$, by MRA property (v). Equation (11.152) works both ways, and we see that $F = \{g(t-k) \mid k \in \mathbb{Z}\}$—which is dense in V_0—is in the closure of $\{\phi(t-k\} \mid k \in \mathbb{Z}\}$. Thus, $\overline{\text{Span}\{\phi(t-k)\}} = V_0$. It remains to show that the $\phi(t-k)$ are orthonormal. We calculate

$$\sum_{k=-\infty}^{\infty} |\Phi(\omega + 2\pi k)|^2 = \sum_{k=-\infty}^{\infty} \left| \frac{G(\omega + 2\pi k)}{\sqrt{P_G(\omega + 2\pi k)}} \right|^2 = \sum_{k=-\infty}^{\infty} \frac{|G(\omega + 2\pi k)|^2}{P_G(\omega)} = 1. \tag{11.153}$$

By the Orthonormal Translates criterion (11.129), $\{\phi(t-k\} \mid k \in \mathbb{Z}\}$ is an orthonormal set.

Corollary. Let $\{V_i: i \in \mathbb{Z}\}$ be an MRA, and $\phi(t) \in V_0$ be given by the theorem. Then, $\{2^{i/2}\phi(2^i t - k): k \in \mathbb{Z}\}$ is an orthonormal basis of V_i.

Proof: By properties (iv) and (v) of the MRA, the scaled versions of an orthonormal basis for V_0 will constitute an orthonormal basis for V_i. ■

Definition (Scaling Function). Let $V = \{V_i: i \in \mathbb{Z}\}$ be an MRA and $\phi(t) \in V_0$ such that $\{\phi(t-k): k \in \mathbb{Z}\}$ is an orthonormal basis of V_0. Then $\phi(t)$, known from the theorem, is called a *scaling function* of the MRA.

Any translate $\phi(t-k)$ of a scaling function $\phi(t)$ is still a scaling function. The next corollary [9] characterizes scaling functions for an MRA.

Corollary (Uniqueness). Let $V = \{V_i: i \in \mathbb{Z}\}$ be an MRA and $\phi(t) \in V_0$ be the scaling function found in the proof. Then $\theta(t) \in V_0$ is a scaling function for V if and only if there is a 2π-periodic function $P(\omega)$ such that

(i) $\Theta(\omega) = P(\omega)\Phi(\omega)$;
(ii) $|P(\omega)| = 1$ almost everywhere on $[0, 2\pi]$.

Proof: Exercise. ■

11.4.2.2 Examples. Let us look at some examples of scaling functions for the three multiresolution analyses that we know.

Example (Step Functions). The scaling function for the Haar MRA, for which V_0 consists of constant functions on unit intervals $(n, n-1)$, is just the unit square pulse $\phi(t) = u(t) - u(t-1)$.

Example (Piecewise Continuous Functions). Finding this scaling function is not so direct. The continuity of V_0 elements forces us to reflect on how a possible scaling function $\phi(t)$ might be orthogonal to its translates $\phi(t-k)$. It is clear that $\phi(t)$ cannot be finitely supported. For then we could take the last interval $(n, n+1)$ to the right over which $\phi(t)$ is nonzero, the last interval $(m, m+1)$ proceeding to the left over which $\phi(t)$ is nonzero, and compute the inner product $\langle\phi(t), \phi(t-(n-m))\rangle$. A simple check of cases shows that it is never zero. Evidently, $\phi(t) \neq 0$ on $(n, n+1)$ for arbitrarily large $|n|$, and the inner products $\langle\phi(t), \phi(t-k)\rangle$ involve an infinite number of terms.

But rather than stipulating from the start that V_0 must have a scaling function, we have elected to define our MRAs using the apparently weaker isomorphism condition (vi). The existence of this isomorphism $I: V_0 \to l^2$, which commutes with translations by integral amounts, is equivalent to V_0 having a Riesz basis. This facilitates our study of the Strömberg MRA. If we can find a Riesz basis $F = \{g(t-k) \mid k \in \mathbb{Z}\}$ for V_0, then the Scaling Function Theorem (11.152) readily gives the Fourier transform $\Phi(\omega)$ of $\phi(t)$. Let $g(t) \in V_0$ be the simple triangular pulse with $g(0) = 1$, $g(t) = t+1$ on $(-1, 0)$, $g(t) = 1-t$ on $(0, 1)$, and $g(t) = 0$ otherwise. Then $x(t) = \Sigma a_k g(t-k)$ is piecewise linear, continuous, and $x(k) = a_k$ for all $k \in \mathbb{Z}$. We can define the isomorphism I by $(Ix)(k) = a_k$. This map commutes with integer translations. The Riesz Basis Characterization implies that $\{g(t-k) \mid k \in \mathbb{Z}\}$ is a Riesz basis. In fact, $g(t) = I^{-1}(\delta(n))$, where I is the isomorphism from V_0 to l^2, and $\delta(n)$ is the discrete delta signal. We compute

$$G(\omega) = \int_{-\infty}^{\infty} g(t)e^{-j\omega t}dt = \left[\frac{\sin\left(\frac{\omega}{2}\right)}{\frac{\omega}{2}}\right]^2 = \operatorname{sinc}^2\left(\frac{\omega}{2}\right). \tag{11.154}$$

Define

$$\Phi(\omega) = \frac{G(\omega)}{\sqrt{P_G(\omega)}} = \frac{\mathrm{sinc}^2\left(\frac{\omega}{2}\right)}{\sqrt{\sum\limits_{k=-\infty}^{\infty} \mathrm{sinc}^4\left(\frac{\omega+2\pi k}{2}\right)}} = \frac{\mathrm{sinc}^2\left(\frac{\omega}{2}\right)}{4\sin^2(\omega)\sqrt{\sum\limits_{k=-\infty}^{\infty} (\omega+2\pi k)^{-4}}}. \tag{11.155}$$

We define the utility the function $\Sigma_n(\omega)$ as follows:

$$\Sigma_n(\omega) = \sum_{k=-\infty}^{\infty} (\omega+2\pi k)^{-n}. \tag{11.156}$$

Though it tests our competence in differential calculus, it is possible [23] to develop closed form expressions for the $\Sigma_n(\omega)$, beginning with the standard summation [36]:

$$\Sigma_2(\omega) = \sum_{k=-\infty}^{\infty} (\omega+2\pi k)^{-2} = \frac{1}{4}\sin^{-2}\left(\frac{\omega}{2}\right). \tag{11.157}$$

Twice differentiating (11.157) gives

$$\frac{d^2}{d\omega^2}\Sigma_2(\omega) = 6\sum_{k=-\infty}^{\infty} (\omega+2\pi k)^{-4} = 6\Sigma_4(\omega) = \frac{1}{4}\cot^2\left(\frac{\omega}{2}\right)\csc^2\left(\frac{\omega}{2}\right) + \frac{1}{8}\csc^4\left(\frac{\omega}{2}\right). \tag{11.158}$$

Finally,

$$\Phi(\omega) = \frac{G(\omega)}{\sqrt{P_G(\omega)}} = \frac{\sqrt{6}}{\omega^2\sqrt{\frac{1}{4}\cot^2\left(\frac{\omega}{2}\right)\csc^2\left(\frac{\omega}{2}\right) + \frac{1}{8}\csc^4\left(\frac{\omega}{2}\right)}}. \tag{11.159}$$

Taking the inverse Fourier transform gives $\phi(t)$ (Figure 11.10). Notice from its magnitude spectrum that $\phi(t)$ is an analog low-pass filter.

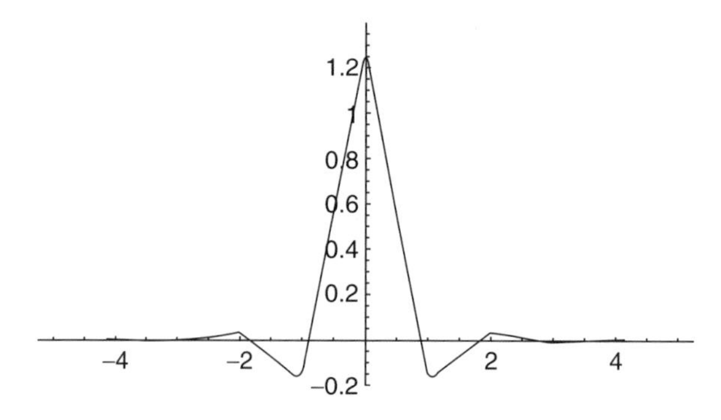

Fig. 11.10. Scaling function for the MRA consisting of continuous piecewise signals.

Example (Spline Functions). The procedure for finding this scaling function is similar to the one we use for the piecewise continuous MRA. The root space V_0 contains the $L^2(\mathbb{R})$ signals that are continuously differentiable and equal to cubic polynomials on each interval $[n, n+1]$, where $n \in \mathbb{Z}$ [37, 38]. To find a scaling function we need to find a Riesz basis. Let $g(t) \in V_0$ be the cubic spline that satisfies $g(0) = 1$ and $g(n) = 0$ otherwise. This is a rounded tent function. Then $x(t) = \Sigma a_k g(t-k)$ is a cubic spline on intervals $[n, n+1]$, continuously differentiable, and $x(k) = a_k$ for all $k \in \mathbb{Z}$. Once again we set $(Ix)(k) = a_k$ and invoke the Riesz Basis Characterization toward showing $\{g(t-k) \mid k \in \mathbb{Z}\}$ to be a Riesz basis. We compute the radial Fourier transform of $g(t)$:

$$G(\omega) = \left(1 - \frac{2}{3}\sin^2\left(\frac{\omega}{2}\right)\right)^{-1} \operatorname{sinc}^4\left(\frac{\omega}{2}\right). \tag{11.160}$$

We can derive a cubic spline scaling function by the standard formula (11.152):

$$\Phi(\omega) = \frac{G(\omega)}{\sqrt{P_G(\omega)}} = \frac{[\Sigma_8(\omega)]^{-\frac{1}{2}}}{\omega^4}, \tag{11.161}$$

where $\Sigma_n(\omega)$ is given by (11.156). Again, we can compute the $\Sigma_n(\omega)$ by taking successive derivatives—six actually—of $\Sigma_2(\omega)$. With either resolute patience or a symbolic computation software package, we calculate

$$\begin{aligned}\frac{d^6}{d\omega^6}\Sigma_2(\omega) = {} & \frac{1}{4}\cot^6\left(\frac{\omega}{2}\right)\csc^2\left(\frac{\omega}{2}\right) + \frac{57}{8}\cot^4\left(\frac{\omega}{2}\right)\csc^4\left(\frac{\omega}{2}\right) \\ & + \frac{45}{4}\cot^2\left(\frac{\omega}{2}\right)\csc^6\left(\frac{\omega}{2}\right) + \frac{17}{16}\csc^8\left(\frac{\omega}{2}\right).\end{aligned} \tag{11.162}$$

Consequently,

$$\Sigma_8(\omega) = \frac{1}{5040}\frac{d^6}{d\omega^6}(\Sigma_2(\omega)) \tag{11.163}$$

and

$$\Phi(\omega) = \frac{1}{\omega^4\sqrt{\Sigma_8(\omega)}}. \tag{11.164}$$

So once again, we find a scaling function using frequency-domain analysis. The key relationship is (11.152). Inverse Fourier transformation gives $\phi(t)$ (Figure 11.11). For the cubic spline MRA too, the scaling function is a low-pass filter. In comparison to the MRA for piecewise linear signals, observe the flatter reject band of $\Phi(\omega)$ for the spline MRA.

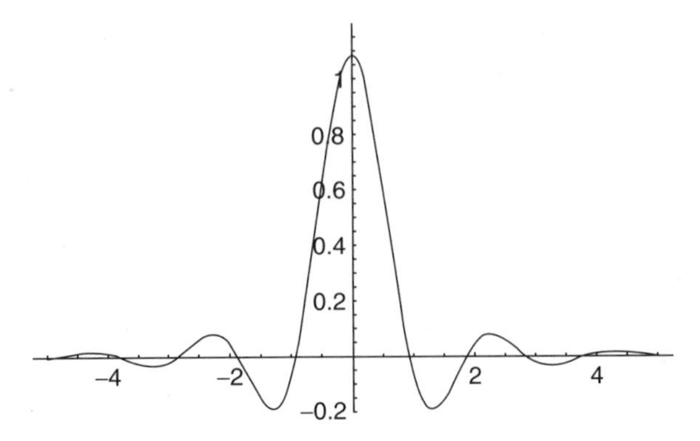

Fig. 11.11. Scaling function for the MRA consisting of cubic splines.

11.4.3 Discrete Low-Pass Filter

The scaling function is not the only special function connected with the multiresolution analysis structure. This section shows that for every MRA of finite energy signals we can find a special discrete filter [10, 23, 29]. This filter will prove useful when we discretize our theory and use it in signal analysis applications. In fact, we shall show that it is a low-pass filter. Mathematically thorough introductions to this material include [8, 9].

Suppose $V = \{V_i: i \in \mathbb{Z}\}$ is an MRA and $\phi(t) \in V_0$ is its scaling function. Because $\phi(t/2) \in V_{-1} \subset V_0$ and since integral translates of $\phi(t)$ span V_0, we see that

$$\frac{1}{2}\phi\left(\frac{t}{2}\right) = \sum_{n=-\infty}^{\infty} h_n \phi(t-n), \tag{11.165}$$

where the sequence $\{h_n \mid n \in \mathbb{Z}\}$ is square-summable. Hilbert space theory tells us that

$$h_n = \left\langle \frac{1}{2}\phi\left(\frac{t}{2}\right), \phi(t-n) \right\rangle. \tag{11.166}$$

These observations lead to the following definition.

Definition (Associated Filter). The $\phi(t)$ be the scaling function of an MRA, $V = \{V_i: i \in \mathbb{Z}\}$. If H_ϕ is the discrete filter with impulse response $h_\phi(n) = h_n$, where h_n is given by (11.166), then H_ϕ is called the *associated filter* to V (and to $\phi(t)$).

As we develop the properties of the discrete filter associated to an MRA, we shall see that it is indeed a low-pass filter. When there is no ambiguity, we drop the subscript: $H = H_\phi$. The following proposition gives a formula for the discrete-time Fourier transform $H(\omega)$ of the associated filter [23].

Proposition. Let $\phi(t)$ be the scaling function of an MRA, $V = \{V_i: i \in \mathbb{Z}\}$; let $\Phi(\omega)$ be its Fourier transform; and let H be the associated discrete filter with impulse response, $h(n) = h_n$, given by (11.166). Then,

$$\Phi(2\omega) = \Phi(\omega)H(\omega), \tag{11.167}$$

where $H(\omega)$ is the DTFT of $h(n)$: $H(\omega) = \sum_n h(n)e^{-j\omega n}$.

Proof: Apply the radial Fourier transform to both sides of (11.165). ■

Remark. The relation shows that $H(\omega)$ has a low-pass filter characteristic (Figure 11.12). The dilation $\Phi(2\omega)$ looks just like $\Phi(\omega)$, except that it is contracted with respect to the independent frequency-domain variable ω by a factor of two. The relation (11.167) shows that a multiplication by $H(\omega)$ accomplishes this, and the

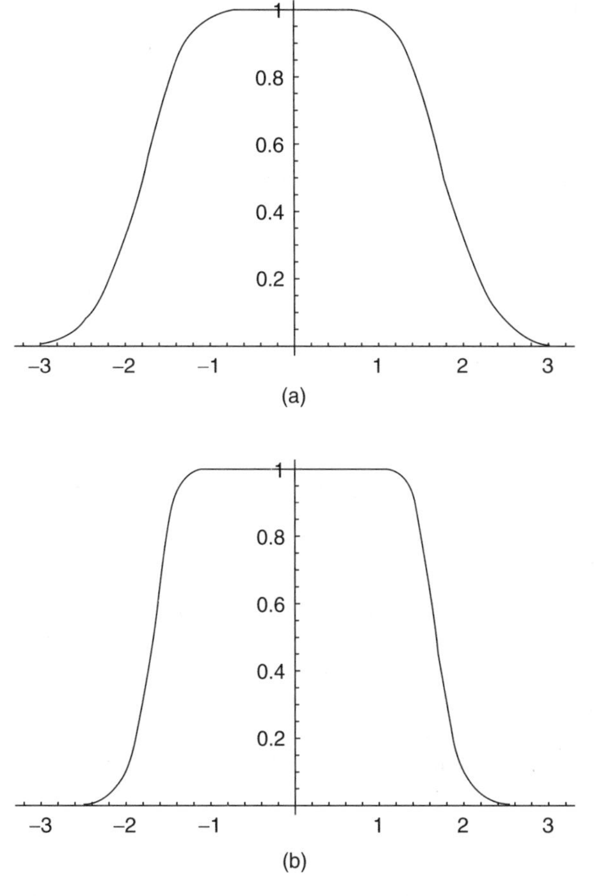

Fig. 11.12. Associated low-pass filters for the MRAs consisting of piecewise linear functions (a) and cubic splines (b). Note the better cutoff behavior of the filter based on cubic spline approximations.

only this can be the case, intuitively, is that $H(\omega)$ is approximately unity around the DC value $\omega = 0$, and it falls off to essentially zero halfway through the passband of the spectrum of $\Phi(\omega)$. This is not a rigorous argument, but it can be made so by assuming some decay constraints on the scaling function $\phi(t)$ [23, 29].

Proposition. Let $\phi(t)$ be the scaling function of an MRA, $V = \{V_i: i \in \mathbb{Z}\}$; $\Phi(\omega)$ its Fourier transform; $h(n)$ the associated discrete low-pass filter (11.166), and $H(\omega)$ its DTFT. Then,

$$|H(\omega)|^2 + |H(\omega+\pi)|^2 = 1, \tag{11.168}$$

for almost all $\omega \in \mathbb{R}$.

Proof: By the Orthonormal Translates Theorem, $\sum_{k=-\infty}^{\infty} |\Phi(2\omega + 2\pi k)|^2 = 1$. Inserting (11.167) into this identity gives

$$\sum_{k=-\infty}^{\infty} |\Phi(\omega+\pi k)|^2 |H(\omega+\pi k)|^2 = 1 \tag{11.169}$$

almost everywhere. Next, we split left-hand side of (11.169) into a sum over even integers and odd integers:

$$|H(\omega)|^2 \sum_{k=-\infty}^{\infty} |\Phi(\omega+2\pi k)|^2 + |H(\omega+\pi)|^2 \sum_{k=-\infty}^{\infty} |\Phi(\omega+2\pi k+\pi)|^2 = 1, \tag{11.170}$$

where we have used $H(\omega) = H(\omega + 2\pi)$. The Orthonormal Translates Theorem tells us that the infinite sums in (11.170) are unity, and (11.168) follows. ■

Remarks. Discrete filters satisfying (11.168) are *conjugate* filters, familiar from Chapter 9. Conjugate filters are used to filter a signal in such a way that it can be subsampled and exactly reconstructed later. The Laplacian pyramid technique provides decomposition by filtering and subsampling as well as exact reconstruction [24]. Various efficient signal compression and transmission techniques rely on this idea [25, 39, 40]. What is particularly striking about the MRA structure is that quite disparate signal theoretic techniques such as scale space analysis [26–28], orthogonal bases, and pyramid decompositions all meet together in one place.

The following results [8] link the low-pass filter H of an MRA, $V = \{V_i: i \in \mathbb{Z}\}$, with the low-resolution subspace $V_{-1} \subset V_0 \in V$.

Lemma. Let $\phi(t)$ be the scaling function of an MRA, let $\Phi(\omega)$ be its Fourier transform, and let $C(\omega) \in L^2[0, 2\pi]$ be 2π-periodic. Then $C(\omega)\Phi(\omega) \in L^2(\mathbb{R})$.

Proof: We compute

$$\int_{-\infty}^{\infty} |\Phi(\omega)|^2 |C(\omega)|^2 d\omega = \sum_{n=-\infty}^{\infty} \int_0^{2\pi} |\Phi(\omega + 2\pi n)|^2 |C(\omega + 2\pi n)|^2 d\omega$$
$$= \sum_{n=-\infty}^{\infty} \int_0^{2\pi} |\Phi(\omega + 2\pi n)|^2 |C(\omega)|^2 d\omega, \qquad (11.171)$$

where we have invoked $C(\omega) = C(\omega + 2\pi)$. Interchanging summation and integration in (11.171), and again using $\sum |\Phi(\omega + 2\pi n)|^2 = 1$, we get

$$\int_{-\infty}^{\infty} |\Phi(\omega)|^2 |C(\omega)|^2 d\omega = \int_0^{2\pi} \sum_{n=-\infty}^{\infty} |\Phi(\omega + 2\pi n)|^2 |C(\omega)|^2 d\omega$$
$$= \int_0^{2\pi} |C(\omega)|^2 d\omega = \|C\|^2_{2, L^2[0, 2\pi]}. \qquad (11.172)$$

■

Proposition ($\mathbf{V_0}$ Characterization). Let $V = \{V_i : i \in \mathbb{Z}\}$ be an MRA, let $\phi(t)$ be its scaling function, and let $\Phi = \mathcal{F}(\phi)$ be its (radial) Fourier transform. Then the root space $V_0 \in V$ contains precisely those $x(t) \in L^2(\mathbb{R})$ such that $X(\omega) = C(\omega)\Phi(\omega)$ for some 2π-periodic $C(\omega) \in L^2[0, 2\pi]$, where $X(\omega) = \mathcal{F}(x)$.

Proof: Let $x(t) \in V_0$. Then $x(t) = \Sigma c_k \phi(t-k)$ for some $c(k) = c_k$, where $c \in l^2$. We compute $X(\omega)$ as follows:

$$\int_{-\infty}^{\infty} x(t)e^{-j\omega t}dt = \int_{-\infty}^{\infty} \sum_{k=-\infty}^{\infty} c_k \phi(t-k)e^{-j\omega t}dt = \sum_{k=-\infty}^{\infty} c_k \int_{-\infty}^{\infty} \phi(t-k)e^{-j\omega t}dt$$
$$= \sum_{k=-\infty}^{\infty} c_k e^{-j\omega k} \int_{-\infty}^{\infty} \phi(t)e^{-j\omega t}dt = C(\omega)\Phi(\omega). \qquad (11.173)$$

Now suppose $X(\omega) = C(\omega)\Phi(\omega)$ for some 2π-periodic $C(\omega) \in L^2[0, 2\pi]$. By the lemma, $X(\omega) \in L^2(\mathbb{R})$, and we can write $C(\omega) = \Sigma c_k e^{-j\omega k}$, where $c(k)$ is the inverse DTFT of $C(\omega)$. Whence the computation (11.173) shows that $x(t)$ is in the closure of the span of $\{\phi(t-k) \mid k \in \mathbb{Z}\}$; so $x \in V_0$. ■

Corollary. With the same notation, define the operator T: $V_0 \to L^2[0, 2\pi]$ by $Tx = C$, where $C(\omega)$ is the 2π-periodic function with $X(\omega) = C(\omega)\Phi(\omega)$ guaranteed by the proposition. Then:

(i) T is linear.

(ii) If $x \in V_0$, then $2\pi\|x\|_2^2 = \|C\|^2_{2, L^2[0, 2\pi]}$.

(iii) If $x, y \in V_0$, $C = Tx$, and $D = Ty$, then $2\pi\langle x, y\rangle_{L^2(\mathbb{R})} = \langle C, D\rangle_{L^2[0, 2\pi]}$.

Proof: Linearity (i) is left as an exercise. For (ii), let $x(t) \in V_0$, so that $X(\omega) = C(\omega)\Phi(\omega)$. Then, the Plancherel's formula for $L^2(\mathbb{R})$ and (11.172) entail

$$\|X\|^2_{2, L^2(\mathbb{R})} = \int_{-\infty}^{\infty} |\Phi(\omega)|^2|C(\omega)|^2 d\omega = \|C\|^2_{2, L^2[0, 2\pi]} = 2\pi\|x\|^2_{2, L^2(\mathbb{R})}. \tag{11.174}$$

From (11.174) and the polarization identity [8, 30] for inner product spaces (Chapter 2), $4\langle x, y\rangle = \|x+y\|^2 - \|x-y\|^2 + j\|x+jy\|^2 - j\|x-jy\|^2$, (iii) follows. ∎

Definition (Canonical Linear Map on V_0). The linear map $Tx = C$, where $C(\omega)$ is the 2π-periodic function with $X(\omega) = C(\omega)\Phi(\omega)$ guaranteed by the corollary, is called the *canonical map* from V_0 to $L^2[0, 2\pi]$.

Proposition (V_{-1} Characterization). Let $V = \{V_i : i \in \mathbb{Z}\}$ be an MRA, let $\phi(t)$ be its scaling function, $\Phi = \mathcal{F}(\phi)$, and let $H = H_\phi$ the associated low-pass filter. Then the first low resolution subspace $V_{-1} \in V$ contains precisely those $x(t) \in L^2(\mathbb{R})$ such that $X(\omega) = C(2\omega)H(\omega)\Phi(\omega)$ for some 2π-periodic $C(\omega) \in L^2[0, 2\pi]$, where $X(\omega) = \mathcal{F}(x)$.

Proof: Let $x(t) \in V_{-1}$, so that $2x(2t) \in V_0$. Then $2x(2t) = \Sigma c_k\phi(t-k)$ for some $c_k = c(k) \in l^2$. Thus,

$$2x(t) = \sum_{k=-\infty}^{\infty} c(k)\phi\left(\frac{t}{2} - k\right). \tag{11.175}$$

Taking Fourier transforms again [8]:

$$\begin{aligned} 2X(\omega) &= 2\sum_{k=-\infty}^{\infty} c_k \int_{-\infty}^{\infty} \phi(s)e^{-j\omega(2s+2k)}ds \\ &= 2\sum_{k=-\infty}^{\infty} c_k e^{-2j\omega k}\int_{-\infty}^{\infty} \phi(s)e^{-j(2\omega)s}ds = 2C(2\omega)\Phi(2\omega), \end{aligned} \tag{11.176}$$

where we have made the substitution $s = t/2 - k$, and $C(\omega)$ is the DTFT of $c(k)$. From (11.167), $\Phi(2\omega) = H(\omega)\Phi(\omega)$; thus, $X(\omega) = C(2\omega)H(\omega)\Phi(\omega)$. For the converse, let $X(\omega) = C(2\omega)H(\omega)\Phi(\omega)$ for some 2π-periodic $C(\omega) \in L^2[0, 2\pi]$. Since $C(2\omega)H(\omega)$ is still 2π-periodic, the Lemma applies, and $X(\omega) \in L^2(\mathbb{R})$. Finally, (11.176) reverses to show $x(t) \in V_0$. ∎

It is possible to generalize this result (exercise). The next section explains a mathematical surprise that arises from MRA theory.

11.4.4 Orthonormal Wavelet

Besides the scaling function and the associated discrete low-pass filter, a third special function accompanies any multiresolution approximation of $L^2(\mathbb{R})$: the *orthonormal wavelet* [23, 29]. Our presentation has been guided by the mathematically complete introductions [8, 9].

Definition (Orthonormal Wavelet). Let $\psi(t) \in L^2(\mathbb{R})$. If its dilations and translations $\{2^{n/2}\psi(2^n t - m)\colon m, n \in \mathbb{Z}\}$ are an orthogonal basis of $L^2(\mathbb{R})$, then ψ is an *orthogonal wavelet*. If $\|\psi\| = 1$, then ψ is an *orthonormal* wavelet.

At the beginning of this chapter we considered an extension of the Fourier transform based on scale and location as transform parameters. The transform inversion required a special signal, the *admissible wavelet*, in order to succeed, and we found that admissible wavelets had to be analog band-pass filters. Now, for MRAs it turns out that the special associated orthogonal wavelet too is a band-pass filter. To discover how it is that an MRA supports this extraordinary function, we examine the orthogonal complements of the component spaces V_i of the multiresolution analysis $V = \{V_i : i \in \mathbb{Z}\}$.

11.4.4.1 Existence. Consider first $V_{-1} \subset V_0$. From Hilbert space theory (Chapters 2 and 3), we know that every element of V_0 can be written uniquely as a sum $x = v + w$, where $v \in V_{-1}$ and $w \perp v$. The set of all such $w \in V_0$ constitute a Hilbert subspace of V_0; let us denote it by W_{-1}. We say that V_0 is the *direct sum* of V_{-1} and W_{-1}: $V_0 = V_{-1} \oplus W_{-1}$. In general, every V_{i+1} is the direct sum of V_i and W_i, where W_i is the *orthogonal complement* of V_i in V_{i+1}. We know already that the V_i have orthonormal bases made up of translations and dilations of the scaling function $\phi(t)$. We can also find an orthonormal basis of W_i by the Gram–Schmidt orthonormalization procedure, of course [31]. But this does not imply that the basis elements are translates of one another, and it exposes no relation between the basis elements so found and the rest of the MRA structure. We want a more enlightening theory.

Lemma (First W_{-1} Characterization). Let $V = \{V_i\colon i \in \mathbb{Z}\}$ be an MRA; let $\phi(t)$ be its scaling function; let $\Phi = \mathcal{F}(\phi)$ be the (radial) Fourier transform of $\phi(t)$; let $H = H_\phi$ be the associated low-pass filter; and let $Tx = C$ be the canonical linear map from V_0 to $L^2[0, 2\pi]$. Then $x(t) \in W_{-1} \subset V_0$ if and only if

$$C(\omega)\overline{H(\omega)} + C(\omega+\pi)\overline{H(\omega+\pi)} = 0 \tag{11.177}$$

for almost all $\omega \in [0, 2\pi]$.

Proof: Let $y(t) \in V_{-1}$ and let $Ty = D$. Then by the V_{-1} characterization, $Y(\omega) = A(2\omega)H(\omega)\Phi(\omega)$ for some 2π-periodic $A \in L^2[0, 2\pi]$. So $D(\omega) = A(2\omega)H(\omega)$ almost everywhere on $[0, 2\pi]$. By the corollary to the V_0 characterization, $\langle x, y\rangle = \langle C, D\rangle$. Thus, $x(t) \in W_{-1}$ if and only if $\langle C(\omega), A(2\omega)H(\omega)\rangle = 0$, for almost all $\omega \in [0, 2\pi]$. Writing out the inner product integral [8], we see that this is further equivalent to

$$\int_0^{2\pi} C(\omega)\overline{A(2\omega)H(\omega)}\,d\omega = \int_0^{\pi} \overline{A(2\omega)}[C(\omega)\overline{H(\omega)} + C(\omega+\pi)\overline{H(\omega+\pi)}]\,d\omega = 0 \tag{11.178}$$

for almost all $\omega \in [0, 2\pi]$. Since $y(t)$ is any element of V_{-1}, the $A(2\omega)$ in the integrand on the right-hand side of (11.178) is an arbitrary π-periodic signal; evidently, $x(t) \in W_{-1}$ if and only if the π-periodic factor $C(\omega)\overline{H(\omega)} + C(\omega+\pi)\overline{H(\omega+\pi)} = 0$ almost everywhere on $[0, \pi]$. Finally, by the 2π-periodicity of $C(\omega)$ and $H(\omega)$, this same expression must be almost everywhere zero on $[0, 2\pi]$.

Lemma (Second $\mathbf{W_{-1}}$ Characterization). Let $V = \{V_i: i \in \mathbb{Z}\}$ be an MRA, let $\phi(t)$ be its scaling function, let $\Phi = \mathcal{F}(\phi)$ be the Fourier transform of ϕ, let $H = H_\phi$ be the associated low-pass filter, and let $Tx = C$ be the canonical map from V_0 to $L^2[0, 2\pi]$. Then, $x(t) \in W_{-1}$ if and only if $X(\omega) = e^{-j\omega}S(2\omega)\overline{H(\omega+\pi)}\Phi(\omega)$ for some 2π-periodic $S(\omega) \in L^2[0, 2\pi]$.

Proof: Resorting to some linear algebra tricks, we formulate the previous lemma's criterion as a determinant. Thus, $x(t) \in W_{-1}$ is equivalent to

$$\det\begin{bmatrix} \overline{H(\omega+\pi)} & C(\omega) \\ -\overline{H(\omega)} & C(\omega+\pi) \end{bmatrix} = 0 \tag{11.179}$$

for almost all $\omega \in [0, 2\pi]$. This means that the columns of the matrix (11.179) are linearly dependent—that is, they are proportional via a 2π-periodic function $R(\omega)$:

$$\begin{bmatrix} C(\omega) \\ C(\omega+\pi) \end{bmatrix} = R(\omega)\begin{bmatrix} \overline{H(\omega+\pi)} \\ -\overline{H(\omega)} \end{bmatrix} \tag{11.180}$$

for just as many $\omega \in [0, 2\pi]$. Now substitute $\omega + \pi$ for ω in (11.180) to see

$$\begin{bmatrix} C(\omega+\pi) \\ C(\omega) \end{bmatrix} = R(\omega+\pi)\begin{bmatrix} \overline{H(\omega)} \\ -\overline{H(\omega+\pi)} \end{bmatrix} \tag{11.181}$$

whence $C(\omega) = -R(\omega+\pi)\overline{H(\omega+\pi)}$. Further putting $\omega + \pi$ for ω in (11.181) gives $C(\omega) = R(\omega)\overline{H(\omega+\pi)}$. Evidently, $R(\omega) = -R(\omega+\pi)$ for almost all $\omega \in [0, 2\pi]$. Hence we have shown that $x(t) \in W_{-1}$ if and only if $X(\omega) = C(\omega)\Phi(\omega)$, where

$C(\omega) = R(\omega)\overline{H(\omega+\pi)}$ for some 2π-periodic $R(\omega) \in L^2[0, 2\pi]$ with $R(\omega) = -R(\omega + \pi)$. We define $S(\omega) = \exp(j\omega/2)R(\omega/2)$. Then $S(\omega + 2\pi) = S(\omega)$ almost everywhere, and

$$X(\omega) = C(\omega)\Phi(\omega) = R(\omega)\overline{H(\omega+\pi)}\Phi(\omega) = e^{-j\omega}S(2\omega)\overline{H(\omega+\pi)}\Phi(\omega). \tag{11.182}$$

■

Lemma (W_0 Characterization). Let $V = \{V_i: i \in \mathbb{Z}\}$ be an MRA, $\phi(t)$ its scaling function, $\Phi = \mathcal{F}(\phi)$ the Fourier transform of ϕ, $H = H_\phi$ the associated low-pass filter, and $Tx = C$ the canonical map from V_0 to $L^2[0, 2\pi]$. Then, $x(t) \in W_0$ if and only if $X(2\omega) = e^{-j\omega}S(2\omega)\overline{H(\omega+\pi)}\Phi(\omega)$ for some 2π-periodic $S(\omega) \in L^2[0, 2\pi]$.

Proof: We note that $x(t) \in W_0$ if and only if $\langle x(t), v(t)\rangle = 0$ for all $v(t) \in V_0$, which is equivalent to $\langle x(t/2), v(t/2)\rangle = 0$. But any $f(t) \in V_{-1}$ is of the form $v(t/2)$, so this is also equivalent to $x(t/2) \perp V_{-1}$; in other words, $y(t) = x(t/2) \in W_{-1}$. The previous lemma says $Y(\omega) = e^{-j\omega}S(2\omega)\overline{H(\omega+\pi)}\Phi(\omega)$ for some 2π-periodic $S(\omega) \in L^2[0, 2\pi]$. But $Y(\omega) = 2X(2\omega)$. ■

The main result of this section—very probably the main result of this chapter, likely the main result of this book, and arguably the main result of Fourier analysis in the latter half of the twentieth century—is expressed in the following theorem and its corollary [23, 29].

Theorem (Orthonormal Basis of W_0). Suppose $V = \{V_i: i \in \mathbb{Z}\}$ is an MRA, $\phi(t)$ its scaling function, $\Phi = \mathcal{F}(\phi)$ the Fourier transform of ϕ, and $H = H_\phi$ is the associated low-pass filter. If $\psi(t)$ is the finite energy signal whose Fourier transform is given by

$$\Psi(2\omega) = e^{-j\omega}\overline{H(\omega+\pi)}\Phi(\omega), \tag{11.183}$$

then $\{\psi(t-k) \mid k \in \mathbb{Z}\}$ is an orthonormal basis for W_0.

Proof: We know that the W_0 characterization lemma entails $\psi(t) \in W_0$. (In fact, $\psi(t)$ represents the very straightforward case where the lemma's $S(\omega) = 1$ almost everywhere on $[0, 2\pi]$.) Let $x(t) \in W_0$. Let us show that a linear combination of translates of $\psi(t)$ is arbitrarily close to $x(t)$. The previous lemma says that $X(2\omega) = e^{-j\omega}S(2\omega)\overline{H(\omega+\pi)}\Phi(\omega)$ for some 2π-periodic $S(\omega) \in L^2[0, 2\pi]$. Thus, $X(\omega) = S(\omega)\Psi(\omega)$, almost everywhere on $[0, 2\pi]$. But we know this condition already from the Spanning Translates Proposition of Section 11.4.2.1: It means that

$$x(t) = \sum_{n=-\infty}^{\infty} s(n)\psi(t-n), \tag{11.184}$$

where $s(n)$ is the inverse discrete-time Fourier transform of $S(\omega)$. Verily, the closure of $\{\psi(t-k) \mid k \in \mathbb{Z}\}$ is all of W_0.

What about orthonormality? If we attempt to apply the Orthonormal Translates criterion (11.129), then we have

$$\begin{aligned}\sum_{k=-\infty}^{\infty} |\Psi(\omega+2k\pi)|^2 &= \sum_{k=-\infty}^{\infty} \left|\Phi\left(\frac{\omega}{2}+k\pi\right)\right|^2 \left|H\left(\frac{\omega}{2}+(k+1)\pi\right)\right|^2 \\ &= \sum_{k=-\infty}^{\infty} \left|\Phi\left(\frac{\omega}{2}+2k\pi\right)\right|^2 \left|H\left(\frac{\omega}{2}+2k\pi+\pi\right)\right|^2 \\ &+ \sum_{k=-\infty}^{\infty} \left|\Phi\left(\frac{\omega}{2}+2k\pi+\pi\right)\right|^2 \left|H\left(\frac{\omega}{2}+2k\pi+2\pi\right)\right|^2 .\end{aligned} \tag{11.185}$$

By the orthonormality of $\{\phi(t-k)\}$, we have

$$\begin{aligned}\sum_{k=-\infty}^{\infty} |\Psi(\omega+2k\pi)|^2 &= \sum_{k=-\infty}^{\infty} \left|\Phi\left(\frac{\omega}{2}+k\pi\right)\right|^2 \left|H\left(\frac{\omega}{2}+(k+1)\pi\right)\right|^2 \\ &= \sum_{k=-\infty}^{\infty} \left|\Phi\left(\frac{\omega}{2}+2k\pi\right)\right|^2 \left|H\left(\frac{\omega}{2}+2k\pi+\pi\right)\right|^2 \\ &+ \sum_{k=-\infty}^{\infty} \left|\Phi\left(\frac{\omega}{2}+2k\pi+\pi\right)\right|^2 \left|H\left(\frac{\omega}{2}+2k\pi+2\pi\right)\right|^2 .\end{aligned} \tag{11.186}$$

By the 2π-periodicity of $H(\omega)$, this becomes

$$\begin{aligned}\sum_{k=-\infty}^{\infty} |\Psi(\omega+2k\pi)|^2 &= \left|H\left(\frac{\omega}{2}+\pi\right)\right|^2 \sum_{k=-\infty}^{\infty} \left|\Phi\left(\frac{\omega}{2}+2k\pi\right)\right|^2 \\ &+ \left|H\left(\frac{\omega}{2}\right)\right|^2 \sum_{k=-\infty}^{\infty} \left|\Phi\left(\frac{\omega}{2}+2k\pi+\pi\right)\right|^2\end{aligned} \tag{11.187}$$

and by Orthonormal Translates applied to the Φ summations on the bottom of (11.187), we get

$$\sum_{k=-\infty}^{\infty} |\Psi(\omega+2k\pi)|^2 = \left|H\left(\frac{\omega}{2}+\pi\right)\right|^2 + \left|H\left(\frac{\omega}{2}\right)\right|^2. \tag{11.188}$$

But this last expression is unity by (11.168), and thus $\{\psi(t-k) \mid k \in \mathbb{Z}\}$ is orthonormal.

Corollary (Existence of Orthonormal Wavelet). With the theorem's assumptions and notation, let $\psi(t) \in L^2(\mathbb{R})$ be defined as above. Then the dilations and translations $\{2^{n/2}\psi(2^n t - m)\colon m, n \in \mathbb{Z}\}$ are an orthonormal basis of $L^2(\mathbb{R})$, and so ψ is an orthonormal wavelet [8–10].

Proof: Since $V_1 = V_0 \oplus W_0$, dilations of $x(t) \in W_0$ by 2^i are in W_i: $x(2^i t) \in W_i$. In fact, $\{2^{i/2}\psi(2^i t - m)\colon m \in \mathbb{Z}\}$ is an orthonormal basis for W_i. But, $V_{i+1} = V_i \oplus W_i$, so

$B_{i+1} = \{2^{n/2}\psi(2^n t - m): n < i+1 \text{ and } m \in \mathbb{Z}\}$ is an orthonormal set inside V_{i+1}. By the intersection property (iii) of the MRA, though, $\bigcap_{i=-\infty}^{\infty} V_i = 0$; B_{i+1} must be dense in V_{i+1}. By the MRA union property (ii), $\overline{\bigcup_{n=-\infty}^{\infty} V_i} = L^2(\mathbb{R})$; $L^2(\mathbb{R})$ must be the Hilbert space direct sum of the W_i:

$$\bigoplus_{i=-\infty}^{\infty} W_i = L^2(\mathbb{R}). \tag{11.189}$$

■

The wavelet $\psi(t)$ that we have found is essentially unique. The exercises outline an argument that any other W_0 function that is an orthogonal wavelet for square-integrable signals must have a Fourier transform that differs from the formula (11.183) by a factor that is unity almost everywhere on $[0, 2\pi]$.

11.4.4.2 Examples.

Let us show some examples of orthonormal wavelets from the multiresolution analyses of square-integrable signals that we already know.

Example (Step Functions). In the Haar MRA [20], the root space V_0 consists of constant functions on unit intervals $(n, n-1)$, and so $\phi(t) = u(t) - u(t-1)$. We compute its Fourier transform by

$$\Phi(\omega) = \int_{-\infty}^{\infty} \phi(t) e^{-j\omega t} dt = e^{-\frac{j\omega}{2}} \frac{\sin(\omega/2)}{\omega/2} \tag{11.190}$$

and the relation $\Phi(2\omega) = \Phi(\omega)H(\omega)$ gives the associated low-pass filter:

$$H(\omega) = e^{-\frac{j\omega}{2}} \cos(\omega/2). \tag{11.191}$$

From $\Psi(\omega) = e^{-j\omega}\overline{H(\omega+\pi)}\Phi(\omega)$, we calculate

$$\Psi(\omega) = -je^{\frac{j\omega}{2}} \frac{\sin^2(\omega/4)}{\omega/4}. \tag{11.192}$$

But (11.192) is the radial Fourier transform of the function

$$\psi(t) = \begin{cases} -1 & \text{if } -1 \le t < \frac{1}{2}, \\ 1 & \text{if } -\frac{1}{2} \le t < 0, \\ 0 & \text{if otherwise.} \end{cases} \tag{11.193}$$

so $\psi(t)$ above is the orthogonal wavelet for the step function MRA.

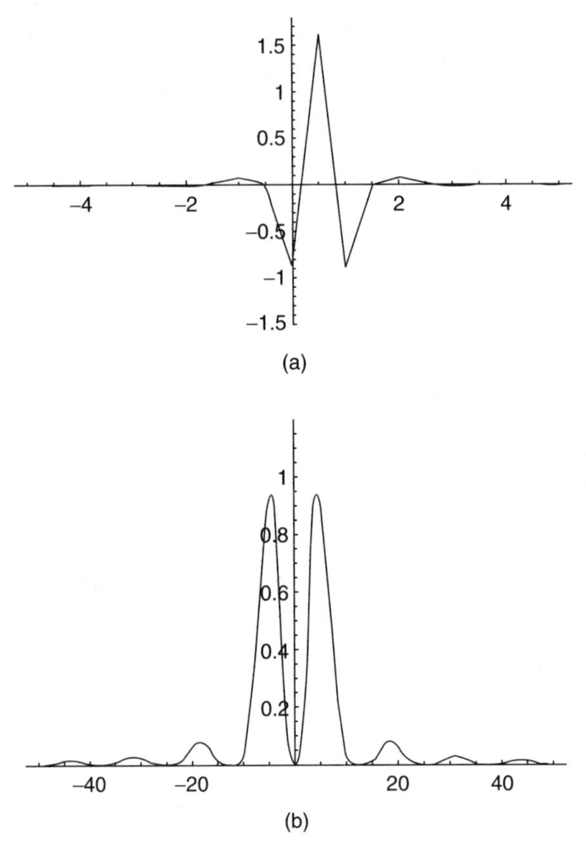

Fig. 11.13. For the Strömberg MRA: The orthogonal wavelet, (a) which we compute from a messy, but exact expression for its Fourier transform (b).

Example (Piecewise Continuous Functions). We found the scaling function for the Strömberg [22] MRA in Section 11.4.2.2 and the associated low-pass filter in Section 11.4.3. The expression $\Psi(2\omega) = e^{j\omega}\overline{H(\omega + \pi)}\,\Phi(\omega)$ as well as (11.159) and (11.167) give us the Fourier transform for the piecewise continuous MRA's wavelet. Then, we can compute $\psi(t)$ via the inverse transform (Figure 11.13).

Example (Spline Functions). In the third MRA we have studied, the root space V_0 contains continuously differentiable, finite-energy signals that are cubic polynomials on unit intervals $[n, n+1]$. This MRA was developed by Lemarié [37] and Battle [38]. The same strategy works once more. We know the scaling function's Fourier transform $\Phi(\omega)$ from (11.164). The discrete-time Fourier transform for the associated low-pass filter is the ratio $H(\omega) = \Phi(2\omega)/\Phi(\omega)$. Hence, we find $\Psi(\omega)$, and inverse transforming gives the orthogonal wavelet (Figure 11.14).

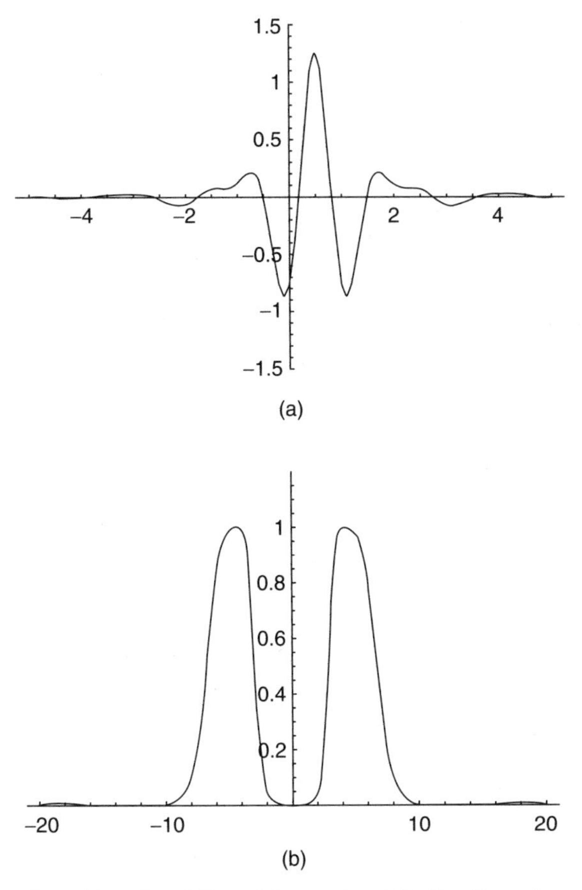

Fig. 11.14. For the cubic spline MRA of Lemarié and Battle: The orthogonal wavelet (a) and its Fourier transform (b).

The next chapter provides various examples of mixed domain signal analysis. In particular, it covers the use of the multiresolution analysis structure for combining time- and scale-domain analysis of signals that arise in typical applications.

11.5 SUMMARY

For signal analysis, both the continuous wavelet transform and the multiresolution analysis that leads to orthogonal wavelets are attractive and feasible. This chapter covered the signal analytic essentials of wavelet theory. The continuous transform finds more applications in signal understanding, as long as basic facts about input signals—location, scale, spectrum—are known beforehand. The orthogonal transforms, based on multiresolution analysis, tend to find more use in compression. We shall see in the last chapter that there are efficient algorithms for decomposition using MRAs that lend themselves to efficient signal description, compression, as well as pattern recognition applications.

We omitted the proof of the theorem on the necessity of admissibility for frames based on translation and dilation (Section 11.3.2). The proof itself is somewhat technical, and, besides, in compensation, we later showed that an elegant construction, the MRA, leads to orthogonal wavelets from which efficient algorithms and straightforward frame constructions proceed.

There are three special functions that arise from an MRA, $V = \{V_i: i \in \mathbb{Z}\}$:

- The scaling function $\phi(t)$ whose translates form an orthonormal basis for the root space V_0;
- The associated low-pass filter H_ϕ;
- And, finally, the orthonormal wavelet $\psi(t)$ whose translations and dilations constitute and orthonormal basis for finite-energy signals.

Some conditions in the definition we give for an MRA are consequences of the others. For example, the third criterion in Section 11.4.1.1—that only the zero signal should appear in all V_i—follows from the rest. This and other interdependencies were only noted some years later [8, 41].

We provided only a few examples of MRAs, but these suffice for introductory signal analysis applications. Important extensions include compactly supported wavelets [3], approximation theory using wavelets [2], and multidimensional signal analysis using wavelets [10, 23].

Let us remark about how wavelets can be used in image processing and analysis. The principal focus of this book notwithstanding, we should also note that our MRA definition extends from one-dimensional functions (signals) to two-dimensional functions (images). Indeed there are MRA structures for the n-dimensional Hilbert space $L^2(\mathbb{R}^n)$. The technique is identical to that used to extend the Gaussian and Laplacian pyramid constructions to images [24].

We just repeat the definition of Section 11.4.1.1 for functions of two variables, $x(s, t) \in L^2(\mathbb{R}^2)$, denoting the subspace chain $V_{2,n}$. We can then derive a two-dimensional scaling function $\phi(s, t)$ as before (exercise). The most common approach, however, is to use a tensor product of one-dimensional MRAs to get an MRA for $L^2(\mathbb{R}^2)$: $V_{2,n} = V_{1,n} \otimes V_{1,n}$, where $\{V_n: n \in \mathbb{Z}\}$ is an MRA in $L^2(\mathbb{R})$. Then, there is a scaling function for the two-dimensional MRA of $L^2(\mathbb{R}^2)$: $\Phi(x, y) = \phi(x)\phi(y)$. While there is a single scaling function for the two-dimensional case, there are now three wavelets: $\Psi_1(x, y) = \phi(x)\psi(y)$, sensitive to *high vertical frequencies*; $\Psi_2(x, y) = \psi(x)\phi(y)$, *horizontal*; $\Psi_3(x, y) = \psi(x)\psi(y)$, *diagonal* or *corners*.

This gives an orientation selective decomposition, especially suitable for images with large x- and y-direction edge components. A number of applications involve such images: inspection of manufactured materials, remote sensing applications, and seismic signal processing among others.

For a history of wavelets, see Ref. [1]. References 10 and 42 list software resources and toolkits for signal analysis using wavelets.

REFERENCES

1. B. B. Hubbard, *The World According to Wavelets*, Wellesley, MA: A. K. Peters, 1996.
2. C. Chui, *Introduction to Wavelets*, San Diego, CA: Academic, 1992.
3. I. Daubechies, *Ten Lectures on Wavelets*, Philadelphia: SIAM, 1992.
4. Y. Meyer, *Wavelets: Algorithms and Applications*, Philadelphia: SIAM, 1993.
5. T. H. Koornwinder, ed., *Wavelets: An Elementary Treatment of Theory and Applications*, Singapore: World Scientific, 1993.
6. G. Kaiser, *A Friendly Guide to Wavelets*, Boston: Birkhauser, 1994.
7. M. Holschneider, *Wavelets: An Analysis Tool*, Oxford: Oxford University Press, 1995.
8. E. Hernandez and G. Weiss, *A First Course on Wavelets*, Boca Raton, FL: CRC Press, 1996.
9. P. Wojtaszczyk, *A Mathematical Introduction to Wavelets*, Cambridge: Cambridge University Press, 1997.
10. S. Mallat, *A Wavelet Tour of Signal Processing*, San Diego, CA: Academic, 1998.
11. H. L. Resnikoff and R. O. Wells, Jr., *Wavelet Analysis: The Scalable Structure of Information*, New York: Springer-Verlag, 1998.
12. C. Blatter, *Wavelets: A Primer*, Natick, MA: A. K. Peters, 1999.
13. A. Grossmann and J. Morlet, Decompositions of Hardy functions into square integrable wavelets of constant shape, *SIAM Journal of Mathematical Analysis*, vol. 15, pp. 723–736, 1984.
14. I. Daubechies, A. Grossmann, and Y. Meyer, Painless non-orthogonal expansions, *Journal of Mathematical Physics*, vol. 27, pp. 1271–1283, 1986.
15. I. Daubechies, The wavelet transform, time-frequency localization and signal analysis, *IEEE Transactions on Information Theory*, vol. 36, no. 5, pp. 961–1005, September 1990.
16. P. Goupillaud, A. Grossmann, and J. P. Morlet, Cycle-octave and related transforms in seismic signal analysis, *Geoexploration*, vol. 23, pp. 85–102, 1984.
17. R. Kronland-Martinet, J. P. Morlet, and A. Grossmann, Analysis of sound patterns through wavelet transforms, *International Journal of Pattern Recognition and Artificial Intelligence*, vol. 1, no. 2, pp. 97–126, 1987.
18. A. Grossmann, Wavelet transforms and edge detection, in *Stochastic Processes in Physics and Engineering*, S. Albeverio et al., ed., Dordrecht, Holland: D. Reidel Publishing Company, pp. 149–157, 1988.
19. Y. Meyer, Principe d'incertitude, bases hilbertiennes et algèbres d'opérateurs, *Séminaire Bourbaki*, no. 662, 1985–1986.
20. A. Haar, Zur theorie der orthogonalen Functionensysteme, *Mathematische Annalen*, vol. 69, pp. 331–371, 1910.
21. C. E. Shannon, Communication in the presence of noise, *Proceedings of the Institute of Radio Engineers*, vol. 37, pp. 10–21, 1949.
22. J.-O. Strömberg, A modified Franklin system and higher order spline systems on $\mathbb{R}^n$ as unconditional bases for Hardy spaces, in *Proceedings of the Conference in Honor of Antoni Zygmund*, vol. II, W. Becker, A. P. Calderon, R. Fefferman, and P. W. Jones, eds., New York: Wadsworth, pp. 475–493, 1981.

23. S. Mallat, "A theory for multiresolution signal decomposition: The wavelet representation," *IEEE Transactions on Pattern Analysis and Machine Intelligence*, pp. 674–693, July 1989.

24. P. J. Burt and E. H. Adelson, The Laplacian pyramid as a compact image code, *IEEE Transactions on Communications*, vol. COM-31, no. 4, pp. 532–540, April 1983.

25. R. E. Crochiere and L. R. Rabiner, *Multirate Digital Signal Processing*, Englewood Cliffs, NJ: Prentice-Hall, 1983.

26. A. P. Witkin, Scale-space filtering, *Proceedings of the 8th International Joint Conference on Artificial Intelligence*, Karlsruhe, W. Germany, 1983. Also in *From Pixels to Predicates*, A. P. Pentland, ed., Norwood, NJ: Ablex, 1986.

27. J. J. Koenderink, The structure of images, *Biological Cybernetics*, vol. 50, pp. 363–370, 1984.

28. T. Lindeberg, Scale space for discrete signals, *IEEE Transactions on Pattern Analysis and Machine Intelligence*, vol. 12, no. 3, pp. 234–254, March 1990.

29. S. G. Mallat, Multiresolution approximations and wavelet orthonormal bases of $L^2(\mathbb{R}^n)$, *Transactions of the American Mathematical Society*, vol. 315, no. 1, pp. 69–87, September 1989.

30. E. Kreyszig, *Introductory Functional Analysis with Applications*, New York: John Wiley & Sons, 1989.

31. J. Stoer and R. Bulirsch, *Introduction to Numerical Analysis*, 2nd ed., New York: Springer-Verlag, 1992.

32. A. Rosenfeld and M. Thurston, Edge and curve detection for visual scene analysis, *IEEE Transactions on Computers*, vol. 20, no. 5, pp. 562–569, May 1971.

33. S. Tanimoto and T. Pavlidis, A hierarchical data structure for picture processing, *Computer Graphics and Image Processing*, vol. 4, no. 2, pp. 104–119, June 1975.

34. R. Y. Wong and E. L. Hall, Sequential hierarchical scene matching, *IEEE Transactions on Computers*, vol. C-27, no. 4, pp. 359–366, April 1978.

35. D. Marr, *Vision*, New York: W. H. Freeman and Company, 1982.

36. I. S. Gradshteyn and I. M. Ryzhik, *Table of Integrals, Series, and Products*, Orlando, FL: Academic, 1980.

37. P.-G. Lemarié, Ondelettes à localisation exponentielle, *Journal de Mathématiques Pures et Appliquées*, vol. 67, pp. 227–236, 1988.

38. G. Battle, A block spin construction of ondelettes. Part 1: Lemarié functions, *Communications in Mathematical Physics*, vol. 110, pp. 601–615, 1987.

39. P. P. Vaidyanathan, Multirate digital filters, filter banks, polyphase networks, and applications: A tutorial, *Proceedings of the IEEE*, vol. 78, pp. 56–93, January 1990.

40. O. Rioul and M. Vetterli, Wavelets and signal processing, *IEEE SP Magazine*, pp. 14–38, October 1991.

41. W. R. Madych, Some elementary properties of multiresolution analyses of $L^2(\mathbb{R}^n)$, in *Wavelets: A Tutorial in Theory and Applications*, C. K. Chui, ed., Boston: Academic, pp. 259–294, 1992.

42. A. Bruce, D. Donoho, and H. Y. Gao, Wavelet analysis, *IEEE Spectrum*, pp. 26–35, October 1996.

PROBLEMS

1. Let $\psi(t)$ be an analyzing wavelet.
   (a) Show that $\Psi(0) = 0$.
   (b) Also, $\int_{-\infty}^{\infty} \psi(t)\, dt = 0$.
   (c) Show that $\psi(t)$ is a bandpass filter.
2. Provide an example of a bandpass filter that is not an analyzing wavelet.
3. Let $V = \{V_i: i \in \mathbb{Z}\}$ be a multiresolution analysis of $L^2(\mathbb{R})$ signals; let $\phi(t)$ be the scaling function for V; let $\Phi(\omega)$ be the (radial) Fourier transform of $\phi(t)$; and let H be the associated discrete filter with impulse response, $h(n) = h_n$, given by (11.166). Show in detail that $\Phi(2\omega) = \Phi(\omega)H(\omega)$, where $H(\omega)$ is the DTFT of $h(n)$.
4. Let $\alpha, \beta \in \mathbb{C}$ and $\psi(t),\phi(t)$ be (analyzing) wavelets. If $\theta(t) = \alpha\psi(t) + \beta\phi(t)$, then show

$$W_\theta[f(t)](a, b) = \bar{\alpha}W_\psi[f(t)](a, b) + \bar{\beta}W_\phi[f(t)](a, b). \tag{11.194}$$

5. Let $\eta > 0$, let $\psi(t)$ be a wavelet, and let $\theta(t) = \frac{1}{\eta}\psi\left(\frac{t}{\eta}\right)$. Show that

$$W_\theta[f(t)](a, b) = \frac{1}{\sqrt{\eta}}W_\psi[f(t)](a\eta, b). \tag{11.195}$$

6. Let $\alpha, \beta \in \mathbb{C}$ and define $\theta(t) = \alpha\psi(t) + \beta\phi(t)$. Show that

$$W_\theta[\alpha f(t) + \beta g(t)](a, b) = \alpha W_\psi[f(t)](a, b) + \beta W_\phi[f(t)](a, b) \tag{11.196}$$

7. If $\gamma \in \mathbb{R}$, then show

$$W[f(t-\gamma)](a, b) = W[f(t)](a, b-\gamma). \tag{11.197}$$

8. Let $m, n \in \mathbb{Z}$; $a_0, b_0 > 0$; and suppose that

$$\psi_{m,n}(t) = a_0^{-\frac{m}{2}}\psi\left(\frac{t - nb_0a_0^m}{a_0^m}\right) = a_0^{-\frac{m}{2}}\psi(a_0^{-m}t - nb_0). \tag{11.198}$$

   (a) Find the radial Fourier transform $\Psi_{m,n}(\omega)$ of $\psi_{m,n}(t)$.
   (b) Defining $\Psi^-(\omega) = \Psi^+(-\omega)$, where $\Psi^+(\omega)$ is given by (11.108), and using the arguments of Section 11.3, show that

$$\sum_{m,n=-\infty}^{\infty} \left|\langle x, \psi^{-}{}_{m,n}\rangle\right|^2 = \frac{1}{2\pi b_0 \ln a_0}\sum_{m}\int_{-\infty}^{0}|X(\omega)|^2 d\omega, \tag{11.199}$$

where $\psi^{-}(t)$ is the inverse (radial) Fourier transform of $\Psi^{-}(\omega)$.

9. Let H be a Hilbert space.

(a) Show that an orthonormal basis in H is also a Riesz basis with unit bounds.

(b) Show that a Riesz basis is linearly independent.

(c) Give an example of a frame in $L^2(\mathbb{R})$ that is not a Riesz basis.

10. Let $V = \{V_i\}$ be the Haar MRA which we defined in Section 11.4.1.1. Signals in V_i are constant on intervals $(n2^{-i}, (n+1)2^{-i})$, where $n \in \mathbb{Z}$. If $x(t) \in V_0$, and $x(t) = c_n[u(t-n) - u(t-n-1)]$ for $t \in [n, n+1)$, then we set $I(x(t)) = s(n)$, where $s(n) = c_n$ for all $n \in \mathbb{Z}$.

(a) Show that I is an isomorphism, a bounded linear operator that is one-to-one and onto.

(b) Show that if $I(x) = s(n)$ and $k \in \mathbb{Z}$, then $I(x(t-k)) = s(n-k)$.

(c) Is I an isometry? Explain.

(d) Are the elements of V_0 compactly supported? Explain.

11. Let W_0 be the continuous $L^2(\mathbb{R})$ signals that are piecewise linear on $[n, n+1]$, $n \in \mathbb{Z}$. Define $x(t) \in W_i \Leftrightarrow x(2t) \in W_{i+1}$.

(a) Verify MRA properties (i), (iii)–(v) for $W = \{W_i\}$.

(b) Let $V = \{V_i\}$ be the Haar MRA of the previous problem. Assuming that step functions are dense in $L^2(\mathbb{R})$, argue likewise for W by showing that given $v(t) \in V_0$, then some $w(t) \in W$ is arbitrarily close to $v(t)$. *Moral*: An approximation of an approximation is still an approximation.

(c) Let $w \in W_0$ and set $Iw = s$, where $s(n) = w(n)$. Show that I is an isomorphism.

12. Let $l^{\infty}(\mathbb{Z})$ be the normed linear space of bounded sequences, with $\|x\| = \sup\{|x(n)|: n \in \mathbb{Z}\}$. Define an operator $Tx = y$ as follows: $y(n) = x(n)(|n|+1)^{-1}$. Show that

(a) T is linear.

(b) T is one-to-one.

(c) T is onto.

(d) T is bounded.

(e) T^{-1} is not bounded [30].

13. Let $I: H \to K$ be a bounded Hilbert space isomorphism and let I^* be the Hilbert space adjoint of I.

(a) Show that $\|I^*\| = \|I\|$ [30].

(b) Supposing that I is onto, show that I^* is an isomorphism.

(c) Show that if $E = \{e_k \mid k \in \mathbb{Z}\}$ is a Riesz basis in H, then there is $F = \{f_k \mid k \in \mathbb{Z}\}$ such that F is a Riesz basis and [9]

$$\langle e_m, f_n \rangle = \begin{cases} 1 & \text{if } m = n, \\ 0 & \text{if } m \neq n. \end{cases} \tag{11.200}$$

14. Let $\phi(t) \in L^2(\mathbb{R})$, let $\Phi(\omega) = \mathcal{F}[\phi(t)](\omega)$ be its radial Fourier transform, let $s \in l^2$, $S(\omega)$ be its discrete-time Fourier transform, and set $P(\omega) = \sum_{n=-\infty}^{\infty} |\Phi(\omega + 2\pi n)|^2$. Show that

(a) $\sum\limits_{k=-\infty}^{\infty} s(k)\Phi(t-k) \in L^2(\mathbb{R})$;

(b) $\left\| \sum\limits_{k=-\infty}^{\infty} s(k)\phi(t-k) \right\|_2^2 = \frac{1}{2\pi} \int\limits_0^{2\pi} |S(\omega)|^2 P(\omega)\, d\omega.$

15. This problem uses the concept of Lebesgue measure. Let $P(\omega)$ be defined as above. Define $P_a = \{\omega \in [0, 2\pi] : P(\omega) > a\}$ and $Q_a = \{\omega \in [0, 2\pi]: P(\omega) < a\}$. Referring to Ref. 9, show that:

(a) If the Lebesgue measure of P_a, $\mu(P_a)$, is zero for almost all $a \in \mathbb{R}$, then $P(\omega) = 0$ almost everywhere. *Hint*: Suppose not, so that $P(\omega) > 0$ on $U \subseteq \mathbb{R}$. Then $U = \cup P_{1/n}$, where $n > 0$ is a natural number. What is $\mu(P_{1/n})$? Apply the Lebesgue measure to the countable union.

(b) Similarly, if $\mu(Q_a) = 0$ for almost all $a \in \mathbb{R}$, then $P(\omega) = \infty$ almost everywhere.

16. Complete the proof of the second Riesz basis characterization theorem [9]. Suppose $\phi(t) \in L^2(\mathbb{R})$, $\Phi(\omega) = \mathcal{F}[\phi(t)](\omega)$ is its Fourier transform, $F = \{\phi(t-k) \mid k \in \mathbb{Z}\}$, $0 < A \leq B < \infty$, $P(\omega) = \Sigma|\Phi(\omega + 2\pi n)|^2$, and F is a Riesz basis with lower and upper bounds $\sqrt{A}$ and $\sqrt{B}$, respectively. Show that $P(\omega) \leq B$ for almost all $\omega \in \mathbb{R}$.

(a) Define $P_a = \{\omega \in [0, 2\pi]: P(\omega) > a\}$. Show that if $\mu(P_a) = 0$ almost everywhere, then $P(\omega) = 0$ for almost all $\omega \in [0, 2\pi]$, and $P(\omega) \leq B$. *Hint*: Consider the case $a = 1/(n+1)$ for $n \in \mathbb{N}$.

(b) Hence, assume that for some $a > 0$, $\mu(P_a) > 0$. Let χ_a be the characteristic function on P_a and $x_a(n)$ be the inverse discrete-time Fourier transform of $\chi_a(\omega)$. Show $\|x_a\|^2 = (2\pi)^{-1}\mu(P_a)$.

(c) Show $\|\Sigma x_a(k)\phi(t-k)\|^2 \geq a\|x_a\|^2$.

(d) Show $B \geq a$.

(e) Conclude that, unless $B \geq \Sigma|\Phi(\omega + 2\pi k)|^2$ almost everywhere on $[0, 2\pi]$, a contradiction arises.

17. This problem uses Lebesgue measure, but is straightforward. Suppose $\phi(t) \in L^2(\mathbb{R})$, $\Phi(\omega) = \mathcal{F}[\phi(t)](\omega)$ is its (radial) Fourier transform, and $F = \{\phi(t-k) \mid k \in \mathbb{Z}\}$ is an orthonormal set.

(a) Show that $\|\Phi\|_2 = (2\pi)^{1/2}$.

(b) Show that $|\Phi(\omega)| \le 1$ for almost all $\omega \in \mathbb{R}$.

18. This problem studies Strömberg MRA [9, 22], wherein V_0 consists of continuous signals that are linear on integral segments $[n, n+1]$.

(a) Let $\phi(t)$ be a scaling function for $V = \{V_i\}$. Prove that $\phi(t)$ cannot be finitely supported. Assume that it is finitely supported and derive a contradiction as follows. Take the last interval $(n, n+1)$ to the right over which $\phi(t)$ is nonzero, take the last interval $(m, m+1)$ proceeding to the left over which $\phi(t)$ is nonzero, and compute the inner product $\langle \phi(t), \phi(t-(n-m)) \rangle$.

(b) Enumerate all of the cases for the inner product, and show that it is never zero.

(c) Show that $\phi(t) \ne 0$ on $(n, n+1)$ for arbitrarily large $|n|$.

(d) Conclude that the inner products $\langle \phi(t), \phi(t-k) \rangle$ involve an infinite number of terms.

19. Show that the scaling function for an MRA is essentially unique. More precisely, let $V = \{V_i: i \in \mathbb{Z}\}$ be an MRA and let $\phi(t) \in V_0$ be its scaling function. Show that $\theta(t) \in V_0$ is a scaling function for V if and only if there is a 2π-periodic function $P(\omega)$ such that $\Theta(\omega) = P(\omega)\Phi(\omega)$ and $|P(\omega)| = 1$ almost everywhere on $[0, 2\pi]$.

20. Let $V = \{V_i: i \in \mathbb{Z}\}$ be an MRA, $\phi(t)$ its scaling function, $\Phi = \mathcal{F}(\phi)$, and $H = H_\phi$ the associated low-pass filter. State and prove a generalization of the V_{-1} Characterization of Section 11.4.3 for any $V_N \subset V_0$, where $N < 0$.

21. With the same notation as in the previous problem, define the operator T: $V_0 \to L^2[0, 2\pi]$ by $Tx = C$, where $C(\omega)$ is the 2π-periodic function with $X(\omega) = C(\omega)\Phi(\omega)$ guaranteed by the V_0 characterization (Section 11.4.3).

(a) Show that T is linear: $T(x + y) = Tx + Ty$ and $T(ax) = aTx$ for $a \in \mathbb{C}$.

(b) If $c(n) \in l^2$ is the inverse DTFT of $C(\omega) = Tx$, show that $\|c\| = \|x\|$.

(c) Show that T is a bounded linear map.

22. This problem uses Lebesgue measure to show that the orthogonal wavelet for an MRA is essentially unique [8]. Suppose $V = \{V_i: i \in \mathbb{Z}\}$ is an MRA, $\phi(t)$ is its scaling function, $\Phi = \mathcal{F}(\phi)$ is the Fourier transform of ϕ, and $H = H_\phi$ is the associated low-pass filter. Let $\psi(t) \in W_0$ be an orthogonal wavelet for $L^2(\mathbb{R})$.

(a) Show that there is a 2π-periodic $v(\omega) \in L^2[0, 2\pi]$ such that

$$\Psi(2\omega) = v(2\omega)e^{-j\omega}\overline{H(\omega + \pi)}\,\Phi(\omega). \tag{11.201}$$

(b) Show that

$$\sum_{k=-\infty}^{\infty} |\Psi(\omega + 2\pi k)|^2 = 1 = |v(\omega)|^2 \sum_{k=-\infty}^{\infty} \left|H\left(\frac{\omega}{2} + \pi k + \pi\right)\right|^2 \left|\Phi\left(\frac{\omega}{2} + \pi k\right)\right|^2. \tag{11.202}$$

(c) Summing over even and odd integers separately, show that the final expression above becomes

$$|v(\omega)|^2 \left(\sum_{k=-\infty}^{\infty} \left|H\left(\frac{\omega}{2} + \pi\right)\right|^2 \left|\Phi\left(\frac{\omega}{2} + 2\pi k\right)\right|^2 + \sum_{k=-\infty}^{\infty} \left|H\left(\frac{\omega}{2}\right)\right|^2 \left|\Phi\left(\frac{\omega}{2} + 2\pi k + \pi\right)\right|^2 \right). \tag{11.203}$$

(d) Prove that

$$1 = |v(\omega)|^2 \left(\left|H\left(\frac{\omega}{2} + \pi\right)\right|^2 + \left|H\left(\frac{\omega}{2}\right)\right|^2 \right) = |v(\omega)|^2 \tag{11.204}$$

for almost all $\omega \in [0, 2\pi]$.

(e) Conclude that the Fourier transform of $\psi(t)$ has the form

$$\Psi(2\omega) = v(2\omega) e^{-j\omega} \overline{H(\omega + \pi)} \Phi(\omega), \tag{11.205}$$

where $v(\omega)$ is measurable, and has period 2π, and $v(\omega) = 1$ almost everywhere on $[0, 2\pi]$.

23. Show that to get an orthonormal basis for W_0 an alternative definition for the Fourier transform of $\psi(t)$ is

$$\Psi(2\omega) = e^{j\omega} \overline{H(\omega + \pi)} \Phi(\omega). \tag{11.206}$$

Show that with this change of the exponent's sign $\{\psi(t - k) \mid k \in \mathbb{Z}\}$ is still an orthonormal basis for W_0.

The following problems involve some extension of concepts in the text, may require some exploration of the research literature, and are generally more difficult than the preceding exercises.

24. Expand the construction of tight wavelet frames in Section 11.3.3 to include the case $0 < a_0 < 1$. Show that (11.109) continues to hold.

25. Investigate the application of frames of translations and dilations as constructed in Section 11.3.3. Assume that $a_0 = 2$ and $b_0 = 1$.

(a) Using a mathematical software package such as Mathematica or Matlab, or by developing your own Fourier transform software in a high-level programming language, find the inverse Fourier transforms for $\Psi^{+}(\omega)$ and $\Psi^{-}(\omega)$.

(b) As Daubechies remarks [3], this frame is not well-localized, and this becomes a problem for certain signal analysis tasks. By experiments of your own design, justify this claim.

(c) Continue your critique of this frame by considering the fact that it consists of translations and dilations of two distinct root elements, $\psi^{+}(t)$ and $\psi^{-}(t)$. In particular, explore the consequences of the definition $\Psi^{-}(\omega) = \Psi^{+}(-\omega)$. What difficulties does this impose on signal analysis applications? Develop experiments that justify your contention.

(d) Develop experiments using translations and dilations of the Mexican hat wavelet and compare the performance to the frame in part (c) based on $\psi^{+}(t)$ and $\psi^{-}(t)$.

26. Extend the idea of a multiresolution analysis to $L^2(\mathbb{R}^2)$.

(a) Reformulate the definition of Section 11.4.1.1 for functions of two variables, $x(s, t) \in L^2(\mathbb{R}^2)$.

(b) Let $\{V_{2,k}: k \in \mathbb{Z}\}$ be an MRA for $L^2(\mathbb{R}^2)$. Show that there is a unique image $\phi(s, t)$ such that $\{2^k\phi(2^k s - n, 2^k t - m): m, n \in \mathbb{Z}\}$ constitutes an orthonormal basis for $V_{2,k}$.

CHAPTER 12

Mixed-Domain Signal Analysis

This final chapter explains the methods for using time-frequency or time-scale transforms to segment, classify, and interpret signals. The previous two chapters introduced these *mixed-domain* transforms and their application to elementary analysis tasks. The short-time Fourier (or Gabor) transform (Chapter 10) and the wavelet transform (Chapter 11) are the main tools for the applications we discuss. The applications explain their practical and efficient use, spotlight their strengths and weaknesses, and contrast them with pure time- and frequency-domain techniques.

This chapter covers three methods that, together with the local frequency or scale information given by the transforms, are capable of elucidating signal structure:

- A type of structured neural network, which we call the *pattern recognition network*;
- The *hidden Markov model* (HMM), which has become very popular for speech, text, and biological sequence analysis;
- The *matching pursuit*, a Hilbert space search technique for efficient signal description using a dictionary of signal models.

In place of a summary, there is an afterword to the entire book.

12.1 WAVELET METHODS FOR SIGNAL STRUCTURE

This section follows up on the theoretical work of the previous chapter. There we discovered a special tool for describing multiscale signal structure, the multiresolution analysis (MRA) of finite-energy signals. Now we want to explain how, working within the framework of a chosen MRA, we can:

(i) Develop a discrete version of the wavelet transform.

(ii) Show how an efficient algorithm for signal decomposition arises from the MRA of finite-energy signals.

Signal Analysis: Time, Frequency, Scale, and Structure, by Ronald L. Allen and Duncan W. Mills
ISBN: 0-471-23441-9

(iii) Link this result to the perfect reconstruction filter banks covered in Chapter 9.

(iv) And, finally, show how to employ these methods for analyzing signal shape across many scales.

12.1.1 Discrete Wavelet Transform

Let us assume that we have selected a multiresolution analysis of square-integrable analog signals. Section 11.4 introduced this theory. To recapitulate, an MRA [1] is a chain of subspaces $\{V_i: i \in \mathbb{Z}\}$ in $L^2(\mathbb{R})$ such that:

(i) The V_i are closed and nested: $... \subset V_{-1} \subset V_0 \subset V_1 \subset V_2 \subset ....$

(ii) Their union is dense in $L^2(\mathbb{R})$: $\overline{\bigcup_{n=-\infty}^{\infty} V_i} = L^2(\mathbb{R})$.

(iii) The only signal common to all the V_i is the signal that is zero almost everywhere: $\bigcap_{i=-\infty}^{\infty} nV_i = 0$.

(iv) Dilation by a factor of two links the closed subspaces: $x(t) \in V_i \Leftrightarrow x(2t) \in V_{i+1}$.

(v) The root space V_0 is closed under integral translation: If $x(t) \in V_0$ and $k \in \mathbb{Z}$, then $x(t-k) \in V_0$.

(vi) There is a bounded, one-to-one, linear map, with a bounded inverse I: $V_0 \to l^2$ that commutes with integral translation: If $k \in \mathbb{Z}$, and $I(x(t)) = s(n) \in l^2$, then $I(x(t-k)) = s(n-k)$.

Property (vi) is equivalent to the existence of a Riesz basis within V_0 (Section 11.4.1.4). The previous chapter provided examples of MRAs: step functions, piecewise linear functions, and cubic spline functions. Most importantly, associated with an MRA $\{V_i: i \in \mathbb{Z}\}$ are three special signals:

(i) An analog *scaling function*, $\phi(t) \in V_0$, such that $\{\phi(t-k): k \in \mathbb{Z}\}$ is an orthonormal basis of V_0.

(ii) A discrete *associated lowpass filter* H_ϕ with impulse response, $h(n) = h_n$, given by $h_n = \langle \frac{1}{2}\phi(\frac{t}{2}), \phi(t-n) \rangle$, where $\phi(t)$ is the scaling function in (i).

(iii) An analog *orthogonal wavelet* $\psi(t)$, defined by its Fourier transform as follows: $\Psi(2\omega) = e^{-j\omega}\overline{H(\omega+\pi)}\,\Phi(\omega)$, where $\phi(t)$ is the scaling function of (i), $\Phi = \mathcal{F}(\phi)$ is the Fourier transform of ϕ, and $H = H_\phi$ is the associated low-pass filter of (ii).

To make the MRA signal decomposition discrete, we assume that the analog source signals reside in root space $x(t) \in V_0$ and that we sample them at unit distance to get $x(n)$. The scaling function is a lowpass filter, and its expanding dilations—for example, $\phi(t/2)$, $\phi(t/4)$, $\phi(t/8)$, and so on—have successively narrower passbands. Filtering $x(t)$ by these dyadic dilations of $\phi(t)$ produces approximate versions of $x(t)$

which are increasingly smooth. Next, since some high frequency detail has been removed from $x(t)$ by the filtering, we may select samples that are further apart. For example, after the convolution $x * \phi(t/4)$ removes noise and sharp transitions from $x(t)$, sampling occurs on quadruple unit intervals. This idea comes from the Laplacian pyramid decomposition, covered at the end of Chapter 9 [2].

Let us formalize these ideas for the case of the MRA. Following the notation of Ref. 1, let $x_a(t) = ax(at)$ be the scaled dilation of $x(t)$ by factor a. Typically, $a = 2^i$ for $i \in \mathbb{Z}$. Then the *discrete approximate* representation of signal $x(t)$ at resolution 2^i is

$$(A_i^d x)(n) = (x(t) * \phi_{2^i}(-t))(2^{-i}n) . \tag{12.1a}$$

On the other hand, the orthogonal wavelet $\psi(t)$ is a bandpass filter. Its dilation by various dyadic factors results in filters with narrower passbands and lower center frequencies. Thus, we define the *discrete detail* representation of $x(t)$ at resolution 2^i:

$$(D_i^d x)(n) = (x(t) * \psi_{2^i}(-t))(2^{-i}n) . \tag{12.1b}$$

Although (12.1a) and (12.1b) discretize the decomposition of a square-integrable signal $x(t)$, it remains to see how to compute the various analog convolutions that are required.

12.1.2 Wavelet Pyramid Decomposition

The *discrete orthogonal wavelet representation* or *wavelet pyramid decomposition* consists of the following filtered and coarsely sampled discrete signals:

$$A_{-J}^d x, D_{-J}^d x, D_{-J+1}^d x, \ldots, D_{-1}^d x . \tag{12.2}$$

Notice in (12.2) that only the pyramid maintains all of the detail signals, up to the decomposition level $-J$, but only the coarsest approximate representation of $x(t)$. Let us turn our attention to the convolution operations needed to derive this special structural description.

12.1.2.1 Coarse Signal Structure: The Approximate Signal. We concentrate on deriving the coarse signal approximation, $A_{-J}^d x$ in the pyramid decomposition (12.2). The next section explains how to derive the detailed structural descriptions.

Again, let $\{V_i\}_{i \in \mathbb{Z}}$ be a multiresolution analysis of $L^2(\mathbb{R})$, $\phi \in V_0$ be its scaling function, and $\psi \in V_1$ be its orthonormal wavelet (Section 11.4.4). We define $\tilde{h}(n) = h(-n)$ to be the reflection of $h(n)$, the impulse response of the associated low-pass filter $H = H_\phi$. (Since a scaling function ϕ is known for the MRA, we drop the subscript.)

We rewrite the convolution (12.1a) as an inner product:

$$(A_i^d x)(n) = 2^i \langle x(t), \phi(2^i t - n) \rangle , \tag{12.3}$$

Now let $p > 0$ and i be integers, let $x \in L^2(\mathbb{R})$, and let H_p be the discrete filter with impulse response h_p:

$$h_p(n) = 2^{-p} \langle \phi(2^{-p} t), \phi(t - n) \rangle = \langle \phi_{2^{-p}}(t), \phi(t - n) \rangle . \tag{12.4}$$

Note that $h_1(n) = h(n)$, the impulse response of the quadrature mirror filter associated to the multiresolution analysis $\{V_i\}$.

Then we claim that the decomposition for the discrete approximate representation of signal $x(n)$ at level i is given in terms of the approximate representation of $x(n)$ at level $i + p$ by

$$(A_i^d x)(n) = \sum_{k = -\infty}^{\infty} \tilde{h}_p(2^p n - k)(A_{i+p}^d x)(k) . \tag{12.5}$$

This means that we can get completely rid of the analog convolutions through which we originally defined the pyramid. Indeed, if we take $p = 1$ in (12.5), then each approximate level of representation comes from convolution with $\tilde{h}_p(n)$ followed by dyadic subsampling. This continues recursively for p levels to produce the approximate signal structure of $x(n)$ at level i from the approximate representation at level $i + p$.

To show how this works, we consider the expansion of V_i signals on the orthogonal basis elements of V_{i+p}. For any i, the signals $\{\phi_{2^i} t - 2^{-i} n\}_{n \in \mathbb{Z}}$ span $V_i \subset V_{i+p}$. Indeed, an orthonormal basis of V_{i+p} is $\{2^{-(i+p)/2} \phi_{2^{i+p}}(t - 2^{-i-p} n)\}_{n \in \mathbb{Z}}$. Consequently,

$$\phi(2^i t - n) = 2^{i+p} \sum_{k = -\infty}^{\infty} \langle \phi(2^i s - n), \phi(2^{i+p} s - k) \rangle \phi(2^{i+p} t - k) . \tag{12.6}$$

With a change of variables $s = 2^{-i-p}(t + 2^p n)$, the inner product in (12.6) is

$$\int_{-\infty}^{\infty} \phi(2^i s - n) \phi(2^{i+p} s - k)\, ds = 2^{-i-p} \int_{-\infty}^{\infty} \phi(2^{-p} t) \phi(t + 2^p n - k)\, dt$$

$$= 2^{-i-p} \int_{-\infty}^{\infty} \phi(2^{-p} s) \phi(s - (k - 2^p n))\, ds. \tag{12.7}$$

Putting (12.6) back into (12.7), it follows that

$$\langle x(t), \phi(2^i t - n)\rangle = \sum_{k=-\infty}^{\infty} \int_{-\infty}^{\infty} \phi(2^{-p}s)\phi(s-(k-2^p n))\, ds \langle x(t), \phi(2^{i+p}t-k)\rangle. \tag{12.8}$$

From the definition of the impulse response h_p (12.4), we get

$$\langle x(t), \phi(2^i t - n)\rangle = 2^p \sum_{k=-\infty}^{\infty} \tilde{h}_p(2^p n - k)\langle x(t), \phi(2^{i+p}t-k)\rangle. \tag{12.9}$$

But the inner products in (12.9) are in fact the discrete representations of signal $x(t)$ at levels i and $i + p$. So (12.5) follows directly.

Let us find the impulse response of the discrete filter H_p. Since $\{\phi(t-k)\}_{k \in \mathbb{Z}}$ is a basis for V_0,

$$\phi(2^{-p}t) = \sum_{k=-\infty}^{\infty} \langle \phi(2^{-p}s), \phi(s-k)\rangle \phi(t-k). \tag{12.10}$$

Taking radial Fourier transforms on both sides of (12.10) and simplifying,

$$\Phi(2^p\omega) = \sum_{k=-\infty}^{\infty} h_p(k) \int_{-\infty}^{\infty} \phi(t-k)e^{-j\omega t}dt = H_p(\omega)\Phi(\omega). \tag{12.11}$$

The discrete-time Fourier transform of the filter H_p is

$$H_p(\omega) = \frac{\Phi(2^p\omega)}{\Phi(2\omega)}, \tag{12.12}$$

where $\Phi(\omega)$ is the radial Fourier transforms of scaling function ϕ (t). Applying the inverse discrete time Fourier transform to (12.12), gives $h_p(n)$ (Figure 12.1).

Figure 12.2 shows the $H_p(\omega)$. (12.18) lists filter values for the cubic spline MRA. Note that the $h_p(n)$ are even.

To extract a coarse approximation of a signal's structure using a given MRA, then, the steps are:

(i) Select a resolution step factor $p > 0$.

(ii) Compute the impulse response $h_p(n)$ of the filter with discrete time Fourier transform given by (12.12).

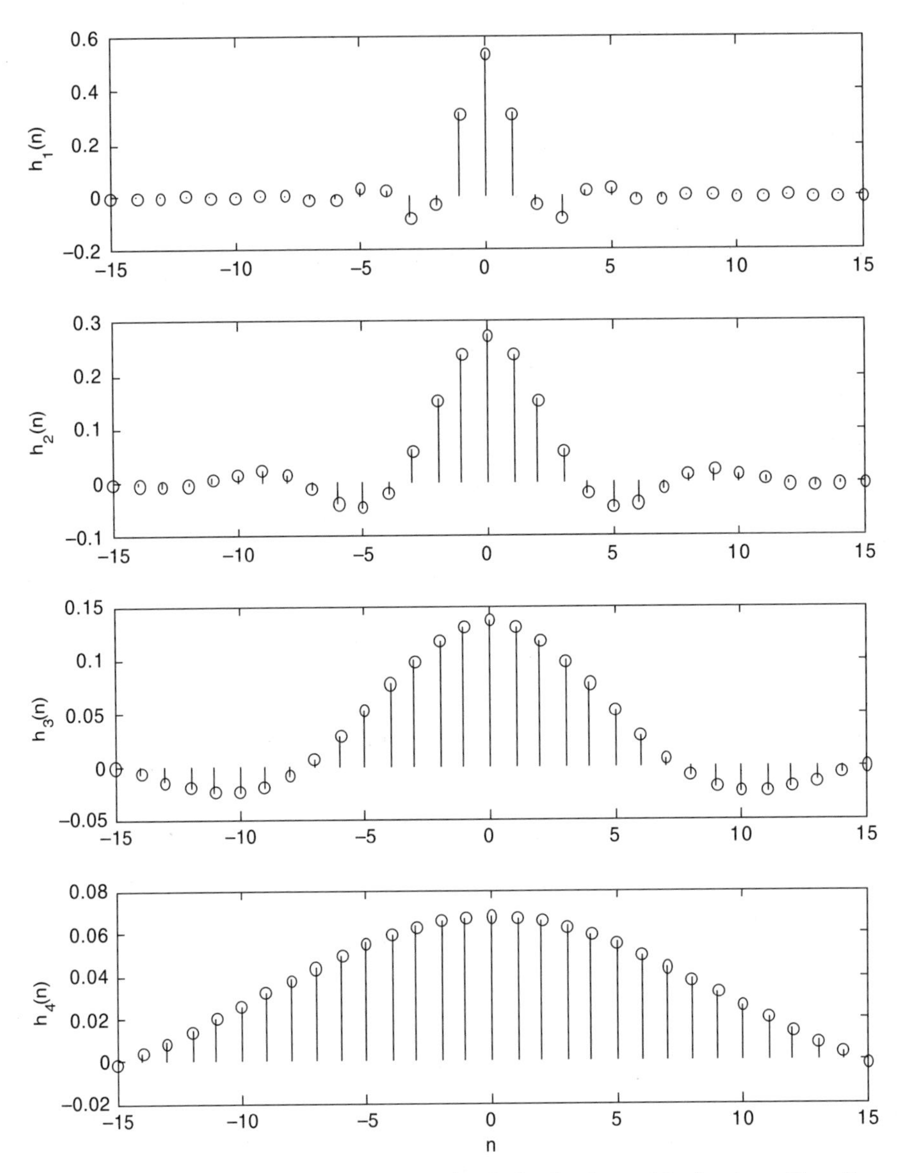

Fig. 12.1. Discrete filter impulse response signals h_1, h_2, h_3, and h_4 for the cubic spline MRA.

(iii) Compute the convolution (12.5) on 2^p-wide intervals to get a coarse approximation $(A_i^d x)(n)$ from $(A_{i+p}^d x)(n)$.

(iv) Employ one of Chapter 4's thresholding methods to the magnitude of the decomposition coefficients, identifying large and small coefficients with significant signal features and background noise, respectively.

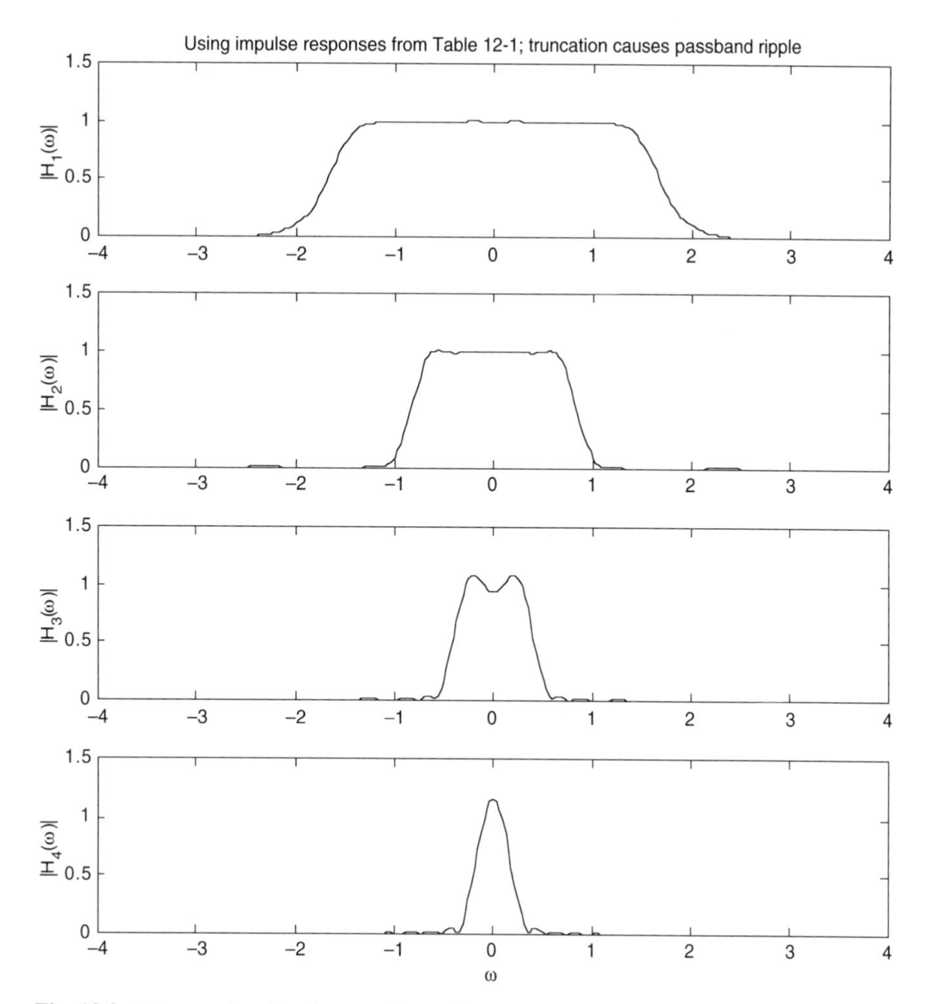

Fig. 12.2. Discrete-time Fourier transforms H_1, H_2, H_3, and H_4 of h_1, h_2, h_3, and h_4 for the cubic spline MRA.

12.1.2.2 *Fine Signal Structure: The Detail Signal.*

A similar derivation gives the signal details at various resolutions. Let G_p be the discrete filter with impulse response g_p,

$$g_p(n) = 2^{-p}\langle\psi(2^{-p}t), \phi(t-n)\rangle, \tag{12.13}$$

and note again that (12.1b) expresses an inner product:

$$(D_i^d x)(n) = 2^i\langle x(t), \psi(2^i t - n)\rangle. \tag{12.14}$$

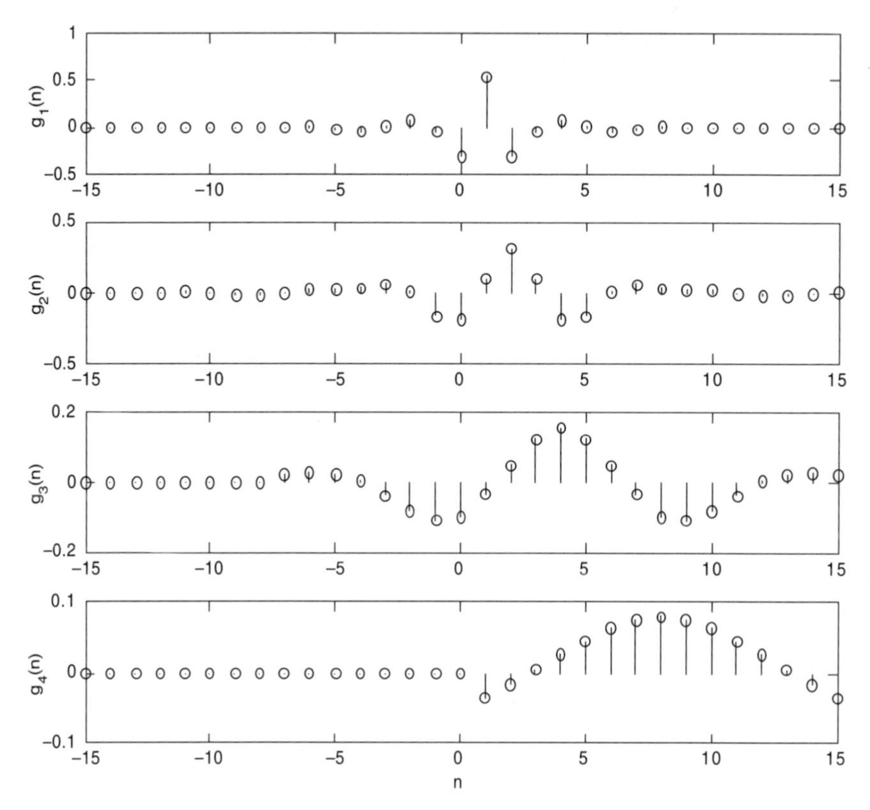

Fig. 12.3. Discrete filter impulse respone signals g_1, g_2, g_3, and g_4 for the cubic spline MRA.

We claim that $(D_i^d x)(n)$, the detail signal at level i, is given in terms of the approximate signal at level $i + p$ by

$$(D_i^d x)(n) = \sum_{k=-\infty}^{\infty} \tilde{g}_p(2^p n - k)(A_{i+p}^d x)(k). \tag{12.15}$$

To verify this, let O_i be the orthogonal complement of V_i inside V_{i+1}: $V_i \perp O_i$ and $V_{i+1} = V_i \oplus O_i$. The shifted, dilated orthogonal wavelets $\{\psi_{2^i} t - 2^{-i} n\}_{n \in \mathbb{Z}}$ span $O_i \subset V_{i+p}$. Since $\{2^{-(i+p)/2}\phi_{2^{i+p}}(t - 2^{-i-p}n)\}_{n \in \mathbb{Z}}$ is an orthonormal basis of V_{i+p},

$$\psi(2^i t - n) = 2^{i+p} \sum_{k=-\infty}^{\infty} \langle \psi(2^i s - n), \phi(2^{i+p} s - k) \rangle \phi(2^{i+p} t - k). \tag{12.16}$$

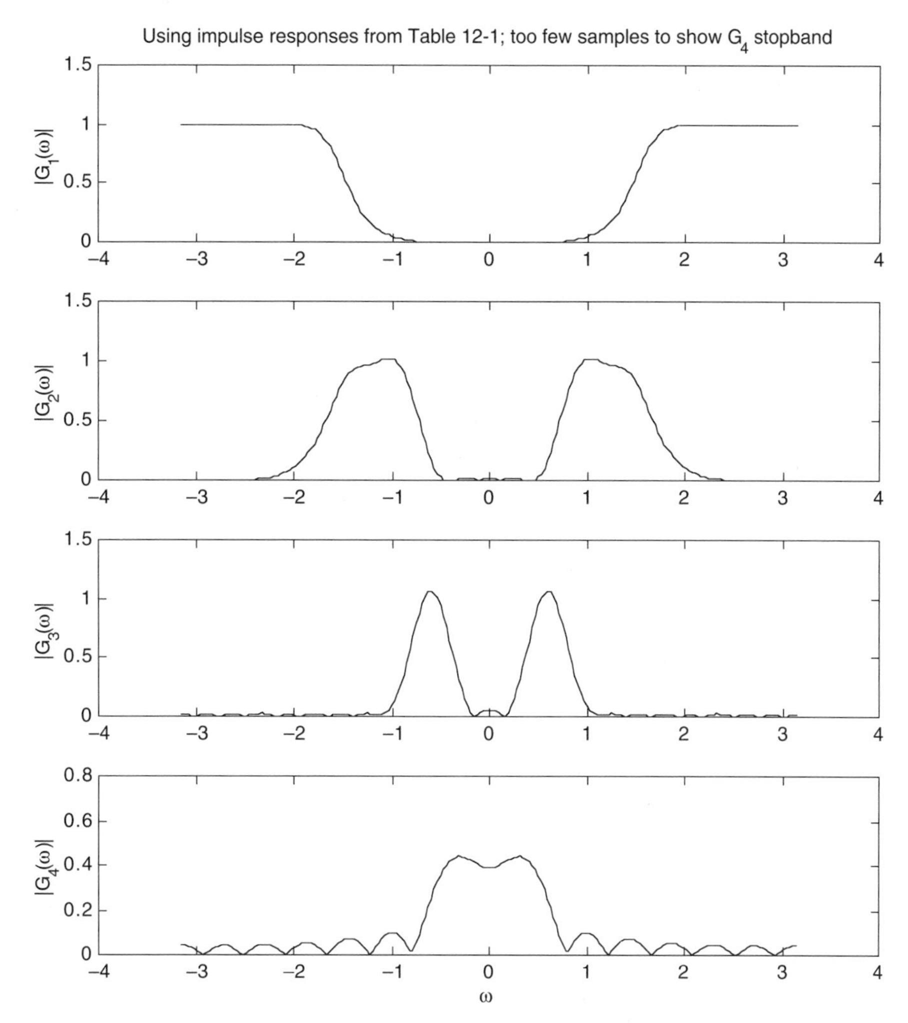

Fig. 12.4. Discrete-time Fourier transforms G_1, G_2, G_3, and G_4 of g_1, g_2, g_3, and g_4 for the cubic spline MRA.

Then, following the argument about approximate signals from the previous section (exercise) gives

$$\langle x(t), \psi(2^i t - n)\rangle = 2^p \sum_{k=-\infty}^{\infty} \tilde{g}_p(2^p n - k)\langle x(t), \phi(2^{i+p} t - k)\rangle , \qquad (12.17)$$

and, consequently,

$$(D_i^d x)(n) = \sum_{k=-\infty}^{\infty} \tilde{g}_p(2^p n - k)(A_{i+p}^d x)(k) . \qquad (12.18)$$

TABLE 12.1. Cubic Spline MRA Orthogonal Wavelet Pyramid Filters

n	$h_1(n)$	$h_2(n)$	$h_3(n)$	$h_4(n)$	$g_2(n)$	$g_3(n)$	$g_4(n)$
0	0.542	0.272	0.136	0.068	–0.189	–0.095	–0.048
1	0.307	0.237	0.131	0.067	0.099	–0.035	–0.035
2	–0.035	0.153	0.118	0.066	0.312	0.049	–0.018
3	–0.078	0.057	0.099	0.063	0.099	0.125	0.003
4	0.023	–0.019	0.077	0.059	–0.189	0.157	0.025
5	0.030	–0.047	0.052	0.055	–0.161	0.125	0.045
6	–0.012	–0.039	0.028	0.050	0.005	0.049	0.062
7	–0.013	–0.013	0.007	0.044	0.054	–0.035	0.074
8	0.006	0.012	–0.009	0.038	0.027	–0.095	0.079
9	0.006	0.020	–0.019	0.032	0.018	–0.107	0.074
10	–0.003	0.015	–0.024	0.026	0.017	–0.080	0.062
11	–0.003	0.004	–0.023	0.020	0.000	–0.037	0.045
12	0.002	–0.006	–0.019	0.014	–0.018	0.003	0.025
13	0.001	–0.009	–0.013	0.009	–0.016	0.023	0.003
14	–0.001	–0.006	–0.006	0.004	–0.004	0.027	–0.018
15	–0.001	–0.001	0.000	–0.001	0.003	0.021	–0.035

Seeking the impulse response of the filter G_p, we expand on $\{\phi(t-k)\}_{k \in \mathbb{Z}}$, an orthongormal basis for V_0:

$$\psi(2^{-p}t) = \sum_{k=-\infty}^{\infty} \langle \psi(2^{-p}s), \phi(s-k) \rangle \phi(t-k) . \tag{12.19}$$

Fourier transformation of (12.19) produces

$$G_p(\omega) = \frac{\Psi(2^p \omega)}{\Phi(2\omega)} . \tag{12.20}$$

Filters for generating the detail structure of signals via the orthogonal wavelet decomposition are shown in Table 12.1(12.18). We set $g(n) = g_1(n) = (-1)^{1-n} h(1-n)$, so it is not shown. Observe that $g_p(n)$ is symmetric about 2^{p-1}.

12.1.2.3 Quadrature Mirror Filters. We have shown that discrete filters, H_p and G_p, with impulse responses h_p and g_p, respectively, are all that we need for the wavelet pyramid decomposition (12.2). Since we know the Fourier transforms of the wavelet and scaling function, we can compute these impulse responses from the inverse discrete-time Fourier transforms of (12.12) and (12.20). Figure 12.5 shows how the pyramid decomposition occurs by successive filtering and subsampling operations.

From Chapter 11's theoretical development, we know that the discrete low-pass filter $H(\omega)$ associated to an MRA satisfies $|H(\omega)|^2 + |H(\omega + \pi)|^2 = 1$. Within an amplification factor, this is precisely the perfect reconstruction criterion of Chapter 9. In fact, $\sqrt{2}h(n)$ is a quadrature mirror filter. We can decompose the signal using

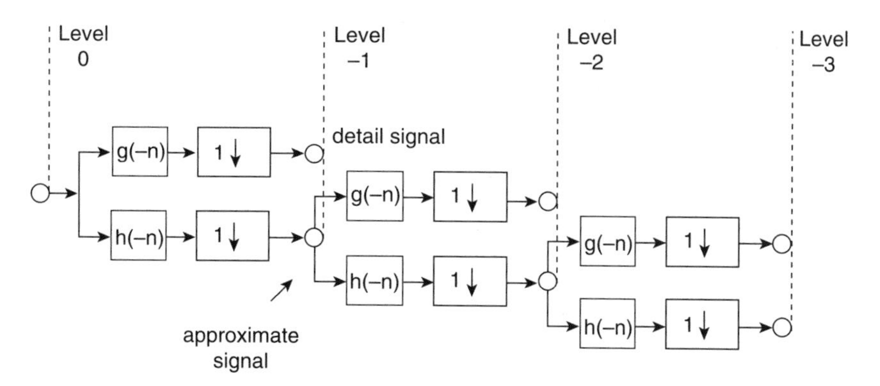

Fig. 12.5. Signal decomposition using the orthogonal wavelet pyramid.

this scaled filter, or we can slightly modify the perfect reconstruction scheme of Chapter 9 by supplying an additional amplification factor upon reconstruction [1, 3].

Consider the QMF pyramid decomposition in Figure 12.5. Let $\tilde{h}(n) = h(-n)$ be the reflection of $h(n)$ and

$$\tilde{H}(z) = \sum_{n=-\infty}^{\infty} \tilde{h}(n) z^{-n} \tag{12.21}$$

be the z-transform of $\tilde{h}(n)$. Subsampling by two followed by $\tilde{H}(z)$ filtering is the same discrete system as $\tilde{H}(z^2)$ filtering followed by subsampling [4] (exercise). Applying the same idea to $\tilde{g}(n)$, note that we can directly obtain level −2 coefficients by filtering with $\tilde{H}(z)\tilde{H}(z^2)$ and $\tilde{H}(z)\tilde{G}(z^2)$ and subsampling by four. We can compute the impulse response of the filter with transfer function $\tilde{H}(z)\tilde{H}(z^2)$ by convolving $\tilde{h}(n)$ with the filter obtained by putting a zero between every $\tilde{h}(n)$ value.

12.1.3 Application: Multiresolution Shape Recognition

This section shows how to use the orthogonal wavelet pyramid decomposition to recognize signal patterns at varying scales. This is one type of pattern recognition application [5]. We seek a known signal pattern, the *model*, in a *sample* signal that—perhaps along with background shapes and noise—contains a dilated version of the model. Moreover, if our pattern comparison algorithm gives localized information, we can attempt to *register* the time-domain position of the model as well. *Registration* is the process of finding the position of a *prototype*, or model, signal within a *candidate*, or sample, signal. For these tasks, signal analysts have relied upon multiple resolution methods for the following reasons.

- If the comparisons between the model and sample proceed pointwise, then the number of computations may become prohibitive—an especially acute problem in multiple dimensions (image analysis, computer vision, video analysis). Hierarchical structures that analyze signals at several resolutions can make the number of computations tractable [6]. Comparisons at coarse scales are iteratively improved in transition to fine scales.
- Coarse representations of signal structure can isolate significant features that are apparent only at certain resolutions [7–9].

When applying pyramid techniques to register a prototype object in a candidate signal, we first decompose both the model pattern and the sample. At the coarsest scale of decomposition, the algorithm compares the model and sample at all possible relative positions. The decomposition coefficients should not change as the model's offset into the sample varies. For otherwise, the decomposition would need to be recomputed for each location; computation time then increases drastically. The model's coefficients will not change if the decomposition procedure is translation-invariant. Of course, the coefficients could change in some simple way that is comparatively inexpensive to compute. Eventually, this produces a set of suffiently good—or, *feasible*—comparison locations between prototype and candidate. The search continues at the next higher resolution with—hopefully—a greatly confined set of feasible registrations. The best acceptable match at the finest scale gives the final result.

We apply the multiresolution analysis (MRA) of $L^2(\mathbb{R})$ to the multiscale shape recognition problem [1]. Once a particular MRA is chosen, it leads to simple, compact, and efficient pyramid decompositions using quadrature mirror filter (QMF) banks (Figure 12.5). The algorithms do not increase the amount of memory space required for storing the representations, yet exactly reconstruct the original signal.

We note that the MRA concept extends to two (or more) dimensions for image analysis. In this case, the separable two-dimensional pyramid decomposition distinguishes between horizontal and vertical spatial frequencies, which is useful for texture and image analysis in artificial environments [1]. Supplementary orientation tunings are possible too [10].

Orthogonal wavelet pyramids suffer from the following difficulties in registration and matching applications:

(i) A registration problem is that the lower resolution coefficients do not translate as the original signal is shifted; in fact, the decomposition coefficients change drastically (Figure 12.6). This greatly complicates the tasks of matching and registration and has inspired research into alternative representations that support pattern matching [11].

(ii) A second registration difficulty arises from the orthogonality of the representation, a consequence of its derivation from an MRA of $L^2(\mathbb{R})$. A registration between prototype and candidate at one resolution may not indicate any correlation between them at a finer scale. Whether this second difficulty appears depends on the nature of the signals acquired by the processing

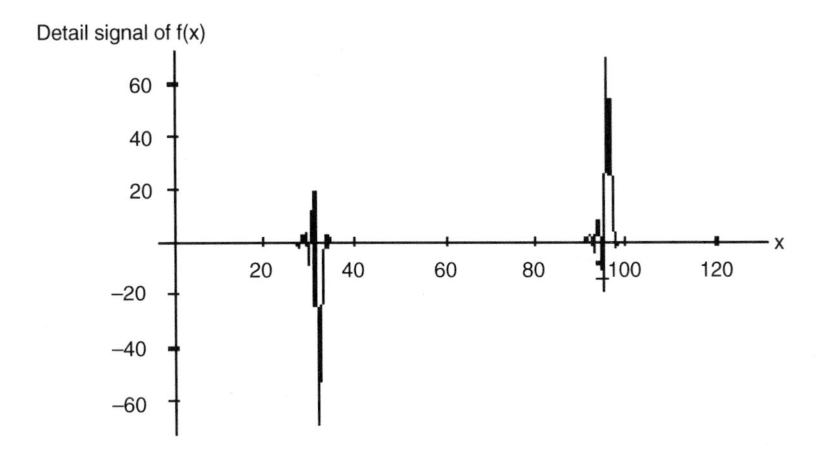

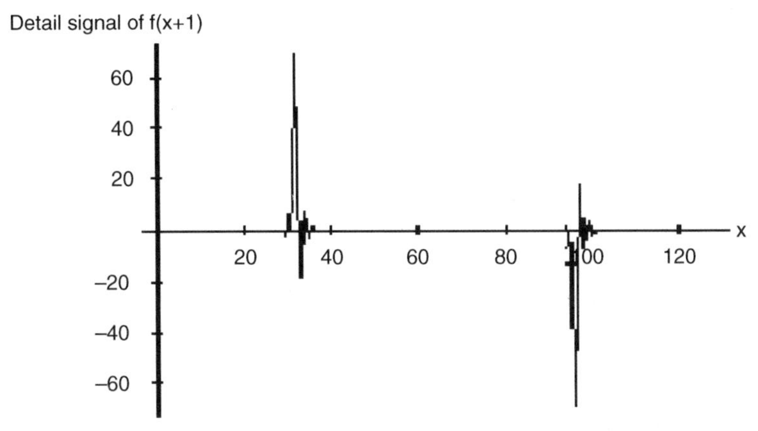

Fig. 12.6. Decomposing a prototype square pulse one resolution level using the orthogonal wavelet pyramid decomposition (a). The same pulse, shifted by a unit distance, represents the candidate. Its first coarse representation is shown in (b). This confounds the basic pyramid registration algorithm when it uses raw detail signals. The best registration position of candidate with respect to prototype is to align the left edge of one with the right edge of the other. On the other hand, the registration algorithm succeeds when using the magnitudes to measure of local energy.

system. Below, we describe an algorithm for coarse-to-fine tracking of registrations in orthogonal wavelet pyramids. The exercises suggest a comparison using the Laplacian pyramid [2] representation with the same registration algorithm. We report some earlier results here [5].

(iii) A problem in multiscale matching is the dyadic dilation factor between pyramid levels. If the modeled object does not happen to be a dyadically scaled version of the candidate's pattern, then it is possible to overlook a match.

In matching applications, we seek instances of model patterns in an acquired signal. Suppose that prototypes are decomposed and stored in a phoneme recognition system, for example. The large number of prototypes needs a compact representation for the model pyramids to trim memory requirements. Applying orthogonal wavelets for such recognition systems is attractive and is a motivation for trying to circumvent the translation of coefficients problem. It is desirable to quickly reject candidate patterns that are not represented by any model. Decomposition of a full pyramid for a candidate pattern is costly when, for instance, only the coefficients at the fourth level of decomposition (1/16 of the total values) are used for comparison with prototypes. If the pattern is accepted, then the time constructing a full pyramid is not lost. It helps obtain a precise registration. But if the pattern is rejected, a full pyramid is built for a candidate even though a tiny fraction of the coefficients find use. One need not derive all pyramid levels, in sequence from finest to coarsest scale of representation, in order to reject or tentatively accept a candidate pattern. We derived formulas for filters that allow us to directly compute coarse pyramid levels in the orthogonal wavelet representation in Sections 12.1.2.1.2.

We use the well-known cubic spline MRA for registration experiments [12]. If $\{V_i : i \in \mathbb{Z}\}$ is the MRA, then the root space V_0 is all finite-energy continuously differentiable functions that are cubic polynomials on intervals $[k, k+1]$. We studied this example of an MRA in Section 11.4. It has a particularly suitable scaling function $\phi(t) \in V_0$ for signal analysis, with exponential decay in the time domain and polynomial decay of ω^{-4} in the frequency domain. We recall that the associated discrete low-pass filter is $h(n) = \langle \frac{1}{2}\phi(\frac{t}{2}), \phi(t-n)\rangle$. We set $g(n) = (-1)^{1-n}h(1-n)$, which is a discrete high-pass filter.

The registration algorithm begins with a set of *feasible points*, where sufficient correlation exists between the candidate and prototype signals, at the coarsest level of representation. Feasible points at finer scales are found, furnishing *feasible paths* up the pyramid. In order for a feasible point to continue a feasible path from a lower level, it must be close to a previous, coarser registration value. The best complete feasible path to the finest level of representation gives the registration between candidate and prototype patterns.

The algorithm uses the limited shift-invariance in the orthogonal wavelet pyramid representation. The coefficients of the wavelet representation at level $l < 0$ translate by amount k when the original signal is translated by amount $k2^{-l}$. The steps are:

(i) The candidate signal is decomposed with the wavelet pyramid (Figure 12.5).

(ii) The minimum of the registration cost function m over all registrations r,

$$m(r, l) = \sum_i [X_{l,c}(i) - X_{l,p}(i-r)]^2 , \tag{12.22}$$

is computed for level $l = -L$. $X_{l,c}$ is the candidate signal decomposition at level l, $X_{l,p}$ is prototype signal decomposition at level l, and i varies over candidate signal values. Let r_{-L} be a registration at which the minimum occurs, and call the minimum $M_{-L} = m(r_{-L}, -L)$. We pad arrays with zeros when endpoints overlap.

(iii) All registrations s such that $m(s, -L) \le Tm(r_{-L}, -L)$, where $T > 1$ is a threshold, are the feasible points at level $-L$. Call this set of registrations FR_{-L}, the starting points of feasible paths up the pyramid levels.

(iv) Steps (ii) and (iii) repeat at higher levels $-L + k$, $1 \le k < L$.

(v) We prune the feasible point sets at the higher levels to retain only feasible points that continue, within an offset tolerance factor, a feasible path coming from a lower level. Thus, $s \in FR_{-k+1}$ only if $m(s, -k+1) \le TM_{-k+1}$ and $t - \tau \le s \le t + \tau$ for some t in FR_{-k} and offset tolerance τ.

(vi) Finally, if at level -1 no full feasible path has been found, then registration failed. If at least one feasible path is found, the best is selected as the final registration. In a local neighborhood of the best registration found at level -1, the original prototype signal and original candidate signal are examined for the best correspondence value.

(vii) To extend the algorithm to matching, where the scale of the candidate object is unknown, then we allow feasible paths to start and stop at intermediate levels.

(viii) For matching applications based on coarse signal structure, it is useful to generate low-resolution pyramid levels directly rather than iteratively. The wide filters we developed above (Table 12.1(12.18)) allow us to jump many scales, quickly compare, and tentatively accept or reject candidate signals. This saves the time and memory cost of performing full pyramid decompositions. (12.18) Table 12.1 gives sample coefficients for $h_p(n)$ and $g_p(n)$. We use filters $g_1(n) = g(n)$, $g_2(n)$, ..., $g_p(n)$, and $h_p(n)$ for an orthogonal wavelet pyramid to level $-p$.

For experimentation, we register sections of the Australian coastline. Digitized, it contains 1090 points for the experiments herein. The scale-space representation has been studied in similar experiments [13]. Beginning from a reference zero position, we plot the outline of the Australian coast (Figure 12.7) as ordered pairs $(x(t), y(t))$, where t is the distance along the coastline to the reference position.

To find derivatives, we approximate $x(t)$ and $y(t)$ to a quadratic using Lagrange's interpolation formula. A signed curvature function [14],

$$\kappa(t) = \frac{(x'y'' - y'x'')}{[(x')^2 + (y')^2]^{3/2}}, \tag{12.23}$$

gives the coastline curvature at $(x(t), y(t))$ (cf. Figure 12.8a). Prototype signal $\kappa(t)$ is decomposed using the orthogonal wavelet decomposition (Figure 12.9a). The

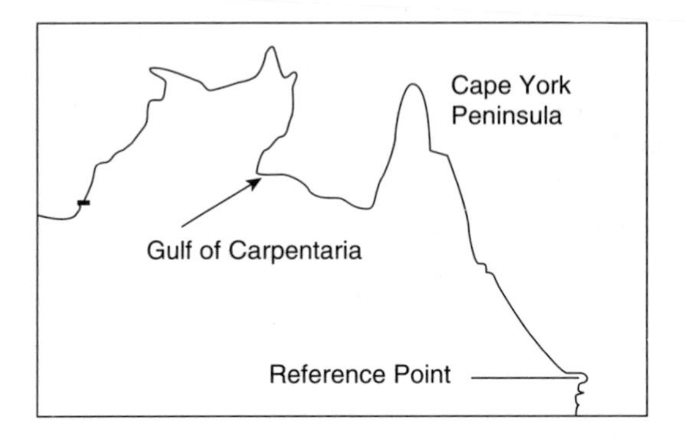

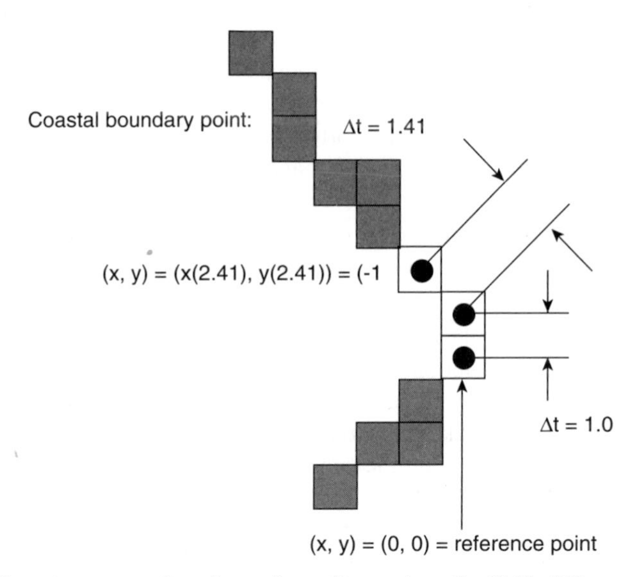

Fig. 12.7. Curvature maps of sections of coastline such as the Gulf of Carpentaria and the Cape York Peninsula are registered within an entire boundary curvature map of Australia. The independent variable of the curvature map is the arc-length distance from the reference point.

candidate signals for the experiments are partial curvature maps of coastal sections, for example the Gulf of Carpentaria (Figure 12.8b, Figure 12.9b).

The registration algorithm generates feasible paths through the pyramid (Figure 12.10). Thresholds T and τ depend on the application. For the boundary matching shown here, $T = 1.2$, and τ in Step (v) was chosen to be 2^{k+1} at level $-k$. If the value of T is too small, the registration can fail at the coarsest level; the r_{-L} is incorrect, and the neighborhood of the correct registration holds no feasible point. When τ is too large, the number of feasible paths to check increases, slowing the coarse-to-fine

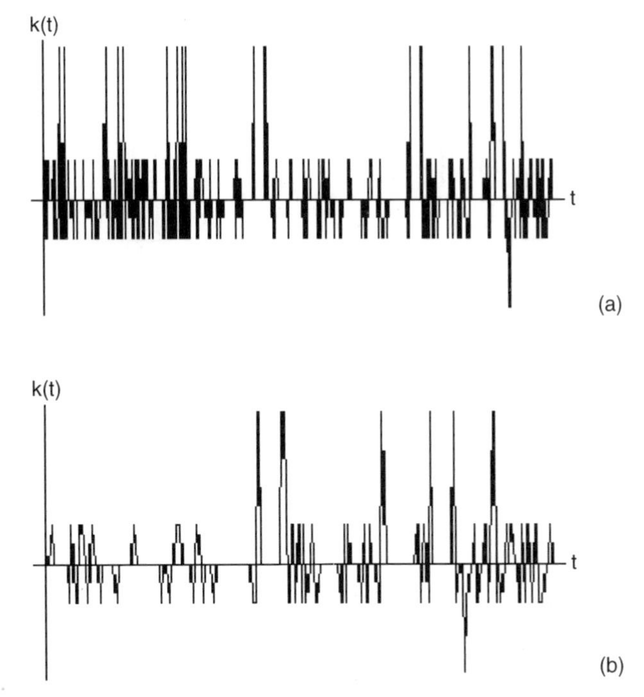

Fig. 12.8. Curvature maps. (a) Partial curvature map of Australian coastline. (b) Curvature map of Gulf of Carpentaria. The correct registration is evident.

algorithm. If τ is too small, initially correct feasible paths terminate at coarse resolutions. It turns out that this holds for both the Laplacian- and wavelet-based registration schemes [5]. The exercises suggest a comparison of these methods along with the piecewise continuous MRA [15].

Table 12.2 shows experimental results in registering continental boundaries. Over all but the lowest resolution level we rely on the detail signal magnitudes of the candidate and prototype signals. For the coarsest comparisons of structure, the approximate signals are used for matching. We add noise to the candidate curvature maps in some experiments. Note that the mean-square signal-to-noise ratio (SNR) employed here is $SNR = \sum s^2(t)/\sum N^2(t)$. Experiments reported in Ref. 5 were performed in which the signals $X_{l,c}$ and $X_{l,p}$ in (12.22) were taken to be either the approximate signal, the detail signal, or the absolute value of the detail signal coefficients. The outcome is problematic registration with the raw detail signals, but satisfacory convergence if $X_{l,c}$ and $X_{l,p}$ are approximate signals [5]. This is not unexpected, as the algorithm then compares successively low-pass filtered signals. The approximate signals are always used at the coarsest level (here, level -4) to generate the initial list of feasible points FR_{-4}. The feasible paths are robust when the candidate has large support, but smaller candidates of nearly straight coastline can fail to correctly register [5].

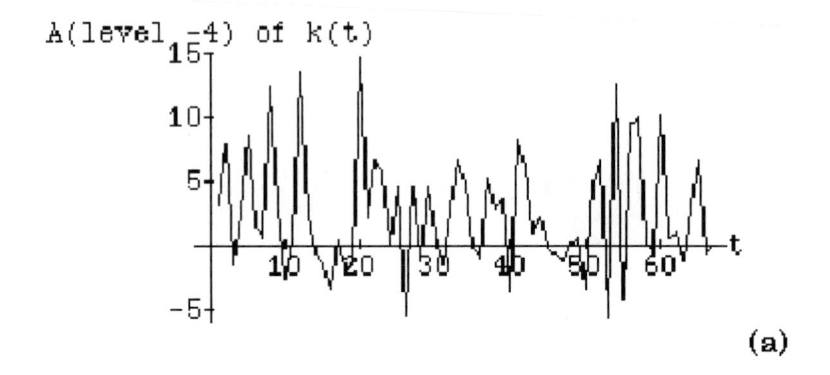

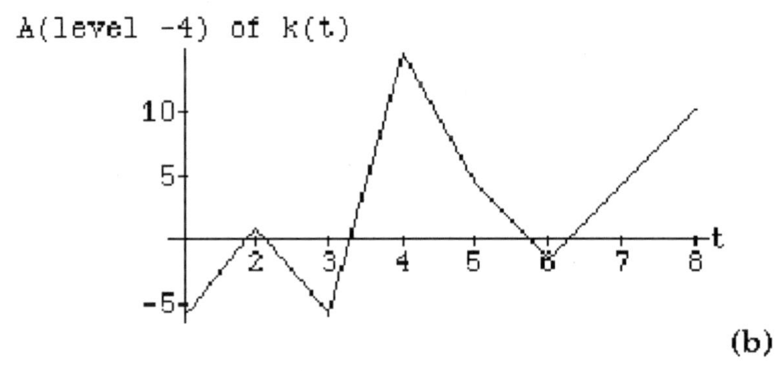

Fig. 12.9. Beginning registration of Gulf of Carpentaria at level –4. (a) Australian coastline curvature at resolution 1/16. (b) Gulf of Carpentaria curvature at resolution 1/16. Feasible points are not so evident.

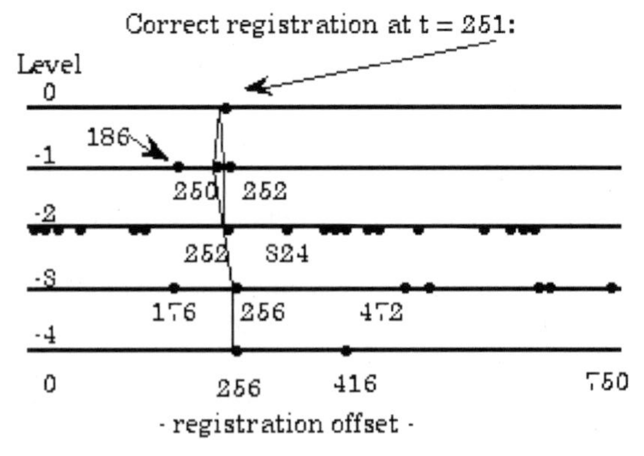

Approximate signals compared at level -4.
Detail signal magnitudes compared at levels -3, -2, and -1.

Fig. 12.10. Registering Gulf segment against Australia boundary: two feasible paths, both ending at the correct registration offset of 251, are found.

TABLE 12.2. Orthogonal Wavelet Pyramid Algorithm Performance

Run	Candidate	Correct Offset	Match Measure	Pyramid Levels	Feasible Points	Success
1.0W	128 point segment of Gulf of Carpentaria	250	Detail signal magnitude	4	19	Yes
1.1W	Same as 1.0W, except shifted	251	Same	4	90	Yes
2.0W	Same as 1.0W, except uniform noise with mean-square SNR of 10.0 added to candidate signal	250	Same	4	87	Yes
2.1W	Same as 2.0W, except shifted	251	Same	4	111	Yes
2.2W	Same as 2.1W, except shifted	252	Same	4	79	Yes

Now, consider the Laplacian pyramid. It analyzes N-point signals into approximately $2N$ coefficients and images into $4N/3$ coefficients. Correlation between levels causes the larger pyramid sizes. Since the orthogonal wavelet and Laplacian pyramid representations are computationally quite alike [1], it is natural to study the registration algorithm using the Laplacian pyramid decomposition. For these experiments, we implemented the Laplacian pyramid using the approximately Gaussian [2] low-pass filter $\{0.05, 0.25, \underline{0.4}, 0.25, 0.05\}$. We find that raw difference signals of the pyramid (not their magnitudes) suffice for good registration. Figure 12.11 shows the result of the experiment of Figure 12.10 using the Laplacian pyramid decomposition.

Table 12.3 shows the results of the Laplacian-based registration algorithm on the same battery of experiments for which we used orthogonal wavelets.

To summarize the results, registering curvature maps with wavelet pyramids produces many more feasible points and paths. However, some Laplacian pyramid runs produce many feasible points too. Both methods are robust in the presence of noise, although the Laplacian pyramid suffers from very large feasible point counts at coarse resolutions. It turns out that candidate signals with small support often make the wavelet registration fail, whereas the Laplacian pyramid algorithm withstands these same small structures [5]. Overall, the Laplacian pyramid decomposition is somewhat better, since the correlation between levels stabilizes the coarse-to-fine tracking.

It appears that these results do not depend substantially on the type of MRA—and hence the discrete pyramid decomposition filters—chosen for deriving the coarse resolution signal structures. Both the Haar [16] and the Daubechies [17] compactly supported wavelets were used in similar registration experiments [18].

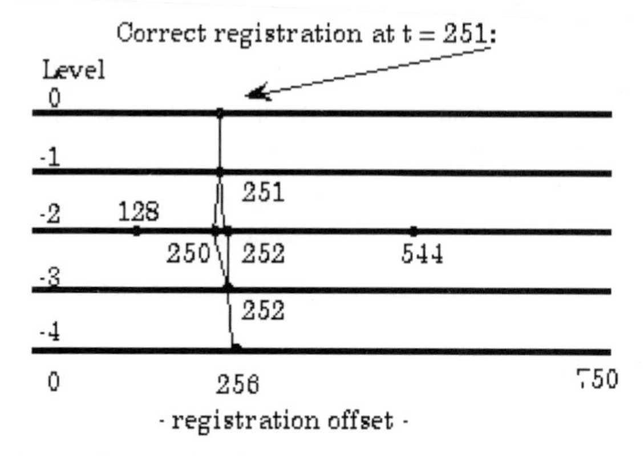

Fig. 12.11. Registering Gulf segment against Australia boundary using the Laplacian pyramid decomposition. Again, two feasible paths arise. The correct registration is found very quickly, at level −1, where the unit distance of the difference signal is the same as the original curvature maps.

These researchers concluded that as long as two or more samples support the signal structure of interest, the approximate signal structures suffice for coarse-to-fine matching and registration. The same authors advise caution with the pyramid's detail signals. Neither method—Haar or Daubechies wavelet pyramids—significantly outperformed the other, although the Haar structures were slightly better when using detail signals [18].

TABLE 12.3. Laplacian Pyramid Algorithm Performance

Run	Candidate	Correct Offset	Match Measure	Pyramid Levels	Feasible Points	Success
1.0L	128-point segment of Gulf of Carpentaria	250	Detail signal	4	17	Yes
1.1L	Same as 1.0L, except shifted	251	Same	4	7	Yes
2.0L	Same as 1.0L, except uniform noise with SNR = 10.0 added to candidate	250	Same	4	954	Yes
2.1L	Same as 2.0L, except shifted	251	Same	4	1090	Yes
2.2L	Same as 2.1L, except shifted	252	Same	4	942	Yes

12.2 MIXED-DOMAIN SIGNAL PROCESSING

Mixed-domain signal transformations provide some new insights into signal processing tasks. Although this section concerns applications that refine rather than interpret a signal, we observe that these steps are often important ancillary feature of a signal analysis application. Here, we confine our remarks to three areas:

- Compression;
- Filter methods and filter banks;
- Enhancement.

Compression is necessary for constructing large signal databases, such as a model-based recognition system might employ. Filtering is important at the front end of an analysis application. Enhancement can essential for building signal prototypes. In fact, although it is a purely signal-in, signal-out technology, compression is perhaps the most important commercial application of the orthogonal wavelet transform. We hasten to add that a very large research literature continues to flourish in all of these areas; our present assessment by no means constitutes the final word.

Good compression methods currently employ either the discrete cosine transform (DCT) or orthogonal wavelet transform coding. The idea is that the transform coefficients are statistically far simpler than the original signals or images, and therefore the transformed data can be described with fewer numerical values. Of course, the orthogonality of the wavelet transform and the efficient, perfect reconstruction filter banks that it provides promote its use in compression. After compression, a handful of transform coefficients nonlinearly encode complex signal and image patterns. Perfect reconstruction is, in principle, possible. Even with lossy compression, ratios of one-bit compressed versus one-byte (8-bit) original signal gives excellent reconstruction. When more decomposition coefficients are discarded in compressing the signal and there remains only a single bit versus 4 bytes of source signal, the reconstruction is still fairly good for human perception.

Digital sound, image, and video databases are huge. Whether they support a signal analysis system or not, compression is essential for economy of storage, retrieval, and transmission. The earliest wavelet compression methods were based on the orthogonal wavelet pyramid decomposition (12.2), shown in Figure 12.5 [1]. New compression methods—some of which provide compression ratios of two orders of magnitude—based on wavelet transforms have been reported in the research literature [19–22]. The basic idea is that many detail coefficients carry no useful signal information and can be set to zero without appreciably affecting the result of the pyramid reconstruction algorithm. There are many variations, but typically the steps are as follows:

(i) Select a multiresoluton analysis and a final level of pyramid decomposition $-L$, where $L > 0$.

(ii) Decompose the signal $x(n)$ into its pyramid decomposition (12.2), producing detail coefficients for levels $-L \le l \le -1$ and approximate coefficients for level $l = -L$.

(iii) Apply a threshold to the fine structure signals $D^d_{-L}x, D^d_{-L+1}x, \ldots, D^d_{-1}x$, so that small magnitude coefficients are set to zero. This is typically a *hard threshold*: If $s(t)$ is a signal, then its hard threshold by $T > 0$ is given by

$$s_{T,h}(t) = \begin{cases} s(t) & \text{if } |s(t)| > T, \\ 0 & \text{if } |s(t)| \leq T. \end{cases} \tag{12.24}$$

(iv) Apply a standard compression technique to the small coarse-resolution trend signal $A^d_{-L}x$. Examples include the Karhunen–Loeve compression [23] or—for images, especially—the Joint Photographic Experts Group (JPEG) standard algorithm [24], which derives from the discrete cosine transform (DCT) [25].

(v) Apply an entropy coding technique, such as simple run-length encoding [26] to the detail signals.

(vi) Decode the compressed pyramid levels from (iv) and (v) and reconstruct the original signal (with some loss, principally due to the thresholding operations) using the exact reconstruction afforded by the MRA's quadrature mirror filters;

(vii) A conservative guideline for hard threshold selection is

$$T = \sigma\sqrt{2\frac{\log(N)}{N}}, \tag{12.25}$$

where σ^2 is the variance of $x(n)$ at level $l = 0$ and N is the number of samples.

Compression ratios of about 25:1 on natural images are possible with the above method. The extension of the orthogonal wavelet pyramid decomposition to two dimensions (images) is necessary for this technique [1], but is unfortunately beyond our present scope. However, Figure 12.12 gives the idea. As with signals, the detail

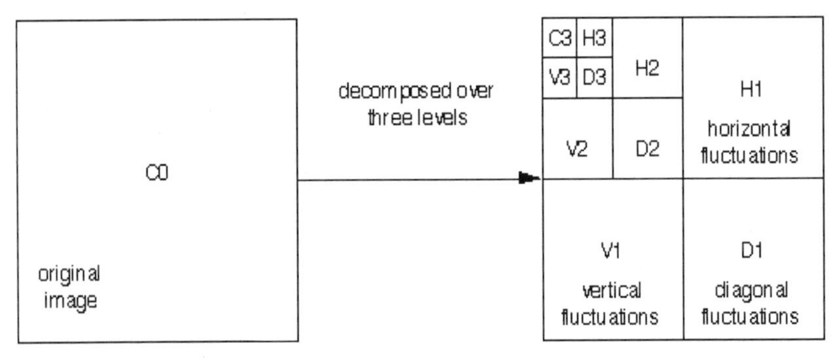

Fig. 12.12. Orthogonal wavelet image compression decomposes the original into four uncorrelated subimages. The trend image C_1 is analogous to the one-dimensional approximate signal. The three detail images contain the direction-sensitive fluctuations of the image C_0. These include vertical details, horizontal details, and diagonal (corner) details. The image pyramid decomposition applies the same algorithm to each coarse structure trend image, C_1, C_2, and so on. This produces more detail images and a final trend image, C_3 above.

images tend to have simple statistics. They can be modeled, quantized, and individually compressed to impressive ratios. The decomposition allows the algorithm designer to tune the vertical, horizontal, and diagonal quantizations so that the direction-sensitive human viusal system perceives minimum distortion in the reconstructed image [1, 27].

More recent approaches to signal compression are as follows:

- *Malvar "wavelets"* are in fact a time-frequency signal decomposition [28]. They are an alternative to the time-scale decomposition using wavelet pyramids. This time-frequency compression technique breaks down an image into a sequence of sinusoidally textured atoms, with adjustable leading and trailing borders. The overall size of the atoms is also tunable. Lastly, as we noted briefly in Chapter 10, as an atomic signal decomposition that uses sinusoids instead of complex exponentials may avoid the limitation of the Balian–Low theorem. This flexibility allows Malvar wavelet compression schemes to beat others when tested in constrained problem domains, such as fingerprint images [29].
- *Wavelet packets* are functions of the form $2^{m/2}W_n(2^m t - k)$, where m, n, and k are integers and $n > 0$ [30,31]. W_n extends only over a finite interval $[0, N]$, and it contains the root frequency of its family of atoms. The decomposition scheme is similar to the orthogonal wavelet pyramid, except that the detail signal structures are also composed at every level. Image decomposition uses tensor products of the translations and dilations of the W_n. A distinct advantage of wavelet packets over Malvar wavelets is that each set of atoms is generated by translation, dilation, and modulation of a single function. This simplifies the construction of algorithms and special-purpose compression hardware. Wavelet packets offer excellent compression ratios, in the realm of 100:1 [19].
- *Structural approaches* to wavelet-based image compression take a two-stage approach. This scheme first extracts edges across several scales. It then encodes the texture representing the difference between the original and the reconstruction from edge information. Combining the texture-coded error image with the edge-coded image gives a perceptually acceptable rendition of the original [20]. Closely related to transform signal compression are mixed-domain processing techniques for noise removal and enhancement.

12.2.1 Filtering Methods

Wavelet decompositions provide for methods that remove background noise from a signal but preserve its sharp edges. This can be especially valuable in signal analysis applications where it is necessary to identify local shapes, for example, that may be corrupted by noise, but still obtain a precice registration [29]. We recall from Chapter 9 that low-pass and bandpass filters removed high-frequency components from signals, but as a rule, these convolutional systems blur the edges as well. Once again, there are many alternatives for wavelet-based noise removal. The typical approach follows the compression algorithm, with a twist at the thresholding step [32–34]:

(i) Select an MRA and final level of decomposition $-L$, where $L > 0$.

(ii) Decompose $x(n)$ according to (12.2).

(iii) Retain the coarse structure approximate coefficients at $l = -L$, but apply a *soft threshold* to $D^d_{-L}x, D^d_{-L+1}x, \ldots, D^d_{-1}x$: If s(t) is a signal, then its *soft threshold* by $T > 0$ is given by

$$s_{T,s}(t) = \begin{cases} \operatorname{sgn}[s(t)](|s(t)| - T) & \text{if } |s(t)| > T, \\ 0 & \text{if } |s(t)| \le T. \end{cases} \tag{12.26}$$

Soft thresholding results in a continuous signal (Figure 12.13).

(iv) Reconstruct the original signal.

(v) Soft threshold selection is either heuristic, based on the hard threshold selection (12.25), or extracted via the Stein unbiased risk estimate (SURE) [33, 35].

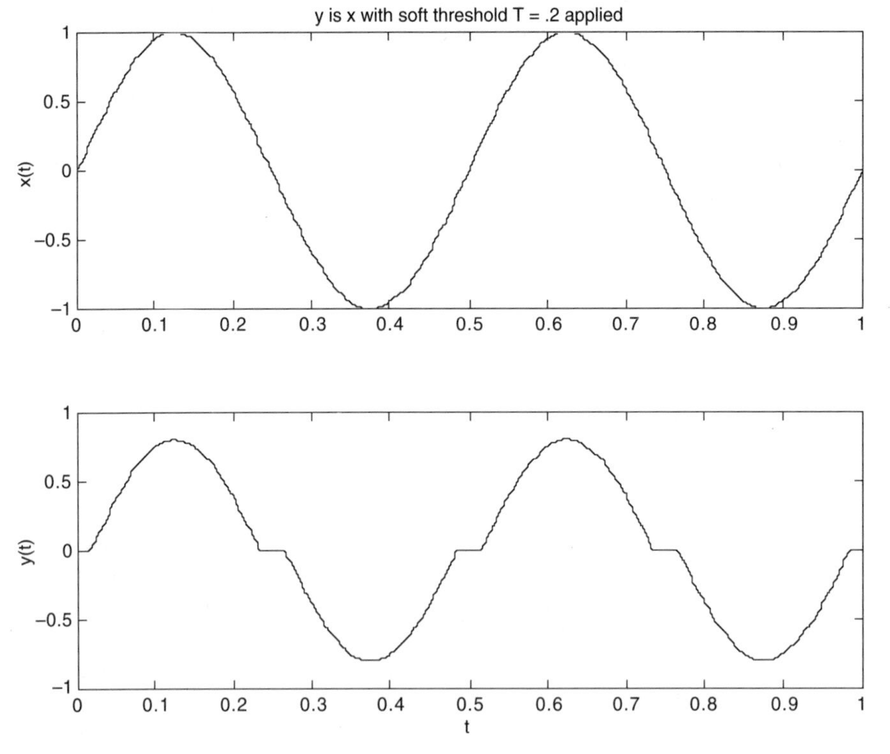

Fig. 12.13. Soft thresholding a sinusoidal signal.

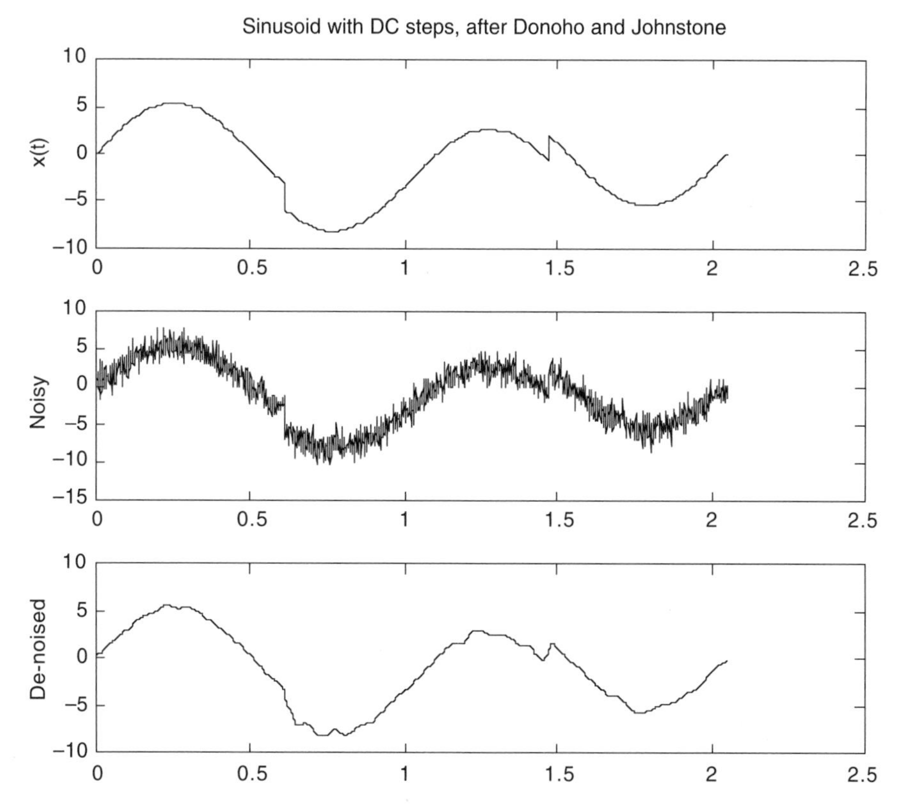

Fig. 12.14. Noise removal filtering using wavelets, soft thresholding, and the SURE threshold selection.

Figure 12.14 provides an example.

12.2.2 Enhancement Techniques

Let us consider a third processing technique using wavelet transforms. Many signal acquistion and imaging systems use photon detectors, such as the popular charge-coupled device (CCD) or the sensitive photomultiplier tube (PMT), as their data source. Examples are spectrometers used to control plasma etchers in the semiconductor manufacturing industry, an application we considered in Chapter 4; astronomical instruments; remote sensing devices; and photo-optical sensors in general.

CCDs now have become the digital image acquistion device of choice [36]. A *charge well* corresponds to a single pixel. They are small, holding perhaps 800 electrons per micron^2 (μ, 10^{-6}m). This is called the *well capacity*, W_c. Factors affecting CCD sensor performance include the following:

- Wells have finite capactiy, so if exposure time is too long, electrons spill over to adjacent wells, causing *blooming* of the image (bright blurry spots).
- There are three noise sources. These are due to thermal effects, N_θ, (also called *dark current*); the quantum nature of light, N_ϕ; and logic noise during readout, N_ρ.
- Total noise within the image at a pixel is therefore $N = N_\phi + N_\theta + N_\rho$.
- *Dynamic range* of the well is defined as (capacity)/(readout noise level) = W_c/N_ρ.
- Thermal effects may be mitigated by cooling the sensor; typically, 6 degrees C warmer means twice as much thermal noise; in other words,

$$N_\theta(t) = \int_{t_0}^{t} K_\theta 2^{\frac{T}{6}} dt, \tag{12.27}$$

  where T is the temperature in degrees Celsius, K_θ is a constant, and i_θ is the dark current.
- Readout noise rate increases with readout frequency.
- Light flux striking sensor obeys a Poisson distribution, where α is the parameter of the distribution:

$$p_I(k) = e^{-\alpha}\frac{\alpha^k}{k!} = Prob(I = k). \tag{12.28}$$

- The mean of a Poisson distributed random variable is $\mu = \alpha$ and its standard deviation is $\sigma = \alpha^{1/2}$ in (12.28).

Of the diverse image noise sources, the most troublesome is *quantum noise*, N_ϕ, which arises from the discrete nature of light quanta detection. Its magnitude changes with the light intensity and is thus image-dependent. Toward eliminating this pernicious source of noise in photon imaging systems—so-called *photon noise*—Nowak and Baraniuk [37] have applied an adaptive filtering method, based on the wavelet pyramid decomposition.

We have already observed that in some applications, wavelet pyramids furnish an especially convenient and powerful tool for suppressing noise in signals and images. The decomposition of a signal into a pyramid allows us to design algorithms that eliminate apparent noise in certain frequency bands by simply attenuating (or even zeroing) the coefficients in the suspect pyramid level. When the signal is reconstructed from the modified pyramid, troublesome noise is then absent. Better compression ratios are also obtained, without significant loss in perceptual signal quality.

In Ref. 37, the idea is to adjust pyramid level intensities according to the energy in other parts of the representation. For example, we might decompose to two coarse levels of representation, then attenuate the detail coefficients selectively, where the approximate signals have large magnitudes. Detail signals corresponding to regions where the approximate signal is weak remain the same or are amplified. Then, we reconstruct the profile using the pyramid scheme. A number of the computational experiments in Ref. 37 markedly improve faint, low-constrast medical

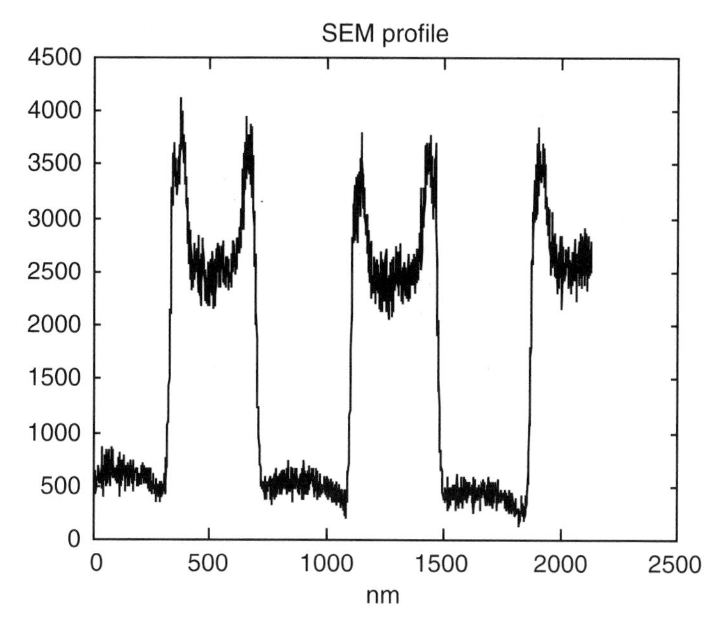

Fig. 12.15. Example of scanning electron microscope profile of lines on a test wafer. Noise magnitudes are roughly equal on tops of the high-magnitude lines and at the bottoms of the low-magnitude spaces between them.

and astronomical images. There is also a promise of better histogramming, edge detection, and higher-level image interpretation results. Figure 12.15 and Figure 12.16 show an application of this technique to raster lines from a scanning electron microscope (SEM) image.

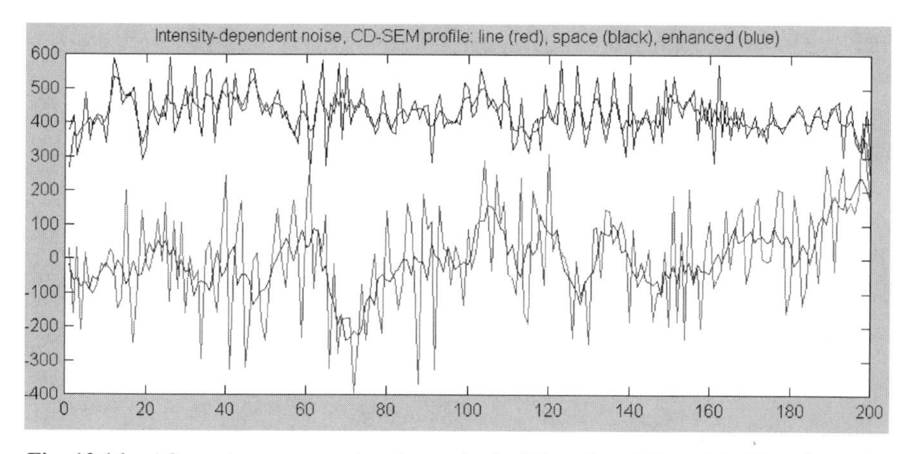

Fig. 12.16. After enhancement using the method of Nowak and Baraniuk. Note that noise on the wafer lines (*lower trace*) is reduced, whereas the details on the low magnitude spaces (*upper trace*) tends to be preserved.

12.3 BIOPHYSICAL APPLICATIONS

There was a surge of interest in these time-frequency and time-scale transforms in the mid-1980s. Psychophysicists noticed that Gabor elementary signals (Chapter 10) could model some aspects of the brain's visual processing. In particular, the *receptive fields* of adjacent neurons in the visual cortex seem to have profiles that resemble the real and imaginary parts of the Gabor elementary function. A controversy ensued, and researchers—electrical engineers, computer scientists, physiologists, and psychologists—armed with the techniques of mixed-domain signal decomposition continue to investigate and debate the mechanisms of animal visual perception [38, 39].

12.3.1 David Marr's Program

Signal concavity remains an important concept in analysis applications. Years ago, the psychologist Attneave [40] noted that a scattering of simple curves suffices to convey the idea of a complex shape, for instance, a cat. Later, computer vision researchers developed the idea of assemblages of simple, oriented edges into complete theories of low-level image understanding [41, 42]. Perhaps the most influential among them was Marr, who conjectured that understanding a scene depends upon the extraction of edge information over a range of visual resolutions [7]. Marr challenged computer vision researchers to find processing and analysis paradigms within biological vision and apply them to machine vision. Researchers investigated concavity and convexity descriptions as well as finding their boundaries at multiple scales. Thus, we might resolve an image into an intricately patterned structure at a fine scale, but coarse representation reveals just a butterfly wing. Marr speculated, but could not prove, that multiscale edges could uniquely describe signals and images. This would imply that the ultimate structural description of a signal would consist of its edge maps across all scales.

Two important early outcomes from Marr's program were scale space theory [8, 9, 43, 44] and optimal multiscale edge detectors [45–48] (Chapter 4). These theoretical results and the practical success of edge-based analysis and description of signal structure, bolstered Marr's conjecture. But wavelets weighed in as well. Mallat tried to use wavelet transform *zerocrossings* [49] as a multiscale structural signal characterization, and showed how the technique could be used for stereometry, but the method suffered from instability. Mallat and Zhong [50] changed strategies and showed that finding wavelet transform *maxima* across scales was equivalent to the Canny edge detector [45]. Then both Berman and Baras [51] and Meyer [29] found counterexamples to Marr's conjecture. In fact, Meyer's example gives a wavelet and a collection of sinusoidal sums that have the same zero crossings when convolved with the wavelet.

12.3.2 Psychophysics

Among the research efforts Marr's work inspired are comparisons between biological and computer vision. Such comparisons tend to support the notion that the particular computer vision innovation being considered is more than an ad-hoc technological trick. When the biological analogies are clear, in fact, it is plausible that the technology is taking advantage of some fundamental physical properties of objects in the world and their possible understanding from irradiance patterns.

While most physiological studies of the visual brain have concentrated on cats and monkeys as experimental subjects [52], the evidence shows many similarities between the function of the cat's visual cortex and that of the monkey. It is therefore reasonable to assume—and the psychophysical studies done on human subjects support this—that the human visual system implements these same principles of cortical organization. Visual information arrives through the retina, and then passes down the optic nerve to the lateral geniculate nucleus (LGN), from which it is relayed to the visual part of the brain, variously known as V1, *area* 17, the *striate cortex*, or the *primary visual cortex*. Some two or three dozen separate visual areas of the brain have been identified according to their visual function [52].

Light impinging on the retina directly stimulates V1 neurons, as well as some other cortical areas. The area of the retina upon which a pattern of irradiance may stimulate a neuron is called the neuron's *receptive field* (RF). In their pioneering work, Hubel and Wiesel [53] differentiated between *simple* cortical neurons and *complex* cortical neurons. A simple cell tends to assume one of two states, "on" or "off," according to whether special light patterns were directed within its RF. Complex cells, on the other hand, do not exhibit this binary behavior, are prone to have larger RFs than the simple cells, and can be stimulated over a much broader RF area [53]. Studying the cat's visual cortex the researchers further demonstrated that both the simple and complex cells within area V1 have a very high orientation selectivity. Monitoring the responses of cells while slits of light at various angles were flashed onto the RFs of the neurons demonstrated this. Interestingly, such orientation specificity is not shown in the retinal area of the eye or in the LGN, but only appears when the visual information finally reaches the cortex [53]. Campbell and Robson [54] confirmed this property for human subjects through a series of pyschophysical experiments.

Campbell and Kulikowski [55] and Blakemore and Campbell [56] described another property showing independent vision channels to exist. These channels have an orientation selectivity in addition to the spatial frequency selectivity. The orientation selectivity exhibited by the independent channels is not well accounted for either in Marr's system [7] or in the Laplacian pyramid algorithms [2]. However, orientation selectivity as well as spatial frequency tuning is a feature of the channels in the two-dimensional wavelet multiresolution representation [1] as shown in Figure 12.12.

Originally, researchers in the animal vision physiology thought that cortical cells were *feature detectors*, activated by the presence of a dot, bar, edge, or corner. Orban [52] emphasizes that it is now clear that cortical cells are actually *filter*s and not feature detectors. It is also now possible to identify some visual cortical areas with the animal's behavior. For cats, nearly all neurons of cortical areas 17 (V1), 18, and 19 are orientation-sensitive bandpass filters covering all orientations [52]. Cells in area 17 have the smallest orientation bandwidth and show strong preference for

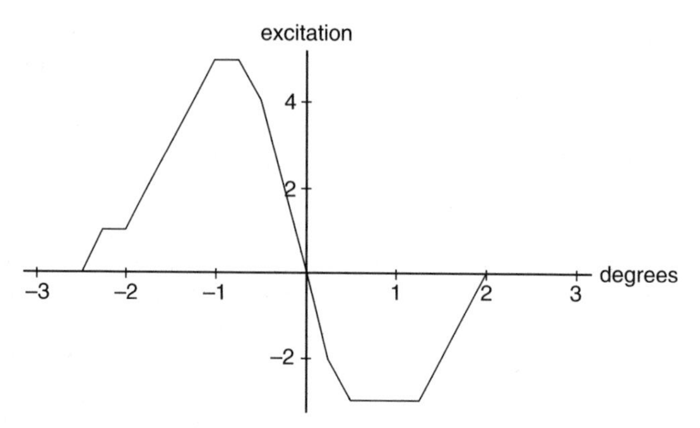

Fig. 12.17. RF profile of cortical simple cell #1316 of Ref. 57, odd symmetry.

horizontal and vertical directions. They are therefore most useful for observing stationary objects. The situation in area 19 is less clear. These cells have large receptive fields, broad orientation tuning, and little motion sensitivity. Area 18 cells have very large RFs, are sensitive only to low spatial frequencies, maintain a high orientation bandwidth, and have some velocity sensitivity. These cells work together to provide the animal with motion analysis [52].

To obtain an RF profile for a cortical cell, Jones and Palmer [57, 58] and Jones, Stepnoski, and Palmer [59] plotted neuron firing rate—the cell *activation level*—against the position of a spot of light within a small 16 × 16 grid. According to the widely held view that the simple cells are filters of varying orientation and spatial frequency sensitivity, it should be possible to model the impulse response of the simple cell filter by mapping firing rate versus stimulus position in the RF. In fact, these researchers were able to obtain plots of the spatial and spectral structure of simple RFs in cats. Figure 12.17 and Figure 12.18 are based on cross sections of contour plots of typical cat simple receptive fields provided in Ref. 57.

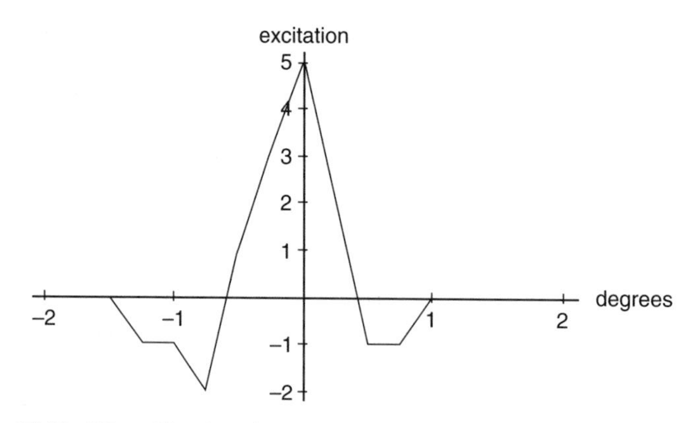

Fig. 12.18. RF profile of cortical simple cell #0219 of Ref. 57, even symmetry.

Many cortical simple cells have either an odd or even symmetry. Cells with odd symmetry, such as in Figure 12.17, have areas of excitation and inhibition on opposite sides of the RF center. On the other hand, those cells with even symmetry, such as in Figure 12.18, have a central excitory (or inhibitory) region that is surrounded by an inhibitory (or excitory) areas. The two basic types are illustrated in the above figures. In Figure 12.17 and Figure 12.18 the regions of positive excitation represent RF areas where responses to *bright stimuli* were obtained. The regions of negative excitation represent RF positions in which the response was found to correlate with *dark stimuli.* All of their excitation frequency measurements were made by microelectrodes inserted into cortical area 17.

Not long after the profiles of cortical neuron receptive fields became more widely understood, Marcelja [60] showed that the RF profiles were strikingly similar to the graphs of Gabor elementary functions (GEF). Marcelja's paper focused the attention of the computer and biological vision research communities onto the potential applications of these functions in vision research. We recall from Chapter 10 that a GEF is a sinusoid multiplied by a Gaussian. Gabor [61] showed that these functions are optimally localized in the time and frequency domains. The product of their spatial extent and bandwidth is minimal. Further, the functions form a complete mathematical set [62] for expansions of other signals.

When we choose the parameters appropriately and graph the GEF's real and imaginary parts separately, the functions resemble the RF profiles of cortical neurons. More remarkably, Pollen and Ronner [63] discovered that adjacent simple cells are often tuned to similar spatial frequencies, and have similar orientation selectivities, but appear to have phase difference of 90 degrees. That is, it appears that the cosine (even) and sine (odd) components of the GEFs are implemented in the visual cortex by pairs of adjacent simple cells [63]. Not surprisingly, these discoveries in physiological and psychophysical research aroused intense new interest in the study and application of the Gabor functions for computational vision.

Daugman [64] extended Gabor's results [61] to the case of two-dimensions, showing that the resulting two-dimensional elliptical Gaussians were optimally localized in the spatial and spatial frequency domains. Daugman suggested that filters based on the elliptical Gaussians modulated by sinusoids represent a necessary evolutionary compromise for a biological vision system with incompatible high-level goals. The organism must find both spatial information and spatial frequency information from its visual space. The way to accomplish this is to implement in the visual cortex the transform that has the best joint specificity in the spatial and frequency domains. This transform is the Gabor transform [65]. Further, the correspondence remains between the shapes of these two-dimensional Gabor functions and the cortical RFs considered as two-dimensional filters. Thus, by basing themselves on the GEFs as models, vision physiologists are evidently equipped with a formalism that explains the properties of orientation selectivity, spatial frequency selectivity, and the empirically observed quadrature relationship for pairs of cortical simple cells.

Does the visual cortex implement some kind of Gabor transform? The physiological experiments on animals and the psychophysical experiments on humans seem to overwhelmingly support the view that the simple RFs are filters selective to orientation and frequency. Further, the spatial structure of the simple RFs closely resembles the GEFs. It is no wonder that the Gabor representation, with its optimal localization properties, was seized upon as a candidate model for the functioning of the visual cortex. Mallat questions the choice of the Gabor transform as a model for cortical functions, however [1]. He points out that the simple cells of the visual cortex do not have impulse responses which contain more cycles when the tuning is for a higher frequency. This would be the case if the cells were organized in the form of the logons of the Gabor representation. Instead, Mallat argues, the simple cells have larger RFs when the frequency tuning is lower and smaller RFs when the frequency tuning is correspondingly higher.

The experimental evidence is not completely clear on this issue however. The question at hand is whether simple cells with high-frequency tuning exhibit more cycles within their fields than cells selective of lower spatial frequencies. This is equivalent to saying that the bandwidth in octaves varies with the particular spatial frequency preferred by a cortical simple cell. Pollen and Ronner stress that a variety of bandwidths are typically associated with cells of a given frequency tuning [63].

Nevertheless, the correlations between preferred spatial frequency and bandwidth tend not to support the contention that a full Gabor transform, with its specific requirement of increased cycles in RFs with higher frequency tunings, is implemented in the visual cortex. A model counterposed to the Gabor model of cortical architecture, wherein the RF sizes vary inversely with the frequency tuning of the simple cells, is presented by Kulikowski, Marcelja, and Bishop [39].

Finally, some more recent studies of the visual cortex support the viewpoint that the receptive field sizes vary inversely with preferred spatial frequency. This would be the case if the visual cortex implements a kind of time-scale transform. A case in point is Anderson and Burr's investigations of human motion detection neurons in the visual cortex [66]. The authors discover a regular decrease in RF size as the observed preferred frequency tuning of these cells increases. The RF size was found to progressively diminish from as high as 7 degrees at low spatial frequencies to 2 minutes of arc for cells with the highest preferred frequencies.

As Gabor—and perhaps wavelet—transforms have proven useful in modeling aspects of the human visual system, so have they found applications in studying the auditory system. One can think of the cochlea as a bandpass filter bank (Chapter 9). From measurements of sound sensitivity above 800 Hz, it then turns out that the filter impulse responses are approximately *dilations* of one another [67]. Thus, cochlear sound processing roughly implements a wavelet transform.

12.4 DISCOVERING SIGNAL STRUCTURE

Time-frequency and time-scale transforms provide alternative tools for local signal description. The local descriptions can be merged and split, according to application design, resulting in a structural description of a signal. The motivation for this is that the physical processes that produced the signal changed over time and that the structure of the signal, properly extracted, provides a means for identifying and understanding the the mechanism that generated it. Structures may be simple time-ordered chains of descriptors. When there are long-term and short-term variations in the signal's generation, then it may be effective to build a hierarchical graph structure for describing the data.

12.4.1 Edge Detection

Let us examine how well the windowed Fourier and wavelet transforms can support basic edge detection signal analysis tasks. Signal edges represent abrupt changes in signal intensity and are a typical initial step to segmenting the signal.

12.4.1.1 Time-Frequency Strategies. A simple sawtooth edge experiment demonstrates that the windowed Fourier transform is a problematic edge detector.

The Gabor transform responses indicate local frequencies, and there are indeed high frequency components in the neighborhood of signal edges (Figure 12.19). Locality is poor, however. Shrinking the time width of the Gabor elementary functions provides better resolution. The problem is that this essentially destroys the nature of the transform. Using different window widths makes it more resemble the wavelet transform. Perhaps the most effective application for the short-time Fourier transforms is to indirectly detect edges by locating regions of distinct texture. Indeed, the windowed Fourier transforms are very effective for this purpose, and a number of research efforts have successfully applied them for texture segmentation [68–70]. The filter banks and time-frequency maps of Chapter 9 provide starting points for the spectral analysis of signal texture. The edges between differently textured regions are inferred as part of the higher-level interpretation steps.

12.4.1.2 Time-Scale Strategies. Better suited to edge detection are the time-scale transforms. Not only does the wavelet transform provide for narrowing the time-domain support of the analyzing wavelet, allowing it to zoom in on signal discontinuities, there are two theoretical results that support wavelet-based edge detection:

(i) For certain continuous wavelet transforms, finding maximal response is identical to applying the optimal Canny edge detector [45].

(ii) The decay of the wavelet transform maxima across scales determines the local regularity of the analyzed signal.

However, we also now know that edge-based descriptions of signal structure are not the final answer:

(iii) Marr's conjecture is false [29, 51].

Yet, structural description by edges and extrema (ridge edges) remains a powerful tool for understanding signals. Both continuous and discrete wavelet transforms are closely related. For example, a wavelet transform can be built around spline wavelets

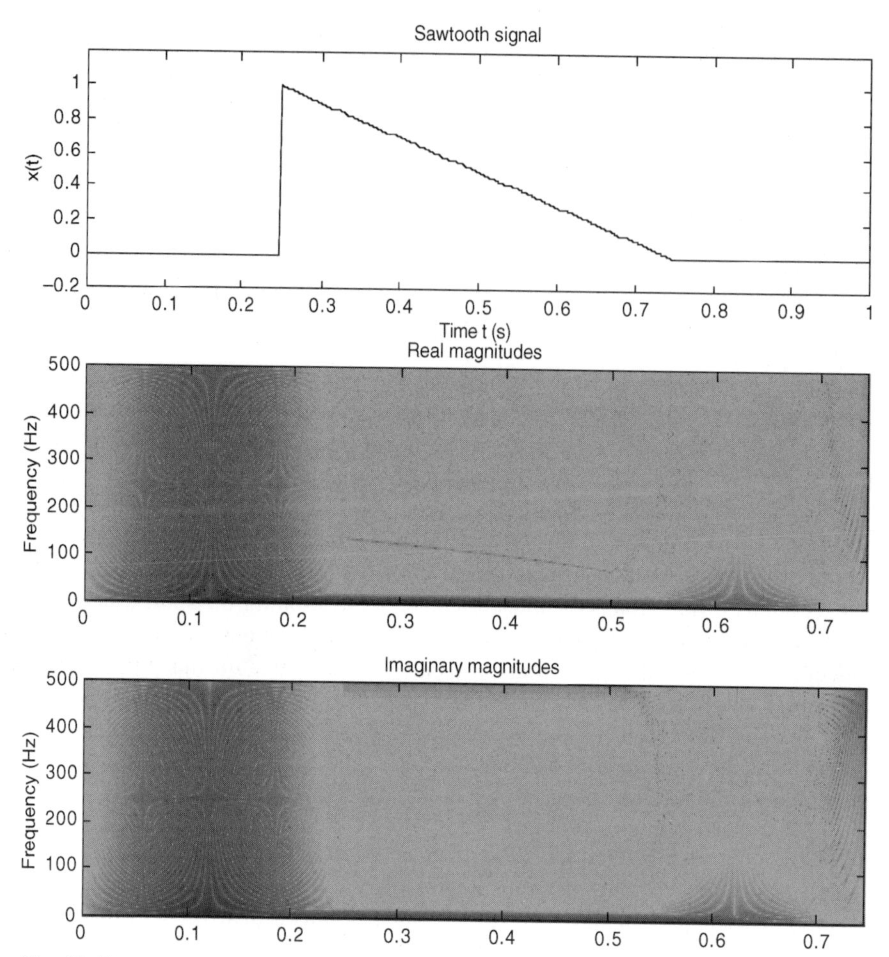

Fig. 12.19. A sawtooth edge (*top*); real and imaginary Gabor transform coefficient magnitudes (*middle*); and convolution of the discrete high-pass filter associated with the cubic spline orthogonal wavelet representation.

that mimic the optimal derivative of Gaussian [48]. The multiresolution decomposition of the signal supports a multiscale edge detector [50]. This is useful for discriminating background noise from substantive signal features according to perceptual criteria [71]. The discrete high-pass filters $g_p(n)$ given in Figure 12.3 and Figure 12.4 function as edge detectors for the orthogonal wavelet pyramid (Figure 12.5).

Let us turn to the continuous wavelet transform. Suppose $g(t) = Ae^{-Bt^2}$ is a Gaussian of zero mean. Set $\psi(t) = \frac{d}{dt}g(t)$. Then $\psi(t)$ is a wavelet because it has zero mean, is integrable, and has finite energy. Let $x_a(t) = ax(at)$ be the scaled dilation of $x(t)$ by factor a. Typically, $a = 2^i$ for $i \in Z$, with $i < 0$ in our notation [1].

Then, changing notations from Chapter 11 slightly, the wavelet transform of an analog signal x is

$$(W_{-i}x)(t) = (\psi_{2^{-i}} * x)(t) \tag{12.29}$$

at time instant t. Consequently,

$$(W_{-i}x)(t) = (\psi_{2^{-i}} * x)(t) = \left(2^{-i}\frac{d}{ds}g(2^{-i}s) * x\right)(t) = 2^{-i}\frac{d}{dt}(g_{2^{-i}} * x)(t). \tag{12.30}$$

So where $|W_{-i}x|$ is large, the version of x, smoothed to resolution 2^{-i}, is changing rapidly [72].

Let us recount the result (ii) above [72]. Suppose $x(t)$ is a signal defined in a neighborhood of t_0, $0 \le \alpha \le 1$, and there is a constant c such that for all t in an interval about t_0, we have

$$|x(t) - x(t_0)| \le c|t - t_0|^{\alpha}. \tag{12.31}$$

Then $x(t)$ is *Lipschitz*[1] α at $t = t_0$. The *Lipschitz regularity* of $x(t)$ at $t = t_0$ is the least upper bound of all α such that (12.31) holds. If there is an $0 \le \alpha \le 1$ and an open interval such that (12.31) holds for all $t \in (a, b)$, then the signal $x(t)$ is *uniformly Lipschitz* α on (a, b). In other words, $x(t)$ is uniformly Lipschitz if it is as tame as an exponential function in some region.

Now suppose we have a continuous wavelet $\psi(t)$ that decays at infinity as $1/(1 + t^2)$ and a square-integrable signal $x(t)$. Then it can be shown [73] that $x(t)$ is uniformly Lipschitz α on (a, b) if and only if there is a $c > 0$ such that for all $t \in (a, b)$

$$|(W_{-i}x)(t)| \le c2^{i\alpha}. \tag{12.32}$$

The decay of wavelet transform maxima over many resolutions is essentially a study of the degree of singularity of the original signal. An extensive study of continuous wavelet transformation as a characterization of regularity is Ref. 74.

12.4.1.3 Application: The Electrocardiogram.

Biomedical technology has investigated almost every avenue of signal analysis in order to improve electrocardiogram (ECG) interpretation. Researchers have experimented with time-domain, frequency domain, time-frequency domain, and now time-scale domain methods [75, 76]. Chapter 1 introduced ECG signal processing and analysis.

[1]Analyst Rudolf Lipschitz (1832–1903) was professor at the University of Bonn.

Frequency-domain methods are effective for many important tasks in computerized electrocardiography, such as convolutional noise removal and band rejection of noise from (50 or 60 Hz, for example) alternating current power sources [77]. Edges and transients in the ECG are crucial to interpreting abnormalities. In order to preserve these features, yet remove noise, research has turned to mixed-domain filtering techniques, such as we covered in Section 12.2 [78]. Compression techniques using wavelet transforms are also known [79]. The most important task in automated ECG analysis is QRS complex detection [80], essentially a ridge edge detection problem. The foundation of these application is the characterization of signal regularity by wavelet transform maxima across scales [74]. Algorithms for QRS detection and time-scale decomposition of ECGs using the orthogonal wavelet decomposition are shown in Ref. 81. The continuous wavelet transform is studied for QRS characterization in Ref. 82. The wavelet transform is effective in revealing abnormalities, such as the ventricular late potential (VLP) [83]. For example, in Ref. 84 a synthetic VLP is introduced into the ECG. The late potential is difficult to discern in the time-domain trace. However, wavelet transformation reveals that the defect is as an enlargement in the time-domain support of the QRS complex at certain scales.

12.4.2 Local Frequency Detection

Both the short-time Fourier and wavelet transforms perform local frequency detection. The STFT or Gabor transform relies on time-frequency cells of fixed size (Chapter 10). The wavelet transform adapts the time domain extent according to the frequency tuning (Chapter 11).

12.4.2.1 Mixed-Domain Strategies. The fixed window width of the short-time Fourier transform is useful when the range of frequencies in the analyzed signal is known to remain within fixed bounds (Figure 12.20). An example of this is in texture analysis, where the local frequencies of the signal pattern are expected within given spectral ranges. Small defects in the texture are not readily detected, but the time-frequency map displays the overall local pattern quite well.

On the other hand, the wavelet pyramid decomposition tends to mimic the coarse structure of the signal in the approximate coefficients and provides a range of highpass filters sensitive to local textures in the detail coefficients (Figure 12.21).

Finally, the continuous wavelet transform clearly shows the scale of the underlying pattern features in its amplitude (Figure 12.22).

12.4.2.2 Application: Echo Cancellation. One application of wavelet transform-based filter banks has been to improve echo canceller performance. Chapter 2 (Section 2.4.4) explained the need for echo cancellation in digital telephony. The echo arises from an impedance mismatch in the four-wire to two-wire hybrid transformer. This causes an echo, audible to the far-end listener, to pass into the speech signal from the near-end speaker. The classical time-domain approach for reducing the echo is to remove the echo by an adaptive convolutional filter [85]. One problem is getting the canceller to converge quickly to an accurate echo model when the

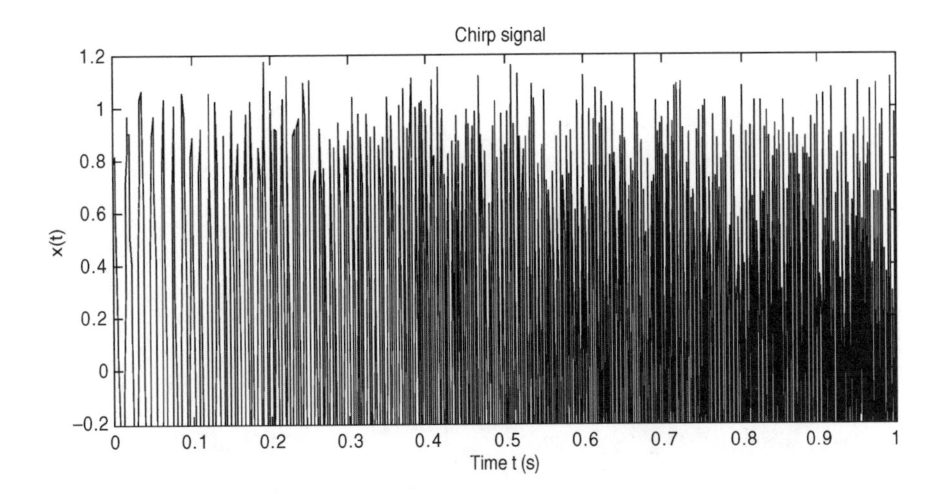

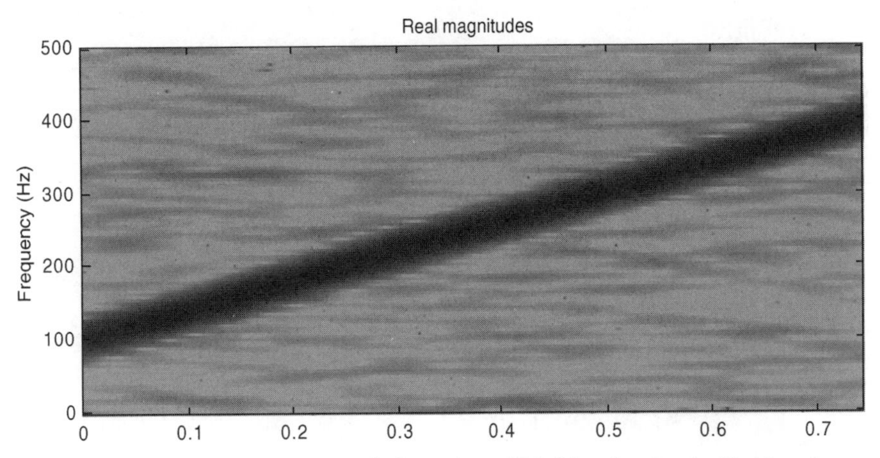

Fig. 12.20. A Gabor transform applied to a sinusoidal chirp signal embedded in noise.

echo path length varies. This can happen in digital telephony, and it is a problem in other applications, such as acoustic echo cancellers employed in teleconferencing systems. Both conventional quadrature mirror filter bank decompositions [86] and wavelet packet decompositions [87] have been used to replace the adaptive time-domain convolution in the classical echo canceller.

12.4.2.3 Application: Seismic Signal Interpretation. The continuous wavelet transform arose out of problematic attempts to use time-frequency methods in seismic data analysis [88]. Both the continuous wavelet transform and the discrete orthogonal pyramid decomposition are applicable to seismic signal interpretation. Early applications used the wavelet transform to improve visualization and interpretation of seismic sections [88–90].

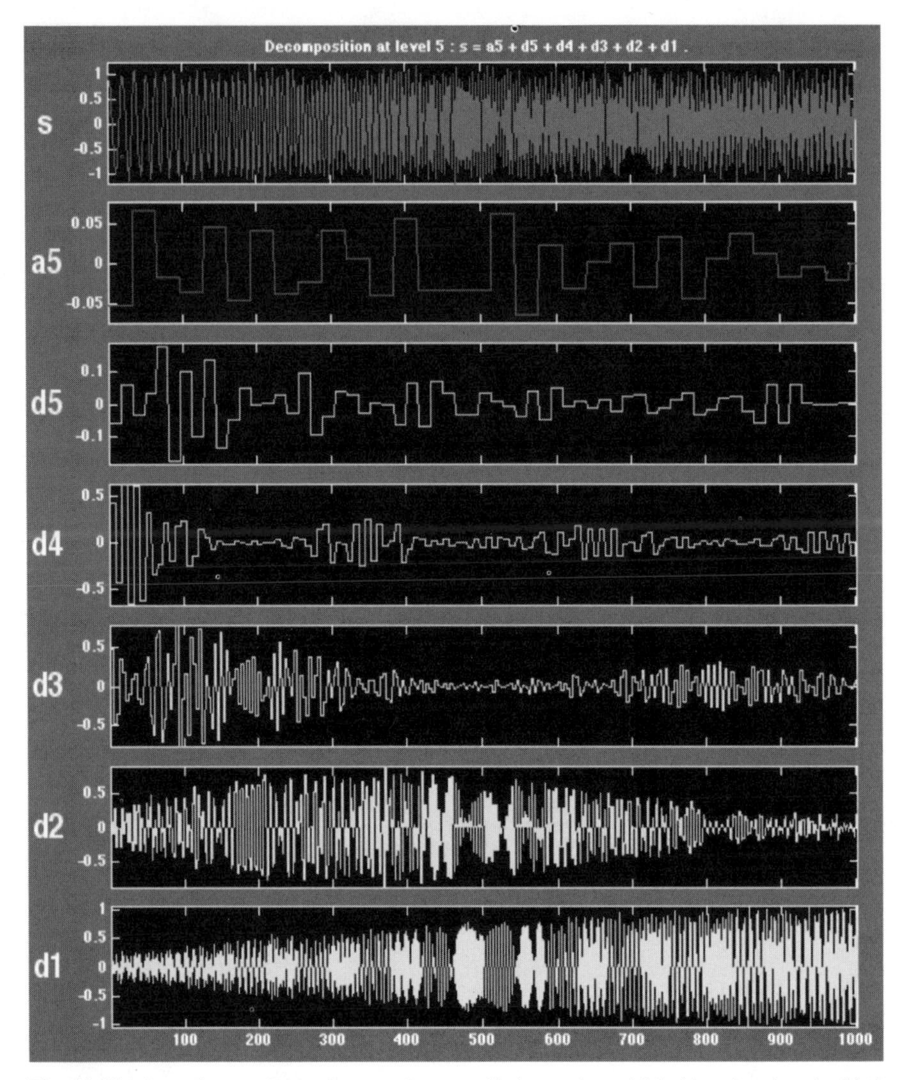

Fig. 12.21. An orthogonal wavelet transform applied to a sinusoidal chirp signal embedded in noise. Decomposition to five levels using the Haar wavelets.

Both the continuous and orthogonal wavelet transforms have been applied to seismic signal analysis. In Ref. 91, for example, researchers recommend the Morlet wavelet (12.33) for removing correlated ground roll noise from seismic exploration data sets.

$$\psi(t) = e^{-\frac{t^2}{2}}\cos(\omega_0 t). \tag{12.33}$$

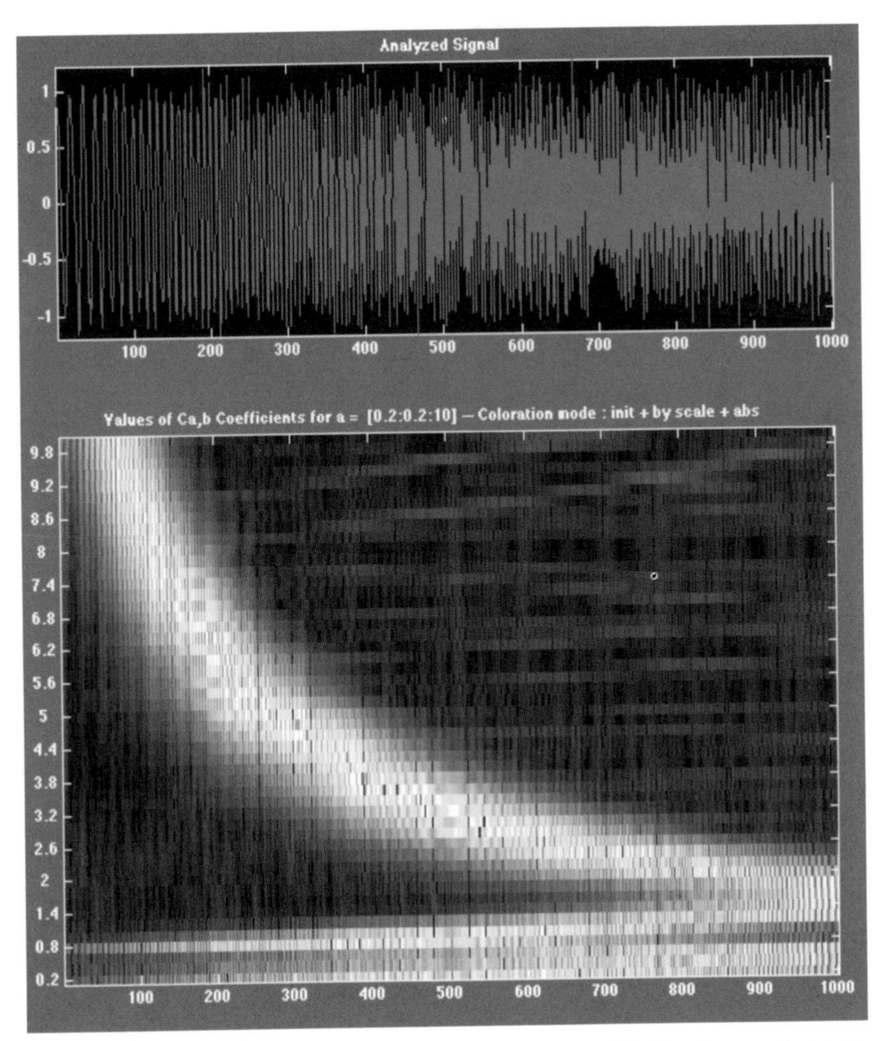

Fig. 12.22. A continuous wavelet transform applied to a sinusoidal chirp signal embedded in noise. Decomposition over 50 scale values using the Morlet wavelet.

The authors perform the continuous transformation by convolving the wavelet $\psi(t)$ with the raw data, exclude the bands containing the ground roll noise, and reconstruct the signal using the inverse transform.

In Ref. 92 the wavelet pyramid transform is considered for analyzing arrival times estimation in seismic traces. The researchers conclude that preliminary denoising is essential. The wavelet-based filtering preserves high-frequency components necessary for finding the boundaries between oscillatory components.

12.4.2.4 Application: Phoneme Recognition. The mixed-domain transforms offer different structural decomposition tools for the speech analyst. In Chapter 9, we considered two types of algorithm for localized frequency analysis: the filter bank and the time-frequency map. The wavelet pyramid decomposition (Figure 12.5) furnishes a filter bank scheme where the outputs are independent of one another and perfect reconstruction of the original signal is possible. In addition, both discrete time-frequency and time-scale transforms support a time-frequency map methodology. The difference between the two is that the time-frequency cells of the Gabor transform, for example, have a fixed time-domain extent (Chapter 10). The wavelet transform cells vary their time spans; cells tuned to higher frequencies have narrower time domain support. Local-frequency estimation, as a preliminary step for recognizing phonemes, remains an active area of research.

It is difficult to design a pitch detector that adapts to both high and low speech frequencies while maintaining adequate noise immunity [93–95]. Recently, the dyadic continuous wavelet transform, given by

$$F_{\psi}(a, b) = W[x(t)](a, b) = \int_{-\infty}^{\infty} x(t)\overline{\psi_{a,b}(t)}\,dt \equiv \langle x(t), \psi_{a,b}(t)\rangle , \tag{12.34a}$$

where

$$\psi_{a,b}(t) = \frac{1}{\sqrt{|a|}}\psi\left(\frac{t-b}{a}\right), \tag{12.34b}$$

$x(t)$ has finite energy, and $a = 2^i$ for some integer i, has been applied to this problem [96]. One advantage is that the the analysis then corresponds to the apparent time-scale operation of the human auditory system [67]. Surprisingly, the researchers report that only a few scales $a = 2^i$ are necessary for accurate detection [96]. Compared to conventional time- and frequency-domain methods, the dyadic wavelet pitch detector:

(i) Is robust to nonstationary signals within its analysis window;

(ii) Works on a wide range of pitch signals, such as from male and female speakers;

(iii) Can detect the beginning of the voiced segment of the speech sample, making it the possible basis for a pitch detection algorithm that operates synchronously with the pitch bearing event;

(iv) Is superior within low frequencies to pitch determination by the time-domain autocorrelation method [95];

(v) Is superior within high frequencies to the frequency-domain cepstrum method [97].

12.4.3 Texture Analysis

Until recently, texture has been a persistently problematic area for signal and image analysis. Although the human subject readily distinguishes visual textures, it has hitherto not been possible to classify them with computerized algorithms, let alone provide theoretical models for synthesizing visually realistic textures.

12.4.3.1 Mixed-Domain Strategies. Some promising early applications of the wavelet transform were to texture analysis [98]. An important contribution to texture analysis and synthesis has come from applying overcomplete wavelet pyramids to the problem [99]. A variety of statistics on the transform coefficients are used to characterize textures. Deletion of certain groups of statistical parameters and subsequent flawed reconstruction of the original image demonstrates that the necessity of the chosen statistics.

12.4.3.2 Application: Defect Detection and Classification. It is possible to apply the statistics of overcomplete pyramids to the problem of texture flaw detection. This application is important in manufacturing defect detection systems, for example. The algorithm of Ref. 99 is capable of synthesizing textures that appear to lie in between two others and offers the promise of a statistical divergence measure for textures. Defects in local regions can be detected by developing the statistical parameters from a prototype sample and and comparing them to statistics extracted from candidate textures.

12.5 PATTERN RECOGNITION NETWORKS

This section explains pattern recognition methods that are useful for analyzing signals that have been decomposed through mixed doman transforms.

12.5.1 Coarse-to-Fine Methods

Pattern recognition where the time-domain size of the recognized signal structures are unknown present a variety of problems for the algorithm designer. In particular, the shape recognition computations can require more time than is available in real-time. We have already reviewed a variety of multiresolution methods for this purpose:

- Multiscale signal edge operators [45–48];
- Time-scale representations such as the wavelet multiresolution analysis [1] and the Laplacian pyramid decomposition [2];
- Scale-space smoothing with a range of kernel sizes [5].

Such decompositions demand large numbers of floating-point multiplications and additions. However, online process control and speech recognition software must keep a real-time pace and make a recognition decision with a fraction of a second. Sometimes, cost constrains the type of processor. One way to stay within the paradigms provided by time-scale signal analysis, yet achieve a real-time recognition time is to employ the classic Haar wavelet approximation [16–18] (Figure 12.23). How to do this is described in process control applications [100, 101].

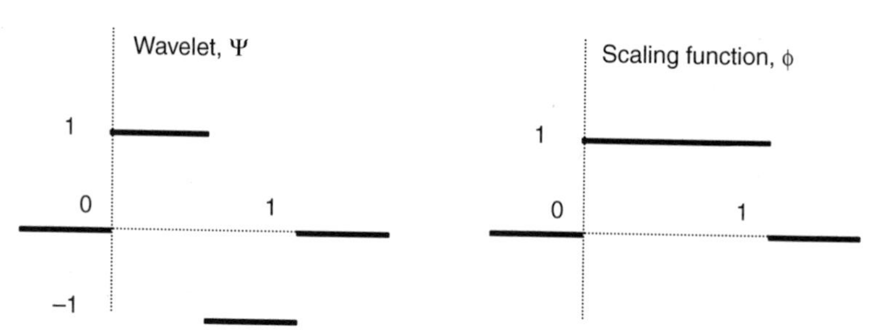

Fig. 12.23. The Haar MRA wavelet and scaling function. Translations and dilations of the wavelet form an orthonormal basis for all finite-energy signals, so the Haar representation can accommodate any signal shape. The scaling function can be used to develop step function approximations to signal shapes. The Fourier characteristics of these approximations are not as attractive as other decompositions, such as smooth spline functions. However, with this technique it is possible to implement a real-time signal decomposition with small industrial control computers.

It is possible to approximate signals by decomposing them into steps or to model them using the signal envelopes [101]. Figure 12.24 shows an example of an endpoint signal from the optical emission monitoring application in [101].

The idea behind using the Haar MRA for signal pattern recognition is that there is a simple relation between certain coarse and fine resolution patterns that allows the application to economize on matching.

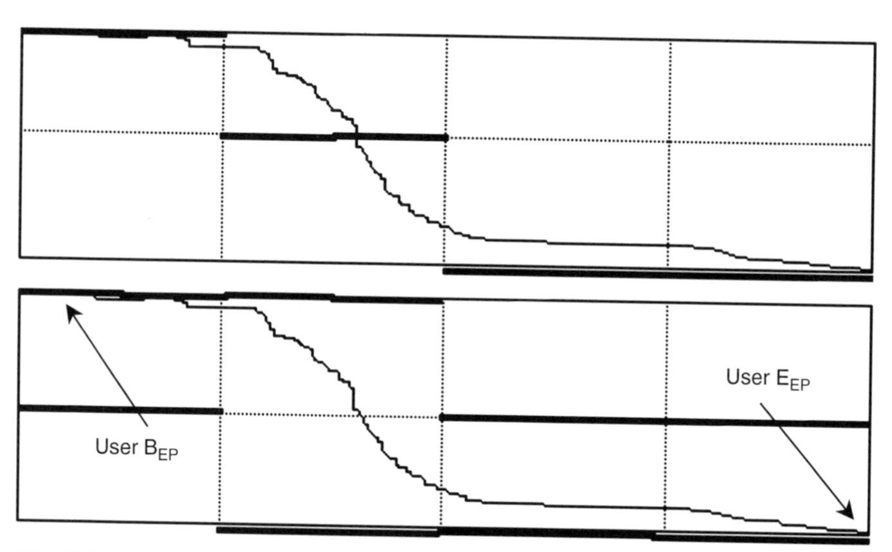

Fig. 12.24. Two methods to find Haar approximations for signal regions: projection to the mean (*top*) and the Haar envelope (*bottom*). In this figure, the shapes represent optical emission endpoint traces selected by users.

12.5.2 Pattern Recognition Networks

A pattern recognition network is a set of pattern detector elements or *neurons* tuned to different resolutions and different signal shapes [101].

Network *training* creates multiple resolution models of the signal shapes which are stored inside the neurons. The multiple resolution matching provides noise immunity during recognition, although linear and nonlinear smoothing operations help to clean acquired signals. To trim the number of pattern detectors in the network, the finest resolutions necessary for the model time and dynamic range divisions can be estimated as follows [101].

Given an input data file $x(n)$ with N values, one computes the discrete Fourier transform:

$$X(k) = \sum_{n=0}^{N-1} x(n)e^{-2\pi jnk}. \tag{12.35}$$

Then, the non-DC coefficients representing the bulk of the signal's energy are selected: $k = 1, 2, \ldots, k_c, N-1, N-2, \ldots, N-k_c$. This selection is based on an energy percentage threshold. Then the signal detectors only need enough time-domain resolution so as to capture the shape of a single cycle of wavelength NT/k_c seconds, where T is the real-time application sampling rate. The length of a signal shape model, together with this minimum resolution value, determines the maximum length in samples of the discrete patterns stored in the detector elements.

The scheme of Ref. 101 also limits the dynamic range of signal models. Again using heuristic thresholds, typically modified by the application user, the amount of noise in the signal models is estimated. Then using a DFT approach again, the necessary dynamic range resolution in the step-shaped signal models is found. The result is a rectangular array of pattern detectors, from the lowest time resolution to the highest and from the lowest dynamic range division to the highest.

The projection of an acquired signal onto the step functions of the Haar representation are the coarse resolution representations used as models for pattern detection. Each "neuron" is a step function pattern detector. Before the network runs on real data, the patterns are checked against previously acquired data sets. One criterion is *stability*—how long a signal pattern persists in the data stream. Another criterion is the tendency to make false detections in the input data. Any neurons that fail to meet these criteria are disabled and not used by the network on real data.

The neurons of the network are interconnected with enabling and disabling links. When a coarse resolution node does not detect its established block shape pattern, it may disable certain finer resolution nodes. An example is shown in Figure 12.25.

At run time, the network presents the current and past signal data to each neuron. Each node computes the step pattern according to its particular resolutions. It compares the candidate pattern to its training pattern. When a node shows no match, higher-resolution nodes whose time or range resolutions are multiples of the non-matching unit cannot possibly activate and are disabled (Figure 12.25). When a node actually finds its pattern, all nodes of lesser resolution are disabled. The network continues to seek a more precise registration of the signal pattern. Any

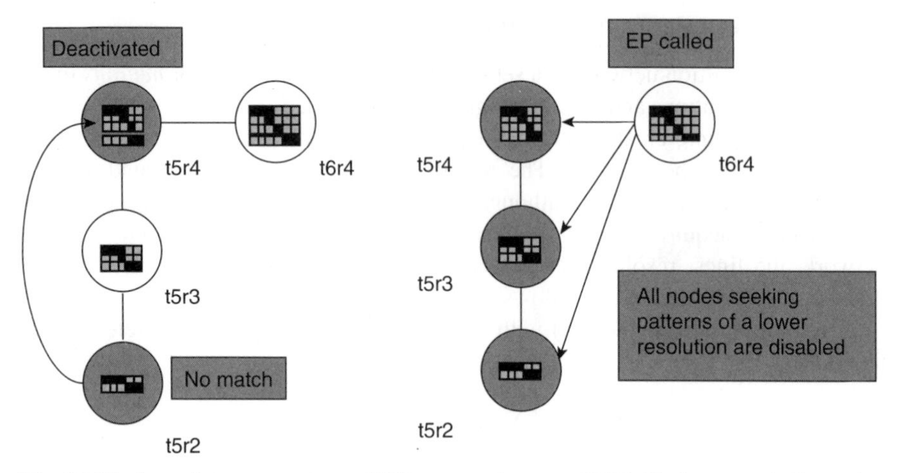

Fig. 12.25. Some interconnections of EP pattern detectors [101]. Node names indicate the time-domain and dynamic range resolutions of the Haar decomposition step functions that approximate the EP region. When a node shows no match, certain higher-resolution nodes cannot possibly activate and are disabled (*left*). When a node detects its pattern, the network disables all nodes of lesser resolution while it seeks a more precise registration of the signal shape. There is a single output node. Any enabled neuron that detects its own pattern can activate the output node to signal the detection of the prototype pattern.

enabled neuron that detects its own pattern can activate the output node to indicate a recognition success.

This is the key idea behind using the Haar step functions to model and compare signals. With a faster computer or freedom from the real-time processing requirement, other multiresolution approximations can be used. Of course, in this situation, the relation between detectors of different resolutions is not so easy to characterize and remains a potential problem for the algorithm design.

12.5.3 Neural Networks

Neural networks are an alternative to the structured design of the pattern recognition network above. Both supervised and unsupervised neural networks have been intensively studied in the last 20 years. An advantage of neural networks is that their training can be conditioned by training data, learning, as it were, the salient patterns present in the raw data [102]. The problem is that large amounts of data are sometimes necessary to train such networks. An example of applying neural networks to semiconductor process control is Ref. 103.

12.5.4 Application: Process Control

In semiconductor integrated circuit fabrication, plasma etch processes selectively remove materials from silicon wafers in a reactor [104]. The chemical species in the

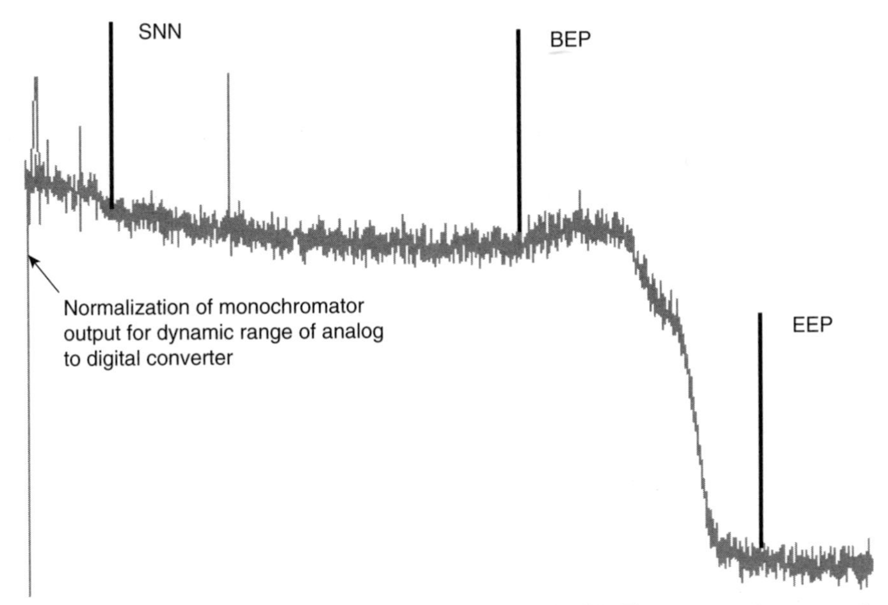

Fig. 12.26. Plasma etch optical emission trace from Ref. 101. The user acquires a sample trace and must indicate the points at which valid data begin (SNN), the endpoint is beginning (BEP), and the endpoint event is over (EEP). A Haar model can be made out of the data for training a pattern recognition network.

reaction emit characteristic wavelengths of light. It is typical to employ a monochromator and digitizer pass the light intensity signal to computer software algorithms to monitor and control the etch progress. When the target layer disappears, process endpoint occurs; the control computer extinguishes the plasma. For signal analysis, the problem is that process endpoints vary from wafer to wafer over a run. Both traditional neural networks [103] and the structured pattern recognition network [100, 101] have been used for this application. Extensive testing is reported in Ref. 101.

12.6 SIGNAL MODELING AND MATCHING

This final section mentions methods for extracting signal structure that have been particularly popular in conjunction with time-frequency and time-scale transforms.

12.6.1 Hidden Markov Models

The hidden Markov model (HMM) is a stochastic state machine that is especially useful for sequential analysis of data. Thus, it has been applied widely in speech recognition [105], handwriting recognition [106], and biological sequence analysis [107].

12.6.2 Matching Pursuit

The matching pursuit is a greedy, iterative algorithm [108]. In Ref. 109 the method is applied with overcomplete dictionaries of damped sinusoids. The method typically uses an overcomplete dictionary for numerical stability. It improves upon traditional techniques such as least squares, singular value decomposition, and orthonormal basis decompositions—for example, the orthogonal wavelet pyramid [1].

12.6.3 Applications

Two of the most important applications of hidden Markov models have been in speech recognition and biological sequence analysis.

12.6.3.1 Speech Analysis. Speech recognition applications are described in the classic tutorial [105].

12.6.3.2 Protein Analysis. A tutorial on protein analysis is Ref. 110.

12.7 AFTERWORD

There is no doubt that mixed-domain signal transforms, combining both time and either frequency or scale information, have altered the signal processing landscape. It is almost impossible today to browse an academic journal in the discipline and not find a contribution that concentrates on the theoretical or practical implications of these techniques. This text introduces the new methods into the mainstream of signal processing education.

Some of the trends we identified when contemplating the task of writing this book have become clearer and stronger. It is still true that learning windowed Fourier and wavelet transforms has its mathematical challenges. The entire signal processing research community has embraced the underlying mathematical tools—especially Hilbert space theory—even though they may entail a steep learning curve. Here we have developed the mathematics incrementally and colored it with terminology, notations, and concepts directly relevant to signal theory. This might relieve some anxiety and make the climb less daunting. Also, the selection of algorithms and applications in *signal analysis* not only reflects the modern mathematical slant but also emphasizes signal understanding as opposed to pure processing. We think that this too is timely, as more and more automated signal recognition technologies have intruded into our lives.

Of the new mixed-domain transforms, probably the most surprises came from orthogonal wavelets. These functions captivated researchers in so many areas. Today, however, the associated signal decomposition seem to be retreating into compression applications, with its analysis powers having been tested and found lacking for pattern recognition. But it does offer insights into texture characterization. On the other hand, the continuous wavelet transform rises up to be the better tool for transient signals. The exponential short-time Fourier methods cannot be sparse and complete, as we now know, but this has not prevented them from becoming the tool of choice in a number of early processing applications. The time-frequency and time-scale tools are just alternative ways to break a signal into pieces and sort them out into a structural description. Beyond that, their importance is that they highlight the special global nature of the wellspring of them all—Fourier's transform.

REFERENCES

1. S. Mallat, A theory for multiresolution signal decomposition: The wavelet representation, *IEEE Transactions on Pattern Analysis and Machine Intelligence*, pp. 674–693, July 1989.
2. P. J. Burt and E. H. Adelson, The Laplacian pyramid as a compact image code, *IEEE Transactions on Communication*, vol. 31, no. 4, pp. 532–540, April 1983.
3. S. Mallat, *A Wavelet Tour of Signal Processing*, San Diego, CA: Academic Press, 1998.
4. O. Rioul and M. Vetterli, Wavelets and signal processing, *IEEE SP Magazine*, pp. 14–38, October 1991.
5. R. L. Allen, F. A. Kamangar, and E. M. Stokely, Laplacian and orthogonal wavelet pyramid decomposition in coarse-to-fine registration, *IEEE Transactions on Signal Processing*, vol. 41, no. 12, pp. 3536-3543, December 1993.
6. R. Y. Wong and E. L. Hall, Sequential hierarchical scene matching, *IEEE Transactions on Computers*, vol. C-27, no. 4, pp. 359–366, April 1978.
7. D. Marr, *Vision*, New York: W. H. Freeman, 1982.
8. T. Lindeberg, *Scale-Space Theory in Computer Vision*, Hingham, MA: Kluwer, 1994.
9. B. ter Haar Romeny, L. Florack, J. Koenderink, M. Viergever, eds., *Scale-Space Theory in Computer Vision* (Proceedings, First International Conference, Scale-Space '97, Utrecht, The Netherlands), Berlin: Springer-Verlag, 1997.
10. R. H. Bamberger and M. J. T. Smith, A filter bank for the directional decomposition of images: Theory and design, *IEEE Transactions on Signal Processing*, vol. 40, no. 4, pp. 882–893, April 1992.
11. E. P. Simoncelli, W. T. Freeman, E. H. Adelson, and D. J. Heeger, Shiftable multiscale transforms, *IEEE Transactions on Information Theory*, vol. 38, no. 2, pp. 587–607, March 1992.
12. P.-G. Lemarié, Ondelettes à localisation exponentielle, *Journal de Mathématiques Pures et Appliquées*, vol. 67, pp. 227–236, 1988.
13. F. Mokhtarian and A. Mackworth, Scale-based description and recognition of planar curves and two-dimensional shapes, *IEEE Transactions on Pattern Analysis and Machine Intelligence*, vol. PAMI-8, no. 1, pp. 34–43, January 1986.
14. M. P. do Carmo, *Differential Geometry of Curves and Surfaces*, Englewood Cliffs, NJ: Prentice-Hall, 1976.
15. J.-O. Strömberg, A modified Franklin system and higher order spline systems on R^n as unconditional bases for Hardy spaces, in W. Becker, A. P. Calderon, R. Fefferman, and P.W. Jones, eds., *Proceedings of the Conference in Honor of Antoni Zygmund*, vol. II, New York: Wadsworth, pp. 475–493, 1981.

16. A. Haar, Zur theorie der orthogonalen Functionensysteme, *Mathematische Annalen*, vol. 69, pp. 331–371, 1910.
17. I. Daubechies, *Ten Lectures on Wavelets*, Philadelphia: SIAM, 1992.
18. H. S. Stone, J. Le Moigne, and M. McGuire, The translation sensitivity of wavelet-based registration, *IEEE Transactions on Pattern Analysis and Machine Intelligence*, vol. 21, no. 10, pp. 1074–1081, October 1999.
19. M. Antonini, M. Barlaud, I. Daubechies, and P. Mathieu, Image coding using vector quantization in the wavelet transform domain, *Proceedings of the IEEE Conference on Acoustics, Speech, and Signal Processing*, pp. 2297–2300, April 1990.
20. S. G. Mallat and J. Froment, Second generation compact image coding with wavelets, in C. K. Chui, ed., *Wavelets: A Tutorial in Theory and Applications*, San Diego, CA: Academic Press, 1992.
21. D. Sinha, and A. H. Tewfik, Low bit rate transparent audio compression using adapted wavelets, *IEEE Transactions on Signal Processing*, vol. 41, no. 12, pp. 3463–3479, December 1993.
22. M. Vetterli and J. Kovacevic, *Wavelets and Subband Coding*, Upper Saddle River, NJ: Prentice-Hall, 1995.
23. A. Rosenfeld and A. C. Kak, *Digital Picture Processing*, San Diego, CA: Academic Press, 1982.
24. G. K. Wallace, The JPEG still picture compression standard, *Communications of the ACM*, vol. 34, no. 4, pp. 31–44, April 1991.
25. N. Ahmed, T. Natarajan, and K. R. Rao, Discrete cosine transform, *IEEE Transactions on Computers*, vol. C-23, no. 1, pp. 90–93, January 1974.
26. A. Gersho and A. M. Gray, *Vector Quantization and Signal Compression*, Boston: Kluwer, 1992.
27. F. W. C. Campbell and J. J. Kulikowski, Orientation selectivity of the human visual system, *Journal of Physiology*, vol. 197, pp. 437–441, 1966.
28. H. Malvar, Lapped transforms for efficient transform/subband coding, *IEEE Transactions on Acoustics, Speech, and Signal Processing*, vol. 38, no. 6, pp. 969–978, June 1990.
29. Y. Meyer: *Wavelets: Algorithms and Applications*, Philadelphia: Society for Industrial and Applied Mathematics, 1993.
30. R. R. Coifman, Y. Meyer, and V. Wickerhauser, Wavelet analysis and signal processing, in M. B. Ruskai, et al., eds., *Wavelets and Their Applications*, Boston: Jones and Bartlett, pp. 153–178, 1992.
31. R. R. Coifman, Y. Meyer, and V. Wickerhauser, Size properties of wavelet-packets, in M. B. Ruskai et al., eds., *Wavelets and Their Applications*, Boston: Jones and Bartlett, pp. 453–470, 1992.
32. D. L. Donoho and I. M. Johnstone, Ideal spatial adaptation by wavelet shrinkage, *Biometrika*, vol. 81, pp. 425–455, 1994.
33. D. L. Donoho, De-noising by soft-thresholding, *IEEE Transactions on Information Theory*, vol. 41, no. 3, pp. 613–627, March 1995.
34. D. L. Donoho, I. M. Johnstone, G. Kerkyacharian, and D. Picard, Density estimation by wavelet thresholding, *Annals of Statistics*, vol. 24, pp. 508–539, 1996.

35. C. M. Stein, Estimation of the mean of a multivariate normal distribution, *Annals of Statistics*, vol. 9, pp. 1135–1151, 1981.
36. K. R. Castleman, *Digital Image Processing*, Upper Saddle River, NJ: Prentice-Hall, 1996.
37. Nowak and Baraniuk, Wavelet-domain filtering for photon imaging systems, *IEEE Trans. IP*, pp. 666–678, May 1999.
38. D. A. Pollen and S. F. Ronner, Visual cortical neurons as localized spatial frequency filters, *IEEE Transactions on Systems, Man, and Cybernetics*, vol. SMC–13, no. 5, pp. 907–916, September–October 1983.
39. J. J. Kulikowski, S. Marcelja, and P. O. Bishop, Theory of spatial position and spatial frequency relations in the receptive fields of simple cells in the visual cortex, *Biological Cybernetics*, vol. 43, pp. 187–198, 1982.
40. F. Attneave, Some informational aspects of visual perception, *Psychological Review*, vol. 61, pp. 183–193, 1954.
41. H. Asada and M. Brady, The curvature primal sketch, *IEEE Transactions on Pattern Analysis and Machine Intelligence*, vol. PAMI-8, no. 1, pp. 2–14, January 1986.
42. I. Biedermann, Human image understanding: Recent research and a theory, *Computer Vision, Graphics, and Image Processing*, vol. 32, pp. 29–73, 1985.
43. A. P. Witkin, Scale-space filtering, *Proceedings of the 8th International Joint Conference on Artificial Intelligence*, Karlsruhe, W. Germany, 1983. See also A. P. Witkin, Scale-space filtering, in *From Pixels to Predicates*, A. P. Pentland, ed., Norwood, NJ: Ablex, 1986.
44. T. Lindeberg, Scale space for discrete signals, *IEEE Transactions on Pattern Analysis and Machine Intelligence*, vol. 12, no. 3, pp. 234–254, March 1990.
45. J. Canny, A computational approach to edge detection, *IEEE Transactions on Pattern Analysis and Machine Intelligence*, vol. PAMI-8, no. 6, pp. 679–698, November 1986.
46. H. D. Tagare and R. J. P. deFigueiredo, On the localization performance measure and optimal edge detection, *IEEE Transactions on Pattern Analysis and Machine Intelligence*, vol. 12, no. 12, pp. 1186–1190, 1990.
47. R. J. Qian and T. S. Huang, Optimal edge detection in two-dimensional images, *IEEE Transactions on Image Processing*, vol. 5, no. 7, pp. 1215–1220, 1996.
48. M. Gökmen and A. K. Jain, $\lambda\tau$-space representation of images and generalized edge detector, *IEEE Transactions on Pattern Analysis and Machine Intelligence*, vol. 19, no. 6, pp. 545–563, June 1997.
49. S. Mallat, Zero-crossings of a wavelet transform, *IEEE Transactions on Information Theory*, vol. 37, no. 4, pp. 1019–1033, July 1991.
50. S. Mallat and S. Zhong, Characterization of signals from multiscale edges, *IEEE Transactions on Pattern Analysis and Machine Intelligence*, vol. 14, no. 7, pp. 710–732, July 1992.
51. Z. Berman and J. S. Baras, Properties of the multiscale maxima and sero-crossings representations, *IEEE Transactions on Signal Processing*, vol. 41, no. 12, pp. 3216–3231, 1993.
52. G. A. Orban, *Neuronal Operations of the Visual Cortex*, Berlin: Springer-Verlag, 1984.
53. D. H. Hubel and T. N. Wiesel, Receptive fields, binocular interaction and functional architecture in the cat's visual cortex, *Journal of Physiology*, vol. 160, pp. 106–154, 1962.

54. F. W. C. Campbell and J. Robson, Application of Fourier analysis to the visibility of gratings, *Journal of Physiology*, vol. 197, pp. 551–566, 1968.
55. F. W. C. Campbell and J. J. Kulikowski, Orientation selectivity of the human visual system, *Journal of Physiology*, vol. 195, pp. 437–441, 1966.
56. C. Blakemore and F. W. C. Campbell, On the existence in the human visual system of neurons selectively sensitive to the orientation and size of retinal images, *Journal of Physiology*, vol. 203, pp. 237–260, 1969.
57. J. P. Jones and L. A. Palmer, The two-dimensional spatial structure of simple receptive fields in cat striate cortex, *Journal of Neurophysiology*, vol. 58, pp. 1187–1211, 1987.
58. J. P. Jones and L. A. Palmer, An evaluation of the two-dimensional Gabor filter model of simple receptive fields in cat striate cortex, *Journal of Neurophysiology*, vol. 58, pp. 1233–1258, 1987.
59. J. P. Jones, A. Stepnoski, and L. A. Palmer, The two-dimensional spectral structure of simple receptive fields in cat striate cortex, *Journal of Neurophysiology*, vol. 58, pp. 1212–1232, 1987.
60. S. Marcelja, Mathematical description of the responses of simple cortical cells, *Journal of the Optical Society of America*, vol. 70, pp. 1297–1300, 1980.
61. D. Gabor, Theory of communication, *Journal of the Institute of Electrical Engineers*, vol. 93, pp. 429–459, 1946.
62. M. Porat and Y. Y. Zeevi, The generalized Gabor scheme of image representation in biological and machine vision, *IEEE Transactions on Pattern Analysis and Machine Intelligence*, vol. 10, no. 4, pp. 452–468, July 1988.
63. D. A. Pollen and S. F. Ronner, Visual cortical neurons as localized spatial frequency filters, *IEEE Transactions on Systems, Man, and Cybernetics*, vol. SMC-13, no. 5, pp. 907–916, September/October 1983.
64. J. G. Daugman, Uncertainty relation for resolution in space, spatial frequency, and orientation optimized by two-dimensional visual cortical filters, *Journal of the Optical Society of America A*, vol. 2, no. 7, pp. 1160–1169, July 1985.
65. J. G. Daugman, Spatial visual channels in the Fourier plane, *Vision Research*, vol. 24, no. 9, pp. 891–910, 1984.
66. S. J. Anderson and D. C. Burr, Receptive field size of human motion detection units, *Vision Research*, vol. 27, no. 4, pp. 621–635, 1987.
67. X. Yang, K. Wang, and S.A. Shamma, Auditory representation of acoustic signals, *IEEE Transactions on Information Theory*, vol. 38, no. 2, pp. 824–839, March 1992.
68. M. R. Turner, Texture discrimination by Gabor functions, *Biological Cybernetics*, vol. 55, pp. 71–82, 1986.
69. A. C. Bovik, M. Clark, and W. S. Geisler, Multichannel texture analysis using localized spatial filters, *IEEE Transactions on Pattern Analysis and Machine Intelligence*, vol. 12, no. 1, pp. 55–73, January 1990.
70. T. Weldon and W. E. Higgins, Designing multiple Gabor filters for multitexture image segmentation, *Optical Engineering*, vol. 38, no. 9, pp. 1478–1489, September 1999.
71. J. Lu, J. B. Weaver, D. M. Healy, Jr., and Y. Xu, Noise reduction with a multiscale edge representation and perceptual criteria, *Proceedings of the IEEE-SP International Symposium on Time-Frequency and Time-Scale Analysis*, Victoria, BC, Canada, pp. 555–558, October 4–6, 1992.

72. S. Mallat and S. Zhong, Wavelet transform maxima and multiscale edges, in M. B. Ruskai, et al., eds., *Wavelets and Their Applications*, Boston: Jones and Bartlett, pp. 67–104, 1992.
73. M. Holschneider, *Wavelets: An Analysis Tool*, New York: Oxford University Press, 1995.
74. S. Mallat and W. L. Hwang, Singularity detection and processing with wavelets, *IEEE Transactions on Information Theory*, vol. 38, no. 2, pp. 617–643, March 1992.
75. M. Akay, ed., *Time-Frequency and Wavelets in Biomedical Signal Processing*, New York: Wiley-IEEE Press, 1997.
76. M. Unser and A. Aldroubi, A review of wavelets in biomedical applications, *Proceedings of the IEEE*, vol. 84, no. 4, pp. 626–638, April 1996.
77. J. R. Cox, Jr., F. M. Nolle, and R. M. Arthur, Digital analysis of electroencephalogram, the blood pressure wave, and the electrocardiogram, *Proceedings of the IEEE*, vol. 60, pp. 1137–1164, 1972.
78. P. E. Tikkanen, Nonlinear wavelet and wavelet packet denoising of electrocardiogram signal, *Biological Cybernetics*, vol. 80, no. 4, pp. 259–267, April 1999.
79. B. A. Rajoub, An efficient coding algorithm for the compression of ECG signals using the wavelet transform, *IEEE Transactions on Biomedical Engineering*, vol. 49, no. 4, pp. 255–362, April 2002.
80. B.-U. Kohler, C. Hennig, and R. Orglmeister, The principles of software QRS detection, *IEEE Engineering in Medicine and Biology Magazine*, vol. 21, no. 1, pp. 42–57, January/February 2002.
81. C. Li, C. Zheng, and C. Tai, Detection of ECG characteristic points using wavelet transforms, *IEEE Transactions on Biomedical Engineering*, vol. 42, no. 1, pp. 21–28, January 1995.
82. S. Kadambe, R. Murray, and G. F. Boudreaux-Bartels, Wavelet transformbased QRS complex detector, *IEEE Transactions on Biomedical Engineering*, vol. 46, no. 7, pp. 838–848, July 1999.
83. L. Khadra, M. Matalgah, B. El_Asir, and S. Mawagdeh, Representation of ECG-late potentials in the time frequency plane, *Journal of Medical Engineering and Technology*, vol. 17, no. 6, pp. 228–231, 1993.
84. F. B. Tuteur, Wavelet transformations in signal detection, in *Wavelets: Time-Frequency Methods and Phase Space*, J. M. Combes, A. Grossmann, and P. Tchamitchian, eds., 2nd ed., Berlin: Springer-Verlag, pp. 132–138, 1990.
85. K. Murano, S. Unagami, and F. Amano, Echo cancellation and applications, *IEEE Communications Magazine*, vol. 28, no. 1, pp. 49–55, January 1990.
86. A. Gilloire and M. Vetterli, Adaptive filtering in sub-bands with critical sampling: Analysis, experiments and applications to acoustic echo cancellation, *IEEE Transactions on Signal Processing*, vol. 40, no. 8, pp. 1862–1875, August 1992.
87. O. Tanrikulu, B. Baykal, A. G. Constantinides, and J. A. Chambers, Residual echo signal in critically sampled subband acoustic echo cancellers based on IIR and FIR filter banks, *IEEE Transactions on Signal Processing*, vol. 45, no. 4, pp. 901–912, April 1997.
88. A. Grossmann and J. Morlet, Decomposition of Hardy functions into iquare Integrable wavelets of constant shape, *SIAM Journal of Mathematical Analysis*, vol. 15, pp. 723–736, July 1984.

89. P. Goupillaud, A. Grossmann, and J. Morlet, Cycle-octave and related transforms in seismic signal analysis, *Geoexploration*, vol. 23, pp. 85–102, 1984–1985.
90. J. L. Larsonneur and J. Morlet, Wavelets and seismic interpretation, in J. M. Combes, A. Grossmann, and P. Tchamitchian, eds., *Wavelets: Time-Frequency Methods and Phase Space*, 2nd ed., Berlin: Springer-Verlag, pp. 126–131, 1990.
91. X.-G. Miao and W. M. Moon, Application of wavelet transform in reflection seismic data analysis, *Geosciences Journal*, vol. 3, no. 3, pp. 171–179, September 1999.
92. G. Olmo and L. Lo Presti, Applications of the wavelet transform for seismic activity monitoring, in *Wavelets: Theory, Applications, and Applications*, C. K. Chui, L. Montefusco, and L. Puccio, eds., San Diego, CA: Academic Press, pp. 561–572, 1994.
93. M. Cooke, S. Beet, and M. Crawford, eds., *Visual Representations of Speech Signals*, Chichester: Wiley, 1993.
94. T. Parsons, *Voice and Speech Processing*, New York: McGraw-Hill, 1987.
95. W. Hess, *Pitch Determination of Speech Signals: Algorithms and Devices*, New York: Springer-Verlag, 1983.
96. S. Kadambe and G. F. Boudreaux-Bartels, Application of the wavelet transform for pitch detection of speech signals, *IEEE Transactions on Information Theory*, vol. 38, no. 2, pp. 917–924, March 1992.
97. A. M. Noll, Cepstrum pitch determination, *Journal of the Acoustical Society of America*, vol. 41, pp. 293–309, February 1967.
98. T. Chang and C.-C. J. Kuo, Texture analysis and classification with tree-structured wavelet transform, *IEEE Transactions on Image Processing*, vol. 2, no. 4, pp. 429–441, October 1993.
99. J. Portilla and E. P. Simoncelli, A parametric texture model based on joint statistics of complex wavelet coefficients, *International Journal of Computer Vision*, vol. 40, no. 1, pp. 49–71, October 2000.
100. R. L. Allen, R. Moore, and M. Whelan, Multiresolution pattern detector networks for controlling plasma etch reactors, *Process, Equipment, and Materials Control in Integrated Circuit Manufacturing*, Proceedings SPIE 2637, pp. 19–30, 1995.
101. R. L. Allen, R. Moore, and M. Whelan, Application of neural networks to plasma etch endpoint detection, *Journal of Vacuum Science and Technology (B)*, pp. 498–503, January–February 1996.
102. J. Hertz, A. Krogh, and R. G. Palmer, *Introduction to the Theory of Neural Computation*, Redwood City, CA: Addison-Wesley, 1991.
103. E. A. Rietman, R. C. Frye, E. R. Lory, and T. R. Harry, Active neural network control of wafer attributes in a plasma etch process, *Journal of Vacuum Science and Technology*, vol. 11, p. 1314, 1993.
104. D. M. Manos and G. K. Herb, Plasma etching technology—An overview, in D. M. Manos and D. L. Flamm, eds., *Plasma Etching: An Introduction*, Boston: Academic Press, 1989.
105. L. R. Rabiner, A tutorial on hidden Markov models and selected applications in speech recognition, *Proceedings of the IEEE*, vol. 77, no. 2, pp. 257–286, February 1989.
106. N. Arica and F. T. Yarman-Vural, Optical character recognition for cursive handwriting, *IEEE Transactions on Pattern Analysis and Machine Intelligence*, vol. 24, no. 6, pp. 801–813, June 2002.

107. R. Durbin, S. Eddy, A. Krogh, and G. Mitchison, *Biological Sequence Analysis*, Cambridge: Cambridge University Press, 1998.

108. S. Mallat and S. Zhang, Matching pursuits with time-frequency dictionaries, *IEEE Transactions on Signal Processing*, pp. 3397–3415, December 1993.

109. M. M. Goodwin and M. Vetterli, Matching pursuit and atomic signal models based on recursive filter banks, *IEEE Transactions on Signal Processing*, pp. 1890–1902, July 1999.

110. R. Karchin, *Hidden Markov Models and Protein Sequence Analysis*, Honors Thesis, Computer Engineering Department, University of California, Santa Cruz, June 1998.

PROBLEMS

1. Using material from Chapters 9 and 11, suppose we are given a multiresolution analysis of finite-energy signals.

(a) Show that the discrete lowpass filter $H(\omega)$ associated to the MRA satisfies $|H(\omega)|^2 + |H(\omega+\pi)|^2 = 1$;

(b) Let $g(n) = g_1(n) = (-1)^{1-n}h(1-n)$ and $G(\omega) = e^{-j\omega}\overline{H(\omega+\pi)}$. Show that, indeed, $g(n)$ is the inverse discrete-time Fourier transform of $G(\omega)$.

(c) Show that $|H(\omega)|^2 + |G(\omega)|^2 = 1$.

(d) Using the perfect reconstruction criterion of Chapter 9, show that $\sqrt{2}h(n)$ is a quadrature mirror filter (QMF).

(e) Sketch a reconstruction diagram using $h(n)$ and $g(n)$ for the reconstruction of the original signal decomposed on the pyramid [1].

2. In the QMF pyramid decomposition (Figure 12.5), let $\tilde{h}(n) = h(-n)$ be the reflection of $h(n)$ and $\tilde{H}(z)$ be its z-transform. Similarly, let $\tilde{g}(n) = g(-n)$ and $\tilde{G}(z)$ be the transfer function of the filter with impulse response $\tilde{g}(n)$.

(a) Show that subsampling a signal $x(n)$ by two followed by $\tilde{H}(z)$ filtering is the same discrete system as $\tilde{H}(z^2)$ filtering followed by subsampling [4].

(b) Applying the same idea to $\tilde{g}(n)$, prove filtering with $\tilde{H}(z)\tilde{H}(z^2)$ and $\tilde{H}(z)\tilde{G}(z^2)$ and subsampling by four produces the level -2 approximate and detail coefficients, respectively.

(c) Show that we can compute the impulse response of the filter with transfer function $\tilde{H}(z)\tilde{H}(z^2)$ by convolving $\tilde{h}(n)$ with the filter obtained by putting a zero between every $\tilde{h}(n)$ value.

(d) State and prove a property similar to (c) for $\tilde{H}(z)\tilde{G}(z^2)$.

(e) State and prove properties for level $l = -L$, where $L > 0$, that generalize these results.

3. Suppose $p > 0$ and define the filter G_p as in (12.13). Let O_i be the orthogonal complement of V_i inside V_{i+1}: $V_i \perp O_i$ and $V_{i+1} = V_i \oplus O_i$.

(a) Show (12.16).

(b) Show (12.17).

(c) Show (12.18).

(d) Since $\{\phi(t-k)\}_{k \in Z}$ is an orthongormal basis for V_0, explain the expansion (12.19).

(e) By Fourier transformation of (12.19), show that $G_p(\omega) = \dfrac{\Psi(2^p\omega)}{\Phi(2\omega)}$.

4. Suppose that $y(n) = x(n-2)$ and both signal $x(n)$ and $y(n)$ are decomposed using the orthogonal wavelet pyramid.

(a) How do the first-level $L = -1$ coefficients for $y(n)$ differ from the first-level coefficients for $x(n)$?

(b) Generalize this result to delays that are higher powers of 2.

5. Show by simple convolutions on discrete steps and ridge edges that discrete highpass filters $g_p(n)$ given in Figure 12.3 and Figure 12.4 function as edge detectors for the orthogonal wavelet pyramid.

6. Suppose $g(t) = Ae^{-Bt^2}$ is a Gaussian of zero mean and $\psi(t) = \frac{d}{dt}g(t)$.

(a) Show that $\psi(t)$ is a wavelet.

(b) Let $x_a(t) = ax(at)$ be the scaled dilation of $x(t)$ by factor $a = 2^{-i}$ for $i \in \mathbb{Z}$, with $i > 0$. Define the wavelet transform $(W_{-i}x)(t) = (\psi_a * x)(t)$. Show that

$$(W_{-i}x)(t) = a\frac{d}{dt}(g_a * x)(t). \tag{12.36}$$

(c) Explain the significance of $|W_{-i}x|$ being large.

(d) Explain the significance of large $|W_{-i}x|$ when a is large. What if a is small?

7. Suppose $x(t)$ is discontinuous at $t = t_0$. Show that its Libschitz regularity at t_0 is zero.

Advanced problems and projects.

8. Implement the multiscale matching and registration algorithm of Section 12.1.3.

(a) Use the cubic spline MRA as described in the text.

(b) Use the Laplacian pyramid.

(c) Use the MRA based on piecewise continuous functions.

(d) Develop matching and registration expreriments using object boundaries or signal envelopes.

(e) Compare the performance of the above algorithms based on your chosen applications.

(f) Explore the effect of target shape support in the candidate signal data.

9. Derive the impulse responses for the $h_p(n)$ and $g_p(n)$ for the case where the MRA is

(a) Based on the Haar functions;

(b) The Stromberg MRA.

10. Compare linear and nonlinear filtering of the electrocardiogram to the wavelet de-noising algorithms.

(a) Obtain and plot an ECG trace (for example, from the signal processing information base; see Section 1.9.2.2).

(b) Develop algorithms based on wavelet noise removal as in Section 12.2.1. Compare hard and soft thresholding methods.

(c) Compare your results in (b) to algorithms based on edge-preserving nonlinear filters, such as the median filter.

(d) Compare your results in (b) and (c) to algorithms based on linear filters, such as the Butterworth, Chebyshev, and elliptic filters of Chapter 9.

(e) Consider the requirements of real-time processing and analysis. Reevaluate your comparisons with this in mind.

11. Compare discrete and continuous wavelet transforms for QRS complex detection [81, 82].

(a) Using your data set from the previous problem, apply a nonlinear filter to remove impulse noise and a convolutional bandpass filter to further smooth the signal.

(b) Decompose the filtered ECG signal using one of the discrete wavelet pyramid decompositions discussed in the text (the cubic spline multiresolution analysis, for instance). Describe the evolution of the QRS complexes across multiple scales [81]. Develop a threshold-based QRS detector and assess its usefulness with regard to changing scale and QRS pulse offset within the filtered data.

(c) Select a scale for decomposition based on a continuous wavelet transform [82]. Compare this method of analysis to the discrete decomposition in (b).

(d) Consider differentiating the smoothed ECG signals to accentuate the QRS peak within the ECG. Does this improve either the discrete or continuous algorithms?

(e) Consider squaring the signal after smoothing to accentuate the QRS complex. Does this offer any improvement? Explain.

(f) Do soft or hard thresholding with wavelet de-noising help in detecting the QRS complexes?

(g) Synthesize some defects in the QRS pulse, such as ventricular late potentials, and explore how well the two kinds of wavelet transform perform in detecting this anomaly.

INDEX

Signal Analysis: Time, Frequency, Scale, and Structure, by Ronald L. Allen and Duncan W. Mills
ISBN: 0-471-23441-9

G